## SOLIDS (SPACE FIGURES):

$L$ = Lateral Area; $T$ (or $S$) = Total (Surface) Area;
$V$ = Volume

Parallelepiped (box):

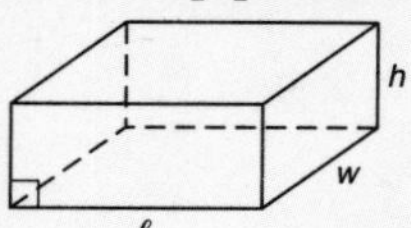


$T = 2\ell w + 2\ell h + 2wh$
$V = \ell wh$

Right Prism:

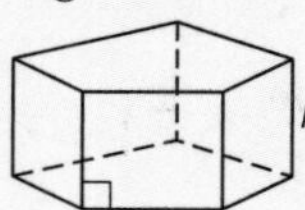


$L = hP$
$T = L + 2B$
$V = Bh$

Regular Pyramid:

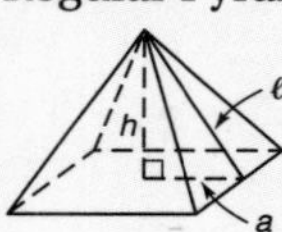


$L = \frac{1}{2}\ell P$
$\ell^2 = a^2 + h^2$
$T = L + B$
$V = \frac{1}{3}Bh$

Right Circular Cylinder:

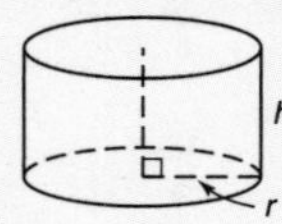


$L = 2\pi rh$
$T = 2\pi rh + 2\pi r^2$
$V = \pi r^2 h$

Right Circular Cone:

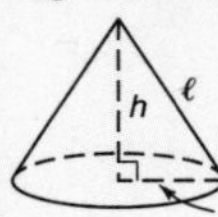


$L = \pi r\ell$
$\ell^2 = r^2 + h^2$
$T = \pi r\ell + \pi r^2$
$V = \frac{1}{3}\pi r^2 h$

Sphere:

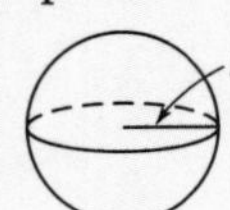


$S = 4\pi r^2$
$V = \frac{4}{3}\pi r^3$

Miscellaneous:
Euler's Equation: $V + F = E + 2$

## ANALYTIC GEOMETRY:

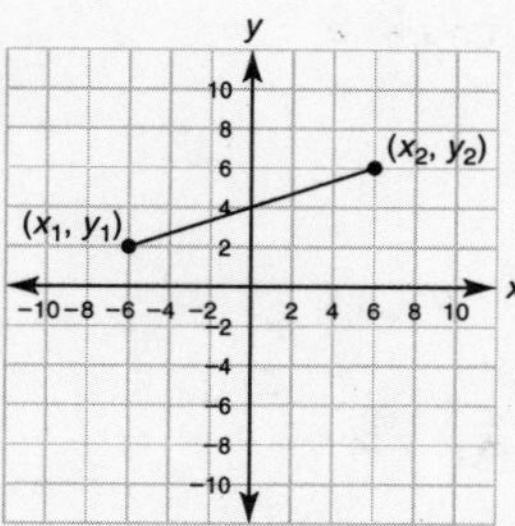


Distance:
$d = \sqrt{(x_2 - x_1)^2 + (y_2 - y_1)^2}$

Midpoint:
$M = \left(\frac{x_1 + x_2}{2}, \frac{y_1 + y_2}{2}\right)$

Slope: $m = \frac{y_2 - y_1}{x_2 - x_1}, x_1 \neq x_2$

Parallel Lines:
$\ell_1 \parallel \ell_2 \leftrightarrow m_1 = m_2$

Perpendicular Lines:
$\ell_1 \perp \ell_2 \leftrightarrow m_1 \cdot m_2 = -1$

Equations of a Line:
Slope-Intercept: $y = mx + b$
Point-Slope: $y - y_1 = m(x - x_1)$
General: $Ax + By = C$

## TRIGONOMETRY:

Right Triangle:

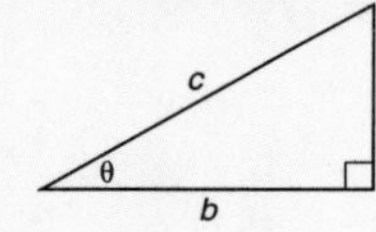


$\sin\theta = \frac{opposite}{hypotenuse} = \frac{a}{c}$
$\cos\theta = \frac{adjacent}{hypotenuse} = \frac{b}{c}$
$\tan\theta = \frac{opposite}{adjacent} = \frac{a}{b}$
$\sin^2\theta + \cos^2\theta = 1$

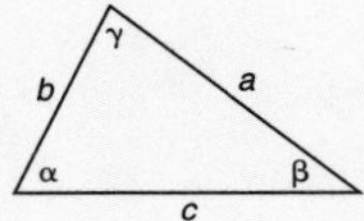


$A = \frac{1}{2}bc\sin\alpha$
$\frac{\sin\alpha}{a} = \frac{\sin\beta}{b} = \frac{\sin\gamma}{c}$
$c^2 = a^2 + b^2 - 2ab\cos\nu$

# Elementary Geometry

## FOR COLLEGE STUDENTS

Fourth Edition

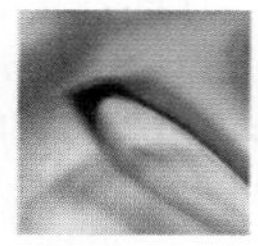

Daniel C. Alexander
*Parkland College*

Geralyn M. Koeberlein
*Mahomet-Seymour High School*

Houghton Mifflin Company
*Boston New York*

*Dan would like to dedicate this edition to the students he has taught at Parkland College in Champaign, Illinois, while Geralyn would like to dedicate this edition to the students she has taught at Mahomet-Seymour High School in Mahomet, Illinois.*

*Math Publisher:* Richard Stratton
*Senior Sponsoring Editor:* Lynn Cox
*Associate Editor:* Melissa Parkin
*Assistant Editor:* Noel Kamm
*Senior Project Editor:* Carol Merrigan
*Editorial Assistant:* Eric Moore
*Senior Manufacturing Coordinator:* Renée Ostrowski
*Executive Marketing Manager:* Brenda Bravener
*Senior Marketing Manager:* Katherine Greig
*Marketing Assistant:* Naveen Hariprasad

*Cover photo credit:* © Shigeru Tanaka/Photonica/Getty Images

PHOTO CREDITS:
**Chapter 1:** p. 1: M.C. Escher's "Concave and Convex," © 2005. The M.C. Escher Company—Holland. All rights reserved; p. 26: Burke/Triolo Productions/Getty Images; p. 48: Jonathan Hogue/www.burningwell.org; **Chapter 2:** p. 71: PhotoDisc/US Landmarks and Travel2/Getty Images; p. 73: Luc Beziat/Getty Images; p. 108: Mike Reese/www.wisconsinbutterflies.org; **Chapter 3:** p. 127: Keith Macgregor/Getty Images; **Chapter 4:** p. 177: CORBIS; p. 182: PhotoDisc/Shipping and Handling/Getty Images; p. 188: Thinkstock/Getty Images; **Chapter 5:** p. 217: www.burningwell.org; p. 227: PhotoDisc/Objects of Nature/Getty Images; p. 248: CORBIS; **Chapter 6:** p. 275: Barbara Stitzer/PhotoEdit; p. 289: NSSDC/NASA; **Chapter 7:** p. 335: Yann Arthus-Bertrand/CORBIS; p. 369: Dennis MacDonald/PhotoEdit; p. 372: PhotoDisc/European People, Business and Lifestyles/Getty Images; **Chapter 8:** p. 389: CORBIS; p. 423: Richard Price/Getty Images; p. 427: Stockdisc/Getty Images; **Chapter 9:** p. 435: Bill Aron/PhotoEdit; **Chapter 10:** p. 481: Gandee Vasan/Getty Images; p. 511: Elena Rooriad/PhotoEdit.

***Company logos:*** Permission was obtained for use of logos on pp. 110, 114, and 117 from the following sources: Hilton Corporation, Health Alliance Corporation, Toyota Motor Corporation, Ford Motor Company, Volkswagen of America, Inc., Target Corporation, and Saks, Inc.



Printed in the U.S.A.
Library of Congress Control Number: 2005934923

Instructor's exam copy:
ISBN 13: 978-0-618-73068-1
ISBN 10: 0-618-73068-0

For orders, use student text ISBNs:
ISBN 13: 978-0-618-64525-1
ISBN 10: 0-618-64525-X
6789 – DOW – 10 09 08 07

# Contents

# Preface

*Elementary Geometry for College Students,* Fourth Edition, shows students how to apply the principles of geometry, as well as how to recognize geometry's relevance to the real world. The text has been written for students who have not completed a course in geometry and are studying these topics for the first time, and for those who need to take a fresh look at the subject. Some background in elementary algebra is helpful for the students' success with this geometry book. Previous editions of this text have been well received and used successfully in the classroom. Many of the book's attractive features have been maintained, while new additions have been included to enhance this edition.

## AUTHORS' PHILOSOPHY AND APPROACH

Our philosophy toward teaching geometry has been the driving force in both the development and the revision of this work. We believe the complete development of geometry begins with an idea, followed by examination and development of a theory, verification of the theory through deduction, and the application of resulting principles in the real world. Our approach to college geometry is largely visual, as it should be.

We present material in a manner and order similar to what we have found effective in our classrooms. We introduce the concept, develop a number of insights into its meaning and application, and establish as well as verify the relationships between these insights. In establishing the deductive basis for many of our principles (theorems), we have used two-column proofs, paragraph proofs, and less formal explanations. This method of proof can have far-reaching effects, including expanding the student's ability to reason, to write a better paragraph, and to order the lines and subroutines in a computer code.

This textbook, which parallels the goals of secondary-level geometry textbooks, is heavily influenced by the standards recommended by both the National Council of Teachers of Mathematics (NCTM) and the American Mathematical Association of Two-Year Colleges (AMATYC). For those interested in building a solid foundation in geometry, we believe that this is the most comprehensive textbook written for the college level. It is suitable for both the student of geometry and the future teacher of its content.

## WRITING STYLE

Geometry teachers, who are aware of the challenges students face in this course, wrote this textbook. We explain what we are doing, clearly indicate where we are going, and demonstrate relationships between topics. We use many paragraph proofs (common at the college level). This method of proof often improves the student's writing style because each paragraph must be ordered and justified.

## OUTCOMES FOR THE STUDENT

- Mastery of essentials for the use of geometry in a vocation
- Preparation of the transfer student for further study of mathematics and related disciplines
- Understanding of the step-by-step development of a logical mathematical system
- Enhancement of an interest in geometry through discovery activities and exercises

## NEW! CHANGES FOR THIS EDITION

- A more streamlined and uniform approach is taken in the outline of steps leading to basic constructions in Chapters 1 and 2.
- An introduction to set concepts appears in Section 1.1 to provide more references to sets and geometric figures as point sets.
- A new section on symmetry and transformations, Section 2.6, gives a more focused treatment of these topics.
- Coverage of similar polygons and applications has been split in order to emphasize these topics individually.
- A more streamlined presentation of trigonometric applications to geometry has been accomplished by treating just the Law of Sines and the Law of Cosines in Section 10.4.
- There are more comparisons provided in Chapters 4 to 6 to help students with the transition from the two-column proof to the paragraph proof.
- A number of Picture Proofs have been included to provide a more visual representation of the material.

- A Chapter Test has been added to the end of every chapter in order to simulate an actual exam; the answer key includes a section reference for every answer. These resources will better prepare students and help them gauge where they need additional practice.
- Included in each section are references to the Student Study Guide, which contains new practice exercise sets so students can work solutions to sample problems immediately after the material has been introduced.
- There has been a 5% increase in the number of exercises in order to provide more practice for students.
- There has been an increase in the number of Geometry in the Real World, Geometry in Nature, Technology Exploration, and Discover! margin features.
- A list of Key Concepts now appears at the beginning of each section to be sure students know which topics they will need to have mastered after completing that section.
- Twice as many end-of-chapter Perspective on History and Perspective on Application features have been included to enable interested students to study geometry in greater depth.
- New chapter-opening photos and interior photos have been incorporated to make more vivid the real-life applications of the concepts presented.
- Tables have been included at the end of the first eight chapters—these are visual summaries and side-by-side comparisons of the essential material within the chapter.

## FEATURES MAINTAINED FOR THE FOURTH EDITION

- **Reminders** are located at points in the textbook that require students to recall earlier learning.
- **Discover!** activities are used to call attention to the inductive method (recognizable logos are used in Chapter 4 as examples of geometric features used in everyday life).
- The margin notes **Geometry in Nature** and **Geometry in the Real World.**
- **Tables** summarize the properties of triangles and quadrilaterals.
- A **Glossary of Terms** and an **Index of Applications** are provided.
- Interesting **chapter-opening** photographs are provided to introduce the principal notion of the chapter.
- **Warnings** are provided as needed to keep students focused.
- **Chapter summaries** review the chapter, preview the following chapter, and provide a list of the most important concepts of the chapter.
- **Chapter reviews** offer numerous review problems.
- **Color** is used to emphasize and highlight important features of the textbook.
- **Front sheets** list important formulas.
- **End sheets** summarize abbreviations and symbols used in the textbook.
- **Summaries** are given of the constructions, postulates, and theorems.
- **Selected answers** are provided in the back of the textbook.
- More difficult homework problems are indicated by an **asterisk (*)** printed in front of a problem number.
- **Definitions** of primary importance are boxed, and other terms are printed in boldface type.
- A convenient **numbering system** for postulates and theorems is used (for example, Theorem 5.3.2 is the second theorem of Section 5.3).
- References to **calculator** usage are made as needed.

## ANCILLARIES FOR THE STUDENT

SSG A ***Student Study Guide*** with suggestions for success in the study of geometry is available. This tool provides a solutions manual for the student, containing selected solutions, and an exciting new component that students will use to solve sample problems immediately after the relevant material has been presented in class. Margin notes throughout the text reference these exercises.

An ***Online Study Center*** is accessible as an additional resource. Included on this website are extra proofs, quizzes, and discovery exercises. Visit this textbook's website at http://college.hmco.com/PIC/alexander4e.

Unique ***text-specific videos*** are available to accompany this text. Professionally produced for this text and hosted by Dana Mosely, these videotapes cover essential topics of the text and offer a valuable resource for further instruction and review.

## SMARTHINKING®

Houghton Mifflin's unique partnership with SMARTHINKING brings students real-time, online tutorial support when they need it most.

This partnership offers a range of tutorial services exclusively for students using Houghton Mifflin textbooks. Employing state-of-the-art whiteboard technology and feedback tools, students interact and communicate with "e-structors." These specially trained tutors guide students through the learning and problem-solving process without providing answers or rewriting a student's work.

SMARTHINKING offers three levels of service.*

**Live Tutorial Help** provides real-time, one-on-one instruction.

**Questions Any Time** allows students to e-mail questions to a tutor outside of the scheduled tutorial sessions and receive a reply, usually within 24 hours.

**Independent Study Resources** connects students around-the-clock to additional educational resources, ranging from interactive websites to Frequently Asked Questions.

**Visit smarthinking.com for more information.**

## ANCILLARIES FOR THE INSTRUCTOR

An ***Online Instructor's Solutions Manual*** is available for the instructor. The instructor's manual provides solutions to all the exercises in the book, alternatives for order of presentation of the topics included, transparency masters, and suggestions for teaching each topic.

There are ***Online Test and Quiz Banks*** available to the instructor.

To access these online components, visit this textbook's online teaching center at http://college.hmco.com/pic/alexander4e.

***HMTesting*** also offers online testing and gradebook functions.

## ACKNOWLEDGEMENTS

We wish to thank Lynn Cox, Senior Sponsoring Editor; Melissa Parkin, Associate Editor; Noel Kamm, Assistant Editor; Carol Merrigan, Senior Project Editor; Eric Moore, Editorial Assistant; and all others at Houghton Mifflin who helped to bring the Fourth Edition of *Elementary Geometry for College Students* to completion. Lest we forget, we wish to express our gratitude to Maureen O'Connor, Dawn Nuttall, Florence Powers, Theresa Grutz, and Beth Dahlke, who were most helpful in the preparation of earlier editions of this book.

We would like to thank reviewers of previous editions, including

**Paul Allen,** *University of Alabama*
**Jane C. Beatie,** *University of South Carolina at Aiken*
**Steven Blasberg,** *West Valley College*
**Barbara Brown,** *Anoka Ramsey Community College*
**Patricia Clark,** *Indiana State University*
**Joyce Cutler,** *Framingham State College*
**Walter Czarnec,** *Framingham State College*
**William W. Durand,** *Henderson State University*
**Zoltan Fischer,** *Minneapolis Community and Technical College*
**Kathryn E. Godshalk,** *Cypress College*
**Chris Graham,** *Mt. San Antonio Community College*
**Sharon Gronberg,** *Southwest Texas State University*
**Geoff Hagopian,** *College of the Desert*
**Edith Hays,** *Texas Woman's University*
**Ben L. Hill,** *Lane Community College*
**George L. Holloway,** *Los Angeles Valley College*
**Tracy Hoy,** *College of Lake County*
**Josephine G. Lane,** *Eastern Kentucky University*
**John C. Longnecker,** *University of Northern Iowa*
**Nicholas Martin,** *Shepherd College*
**Jill McKenney,** *Lane Community College*
**James R. McKinney,** *Cal Poly at Pomona*
**Iris C. McMurtry,** *Motlow State Community College*
**Michael Naylor,** *Western Washington University*
**Maurice Ngo,** *Chabot College*
**Ellen L. Rebold,** *Brookdale Community College*
**Lauri Semarne**
**Patty Shovanec,** *Texas Technical University*
**Joseph F. Stokes,** *Western Kentucky University*
**Kay Stroope,** *Phillips Community College–University of Arkansas*
**Karen R. Swick,** *Palm Beach Atlantic College*
**Steven L. Thomassin,** *Ventura College*
**Jean A. Vrechek,** *Sacramento City College*

We are most grateful to our reviewers of the Fourth Edition:

**Darwin G. Dorn,** *University of Wisconsin–Washington County*
**Erin C. Martin,** *Parkland College*
**Patty Schovanec,** *Texas Technical University*
**Lauri Semarne**
**Marvin Stick,** *University of Massachusetts–Lowell*
**Dr. John Stroyls,** *Georgia Southwestern State University*
**Bettie A. Truitt, Ph.D.,** *Black Hawk College*
**Tom Zerger,** *Saginaw Valley State University*

* Limits apply; terms and hours of SMARTHINKING service are subject to change.

# Foreword

In the Fourth Edition of *Elementary Geometry for College Students,* the topics that comprise a basic course in plane geometry are found in Chapters 1 to 7. To enhance that material, the final chapters focus on these topics:

Chapter 8: Solid Geometry
Chapter 9: Analytic Geometry
Chapter 10: Trigonometry

The order in which the chapters of the book can be studied is depicted in the following flow chart. It may be necessary to exclude parts of sections if a non-standard sequence is chosen.

$$\begin{array}{c} \qquad\qquad\qquad\qquad\qquad\qquad 8 \\ \qquad\qquad\qquad\qquad\qquad\qquad \uparrow \\ 1 \rightarrow 2 \rightarrow 3 \rightarrow 4 \rightarrow 5 \rightarrow 6 \rightarrow 7 \rightarrow 9 \\ \qquad\qquad\qquad\qquad\qquad\qquad \downarrow \\ \qquad\qquad\qquad\qquad\qquad\qquad 10 \end{array}$$

For students who wish to review algebra topics, these are found in Appendix A.

A.1 Algebraic Expressions
A.2 Formulas and Equations
A.3 Inequalities
A.4 Quadratic Equations

Section A.4 also reviews factoring, the Square Roots method, and the Quadratic Formula.

Students who want a more complete background in the elements of logic should visit this textbook's website at http://college.hmco.com/pic/alexander4ed.

Logic Appendix 1 Truth Tables
Logic Appendix 2 Valid Arguments

For a minimal course in geometry, the following sections can be treated as optional:

Section 2.6 Symmetry and Transformations
Section 3.4 Basic Constructions Justified
Section 3.5 Inequalities in a Triangle
Section 5.6 Segments Divided Proportionally
Section 6.4 Some Constructions and Inequalities for the Circle
Section 7.5 More Area Relationships in the Circle

*Daniel C. Alexander*
*Geralyn M. Koeberlein*

# Index of Applications

# Chapter 1

# Line and Angle Relationships

In geometry, figures can be drawn that create an illusion. Careful inspection of this Escher print reveals that the artist created numerous perceptual flaws; compare the staircases, ladders, and windows in the print in order to discover the wrong-doing. This chapter opens with a discussion of statements and of the types of reasoning used in geometry. Section 1.2 focuses on the tools of geometry, such as the ruler and protractor. The remainder of the chapter begins the formal development of geometry by considering the relationships between lines and angles. For any student who needs an algebra refresher, selected topics can be found in the appendices. Other techniques from algebra are reviewed or developed in the textbook as needed. An introduction to logic can be found on our website.

## Chapter Outline



**For online student resources, visit this textbook's website at math.college.hmco.com/students.**

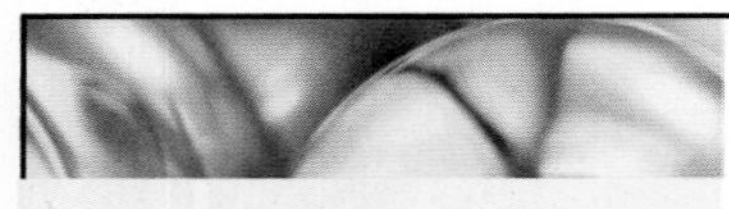

# 1.1 Sets, Statements, and Reasoning

**KEY CONCEPTS**

Statement • Conjunction • Disjunction • Negation • Implication (Conditional) • Hypothesis • Conclusion • Intuition • Induction • Deduction • Law of Detachment • Sets • Subset • Intersection • Union • Venn Diagram

A **set** is any collection of objects; in particular, the objects are known as the *elements* of the set. The statement $A = \{1, 2, 3\}$ is read, "A is the set of elements 1, 2, and 3." In geometry, geometric figures such as lines and angles are actually sets of points.

Where $A = \{1, 2, 3\}$ and $B = \{\text{counting numbers}\}$, $A$ is a *subset* of $B$ because each element in $A$ is also in $B$; in symbols, $A \subseteq B$. In Chapter 2, we will discover that $T = \{\text{all triangles}\}$ is a subset of $P = \{\text{all polygons}\}$.

## STATEMENTS

DEFINITION: A **statement** is a set of words and symbols that collectively make a claim that can be classified as true or false.

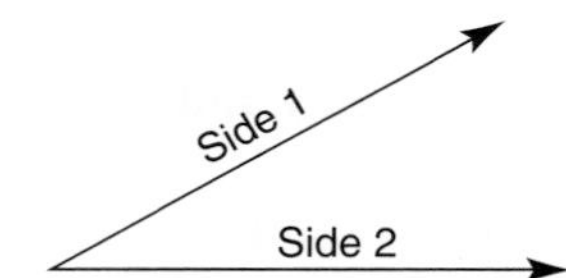


**FIGURE 1.1**

### EXAMPLE 1

Classify each of the following as a true statement, a false statement, or neither.

1. $4 + 3 = 7$
2. An angle has two sides. (See Figure 1.1.)
3. Robert E. Lee played shortstop for the Yankees.
4. $7 < 3$ (This is read "7 is less than 3.")
5. Look out!

***Solution***

1 and 2 are true statements; 3 and 4 are false statements; 5 is not a statement.

Some statements contain one or more *variables;* a **variable** is a letter that represents a number. The claim "$x + 5 = 6$" is called an *open sentence* or *open statement* because it can be classified as true or false, depending on the replacement value of $x$. For instance, $x + 5 = 6$ is true if $x = 1$; for $x$ not equal to 1, $x + 5 = 6$ is false. Some statements containing variables are classified as true because they are true for all replacements. Consider the Commutative Property of Addition, usually stated in the form $a + b = b + a$. In words, this property states that the same result is obtained when two numbers are added in either order; for instance, when $a = 4$ and $b = 7$, it follows that $4 + 7 = 7 + 4$.

The **negation** of a given statement $P$ makes a claim opposite that of the original statement. If the given statement is true, its negation is false, and vice versa. If $P$ is a statement, we use $\sim P$ (which is read "not $P$") to indicate its negation.

### EXAMPLE 2

Give the negation of each statement.

a) $4 + 3 = 7$ b) All fish can swim.

***Solution***

a) $4 + 3 \neq 7$ ($\neq$ means "is not equal to.")
b) Some fish cannot swim. (To negate "All fish can swim," we say that at least one fish cannot swim.)

Sometimes we form a *compound* statement by combining other statements used as "building blocks." In such cases, we may use letters such as $P$ and $Q$ to represent simple statements. For example, the letter $P$ may refer to the statement "$4 + 3 = 7$" and the letter $Q$ to the statement "Babe Ruth was a U.S. president." The statement "$4 + 3 = 7$ *and* Babe Ruth was a U.S. president" has the form $P$ *and* $Q$ and is known as the **conjunction** of $P$ and $Q$. The statement "$4 + 3 = 7$ *or* Babe Ruth was a U.S. president" has the form $P$ *or* $Q$ and is known as the **disjunction** of $P$ and $Q$. A conjunction is true only when $P$ and $Q$ are *both* true. A disjunction is false only when $P$ and $Q$ are *both* false.

The statement "If $P$, then $Q$," known as a **conditional statement** (or **implication**), is classified as true or false as a whole. A statement of this form can be written in equivalent forms; for instance, the conditional statement "If an angle is a right angle, then it measures 90 degrees" is equivalent to the statement "All right angles measure 90 degrees."

EXAMPLE 3

Classify each conditional statement as true or false.

1. If an animal is a fish, then it can swim. (States, "All fish can swim.")
2. If two sides of a triangle are equal in length, then two angles of the triangle are equal in measure. (See Figure 1.2.)

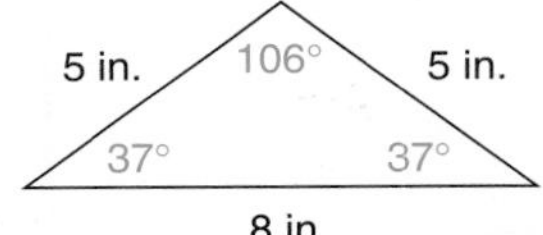


**FIGURE 1.2**

3. If Wendell studies, then he will receive an A on the test.

***Solution***

Statements 1 and 2 are true. Statement 3 is false; Wendell may study, yet not receive an A.

In the conditional statement "If $P$, then $Q$," $P$ is the **hypothesis** and $Q$ is the **conclusion.** In Example 3, statement 2, we have

*Hypothesis:* Two sides of a triangle are equal in length.

*Conclusion:* Two angles of the triangle are equal in measure.

*Exs. 1–7*

For the true statement "If $P$, then $Q$," the hypothetical situation described in $P$ implies the conclusion described in $Q$. This type of statement suggests some form of reasoning, so we turn our attention to this matter.

## REASONING

Success in the study of geometry requires vocabulary development, attention to detail and order, supporting claims, and thinking. **Reasoning** is a process based on experience and principles that allow one to arrive at a conclusion.The following types of reasoning are used to develop mathematical principles.

| | | |
|---|---|---|
| 1. | Intuition | An inspiration leading to the statement of a theory |
| 2. | Induction | An organized effort to test and validate the theory |
| 3. | Deduction | A formal argument that proves the tested theory |

### *Intuition*

We are often inspired to think and say, "It occurs to me that . . . ." With **intuition,** a sudden insight allows one to make a statement without applying any formal reasoning. When intuition is used, we sometimes err by "jumping" to conclusions. In a cartoon, the character having the "bright idea" (using intuition) is shown with a light bulb next to her or his head.

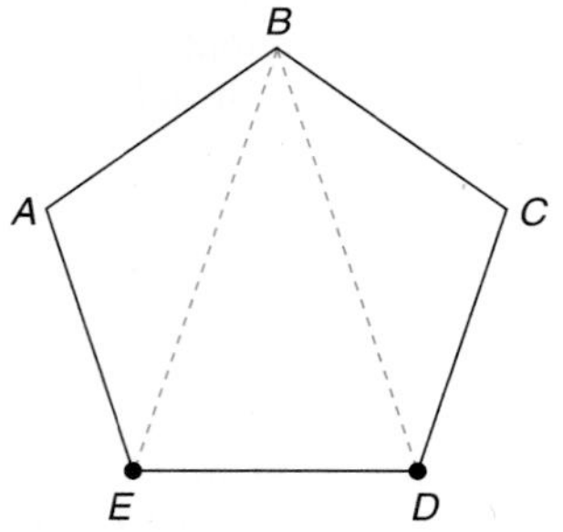

B

A C

E D

**FIGURE 1.3**

EXAMPLE 4

Figure 1.3 is called a *regular pentagon* because its five sides have equal lengths and its angles have equal measures. What do you suspect is true of the lengths of the dashed parts of lines from $B$ to $E$ and from $B$ to $D$?

***Solution***

Intuition suggests that the lengths of the dashed parts of lines (known as *diagonals* of the pentagon) are the same.

NOTE 1: A *ruler* can be used to verify that this claim is true. We will discuss measurement with the ruler in more detail in Section 1.2.

NOTE 2: Using methods found in Chapter 3, we could use deduction to prove that the two diagonals do indeed have the same length.

The role intuition plays in formulating mathematical thoughts is truly significant. But to have an idea is not enough! Testing a theory may lead to a revision of the theory or even to its total rejection. If a theory stands up to testing, it moves one step closer to becoming mathematical law.

### *Induction*

We often use specific observations and experiments to draw a general conclusion. This type of reasoning is called **induction.** As you would expect, the observation/experimentation process is common in laboratory and clinical settings. Chemists, physicists, doctors, psychologists, weather forecasters, and many others use collected data as a basis for drawing conclusions . . . and so will we!

EXAMPLE 5

While in a grocery store, you examine several 8-oz cartons of yogurt. Although the flavors and brands differ, each carton is priced at 75 cents. What do you conclude?

***Conclusion***

Every 8-oz carton of yogurt in the store costs 75 cents.

As you may already know (see Figure 1.2), a figure with three straight sides is called a *triangle.*

EXAMPLE 6

In a geometry class, you have been asked to measure the three interior angles of each triangle in Figure 1.4. You discover that triangles I, II, and IV have two angles (as marked) that have equal measures. What may you conclude?

***Conclusion***

The triangles that have two sides of equal length also have two angles of equal measure.

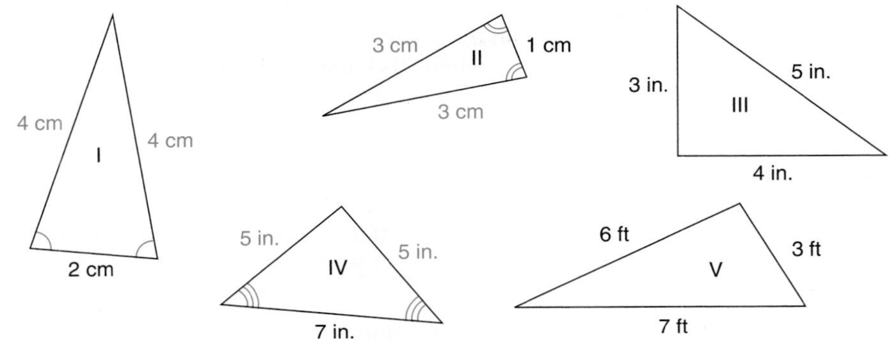


**FIGURE 1.4**

NOTE: A *protractor* can be used to support the conclusion found in Example 6. We will discuss the protractor in Section 1.2.

## *Deduction*

DEFINITION: **Deduction** is the type of reasoning in which the knowledge and acceptance of selected assumptions guarantee the truth of a particular conclusion.

In Example 7, we will illustrate the form of deductive reasoning used most frequently in the development of geometry. In this form, known as a **valid argument,** at least two statements are treated as facts; these assumptions are called the *premises* of the argument. On the basis of the premises, a particular *conclusion* must follow. This form of deduction is called the Law of Detachment.

EXAMPLE 7

If you accept the following statements 1 and 2 as true, what must you conclude?

1. If a student plays on the Rockville High School boys' varsity basketball team, then he is a talented athlete.
2. Todd plays on the Rockville High School boys' varsity basketball team.

***Conclusion***

Todd is a talented athlete.

To more easily recognize this pattern for deductive reasoning, we use letters to represent statements in the following generalization.

**LAW OF DETACHMENT**

Let $P$ and $Q$ represent simple statements, and assume that statements 1 and 2 are true. Then a valid argument having conclusion C has the form

1. If $P$, then $Q$ } premises
2. $P$

C. $\therefore Q$ } conclusion

NOTE: The symbol $\therefore$ means "therefore."

In the preceding form, the statement "If $P$, then $Q$" is often read "$P$ implies $Q$." That is, when $P$ is known to be true, $Q$ must follow.

### EXAMPLE 8

Is the following argument valid? Assume that premises 1 and 2 are true.

1. If it is raining, then Tim will stay in the house.
2. It is raining.

C. $\therefore$ Tim will stay in the house.

***Conclusion***

The argument is valid because the form of the argument is

1. If $P$, then $Q$
2. $P$

C. $\therefore Q$

with $P$ = "It is raining," and $Q$ = "Tim will stay in the house."

### EXAMPLE 9

Is the following argument valid? Assume that premises 1 and 2 are true.

1. If a man lives in London, then he lives in England.
2. William lives in England.

C. $\therefore$ William lives in London.

***Conclusion***

The argument is not valid. Here, $P$ = "A man lives in London," and $Q$ = "A man lives in England." Thus the form of this argument is

1. If $P$, then $Q$
2. $Q$

C. $\therefore P$

But the Law of Detachment does not handle the question "If $Q$, then what?" Even though statement $Q$ is true, it does not enable us to draw a valid conclusion about $P$. Of course, if William lives in England, he *might* live in London; but he might instead live in Liverpool, Manchester, Coventry, or any of countless other places in England. Each of these possibilities is a **counterexample** disproving the validity of the argument. Remember that deductive reasoning is concerned with reaching conclusions that *must be true,* given the truth of the premises.

Warning (!)
In the box, the argument on the left is valid and patterned after Example 8. The argument on the right is invalid; this form was given in Example 9.

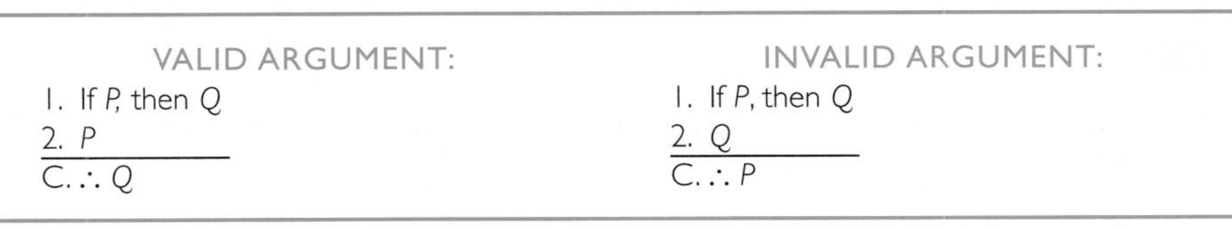

| VALID ARGUMENT: | INVALID ARGUMENT: |
|---|---|
| 1. If $P$, then $Q$ | 1. If $P$, then $Q$ |
| 2. $P$ | 2. $Q$ |
| C. $\therefore Q$ | C. $\therefore P$ |

We will use deductive reasoning throughout our work in geometry. For example, suppose that you know these two facts:

1. If an angle is a right angle, then it measures 90°.
2. Angle $A$ is a right angle.

*Exs. 8–12*

Then you may conclude

C. Angle A measures 90°.

## VENN DIAGRAMS

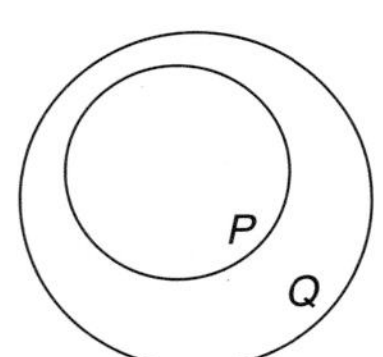


If $P$, then $Q$.

**FIGURE 1.5**

Sets of objects are often represented by geometric figures known as *Venn Diagrams.* Their creator, John Venn, was an Englishman who lived from 1834 to 1923. In a Venn Diagram, each set is represented by a closed (bounded) figure such as a circle or rectangle. If statements $P$ and $Q$ of the conditional statement "If $P$, then $Q$" are represented by sets of objects $P$ and $Q$, respectively, then the Law of Detachment can be justified by a geometric argument. When a Venn Diagram is used to represent the statement "If $P$, then $Q$," it is absolutely necessary that circle $P$ lies in circle $Q$; that is, $P$ is a *subset* of $Q$. (See Figure 1.5.)

### EXAMPLE 10

Use Venn Diagrams to verify Example 7.

***Solution***

Let $B$ = students on the Rockville High varsity boys' basketball team.
Let $A$ = people who are talented athletes.
To represent the statement "If a basketball player ($B$), then a talented athlete ($A$)," we show $B$ within $A$. In Figure 1.6 we use point $T$ to represent Todd, a person on the basketball team ($T$ in $B$). With point $T$ also in circle $A$, we conclude that "Todd is a talented athlete."

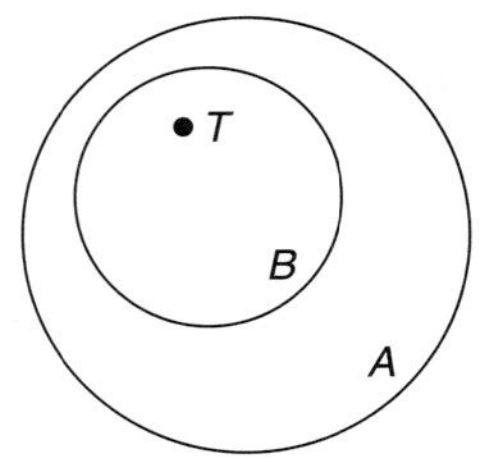


**FIGURE 1.6**

The statement "If $P$, then $Q$" is sometimes expressed in the form "All $P$ are $Q$." For instance, the conditional statement of Examples 7 and 10 can be written "All Rockville high school players are talented athletes." Venn Diagrams can also be used to demonstrate that the argument of Example 9 is not valid. In order to show the invalidity of this argument, one must show that an object in $Q$ may *not* lie in circle $P$. (See Figure 1.5.)

The compound statements known as the conjunction and the disjunction can also be related to the intersection and union of sets, relationships that can be illustrated by the use of Venn Diagrams. For the Venn Diagram, we assume that the sets $P$ and $Q$ may have elements in common.

**Discover!**

In the St. Louis area, an interview of 100 sports enthusiasts shows that 74 support the Cardinals baseball team and 58 support the Rams football team. All of those interviewed support one team or the other or both. How many support both teams?

ANSWER

32 (74 + 58 − 100)

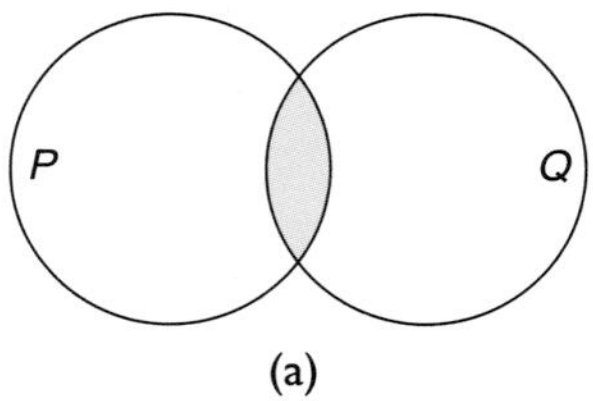


(a)

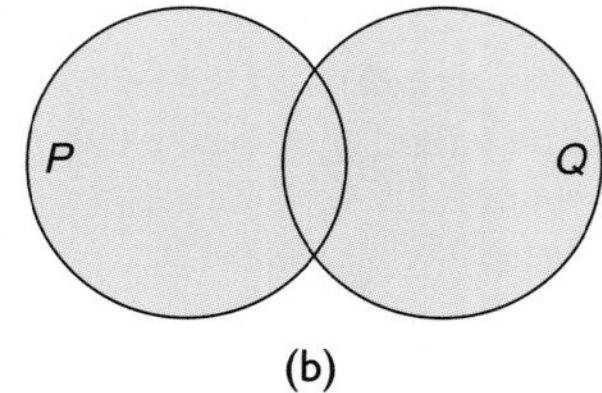


(b)

**FIGURE 1.7**

The elements common to $P$ and $Q$ form the **intersection** of $P$ and $Q$, which is written $P \cap Q$. This set, $P \cap Q$, is the set of all elements in *both P* **and** $Q$.

The elements that are in $P$, in $Q$, or in both form the **union** of $P$ and $Q$, which is written $P \cup Q$. This set, $P \cup Q$, is the set of elements in $P$ **or** $Q$.

*Exs. 13–15*

## 1.1 Exercises

*In Exercises 1 and 2, which sentences are statements? If a sentence is a statement, classify it as true or false.*

1. a) Where do you live?
   b) $4 + 7 \neq 5$.
   c) Washington was the first U.S. president.
   d) $x + 3 = 7$ when $x = 5$.

2. a) Chicago is located in the state of Illinois.
   b) Get out of here!
   c) $x < 6$ (read as "x is less than 6") when $x = 10$.
   d) Babe Ruth is remembered as a great football player.

*In Exercises 3 and 4, give the negation of each statement.*

3. a) Christopher Columbus crossed the Atlantic Ocean.
   b) All jokes are funny.

4. a) No one likes me.
   b) Angle 1 is a right angle.

*In Exercises 5 to 10, classify each statement as simple, conditional, a conjunction, or a disjunction.*

5. If Alice plays, the volleyball team will win.
6. Alice played and the team won.
7. The first-place trophy is beautiful.
8. An integer is odd or it is even.
9. Matthew is playing shortstop.
10. You will be in trouble if you don't change your ways.

*In Exercises 11 to 18, state the hypothesis and the conclusion of each statement.*

11. If you go to the game, then you will have a great time.
12. If two chords of a circle have equal lengths, then the arcs of the chords are congruent.
13. If the diagonals of a parallelogram are perpendicular, then the parallelogram is a rhombus.
14. If $\frac{a}{b} = \frac{c}{d}$ where $b \neq 0$ and $d \neq 0$, then $a \cdot d = b \cdot c$.
15. Corresponding angles are congruent if two parallel lines are cut by a transversal.
16. Vertical angles are congruent when two lines intersect.
17. All squares are rectangles.
18. Base angles of an isosceles triangle are congruent.

*In Exercises 19 to 24, classify each statement as true or false.*

19. If a number is divisible by 6, then it is divisible by 3.
20. Rain is wet and snow is cold.
21. Rain is wet or snow is cold.
22. If Jim lives in Idaho, then he lives in Boise.
23. Triangles are round or circles are square.
24. Triangles are square or circles are round.

*In Exercises 25 to 32, name the type of reasoning (if any) used.*

25. While participating in an Easter egg hunt, Sarah notices that each of the seven eggs she has found is numbered. Sarah concludes that all eggs used for the hunt are numbered.

26. You walk into your geometry class, look at the teacher, and conclude that you will have a quiz today.

27. Albert knows the rule "If a number is added to each side of an equation, then the new equation has the same solution set as the given equation." Given the equation $x - 5 = 7$, Albert concludes that $x = 12$.

28. You believe that "Anyone who plays major league baseball is a talented athlete." Knowing that Duane Gibson has just been called up to the major leagues, you conclude that Duane Gibson is a talented athlete.

29. As a handcuffed man is brought into the police station, you glance at him and say to your friend, "That fellow looks guilty to me."

30. While judging a science fair project, Mr. Cange finds that each of the first 5 projects is outstanding and concludes that all 10 will be outstanding.

31. You know the rule "If a person lives in the Santa Rosa Junior College district, then he or she will receive a tuition break at Santa Rosa." Candace tells you that she has received a tuition break. You conclude that she resides in the Santa Rosa Junior College district.

32. As Mrs. Gibson enters the doctor's waiting room, she concludes that it will be a long wait.

*In Exercises 33 to 36, use intuition to state a conclusion.*

33. You are told that the opposite angles formed when two lines cross are **vertical angles.** In the figure, angles 1 and 2 are vertical angles. Conclusion?

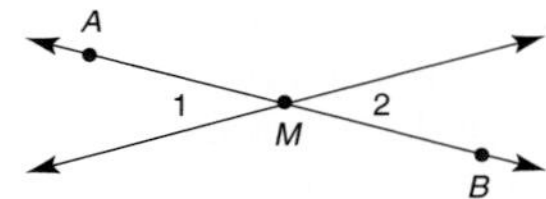


*Exercises 33, 34*

34. In the figure, point $M$ is called the **midpoint** of line segment $AB$. Conclusion?

35. The two triangles shown are **similar** to each other. Conclusion?

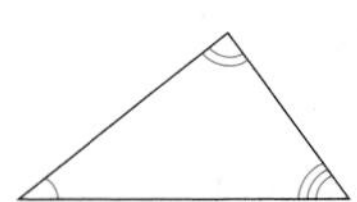

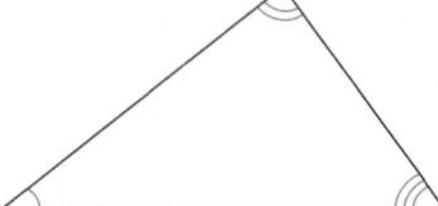

36. Observe (but do not measure) the following angles. Conclusion?

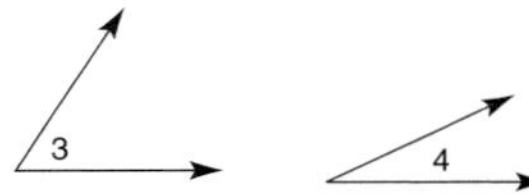


*In Exercises 37 to 40, use induction to state a conclusion.*

37. Several movies directed by Lawrence Garrison have won Academy Awards, and many others have received nominations. His latest work, *A Prisoner of Society,* is to be released next week. Conclusion?

38. On Monday, Matt says to you, "Andy hit his little sister at school today." On Tuesday, Matt informs you, "Andy threw his math book into the wastebasket during class." On Wednesday, Matt tells you, "Because Andy was throwing peas in the school cafeteria, he was sent to the principal's office." Conclusion?

39. While searching for a classroom, Tom stopped at an instructor's office to ask directions. On the office bookshelves are books titled *Intermediate Algebra, Calculus, Modern Geometry, Linear Algebra,* and *Differential Equations.* Conclusion?

40. At a friend's house, you see several food items, including apples, pears, grapes, oranges, and bananas. Conclusion?

*In Exercises 41 to 50, use deduction to state a conclusion, if possible.*

41. If the sum of the measures of two angles is 90°, then these angles are called "complementary." Angle 1 measures 27° and angle 2 measures 63°. Conclusion?

42. If a person attends college, then he or she will be a success in life. Kathy Jones attends Dade County Community College. Conclusion?

43. All mathematics teachers have a strange sense of humor. Alex is a mathematics teacher. Conclusion?

44. All mathematics teachers have a strange sense of humor. Alex has a strange sense of humor. Conclusion?

45. If Stewart Powers is elected president, then every family will have an automobile. Every family has an automobile. Conclusion?

46. If Tabby is meowing, then she is hungry. Tabby is hungry. Conclusion?

47. If a person is involved in politics, then that person will be in the public eye. June Jesse has been elected to the Missouri state senate. Conclusion?

48. If a student is enrolled in a literature course, then he or she will work very hard. Bram Spiegel digs ditches by hand 6 days a week. Conclusion?

49. If a person is rich and famous, then he or she is happy. Marilyn is wealthy and well-known. Conclusion?

50. If you study hard and hire a tutor, then you will make an A in this course. You make an A in this course. Conclusion?

*In Exercises 51 to 54, use Venn Diagrams to determine whether the argument is valid or not valid.*

51. (1) If an animal is a cat, then it makes a "meow" sound.
(2) Tipper is a cat.
(C) Then Tipper makes a "meow" sound.

52. (1) If an animal is a cat, then it makes a "meow" sound.
(2) Tipper makes a "meow" sound.
(C) Then Tipper is a cat.

53. (1) All Boy Scouts serve the United States of America.
(2) Sean serves the United States of America.
(C) Sean is a Boy Scout.

54. (1) All Boy Scouts serve the United States of America.
(2) Sean is a Boy Scout.
(C) Sean serves the United States of America.

55. Where $A = \{1,2,3\}$ and $B = \{2,4,6,8\}$, classify each of the following as true or false.
(a) $A \cap B = \{2\}$
(b) $A \cup B = \{1,2,3,4,6,8\}$
(c) $A \subseteq B$

# 1.2 Informal Geometry and Measurement

**KEY CONCEPTS**

Point • Line • Plane • Collinear Points • Line Segment • Midpoint • Congruence • Protractor • Parallel • Bisect • Intersect • Perpendicular • Compass • Constructions • Circle • Arc • Radius

In geometry, the terms *point, line,* and *plane* are described but not defined. Other concepts that are accepted intuitively, but never defined, include the *straightness* of a line, the *flatness* of a plane, the notion that a point on a line lies *between* two other points on the line, and the notion that a point lies in the *interior* or *exterior* of an angle. Some of the terms found in this section are formally defined in later sections of Chapter 1. The following are descriptions of some of the undefined terms.

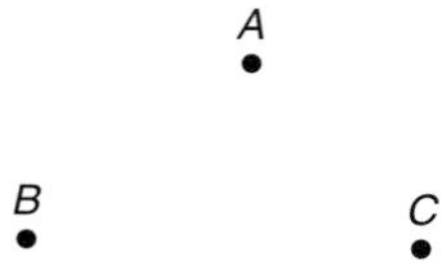

A
B C

**FIGURE 1.8**

A **point,** which is represented by a dot, has location but not size; that is, a point has no dimensions. An uppercase italic letter is used to name a point. Figure 1.8 shows points *A, B,* and *C.* ("Point" may be abbreviated "pt." for convenience.)

The second undefined term is **line.** A line is an infinite set of points. Given any two points on a line, there is always a point that lies between them on that line. Lines have a quality of "straightness" that is not defined but assumed. Given several points on a line, these points form a straight path. Whereas a point has no dimensions, a line is one-dimensional; that is, the distance between any two points on a given line can be measured. Line *AB*, represented symbolically by $\overleftrightarrow{AB}$, extends infinitely far in opposite directions, as suggested by the arrows on the line. A line may also be represented by a single lowercase letter. Figures 1.9(a) and (b) show the lines *AB* and *m*. When a lowercase letter is used to name a line, the line symbol is omitted.

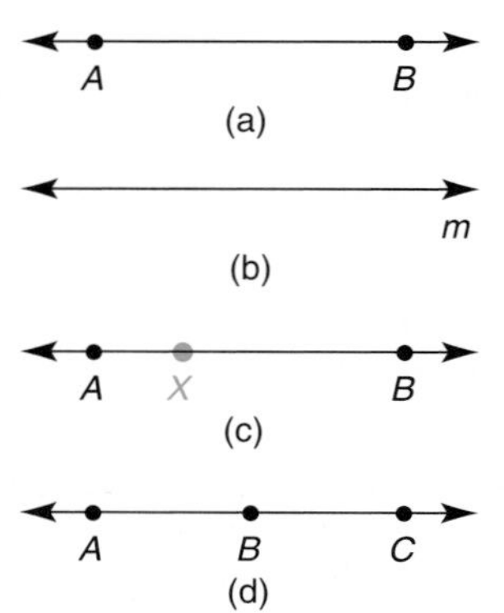


**FIGURE 1.9**

Note the position of point $X$ on $\overleftrightarrow{AB}$ in Figure 1.9(c) on page 10. When three points such as $A$, $X$, and $B$ are on the same line, they are said to be **collinear.** In the order shown, which is symbolized $A$-$X$-$B$ or $B$-$X$-$A$, point $X$ is said to be *between* $A$ and $B$.

When no drawing is provided, the notation $A$-$B$-$C$ means that these points are collinear, with $B$ *between* $A$ and $C$. When a drawing is provided, we assume that all points in the drawing that appear to be collinear *are* collinear, *unless otherwise stated.* Figure 1.9(d) shows that $A$, $B$, and $C$ are collinear, with $B$ between $A$ and $C$.

At this time, we informally introduce some terms that will be formally defined later. You have probably encountered the terms *angle, triangle,* and *rectangle* many times. An example of each is shown in Figure 1.10.

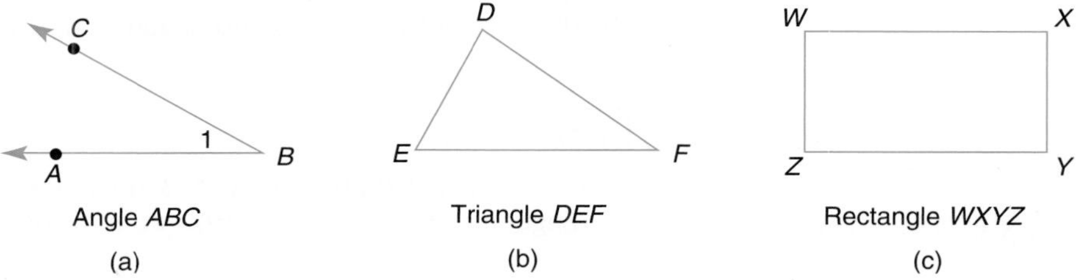


**FIGURE 1.10**

Using symbols and abbreviations, we refer to Figures 1.10(a), (b), and (c) as $\angle ABC$, $\triangle DEF$, and rect. $WXYZ$, respectively. Some caution must be used when naming figures; for instance, it is incorrect to describe the angle in Figure 1.10 (a) as $\angle ACB$ because that order implies a path from point $A$ to point $C$ to point $B$ . . . a different angle! In $\angle ABC$, the point $B$ at which the sides meet is called the **vertex** of the angle. Because there is no confusion regarding the angle described, $\angle ABC$ is also known as $\angle B$ (using only the vertex) or as $\angle 1$. The points $D$, $E$, and $F$ at which the sides of $\triangle DEF$ meet are called the *vertices* (plural of *vertex*) of the triangle. Similarly, $W$, $X$, $Y$, and $Z$ are the vertices of the rectangle.

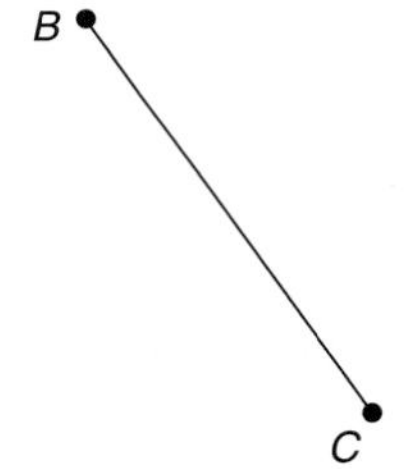


**FIGURE 1.11**

A **line segment** is part of a line. It consists of two distinct points on the line and all points between them. (See Figure 1.11.) Using symbols, we indicate the line segment by $\overline{BC}$; note that $\overline{BC}$ is a set of points but is not a number. We use $BC$ (omitting the segment symbol) to indicate the *length* of this line segment; thus $BC$ is a number. The sides of a triangle or rectangle are line segments. The vertices of a rectangle are named in an order that identifies its line segment sides.

EXAMPLE 1

Can the rectangle in Figure 1.10(c) be named a) $XYZW$? b) $WYXZ$?

***Solution***

a) Yes, because the points taken in this order trace the figure.
b) No; for example, $\overline{WY}$ is not a side of the rectangle.

## MEASURING LINE SEGMENTS

The instrument used to measure a line segment is a scaled straightedge such as a *ruler,* a *yardstick,* or a *meter stick.* Generally, we place the "0 point" of the ruler at one end of the line segment and find the numerical length as the number at the other end. Line segment $RS$ ($\overline{RS}$ in symbols) in Figure 1.12 measures 5 centimeters. Because we express the length of $\overline{RS}$ by $RS$ (with no bar), we write $RS = 5$ cm.

Because manufactured measuring devices such as the ruler, yardstick, and meter stick may lack perfection or be misread, there is a margin of error each time one is

> **Discover!**
>
> In converting from U.S. units to the metric system, a known conversion is the fact that 1 inch = 2.54 cm. What is the "cm" equivalent of 3.7 inches?
>
> ANSWER
>
> 9.4 cm

used. In Figure 1.12, for instance, $RS$ may actually measure 5.02 cm (and that could be rounded from 5.023 cm, etc.). Measurements are approximate, not perfect.

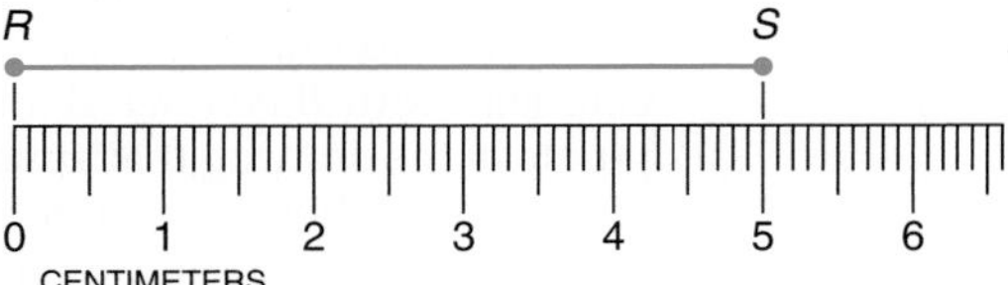


**FIGURE 1.12**

In Example 2, a ruler (not drawn to scale) is shown in Figure 1.13. In the drawing, the distance between consecutive marks on the ruler corresponds to 1 inch.

### EXAMPLE 2

In rectangle $ABCD$ of Figure 1.13, the line segments $\overline{AC}$ and $\overline{BD}$ shown are the diagonals of the rectangle. How do the lengths of the diagonals compare?

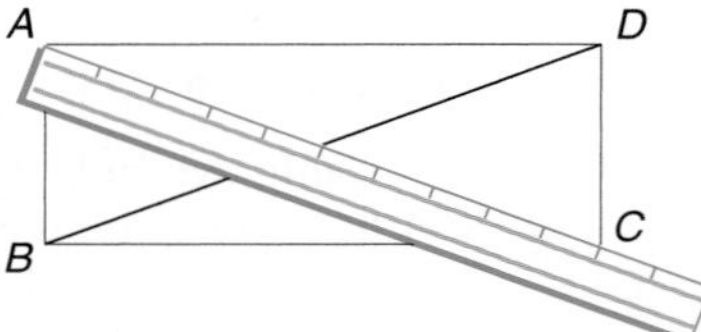


**FIGURE 1.13**

***Solution***

As intuition suggests, the lengths of the diagonals are the same. As shown, $AC = 10''$ and $BD = 10''$.

NOTE: 10″ means 10 inches, and 10′ means 10 feet.

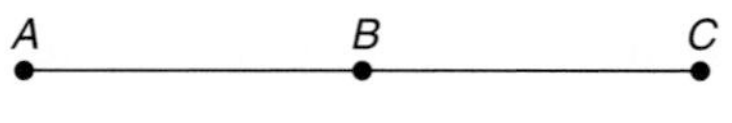

**FIGURE 1.14**

In Figure 1.14, point $B$ lies between $A$ and $C$ on $\overline{AC}$. If $AB = BC$, then $B$ is the **midpoint** of $\overline{AC}$. When $AB = BC$, the geometric figures $\overline{AB}$ and $\overline{BC}$ are said to be **congruent.** Numerical lengths may be equal, but the actual line segments (geometric figures) are congruent. The symbol for congruence is $\cong$; thus $\overline{AB} \cong \overline{BC}$ if $B$ is the midpoint of $\overline{AC}$. Example 3 emphasizes the relationship between $\overline{AB}$, $\overline{BC}$, and $\overline{AC}$ when $B$ lies between $A$ and $C$.

*Exs. 1–8*

### EXAMPLE 3

In Figure 1.15, the lengths of $\overline{AB}$ and $\overline{BC}$ are $AB = 4$ and $BC = 8$. What is $AC$, the length of $\overline{AC}$?

**FIGURE 1.15**

***Solution***

As intuition suggests, the length of $\overline{AC}$ equals $AB + BC$.
Thus $AC = 4 + 8 = 12$.

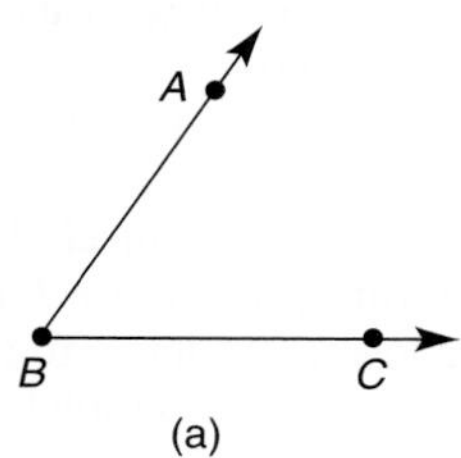


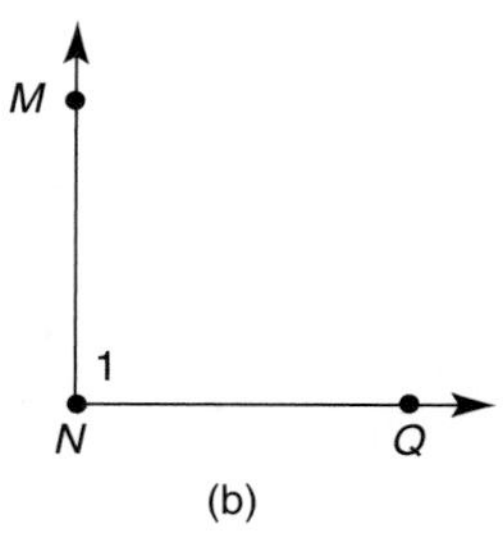


**FIGURE 1.16**

## MEASURING ANGLES

Although we formally define an angle in Section 1.4, we consider it intuitively at this time.

An angle's measure does not depend on the lengths of its sides but on the amount of opening between its sides. In Figure 1.16, the arrows on the angles' sides suggest that the sides extend indefinitely.

You cannot measure an angle with a ruler! The instrument shown in Figure 1.17 (and used in the measurement of angles) is a **protractor.** For example, you would express the measure of $\angle RST$ by writing $m\angle RST = 50°$. When a lowercase m is used before the angle symbol $\angle$, it means the measure of the angle (in degrees). Measuring the angles in Figure 1.16 with a protractor, we find that $m\angle B = 55°$ and $m\angle 1 = 90°$. If the degree symbol is missing, the measure is understood to be in degrees; thus $m\angle 1 = 90$.

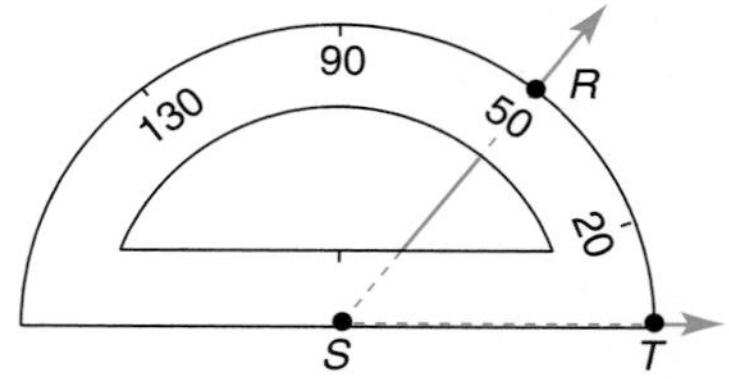


**FIGURE 1.17**

In practice, the protractor shown will measure an angle that is greater than 0° but less than or equal to 180°. To measure an angle with a protractor:

1. Place the notch of the protractor at the point where the sides of the angle meet (the vertex). See point *S* in Figure 1.18.
2. Place the edge of the protractor along a side of the angle so that the scale reads "0." See point *T* in Figure 1.18.
3. Read the angle size by reading the degree measure that corresponds to the second side of the angle. CAUTION: Many protractors show dual scales. See point *R* in Figure 1.18.

EXAMPLE 4

For Figure 1.18, find the measure of $\angle RST$.

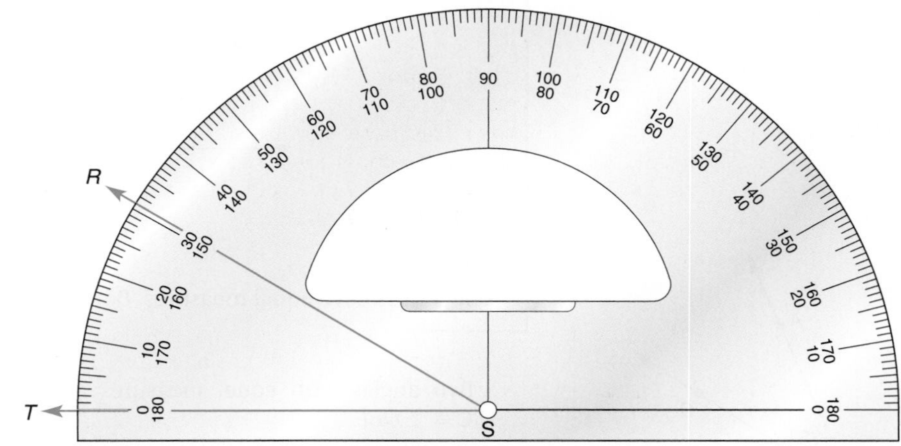


**FIGURE 1.18**

***Solution***

Using the protractor, we find that the measure of angle *RST* is 31°. (In symbols, $m\angle RST = 31°$ or $m\angle RST = 31$.)

Some protractors show a full 360° and are used to measure an angle whose measure is greater than 180°; this type of angle is known as a reflex angle.

Like measurement with a ruler, measurement with a protractor will not be perfect.

The lines on a sheet of paper in a notebook are *parallel.* Informally, **parallel** lines won't cross over each other even if they are extended indefinitely. In Figure 1.19(a), we say that lines $\ell$ and $m$ are parallel; note here the use of a lowercase letter to name a line. We may say that line segments are parallel if they are parts of parallel lines; thus $\overline{RS}$ is parallel to $\overline{MN}$ in Figure 1.19(b).

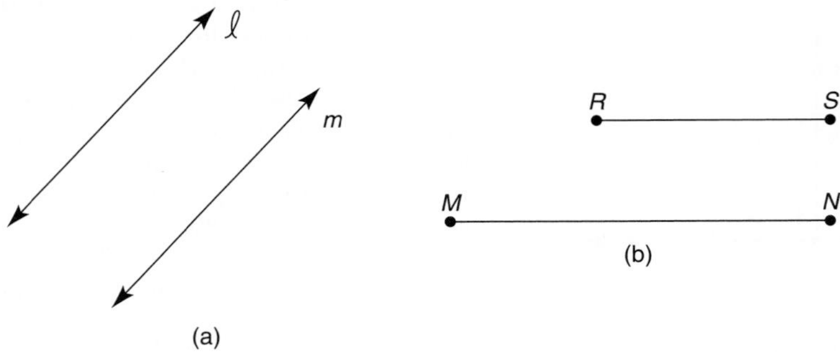


**FIGURE 1.19**

For $A = \{1, 2, 3\}$ and $B = \{6, 8, 10\}$, there are no common elements; for this reason, we say the intersection of $A$ and $B$ is the **empty set** (symbol is $\varnothing$). Just as $A \cap B = \varnothing$, the parallel lines in Figure 1.19(a) are characterized by $\ell \cap m = \varnothing$.

EXAMPLE 5

In Figure 1.20 the sides of angles $ABC$ and $DEF$ are parallel ($\overline{AB}$ to $\overline{DE}$ and $\overline{BC}$ to $\overline{EF}$). Use a protractor to decide whether these angles have equal measures.

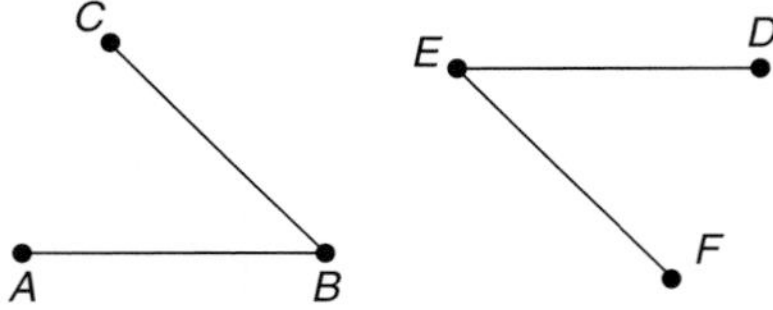


**FIGURE 1.20**

***Solution***

The angles have equal measures. Both measure 44°.

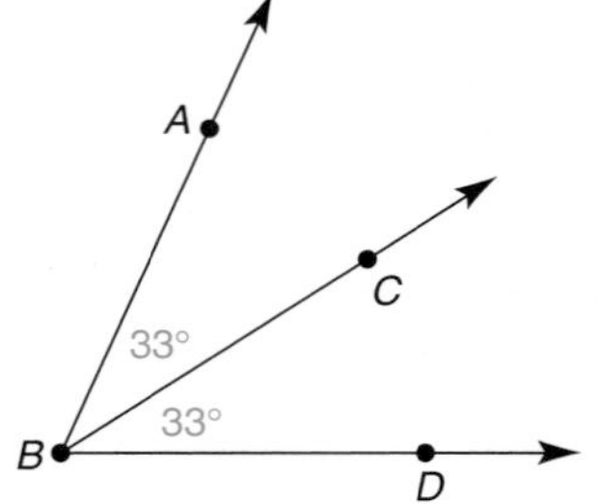


**FIGURE 1.21**

Two angles with equal measures are said to be *congruent.* In Figure 1.20, $\angle ABC \cong \angle DEF$.

In Figure 1.21, angle $ABD$ has been separated into smaller angles $ABC$ and $CBD$; if the two smaller angles are congruent (have equal measures), then angle $ABD$ has been *bisected.* In general, the word **bisect** means to separate into two parts of equal measure. Any angle having a 180° measure is called a **straight angle,** an angle whose sides are in opposite directions. See straight angle $RST$ in Figure 1.22(a). When a straight angle is bisected, as shown in Figure 1.22(b), the two angles formed are **right angles** (each measures 90°).

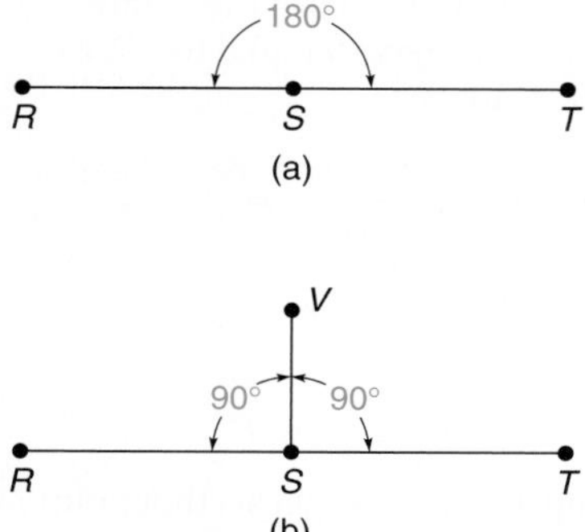


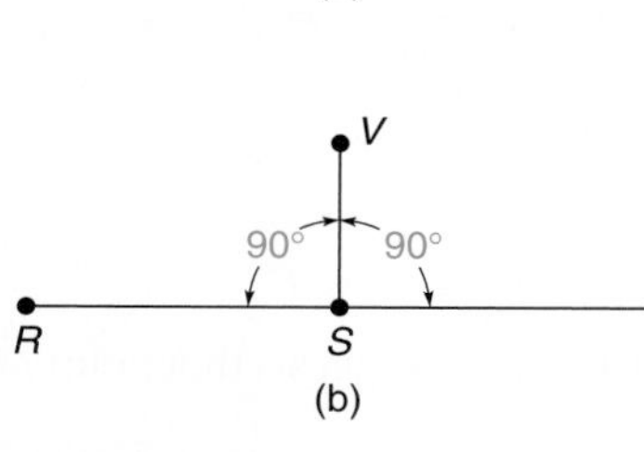


**FIGURE 1.22**

**FIGURE 1.23**

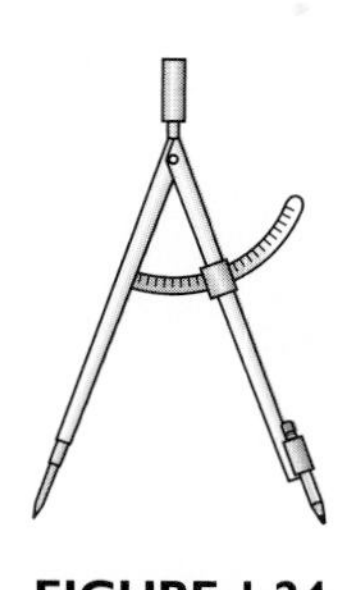

**FIGURE 1.24**

When two lines have a point in common, as in Figure 1.23, they are said to **intersect.** When two lines intersect and form congruent adjacent angles, they are said to be **perpendicular.**

EXAMPLE 6

In Figure 1.23, suppose lines $r$ and $t$ are perpendicular. What is the measure of each of the angles formed?

***Solution***

Each of the marked angles (numbered 1, 2, 3, and 4) is a right angle and measures 90°.

*Exs. 9–13*

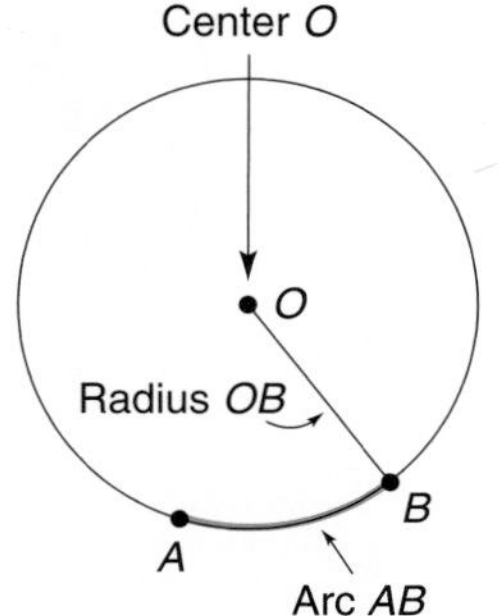


**FIGURE 1.25**

## CONSTRUCTIONS

Another tool used in geometry is the **compass.** This instrument, shown in Figure 1.24, is used to construct circles and parts of circles known as arcs. The compass and circle are discussed in the following paragraphs.

The ancient Greeks insisted that only two tools (a compass and a straightedge) be used for geometric **constructions,** which were idealized drawings assuming perfection in the use of these tools. The compass was used to create "perfect" circles and for marking off segments of "equal" length. The straightedge could be used to pass a line through two designated points.

A **circle** is the set of all points in a plane that are at a given distance from a particular point (known as the "center" of the circle). The part of a circle between any two of its points is known as an **arc.** Any line segment joining the center to a point on the circle is a **radius** (plural, *radii*) of the circle. See Figure 1.25.

Construction 1, which follows, is quite basic and depends only on using arcs of the same radius length to construct line segments of the same length. The arcs are created by using a compass. Construction 2 is more difficult to perform and explain, so we will delay its explanation to a later chapter (see Section 3.4).

CONSTRUCTION 1:

To construct a segment congruent to a given segment.

*Given:* $\overline{AB}$ in Figure 1.26(a), as shown on page 16

*Construct:* $\overline{CD}$ on line $m$ so that $\overline{CD} \cong \overline{AB}$ (or $CD = AB$)

*Construction:* With your compass open to the length of $\overline{AB}$, place the stationary point of the compass at $C$ and mark off a length equal to $AB$, as shown in Figure 1.26(b). Then $CD = AB$.

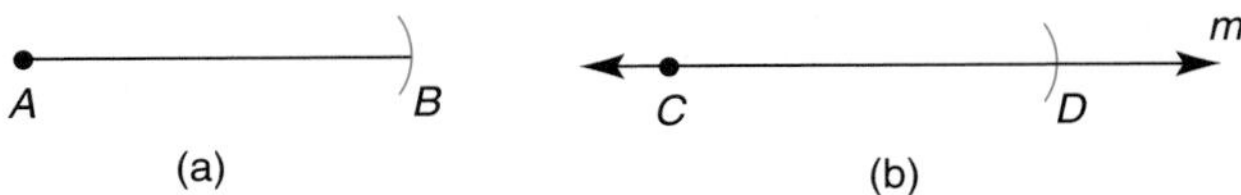


**FIGURE 1.26**

The following construction is shown step by step. Intuition suggests that point $M$ in Figure 1.27(c) is the midpoint of $\overline{AB}$.

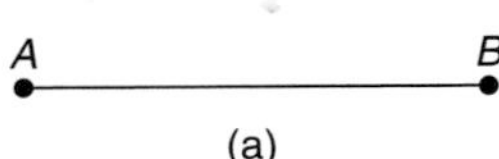


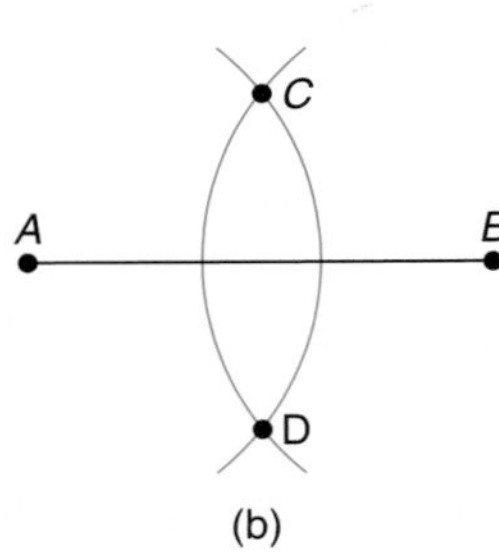


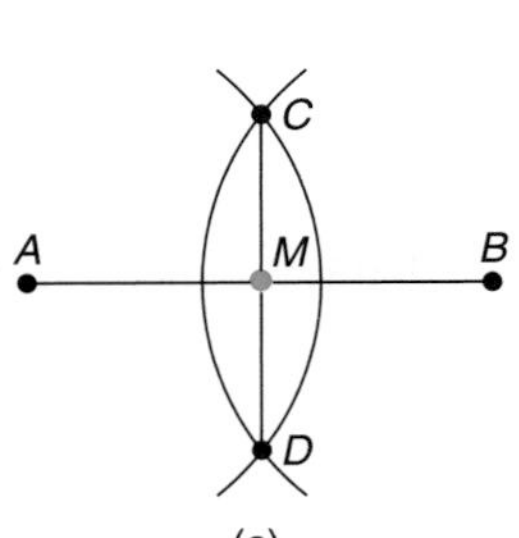


**FIGURE 1.27**

CONSTRUCTION 2:
To construct the midpoint $M$ of a given line segment $AB$.

*Given:* $\overline{AB}$ in Figure 1.27(a)

*Construct:* $M$ so that $AM = MB$ ($M$ is the midpoint of $\overline{AB}$.)

*Construction:* Figure 1.27(a): Open your compass to a length greater than one-half of $\overline{AB}$.

Figure 1.27(b): Using $A$ as the center of the arc, mark off an arc that extends both above and below segment $AB$. With $B$ as the center, and keeping the same length of radius, mark off an arc that extends above and below $\overline{AB}$ so that two points ($C$ and $D$) are determined where the arcs cross.

Figure 1.27(c): Now draw $\overline{CD}$. The point where $\overline{CD}$ crosses $\overline{AB}$ is the midpoint $M$.

*Exs. 14–17*

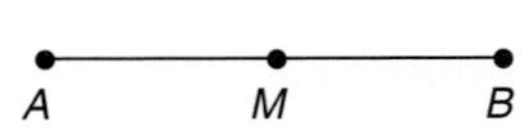


**FIGURE 1.28**

EXAMPLE 7

In Figure 1.28, $M$ is the midpoint of $\overline{AB}$.

a) Find $AM$ if $AB = 15$.
b) Find $AB$ if $AM = 4.3$.
c) Find $AB$ if $AM = 2x + 1$.

***Solution***

a) $AM$ is one-half of $AB$, so $AM = 7\frac{1}{2}$.
b) $AB$ is twice $AM$, so $AB = 2(4.3)$ or $AB = 8.6$.
c) $AB$ is twice $AM$, so $AB = 2(2x + 1)$ or $AB = 4x + 2$.

The technique from algebra used in Example 8 and also needed for Exercises 47 and 48 of this section depends on the following properties of addition and subtraction.

If $a = b$ and $c = d$, then $a + c = b + d$.

*Words:* Equals added to equals provide equal sums.

*Illustration:* Since $0.5 = \frac{5}{10}$ and $0.2 = \frac{2}{10}$, it follows that $0.5 + 0.2 = \frac{5}{10} + \frac{2}{10}$.

If $a = b$ and $c = d$, then $a - c = b - d$.

*Words:* Equals subtracted from equals provide equal differences.

*Illustration:* Since $0.5 = \frac{5}{10}$ and $0.2 = \frac{2}{10}$, it follows that $0.5 - 0.2 = \frac{5}{10} - \frac{2}{10}$.

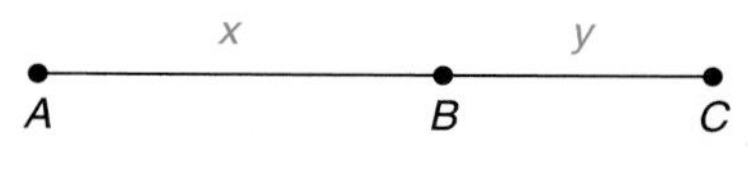


**FIGURE 1.29**

EXAMPLE 8

In Figure 1.29, point $B$ lies on $\overline{AC}$ between $A$ and $C$. If $AC = 10$ and $AB$ is 2 units longer than $BC$, find the length $x$ of $\overline{AB}$ and the length $y$ of $\overline{BC}$.

***Solution***

Because $AB + BC = AC$, we have $x + y = 10$.
Because $AB - BC = 2$, we have $x - y = 2$.
Adding the left and right sides of these equations,

$$\begin{array}{rl} x + y &= 10 \\ x - y &= \ \ 2 \\ \hline 2x \qquad &= 12 \end{array} \qquad \text{so } x = 6.$$

If $x = 6$, then $x + y = 10$ becomes $6 + y = 10$ and $y = 4$.
Thus $AB = 6$ and $BC = 4$.

*Exs. 18, 19*

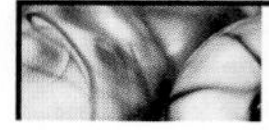

## 1.2 Exercises

1. If line segment $AB$ and line segment $CD$ are drawn to scale, what does intuition tell you about the lengths of these segments?

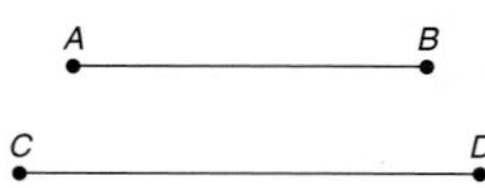


2. If angles $ABC$ and $DEF$ were measured with a protractor, what does intuition tell you about the degree measures of these angles?

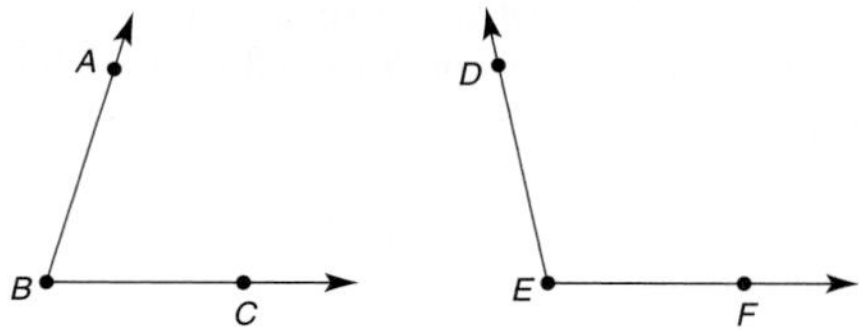


3. How many endpoints does a line segment have? How many midpoints does a line segment have?

4. Do the points $A$, $B$, and $C$ appear to be collinear?

*Exercises 4–6*

5. How many lines can be drawn to contain both points $A$ and $B$? How many lines can be drawn to contain points $A$, $B$, and $C$?

6. Consider noncollinear points $A$, $B$, and $C$. If each line must contain two of the points, what is the total number of lines that are determined by these points?

7. Name all the angles in the figure.

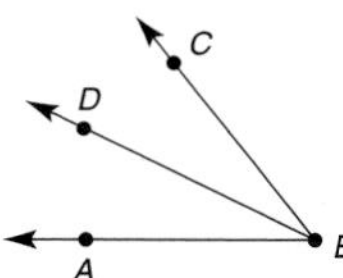


8. Which of the following measures can an angle have? 23°, 90°, 200°, 110.5°, −15°

9. Must two different points be collinear? Must three or more points be collinear? Can three or more points be collinear?

10. Which symbol(s) correctly expresses the order in which the points $A$, $B$, and $X$ lie on the given line, $A$-$X$-$B$ or $A$-$B$-$X$?

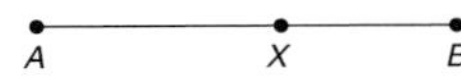


11. Which symbols correctly name the angle shown? $\angle ABC$, $\angle ACB$, $\angle CBA$

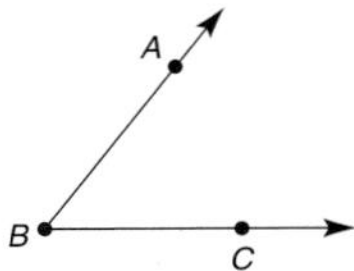


12. A triangle is named $\triangle ABC$. Can it also be named $\triangle ACB$? Can it be named $\triangle BAC$?

13. Consider rectangle $MNPQ$. Can it also be named rectangle $PQMN$? Can it be named rectangle $MNQP$?

14. Suppose $\angle ABC$ and $\angle DEF$ have the same measure. Which statements are expressed correctly?

a. $m\angle ABC = m\angle DEF$ b. $\angle ABC = \angle DEF$
c. $m\angle ABC \cong m\angle DEF$ d. $\angle ABC \cong \angle DEF$

15. Suppose $\overline{AB}$ and $\overline{CD}$ have the same length. Which statements are expressed correctly?

a. $AB = CD$ b. $\overline{AB} = \overline{CD}$
c. $AB \cong CD$ d. $\overline{AB} \cong \overline{CD}$

16. When two lines cross (intersect), they share exactly one point in common. In the drawing, what is the point of intersection? How do the measures of $\angle 1$ and $\angle 2$ compare?

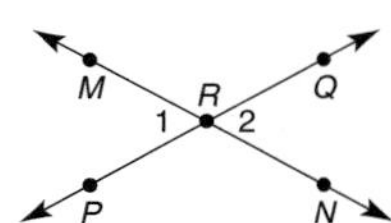


17. Judging from the ruler shown (not to scale), estimate the measure of each line segment.

a) $AB$ b) $CD$

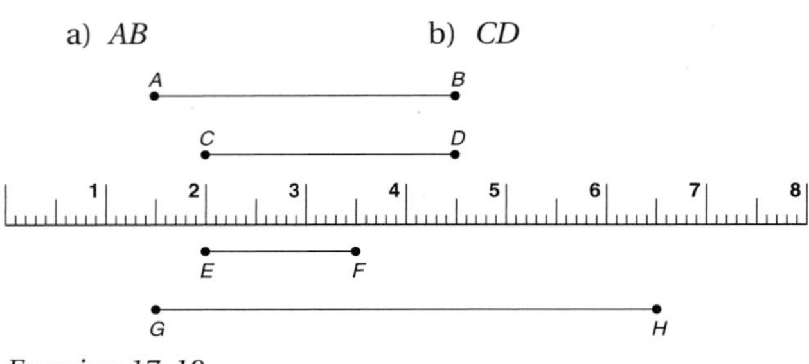


*Exercises 17, 18*

18. Judging from the ruler, estimate the measure of each line segment.

a) $EF$ b) $GH$

19. Judging from the protractor provided, estimate the measure of each angle to the nearest multiple of 5° (e.g., 20°, 25°, 30°, etc.).

a) $m\angle 1$ b) $m\angle 2$

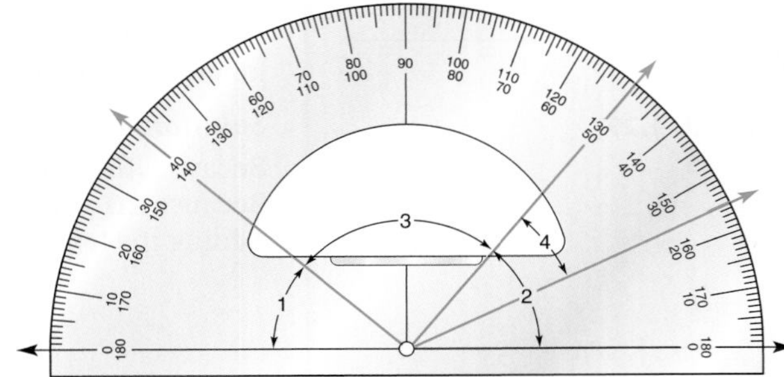


*Exercises 19, 20*

20. Judging from the protractor, estimate the measure of each angle to the nearest multiple of 5° (e.g., 20°, 25°, 30°, etc.).

a) $m\angle 3$ b) $m\angle 4$

21. Consider the square at the right, $RSTV$. It has four right angles and four sides of the same length. How are sides $\overline{RS}$ and $\overline{ST}$ related? How are sides $\overline{RS}$ and $\overline{VT}$ related?

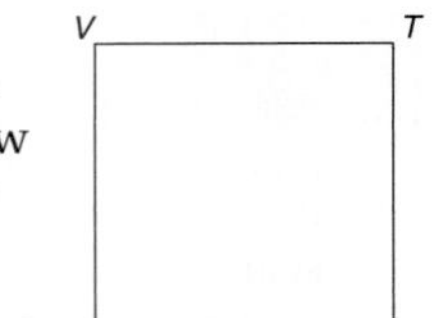


22. Square $RSTV$ has diagonals $\overline{RT}$ and $\overline{SV}$ (not shown). If the diagonals are drawn, how will their lengths compare? Do the diagonals of a square appear to be perpendicular?

23. Use a compass to draw a circle. Draw a radius, a line segment that connects the center to a point on the circle. Measure the length of the radius. Draw other radii and find their lengths. How do the lengths of the radii compare?

24. Use a compass to draw a circle of radius 1 inch. Draw a chord, a line segment that joins two points on the circle. Draw other chords and measure their lengths. What is the largest possible length of a chord in this circle?

25. The sides of the pair of angles are parallel. Are $\angle 1$ and $\angle 2$ congruent?

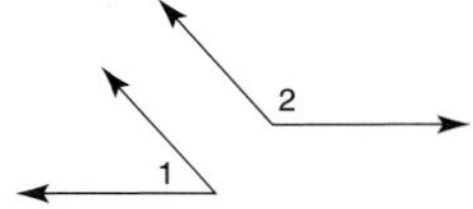

26. The sides of the pair of angles are parallel. Are $\angle 3$ and $\angle 4$ congruent?

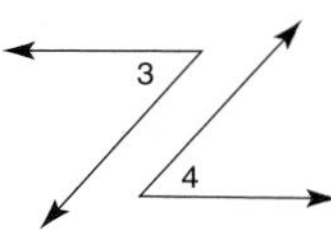


27. The sides of the pair of angles are perpendicular. Are $\angle 5$ and $\angle 6$ congruent?

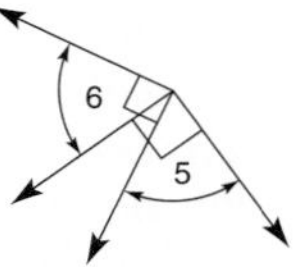


28. The sides of the pair of angles are perpendicular. Are $\angle 7$ and $\angle 8$ congruent?

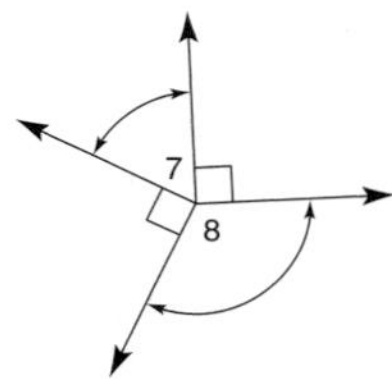


29. On a piece of paper, use your compass to construct a triangle that has two sides of the same length. Cut the triangle out of the paper and fold the triangle in half so that the congruent sides coincide (one lies over the other). What seems to be true of two angles of that triangle?

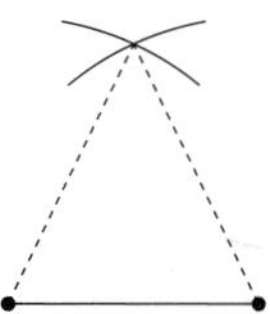

30. On a piece of paper, use your protractor to draw a triangle that has two angles of the same measure. Cut the triangle out of the paper and fold the triangle in half so that the angles of equal measure coincide (one lies over the other). What seems to be true of two of the sides of that triangle?

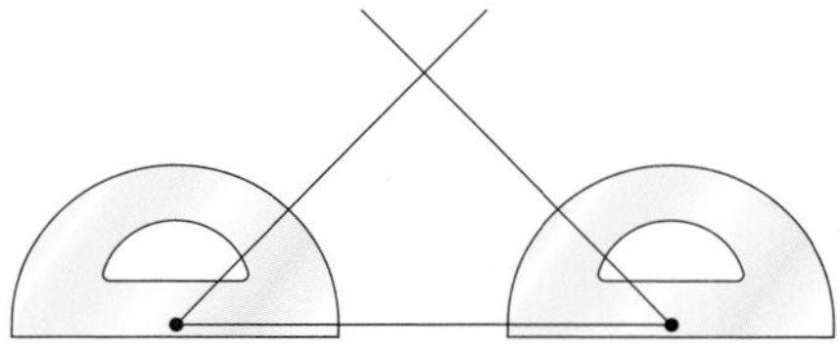

31. A trapezoid is a four-sided figure that contains one pair of parallel sides. Which sides of the trapezoid *MNPQ* appear to be parallel?

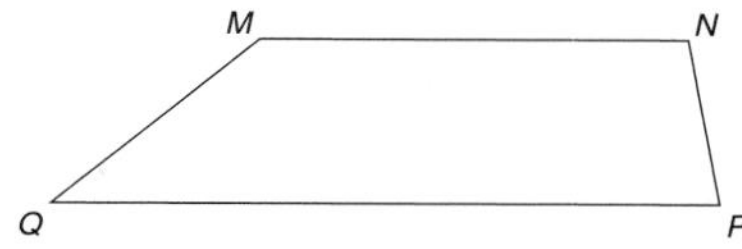


32. In the rectangle shown, what is true of the lengths of each pair of opposite sides?

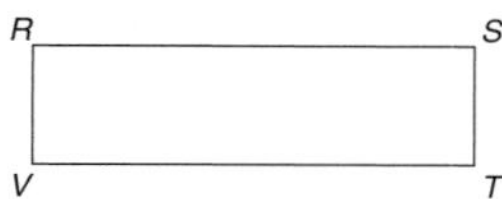


33. A line segment is bisected if its two parts have the same length. Which line segment, $\overline{AB}$ or $\overline{CD}$, is bisected at point *X*?

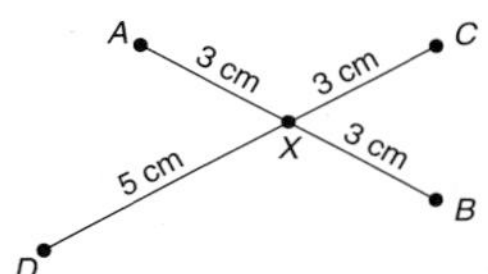


34. An angle is bisected if its two parts have the same measure. Use three letters to name the angle that is bisected.

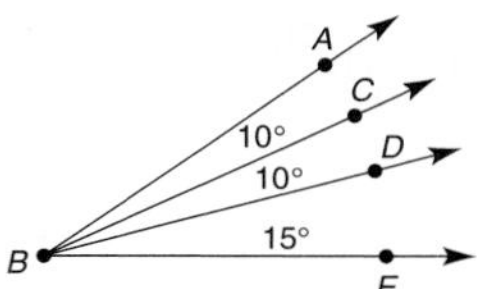


*In Exercises 35 to 38, where A-B-C on $\overline{AC}$, it follows that $AB + BC = AC$.*

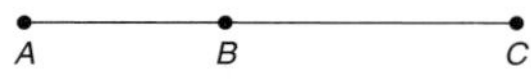


*Exercises 35–38*

35. Find $AC$ if $AB = 9$ and $BC = 13$.
36. Find $AB$ if $AC = 25$ and $BC = 11$.
37. Find $x$ if $AB = x$, $BC = x + 3$, and $AC = 21$.
38. Find an expression for $AC$ (the length of $\overline{AC}$) if $AB = x$ and $BC = y$.
39. $\angle ABC$ is a straight angle. Using your protractor, you can show that $m\angle 1 + m\angle 2 = 180°$. Find $m\angle 1$ if $m\angle 2 = 56°$.

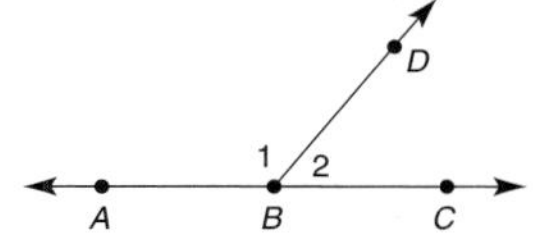


*Exercises 39, 40*

40. Find $m\angle 1$ if $m\angle 1 = 2x$ and $m\angle 2 = x$.
(HINT: See Exercise 39 on page 19.)

*In Exercises 41 to 44, $m\angle 1 + m\angle 2 = m\angle ABC$.*

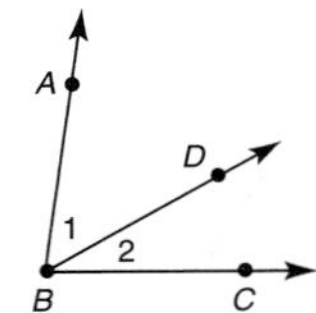


*Exercises 41–44*

41. Find $m\angle ABC$ if $m\angle 1 = 32°$ and $m\angle 2 = 39°$.

42. Find $m\angle 1$ if $m\angle ABC = 68°$ and $m\angle 1 = m\angle 2$.

43. Find $x$ if $m\angle 1 = x$, $m\angle 2 = 2x + 3$, and $m\angle ABC = 72°$.

44. Find an expression for $m\angle ABC$ if $m\angle 1 = x$ and $m\angle 2 = y$.

45. A compass was used to mark off three congruent segments, $\overline{AB}$, $\overline{BC}$, and $\overline{CD}$. Thus $\overline{AD}$ has been trisected at points $B$ and $C$. If $AD = 32.7$, how long is $\overline{AB}$?

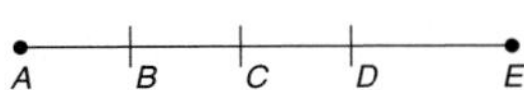


46. Use your compass and straightedge to bisect $\overline{EF}$.

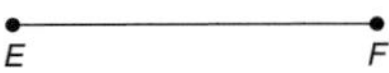


*47. In the figure, $m\angle 1 = x$ and $m\angle 2 = y$. If $x - y = 24°$, find $x$ and $y$.
(HINT: $m\angle 1 + m\angle 2 = 180°$.)

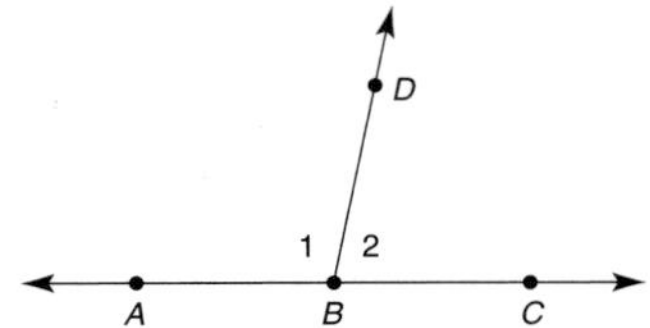


*48. In the drawing, $m\angle 1 = x$ and $m\angle 2 = y$. If $m\angle RSV = 67°$ and $x - y = 17°$, find $x$ and $y$.
(HINT: $m\angle 1 + m\angle 2 = m\angle RSV$.)

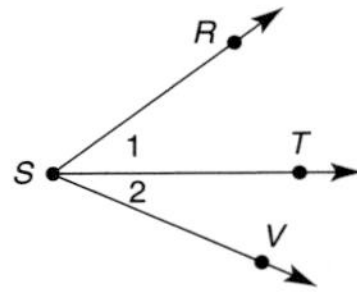


*For Exercises 49 and 50, use the following information. Relative to its point of departure or some other point of reference, the angle that is used to locate the position of a ship or airplane is called its bearing. The bearing may also be used to describe the direction in which the airplane or ship is moving. By using an angle between 0° and 90°, a bearing is measured from the North-South line toward the East or West. In the diagram, airplane A (which is 250 miles from Chicago's O'Hare airport's control tower) has a bearing of S 53° W.*

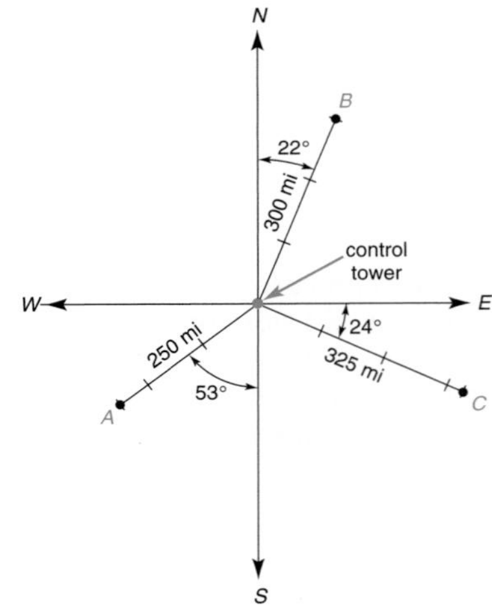


*Exercises 49, 50*

49. Find the bearing of airplane $B$ relative to the control tower.

50. Find the bearing of airplane $C$ relative to the control tower.

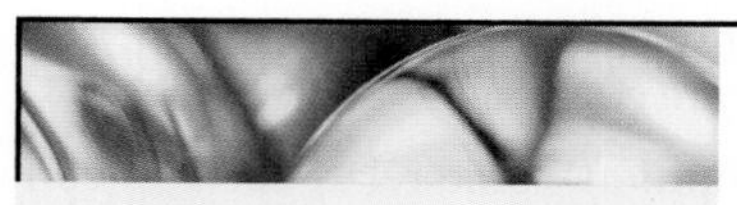

# 1.3 Early Definitions and Postulates

**KEY CONCEPTS**

Mathematical System • Postulate • Theorem • Ruler Postulate • Distance • Segment-Addition Postulate • Ray • Opposite Rays • Coplanar Points • Space

## A MATHEMATICAL SYSTEM

Like algebra, the branch of mathematics called geometry determines a **mathematical system.** Each system has its own vocabulary and properties. In the formal study of a mathematical system, we begin with undefined terms. Building on this foundation, we can then define additional terms. Once the terminology is sufficiently developed, certain properties (characteristics) of the system become apparent. These properties are known as **axioms** or **postulates** of the system; more generally, such statements are called **assumptions.** Once we have developed a vocabulary and accepted certain postulates, many principles follow logically when we apply deductive methods. These statements can be proved and are called **theorems.** The following box summarizes the components of a mathematical system (sometimes called a logical system or deductive system).

FOUR PARTS OF A MATHEMATICAL SYSTEM

1. Undefined terms
2. Defined terms } vocabulary
3. Axioms or postulates
4. Theorems } principles

## CHARACTERISTICS OF A GOOD DEFINITION

Terms such as *point, line,* and *plane* are classified as undefined because they do not fit into any set or category that has been previously determined. Terms that *are* defined, however, should be described precisely. *But what is a good definition?* A good definition is like a mathematical equation written using words. A good definition must possess four characteristics. We illustrate this with a term that we will redefine at a later time.

**Discover!**

Although we cannot actually define *line* and *plane,* we can compare them in the following analogy. Please complete: A ? is to *straight* as a ? is to *flat.*

ANSWERS

Line; plane

DEFINITION: An **isosceles triangle** is a triangle that has two congruent sides.

In the definition, notice that: (1) The term being defined—*isosceles triangle*—is named. (2) The term being defined is placed into a larger category (a type of *triangle*). (3) The distinguishing quality (that two sides of the triangle are congruent) is included. (4) The *reversibility* of the definition is illustrated by these statements:

"If a triangle is isosceles, then it has two congruent sides."
"If a triangle has two congruent sides, then it is an isosceles triangle."

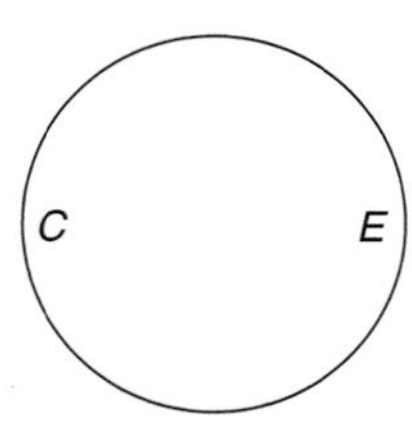


**FIGURE 1.30**

In summary, a good definition will possess these qualities:

1. It names the term being defined.
2. It places the term into a set or category.
3. It distinguishes the defined term from other terms without providing unnecessary facts.
4. It is reversible.

In many textbooks, it is common to use the phrase "if and only if" in expressing the definition of a term. For instance, we could define *congruent angles* by saying that two angles are congruent if and only if these angles have equal measures. The "if and only if" statement has the following dual meaning:

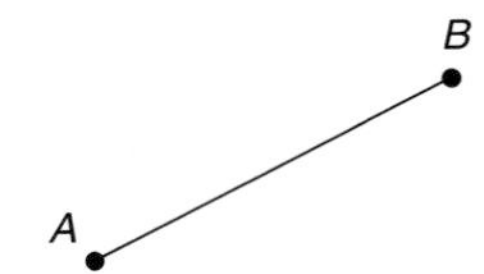


FIGURE 1.31

"If two angles are congruent, then they have equal measures."
"If two angles have equal measures, then they are congruent."

When represented by a Venn Diagram, this definition would relate set $C$ = {congruent angles} to set $E$ = {angles with equal measures} as shown in Figure 1.30 on page 21. The sets $C$ and $E$ are identical and are known as **equivalent sets.**

Once undefined terms have been described, they become the building blocks for other terminology. In this textbook, primary terms are defined within boxes, whereas related terms are often boldfaced and defined within statements. Consider the following definition (see Figure 1.31).

*Exs. 1–4*

DEFINITION: A **line segment** is the part of a line that consists of two points, known as *endpoints*, and all points between them.

Considering this definition, we see that

1. The term being defined, *line segment*, is clearly present in the definition.
2. A line segment is defined as part of a line (a category).
3. The definition distinguishes the line segment as a specific part of a line.
4. The definition is reversible.
   i) A line segment is the part of a line between and including two points.
   ii) The part of a line between and including two points is a line segment.

*Geometry in the Real World*

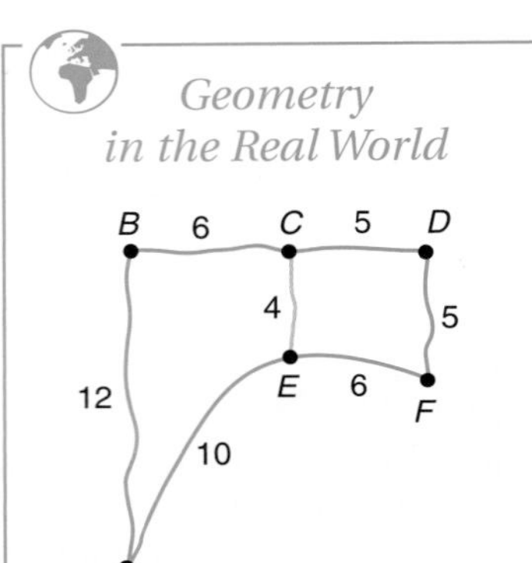


On the road map, driving distances between towns are shown. In traveling from town $A$ to town $D$, which path traverses the least distance?

SOLUTION $A$ to $E$, $E$ to $C$, $C$ to $D$: $10 + 4 + 5 = 19$

## INITIAL POSTULATES

POSTULATE 1:
Through two distinct points, there is exactly one line.

Postulate 1 is sometimes stated in the form "Two points determine a line." See Figure 1.32, in which points $C$ and $D$ determine exactly one line, namely $\overleftrightarrow{CD}$. Of course, Postulate 1 also implies that there is a unique line segment determined by two distinct points used as endpoints. Recall Figure 1.31, in which points $A$ and $B$ determine $\overline{AB}$.

NOTE: In geometry, the reference numbers used with postulates need not be memorized.

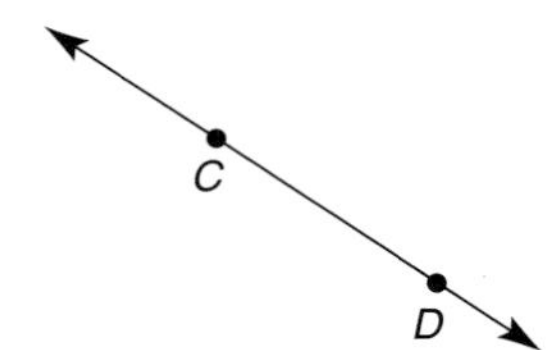


FIGURE 1.32

EXAMPLE 1

In Figure 1.33, how many distinct lines can be drawn through

a) point $A$?
b) both points $A$ and $B$ at the same time?
c) all points $A$, $B$, and $C$ at the same time?

***Solution***

a) An infinite (countless) number
b) Exactly one
c) No line contains all three points.

A
B
C

FIGURE 1.33

Recall from Section 1.2 that the symbol for line segment $AB$, named by its endpoints, is $\overline{AB}$. Omission of the bar from $\overline{AB}$, as in $AB$, means that we are considering the *length* of the segment. These symbols are summarized in Table 1.1.

**TABLE 1.1**

| *Symbol* | *Words for Symbol* | *Geometric Figure* |
|---|---|---|
| $\overleftrightarrow{AB}$ | Line $AB$ |  |
| $\overline{AB}$ | Line segment $AB$ |  |
| $AB$ | Length of segment $AB$ | A number |

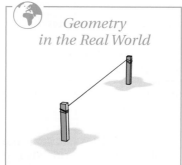

A ruler is used to measure the length of any line segment like $\overline{AB}$. This length may be represented by $AB$ or $BA$ (the order of $A$ and $B$ is not important). $AB$ must be a positive number.

*Geometry in the Real World*

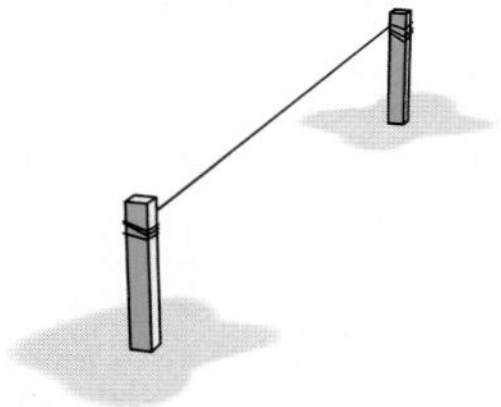

In construction, a string joins two stakes. The line determined is described in Postulate 1 on the previous page.

POSTULATE 2: **(Ruler Postulate)**
The measure of any line segment is a unique positive number.

We wish to call attention to the term *unique* and to the general notion of uniqueness. Postulate 2 implies the following:

1. There exists a number measure for each line segment.
2. Only *one* measure is permissible.

Characteristics 1 and 2 are both necessary for uniqueness! Other phrases that may replace the term *unique* include

One and only one
Exactly one
One and no more than one

A more accurate claim than the commonly heard statement "The shortest distance between two points is a straight line" is found in the following definition.

DEFINITION: The **distance** between two points $A$ and $B$ is the length of the line segment $\overline{AB}$ that joins the two points.

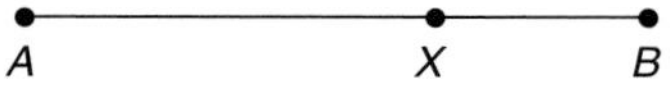

**FIGURE 1.34**

As we saw in Section 1.2, there is a relationship between the lengths of the line segments determined in Figure 1.34. This relationship is stated in the third postulate. Note that postulate numbers need not be memorized. It is the title and meaning that are important!

POSTULATE 3: **(Segment-Addition Postulate)**
If $X$ is a point of $\overline{AB}$ and $A$-$X$-$B$, then $AX + XB = AB$.

*Technology Exploration*

Use software if available.

1) Draw line segment $\overline{XY}$.
2) Choose point $P$ on $\overline{XY}$.
3) Measure $\overline{XP}$, $\overline{PY}$, and $\overline{XY}$.
4) Show that $XP + PY = XY$.

EXAMPLE 2

In Figure 1.34, find $AB$ if

a) $AX = 7.32$ and $XB = 6.19$. b) $AX = 2x + 3$ and $XB = 3x - 7$.

***Solution***

a) $AB = 7.32 + 6.19$, so $AB = 13.51$.
b) $AB = (2x + 3) + (3x - 7)$, so $AB = 5x - 4$.

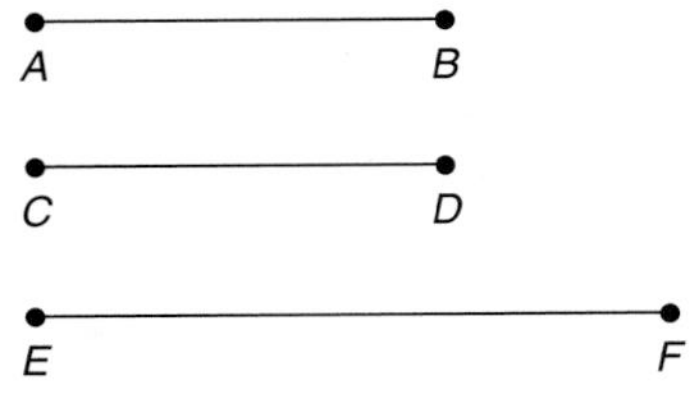


**FIGURE 1.35**

DEFINITION: **Congruent (≅) segments** are two segments that have the same length.

In general, geometric figures that can be made to coincide (fit perfectly one on top of the other) are said to be **congruent.** The symbol ≅ is a combination of the symbol ~, which means that the figures have the same shape, and =, which means that the corresponding parts of the figures have the same measure. In Figure 1.35, $\overline{AB} \cong \overline{CD}$, but $\overline{AB} \ncong \overline{EF}$. Does it appear that $\overline{CD} \cong \overline{EF}$?

EXAMPLE 3

In the U.S. system of measures, 1 foot = 12 inches. If $AB = 2.5$ feet and $CD = 2$ feet 6 inches, are $\overline{AB}$ and $\overline{CD}$ congruent?

***Solution***

Yes, $\overline{AB} \cong \overline{CD}$ because 2.5 feet = 2 feet + 0.5 feet or 2 feet + 0.5(12 inches) or 2 feet 6 inches.

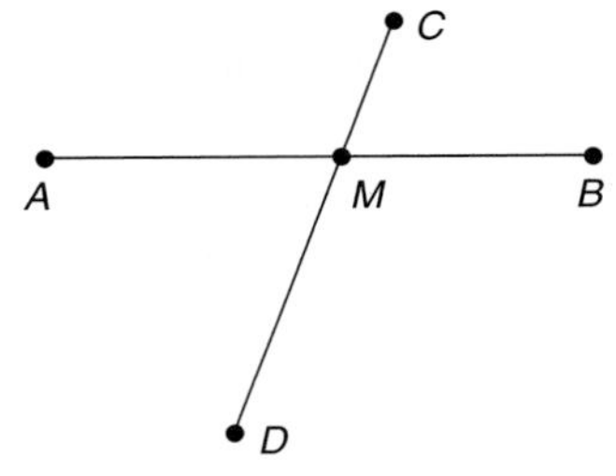


**FIGURE 1.36**

DEFINITION: The **midpoint** of a line segment is the point that separates the line segment into two congruent parts.

In Figure 1.36, if *A*, *M*, and *B* are collinear and $\overline{AM} \cong \overline{MB}$, then *M* is the **midpoint** of $\overline{AB}$. Equivalently, *M* is the midpoint of $\overline{AB}$ if $AM = MB$. Also, if $\overline{AM} \cong \overline{MB}$, then $\overline{CD}$ is described as a **bisector** of $\overline{AB}$. Under what condition would $\overline{AB}$ be a bisector of $\overline{CD}$?

Given that *M* is the midpoint of $\overline{AB}$ in Figure 1.36, we can draw these conclusions:

$AM = MB \qquad AB = 2\,(AM)$

$AM = \frac{1}{2}(AB) \qquad AB = 2\,(MB)$

$MB = \frac{1}{2}(AB)$

EXAMPLE 4

*Given:* *M* is the midpoint of $\overline{EF}$ (not shown). $EM = 3x + 9$ and $MF = x + 17$

*Find:* *x* and *EM*

***Solution***

Because *M* is the midpoint of $\overline{EF}$, $EM = MF$. Then

$$\begin{aligned} 3x + 9 &= x + 17 \\ 2x + 9 &= 17 \\ 2x &= 8 \\ x &= 4 \end{aligned}$$

By substitution, $EM = 3(4) + 9 = 12 + 9 = 21$.

**Discover!**

Assume that *M* is the midpoint of $\overline{AB}$ in Figure 1.36. Can you also conclude that *M* is the midpoint of $\overline{CD}$?

ANSWER

No

In geometry, the word **union** is used to describe the joining or combining of two figures or sets of points.

DEFINITION: **Ray** *AB*, denoted by $\overrightarrow{AB}$, is the union of $\overline{AB}$ and all points *X* on $\overleftrightarrow{AB}$ such that *B* is between *A* and *X*.

In Figure 1.37, $\overleftrightarrow{AB}$, $\overrightarrow{AB}$, and $\overrightarrow{BA}$ are shown; note that $\overrightarrow{AB}$ and $\overrightarrow{BA}$ are not the same ray.

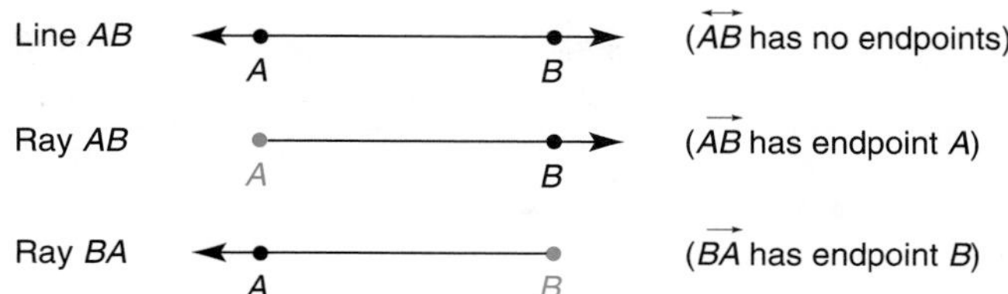


**FIGURE 1.37**

**Opposite rays** are two rays that share a common endpoint; the union of opposite rays is a straight line. In Figure 1.39(a), $\overrightarrow{BA}$ and $\overrightarrow{BC}$ are opposite rays.

The **intersection** of two geometric figures is the set of points that the two figures have in common. In everyday life, the intersection of Bradley Avenue and Neil Street is the part of the roadway that the two roads have in common (Figure 1.38).

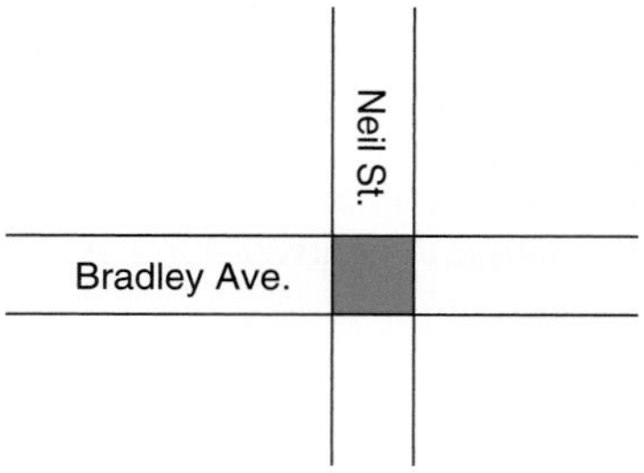

**FIGURE 1.38**

POSTULATE 4:
If two lines intersect, they intersect at a point.

When two lines share two (or more) points, the lines coincide; in this situation, we say there is only one line. In Figure 1.39(a), $\overleftrightarrow{AB}$ and $\overleftrightarrow{BC}$ are the same as $\overleftrightarrow{AC}$. In Figure 1.39(b), lines $\ell$ and $m$ intersect at point $P$.

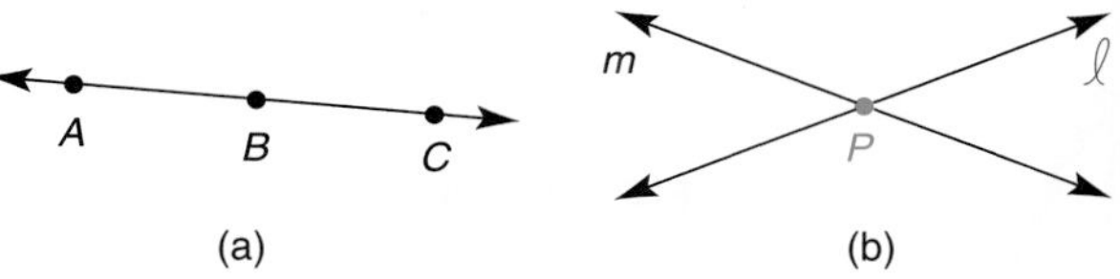


**FIGURE 1.39**

DEFINITION: **Parallel lines** are lines that lie in the same plane but do not intersect.

In Figure 1.40, $\ell$ and $n$ are parallel; in symbols, $\ell \parallel n$ and $\ell \cap n = \emptyset$. However, $\ell$ and $m$ are not parallel and $\ell \cap m = A$; so, $\ell \nparallel m$.

*Exs. 5–12*

EXAMPLE 5

In Figure 1.40, what is the intersection of

a) lines $\ell$ and $m$? b) line $\ell$ and line $n$?

***Solution***

a) Point $A$ b) Parallel lines do not intersect.

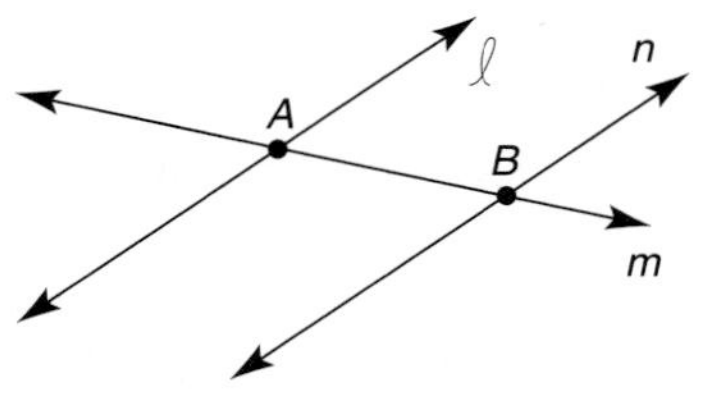


**FIGURE 1.40**

Another undefined term in geometry is **plane.** A plane is two-dimensional; that is, it has infinite length and infinite width, but no thickness. Except for its limited size, a flat surface such as the top of a table could be used as an example of a plane. An uppercase letter can be used to name a plane. Because a plane (like a line) is infinite, we can show only a portion of the plane or planes, as in Figure 1.41 on page 26.

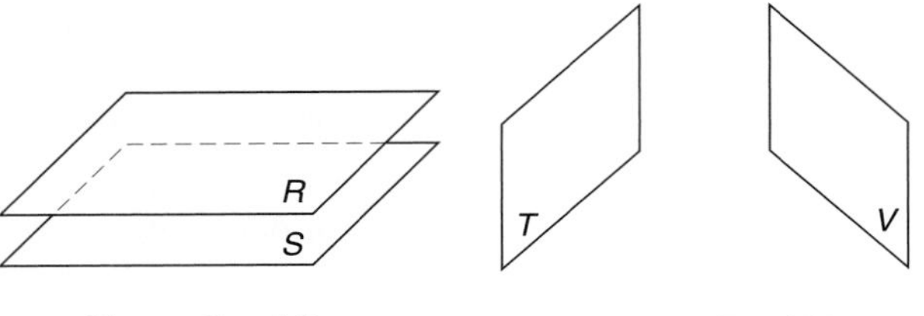


**FIGURE 1.41**

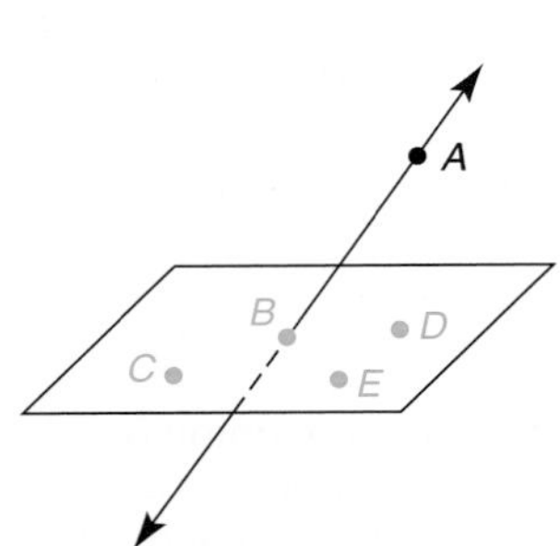


**FIGURE 1.42**

A plane is two-dimensional, consists of an infinite number of points, and contains an infinite number of lines. Two distinct points may determine (or "fix") a line; likewise, exactly three noncollinear points determine a plane. Just as collinear points lie on the same line, **coplanar points** lie in the same plane. In Figure 1.42, points *B*, *C*, *D*, and *E* are coplanar, whereas *A*, *B*, *C*, and *D* are noncoplanar.

In this book, points shown in figures are assumed to be coplanar unless otherwise stated. For instance, points *A*, *B*, *C*, *D*, and *E* are coplanar in Figure 1.43(a), as are points *F*, *G*, *H*, *J*, and *K* in Figure 1.43(b).

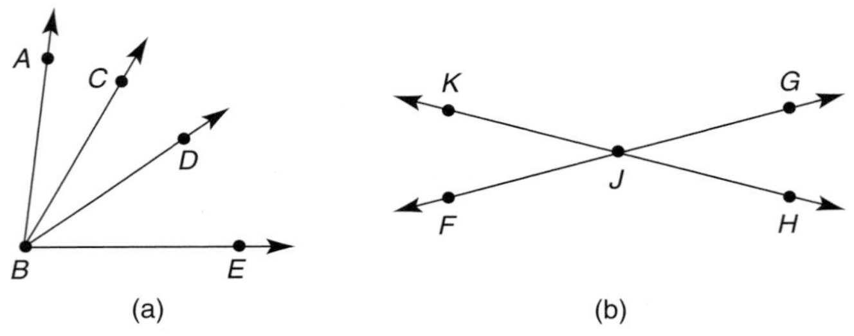


**FIGURE 1.43**

Geometry in the Real World

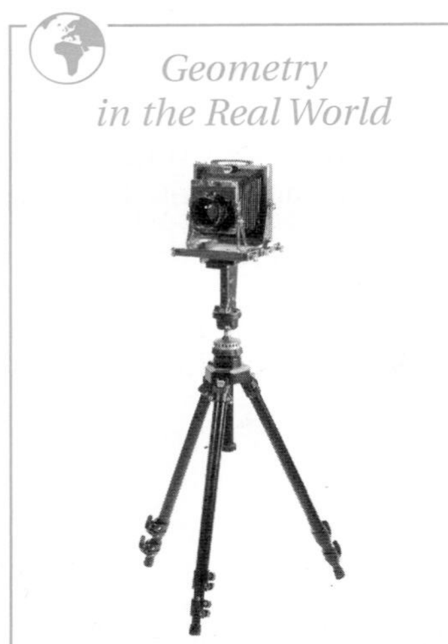

The tripod illustrates Postulate 5 in that the three points at the base enable the unit to sit level.

POSTULATE 5:
Through three noncollinear points, there is exactly one plane.

On the basis of Postulate 5, we can see why a three-legged table sits evenly, but a four-legged table would "wobble" if the legs were of unequal length.

**Space** is the set of all possible points. It is three-dimensional, having qualities of length, width, and depth. When two planes intersect in space, their intersection is a line. An opened greeting card suggests this relationship, as does Figure 1.44(a). This notion gives rise to our next postulate.

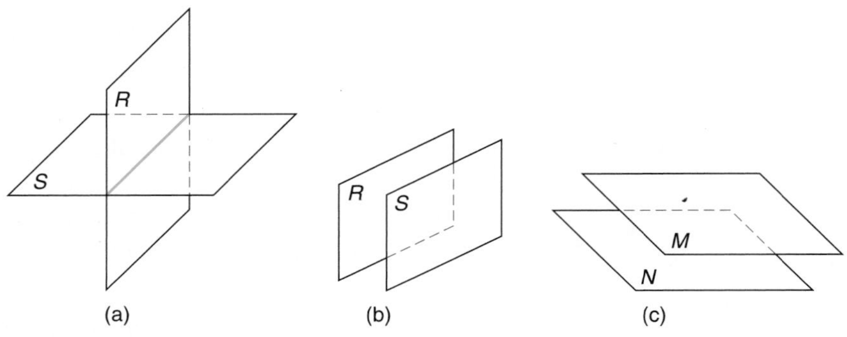


**FIGURE 1.44**

**POSTULATE 6:**
If two distinct planes intersect, then their intersection is a line.

The intersection of two planes is infinite because it is a line. [See Figure 1.44(a).] If two planes do not intersect, then they are **parallel.** The parallel **vertical** planes $R$ and $S$ in Figure 1.44(b) may remind you of the opposite walls of your classroom. The parallel horizontal planes $M$ and $N$ in Figure 1.44(c) suggest the relationship between ceiling and floor.

Imagine a plane and two points of that plane, points $A$ and $B$. Now think of the line containing the two points and the relationship of $\overleftrightarrow{AB}$ to the plane. Perhaps your conclusion can be summed up as follows.

**POSTULATE 7:**
Given two distinct points in a plane, the line containing these points also lies in the plane.

SSG *Exs. 13–16*

Because the uniqueness of the midpoint of a line segment can be justified, we call the following statement a theorem.The "proof" of the theorem is found in Section 2.2.

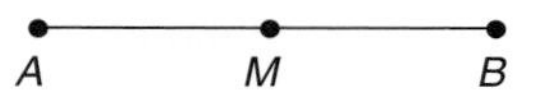


**FIGURE 1.45**

**THEOREM 1.3.1:** The midpoint of a line segment is unique.

If $M$ is the midpoint of $\overline{AB}$ in Figure 1.45, then no other point can separate $\overline{AB}$ into two congruent parts. The proof of this theorem is based on the Ruler Postulate. $M$ is *the* point that is located $\frac{1}{2}(AB)$ units from $A$ (and from $B$).

The numbering system used to identify Theorem 1.3.1 need not be memorized. However, this theorem number may be used in a later reference. The numbering system works as follows:

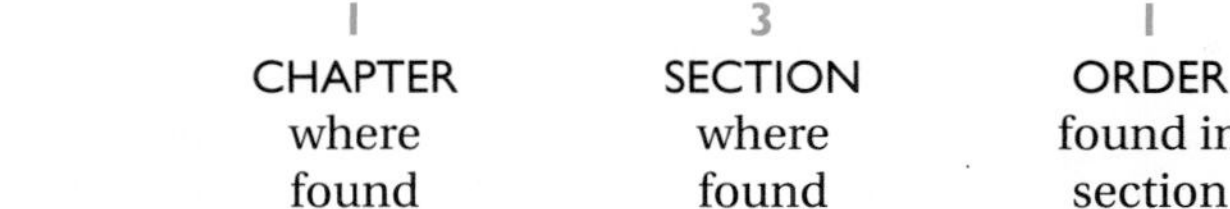

*Exs. 17–20*

A summary of the theorems presented in this textbook appears at the end of the book.

## 1.3 Exercises

*In Exercises 1 and 2, complete the statement.*

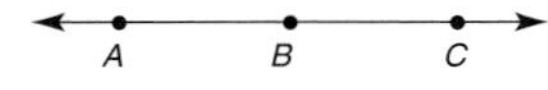


*Exercises 1, 2*

1. $AB + BC = \underline{\ ?\ }$
2. If $AB = BC$, then $B$ is the $\underline{\ ?\ }$ of $\overline{AC}$.

*In Exercises 3 and 4, use the fact that 1 foot = 12 inches.*

3. Convert 6.25 feet to a measure in inches.
4. Convert 52 inches to a measure in feet and inches.

*In Exercises 5 and 6, use the fact that 1 meter ≈ 3.28 feet (measure is approximate).*

5. Convert $\frac{1}{2}$ meter to feet.
6. Convert 16.4 feet to meters.
7. In the figure, the 15-mile road from $A$ to $C$ is under construction. A detour from $A$ to $B$ of 5 miles and then from $B$ to $C$ of 13 miles must be taken. How much farther is the "detour" from $A$ to $C$ than the road from $A$ to $C$?

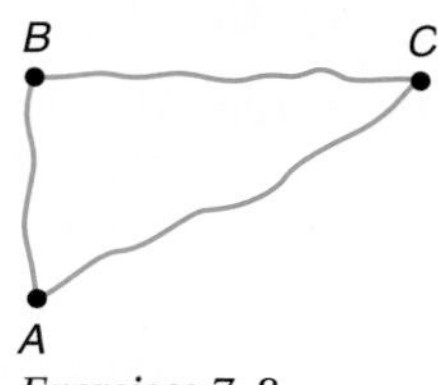


*Exercises 7, 8*

8. A cross-country runner jogs at a rate of 15 meters per second. If she runs 300 meters from $A$ to $B$, 450 meters from $B$ to $C$, and then 600 meters from $C$ back to $A$, how long will it take her to return to point $A$? (HINT: See figure for Exercise 7.)

*In Exercises 9 to 28, use the drawings as needed to answer the following questions.*

9. Name three points that appear to be

   a) collinear. b) noncollinear.

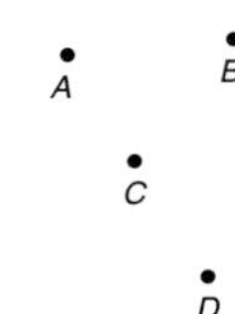


*Exercises 9, 10*

10. How many lines can be drawn through

    a) point $A$?
    b) points $A$ and $B$?
    c) points $A$, $B$, and $C$?
    d) points $A$, $B$, and $D$?

11. Give the meanings of $\overleftrightarrow{CD}$, $\overline{CD}$, $CD$, and $\overrightarrow{CD}$.

12. Explain the difference, if any, between

    a) $\overleftrightarrow{CD}$ and $\overleftrightarrow{DC}$.
    b) $\overline{CD}$ and $\overline{DC}$.
    c) $CD$ and $DC$.
    d) $\overrightarrow{CD}$ and $\overrightarrow{DC}$.

13. Name two lines that appear to be

    a) parallel. b) nonparallel.

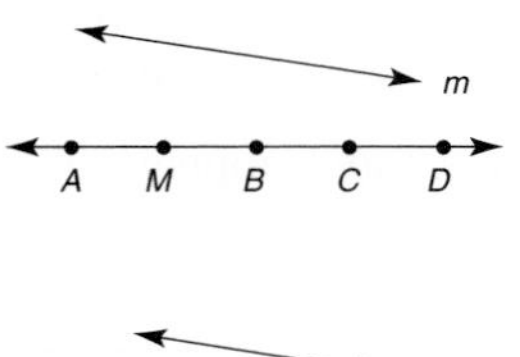


*Exercises 13–17*

14. Classify as true or false:

    a) $AB + BC = AD$
    b) $AD - CD = AB$
    c) $AD - CD = AC$
    d) $AB + BC + CD = AD$
    e) $AB = BC$

15. *Given:* $M$ is the midpoint of $\overline{AB}$
    $AM = 2x + 1$ and $MB = 3x - 2$
    *Find:* $x$ and $AM$

16. *Given:* $M$ is the midpoint of $\overline{AB}$
    $AM = 2(x + 1)$ and $MB = 3(x - 2)$
    *Find:* $x$ and $AB$

17. *Given:* $AM = 2x + 1$, $MB = 3x + 2$, and $AB = 6x - 4$
    *Find:* $x$ and $AB$

18. Can a segment bisect a line? a segment? Can a line bisect a segment? a line?

19. In the figure, name

    a) two opposite rays.
    b) two rays that are not opposite.

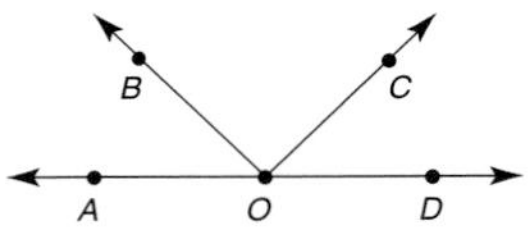


20. Suppose that (a) point $C$ lies in plane $X$ and (b) point $D$ lies in plane $X$. What can you conclude regarding $\overleftrightarrow{CD}$?

21. Make a sketch of

    a) two intersecting lines that are perpendicular.
    b) two intersecting lines that are *not* perpendicular.
    c) two parallel lines.

22. Make a sketch of

    a) two intersecting planes.
    b) two parallel planes.
    c) two parallel planes intersected by a third plane that is not parallel to the first or the second plane.

23. Suppose that (a) planes $M$ and $N$ intersect, (b) point $A$ lies in both planes $M$ and $N$, and (c) point $B$ lies in both planes $M$ and $N$. What can you conclude regarding $\overleftrightarrow{AB}$?

24. Suppose that (a) points $A$, $B$, and $C$ are collinear and (b) $AB > AC$. Which point can you conclude *cannot* lie between the other two?

25. Suppose that points $A$, $R$, and $V$ are collinear. If $AR = 7$ and $RV = 5$, then which point cannot possibly lie between the other two?

26. Points $A$, $B$, $C$, and $D$ are coplanar; $B$, $C$, and $D$ are collinear; point $E$ is not in plane $M$. How many planes contain

    a) points $A$, $B$ and $C$?
    b) points $B$, $C$, and $D$?
    c) points $A$, $B$, $C$, and $D$?
    d) points $A$, $B$, $C$, and $E$?

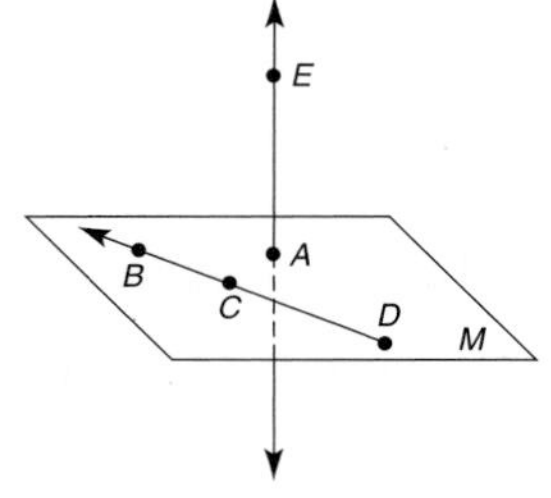

27. Using the number line provided, name the point that

a) is the midpoint of $\overline{AE}$.
b) is the endpoint of a segment of length 4, if the other endpoint is point $G$.
c) has a distance from $B$ equal to $3(AC)$.

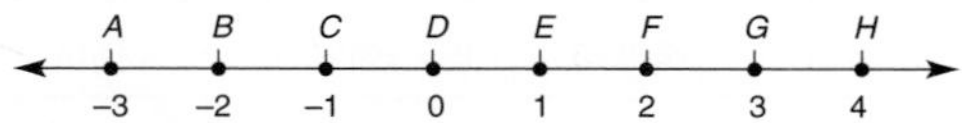

*Exercises 27, 28*

28. Given that $B$ is the midpoint of $\overline{AC}$ and $C$ is the midpoint of $\overline{BD}$, what can you conclude about the lengths of

a) $\overline{AB}$ and $\overline{CD}$?
b) $\overline{AC}$ and $\overline{BD}$?
c) $\overline{AC}$ and $\overline{CD}$?

*In Exercises 29 to 32, use only a compass and a straightedge to complete each construction.*

29. *Given:* $\overline{AB}$ and $\overline{CD}$ $(AB > CD)$
*Construct:* $\overline{MN}$ on line $\ell$ so that $MN = AB + CD$

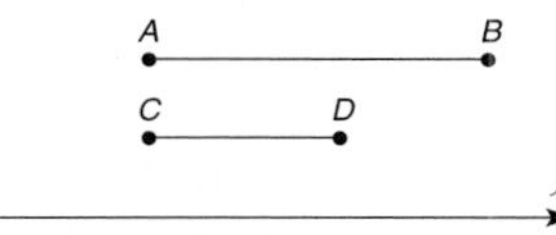

*Exercises 29, 30*

30. *Given:* $\overline{AB}$ and $\overline{CD}$ $(AB > CD)$
*Construct:* $\overline{EF}$ so that $EF = AB - CD$

31. *Given:* $\overline{AB}$ as shown in the figure
*Construct:* $\overline{PQ}$ on line $n$ so that $PQ = 3(AB)$

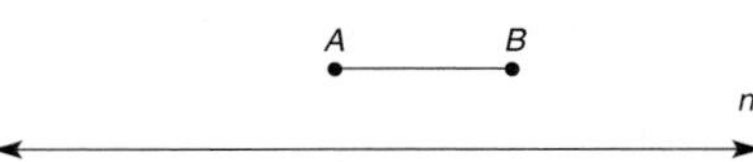

*Exercises 31, 32*

32. *Given:* $\overline{AB}$ as shown in the figure
*Construct:* $\overline{TV}$ on line $n$ so that $TV = \frac{1}{2}(AB)$

33. Can you use the construction for the midpoint of a segment to divide a line segment into

a) three congruent parts?
b) four congruent parts?
c) six congruent parts?
d) eight congruent parts?

34. Generalize your findings in Exercise 33.

35. Consider points $A$, $B$, $C$, and $D$, no three of which are collinear. Using two points at a time (such as $A$ and $B$), how many lines are determined by these points?

36. Consider noncoplanar points $A$, $B$, $C$, and $D$. Using three points at a time (such as $A$, $B$, and $C$), how many planes are determined by these points?

37. Line $\ell$ is parallel to plane $P$ (that is, it will not intersect $P$ even if extended). Line $m$ intersects line $\ell$. What can you conclude about $m$ and $P$?

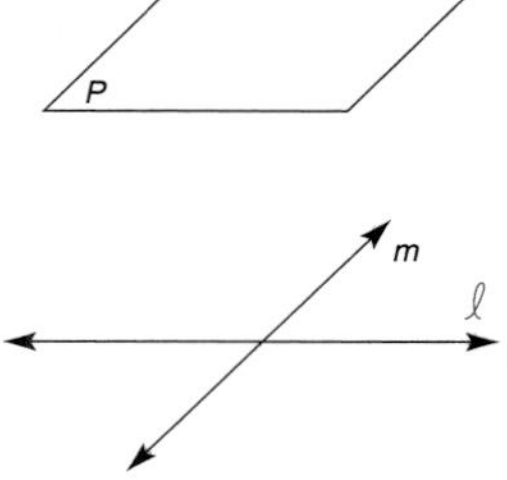


38. $\overleftrightarrow{AB}$ and $\overleftrightarrow{EF}$ are said to be **skew** lines because they neither intersect nor are parallel. How many planes are determined by

a) parallel lines $AB$ and $DC$?
b) intersecting lines $AB$ and $BC$?
c) skew lines $AB$ and $EF$?
d) lines $AB$, $BC$, and $DC$?
e) points $A$, $B$, and $F$?
f) points $A$, $C$, and $H$?
g) points $A$, $C$, $F$, and $H$?

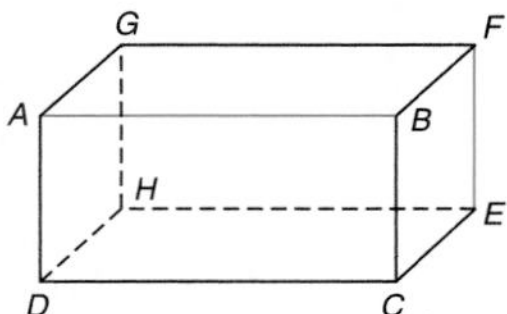

# 1.4 Angles and Their Relationships

**KEY CONCEPTS**

Angle: Sides of Angle, Vertex of Angle • Protractor Postulate • Acute, Right, Obtuse, Straight, and Reflex Angles • Adjacent Angles • Complementary Angles • Supplementary Angles • Vertical Angles

This section introduces you to the language of angles. Recall from Sections 1.1 and 1.3 that the word *union* means that two sets or figures are joined.

> DEFINITION: An **angle** is the union of two rays that share a common endpoint.

In Figure 1.46, the angle is symbolized by $\angle ABC$ or $\angle CBA$. The rays $BA$ and $BC$ are known as the **sides** of the angle. $B$, the common endpoint of these rays, is known as the **vertex** of the angle. When three letters are used to name an angle, the vertex is always named in the middle. In many instances, a single letter or numeral is used to name the angle. The angle in Figure 1.46 may be described as $\angle B$ (the vertex) or as $\angle 1$. In set notation, $\angle B = \overrightarrow{BA} \cup \overrightarrow{BC}$.

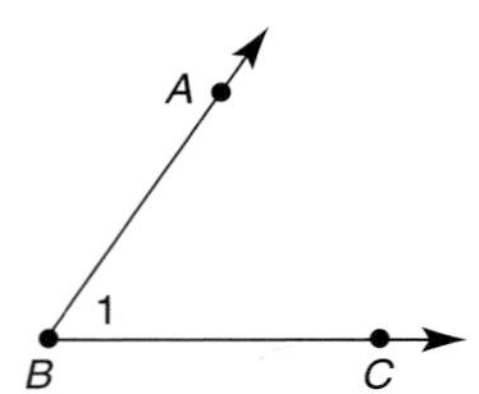


**FIGURE 1.46**

> POSTULATE 8: **(Protractor Postulate)**
> The measure of an angle is a unique positive number.

NOTE: In Chapters 1 to 9, the measures of angles are generally between 0° and 180°, including 180°. Angles with measures between 180° and 360° are introduced in this section; these angles are not used often in the study of geometry.

## TYPES OF ANGLES

An angle whose measure is less than 90° is an **acute angle.** If the angle's measure is exactly 90°, the angle is a **right angle.** If the angle's measure is between 90° and 180°, the angle is **obtuse.** An angle whose measure is exactly 180° is a **straight angle;** alternatively, a straight angle is one whose sides form opposite rays (a straight line). A **reflex angle** is one whose measure is between 180° and 360°.

In Figure 1.47, $\angle ABC$ contains the noncollinear points $A$, $B$, and $C$. These three points, in turn, determine a plane. The plane containing $\angle ABC$ is separated into three subsets by the angle:

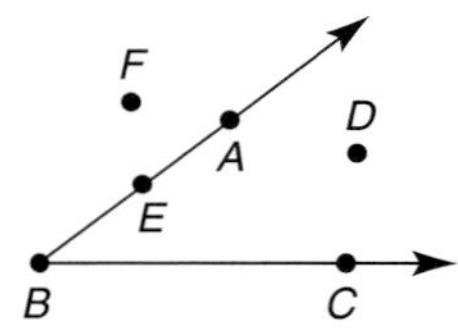


**FIGURE 1.47**

Points like $D$ are in the *interior* of $\angle ABC$.
Points like $E$ are said to be *on* $\angle ABC$.
Points like $F$ are in the *exterior* of $\angle ABC$.

With this description, it is possible to state the counterpart of the Segment-Addition Postulate! Consider Figure 1.48 as you read Postulate 9.

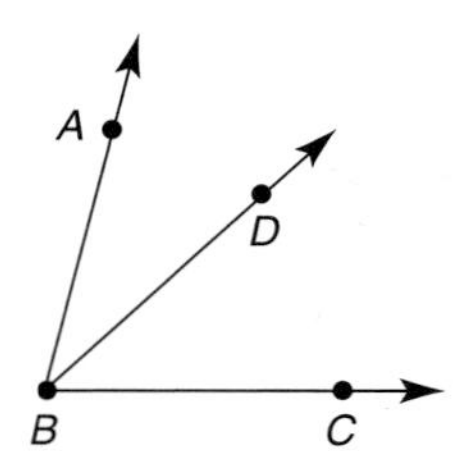


**FIGURE 1.48**

> POSTULATE 9: **(Angle-Addition Postulate)**
> If a point $D$ lies in the interior of an angle $ABC$, then $m\angle ABD + m\angle DBC = m\angle ABC$.

In Table 1.2, we summarize the angle types we will encounter.

**TABLE 1.2**

**Angles**

| *Angle* | *Example* |
|---|---|
| Acute (1) | $m\angle 1 = 23°$ |

*continued*

*Technology Exploration*

Use software if available.

1) Draw $\angle RST$.
2) Through point $V$ in the interior of $\angle RST$, draw $\overrightarrow{SV}$.
3) Measure $\angle RST$, $\angle RSV$, and $\angle VST$.
4) Show that $m\angle RSV + m\angle VST = m\angle RST$.

**TABLE 1.2 CONT.**

**Angles**

| *Angle* | *Example* | |
|---|---|---|
| Right (2) | $m\angle 2 = 90°$ | |
| Obtuse (3) | $m\angle 3 = 112°$ | |
| Straight (4) | $m\angle 4 = 180°$ | |
| Reflex (5) | $m\angle 5 = 337°$ | |

NOTE: The arc is necessary in indicating the reflex angle.

EXAMPLE 1

Use Figure 1.48 to find $m\angle ABC$ if:

a) $m\angle ABD = 27°$ and $m\angle DBC = 42°$
b) $m\angle ABD = x°$ and $m\angle DBC = (2x - 3)°$

***Solution***

a) Using the Angle-Addition Postulate,
$m\angle ABC = m\angle ABD + m\angle DBC$. That is, $m\angle ABC = 27° + 42° = 69°$.
b) $m\angle ABC = m\angle ABD + m\angle DBC = x° + (2x - 3)° = (3x - 3)°$

## Discover!

An index card can be used to categorize each type of angle displayed. In each sketch, an index card is placed over an angle. A dashed ray indicates that a side is hidden. What type of angle is shown in each figure? (Note the placement of the card in each figure.)

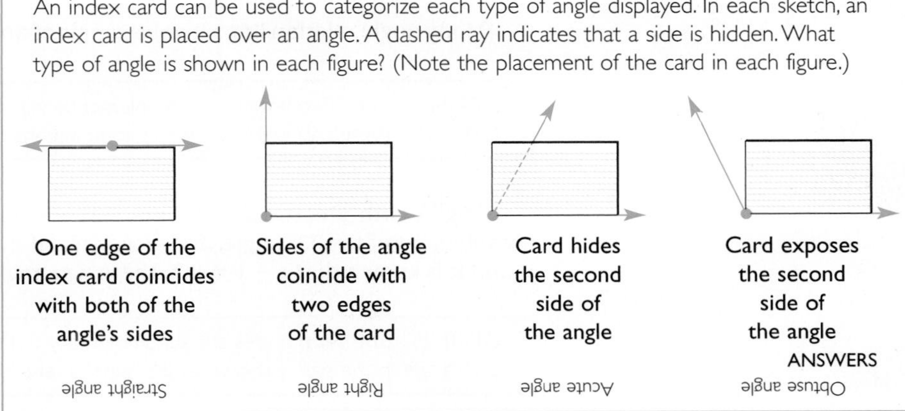

One edge of the index card coincides with both of the angle's sides
Sides of the angle coincide with two edges of the card
Card hides the second side of the angle
Card exposes the second side of the angle
ANSWERS
Straight angle
Right angle
Acute angle
Obtuse angle

*Exs. 1–6*

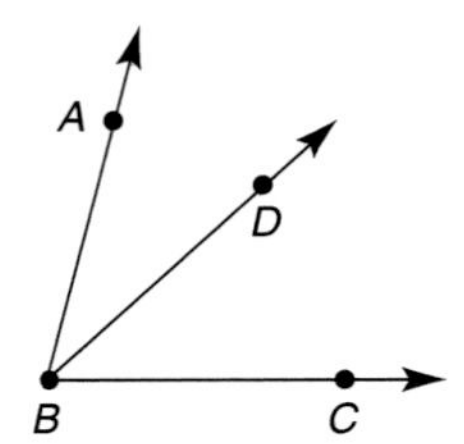


FIGURE 1.48

## CLASSIFYING PAIRS OF ANGLES

In Figure 1.48, $\angle ABD$ and $\angle DBC$ are said to be adjacent. Two angles are **adjacent** if they share a common side and a common vertex but have no interior points in common. In Figure 1.48, $\angle ABC$ and $\angle ABD$ are not adjacent because they do have interior points in common.

> DEFINITION: **Congruent angles** ($\cong \angle$s) are two angles with the same measure.

Congruent angles must coincide when one is placed over the other. (Do not consider that the sides appear to have different lengths; remember that rays are infinite in length!) In symbols, $\angle 1 \cong \angle 2$ if $m\angle 1 = m\angle 2$. In Figure 1.49, similar markings indicate that $\angle 1 \cong \angle 2$.

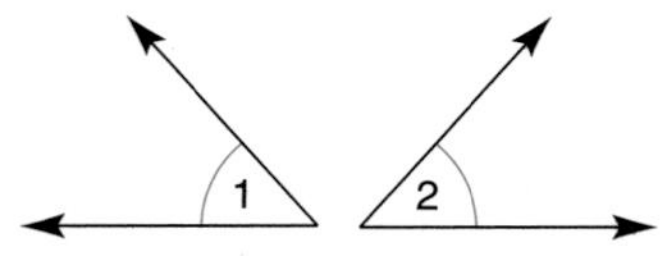


FIGURE 1.49

EXAMPLE 2

*Given:* $\angle 1 \cong \angle 2$
$m\angle 1 = 2x + 15$
$m\angle 2 = 3x - 2$

*Find:* $x$

***Solution***

$\angle 1 \cong \angle 2$ means $m\angle 1 = m\angle 2$. Therefore

$$2x + 15 = 3x - 2$$
$$17 = x \qquad \text{or} \qquad x = 17$$

NOTE: $m\angle 1 = 2(17) + 15 = 49°$ and $m\angle 2 = 3(17) - 2 = 49°$.

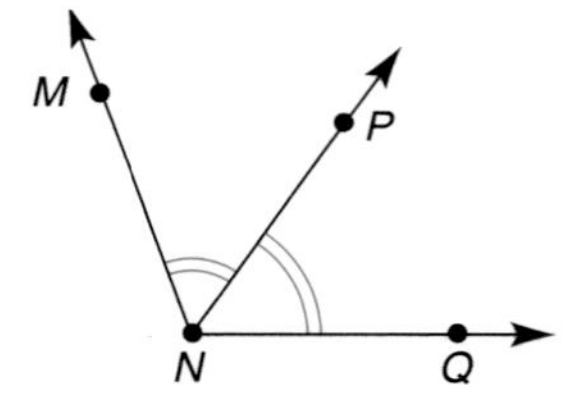


FIGURE 1.50

> DEFINITION: The **bisector** of an angle is the ray that separates the given angle into two congruent angles.

With $P$ in the interior of $\angle MNQ$ so that $\angle MNP \cong \angle PNQ$, $\overrightarrow{NP}$ is said to **bisect** $\angle MNQ$. Equivalently, $\overrightarrow{NP}$ is the bisector or angle-bisector of $\angle MNQ$. Based on Figure 1.50, possible consequences of the definition of bisector of an angle are

$m\angle MNP = m\angle PNQ$ $\qquad$ $m\angle MNP = \frac{1}{2}(m\angle MNQ)$
$m\angle PNQ = \frac{1}{2}(m\angle MNQ)$ $\qquad$ $m\angle MNQ = 2(m\angle MNP)$
$m\angle MNQ = 2(m\angle PNQ)$

Many angle relationships involve a pair of angles.

> DEFINITION: Two angles are **complementary** if the sum of their measures is 90°. Each angle in the pair is known as the **complement** of the other angle.

Angles with measures of 37° and 53° are complementary. The 37° angle is the complement of the 53° angle, and vice versa. If the measures of two angles are $x$ and $y$, and it is known that $x + y = 90°$, these two angles are complementary.

> DEFINITION: Two angles are **supplementary** if the sum of their measures is 180°. Each angle in the pair is known as the **supplement** of the other angle.

EXAMPLE 3

Given that m∠1 = 29°, find:

a) the complement $x$ b) the supplement $y$

***Solution***

a) $x + 29 = 90$, so $x = 61°$; complement = 61°
b) $y + 29 = 180$, so $y = 151°$; supplement = 151°

EXAMPLE 4

*Given:* ∠$P$ and ∠$Q$ are complementary so that

$$m\angle P = \frac{x}{2} \quad \text{and} \quad m\angle Q = \frac{x}{3}$$

*Find:* $x$, m∠$P$, and m∠$Q$

***Solution***

$$m\angle P + m\angle Q = 90$$
$$\frac{x}{2} + \frac{x}{3} = 90$$

Multiplying by 6 (the least common denominator, or LCD, of 2 and 3),

$$6 \cdot \frac{x}{2} + 6 \cdot \frac{x}{3} = 6 \cdot 90$$
$$3x + 2x = 540$$
$$5x = 540$$
$$x = 108$$

$$m\angle P = \frac{x}{2} = \frac{108}{2} = 54°$$
$$m\angle Q = \frac{x}{3} = \frac{108}{3} = 36°$$

NOTE: m∠$P$ = 54° and m∠$Q$ = 36°, so their sum is exactly 90°.

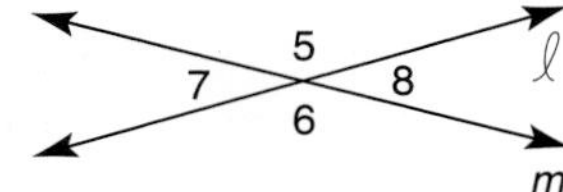


**FIGURE 1.51**

When two straight lines intersect, the pairs of nonadjacent angles formed are each known as **vertical angles.** In Figure 1.51, ∠5 and ∠6 are vertical angles, as are ∠7 and ∠8. In addition, ∠5 and ∠7 can be described as adjacent and supplementary angles, as can ∠5 and ∠8. If m∠7 = 30°, what is m∠5 and what is m∠8°? It is true in general that vertical angles are congruent, and we will prove this in Example 3 of Section 1.6. We apply this property in Example 5 of this section.

*Exs. 7–12*

Recall the Addition and Subtraction Properties of Equality: If $a = b$ and $c = d$, then $a \pm c = b \pm d$. These principles can be used in solving a system of equations such as the following:

$$\begin{aligned} x + y &= 5 \\ 2x - y &= 7 \\ \hline 3x \quad &= 12 \\ x &= 4 \end{aligned}$$ left and right sides are added

We substitute 4 for $x$ in either equation to solve for $y$:

$$\begin{aligned} x + y &= 5 \\ 4 + y &= 5 \\ y &= 1 \end{aligned}$$ by substitution

If $x = 4$ and $y = 1$, then $x + y = 5$ *and* $2x - y = 7$.

When each term in an equation is multiplied by the same nonzero number, the solutions of the equation are not changed. For instance, the equations $2x - 3 = 7$ and $6x - 9 = 21$ (each term multiplied by 3) both have the solution $x = 5$. Likewise, the values of $x$ and $y$ that make the equation $4x + y = 180$ true also make the equation $16x + 4y = 720$ (each term multiplied by 4) true. We use this method in Example 5.

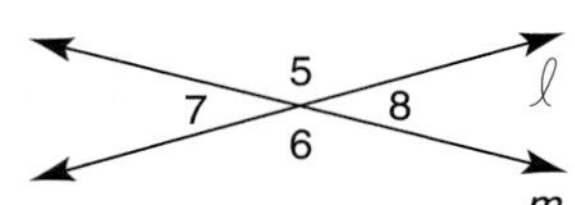
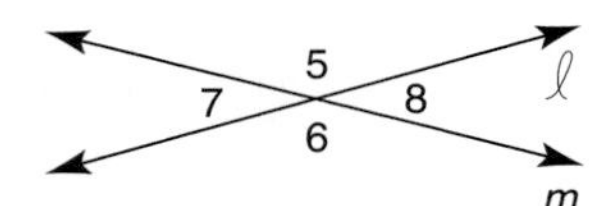


**FIGURE 1.51**

EXAMPLE 5

*Given:* In Figure 1.51, $\ell$ and $m$ intersect so that

$m\angle 5 = 2x + 2y$
$m\angle 8 = 2x - y$
$m\angle 6 = 4x - 2y$

*Find:* $x$ and $y$

***Solution***

$\angle 5$ and $\angle 8$ are supplementary (adjacent and together form a straight angle). Therefore, $m\angle 5 + m\angle 8 = 180$. $\angle 5$ and $\angle 6$ are congruent (vertical). Therefore, $m\angle 5 = m\angle 6$. Consequently, we have

$$(2x + 2y) + (2x - y) = 180 \quad \text{(supplementary } \angle\text{s 5 and 8)}$$
$$2x + 2y = 4x - 2y \quad (\cong \angle\text{s 5 and 6})$$

Simplifying,

$$4x + y = 180$$
$$2x - 4y = 0$$

Using the Multiplication Property of Equality, we multiply the first equation by 4. Then the equivalent system allows us to eliminate variable $y$ by addition.

$$\begin{array}{r} 16x + 4y = 720 \\ 2x - 4y = 0 \\ \hline 18x \qquad = 720 \\ x = 40 \end{array} \quad \text{adding left, right sides}$$

From the equation $4x + y = 180$, it follows that

$$4(40) + y = 180$$
$$160 + y = 180$$
$$y = 20$$

Summarizing, $x = 40$ and $y = 20$.

NOTE: $m\angle 5 = 120°$, $m\angle 8 = 60°$, and $m\angle 6 = 120°$.

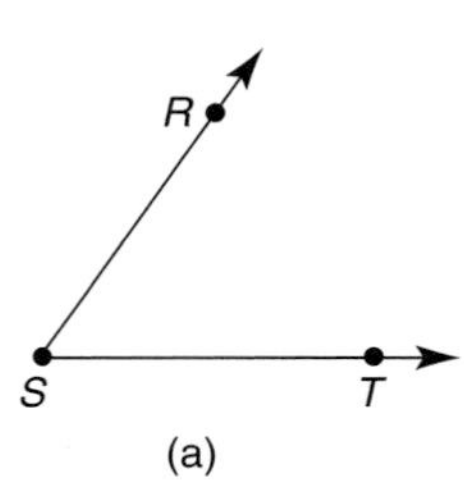


(a)

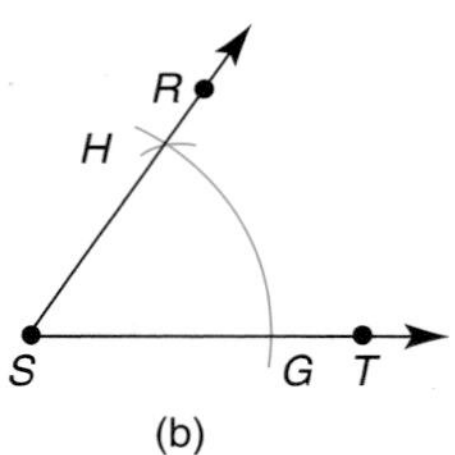


(b)

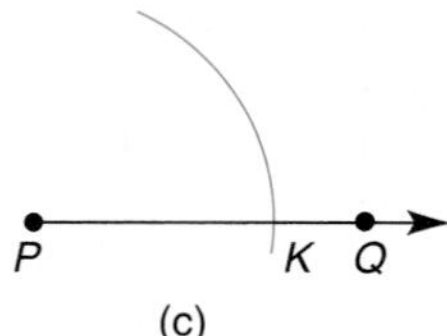


(c)

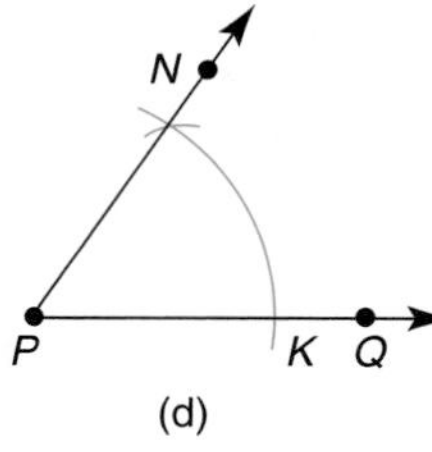


(d)

**FIGURE 1.52**

## CONSTRUCTIONS WITH ANGLES

In Section 1.2, we considered Constructions 1 and 2 with line segments. Now consider two constructions that involve angle concepts. In Section 3.4, it will become clear why these methods are valid. However, intuition suggests that the techniques are appropriate.

CONSTRUCTION 3:

To construct an angle congruent to a given angle.

*Given:* $\angle RST$ in Figure 1.52(a)

*Construct:* With $\overrightarrow{PQ}$ as one side, $\angle NPQ \cong \angle RST$

*Construction:* Figure 1.52(b): With a compass, mark an arc to intersect both sides of $\angle RST$ (at points $G$ and $H$, respectively).

Figure 1.52(c): Without changing the radius, mark an arc to intersect $\overrightarrow{PQ}$ at $K$ and the "would-be" second side of $\angle NPQ$.

Figure 1.52(b): Now mark an arc to measure the distance from $G$ to $H$.

Figure 1.52(d): Using the same radius, mark an arc with $K$ as center to intersect the would-be second side of the desired angle. Now draw the ray from $P$ through the point of intersection of the two arcs.

The resulting angle is the one desired, as we will prove in Section 3.4, Example 1.

Just as a line segment can be bisected, so can an angle. This takes us to a fourth construction method.

CONSTRUCTION 4:
To construct the angle bisector of a given angle.

*Given:* $\angle PRT$ in Figure 1.53(a)

*Construct:* $\overrightarrow{RS}$ so that $\angle PRS \cong \angle SRT$

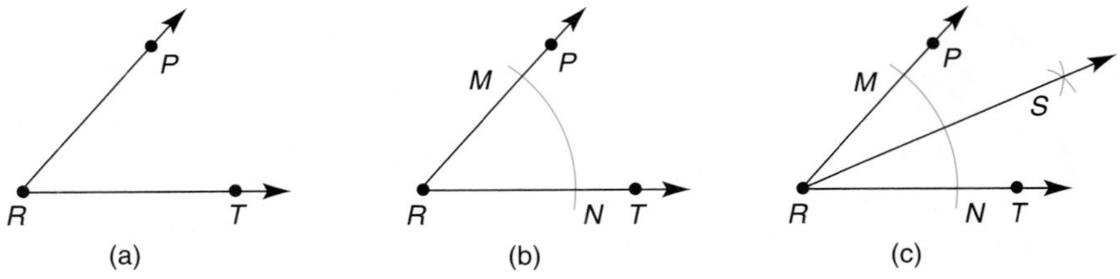


**FIGURE 1.53**

*Construction:* Figure 1.53(b): Using a compass, mark an arc to intersect the sides of $\angle PRT$ at points $M$ and $N$.

Figure 1.53(c): Now, with $M$ and $N$ as centers, mark off two arcs with equal radii to intersect at point $S$ in the interior of $\angle PRT$, as shown. Now draw ray $RS$, the desired angle bisector.

*Exs. 13–20*

Reasoning from the definition of an angle bisector, the Angle-Addition Postulate, and the Protractor Postulate, we can justify the following theorem.

THEOREM 1.4.1: There is one and only one angle bisector for a given angle.

This theorem is often stated, "The angle bisector of an angle is unique." This statement is proved in Example 5 of Section 2.2.

## 1.4 Exercises

1. What *type of angle* is each of the following?

   a) 47° b) 90° c) 137.3°

2. What *type of angle* is each of the following?

   a) 115° b) 180° c) 36°

3. What *relationship*, if any, exists between two angles:

   a) with measures of 37° and 53°?
   b) with measures of 37° and 143°?

4. What *relationship*, if any, exists between two angles:

   a) with equal measures?
   b) that have the same vertex and a common side between them?

*In Exercises 5 to 8, describe in one word the relationship between the angles.*

5. $\angle ABD$ and $\angle DBC$

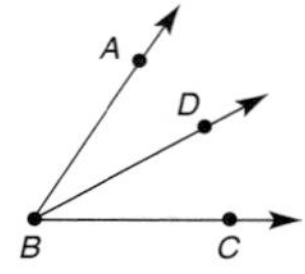


6. $\angle 7$ and $\angle 8$

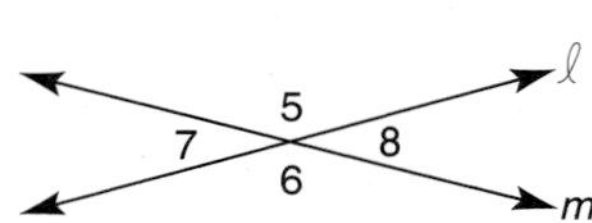


7. $\angle 1$ and $\angle 2$

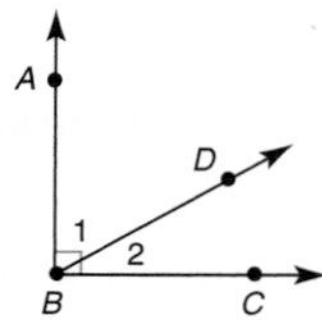


8. $\angle 3$ and $\angle 4$

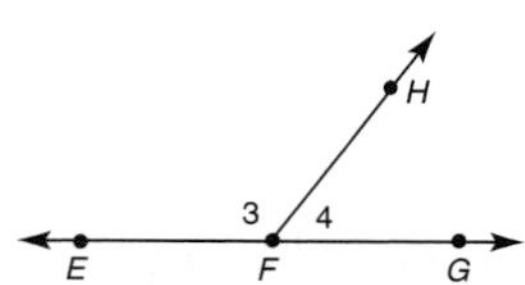


*Use drawings as needed to answer each of the following questions.*

9. Must two rays with a common endpoint be coplanar? Must three rays with a common endpoint be coplanar?

10. Suppose that $\overrightarrow{AB}$, $\overrightarrow{AC}$, $\overrightarrow{AD}$, $\overrightarrow{AE}$, and $\overrightarrow{AF}$ are coplanar.

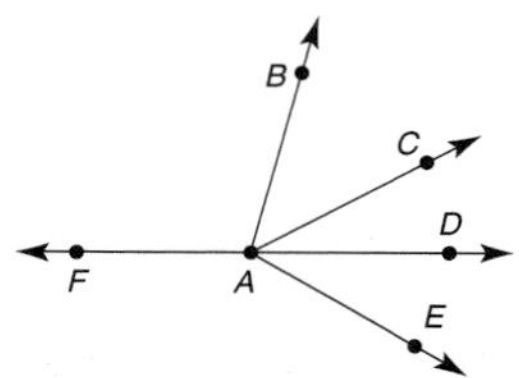


*Exercises 10–13*

Classify the following as true or false:

a) $m\angle BAC + m\angle CAD = m\angle BAD$
b) $\angle BAC \cong \angle CAD$
c) $m\angle BAE - m\angle DAE = m\angle BAC$
d) $\angle BAC$ and $\angle DAE$ are adjacent
e) $m\angle BAC + m\angle CAD + m\angle DAE = m\angle BAE$

11. Without using a protractor, name the type of angle represented by:

    a) $\angle BAE$ b) $\angle FAD$ c) $\angle BAC$ d) $\angle FAE$

12. What, if anything, is wrong with the claim $m\angle FAB + m\angle BAE = m\angle FAE$?

13. $\angle FAC$ and $\angle CAD$ are adjacent and $\overrightarrow{AF}$ and $\overrightarrow{AD}$ are opposite rays. What can you conclude about $\angle FAC$ and $\angle CAD$?

*For Exercises 14 and 15, let $m\angle 1 = x$ and $m\angle 2 = y$.*

14. Using variables $x$ and $y$, write an equation that expresses the fact that $\angle 1$ and $\angle 2$ are:

    a) supplementary b) congruent

15. Using variables $x$ and $y$, write an equation that expresses the fact that $\angle 1$ and $\angle 2$ are:

    a) complementary b) vertical

16. *Given:* $m\angle RST = 39°$
    $m\angle TSV = 23°$
    *Find:* $m\angle RSV$

17. *Given:* $m\angle RSV = 59°$
    $m\angle TSV = 17°$
    *Find:* $m\angle RST$

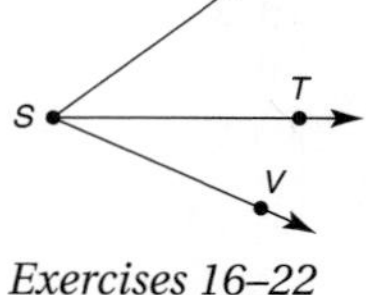


*Exercises 16–22*

18. *Given:* $m\angle RST = 2x + 9$
    $m\angle TSV = 3x - 2$
    $m\angle RSV = 67°$
    *Find:* $x$

19. *Given:* $m\angle RST = 2x - 10$
    $m\angle TSV = x + 6$
    $m\angle RSV = 4(x - 6)$
    *Find:* $x$ and $m\angle RSV$

20. *Given:* $m\angle RST = 5(x + 1) - 3$
    $m\angle TSV = 4(x - 2) + 3$
    $m\angle RSV = 4(2x + 3) - 7$
    *Find:* $x$ and $m\angle RSV$

21. *Given:* $\overrightarrow{ST}$ bisects $\angle RSV$
$m\angle RST = x + y$
$m\angle TSV = 2x - 2y$
$m\angle RSV = 64°$
*Find:* $x$ and $y$

22. *Given:* $\overrightarrow{ST}$ bisects $\angle RSV$
$m\angle RST = 2x + 3y$
$m\angle TSV = 3x - y + 2$
$m\angle RSV = 80°$
*Find:* $x$ and $y$

23. *Given:* $\overleftrightarrow{AB}$ and $\overleftrightarrow{AC}$ in plane $P$ as shown
$\overleftrightarrow{AD}$ intersects $P$ at point $A$
$\angle CAB \cong \angle DAC$
$\angle DAC \cong \angle DAB$

What can you conclude?

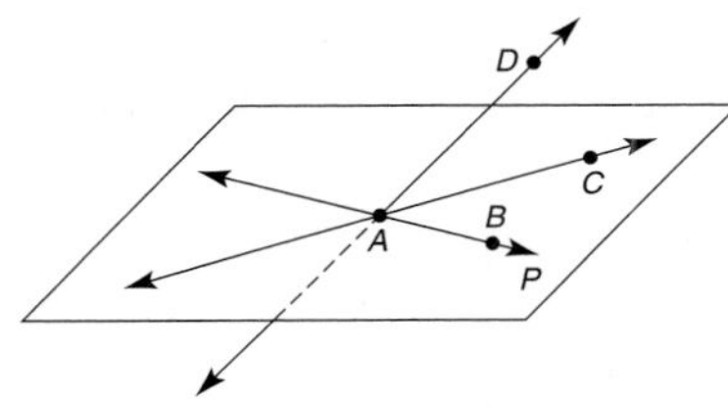


24. Two angles are complementary. One angle is 12° larger than the other. Using two variables $x$ and $y$, find the size of each angle by solving a system of equations.

25. Two angles are supplementary. One angle is 24° more than twice the other. Using two variables $x$ and $y$, find the measure of each angle.

26. For two complementary angles, find an expression for the measure of the second angle if the measure of the first is:

a) $x°$
b) $(3x - 12)°$
c) $(2x + 5y)°$

27. Suppose that two angles are supplementary. Find expressions for the supplements, using the expressions provided in Exercise 26, parts (a) to (c).

28. On the protractor shown, $\overrightarrow{NP}$ bisects $\angle MNQ$. Find $x$.

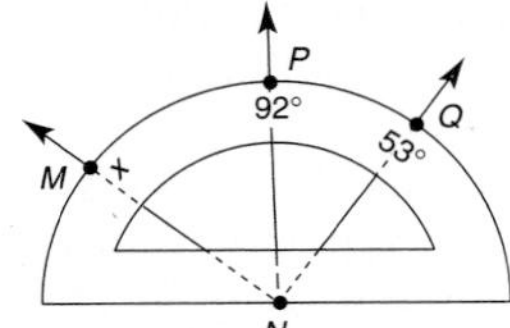


*Exercises 28, 29*

29. On the protractor shown for Exercise 28, $\angle MNP$ and $\angle PNQ$ are complementary. Find $x$.

30. Classify as true or false:

a) If points $P$ and $Q$ lie in the interior of $\angle ABC$, then $\overline{PQ}$ lies in the interior of $\angle ABC$.
b) If points $P$ and $Q$ lie in the interior of $\angle ABC$, then $\overleftrightarrow{PQ}$ lies in the interior of $\angle ABC$.
c) If points $P$ and $Q$ lie in the interior of $\angle ABC$, then $\overrightarrow{PQ}$ lies in the interior of $\angle ABC$.

*In Exercises 31 to 38, use only a compass and a straightedge to perform the indicated constructions.*

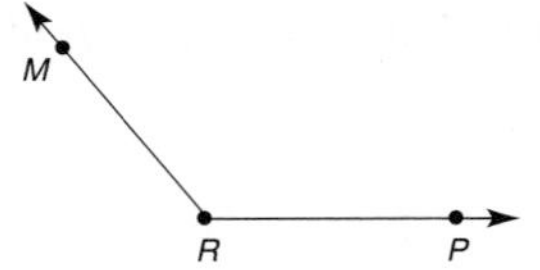


*Exercises 31–33*

31. *Given:* Obtuse $\angle MRP$
*Construct:* With $\overrightarrow{OA}$ as one side, an angle $\cong \angle MRP$

32. *Given:* Obtuse $\angle MRP$
*Construct:* $\overrightarrow{RS}$, the angle bisector of $\angle MRP$

33. *Given:* Obtuse $\angle MRP$
*Construct:* Rays $\overrightarrow{RS}$, $\overrightarrow{RT}$, and $\overrightarrow{RU}$ so that $\angle MRP$ is divided into four $\cong$ angles

34. *Given:* Straight $\angle DEF$
*Construct:* A right angle with vertex at $E$
(HINT: Use Construction 4.)

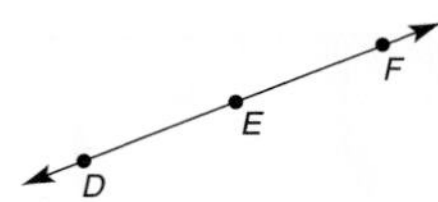


35. Draw a triangle with three acute angles. Construct angle bisectors for each of the three angles. On the basis of the appearance of your construction, what seems to be true?

36. *Given:* Acute $\angle 1$ and $\overline{AB}$
*Construct:* Triangle $ABC$ with $\angle A \cong \angle 1$, $\angle B \cong \angle 1$, and base $\overline{AB}$

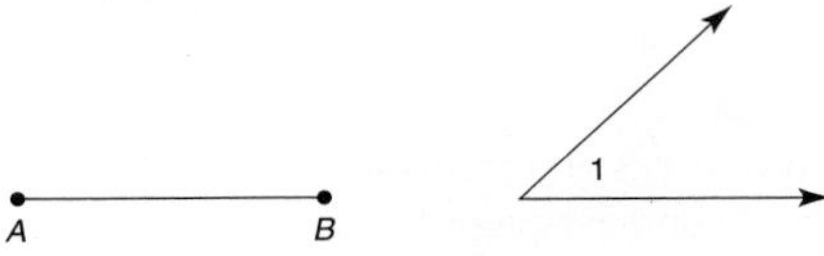


37. What seems to be true of two of the sides in the triangle you constructed in Exercise 36?

38. *Given:* Straight $\angle ABC$ and $\overrightarrow{BD}$
*Construct:* Bisectors of $\angle ABD$ and $\angle DBC$

What type of angle is formed by the bisectors of the two angles?

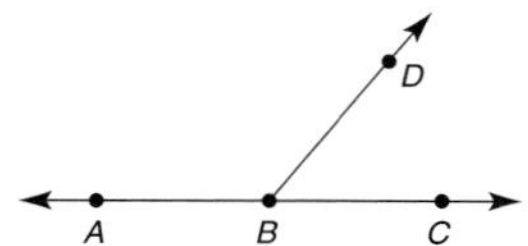


39. Refer to the circle with center $O$.

a) Use a protractor to find m $\angle B$.
b) Use a protractor to find m $\angle D$.
c) Compare results in parts (a) and (b).

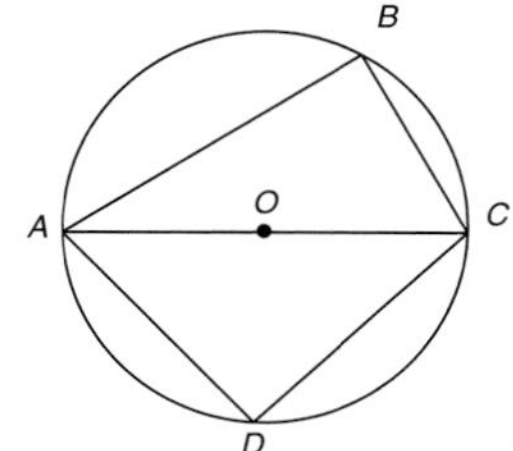


40. Refer to the circle with center $P$.

a) Use a protractor to find m $\angle 1$.
b) Use a protractor to find m $\angle 2$.
c) Compare results in parts (a) and (b).

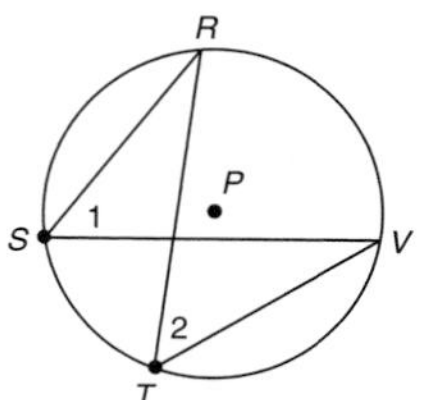


41. On the hanging sign, the three angles ( $\angle ABD$, $\angle ABC$ , and $\angle DBC$ ) at vertex $B$ have the sum of measures 360°. If m $\angle DBC = 90°$ and $\overrightarrow{BA}$ bisects the indicated reflex angle, find m$\angle ABC$ .

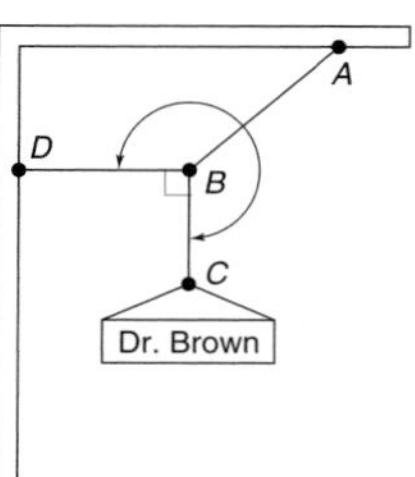


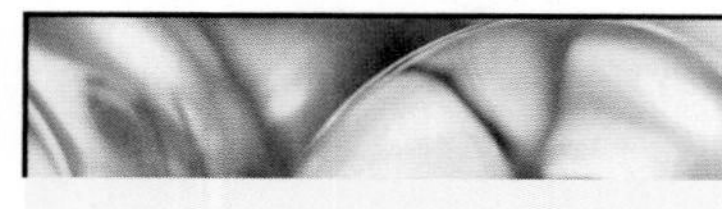

**KEY CONCEPTS**

Proof • Algebraic Properties • Given Problem and Prove Statement • Sample Proofs

# 1.5 Introduction to Geometric Proof

This section introduces some guidelines for proving geometric properties. Several examples are offered to help you develop your own proofs. In the beginning, the form of proof will be a two-column proof, with statements in the left column and reasons in the right column. But where do the statements and reasons come from?

To deal with this question, you must ask "What" it is that is known (Given) and "Why" the conclusion (Prove) should follow from this information. Understanding the why may mean dealing with several related conclusions and thus several intermediate whys. In correctly piecing together a proof, you will usually scratch out several conclusions and reorder them. Of course, each conclusion must be justified by citing the Given (hypothesis), a previously stated definition or postulate, or a theorem previously proved.

Reminder 

Additional properties and techniques of algebra are found in Appendix A.

Selected properties from algebra are often used as reasons to justify statements. For instance, we use the Addition Property of Equality to justify adding the same number to each side of an equation. Reasons found in a proof often include the properties found in Tables 1.3 and 1.4.

**TABLE 1.3**

**Properties of Equality**

| | |
|---|---|
| Addition Property of Equality: | If $a = b$, then $a + c = b + c$. |
| Subtraction Property of Equality: | If $a = b$, then $a - c = b - c$. |
| Multiplication Property of Equality: | If $a = b$, then $a \cdot c = b \cdot c$. |
| Division Property of Equality: | If $a = b$ and $c \neq 0$, then $\frac{a}{c} = \frac{b}{c}$. |

As we discover in Example 1, some properties can be used interchangably.

EXAMPLE 1

Which property of equality justifies each conclusion?

a) If $2x - 3 = 7$, then $2x = 10$. b) If $2x = 10$, then $x = 5$.

***Solution***

a) Addition Property of Equality; added 3 to each side of the equation.
b) Multiplication Property of Equality; multiplied each side of the equation by $\frac{1}{2}$. *OR* Division Property of Equality; divided each side of the equation by 2.

**TABLE 1.4**

**Further Algebraic Properties**

| | |
|---|---|
| Distributive Property: | $a(b + c) = a \cdot b + a \cdot c$. |
| Substitution Property: | If $a = b$, then $a$ replaces $b$ in any equation. |
| Transitive Property: | If $a = b$ and $b = c$, then $a = c$. |

Before considering geometric proof, we study algebraic proof, in which each statement in a sequence of steps is supported by the reason *why* we can make that statement (claim). The first claim in the proof is the *Given* problem; and the sequence of steps must conclude with a final statement representing the claim to be proved (called the *Prove statement*).

*Exs. 1–4*

Study Example 2. Then cover the reasons and try to provide the reason for each statement. With statements covered, find the statement for each reason.

EXAMPLE 2

*Given:* $2(x - 3) + 4 = 10$

*Prove:* $x = 6$

**PROOF**

| Statements | Reasons |
|---|---|
| 1. $2(x - 3) + 4 = 10$ | 1. Given |
| 2. $2x - 6 + 4 = 10$ | 2. Distributive Property |
| 3. $2x - 2 = 10$ | 3. Substitution |
| 4. $2x = 12$ | 4. Addition Property of Equality |
| 5. $x = 6$ | 5. Division Property of Equality |

NOTE 1: Alternatively, Step 5 could use the reason Multiplication Property of Equality $\left(\text{multiply by } \frac{1}{2}\right)$.

NOTE 2: The fifth step is the final step because the Prove statement has been made and justified.

*Exs. 5–7*

## Discover!

In the diagram, the wooden trim pieces are mitered (cut at an angle) to be equal and to form a right angle when placed together. Use the properties of algebra to explain why the measures of $\angle 1$ and $\angle 2$ are both 45°. What you have done is an informal "proof."

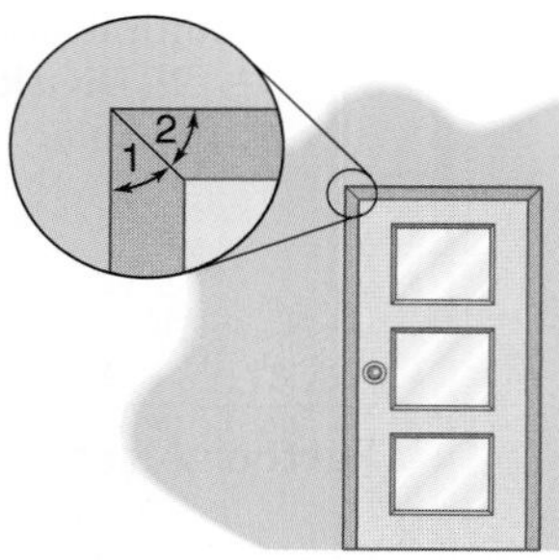


**ANSWER**

$m\angle 1 + m\angle 2 = 90°$. Because $m\angle 1 = m\angle 2$, we see that $m\angle 1 + m\angle 1 = 90°$. Thus $2 \cdot m\angle 1 = 90$ and, dividing by 2, we see that $m\angle 1 = 45°$. Then $m\angle 2 = 45°$ also.

This Discover! activity suggests that a formal geometric proof also exists. The typical format for a problem requiring geometric proof is

*Given:* __________ [Drawing]

*Prove:* __________

Consider this problem:

**FIGURE 1.54**

*Given:* $A$-$P$-$B$ on $\overline{AB}$ (Figure 1.54)

*Prove:* $AP = AB - PB$

First consider the Drawing (Figure 1.54), and relate it to any additional information described by the Given. Then consider the Prove statement. Do you understand the claim, and does it seem reasonable? If it seems reasonable, the intermediate claims must be ordered and supported to form the contents of the proof. Because a proof must begin with the Given and conclude with the Prove, the proof of the preceding problem has this form:

**PROOF**

| Statements | Reasons |
|---|---|
| 1. *A*-*P*-*B* on $\overline{AB}$ | 1. Given |
| 2. ? | 2. ? |
| . | . |
| . | . |
| . | . |
| ?. $AP = AB - PB$ | ?. ? |

To construct the proof, you must glean from the Drawing and the Given that

$$AP + PB = AB$$

You can then deduce (through subtraction) that $AP = AB - PB$. Thus the complete proof problem will have the appearance of Example 3.

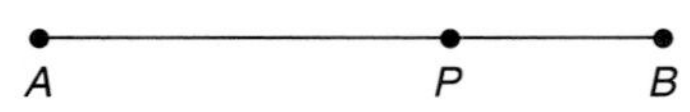


**FIGURE 1.55**

EXAMPLE 3

*Given:* *A*-*P*-*B* on $\overline{AB}$ (Figure 1.55)

*Prove:* $AP = AB - PB$

**PROOF**

| Statements | Reasons |
|---|---|
| 1. *A*-*P*-*B* on $\overline{AB}$ | 1. Given |
| 2. $AP + PB = AB$ | 2. Segment-Addition Postulate |
| 3. $AP = AB - PB$ | 3. Subtraction Property of Equality |

*Exs. 8–10*

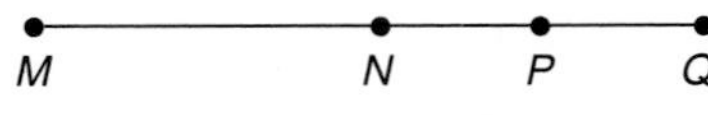


**FIGURE 1.56**

## SAMPLE PROOFS

Consider Figure 1.56 and this problem:

*Given:* $MN > PQ$

*Prove:* $MP > NQ$

To understand the situation, first study the Drawing (Figure 1.56) and the related Given. Then read the Prove with reference to the drawing. Constructing the proof requires that you begin with the Given and end with the Prove. What may be confusing here is that the Given involves *MN* and *PQ*, whereas the Prove involves *MP* and *NQ*. However, this is easily remedied through the addition of *NP* to each side of the inequality $MN > PQ$; see step 2 in the proof of Example 4.

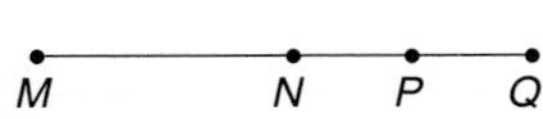


**FIGURE 1.57**

EXAMPLE 4

*Given:* $MN > PQ$ (Figure 1.57)

*Prove:* $MP > NQ$

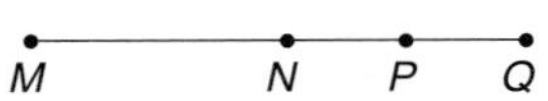


**FIGURE 1.57**

PROOF

| Statements | Reasons |
|---|---|
| 1. $MN > PQ$ | 1. Given |
| 2. $MN + NP > NP + PQ$ | 2. Addition Property of Inequality |
| 3. But $MN + NP = MP$ and $NP + PQ = NQ$ | 3. Segment-Addition Postulate |
| 4. $MP > NQ$ | 4. Substitution |

NOTE: The final reason may come as a surprise. However, the Substitution Axiom of Equality allows you to replace a quantity with its equal in *any* statement—including an inequality! See Appendix A.3 for more information.

Now that you have a better idea of the development of proofs, consider an example that involves an angle bisector.

EXAMPLE 5

Study this proof, noting the order of the statements and reasons.

*Given:* $\overrightarrow{ST}$ bisects $\angle RSU$
$\overrightarrow{SV}$ bisects $\angle USW$ (Figure 1.58)

*Prove:* $m\angle RST + m\angle VSW = m\angle TSV$

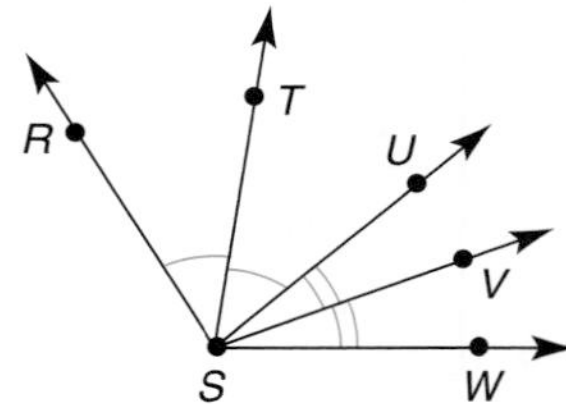


**FIGURE 1.58**

PROOF

| Statements | Reasons |
|---|---|
| 1. $\overrightarrow{ST}$ bisects $\angle RSU$ | 1. Given |
| 2. $m\angle RST = m\angle TSU$ | 2. If an angle is bisected, then the measures of the resulting angles are equal |
| 3. $\overrightarrow{SV}$ bisects $\angle USW$ | 3. Same as reason 1 |
| 4. $m\angle VSW = m\angle USV$ | 4. Same as reason 2 |
| 5. $m\angle RST + m\angle VSW = m\angle TSU + m\angle USV$ | 5. Addition Property of Equality (use the equations from statements 2 and 4) |
| 6. $m\angle TSU + m\angle USV = m\angle TSV$ | 6. Angle-Addition Postulate |
| 7. $m\angle RST + m\angle VSW = m\angle TSV$ | 7. Substitution |

*Exs. 11, 12*

## 1.5 Exercises

*In Exercises 1 to 6, which property justifies the conclusion of the statement?*

1. If $2x = 12$, then $x = 6$.
2. If $x + x = 12$, then $2x = 12$.
3. If $x + 5 = 12$, then $x = 7$.
4. If $x - 5 = 12$, then $x = 17$.
5. If $\frac{x}{5} = 3$, then $x = 15$.
6. If $3x - 2 = 13$, then $3x = 15$.

*In Exercises 7 to 10, state the property or definition that justifies the conclusion (the "then" clause).*

7. Given that $\angle$s 1 and 2 are supplementary, then $m\angle 1 + m\angle 2 = 180°$.
8. Given that $m\angle 3 + m\angle 4 = 180°$, then $\angle s$ 3 and 4 are supplementary.
9. Given $\angle RSV$ and $\overrightarrow{ST}$ as shown, then $m\angle RST + m\angle TSV = m\angle RSV$.
10. Given that $m\angle RST = m\angle TSV$, then $\overrightarrow{ST}$ bisects $\angle RSV$.

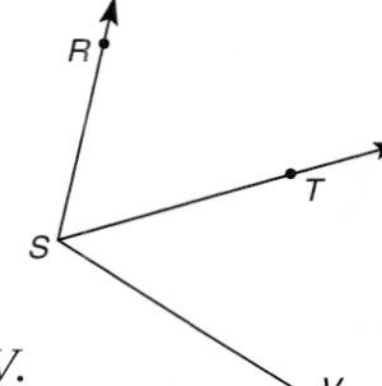


*Exercises 9, 10*

*In Exercises 11 to 22, use the Given information to draw a conclusion based on the stated property or definition.*

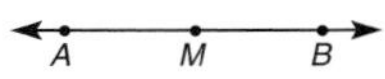


*Exercises 11, 12*

11. *Given:* *A-M-B*; Segment-Addition Postulate
12. *Given:* *M* is the midpoint of $\overline{AB}$; definition of midpoint
13. *Given:* $m\angle 1 = m\angle 2$; definition of angle bisector
14. *Given:* $\overrightarrow{EG}$ bisects $\angle DEF$; definition of angle bisector
15. *Given:* $\angle s$ 1 and 2 are complementary; definition of complementary angles
16. *Given:* $m\angle 1 + m\angle 2 = 90°$; definition of complementary angles

*Exercises 13–16*

17. *Given:* $2x - 3 = 7$; Addition Property of Equality
18. *Given:* $3x = 21$; Division Property of Equality
19. *Given:* $7x + 5 - 3 = 30$; Substitution Property of Equality
20. *Given:* $\frac{1}{2} = 0.5$ and $0.5 = 50\%$; Transitive Property of Equality
21. *Given:* $3(2x - 1) = 27$; Distributive Property
22. *Given:* $\frac{x}{5} = -4$; Multiplication Property of Equality

*In Exercises 23 and 24, fill in the missing reasons for the algebraic proof.*

23. *Given:* $3(x - 5) = 21$
    *Prove:* $x = 12$

**PROOF**

| Statements | Reasons |
|---|---|
| 1. $3(x - 5) = 21$ | 1. ? |
| 2. $3x - 15 = 21$ | 2. ? |
| 3. $3x = 36$ | 3. ? |
| 4. $x = 12$ | 4. ? |

24. *Given:* $2x + 9 = 3$
    *Prove:* $x = -3$

**PROOF**

| Statements | Reasons |
|---|---|
| 1. $2x + 9 = 3$ | 1. ? |
| 2. $2x = -6$ | 2. ? |
| 3. $x = -3$ | 3. ? |

*In Exercises 25 and 26, fill in the missing statements for the algebraic proof.*

25. *Given:* $2(x + 3) - 7 = 11$
    *Prove:* $x = 6$

**PROOF**

| Statements | Reasons |
|---|---|
| 1. ? | 1. Given |
| 2. ? | 2. Distributive Property |
| 3. ? | 3. Substitution (Addition) |
| 4. ? | 4. Addition Property of Equality |
| 5. ? | 5. Division Property of Equality |

26. *Given:* $\frac{x}{5} + 3 = 9$
    *Prove:* $x = 30$

**PROOF**

| Statements | Reasons |
|---|---|
| 1. ? | 1. Given |
| 2. ? | 2. Subtraction Property of Equality |
| 3. ? | 3. Multiplication Property of Equality |

*In Exercises 27 to 30, fill in the missing reasons for each geometric proof.*

*Exercises 27, 28*

27. *Given:* $D$-$E$-$F$ on $\overleftrightarrow{DF}$
    *Prove:* $DE = DF - EF$

**PROOF**

| Statements | Reasons |
|---|---|
| 1. $D$-$E$-$F$ on $\overleftrightarrow{DF}$ | 1. ? |
| 2. $DE + EF = DF$ | 2. ? |
| 3. $DE = DF - EF$ | 3. ? |

28. *Given:* $E$ is the midpoint of $\overline{DF}$
    *Prove:* $DE = \frac{1}{2}(\text{DF})$

**PROOF**

| Statements | Reasons |
|---|---|
| 1. $E$ is the midpoint of $\overline{DF}$ | 1. ? |
| 2. $DE = EF$ | 2. ? |
| 3. $DE + EF = DF$ | 3. ? |
| 4. $DE + DE = DF$ | 4. ? |
| 5. $2(DE) = DF$ | 5. ? |
| 6. $DE = \frac{1}{2}(DF)$ | 6. ? |

29. *Given:* $\overrightarrow{BD}$ bisects $\angle ABC$
    *Prove:* $m\angle ABD = \frac{1}{2}(m\angle ABC)$

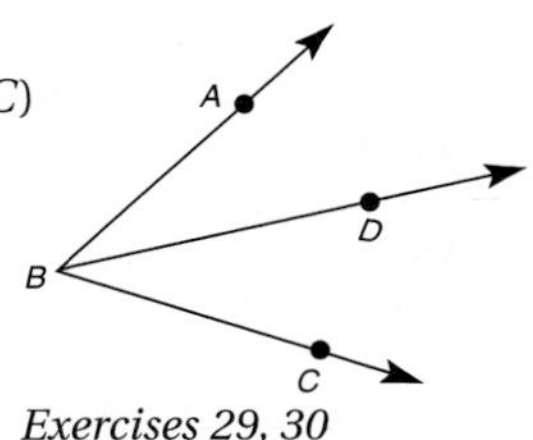

*Exercises 29, 30*

**PROOF**

| Statements | Reasons |
|---|---|
| 1. $\overrightarrow{BD}$ bisects $\angle ABC$ | 1. ? |
| 2. $m\angle ABD = m\angle DBC$ | 2. ? |
| 3. $m\angle ABD + m\angle DBC = m\angle ABC$ | 3. ? |
| 4. $m\angle ABD + m\angle ABD = m\angle ABC$ | 4. ? |
| 5. $2(m\angle ABD) = m\angle ABC$ | 5. ? |
| 6. $m\angle ABD = \frac{1}{2}(m\angle ABC)$ | 6. ? |

30. *Given:* $\angle ABC$ and $\overrightarrow{BD}$
    *Prove:* $m\angle ABD = m\angle ABC - m\angle DBC$

**PROOF**

| Statements | Reasons |
|---|---|
| 1. $\angle ABC$ and $\overrightarrow{BD}$ | 1. ? |
| 2. $m\angle ABD + m\angle DBC = m\angle ABC$ | 2. ? |
| 3. $m\angle ABD = m\angle ABC - m\angle DBC$ | 3. ? |

*In Exercises 31 and 32, fill in the missing statements and reasons.*

31. *Given:* $M$-$N$-$P$-$Q$ on $\overline{MQ}$
*Prove:* $MN + NP + PQ = MQ$

**PROOF**

| Statements | Reasons |
|---|---|
| 1. ? | 1. ? |
| 2. $MN + NQ = MQ$ | 2. ? |
| 3. $NP + PQ = NQ$ | 3. ? |
| 4. ? | 4. Substitution Property of Equality |

32. *Given:* $\angle TSW$ with $\overrightarrow{SU}$ and $\overrightarrow{SV}$
*Prove:* $m\angle TSW = m\angle TSU + m\angle USV + m\angle VSW$

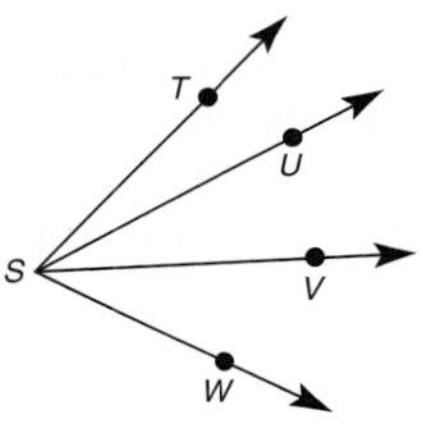


**PROOF**

| Statements | Reasons |
|---|---|
| 1. ? | 1. ? |
| 2. $m\angle TSW = m\angle TSU + m\angle USW$ | 2. ? |
| 3. $m\angle USW = m\angle USV + m\angle VSW$ | 3. ? |
| 4. ? | 4. Substitution Property of Equality |

33. When the Distributive Property is written in its *symmetric* form, it reads $a \cdot b + a \cdot c = a(b + c)$. Use this form to rewrite $5x + 5y$.

34. Another form of the Distributive Property (see Exercise 33) reads $b \cdot a + c \cdot a = (b + c)a$. Use this form to rewrite $5x + 7x$. Then simplify.

35. The Multiplication Property of Inequality requires that we *reverse* the inequality symbol when multiplying by a *negative* number. Given that $-7 < 5$, form the inequality that results when we multiply each side by $-2$.

36. The Division Property of Inequality requires that we *reverse* the inequality symbol when dividing by a *negative* number. Given that $12 > -4$, form the inequality that results when we divide each side by $-4$.

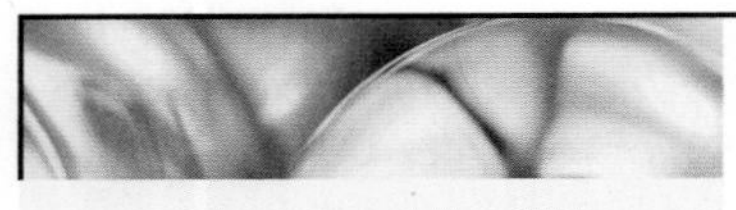

# 1.6 Relationships: Perpendicular Lines

**KEY CONCEPTS**

Vertical Line(s) • Horizontal Line(s) • Perpendicular Lines • Relations: Reflexive, Symmetric, and Transitive Properties • Equivalence Relation • Perpendicular-Bisector of a Line Segment

Informally, a **vertical** line is one that extends up and down, like a flagpole. On the other hand, a line that extends left to right is **horizontal.** In Figure 1.59, which follows, $\ell$ is vertical and $j$ is horizontal. Where lines $\ell$ and $j$ intersect, they appear to form angles of equal measure.

DEFINITION: **Perpendicular lines** are two lines that meet to form congruent adjacent angles.

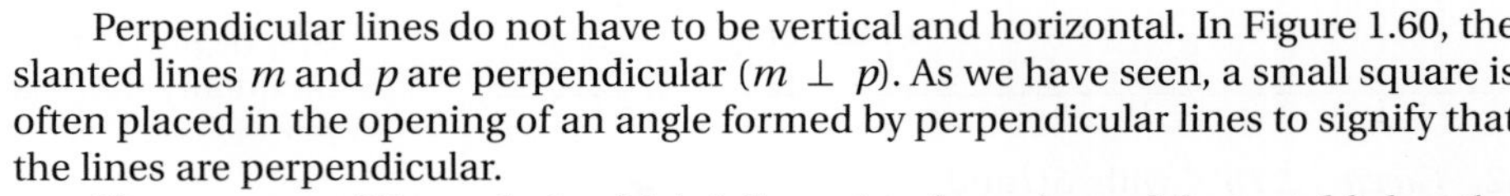

Perpendicular lines do not have to be vertical and horizontal. In Figure 1.60, the slanted lines $m$ and $p$ are perpendicular ($m \perp p$). As we have seen, a small square is often placed in the opening of an angle formed by perpendicular lines to signify that the lines are perpendicular.

The purpose of Example 1, which follows, is a formal proof that establishes the relationship between perpendicular lines and right angles. Study this proof, noting the order of the statements and reasons. The numbers in parentheses to the left of the statements refer to the earlier statement(s) on which the new statement is based.

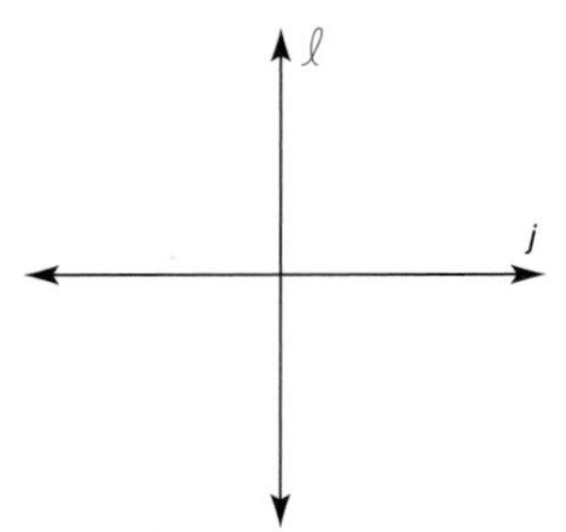


**FIGURE 1.59**

THEOREM 1.6.1: If two lines are perpendicular, then they meet to form right angles.

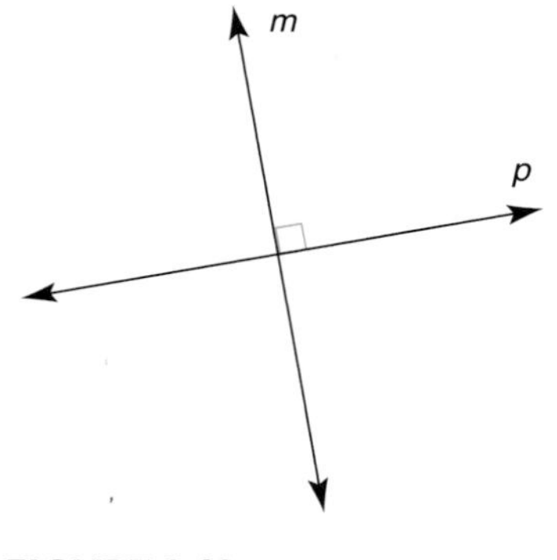


**FIGURE 1.60**

EXAMPLE 1

*Given:* $\overleftrightarrow{AB} \perp \overleftrightarrow{CD}$, intersecting at $E$ (Figure 1.61 on page 47)

*Prove:* $\angle AEC$ is a right angle

**PROOF**

| | Statements | Reasons |
|---|---|---|
| | 1. $\overleftrightarrow{AB} \perp \overleftrightarrow{CD}$, intersecting at $E$ | 1. Given |
| (1) | 2. $\angle AEC \cong \angle CEB$ | 2. Perpendicular lines meet to form congruent adjacent angles (Definition) |
| (2) | 3. $m\angle AEC = m\angle CEB$ | 3. If two angles are congruent, their measures are equal |
| | 4. $\angle AEB$ is a straight angle and $m\angle AEB = 180°$ | 4. Measure of a straight angle equals 180° |
| | 5. $m\angle AEC + m\angle CEB = m\angle AEB$ | 5. Angle-Addition Postulate |
| (4), (5) | 6. $m\angle AEC + m\angle CEB = 180°$ | 6. Substitution |
| (3), (6) | 7. $m\angle AEC + m\angle AEC = 180°$ or $2 \cdot m\angle AEC = 180°$ | 7. Substitution |
| (7) | 8. $m\angle AEC = 90°$ | 8. Division Property of Equality |
| (8) | 9. $\angle AEC$ is a right angle | 9. If the measure of an angle is 90°, then the angle is a right angle |

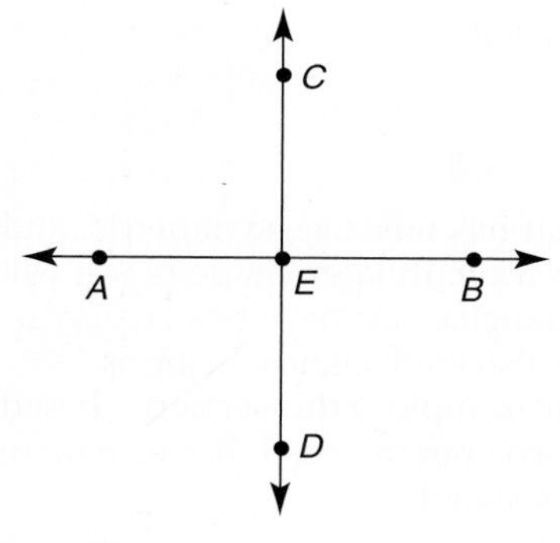


**FIGURE 1.61**

## RELATIONS

The relationship between perpendicular lines suggests the more general, but undefined, mathematical concept of **relation.** In general, a relation "connects" two elements of an associated set of objects. Table 1.5 provides several examples of this concept.

*Exs. 1, 2*

**TABLE 1.5**

| *Relation* | *Objects Related* | *Example of Relationship* |
|---|---|---|
| is equal to | numbers | $2 + 3 = 5$ |
| is greater than | numbers | $7 > 5$ |
| is perpendicular to | lines | $\ell \perp m$ |
| is complementary to | angles | $\angle 1$ is comp. to $\angle 2$ |
| is congruent to | line segments | $\overline{AB} \cong \overline{CD}$ |
| is a brother of | people | Matt is a brother of Phil |

Reminder 

Numbers that measure may be **equal** ($AB = CD$ or $m\angle 1 = m\angle 2$) whereas geometric figures may be **congruent** ($\overline{AB} \cong \overline{CD}$ or $\angle 1 \cong \angle 2$).

There are three special properties that may exist for a given relation R. Where *a, b,* and *c* are associated objects for relation R, the properties consider one object (reflexive), two objects in either order (symmetric), or three objects (transitive). For the properties to exist, it is necessary that the statements be true for all objects selected from the associated set. These properties are generalized and given examples as follows:

**Reflexive property:** *a*R*a* ($5 = 5$; equality of numbers has a reflexive property)
**Symmetric property:** If *a*R*b*, then *b*R*a*. (If $\ell \perp m$, then $m \perp \ell$; perpendicularity of lines has a symmetric property)
**Transitive property:** If *a*R*b* and *b*R*c*, then *a*R*c*. (If $\angle 1 \cong \angle 2$ and $\angle 2 \cong \angle 3$, then $\angle 1 \cong \angle 3$; congruence of angles has a transitive property)

### EXAMPLE 2

Does the relation "is less than" for numbers have a reflexive property? a symmetric property? a transitive property?

***Solution***

Because "$2 < 2$" is false, there is *no* reflexive property.
"If $2 < 5$, then $5 < 2$" is also false; there is *no* symmetric property.
"If $2 < 5$ and $5 < 9$, then $2 < 9$" is true; there *is* a transitive property.

NOTE: The same results are obtained for choices other than 2, 5, and 9.

Congruence of angles (or of line segments) is closely tied to equality of angle measures (or segment measures) by the definition of congruence. The following list gives some properties of the congruence of angles.

*Reflexive:* $\angle 1 \cong \angle 1$; an angle is congruent to itself.

*Symmetric:* If $\angle 1 \cong \angle 2$, then $\angle 2 \cong \angle 1$.

*Transitive:* If $\angle 1 \cong \angle 2$ and $\angle 2 \cong \angle 3$, then $\angle 1 \cong \angle 3$.

*Exs. 3–9*

Any relation (such as congruence of angles) that has reflexive, symmetric, and transitive properties is known as an *equivalence relation.* In later chapters, we will see that congruence of triangles and similarity of triangles also have reflexive, symmetric, and transitive properties; these relations are also equivalence relations.

Returning to the formulation of a proof, the final example in this section is based on the fact that vertical angles are congruent when two lines intersect. Because there are two pairs of congruent angles, the Prove could be stated

*Prove:* $\angle 1 \cong \angle 3$ and $\angle 2 \cong \angle 4$

Such a conclusion is a conjunction and would be proved if both congruences were established. For simplicity, the Prove of Example 3 is stated

*Prove:* $\angle 2 \cong \angle 4$

Study this proof of Theorem 1.6.2, noting the order of the statements and reasons.

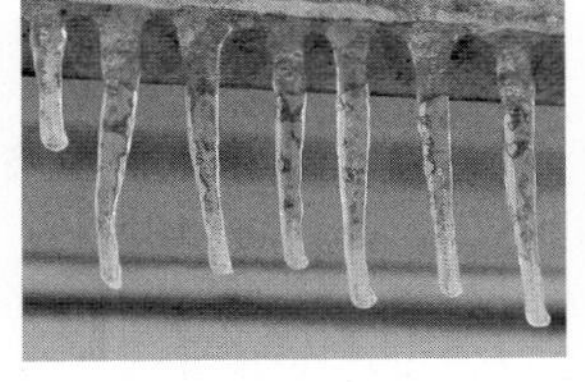

An icicle formed from freezing water assumes a vertical path.

THEOREM 1.6.2: If two lines intersect, then the vertical angles formed are congruent.

EXAMPLE 3

*Given:* $\overleftrightarrow{AC}$ intersects $\overleftrightarrow{BD}$ at $O$ (Figure 1.62)

*Prove:* $\angle 2 \cong \angle 4$

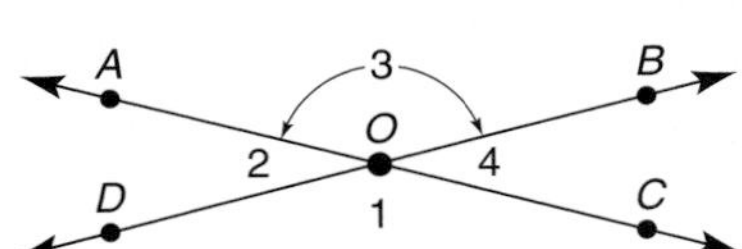


**FIGURE 1.62**

**PROOF**

| Statements | Reasons |
|---|---|
| 1. $\overleftrightarrow{AC}$ intersects $\overleftrightarrow{BD}$ at $O$ | 1. Given |
| 2. $\angle$s $AOC$ and $DOB$ are straight $\angle$s, with $m\angle AOC = 180$ and $m\angle DOB = 180$ | 2. The measure of a straight angle is 180° |
| 3. $m\angle AOC = m\angle DOB$ | 3. Substitution |
| 4. $m\angle 1 + m\angle 4 = m\angle DOB$ and $m\angle 1 + m\angle 2 = m\angle AOC$ | 4. Angle-Addition Postulate |
| 5. $m\angle 1 + m\angle 4 = m\angle 1 + m\angle 2$ | 5. Substitution |
| 6. $m\angle 4 = m\angle 2$ | 6. Subtraction Property of Equality |
| 7. $\angle 4 \cong \angle 2$ | 7. If two angles are equal in measure, the angles are congruent |
| 8. $\angle 2 \cong \angle 4$ | 8. Symmetric Property of Congruence of Angles |

**Technology Exploration**

Use computer software if available.

1) Construct $\overleftrightarrow{AC}$ and $\overleftrightarrow{DB}$ to intersect at point $O$. (See Figure 1.62).

2) Measure $\angle 1$, $\angle 2$, $\angle 3$, and $\angle 4$.

3) Show that $m\angle 1 = m\angle 3$ and $m\angle 2 = m\angle 4$.

In the preceding proof, the degree symbol (°) has been omitted from the statements, and this will continue to be done in future proofs. Moreover, there is no need to reorder the congruent angles from statement 7 to statement 8 because congruence of angles is symmetric; in the later work, statement 7 will be written to match the

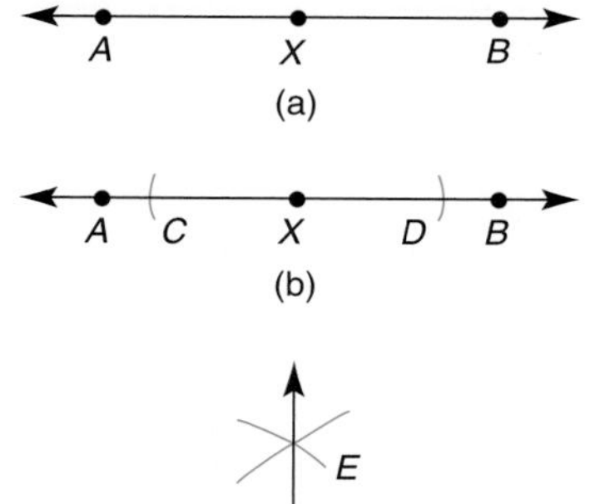


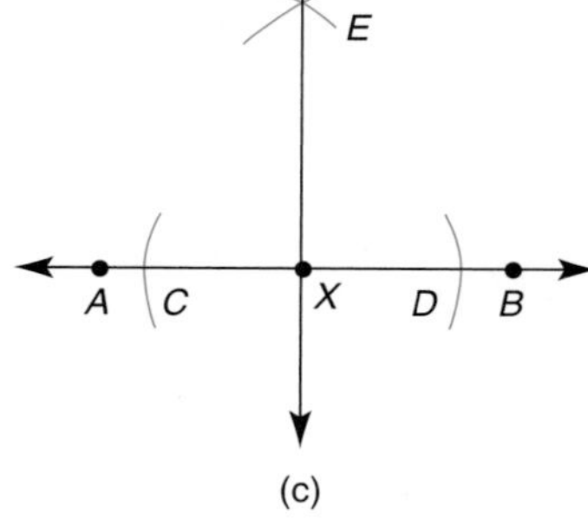


**FIGURE 1.63**

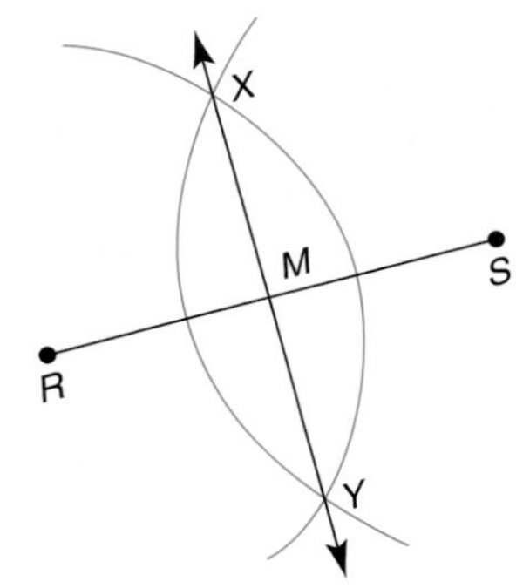


**FIGURE 1.64**

*Exs. 10–14*

Prove statement even if the previous line does not have the same order. The same type of thinking applies to proving lines perpendicular or parallel: the order is simply not important!

## CONSTRUCTIONS LEADING TO PERPENDICULAR LINES

Construction 2 in Section 1.2 determined not only the midpoint of $\overline{AB}$ but also that of the **perpendicular bisector** of $\overline{AB}$. In many instances, we need the perpendicular line at a point other than the midpoint of a segment.

---

CONSTRUCTION 5:

To construct the line perpendicular to a given line at a specified point on the given line.

*Given:* $\overleftrightarrow{AB}$ with point $X$ in Figure 1.63(a)

*Construct:* A line $\overleftrightarrow{EX}$, so that $\overleftrightarrow{EX} \perp \overleftrightarrow{AB}$

*Construction:* *Figure 1.63(b):* Using $X$ as the center, mark off arcs of equal radii on each side of $X$ to intersect $\overleftrightarrow{AB}$ at $C$ and $D$.

*Figure 1.63(c):* Now, using $C$ and $D$ as centers, mark off arcs of equal radii with a length greater than $XD$ so that these arcs intersect either above (as shown) or below $\overleftrightarrow{AB}$.

Calling the point of intersection $E$, draw $\overleftrightarrow{EX}$, which is the desired perpendicular line.

---

The theorem that Construction 5 is based on is a consequence of the Protractor Postulate, and we state it without proof.

> THEOREM 1.6.3: In a plane, there is exactly one line perpendicular to a given line at any point on the line.

Construction 2, which was used to locate the midpoint of a line segment in Section 1.2, is also the method for constructing the perpendicular bisector of a line segment. In Figure 1.64, $\overleftrightarrow{XY}$ is the perpendicular bisector of $\overline{RS}$. The following theorem can be proved by methods developed later in this book.

> THEOREM 1.6.4: The perpendicular bisector of a line segment is unique.

## 1.6 Exercises

*In Exercises 1 and 2, supply reasons.*

1. *Given:* $\angle 1 \cong \angle 3$
   *Prove:* $\angle MOP \cong \angle NOQ$

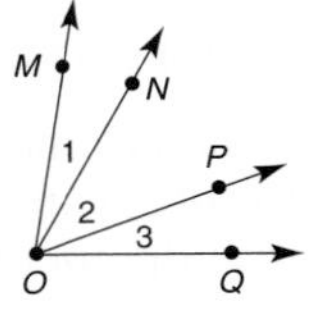


**PROOF**

| Statements | Reasons |
|---|---|
| 1. $\angle 1 \cong \angle 3$ | 1. ? |
| 2. $m\angle 1 = m\angle 3$ | 2. ? |
| 3. $m\angle 1 + m\angle 2 = m\angle MOP$ and $m\angle 2 + m\angle 3 = m\angle NOQ$ | 3. ? |
| 4. $m\angle 1 + m\angle 2 = m\angle 2 + m\angle 3$ | 4. ? |
| 5. $m\angle MOP = m\angle NOQ$ | 5. ? |
| 6. $\angle MOP \cong \angle NOQ$ | 6. ? |

2. *Given:* $\overleftrightarrow{AB}$ intersects $\overleftrightarrow{CD}$ at $O$ so that $\angle 1$ is a right $\angle$ (Use the figure at the top of the next column.)
   *Prove:* $\angle 2$ and $\angle 3$ are complementary

**PROOF**

| Statements | Reasons |
|---|---|
| 1. $\overleftrightarrow{AB}$ intersects $\overleftrightarrow{CD}$ at $O$ | 1. ? |
| 2. $\angle AOB$ is a straight $\angle$, so $m\angle AOB = 180$ | 2. ? |
| 3. $m\angle 1 + m\angle COB = m\angle AOB$ | 3. ? |
| 4. $m\angle 1 + m\angle COB = 180$ | 4. ? |
| 5. $\angle 1$ is a right angle | 5. ? |
| 6. $m\angle 1 = 90$ | 6. ? |
| 7. $90 + m\angle COB = 180$ | 7. ? |
| 8. $m\angle COB = 90$ | 8. ? |
| 9. $m\angle 2 + m\angle 3 = m\angle COB$ | 9. ? |
| 10. $m\angle 2 + m\angle 3 = 90$ | 10. ? |
| 11. $\angle 2$ and $\angle 3$ are complementary | 11. ? |

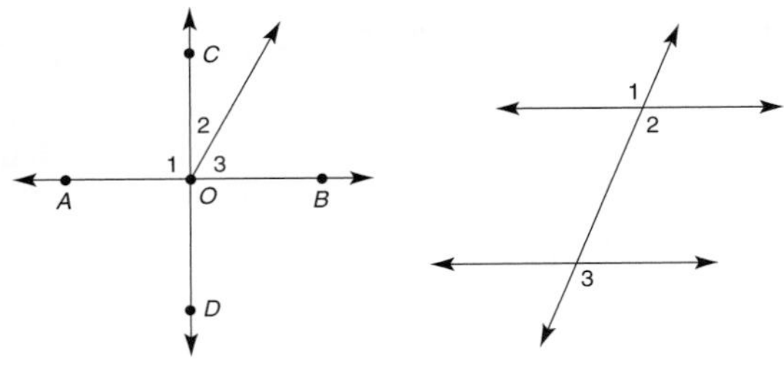


*Exercise 2* *Exercise 3*

*In Exercises 3 and 4, supply statements.*

3. *Given:* $\angle 1 \cong \angle 2$ and $\angle 2 \cong \angle 3$
   *Prove:* $\angle 1 \cong \angle 3$

**PROOF**

| Statements | Reasons |
|---|---|
| 1. ? | 1. Given |
| 2. ? | 2. Transitive Property of Congruence |

4. *Given:* $m\angle AOB = m\angle 1$
   $m\angle BOC = m\angle 1$
   *Prove:* $\overrightarrow{OB}$ bisects $\angle AOC$

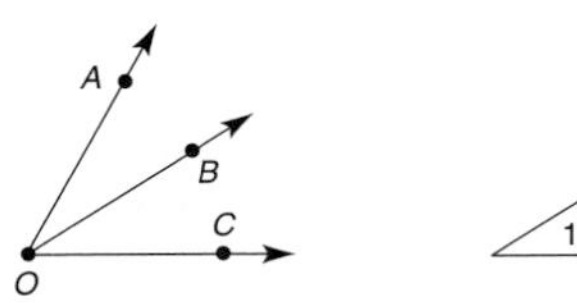


**PROOF**

| Statements | Reasons |
|---|---|
| 1. ? | 1. Given |
| 2. ? | 2. Substitution |
| 3. ? | 3. Angles with equal measures are congruent |
| 4. ? | 4. If a ray divides an angle into two congruent angles, then the ray bisects the angle |

*In Exercises 5 to 9, use a compass and a straightedge to complete the constructions.*

5. *Given:* Point $N$ on line $s$
   *Construct:* Line $m$ through $N$ so that $m \perp s$

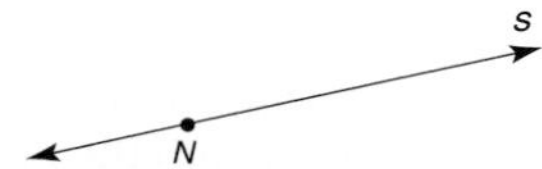


6. *Given:* $\overrightarrow{OA}$
   *Construct:* Right angle $BOA$
   (HINT: Use a straightedge to extend $\overrightarrow{OA}$ to the left.)

7. *Given:* Line $\ell$ containing point $A$
   *Construct:* A 45° angle with vertex at $A$

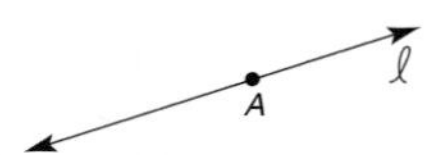


8. *Given:* $\overline{AB}$
   *Construct:* The perpendicular bisector of $\overline{AB}$

9. *Given:* Triangle $ABC$
   *Construct:* The perpendicular bisectors of sides $\overline{AB}$, $\overline{AC}$, and $\overline{BC}$

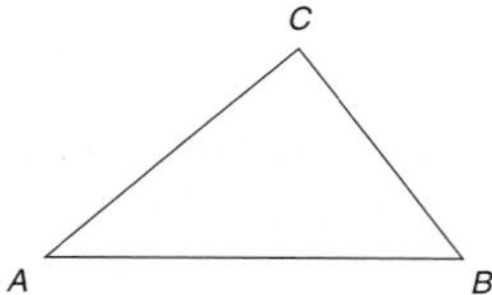


10. Draw a conclusion based on the results of Exercise 9.

*In Exercises 11 and 12, provide the missing statements and reasons.*

11. *Given:* $\angle$s 1 and 3 are complementary
    $\angle$s 2 and 3 are complementary
    *Prove:* $\angle 1 \cong \angle 2$

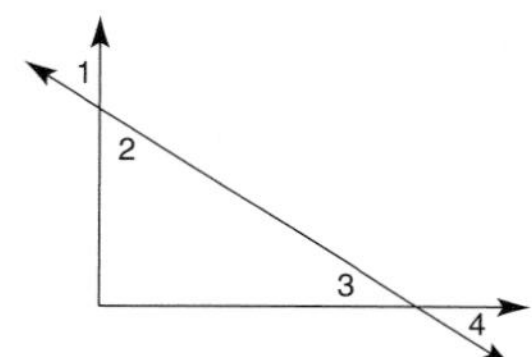


**PROOF**

| | Statements | Reasons |
|---|---|---|
| | 1. $\angle$s 1 and 3 are complementary; $\angle$s 2 and 3 are complementary | 1. ? |
| | 2. $m\angle 1 + m\angle 3 = 90$; $m\angle 2 + m\angle 3 = 90$ | 2. The sum of the measures of complementary $\angle$s is 90 |
| (2) | 3. $m\angle 1 + m\angle 3 = m\angle 2 + m\angle 3$ | 3. ? |
| | 4. ? | 4. Subtraction Property of Equality |
| (4) | 5. ? | 5. If two $\angle$s are = in measure, they are $\cong$ |

12. *Given:* $\angle 1 \cong \angle 2$; $\angle 3 \cong \angle 4$
    $\angle$s 2 and 3 are complementary
    *Prove:* $\angle$s 1 and 4 are complementary

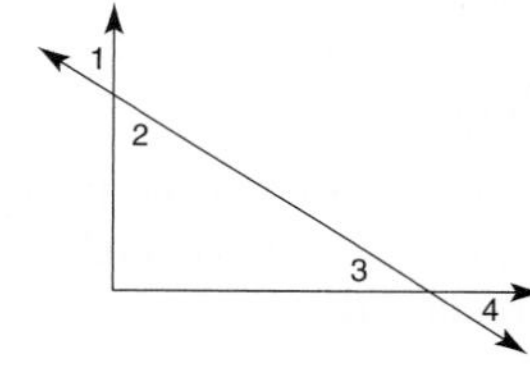


**PROOF**

| | Statements | Reasons |
|---|---|---|
| | 1. $\angle 1 \cong \angle 2$ and $\angle 3 \cong \angle 4$ | 1. ? |
| | 2. ? and ? | 2. If two $\angle$s are $\cong$, then their measures are equal |
| | 3. $\angle$s 2 and 3 are complementary | 3. ? |
| (3) | 4. ? | 4. The sum of the measures of complementary $\angle$s is 90 |
| (2), (4) | 5. $m\angle 1 + m\angle 4 = 90$ | 5. ? |
| | 6. ? | 6. If the sum of the measures of two angles is 90, then the angles are complementary |

13. Does the relation "is perpendicular to" have a reflexive property (consider line $\ell$)? a symmetric property (consider lines $\ell$ and $m$)? a transitive property (consider lines $\ell$, $m$, and $n$)?

14. Does the relation "is greater than" have a reflexive property (consider real number $a$)? a symmetric property (consider real numbers $a$ and $b$)? a transitive property (consider real numbers $a$, $b$, and $c$)?

15. Does the relation "is complementary to" for angles have a reflexive property (consider one angle)? a symmetric property (consider two angles)? a transitive property (consider three angles)?

16. Does the relation "is less than" for numbers have a reflexive property (consider one number)? a symmetric property (consider two numbers)? a transitive property (consider three numbers)?

17. Does the relation "is a brother of" have a reflexive property (consider one male)? a symmetric property (consider two males)? a transitive property (consider three males)?

18. Does the relation "is in love with" have a reflexive property (consider one person)? a symmetric property (consider two people)? a transitive property (consider three people)?

19. By this time, the text has used numerous symbols and abbreviations. In this exercise, indicate what word is represented or abbreviated by each of the following:

a) $\perp$
b) $\angle$s
c) supp.
d) rt.
e) m$\angle 1$
f) adj.
g) comp.
h) $\overrightarrow{AB}$
i) $\cong$
j) vert.

20. If there were no understood restriction to lines in a plane in Theorem 1.6.3, the theorem would be false. Explain why the following statement is false: "In space, there is exactly one line perpendicular to a given line at any point on the line."

21. Prove the Extended Segment Addition Property by using the Drawing, the Given, and the Prove that follow.

*Given:* $M$-$N$-$P$-$Q$ on $\overline{MQ}$
*Prove:* $MN + NP + PQ = MQ$

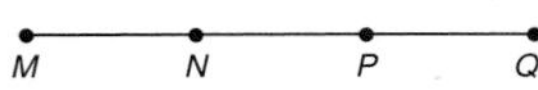


22. The Segment Addition Postulate can be generalized as follows: "The length of a line segment equals the sum of the lengths of its parts." State a general conclusion about $AE$ based on the following figure.

23. Prove the Extended Angle Addition Property by using the Drawing, the Given, and the Prove that follow.

*Given:* $\angle TSW$ with $\overrightarrow{SU}$ and $\overrightarrow{SV}$
*Prove:* m$\angle TSW$ = m$\angle TSU$ + m$\angle USV$ + m$\angle VSW$

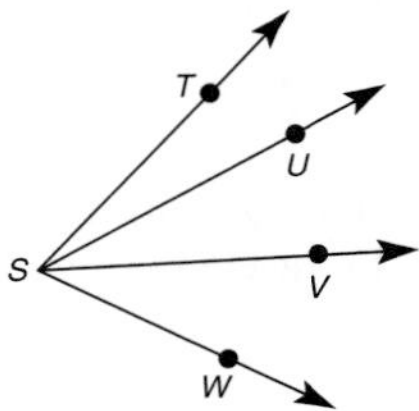


24. The Angle Addition Postulate can be generalized as follows: "The measure of an angle is equal the sum of the measures of its parts." State a general conclusion about m$\angle GHK$ based on the following figure.

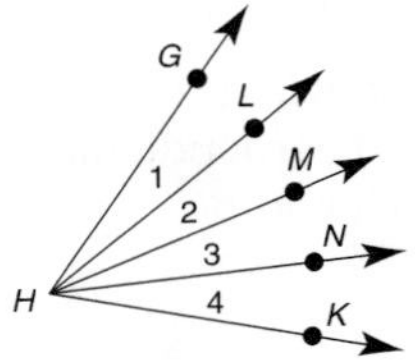


25. If there were no understood restriction to lines in a plane in Theorem 1.6.4, the theorem would be false. Explain why the following statement is false: "In space, the perpendicular bisector of a line segment is unique."

* 26. In the proof that follows on page 53, provide the missing reasons.

*Given:* $\angle 1$ and $\angle 2$ are complementary
$\angle 1$ is acute
*Prove:* $\angle 2$ is also acute

PROOF

| | Statements | Reasons |
|---|---|---|
| | 1. $\angle 1$ and $\angle 2$ are complementary | 1. ? |
| (1) | 2. $m\angle 1 + m\angle 2 = 90$ | 2. ? |
| | 3. $\angle 1$ is acute | 3. ? |
| (3) | 4. Where $m\angle 1 = x$, $0 < x < 90$ | 4. ? |
| (2) | 5. $x + m\angle 2 = 90$ | 5. ? |
| (5) | 6. $m\angle 2 = 90 - x$ | 6. ? |
| (4) | 7. $-x < 0 < 90 - x$ | 7. ? |
| (7) | 8. $90 - x < 90 < 180 - x$ | 8. ? |
| (7, 8) | 9. $0 < 90 - x < 90$ | 9. ? |
| (6, 9) | 10. $0 < m\angle 2 < 90$ | 10. ? |
| (10) | 11. $\angle 2$ is acute | 11. ? |

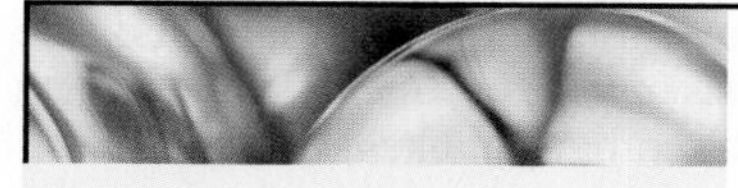

**KEY CONCEPTS**

Formal Proof of a Theorem • Converse of a Theorem • Picture Proof (Informal) of a Theorem

# 1.7 The Formal Proof of a Theorem

Recall from Section 1.3 that statements that follow logically from known undefined terms, definitions, and postulates are called theorems, statements that can be proved. The formal proof of a theorem has several parts. To begin to understand how these parts are related, you need to consider carefully the terms *hypothesis* and *conclusion.* The hypothesis of a statement describes the given situation (Given), whereas the conclusion describes what you need to establish (Prove). When a statement has the form "If H, then C," the hypothesis is the H statement and the conclusion is the C statement. Some theorems must be reworded to fit into "If . . . , then . . ." form so that the hypothesis and conclusion are easy to recognize.

EXAMPLE 1

Give the hypothesis H and conclusion C for each of these statements.

a) If two lines intersect, then the vertical angles formed are congruent.
b) All right angles are congruent.
c) Parallel lines do not intersect.
d) Lines are perpendicular when they meet to form congruent adjacent angles.

***Solution***

a) As is — H: Two lines intersect.
C: The vertical angles formed are congruent.

b) Reworded — If two angles are right angles, then these angles are congruent.
H: Two angles are right angles.
C: The angles are congruent.

c) Reworded — If two lines are parallel, then these lines do not intersect.
H: Two lines are parallel.
C: The lines do not intersect.

d) Reordered — When (if) two lines meet to form congruent adjacent angles, these lines are perpendicular.
H: Two lines meet to form congruent adjacent angles.
C: The lines are perpendicular.

*Exs. 1–3*

Why do we need to distinguish between the hypothesis and the conclusion? For a theorem, the hypothesis determines the Drawing and the Given, providing a description of the Drawing's known characteristics. The conclusion determines the relationship (the Prove) that you wish to establish in the Drawing.

## THE WRITTEN PARTS OF A FORMAL PROOF

The five necessary parts of a formal proof are listed in the accompanying box in the order in which they should be developed.

> ESSENTIAL PARTS OF THE FORMAL PROOF OF A THEOREM
> 1. *Statement:* States the theorem to be proved.
> 2. *Drawing:* Represents the hypothesis of the theorem.
> 3. *Given:* Describes the Drawing according to the information found in the hypothesis of the theorem.
> 4. *Prove:* Describes the Drawing according to the claim made in the conclusion of the theorem.
> 5. *Proof:* Orders a list of claims (Statements) and justifications (Reasons), beginning with the Given and ending with the Prove; there must be a logical flow in this Proof.

The most difficult aspect of a formal proof is the thinking process that must take place between parts 4 and 5. This game plan or analysis involves deducing and ordering conclusions based on the given situation. One must be somewhat like a lawyer—selecting the claims that help prove the case, while discarding those that are superfluous. In the process of ordering the statements, it may be beneficial to think in reverse order, like so:

*The Prove statement would be true if what else were true?*

The final proof must be arranged in an order that allows one to reason from an earlier statement to a later claim by using deduction (perhaps several times).

| | | |
|---|---|---|
| H: hypothesis | $\longleftarrow$ | statement of proof |
| P: principle | $\longleftarrow$ | reason of proof |
| $\therefore$ C: conclusion | $\longleftarrow$ | next statement in proof |

Consider the following theorem, which was proved in Example 1 of Section 1.6.

> THEOREM 1.6.1: If two lines are perpendicular, then they meet to form right angles.

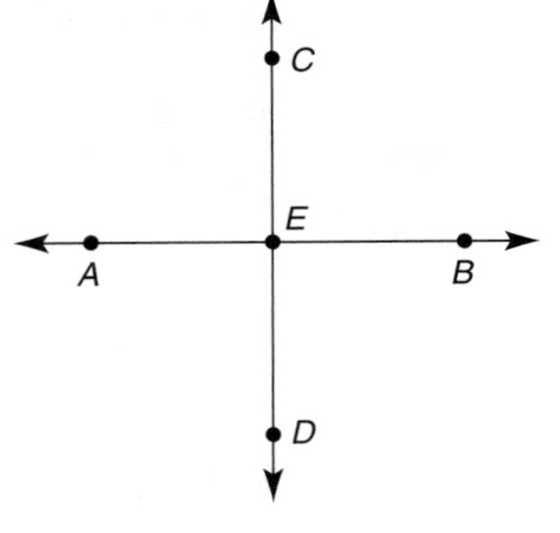

**FIGURE 1.65**

### EXAMPLE 2

Write the parts of the formal proof of Theorem 1.6.1.

***Solution***

1. State the theorem.
   *If two lines are perpendicular, then they meet to form right angles.*
2. The hypothesis is H: Two lines are perpendicular.
   Make a Drawing to fit this description. (See Figure 1.65.)
3. Write the Given statement, using the Drawing and based on the hypothesis H: Two lines are $\perp$.
   *Given:* $\overleftrightarrow{AB} \perp \overleftrightarrow{CD}$ intersecting at E
4. Write the Prove statement, using the Drawing and based on the conclusion C: They meet to form right angles.
   *Prove:* $\angle AEC$ is a right angle.
5. Construct the Proof. This formal proof is found in Example 1, Section 1.6.

*Exs. 4, 5*

## CONVERSE OF A STATEMENT

The converse of the statement "If $P$, then $Q$" is "If $Q$, then $P$." That is, the converse of a given statement interchanges its hypothesis and conclusion. Consider the following:

*Statement:* If a person lives in London, then that person lives in England.

*Converse:* If a person lives in England, then that person lives in London.

In this case, the given statement is true, whereas its converse is false. For more information on converse statements, see Section 2.2. Sometimes the converse of a true statement is also true. In fact, Example 3 presents the formal proof of a theorem that is the converse of Theorem 1.6.1.

Once a theorem has been proved, it may be cited thereafter as a reason in future proofs. Thus any theorem found in this section can be used for justification in later sections.

Warning (!)
You should not make a drawing that embeds qualities beyond those described in the hypothesis; neither should your drawing indicate fewer qualities than the hypothesis prescribes!

The proof that follows is nearly complete! It is difficult to provide a complete formal proof that explains the "how to" and simultaneously presents the final polished form. Example 2 aims only at the "how to", whereas Example 3 illustrates the polished form. What you do not see in Example 3 are the thought process and the scratch paper needed to piece this puzzle together.

The proof of a theorem is not unique! From the start, students' Drawings need not match, although the same relationships should be indicated. Certainly, different letters are likely to be chosen in illustrating the hypothesis.

THEOREM 1.7.1: If two lines meet to form a right angle, then these lines are perpendicular.

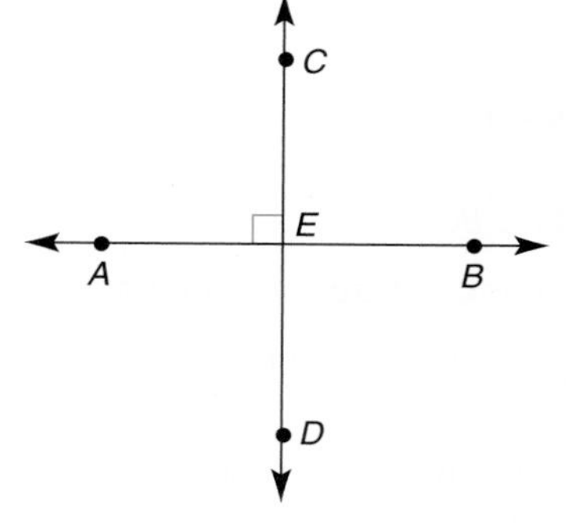


**FIGURE 1.66**

EXAMPLE 3

Give a formal proof for Theorem 1.7.1.

*If two lines meet to form a right angle, then these lines are perpendicular.*

*Given:* $\overleftrightarrow{AB}$ and $\overleftrightarrow{CD}$ intersect at $E$ so that $\angle AEC$ is a right angle (Figure 1.66)

*Prove:* $\overleftrightarrow{AB} \perp \overleftrightarrow{CD}$

**PROOF**

| | Statements | Reasons |
|---|---|---|
| | 1. $\overleftrightarrow{AB}$ and $\overleftrightarrow{CD}$ intersect so that $\angle AEC$ is a right angle | 1. Given |
| | 2. $m\angle AEC = 90$ | 2. If an $\angle$ is a right $\angle$, its measure is 90 |
| | 3. $\angle AEB$ is a straight $\angle$, so $m\angle AEB = 180$ | 3. If an $\angle$ is a straight $\angle$, its measure is 180 |
| | 4. $m\angle AEC + m\angle CEB = m\angle AEB$ | 4. Angle-Addition Postulate |
| (2), (3), (4) | 5. $90 + m\angle CEB = 180$ | 5. Substitution |
| (5) | 6. $m\angle CEB = 90$ | 6. Subtraction Property of Equality |
| (2), (6) | 7. $m\angle AEC = m\angle CEB$ | 7. Substitution |
| | 8. $\angle AEC \cong \angle CEB$ | 8. If two $\angle$s have = measures, the $\angle$s are $\cong$ |
| | 9. $\overleftrightarrow{AB} \perp \overleftrightarrow{CD}$ | 9. If two lines form $\cong$ adjacent $\angle$s, these lines are $\perp$ |

*Exs. 6–8*

Several other theorems are now stated, the proofs of which are left as exercises. This list contains theorems that are quite useful when cited as reasons in later proofs. Only a formal proof of Theorem 1.7.6 is provided.

> THEOREM 1.7.2: If two angles are complementary to the same angle (or to congruent angles), then these angles are congruent.

See Exercise 21 for a drawing describing Theorem 1.7.2.

> THEOREM 1.7.3: If two angles are supplementary to the same angle (or to congruent angles), then these angles are congruent.

See Exercise 22 for a drawing describing Theorem 1.7.3.

> THEOREM 1.7.4: Any two right angles are congruent.

> THEOREM 1.7.5: If the exterior sides of two adjacent acute angles form perpendicular rays, then these angles are complementary.

For Theorem 1.7.5, we create an informal proof called a picture proof. Although such a proof is less detailed, the impact of the explanation is the same! This is the first of numerous "picture proofs" used in this textbook.

### *Technology Exploration*

Use computer software if available.

1) Draw $\overleftrightarrow{EG}$ containing point $F$. Also draw $\overrightarrow{FH}$ as in Figure 1.68.
2) Measure $\angle 3$ and $\angle 4$.
3) Show that $m\angle 3 + m\angle 4 = 180°$. (Answer may not be perfect.)

### PICTURE PROOF OF THEOREM 1.7.5

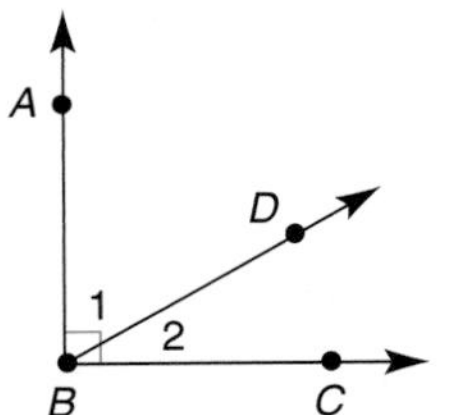


**FIGURE 1.67**

*Given:* $\overrightarrow{BA} \perp \overrightarrow{BC}$

*Prove:* $\angle 1$ and $\angle 2$ are complementary.

*Proof:* We see that $\angle 1$ and $\angle 2$ are parts of a right angle.

Then $m\angle 1 + m\angle 2 = 90°$, so $\angle 1$ and $\angle 2$ are complementary.

> EXAMPLE 4
>
> Study the formal proof of Theorem 1.7.6.

> THEOREM 1.7.6: If the exterior sides of two adjacent angles form a straight line, then these angles are supplementary.

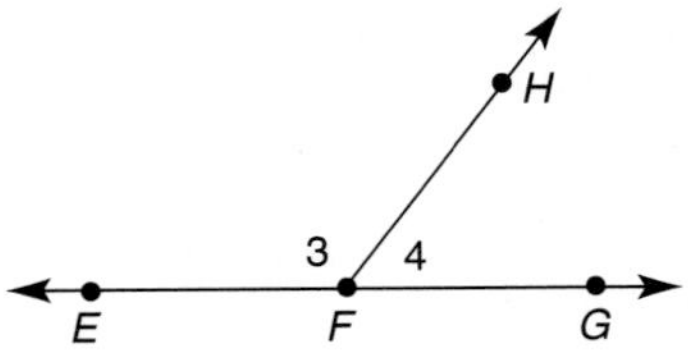

**FIGURE 1.68**

*Given:* $\angle 3$ and $\angle 4$ and $\overleftrightarrow{EG}$ (Figure 1.68)

*Prove:* $\angle 3$ and $\angle 4$ are supplementary

**PROOF**

| Statements | Reasons |
|---|---|
| 1. $\angle 3$ and $\angle 4$ and $\overleftrightarrow{EG}$ | 1. Given |
| 2. $m\angle 3 + m\angle 4 = m\angle EFG$ | 2. Angle-Addition Postulate |
| 3. $\angle EFG$ is a straight angle | 3. If the sides of an $\angle$ are opposite rays, it is a straight $\angle$ |
| 4. $m\angle EFG = 180$ | 4. The measure of a straight $\angle$ is 180 |
| 5. $m\angle 3 + m\angle 4 = 180$ | 5. Substitution |
| 6. $\angle 3$ and $\angle 4$ are supplementary | 6. If the sum of the measures of two $\angle$s is 180, the $\angle$s are supplementary |

*Exs. 9–12*

The final two theorems in this section are stated for convenience. When cited as "reasons," they will make later proofs easier to complete. We suggest that the student make drawings to illustrate Theorem 1.7.7 and Theorem 1.7.8.

THEOREM 1.7.7: If two line segments are congruent, then their midpoints separate these segments into four congruent segments.

*Exs. 13, 14*

THEOREM 1.7.8: If two angles are congruent, then their bisectors separate these angles into four congruent angles.

## 1.7 Exercises

*In Exercises 1 to 6, state the hypothesis H and the conclusion C for each statement.*

1. If a line segment is bisected, then each of the equal segments has half the length of the original segment.
2. If two sides of a triangle are congruent, then the triangle is isosceles.
3. All squares are quadrilaterals.
4. Every regular polygon has congruent interior angles.
5. Two angles are congruent if each is a right angle.
6. The lengths of corresponding sides of similar polygons are proportional.
7. Name, in order, the five parts of the formal proof of a theorem.
8. Which part (hypothesis or conclusion) of a theorem determines the

   a) Drawing? b) Given? c) Prove?

*For each theorem stated in Exercises 9 to 14, make a Drawing. On the basis of your Drawing, write a Given and a Prove for the theorem.*

9. If two lines are perpendicular, then these lines meet to form a right angle.
10. If two lines meet to form a right angle, then these lines are perpendicular.
11. If two angles are complementary to the same angle, then these angles are congruent.
12. If two angles are supplementary to the same angle, then these angles are congruent.
13. If two lines intersect, then the vertical angles formed are congruent.
14. Any two right angles are congruent.

*In Exercises 15 to 20, use the drawing at the right and apply the theorems of this section.*

15. If $m\angle 1 = 125°$, find $m\angle 2$, $m\angle 3$, and $m\angle 4$.

16. If $m\angle 2 = 47°$, find $m\angle 1$, $m\angle 3$, and $m\angle 4$.

17. If $m\angle 1 = 3x + 10$ and $m\angle 3 = 4x - 30$, find $x$ and $m\angle 1$.

18. If $m\angle 2 = 6x + 8$ and $m\angle 4 = 7x$, find $x$ and $m\angle 2$.

19. If $m\angle 1 = 2x$ and $m\angle 2 = x$, find $x$ and $m\angle 1$.

20. If $m\angle 2 = x + 15$ and $m\angle 3 = 2x$, find $x$ and $m\angle 2$.

*In Exercises 21 to 29, complete the formal proof of each theorem.*

21. If two angles are complementary to the same angle, then these angles are congruent.

*Given:* $\angle 1$ is comp. to $\angle 3$
$\angle 2$ is comp. to $\angle 3$
*Prove:* $\angle 1 \cong \angle 2$

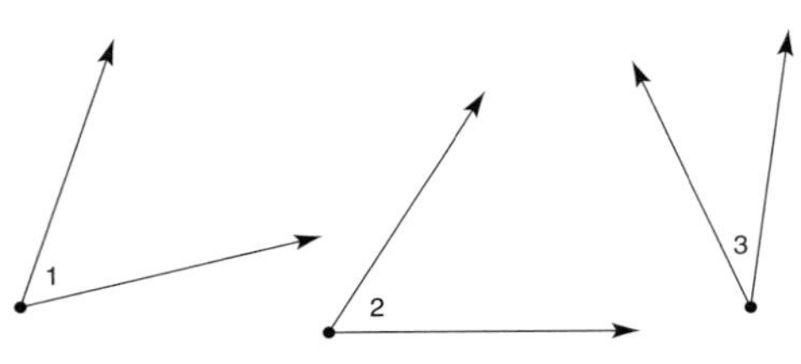


**PROOF**

| Statements | Reasons |
|---|---|
| 1. $\angle 1$ is comp. to $\angle 3$<br>$\angle 2$ is comp. to $\angle 3$ | 1. ? |
| 2. $m\angle 1 + m\angle 3 = 90$<br>$m\angle 2 + m\angle 3 = 90$ | 2. ? |
| 3. $m\angle 1 + m\angle 3 =$<br>$m\angle 2 + m\angle 3$ | 3. ? |
| 4. $m\angle 1 = m\angle 2$ | 4. ? |
| 5. $\angle 1 \cong \angle 2$ | 5. ? |

22. If two angles are supplementary to the same angle, then these angles are congruent.

*Given:* $\angle 1$ is supp. to $\angle 2$
$\angle 3$ is supp. to $\angle 2$
*Prove:* $\angle 1 \cong \angle 3$
(HINT: See Exercise 21 for help. The figure is in the next column.)

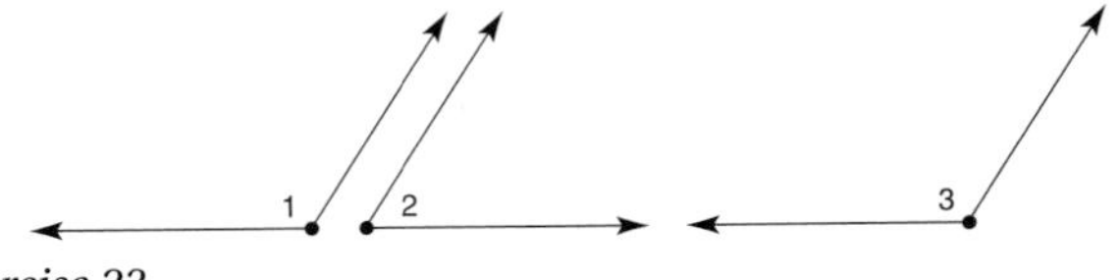


*Exercise 22*

23. If two lines intersect, the vertical angles formed are congruent.

24. Any two right angles are congruent.

25. If the exterior sides of two adjacent acute angles form perpendicular rays, then these angles are complementary.

*Given:* $\overrightarrow{BA} \perp \overrightarrow{BC}$
*Prove:* $\angle 1$ is comp. to $\angle 2$

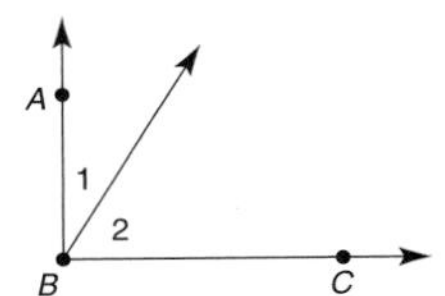


**PROOF**

| Statements | Reasons |
|---|---|
| 1. $\overrightarrow{BA} \perp \overrightarrow{BC}$ | 1. ? |
| 2. ? | 2. If two rays are $\perp$, then they meet to form a rt. $\angle$ |
| 3. $m\angle ABC = 90$ | 3. ? |
| 4. $m\angle ABC = m\angle 1 + m\angle 2$ | 4. ? |
| 5. $m\angle 1 + m\angle 2 = 90$ | 5. Substitution |
| 6. ? | 6. If the sum of the measures of two angles is 90, then the angles are complementary |

26. If two line segments are congruent, then their midpoints separate these segments into four congruent segments.

*Given:* $\overline{AB} \cong \overline{DC}$
$M$ is the midpoint of $\overline{AB}$
$N$ is the midpoint of $\overline{DC}$
*Prove:* $\overline{AM} \cong \overline{MB} \cong \overline{DN} \cong \overline{NC}$

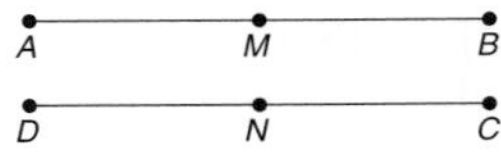

27. If two angles are congruent, then their bisectors separate these angles into four congruent angles.

Given: $\angle ABC \cong \angle EFG$
$\overrightarrow{BD}$ bisects $\angle ABC$
$\overrightarrow{FH}$ bisects $\angle EFG$
Prove: $\angle 1 \cong \angle 2 \cong \angle 3 \cong \angle 4$

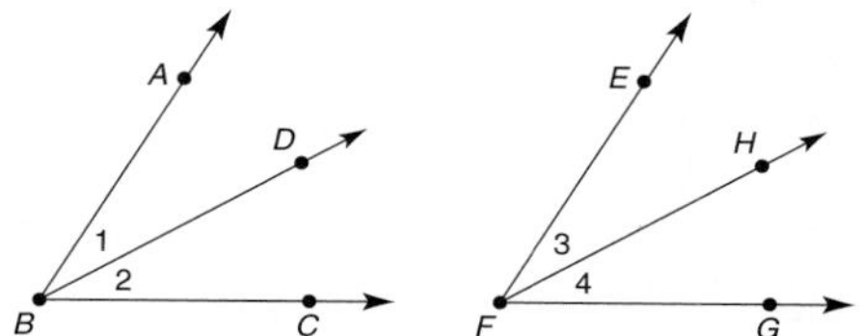


28. The bisectors of two adjacent supplementary angles form a right angle.

Given: $\angle ABC$ is supp. to $\angle CBD$
$\overrightarrow{BE}$ bisects $\angle ABC$
$\overrightarrow{BF}$ bisects $\angle CBD$
Prove: $\angle EBF$ is a right angle

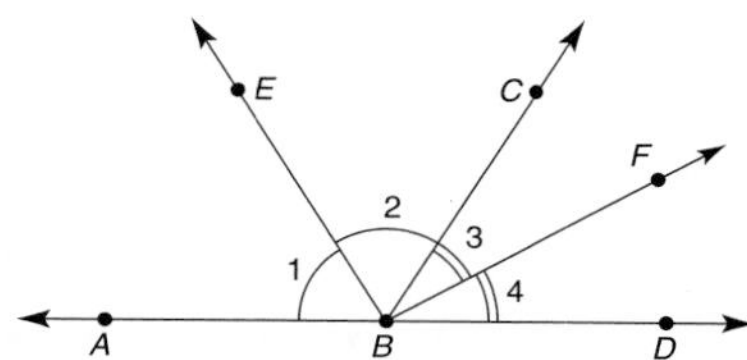


29. The supplement of an acute angle is an obtuse angle.

(HINT: Use Exercise 26 of Section 1.6 as a guide.)

# Perspective on History

## THE DEVELOPMENT OF GEOMETRY

One of the first written accounts of geometric knowledge appears in the Rhind papyrus, a collection of documents that date back to more than 1000 years before Christ. In this document, Ahmes (an Egyptian scribe) describes how north-south and east-west lines were redrawn following the overflow of the Nile River. Astronomy was used to lay out the north-south line. The rest was done by people known as "rope-fasteners." By tying knots in a rope, it was possible to separate the rope into segments with lengths that were in the ratio 3 to 4 to 5. The knots were fastened at stakes in such a way that a right triangle would be formed. In Figure 1.69, the right angle is formed so that one side (of length 4, as shown) lies in the north-south line, and the second side (of length 3, as shown) lies in the east-west line.

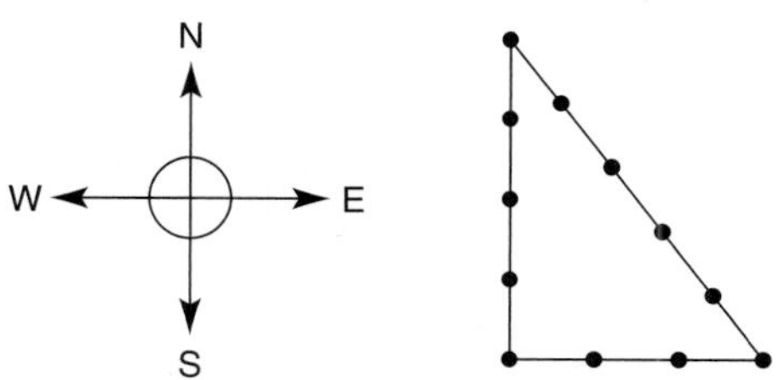


**FIGURE 1.69**

The principle that was used by the rope-fasteners is known as the Pythagorean Theorem. However, we also know that the ancient Chinese were aware of this relationship. That is, the Pythagorean Theorem was known and applied many centuries before the time of Pythagoras (for whom the theorem is named).

Ahmes describes other facts of geometry that were known to the Egyptians. Perhaps the most impressive of these facts was that their approximation of $\pi$ was 3.1604. To four decimal places of accuracy, we know today that the correct value of $\pi$ is 3.1416.

Like the Egyptians, the Chinese treated geometry in a very practical way. In their constructions and designs, the Chinese used the rule (ruler), the square, the compass, and the level. Unlike the Egyptians and the Chinese, the Greeks formalized and expanded the knowledge base of geometry by pursuing geometry as an intellectual endeavor.

According to the Greek scribe Proclus (about 50 B.C.), Thales (625–547 B.C.) first established deductive proofs for several of the known theorems of geometry. Proclus also notes that it was Euclid (330–275 B.C.) who collected, summarized, ordered, and verified the vast quantity of knowledge of geometry in his time. Euclid's work *Elements* was the first textbook of geometry. Much of what was found in *Elements* is the core knowledge of geometry and thus can be found in this textbook as well.

# Perspective on Application

## PATTERNS

In much of the study of mathematics, we seek patterns related to the set of counting numbers $N = \{1,2,3,4,5, \ldots\}$. Some of these patterns are geometric and are given special names that reflect the configuration of sets of points. For instance, the set of *square numbers* are shown below geometrically and, of course, correspond to the numbers 1, 4, 9, 16, . . . .

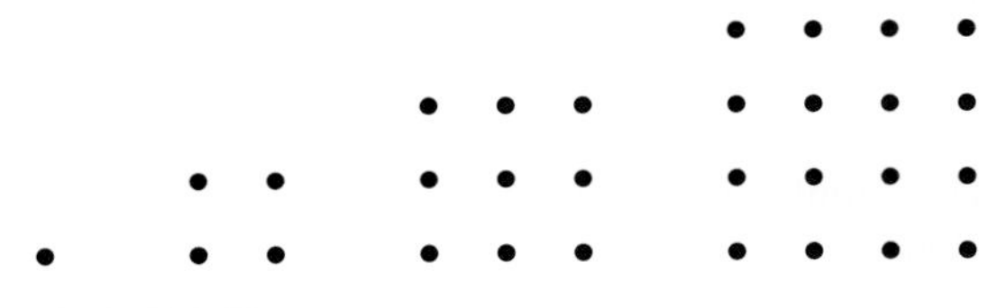

**FIGURE 1.70**

### EXAMPLE 1

Find the fourth number in the pattern of triangular numbers shown in Figure 1.71(a).

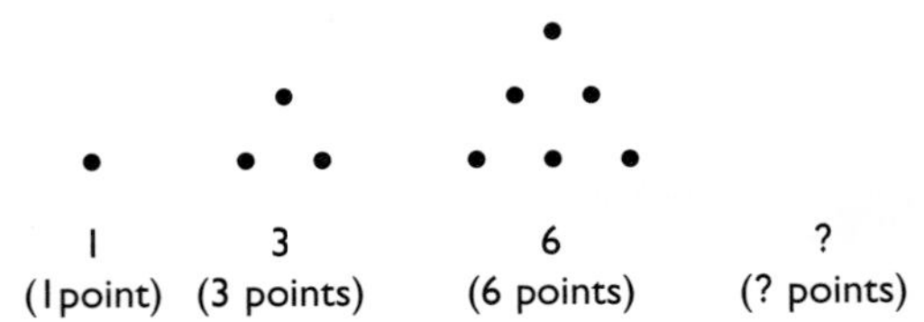


**FIGURE 1.71(a)**

*Solution*

Adding a row of 4 points at the bottom, we have the diagram shown in Figure 1.71(b), which contains 10 points. The fourth triangular number is 10.

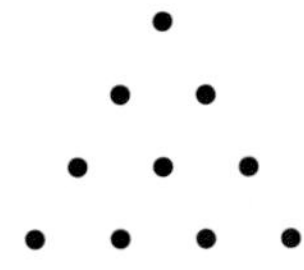

**FIGURE 1.71(b)**

Some patterns of geometry lead to principles known as postulates and theorems. One of the principles that we will explore in the next example is based on the total number of *diagonals* found in a polygon with a given number of sides. A diagonal of a polygon (many-sided figure) joins two non-consecutive vertices of the polygon together. Of course, joining any two vertices of a triangle will determine a side; thus a triangle has no diagonals. In Example 2, both the number of sides of the polygon and the number of diagonals are shown.

### EXAMPLE 2

Find the total number of diagonals for a polygon of 6 sides.

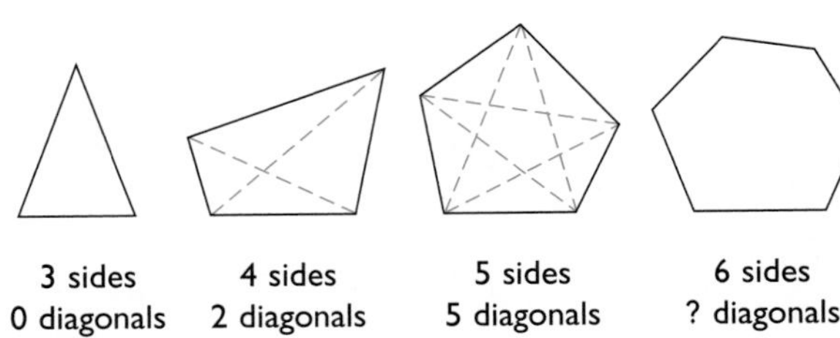


**FIGURE 1.72(a)**

*Solution*

By drawing all possible diagonals as shown in Figure 1.72(b) and counting them, we find that there are a total of 9 diagonals!

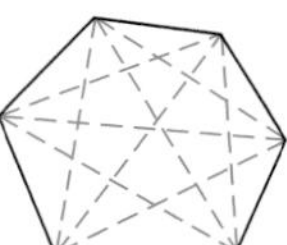

**FIGURE 1.72(b)**

Certain geometric patterns are used to test students, as in testing for intelligence (IQ) or on college admissions tests. A simple example might have you predict the next (fourth) figure in the pattern of squares shown in Figure 1.73(a).

**FIGURE 1.73(a)**

We rotate the square once more to obtain the fourth figure as shown in Figure 1.73(b).

**FIGURE 1.73(b)**

EXAMPLE 3

Midpoints of the sides of a *square* are used to generate new figures in the sequence shown in Figure 1.74(a). Draw the fourth figure.

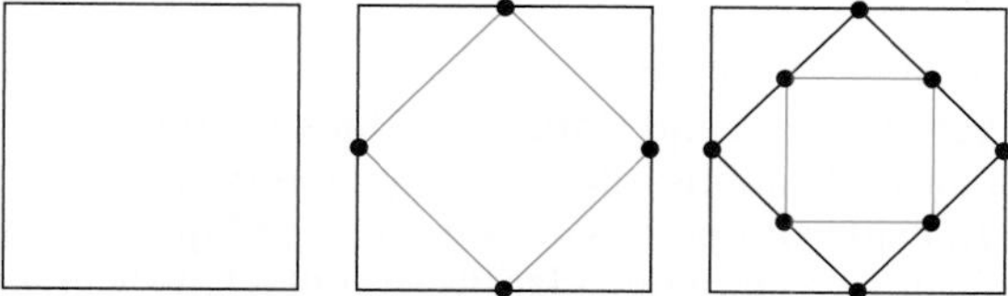

**FIGURE 1.74(a)**

*Solution*

By continuing to add and join midpoints in the third figure, we form a figure like the one shown in Figure 1.74(b).

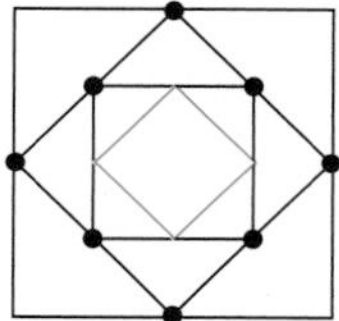

**FIGURE 1.74(b)**

Note that each new figure within the previous figure is also a square!

# Summary

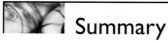

## A LOOK BACK AT CHAPTER 1

Our goal in this chapter has been to introduce geometry. We discussed the types of reasoning that are used to develop geometric relationships. The use of the tools of measurement (ruler and protractor) was described. We encountered the four elements of a mathematical system: undefined terms, definitions, postulates, and theorems. The undefined terms were needed to lay the foundation for defining new terms. The postulates were needed to lay the foundation for the theorems we proved here and for the theorems that lie ahead. Constructions presented in this chapter included the bisector of an angle and the perpendicular to a line at a point on the line.

## A LOOK AHEAD TO CHAPTER 2

The theorems we will prove in the next chapter are based on a postulate known as the Parallel Postulate. A new method of proof, called indirect proof, will be introduced; it will be used in later chapters. Although many of the theorems in Chapter 2 deal with parallel lines, several theorems in the chapter deal with the angles of a polygon. Symmetry and transformations will be discussed.

## KEY CONCEPTS

1.1 Statement • Variable • Conjunction • Disjunction • Negation • Implication (Conditional) • Hypothesis • Conclusion • Intuition • Induction • Deduction • Argument (Valid and Invalid) • Law of Detachment • Counterexample • Venn Diagram • Intersection • Union

1.2 Point • Line • Collinear Points • Vertex • Line Segment • Betweenness of Points • Midpoint • Congruent • Protractor • Parallel Lines • Bisect • Straight Angle • Right Angle • Intersect • Perpendicular • Compass • Constructions • Circle • Arc • Radius

1.3 Mathematical System • Axiom or Postulate • Assumption • Theorem • Ruler Postulate • Distance • Segment-Addition Postulate • Congruent Segments • Midpoint of a Line Segment • Bisector of a Line Segment • Union • Ray • Opposite Rays • Intersection of Two Geometric Figures • Parallel Lines • Plane • Coplanar Points • Space • Parallel, Vertical, Horizontal Planes

1.4 Angle • Sides of an Angle • Vertex of an Angle • Protractor Postulate • Acute, Right, Obtuse, and Straight Angles • Angle-Addition Postulate • Adjacent Angles • Congruent Angles • Bisector of an Angle • Complementary and Supplementary Angles • Vertical Angles

1.5 Algebraic Properties

1.6 Vertical Lines and Horizontal Lines • Perpendicular Lines • Relations • Reflexive, Symmetric, and Transitive Properties of Congruence • Equivalence Relation • Perpendicular Bisector of a Line Segment

1.7 Formal Proof of a Theorem • Converse

## TABLE 1.6 AN OVERVIEW OF CHAPTER ONE

### Line and Line Segment Relationships

| Figure | Relationship | Symbols |
|---|---|---|
| 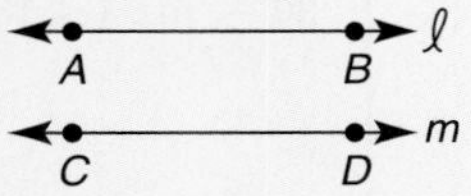 A B ℓ; C D m | Parallel lines (and segments) | $\ell \parallel m$ or $\overleftrightarrow{AB} \parallel \overleftrightarrow{CD}$; $\overline{AB} \parallel \overline{CD}$ |
| 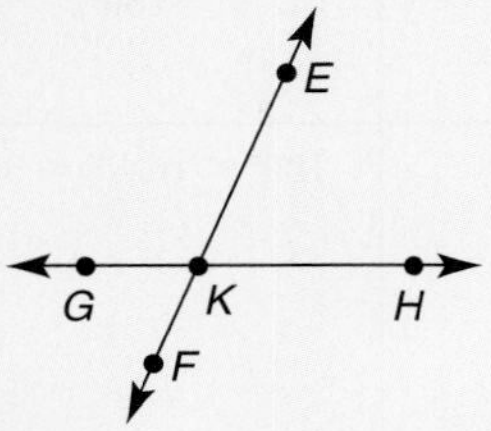 E, G, K, H, F | Intersecting lines | $\overleftrightarrow{EF} \cap \overleftrightarrow{GH} = K$ |
| 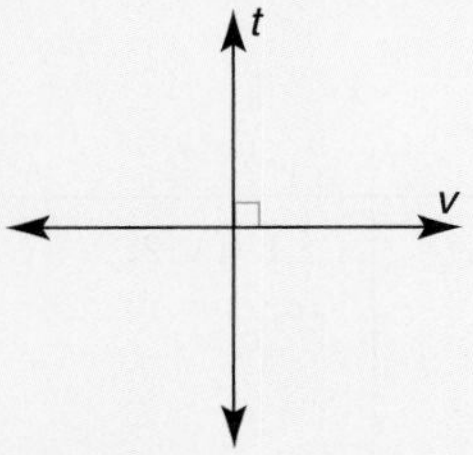 t, v | Perpendicular lines ($t$ shown vertical, $v$ shown horizontal) | $t \perp v$ |
|  M N; P Q | Congruent line segments | $\overline{MN} \cong \overline{PQ}$; $MN = PQ$ |
|  A B C | Point $B$ between $A$ and $C$ on $\overline{AC}$ | $AB + BC = AC$ |
|  P M Q | Point $M$ the midpoint of $\overline{PQ}$ | $\overline{PM} \cong \overline{MQ}$; $PM = MQ$; $PM = \frac{1}{2}(PQ)$ |

### Angle Classification (One Angle)

| Figure | Type | Angle Measure |
|---|---|---|
| 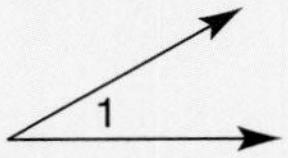 1 | Acute angle | $0° < m\angle 1 < 90°$ |
| 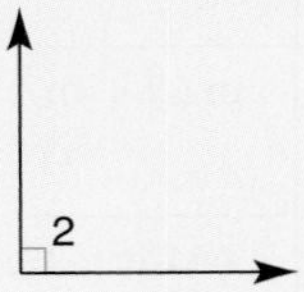 2 | Right angle | $m\angle 2 = 90°$ |

*continued*

**TABLE 1.6 CONT.**

## Angle Classification (One Angle)

| *Figure* | *Type* | *Angle Measure* |
|---|---|---|
| 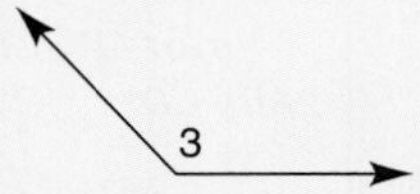 3 | Obtuse angle | $90° < m\angle 3 < 180°$ |
|  4 | Straight angle | $m\angle 4 = 180°$ |
| 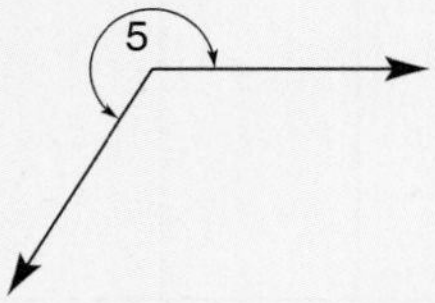 5 | Reflex angle | $180° < m\angle 5 < 360°$ |

## Angle Relationships (Two Angles)

| *Figure* | *Relationship* | *Symbols* |
|---|---|---|
| 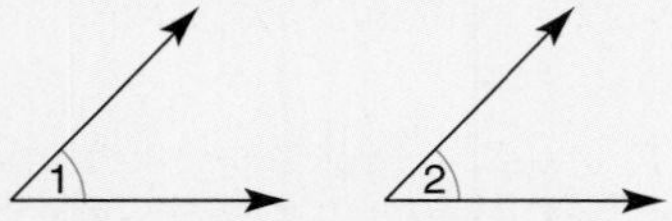 1, 2 | Congruent angles | $\angle 1 \cong \angle 2$;<br>$m\angle 1 = m\angle 2$ |
| 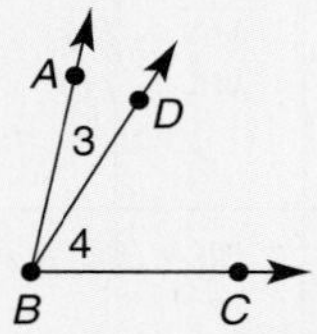 A, D, 3, 4, B, C | Adjacent angles | $m\angle 3 + m\angle 4 = m\angle ABC$ |
| 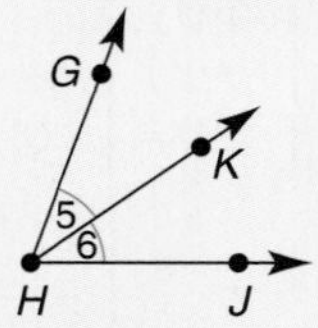 G, K, 5, 6, H, J | Bisector of angle<br>($\overrightarrow{HK}$ bisects $\angle GHJ$) | $\angle 5 \cong \angle 6$;<br>$m\angle 5 = m\angle 6$;<br>$m\angle 5 = \frac{1}{2}(m\angle GHJ)$ |
| 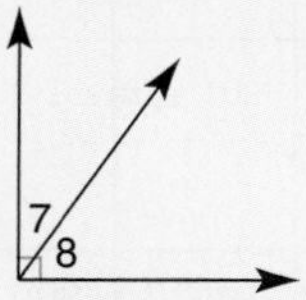 7, 8 | Complementary angles | $m\angle 7 + m\angle 8 = 90°$ |
| 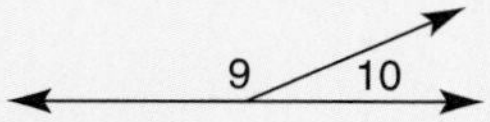 9, 10 | Supplementary angles | $m\angle 9 + m\angle 10 = 180°$ |
| 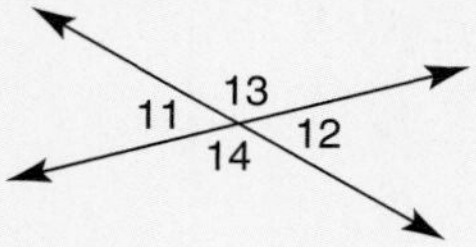 13, 11, 12, 14 | Vertical angles<br>($\angle 11$ and $\angle 12$;<br>$\angle 13$ and $\angle 14$) | $\angle 11 \cong \angle 12$; $\angle 13 \cong \angle 14$ |

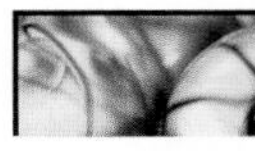

## Chapter 1 Review Exercises

1. Name the four components of a mathematical system.

2. Name three types of reasoning.

3. Name the four characteristics of a good definition.

*In Review Exercises 4 to 6, name the type of reasoning illustrated.*

4. While watching the pitcher warm up, Phillip thinks, "I'll be able to hit against him."

5. Laura is away at camp. On the first day, her mother brings her additional clothing. On the second day, her mother brings her another pair of shoes. On the third day, her mother brings her cookies. Laura concludes that her mother will bring her something on the fourth day.

6. Sarah knows the rule "A number (not 0) divided by itself equals 1." The teacher asks Sarah, "What is 5 divided by 5?" Sarah says, "The answer is 1."

*In Review Exercises 7 and 8, state the hypothesis and conclusion for each statement.*

7. If the diagonals of a trapezoid are equal in length, then the trapezoid is isosceles.

8. The diagonals of a parallelogram are congruent if the parallelogram is a rectangle.

*In Review Exercises 9 to 11, draw a valid conclusion where possible.*

9. 1. If a person has a good job, then that person has a college degree.
   2. Billy Fuller has a college degree.
   C. $\therefore$ ?

10. 1. If a person has a good job, then that person has a college degree.
    2. Jody Smithers has a good job.
    C. $\therefore$ ?

11. 1. If the measure of an angle is 90°, then that angle is a right angle.
    2. Angle A has a measure of 90°.
    C. $\therefore$ ?

12. $A$, $B$, and $C$ are three points on a line. $AC = 8$, $BC = 4$, and $AB = 12$. Which point must be between the other two points?

13. Use three letters to name the angle shown. Also use one letter to name the same angle. Decide whether the angle is less than 90°, equal to 90°, or greater than 90°.

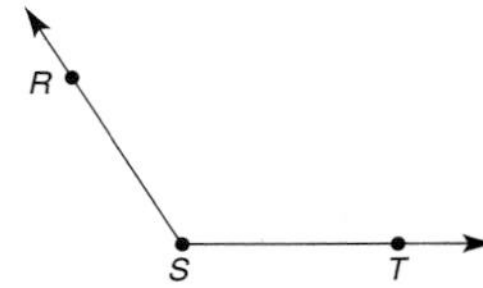


14. Figure $MNPQ$ is a rhombus. Draw diagonals $\overline{MP}$ and $\overline{QN}$ of the rhombus. How do $\overline{MP}$ and $\overline{QN}$ appear to be related?

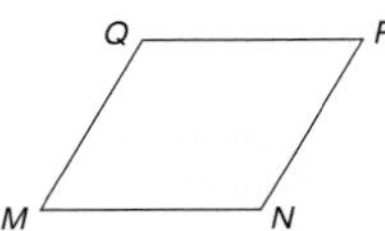


*In Review Exercises 15 to 17, sketch and label the figures described.*

15. Points $A$, $B$, $C$, and $D$ are coplanar. $A$, $B$, and $C$ are the only three of these points that are collinear.

16. Line $\ell$ intersects plane $X$ at point $P$.

17. Plane $M$ contains intersecting lines $j$ and $k$.

18. *Given:* $\overrightarrow{BD}$ bisects $\angle ABC$
    $m\angle ABD = 2x + 15$
    $m\angle DBC = 3x - 2$
    *Find:* $m\angle ABC$

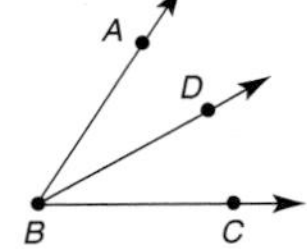


*Exercises 18, 19*

19. *Given:* $m\angle ABD = 2x + 5$
    $m\angle DBC = 3x - 4$
    $m\angle ABC = 86°$
    *Find:* $m\angle DBC$

20. *Given:* $AM = 3x - 1$
    $MB = 4x - 5$
    $M$ is the midpoint of $\overline{AB}$
    *Find:* $AB$

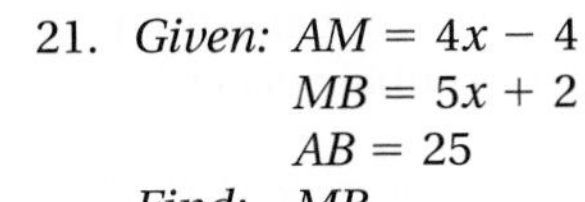


*Exercises 20, 21*

21. *Given:* $AM = 4x - 4$
    $MB = 5x + 2$
    $AB = 25$
    *Find:* $MB$

22. *Given:* $D$ is the midpoint of $\overline{AC}$
    $\overline{AC} \cong \overline{BC}$
    $CD = 2x + 5$
    $BC = x + 28$
    *Find:* $AC$

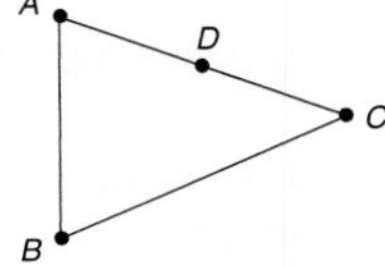

23. *Given:* $m\angle 3 = 7x - 21$
$m\angle 4 = 3x + 7$
*Find:* $m\angle FMH$

24. *Given:* $m\angle FMH = 4x + 1$
$m\angle 4 = x + 4$
*Find:* $m\angle 4$

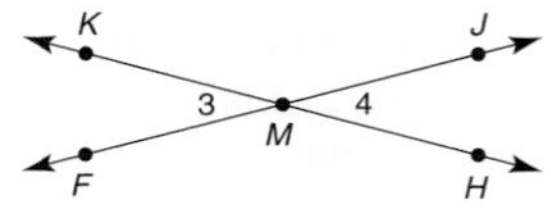

*Exercises 23–25*

25. In the figure, find:

a) $\overleftrightarrow{KH} \cap \overleftrightarrow{FJ}$
b) $\overrightarrow{MJ} \cup \overrightarrow{MH}$
c) $\angle KMJ \cap \angle JMH$
d) $\overrightarrow{MK} \cup \overrightarrow{MH}$

26. *Given:* $\angle EFG$ is a right angle
$m\angle HFG = 2x - 6$
$m\angle EFH = 3 \cdot m\angle HFG$
*Find:* $m\angle EFH$

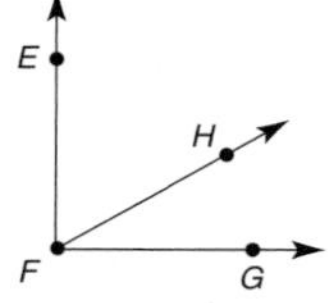


27. Two angles are supplementary. One angle is 40° more than four times the other. Find the measures of the two angles.

28. a) Write an expression for the perimeter of the triangle shown. (HINT: Add the lengths of the sides.)

b) If the perimeter is 32 centimeters, find the value of $x$.

c) Find the length of each side of the triangle.

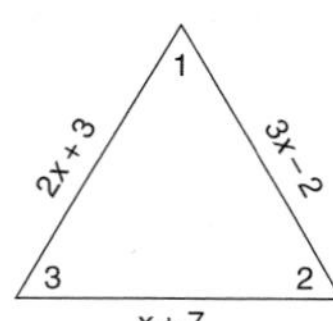


29. The sum of the measures of all three angles of the triangle in Review Exercise 28 is 180°. If the sum of the measures of angles 1 and 2 is more than 130°, what can you conclude about the measure of angle 3?

30. Susan wants to have a 4-ft board with some pegs on it. She wants to leave 6 in. on each end and 4 in. between pegs. How many pegs will fit on the board? [HINT: If $n$ represents the number of pegs, then $(n - 1)$ represents the number of equal spaces.]

*State whether the sentences in Review Exercises 31 to 35 are always true (A), sometimes true (S), or never true (N).*

31. If $AM = MB$, then $A$, $M$, and $B$ are collinear.

32. If two angles are congruent, then they are right angles.

33. The bisectors of vertical angles are opposite rays.

34. Complementary angles are congruent.

35. The supplement of an obtuse angle is another obtuse angle.

36. Fill in the missing statements or reasons.

*Given:* $\angle 1 \cong \angle P$
$\angle 4 \cong \angle P$
$\overrightarrow{VP}$ bisects $\angle RVO$
*Prove:* $\angle TVP \cong \angle MVP$

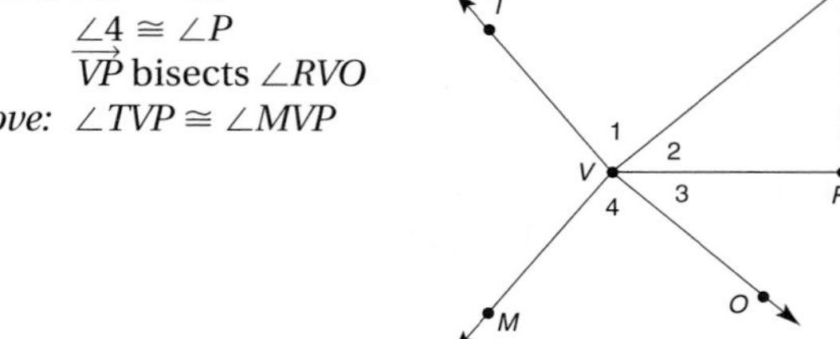


**PROOF**

| | Statements | Reasons |
|---|---|---|
| | 1. $\angle 1 \cong \angle P$ | 1. Given |
| | 2. ? | 2. Given |
| (1), (2) | 3. ? | 3. Transitive Prop. of $\cong$ |
| (3) | 4. $m\angle 1 = m\angle 4$ | 4. ? |
| | 5. $\overrightarrow{VP}$ bisects $\angle RVO$ | 5. ? |
| | 6. ? | 6. If a ray bisects an $\angle$, it forms two $\angle$s of equal measure |
| (4), (6) | 7. ? | 7. Addition Prop. of Equality |
| | 8. $m\angle 1 + m\angle 2 = m\angle TVP$; $m\angle 4 + m\angle 3 = m\angle MVP$ | 8. ? |
| (7), (8) | 9. $m\angle TVP = m\angle MVP$ | 9. ? |
| | 10. ? | 10. If two $\angle$s are = in measure, then they are $\cong$ |

*Write two-column proofs for Review Exercises 37 to 44.*

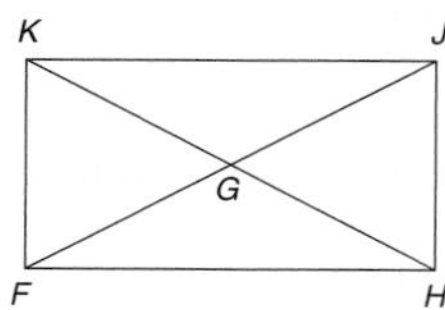

*Exercises 37–39*

37. *Given:* $\overline{KF} \perp \overline{FH}$
$\angle JHF$ is a right $\angle$
*Prove:* $\angle KFH \cong \angle JHF$

*For Review Exercises 38 and 39, see the figure on page 66.*

38. *Given:* $\overline{KH} \cong \overline{FJ}$
    $G$ is the midpoint of both $\overline{KH}$ and $\overline{FJ}$
    *Prove:* $\overline{KG} \cong \overline{GJ}$

39. *Given:* $\overline{KF} \perp \overline{FH}$
    *Prove:* $\angle KFJ$ is comp. to $\angle JFH$

40. *Given:* $\angle 1$ is comp. to $\angle M$
    $\angle 2$ is comp. to $\angle M$
    *Prove:* $\angle 1 \cong \angle 2$

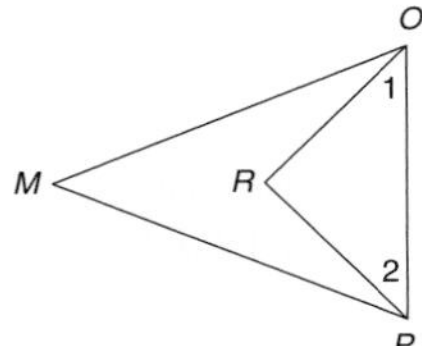


*Exercises 40, 41*

41. *Given:* $\angle MOP \cong \angle MPO$
    $\overrightarrow{OR}$ bisects $\angle MOP$
    $\overrightarrow{PR}$ bisects $\angle MPO$
    *Prove:* $\angle 1 \cong \angle 2$

*For Review Exercise 42, see the figure that follows Review Exercise 43.*

42. *Given:* $\angle 4 \cong \angle 6$
    *Prove:* $\angle 5 \cong \angle 6$

43. *Given:* Figure as shown
    *Prove:* $\angle 4$ is supp. to $\angle 2$

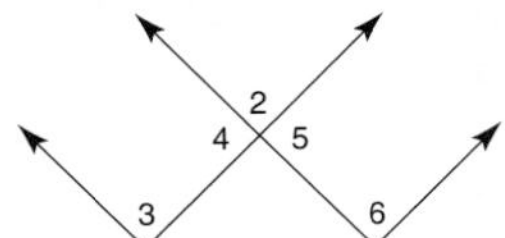


*Exercises 42–44*

44. *Given:* $\angle 3$ is supp. to $\angle 5$
    $\angle 4$ is supp. to $\angle 6$
    *Prove:* $\angle 3 \cong \angle 6$

45. *Given:* $\overline{VP}$
    *Construct:* $\overline{VW}$ such that
    $VW = 4 \cdot VP$

46. Construct a 135° angle.

47. *Given:* Triangle $PQR$
    *Construct:* The three angle bisectors

    What did you discover about the three angle bisectors of this triangle?

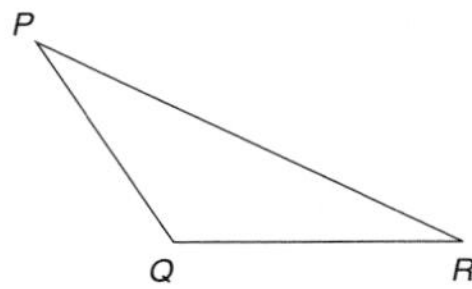


48. *Given:* $\overline{AB}$, $\overline{BC}$, and $\angle B$ as shown in Review Exercise 49
    *Construct:* Triangle $ABC$

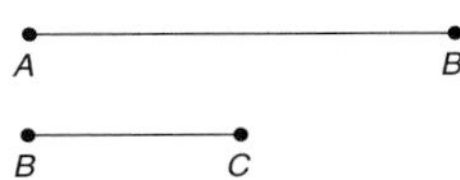


49. *Given:* $m\angle B = 50°$
    *Construct:* An angle whose measure is 20°

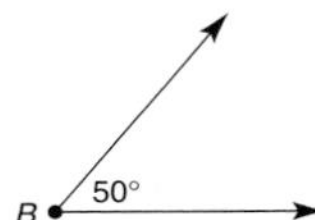


50. If $m\angle 1 = 90°$, find the measure of reflex angle 2.

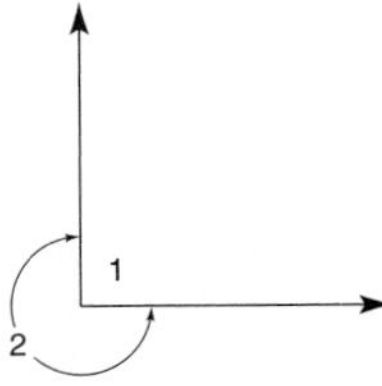

# Chapter 1 Test

1. Which type of reasoning is illustrated below? ________

   Because it has rained the previous four days, Annie concludes that it will rain again today.

2. Given $\angle ABC$ (as shown), provide a second correct method for naming this angle. ________

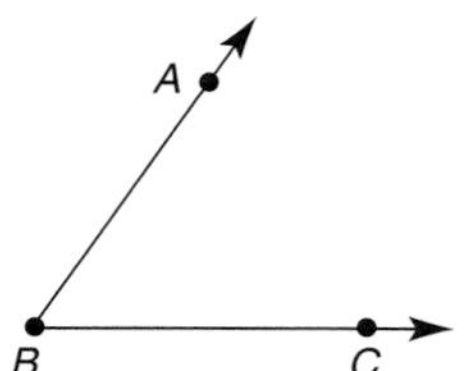


3. Using the Segment-Addition Postulate, state a conclusion regarding the accompanying figure. ________

4. Complete each postulate:
   a) If two lines intersect, they intersect in a ______
   b) If two planes intersect, they intersect in a ______

5. Given that $x$ is the measure of an angle, name the type of angle when:
   a) $x = 90°$ ______ b) $90° < x < 180°$ ______

6. What would describe two angles
   a) whose sum of measures is equal to 180°? ______
   b) that have equal measures? ______

7. Given that $\overrightarrow{NP}$ bisects $\angle MNQ$, state a conclusion involving $m\angle MNP$ and $m\angle PNQ$. ________________

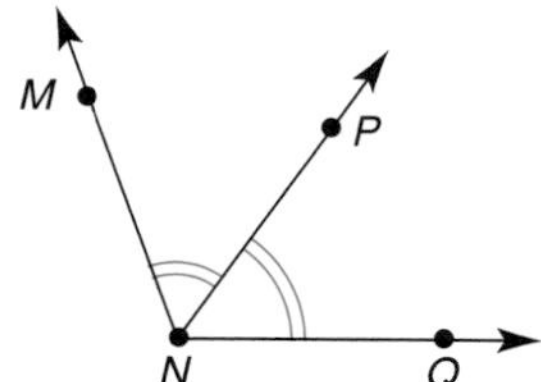


8. Complete each theorem:
   a) If two lines are perpendicular, they meet to form ______ angles.
   b) If the exterior sides of two adjacent angles form a straight line, these angles are ______

9. State the conclusion for the following deductive argument.
   (1) If you study geometry, then you will develop reasoning skills.
   (2) Kianna is studying geometry this semester.
   (C) ________________

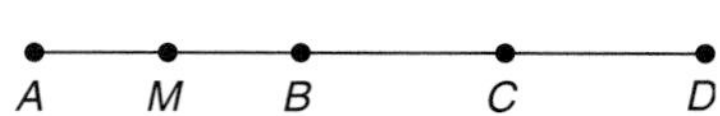


*Questions 10, 11*

10. In the figure $A$-$B$-$C$-$D$, $M$ is the midpoint of $\overline{AB}$. If $AB = 6.4$ inches and $BD = 7.2$ inches, find $MD$. ______

11. In the figure, $AB = x$, $BD = x + 5$, and $AD = 27$. Find: a) $x$ ______ b) $BD$ ______

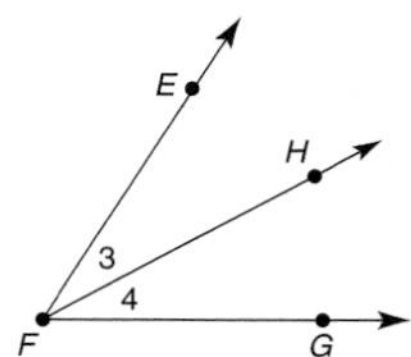


*Questions 12, 13*

12. In the figure, $m\angle EFG = 68°$ and $m\angle 3 = 33°$. Find $m\angle 4$. ______

13. In the figure, $m\angle 3 = x$ and $m\angle 4 = 2x - 3$. If $m\angle EFG = 69°$, find:
    a) $x$ ______ b) $m\angle 4$ ______

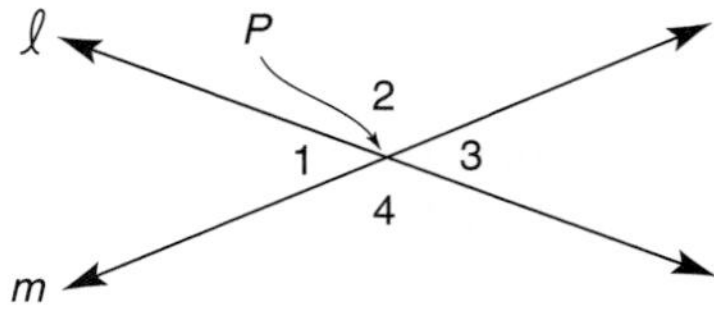


*Questions 14–16*

14. Lines $\ell$ and $m$ intersect at point $P$. If $m\angle 1 = 43°$, find:
    a) $m\angle 2$ ______ b) $m\angle 3$ ______

15. If $m\angle 1 = 2x - 3$ and $m\angle 3 = 3x - 28$, find: a) $x$ ______ b) $m\angle 1$ ______

16. If $m\angle 1 = 2x - 3$ and $m\angle 2 = 6x - 1$, find: a) $x$ ______ b) $m\angle 2$ ______

17. $\angle$s 3 and 4 (not shown) are complementary. Where m$\angle 3 = x$ and m$\angle 4 = y$, write an equation using variables $x$ and $y$. ________________

18. Construct the angle-bisector of obtuse angle $RST$.

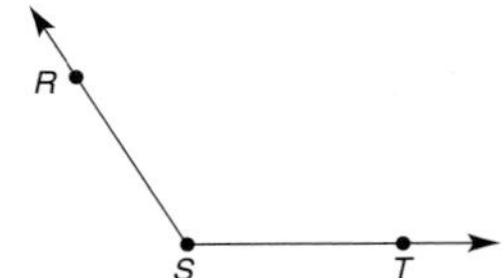


19. Construct the perpendicular-bisector of $\overline{AB}$.

*In Exercises 20 to 22, complete the missing statements/reasons for each proof.*

20. *Given:* $M$-$N$-$P$-$Q$ on $\overline{MQ}$
*Prove:* $MN + NP + PQ = MQ$

**PROOF**

| Statements | Reasons |
|---|---|
| 1. $M$-$N$-$P$-$Q$ on $\overline{MQ}$ | 1. ________ |
| 2. $MN + NQ = MQ$ | 2. ________ |
| 3. $NP + PQ = NQ$ | 3. ________ |
| 4. $MN + NP + PQ = MQ$ | 4. ________ |

21. *Given:* $2x - 3 = 17$
*Prove:* $x = 10$

**PROOF**

| Statements | Reasons |
|---|---|
| 1. ________ | 1. Given |
| 2. ________ | 2. Addition Property of Equality |
| 3. ________ | 3. Division Property of Equality |

22. *Given:* $\angle ABC$ is a right angle;
$\overrightarrow{BD}$ bisects $\angle ABC$
*Prove:* m$\angle 1 = 45°$

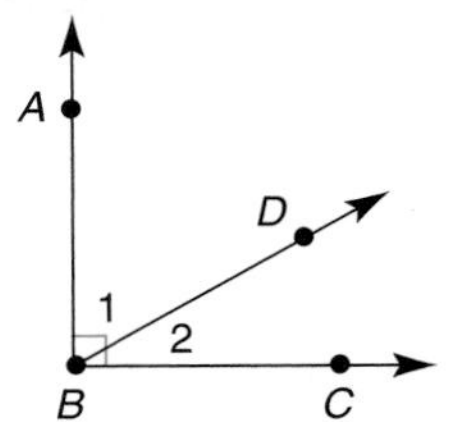


**PROOF**

| Statements | Reasons |
|---|---|
| 1. $\angle ABC$ is a right angle | 1. ________ |
| 2. m$\angle ABC$ = ________ | 2. Definition of a right angle |
| 3. m$\angle 1$ + m$\angle 2$ = m$\angle ABC$ | 3. ________ |
| 4. m$\angle 1$ + m$\angle 2$ = ________ | 4. Substitution Property of Equality |
| 5. $\overrightarrow{BD}$ bisects $\angle ABC$ | 5. ________ |
| 6. m$\angle 1$ = m$\angle 2$ | 6. ________ |
| 7. m$\angle 1$ + m$\angle 1 = 90°$ or $2 \cdot$ m$\angle 1 = 90°$ | 7. ________ |
| 8. ________ | 8. Division Property of Equality |

# Chapter 2

# Parallel Lines

Over 1 mile in length, the Manhattan Bridge in New York City joins the burroughs of Brooklyn and Manhattan. Although construction of the bridge began in 1901, it was not open to traffic until late 1909. From the cable that is suspended from the uppermost points of the bridge towers, vertical supports are dropped perpendicular to the bridge floor. The vertical supports appear to be parallel, which is not by chance! In fact, the parallel relationship between the vertical supports illustrates Theorem 2.3.7 of this chapter.

## Chapter Outline



**For online student resources, visit this textbook's website at math.college.hmco.com/students.**

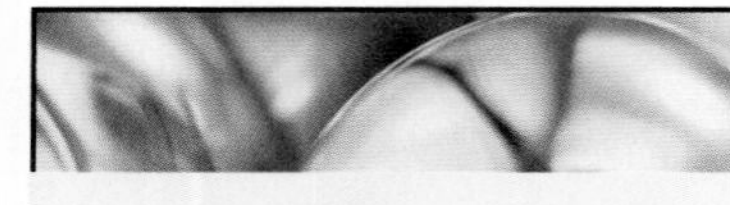

**KEY CONCEPTS**

Perpendicular Lines • Perpendicular Planes • Parallel Lines • Parallel Planes • Parallel Postulate • Transversal • Interior Angles • Exterior Angles • Corresponding Angles • Alternate Interior Angles • Alternate Exterior Angles

# 2.1 The Parallel Postulate and Special Angles

## PERPENDICULAR LINES

By definition, two lines (or segments or rays) are perpendicular if they meet to form congruent adjacent angles. Using this definition, we proved the theorem stating that "perpendicular lines meet to form right angles." We can also say that two rays or line segments are perpendicular if they are parts of perpendicular lines. We now consider a method for constructing a line perpendicular to a given line.

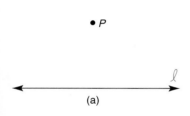

CONSTRUCTION 6:
To construct the line that is perpendicular to a given line from a point not on the given line.

*Given:* In Figure 2.1(a), line $\ell$ and point $P$ not on $\ell$

*Construct:* $\overleftrightarrow{PQ} \perp \ell$

*Construction:* Figure 2.1(b): With $P$ as the center, open the compass to a length great enough to intersect $\ell$ in two points $A$ and $B$.

Figure 2.1(c): With $A$ and $B$ as centers, mark off arcs of equal radii (using the same compass opening) to intersect at a point $Q$, as shown.

Draw $\overleftrightarrow{PQ}$ to complete the desired line.

(a)

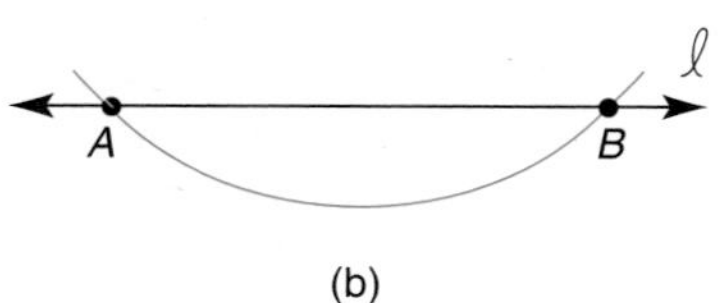


(b)

In this construction, $\angle PRA$ and $\angle PRB$ are right angles. The arcs drawn from $A$ and $B$ to intersect below line $\ell$ provide greater accuracy than would arcs intersecting above $\ell$.

Construction 6 suggests a uniqueness relationship that can be proved.

THEOREM 2.1.1: From a point not on a given line, there is exactly one line perpendicular to the given line.

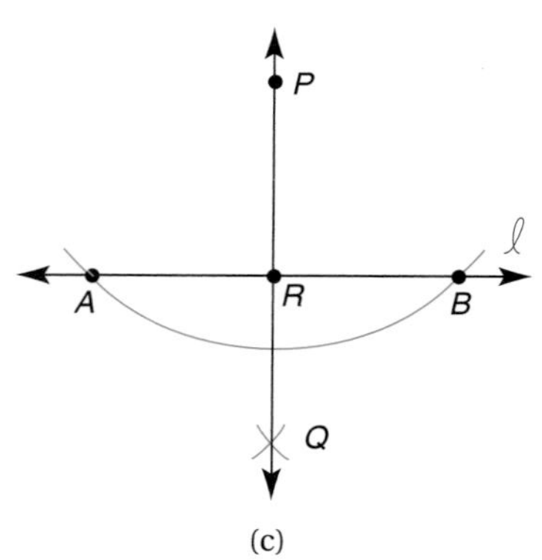


(c)

**FIGURE 2.1**

The term *perpendicular* includes line-ray, line-plane, and plane-plane relationships. The drawings in Figure 2.2 indicate two perpendicular lines, a line perpendicular to a plane, and two perpendicular planes. In Figure 2.1(c), $\overrightarrow{RP} \perp \ell$.

## PARALLEL LINES

Just as the word *perpendicular* can relate lines and planes, the word *parallel* can be used to describe possible relationships among lines and planes. However, parallel lines must lie in the same plane, as the following definition emphasizes.

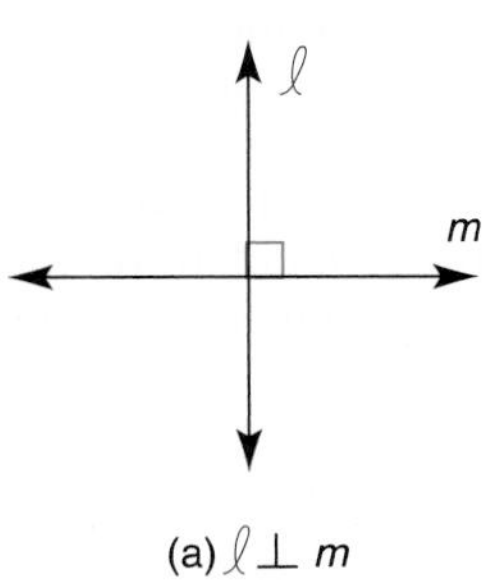


(a) $\ell \perp m$

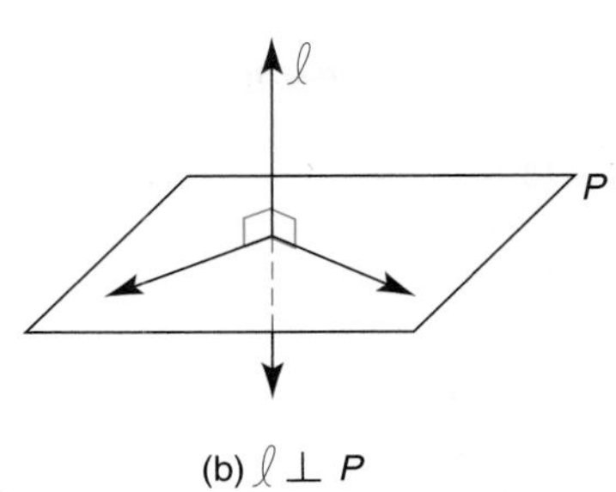


(b) $\ell \perp P$

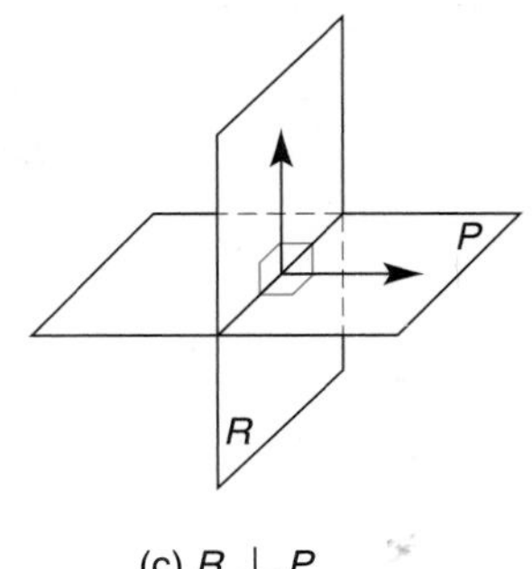


(c) $R \perp P$

**FIGURE 2.2**

DEFINITION: **Parallel lines** are lines in the same plane that do not intersect.

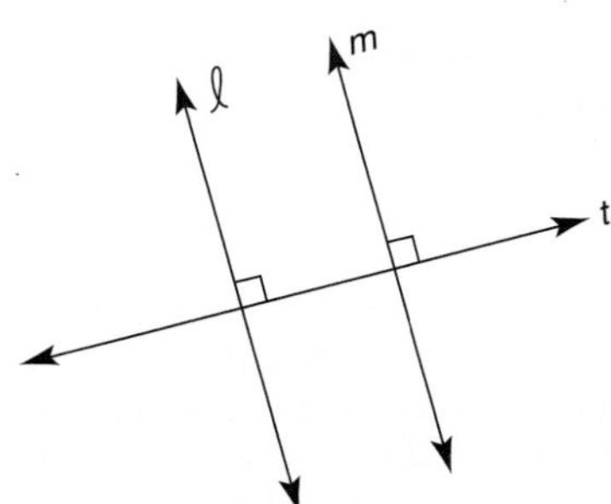


## Discover!

In the sketch at the left, lines $\ell$ and $m$ lie in the same plane with line $t$ and are perpendicular to line $t$. How are the lines $\ell$ and $m$ related to each other?

ANSWER

These lines are said to be *parallel*. They will not intersect.

More generally, two lines in a plane, a line and a plane, or two planes are parallel if they do not intersect (see Figure 2.3). Figure 2.3 illustrates possible applications of the word *parallel.* In Figure 2.4, two parallel planes $M$ and $N$ are intersected by a third plane $G$. How must the lines of intersection $a$ and $b$ be related?

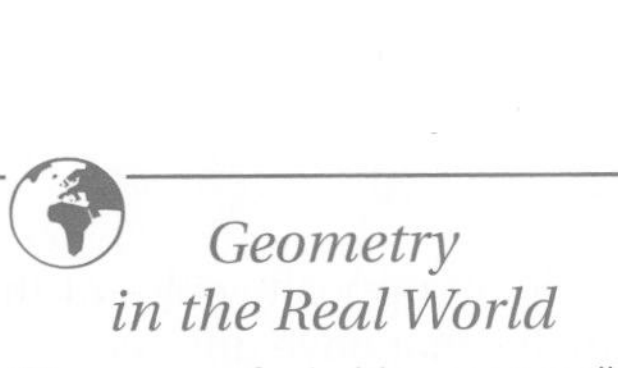

Geometry in the Real World

The rungs of a ladder are parallel line segments.

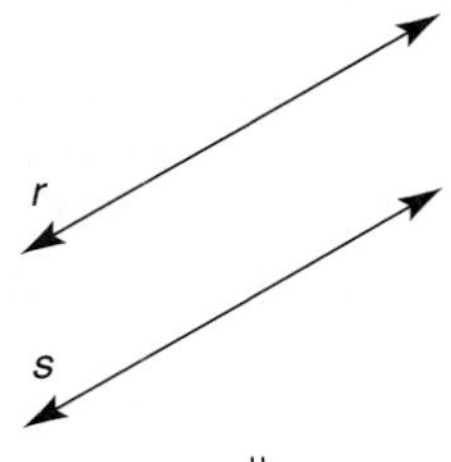


(a) $r \parallel s$

$r \cap s = \varnothing$

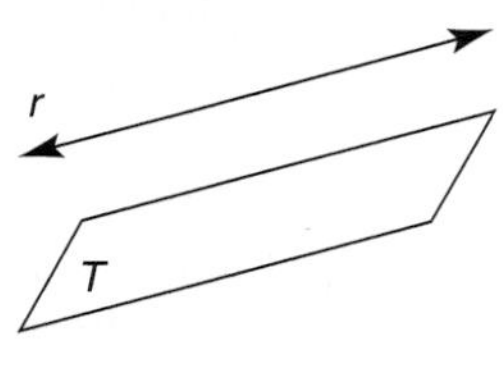


(b) $r \parallel T$

$r \cap T = \varnothing$

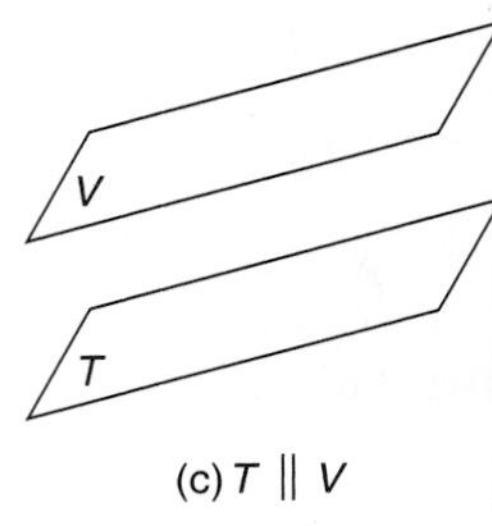


(c) $T \parallel V$

$T \cap V = \varnothing$

**FIGURE 2.3**

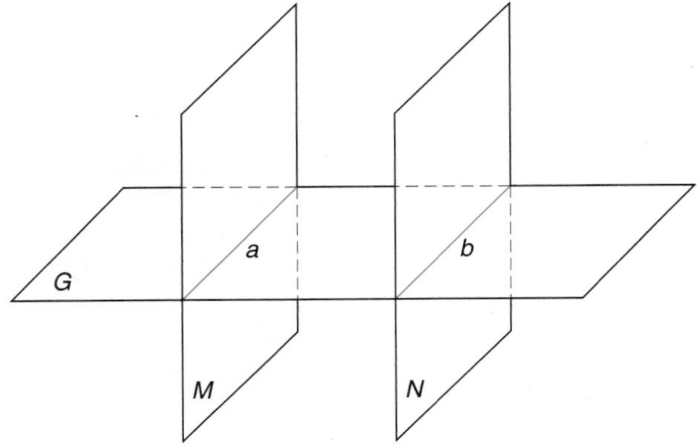

*Exs. 1–3*

**FIGURE 2.4**

## EUCLIDEAN GEOMETRY

The type of geometry found in this textbook is known as Euclidean geometry. In this geometry, a plane is a flat, two-dimensional surface in which the line segment joining any two points of the plane lies entirely within the plane. Whereas the postulate that follows characterizes Euclidean geometry, the Perspective on Application section near the end of this chapter discusses alternative geometries. Postulate 10, the Euclidean Parallel Postulate, is easy to accept because of the way we perceive a plane.

> POSTULATE 10: **(Parallel Postulate)**
> Through a point not on a line, exactly one line is parallel to the given line.

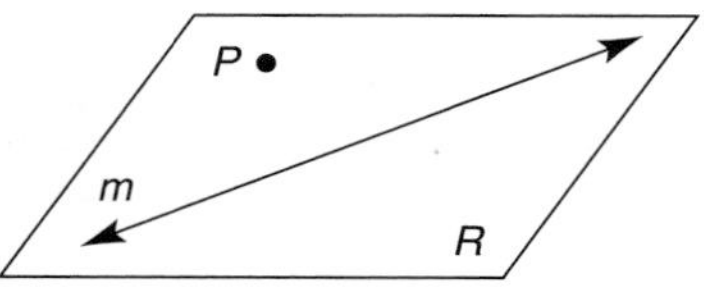


**FIGURE 2.5**

Consider Figure 2.5, in which line $m$ and point $P$ (with $P$ not on $m$) both lie in plane $R$. It seems reasonable that exactly one line can be drawn through $P$ parallel to line $m$. The method of construction for the unique line through $P$ parallel to $m$ is provided in Section 2.3.

A **transversal** is a line that intersects two (or more) other lines at distinct points; all of the lines lie in the same plane. In Figure 2.6, $t$ is a transversal for lines $r$ and $s$. Angles that are formed between $r$ and $s$ are **interior angles;** those outside $r$ and $s$ are **exterior angles.**

*Interior angles:* $\angle 3$, $\angle 4$, $\angle 5$, $\angle 6$

*Exterior angles:* $\angle 1$, $\angle 2$, $\angle 7$, $\angle 8$

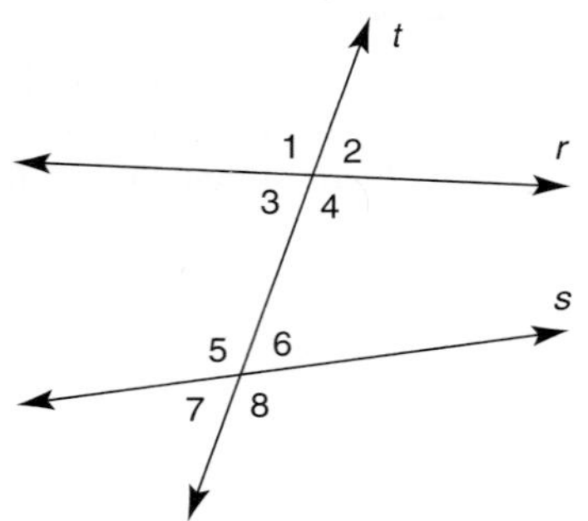


**FIGURE 2.6**

Consider the angles in Figure 2.6 that are formed when lines are cut by a transversal. Two angles that lie in the same relative positions (such as *above* and *left*) are called **corresponding angles** for these lines. In Figure 2.6, $\angle 1$ and $\angle 5$ are corresponding angles; each angle is *above* the line and to the *left* of the transversal that helps form the angle.

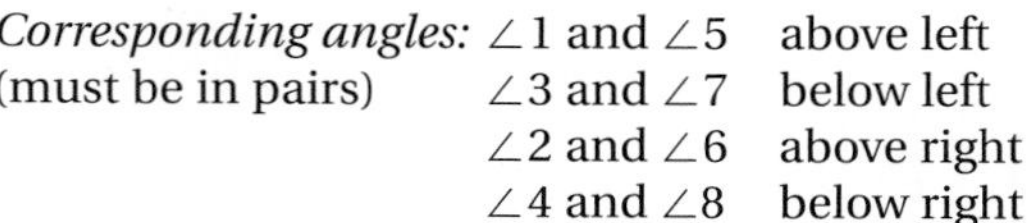

| | | |
|---|---|---|
| *Corresponding angles:* | $\angle 1$ and $\angle 5$ | above left |
| (must be in pairs) | $\angle 3$ and $\angle 7$ | below left |
| | $\angle 2$ and $\angle 6$ | above right |
| | $\angle 4$ and $\angle 8$ | below right |

Two interior angles that have different vertices and lie on opposite sides of the transversal are **alternate interior angles.** Two exterior angles that have different vertices and lie on opposite sides of the transversal are **alternate exterior angles.** Both types of alternate angles must occur in pairs; in Figure 2.6, these pairs of angles are numbered as follows:

*Alternate interior angles:* $\angle 3$ and $\angle 6$
$\angle 4$ and $\angle 5$
*Alternate exterior angles:* $\angle 1$ and $\angle 8$
$\angle 2$ and $\angle 7$

*Exs. 4–6*

## PARALLEL LINES AND CONGRUENT ANGLES

Now suppose that *parallel* lines $\ell$ and $m$ in Figure 2.7 are cut by transversal $v$. If a protractor were used to measure $\angle 1$ and $\angle 5$, these corresponding angles would be found to have equal measures; that is, they are congruent. Similarly, any other pair of corresponding angles will be congruent as long as $\ell \parallel m$.

> POSTULATE 11:
> If two parallel lines are cut by a transversal, then the corresponding angles are congruent.

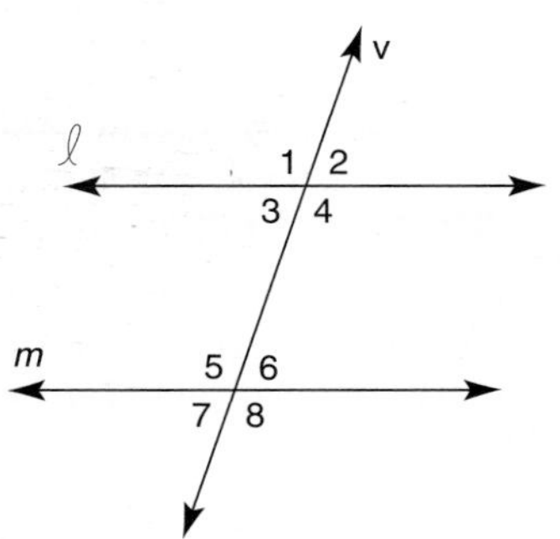


**FIGURE 2.7**

EXAMPLE 1

In Figure 2.7, $\ell \parallel m$ and $m\angle 1 = 117°$. Find:

a) $m\angle 2$ c) $m\angle 4$
b) $m\angle 5$ d) $m\angle 8$

***Solution***

a) $m\angle 2 = 63°$ supplementary to $\angle 1$
b) $m\angle 5 = 117°$ corresponding to $\angle 1$
c) $m\angle 4 = 117°$ vertical $\angle$ to $\angle 1$
d) $m\angle 8 = 117°$ corresponding to $\angle 4$ [found in part (c)]

Several theorems follow from Postulate 11; for some of these theorems, formal proofs are provided. Study the proofs and be able to state all the theorems. Later, you can cite the theorems as reasons in subsequent proofs.

THEOREM 2.1.2: If two parallel lines are cut by a transversal, then the alternate interior angles are congruent.

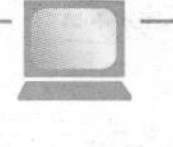

*Technology Exploration*

Use computer software if available.

1) Draw $\overleftrightarrow{AB} \parallel \overleftrightarrow{CD}$.
2) Draw transversal $\overleftrightarrow{EF}$.
3) By numbering the angles as in Figure 2.8, find the measures of all eight angles.
4) Show that pairs of corresponding angles are congruent.

*Given:* $a \parallel b$ in Figure 2.8
Transversal $k$

*Prove:* $\angle 3 \cong \angle 6$

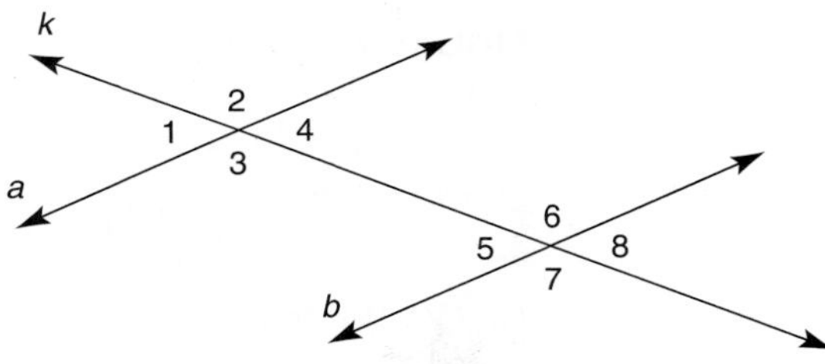


**FIGURE 2.8**

**PROOF**

| Statements | Reasons |
| --- | --- |
| 1. $a \parallel b$; transversal $k$ | 1. Given |
| 2. $\angle 2 \cong \angle 6$ | 2. If two $\parallel$ lines are cut by a transversal, corresponding $\angle$s are $\cong$ |
| 3. $\angle 3 \cong \angle 2$ | 3. If two lines intersect, vertical $\angle$s formed are $\cong$ |
| 4. $\angle 3 \cong \angle 6$ | 4. Transitive (of $\cong$) |

Although we did not establish that alternate interior angles 4 and 5 are congruent, it is easy to prove that these are congruent because they are supplements to $\angle 3$ and $\angle 6$. Another theorem that is similar to Theorem 2.1.2 follows, but the proof is left as Exercise 26.

> THEOREM 2.1.3: If two parallel lines are cut by a transversal, then the alternate exterior angles are congruent.

## PARALLEL LINES AND SUPPLEMENTARY ANGLES

Whenever two parallel lines are cut by a transversal, an interesting relationship exists between the two interior angles on the same side of the transversal. The following proof establishes that these interior angles are supplementary; a similar claim can be made for the pair of exterior angles on the same side of the transversal.

> THEOREM 2.1.4: If two parallel lines are cut by a transversal, then the interior angles on the same side of the transversal are supplementary.

*Given:* In Figure 2.9, $\overleftrightarrow{TV} \parallel \overleftrightarrow{WY}$ with transversal $\overleftrightarrow{RS}$

*Prove:* $\angle 1$ and $\angle 3$ are supplementary

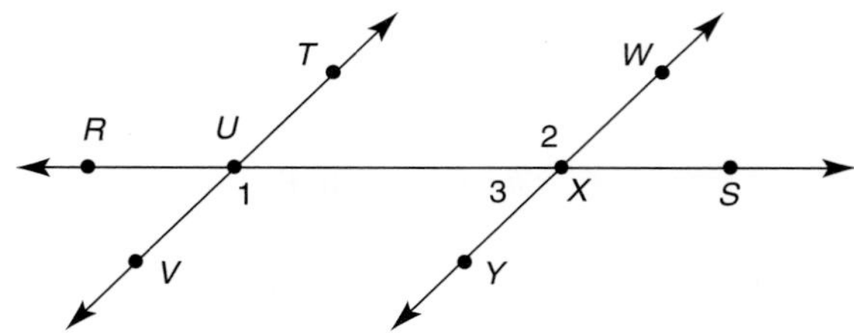


**FIGURE 2.9**

**PROOF**

| Statements | Reasons |
|---|---|
| 1. $\overleftrightarrow{TV} \parallel \overleftrightarrow{WY}$; transversal $\overleftrightarrow{RS}$ | 1. Given |
| 2. $\angle 1 \cong \angle 2$ | 2. If two $\parallel$ lines are cut by a transversal, alternate interior $\angle$s are $\cong$ |
| 3. $m\angle 1 = m\angle 2$ | 3. If two $\angle$s are $\cong$, their measures are $=$ |
| 4. $\angle WXY$ is a straight $\angle$, so $m\angle WXY = 180°$ | 4. If an $\angle$ is a straight $\angle$, its measure is 180° |
| 5. $m\angle 2 + m\angle 3 = m\angle WXY$ | 5. Angle-Addition Postulate |
| 6. $m\angle 2 + m\angle 3 = 180°$ | 6. Substitution |
| 7. $m\angle 1 + m\angle 3 = 180°$ | 7. Substitution |
| 8. $\angle 1$ and $\angle 3$ are supplementary | 8. If the sum of measures of two $\angle$s is 180°, the $\angle$s are supplementary |

The proof of the following theorem is left as an exercise.

> THEOREM 2.1.5: If two parallel lines are cut by a transversal, then the exterior angles on the same side of the transversal are supplementary.

*Exs. 7–11*

The remaining examples in this section illustrate methods from algebra and deal with the angles formed when two parallel lines are cut by a transversal.

EXAMPLE 2

*Given:* $\overleftrightarrow{TV} \parallel \overleftrightarrow{WY}$ with transversal $\overleftrightarrow{RS}$

$m\angle RUV = (x + 4)(x - 3)$
$m\angle WXS = x^2 - 3$

*Find:* $x$

***Solution***

$\angle RUV$ and $\angle WXS$ are alternate exterior angles, so they are congruent. Then $m\angle RUV = m\angle WXS$. Therefore,

$$\begin{aligned}(x + 4)(x - 3) &= x^2 - 3\\ x^2 + x - 12 &= x^2 - 3\\ x - 12 &= -3\\ x &= 9\end{aligned}$$

NOTE: Both angles measure 78° when $x = 9$.

In Figure 2.10, lines $r$ and $s$ are parallel. For $\ell$ and $m$ to be parallel, which angle, $\angle 1$ or $\angle 9$, must be congruent to $\angle 5$? To answer the question, it might be helpful to use a second color to draw in the lines that are known to be parallel. Now decide which transversal to use, $\ell$ or $m$. In this case, $\ell$ is the transversal for $r$ and $s$ that forms corresponding angles 5 and 1. Therefore, $\angle 1$ must be congruent to $\angle 5$ if $r \parallel s$.

*Exs. 12, 13*

For Example 3, recall that two equations are necessary to solve a problem in two variables.

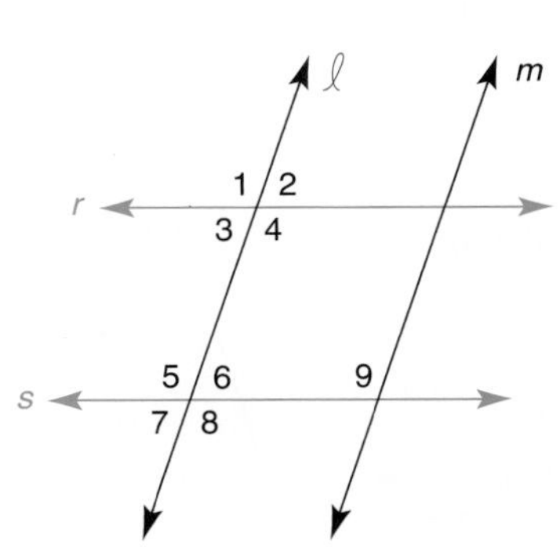

**FIGURE 2.10**

EXAMPLE 3

*Given:* In Figure 2.10, $r \parallel s$ and transversal $\ell$

$m\angle 3 = 4x + y$
$m\angle 5 = 6x + 5y$
$m\angle 6 = 5x - 2y$

*Find:* $x$ and $y$

***Solution***

$\angle 3$ and $\angle 6$ are congruent alternate interior angles; also, $\angle 3$ and $\angle 5$ are supplementary angles according to Theorem 2.1.4. These facts lead to the following system of equations:

$$\begin{aligned}4x + y &= 5x - 2y\\ (4x + y) + (6x + 5y) &= 180\end{aligned}$$

These equations can be simplified to

$$\begin{aligned}x - 3y &= 0\\ 10x + 6y &= 180\end{aligned}$$

After we divide each term of the second equation by 2, the system becomes

$$\begin{aligned}x - 3y &= 0\\ 5x + 3y &= 90\end{aligned}$$

Addition leads to the equation $6x = 90$, so $x = 15$. Substituting 15 for $x$ into the equation $x - 3y = 0$, we have

$$\begin{aligned}15 - 3y &= 0\\ -3y &= -15\\ y &= 5\end{aligned}$$

Our solution, $x = 15$ and $y = 5$, yields the following angle measures:

$$m\angle 3 = 65°$$
$$m\angle 5 = 115°$$
$$m\angle 6 = 65°$$

NOTE: For an alternative solution, the equation $x - 3y = 0$ could be multiplied by 2 to obtain $2x - 6y = 0$. Then the equations $2x - 6y = 0$ and $10x + 6y = 180$ could be added.

Note that the angle measures determined in Example 3 are consistent with Figure 2.10 and the required relationships for the angles named. For instance, $m\angle 3 + m\angle 5 = 180°$, and we see that interior angles on the same side of the transversal are indeed supplementary.

## 2.1 Exercises

*For Exercises 1 to 4, $\ell \parallel m$ with transversal $v$.*

1. If $m\angle 1 = 108°$, find: a) $m\angle 5$ b) $m\angle 7$
2. If $m\angle 3 = 71°$, find: a) $m\angle 5$ b) $m\angle 6$
3. If $m\angle 2 = 68.3°$, find: a) $m\angle 3$ b) $m\angle 6$
4. If $m\angle 4 = 110.8°$, find: a) $m\angle 5$ b) $m\angle 8$

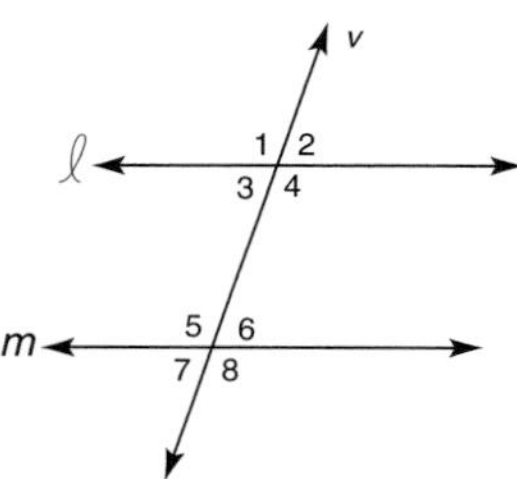


*Use drawings, as needed, to answer each question.*

5. Does the relation "is parallel to" have a
   a) reflexive property? (consider a line $m$)
   b) symmetric property? (consider lines $m$ and $n$ in a plane)
   c) transitive property? (consider coplanar lines $m$, $n$, and $q$)
6. In a plane, $\ell \perp m$ and $t \perp m$. By appearance, how are $\ell$ and $t$ related?
7. Suppose that $r \parallel s$. Each interior angle on the right side of the transversal $t$ has been bisected. Using intuition, what appears to be true of $\angle 9$ formed by the bisectors?

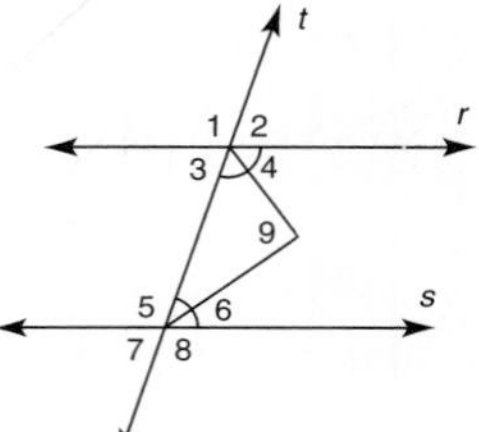


8. Make a sketch to represent two planes that are
   a) parallel.
   b) perpendicular.
9. Suppose that $r$ is parallel to $s$ and $m\angle 2 = 87°$. Find:
   a) $m\angle 3$ c) $m\angle 1$
   b) $m\angle 6$ d) $m\angle 7$

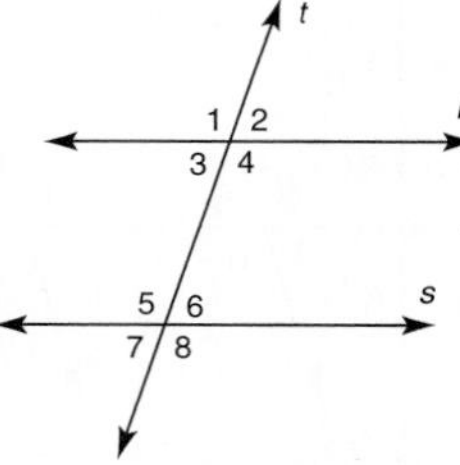


10. In Euclidean geometry, how many lines can be drawn through a point $P$ not on a line $\ell$ that are
    a) parallel to line $\ell$?
    b) perpendicular to line $\ell$?
11. Lines $r$ and $s$ are cut by transversal $t$. Which angle
    a) corresponds to $\angle 1$?
    b) is the alternate interior $\angle$ for $\angle 4$?

    c) is the alternate exterior $\angle$ for $\angle 1$?
    d) is the other interior angle on the same side of transversal $t$ as $\angle 3$?

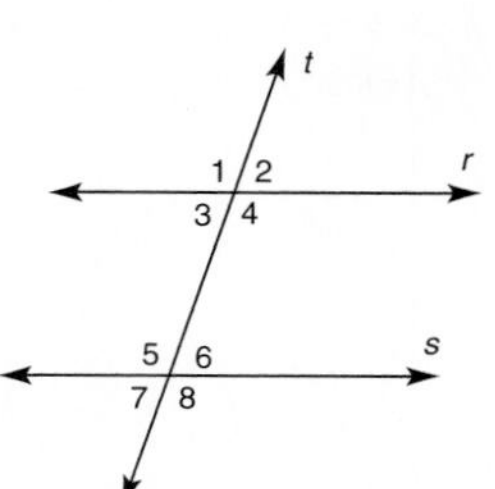

12. $\overline{AD} \parallel \overline{BC}$, $\overline{AB} \parallel \overline{DC}$, and $m\angle A = 92°$.

 a) $m\angle B$
 b) $m\angle C$
 c) $m\angle D$

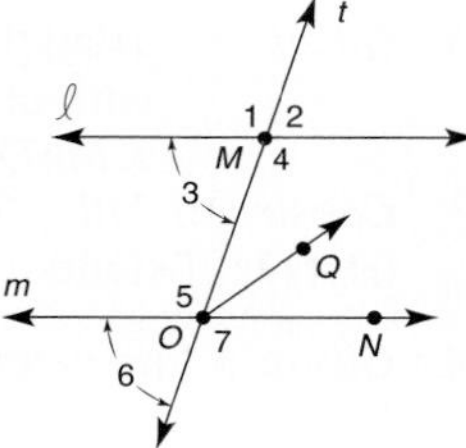


13. $\ell \parallel m$, with transversal $t$, and $\overrightarrow{OQ}$ bisects $\angle MON$. If $m\angle 1 = 112°$, find the following:

 a) $m\angle 2$
 b) $m\angle 4$
 c) $m\angle 5$
 d) $m\angle MOQ$

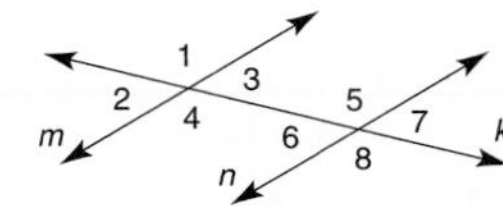


*Exercises 13, 14*

14. *Given:* $\ell \parallel m$
 Transversal $t$
 $m\angle 1 = 4x + 2$
 $m\angle 6 = 4x - 2$
 *Find:* $x$ and $m\angle 5$

15. *Given:* $m \parallel n$
 Transversal $k$
 $m\angle 3 = x^2 - 3x$
 $m\angle 6 = (x + 4)(x - 5)$
 *Find:* $x$ and $m\angle 4$

16. *Given:* $m \parallel n$
 Transversal $k$
 $m\angle 1 = 5x + y$
 $m\angle 2 = 3x + y$
 $m\angle 8 = 3x + 5y$
 *Find:* $x$, $y$, and $m\angle 8$



*Exercises 15–17*

17. *Given:* $m \parallel n$
 Transversal $k$
 $m\angle 3 = 6x + y$
 $m\angle 5 = 8x + 2y$
 $m\angle 6 = 4x + 7y$
 *Find:* $x$, $y$, and $m\angle 7$

18. In the three-dimensional figure, $\overline{CA} \perp \overline{AB}$ and $\overline{BE} \perp \overline{AB}$. Are $\overleftrightarrow{CA}$ and $\overleftrightarrow{BE}$ parallel to each other? (Compare with Exercise 6.)

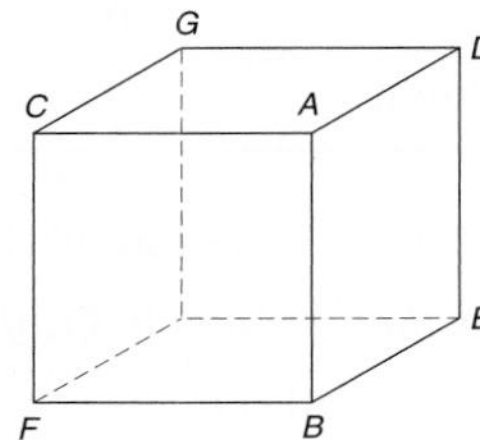


19. *Given:* $\ell \parallel m$ and $\angle 3 \cong \angle 4$
 *Prove:* $\angle 1 \cong \angle 4$
 (See figure in second column.)

**PROOF**

| Statements | Reasons |
|---|---|
| 1. $\ell \parallel m$ | 1. ? |
| 2. $\angle 1 \cong \angle 2$ | 2. ? |
| 3. $\angle 2 \cong \angle 3$ | 3. ? |
| 4. ? | 4. Given |
| 5. ? | 5. Transitive of $\cong$ |

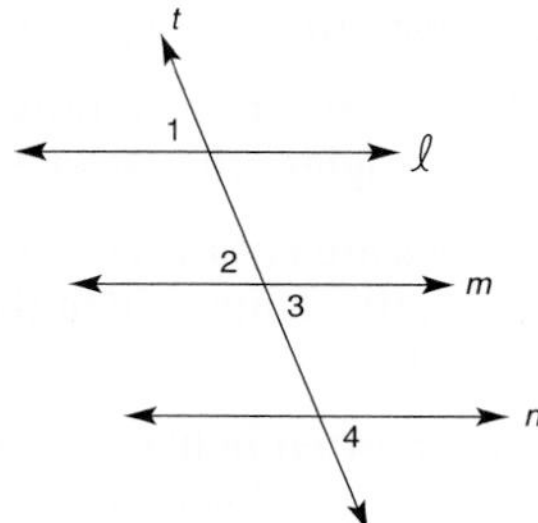


*Exercises 19, 20*

20. *Given:* $\ell \parallel m$ and $m \parallel n$
 *Prove:* $\angle 1 \cong \angle 4$

**PROOF**

| Statements | Reasons |
|---|---|
| 1. $\ell \parallel m$ | 1. ? |
| 2. $\angle 1 \cong \angle 2$ | 2. ? |
| 3. $\angle 2 \cong \angle 3$ | 3. ? |
| 4. ? | 4. Given |
| 5. $\angle 3 \cong \angle 4$ | 5. ? |
| 6. ? | 6. ? |

21. *Given:* $\overleftrightarrow{CE} \parallel \overleftrightarrow{DF}$
 Transversal $\overleftrightarrow{AB}$
 $\overrightarrow{CX}$ bisects $\angle ACE$
 $\overrightarrow{DE}$ bisects $\angle CDF$
 *Prove:* $\angle 1 \cong \angle 3$

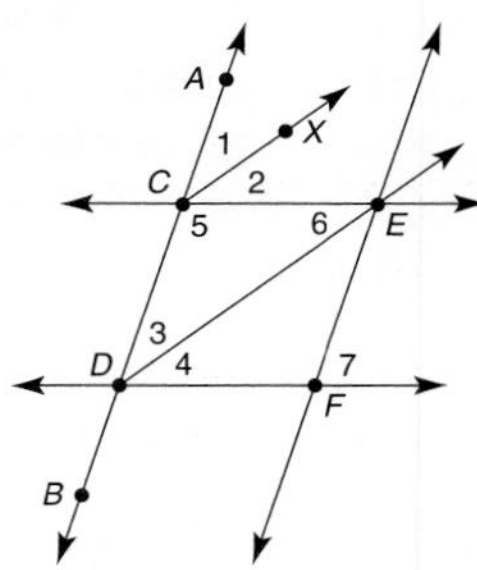


*Exercises 21, 22*

22. *Given:* $\overleftrightarrow{CE} \parallel \overleftrightarrow{DF}$
 Transversal $\overleftrightarrow{AB}$
 $\overrightarrow{DE}$ bisects $\angle CDF$
 *Prove:* $\angle 3 \cong \angle 6$

23. *Given:* $r \parallel s$
 Transversal $t$
 $\angle 1$ is a right $\angle$
 *Prove:* $\angle 2$ is a right $\angle$

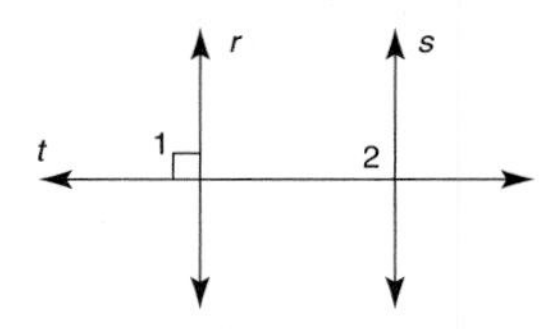


*Exercises 23, 24*

24. *Given:* $r \parallel s$, $r \perp t$ (See figure on page 79.)
*Prove:* $s \perp t$

25. In triangle $ABC$, line $t$ is drawn through vertex $A$ in such a way that $t \parallel \overline{BC}$.

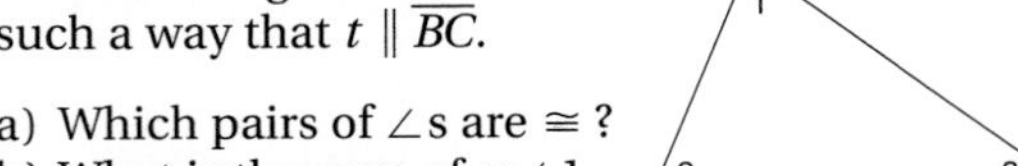

a) Which pairs of $\angle$s are $\cong$?
b) What is the sum of $m\angle 1$, $m\angle 4$, and $m\angle 5$?
c) What is the sum of measures of the $\angle$s of $\triangle ABC$?

*In Exercises 26 to 28, write a formal proof of each theorem.*

26. If two parallel lines are cut by a transversal, then the alternate exterior angles are congruent.

27. If two parallel lines are cut by a transversal, then the exterior angles on the same side of the transversal are supplementary.

28. If a transversal is perpendicular to one of two parallel lines, then it is also perpendicular to the other line.

29. Suppose that two lines are cut by a transversal in such a way that corresponding angles are not congruent. Can those two lines be parallel?

30. *Given:* Line $\ell$ and point $P$ not on $\ell$
*Construct:* $\overleftrightarrow{PQ} \perp \ell$

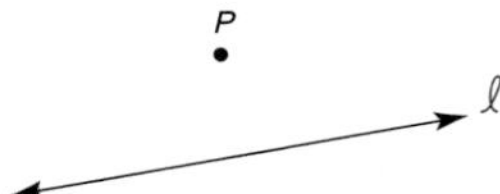


31. *Given:* Triangle $ABC$ with three acute angles
*Construct:* $\overline{BD} \perp \overline{AC}$

32. *Given:* Triangle $MNQ$ with obtuse $\angle MNQ$
*Construct:* $\overline{NE} \perp \overline{MQ}$

33. *Given:* Triangle $MNQ$ with obtuse $\angle MNQ$
*Construct:* $\overline{MR} \perp \overline{NQ}$
(HINT: Extend $\overline{NQ}$.)

*Exercises 32, 33*

34. *Given:* A line $m$ and a point $T$ not on $m$

Suppose that you do the following:

i) Construct a perpendicular line $r$ from $T$ to line $m$.
ii) Construct a line $s$ perpendicular to line $r$ at point $T$.

What relationship holds between lines $s$ and $m$?

## 2.2 Indirect Proof

**KEY CONCEPTS**
Conditional • Converse • Inverse • Contrapositive • Law of Negative Inference • Indirect Proof

Let $P \rightarrow Q$ represent the statement "If $P$, then $Q$." The following statements are related to this conditional statement.

NOTE: Recall that $\sim P$ represents the negation of $P$.

| | | |
|---|---|---|
| Conditional (or Implication) | $P \rightarrow Q$ | If $P$, then $Q$. |
| Converse of Conditional | $Q \rightarrow P$ | If $Q$, then $P$. |
| Inverse of Conditional | $\sim P \rightarrow \sim Q$ | If not $P$, then not $Q$. |
| Contrapositive of Conditional | $\sim Q \rightarrow \sim P$ | If not $Q$, then not $P$. |

For example, consider the following conditional statement.

*If Tom lives in San Diego, then he lives in California.*

This true statement has these related statements:

*Converse:* If Tom lives in California, then he lives in San Diego. (false)
*Inverse:* If Tom does not live in San Diego, then he does not live in California. (false)
*Contrapositive:* If Tom does not live in California, then he does not live in San Diego. (true)

In general, the conditional statement and its contrapositive are either both true or both false! In advanced courses, a statement of the form "If $P$, then $Q$" may be established by proving that its contrapositive is true, if that seems the easier proof to complete. Similarly, the converse and the inverse are also either both true or both false. See our textbook website for more information about the conditional and related statements.

Venn Diagrams can also be used to explain why the conditional statement $P \rightarrow Q$ and its contrapositive $\sim Q \rightarrow \sim P$ are equivalent. The relationship "If $P$, then $Q$" is represented at the right. Note that if any point is selected outside of $Q$ ($\therefore \sim Q$), then it cannot possibly lie in set $P$. Thus the statement "If not $Q$, then not $P$" relates statements $P$ and $Q$ in a manner equivalent to that of "If $P$, then $Q$."

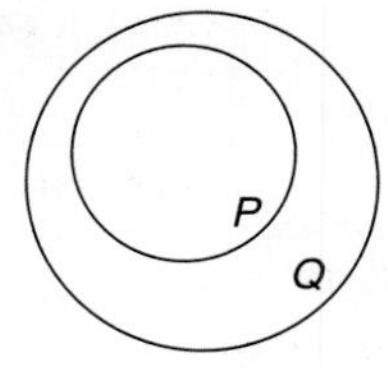


**FIGURE 2.11**

EXAMPLE 1

For the conditional statement that follows, give the converse, the inverse, and the contrapositive. Then classify each as true or false.

*If two angles are vertical angles, then they are congruent angles.*

***Solution***

*Converse:* If two angles are congruent angles, then they are vertical angles. (false)

*Inverse:* If two angles are not vertical angles, then they are not congruent angles. (false)

*Contrapositive:* If two angles are not congruent angles, then they are not vertical angles. (true)

## THE LAW OF NEGATIVE INFERENCE (CONTRAPOSITION)

*Exs. 1, 2*

Consider the following circumstances, and accept each premise as true:

1. If Matt cleans his room, then he will go to the movie. ($P \rightarrow Q$)
2. Matt does not get to go to the movie. ($\sim Q$)

What can you conclude? You should have deduced that Matt did not clean his room; if he had, he would have gone to the movie. This "backdoor" reasoning is based on the fact that the truth of $P \rightarrow Q$ implies the truth of $\sim Q \rightarrow \sim P$.

LAW OF NEGATIVE INFERENCE (CONTRAPOSITION)

$$\begin{array}{l} P \rightarrow Q \\ \underline{\quad \sim Q} \\ \therefore \sim P \end{array}$$

Like the Law of Detachment from Section 1.1, the Law of Negative Inference (Law of Contraposition) is a form of deduction. Whereas the Law of Detachment characterizes the method of "direct proof" encountered in preceding sections, the Law of Negative Inference is the backbone of the method of proof known as **indirect proof.**

## INDIRECT PROOF

*Exs. 3, 4*

You will need to know when to use the indirect method of proof. Often the theorem to be proved has the form $P \rightarrow Q$, in which $Q$ is a negation and denies some claim. For instance, an indirect proof might be best if $Q$ reads in one of these ways:

$c$ is *not* equal to $d$
$\ell$ is *not* perpendicular to $m$

However, we will see in Example 4 of this section that the indirect method can be used to prove that line $\ell$ is parallel to line $m$. Indirect proof is also used for proving existence and uniqueness theorems; see Example 5.

The method of indirect proof is illustrated in Example 2. All indirect proofs in this book are given in paragraph form (as are many of the direct proofs).

In any paragraph proof, each statement must still be justified. Because of the need to order your statements properly, writing this type of proof may have a positive impact on the essays you write for your other classes!

*Geometry in the Real World*

When the bubble displayed on the level is not centered, the board used in construction is neither vertical nor horizontal.

EXAMPLE 2

*Given:* In Figure 2.12, $\overrightarrow{BA}$ is *not* perpendicular to $\overrightarrow{BD}$

*Prove:* $\angle 1$ and $\angle 2$ are *not* complementary

*Proof:* Suppose that $\angle 1$ and $\angle 2$ *are* complementary. Then $m\angle 1 + m\angle 2 = 90°$ because the sum of the measures of two complementary $\angle$s is 90. We also know that $m\angle 1 + m\angle 2 = m\angle ABD$, by the Angle-Addition Postulate. In turn, $m\angle ABD = 90°$ by substitution. Then $\angle ABD$ is a right angle. In turn, $\overrightarrow{BA} \perp \overrightarrow{BD}$. But this contradicts the given hypothesis; therefore, the supposition must be false, and it follows that $\angle 1$ and $\angle 2$ are not complementary.

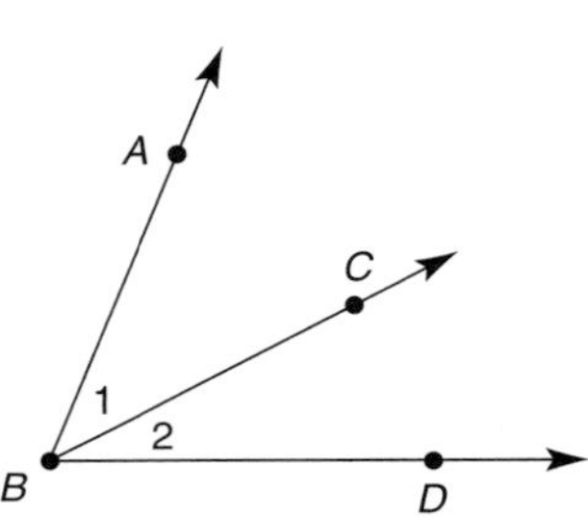


**FIGURE 2.12**

In Example 2 and in all indirect proofs, the first statement takes the form

*"Suppose that . . ."* or *"Assume that . . ."*

By its very nature, this statement cannot be supported even though every other statement in the proof can be justified; thus, when a contradiction is reached, the finger of blame points to the supposition. At this stage of the proof, we may say that the claim involving $\sim Q$ has failed and is false; thus, our only recourse is to conclude that $Q$ is true. Following is an outline of this technique.

*Exs. 5–7*

METHOD OF INDIRECT PROOF

To prove the statement $P \rightarrow Q$ or to complete the proof problem of the form
*Given:* $P$
*Prove:* $Q$
where $Q$ may be a negation, use the following steps:

1. Suppose that $\sim Q$ is true.
2. Reason from the supposition until you reach a contradiction.
3. Note that the supposition claiming that $\sim Q$ is true must be false and that $Q$ must therefore be true.

Step 3 completes the proof.

The contradiction that is discovered in an indirect proof often has the form $\sim P$. Thus the assumed statement $\sim Q$ has forced the conclusion $\sim P$, asserting that $\sim Q \rightarrow \sim P$ is true. Then the desired theorem $P \rightarrow Q$ (the contrapositive of $\sim Q \rightarrow \sim P$) is also true.

### EXAMPLE 3

Complete a formal proof of the following theorem:

*If two lines are cut by a transversal so that corresponding angles are not congruent, then the two lines are not parallel.*

*Given:* In Figure 2.13, $\ell$ and $m$ are cut by transversal $t$

$\angle 1 \not\cong \angle 5$

*Prove:* $\ell \nparallel m$

SSG *Exs. 8, 9*

*Proof:* Assume that $\ell \parallel m$. When these lines are cut by transversal $t$, the corresponding angles (including $\angle 1$ and $\angle 5$) are congruent. But $\angle 1 \not\cong \angle 5$ by hypothesis. Thus the assumed statement, which claims that $\ell \parallel m$, must be false. It follows that $\ell \nparallel m$.

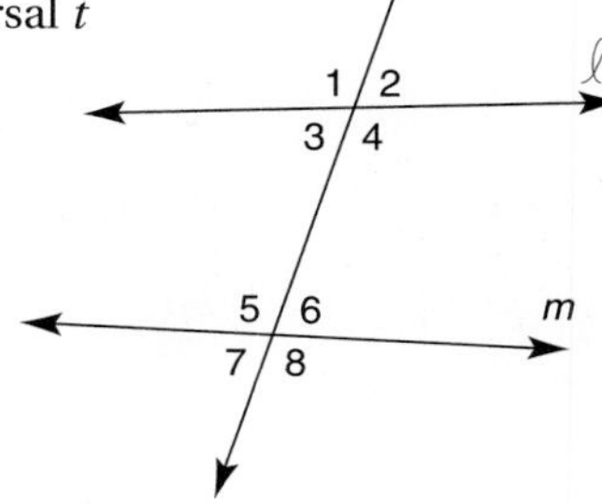


**FIGURE 2.13**

The versatility of the indirect proof is shown in the final examples of this section. The indirect proofs preceding Example 4 all contain a negation in the conclusion (Prove); the proofs in the final illustrations use the indirect method to arrive at a positive conclusion.

### EXAMPLE 4

*Given:* In Figure 2.14, parallel planes $P$ and $Q$ are intersected by plane $T$ in lines $\ell$ and $m$

*Prove:* $\ell \parallel m$

*Proof:* Assume that $\ell$ is not parallel to $m$. Then $\ell$ and $m$ intersect at some point $A$. But if so, point $A$ must be on both planes $P$ and $Q$, which means that planes $P$ and $Q$ intersect; but $P$ and $Q$ are parallel by hypothesis. Therefore, the assumption that $\ell$ and $m$ are not parallel must be false, and it follows that $\ell \parallel m$.

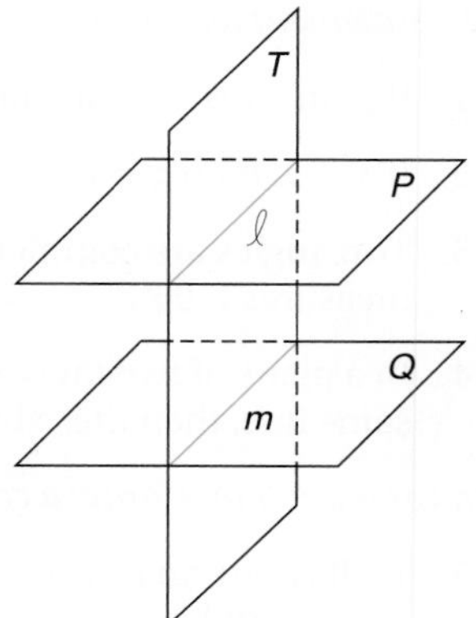


**FIGURE 2.14**

Indirect proofs are also used to establish uniqueness theorems, as Example 5 illustrates.

EXAMPLE 5

Prove the statement "The angle bisector of an angle is unique."

*Given:* In Figure 2.15(a), $\overrightarrow{BD}$ bisects $\angle ABC$

*Prove:* $\overrightarrow{BD}$ is the only angle bisector for $\angle ABC$

*Proof:* $\overrightarrow{BD}$ bisects $\angle ABC$, so $m\angle ABD = \frac{1}{2}m\angle ABC$. Suppose that $\overrightarrow{BE}$ [as shown in Figure 2.15(b)] is also a bisector of $\angle ABC$ and that $m\angle ABE = \frac{1}{2}m\angle ABC$.

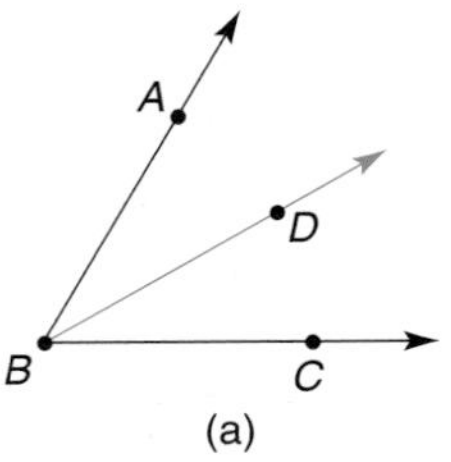


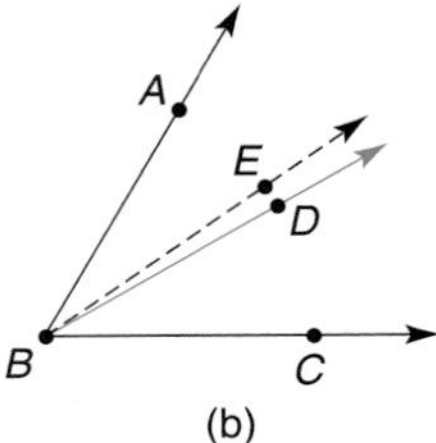


**FIGURE 2.15**

By the Angle-Addition Postulate, $m\angle ABD = m\angle ABE + m\angle EBD$. By substitution, $\frac{1}{2}m\angle ABC = \frac{1}{2}m\angle ABC + m\angle EBD$; but then $m\angle EBD = 0$ by subtraction. An angle with a measure of 0 contradicts the Protractor Postulate, which states that the measure of an angle is a unique positive number. Therefore, the assumed statement must be false, and it follows that the angle bisector of an angle is unique.

*Ex. 10*

## 2.2 Exercises

*In Exercises 1 to 4, write the converse, the inverse, and the contrapositive of each statement. When possible, classify the statement as true or false.*

1. If Juan wins the state lottery, then he will be rich.
2. If $x > 2$, then $x \neq 0$.
3. Two angles are complementary if the sum of their measures is 90°.
4. In a plane, if two lines are not perpendicular to the same line, then these lines are not parallel.

*In Exercises 5 to 8, draw a conclusion where possible.*

5. 1. If two triangles are congruent, then the triangles are similar.
   2. Triangles $ABC$ and $DEF$ are not congruent.
   
   C. $\therefore$ ?

6. 1. If two triangles are congruent, then the triangles are similar.
   2. Triangles $ABC$ and $DEF$ are not similar.
   
   C. $\therefore$ ?

7. 1. If $x > 3$, then $x = 5$.
   2. $x > 3$
   
   C. $\therefore$ ?

8. 1. If $x > 3$, then $x = 5$.
   2. $x \neq 5$
   
   C. $\therefore$ ?

9. Which of the following statements would you prove by the indirect method?
   a) In triangle $ABC$, if $m\angle A > m\angle B$, then $AC \neq BC$.
   b) If alternate exterior $\angle 1 \not\cong$ alternate exterior $\angle 8$, then $\ell$ is not parallel to $m$.
   c) If $(x + 2) \cdot (x - 3) = 0$, then $x = -2$ or $x = 3$.
   d) If two sides of a triangle are congruent, then the two angles opposite these sides are also congruent.
   e) The perpendicular bisector of a line segment is unique.

10. For each statement in Exercise 9 that can be proved by the indirect method, give the first statement in each proof.

11. A periscope uses an indirect method of observation. This instrument allows one to see what would otherwise be obstructed. Mirrors are located (see $\overline{AB}$ and $\overline{CD}$ in the drawing) so that an image is reflected twice. How are $\overline{AB}$ and $\overline{CD}$ related to each other?

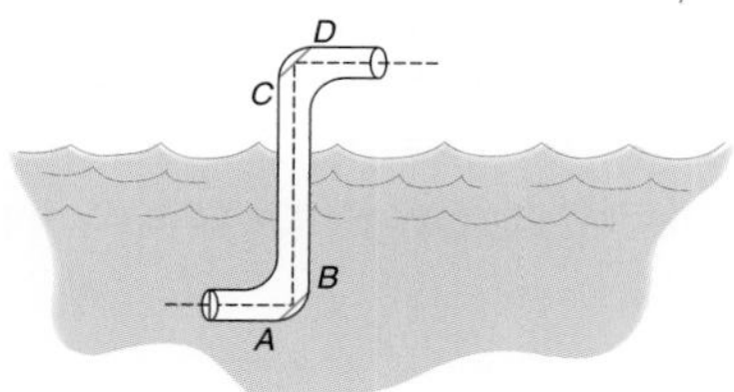


12. Some stores use an indirect method of observation. The purpose may be for safety (to avoid collisions) or to foil the attempts of would-be shoplifters. In this situation, a mirror (see $\overline{EF}$ in the drawing) is placed at the intersection of two aisles as shown. An observer at point $P$ can then see any movement along the indicated aisle. In the sketch, what is the measure of $\angle GEF$?

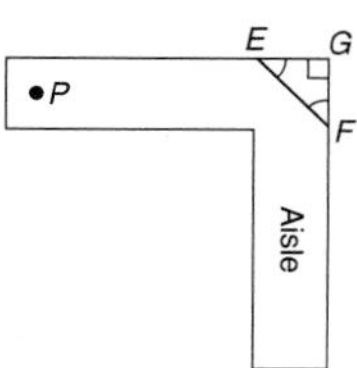


*In Exercises 13 to 24, give the indirect proof for each problem or statement.*

13. *Given:* $\angle 1 \not\cong \angle 5$
*Prove:* $r \nparallel s$

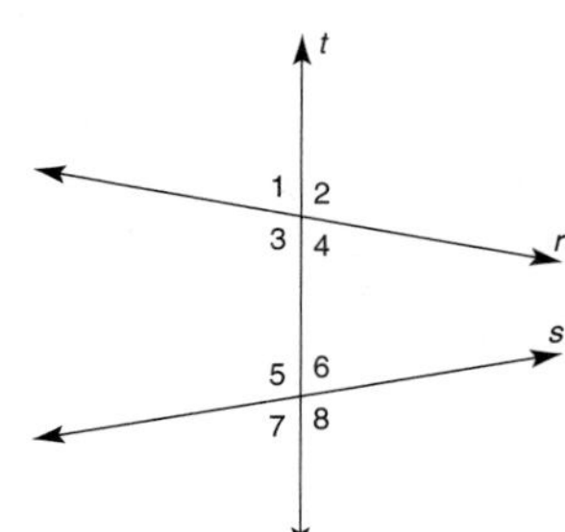


14. *Given:* $\angle ABD \not\cong \angle DBC$
*Prove:* $\overrightarrow{BD}$ does not bisect $\angle ABC$

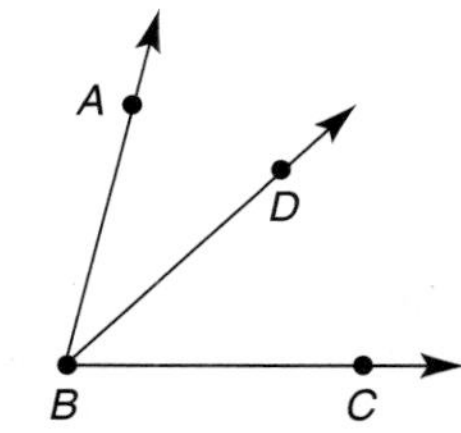


15. *Given:* $m\angle 3 > m\angle 4$
*Prove:* $\overrightarrow{FH}$ is not $\perp$ to $\overleftrightarrow{EG}$

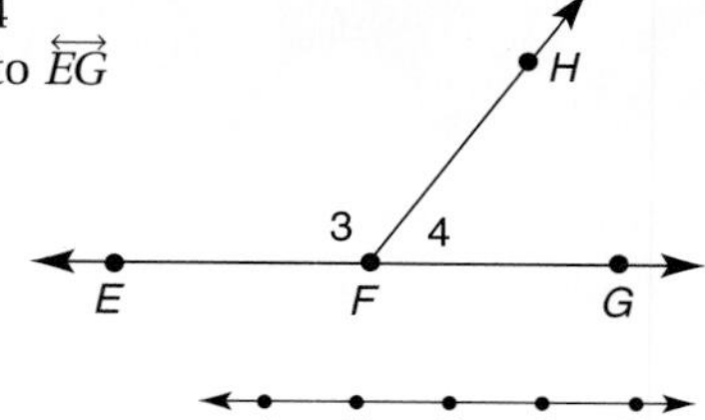


16. *Given:* $MB > BC$
$AM = CD$
*Prove:* $B$ is not the midpoint of $\overline{AD}$

A M B C D

17. If two angles are not congruent, then these angles are not vertical angles.

18. If $x^2 \neq 25$, then $x \neq 5$.

19. If alternate interior angles are not congruent when two lines are cut by a transversal, then the lines are not parallel.

20. If $a$ and $b$ are positive numbers, then $\sqrt{a^2 + b^2} \neq a + b$.

21. The midpoint of a line segment is unique.

22. There is exactly one line perpendicular to a given line at a point on the line.

*23. In a plane, if two lines are parallel to a third line, then the two lines are parallel to each other.

*24. In a plane, if two lines are intersected by a transversal so that the corresponding angles are congruent, then the lines are parallel.

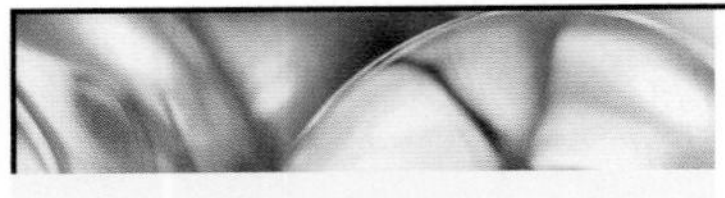

**KEY CONCEPTS**
Proving Lines Parallel

# 2.3 Proving Lines Parallel

In Section 2.1, several methods for proving angles congruent or supplementary were developed by using parallel lines. Here is a quick review of the relevant postulate and theorems. Each has the hypothesis "If two parallel lines are cut by a transversal."

POSTULATE 11: If two parallel lines are cut by a transversal, then the corresponding angles are congruent.

THEOREM 2.1.2: If two parallel lines are cut by a transversal, then the alternate interior angles are congruent.

THEOREM 2.1.3: If two parallel lines are cut by a transversal, then the alternate exterior angles are congruent.

THEOREM 2.1.4: If two parallel lines are cut by a transversal, then the interior angles on the same side of the transversal are supplementary.

THEOREM 2.1.5: If two parallel lines are cut by a transversal, then the exterior angles on the same side of the transversal are supplementary.

Suppose that we wish to prove that two lines are parallel rather than to establish an angle relationship (as the previous statements do). Such a theorem would take the form "If . . . , then these lines are parallel." At present, the only method we have of proving lines parallel is based on the definition of parallel lines. Establishing the conditions of the definition (that coplanar lines do *not* intersect) is virtually impossible! Thus we begin to develop methods for proving that lines in a plane are parallel by proving Theorem 2.3.1 by the indirect method. Counterparts of Theorems 2.1.2–2.1.5, namely Theorems 2.3.2–2.3.5, are proved directly but depend on Theorem 2.3.1. Except for Theorem 2.3.6, the theorems of this section require coplanar lines.

*Exs. 1, 2*

THEOREM 2.3.1: If two lines are cut by a transversal so that the corresponding angles are congruent, then these lines are parallel.

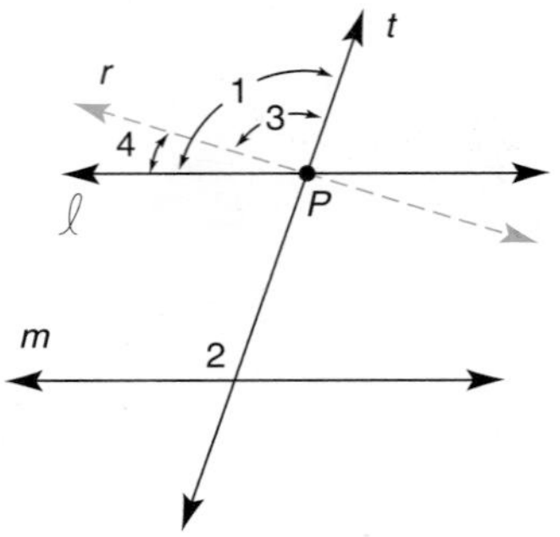

*Given:* $\ell$ and $m$ cut by transversal $t$
$\angle 1 \cong \angle 2$ (Figure 2.16)

*Prove:* $\ell \parallel m$

*Proof:* Suppose that $\ell \nparallel m$. Then a line $r$ can be drawn through point $P$ such that it is parallel to $m$; this follows from the Parallel Postulate. If $r \parallel m$, then $\angle 3 \cong \angle 2$ because these angles correspond. But $\angle 1 \cong \angle 2$ by hypothesis. Now $\angle 3 \cong \angle 1$ by the Transitive Property of Congruence; therefore, $m\angle 3 = m\angle 1$. But $m\angle 3 + m\angle 4 = m\angle 1$. (See Figure 2.16.) Substitution of $m\angle 1$ for $m\angle 3$ leads to $m\angle 1 + m\angle 4 = m\angle 1$; and by subtraction, $m\angle 4 = 0$. This contradicts the Protractor Postulate, which states that the measure of any angle must be a

**FIGURE 2.16**

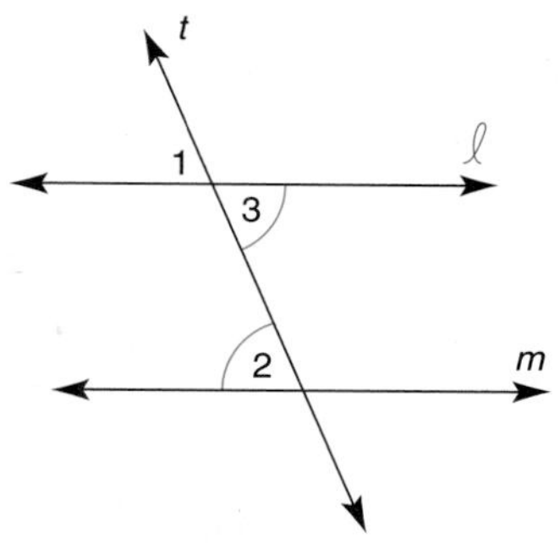


**FIGURE 2.17**

positive number. Consequently, $r$ and $\ell$ must coincide, and it follows that $\ell \parallel m$.

Once proved, Theorem 2.3.1 opens the doors to a host of other methods for proving that lines are parallel. Each claim, Theorem 2.3.2–Theorem 2.3.5, is the converse of its counterpart in Section 2.1.

> THEOREM 2.3.2: If two lines are cut by a transversal so that the alternate interior angles are congruent, then these lines are parallel.

*Given:* Lines $\ell$ and $m$ and transversal $t$
$\angle 2 \cong \angle 3$ (Figure 2.17)

*Prove:* $\ell \parallel m$

*Plan for the proof:* Show that $\angle 1 \cong \angle 2$ (corresponding angles). Then apply Theorem 2.3.1, in which $\cong$ corresponding $\angle$s imply parallel lines.

**PROOF**

| Statements | Reasons |
|---|---|
| 1. $\ell$ and $m$; trans. $t$; $\angle 2 \cong \angle 3$ | 1. Given |
| 2. $\angle 1 \cong \angle 3$ | 2. If two lines intersect, vertical $\angle$s are $\cong$ |
| 3. $\angle 1 \cong \angle 2$ | 3. Transitive Property of Congruence |
| 4. $\ell \parallel m$ | 4. If two lines are cut by a transversal so that corr. $\angle$s are $\cong$, then these lines are parallel |

> **Discover!**
>
> When a staircase is designed, "stringers" are cut for each side of the stairs as shown. How are angles 1 and 3 related? How are angles 1 and 2 related?
>
> 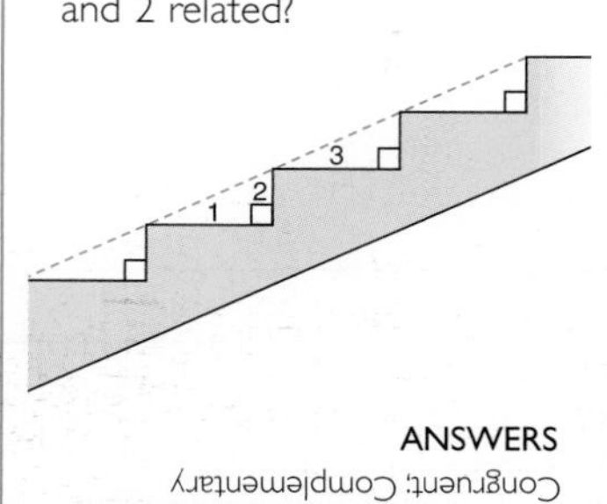
> 
>
> ANSWERS
>
> Congruent; Complementary

The following theorem is proved in a manner much like the proof of Theorem 2.3.2. The proof is left as an exercise.

> THEOREM 2.3.3: If two lines are cut by a transversal so that the alternate exterior angles are congruent, then these lines are parallel.

In a more involved drawing, it may be difficult to decide which lines are parallel because of congruent angles. Consider Figure 2.18. Suppose that $\angle 1 \cong \angle 3$. Which lines must be parallel? The resulting confusion (it appears that $a$ may be parallel to $b$ and $c$ may be parallel to $d$) can be overcome by asking, "Which lines help form $\angle 1$ and $\angle 3$?" In this case, $\angle 1$ and $\angle 3$ are formed by lines $a$ and $b$ with $c$ as the transversal. Thus $a \parallel b$.

**EXAMPLE 1**

In Figure 2.18, which lines must be parallel if $\angle 3 \cong \angle 8$?

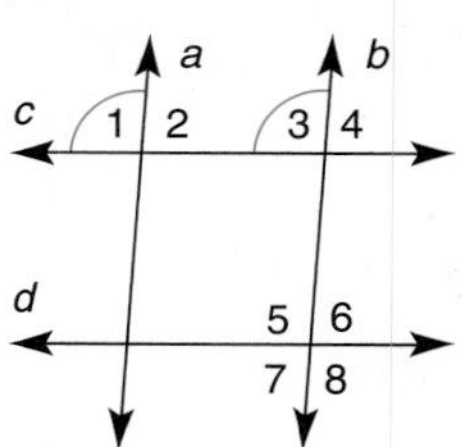


**FIGURE 2.18**

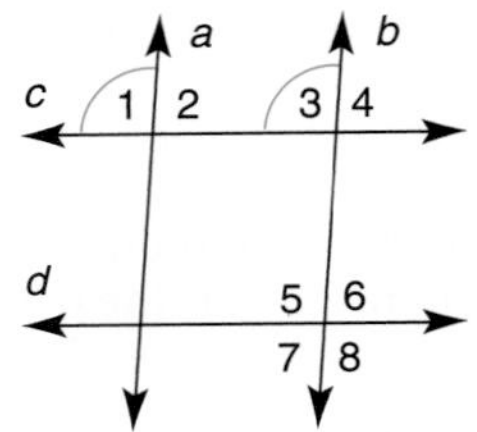


**FIGURE 2.18**

***Solution***

$\angle 3$ and $\angle 8$ are the alternate exterior angles formed when lines $c$ and $d$ are cut by transversal $b$. Thus $c \parallel d$.

EXAMPLE 2

In Figure 2.18, $m\angle 3 = 94°$. Find $m\angle 5$ such that $c \parallel d$.

***Solution***

With $b$ as a transversal for lines $c$ and $d$, $\angle 3$ and $\angle 5$ are corresponding angles. Then $c$ would be parallel to $d$ if $\angle 3$ and $\angle 5$ were congruent. Thus, $m\angle 5 = 94°$.

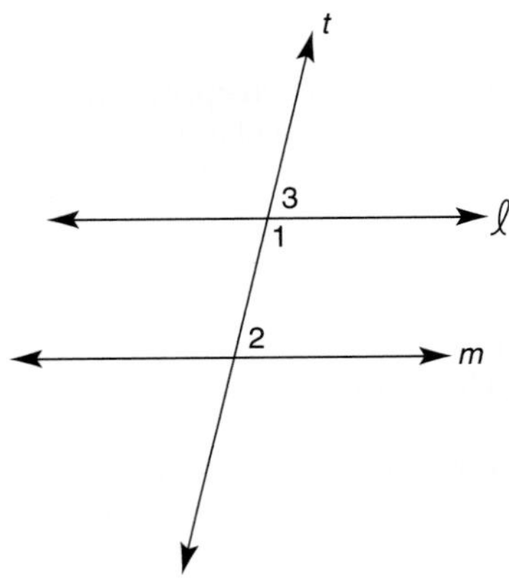


**FIGURE 2.19**

Theorems 2.3.4 and 2.3.5 enable us to prove that lines are parallel when certain pairs of angles are supplementary.

THEOREM 2.3.4: If two lines are cut by a transversal so that the interior angles on the same side of the transversal are supplementary, then these lines are parallel.

EXAMPLE 3

Prove Theorem 2.3.4. (See Figure 2.19.)

*Given:* Lines $\ell$ and $m$; transversal $t$
$\angle 1$ is supplementary to $\angle 2$

*Prove:* $\ell \parallel m$

**PROOF**

| Statements | Reasons |
|---|---|
| 1. $\ell$ and $m$; trans. $t$ <br> $\angle 1$ is supp. to $\angle 2$ | 1. Given |
| 2. $\angle 1$ is supp. to $\angle 3$ | 2. If the exterior sides of two adjacent $\angle$s form a straight line, these $\angle$s are supplementary |
| 3. $\angle 2 \cong \angle 3$ | 3. If two $\angle$s are supp. to the same $\angle$, they are $\cong$ |
| 4. $\ell \parallel m$ | 4. If two lines are cut by a transversal so that corr. $\angle$s are $\cong$, then these lines are parallel |

The proof of Theorem 2.3.5 is similar to that of Theorem 2.3.4. The proof is left as an exercise.

THEOREM 2.3.5: If two lines are cut by a transversal so that the exterior angles on the same side of the transversal are supplementary, then these lines are parallel.

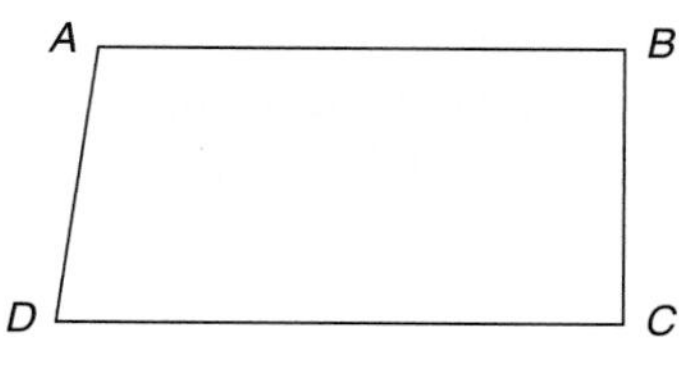

FIGURE 2.20

EXAMPLE 4

In Figure 2.20, which line segments must be parallel if $\angle B$ and $\angle C$ are supplementary?

***Solution***

Again, the solution lies in the question "Which line segments form $\angle B$ and $\angle C$?" With $\overline{BC}$ as a transversal, $\angle B$ and $\angle C$ are formed by $\overline{AB}$ and $\overline{DC}$. It follows that $\overline{AB} \parallel \overline{DC}$, because $\angle$s $B$ and $C$ are supplementary.

We include two final theorems that provide additional means of proving that lines are parallel. The proof of Theorem 2.3.6 (see Exercise 31) requires an auxiliary line (a transversal). Proof of Theorem 2.3.7 is found in Example 5.

THEOREM 2.3.6: If two lines are each parallel to a third line, then these lines are parallel to each other.

Theorem 2.3.6 is true even if the three lines described are not coplanar. In Theorem 2.3.7, the lines must be coplanar.

THEOREM 2.3.7: If two coplanar lines are each perpendicular to a third line, then these lines are parallel to each other.

*Exs. 3–8*

EXAMPLE 5

*Given:* $\overleftrightarrow{AC} \perp \overleftrightarrow{BE}$ and $\overleftrightarrow{DF} \perp \overleftrightarrow{BE}$ (Figure 2.21)

*Prove:* $\overleftrightarrow{AC} \parallel \overleftrightarrow{DF}$

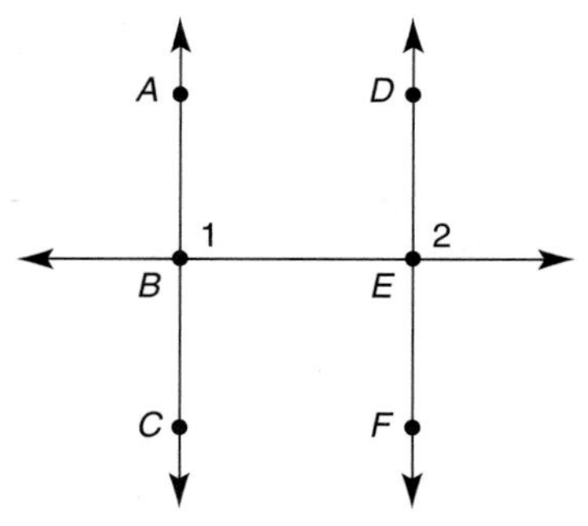

FIGURE 2.21

**PROOF**

| Statements | Reasons |
|---|---|
| 1. $\overleftrightarrow{AC} \perp \overleftrightarrow{BE}$ and $\overleftrightarrow{DF} \perp \overleftrightarrow{BE}$ | 1. Given |
| 2. $\angle$s 1 and 2 are rt. $\angle$s | 2. If two lines are perpendicular, they meet to form right $\angle$s |
| 3. $\angle 1 \cong \angle 2$ | 3. All right angles are $\cong$ |
| 4. $\overleftrightarrow{AC} \parallel \overleftrightarrow{DF}$ | 4. If two lines are cut by a transversal so that corr. $\angle$s are $\cong$, then these lines are parallel |

EXAMPLE 6

*Given:* $m\angle 1 = 7x$ and $m\angle 2 = 5x$ (See Figure 2.22 on the next page.)

*Find:* $x$, so that $\ell$ will be parallel to $m$

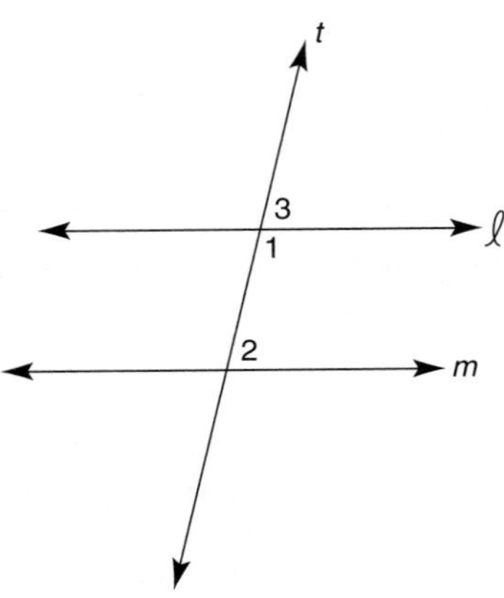


**FIGURE 2.22**

***Solution***

For $\ell$ to be parallel to $m$, $\angle$s 1 and 2 would have to be supplementary. This follows from Theorem 2.3.4 because $\angle$s 1 and 2 are interior angles on the same side of transversal $t$. Then

$$7x + 5x = 180$$
$$12x = 180$$
$$x = 15$$

NOTE: With $m\angle 1 = 105°$ and $m\angle 2 = 75°$, we see that $\angle 1$ and $\angle 2$ are supplementary. Then $\ell \parallel m$.

Construction 7 depends on Theorem 2.3.1, which is restated below.

THEOREM 2.3.1: If two lines are cut by a transversal so that corresponding angles are congruent, then these lines are parallel.

*Exs. 9–16*

CONSTRUCTION 7

To construct the line parallel to a given line from a point not on that line.

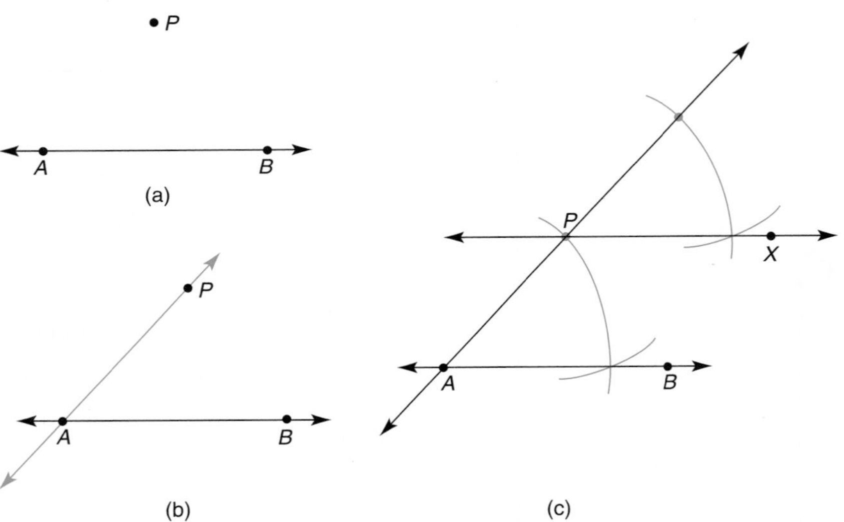


**FIGURE 2.23**

*Given:* $\overleftrightarrow{AB}$ and point $P$ not on $\overleftrightarrow{AB}$, as in Figure 2.23(a)

*Construct:* The line through point $P$ parallel to $\overleftrightarrow{AB}$

*Construction:* Figure 2.23(b): Draw a line (to become a transversal) through point $P$ and some point on $\overleftrightarrow{AB}$. For convenience, we choose point $A$ and draw $\overleftrightarrow{AP}$ as in Figure 2.23(c). Using $P$ as the vertex, construct the angle that corresponds to $\angle PAB$ so that this angle is congruent to $\angle PAB$. It may be necessary to extend $\overleftrightarrow{AP}$ upward to accomplish this. $\overleftrightarrow{PX}$ is the desired line parallel to $\overleftrightarrow{AB}$.

## 2.3 Exercises

*In Exercises 1 to 6, ℓ and m are cut by transversal v. On the basis of the information given, determine whether ℓ must be parallel to m.*

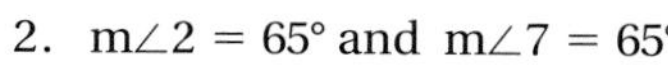

1. m∠1 = 107° and m∠5 = 107°
2. m∠2 = 65° and m∠7 = 65°
3. m∠1 = 106° and m∠7 = 76°
4. m∠1 = 106° and m∠4 = 106°
5. m∠3 = 113.5° and m∠5 = 67.5°
6. m∠6 = 71.4° and m∠7 = 71.4°

*In Exercises 7 to 16, name the lines (if any) that must be parallel under the given conditions.*

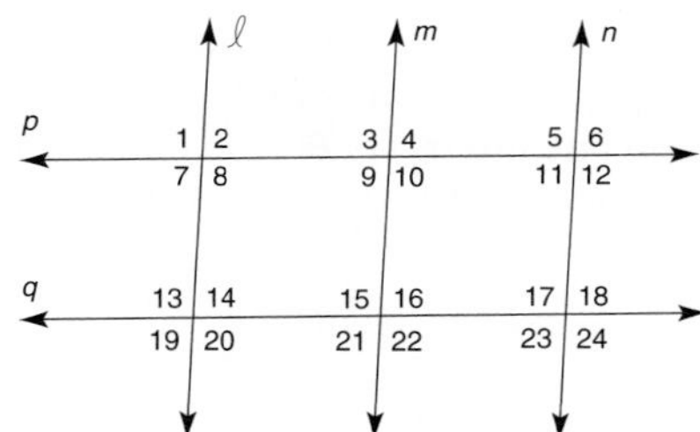


*Exercises 7–16*

7. ∠1 ≅ ∠20
8. ∠3 ≅ ∠10
9. ∠9 ≅ ∠14
10. ∠7 ≅ ∠11
11. ℓ ⊥ p and n ⊥ p
12. ℓ ∥ m and m ∥ n
13. ℓ ⊥ p and m ⊥ q
14. ∠8 and ∠9 are supplementary.
15. m∠8 = 110°, p ∥ q, and m∠18 = 70°
16. The bisectors of ∠9 and ∠21 are parallel.

*In Exercises 17 and 18, complete each proof by filling in the missing statements and reasons.*

17. *Given:* ∠1 and ∠2 are complementary
∠3 and ∠1 are complementary
*Prove:* $\overline{BC} \parallel \overline{DE}$

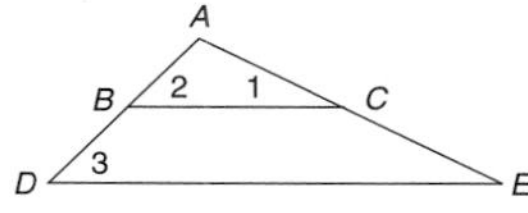


**PROOF**

| Statements | Reasons |
|---|---|
| 1. ∠s 1 and 2 are comp.; ∠s 3 and 1 are comp. | 1. ? |
| 2. ∠2 ≅ ∠3 | 2. ? |
| 3. ? | 3. If two lines are cut by a transversal so that corr. ∠s are ≅, the lines are ∥ |

18. *Given:* ℓ ∥ m
∠3 ≅ ∠4
*Prove:* ℓ ∥ n

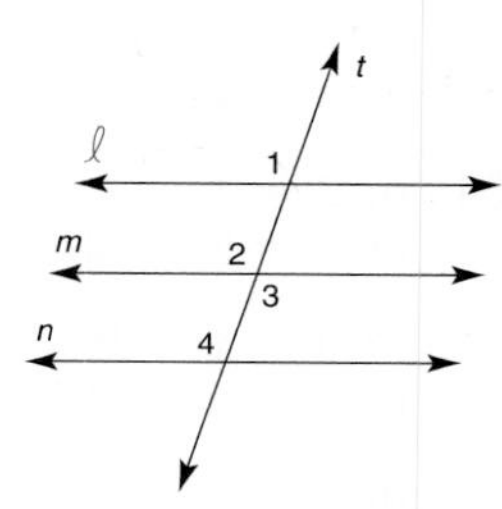


**PROOF**

| Statements | Reasons |
|---|---|
| 1. ℓ ∥ m | 1. ? |
| 2. ∠1 ≅ ∠2 | 2. ? |
| 3. ∠2 ≅ ∠3 | 3. If two lines intersect, the vertical ∠s formed are ≅ |
| 4. ? | 4. Given |
| 5. ∠1 ≅ ∠4 | 5. Transitive Prop. of ≅ |
| 6. ? | 6. ? |

*In Exercises 19 to 22, complete the proof.*

19. *Given:* $\overline{AD} \perp \overline{DC}$
$\overline{BC} \perp \overline{DC}$
*Prove:* $\overline{AD} \parallel \overline{BC}$

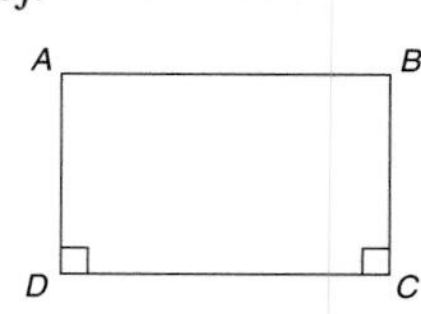


20. *Given:* m∠2 + m∠3 = 90°
$\overrightarrow{BE}$ bisects ∠ABC
$\overrightarrow{CE}$ bisects ∠BCD
*Prove:* ℓ ∥ n

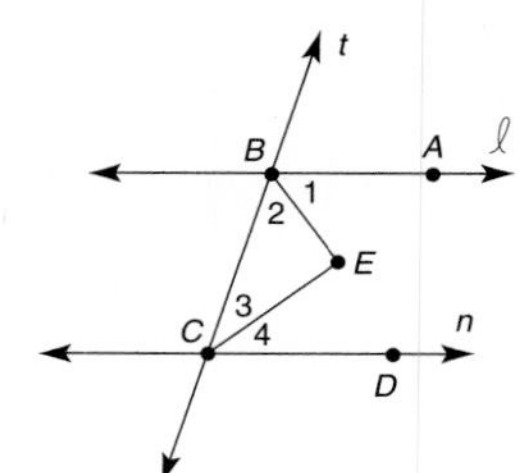

21. *Given:* $\overrightarrow{DE}$ bisects $\angle CDA$
$\angle 3 \cong \angle 1$
*Prove:* $\overline{ED} \parallel \overline{AB}$

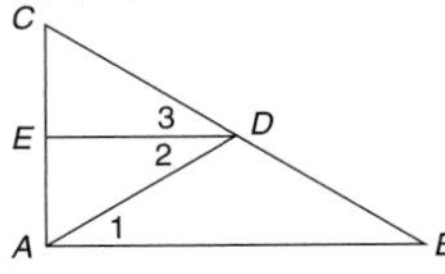


22. *Given:* $\overline{XY} \parallel \overline{WZ}$
$\angle 1 \cong \angle 2$
*Prove:* $\overline{MN} \parallel \overline{XY}$

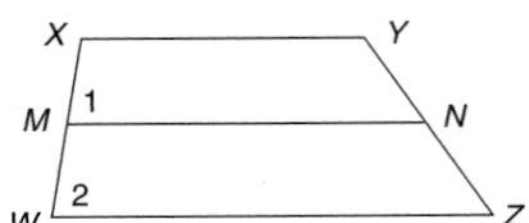


*In Exercises 23 to 28, determine the value of x so that line ℓ will be parallel to line m.*

23. $m\angle 4 = 5x$
$m\angle 5 = 4(x + 5)$

24. $m\angle 2 = 4x + 3$
$m\angle 7 = 5(x - 3)$

25. $m\angle 6 = x^2 - 9$
$m\angle 2 = x(x - 1)$

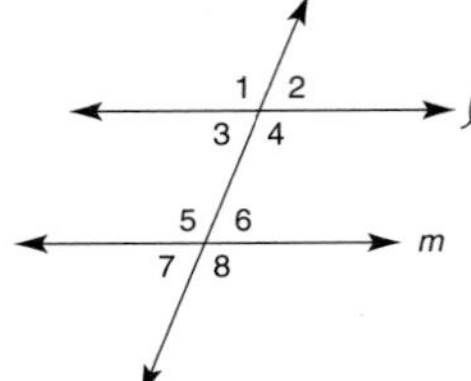


*Exercises 23–28*

26. $m\angle 4 = 2x^2 - 3x + 6$
$m\angle 5 = 2x(x - 1) - 2$

27. $m\angle 3 = (x + 1)(x + 4)$
$m\angle 5 = 16(x + 3) - (x^2 - 2)$

28. $m\angle 2 = (x^2 - 1)(x + 1)$
$m\angle 8 = 185 - x^2(x + 1)$

*In Exercises 29 to 31, give a formal proof for each theorem.*

29. If two lines are cut by a transversal so that the alternate exterior angles are congruent, then these lines are parallel.

30. If two lines are cut by a transversal so that the exterior angles on the same side of the transversal are supplementary, then these lines are parallel.

31. If two lines are parallel to the same line, then these lines are parallel to each other. (Assume three coplanar lines.)

32. Explain why the statement in Exercise 31 remains true even if the three lines are not coplanar.

33. Given that point $P$ does *not* lie on line $\ell$, construct the line through point $P$ that is parallel to line $\ell$.

•P

34. Given that point $Q$ does *not* lie on $\overline{AB}$, construct the line through point $Q$ that is parallel to $\overline{AB}$.

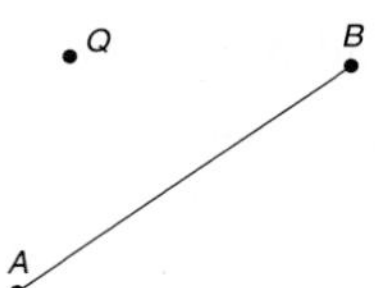


35. A carpenter drops a plumb line from point $A$ to $\overline{BC}$. Assuming that $\overline{BC}$ is horizontal, the point $D$ at which the plumb line intersects $\overline{BC}$ will determine the vertical line segment $\overline{AD}$. Use a construction to locate point $D$.

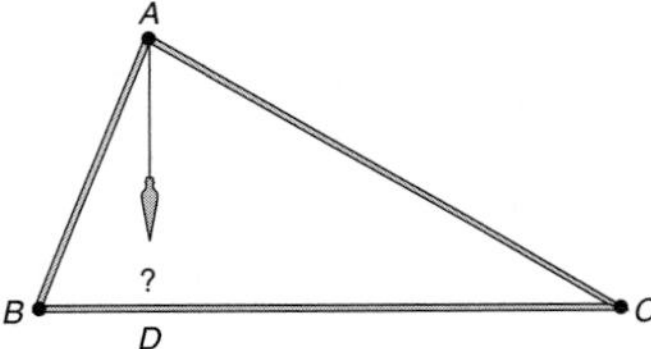

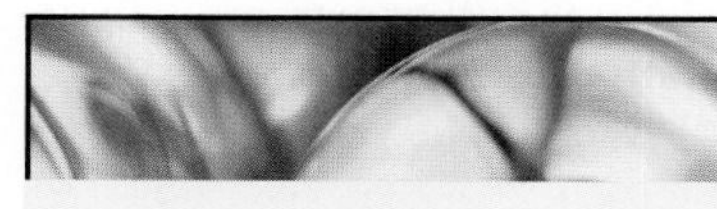

**KEY CONCEPTS**

Triangles • Vertices • Sides of a Triangle • Interior and Exterior of a Triangle • Scalene Triangle • Isosceles Triangle • Equilateral Triangle • Acute Triangle • Obtuse Triangle • Right Triangle • Equiangular Triangle • Auxiliary Line • Determined • Underdetermined • Overdetermined • Corollary • Exterior Angle of a Triangle

# 2.4 The Angles of a Triangle

Recall that in geometry, the word *union* means that figures are joined or combined. Picture three noncollinear points. Now read the definition below.

> DEFINITION: A **triangle** is the union of three line segments that are determined by three noncollinear points.

Symbolized by $\triangle$, the triangle is a figure you have encountered many times. In the triangle in Figure 2.24, known as $\triangle ABC$, or $\triangle BCA$, etc. (order of letters $A$, $B$, and $C$ being unimportant), each point $A$, $B$, and $C$ is a **vertex** of the triangle; collectively, these three points are the **vertices** of the triangle. $\overline{AB}$, $\overline{BC}$, and $\overline{AC}$ are the **sides** of the triangle. Point $D$ is in the **interior** of the triangle; point $E$ is on the triangle; and point $F$ is in the **exterior** of the triangle.

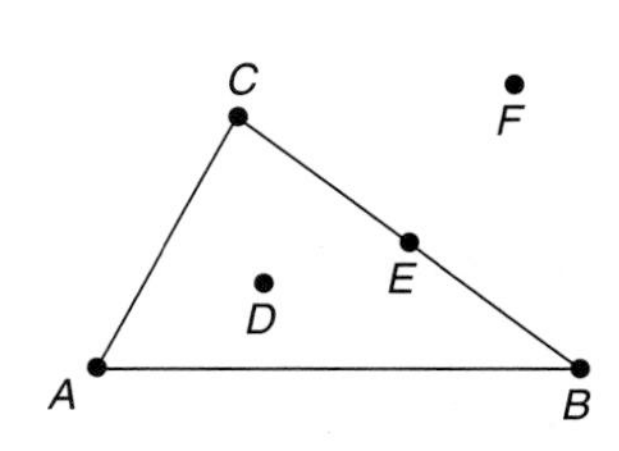


**FIGURE 2.24**

Triangles may be categorized by the lengths of their sides. Table 2.1 presents each type of triangle, the relationship among its sides, and a drawing in which congruent parts are indicated.

**TABLE 2.1**

**Triangles Classified by Congruent Sides**

| *Type* | | *Number of Congruent Sides* |
|---|---|---|
| Scalene | | None |
| Isosceles | | At least two congruent sides |
| Equilateral | | All three sides congruent |

Triangles may also be classified according to their angles (see Table 2.2).

In an earlier exercise, it was suggested that the sum of the measures of the three interior angles of a triangle is 180°. This is now stated as a theorem and proved through the use of an **auxiliary** (or helping) **line.** When an auxiliary line is added to the drawing for a proof, a justification must be given for the existence of that line. Justifications include statements such as

*There is exactly one line through two distinct points.*
*An angle has exactly one bisector.*
*There is only one line perpendicular to another line at a point on that line.*

When an auxiliary line is introduced into a proof, the original drawing is sometimes redrawn for the sake of clarity. Each auxiliary figure must be **determined,** but it must not be **underdetermined** or **overdetermined.** A figure is underdetermined when there is more than one possible figure described. On the other extreme, a figure is overdetermined when it is impossible for *all* conditions described to be satisfied.

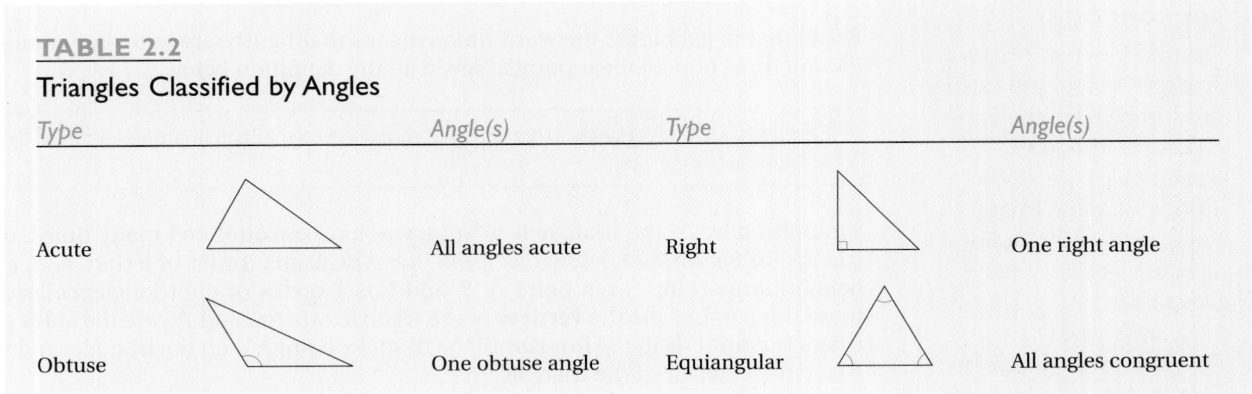

**TABLE 2.2**

**Triangles Classified by Angles**

| *Type* | | *Angle(s)* | *Type* | | *Angle(s)* |
|---|---|---|---|---|---|
| Acute | | All angles acute | Right | | One right angle |
| Obtuse | | One obtuse angle | Equiangular | | All angles congruent |

*Exs. 1–7*

EXAMPLE 1

In $\triangle HJK$ (not shown), $HJ = 4$, $JK = 4$, and $m\angle J = 90°$. Describe completely the type of triangle represented.

***Solution***

$\triangle HJK$ is a right isosceles triangle, or $\triangle HJK$ is an isosceles right triangle.

**Discover!**

From a paper triangle, cut the angles from the "corners." Now place the angles together at the same vertex as shown. What is the sum of the measures of the three angles?

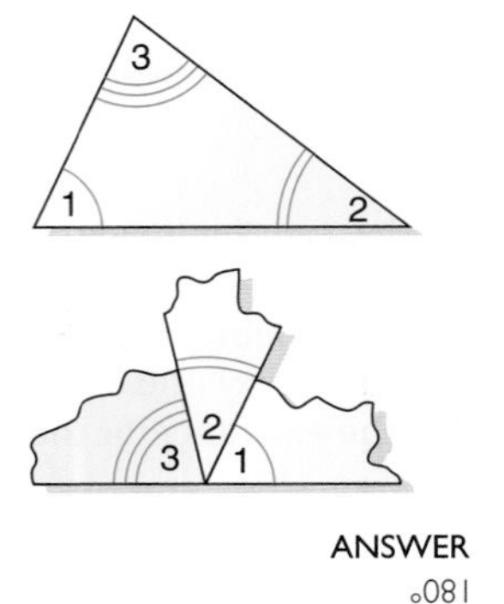


ANSWER

180°

THEOREM 2.4.1: In a triangle, the sum of the measures of the interior angles is 180°.

The first statement in the following "picture proof" establishes and justifies the uniqueness of the auxiliary line that is used.

**PICTURE PROOF OF THEOREM 2.4.1**

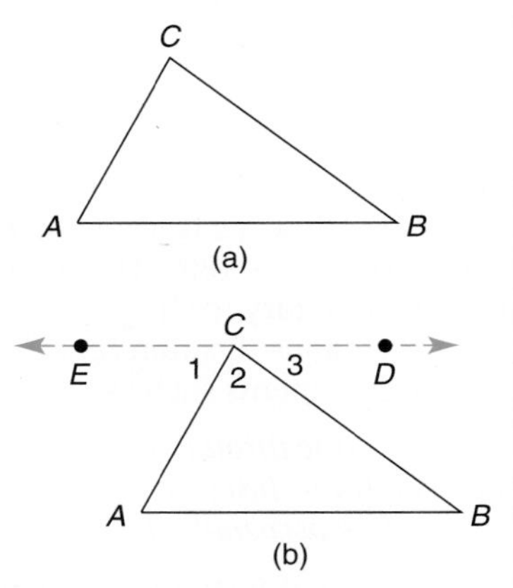


**FIGURE 2.25**

*Given:* $\triangle ABC$ in Figure 2.25(a)
*Prove:* $m\angle A + m\angle B + m\angle C = 180°$

*Proof:* Through $C$, draw $\overleftrightarrow{ED} \parallel \overline{AB}$.
We see that $m\angle 1 + m\angle 2 + m\angle 3 = 180°$.
But $m\angle 1 = m\angle A$ and $m\angle 3 = m\angle B$ (alternate interior angles are congruent).
Then $m\angle A + m\angle B + m\angle C = 180°$ in Figure 2.25(a).

At times, we use the notions of the equality and congruence of angles interchangeably within a proof, without stating both. See the preceding "picture proof."

*Technology Exploration*

Use computer software, if available.

1) Draw $\triangle ABC$.

2) Measure $\angle A$, $\angle B$, and $\angle C$.

3) Show that $m\angle A + m\angle B + m\angle C = 180°$

(Answer may not be "perfect.")

EXAMPLE 2

In $\triangle RST$ (not shown), $m\angle R = 45°$ and $m\angle S = 64°$. Find $m\angle T$.

***Solution***

In $\triangle RST$, $m\angle R + m\angle S + m\angle T = 180°$, so $45° + 64° + m\angle T = 180°$. Thus $109° + m\angle T = 180°$ and $m\angle T = 71°$.

A theorem that follows directly from a previous theorem is known as a **corollary** of that theorem. Corollaries, like theorems, must be proved before they can be used. These proofs are often brief, but they depend on the related theorem. Here are some corollaries of Theorem 2.4.1. We suggest that the student make a drawing to illustrate each corollary.

COROLLARY 2.4.2: Each angle of an equiangular triangle measures 60°.

*Exs. 8–12*

COROLLARY 2.4.3: The acute angles of a right triangle are complementary.

EXAMPLE 3

*Given:* $\angle M$ is a right $\angle$ in $\triangle NMQ$ (not shown)
$m\angle N = 57°$

*Find:* $m\angle Q$

Because the acute $\angle$s of a right triangle are complementary,

$$m\angle N + m\angle Q = 90°$$
$$\therefore 57° + m\angle Q = 90°$$
$$m\angle Q = 33°$$

COROLLARY 2.4.4: If two angles of one triangle are congruent to two angles of another triangle, then the third angles are also congruent.

The following example illustrates Corollary 2.4.4.

EXAMPLE 4

In $\triangle RST$ and $\triangle XYZ$ (triangles not shown), $m\angle R = m\angle X = 52°$. Also, $m\angle S = m\angle Y = 59°$.

a) Find $m\angle T$. b) Find $m\angle Z$. c) Is $\angle T \cong \angle Z$?

***Solution***

a) $m\angle R + m\angle S + m\angle T = 180°$
$52° + 59° + m\angle T = 180°$
$111° + m\angle T = 180°$
$m\angle T = 69°$

b) Using $m\angle X + m\angle Y + m\angle Z = 180°$, we repeat part (a) to find $m\angle Z = 69°$.

c) Yes, $\angle T \cong \angle Z$ (both measure 69°).

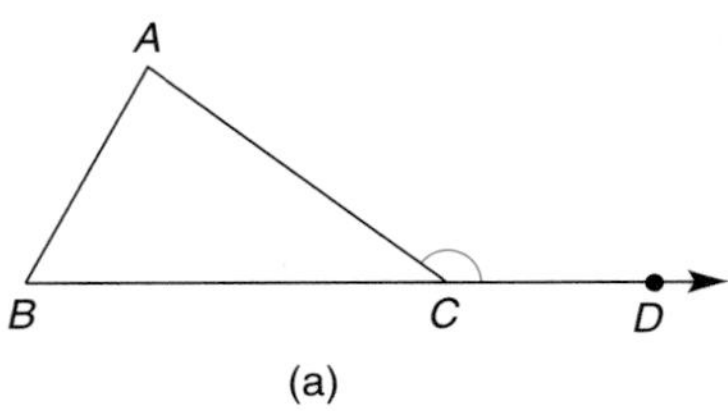

(a)

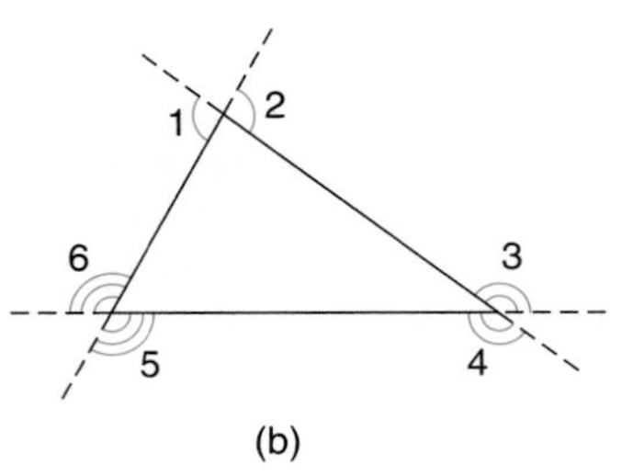

(b)

**FIGURE 2.26**

When the sides of a triangle are extended, each angle that is formed by a side and an extension of the adjacent side is an **exterior angle** of the triangle. In Figure 2.26(a), $\angle ACD$ is an exterior angle of $\triangle ABC$; for a triangle, there are a total of six exterior angles—two at each vertex. [See Figure 2.26(b).]

> COROLLARY 2.4.5: The measure of an exterior angle of a triangle equals the sum of the measures of the two nonadjacent interior angles.

In Figure 2.26(a), $\angle A$ and $\angle B$ are the two *nonadjacent* interior angles for exterior $\angle ACD$. These angles ($A$ and $B$) are sometimes called *remote* interior angles for exterior $\angle ACD$.

EXAMPLE 5

*Given:* In Figure 2.27,

$$m\angle 1 = x^2 + 2x$$
$$m\angle S = x^2 - 2x$$
$$m\angle T = 3x + 10$$

*Find:* $x$

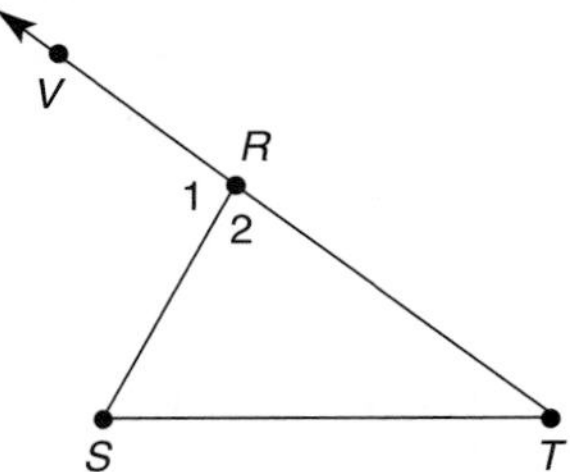

**FIGURE 2.27**

***Solution***

By Corollary 2.4.5,

$$m\angle 1 = m\angle S + m\angle T$$
$$x^2 + 2x = (x^2 - 2x) + (3x + 10)$$
$$2x = x + 10$$
$$x = 10$$

*Exs. 13–19*

*Check:* $m\angle 1 = 120°$, $m\angle S = 80°$, and $m\angle T = 40°$; so $120 = 80 + 40$, which satisfies the conditions of Corollary 2.4.5.

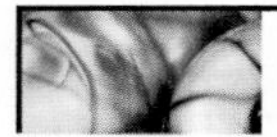

## 2.4 Exercises

*In Exercises 1 to 4, refer to $\triangle ABC$. On the basis of the information given, determine the measure of the remaining angle(s) of the triangle.*

1. $m\angle A = 63°$ and $m\angle B = 42°$
2. $m\angle B = 39°$ and $m\angle C = 82°$
3. $m\angle A = m\angle C = 67°$
4. $m\angle B = 42°$ and $m\angle A = m\angle C$

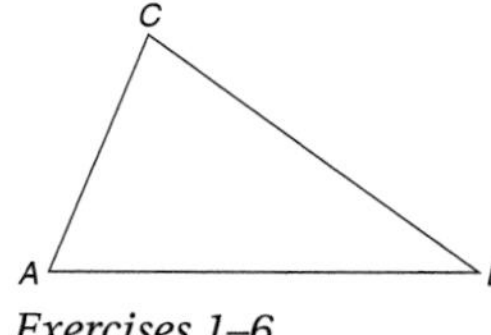

*Exercises 1–6*

5. Describe the auxiliary line (segment) as determined, overdetermined, or underdetermined.
   a) Draw the line through vertex $C$ of $\triangle ABC$.
   b) Through vertex $C$, draw the line parallel to $\overline{AB}$.
   c) With $M$ the midpoint of $\overline{AB}$, draw $\overline{CM}$ perpendicular to $\overline{AB}$.
6. Describe the auxiliary line (segment) as determined, overdetermined, or underdetermined.
   a) Through vertex $B$ of $\triangle ABC$, draw $\overleftrightarrow{AB} \perp \overline{AC}$.
   b) Draw the line that contains $A$, $B$, and $C$.
   c) Draw the line that contains $M$, the midpoint of $\overline{AB}$.

*In Exercises 7 and 8, classify the triangle (not shown) by considering the lengths of its sides.*

7. a) All sides of $\triangle ABC$ are of the same length.
   b) In $\triangle DEF$, $DE = 6$, $EF = 6$, and $DF = 8$.
8. a) In $\triangle XYZ$, $\overline{XY} \cong \overline{YZ}$.
   b) In $\triangle RST$, $RS = 6$, $ST = 7$, and $RT = 8$.

*In Exercises 9 and 10, classify the triangle (not shown) by considering the measures of its angles.*

9. a) All angles of $\triangle ABC$ measure 60°.
   b) In $\triangle DEF$, $m\angle D = 40°$, $m\angle E = 50°$, and $m\angle F = 90°$.

10. a) In $\triangle XYZ$, $m\angle X = 123°$.
    b) In $\triangle RST$, $m\angle R = 45°$, $m\angle S = 65°$, and $m\angle T = 70°$.

*In Exercises 11 and 12, make drawings as needed.*

11. Suppose that for $\triangle ABC$ and $\triangle MNQ$, you know that $\angle A \cong \angle M$ and $\angle B \cong \angle N$. Explain why $\angle C \cong \angle Q$.

12. Suppose that $T$ is a point on side $\overline{PQ}$ of $\triangle PQR$. Also, $\overrightarrow{RT}$ bisects $\angle PRQ$, and $\angle P \cong \angle Q$. If $\angle 1$ and $\angle 2$ are the angles formed when $\overrightarrow{RT}$ intersects $\overline{PQ}$, explain why $\angle 1 \cong \angle 2$.

*In Exercises 13 to 15, $j \parallel k$ and $\triangle ABC$.*

13. *Given:* $m\angle 3 = 50°$
    $m\angle 4 = 72°$
    *Find:* $m\angle 1$, $m\angle 2$, and $m\angle 5$

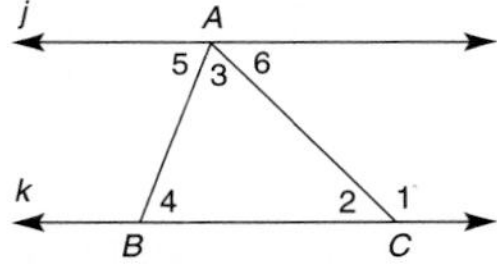


Exercises 13–15

14. *Given:* $m\angle 3 = 55°$
    $m\angle 2 = 74°$
    *Find:* $m\angle 1$, $m\angle 4$, and $m\angle 5$

15. *Given:* $m\angle 1 = 122.3°$, $m\angle 5 = 41.5°$
    *Find:* $m\angle 2$, $m\angle 3$, and $m\angle 4$

16. *Given:* $\overline{MN} \perp \overline{NQ}$ and $\angle$s as shown
    *Find:* $x$, $y$, and $z$

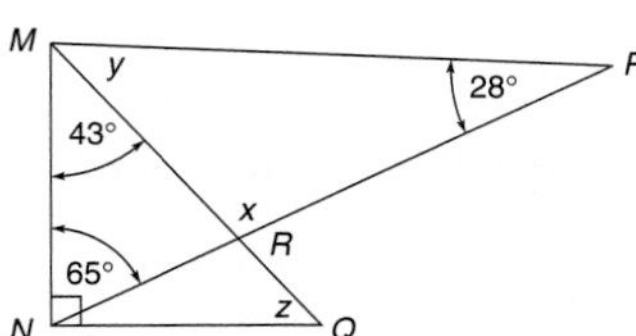


17. *Given:* $\overline{AB} \parallel \overline{DC}$
    $\overrightarrow{DB}$ bisects $\angle ADC$
    $m\angle A = 110°$
    *Find:* $m\angle 3$

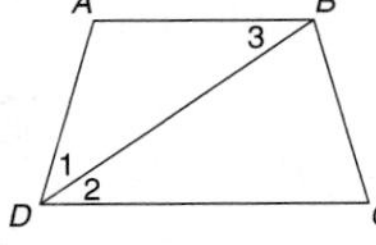


Exercises 17, 18

18. *Given:* $\overline{AB} \parallel \overline{DC}$
    $\overrightarrow{DB}$ bisects $\angle ADC$
    $m\angle 1 = 36°$
    *Find:* $m\angle A$

19. *Given:* $\triangle ABC$ with $B$-$D$-$E$-$C$
    $m\angle 3 = m\angle 4 = 30°$
    $m\angle 1 = m\angle 2 = 70°$
    *Find:* $m\angle B$

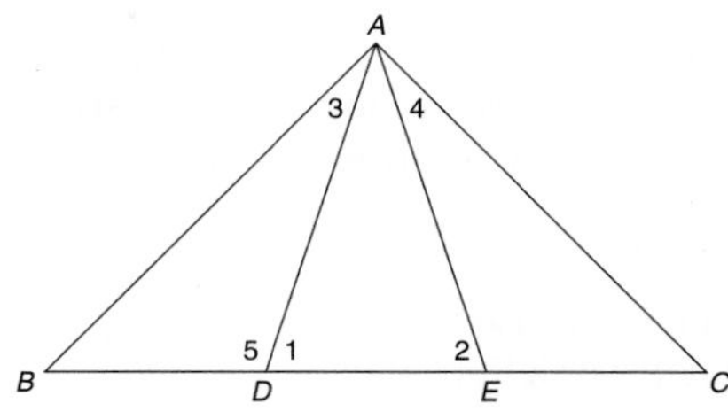


Exercises 19, 20

20. *Given:* $\triangle ABC$ with $B$-$D$-$E$-$C$
    $m\angle 1 = 2x$
    $m\angle 3 = x$
    *Find:* $m\angle B$ in terms of $x$

21. Consider any triangle and one exterior angle at each vertex. What is the sum of the measures of the three exterior angles of the triangle?

22. *Given:* Right $\triangle ABC$ with right $\angle C$
    $m\angle 1 = 7x + 4$
    $m\angle 2 = 5x + 2$
    *Find:* $x$

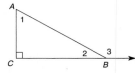

Exercises 22, 23

23. *Given:* $m\angle 1 = x$
    $m\angle 2 = y$
    $m\angle 3 = 3x$
    *Find:* $x$ and $y$

24. *Given:* $m\angle 1 = 8(x + 2)$
    $m\angle 3 = 5x - 3$
    $m\angle 5 = 5(x + 1) - 2$
    *Find:* $x$

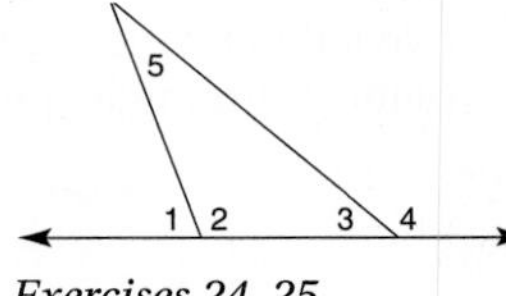


Exercises 24, 25

25. *Given:* $m\angle 1 = x$
    $m\angle 2 = 4y$
    $m\angle 3 = 2y$
    $m\angle 4 = 2x - y - 40$
    *Find:* $x$, $y$, and $m\angle 5$

26. *Given:* Equiangular $\triangle RST$
    $\overrightarrow{RV}$ bisects $\angle SRT$
    *Prove:* $\triangle RVS$ is a right $\triangle$

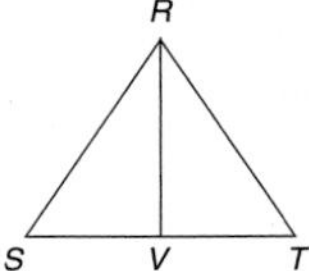

27. *Given:* $\overline{MN}$ and $\overline{PQ}$ intersect at $K$; $\angle M \cong \angle Q$
*Prove:* $\angle P \cong \angle N$

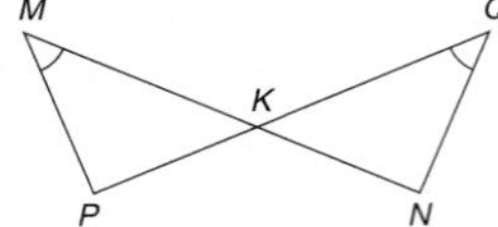


28. The sum of the measures of two angles of a triangle equals the measure of the third (largest) angle. What type of triangle is described?

29. Draw, if possible, an
   a) isosceles obtuse triangle.
   b) equilateral right triangle.

30. Draw, if possible, a
   a) right scalene triangle.
   b) triangle having both a right angle and an obtuse angle.

31. Along a straight shoreline, two houses are located at points $H$ and $M$. The houses are 5000 feet apart. A small island lies in view of both houses, with angles as indicated. Find $m\angle I$.

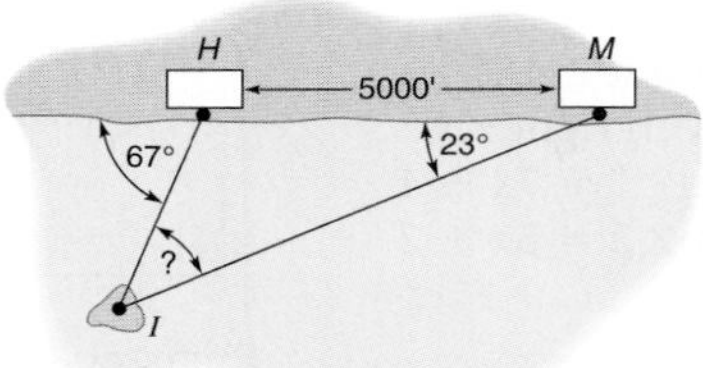


32. An airplane has leveled off (is flying horizontally) at an altitude of 12,000 feet. Its pilot can see each of two small towns at points $R$ and $T$ in front of the plane. With angle measures as indicated, find $m\angle R$.

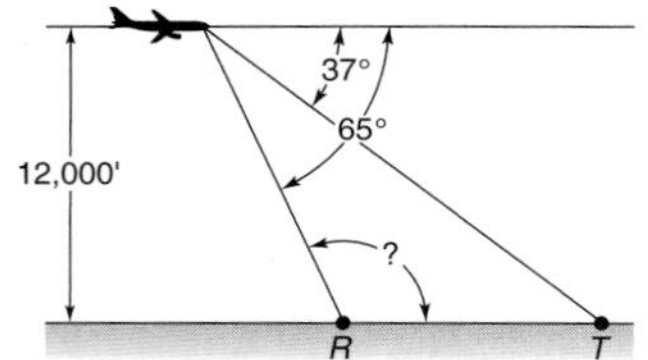


33. On a map, three Los Angeles suburbs are located at points $N$ (Newport Beach), $P$ (Pomona), and $B$ (Burbank). With angle measures as indicated, determine $m\angle N$ and $m\angle P$.

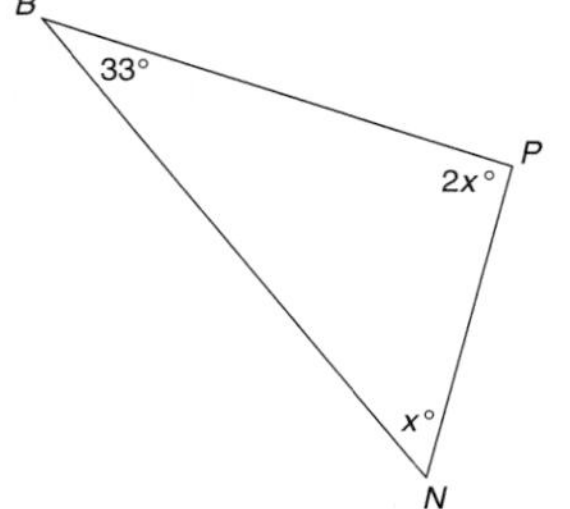


34. The roofline of a house shows the shape of right triangle $ABC$ with $m\angle C = 90°$. If the measure of $\angle CAB$ is 24° larger than the measure of $\angle CBA$, then how large is each angle?

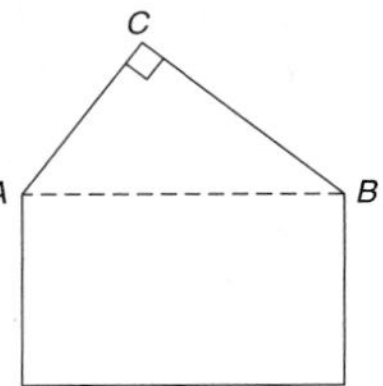


35. A lamppost has a design such that $m\angle C = 110°$ and $\angle A \cong \angle B$. Find $m\angle A$ and $m\angle B$.

36. For the lamppost of Exercise 35, suppose that $m\angle A = m\angle B$ and that $m\angle C = 3(m\angle A)$. Find $m\angle A$, $m\angle B$, and $m\angle C$.

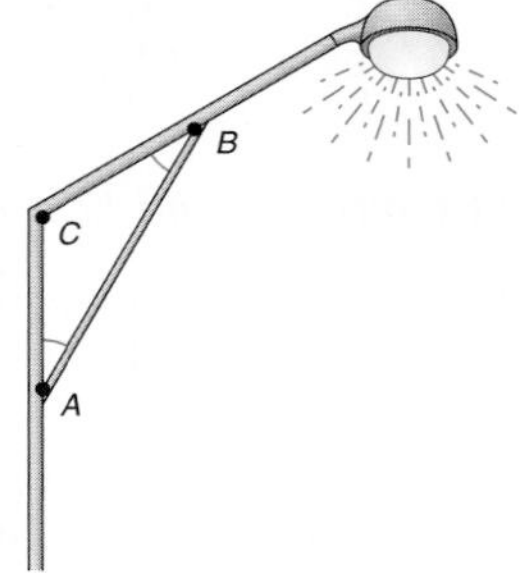


37. The triangular symbol on the "PLAY" button of a VCR has congruent angles at $M$ and $N$. If $m\angle P = 30°$, what are the measures of angle $M$ and angle $N$?

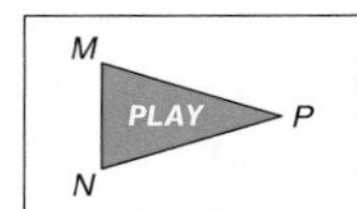


38. A polygon with four sides is called a *quadrilateral*. Consider the figure and the dashed auxiliary line. What is the sum of the measures of the four interior angles of this (or any other) quadrilateral?

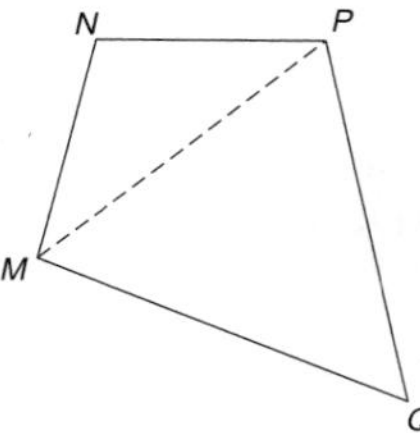


39. Explain why the following statement is true.

   *Each interior angle of an equiangular triangle measures 60°.*

40. Explain why the following statement is true.

   *The acute angles of a right triangle are complementary.*

*In Exercises 41 to 43, write a formal proof for each corollary.*

41. The measure of an exterior angle of a triangle equals the sum of the measures of the two nonadjacent interior angles.

42. If two angles of one triangle are congruent to two angles of another triangle, then the third angles are also congruent.

43. Use an indirect proof to establish the following theorem: A triangle cannot have more than one right angle.

44. *Given:* $\overleftrightarrow{AB}$, $\overleftrightarrow{DE}$, and $\overleftrightarrow{CF}$
$\overleftrightarrow{AB} \parallel \overleftrightarrow{DE}$
$\overrightarrow{CG}$ bisects $\angle BCF$
$\overrightarrow{FG}$ bisects $\angle CFE$
*Prove:* $\angle G$ is a right angle

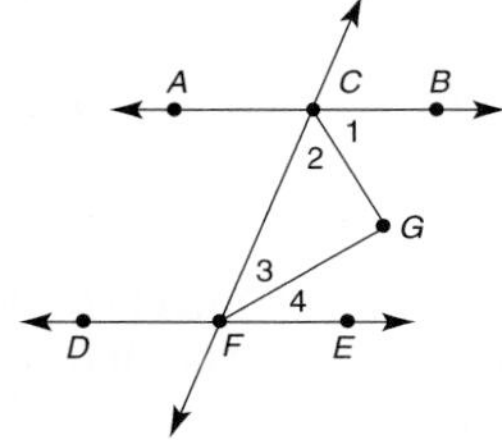


*45. *Given:* $\overrightarrow{NQ}$ bisects $\angle MNP$
$\overrightarrow{PQ}$ bisects $\angle MPR$
$m\angle Q = 42°$
*Find:* $m\angle M$

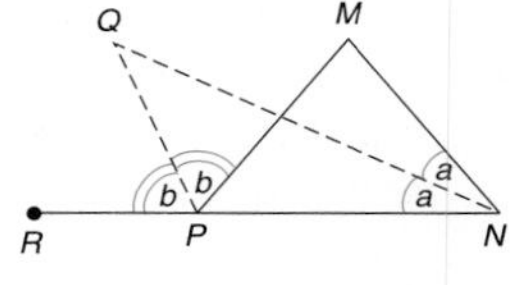


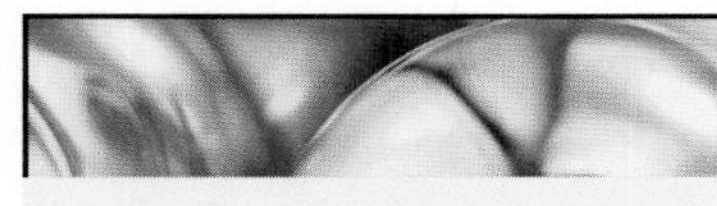

## 2.5 Convex Polygons

**KEY CONCEPTS**

Convex Polygons (Triangle, Quadrilateral, Pentagon, Hexagon, Heptagon, Octagon, Nonagon, Decagon) • Concave Polygon • Diagonals of a Polygon • Regular Polygon • Equilateral Polygon • Equiangular Polygon • Polygram

DEFINITION: A **polygon** is a closed plane figure whose sides are line segments that intersect only at the endpoints.

The polygons we generally consider in this textbook are **convex;** the angle measures of convex polygons are between 0° and 180°. Convex polygons are shown in Figure 2.28; those in Figure 2.29 are **concave.** A line segment joining two points of a concave polygon can contain points in the exterior of the polygon. Thus, a concave polygon always has at least one reflex angle. Figure 2.30 shows some figures that aren't polygons at all!

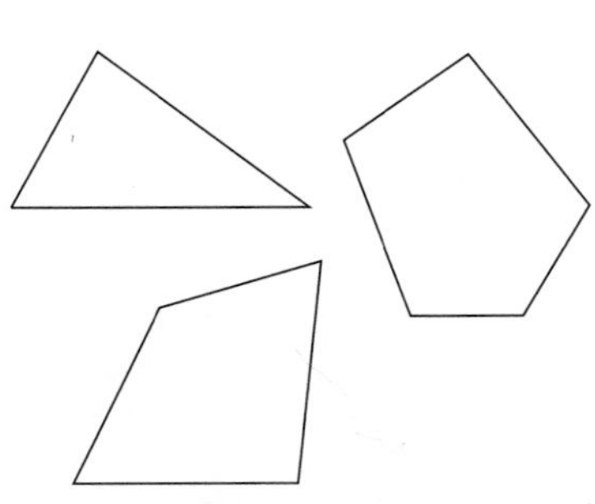

Convex Polygons

**FIGURE 2.28**

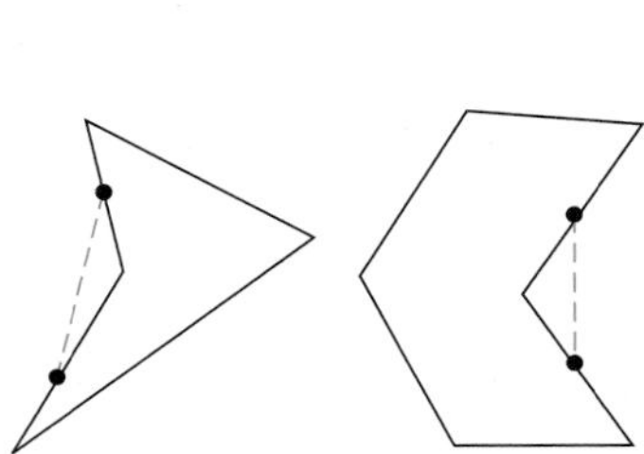

Concave Polygons

**FIGURE 2.29**

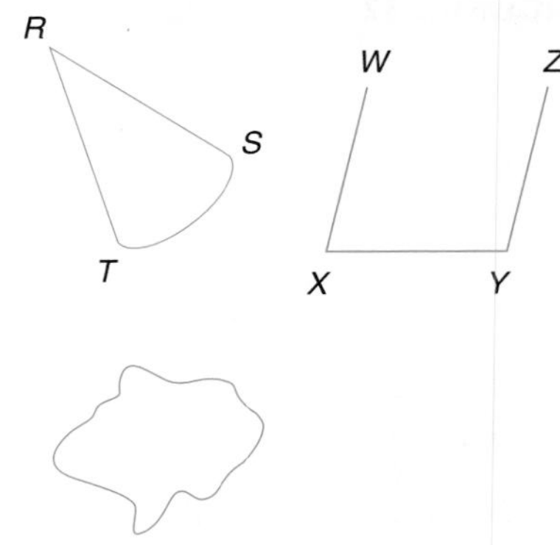


Not Polygons

**FIGURE 2.30**

In Figure 2.29, each concave polygon contains one reflex angle.

Table 2.3 shows some special names for polygons with fixed numbers of sides.

**TABLE 2.3**

| *Polygon* | *Number of Sides* | *Polygon* | *Number of Sides* |
|---|---|---|---|
| Triangle | 3 | Heptagon | 7 |
| Quadrilateral | 4 | Octagon | 8 |
| Pentagon | 5 | Nonagon | 9 |
| Hexagon | 6 | Decagon | 10 |

With Venn Diagrams, the set of all objects under consideration is called the **universe.** If $P$ = {all polygons} is the universe, then we can describe sets $T$ = {triangles} and $Q$ = {quadrilaterals} as subsets that lie within universe $P$. Sets $T$ and $Q$ are described as **disjoint** because they have no elements in common. See Figure 2.31.

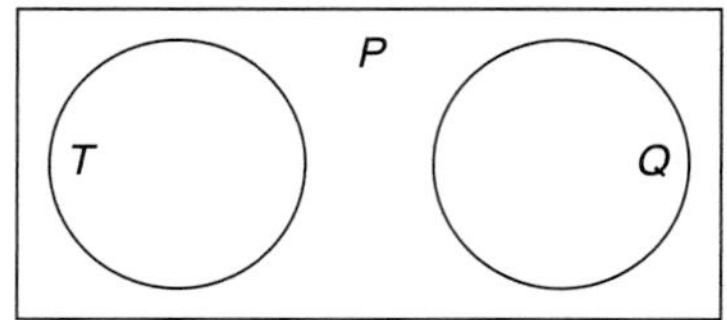


**FIGURE 2.31**

## DIAGONALS OF A POLYGON

A **diagonal** of a polygon is a line segment that joins two nonconsecutive vertices.

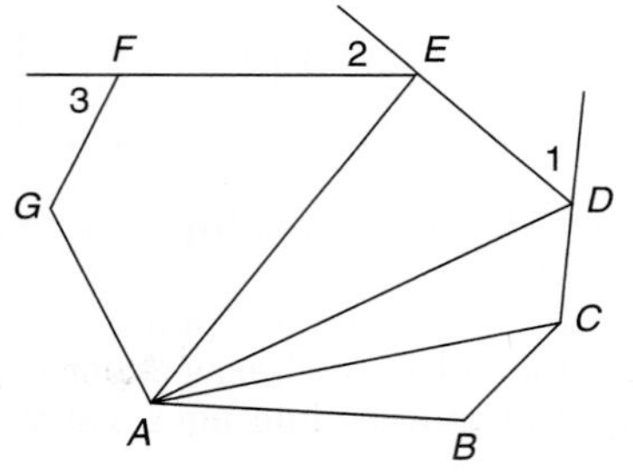


**FIGURE 2.32**

Figure 2.32 shows heptagon $ABCDEFG$ for which $\angle GAB$, $\angle B$, and $\angle BCD$ are some of the interior angles and $\angle 1$, $\angle 2$, and $\angle 3$ are some of the exterior angles. $\overline{AB}$, $\overline{BC}$, and $\overline{CD}$ are some of the sides of the heptagon, because these join consecutive vertices. Because a diagonal joins nonconsecutive vertices of $ABCDEFG$, $\overline{AC}$, $\overline{AD}$, and $\overline{AE}$ are among the many diagonals of the polygon.

Table 2.4 illustrates polygons by numbers of sides and their corresponding numbers of diagonals.

When the number of sides of a polygon is small, we can list all diagonals by name. In pentagon $ABCDE$ of Table 2.4, we see diagonals $\overline{AC}$, $\overline{AD}$, $\overline{BD}$, $\overline{BE}$, and $\overline{CE}$, a total of five. As the number of sides increases, it becomes more difficult to count all the diagonals. In such a case, the formula of Theorem 2.5.1 is most convenient to use. Although this theorem is given without proof, Exercise 39 of this section provides some insight for the proof.

**TABLE 2.4**

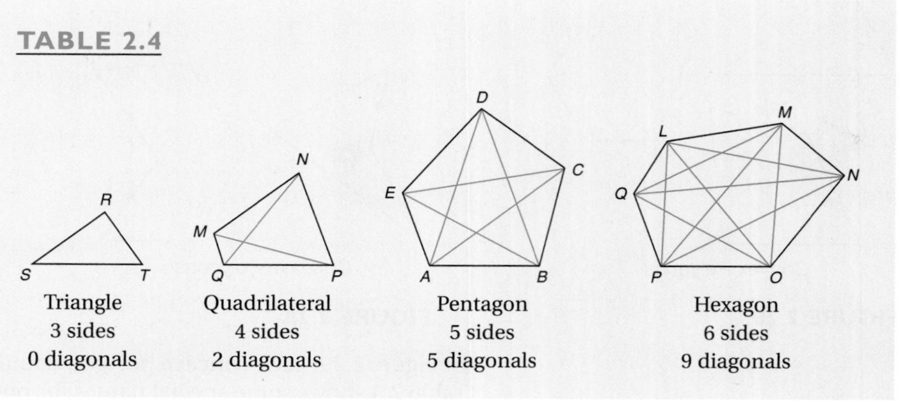

> THEOREM 2.5.1: The total number of diagonals $D$ in a polygon of $n$ sides is given by the formula $D = \frac{n(n-3)}{2}$.

This theorem reminds us that a triangle has no diagonals; when $n = 3$, $D = \frac{3(3-3)}{2} = 0$.

*Exs. 1–5*

**EXAMPLE 1**

Use the formula of Theorem 2.5.1 to find the number of diagonals for any pentagon.

***Solution***

To use the formula of Theorem 2.5.1, we note that $n = 5$ in a pentagon. Then $D = \frac{5(5-3)}{2} = \frac{5(2)}{2} = 5$.

**Reminder**

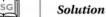

The sum of the interior angles of a triangle is 180°.

## SUM OF THE INTERIOR ANGLES OF A POLYGON

The following theorem provides the formula for the sum of the interior angles of any polygon.

> THEOREM 2.5.2: The sum $S$ of the measures of the interior angles of a polygon with $n$ sides is given by $S = (n - 2) \cdot 180°$. Note that $n > 2$ for any polygon.

Let us consider an informal proof of Theorem 2.5.2 for the special case of a pentagon. The proof would change for a polygon of a different number of sides, but only by the number of triangles into which the polygon can be separated. Although Theorem 2.5.2 is also true for concave polygons, we consider the proof only for the case of the convex polygon.

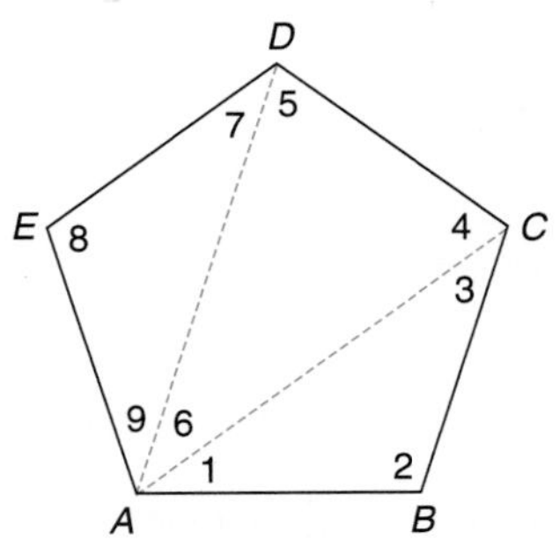


**FIGURE 2.33**

***Proof***

Consider the pentagon $ABCDE$ in Figure 2.33 with auxiliary segments (diagonals from one vertex) as shown.

With angles marked as shown in triangles $ABC$, $ACD$, and $ADE$,

$$\begin{array}{rl} m\angle 1 + m\angle 2 + m\angle 3 &= 180° \\ m\angle 6 + m\angle 5 + m\angle 4 &= 180° \\ m\angle 8 + m\angle 9 + m\angle 7 &= 180° \\ \hline m\angle E + m\angle A + m\angle D + m\angle B + m\angle C &= 540° \end{array} \quad \text{adding}$$

For pentagon $ABCDE$, in which $n = 5$, the sum of the measures of the interior angles is $(5 - 2) \cdot 180°$, which equals 540°.

When drawing diagonals from one vertex of a polygon of $n$ sides, we always form $(n - 2)$ triangles. The sum of the measures of the interior angles always equals $(n - 2) \cdot 180°$.

EXAMPLE 2

Find the sum of the measures of the interior angles of a hexagon. Then find the measure of each interior angle of an equiangular hexagon.

***Solution***

For the hexagon, $n = 6$, so the sum of the measures of the interior angles is $S = (6 - 2) \cdot 180°$ or $4(180°)$ or $720°$.

In an equiangular hexagon, each of the six interior angles measures $\frac{720°}{6}$ or $120°$.

EXAMPLE 3

Find the number of sides in a polygon whose sum of interior angles is 2160°.

***Solution***

Here $S = 2160$ in the formula of Theorem 2.5.2. Because $(n - 2) \cdot 180 = 2160$, we have $180n - 360 = 2160$.

Then $180n = 2520$

$n = 14$

The polygon has 14 sides.

*Exs. 6–9*

## REGULAR POLYGONS

Figure 2.34 shows polygons that are, respectively, (a) **equilateral,** (b) **equiangular,** and (c) **regular** (both sides and angles are congruent). Note the parts that are marked congruent in each figure.

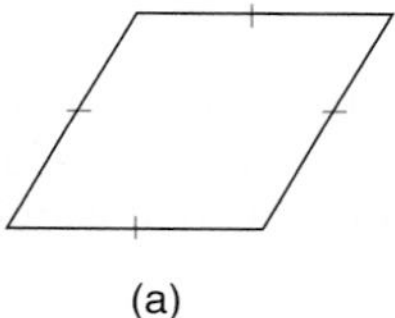
(a)

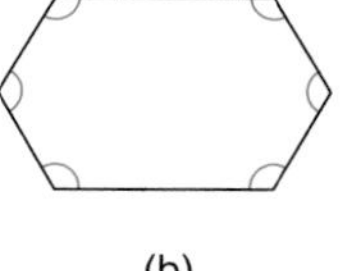
(b)

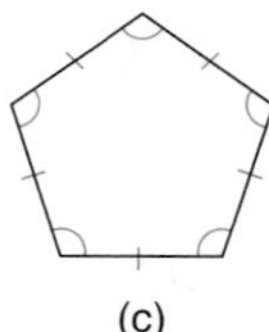
(c)

**FIGURE 2.34**

DEFINITION: A **regular polygon** is a polygon that is both equilateral and equiangular.

The polygon in Figure 2.34(c) is a *regular pentagon.* Other examples of regular polygons include the equilateral triangle and the square.

Based upon the formula $S = (n - 2) \cdot 180°$ from Theorem 2.5.2, there is also a formula for the measure of each interior angle of a regular polygon. It applies to equiangular polygons as well.

COROLLARY 2.5.3: The measure $I$ of each interior angle of a regular polygon or equiangular polygon of $n$ sides is $I = \frac{(n - 2) \cdot 180°}{n}$.

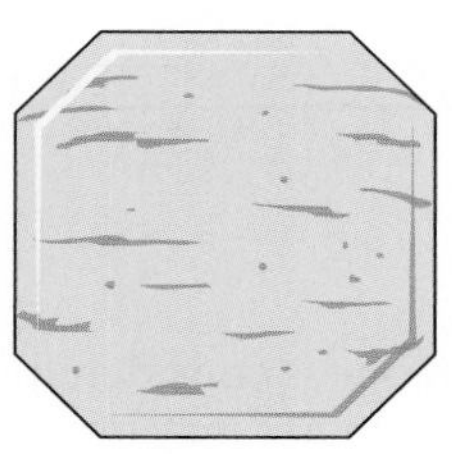

**FIGURE 2.35**

EXAMPLE 4

Find the measure of each interior angle of a ceramic floor tile in the shape of an equiangular octagon (Figure 2.35).

***Solution***

For an octagon, $n = 8$.

Then
$$I = \frac{(8-2)\cdot 180}{8}$$
$$= \frac{6 \cdot 180}{8}$$
$$= \frac{1080}{8}, \quad \text{so} \quad I = 135^\circ$$

Each interior angle of the tile measures 135°.

NOTE: For the octagonal tiles of Example 4, small squares are used as "fillers" to cover the floor. The pattern, a tessellation, is shown on page 363.

EXAMPLE 5

Each interior angle of a certain regular polygon has a measure of 144°. Find its number of sides, and identify the type of polygon it is.

***Solution***

Let $n$ be the number of sides the polygon has. All $n$ of the interior angles are equal in measure.

The measure of each interior angle is given by

$$I = \frac{(n-2)\cdot 180}{n} \quad \text{where } I = 144$$

Then
$$\frac{(n-2)\cdot 180}{n} = 144$$
$$(n-2)\cdot 180 = 144n$$
$$180n - 360 = 144n$$
$$36n = 360$$
$$n = 10$$

The polygon is a regular decagon.

*Exs. 10–12*

**Discover!**

From a paper quadrilateral, cut the angles from the "corners." Now place the angles so that they have the same vertex and do *not* overlap. What is the sum of measures of the four angles?

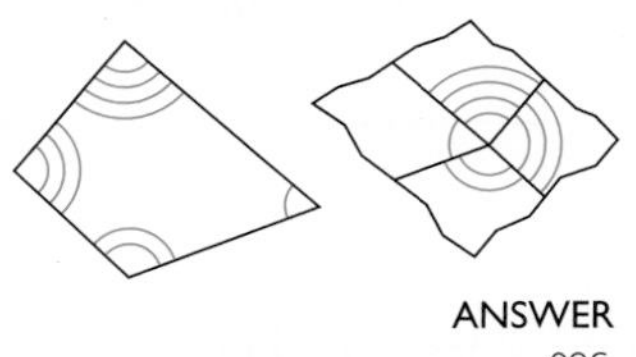

ANSWER

360°

A second corollary to Theorem 2.5.2 concerns the sum of the interior angles of any quadrilateral. For the proof, we simply let $n = 4$ in the formula $S = (n-2)\cdot 180°$. Then $S = (4-2)\cdot 180° = 2 \cdot 180° = 360°$. Also, see the Discover! at the left.

COROLLARY 2.5.4: The sum of the four interior angles of a quadrilateral is 360°.

Based upon Corollary 2.5.4, it is clearly the case that each interior angle of a square or rectangle measures 90°.

The following interesting corollary to Theorem 2.5.2 can be established through algebra.

> COROLLARY 2.5.5: The sum of the measures of the exterior angles of a polygon, one at each vertex, is 360°.

We now consider an algebraic proof for Corollary 2.5.5.

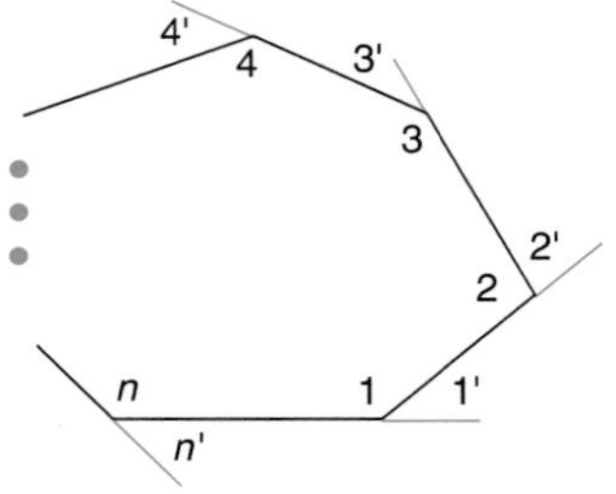


**FIGURE 2.36**

*Proof*

A polygon of $n$ sides has $n$ interior angles and $n$ exterior angles, if one is considered at each vertex. As shown in Figure 2.36, these interior and exterior angles may be grouped into pairs of supplementary angles. Because there are $n$ pairs of angles, the sum of the measures of all pairs is $180 \cdot n$ degrees.

Of course, the sum of the measures of the interior angles is $(n - 2) \cdot 180°$.

In words, we have

$$\text{Sum of Measures of Interior Angles} + \text{Sum of Measures of Exterior Angles} = \text{Sum of Measures of All Supplementary Pairs}$$

Let $S$ represent the sum of the measures of the exterior angles.

$$\begin{aligned} (n - 2) \cdot 180 + S &= 180n \\ 180n - 360 + S &= 180n \\ -360 + S &= 0 \\ \therefore S &= 360 \end{aligned}$$

---

**EXAMPLE 6**

Use Corollary 2.5.5 to find the number of sides of a regular polygon if each interior angle measures 144°. (Note that we are repeating Example 5.)

***Solution***

If each interior angle measures 144°, then each exterior angle measures 36° (they are supplementary, because exterior sides of these adjacent angles form a straight line).

Now each of the $n$ exterior angles has the measure

$$\frac{360°}{n}$$

In this case, $\frac{360}{n} = 36$, and it follows that $36n = 360$, so $n = 10$. The polygon (a decagon) has 10 sides.

The next corollary follows from Corollary 2.5.5. The claim made in Corollary 2.5.6 was illustrated in Example 6.

> COROLLARY 2.5.6: The measure $E$ of each exterior angle of a regular polygon or equiangular polygon of $n$ sides is $E = \frac{360°}{n}$.

## POLYGRAMS

*Exs. 13, 14*

A **polygram** is the star-shaped figure that results when the sides of certain polygons are extended. The polygon must be convex with five or more sides. When the polygon is regular, the resulting polygram is also regular—that is, the interior acute angles

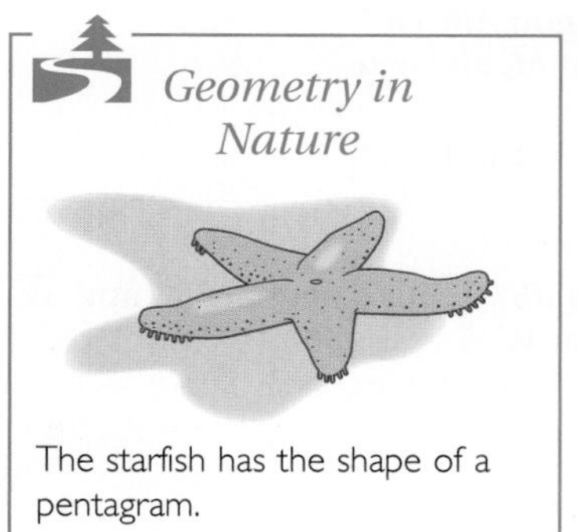

The starfish has the shape of a pentagram.

are congruent, the interior reflex angles are congruent, and all sides are congruent. The names of polygrams come from the names of the polygons whose sides were extended. Figure 2.37 shows a pentagram, a hexagram, and an octagram. With congruent angles and sides indicated, these figures are **regular polygrams.**

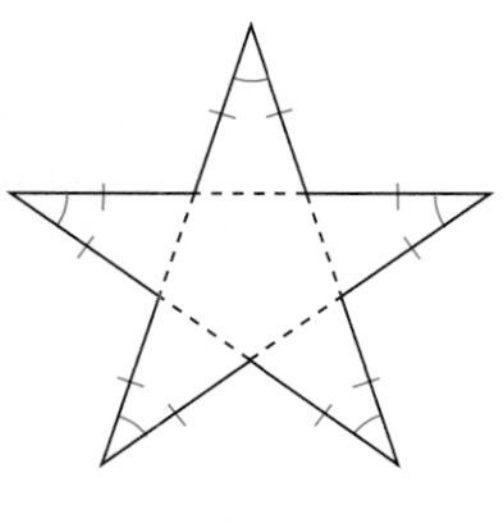
Pentagram

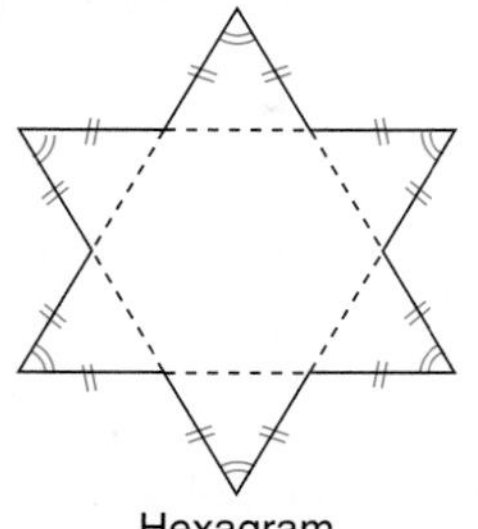
Hexagram

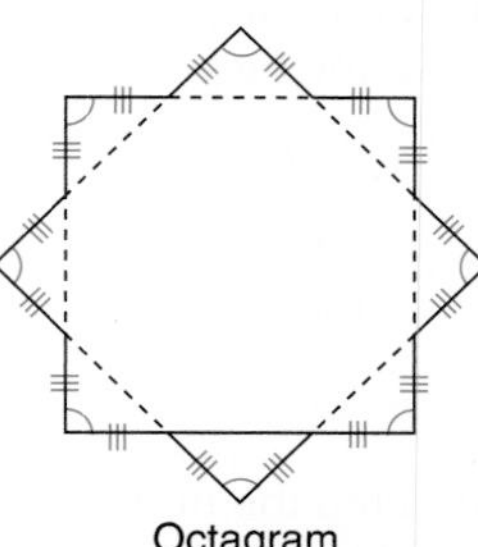
Octagram

*Exs. 15, 16*

**FIGURE 2.37**

## 2.5 Exercises

1. As the number of sides of a regular polygon increases, does each interior angle increase or decrease in measure?

2. As the number of sides of a regular polygon increases, does each exterior angle increase or decrease in measure?

3. *Given:* $\overline{AB} \parallel \overline{DC}$, $\overline{AD} \parallel \overline{BC}$, $\overline{AE} \parallel \overline{FC}$, with angle measures as indicated
*Find:* $x$, $y$, and $z$

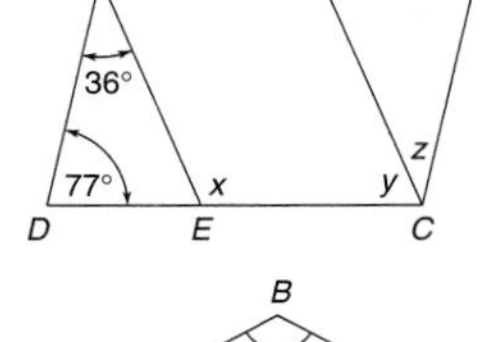


4. In pentagon $ABCDE$ with $\angle B \cong \angle D \cong \angle E$, find the measure of interior angle $EDC$.

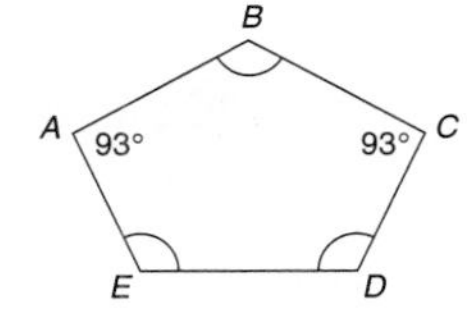


5. Find the total number of diagonals for a polygon of $n$ sides if:

   a) $n = 5$ b) $n = 10$

6. Find the total number of diagonals for a polygon of $n$ sides if:

   a) $n = 6$ b) $n = 8$

7. Find the sum of the measures of the interior angles of a polygon of $n$ sides if:

   a) $n = 5$ b) $n = 10$

8. Find the sum of the measures of the interior angles of a polygon of $n$ sides if:

   a) $n = 6$ b) $n = 8$

9. Find the measure of each interior angle of a regular polygon of $n$ sides if:

   a) $n = 4$ b) $n = 12$

10. Find the measure of each interior angle of a regular polygon of $n$ sides if:

    a) $n = 6$ b) $n = 10$

11. Find the measure of each exterior angle of a regular polygon of $n$ sides if:

    a) $n = 4$ b) $n = 12$

12. Find the measure of each exterior angle of a regular polygon of $n$ sides if:

    a) $n = 6$ b) $n = 10$

13. Find the number of sides that a polygon has if the sum of the measures of its interior angles is:

    a) 900° b) 1260°

14. Find the number of sides that a polygon has if the sum of the measures of its interior angles is:

    a) 1980° b) 2340°

15. Find the number of sides that a regular polygon has if the measure of each interior angle is:

a) 108° b) 144°

16. Find the number of sides that a regular polygon has if the measure of each interior angle is:

a) 150° b) 168°

17. Find the number of sides in a regular polygon whose exterior angles each measure:

a) 24° b) 18°

18. Find the number of sides in a regular polygon whose exterior angles each measure:

a) 45° b) 9°

19. What is the measure of each interior angle of a stop sign?

20. Lug bolts are equally spaced about the wheel to form the equal angles shown in the figure. What is the measure of each of the equal acute angles?

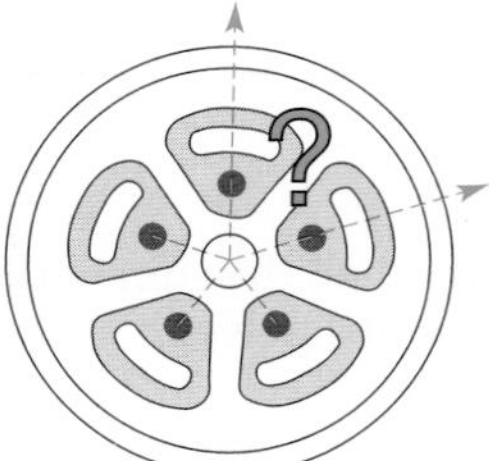

*In Exercises 21 to 26, with P = {all polygons} as the universe, draw a Venn Diagram to represent the relationship between these sets. Describe a subset relationship, if one exists. Are the sets described disjoint or equivalent? Do the sets intersect?*

21. $T$ = {triangles}; $I$ = {isosceles triangles}

22. $R$ = {right triangles}; $S$ = {scalene triangles}

23. $A$ = {acute triangles}; $S$ = {scalene triangles}

24. $Q$ = {quadrilaterals}; $L$ = {equilateral polygons}

25. $H$ = {hexagons}; $O$ = {octagons}

26. $T$ = {triangles}; $Q$ = {quadrilaterals}

27. *Given:* Quadrilateral $RSTQ$ with exterior $\angle$s at $R$ and $T$
*Prove:* $m\angle 1 + m\angle 2 = m\angle 3 + m\angle 4$

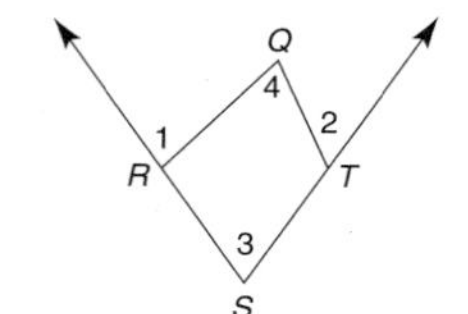


28. *Given:* Regular hexagon $ABCDEF$ with diagonal $\overline{AC}$ and exterior $\angle 1$
*Prove:* $m\angle 2 + m\angle 3 = m\angle 1$

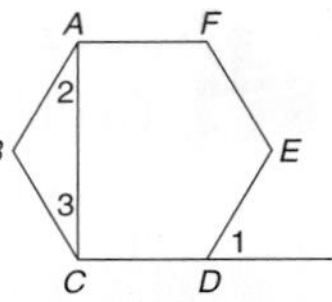


29. *Given:* Quadrilateral $RSTV$ with diagonals $\overline{RT}$ and $\overline{SV}$ intersecting at $W$
*Prove:* $m\angle 1 + m\angle 2 = m\angle 3 + m\angle 4$

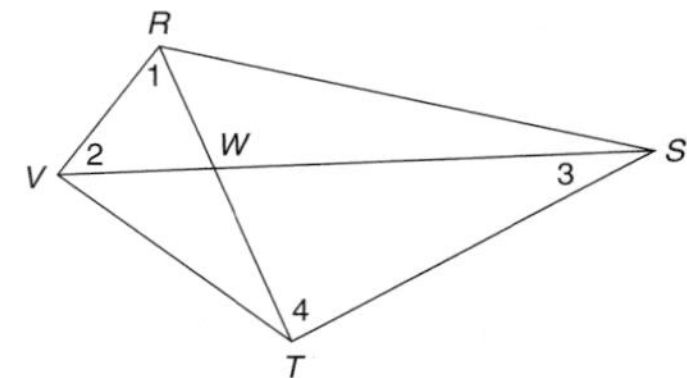


30. *Given:* Quadrilateral $ABCD$ with $\overline{BA} \perp \overline{AD}$ and $\overline{BC} \perp \overline{DC}$
*Prove:* $\angle$s $B$ and $D$ are supplementary

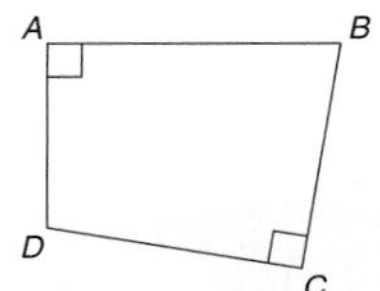


31. A father wishes to make a home plate for his son to use in practicing baseball. Find the size of each of the equal angles if the home plate is modeled on the one in (a) and if it is modeled on the one in (b).

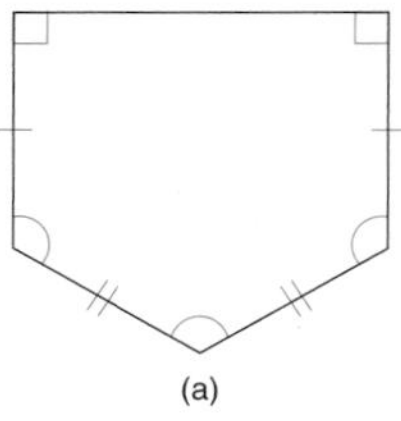
(a)

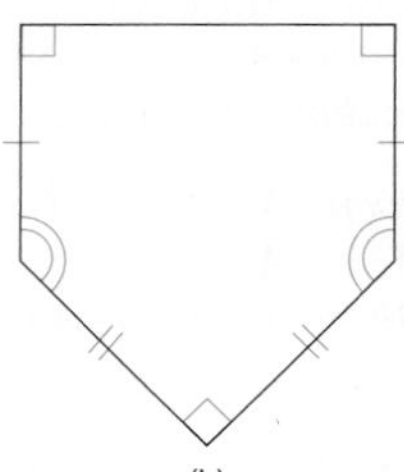
(b)

32. The adjacent interior and exterior angles of a certain polygon are supplementary, as indicated in the drawing. Assume that you know that the measure of each interior angle of a regular polygon is $\frac{(n-2)180}{n}$.

a) Express the measure of each exterior angle as the supplement of the interior angle.
b) Simplify the expression in part (a) to show that each exterior angle has a measure of $\frac{360}{n}$.

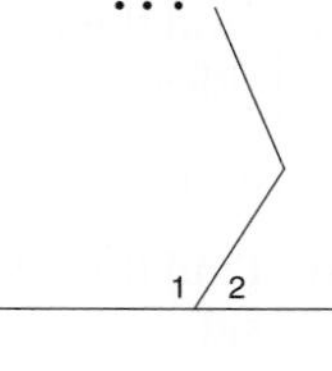


33. Find the measure of each acute interior angle of a regular pentagram.

34. Find the measure of each acute interior angle of a regular octagram.

35. Consider any regular polygon; find and join (in order) the midpoints of the sides. What does intuition tell you about the resulting polygon?

36. Consider a regular hexagon *RSTUVW*. What does intuition tell you about $\triangle RTV$, the result of drawing diagonals $\overline{RT}$, $\overline{TV}$, and $\overline{VR}$?

37. The face of a clock has the shape of a regular polygon with 12 sides. What is the measure of the angle formed by two consecutive sides?

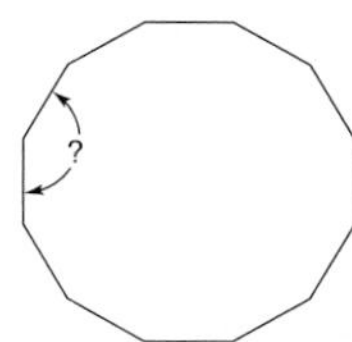


38. The top surface of a picnic table is in the shape of a regular hexagon. What is the measure of the angle formed by two consecutive sides?

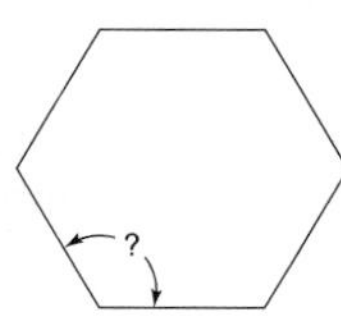


* 39. Consider a polygon of $n$ sides determined by the $n$ noncollinear vertices $A$, $B$, $C$, $D$, and so on.

a) Choose any vertex of the polygon. To how many of the remaining vertices of the polygon can the selected vertex be joined to form a diagonal?

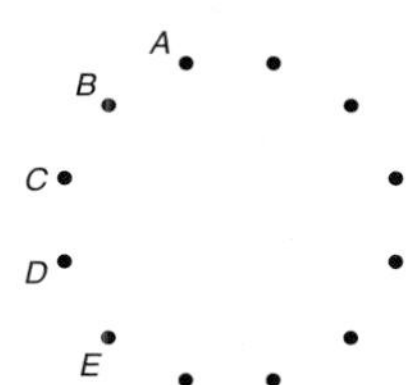


b) Considering that each of the $n$ vertices in (a) can be joined to any one of the remaining $(n - 3)$ vertices to form diagonals, the product $n(n - 3)$ appears to represent the total number of diagonals possible. However, this number includes duplications, such as $\overline{AC}$ and $\overline{CA}$. What expression actually represents $D$, the total number of diagonals in a polygon of $n$ sides?

40. For the concave quadrilateral *ABCD*, explain why the sum of the interior angles is 360°. (HINT: Draw $\overline{BD}$.)

41. If $m\angle A = 20°$, $m\angle B = 88°$, and $m\angle C = 31°$, find the measure of the reflex angle at vertex $D$. (HINT: See Exercise 40.)

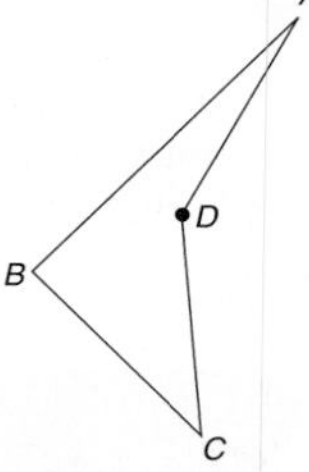


*Exercises 40, 41*

# 2.6 Symmetry and Transformations

**KEY CONCEPTS**

Symmetry • Line of Symmetry • Axis of Symmetry • Point Symmetry • Transformations • Slides • Translations • Reflections • Rotations

## LINE SYMMETRY

In the figure below, rectangle *ABCD* is said to have *symmetry with respect to line* $\ell$. Notice that each point to the left of the *line of symmetry* or *axis of symmetry* has a corresponding point to the right; for instance, $X$ and $Y$ are *corresponding points.*

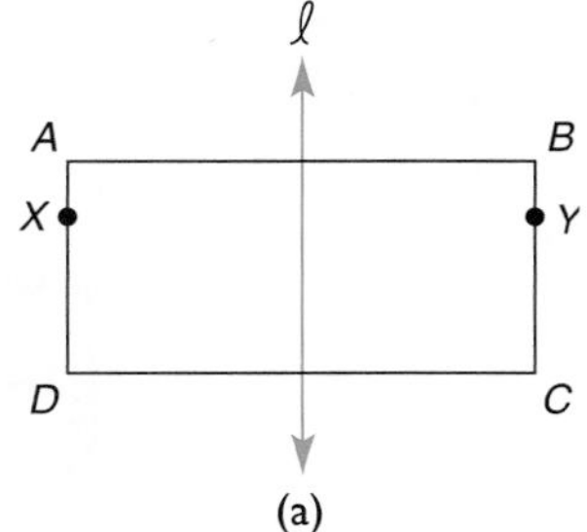


(a)

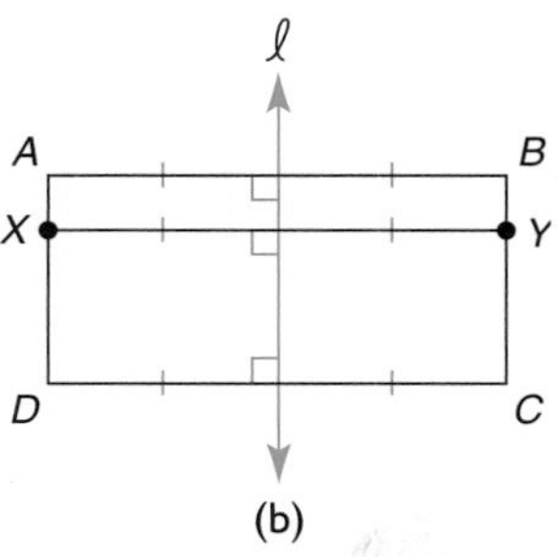


(b)

**FIGURE 2.38**

**DEFINITION:** A figure has *symmetry with respect to a line* $\ell$ if for every point $A$ on the figure, there is a second point $B$ on the figure for which $\ell$ is the perpendicular bisector of $\overline{AB}$.

In particular, $ABCD$ of Figure 2.38 has *horizontal symmetry* with respect to line $\ell$. That is, a vertical axis of symmetry leads to a pairing of corresponding points on a horizontal line. In Example 1, we see that a horizontal axis leads to *vertical* symmetry for points.

*Geometry in Nature*

Like many of nature's creations, the butterfly displays line symmetry.

## EXAMPLE 1

Rectangle $ABCD$ (Figure 2.39) has a second line of symmetry. Draw this line (or axis) for which there is *vertical symmetry.*

***Solution***

Line $m$ (determined by the midpoints of $\overline{AD}$ and $\overline{BC}$) is the desired line of symmetry. As shown in Figure 2.39 (b), $R$ and $S$ are located symmetrically with respect to line $m$.

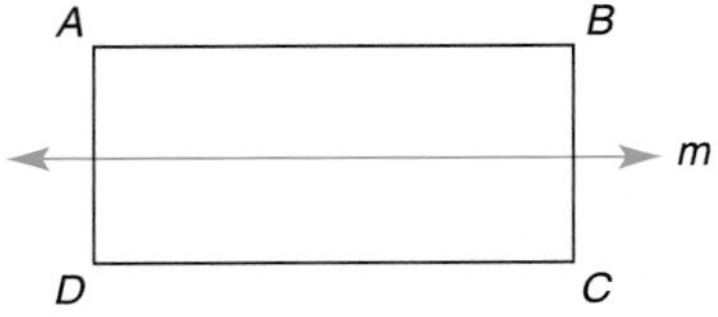


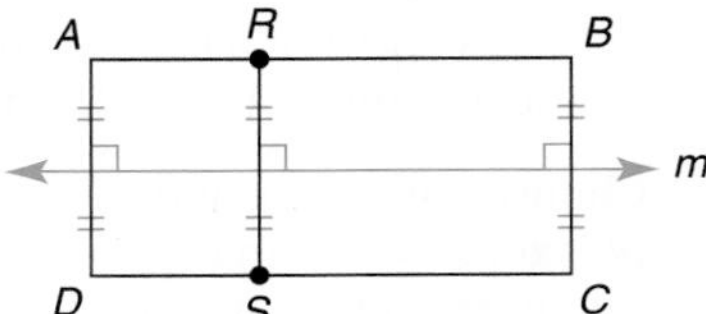


**FIGURE 2.39**

## Discover!

The uppercase block form of the letter A is shown at the right. Does it have symmetry with respect to a line?

ANSWER

Yes, line $\ell$ as shown at the right is a line of symmetry. This vertical line $\ell$ is the only line of symmetry for the uppercase A.

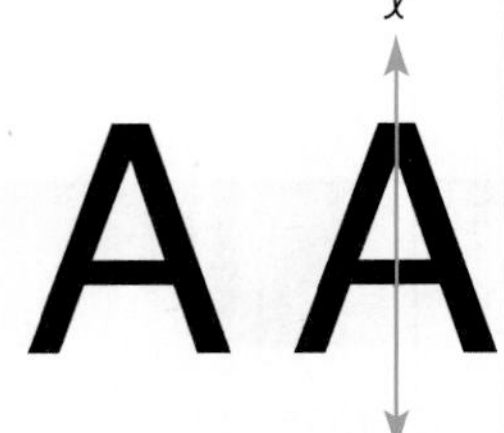


## EXAMPLE 2

a) Which letter(s) shown below has (have) a line of symmetry?
b) Which letter(s) has (have) more than one line of symmetry?

**B D F G H**

***Solution***

a) B, D, and H as shown

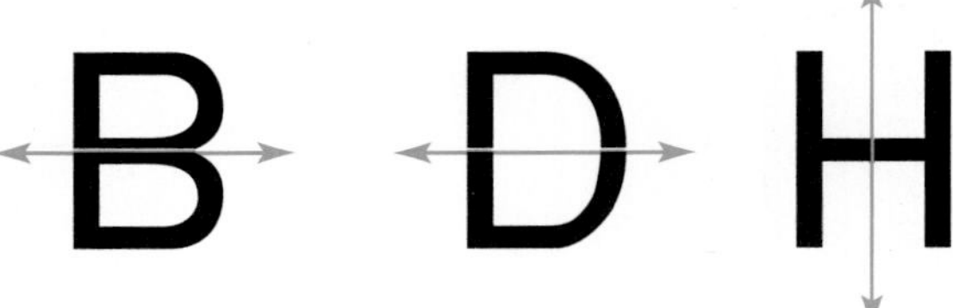

b) H as shown

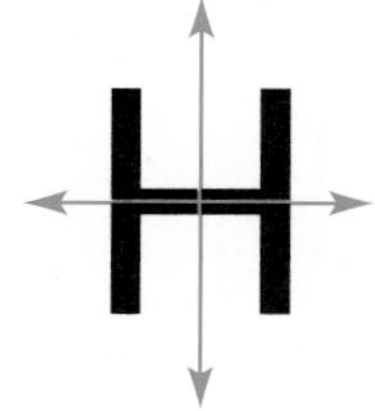


In Chapter 4, we will discover formal definitions of the types of quadrilaterals known as the parallelogram, square, rectangle, kite, rhombus, and rectangle. Some of these are included in Examples 3 and 5.

EXAMPLE 3

a) Which figures have at least one line of symmetry?
b) Which figures have more than one line of symmetry?

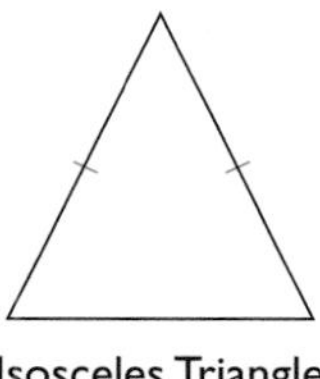




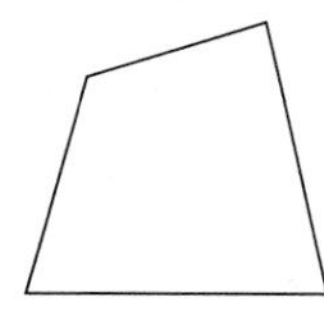


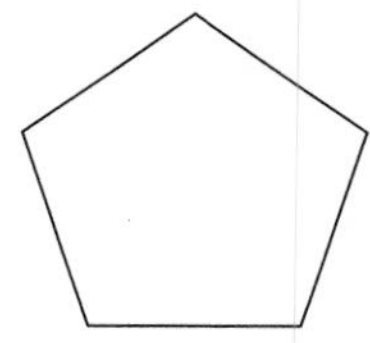


**FIGURE 2.40(a)**

***Solution***

a) The isosceles triangle, square, and the regular pentagon all have a line of symmetry.
b) The square and regular pentagon have more than one line of symmetry, so these figures are shown with two lines of symmetry. (There are actually more than two lines of symmetry.)

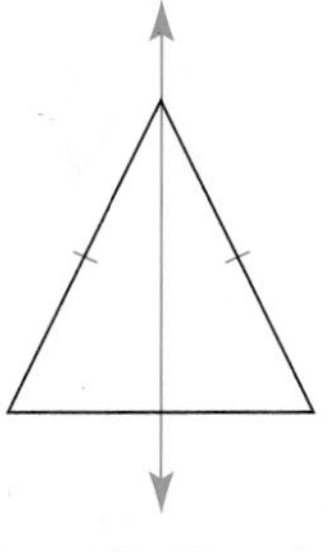


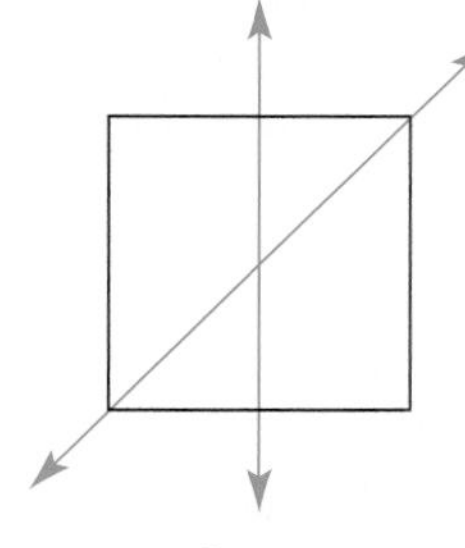


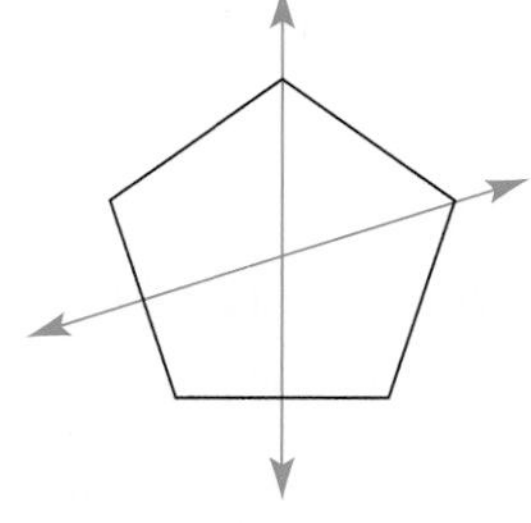


**FIGURE 2.40(b)**

*Exs. 1–4*

## POINT SYMMETRY

The rectangle *ABCD* also is said to have *symmetry with respect to a point.* In Figure 2.41, point *P* is determined by the intersection of the diagonals of rectangle *ABCD.*

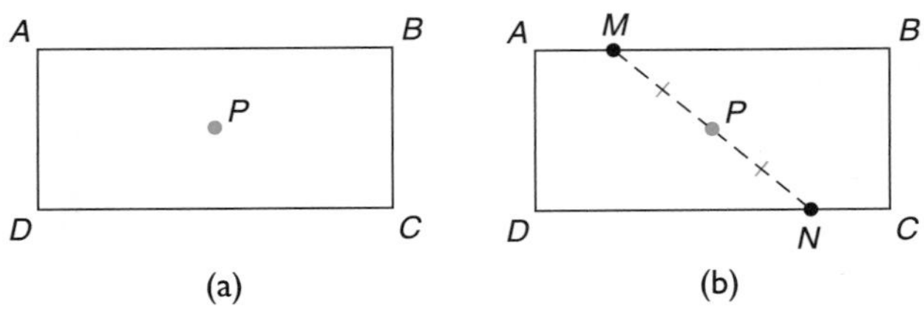


**FIGURE 2.41**

> DEFINITION: A figure has **symmetry with respect to point *P*** if for every point *A* on the figure, there is a second point *C* for which point *P* is the midpoint of $\overline{AC}$.

Based upon this definition, each point on rectangle *ABCD* in Figure 2.41(a) has a corresponding point that is the same distance from *P* but lies in the opposite direction from *P*. In Figure 2.41(b), *M* and *N* are a pair of corresponding points. Even though a figure may have multiple lines of symmetry, a figure can have only one point of symmetry. Thus, the point of symmetry (when one exists) is unique.

### Discover!

The uppercase block form of the letter O is shown at the right. Does it have symmetry with respect to a point?

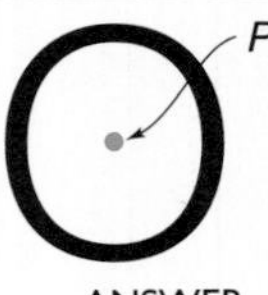


ANSWER

Yes, point *P* (centered) is the point of symmetry.
This point *P* is the only point of symmetry for the uppercase O.

### EXAMPLE 4

Which letter(s) shown below have a point of symmetry?

**M N P S X**

***Solution***

N, S, and X as shown all have point symmetry.

N S X

*Geometry in the Real World*

Taking a good look at the hexagonal shape used in the Hampton Inn logo reveals both point symmetry and line symmetry.

### EXAMPLE 5

Which figures have point symmetry?

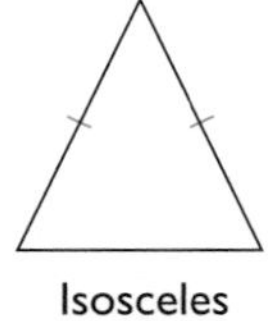

Isosceles Triangle

Square

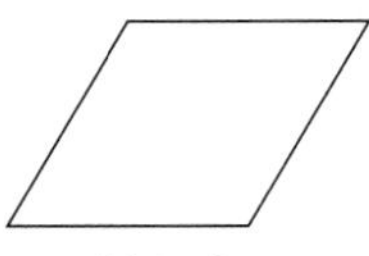

Rhombus

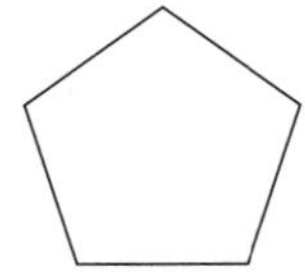

Regular Pentagon

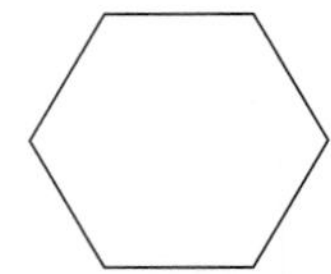

Regular Hexagon

**FIGURE 2.42(a)**

***Solution***

Only the square, the rhombus, and the regular hexagon have a point of symmetry. In the regular pentagon, consider the centrally located point $P$ and note that $AP \neq PM$.

Square
YES

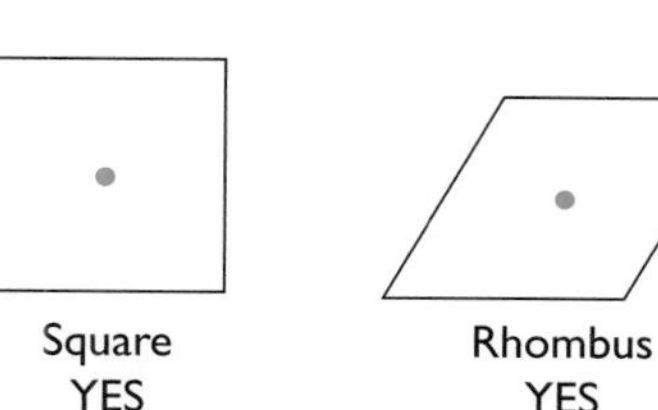


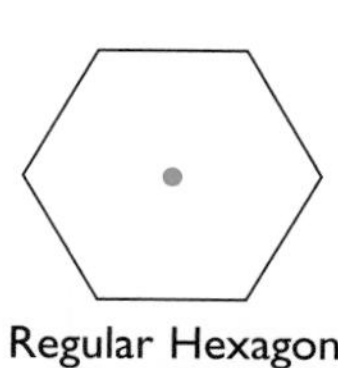


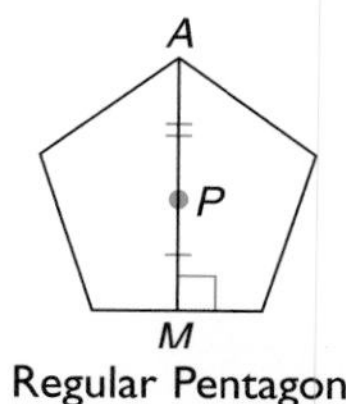


SSG *Exs. 5–8*

**FIGURE 2.42(b)**

## TRANSFORMATIONS

In the following material, we will generate new figures from old figures by association of points. In particular, the transformations included in this textbook will preserve the shape and size of the given figure; in other words, these transformations lead to a second figure that is *congruent* to the given figure. The types of transformations included are (1) the *slide* or *translation,* (2) the *reflection,* and (3) the *rotation.*

### *Slides (Translations)*

With this type of transformation, every point of the original figure is associated with a new point by using a fixed length and direction. In Figure 2.43, $\triangle ABC$ is translated to the second triangle location ($\triangle DEF$) by sliding each point through the distance and in the direction that takes point $A$ to point $D$. The background grid is not necessary to demonstrate the slide, but it lends credibility to our claim that length and direction are preserved!

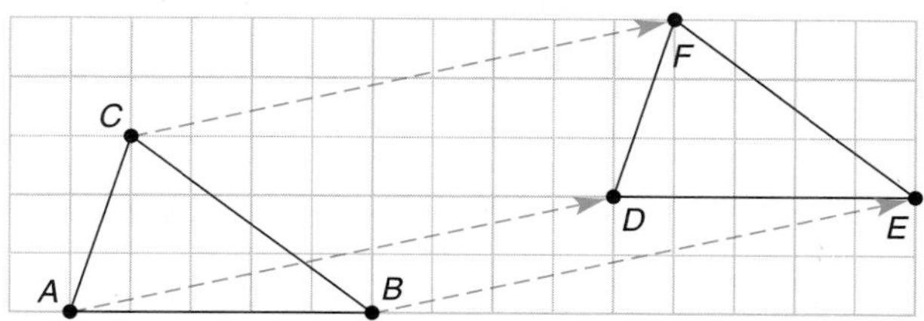


**FIGURE 2.43**

EXAMPLE 6

Slide $\triangle XYZ$ horizontally to form $\triangle RST$. In this example, the distance (length) of the slide is $XR$.

***Solution***

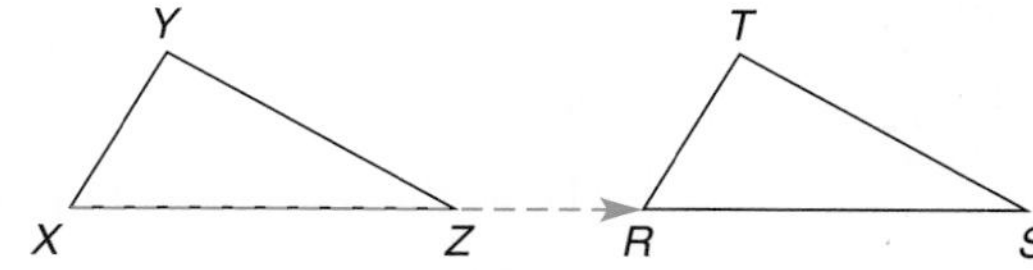


**FIGURE 2.44**

In Example 6, $\triangle XYZ \cong \triangle RTS$ and also $\triangle RTS \cong \triangle XYZ$. In every slide, the given figure and the produced figure (its *image*) are necessarily congruent. In Example 6, the correspondence of vertices is given by $X \leftrightarrow R$, $Y \leftrightarrow T$, and $Z \leftrightarrow S$.

EXAMPLE 7

Where $A \leftrightarrow E$, complete the slide of quadrilateral $ABCD$ to form quadrilateral $EFGH$. Indicate the correspondence among the remaining vertices.

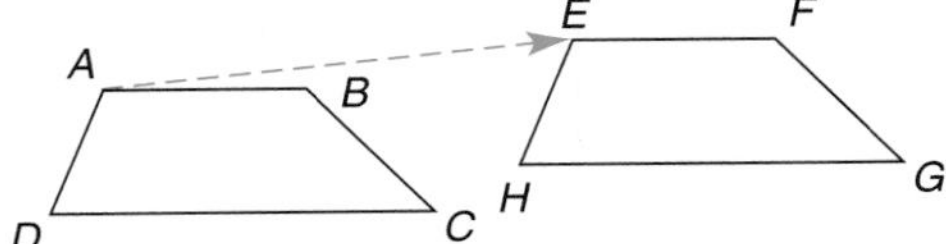


**FIGURE 2.45**

***Solution***

$B \leftrightarrow F$, $C \leftrightarrow G$, and $D \leftrightarrow H$.

### *Reflections*

With the reflection, every point of the original figure is reflected across a line in such a way as to make the given line a line of symmetry. Each pair of corresponding points will lie on opposite sides of the line of reflection and at like distances. In Figure 2.46, obtuse triangle $MNP$ is reflected across the vertical line $\overleftrightarrow{AB}$ to produce the image $\triangle GHK$. The vertex $N$ of the given obtuse angle corresponds to the vertex $H$ of the obtuse angle in the image triangle. It is possible for the line of reflection to be horizontal or oblique (slanted). With the vertical line as the axis of reflection, a drawing such as Figure 2.46 is sometimes called a *horizontal reflection* since the image lies to the right of the given figure.

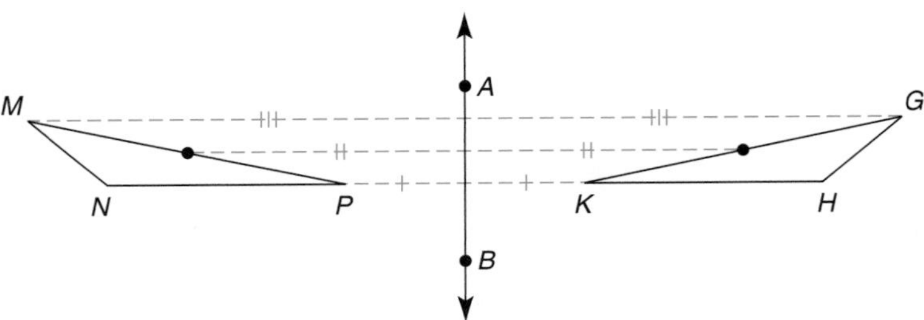


**FIGURE 2.46**

EXAMPLE 8

Draw the reflection of right $\triangle ABC$ (shown on page 113)

a) across line $\ell$ to form $\triangle XYZ$.
b) across line $m$ to form $\triangle PQR$.

***Solution***

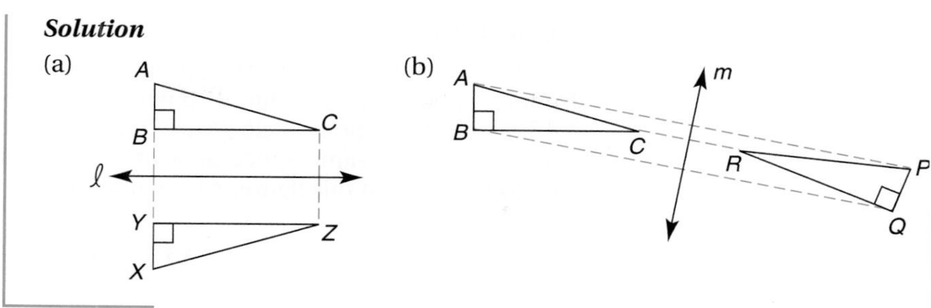


**FIGURE 2.47**

With the horizontal axis (line) of reflection, the reflection in Example 8(a) is often called a *vertical reflection.* In the vertical reflection of Figure 2.47(a), the image lies below the given figure. In Example 9, we use a side of the given figure as the line (line segment) of reflection. This reflection is neither horizontal nor vertical.

EXAMPLE 9

Draw the reflection of $\triangle ABC$ across side $\overline{BC}$ to form $\triangle DBC$. How are $\triangle ABC$ and $\triangle DBC$ related?

***Solution***

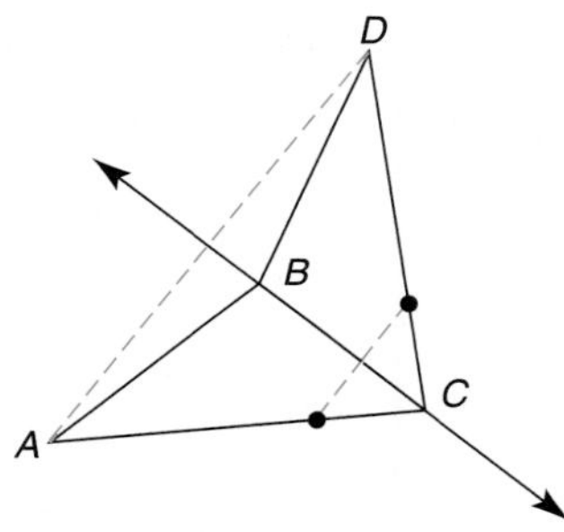


**FIGURE 2.48**

The triangles are congruent; also, notice that $D \leftrightarrow A$.

EXAMPLE 10

Complete the figure produced by a reflection across the given line.

***Solution***

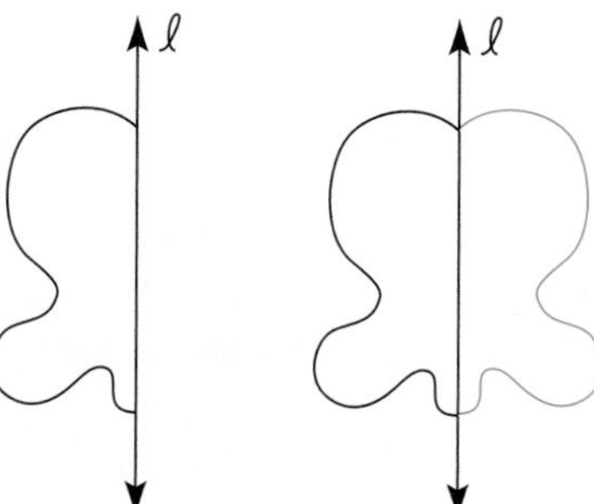


**FIGURE 2.49**

### *Rotations*

In this transformation, every point of the given figure leads to a point (its image) by rotation about a given point through a prescribed angle measure. In Figure 2.50, ray *AB* rotates about point *A* clockwise through an angle of 30° to produce the image ray *AC*. This has the same appearance as the second hand of a clock over a five-second period of time. In this figure, $A \leftrightarrow A$ and $B \leftrightarrow C$.

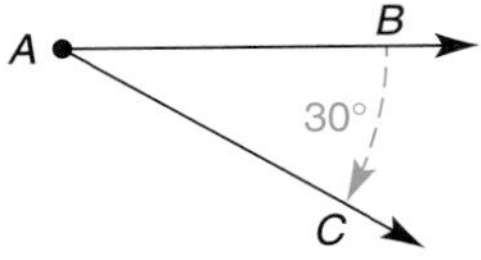


**FIGURE 2.50**

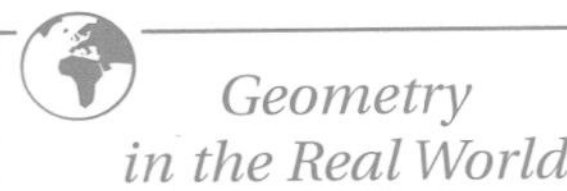


The logo that identifies the Health Alliance Corporation begins with a figure that consists of a rectangle and an adjacent square. The logo is completed by rotating this basic unit through angles of 90°.

EXAMPLE 11

In Figure 2.51, square *WXYZ* is rotated counterclockwise about its center (intersection of diagonals) through an angle of 45° to form congruent square *QMNP*. What is the name of the eight-pointed geometric figure that is formed by the two intersecting squares?

***Solution***

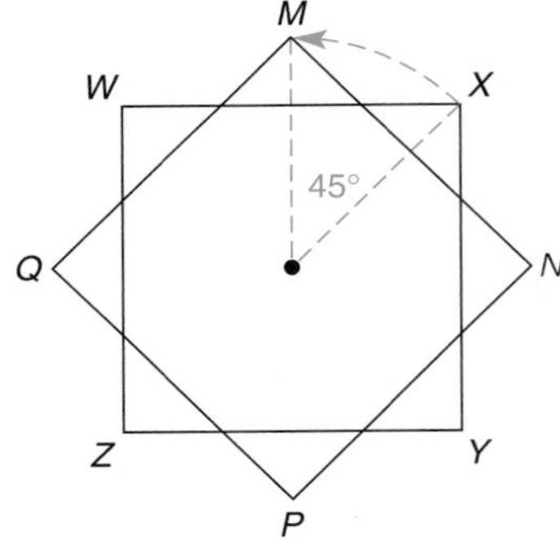


**FIGURE 2.51**

The eight-pointed figure formed is a regular octagram.

EXAMPLE 12

In Figure 2.52 are the uppercase A, line $\ell$, and point *O*. Which of the pairs of transformations produce the original figure?

a) The letter A is reflected across $\ell$, and that image is reflected across $\ell$ again.
b) The letter A is reflected across $\ell$, and that image is rotated clockwise 60° about point *O*.
c) The letter A is rotated 180° about *O*, followed by another 180° rotation about *O*.

**Solution**
(a) and (c)

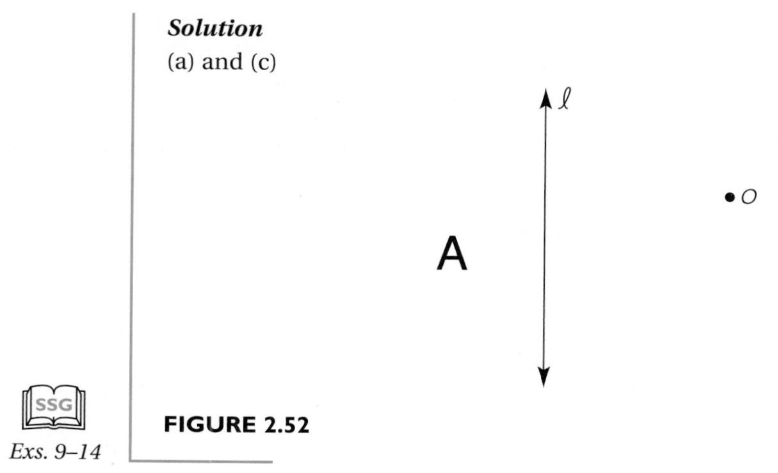

*Exs. 9–14*

**FIGURE 2.52**

## 2.6 Exercises

1. Which letters have symmetry with respect to a line?

   M N P T X

2. Which letters have symmetry with respect to a line?

   I K S V Z

3. Which letters have symmetry with respect to a point?

   M N P T X

4. Which letters have symmetry with respect to a point?

   I K S V Z

5. Which geometric figures have symmetry with respect to at least one line?

   a) b) c)

6. Which geometric figures have symmetry with respect to at least one line?

   a) b) c)

7. Which geometric figures have symmetry with respect to a point?

   a) b) c)

8. Which geometric figures have symmetry with respect to a point?

   a) b) c)

9. Which words have a vertical line of symmetry?

   DAD MOM NUN EYE

10. Which words have a vertical line of symmetry?

    WOW BUB MAM EVE

11. Complete each figure so that it has symmetry with respect to line $\ell$.

    a) b)

12. Complete each figure so that it has symmetry with respect to line $m$.

a) b)

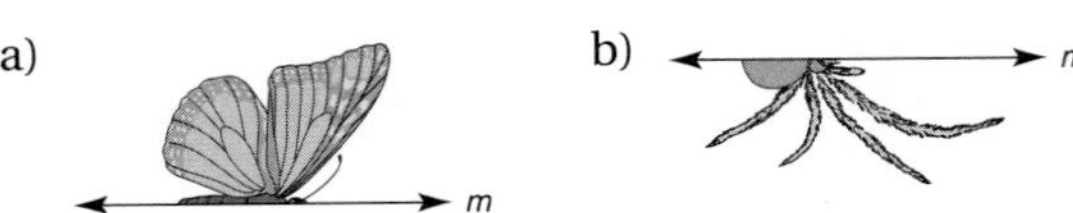


13. Complete each figure so that it reflects across line $\ell$.

a) b)

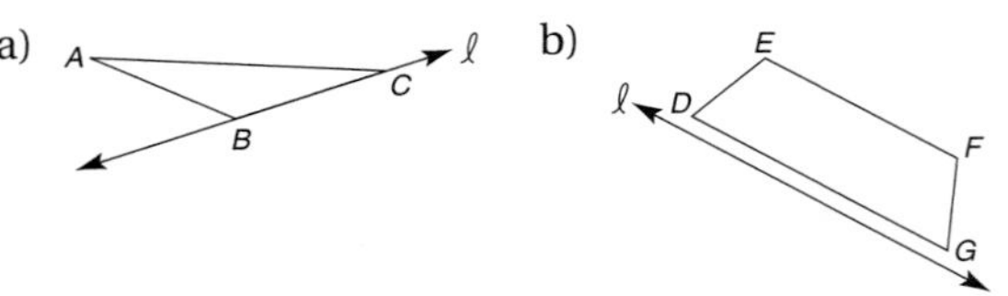


14. Complete each figure so that it reflects across line $m$.

a) b)

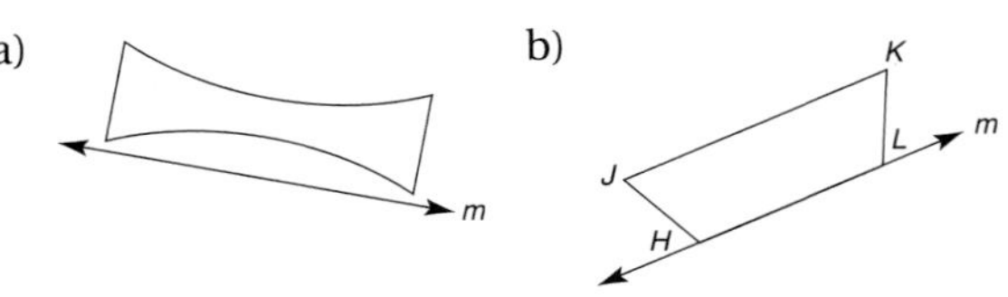


15. Suppose that $\triangle ABC$ slides to the right to the position of $\triangle DEF$.

a) If $m\angle A = 63°$, find $m\angle D$. b) Is $\overline{AC} \cong \overline{DF}$?
c) Is $\triangle ABC$ congruent to $\triangle DEF$?

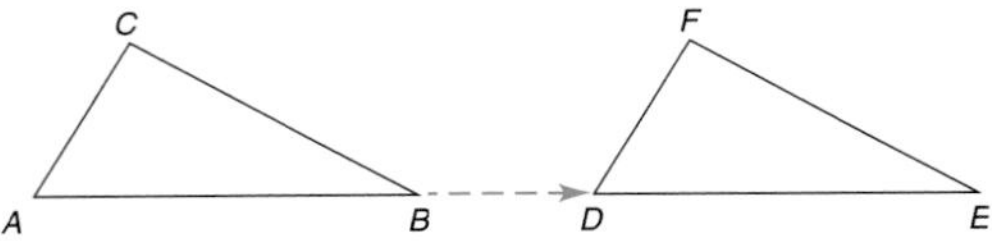


16. Suppose that square $RSTV$ slides point for point to form quadrilateral $WXYZ$.

a) Is $WXYZ$ a square? b) Is $RSTV \cong WXYZ$?
c) If $RS = 1.8$ cm, find $WX$.

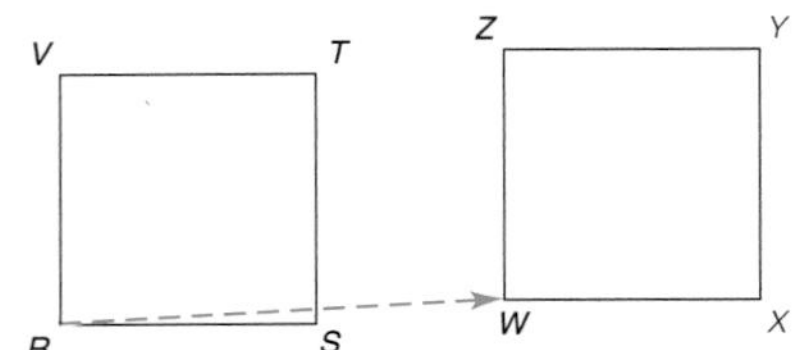


17. Given that the vertical line is a line of symmetry, complete each letter to discover the hidden word.

18. Given that the horizontal line is a line of symmetry, complete each letter to discover the hidden word.

19. Given that each letter has symmetry with respect to the indicated point, complete each letter to discover the hidden word.

20. What word is produced by a 180° rotation about the point?

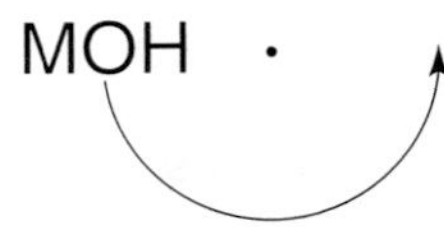


21. What word is produced by a 180° rotation about the point?

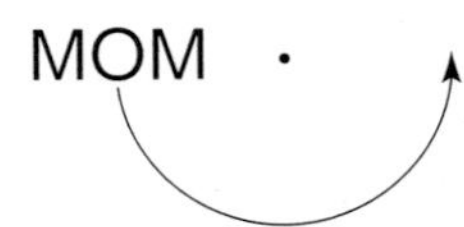


22. What word is produced by a 360° rotation about the point?

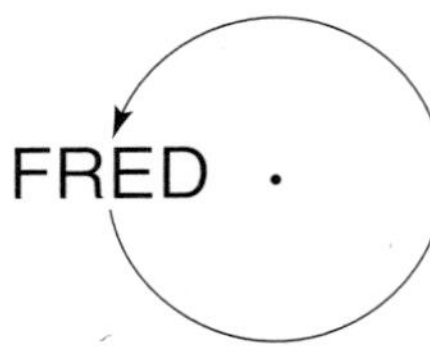


23. In which direction (clockwise or counterclockwise) will pulley 1 rotate if pulley 2 rotates in the clockwise direction?

a) b)

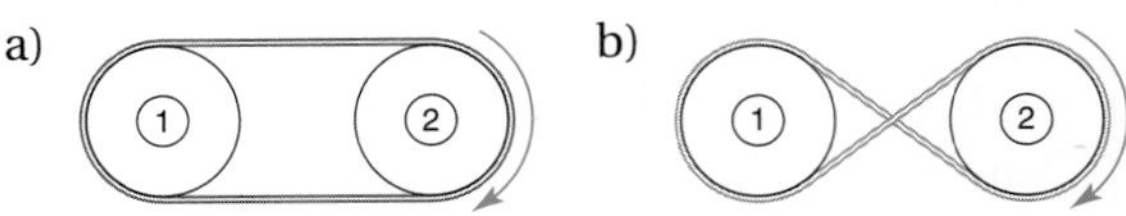


24. In which direction (clockwise or counterclockwise) will gear 1 rotate if gear 2 rotates in the clockwise direction?

a) b)

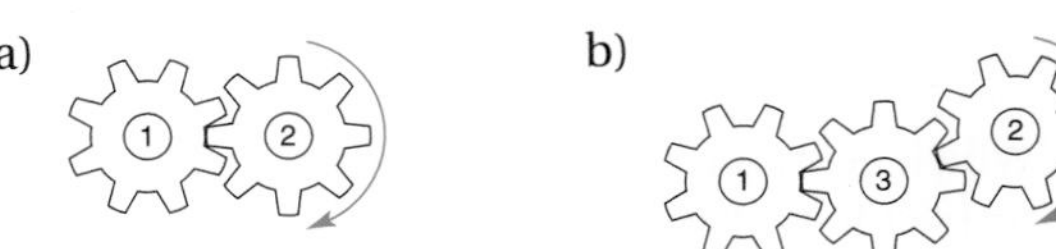

25. Considering that the consecutive dials on the electric meter rotate in opposite directions, what is the current reading in kilowatt hours of usage?

KWH

26. Considering that the consecutive dials on the natural gas meter rotate in opposite directions, what is the current reading in cubic feet of usage?

CU FT

27. Describe the type(s) of symmetry displayed by each of these automobile logos.

a) Toyota b) Mercury c) Volkswagen

28. Describe the type(s) of symmetry displayed by each of these department store logos.

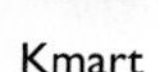

a) Kmart b) Target c) Bergner's

29. Given a figure, which of the following pairs of transformations leads to an image that repeats the original figure?

a) Figure slides 10 cm to the right *twice.*
b) Figure is reflected about a vertical line *twice.*
c) Figure is rotated clockwise about a point 180° *twice.*
d) Figure is rotated clockwise about a point 90° *twice.*

30. Given a figure, which of the following pairs of transformations leads to an image that repeats the original figure?

a) Figure slides 10 cm to the right, followed by slide of 10 cm to the left.
b) Figure is reflected about the same horizontal line *twice.*
c) Figure is rotated clockwise about a point 120° *twice.*
d) Figure is rotated clockwise about a point 360° *twice.*

31. A regular hexagon is rotated about a centrally located point (as shown). How many rotations are needed to repeat the given hexagon vertex for vertex if the angle of rotation is

a) 30°? b) 60°? c) 90°? d) 240°?

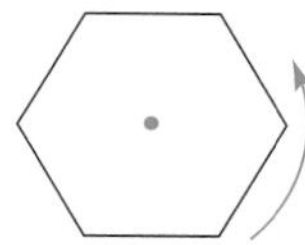

32. A regular octagon is rotated about a centrally located point (as shown). How many rotations are needed to repeat the given octagon vertex for vertex if the angle of rotation is

a) 10°? b) 45°? c) 90°? d) 120°?

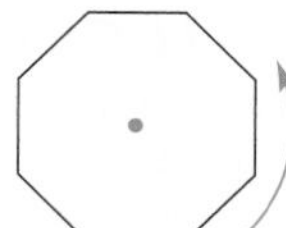

# Perspective on History

## SKETCH OF EUCLID

Names often associated with the early development of Greek mathematics, beginning in approximately 600 B.C., include Thales, Pythagoras, Archimedes, Appolonius, Diophantus, Eratosthenes, and Heron. However, the name most often associated with traditional geometry is that of Euclid, who lived around 300 B.C.

Euclid, himself a Greek, was asked to head the mathematics department at the University of Alexandria (in Egypt), which was the center of Greek learning. It is believed that Euclid told Ptolemy (the local ruler) that "There is no royal road to geometry," in response to Ptolemy's request for a quick and easy knowledge of the subject.

Euclid's best-known work is the *Elements*, a systematic treatment of geometry with some algebra and number theory. That work, which consists of 13 volumes, has dominated the study of geometry for more than 2000 years. Most secondary-level geometry courses, even today, are based on Euclid's *Elements* and in particular on these volumes:

**Book I:** Triangles and congruence, parallels, quadrilaterals, the Pythagorean theorem, and area relationships

**Book III:** Circles, chords, secants, tangents, and angle measurement

**Book IV:** Constructions and regular polygons

**Book VI:** Similar triangles, proportions, and the Angle Bisector theorem

**Book XI:** Lines and planes in space, and parallelepipeds

One of Euclid's theorems was a forerunner of the theorem of trigonometry known as the Law of Cosines. Although it is difficult to understand now, it will make sense to you later. As stated by Euclid, "In an obtuse-angled triangle, the square of the side opposite the obtuse angle equals the sum of the squares of the other two sides and the product of one side and the projection of the other upon it."

While it is believed that Euclid was a great teacher, he is also recognized as a great mathematician and as the first author of an elaborate textbook. In Chapter 2 of *this* textbook, Euclid's Parallel Postulate has been central to our study of plane geometry.

# Perspective on Application

## NON-EUCLIDEAN GEOMETRIES

The geometry we present in this book is often described as Euclidean geometry. A non-Euclidean geometry is a geometry characterized by the existence of at least one contradiction of a Euclidean geometry postulate. To appreciate this subject, you need to realize the importance of the word *plane* in the Parallel Postulate. Thus the Parallel Postulate is now restated.

> PARALLEL POSTULATE:
> *In a plane*, through a point not on a line, exactly one line is parallel to the given line.

The Parallel Postulate characterizes a course in plane geometry; it corresponds to the theory that "the earth is flat." On a small scale (most applications aren't global), the theory works well and serves the needs of carpenters, designers, and most engineers.

To begin the move to a different geometry, consider the surface of a sphere (like Earth). (See Figure 2.53.) By definition, a **sphere** is the set of all points in space that are at a fixed distance from a given point. If a line segment on the surface of the sphere is extended to form a line, it becomes a great circle (like the equator of the earth). Each line in this geometry, known as "spherical geometry," is the intersection of a plane containing the center of the sphere with the sphere.

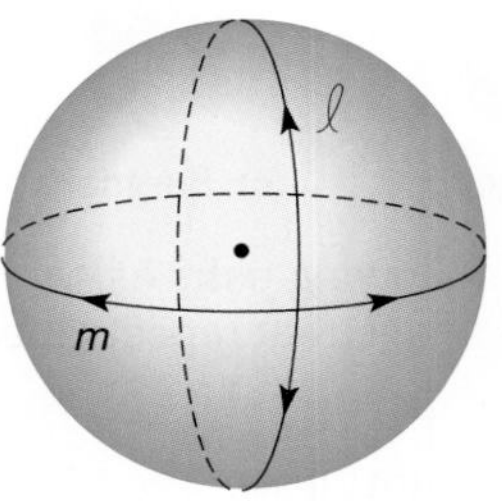


(a) ℓ and *m* are lines in spherical geometry

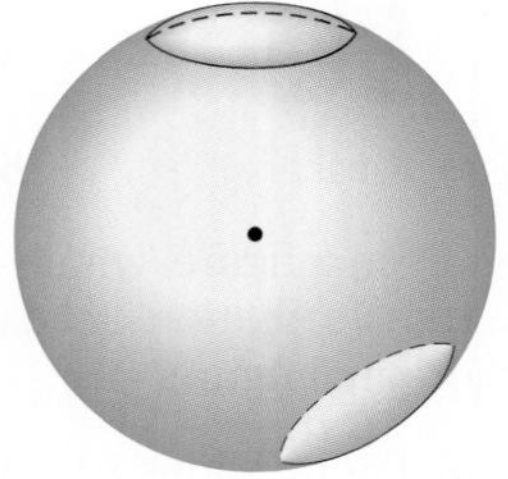

(b) These circles are *not* lines in spherical geometry

**FIGURE 2.53**

Spherical geometry (or elliptic geometry) is actually a model of Riemannian geometry, named in honor of Georg F. B. Riemann (1826–1866), the German mathematician responsible for the next postulate. The Reimannian Postulate is not numbered in this book, because it does not characterize Euclidean geometry.

> RIEMANNIAN POSTULATE:
> Through a point not on a line, there are no lines parallel to the given line.

To understand the Reimannian Postulate, consider a sphere (Figure 2.54) containing line $\ell$ and point $P$ not on $\ell$. Any line drawn through point $P$ must intersect $\ell$ in two points. To see this develop, follow the frames in Figure 2.55, which depict an attempt to draw a line parallel to $\ell$ through point $P$.

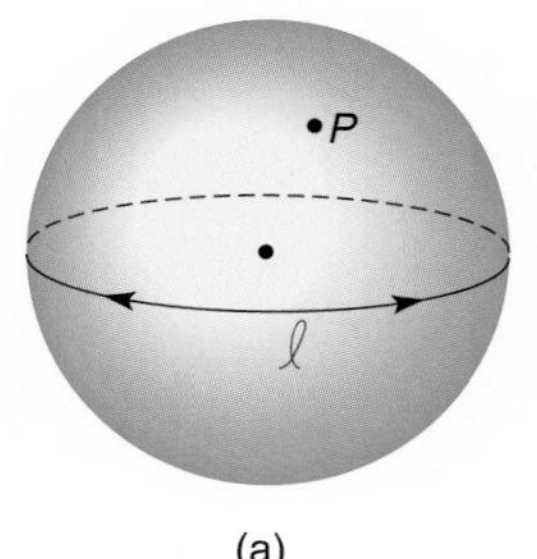


(a)

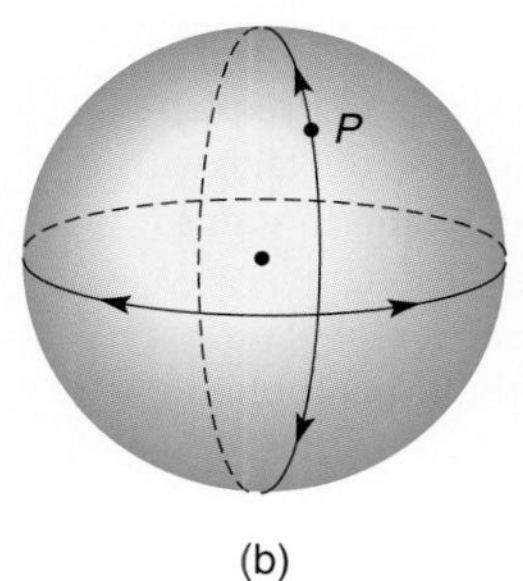


(b)

**FIGURE 2.54**

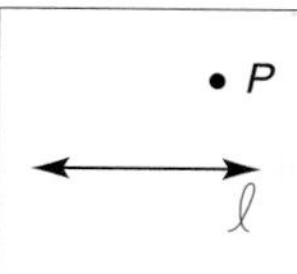


(a) Small part of surface of the sphere

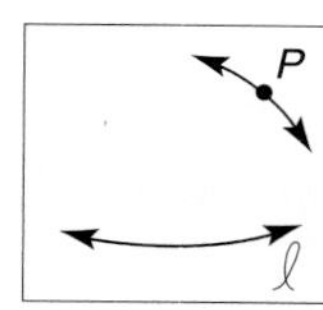


(b) Line through $P$ "parallel" to $\ell$ on larger part of surface

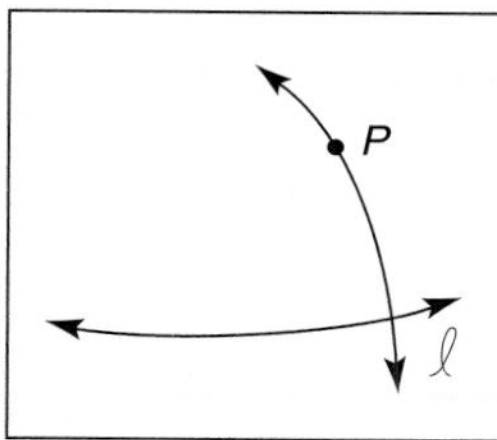


(c) Line through $P$ shown to intersect $\ell$ on larger portion of surface

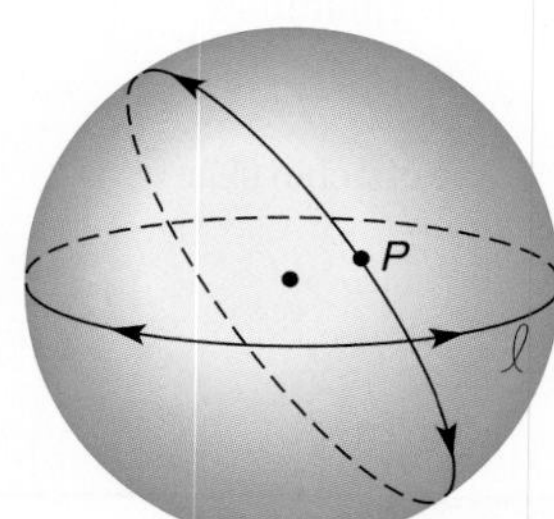


(d) All of line $\ell$ and the line through $P$ shown on entire sphere

**FIGURE 2.55**

Consider the natural extension to Riemannian geometry of the claim that the shortest distance between two points is a straight line. For the sake of efficiency and common sense, a person traveling from New York City to London will follow the path of a line as it is known in spherical geometry. As you might guess, this concept is used to chart international flights between cities. In Euclidean geometry, the claim suggests that a person tunnel under Earth's surface from one city to the other.

A second type of non-Euclidean geometry is attributed to the works of a German, Karl F. Gauss (1777–1855); a Russian, Nikolai Lobachevski (1793–1856); and a Hungarian, Johann Bolyai (1802–1862). The postulate for this system of non-Euclidean geometry is as follows:

LOBACHEVSKIAN POSTULATE:
Through a point not on line, there are infinitely many lines parallel to the given line.

This form of non-Euclidean geometry is termed "hyperbolic geometry." Rather than use the plane or sphere as the surface for study, mathematicians use a saddle-like surface known as a **hyperbolic paraboloid.** (See Figure 2.56.) A line $\ell$ is the intersection of a plane with this surface. Clearly, more than one plane can intersect this surface to form a line containing $P$ that does not intersect $\ell$. In fact, an infinite number of planes intersect the surface in an infinite number of lines parallel to $\ell$ and containing $P$. Table 2.5 compares the three types of geometry.

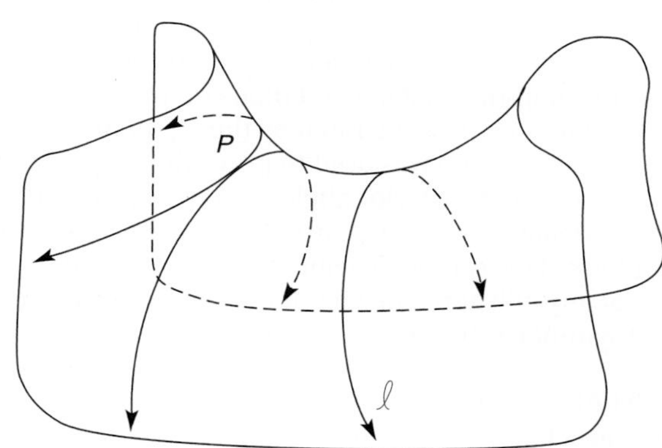


**FIGURE 2.56**

### Comparison of Types of Geometry

| *Postulate* | *Model* | *Line* | *Number of Lines Through P Parallel to ℓ* |
|---|---|---|---|
| Parallel (Euclidean) | Plane geometry | Intersection of two planes | One |
| Riemannian | Spherical geometry | Intersection of plane with sphere (plane contains center of sphere) | None |
| Lobachevskian | Hyperbolic geometry | Intersection of plane with hyperbolic paraboloid | Infinitely many |

## Summary

### A LOOK BACK AT CHAPTER 2

The goal of this chapter has been to prove several theorems based on the postulate "If two parallel lines are cut by a transversal, then the corresponding angles are congruent." The method of indirect proof was introduced as a basis for proving lines parallel if the corresponding angles are congruent. Several methods of proving lines parallel were then demonstrated by the direct method. The Parallel Postulate was used to prove that the sum of the measures of the interior angles of a triangle is 180°. Several corollaries followed naturally from this theorem. A sum formula was then developed for the interior angles of any polygon. The chapter closed with a discussion of symmetry and transformations.

### A LOOK AHEAD TO CHAPTER 3

In the next chapter, the concept of congruence will be extended to triangles, and several methods of proving triangles congruent will be developed. Several theorems dealing with the inequalities of a triangle will also be proved. The Pythagorean Theorem will be introduced.

### KEY CONCEPTS

2.1 Perpendicular Lines • Perpendicular Planes • Parallel Lines • Parallel Planes • Parallel Postulate • Transversal • Interior Angles • Exterior Angles • Corresponding Angles • Alternate Interior Angles • Alternate Exterior Angles

2.2 Conditional • Converse • Inverse • Contrapositive • Law of Negative Inference • Indirect Proof

2.3 Proving Lines Parallel

2.4 Triangle • Vertices • Sides of a Triangle • Interior and Exterior of a Triangle • Scalene Triangle • Isosceles Triangle • Equilateral Triangle • Acute Triangle • Obtuse Triangle • Right Triangle • Equiangular Triangle • Auxiliary Line • Corollary • Exterior Angle of a Triangle

2.5 Convex Polygons (Triangle, Quadrilateral, Pentagon, Hexagon, Heptagon, Octagon, Nonagon, Decagon) • Concave Polygon • Diagonals of a Polygon • Regular Polygon • Equilateral Polygon • Equiangular Polygon • Polygram

2.6 Symmetry • Axis of Symmetry • Line Symmetry • Point Symmetry • Transformations • Slides • Translations • Reflections • Rotations

## TABLE 2.6 AN OVERVIEW OF CHAPTER TWO

### Parallel Lines and Transversal

| *Figure* | *Relationship* | *Symbols* |
|---|---|---|
| 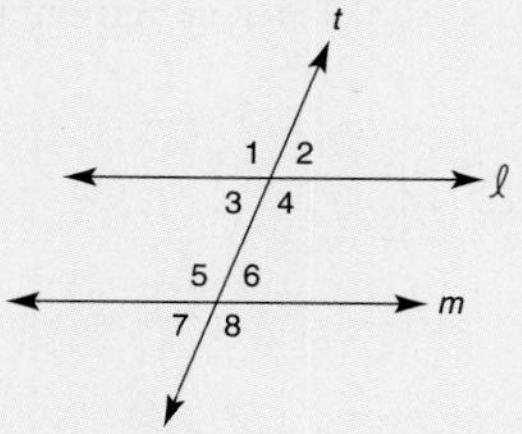  | $\ell \parallel m$ | Corresponding ∠s ≅;<br>∠1 ≅ ∠5, ∠2 ≅ 6, etc.<br>Alternate interior ∠s ≅;<br>∠3 ≅ ∠6 and ∠4 ≅ ∠5<br>Alternate exterior ∠s ≅;<br>∠1 ≅ ∠8 and ∠2 ≅ ∠7<br>Supplementary ∠s;<br>m∠3 + m∠5 = 180°;<br>m∠1 + m∠7 = 180°, etc. |

### Triangles Classified by Sides

| *Figure* | *Type* | *Number of Congruent Sides* |
|---|---|---|
| 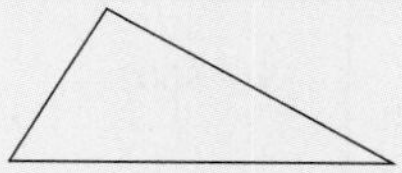 | Scalene | None |
| 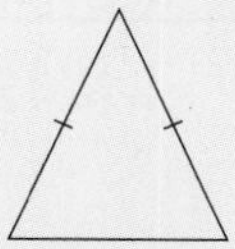 | Isosceles | Two |
| 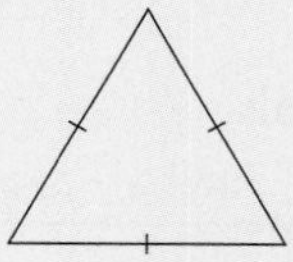 | Equilateral | Three |

### Triangles Classified by Angles

| *Figure* | *Type* | *Angle(s)* |
|---|---|---|
| 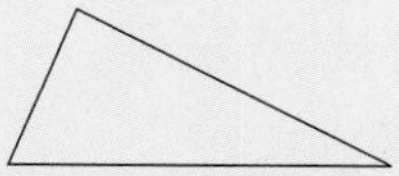 | Acute | Three acute angles |
| 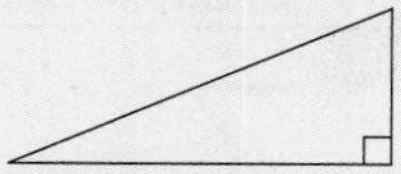 | Right | One right angle |
| 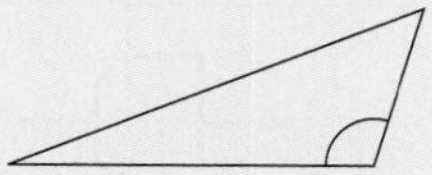 | Obtuse | One obtuse angle |

*continued*

**TABLE 2.6 CONT.**

### Triangles Classified by Angles

| *Figure* | *Type* | *Angle(s)* |
|---|---|---|
| 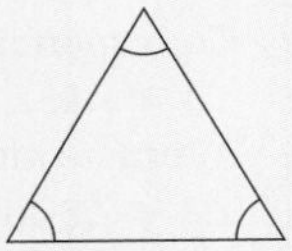 | Equiangular | Three congruent angles |

### Polygons: Sum $S$ of All Interior Angles

| *Figure* | *Type of Polygon* | *Sum of Interior Angles* |
|---|---|---|
| 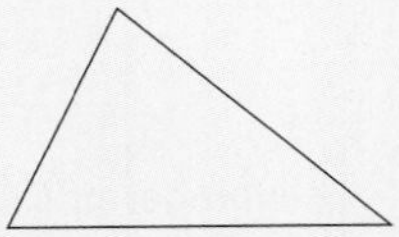 | Triangle | $S = 180°$ |
| 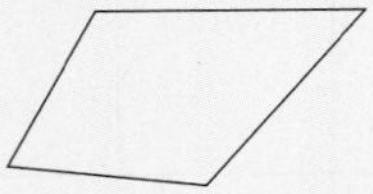 | Quadrilateral | $S = 360°$ |
| 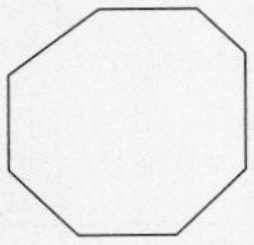 | Polygon with $n$ sides | $S = (n - 2) \cdot 180°$ |

### Polygons: Sum $S$ of All Exterior Angles; $D$ Is the Total Number of Diagonals

| *Figure* | *Type of Polygon* | *Relationships* |
|---|---|---|
| 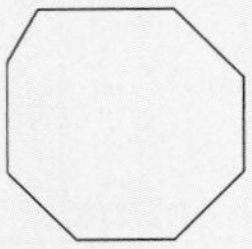 | Polygon with $n$ sides | $S = 360°$<br>$D = \frac{n(n-3)}{2}$ |

### Symmetry

| *Figure* | *Type of Symmetry* | *Figure Redrawn to Display Symmetry* |
|---|---|---|
| Z | Point |  Z |
| D | Line | 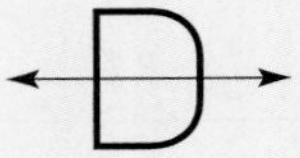 D |

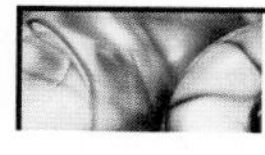

# Chapter 2 Review Exercises

1. If m∠1 = m∠2, which lines are parallel?

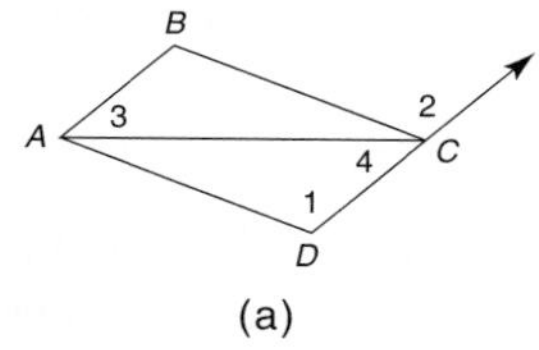

(a)

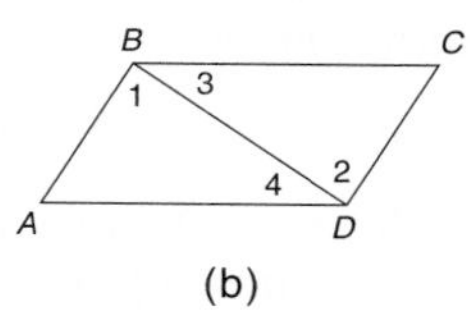

(b)

2. *Given:* m∠13 = 70°
   *Find:* m∠3

3. *Given:* m∠9 = $2x + 17$
   m∠11 = $5x - 94$
   *Find:* $x$

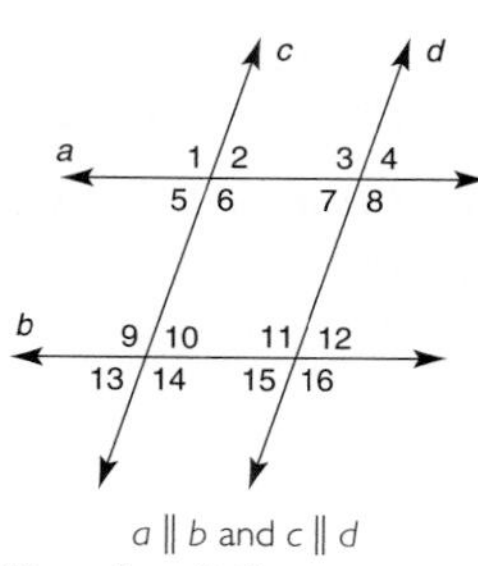

*a* ∥ *b* and *c* ∥ *d*
*Exercises 2, 3*

4. *Given:* m∠*B* = 75°, m∠*DCE* = 50°
   *Find:* m∠*D* and m∠*DEF*

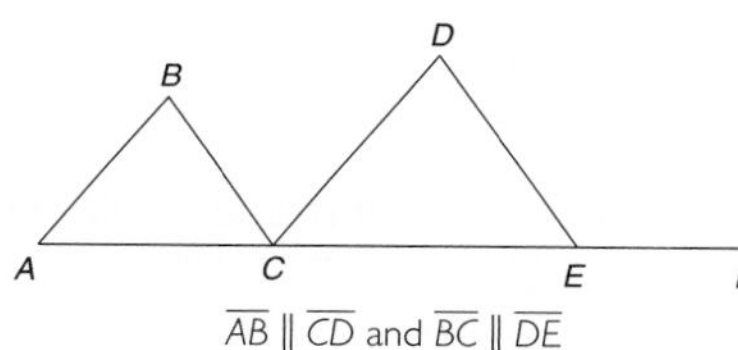

$\overline{AB} \parallel \overline{CD}$ and $\overline{BC} \parallel \overline{DE}$
*Exercises 4, 5*

5. *Given:* m∠*DCA* = 130°
   m∠*BAC* = $2x + y$
   m∠*BCE* = 150°
   m∠*DEC* = $2x - y$
   *Find:* $x$ and $y$

6. *Given:* In the drawing for Review Exercises 6 to 11,
   $\overline{AC} \parallel \overline{DF}$
   $\overline{AE} \parallel \overline{BF}$
   m∠*AEF* = $3y$
   m∠*BFE* = $x + 45$
   m∠*FBC* = $2x + 15$
   *Find:* $x$ and $y$

*Exercises 6–11*

*Use the given information to name the segments that must be parallel. If there are no such segments, write "none." Assume A-B-C and D-E-F. (Use the drawing from Exercise 6.)*

7. ∠3 ≅ ∠11

8. ∠4 ≅ ∠5

9. ∠7 ≅ ∠10

10. ∠6 ≅ ∠9

11. ∠8 ≅ ∠5 ≅ ∠3

*For Review Exercises 12 to 15, find the values of x and y.*

12. 13.

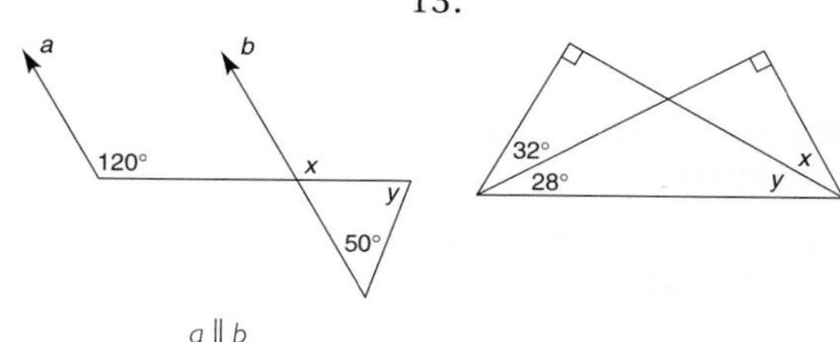

*a* ∥ *b*

14. 15.

*a* ∥ *b*

16. *Given:* m∠1 = $x^2 - 12$
    m∠4 = $x(x - 2)$
    *Find:* $x$ so that $\overline{AB} \parallel \overrightarrow{CD}$

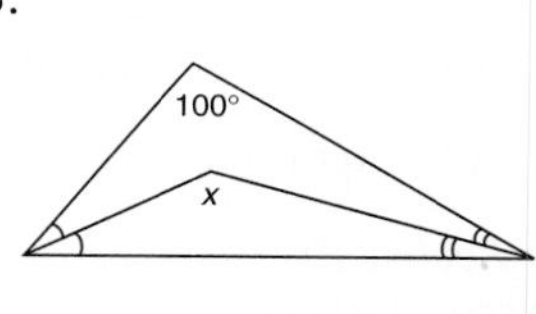

*Exercises 16, 17*

17. *Given:* $\overline{AB} \parallel \overrightarrow{CD}$
    m∠2 = $x^2 - 3x + 4$
    m∠1 = $17x - x^2 - 5$
    m∠*ACE* = 111°
    *Find:* m∠3, m∠4, and m∠5

18. *Given:* $\overline{DC} \parallel \overline{AB}$
    ∠*A* ≅ ∠*C*
    m∠*A* = $3x + y$
    m∠*D* = $5x + 10$
    m∠*C* = $5y + 20$
    *Find:* m∠*B*

*For Review Exercises 19 to 24, decide whether the statements are always true (A), sometimes true (S), or never true (N).*

19. An isosceles triangle is a right triangle.

20. An equilateral triangle is a right triangle.

21. A scalene triangle is an isosceles triangle.

22. An obtuse triangle is an isosceles triangle.

23. A right triangle has two congruent angles.

24. A right triangle has two complementary angles.

25. Complete the following table for regular polygons.

| Number of sides | 8 | 12 | 20 | | | | |
|---|---|---|---|---|---|---|---|
| Measure of each exterior ∠ | | | | 24 | 36 | | |
| Measure of each interior ∠ | | | | | | 157.5 | 178 |
| Number of diagonals | | | | | | | |

*For Review Exercises 26 to 29, sketch, if possible, the polygon described.*

26. A quadrilateral that is equiangular but not equilateral

27. A quadrilateral that is equilateral but not equiangular

28. A triangle that is equilateral but not equiangular

29. A hexagon that is equilateral but not equiangular

*For Review Exercises 30 and 31, write the converse, inverse, and contrapositive of each statement.*

30. If two angles are right angles, then the angles are congruent.

31. If it is not raining, then I am happy.

32. Which statement—the converse, the inverse, or the contrapositive—always has the same truth or falsity as a given implication?

33. *Given:* $\overline{AB} \parallel \overline{CF}$
$\angle 2 \cong \angle 3$
*Prove:* $\angle 1 \cong \angle 3$

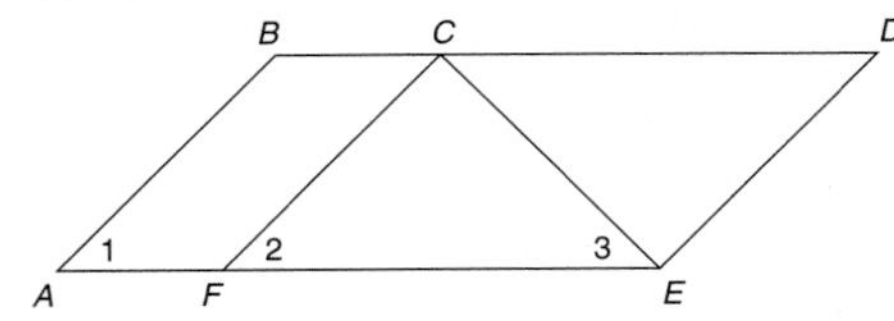


34. *Given:* $\angle 1$ is complementary to $\angle 2$; $\angle 2$ is complementary to $\angle 3$
*Prove:* $\overline{BD} \parallel \overline{AE}$

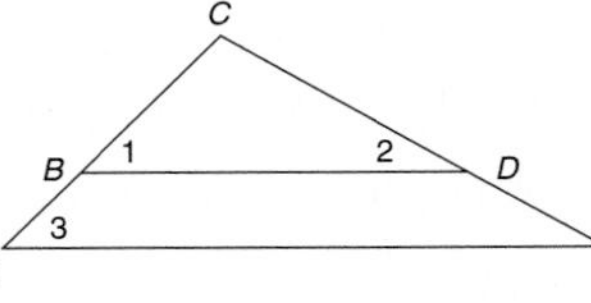


35. *Given:* $\overline{BE} \perp \overline{DA}$
$\overline{CD} \perp \overline{DA}$
*Prove:* $\angle 1 \cong \angle 2$

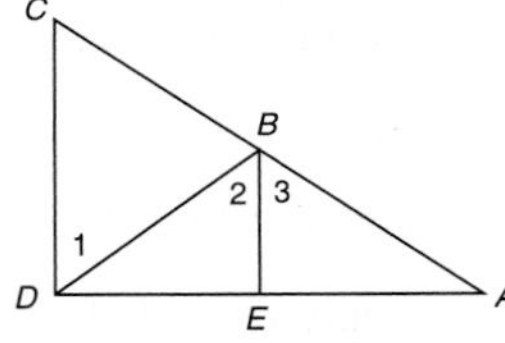


36. *Given:* $\angle A \cong \angle C$
$\overline{DC} \parallel \overrightarrow{AB}$
*Prove:* $\overline{DA} \parallel \overline{CB}$

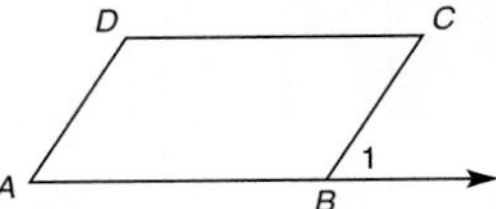


*For Exercises 37 and 38, give the first statement for an indirect proof.*

37. If $x^2 + 7x + 12 \neq 0$, then $x \neq -3$.

38. If two angles of a triangle are not congruent, then the sides opposite those angles are not congruent.

39. *Given:* $m \nparallel n$
*Prove:* $\angle 1 \ncong \angle 2$

40. *Given:* $\angle 1 \ncong \angle 3$
*Prove:* $m \nparallel n$

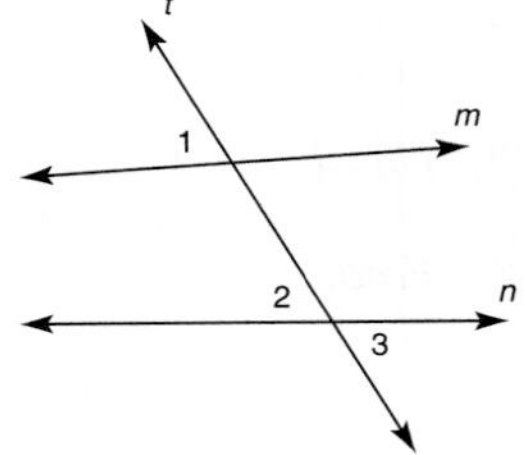


*Exercises 39, 40*

41. Construct the line through $C$ parallel to $\overline{AB}$.

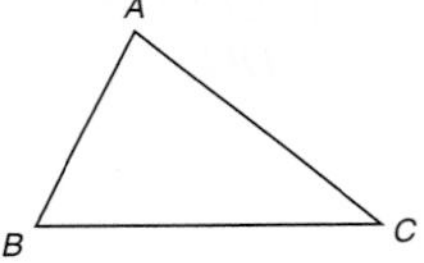


42. Construct an equilateral triangle $ABC$ with side $\overline{AB}$.

A B

43. Which block letters have

a) line symmetry (at least one axis)?
b) point symmetry?

**B H J S W**

44. Which figures have

a) line symmetry (at least one axis)?
b) point symmetry?

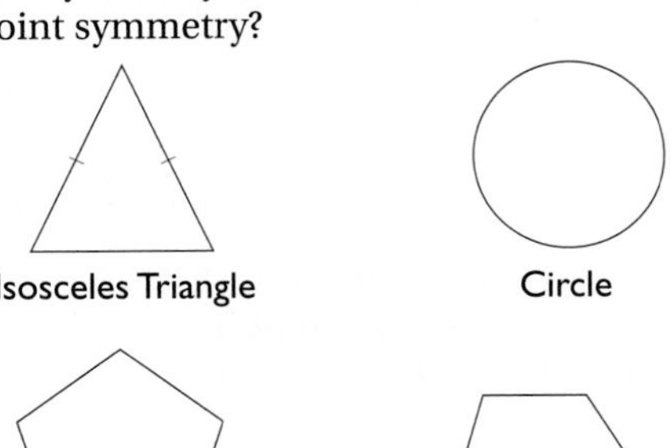


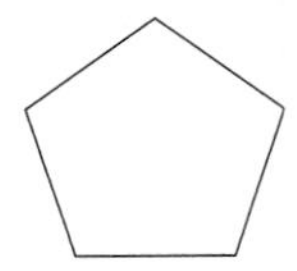


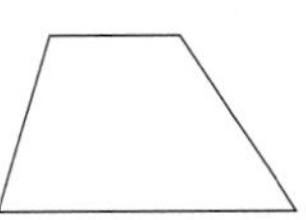

45. When $\triangle ABC$ slides to its image $\triangle DEF$, how are $\triangle ABC$ and $\triangle DEF$ related?

46. Complete the drawing so that the figure is reflected across a) line $\ell$. b) line $m$.

a) 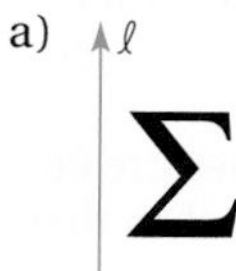


b) 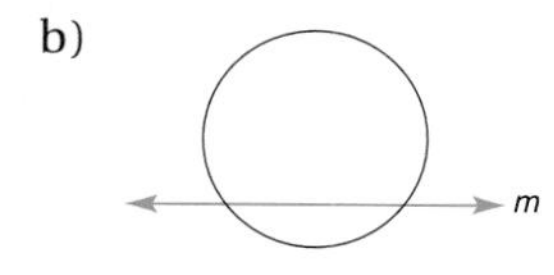


47. Through what approximate angle of rotation must a baseball pitcher turn when throwing to first base rather than home plate?

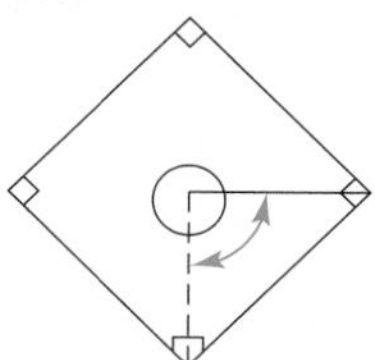

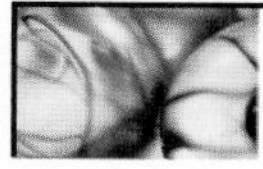

## Chapter 2 Test

1. Consider the figure shown at the right.

   a) Name the angle that corresponds to $\angle 1$.

   ________

   b) Name the alternate interior angle for $\angle 6$.

   ________

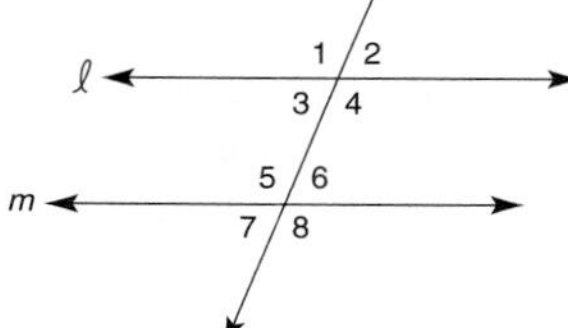


2. In the accompanying figure, $m\angle 2 = 68°$, $m\angle 8 = 112°$, and $m\angle 9 = 110°$.

   a) Which lines ($r$ and $s$ OR $\ell$ and $m$) must be parallel?

   ________

   b) Which pair of lines ($r$ and $s$ OR $\ell$ and $m$) cannot be parallel? ________

3. To prove a theorem of the form "If $P$, then $Q$" by the indirect method, the first line of the proof should read:
   Suppose that ________ is true.

4. Assuming that statements 1 and 2 are true, draw a valid conclusion if possible.

   1. If two angles are both right angles, then the angles are congruent.
   2. $\angle R$ and $\angle S$ are not congruent.

   C. $\therefore$?

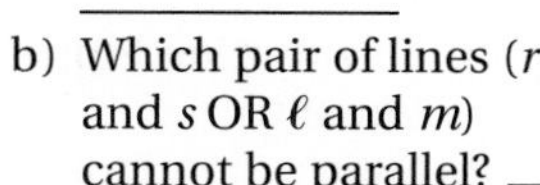

5. Let all of the lines named be coplanar. Make a drawing to reach a conclusion.

   a) If $r \parallel s$ and $s \parallel t$, then ________.
   b) If $a \perp b$ and $b \perp c$, then ________.

6. Through point $A$, construct the line that is perpendicular to line $\ell$.

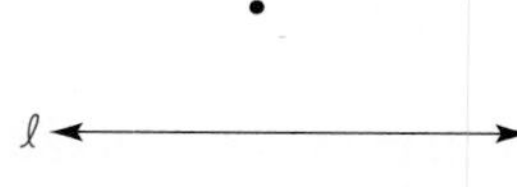


7. For $\triangle ABC$, find $m\angle B$ if

   a) $m\angle A = 65°$ and $m\angle C = 79°$.

   ________

   b) $m\angle A = 2x$, $m\angle B = x$, and $m\angle C = 2x + 15$. ________

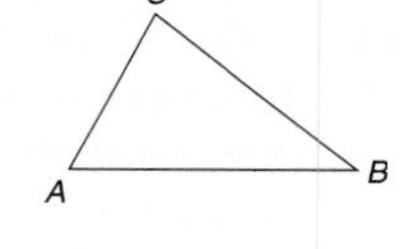

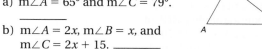

8. a) What word describes a polygon with five sides.

   ________

   b) How many diagonals does a polygon with five sides have? ________

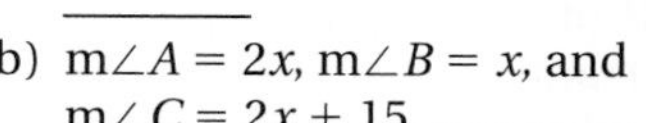

9. a) Given that the polygon shown has six congruent angles, this polygon is known as a(n) ________ ________.

   b) What is the measure of each of the congruent interior angles? ________

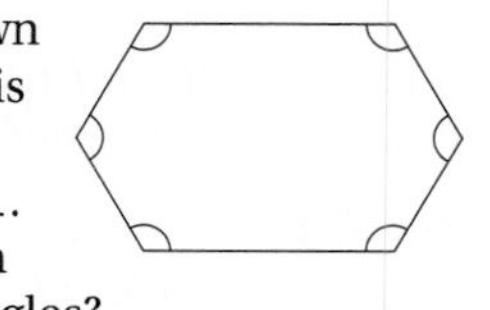

10. Consider the block letters A, D, N, O, and X.

    Which type of symmetry (line symmetry, point symmetry, both types, or neither type) is illustrated by each letter?

    A ________ D ________
    N ________ O ________
    X ________

11. Which type of transformation (slide, reflection, or rotation) is illustrated?

a) ________ b) ________ c) ________

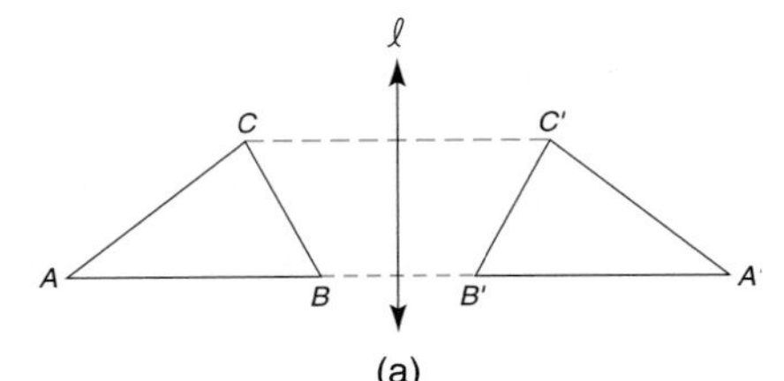

(a)

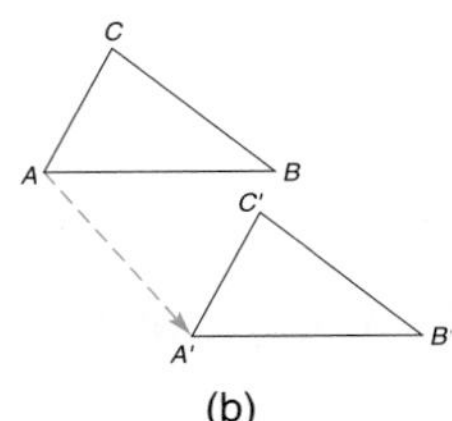

(b)

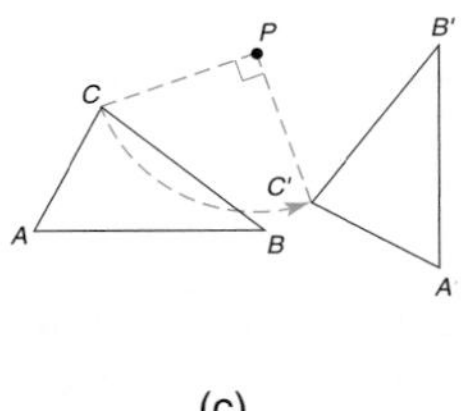

(c)

12. In the figure shown, suppose that $\overline{AB} \parallel \overline{DC}$ and $\overline{AD} \parallel \overline{BC}$. If $m\angle 1 = 82°$ and $m\angle 4 = 37°$, find $m\angle C$. ________

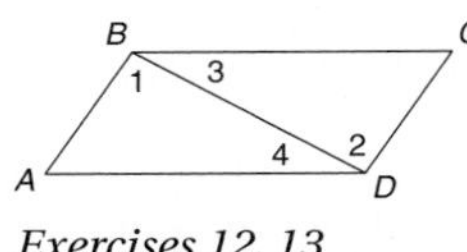

*Exercises 12, 13*

13. If $m\angle 1 = x + 28$ and $m\angle 2 = 2x - 26$, find the value $x$ for which it follows that $\overline{AB} \parallel \overline{DC}$. ________

14. In the figure shown, suppose that ray $CD$ bisects exterior angle $\angle ACE$ of $\triangle ABC$. If $m\angle 1 = 70°$ and $m\angle 2 = 30°$, find $m\angle 4$.

________

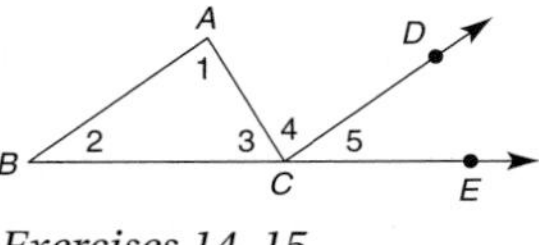

*Exercises 14, 15*

15. In the figure shown, $\angle ACE$ is an exterior angle of $\triangle ABC$. If $\overline{CD} \parallel \overline{BA}$, $m\angle 1 = 2(m\angle 2)$, and $m\angle ACE = 117°$, find the measure of $\angle 1$. ________

*In Exercises 16 and 18, complete the missing statements or reasons for each proof.*

16. *Given:* $\angle 1 \cong 2$
$\angle 3 \cong \angle 4$
*Prove:* $\ell \parallel n$

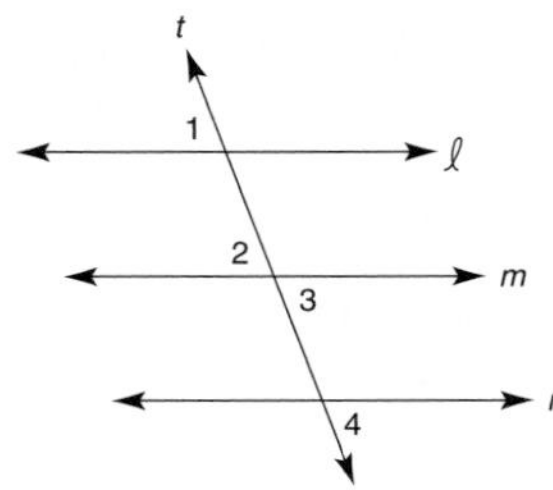


**PROOF**

| Statements | Reasons |
|---|---|
| 1. $\angle 1 \cong \angle 2$ and $\angle 3 \cong \angle 4$ | 1. ________ |
| 2. ________ | 2. If two lines intersect, the vertical $\angle$s are $\cong$ |
| 3. $\angle 1 \cong 4$ | 3. ________ |
| 4. ________ | 4. If two lines are cut by a transversal so that alternate exterior $\angle$s are $\cong$, the lines are $\parallel$ |

17. Use an indirect proof to complete the following proof.
*Given:* $\triangle MNQ$ with $m\angle N = 120°$
*Prove:* $\angle M$ and $\angle Q$ are not complementary
*Proof:*

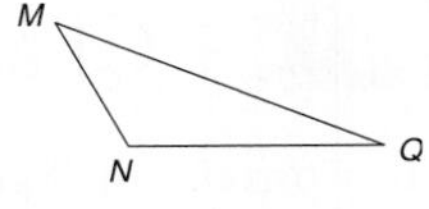

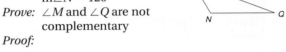

18. *Given:* In $\triangle ABC$, $m\angle C = 90°$
*Prove:* $\angle 1$ and $\angle 2$ are complementary

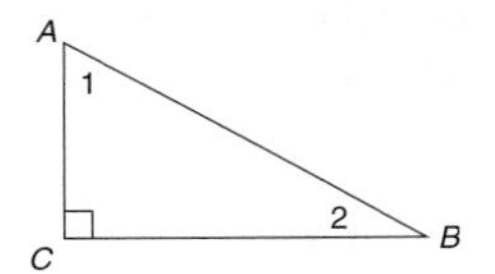


**PROOF**

| Statements | Reasons |
|---|---|
| 1. $\triangle ABC$, $m\angle C = 90°$ | 1. ________ |
| 2. $m\angle 1 + m\angle 2 + m\angle C =$ ________ | 2. The sum of $\angle$s of a $\triangle$ is 180° |
| 3. ________ | 3. Substitution Prop. of Equality |
| 4. $m\angle 1 + m\angle 2 =$ ______ | 4. Subtraction Prop. of Equality |
| 5. ________ | 5. ________ |

# Chapter 3

# Triangles

The building at the center of this breathtaking skyscape of Hong Kong is the Bank of China. The structure displays many triangles that are the same in shape and size. Such triangles, known as congruent triangles, are also seen in the Ferris wheel displayed in Exercise 41 of Section 3.3.

As we shall see, the properties of triangles found in this chapter are also applied in the development of the characteristics of quadrilaterals that we study in Chapter 4.

## Chapter Outline



**For online student resources, visit this textbook's website at math.college.hmco.com/students.**

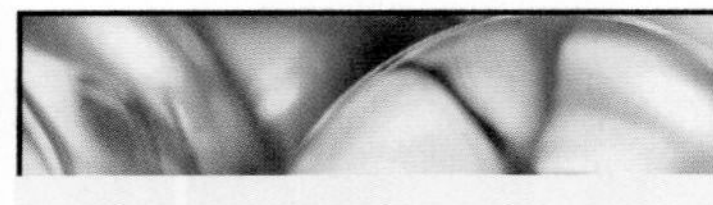

**KEY CONCEPTS**

Congruent Triangles • SSS • SAS • ASA • AAS • Included Side • Included Angle • Reflexive Property of Congruence (Identity) • Symmetric and Transitive Properties of Congruence

# 3.1 Congruent Triangles

Two triangles are **congruent** if one coincides with (fits perfectly over) the other. In Figure 3.1, we say that $\triangle ABC \cong \triangle DEF$ if these congruences hold:

$$\begin{array}{ll} \angle A \cong \angle D & \overline{AB} \cong \overline{DE} \\ \angle B \cong \angle E & \overline{BC} \cong \overline{EF} \\ \angle C \cong \angle F & \overline{AC} \cong \overline{DF} \end{array}$$

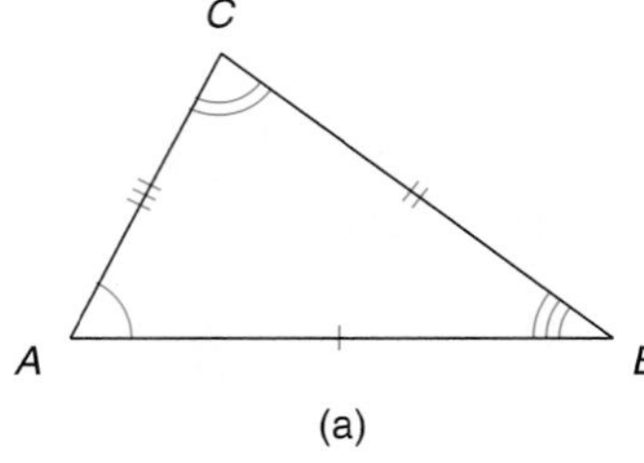


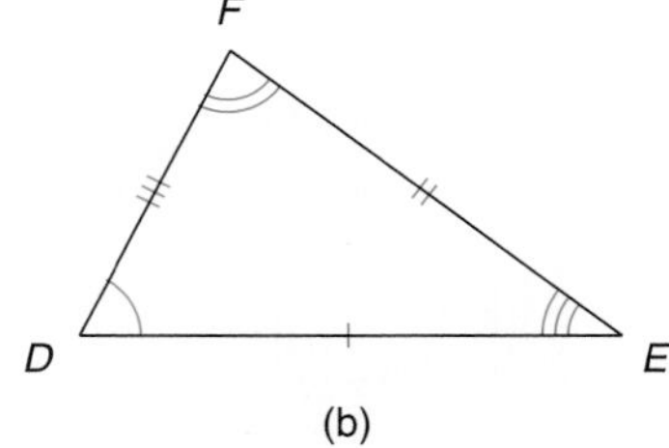


**FIGURE 3.1**

From the indicated congruences, we also say that vertex $A$ corresponds to vertex $D$, as does $B$ to $E$ and $C$ to $F$. In symbols, the correspondences are represented by

$$A \leftrightarrow D \qquad B \leftrightarrow E \qquad C \leftrightarrow F$$

Recall that in Section 2.6, we used a slide transformation on $\triangle ABC$ to form its image $\triangle DEF$.

The claim $\triangle MNQ \cong \triangle RST$ orders corresponding vertices of the triangles (not shown), so we can conclude from this statement that

$$M \leftrightarrow R, \qquad N \leftrightarrow S, \qquad \text{and} \qquad Q \leftrightarrow T$$

This correspondence of vertices implies the congruence of corresponding parts such as $\angle M \cong \angle R$ and $\overline{NQ} \cong \overline{ST}$.

Conversely, if the correspondence of vertices of two congruent triangles is $M \leftrightarrow R$, $N \leftrightarrow S$, and $Q \leftrightarrow T$, we can write $\triangle MNQ \cong \triangle RST$, $\triangle NQM \cong \triangle STR$, and so on.

> DEFINITION: Two triangles are **congruent** when the six parts of the first triangle are congruent to the six corresponding parts of the second triangle.

As always, any definition is reversible! If two triangles are known to be congruent, we may conclude that the corresponding parts are congruent. Moreover, if the six pairs of parts are known to be congruent, then so are the triangles! From the congruent parts indicated in Figure 3.2, we can conclude that $\triangle MNQ \cong \triangle RST$.

Again using the terminology introduced in Section 2.6, we note that in Figure 3.2, $\triangle TSR$ is the reflection of $\triangle QNM$ across a vertical line (not shown) midway between the two triangles.

On page 129 are some of the properties of congruent triangles that are useful in later proofs and explanations.

**Discover!**

Holding two sheets of construction paper together, use scissors to cut out two triangles. How do the triangles compare?

ANSWER

The triangles are congruent.

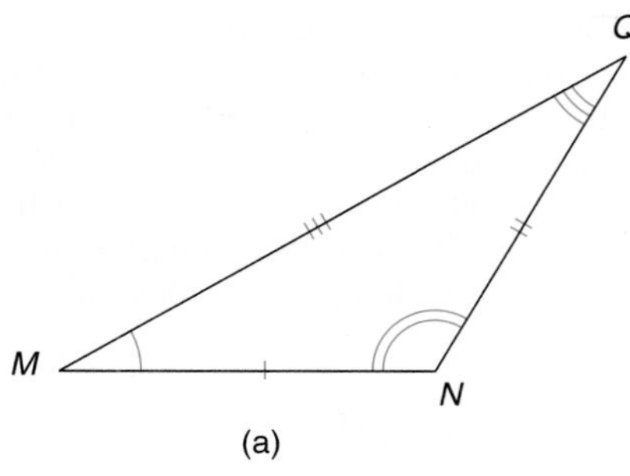


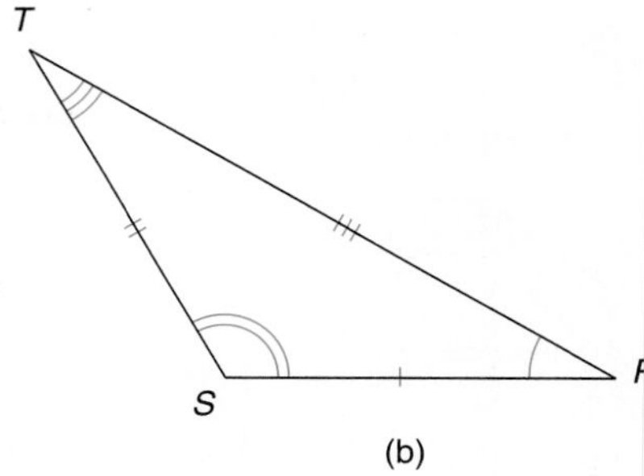


**FIGURE 3.2**

1. $\triangle ABC \cong \triangle ABC$ (Reflexive Property of Congruence)
2. If $\triangle ABC \cong \triangle DEF$, then $\triangle DEF \cong \triangle ABC$. (Symmetric Property of Congruence)
3. If $\triangle ABC \cong \triangle DEF$ and $\triangle DEF \cong \triangle GHI$, then $\triangle ABC \cong \triangle GHI$. (Transitive Property of Congruence)

*Exs. 1, 2*

On the basis of the properties above, we see that the "congruence of triangles" is an equivalence relation.

It would be difficult to establish that triangles were congruent if six pairs of congruent parts had to be verified first. Fortunately, it is possible to prove triangles congruent by establishing fewer than six pairs of congruences. To suggest a first method, consider the construction in Example 1.

EXAMPLE 1

Construct a triangle whose sides have the lengths of the segments provided in Figure 3.3(a).

***Solution***

Figure 3.3(b): Choose $\overline{AB}$ as the first side of the triangle and mark its length as shown.

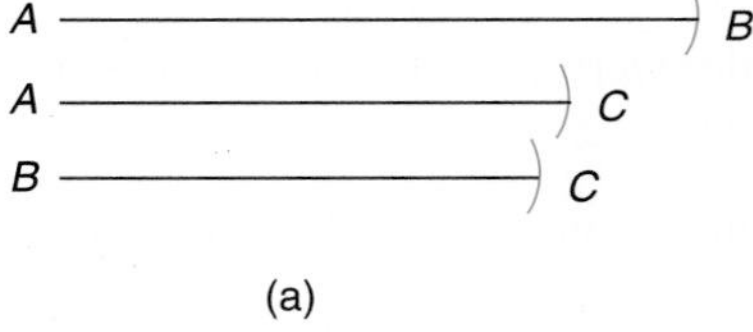


**FIGURE 3.3**

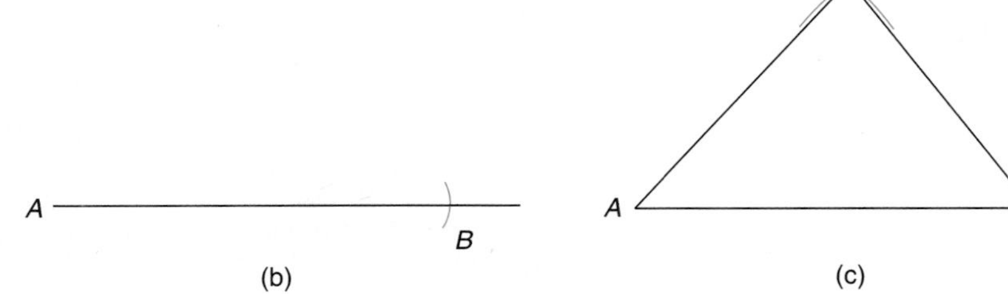


Figure 3.3(c): Using the left endpoint $A$, mark off an arc of length equal to that of $\overline{AC}$. Now mark off an arc the length of $\overline{BC}$ from the right endpoint $B$ so that these arcs intersect at $C$, the third vertex of the triangle. Joining point $C$ to $A$ and then to $B$ completes the desired triangle.

Geometry in the Real World

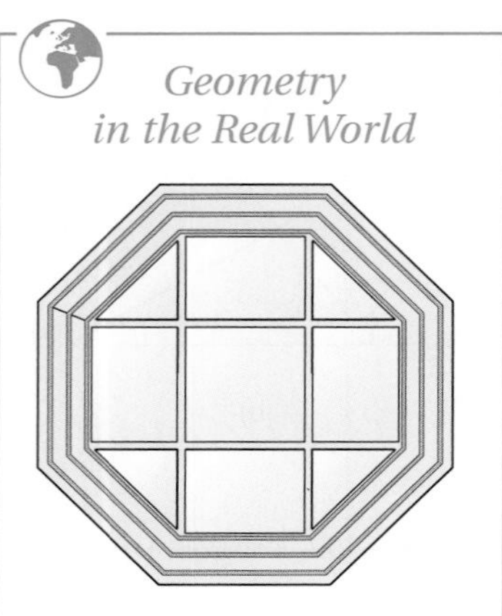

The four triangular panes in the octagonal window are congruent triangles.

Consider Example 1 once more. If a "different" triangle were constructed by choosing $\overline{AC}$ to be the first side, it would be congruent to the one shown. It might be necessary to flip or rotate it to have corresponding vertices match, but that is perfectly acceptable! The point of Example 1 is that it does provide a method for establishing the congruence of triangles, using only three pairs of parts. If corresponding angles are measured in the previous triangle or in any other triangle constructed with the same lengths for sides, these pairs of angles will also be congruent!

## SSS (METHOD FOR PROVING TRIANGLES CONGRUENT)

POSTULATE 12:
If the three sides of one triangle are congruent to the three sides of a second triangle, then the triangles are congruent (SSS).

The designation SSS will be cited as a reason in the proof that follows. The three S letters refer to the three *pairs* of congruent sides.

EXAMPLE 2

*Given:* $\overline{AB}$ and $\overline{CD}$ bisect each other at $M$
$\overline{AC} \cong \overline{DB}$
(See Figure 3.4.)

*Prove:* $\triangle AMC \cong \triangle BMD$

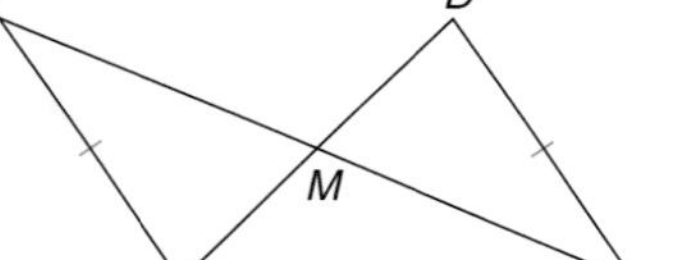


**FIGURE 3.4**

**PROOF**

| Statements | Reasons |
|---|---|
| 1. $\overline{AB}$ and $\overline{CD}$ bisect each other at $M$ | 1. Given |
| 2. $\overline{AM} \cong \overline{MB}$<br>$\overline{CM} \cong \overline{MD}$ | 2. If a segment is bisected, the segments formed are $\cong$ |
| 3. $\overline{AC} \cong \overline{DB}$ | 3. Given |
| 4. $\triangle AMC \cong \triangle BMD$ | 4. SSS |

NOTE 1: In steps 2 and 3, the three pairs of sides were shown to be congruent; thus SSS is cited to justify that $\triangle AMC \cong \triangle BMD$.

NOTE 2: $\triangle BMD$ is the image determined by the rotation of $\triangle AMC$ about point $M$ through a 180° angle.

The two sides that form an angle of a triangle are said to **include that angle** of the triangle. In $\triangle TUV$ in Figure 3.5(a), sides $\overline{TU}$ and $\overline{TV}$ form $\angle T$; therefore, $\overline{TU}$ and $\overline{TV}$ include $\angle T$. In turn, $\angle T$ is said to be the included angle for $\overline{TU}$ and $\overline{TV}$. Similarly, any two angles of a triangle must have a common side, and these two angles are said to **include that side.** In $\triangle TUV$, $\angle U$ and $\angle T$ share the common side $\overline{UT}$; therefore, $\angle U$ and $\angle T$ include the side $\overline{UT}$. $\overline{UT}$ is the side included by $\angle U$ and $\angle T$.

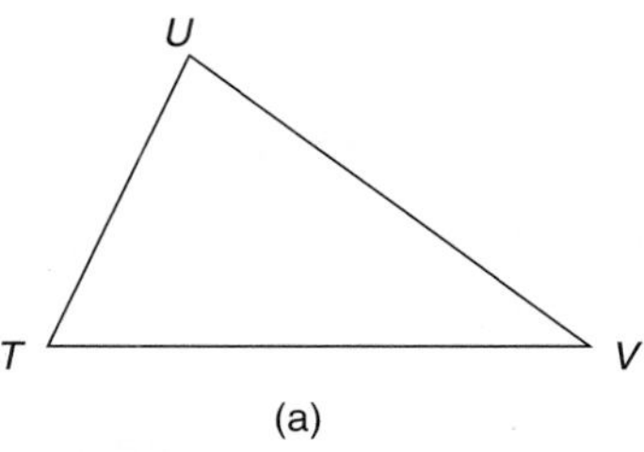

(a)

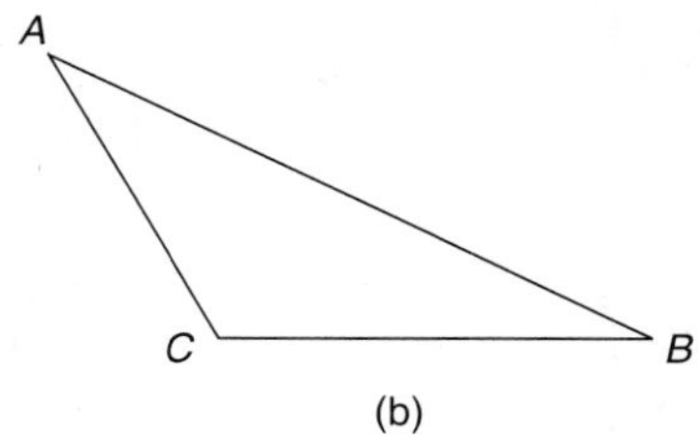

(b)

**FIGURE 3.5**

EXAMPLE 3

In $\triangle ABC$ of Figure 3.5(b):

a) Which angle is included by $\overline{AC}$ and $\overline{CB}$?
b) Which sides include $\angle B$?
c) What is the included side for $\angle A$ and $\angle B$?
d) Which angles include $\overline{CB}$?

***Solution***

a) $\angle C$ (because it is formed by $\overline{AC}$ and $\overline{CB}$)
b) $\overline{AB}$ and $\overline{BC}$ (because these form $\angle B$)
c) $\overline{AB}$ (because it is the common side for $\angle A$ and $\angle B$)
d) $\angle C$ and $\angle B$ (because $\overline{CB}$ is a side of each angle)

## SAS (METHOD FOR PROVING TRIANGLES CONGRUENT)

A second way of establishing that two triangles are congruent involves showing that two sides and the included angle of one triangle are congruent to two sides and the included angle of a second triangle. If two people each draw a triangle so that two of the sides measure 2 cm and 3 cm and the included angle measures 54°, then those triangles are congruent. (See Figure 3.6.)

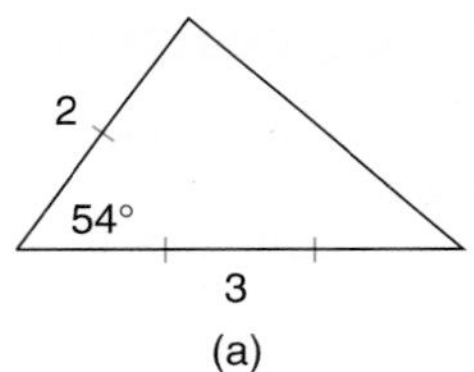

(a)

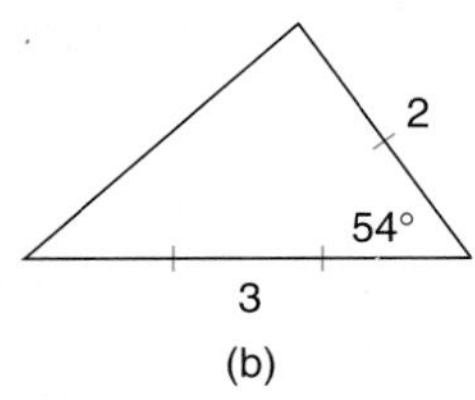

(b)

**FIGURE 3.6**

POSTULATE 13:
If two sides and the included angle of one triangle are congruent to two sides and the included angle of a second triangle, then the triangles are congruent (SAS).

The order of the letters SAS in Postulate 13 helps us remember that the two sides that are named have the angle "between" them. That is, in each triangle, the two sides form the angle.

In Example 4, which follows, the two triangles to be proved congruent share a common side; the statement $\overline{PN} \cong \overline{PN}$ is justified by the Reflexive Property of Congruence, which is conveniently expressed as **Identity.** However stated, this justification applies when triangles (or perhaps other polygons) have a part in common.

DEFINITION: In this context, **Identity** is the reason we cite when verifying that a line segment (or an angle) is congruent to itself; also known as the Reflexive Property of Congruence.

In Example 4, note the use of Identity and SAS as the final reasons.

EXAMPLE 4

*Given:* $\overline{PN} \perp \overline{MQ}$
$\overline{MN} \cong \overline{NQ}$
(See Figure 3.7.)

*Prove:* $\triangle PNM \cong \triangle PNQ$

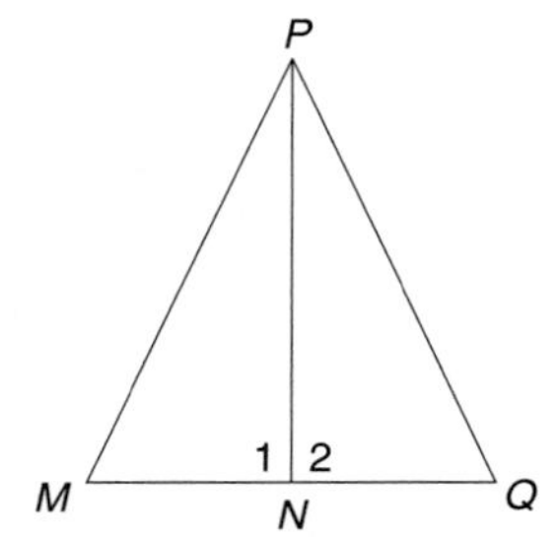


FIGURE 3.7

PROOF

| Statements | Reasons |
|---|---|
| 1. $\overline{PN} \perp \overline{MQ}$ | 1. Given |
| 2. $\angle 1 \cong \angle 2$ | 2. If two lines are $\perp$, they meet to form $\cong$ adjacent $\angle$s |
| 3. $\overline{MN} \cong \overline{NQ}$ | 3. Given |
| 4. $\overline{PN} \cong \overline{PN}$ | 4. Identity (or Reflexive) |
| 5. $\triangle PNM \cong \triangle PNQ$ | 5. SAS |

NOTE: In $\triangle PNM$, $\overline{MN}$ (step 3) and $\overline{PN}$ (step 4) include $\angle 1$; similarly, $\overline{NQ}$ and $\overline{PN}$ include $\angle 2$ in $\triangle PNQ$. Thus SAS is used to verify that $\triangle PNM \cong \triangle PNQ$ in reason 5.

*Exs. 3–6*

## ASA (METHOD FOR PROVING TRIANGLES CONGRUENT)

The next method for proving triangles congruent requires a combination of two angles and the included side. If two people each draw a triangle for which two of the angles measure 33° and 47° and the included side measures 5 centimeters, then those triangles are congruent. See Figure 3.8.

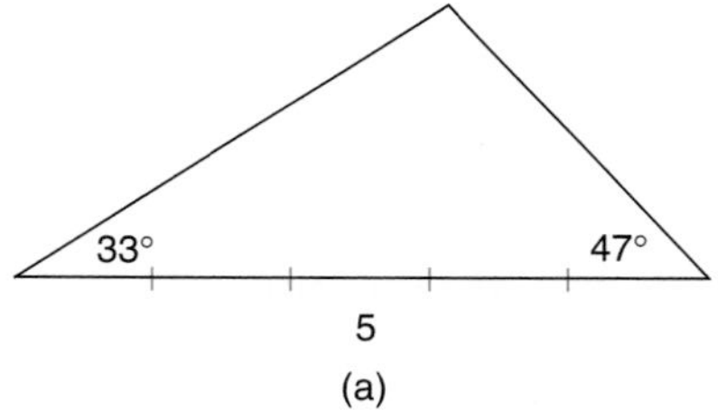


(a)

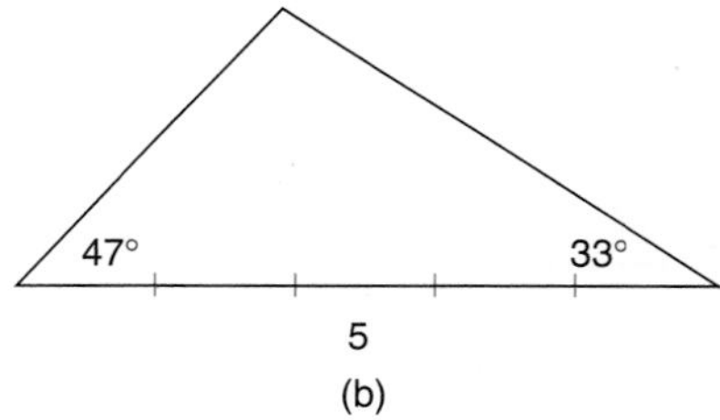


(b)

FIGURE 3.8

POSTULATE 14:
If two angles and the included side of one triangle are congruent to two angles and the included side of a second triangle, then the triangles are congruent (ASA).

Although this method is written compactly as ASA, you must be careful as you write these abbreviations! For example, ASA refers to two angles and the included side, whereas SAS refers to two sides and the included angle. For us to use either postulate, the specific conditions described in it must be satisfied.

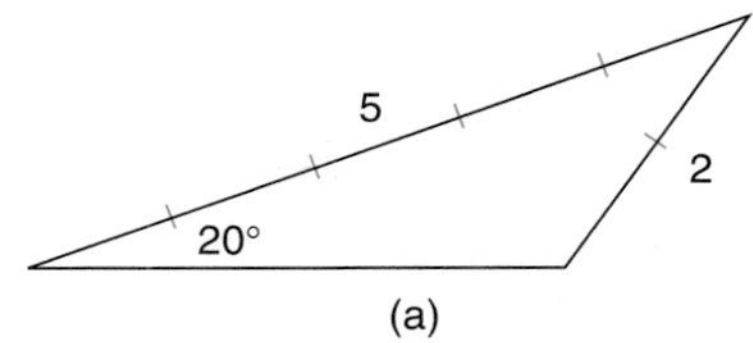


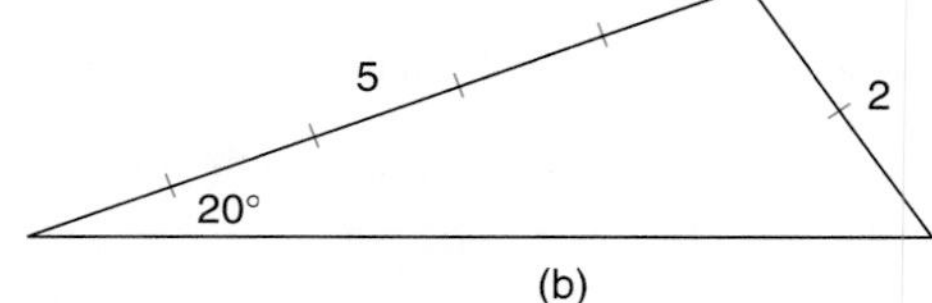


**FIGURE 3.9**

SSS, SAS, and ASA are all valid methods of proving triangles congruent, but SSA is *not* a method and *cannot* be used. In Figure 3.9, the two triangles are marked to show SSA, and yet the two triangles are *not* congruent.

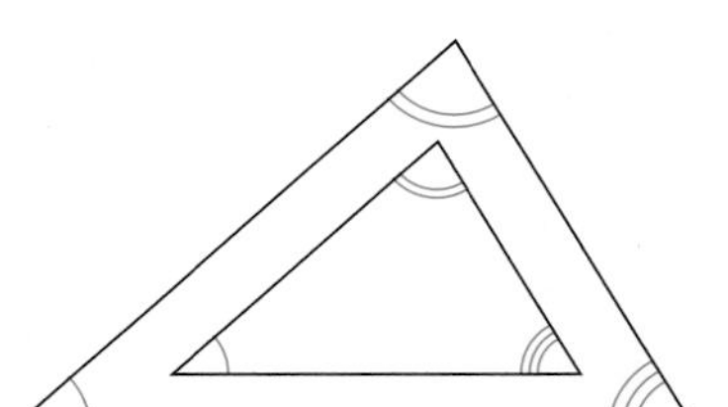

**FIGURE 3.10**

Another combination that cannot be used to prove triangles congruent is AAA. See Figure 3.10. Three congruent pairs of angles in two triangles do not guarantee congruent pairs of sides!

*Exs. 7–11*

In Example 5, the triangles to be proved congruent overlap (see Figure 3.11). To clarify relationships, the triangles have been redrawn separately in Figure 3.12. Note the parts marked congruent as established in the proof. Note that Identity (or Reflexive) can also be used to say that an angle is congruent to itself.

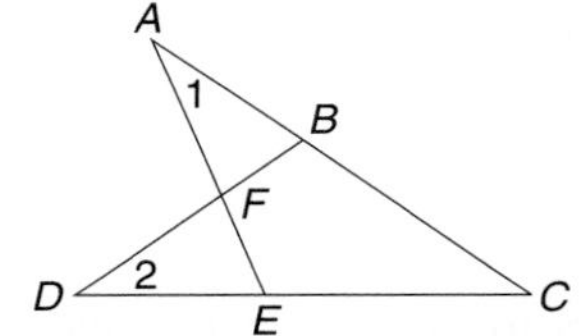


**FIGURE 3.11**

EXAMPLE 5

*Given:* $\overline{AC} \cong \overline{DC}$
$\angle 1 \cong \angle 2$
(See Figure 3.11.)

*Prove:* $\triangle ACE \cong \triangle DCB$

**PROOF**

| Statements | Reasons |
|---|---|
| 1. $\overline{AC} \cong \overline{DC}$ (See Figure 3.12.) | 1. Given |
| 2. $\angle 1 \cong \angle 2$ | 2. Given |
| 3. $\angle C \cong \angle C$ | 3. Identity |
| 4. $\triangle ACE \cong \triangle DCB$ | 4. ASA |

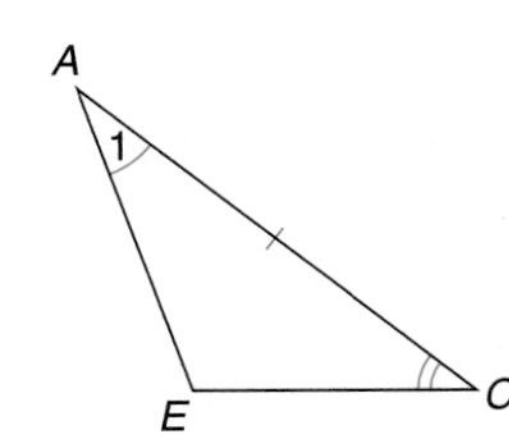


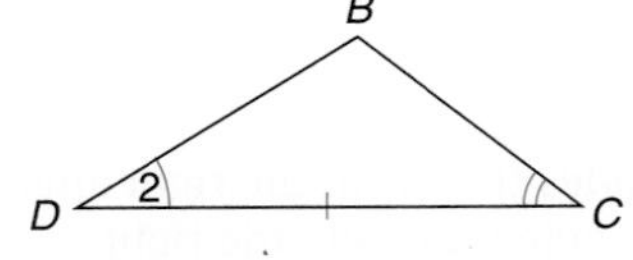


**FIGURE 3.12**

Next we consider a theorem (proved by the ASA postulate) that is convenient as a reason in many proofs.

## AAS (METHOD FOR PROVING TRIANGLES CONGRUENT)

THEOREM 3.1.1: If two angles and a nonincluded side of one triangle are congruent to two angles and a nonincluded side of a second triangle, then the triangles are congruent (AAS).

*Given:* $\angle T \cong \angle K$, $\angle S \cong \angle J$, and $\overline{SR} \cong \overline{HJ}$ (See Figure 3.13 on page 134.)

*Prove:* $\triangle TSR \cong \triangle KJH$

Warning (!)
Do not use AAA or SSA, because they are simply not valid for proving triangles congruent; with AAA the triangles have the same shape but are not necessarily congruent.

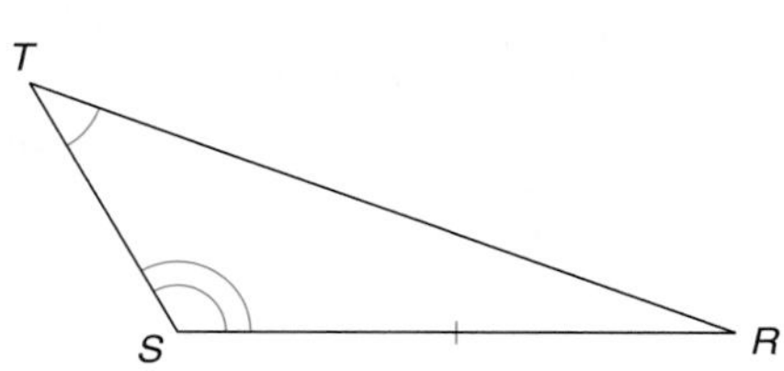

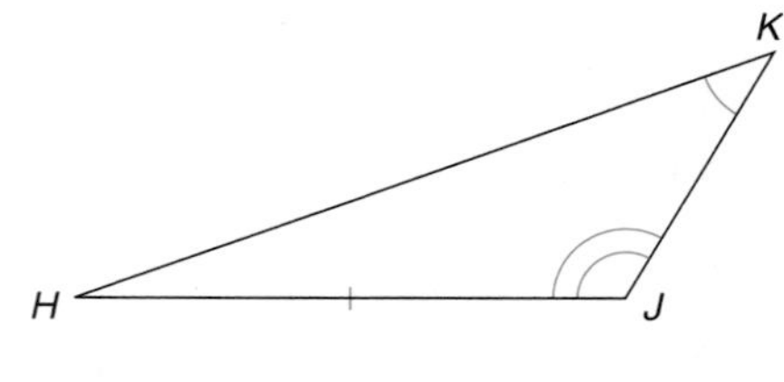


**FIGURE 3.13**

**PROOF**

| Statements | Reasons |
|---|---|
| 1. $\angle T \cong \angle K$<br>$\angle S \cong \angle J$ | 1. Given |
| 2. $\angle R \cong \angle H$ | 2. If two $\angle$s of one $\triangle$ are $\cong$ to two $\angle$s of another $\triangle$, then the third $\angle$s are also congruent |
| 3. $\overline{SR} \cong \overline{HJ}$ | 3. Given |
| 4. $\triangle TSR \cong \triangle KJH$ | 4. ASA |

*Exs. 12–14*

In summary, you may use SSS, SAS, ASA, or AAS to prove that triangles are congruent. Identity (or Reflexive) may be used to state a self-congruence when a side or angle is common to two triangles.

## 3.1 Exercises

*In Exercises 1 to 8, use the drawings provided to answer each question.*

1. Name a common angle and a common side for $\triangle ABC$ and $\triangle ABD$. If $\overline{BC} \cong \overline{BD}$, can you conclude that $\triangle ABC$ and $\triangle ABD$ are congruent? Can SSA be used as a reason for proving triangles congruent?

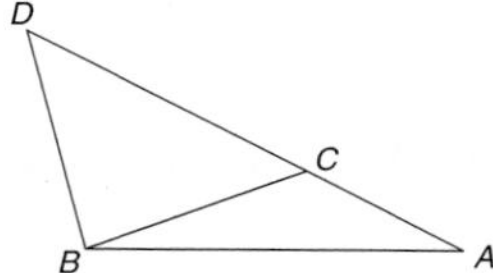


*For Exercises 2 and 3, see the figure in the second column.*

2. With corresponding angles indicated, the triangles are congruent. Find values for *a*, *b*, and *c*.

3. With corresponding angles indicated, find m$\angle A$ if m$\angle F = 72°$.

4. With corresponding angles indicated, find m$\angle E$ if m$\angle A = 57°$ and m$\angle C = 85°$.

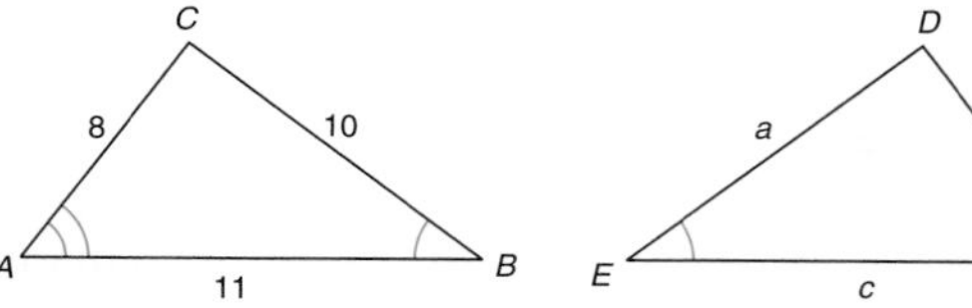


*Exercises 2–4, 6*

5. In a right triangle, the sides that form the right angle are the **legs;** the longest side (opposite the right angle) is the **hypotenuse.** Some textbooks say that when two right triangles have congruent pairs of legs, the right triangles are congruent by the reason LL. In our work, LL is just a special case of one of the postulates in this section. Which postulate is that?

6. In the figure for Exercise 2, write a statement that the triangles are congruent, paying due attention to the order of corresponding vertices.

7. In $\triangle ABC$, the midpoints of the sides are joined. What does intuition tell you about the relationship between $\triangle AED$ and $\triangle FDE$? (We will prove this relationship later.)

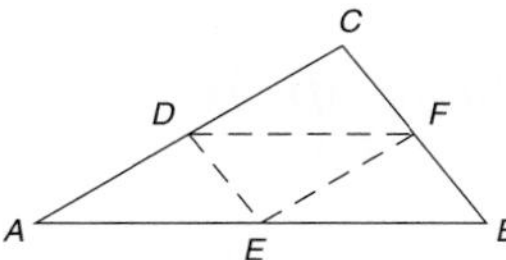


8. Suppose that you wish to prove that $\triangle RST \cong \triangle SRV$. Using the reason Identity, name one pair of corresponding parts that are congruent.

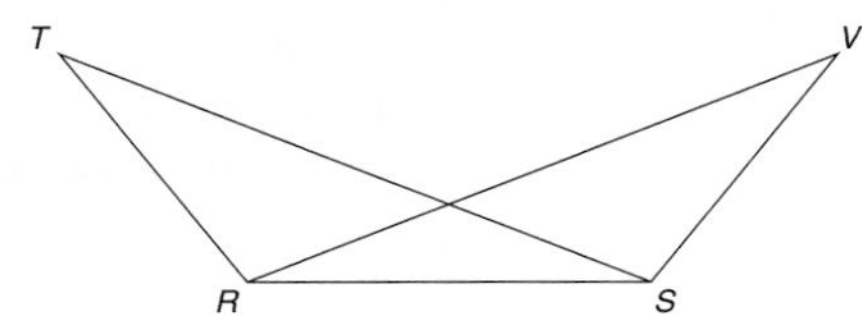


*In Exercises 9 to 12, congruent parts are indicated by like dashes (sides) or arcs (angles). State which method (SSS, SAS, ASA, or AAS) would be used to prove the two triangles congruent.*

9.

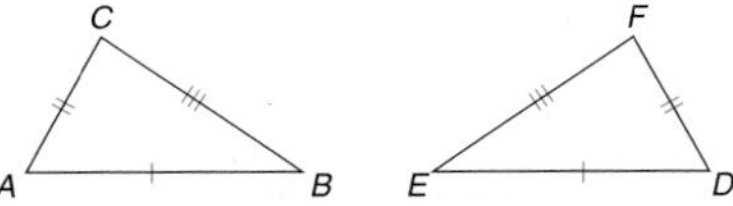


10.

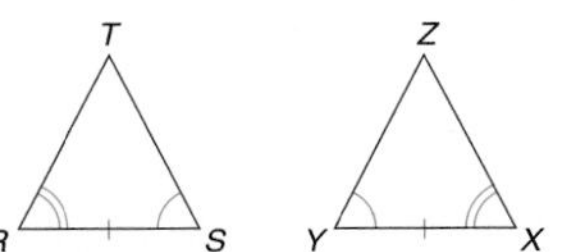


11.

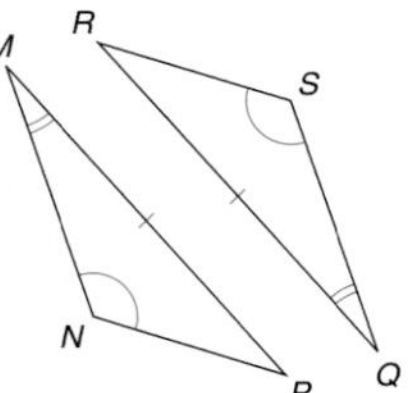


12.

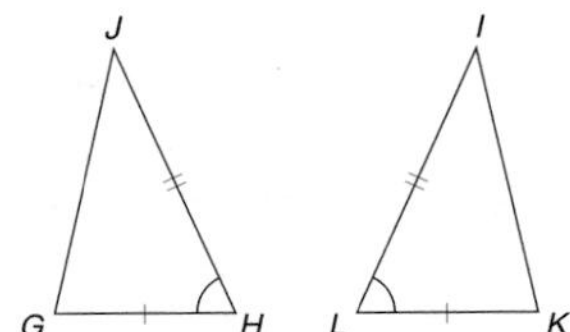


*In Exercises 13 to 18, use only the given information to state the reason why $\triangle ABC \cong \triangle DBC$. Redraw the figure and use marks like those used in Exercises 9 to 12.*

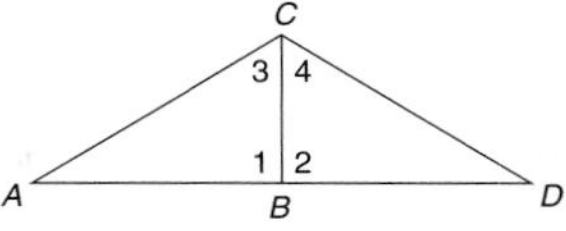


*Exercises 13–18*

13. $\angle A \cong \angle D$, $\overline{AB} \cong \overline{BD}$, and $\angle 1 \cong \angle 2$

14. $\angle A \cong \angle D$, $\overline{AC} \cong \overline{CD}$, and $B$ is the midpoint of $\overline{AD}$

15. $\angle A \cong \angle D$, $\overline{AC} \cong \overline{CD}$, and $\overrightarrow{CB}$ bisects $\angle ACD$

16. $\angle A \cong \angle D$, $\overline{AC} \cong \overline{CD}$, and $\overline{AB} \cong \overline{BD}$

17. $\overline{AC} \cong \overline{CD}$, $\overline{AB} \cong \overline{BD}$, and $\overline{CB} \cong \overline{CB}$ (by Identity)

18. $\angle 1$ and $\angle 2$ are right $\angle$s, $\overline{AB} \cong \overline{BD}$, and $\angle A \cong \angle D$

*In Exercises 19 and 20, the triangles to be proved congruent have been redrawn separately. Congruent parts are marked.*

a) *Name an additional pair of parts that are congruent by Identity.*

b) *Considering the congruent parts, state the reason why the triangles must be congruent.*

19. $\triangle ABC \cong \triangle AED$

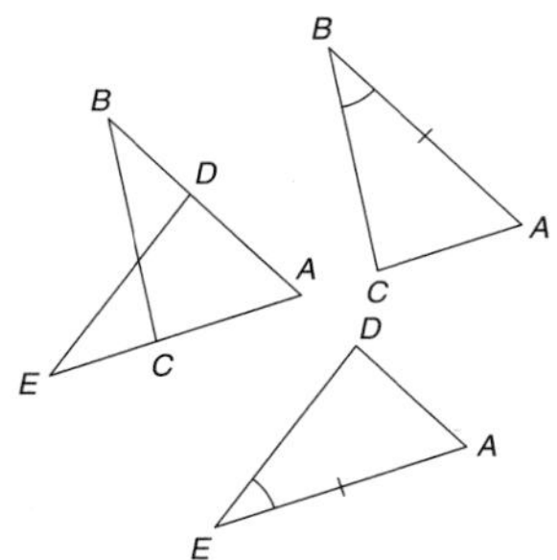


20. $\triangle MNP \cong \triangle MQP$

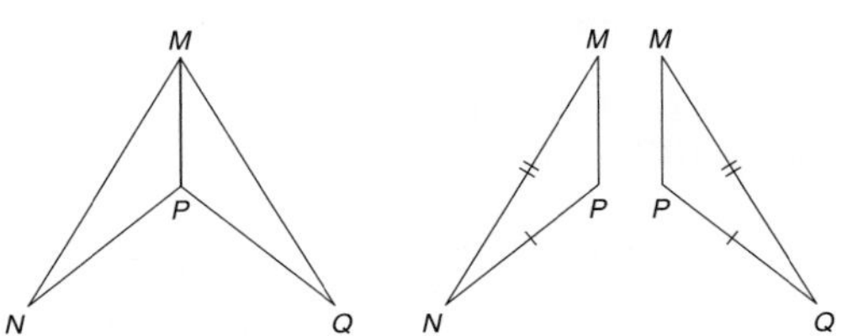

*In Exercises 21 to 24, the triangles named can be proven congruent. Considering the congruent pairs marked, name the additional pair of parts that must be congruent for us to use the method named.*

21. SAS

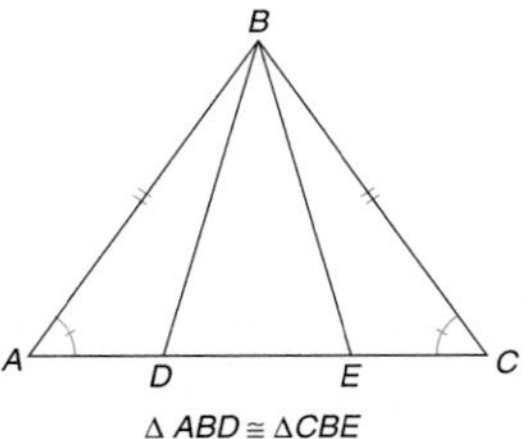


22. ASA

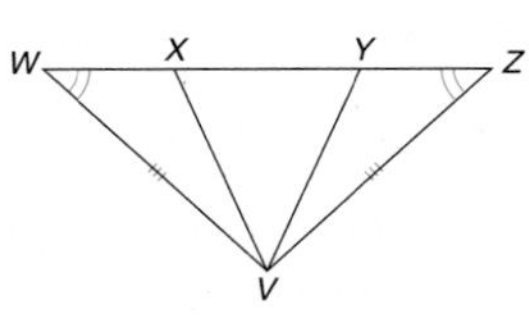


23. SSS

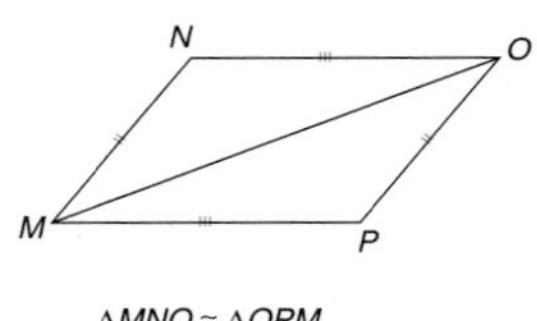


24. AAS

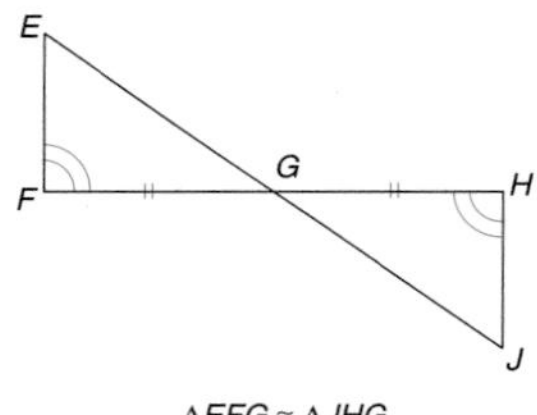


*In Exercises 25 and 26, complete each proof. Use the figure in the second column.*

25. *Given:* $\overline{AB} \cong \overline{CD}$ and $\overline{AD} \cong \overline{CB}$
*Prove:* $\triangle ABC \cong \triangle CDA$

**PROOF**

| Statements | Reasons |
|---|---|
| 1. $\overline{AB} \cong \overline{CD}$ and $\overline{AD} \cong \overline{CB}$ | 1. ? |
| 2. ? | 2. Identity |
| 3. $\triangle ABC \cong \triangle CDA$ | 3. ? |

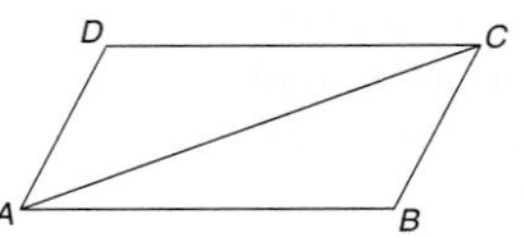


*Exercises 25, 26*

26. *Given:* $\overline{DC} \parallel \overline{AB}$ and $\overline{AD} \parallel \overline{BC}$
*Prove:* $\triangle ABC \cong \triangle CDA$

**PROOF**

| Statements | Reasons |
|---|---|
| 1. $\overline{DC} \parallel \overline{AB}$ | 1. ? |
| 2. $\angle DCA \cong \angle BAC$ | 2. ? |
| 3. ? | 3. Given |
| 4. ? | 4. If two $\parallel$ lines are cut by a transversal, alt. int. $\angle$s are $\cong$ |
| 5. $\overline{AC} \cong \overline{AC}$ | 5. ? |
| 6. ? | 6. ASA |

*In Exercises 27 to 32, use SSS, SAS, ASA, or AAS to prove that the triangles are congruent.*

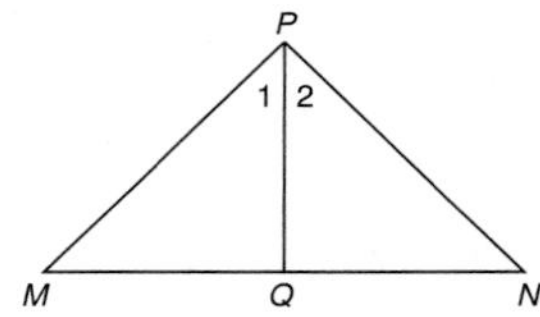


*Exercises 27, 28*

27. *Given:* $\overrightarrow{PQ}$ bisects $\angle MPN$
$\overline{MP} \cong \overline{NP}$
*Prove:* $\triangle MQP \cong \triangle NQP$

28. *Given:* $\overline{PQ} \perp \overline{MN}$ and $\angle 1 \cong \angle 2$
*Prove:* $\triangle MQP \cong \triangle NQP$

29. *Given:* $\overline{AB} \perp \overline{BC}$ and $\overline{AB} \perp \overline{BD}$
$\overline{BC} \cong \overline{BD}$
*Prove:* $\triangle ABC \cong \triangle ABD$

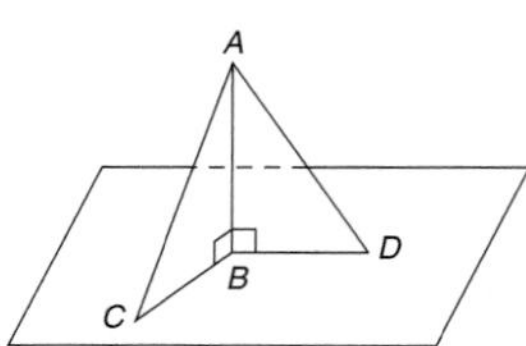

30. *Given:* $\overline{PN}$ bisects $\overline{MQ}$
$\angle M$ and $\angle Q$ are right angles
*Prove:* $\triangle PQR \cong \triangle NMR$

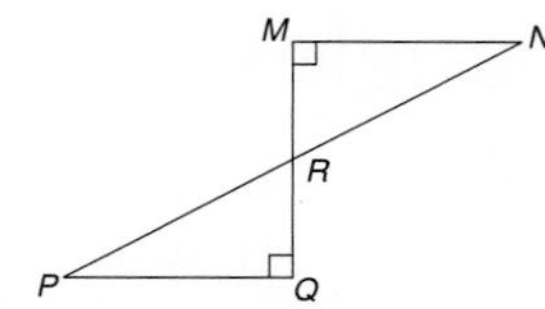


31. *Given:* $\angle VRS \cong \angle TSR$ and $\overline{RV} \cong \overline{TS}$
*Prove:* $\triangle RST \cong \triangle SRV$

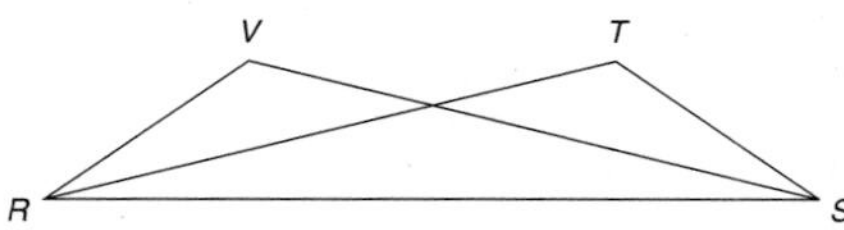


*Exercises 31, 32*

32. *Given:* $\overline{VS} \cong \overline{TR}$ and $\angle TRS \cong \angle VSR$
*Prove:* $\triangle RST \cong \triangle SRV$

*In Exercises 33 to 36, the methods to be used are SSS, SAS, ASA, and AAS.*

33. Given that $\triangle RST \cong \triangle RVU$, does it follow that $\triangle RSU$ is also congruent to $\triangle RVT$? Name the method, if any, used in arriving at this conclusion.

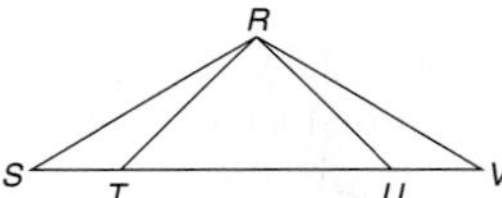


*Exercises 33, 34*

34. Given that $\angle S \cong \angle V$ and $\overline{ST} \cong \overline{UV}$, does it follow that $\triangle RST \cong \triangle RVU$? Which method, if any, did you use?

35. Given that $\angle A \cong \angle E$ and $\angle B \cong \angle D$, does it follow that $\triangle ABC \cong \triangle EDC$? If so, cite the method used in arriving at this conclusion.

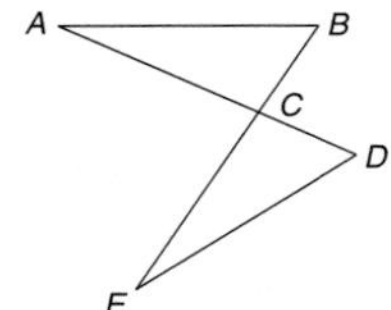


*Exercises 35, 36*

36. Given that $\angle A \cong \angle E$ and $\overline{BC} \cong \overline{DC}$, does it follow that $\triangle ABC \cong \triangle EDC$? Cite the method, if any, used in reaching this conclusion.

37. In quadrilateral $ABCD$, $\overline{AC}$ and $\overline{BD}$ are perpendicular bisectors of each other. Name all triangles that are congruent to:

a) $\triangle ABE$ b) $\triangle ABC$ c) $\triangle ABD$

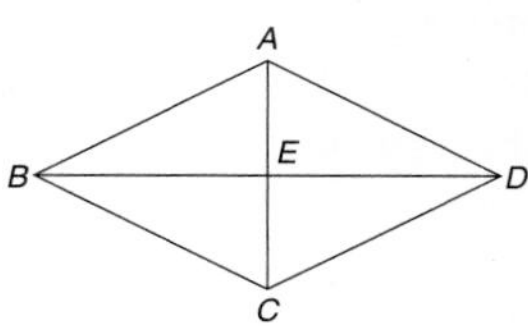


38. In $\triangle ABC$ and $\triangle DEF$, you know that $\angle A \cong \angle D$, $\angle C \cong \angle F$, and $\overline{AB} \cong \overline{DE}$. Before concluding that the triangles are congruent by ASA, you need to show that $\angle B \cong \angle E$. State the postulate or theorem that allows you to confirm this statement ($\angle B \cong \angle E$).

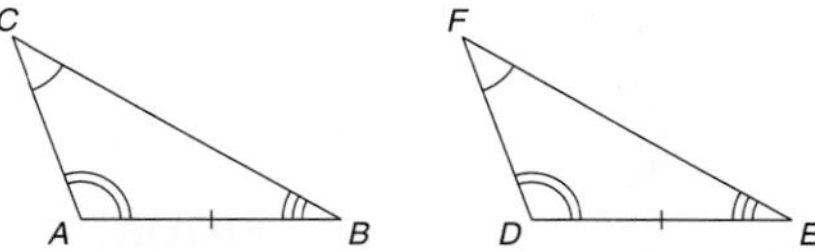


*In Exercises 39 and 40, complete each proof.*

39. *Given:* Plane $M$
$C$ is the midpoint of $\overline{EB}$
$\overline{AD} \perp \overline{BE}$ and $\overline{AB} \parallel \overline{ED}$
*Prove:* $\triangle ABC \cong \triangle DEC$

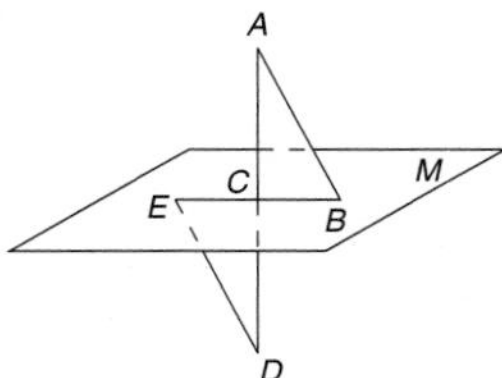


40. *Given:* $\overline{SP} \cong \overline{SQ}$ and $\overline{ST} \cong \overline{SV}$
*Prove:* $\triangle SPV \cong \triangle SQT$ and $\triangle TPQ \cong \triangle VQP$

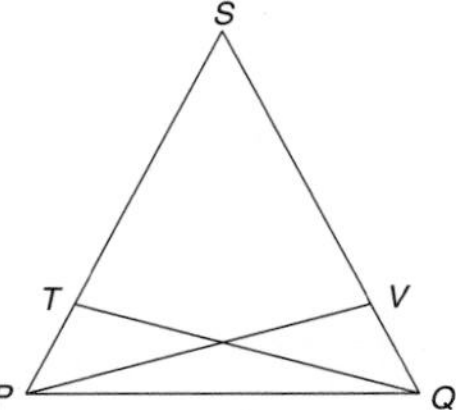

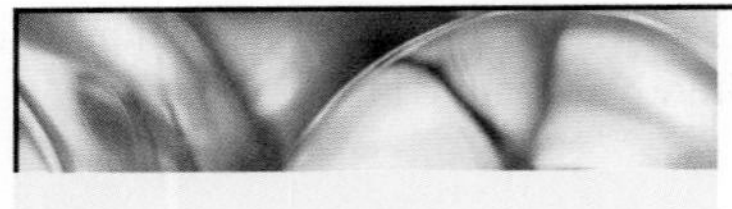

**KEY CONCEPTS**

CPCTC • Hypotenuse and Legs of a Right Triangle • HL • Pythagorean Theorem • Square Roots Property

# 3.2 Corresponding Parts of Congruent Triangles

Recall that the definition of congruent triangles states that *all* six parts (three sides and three angles) of one triangle are congruent respectively to the six corresponding parts of the second triangle. If we have proved that $\triangle ABC \cong \triangle DEF$ by SAS (the congruent parts are marked in Figure 3.14), then we can draw conclusions such as $\angle C \cong \angle F$ and $\overline{AC} \cong \overline{DF}$. The following reason is often cited for drawing such conclusions and is based on the definition of congruent triangles.

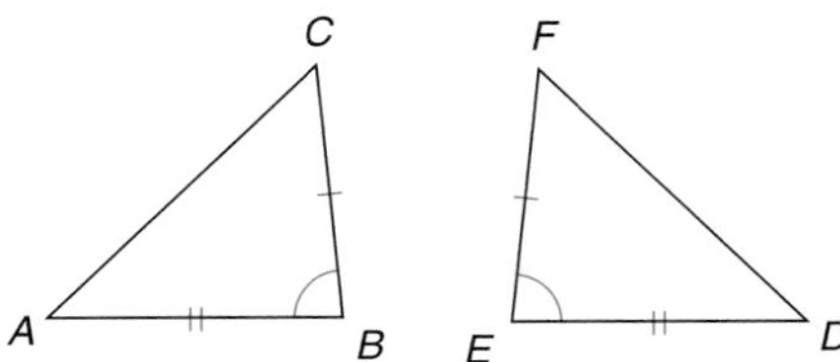


**FIGURE 3.14**

*Exs. 1–3*

> CPCTC: Corresponding parts of congruent triangles are congruent.

For triangles that have been proved congruent, CPCTC may be used to establish that either two line segments (corresponding sides) or two angles (corresponding angles) are congruent.

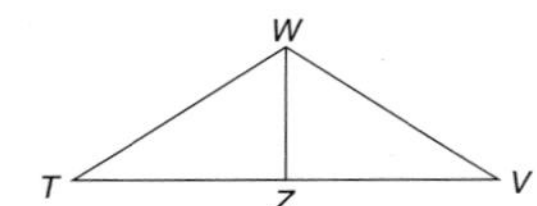


**FIGURE 3.15**

EXAMPLE 1

*Given:* $\overrightarrow{WZ}$ bisects $\angle TWV$
$\overline{WT} \cong \overline{WV}$
(See Figure 3.15.)

*Prove:* $\overline{TZ} \cong \overline{VZ}$

**PROOF**

| Statements | Reasons |
|---|---|
| 1. $\overrightarrow{WZ}$ bisects $\angle TWV$ | 1. Given |
| 2. $\angle TWZ \cong \angle VWZ$ | 2. The bisector of an angle separates it into two $\cong$ $\angle$s |
| 3. $\overline{WT} \cong \overline{WV}$ | 3. Given |
| 4. $\overline{WZ} \cong \overline{WZ}$ | 4. Identity |
| 5. $\triangle TWZ \cong \triangle VWZ$ | 5. SAS |
| 6. $\overline{TZ} \cong \overline{VZ}$ | 6. CPCTC |

**Reminder**

CPCTC means "corresponding parts of congruent triangles are congruent."

In Example 1, we could just as easily have used CPCTC to prove that two angles are congruent. If we had been asked to prove that $\angle T \cong \angle V$, then the final statement would have read

| | |
|---|---|
| 6. $\angle T \cong \angle V$ | 6. CPCTC |

We can take the proof in Example 1 a step further by proving triangles congruent and then using CPCTC to reach another conclusion, such as parallel or perpendicular lines. In Example 1, suppose we had been asked to prove that $\overline{WZ}$ bisects $\overline{TV}$. Then steps 1–6 would have remained as is, and a seventh step would have read

| | |
|---|---|
| 7. $\overline{WZ}$ bisects $\overline{TV}$ | 7. If a line segment is divided into two $\cong$ parts, then it has been bisected |

In our study of triangles, we will establish three types of conclusions:

1. *Proving triangles congruent,* such as $\triangle TWZ \cong \triangle VWZ$
2. *Proving corresponding parts of congruent triangles congruent,* like $\overline{TZ} \cong \overline{VZ}$ (Note that two $\triangle$s have to be proved $\cong$ before CPCTC can be used.)
3. *Establishing a further relationship,* like $\overline{WZ}$ bisects $\overline{TV}$ (Note that we must establish that two $\triangle$s are $\cong$ and also apply CPCTC before this goal can be reached.)

Little is said in this book about a "plan for proof," but every geometry student and teacher must have a plan before a proof can be completed. We demonstrate this "plan" in Example 2.

EXAMPLE 2

*Given:* $\overline{ZW} \cong \overline{YX}$
$\overline{ZY} \cong \overline{WX}$
(See Figure 3.16.)

*Prove:* $\overline{ZY} \parallel \overline{WX}$

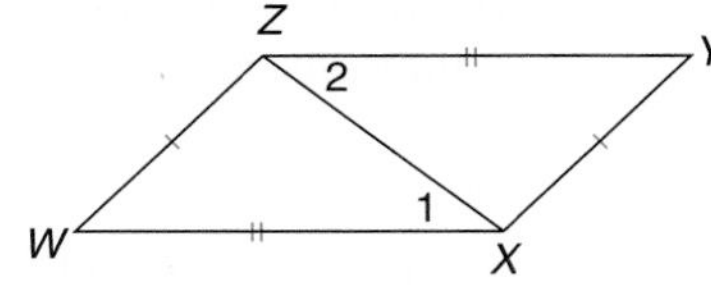


**FIGURE 3.16**

*Plan for Proof:* By showing that $\triangle ZWX \cong \triangle XYZ$, we can show that $\angle 1 \cong \angle 2$ by CPCTC. Then $\angle$s 1 and 2 are congruent alternate interior angles for $\overline{ZY}$ and $\overline{WX}$, which must be parallel.

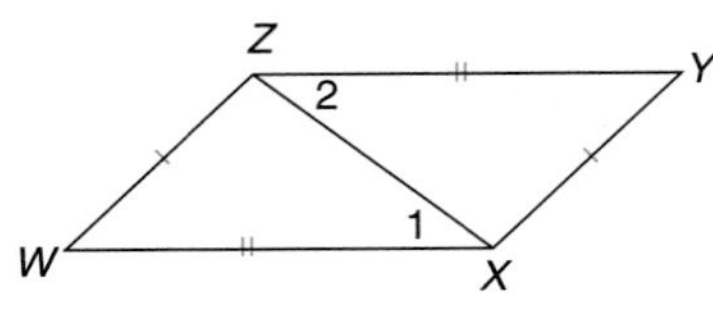


**FIGURE 3.16**

*Exs. 4–6*

PROOF

| Statements | Reasons |
|---|---|
| 1. $\overline{ZW} \cong \overline{YX}$; $\overline{ZY} \cong \overline{WX}$ | 1. Given |
| 2. $\overline{ZX} \cong \overline{ZX}$ | 2. Identity |
| 3. $\triangle ZWX \cong \triangle XYZ$ | 3. SSS |
| 4. $\angle 1 \cong \angle 2$ | 4. CPCTC |
| 5. $\overline{ZY} \parallel \overline{WX}$ | 5. If two lines are cut by a transversal so that the alt. int. $\angle$s are $\cong$, these lines are $\parallel$ |

## SUGGESTIONS FOR PROVING TRIANGLES CONGRUENT

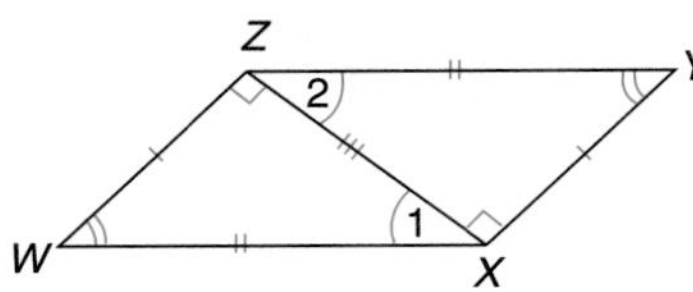


**FIGURE 3.17**

*Exs. 7–9*

Because many proofs depend on establishing congruent triangles, we offer the following suggestions.

Suggestions for a proof that involves congruent triangles:

1. Mark the figures systematically, using:
   a) A *square* in the opening of each right angle
   b) The same number of *dashes* on congruent sides
   c) The same number of *arcs* on congruent angles
2. Trace the triangles to be proved congruent in different colors.
3. If the triangles overlap, draw them separately.

NOTE: See Figure 3.17 for reference.

## RIGHT TRIANGLES

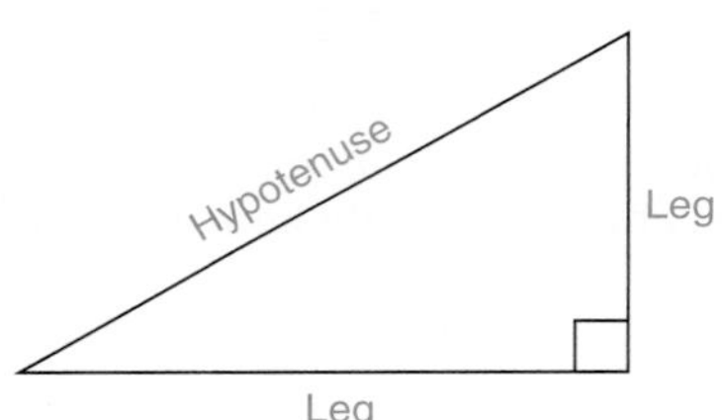


**FIGURE 3.18**

In a right triangle, the side opposite the right angle is the **hypotenuse** of the triangle, and the sides of the right angle are the **legs** of the triangle. These parts of a right triangle are illustrated in Figure 3.18.

In addition to the methods discussed earlier for proving triangles congruent, we also have the HL method, which applies exclusively to right triangles. In HL, H refers to hypotenuse and L refers to leg. The proof of this method will be delayed until Section 5.4.

## HL (METHOD FOR PROVING TRIANGLES CONGRUENT)

THEOREM 3.2.1: If the hypotenuse and a leg of one right triangle are congruent to the hypotenuse and a leg of a second right triangle, then the triangles are congruent (HL).

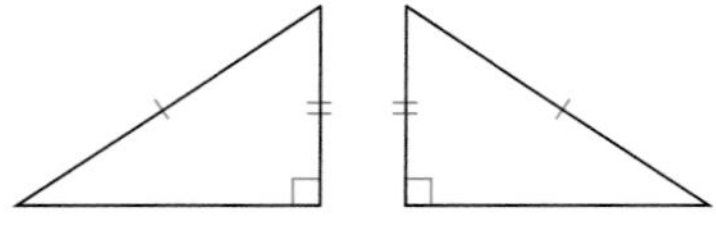

**FIGURE 3.19**

The relationship described in Theorem 3.2.1 is illustrated in Figure 3.19. In Example 3, the construction leads to a unique right triangle.

EXAMPLE 3

*Given:* $\overline{AB}$ and $\overline{CA}$ in Figure 3.20(a); note that $AB > CA$. (See page 141.)

*Construct:* The right triangle with hypotenuse of length equal to $AB$ and one leg of length equal to $CA$

***Solution***

Figure 3.20(b): Construct $\overleftrightarrow{CQ}$ perpendicular to $\overleftrightarrow{EF}$ at point $C$.
Figure 3.20(c): Now mark off the length of $\overline{CA}$ on $\overleftrightarrow{CQ}$.

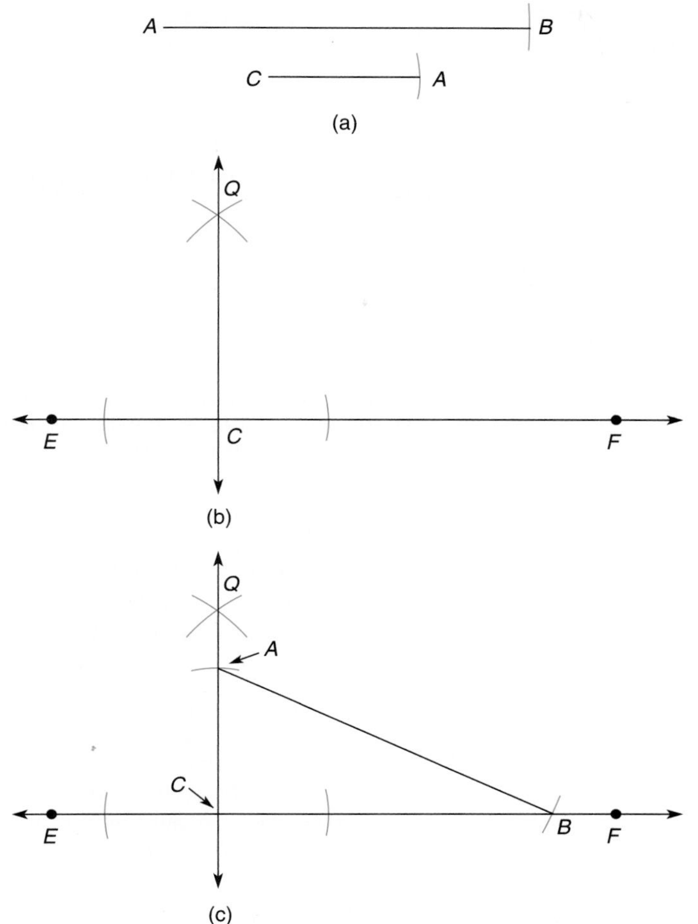


**FIGURE 3.20**

Finally, with point $A$ as center, mark off a length equal to that of $\overline{AB}$ as shown. $\triangle ABC$ is the desired right $\triangle$.

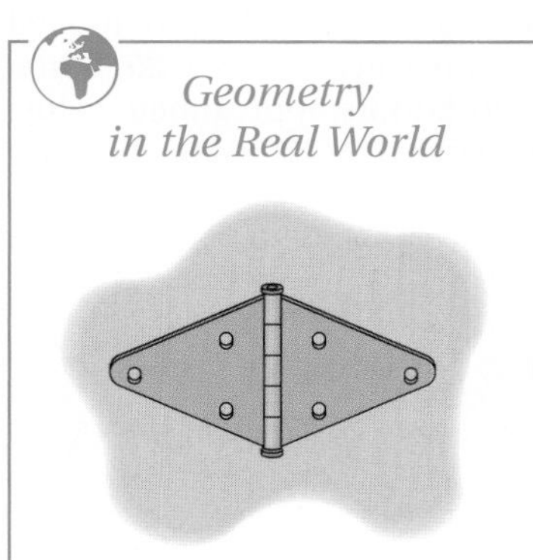


In the manufacturing process, the parts of many machines must be congruent. The two sides of the hinge shown are congruent.

EXAMPLE 4

Cite the reason why the right triangles $\triangle ABC$ and $\triangle ECD$ in Figure 3.21 are congruent if:

a) $\overline{AB} \cong \overline{EC}$ and $\overline{AC} \cong \overline{ED}$
b) $\angle A \cong \angle E$ and $C$ is the midpoint of $\overline{BD}$
c) $\overline{BC} \cong \overline{CD}$ and $\angle 1 \cong \angle 2$
d) $\overline{AB} \cong \overline{EC}$ and $\overline{EC}$ bisects $\overline{BD}$

***Solution***

a) HL b) AAS c) ASA d) SAS

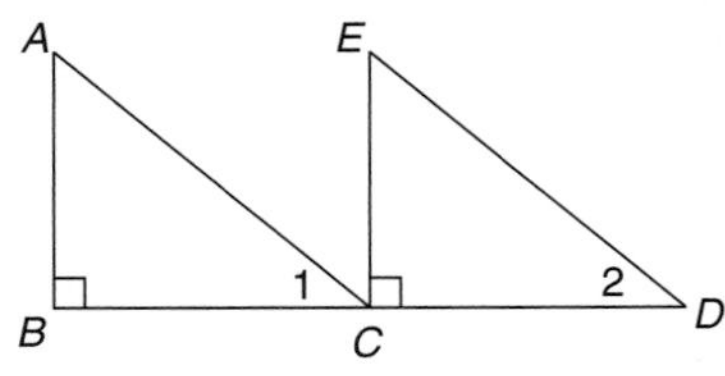


**FIGURE 3.21**

SSG *Exs. 10–11*

The following theorem can be applied only when a triangle is a right triangle (contains a right angle). Proof of the theorem is delayed until Section 5.4.

**Technology Exploration**

Computer software and a calculator are needed.

1. Form a right $\triangle ABC$ with $m\angle C = 90°$.
2. Measure $AB$, $AC$, and $BC$.
3. Show that $(AC)^2 + (BC)^2 = (AB)^2$.

(Answer will probably not be "perfect.")

PYTHAGOREAN THEOREM: The square of the length ($c$) of the hypotenuse of a right triangle equals the sum of squares of the lengths ($a$ and $b$) of the legs of the right triangle; that is, $c^2 = a^2 + b^2$.

In applications of the Pythagorean Theorem, we often arrive at statements such as $c^2 = 25$. Using the following property, we see that $c = \sqrt{25}$ *or* $c = 5$.

SQUARE ROOTS PROPERTY: Let $x$ represent the length of a line segment, and let $p$ represent a positive number. If $x^2 = p$, then $x = \sqrt{p}$.

The *square root* of $p$, symbolized $\sqrt{p}$, represents the number that when multiplied times itself equals $p$. As we indicated earlier, $\sqrt{25} = 5$ because $5 \times 5 = 25$. When a square root is not exact, a calculator can be used to find its approximate value; where the symbol $\approx$ means "is equal to approximately," $\sqrt{22} \approx 4.69$ because $4.69 \times 4.69 = 21.9961 \approx 22$.

EXAMPLE 5

Find the length of the third side of the right triangle. (See the figure below.)

a) Find $c$ if $a = 6$ and $b = 8$.
b) Find $b$ if $a = 7$ and $c = 10$.

***Solution***

a) $c^2 = a^2 + b^2$, so $c^2 = 6^2 + 8^2$ or $c^2 = 36 + 64 = 100$. Then $c = \sqrt{100} = 10$.
b) $c^2 = a^2 + b^2$, so $10^2 = 7^2 + b^2$ or $100 = 49 + b^2$. Subtracting yields $b^2 = 51$, so $b = \sqrt{51} \approx 7.14$.

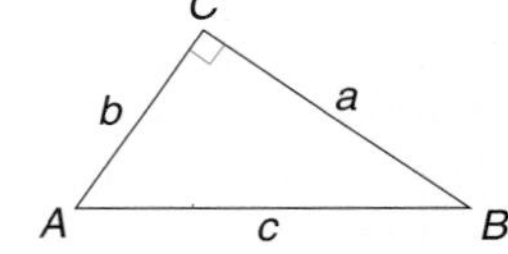

*Exs. 12–14*

## 3.2 Exercises

*In Exercises 1 to 8, plan and write the two-column proof for each problem.*

1. *Given:* $\angle 1$ and $\angle 2$ are right $\angle$s
   $\overline{CA} \cong \overline{DA}$
   *Prove:* $\triangle ABC \cong \triangle ABD$

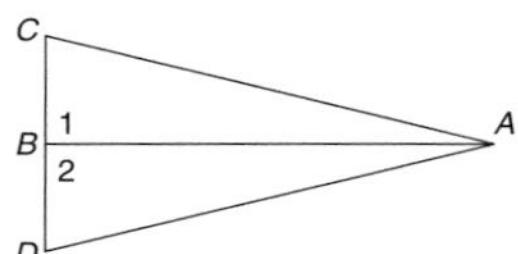


*Exercises 1, 2*

2. *Given:* $\angle 1$ and $\angle 2$ are right $\angle$s
   $\overrightarrow{AB}$ bisects $\angle CAD$
   *Prove:* $\triangle ABC \cong \triangle ABD$

3. *Given:* $P$ is the midpoint of both $\overline{MR}$ and $\overline{NQ}$
   *Prove:* $\triangle MNP \cong \triangle RQP$

4. *Given:* $\overline{MN} \parallel \overline{QR}$ and $\overline{MN} \cong \overline{QR}$
   *Prove:* $\triangle MNP \cong \triangle RQP$

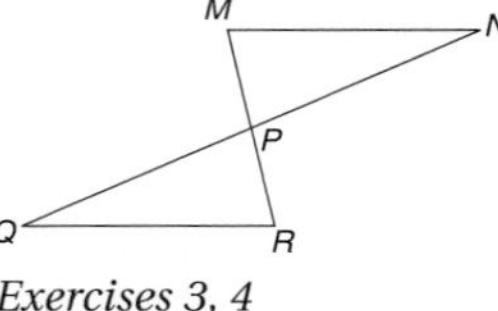


*Exercises 3, 4*

5. *Given:* $\angle R$ and $\angle V$ are right $\angle$s
   $\angle 1 \cong \angle 2$
   *Prove:* $\triangle RST \cong \triangle VST$

6. *Given:* $\angle 1 \cong \angle 2$ and $\angle 3 \cong \angle 4$
   *Prove:* $\triangle RST \cong \triangle VST$

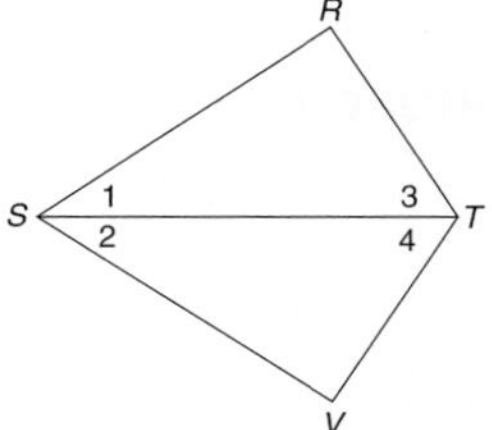


*Exercises 5–8*

*For Exercises 7 and 8, use the figure on page 142.*

7. *Given:* $\overline{SR} \cong \overline{SV}$ and $\overline{RT} \cong \overline{VT}$
   *Prove:* $\triangle RST \cong \triangle VST$

8. *Given:* $\angle R$ and $\angle V$ are right $\angle$s
   $\overline{RT} \cong \overline{VT}$
   *Prove:* $\triangle RST \cong \triangle VST$

9. *Given:* $\overline{UW} \parallel \overline{XZ}$, $\overline{VY} \perp \overline{UW}$, and $\overline{VY} \perp \overline{XZ}$
   $m\angle 1 = m\angle 4 = 42°$
   *Find:* $m\angle 2$, $m\angle 3$, $m\angle 5$, and $m\angle 6$

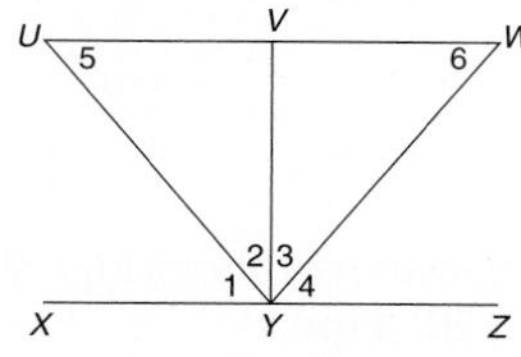


*Exercises 9, 10*

10. *Given:* $\overline{UW} \parallel \overline{XZ}$, $\overline{VY} \perp \overline{UW}$, and $\overline{VY} \perp \overline{XZ}$
    $m\angle 1 = m\angle 4 = 4x + 3$
    $m\angle 2 = 6x - 3$
    *Find:* $m\angle 1$, $m\angle 2$, $m\angle 3$, $m\angle 4$, $m\angle 5$, and $m\angle 6$

*In Exercises 11 and 12, complete each proof.*

11. *Given:* $\overline{HJ} \perp \overline{KL}$ and $\overline{HK} \cong \overline{HL}$
    *Prove:* $\overline{KJ} \cong \overline{JL}$

**PROOF**

| Statements | Reasons |
|---|---|
| 1. $\overline{HJ} \perp \overline{KL}$ and $\overline{HK} \cong \overline{HL}$ | 1. ? |
| 2. $\angle$s $HJK$ and $HJL$ are rt. $\angle$s | 2. ? |
| 3. $\overline{HJ} \cong \overline{HJ}$ | 3. ? |
| 4. ? | 4. HL |
| 5. ? | 5. CPCTC |

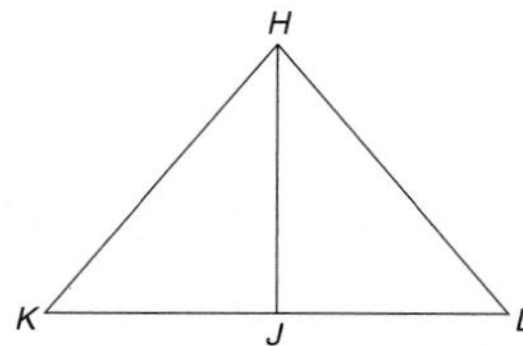


*Exercises 11, 12*

12. *Given:* $\overrightarrow{HJ}$ bisects $\angle KHL$
    $\overline{HJ} \perp \overline{KL}$
    *Prove:* $\angle K \cong \angle L$

**PROOF**

| Statements | Reasons |
|---|---|
| 1. ? | 1. Given |
| 2. $\angle JHK \cong \angle JHL$ | 2. ? |
| 3. $\overline{HJ} \perp \overline{KL}$ | 3. ? |
| 4. $\angle HJK \cong \angle HJL$ | 4. ? |
| 5. ? | 5. Identity |
| 6. ? | 6. ASA |
| 7. $\angle K \cong \angle L$ | 7. ? |

*In Exercises 13 to 16, first prove that triangles are congruent, and then use CPCTC.*

13. *Given:* $\angle P$ and $\angle R$ are right $\angle$s
    $M$ is the midpoint of $\overline{PR}$
    *Prove:* $\angle N \cong \angle Q$

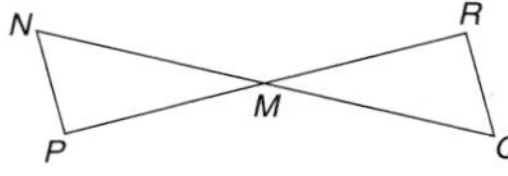


*Exercises 13, 14*

14. *Given:* $M$ is the midpoint of $\overline{NQ}$
    $\overline{NP} \parallel \overline{RQ}$ with transversals $\overline{PR}$ and $\overline{NQ}$
    *Prove:* $\overline{NP} \cong \overline{QR}$

15. *Given:* $\angle 1$ and $\angle 2$ are right $\angle$s
    $H$ is the midpoint of $\overline{FK}$
    $\overline{FG} \parallel \overline{HJ}$
    *Prove:* $\overline{FG} \cong \overline{HJ}$

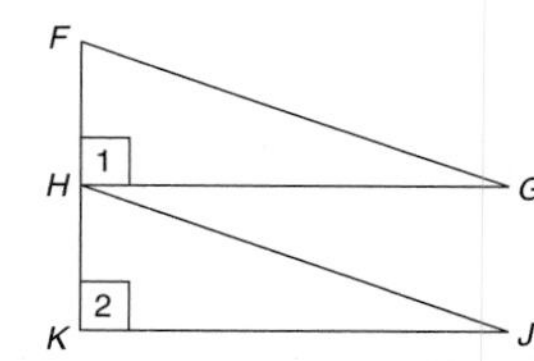


16. *Given:* $\overline{DE} \perp \overline{EF}$ and $\overline{CB} \perp \overline{AB}$
    $\overline{AB} \parallel \overline{FE}$
    $\overline{AC} \cong \overline{FD}$
    *Prove:* $\overline{EF} \cong \overline{BA}$

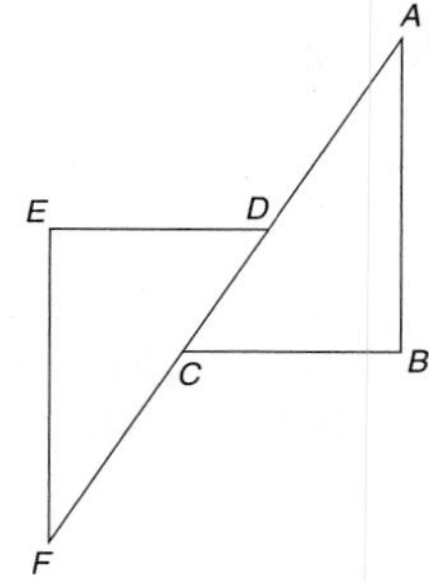


*In Exercises 17 to 22, $\triangle ABC$ is a right triangle. Use the given information to find the length of the third side of the triangle.*

17. $a = 4$ and $b = 3$

18. $a = 12$ and $b = 5$

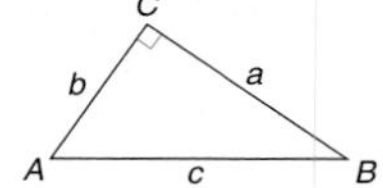

19. $a = 15$ and $c = 17$

20. $b = 6$ and $c = 10$

21. $a = 5$ and $b = 4$

22. $a = 7$ and $c = 8$

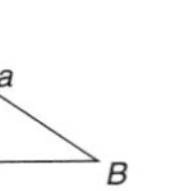


*In Exercises 23 to 25, prove the indicated relationship.*

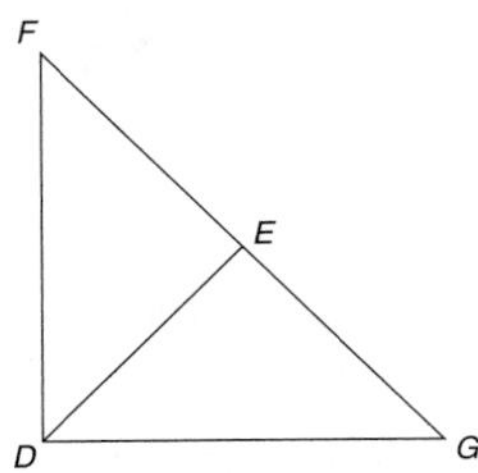


*Exercises 23–25*

23. *Given:* $\overline{DF} \cong \overline{DG}$ and $\overline{FE} \cong \overline{EG}$
*Prove:* $\overrightarrow{DE}$ bisects $\angle FDG$

24. *Given:* $\overrightarrow{DE}$ bisects $\angle FDG$
$\angle F \cong \angle G$
*Prove:* $E$ is the midpoint of $\overline{FG}$

25. *Given:* $E$ is the midpoint of $\overline{FG}$
$\overline{DF} \cong \overline{DG}$
*Prove:* $\overline{DE} \perp \overline{FG}$

*In Exercises 26 to 28, draw the triangles that are to be shown congruent separately.*

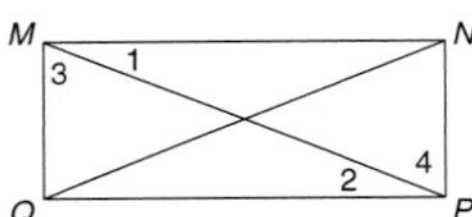


*Exercises 26–28*

26. *Given:* $\angle MQP$ and $\angle NPQ$ are rt. $\angle$s
$\overline{MQ} \cong \overline{NP}$
*Prove:* $\overline{MP} \cong \overline{NQ}$
(HINT: Show $\triangle MQP \cong \triangle NPQ$.)

27. *Given:* $\angle 1 \cong \angle 2$ and $\overline{MN} \cong \overline{QP}$
*Prove:* $\overline{MQ} \parallel \overline{NP}$
(HINT: Show $\triangle NMP \cong \triangle QPM$.)

28. *Given:* $\overline{MN} \parallel \overline{QP}$ and $\overline{MQ} \parallel \overline{NP}$
*Prove:* $\overline{MQ} \cong \overline{NP}$
(HINT: Show $\triangle MQP \cong \triangle PNM$.)

29. *Given:* $\overrightarrow{RW}$ bisects $\angle SRU$
$\overline{RS} \cong \overline{RU}$
*Prove:* $\triangle TRU \cong \triangle VRS$
(HINT: First show that $\triangle RSW \cong \triangle RUW$.)

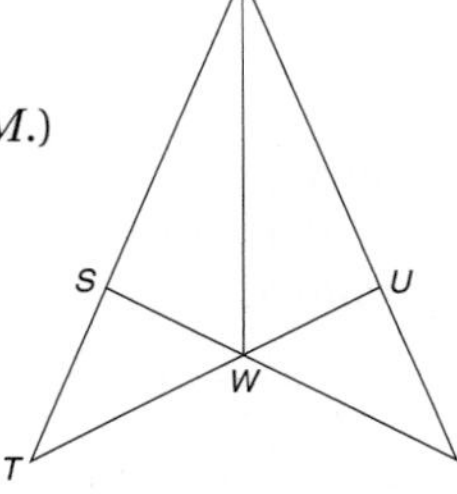


*Exercise 29*

30. *Given:* $\overline{DB} \perp \overline{BC}$ and
$\overline{CE} \perp \overline{DE}$
$\overline{AB} \cong \overline{AE}$
*Prove:* $\triangle BDC \cong \triangle ECD$
(HINT: First show that $\triangle ACE \cong \triangle ADB$.)

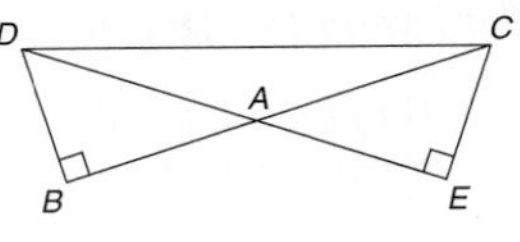


31. In the roof truss shown, $AB = 8$ and $m\angle HAF = 37°$. Find:

a) $AH$ b) $m\angle BAD$ c) $m\angle ADB$

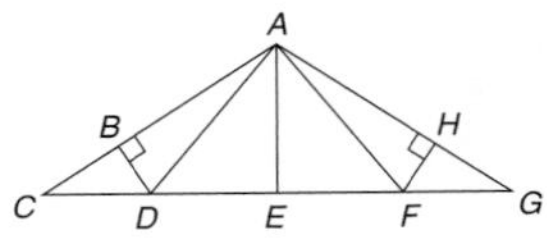


32. In the support system of the bridge shown, $AC = 6$ ft and $m\angle ABC = 28°$. Find:

a) $m\angle RST$ b) $m\angle ABD$ c) $BS$

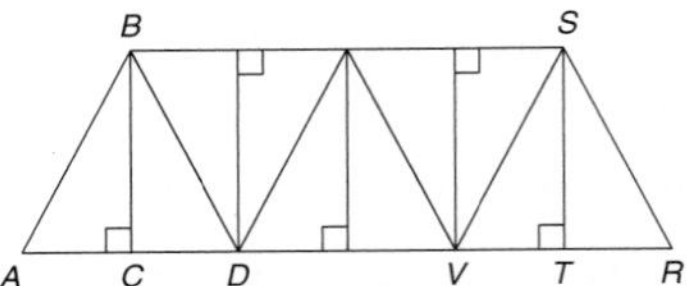


33. As a car moves along the roadway in a mountain pass, it passes through a horizontal run of 750 feet and through a vertical rise of 45 feet. To the nearest foot, how far does the car move along the roadway?

34. Because of construction along the road from $A$ to $B$, Alinna drives 5 miles from $A$ to $C$ and then 12 miles from $C$ to $B$. How much farther did Alinna travel by using the alternative route from $A$ to $B$?

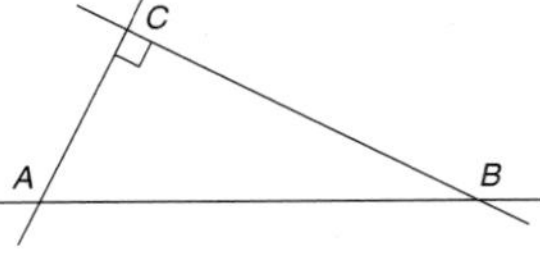


35. *Given:* Regular pentagon $ABCDE$ with diagonals $\overline{BE}$ and $\overline{BD}$
*Prove:* $\overline{BE} \cong \overline{BD}$
(HINT: First prove $\triangle ABE \cong \triangle CBD$.)

36. In the figure with regular pentagon $ABCDE$, do $\overrightarrow{BE}$ and $\overrightarrow{BD}$ trisect $\angle ABC$?
(HINT: $m\angle ABE = m\angle AEB$.)

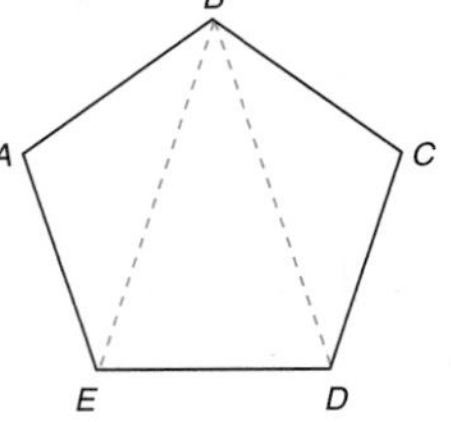


*Exercises 35, 36*

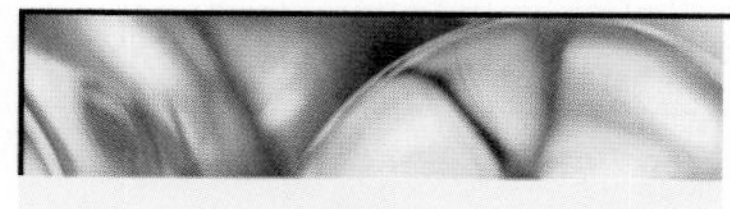

# 3.3 Isosceles Triangles

**KEY CONCEPTS**

Isosceles Triangles • Vertex, Legs, and Base of an Isosceles Triangle • Base Angles • Vertex Angle • Angle Bisector • Median • Altitude • Perpendicular Bisector • Equilateral and Equiangular Triangles • Perimeter

In an isosceles triangle, the two sides of equal length are **legs** and the third side is the **base.** The point at which the two legs meet is the **vertex** of the triangle, so the angle formed by the legs (and opposite the base) is the **vertex angle.** The two remaining angles are **base angles.** (See Figure 3.22.) If $\overline{AC} \cong \overline{BC}$ in Figure 3.23, then $\triangle ABC$ is isosceles with legs $\overline{AC}$ and $\overline{BC}$, base $\overline{AB}$, vertex $C$, vertex angle $C$, and base angles at $A$ and $B$. With $\overline{AC} \cong \overline{BC}$, we see that the base $\overline{AB}$ of this isosceles triangle is not necessarily the "bottom" side. See Figure 3.23.

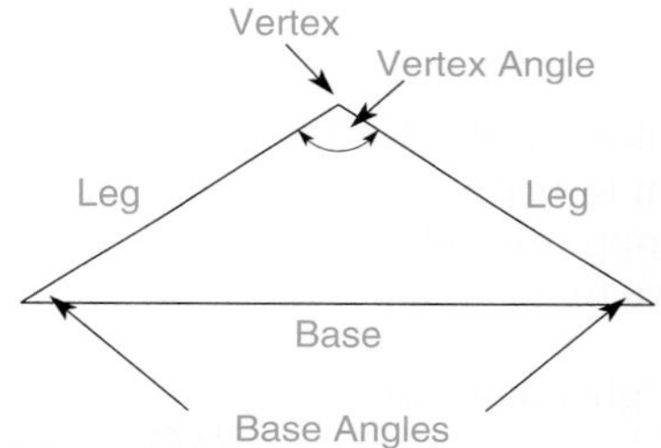


**FIGURE 3.22**

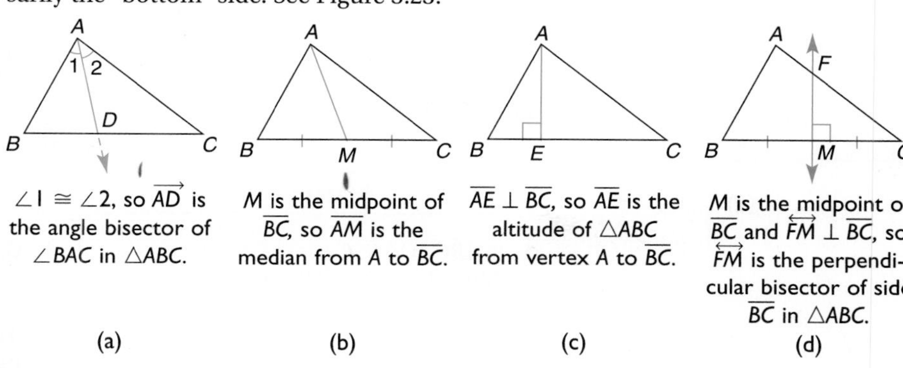

A
1
2
D
B
C
$\angle 1 \cong \angle 2$, so $\overrightarrow{AD}$ is the angle bisector of $\angle BAC$ in $\triangle ABC$.
(a)
A
B
M
C
$M$ is the midpoint of $\overline{BC}$, so $\overline{AM}$ is the median from $A$ to $\overline{BC}$.
(b)
A
B
E
C
$\overline{AE} \perp \overline{BC}$, so $\overline{AE}$ is the altitude of $\triangle ABC$ from vertex $A$ to $\overline{BC}$.
(c)
A
F
B
M
C
$M$ is the midpoint of $\overline{BC}$ and $\overleftrightarrow{FM} \perp \overline{BC}$, so $\overleftrightarrow{FM}$ is the perpendicular bisector of side $\overline{BC}$ in $\triangle ABC$.
(d)


**FIGURE 3.23**

In any triangle, a number of segments, rays, or lines are related to the triangle. (See Figure 3.23.) Each angle of a triangle has a unique **angle bisector,** and this may be indicated by a ray or segment from the vertex of the bisected angle. Just as an angle bisector begins at the vertex of an angle, so does the **median,** which joins a vertex to the midpoint of the opposite side. Generally, the median from a vertex of a triangle is not the same as the angle bisector from that vertex. An **altitude** is a line segment drawn from a vertex to the opposite side such that it is perpendicular to the opposite side. Finally, the **perpendicular bisector** of a side of a triangle is shown as a line in Figure 3.23. A segment or ray could also perpendicularly bisect a side of the triangle. In Figure 3.24, $\overline{AD}$ is the angle bisector of $\angle BAC$; $\overline{AE}$ is the altitude from $A$ to $\overline{BC}$; $M$ is the midpoint of $\overline{BC}$; $\overline{AM}$ is the median from $A$ to $\overline{BC}$; and $\overleftrightarrow{FM}$ is the perpendicular bisector of $\overline{BC}$. Notice that these are four distinct geometric figures.

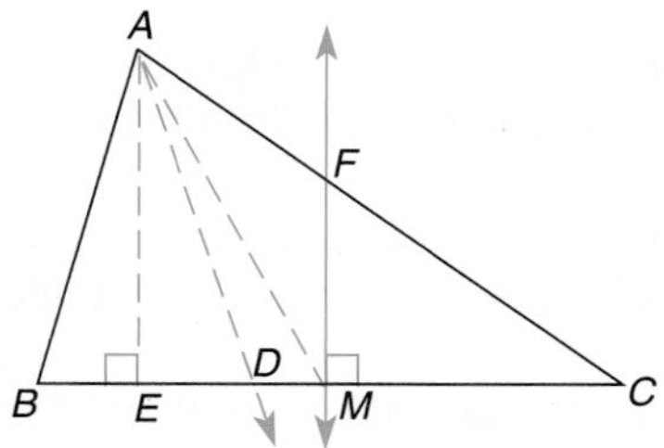


**FIGURE 3.24**

An altitude can actually lie in the exterior of a triangle. In Figure 3.25, which shows the obtuse triangle $\triangle RST$, the altitude from $R$ must be drawn to an extension of side $\overline{ST}$. Later we will use the length of the altitude $\overline{RH}$ in place of $h$ in the following standard formula for the area of a triangle:

$$A = \frac{1}{2}bh$$

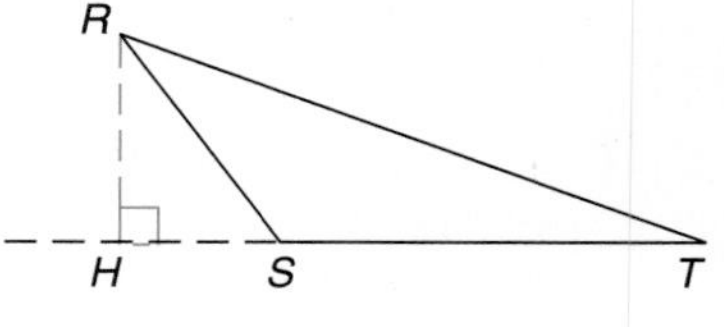


**FIGURE 3.25**

The angle bisector and the median necessarily lie in the interior of the triangle.

Each triangle has three altitudes—one from each vertex. As these are shown for $\triangle ABC$ in Figure 3.26, do the three altitudes seem to meet at a common point?

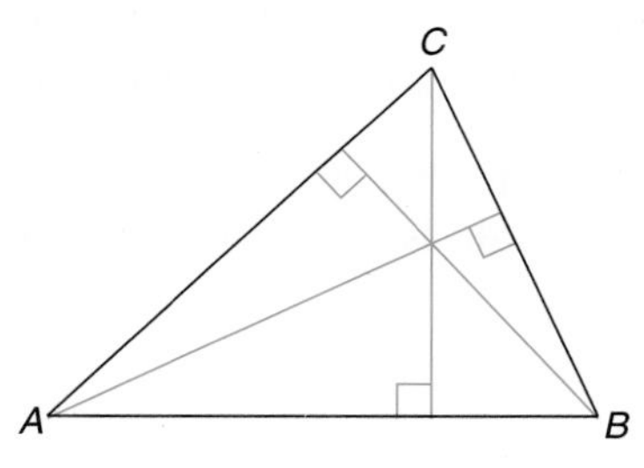


**FIGURE 3.26**

We now consider the proof of a statement that involves the altitudes of congruent triangles.

THEOREM 3.3.1: Corresponding altitudes of congruent triangles are congruent.

*Given:* $\triangle ABC \cong \triangle RST$
Altitudes $\overline{CD}$ to $\overline{AB}$ and $\overline{TV}$ to $\overline{RS}$
(See Figure 3.27.)

*Prove:* $\overline{CD} \cong \overline{TV}$

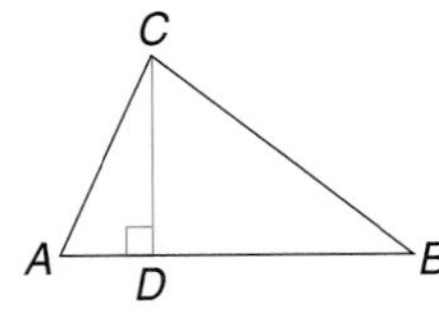

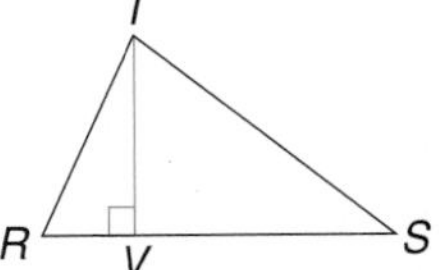


**FIGURE 3.27**

PROOF

| Statements | Reasons |
|---|---|
| 1. $\triangle ABC \cong \triangle RST$ Altitudes $\overline{CD}$ to $\overline{AB}$ and $\overline{TV}$ to $\overline{RS}$ | 1. Given |
| 2. $\overline{CD} \perp \overline{AB}$ and $\overline{TV} \perp \overline{RS}$ | 2. An altitude of a $\triangle$ is the line segment from one vertex drawn $\perp$ to the opposite side |
| 3. $\angle CDA$ and $\angle TVR$ are right $\angle$s | 3. If two lines are $\perp$, they form right $\angle$s |
| 4. $\angle CDA \cong \angle TVR$ | 4. All right angles are $\cong$ |
| 5. $\overline{AC} \cong \overline{RT}$ and $\angle A \cong \angle R$ | 5. CPCTC (from $\triangle ABC \cong \triangle RST$) |
| 6. $\triangle CDA \cong \triangle TVR$ | 6. AAS |
| 7. $\overline{CD} \cong \overline{TV}$ | 7. CPCTC |

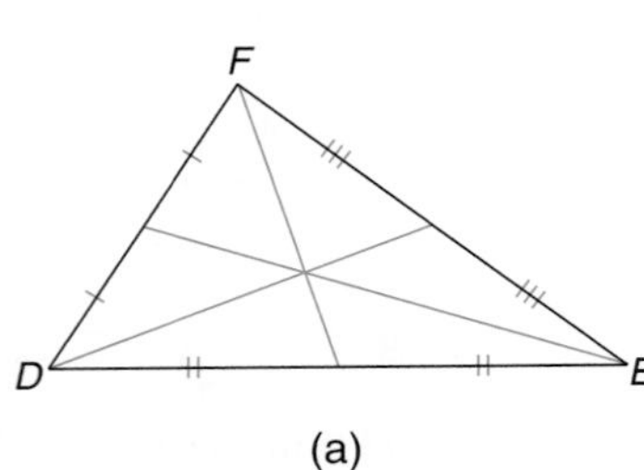

(a)

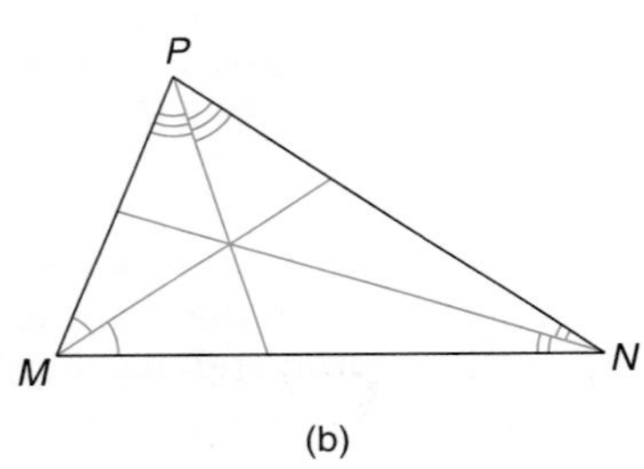

(b)

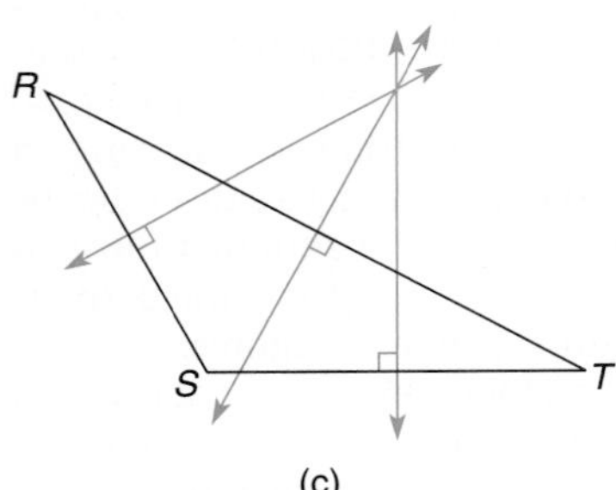

(c)

**FIGURE 3.28**

Each triangle has three medians—one from each vertex to the midpoint of the opposite side. As the medians are drawn for $\triangle DEF$ in Figure 3.28(a), does it appear that the three medians intersect at a point?

Each triangle has three angle bisectors—one for each of the three angles. As these are shown for $\triangle MNP$ in Figure 3.28(b), does it appear that the three angle bisectors have a point in common?

Each triangle has three perpendicular bisectors for its sides; these are shown for $\triangle RST$ in Figure 3.28(c). Like the altitudes, medians, and angle bisectors, the perpendicular bisectors of the sides also meet at a single point.

*Exs. 1–6*

The angle bisectors and the medians of a triangle always meet in the interior of the triangle. However, the altitudes and perpendicular bisectors of the sides [see Figure 3.28(c)] can meet in the exterior of the triangle. These points of intersection will be given more attention in Chapter 6.

The Discover! activity on page 147 opens the doors to further discoveries.

In Figure 3.29, the bisector of the vertex angle of $\triangle ABC$ is a line (segment) of symmetry for $\triangle ABC$.

EXAMPLE 1

Give a formal proof of Theorem 3.3.2.

**Discover!**

Using a sheet of construction paper, cut out an isosceles triangle. Now use your compass to bisect the vertex angle. Fold along the angle bisector to form two smaller triangles. How are the smaller triangles related?

ANSWER

They are congruent.

> THEOREM 3.3.2: The bisector of the vertex angle of an isosceles triangle separates the triangle into two congruent triangles.

*Given:* Isosceles $\triangle ABC$, with $\overline{AB} \cong \overline{BC}$
$\overrightarrow{BD}$ bisects $\angle ABC$
(See Figure 3.29.)

*Prove:* $\triangle ABD \cong \triangle CBD$

**PROOF**

| Statements | Reasons |
|---|---|
| 1. Isosceles $\triangle ABC$ with $\overline{AB} \cong \overline{BC}$ | 1. Given |
| 2. $\overrightarrow{BD}$ bisects $\angle ABC$ | 2. Given |
| 3. $\angle 1 \cong \angle 2$ | 3. The bisector of an $\angle$ separates it into two $\cong$ $\angle$s |
| 4. $\overline{BD} \cong \overline{BD}$ | 4. Identity |
| 5. $\triangle ABD \cong \triangle CBD$ | 5. SAS |

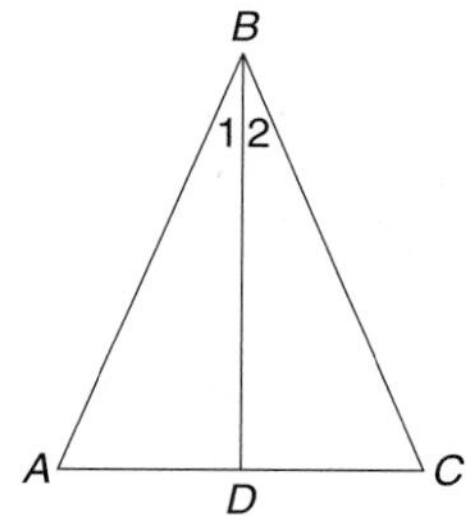


**FIGURE 3.29**

Recall from Section 2.4 that an auxiliary figure must be determined. Consider Figure 3.30 and the following three descriptions, which are coded **D** for determined, **U** for underdetermined, and **O** for overdetermined:

**D:** Draw a line segment from $A$ perpendicular to $\overline{BC}$ so that the terminal point is on $\overline{BC}$. [*Determined* because the line from $A$ perpendicular to $\overline{BC}$ is unique; see Figure 3.30(a).]

**U:** Draw a line segment from $A$ to $\overline{BC}$ so that the terminal point is on $\overline{BC}$. [*Underdetermined* because many line segments are possible; see Figure 3.30(b).]

**O:** Draw a line segment from $A$ perpendicular to $\overline{BC}$ so that it bisects $\overline{BC}$. [*Overdetermined* because the line segment from $A$ drawn perpendicular to $\overline{BC}$ will not contain the midpoint $M$ of $\overline{BC}$; see Figure 3.30(c).]

In Example 2, an auxiliary segment is needed. As you study the proof, note the uniqueness of the segment and its justification (reason 2) in the proof.

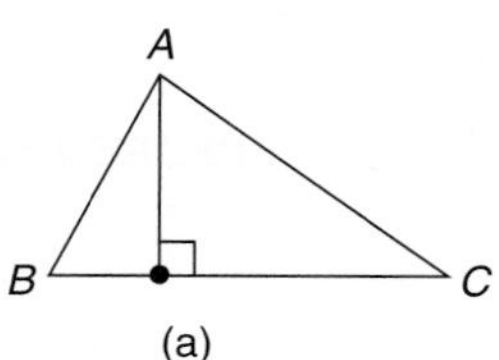


(a)

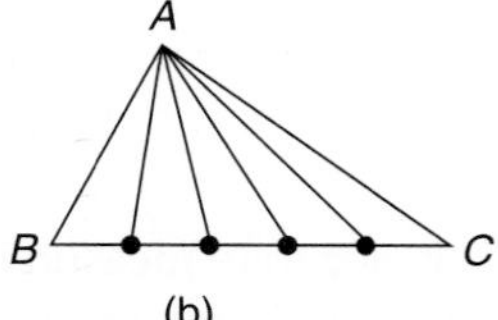


(b)

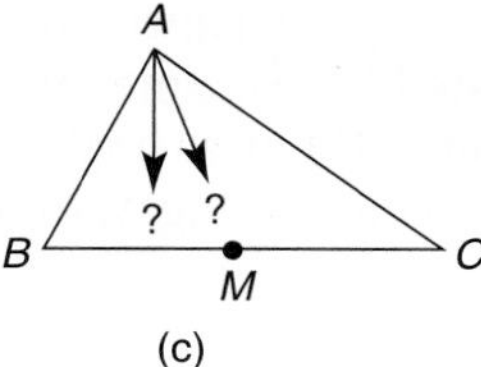


(c)

**FIGURE 3.30**

**EXAMPLE 2**

Give a formal proof of Theorem 3.3.3.

> THEOREM 3.3.3: If two sides of a triangle are congruent, then the angles opposite these sides are also congruent.

*Given:* Isosceles $\triangle MNP$
with $\overline{MP} \cong \overline{NP}$
[See Figure 3.31(a) on page 148.]

*Prove:* $\angle M \cong \angle N$

NOTE: Figure 3.31(b) shows the auxiliary segment.

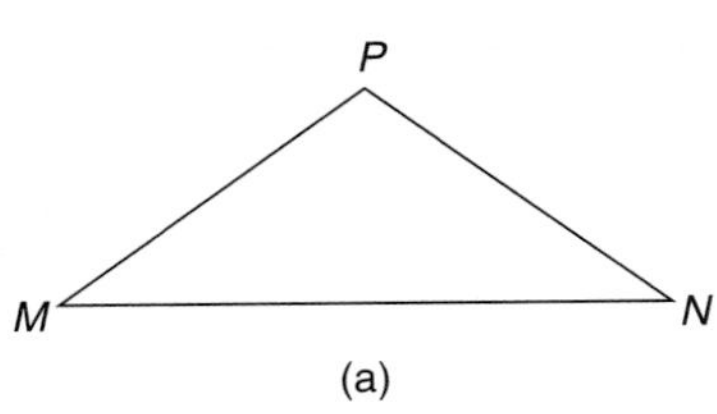

(a)

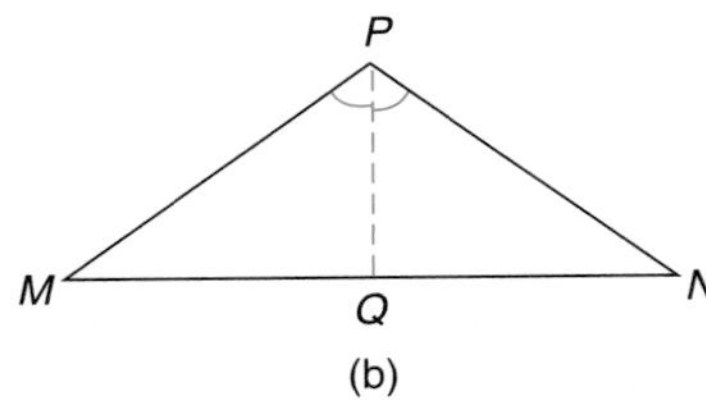

(b)

**FIGURE 3.31**

**PROOF**

| Statements | Reasons |
|---|---|
| 1. Isosceles $\triangle MNP$ with $\overline{MP} \cong \overline{NP}$ | 1. Given |
| 2. Draw $\angle$ bisector $\overrightarrow{PQ}$ from $P$ to $\overline{MN}$ | 2. Every angle has one and only one bisector |
| 3. $\triangle MPQ \cong \triangle NPQ$ | 3. The bisector of the vertex angle of an isosceles $\triangle$ separates it into two $\cong$ $\triangle$s |
| 4. $\angle M \cong \angle N$ | 4. CPCTC |

Theorem 3.3.3 is sometimes stated, "The base angles of an isosceles triangle are congruent." We apply this theorem in Example 3.

> **Discover!**
>
> Using construction paper and scissors, cut out an isosceles triangle $MNP$ with $\overline{MP} \cong \overline{PN}$. Fold it so that $\angle M$ coincides with $\angle N$. What can you conclude?
>
> ANSWER
> $\angle M \cong \angle N$

EXAMPLE 3

Find the size of each angle of the isosceles triangle shown in Figure 3.32 if:

a) $m\angle 1 = 36°$
b) The measure of each base angle is 5° less than twice the measure of the vertex angle

***Solution***

a) $m\angle 1 + m\angle 2 + m\angle 3 = 180°$. Since $m\angle 1 = 36°$ and $\angle 2$ and $\angle 3$ are $\cong$, we have

$$36 + 2(m\angle 2) = 180$$
$$2(m\angle 2) = 144$$
$$m\angle 2 = 72$$

Now $m\angle 1 = 36°$, and $m\angle 2 = m\angle 3 = 72°$.

b) Let the vertex angle measure be given by $x$. Then the size of each base angle is $2x - 5$. Because the sum of the measures is 180°,

$$x + (2x - 5) + (2x - 5) = 180$$
$$5x - 10 = 180$$
$$5x = 190$$
$$x = 38$$
$$2x - 5 = 2(38) - 5 = 76 - 5 = 71$$

Therefore, $m\angle 1 = 38°$ and $m\angle 2 = m\angle 3 = 71°$.

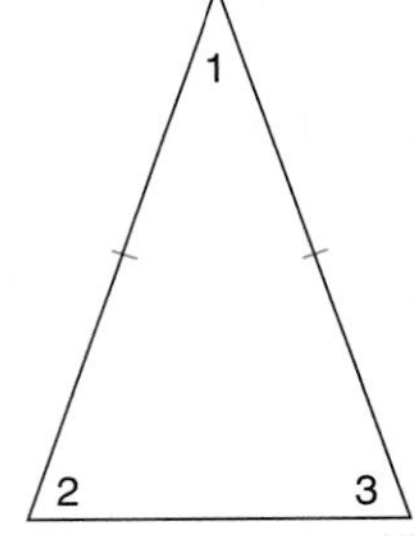

**FIGURE 3.32**

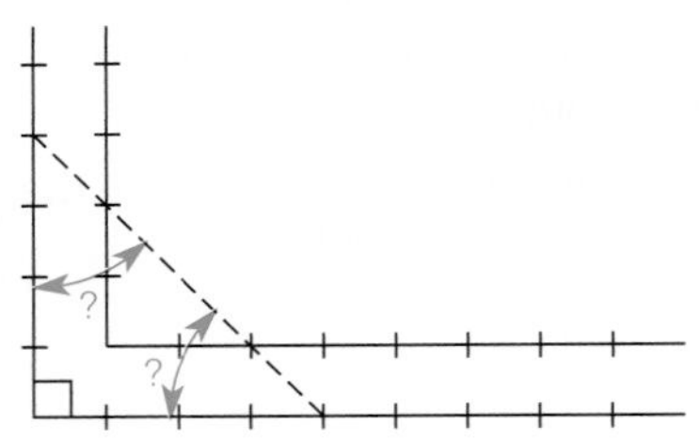

**FIGURE 3.33**

In some instances, a carpenter may want to get a quick, accurate measurement without having to go get his or her tools. Suppose that the carpenter's square shown in Figure 3.33 is handy but that a miter box is not nearby. If two marks are made at lengths of 4 inches from the corner of the square and these are then joined, what size angle is determined? You should see that each angle indicated by an arc measures 45°.

Example 4 shows us that the converse of the theorem "The base angles of an isosceles $\triangle$ are congruent" is also true. However, see the accompanying Warning.

> Warning (!)
> The converse of an "If, then" statement is not always true.

EXAMPLE 4

Study the picture proof of Theorem 3.3.4.

THEOREM 3.3.4: If two angles of a triangle are congruent, then the sides opposite these angles are also congruent.

**PICTURE PROOF OF THEOREM 3.3.4**

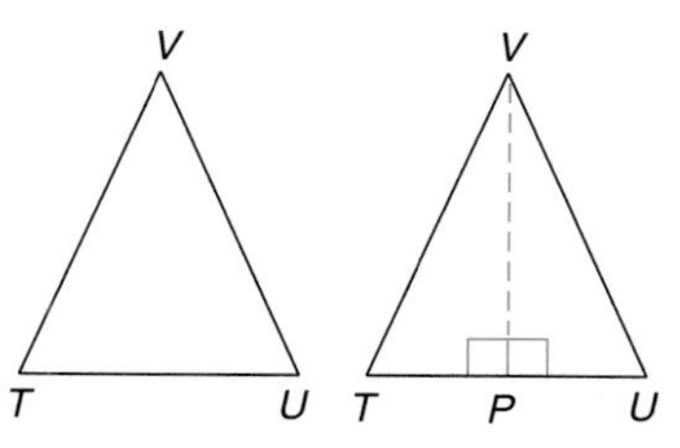


**FIGURE 3.34**

*Given:* $\triangle TUV$ with $\angle T \cong \angle U$

*Prove:* $\overline{VU} \cong \overline{UT}$

*Proof:* Drawing $\overline{VP} \perp \overline{TU}$, we see that $\triangle VPT \cong \triangle VPU$ (by AAS). Now $\overline{VU} \cong \overline{VT}$ (by CPCTC).

*Exs. 7–17*

When all three sides of a triangle are congruent, the triangle is **equilateral.** If all three angles are congruent, then the triangle is **equiangular.** Theorems 3.3.3 and 3.3.4 can be used to prove the sets {equilateral triangles} and {equiangular triangles} are equivalent.

COROLLARY 3.3.5: An equilateral triangle is also equiangular.

COROLLARY 3.3.6: An equiangular triangle is also equilateral.

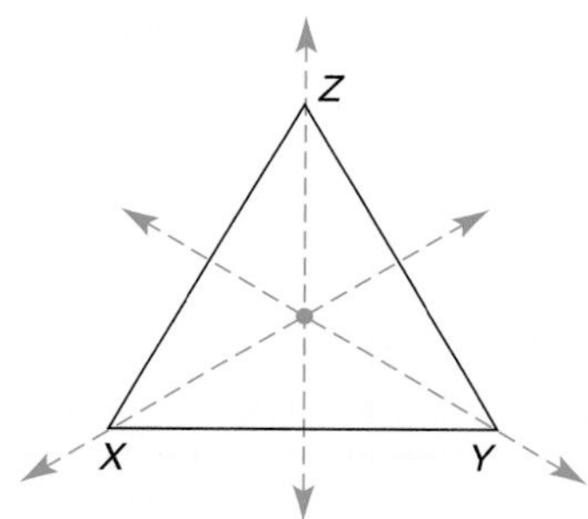


**FIGURE 3.35**

An equilateral (or equiangular) triangle has line symmetry with respect to each of the three axes shown in Figure 3.35.

DEFINITION: The **perimeter** of a triangle is the sum of the lengths of its sides. Thus if $a$, $b$, and $c$ are the lengths of the three sides, then the perimeter $P$ is given by $P = a + b + c$. (See Figure 3.36.)

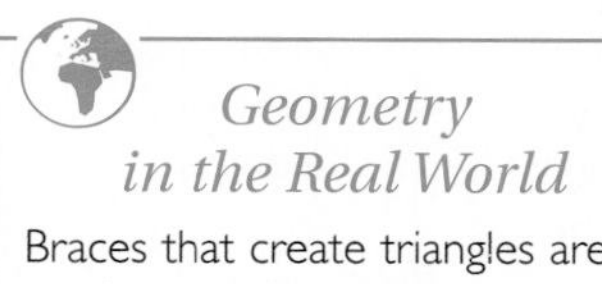

*Geometry in the Real World*

Braces that create triangles are used to provide stability for a bookcase. The triangle is called a rigid figure.

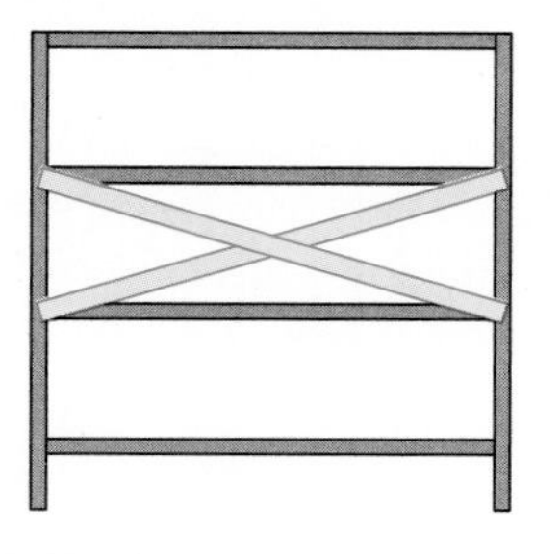

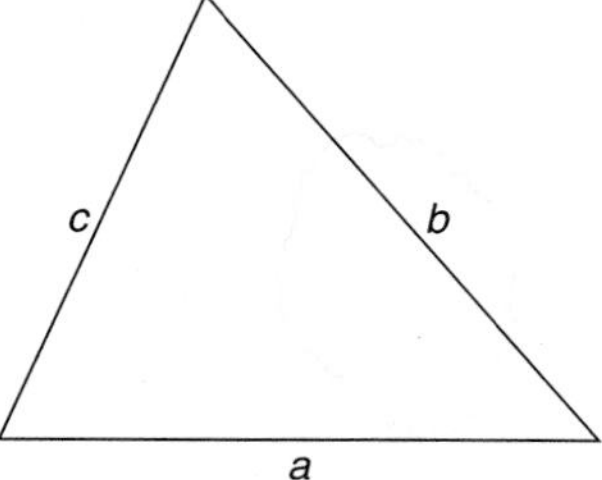


**FIGURE 3.36**

EXAMPLE 5

*Given:* $\angle B \cong \angle C$
$AB = 5.3$ and $BC = 3.6$
(See Figure 3.37 on page 150.)

*Find:* The perimeter of $\triangle ABC$

***Solution***

If $\angle B \cong \angle C$, then $AC = AB = 5.3$. Therefore,

$$P = a + b + c$$
$$P = 3.6 + 5.3 + 5.3$$
$$P = 14.2$$

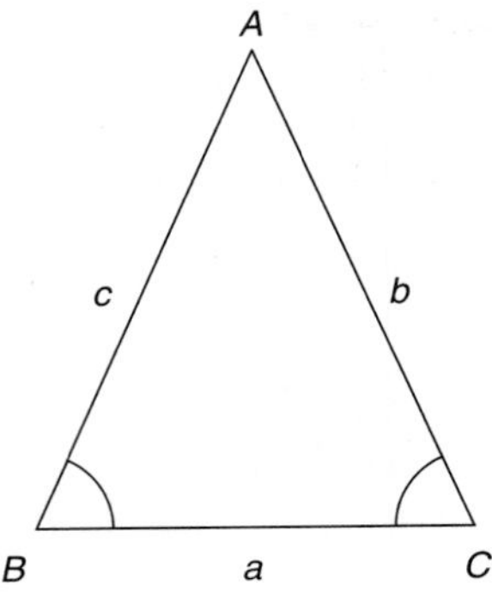


**FIGURE 3.37**

*Exs. 18–22*

Many of the properties of triangles that were investigated in earlier sections are summarized in Table 3.1.

**TABLE 3.1**

**Selected Properties of Triangles**

| | *Scalene* | *Isosceles* | *Equilateral (equiangular)* | *Acute* | *Right* | *Obtuse* |
|---|---|---|---|---|---|---|
| *Sides* | No two are ≅ | Exactly two are ≅ | All three are ≅ | Possibly two or three ≅ sides | Possibly two ≅ sides; $c^2 = a^2 + b^2$ | Possibly two ≅ sides |
| *Angles* | Sum of ∠s is 180° | Sum of ∠s is 180°; two ∠s ≅ | Sum of ∠s is 180°; three ≅ 60° ∠s | All ∠s acute; sum of ∠s is 180°; possibly two or three ≅ ∠s | One right ∠; sum of ∠s is 180°; possibly two ≅ 45° ∠s; acute ∠s are complementary | One obtuse ∠; sum of ∠s is 180°; possibly two ≅ acute ∠s |

## 3.3 Exercises

*For Exercises 1 to 8, use the accompanying drawing.*

1. If $\overline{VU} \cong \overline{VT}$, what type of triangle is $\triangle VTU$?
2. If $\overline{VU} \cong \overline{VT}$, which angles of $\triangle VTU$ are congruent?
3. If $\angle T \cong \angle U$, which sides of $\triangle VTU$ are congruent?
4. If $\overline{VU} \cong \overline{VT}$, $VU = 10$, and $TU = 8$, what is the perimeter of $\triangle VTU$?

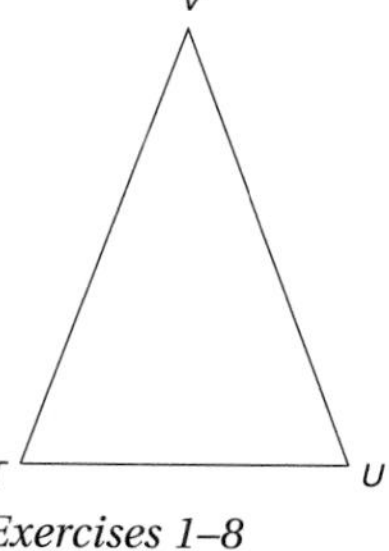

Exercises 1–8

5. If $\overline{VU} \cong \overline{VT}$ and $m\angle T = 69°$, find $m\angle U$.
6. If $\overline{VU} \cong \overline{VT}$ and $m\angle T = 69°$, find $m\angle V$.
7. If $\overline{VU} \cong \overline{VT}$ and $m\angle T = 72°$, find $m\angle V$.
8. If $\overline{VU} \cong \overline{VT}$ and $m\angle V = 40°$, find $m\angle T$.

*In Exercises 9 to 12, determine whether the sets have a subset relationship. Are the two sets disjoint or equivalent? Do the sets intersect?*

9. $L$ = {equilateral triangles}; $E$ = {equiangular triangles}
10. $S$ = {triangles with two ≅ sides}; $A$ = {triangles with two ≅ ∠s}
11. $R$ = {right triangles}; $O$ = {obtuse triangles}
12. $I$ = {isosceles triangles}; $R$ = {right triangles}

*In Exercises 13 to 18, describe the segment as determined, underdetermined, or overdetermined. Use the accompanying drawing for reference.*

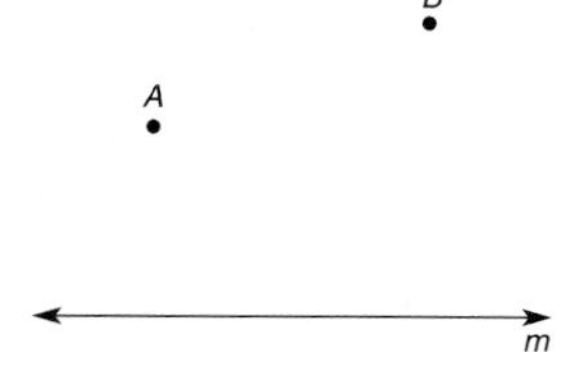

Exercises 13–18

13. Draw a segment through point $A$.
14. Draw a segment with endpoints $A$ and $B$.
15. Draw a segment $\overline{AB}$ parallel to line $m$.
16. Draw a segment $\overline{AB}$ perpendicular to $m$.
17. Draw a segment from $A$ perpendicular to $m$.
18. Draw $\overline{AB}$ so that line $m$ bisects $\overline{AB}$.
19. A surveyor knows that a lot has the shape of an isosceles triangle. If the vertex angle measures 70° and each equal side is 160 ft long, what measure does each of the base angles have?
20. In concave quadrilateral $ABCD$, the angle at $A$ measures 40°. $\triangle ABD$ is isosceles, $\overrightarrow{BC}$ bisects $\angle ABD$, and $\overrightarrow{DC}$ bisects $\angle ADB$. What are the measures of $\angle ABC$, $\angle ADC$, and $\angle 1$?

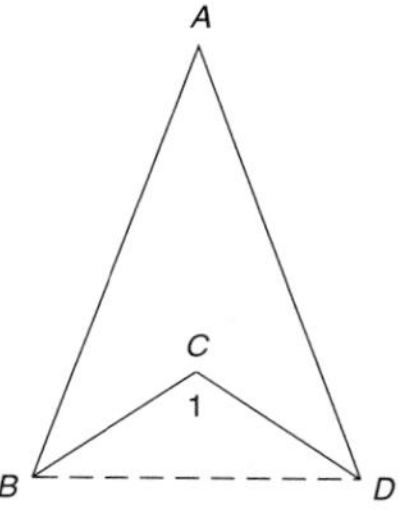


*In Exercises 21 to 26, use arithmetic or algebra as needed to find the measures indicated. Note the use of dashes on equal sides of the given isosceles triangles.*

21. Find $m\angle 1$ and $m\angle 2$ if $m\angle 3 = 68°$.

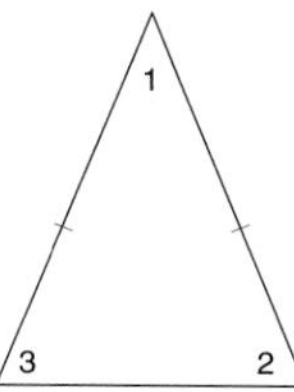


22. If $m\angle 3 = 68°$, find $m\angle 4$, the angle formed by the bisectors of $\angle 3$ and $\angle 2$.
23. Find the measure of $\angle 5$, which is formed by the bisectors of $\angle 1$ and $\angle 3$. Again let $m\angle 3 = 68°$.
24. Find an expression for the measure of $\angle 5$ if $m\angle 3 = 2x$ and the segments shown bisect the angles of the isosceles triangle.

Exercises 22–24

25. In isosceles $\triangle ABC$ with vertex $A$ (not shown), each base angle is 12° larger than the vertex angle. Find the measure of each angle.

26. In isosceles $\triangle ABC$ (not shown), vertex angle $A$ is 5° more than one-half of base angle $B$. Find the size of each angle of the triangle.

*In Exercises 27 to 30, suppose that $\overline{BC}$ is the base of isosceles $\triangle ABC$ (not shown).*

27. Find the perimeter of $\triangle ABC$ if $AB = 8$ and $BC = 10$.

28. Find $AB$ if the perimeter of $\triangle ABC$ is 36.4 and $BC = 14.6$.

29. Find $x$ if the perimeter of $\triangle ABC$ is 40, $AB = x$, and $BC = x + 4$.

30. Find $x$ if the perimeter of $\triangle ABC$ is 68, $AB = x$, and $BC = 1.4x$.

31. Suppose that $\triangle ABC \cong \triangle DEF$. Also, $\overline{AX}$ bisects $\angle CAB$ and $\overline{DY}$ bisects $\angle FDE$. Are the corresponding angle bisectors of congruent triangles congruent?

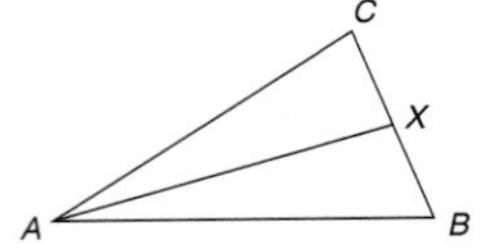

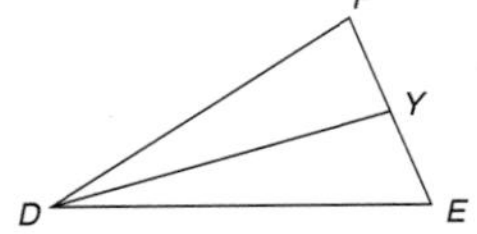


*Exercises 31, 32*

32. Suppose that $\triangle ABC \cong \triangle DEF$, $\overline{AX}$ is the median from $A$ to $\overline{BC}$, and $\overline{DY}$ is the median from $D$ to $\overline{EF}$. Are the corresponding medians of congruent triangles congruent?

*In Exercises 33 and 34, complete each proof using the drawing below.*

33. *Given:* $\angle 3 \cong \angle 1$
*Prove:* $\overline{AB} \cong \overline{AC}$

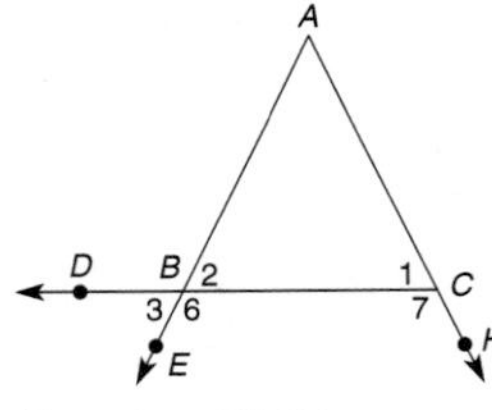


*Exercises 33, 34*

**PROOF**

| Statements | Reasons |
|---|---|
| 1. $\angle 3 \cong \angle 1$ | 1. ? |
| 2. ? | 2. If two lines intersect, the vertical $\angle$s formed are $\cong$ |
| 3. ? | 3. Transitive Property of Congruence |
| 4. $\overline{AB} \cong \overline{AC}$ | 4. ? |

34. *Given:* $\overline{AB} \cong \overline{AC}$
*Prove:* $\angle 6 \cong \angle 7$

**PROOF**

| Statements | Reasons |
|---|---|
| 1. ? | 1. Given |
| 2. $\angle 2 \cong \angle 1$ | 2. ? |
| 3. $\angle 2$ and $\angle 6$ are supplementary; $\angle 1$ and $\angle 7$ are supplementary | 3. ? |
| 4. ? | 4. If two $\angle$s are supplementary to $\cong$ $\angle$s, they are $\cong$ to each other |

*In Exercises 35 to 37, complete each proof.*

35. *Given:* $\angle 1 \cong \angle 3$
$\overline{RU} \cong \overline{VU}$
*Prove:* $\triangle STU$ is isosceles
(HINT: First show that $\triangle RUS \cong \triangle VUT$.)

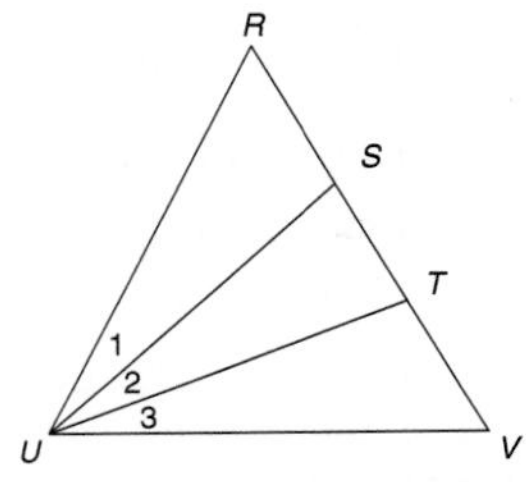


36. *Given:* $\overline{WY} \cong \overline{WZ}$
$M$ is the midpoint of $\overline{YZ}$
$\overline{MX} \perp \overline{WY}$ and $\overline{MT} \perp \overline{WZ}$
*Prove:* $\overline{MX} \cong \overline{MT}$

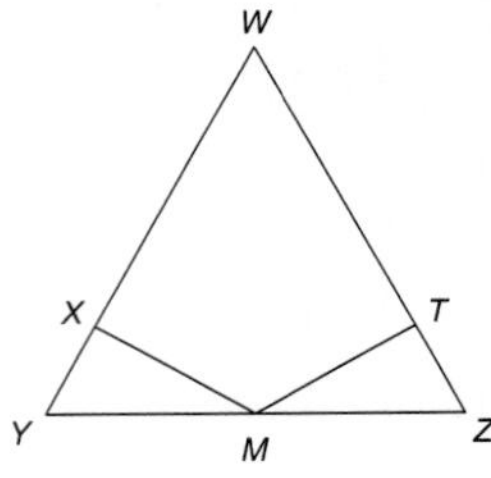


37. *Given:* Isosceles $\triangle MNP$ with vertex $P$
Isosceles $\triangle MNQ$ with vertex $Q$
*Prove:* $\triangle MQP \cong \triangle NQP$

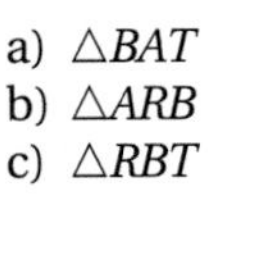


38. In isosceles triangle $BAT$, $\overline{AB} \cong \overline{AT}$. Also, $\overline{BR} \cong \overline{BT} \cong \overline{AR}$. If $AB = 12.3$ and $AR = 7.6$, find the perimeter of:

a) $\triangle BAT$
b) $\triangle ARB$
c) $\triangle RBT$

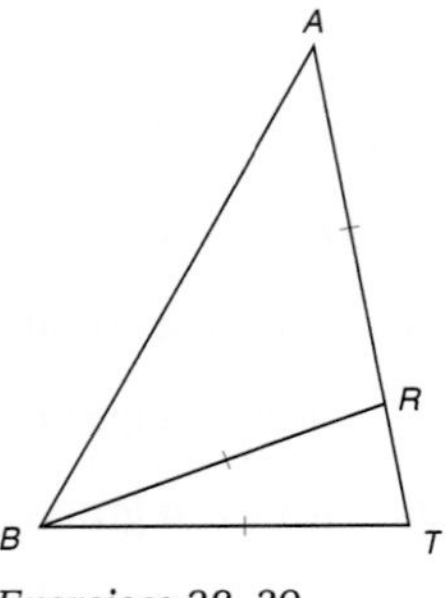


*Exercises 38, 39*

39. In $\triangle BAT$ (on page 152), $\overline{BR} \cong \overline{BT} \cong \overline{AR}$, and $m\angle RBT = 20°$. Find:

a) $m\angle T$
b) $m\angle ARB$
c) $m\angle A$

40. In $\triangle PMN$, $\overline{PM} \cong \overline{PN}$. $\overrightarrow{MB}$ bisects $\angle PMN$, and $\overrightarrow{NA}$ bisects $\angle PNM$. If $m\angle P = 36°$, name all isosceles triangles shown in the drawing.

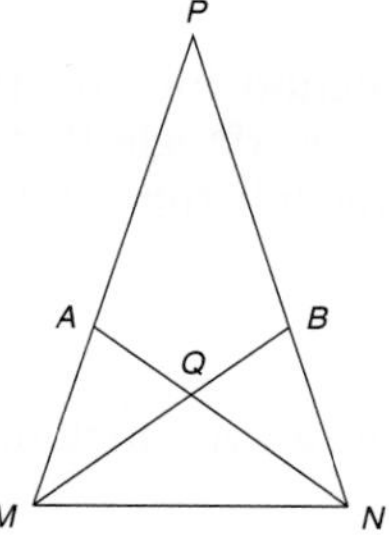


41. $\triangle ABC$ lies in the structural support system of the Ferris wheel. If $m\angle A = 30°$ and $AB = AC = 20$ ft, find the measures of $\angle B$ and $\angle C$.

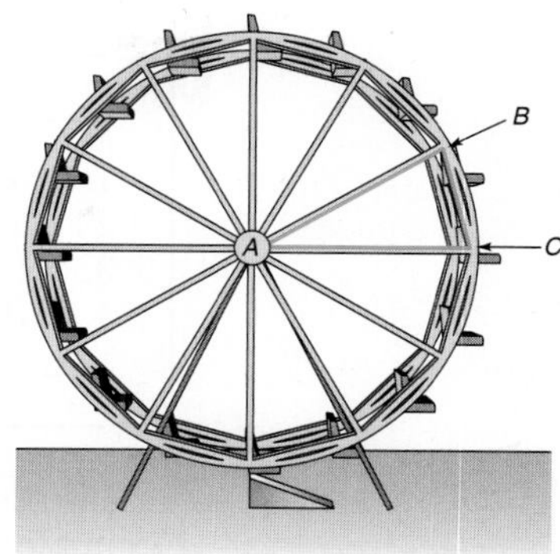


*In Exercises 42 to 44, explain why each statement is true.*

42. The altitude from the vertex of an isosceles triangle is also the median to the base of the triangle.

43. The bisector of the vertex angle of an isosceles triangle bisects the base.

44. The angle bisectors of the base angles of an isosceles triangle, together with the base, form an isosceles triangle.

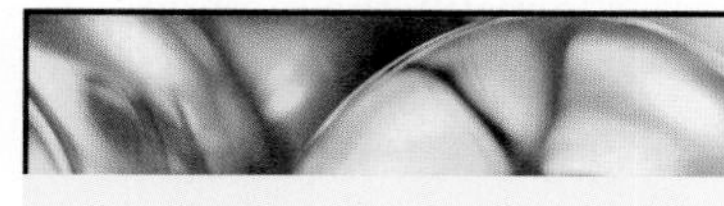

## 3.4 Basic Constructions Justified

**KEY CONCEPTS**
Justification of a Construction

In earlier sections, construction methods were introduced that appeared to achieve their goals; however, the methods were presented intuitively. In this section, we justify the construction methods and apply them in further constructions. The justification of the method is a "proof" that demonstrates that the construction accomplished its purpose. See Example 1.

EXAMPLE 1

Justify the method for constructing an angle congruent to a given angle.

*Given:* $\angle ABC$
$\overline{BD} \cong \overline{BE} \cong \overline{ST} \cong \overline{SR}$ (by construction)
$\overline{DE} \cong \overline{TR}$ (by construction)
(See Figure 3.38.)

*Prove:* $\angle B \cong \angle S$

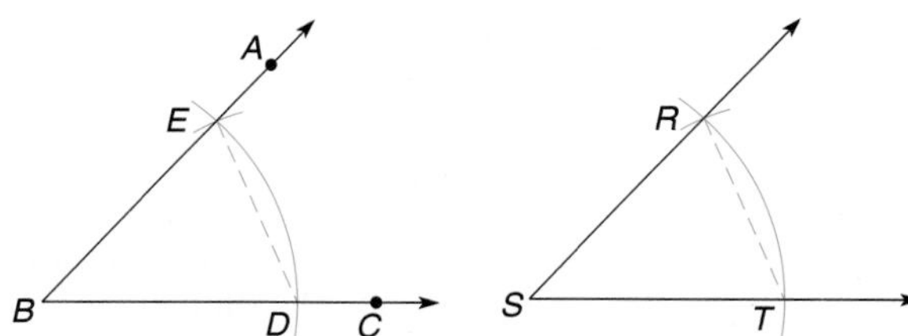


**FIGURE 3.38**

**PROOF**

| Statements | Reasons |
|---|---|
| 1. $\angle ABC$; $\overline{BD} \cong \overline{BE} \cong \overline{ST} \cong \overline{SR}$ | 1. Given |
| 2. $\overline{DE} \cong \overline{TR}$ | 2. Given |
| 3. $\triangle EBD \cong \triangle RST$ | 3. SSS |
| 4. $\angle B \cong \angle S$ | 4. CPCTC |

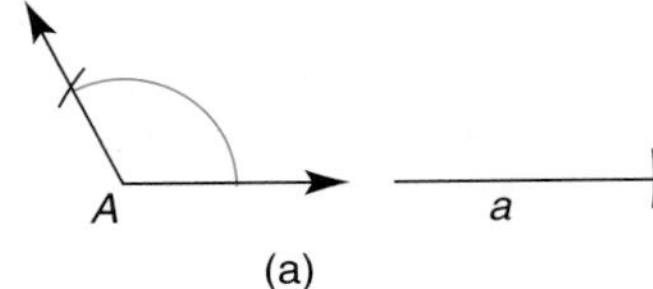


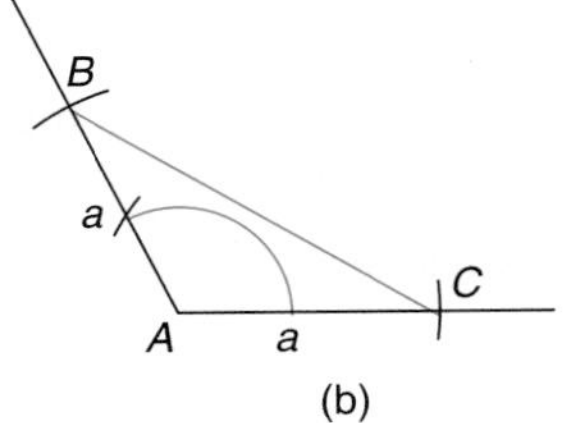


**FIGURE 3.39**

In Example 2, we will apply the construction method that was justified in Example 1. Our goal is to construct an isosceles triangle that contains an obtuse angle. It is necessary that the congruent sides include the obtuse angle.

EXAMPLE 2

Construct an isosceles triangle in which obtuse $\angle A$ is included by two sides of length $a$ [see Figure 3.39(a)].

***Solution***

Construct an angle congruent to $\angle A$. From $A$, mark off arcs of length $a$ at points $B$ and $C$ as shown in Figure 3.39(b). Join $B$ to $C$ to complete $\triangle ABC$.

*Exs. 1–2*

In Example 3, we recall the method of construction used to bisect an angle. Although the technique is illustrated, the purpose here is to justify the method.

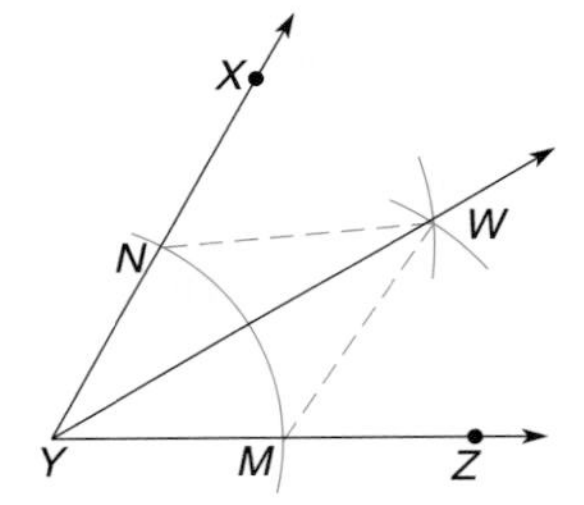


**FIGURE 3.40**

EXAMPLE 3

Justify the method for constructing the bisector of an angle. Provide the missing reasons in the proof.

*Given:* $\angle XYZ$
$\overline{YM} \cong \overline{YN}$ (by construction)
$\overline{MW} \cong \overline{NW}$ (by construction)
(See Figure 3.40.)

*Prove:* $\overrightarrow{YW}$ bisects $\angle XYZ$

**PROOF**

| Statements | Reasons |
|---|---|
| 1. $\angle XYZ$; $\overline{YM} \cong \overline{YN}$ and $\overline{MW} \cong \overline{NW}$ | 1. ? |
| 2. $\overline{YW} \cong \overline{YW}$ | 2. ? |
| 3. $\triangle YMW \cong \triangle YNW$ | 3. ? |
| 4. $\angle MYW \cong \angle NYW$ | 4. ? |
| 5. $\overrightarrow{YW}$ bisects $\angle XYZ$ | 5. ? |

The angle bisector method can be used to construct angles of certain measures. For instance, if a right angle has been constructed, then an angle of measure 45° can be constructed by bisecting the 90° angle. In Example 4, we construct an angle of measure 30°.

EXAMPLE 4

Construct an angle that measures 30°.

***Solution***

We begin by constructing an equilateral (and therefore equiangular) triangle. To accomplish this, mark off a line segment of length $a$ as shown in Figure 3.41(a). From the endpoints of this line segment, mark off arcs using the same radius length $a$. The point of intersection determines the third vertex of this triangle, whose angles measure 60° each [see Figure 3.41(b)]. By constructing the bisector of one angle, we determine an angle that measures 30° [Figure 3.41(c)].

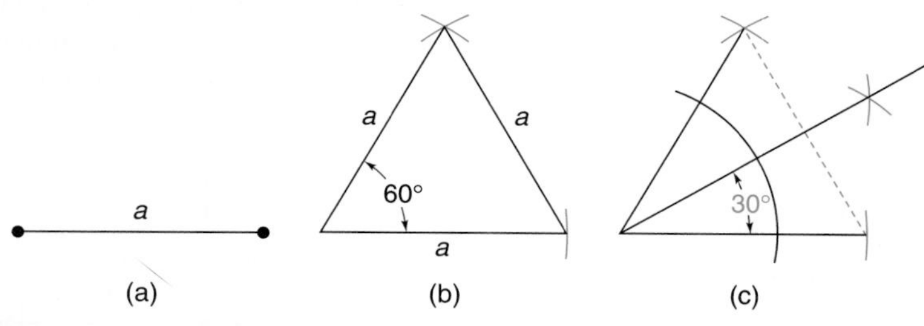


**FIGURE 3.41**

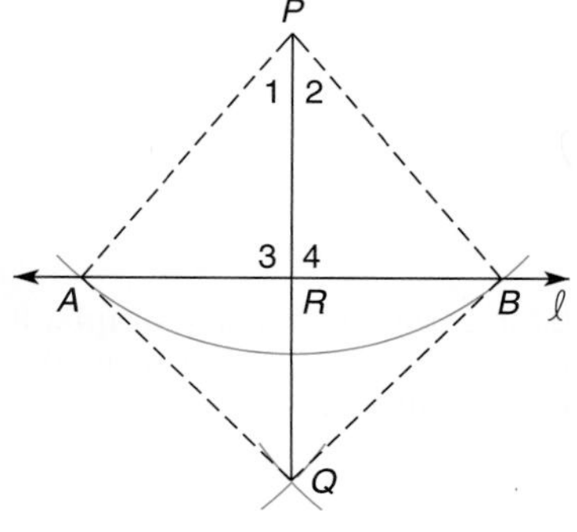


**FIGURE 3.42**

In Example 5, we justify the method for constructing a line perpendicular to a given line from a point not on that line. In the example, point $P$ lies above line $\ell$.

EXAMPLE 5

*Given:* $P$ not on $\ell$
$\overline{PA} \cong \overline{PB}$ (by construction)
$\overline{AQ} \cong \overline{BQ}$ (by construction)
(See Figure 3.42.)

*Prove:* $\overline{PQ} \perp \overline{AB}$

Provide the missing statements and reasons in the proof.

**PROOF**

| Statements | Reasons |
|---|---|
| 1. $P$ not on $\ell$ <br> $\overline{PA} \cong \overline{PB}$ and $\overline{AQ} \cong \overline{BQ}$ | 1. ? given |
| 2. $\overline{PQ} \cong \overline{PQ}$ | 2. ? identity |
| 3. $\triangle PAQ \cong \triangle PBQ$ | 3. ? SSS |
| 4. $\angle 1 \cong \angle 2$ | 4. ? CPCTC |
| 5. $\overline{PR} \cong \overline{PR}$ | 5. ? identity |
| (1), (4), (5) 6. $\triangle PRA \cong \triangle PRB$ | 6. ? SSS |
| 7. $\angle 3 \cong \angle 4$ | 7. ? CPCTC |
| 8. ? | 8. If two lines meet to form $\cong$ adjacent $\angle$s, these lines are $\perp$ |

In Example 6, we recall the method for constructing the line perpendicular to a given line at a point on the line. We illustrate the technique in the example and ask that the student justify the method in Exercise 29. In Example 6, we construct an angle that measures 45°.

EXAMPLE 6

Construct an angle that measures 45°.

***Solution***

We begin by constructing a line segment perpendicular to line $\ell$ at point $P$ [Figure 3.43(a)]. Next we bisect one of the right angles that was determined. The bisector forms an angle whose measure is 45° [Figure 3.43(b)].

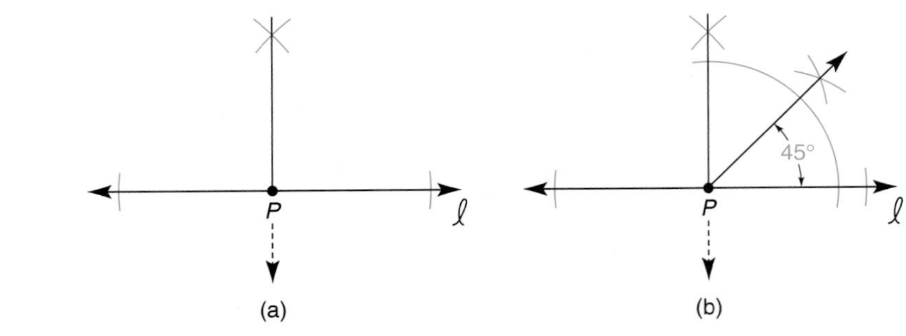

*Exs. 3–5*

**FIGURE 3.43**

As we saw in Example 4, constructing an equilateral triangle is fairly simple. It is also possible to construct other regular polygons, such as a square or a regular hexagon. In the following box, we recall some facts that will help us perform such constructions.

To construct a regular polygon with $n$ sides:

1. Each interior angle must measure $I = \frac{(n-2)180}{n}$ degrees; alternatively, each exterior angle must measure $E = \frac{360}{n}$ degrees.
2. All sides must be congruent.

EXAMPLE 7

Construct a regular hexagon having sides of length $a$.

***Solution***

We begin by marking off a line segment of length $a$, as shown in Figure 3.44(a). Each exterior angle of the hexagon ($n = 6$) must measure $E = \frac{360}{6} = 60°$; then each interior angle measures 120°. We construct an equilateral triangle (all sides measure $a$) so that a 60° exterior angle is formed, as shown in Figure 3.44(b). Again marking off an arc of length $a$ for the second side, we construct another exterior angle of measure 60° as shown in Figure 3.44(c). This procedure is continued until the regular hexagon *ABCDEF* is determined, as shown in Figure 3.44(d).

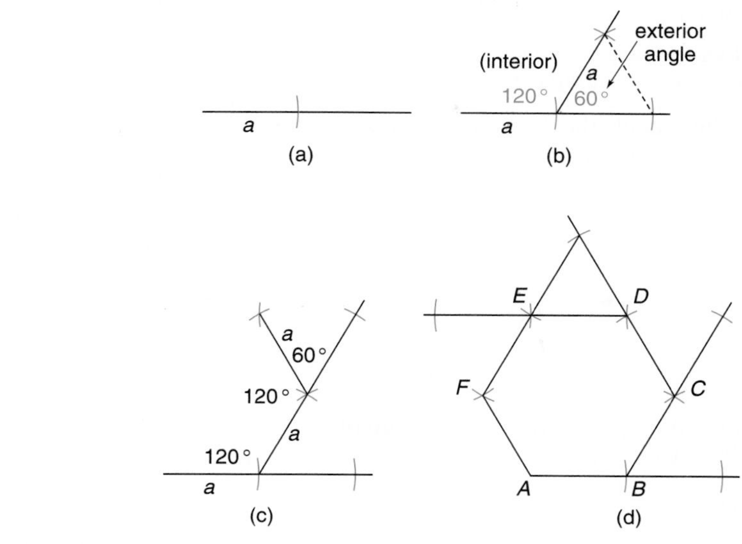

*Exs. 6–7*

**FIGURE 3.44**

## 3.4 Exercises

*In Exercises 1 to 6, use line segments of given lengths a, b, and c to perform the constructions.*

a

b

c

*Exercises 1–6*

1. Construct a line segment of length $2b$.
2. Construct a line segment of length $b + c$.
3. Construct a line segment of length $\frac{1}{2}c$.
4. Construct a line segment of length $a - b$.
5. Construct a triangle with sides of lengths $a$, $b$, and $c$.
6. Construct an isosceles triangle with a base of length $b$ and legs of length $a$.

*In Exercises 7 to 12, use the angles provided to perform the constructions.*

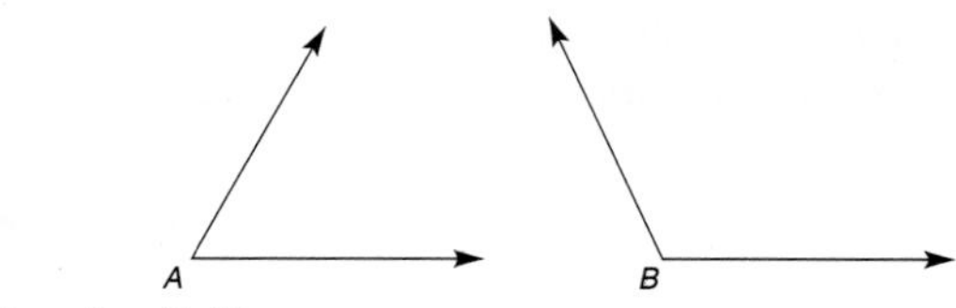


*Exercises 7–12*

7. Construct an angle that is congruent to acute $\angle A$.
8. Construct an angle that is congruent to obtuse $\angle B$.
9. Construct an angle that has one-half the measure of $\angle A$.
10. Construct an angle that has a measure equal to $m\angle B - m\angle A$.

11. Construct an angle that has twice the measure of $\angle A$.

12. Construct an angle whose measure averages the measures of $\angle A$ and $\angle B$.

*In Exercises 13 and 14, use the angles and lengths of sides provided to construct the triangle described.*

13. Construct the triangle that has sides of lengths $r$ and $t$ with included angle $S$.

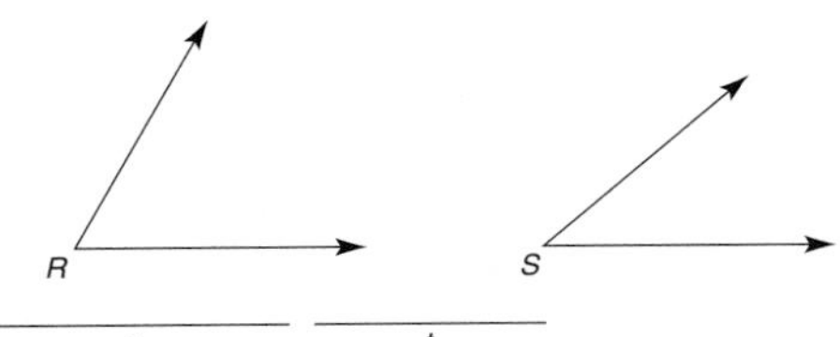


*Exercises 13, 14*

14. Construct the triangle that has a side of length $t$ included by angles $R$ and $S$.

*In Exercises 15 to 18, construct angles having the given measures.*

15. 90° and then 45°

16. 60° and then 30°

17. 30° and then 15°

18. 45° and then 105°
(HINT: 105° = 45° + 60°)

19. Describe how you would construct an angle measuring 22.5°.

20. Describe how you would construct an angle measuring 75°.

21. Construct the complement of the acute angle shown.

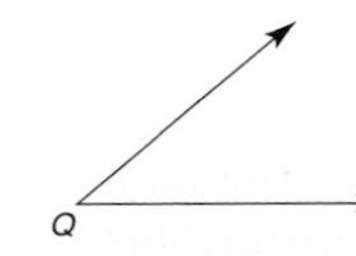


22. Construct the supplement of the obtuse angle shown.

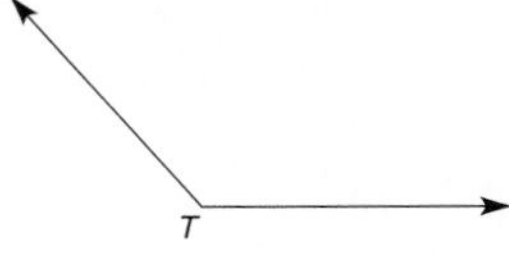


*In Exercises 23 to 26, use line segments of lengths a and c as shown.*

23. Construct the right triangle with hypotenuse of length $c$ and a leg of length $a$.

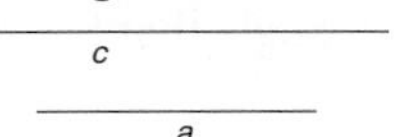


*Exercises 23–26*

24. Construct an isosceles triangle with base of length $c$ and altitude of length $a$.
(HINT: The altitude lies on the perpendicular bisector of the base.)

25. Construct an isosceles triangle with a vertex angle of 30° and each leg of length $c$.

26. Construct a right triangle with base angles of 45° and hypotenuse of length $c$.

*In Exercises 27 and 28, use the given angle and the line segment of length b.*

27. Construct the right triangle in which acute angle $R$ has a side (one leg of the triangle) of length $b$.

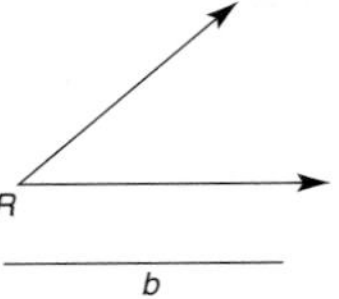


*Exercises 27, 28*

28. Construct an isosceles triangle with base of length $b$ and congruent base angles having the measure of angle $R$.

29. Complete the justification of the construction of the line perpendicular to a given line at a point on that line.

*Given:* Line $m$, with point $P$ on $m$
$\overline{PQ} \cong \overline{PR}$ (by construction)
$\overline{QS} \cong \overline{RS}$ (by construction)
*Prove:* $\overleftrightarrow{SP} \perp m$

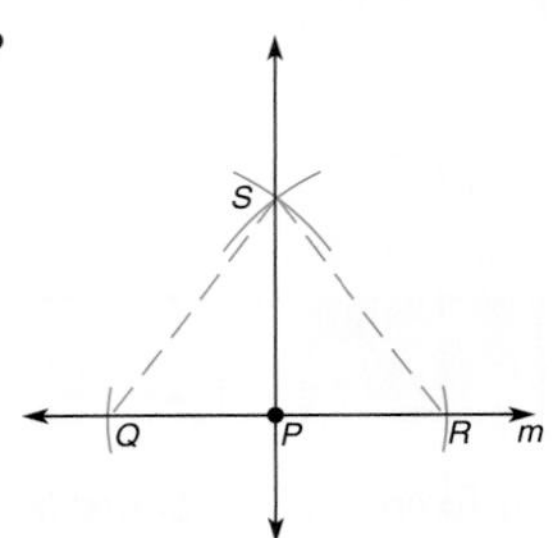


30. Complete the justification of the construction of the perpendicular bisector of a line segment.

*Given:* $\overline{AB}$ with $\overline{AC} \cong \overline{BC} \cong \overline{AD} \cong \overline{BD}$ (by construction)
*Prove:* $\overline{AM} \cong \overline{MB}$ and $\overleftrightarrow{CD} \perp \overline{AB}$

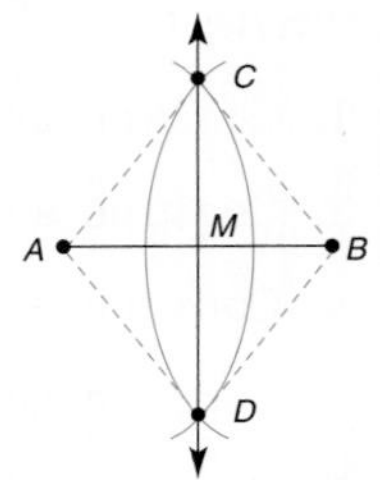

31. To construct a regular hexagon, what measure would be necessary for each interior angle? Construct an angle of that measure.

32. To construct a regular octagon, what measure would be necessary for each interior angle? Construct an angle of that measure.

33. To construct a regular dodecagon (12 sides), what measure would be necessary for each interior angle? Construct an angle of that measure.

34. Draw an acute triangle and construct the three medians of the triangle. Do the medians appear to meet at a common point?

35. Draw an obtuse triangle and construct the three altitudes of the triangle. Do the altitudes appear to meet at a common point?
(HINT: In the construction of two of the altitudes, sides need to be extended.)

36. Draw a right triangle and construct the angle bisectors of the triangle. Do the angle bisectors appear to meet at a common point?

37. Draw an obtuse triangle and construct the three perpendicular bisectors of its sides. Do the perpendicular bisectors of the three sides appear to meet at a common point?

38. Construct an equilateral triangle and its three altitudes. What does intuition tell you about the three medians, the three angle bisectors, and the three perpendicular bisectors of the sides of that triangle?

39. A carpenter has placed a square over an angle in such a manner that $\overline{AB} \cong \overline{AC}$ and $\overline{BD} \cong \overline{CD}$ (see drawing). What can you conclude about the location of point $D$?

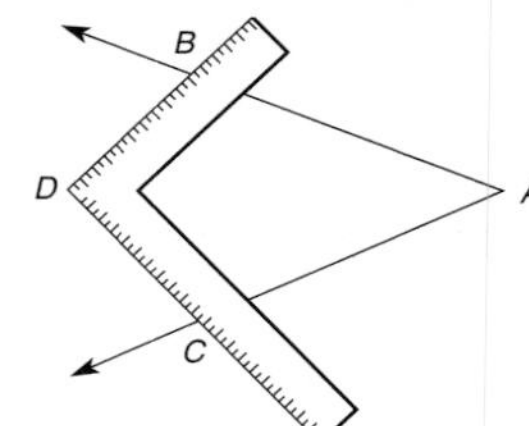


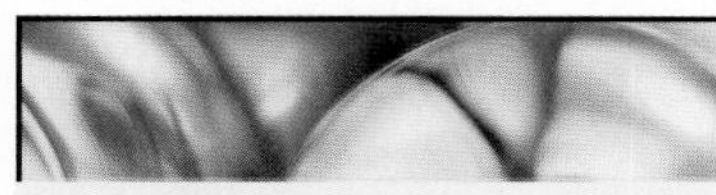

**KEY CONCEPTS**

Lemma • Inequality of Sides and Angles in a Triangle • The Triangle Inequality

# 3.5 Inequalities in a Triangle

Important inequality relationships exist among the measured parts of a triangle. To establish some of these, we recall and apply some facts from both algebra and geometry. A more in-depth review of inequalities can be found in Appendix A, Section A.3.

> DEFINITION: Let $a$ and $b$ be real numbers. $\boldsymbol{a > b}$ (read "a is greater than b") if and only if there is a positive number $p$ for which $a = b + p$.

For instance, $9 > 4$, because there is the positive number 5 for which $9 = 4 + 5$. Because $5 + 2 = 7$, we also know that $7 > 2$ and $7 > 5$. In geometry, let $A$-$B$-$C$ on $\overline{AC}$ so that $AB + BC = AC$; then $AC > AB$, because $BC$ is a positive number.

*Exs. 1–3*

## LEMMAS (HELPING THEOREMS)

We will use the following theorems to help us prove the theorems found later in this section. In their role as "helping" theorems, each of the five boxed statements that follow is called a **lemma.** We will prove the first four lemmas, because their content is geometric.

> LEMMA 3.5.1: If $B$ is between $A$ and $C$ on $\overline{AC}$, then $AC > AB$ and $AC > BC$. (The measure of a line segment is greater than the measure of any of its parts. See Figure 3.45.)

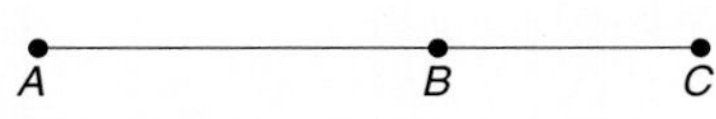


**FIGURE 3.45**

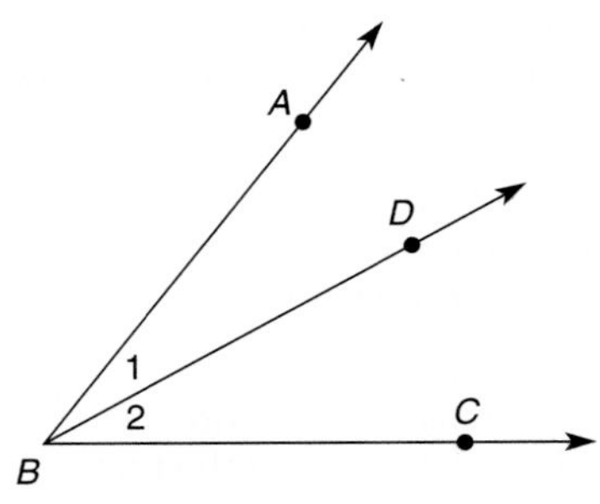


FIGURE 3.46

*Proof*

By the Segment-Addition Postulate, $AC = AB + BC$. According to the Ruler Postulate, $BC > 0$ (meaning $BC$ is positive); it follows that $AC > AB$. Similarly, $AC > BC$. These relationships follow logically from the definition of $a > b$.

---

LEMMA 3.5.2: If $\overrightarrow{BD}$ separates $\angle ABC$ into two parts ($\angle 1$ and $\angle 2$), then $m\angle ABC > m\angle 1$ and $m\angle ABC > m\angle 2$. (The measure of an angle is greater than the measure of any of its parts. See Figure 3.46.)

*Proof*

By the Angle-Addition Postulate, $m\angle ABC = m\angle 1 + m\angle 2$. Using the Protractor Postulate, $m\angle 2 > 0$; it follows that $m\angle ABC > m\angle 1$. Similarly, $m\angle ABC > m\angle 2$.

---

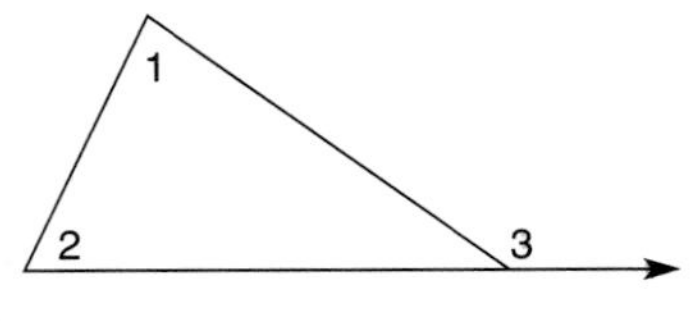


FIGURE 3.47

LEMMA 3.5.3: If $\angle 3$ is an exterior angle of a triangle and $\angle 1$ and $\angle 2$ are the nonadjacent interior angles, then $m\angle 3 > m\angle 1$ and $m\angle 3 > m\angle 2$. (The measure of an exterior angle of a triangle is greater than the measure of either nonadjacent interior angle. See Figure 3.47.)

*Proof*

Because the measure of an exterior angle of a triangle equals the sum of measures of the two nonadjacent interior angles, $m\angle 3 = m\angle 1 + m\angle 2$. It follows that $m\angle 3 > m\angle 1$ and $m\angle 3 > m\angle 2$.

---

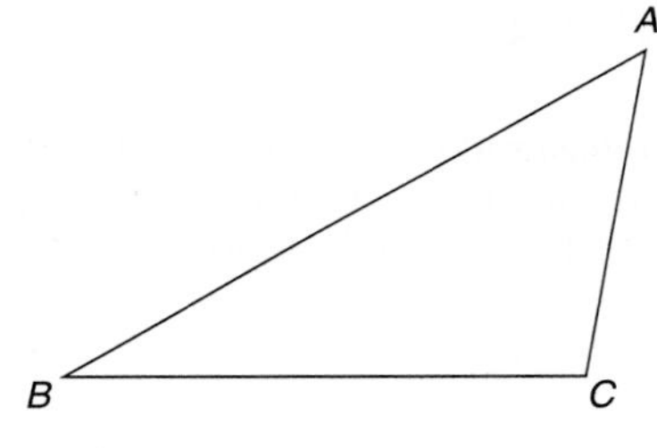


FIGURE 3.48

LEMMA 3.5.4: In $\triangle ABC$, if $\angle C$ is a right angle or an obtuse angle, then $m\angle C > m\angle A$ and $m\angle C > m\angle B$. (If a triangle contains a right or an obtuse angle, then the measure of this angle is greater than the measure of either of the remaining angles. See Figure 3.48.)

*Proof*

In $\triangle ABC$, $m\angle A + m\angle B + m\angle C = 180°$. With $m\angle C \geq 90°$, it follows that $m\angle A + m\angle B \leq 90°$, and each angle ($\angle A$ and $\angle B$) must be acute. Thus $m\angle C > m\angle A$ and $m\angle C > m\angle B$.

---

The following theorem (also a lemma) is used in Example 1. Its proof (not given) depends on the definition of "is greater than," which is found on the previous page.

LEMMA 3.5.5: **(Addition Property of Inequality)** If $a > b$ and $c > d$, then $a + c > b + d$.

EXAMPLE 1

Give a paragraph proof for the following problem.

*Given:* $AB > CD$ and $BC > DE$

*Prove:* $AC > CE$

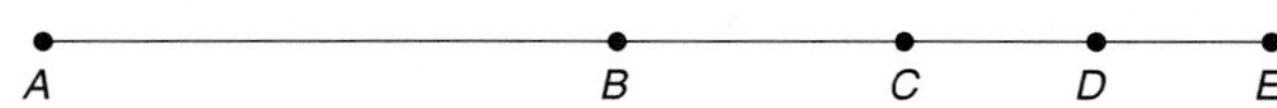


**FIGURE 3.49**

*Proof:* If $AB > CD$ and $BC > DE$, then $AB + BC > CD + DE$ by Lemma 3.5.5. But $AB + BC = AC$ and $CD + DE = CE$ by the Segment-Addition Postulate. Using substitution, it follows that $AC > CE$.

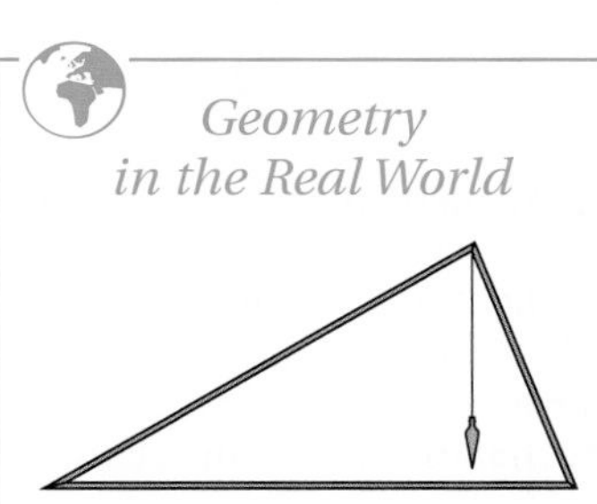


A carpenter's "plumb" determines the shortest distance to a horizontal line. A vertical brace provides structural support for the roof.

The paragraph proof in Example 1 could have been written in this standard format.

**PROOF**

| Statements | Reasons |
|---|---|
| 1. $AB > CD$ and $BC > DE$ | 1. Given |
| 2. $AB + BC > CD + DE$ | 2. Lemma 3.5.5 |
| 3. $AB + BC = AC$ and $CD + DE = CE$ | 3. Segment-Addition Postulate |
| 4. $AC > CE$ | 4. Substitution |

*Exs. 4–8*

The paragraph proof and the two-column proof of Example 1 are equivalent. In either form, statements must be ordered and justified.

The remaining theorems are the "heart" of this section. Before studying the theorem and its proof, it is a good idea to visualize each theorem. Many statements of inequality are intuitive; that is, they are easy to believe even though they may not be easily proved.

Study Theorem 3.5.6 and consider Figure 3.50. It appears that $m\angle C > m\angle B$.

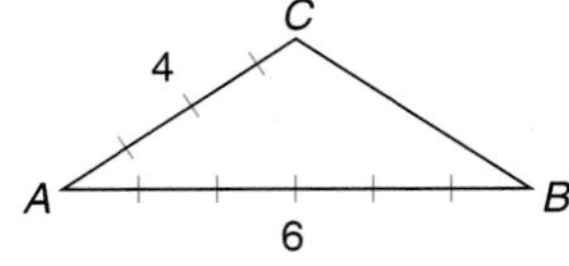


**FIGURE 3.50**

THEOREM 3.5.6: If one side of a triangle is longer than a second side, then the measure of the angle opposite the longer side is greater than the measure of the angle opposite the shorter side.

EXAMPLE 2

Provide a paragraph proof of Theorem 3.5.6.

*Given:* $\triangle ABC$, with $AC > BC$ [See Figure 3.51(a) on page 162.]

*Prove:* $m\angle B > m\angle A$

**Technology Exploration**

Use computer software if available.

1) Draw a $\triangle ABC$ with $\overline{AB}$ as the longest side.
2) Measure $\angle A$, $\angle B$, and $\angle C$.
3) Show that $\angle C$ has the greatest measure.

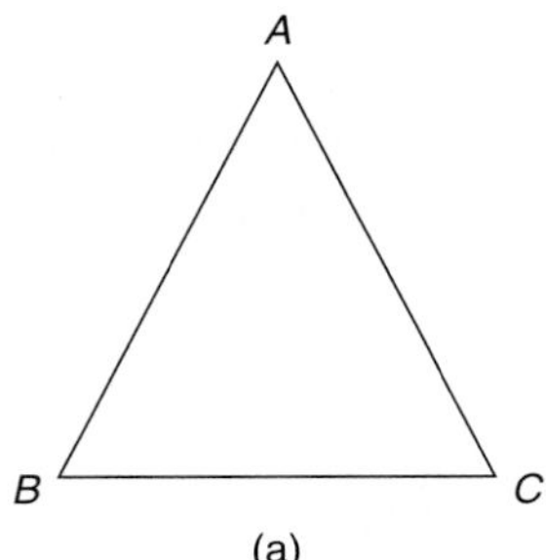


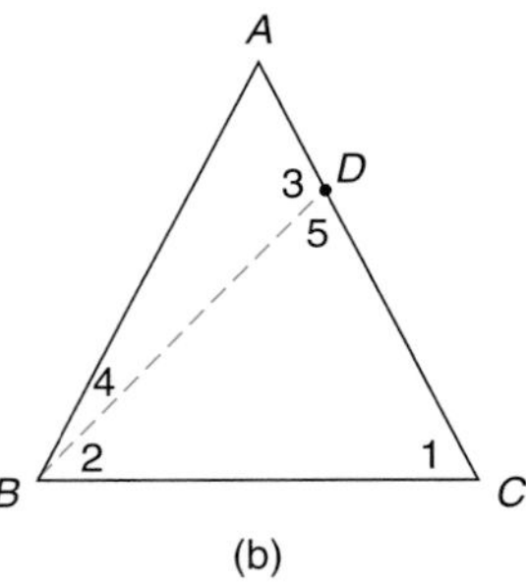


**FIGURE 3.51**

*Proof:* Given $\triangle ABC$ with $AC > BC$, we use the Ruler Postulate to locate point $D$ on $\overline{AC}$ so that $\overline{CD} \cong \overline{BC}$, as in Figure 3.51(b). Now $m\angle 2 = m\angle 5$ in the isosceles triangle $BDC$. By Lemma 3.5.2, $m\angle ABC > m\angle 2$; therefore, $m\angle ABC > m\angle 5$ (*) by substitution. But $m\angle 5 > m\angle A$ (*), because $\angle 5$ is an exterior angle of $\triangle ADB$. Using the two starred statements, we can conclude by the Transitive Property of Inequality that $m\angle ABC > m\angle A$; that is, $m\angle B > m\angle A$ in Figure 3.51(a).

The relationship described in Theorem 3.5.6 extends, of course, to all sides and all angles of a triangle. That is, the largest of the three angles of a triangle is opposite the longest side, and the smallest angle is opposite the shortest side.

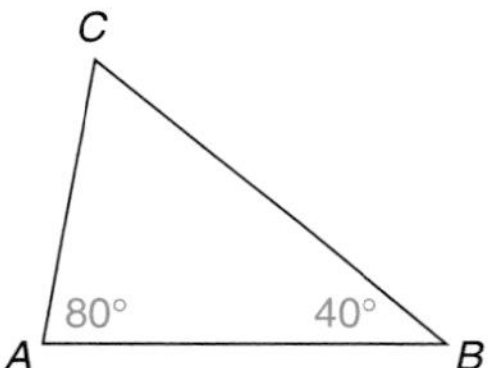


**FIGURE 3.52**

EXAMPLE 3

Given that the three sides of $\triangle ABC$ (not shown) are $AB = 4$, $BC = 5$, and $AC = 6$, arrange the angles by size.

***Solution***

Because $AC > BC > AB$, the largest angle lies opposite $\overline{AC}$, and that is $\angle B$. The angle intermediate in size is $\angle A$, which lies opposite $\overline{BC}$. The smallest angle is $\angle C$, which lies opposite $\overline{AB}$, the shortest side. Thus the order of the angles by size is

$$m\angle B > m\angle A > m\angle C$$

**Discover!**

Using construction paper and a protractor, draw $\triangle RST$ so that $m\angle R = 75°$, $m\angle S = 60°$, and $m\angle T = 45°$. Measure the length of each side.

a) Which side is longest?
b) Which side is shortest?

ANSWERS

(a) $\overline{ST}$ (b) $\overline{RS}$

The converse of Theorem 3.5.6 is also true. It is necessary, however, to use an indirect proof to establish the converse. Recall that this method of proof begins by supposing the opposite of what we want to show. Because this assumption leads to a contradiction, the assumption must be false and the desired claim is therefore true.

Study Theorem 3.5.7 and consider Figure 3.52, in which $m\angle A = 80°$ and $m\angle B = 40°$. It appears that the longer side lies opposite the larger angle; that is, it appears that $BC > AC$.

THEOREM 3.5.7: If the measure of one angle of a triangle is greater than the measure of a second angle, then the side opposite the larger angle is longer than the side opposite the smaller angle.

The proof of Theorem 3.5.7 depends on this fact: Given real numbers $a$ and $b$, only one of the following can be true.

$$a > b, \qquad a = b, \qquad \text{or} \qquad a < b$$

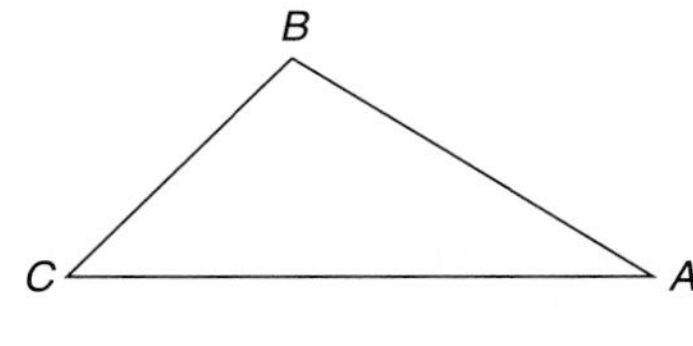

**FIGURE 3.53**

EXAMPLE 4

Prove Theorem 3.5.7 by using an indirect approach.

*Given:* $\triangle ABC$ with $m\angle B > m\angle A$ (See Figure 3.53.)

*Prove:* $AC > BC$

*Proof:* Given $\triangle ABC$ with $m\angle B > m\angle A$, assume that $AC \leq BC$. But if $AC = BC$, then $m\angle B = m\angle A$, which contradicts the hypothesis. Also, if $AC < BC$, then it follows by the previous theorem that $m\angle B < m\angle A$, which also contradicts the hypothesis. Thus the assumed statement must be false, and it follows that $AC > BC$.

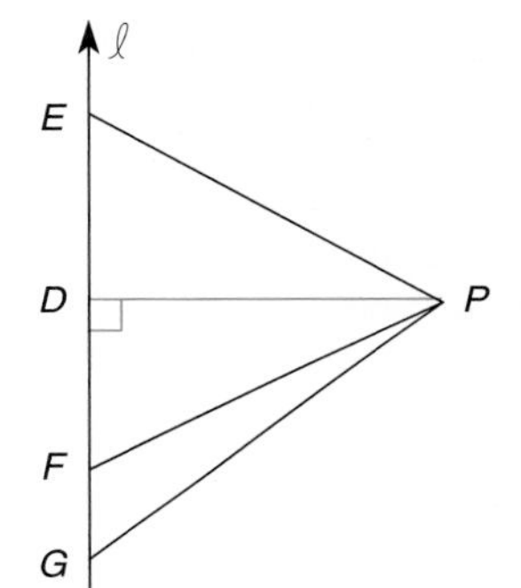

**FIGURE 3.54**

EXAMPLE 5

Given $\triangle RST$ in which $m\angle R = 90°$, $m\angle S = 60°$, and $m\angle T = 30°$, write an extended inequality that compares the lengths of the three sides.

***Solution***

With $m\angle R > m\angle S > m\angle T$, it follows that the sides opposite these $\angle$s are unequal in the same order. That is,

$$ST > RT > SR$$

*Exs. 9–12*

The following corollary is a consequence of Theorem 3.5.7.

COROLLARY 3.5.8: The perpendicular segment from a point to a line is the shortest segment that can be drawn from the point to the line.

In Figure 3.54, $PD < PE$, $PD < PF$, and $PD < PG$. In every case, $\overline{PD}$ is opposite an acute angle of a triangle, whereas the second segment is always opposite a right angle (necessarily the largest angle of the triangle involved). With $\overline{PD} \perp \ell$, we say that $PD$ is the *distance* from $P$ to $\ell$.

Corollary 3.5.8 can easily be extended to three dimensions.

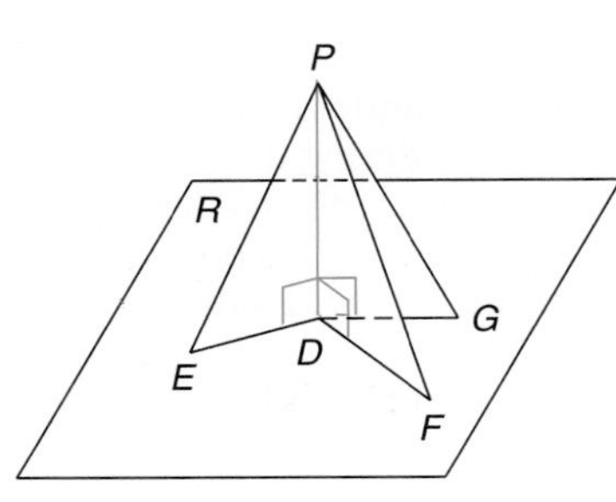

**FIGURE 3.55**

COROLLARY 3.5.9: The perpendicular segment from a point to a plane is the shortest segment that can be drawn from the point to the plane.

In Figure 3.55, $\overline{PD}$ is a leg of each right triangle shown. With $\overline{PE}$ the hypotenuse of $\triangle PDE$, $\overline{PF}$ the hypotenuse of $\triangle PDF$, and $\overline{PG}$ the hypotenuse of $\triangle PDG$, the length of $\overline{PD}$ is less than that of $\overline{PE}$, $\overline{PF}$, $\overline{PG}$, or any other line segment joining point $P$ to a point in plane $R$. The length of $\overline{PD}$ is known as the *distance* from point $P$ to plane $R$.

Our final theorem shows that no side of a triangle can have a length greater than or equal to the sum of the lengths of the other two sides. In the proof, the relationship is validated for only one of three possible inequalities. Theorem 3.5.10 is often called the Triangle Inequality. (See Figure 3.56 on page 164.)

*Exs. 13–14*

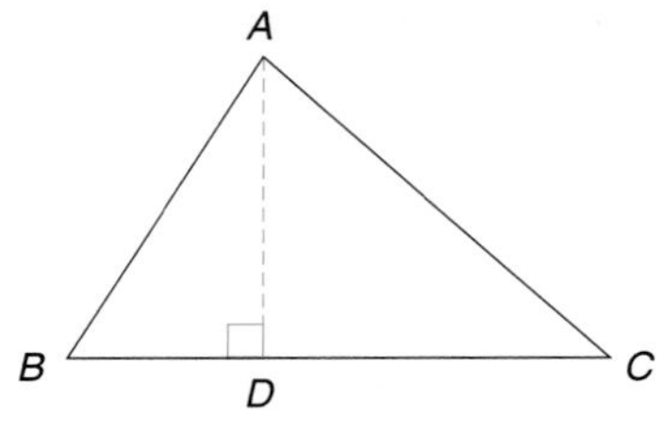


**FIGURE 3.56**

> THEOREM 3.5.10: **(Triangle Inequality)** The sum of the lengths of any two sides of a triangle is greater than the length of the third side.

*Given:* $\triangle ABC$

*Prove:* $BA + CA > BC$

*Proof:* Draw $\overline{AD} \perp \overline{BC}$. Because the shortest segment from a point to $\overline{AD}$ is the perpendicular segment, $BA > BD$ and $CA > CD$. Using Lemma 3.5.5, we add the inequalities; $BA + CA > BD + CD$. By the Segment-Addition Postulate, the sum $BD + CD$ can be replaced by $BC$ to yield $BA + CA > BC$.

The following statement is an alternative and expanded form of Theorem 3.5.10. If $a$, $b$, and $c$ are the lengths of the sides of a triangle and $c$ is the length of any side, then $a - b < c < a + b$.

> THEOREM 3.5.10: **(Triangle Inequality)** The length of any side of a triangle must lie between the sum and difference of the lengths of the other two sides.

### EXAMPLE 6

Can a triangle have sides of the following lengths?

a) 3, 4, and 5
b) 3, 4, and 7
c) 3, 4, and 8
d) 3, 4, and $x$

***Solution***

a) Yes, because no side has a length greater than or equal to the sum of the lengths of the other two sides
b) No, because $7 = 3 + 4$ (need $4 - 3 < 7 < 3 + 4$)
c) No, because $8 > 3 + 4$ (need $4 - 3 < 8 < 3 + 4$)
d) Yes, if $4 - 3 < x < 4 + 3$

From Example 6, you can see that the length of one side cannot be greater than or equal to the sum of the lengths of the other two sides. Considering the alternative form of Theorem 3.5.10, we see that $4 - 3 < 5 < 4 + 3$ in part (a). When 5 [as in part (a)] is replaced by 7 [as in part (b)] or 8 [as in part (c)], this inequality becomes a false statement. Part (d) of Example 6 shows that the length of the third side must be between 1 and 7.

Our final example illustrates a practical application of inequality relationships in triangles.

### EXAMPLE 7

On a map, firefighters are located at points $A$ and $B$. A fire has broken out at point $C$. Which group of firefighters is nearer the location of the fire? (See Figure 3.57 on page 165.)

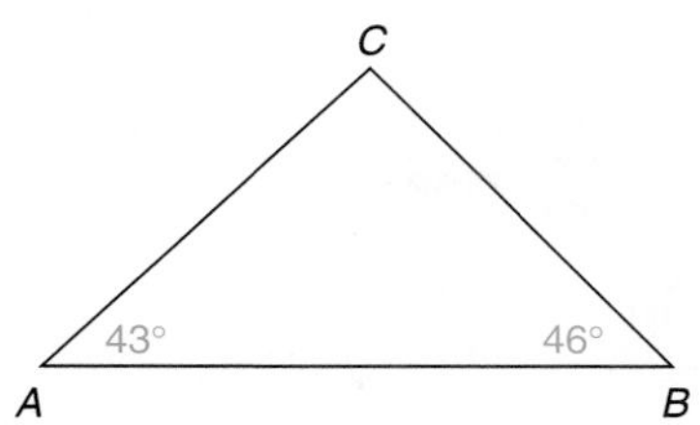


**FIGURE 3.57**

***Solution***

With m∠$A$ = 43° and m∠$B$ = 46°, it follows that $AC > BC$. Because the distance from $B$ to $C$ is less than the distance from $A$ to $C$, the firefighters at site $B$ should be dispatched to the fire located at $C$.

NOTE: In Example 7 we assume that highways from $A$ and $B$ (to $C$) are equally accessible.

*Exs. 15–18*

## 3.5 Exercises

*In Exercises 1 to 10, classify each statement as true or false.*

1. $\overline{AB}$ is the longest side of △$ABC$.

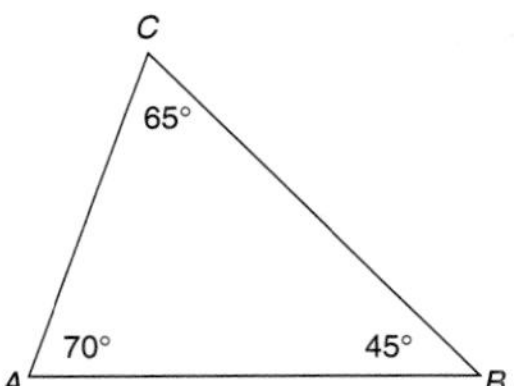


*Exercises 1, 2*

2. $AB < BC$
3. $DB > AB$

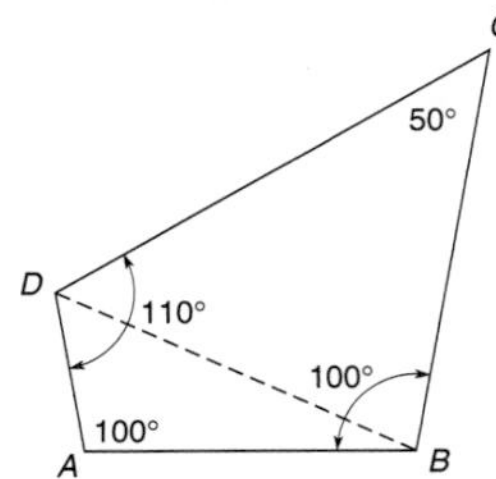


*Exercises 3, 4*

4. Because m∠$A$ = m∠$B$, it follows that $DA = DC$.
5. m∠$A$ + m∠$B$ = m∠$C$

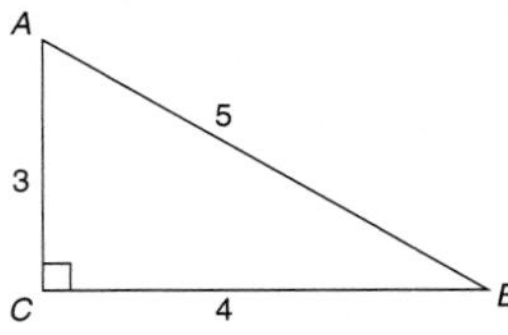


*Exercises 5, 6*

6. m∠$A$ > m∠$B$
7. $DF > DE + EF$

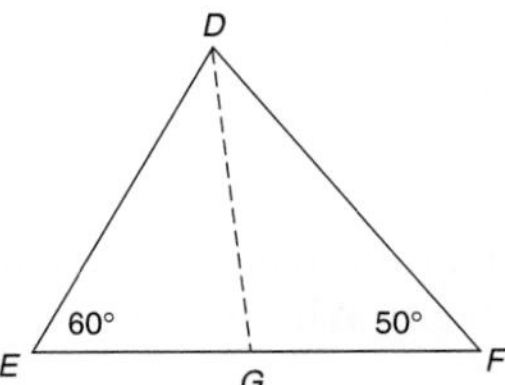


*Exercises 7, 8*

8. If $\overrightarrow{DG}$ is the bisector of ∠$EDF$, then $DG > DE$.

9. $DA > AC$

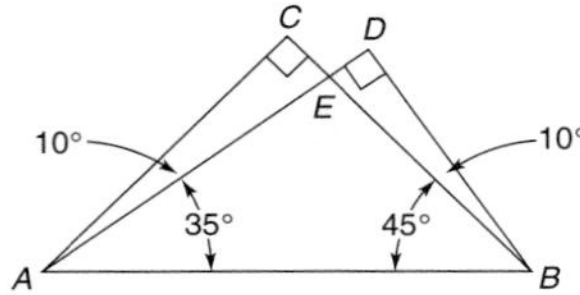


*Exercises 9, 10*

10. $CE = ED$

11. If possible, draw a triangle whose angles measure:

 a) 100°, 100°, and 60°
 b) 45°, 45°, and 90°

12. If possible, draw a triangle whose angles measure:

 a) 80°, 80°, and 50°
 b) 50°, 50°, and 80°

13. If possible, draw a triangle whose sides measure:

 a) 8, 9, and 10
 b) 8, 9, and 17
 c) 8, 9, and 18

14. If possible, draw a triangle whose sides measure:

 a) 7, 7, and 14
 b) 6, 7, and 14
 c) 6, 7, and 8

*In Exercises 15 to 18, describe the triangle ($\triangle XYZ$, not shown) as scalene, isosceles, or equilateral. Also, is the triangle acute, right, or obtuse?*

15. $m\angle X = 43°$ and $m\angle Y = 47°$

16. $m\angle X = 60°$ and $\angle Y \cong \angle Z$.

17. $m\angle X = m\angle Y = 40°$

18. $m\angle X = 70°$ and $m\angle Y = 40°$

19. Two of the sides of an isosceles triangle have lengths of 10 cm and 4 cm. Which length must be the length of the base?

20. The sides of a right triangle have lengths of 6 cm, 8 cm, and 10 cm. Which length is that of the hypotenuse?

21. A triangle is both isosceles *and* acute. If one angle of the triangle measures 36°, what is the measure of the largest angle(s) of the triangle? What is the measure of the smallest angle(s) of the triangle?

22. One of the angles of an isosceles triangle measures 96°. What is the measure of the largest angle(s) of the triangle? What is the measure of the smallest angle(s) of the triangle?

23. NASA in Huntsville, Alabama (at point $H$), has called a manufacturer for parts needed as soon as possible. NASA will, in fact, send a courier for the necessary equipment. The manufacturer has two distribution centers located in nearby Tennessee—one in Nashville (at point $N$) and the other in Jackson (at point $J$). Using the angle measurements indicated on the accompanying map, determine to which town the courier should be dispatched to obtain the needed parts.

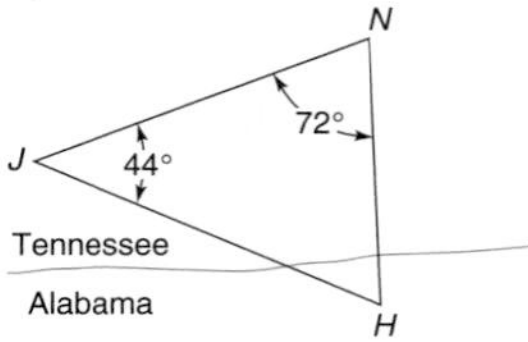


24. A tornado has just struck a small Kansas community at point $T$. There are Red Cross units stationed in both Salina (at point $S$) and Wichita (at point $W$). Using the angle measurements indicated on the accompanying map, determine which Red Cross unit would reach the victims first. (Assume that both units have the same mode of travel and accessible roadways available.)

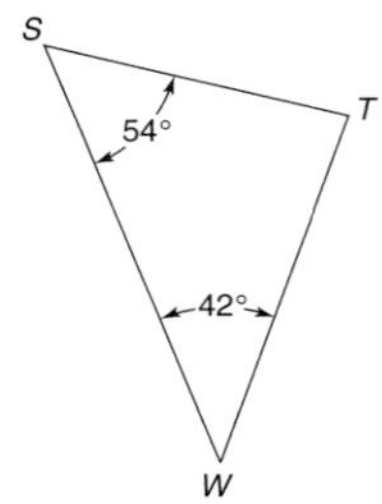


*In Exercises 25 and 26, complete each proof.*

25. *Given:* $m\angle ABC > m\angle DBE$
 $m\angle CBD > m\angle EBF$
 *Prove:* $m\angle ABD > m\angle DBF$

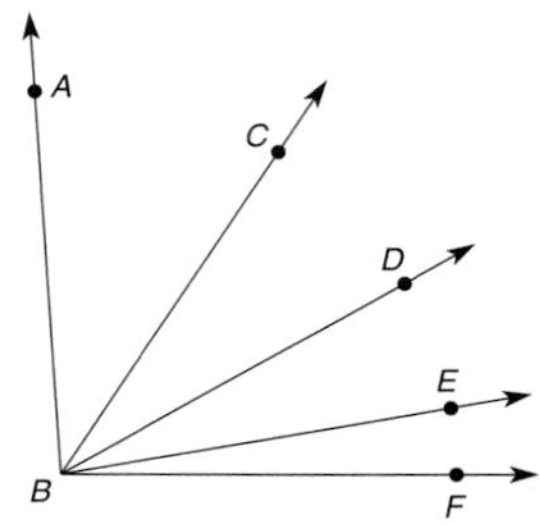

**PROOF**

| Statements | Reasons |
|---|---|
| 1. ? | 1. Given |
| 2. $m\angle ABC + m\angle CBD >$ $m\angle DBE + m\angle EBF$ | 2. Addition Property of Inequality |
| 3. $m\angle ABD = m\angle ABC +$ $m\angle CBD$ and $m\angle DBF =$ $m\angle DBE + m\angle EBF$ | 3. ? |
| 4. ? | 4. Substitution |

26. *Given:* Equilateral $\triangle ABC$ and $D$-$B$-$C$
*Prove:* $DA > AC$

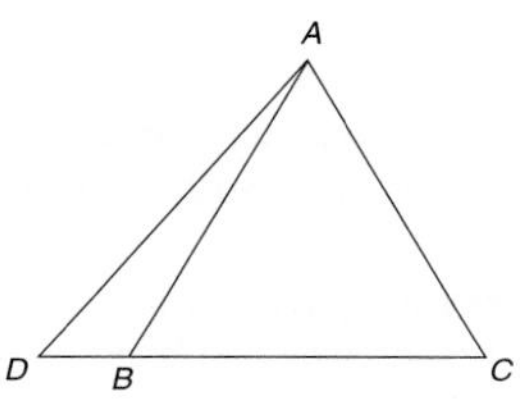


**PROOF**

| Statements | Reasons |
|---|---|
| 1. ? | 1. Given |
| 2. $\triangle ABC$ is equiangular, so $m\angle ABC = m\angle C$ | 2. ? |
| 3. $m\angle ABC > m\angle D$ ($\angle D$ of $\triangle ABD$) | 3. The measure of an ext. $\angle$ of a $\triangle$ is greater than the measure of either nonadjacent int. $\angle$ |
| 4. ? | 4. Substitution |
| 5. ? | 5. ? |

*In Exercises 27 and 28, construct proofs.*

27. *Given:* Quadrilateral $RSTU$ with diagonal $\overline{US}$
$\angle R$ and $\angle TUS$ are right $\angle$s
*Prove:* $TS > UR$

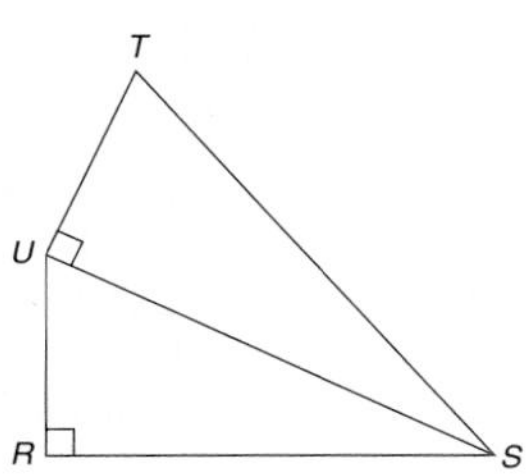


28. *Given:* Quadrilateral $ABCD$ with $\overline{AB} \cong \overline{DE}$
*Prove:* $DC > AB$

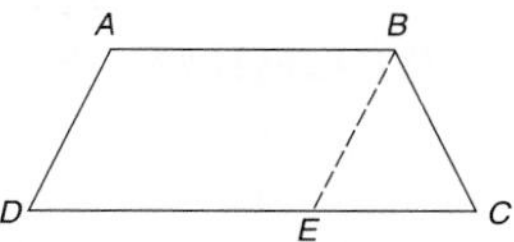


29. For $\triangle ABC$ and $\triangle DEF$ (not shown), suppose that $\overline{AC} \cong \overline{DF}$ and $\overline{AB} \cong \overline{DE}$ but that $m\angle A < m\angle D$. Draw a conclusion regarding the lengths of $\overline{BC}$ and $\overline{EF}$.

30. In $\triangle MNP$ (not shown), $\overrightarrow{MQ}$ bisects $\angle NMP$ and $MN < MP$. Draw a conclusion about the relative lengths of $\overline{NQ}$ and $\overline{QP}$.

*In Exercises 31 to 34, apply a form of Theorem 3.5.10.*

31. The sides of a triangle have lengths of 4, 6, and $x$. Write an inequality that states the possible values of $x$.

32. The sides of a triangle have lengths of 7, 13, and $x$. As in Exercise 31, write an inequality that describes the possible values of $x$.

33. If the lengths of two sides of a triangle are represented by $2x + 5$ and $3x + 7$ (in which $x$ is positive), describe in terms of $x$ the possible lengths of the third side whose length is represented by $y$.

34. Prove by the indirect method: "The length of a diagonal of a square is not equal in length to the length of any of the sides of the square."

35. Prove by the indirect method:
*Given:* $\triangle MPN$ is not isosceles
*Prove:* $PM \neq PN$

*In Exercises 36 and 37, prove each theorem.*

36. The length of the median from the vertex of an isosceles triangle is less than the length of either of the legs.

37. The length of an altitude of an acute triangle is less than the length of either side containing the same vertex as the altitude.

# Perspective on History

## SKETCH OF ARCHIMEDES

Whereas Euclid (see Perspective on History, Chapter 2) was a great teacher and wrote so that the majority might understand the principles of geometry, Archimedes wrote only for the very well educated mathematicians and scientists of his day. Archimedes (287–212 B.C.) wrote on such topics as the measure of the circle, the quadrature of the parabola, and spirals. In his works, Archimedes found a very good approximation of $\pi$. His other geometric works included investigations of conic sections and spirals, and he also wrote about physics. He was a great inventor and is probably remembered more for his inventions than for his writings.

Several historical events concerning the life of Archimedes have been substantiated, and one account involves his detection of a dishonest goldsmith. In that story, Archimedes was called upon to determine whether the crown that had been ordered by the king was constructed entirely of gold. By applying the principle of hydrostatics (which he had discovered), Archimedes established that the goldsmith had not constructed the crown entirely of gold. (The principle of hydrostatics states that an object placed in a fluid displaces an amount of fluid equal in weight to the amount of weight the object loses while submerged.)

One of his inventions is known as Archimedes' screw. This device allows water to flow from one level to a higher level so that, for example, holds of ships can be emptied of water. Archimedes' screw was used in Egypt to drain fields when the Nile River overflowed its banks.

When Syracuse (where Archimedes lived) came under siege by the Romans, Archimedes designed a long-range catapult that was so effective that Syracuse was able to fight off the powerful Roman army for three years before being overcome.

One report concerning the inventiveness of Archimedes has been treated as false, because his result has not been duplicated. It was said that he designed a wall of mirrors that could focus and reflect the sun's heat with such intensity as to set fire to Roman ships at sea. Because recent experiments with concave mirrors have failed to produce such intense heat, this account is difficult to believe.

Archimedes eventually died at the hands of a Roman soldier, even though the Roman army had been given orders not to harm him. After his death, the Romans honored his brilliance with a tremendous monument displaying the figure of a sphere inscribed in a right circular cylinder.

# Perspective on Application

## PASCAL'S TRIANGLE

Blaise Pascal (1623–1662) was a French mathematician who contributed to several areas of mathematics, including conic sections, calculus, and the invention of a calculating machine. But Pascal's name is most often associated with the array of numbers known as Pascal's Triangle, which follows:

```
        1
      1   1
    1   2   1
  1   3   3   1
1   4   6   4   1
```

Each row of entries in Pascal's Triangle begins and ends with the number 1. Intermediate entries in each row are found by the addition of the upper- left and upper-right entries of the preceding row. The row following 1 4 6 4 1 has the form

```
  1   4   6   4   1
↓ ↙ ↓ ↙ ↓ ↙ ↓ ↙ ↓
1   5   10  10  5   1
```

Applications of Pascal's Triangle include the counting of subsets of a given set, which we will consider in the following paragraph. While we do not pursue this notion, Pascal's Triangle is also useful in the algebraic expansion of a binomial to a power such as $(a + b)^2$, which equals $a^2 + 2ab + b^2$. Notice that the multipliers in the product found with exponent 2 are 1 2 1, from a row of Pascal's Triangle. In fact, the expansion $(a + b)^3$ leads to $a^3 + 3a^2b + 3ab^2 + b^3$, in which the multipliers (also known as coefficients) take the form 1 3 3 1, a row of Pascal's Triangle.

### Subsets of a Given Set

A subset of a given set is a set formed from choices of elements from the given set. Because a subset of a set with $n$ elements can have from 0 to $n$ elements, we find that Pascal's Triangle provides a count of the number of subsets containing a given counting number of elements.

| *Pascal's Triangle* | *Set* | *Number of Elements* | *Subsets of the Set* | *Number of Subsets* |
|---|---|---|---|---|
| **1** | ∅ | **0** | ∅ | **1** |
| **1 1** | {a} | **1** | ∅, {a}<br>**1 + 1 subsets** | **2** |
| **1 2 1** | {a, b} | **2** | ∅, {a}, {b}, {a, b}<br>**1 + 2 + 1 subsets** | **4** |
| **1 3 3 1** | {a, b, c} | **3** | ∅, {a}, {b}, {c}, {a, b}, {a, c}, {b, c}, {a, b, c}<br>**1 + 3 + 3 + 1 subsets** | **8** |

**1 subset of 0 elements, 3 subsets of 1 element each, 3 subsets of 2 elements each, 1 subset of 3 elements**

In algebra, it is shown that $2^0 = 1$; not by coincidence, the set $\varnothing$, which has 0 elements, has 1 subset. Just as $2^1 = 2$, the set $\{a\}$ which has 1 element, has 2 subsets, The pattern continues so that a set with 2 elements has $2^2 = 4$ subsets and a set with 3 elements has $2^3 = 8$ subsets. A quick examination suggests this fact:

> The total number of subsets for a set with $n$ elements is $2^n$.

The entries of the fifth row of Pascal's Triangle correspond to the numbers of subsets of the four-element set $\{a, b, c, d\}$; of course, the subsets of $\{a, b, c, d\}$ must have 0 elements, 1 element each, 2 elements each, 3 elements each, or 4 elements each. Based upon the preceding principle, there will be a total of $2^4 = 16$ subsets for $\{a, b, c, d\}$.

**EXAMPLE 1**

List all 16 subsets of the set $\{a, b, c, d\}$ by considering the fifth row of Pascal's Triangle, namely 1 4 6 4 1. Notice also that $1 + 4 + 6 + 4 + 1$ must equal 16.

***Solution***

$\varnothing$, $\{a\}$, $\{b\}$, $\{c\}$, $\{d\}$, $\{a, b\}$, $\{a, c\}$, $\{a, d\}$, $\{b, c\}$, $\{b, d\}$, $\{c, d\}$, $\{a, b, c\}$, $\{a, b, d\}$, $\{a, c, d\}$, $\{b, c, d\}$, $\{a, b, c, d\}$.

**EXAMPLE 2**

Find the number of subsets for a set with six elements.

***Solution***

The number of subsets is $2^6$, or 64.

Looking back at Example 1, we notice that the number of subsets of the four-element set $\{a, b, c, d\}$ is $1 + 4 + 6 + 4 + 1$, which equals 16, or $2^4$. The preceding principle can be restated in the following equivalent form:

> The sum of the entries in row $n$ of Pascal's Triangle is $2^{n-1}$.

**EXAMPLE 3**

The sixth row of Pascal's Triangle is 1 5 10 10 5 1. Use the principle above to find the sum of the entries of this row.

***Solution***

With $n = 6$, it follows that $n - 1 = 5$. Then $1 + 5 + 10 + 10 + 5 + 1 = 2^5$, or 32.

NOTE: There are 32 subsets for a set containing five elements; consider $\{a, b, c, d, e\}$.

In closing, we note that many of the principles based upon Pascal's Triangle have not been explored in this Perspective on Application!

# Summary

### A LOOK BACK AT CHAPTER 3

In this chapter, we considered several methods for proving triangles congruent. We explored properties of isosceles triangles and justified construction methods of earlier chapters. Inequality relationships for the sides and angles of a triangle were also investigated.

### A LOOK AHEAD TO CHAPTER 4

In the next chapter, we use properties of triangles to develop the properties of quadrilaterals. We consider several special types of quadrilaterals, including the parallelogram, kite, rhombus, and trapezoid.

### KEY CONCEPTS

3.1 Congruent Triangles • SSS, SAS, ASA, AAS • Included Angle, Included Side • Reflexive Property of Congruence (Identity) • Symmetric, Transitive Properties of Congruence

3.2 CPCTC • Hypotenuse and Legs of a Right Triangle • HL • Pythagorean Theorem • Square Roots Property

3.3 Isosceles Triangle • Vertex, Legs, and Base of an Isosceles Triangle • Base Angles • Vertex Angle • Angle Bisector • Median • Altitude • Perpendicular Bisector • Auxiliary Line • Determined, Underdetermined, Overdetermined • Equilateral and Equiangular Triangles • Perimeter

3.4 Justifying Constructions

3.5 Lemma • Triangle Inequality

## TABLE 3.2 AN OVERVIEW OF CHAPTER THREE

### Methods of Proving Triangles Congruent: $\triangle ABC \cong \triangle DEF$

| Figure (Note Marks) | Method | Steps Needed in Proof |
|---|---|---|
| 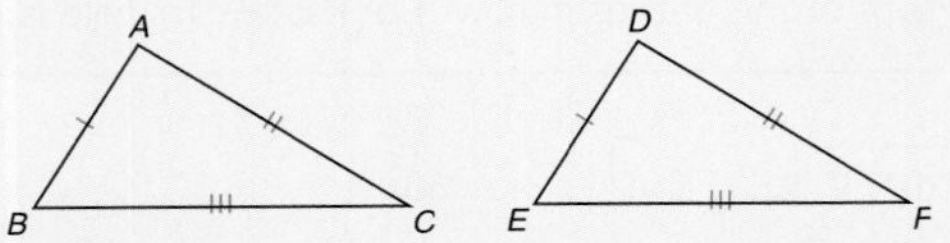 | SSS | $\overline{AB} \cong \overline{DE}$, $\overline{AC} \cong \overline{DF}$, and $\overline{BC} \cong \overline{EF}$ |
| 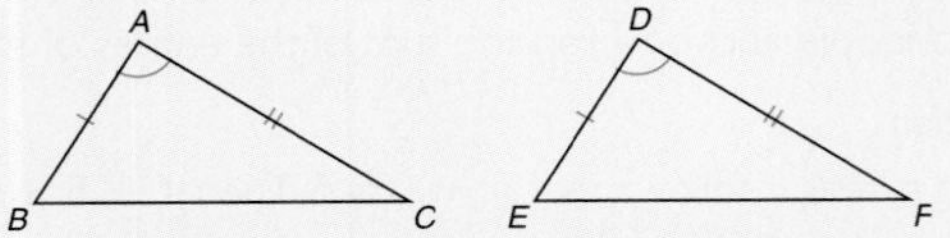 | SAS | $\overline{AB} \cong \overline{DE}$, $\angle A \cong \angle D$, and $\overline{AC} \cong \overline{DF}$ |
| 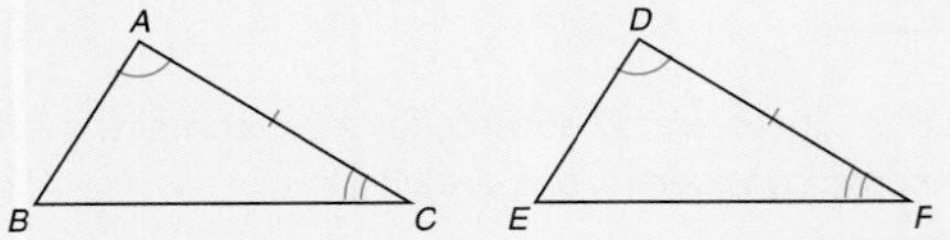 | ASA | $\angle A \cong \angle D$, $\overline{AC} \cong \overline{DF}$, and $\angle C \cong \angle F$ |
| 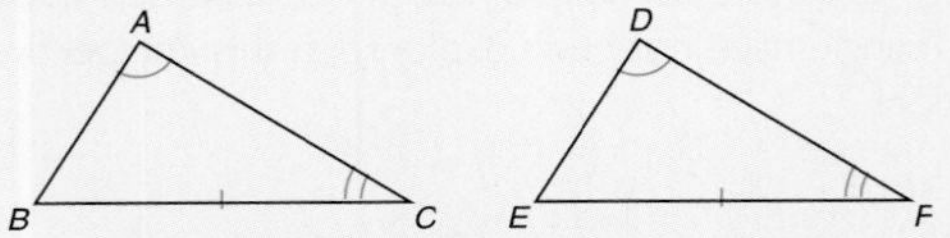 | AAS | $\angle A \cong \angle D$, $\angle C \cong \angle F$, and $\overline{BC} \cong \overline{EF}$ |
| 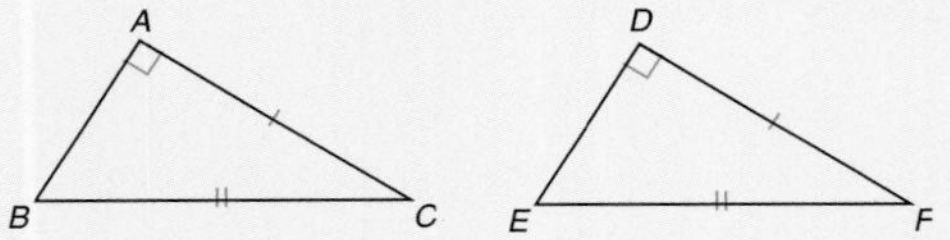 | HL | $\angle A$ and $\angle D$ are rt. $\angle$s, $\overline{AC} \cong \overline{DF}$, and $\overline{BC} \cong \overline{EF}$ |

### Special Relationships

| Figure | Relationship | Conclusion |
|---|---|---|
| 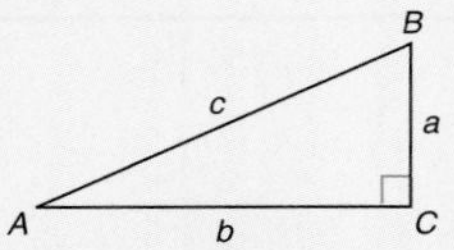 | Pythagorean Theorem | $c^2 = a^2 + b^2$ |
| 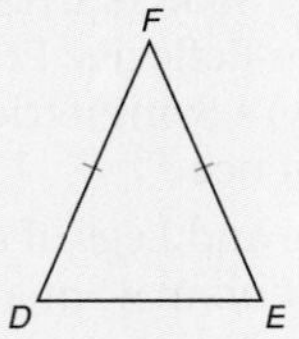 | $\overline{DF} \cong \overline{EF}$ (two $\cong$ sides) | $\angle E \cong \angle D$ (opposite $\angle$s $\cong$) |
| 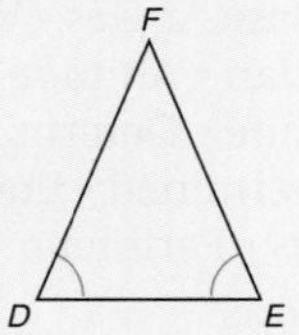 | $\angle D \cong \angle E$ (two $\cong$ angles) | $\overline{EF} \cong \overline{DF}$ (opposite sides $\cong$) |

**TABLE 3.2 CONT.**

**Inequality Relationships in a Triangle**

| *Figure* | *Relationship* | *Conclusion* |
|---|---|---|
| | $ST > RS$ | $m\angle R > m\angle T$ (opposite angles) |
| | $m\angle Y > m\angle X$ | $XZ > YZ$ (opposite sides) |

## Chapter 3 Review Exercises

1. *Given:* $\angle AEB \cong \angle DEC$
   $\overline{AE} \cong \overline{ED}$
   *Prove:* $\triangle AEB \cong \triangle DEC$

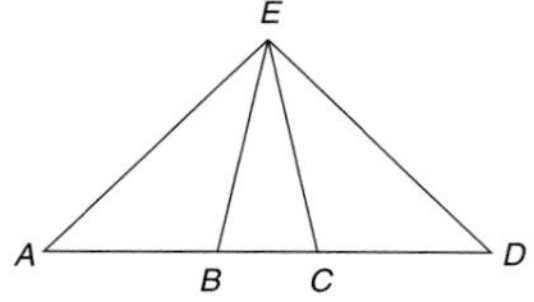


2. *Given:* $\overline{AB} \cong \overline{EF}$
   $\overline{AC} \cong \overline{DF}$
   $\angle 1 \cong \angle 2$
   *Prove:* $\angle B \cong \angle E$

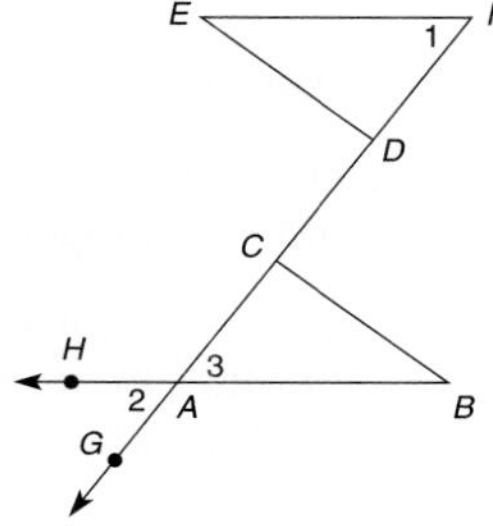


3. *Given:* $\overline{AD}$ bisects $\overline{BC}$
   $\overline{AB} \perp \overline{BC}$
   $\overline{DC} \perp \overline{BC}$
   *Prove:* $\overline{AE} \cong \overline{ED}$

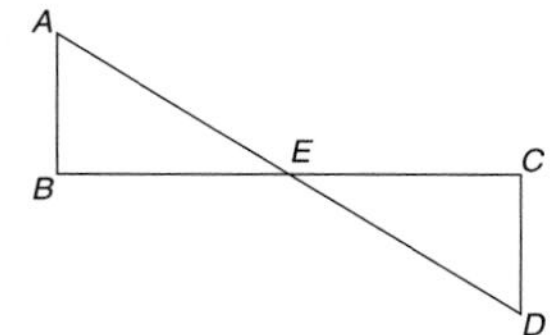


4. *Given:* $\overline{OA} \cong \overline{OB}$
   $\overline{OC}$ is the median to $\overline{AB}$
   *Prove:* $\overline{OC} \perp \overline{AB}$

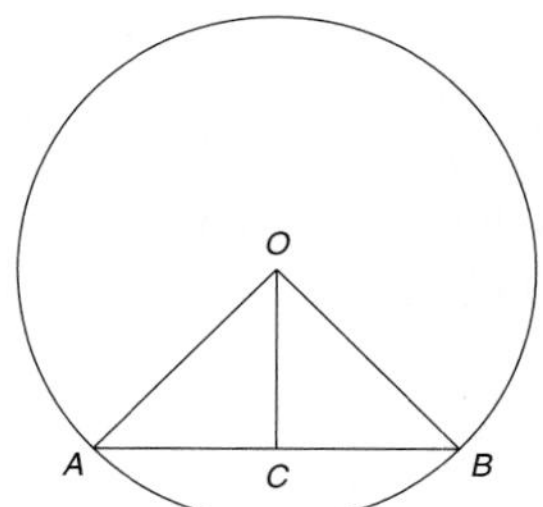


5. *Given:* $\overline{AB} \cong \overline{DE}$
   $\overline{AB} \parallel \overline{DE}$
   $\overline{AC} \cong \overline{DF}$
   *Prove:* $\overline{BC} \parallel \overline{FE}$

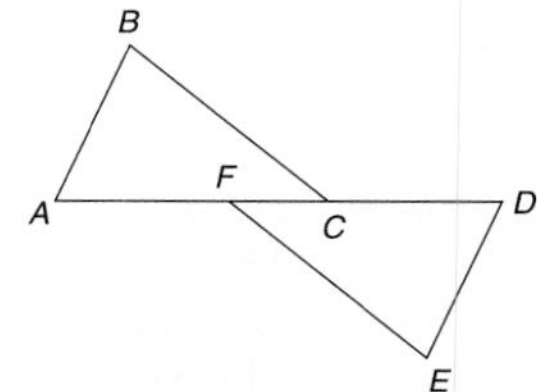


6. *Given:* $B$ is the midpoint of $\overline{AC}$
   $\overline{BD} \perp \overline{AC}$
   Prove: $\triangle ADC$ is isosceles

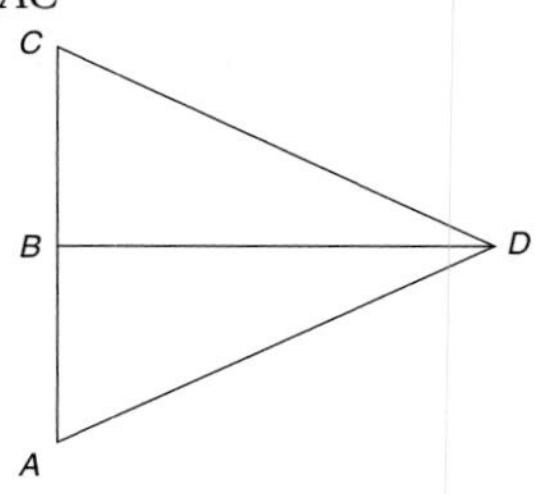

7. *Given:* $\overline{JM} \perp \overline{GM}$ and $\overline{GK} \perp \overline{KJ}$
$\overline{GH} \cong \overline{HJ}$
*Prove:* $\overline{GM} \cong \overline{JK}$

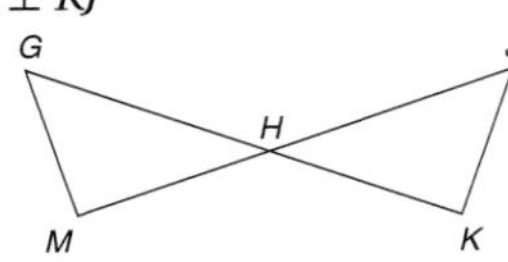


8. *Given:* $\overline{TN} \cong \overline{TR}$
$\overline{TO} \perp \overline{NP}$
$\overline{TS} \perp \overline{PR}$
$\overline{TO} \cong \overline{TS}$
*Prove:* $\angle N \cong \angle R$

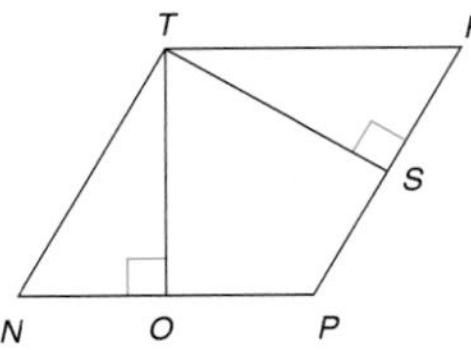


9. *Given:* $\overline{YZ}$ is the base of an isosceles triangle
$\overrightarrow{XA} \parallel \overline{YZ}$
*Prove:* $\angle 1 \cong \angle 2$

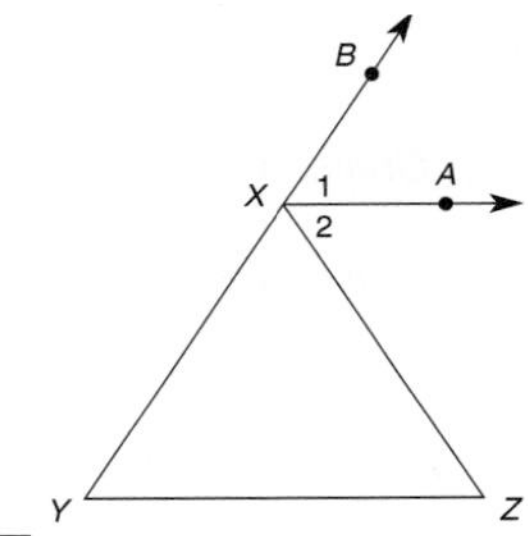


10. *Given:* $\overline{AB} \parallel \overline{DC}$
$\overline{AB} \cong \overline{DC}$
$C$ is the midpoint of $\overline{BE}$
*Prove:* $\overline{AC} \parallel \overline{DE}$

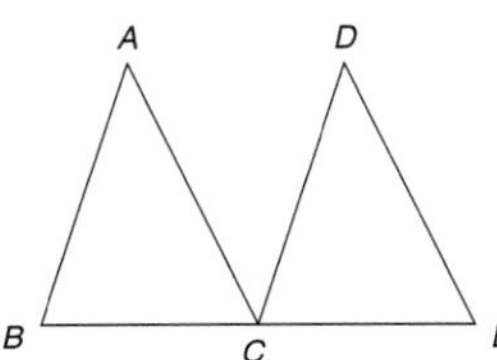


11. *Given:* $\angle BAD \cong \angle CDA$
$\overline{AB} \cong \overline{CD}$
*Prove:* $\overline{AE} \cong \overline{ED}$
(HINT: Prove $\triangle BAD \cong \triangle CDA$ first.)

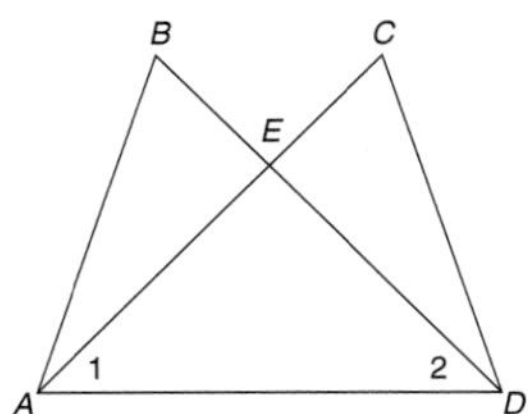


12. Given: $\overline{BE}$ is the altitude to $\overline{AC}$
$\overline{AD}$ is the altitude to $\overline{CE}$
$\overline{BC} \cong \overline{CD}$
*Prove:* $\overline{BE} \cong \overline{AD}$
(HINT: Prove $\triangle CBE \cong \triangle CDA$.)

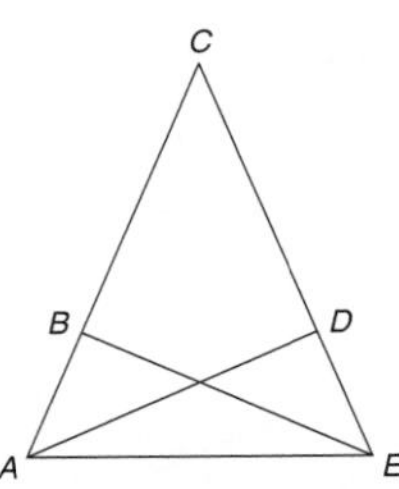


13. *Given:* $\overline{AB} \cong \overline{CD}$
$\angle BAD \cong \angle CDA$
*Prove:* $\triangle AED$ is isosceles
(HINT: Prove $\angle CAD \cong \angle BDA$ by CPCTC.)

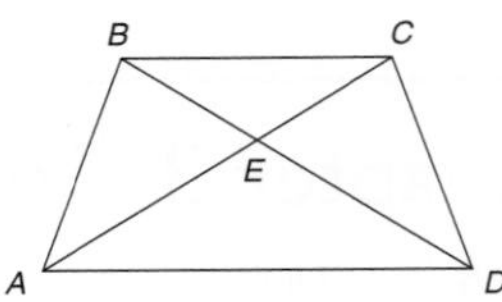


14. *Given:* $\overrightarrow{AC}$ bisects $\angle BAD$
*Prove:* $AD > CD$

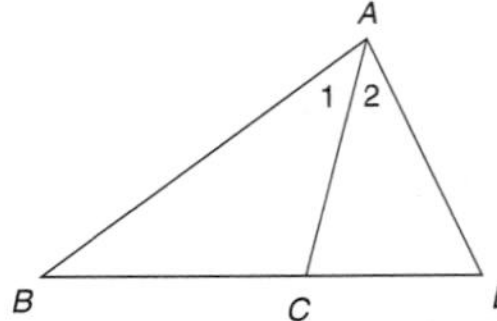


15. In $\triangle PQR$ (not shown), $m\angle P = 67°$ and $m\angle Q = 23°$.
    a) Name the shortest side.
    b) Name the longest side.

16. In $\triangle ABC$ (not shown), $m\angle A = 40°$ and $m\angle B = 65°$. List the sides in order of their lengths, starting with the smallest side.

17. In $\triangle PQR$ (not shown), $PQ = 1.5$, $PR = 2$, and $QR = 2.5$. List the angles in order of size, starting with the smallest angle.

18. Name the longest line segment in the figure.

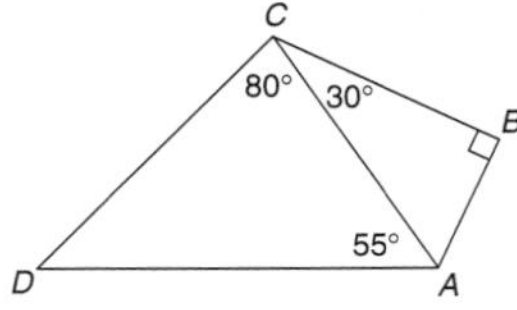

19. Which of the following can be the lengths of the sides of a triangle?

a) 3, 6, 9 b) 4, 5, 8 c) 2, 3, 8

20. Two sides of a triangle have lengths 15 and 20. The length of the third side can be any number between _?_ and _?_.

21. *Given:* $\overline{DB} \perp \overline{AC}$
$\overline{AD} \cong \overline{DC}$
$m\angle C = 70°$
*Find:* $m\angle ADB$

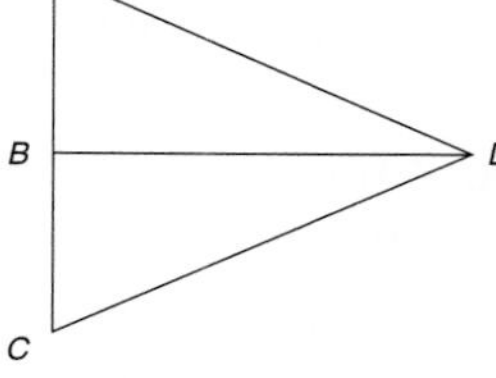


22. *Given:* $\overline{AB} \cong \overline{BC}$
$\angle DAC \cong \angle BCD$
$m\angle B = 50°$
*Find:* $m\angle ADC$

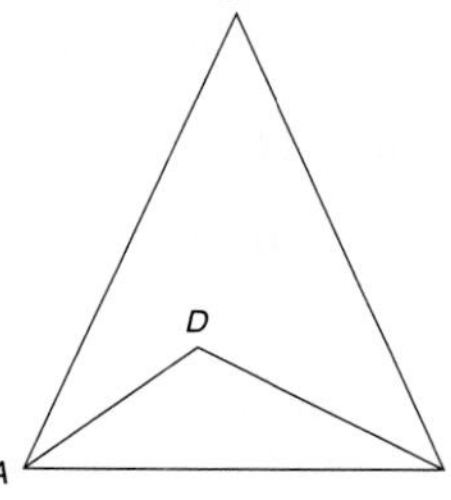


23. *Given:* $\triangle ABC$ is isosceles with base $\overline{AB}$
$m\angle 2 = 3x + 10$
$m\angle 4 = \frac{5}{2}x + 18$
*Find:* $m\angle C$

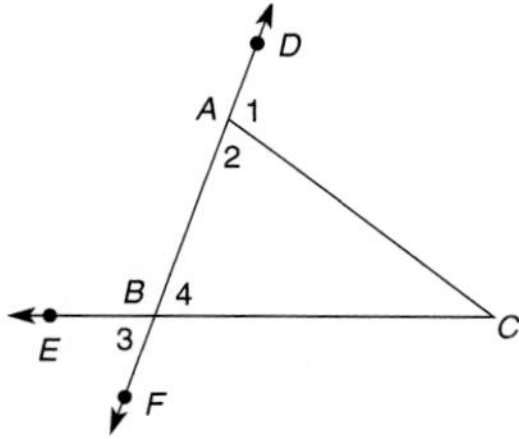


*Exercises 23, 24*

24. *Given:* $\triangle ABC$ with perimeter 40
$AB = 10$
$BC = x + 6$
$AC = 2x - 3$
*Find:* Whether $\triangle ABC$ is scalene, isosceles, or equilateral

25. *Given:* $\triangle ABC$ is isosceles with base $\overline{AB}$
$AB = y + 7$
$BC = 3y + 5$
$AC = 9 - y$
*Find:* Whether $\triangle ABC$ is also equilateral

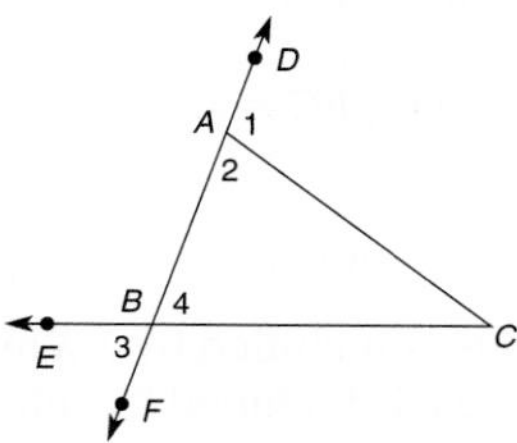


*Exercises 25, 26*

26. *Given:* $\overline{AC}$ and $\overline{BC}$ are the legs of isosceles $\triangle ABC$
$m\angle 1 = 5x$
$m\angle 3 = 2x + 12$
*Find:* $m\angle 2$

27. Construct an angle that measures 75°.

28. Construct a right triangle that has acute angle $A$ and hypotenuse of length $c$.

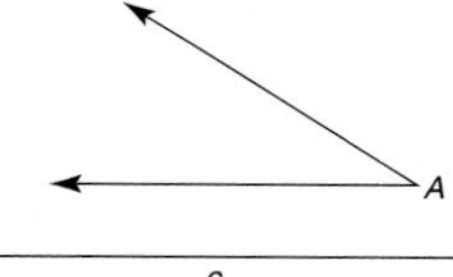


29. Construct a second isosceles triangle in which the base angles are half as large as the base angles of the given isosceles triangle.

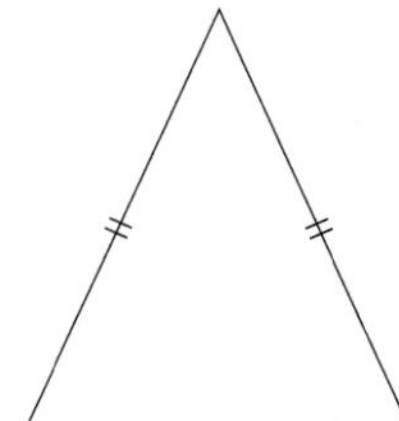

# Chapter 3 Test

1. It is given that $\triangle ABC \cong \triangle DEF$ (triangles not shown).

   a) If $m\angle A = 37°$ and $m\angle E = 68°$, find $m\angle F$. ________

   b) If $AB = 7.3$ cm, $BC = 4.7$ cm, and $AC = 6.3$ cm, find $EF$. ________

2. Consider $\triangle XYZ$ (not shown).

   a) Which side is included by $\angle X$ and $\angle Y$? ________

   b) Which angle is included by sides $\overline{XY}$ and $\overline{YZ}$? ________

3. State the reason (SSS, SAS, ASA, AAS, or HL) why the triangles are congruent. Note the marks that indicate congruent parts.

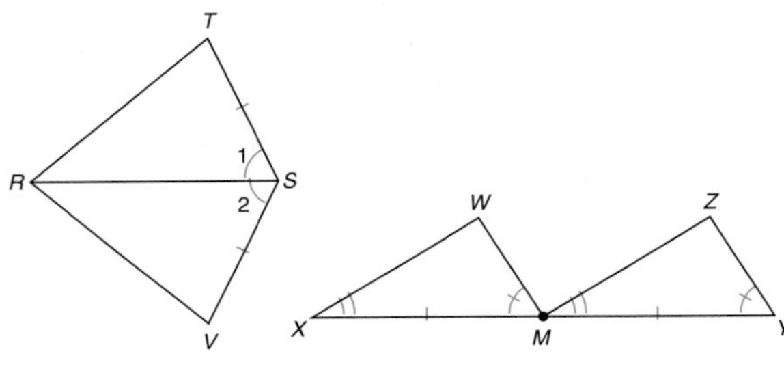


$\triangle RVS \cong \triangle RTS$ ________ $\triangle XMW \cong \triangle MYZ$ ________

4. Write the statement that is represented by the acronym CPCTC. ________

5. With congruent parts marked, are the two triangles congruent? Answer YES or NO.

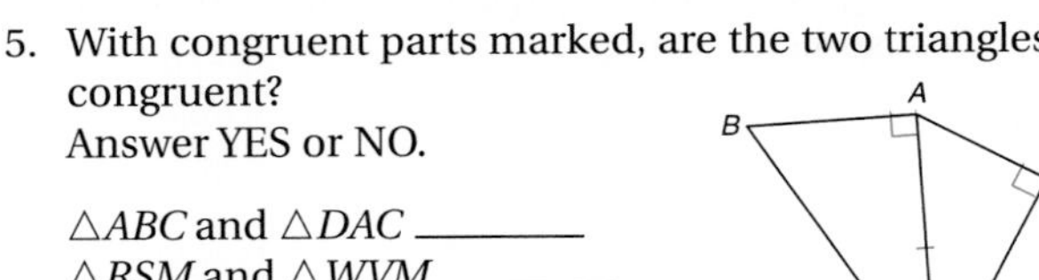

   $\triangle ABC$ and $\triangle DAC$ ________

   $\triangle RSM$ and $\triangle WVM$ ________

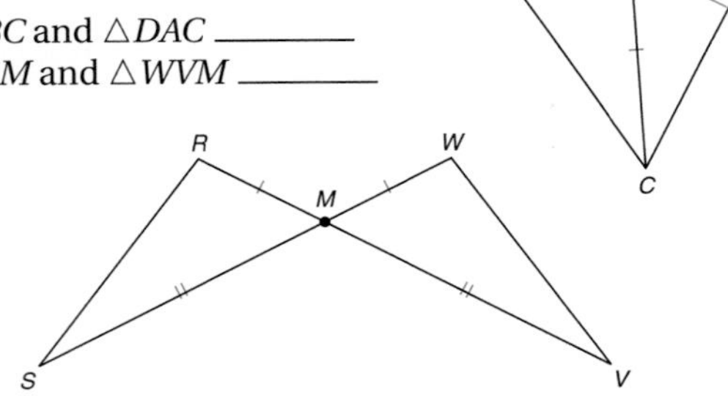


6. With $\triangle ABD \cong \triangle CBE$ and $A\text{-}D\text{-}E\text{-}C$, does it necessarily follow that $\triangle AEB$ and $\triangle CDB$ are congruent? Answer YES or NO. ________

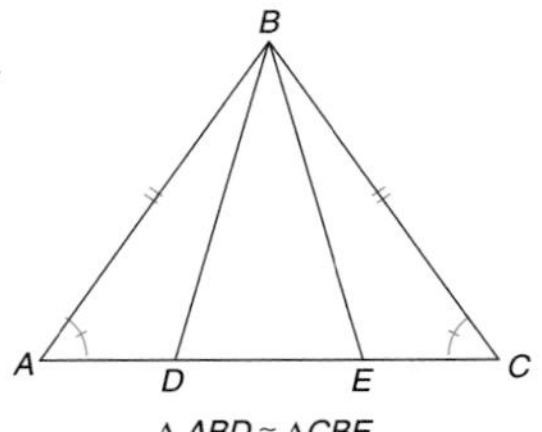


$\triangle ABD \cong \triangle CBE$

7. In $\triangle ABC$, $m\angle C = 90°$. Find:

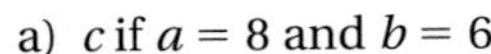

   a) $c$ if $a = 8$ and $b = 6$ ________

   b) $b$ if $a = 6$ and $c = 8$ ________

8. $\overline{CM}$ is the median for $\triangle ABC$ from vertex $C$ to side $\overline{AB}$.

   a) Name two line segments that must be congruent. ________

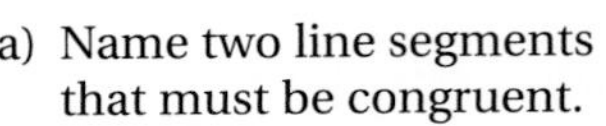

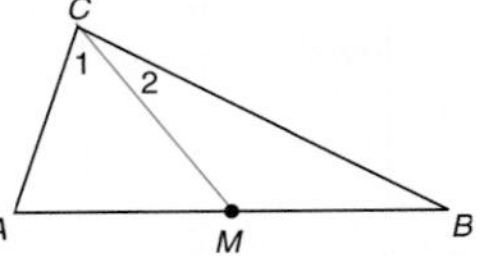


   b) Is $\angle 1$ necessarily congruent to $\angle 2$? ________

9. In $\triangle TUV$, $\overline{TV} \cong \overline{UV}$.

   a) If $m\angle T = 71°$, find $m\angle V$. ________

   b) If $m\angle T = 7x + 2$ and $m\angle U = 9(x - 2)$, find $m\angle V$. ________

10. In $\triangle TUV$, $\angle T \cong \angle U$.

   a) If $VT = 7.6$ inches and $TU = 4.3$ inches, find $VU$. ________

   b) If $VT = 4x + 1$, $TU = 2x$ and $VU = 6x - 10$, find the perimeter of $\triangle TUV$. ________ (HINT: Find the value of $x$.)

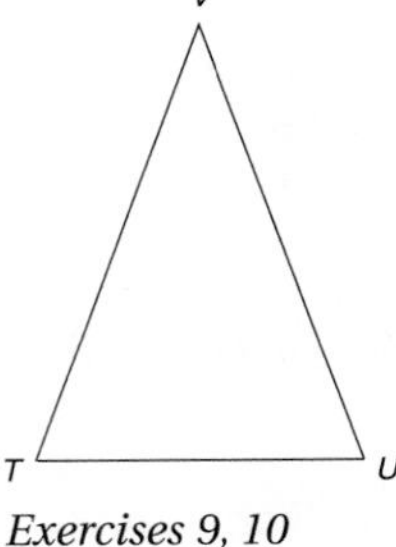


*Exercises 9, 10*

11. Show all arcs in the following construction.

   a) Construct an angle that measures 60°.

   b) Using the result from part (a), construct an angle that measures 30°.

12. Show all arcs in the following construction. Construct an isoceles right triangle in which each leg has the length of line segment $\overline{AB}$.

13. In $\triangle ABC$, $m\angle C = 46°$, and $m\angle B = 93°$.

   a) Name the shortest side of $\triangle ABC$. ________

   b) Name the longest side of $\triangle ABC$. ________

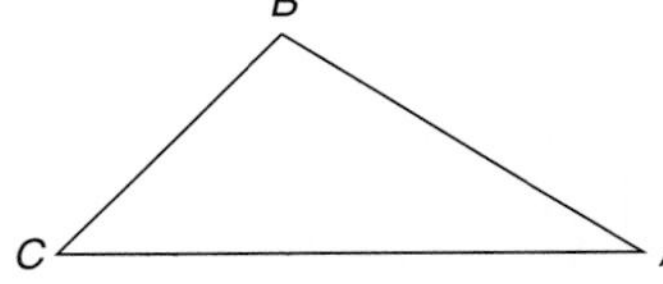

14. In $\triangle TUV$ (not shown), $TU > TV > VU$. Write a three-part inequality that compares the measures of the three angles of $\triangle TUV$.

_______________

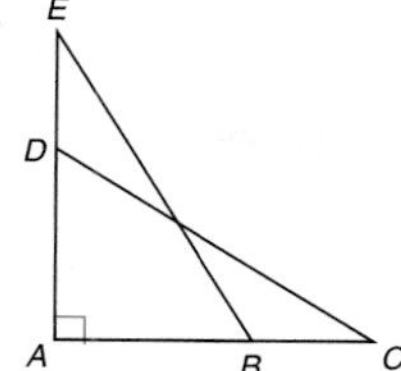


15. In the figure, $\angle A$ is a right angle, $AD = 4$, $DE = 3$, $AB = 5$, and $BC = 2$. Of the two line segments $\overline{DC}$ and $\overline{EB}$, which one is longer? _______

16. Given $\triangle ABC$, draw the triangle that results when $\triangle ABC$ is rotated clockwise 180° about $M$, the midpoint of $\overline{AC}$. Let $D$ name the image of point $B$. In these congruent triangles, which side of $\triangle CDA$ corresponds to side $\overline{BC}$ of $\triangle ABC$?

_______

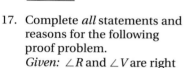

17. Complete *all* statements and reasons for the following proof problem.
*Given:* $\angle R$ and $\angle V$ are right angles; $\angle 1 \cong \angle 2$
*Prove:* $\triangle RST \cong \triangle VST$

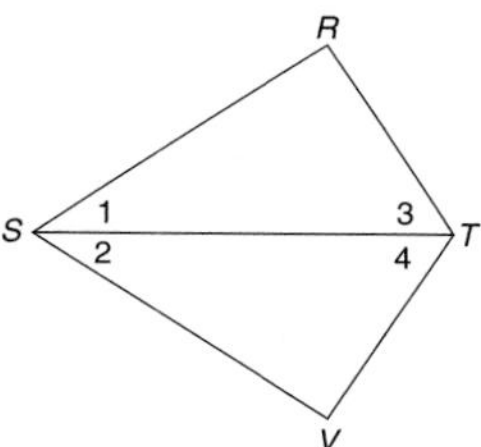


| Statements | Reasons |
|---|---|
| | |

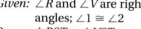

18. Complete the missing statements and reasons in the following proof.
*Given:* $\triangle RUV$; $\angle R \cong \angle V$, and $\angle 1 \cong \angle 3$
*Prove:* $\triangle STU$ is an isosceles triangle

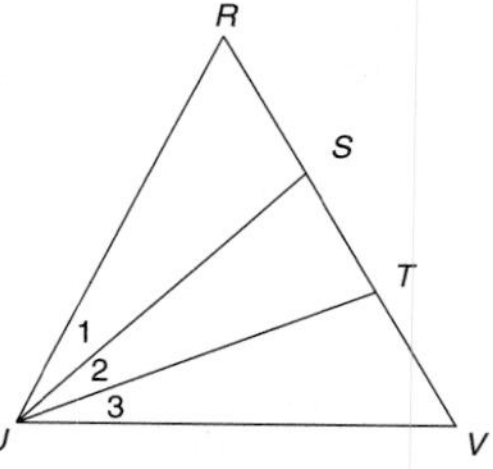


| Statements | Reasons |
|---|---|
| 1. $\triangle RUV$; $\angle R \cong \angle V$ | 1. _______ |
| 2. $\therefore \overline{UV} \cong \overline{UR}$ | 2. _______ |
| 3. _______ | 3. Given |
| 4. $\triangle RSU \cong \triangle VTU$ | 4. _______ |
| 5. _______ | 5. CPCTC |
| 6. _______ | 6. If 2 sides of a $\triangle$ are $\cong$, this triangle is an isosceles triangle. |

# Chapter 4

# Quadrilaterals

Designed by architect Frank Lloyd Wright (1867–1959), this private home in Bear Run, Pennsylvania, was built in 1935. The geometric figure that dominates this house (and many others) is the quadrilateral. In this chapter, we consider several types of quadrilaterals—among them the parallelogram, the rhombus, and the trapezoid—as we develop their properties. Applications based upon the trapezoid will be found in Exercises 37 to 40 of Section 4.4.

## Chapter Outline



**For online student resources, visit this textbook's website at math.college.hmco.com/students.**

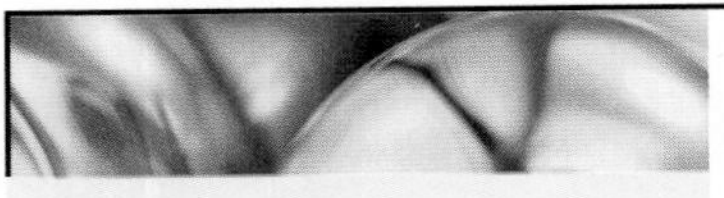

**KEY CONCEPTS**

Quadrilateral • Skew Quadrilateral • Parallelogram • Diagonals of a Parallelogram • Altitudes of a Parallelogram

# 4.1 Properties of a Parallelogram

A **quadrilateral** is a polygon that has four sides. Unless otherwise stated, the term *quadrilateral* refers to a figure such as *ABCD* in Figure 4.1(a), in which the line segment sides lie within a single plane. When the sides of the quadrilateral are not coplanar, as with *MNPQ* in Figure 4.1(b), the quadrilateral is said to be **skew.** Thus *MNPQ* is a skew quadrilateral. In this textbook, we generally consider quadrilaterals whose sides are coplanar.

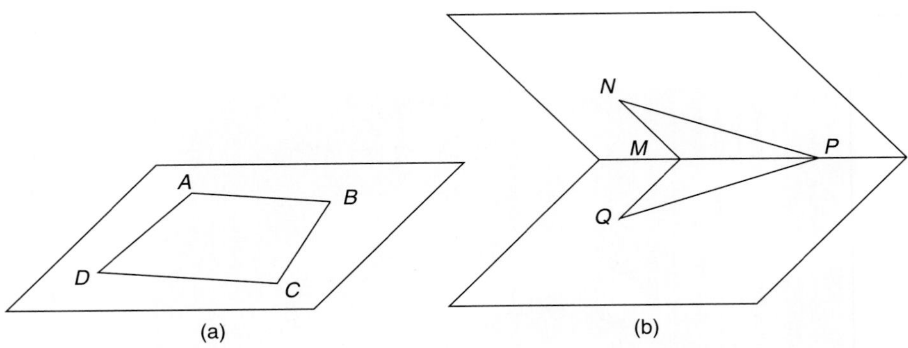


**FIGURE 4.1**

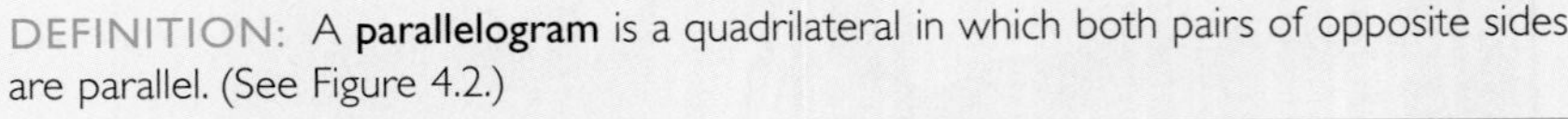

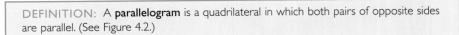

DEFINITION: A **parallelogram** is a quadrilateral in which both pairs of opposite sides are parallel. (See Figure 4.2.)

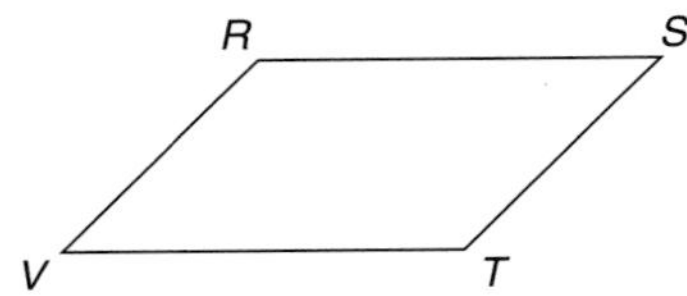


**FIGURE 4.2**

Because the symbol for parallelogram is ▱, the quadrilateral in Figure 4.2 is ▱*RSTV*. The set $P = \{\text{parallelograms}\}$ is a subset of $Q = \{\text{quadrilaterals}\}$.

The related activity guides us toward many of the theorems of this section.

## Discover!

From a standard sheet of construction paper, cut out a parallelogram as shown. Then cut along one diagonal. How are the two triangles that are formed related?

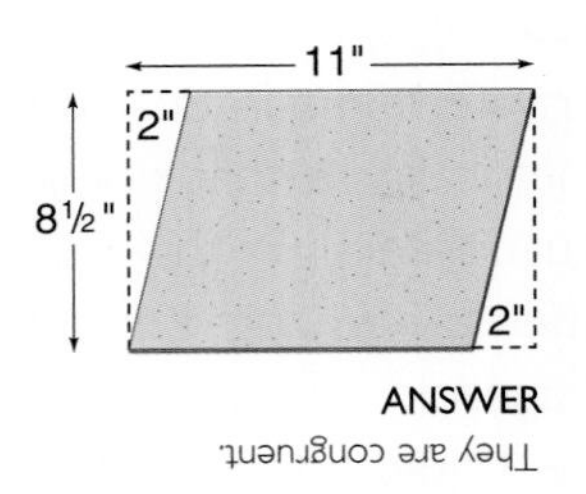


ANSWER

They are congruent.

### EXAMPLE 1

Give a formal proof of Theorem 4.1.1.

THEOREM 4.1.1: A diagonal of a parallelogram separates it into two congruent triangles.

*Given:* ▱$ABCD$ with diagonal $\overline{AC}$ (See Figure 4.3.)

*Prove:* $\triangle ACD \cong \triangle CAB$

**PROOF**

| Statements | Reasons |
|---|---|
| 1. ▱$ABCD$ | 1. Given |
| 2. $\overline{AB} \parallel \overline{CD}$ | 2. The opposite sides of a ▱ are ∥ (definition) |
| 3. $\angle 1 \cong \angle 2$ | 3. If two ∥ lines are cut by a transversal, the alternate interior ∠s are congruent |
| 4. $\overline{AD} \parallel \overline{BC}$ | 4. Same as reason 2 |
| 5. $\angle 3 \cong \angle 4$ | 5. Same as reason 3 |
| 6. $\overline{AC} \cong \overline{AC}$ | 6. Identity |
| 7. $\triangle ACD \cong \triangle CAB$ | 7. ASA |

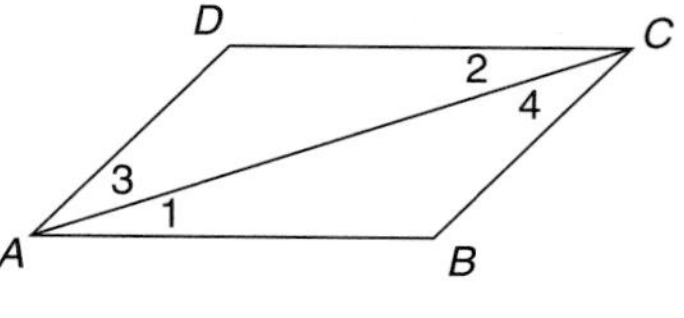


**FIGURE 4.3**

Three corollaries of Theorem 4.1.1 follow. Make a drawing to illustrate each corollary.

**Reminder**
The sum of the measures of the interior angles of a quadrilateral is 360°.

COROLLARY 4.1.2: The opposite angles of a parallelogram are congruent.

COROLLARY 4.1.3: The opposite sides of a parallelogram are congruent.

COROLLARY 4.1.4: The diagonals of a parallelogram bisect each other.

Recall Theorem 2.1.4: "If two parallel lines are cut by a transversal, then the interior angles on the same side of the transversal are supplementary." A corollary of that theorem is stated next.

*Exs. 1–6*

COROLLARY 4.1.5: Two consecutive angles of a parallelogram are supplementary.

EXAMPLE 2

In ▱$RSTV$, $m\angle S = 42°$, $ST = 5.3$ cm, and $VT = 8.1$ cm. Find:

a) $m\angle V$ b) $m\angle T$ c) $RV$ d) $RS$

***Solution***

a) $m\angle V = 42°$; $\angle V \cong \angle S$ because these are opposite ∠s of ▱ $RSTV$.
b) $m\angle T = 138°$; $\angle T$ and $\angle S$ are supplementary because these angles are consecutive angles of ▱$RSTV$.
c) $RV = 5.3$ cm; $\overline{RV} \cong \overline{ST}$ because these are opposite sides of ▱$RSTV$.
d) $RS = 8.1$ cm; $\overline{RS} \cong \overline{VT}$, also a pair of opposite sides of ▱$RSTV$.

Example 3 illustrates Theorem 4.1.6, the fact that two parallel lines are everywhere equidistant. In general, the phrase *distance between two parallel lines* refers to the length of the perpendicular segment between the two parallel lines. These concepts will provide insight into the definition of altitude of a parallelogram.

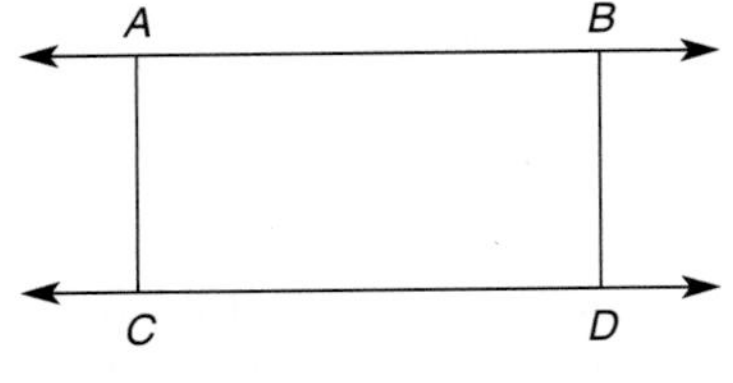


**FIGURE 4.4**

> THEOREM 4.1.6: Two parallel lines are everywhere equidistant.

EXAMPLE 3

*Given:* $\overleftrightarrow{AB} \parallel \overleftrightarrow{CD}$
$\overline{AC} \perp \overleftrightarrow{CD}$ and $\overline{BD} \perp \overleftrightarrow{CD}$
(See Figure 4.4.)

*Prove:* $\overline{AC} \cong \overline{BD}$

**PROOF**

| Statements | Reasons |
|---|---|
| 1. $\overleftrightarrow{AB} \parallel \overleftrightarrow{CD}$ | 1. Given |
| 2. $\overline{AC} \perp \overleftrightarrow{CD}$ and $\overline{BD} \perp \overleftrightarrow{CD}$ | 2. Given |
| 3. $\overline{AC} \parallel \overline{BD}$ | 3. If two lines are ⊥ to the same line, they are parallel |
| 4. *ABDC* is a ▱ | 4. If both pairs of opposite sides of a quadrilateral are ∥, the quadrilateral is a ▱ |
| 5. $\overline{AC} \cong \overline{BD}$ | 5. Opposite sides of a ▱ are congruent |

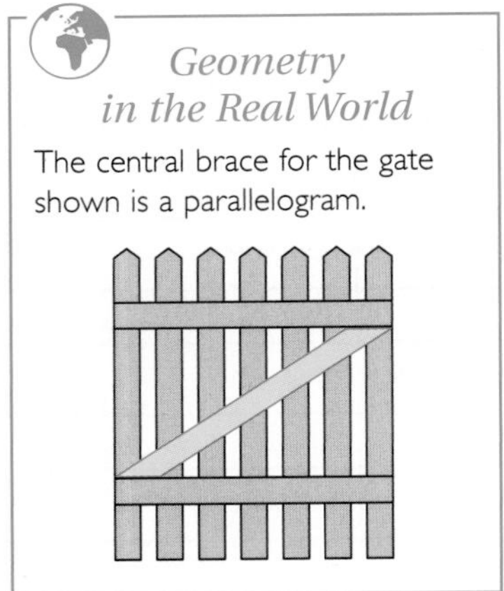

*Geometry in the Real World*

The central brace for the gate shown is a parallelogram.

In Example 3, we used the definition of a parallelogram to prove that a particular quadrilateral was a parallelogram, but there are other ways of establishing that a given quadrilateral is a parallelogram. We will investigate those methods in Section 4.2.

*Exs. 7–11*

> DEFINITION: An **altitude** of a parallelogram is a line segment from one vertex that is perpendicular to a nonadjacent side (or to an extension of that side).

For ▱ *RSTV*, $\overline{RW}$ and $\overline{SX}$ are altitudes to side $\overline{VT}$ (or to side $\overline{RS}$), as shown in Figure 4.5(a). With respect to side $\overline{RS}$, sometimes called base $\overline{RS}$, the length *RW* (or *SX*) is the *height* of *RSTV*. Similarly, $\overline{TY}$ and $\overline{SZ}$ are altitudes to side $\overline{RV}$ (or to side $\overline{ST}$), as shown in Figure 4.5(b). Also, the length *TY* (or *ZS*) is called the *height* of parallelogram *RSTV* with respect to side $\overline{ST}$ (or $\overline{RV}$).

Next we consider an inequality relationship for the parallelogram. In order to develop this relationship, we need to investigate an inequality involving two triangles.

In $\triangle ABC$ and $\triangle DEF$ of Figure 4.6, $\overline{AB} \cong \overline{DE}$ and $\overline{BC} \cong \overline{EF}$. If $m\angle B > m\angle E$, then $AC > DF$.

We will use, but not prove, the following relationship found in Lemma 4.1.7.

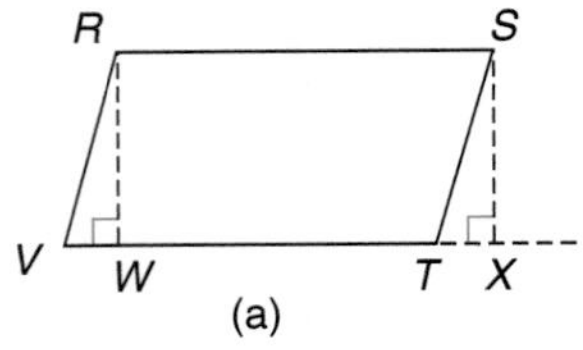


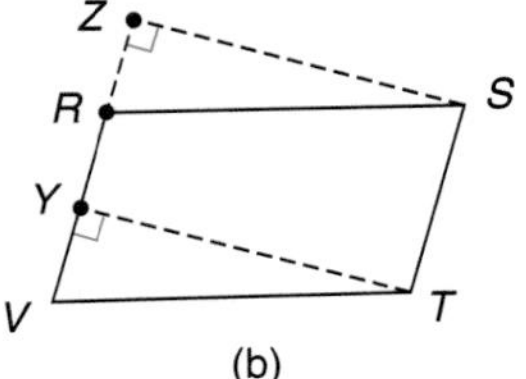


**FIGURE 4.5**

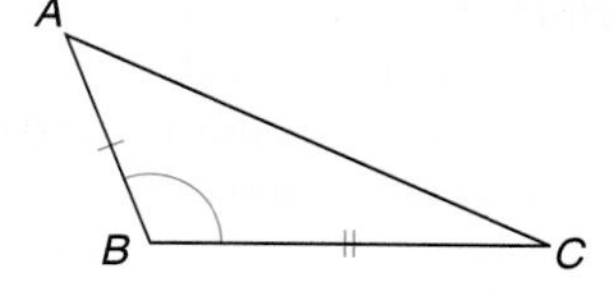

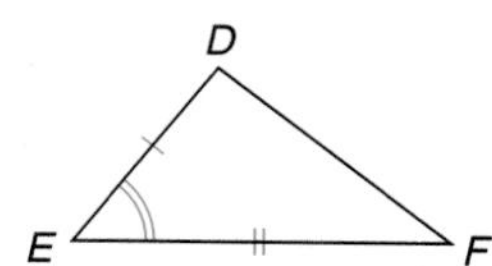


**FIGURE 4.6**

*Given:* $\overline{AB} \cong \overline{DE}$ and $\overline{BC} \cong \overline{EF}$; $m\angle B > m\angle E$ (See Figure 4.6.)

*Prove:* $AC > DF$

The corresponding lemma follows.

**Discover!**

On one piece of paper, draw a triangle ($\triangle ABC$) so that $AB = 3$, $BC = 5$, and $m\angle B = 110°$. Then draw $\triangle DEF$, in which $DE = 3$, $EF = 5$, and $m\angle E = 50°$. Which is longer, $\overline{AC}$ or $\overline{DF}$?

ANSWER
$\overline{AC}$

LEMMA 4.1.7: If two sides of one triangle are congruent to two sides of a second triangle and the included angle of the first triangle is greater than the included angle of the second, then the length of the side opposite the included angle of the first triangle is greater than the length of the side opposite the included angle of the second.

Now we can compare the lengths of the diagonals of a parallelogram. For a parallelogram having no right angles, two consecutive angles are unequal but supplementary; thus one angle of the parallelogram will be acute and the consecutive angle will be obtuse. In Figure 4.7(a), $\square ABCD$ has acute angle $A$ and obtuse angle $D$. Note that the lengths of the two sides of the triangles that include $\angle A$ and $\angle D$ are congruent. In Figure 4.7(b), diagonal $\overline{AC}$ lies opposite the obtuse angle $ADC$ in $\triangle ACD$, and diagonal $\overline{BD}$ lies opposite the acute angle $DAB$ in $\triangle ABD$. In Figures 4.7(c) and (d), we have taken $\triangle ACD$ and $\triangle ABD$ from $\square ABCD$ of Figure 4.7(b). Note that $\overline{AC}$ (opposite obtuse $\angle D$) is longer than $\overline{DB}$ (opposite acute $\angle A$).

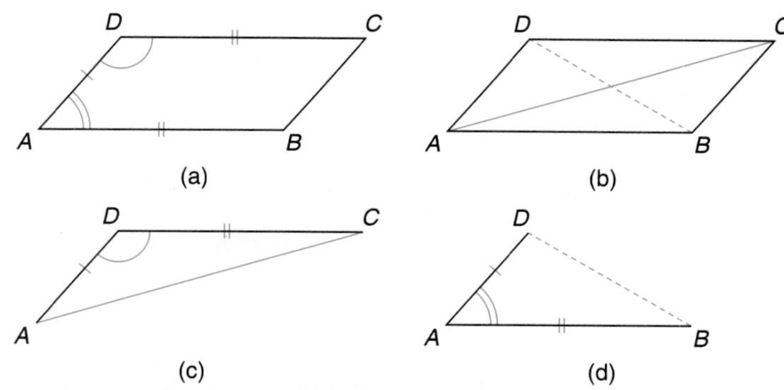


**FIGURE 4.7**

On the basis of Lemma 4.1.7 and the preceding discussion, we have the following theorem.

THEOREM 4.1.8: In a parallelogram with unequal pairs of consecutive angles, the longer diagonal lies opposite the obtuse angle.

**Discover!**

Draw $\square ABCD$ so that $m\angle A > m\angle B$. Which diagonal has the greater length?

ANSWER
$\overline{BD}$

EXAMPLE 4

In parallelogram $RSTV$ (not shown), $m\angle R = 67°$.

a) Find the measure of $\angle S$.
b) Determine which diagonal ($\overline{RT}$ or $\overline{SV}$) has the greater length.

***Solution***

a) $m\angle S = 180° - 67° = 113°$ ($\angle R$ and $\angle S$ are supplementary.)

b) Because $\angle S$ is obtuse, the diagonal opposite this angle is longer; that is, $\overline{RT}$ is the longer diagonal.

We use an indirect approach to solve Example 5.

EXAMPLE 5

In parallelogram *ABCD* (not shown), $\overline{AC}$ and $\overline{BD}$ are diagonals and $AC > BD$. Determine which angles of the parallelogram are obtuse and which angles are acute.

***Solution***

Because the longer diagonal $\overline{AC}$ lies opposite angles *B* and *D*, these angles are obtuse. The remaining angles *A* and *C* are necessarily acute.

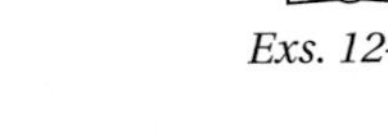
*Exs. 12–15*

Our next example uses algebra to relate angle sizes and diagonal lengths.

EXAMPLE 6

In ▱*MNPQ* in Figure 4.8, $m\angle M = 2(x + 10)$ and $m\angle Q = 3x - 10$. Determine which diagonal would be longer, $\overline{QN}$ or $\overline{MP}$.

**FIGURE 4.8**

***Solution***

Consecutive angles *M* and *Q* are supplementary, so $m\angle M + m\angle Q = 180°$.

$$2(x + 10) + (3x - 10) = 180$$
$$2x + 20 + 3x - 10 = 180$$
$$5x + 10 = 180 \rightarrow 5x = 170 \rightarrow x = 34$$

Then $m\angle M = 2(34 + 10) = 88°$, whereas $m\angle Q = 3(34) - 10 = 92°$. Because $m\angle Q > m\angle M$, diagonal $\overline{MP}$ (opposite $\angle Q$) would be longer than $\overline{QN}$.

## SPEED AND DIRECTION OF AIRCRAFT

For the application to follow in Example 7, we indicate the velocity of an airplane or of the wind by drawing a directed arrow. In each case, a scale is used on a grid in which a north-south line meets an east-west line at right angles. Consider the sketches in Figure 4.9 on page 183 and read their descriptions.

In some scientific applications, such as Example 7, a parallelogram can be used to determine the solution to the problem. For instance, the Parallelogram Law enables us to determine the resulting speed and direction of an airplane when the velocity of the airplane and that of the wind are considered together. In Figure 4.10, also on page 183, the arrows representing the two velocities are placed head-to-tail from the point of origin. Because the order of the two velocities is reversible, the drawing leads to a parallelogram. In the parallelogram, it is the length and direction of the diagonal that solve the problem. In Example 7, accuracy is critical when scaling the drawing. Otherwise, the ruler and protractor will give poor results.

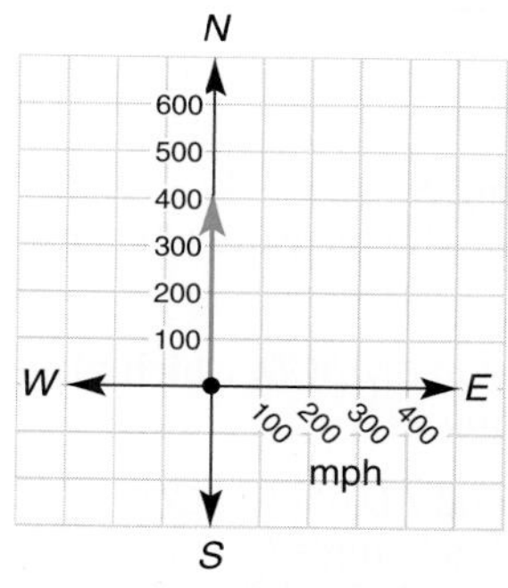


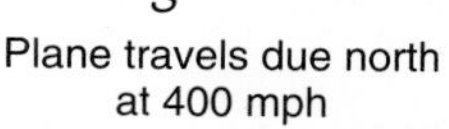
Plane travels due north at 400 mph

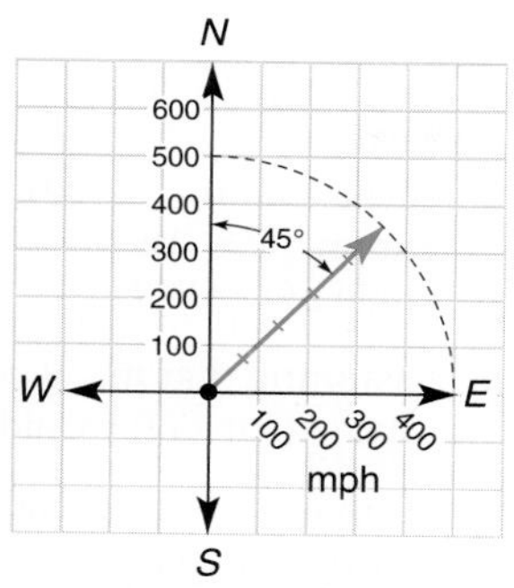


Plane travels at 500 mph in the direction N 45° E

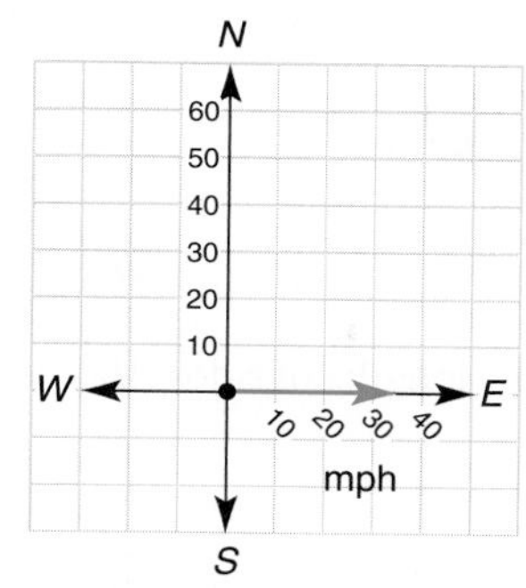


Wind blows at 30 mph in the direction west to east

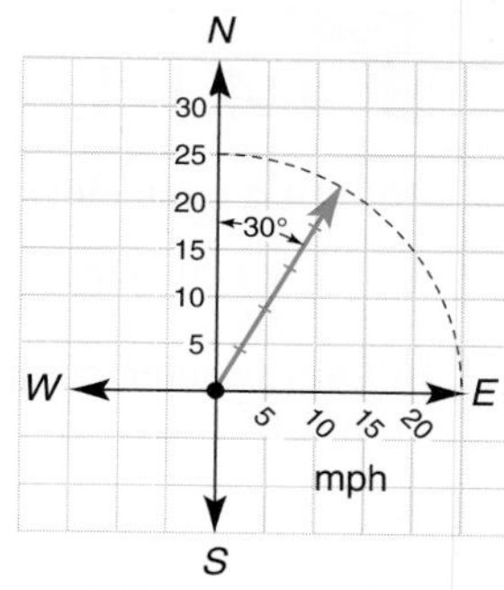


Wind blows at 25 mph in the direction N 30° E

**FIGURE 4.9**

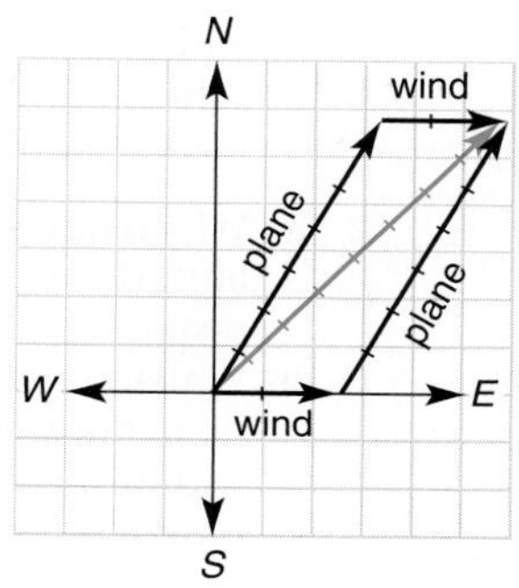


**FIGURE 4.10**

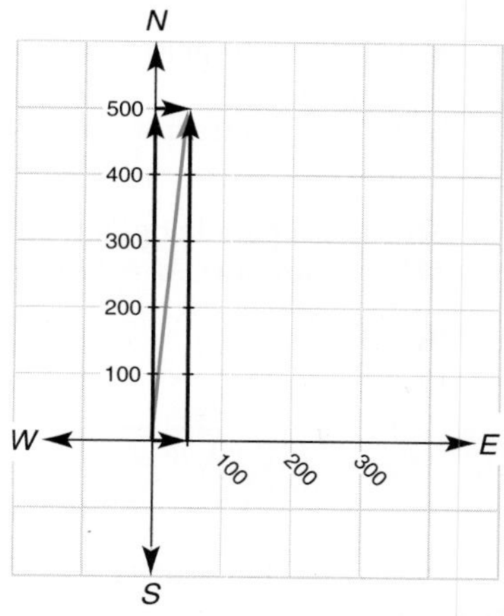


**FIGURE 4.11**

NOTE: In Example 7, kph means *kilometers per hour.*

EXAMPLE 7

An airplane travels due north at 500 kph. If the wind blows at 50 kph from west to east, what are the resulting speed and direction of the plane?

***Solution***

Using a ruler to measure the diagonal of the parallelogram, we find that the length corresponds to a speed of approximately 505 kph. Using a protractor, we find that the direction is approximately N 6° E. (See Figure 4.11.)

NOTE: The actual speed is approximately 502.5 kph while the direction is N 5.7° E.

*Exs. 16–17*

## 4.1 Exercises

1. *ABCD* is a parallelogram.

   a) Using a ruler, compare the lengths of sides $\overline{AB}$ and $\overline{DC}$.
   b) Using a ruler, compare the lengths of sides $\overline{AD}$ and $\overline{BC}$.

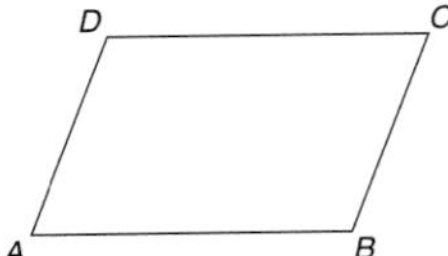


*Exercises 1, 2*

2. *ABCD* is a parallelogram.

   a) Using a protractor, compare the measures of $\angle A$ and $\angle C$.
   b) Using a protractor, compare the measures of $\angle B$ and $\angle D$.

3. *MNPQ* is a parallelogram. Suppose that $MQ = 5$, $MN = 8$, and $m\angle M = 110°$. Find:

   a) $QP$
   b) $NP$
   c) $m\angle Q$
   d) $m\angle P$

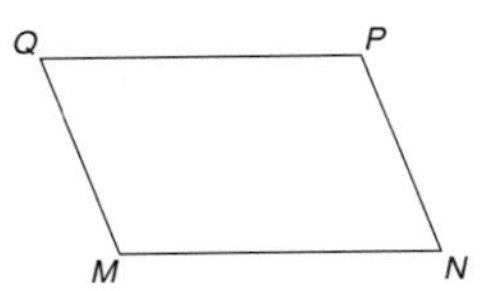


*Exercises 3, 4*

4. *MNPQ* is a parallelogram. Suppose that $MQ = 12.7$, $MN = 17.9$, and $m\angle M = 122°$. Find:

   a) $QP$
   b) $NP$
   c) $m\angle Q$
   d) $m\angle P$

5. Given that $AB = 3x + 2$, $BC = 4x + 1$, and $CD = 5x - 2$, find the length of each side of $\square ABCD$.

6. Given that $m\angle A = 2x + 3$ and $m\angle C = 3x - 27$, find the measure of each angle of $\square ABCD$.

7. Given that $m\angle A = 2x + 3$ and $m\angle B = 3x - 23$, find the measure of each angle of $\square ABCD$.

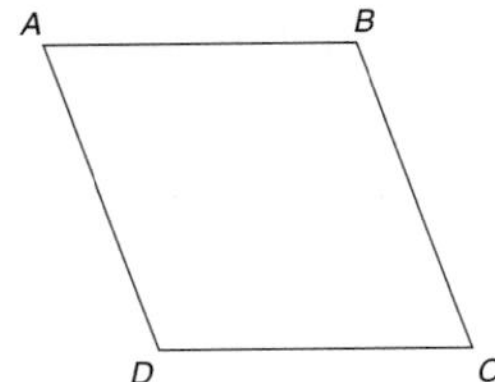


*Exercises 5–10*

8. Given that $m\angle A = 2x + y$, $m\angle B = 2x + 3y - 20$, and $m\angle C = 3x - y + 16$, find the measure of each angle of $\square ABCD$.

9. Assuming that $m\angle B > m\angle A$ in $\square ABCD$, which diagonal ($\overline{AC}$ or $\overline{BD}$) would be longer?

10. Suppose that diagonals $\overline{AC}$ and $\overline{BD}$ of $\square ABCD$ are drawn and that $AC > BD$. Which angle ($\angle A$ or $\angle B$) would have the greater measure?

*In Exercises 11 and 12, consider $\square RSTV$ with $\overline{VX} \perp \overline{RS}$ and $\overline{VY} \perp \overline{ST}$.*

11. a) Which line segment is the altitude of $\square RSTV$ with respect to base $\overline{ST}$?
    b) Which number is the height of $\square RSTV$ with respect to base $\overline{ST}$?

12. a) Which line segment is the altitude of $\square RSTV$ with respect to base $\overline{RS}$?
    b) Which number is the height of $\square RSTV$ with respect to base $\overline{RS}$?

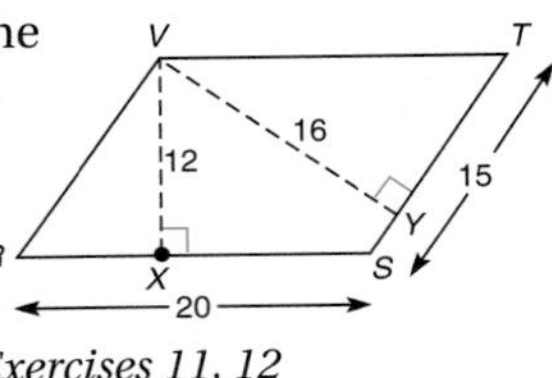


*Exercises 11, 12*

*In Exercises 13 to 16, classify each statement as true or false. In Exercises 13 and 14, recall that the symbol $\subseteq$ means "is a subset of."*

13. Where $Q$ = {quadrilaterals} and $P$ = {polygons}, $Q \subseteq P$.

14. Where $Q$ = {quadrilaterals} and $P$ = {parallelograms}, $Q \subseteq P$.

15. A parallelogram has point symmetry about the point where its two diagonals intersect.

16. A parallelogram has line symmetry and either diagonal is an axis of symmetry.

17. In quadrilateral *RSTV*, the midpoints of consecutive sides are joined in order. Try drawing other quadrilaterals and joining their midpoints. What can you conclude about the resulting quadrilateral in each case?

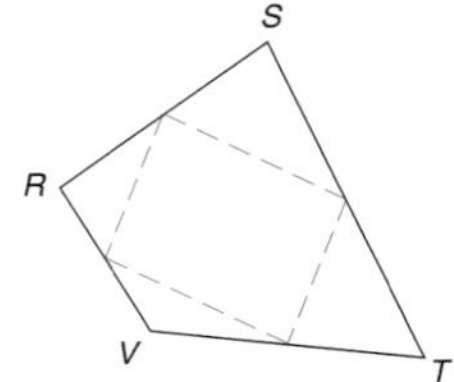

18. In quadrilateral $ABCD$, the midpoints of opposite sides are joined to form two intersecting segments. Try drawing other quadrilaterals and joining their opposite midpoints. What can you conclude about these segments in each case?

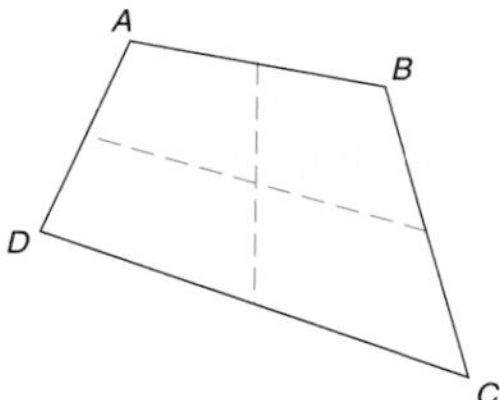


19. Quadrilateral $ABCD$ has $\overline{AB} \cong \overline{DC}$ and $\overline{AD} \cong \overline{BC}$. Using intuition, what type of quadrilateral is $ABCD$?

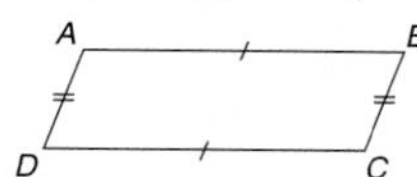


20. Quadrilateral $RSTV$ has $\overline{RS} \cong \overline{TV}$ and $\overline{RS} \parallel \overline{TV}$. Using intuition, what type of quadrilateral is $RSTV$?

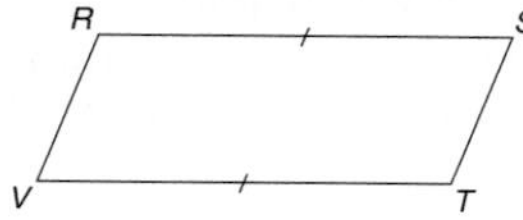


*In Exercises 21 to 24, use the definition of parallelogram to complete each proof.*

21. *Given:* $\overline{RS} \parallel \overline{VT}$, $\overline{RV} \perp \overline{VT}$, and $\overline{ST} \perp \overline{VT}$
*Prove:* $RSTV$ is a parallelogram

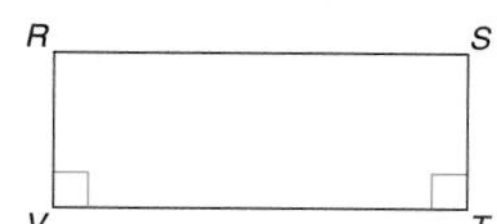


**PROOF**

| Statements | Reasons |
|---|---|
| 1. $\overline{RS} \parallel \overline{VT}$ | 1. ? |
| 2. ? | 2. Given |
| 3. ? | 3. If two lines are ⊥ to the same line, they are ∥ to each other |
| 4. ? | 4. If both pairs of opposite sides of a quadrilateral are ∥, the quad. is a ▱ |

22. *Given:* $\overline{WX} \parallel \overline{ZY}$ and ∠s $Z$ and $Y$ are supplementary
*Prove:* $WXYZ$ is a parallelogram

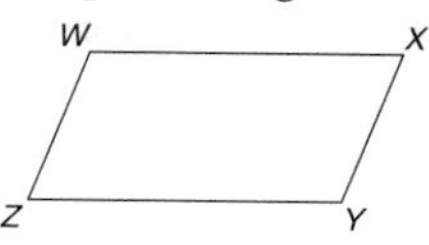


**PROOF**

| Statements | Reasons |
|---|---|
| 1. $\overline{WX} \parallel \overline{ZY}$ | 1. ? |
| 2. ? | 2. Given |
| 3. ? | 3. If two lines are cut by a transversal so that int. ∠s on the same side of the trans. are supplementary, these lines are ∥ |
| 4. ? | 4. If both pairs of opposite sides of a quadrilateral are ∥, the quad. is a ▱ |

23. *Given:* Parallelogram $RSTV$; also $\overline{XY} \parallel \overline{VT}$
*Prove:* $\angle 1 \cong \angle S$
*Plan:* First show that $RSYX$ is a parallelogram.

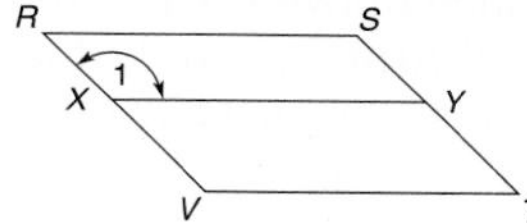


24. *Given:* Parallelogram $ABCD$ with $\overline{DE} \perp \overline{AB}$ and $\overline{FB} \perp \overline{AB}$
*Prove:* $\overline{DE} \cong \overline{FB}$
*Plan:* First show that $DEBF$ is a parallelogram.

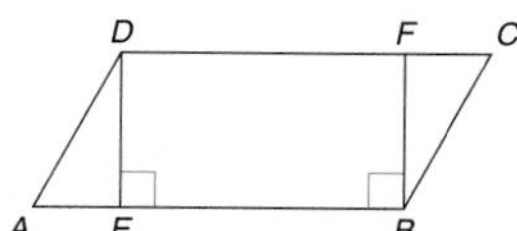


*In Exercises 25 to 28, write a formal proof of each theorem or corollary.*

25. The opposite angles of a parallelogram are congruent.

26. The opposite sides of a parallelogram are congruent.

27. The diagonals of a parallelogram bisect each other.

28. The consecutive angles of a parallelogram are supplementary.

29. The bisectors of two consecutive angles of $\square HJKL$ are shown. What can you conclude about $\angle P$?

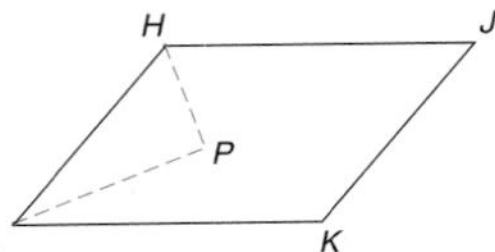


30. When the bisectors of two consecutive angles of a parallelogram meet at a point on the remaining side, what can you conclude about $\triangle DEC$? About $\triangle ADE$? About $\triangle BCE$?

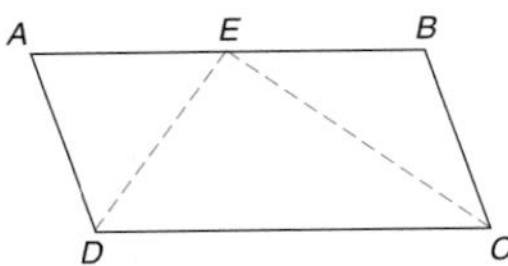


31. Draw parallelogram *RSTV* with $m\angle R = 70°$ and $m\angle S = 110°$. Which diagonal of $\square RSTV$ has the greater length?

32. Draw parallelogram *RSTV* so that the diagonals have the lengths $RT = 5$ and $SV = 4$. Which two angles of $\square RSTV$ have the greater measure?

33. The following problem is based on the Parallelogram Law. In the scaled drawing, each unit corresponds to 50 mph. A small airplane travels due east at 250 mph. The wind is blowing at 50 mph in the direction due north. Using the scale provided, determine the approximate length of the indicated diagonal and use it to determine the speed of the airplane in miles per hour.

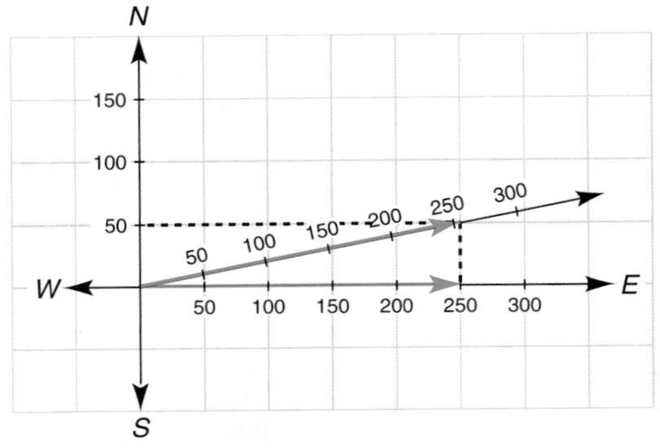


*Exercises 33, 34*

34. In the drawing for Exercise 33, the bearing (direction) in which the airplane travels is described as north $x°$ east, where $x$ is the measure of the angle from the north axis toward the east axis. Using a protractor, find the approximate bearing of the airplane.

35. Two streets meet to form an obtuse angle at point *B*. On that corner, the newly poured foundation for a building takes the shape of a parallelogram. Which diagonal, $\overline{AC}$ or $\overline{BD}$, is longer?

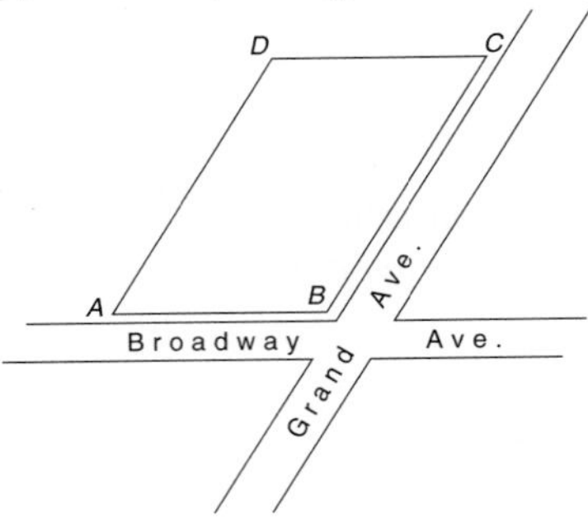


*Exercises 35, 36*

36. To test the accuracy of the foundation's measurements, lines (strings) are joined from opposite corners of the building's foundation. How should the strings that are represented by $\overline{AC}$ and $\overline{BD}$ be related?

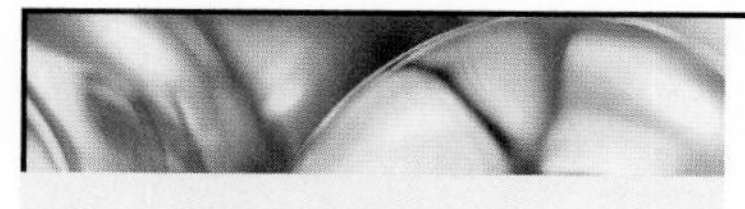

**KEY CONCEPTS**

Quadrilaterals That Are Parallelograms • Rectangle • Kite

# 4.2 The Parallelogram and Kite

The quadrilaterals discussed in this section have two pairs of congruent sides.

## THE PARALLELOGRAM

Because the hypothesis of each theorem in Section 4.1 included a given parallelogram, our goal was to develop the properties of parallelograms. In this section, Theorems 4.2.1–4.2.3 take the form "If . . . , then this quadrilateral is a parallelogram." In this section, we find that quadrilaterals having certain characteristics must be parallelograms. For instance, one set of characteristics that determines a parallelogram is "a quadrilateral has one pair of *opposite* sides that are both congruent and parallel."

### EXAMPLE 1

Give a formal proof of Theorem 4.2.1.

> THEOREM 4.2.1: If two sides of a quadrilateral are both congruent and parallel, then the quadrilateral is a parallelogram.

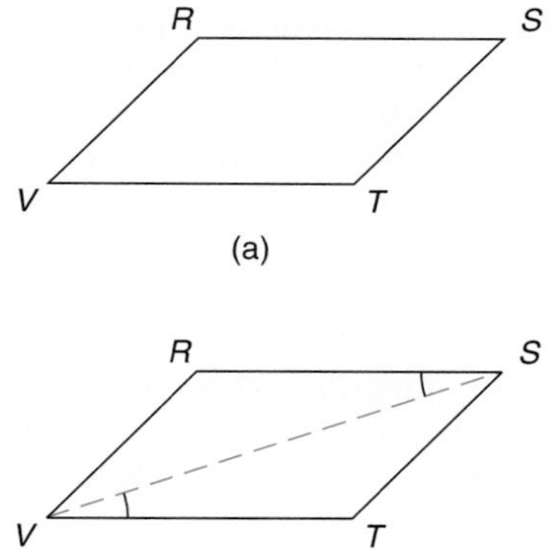


**FIGURE 4.12**

*Given:* In Figure 4.12(a), $\overline{RS} \parallel \overline{VT}$ and $\overline{RS} \cong \overline{VT}$

*Prove:* $RSTV$ is a ▱

**PROOF**

| Statements | Reasons |
|---|---|
| 1. $\overline{RS} \parallel \overline{VT}$ and $\overline{RS} \cong \overline{VT}$ | 1. Given |
| 2. Draw diagonal $\overline{VS}$, as in Figure 4.12(b) | 2. Exactly one line passes through two points |
| 3. $\overline{VS} \cong \overline{VS}$ | 3. Identity |
| 4. $\angle RSV \cong \angle SVT$ | 4. If two ∥ lines are cut by a transversal, alternate interior ∠s are ≅ |
| 5. $\triangle RSV \cong \triangle TVS$ | 5. SAS |
| 6. $\therefore \angle RVS \cong \angle VST$ | 6. CPCTC |
| 7. $\overline{RV} \parallel \overline{ST}$ | 7. If two lines are cut by a transversal so that alternate interior ∠s are ≅, these lines are ∥ |
| 8. $RSTV$ is a ▱ | 8. If both pairs of opposite sides of a quadrilateral are ∥, the quadrilateral is a parallelogram |

Consider the following activity. Through it, we discover another type of quadrilateral that must be a parallelogram.

**Discover!**

Take two straws and cut each straw into two pieces so that the lengths of the pieces of one straw match those of the second. Now form a quadrilateral by placing the pieces end-to-end so that congruent sides lie in opposite positions. What type of quadrilateral is always formed?

ANSWER

A parallelogram

**FIGURE 4.13**

The preceding activity led to a parallelogram and also leads to the following theorem. Proof of the theorem is left to the student.

THEOREM 4.2.2: If both pairs of opposite sides of a quadrilateral are congruent, then it is a parallelogram.

Another quality of quadrilaterals that determines a parallelogram is stated in Theorem 4.2.3. Its proof is also left to the student. To clarify the meaning of Theorem 4.2.3, see the drawing for Exercise 3 on page 192.

THEOREM 4.2.3: If the diagonals of a quadrilateral bisect each other, then the quadrilateral is a parallelogram.

When a figure is drawn to represent the hypothesis of a theorem, we should not include more conditions than the hypothesis states. Relative to Theorem 4.2.3, if we drew two diagonals that not only bisected each other but also were equal in length, then the quadrilateral would be the special type of parallelogram known as a **rectangle.** We will deal with rectangles in the next section.

*Exs. 1–4*

## THE KITE

The next quadrilateral we consider is known as a *kite.* This quadrilateral gets its name from the child's toy pictured in Figure 4.13. In the construction of the kite, there are two pairs of congruent *adjacent* sides. See Figure 4.14(a) on page 189. This leads to the formal definition of a kite.

DEFINITION: A **kite** is a quadrilateral with two distinct pairs of congruent adjacent sides.

The word *distinct* is used in this definition to clarify that the kite does not have four congruent sides.

**Discover!**

Take two straws and cut them into pieces so the lengths match. Now form a quadrilateral by placing congruent pieces together. What type of quadrilateral is always formed?

ANSWER

Kite

THEOREM 4.2.4: In a kite, one pair of opposite angles are congruent.

In Example 2, we verify Theorem 4.2.4 by proving that $\angle B \cong \angle D$. With congruent sides as marked, $\angle A \not\cong \angle C$.

**Discover!**

From a sheet of construction paper, cut out kite $ABCD$ so that $AB = AD$ and $BC = DC$.
a) When you fold kite $ABCD$ along the diagonal $\overline{AC}$, are two congruent triangles formed?
b) When you fold kite $ABCD$ along diagonal $\overline{BD}$, are two congruent triangles formed?

ANSWERS

(a) Yes (b) No

EXAMPLE 2

Complete the proof of Theorem 4.2.4.

*Given:* Kite $ABCD$ with congruent sides as marked. [See Figure 4.14(a).]

*Prove:* $\angle B \cong \angle D$

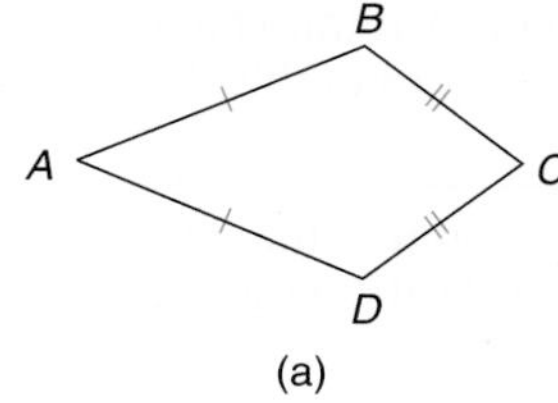

(a)

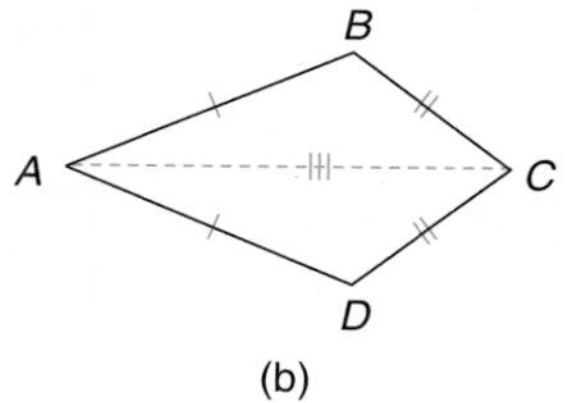

(b)

**FIGURE 4.14**

PROOF

| Statements | Reasons |
|---|---|
| 1. Kite $ABCD$ | 1. ? |
| 2. $\overline{BC} \cong \overline{CD}$ and $\overline{AB} \cong \overline{AD}$ | 2. A kite has two pairs of $\cong$ adjacent sides |
| 3. Draw $\overline{AC}$ [Figure 4.14(b)] | 3. Through two points, there is exactly one line |
| 4. $\overline{AC} \cong \overline{AC}$ | 4. ? |
| 5. $\triangle ACD \cong \triangle ACB$ | 5. ? |
| 6. ? | 6. CPCTC |

*Exs. 5–10*

Two additional theorems involving the kite are found in Exercises 27 and 28 of this section.

When observing an old barn or shed, we often see that it has begun to lean. Unlike a triangle, which is rigid in shape [Figure 4.15(a)] and bends only when broken, a quadrilateral [Figure 4.15(b)] does *not* provide the same level of strength and stability. In the construction of a house, bridge, building, or swing set [Figure 4.15(c)], note the use of wooden or metal triangles as braces.

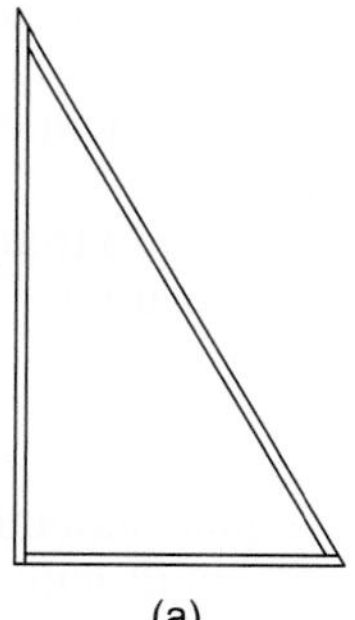
(a)

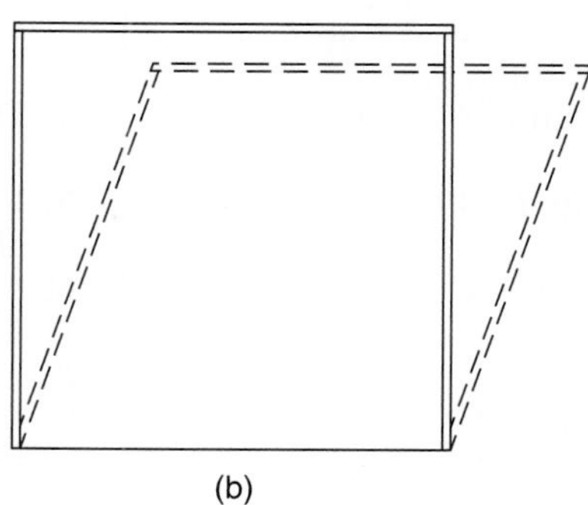
(b)

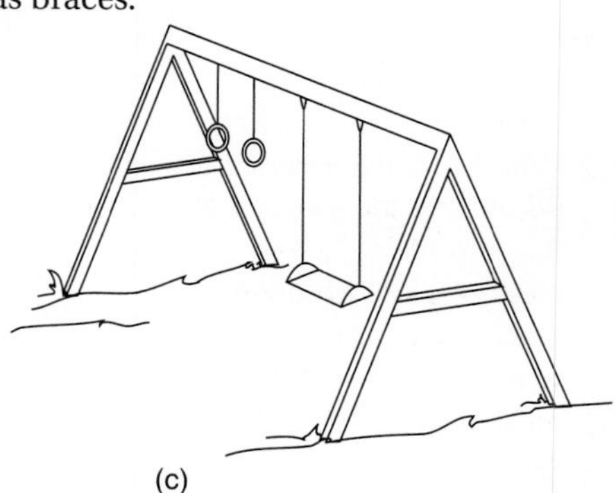
(c)

**FIGURE 4.15**

The brace in the swing set in Figure 4.15(c) suggests the following theorem.

> THEOREM 4.2.5: The segment that joins the midpoints of two sides of a triangle is parallel to the third side and has a length equal to one-half the length of the third side.

Refer to Figure 4.16(a); Theorem 4.2.5 claims that $\overline{MN} \parallel \overline{BC}$ and $MN = \frac{1}{2}(BC)$. We will prove the first part of this theorem but leave the second part as an exercise.

> The line segment that joins the midpoints of two sides of a triangle is parallel to the third side of the triangle.

*Given:* In Figure 4.16(a), $\triangle ABC$ with midpoints $M$ and $N$ of $\overline{AB}$ and $\overline{AC}$, respectively

*Prove:* $\overline{MN} \parallel \overline{BC}$

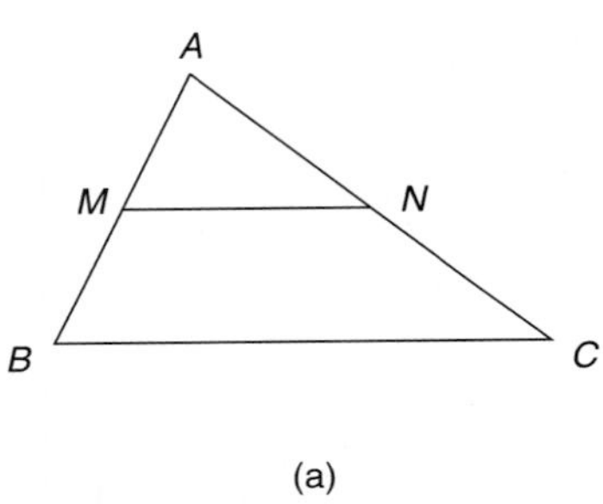

(a)

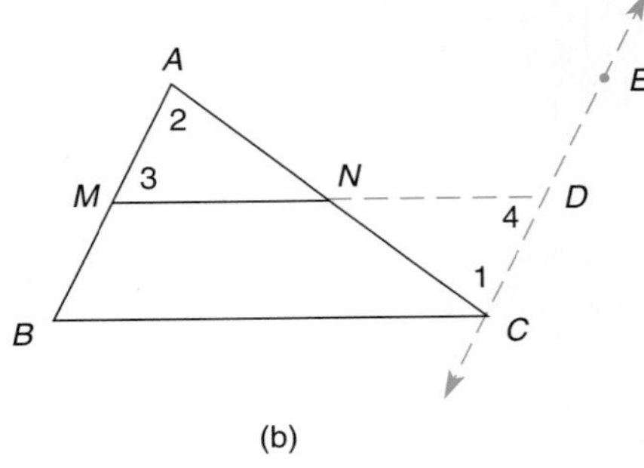

(b)

**FIGURE 4.16**

> **Discover!**
>
> Sketch regular hexagon $ABCDEF$. Draw diagonals $\overline{AE}$ and $\overline{CE}$. What type of quadrilateral is $ABCE$?
>
> ANSWER
>
> Kite

**PROOF**

| Statements | Reasons |
|---|---|
| 1. $\triangle ABC$, with midpoints $M$ and $N$ of $\overline{AB}$ and $\overline{AC}$, respectively | 1. Given |
| 2. Through $C$, construct $\overleftrightarrow{CE} \parallel \overline{AB}$, as in Figure 4.16(b) | 2. Parallel Postulate |
| 3. Extend $\overline{MN}$ to meet $\overleftrightarrow{CE}$ at $D$, as in Figure 4.16(b) | 3. Exactly one line passes through two points |
| 4. $\overline{AM} \cong \overline{MB}$ and $\overline{AN} \cong \overline{NC}$ | 4. The midpoint of a segment divides it into $\cong$ segments |
| 5. $\angle 1 \cong \angle 2$ and $\angle 4 \cong \angle 3$ | 5. If two $\parallel$ lines are cut by a transversal, alternate interior $\angle$s are $\cong$ |
| 6. $\triangle ANM \cong \triangle CND$ | 6. AAS |
| 7. $\overline{AM} \cong \overline{DC}$ | 7. CPCTC |
| 8. $\overline{MB} \cong \overline{DC}$ | 8. Transitive (both are $\cong$ to $\overline{AM}$) |
| 9. Quadrilateral $BMDC$ is a $\square$ | 9. If two sides of a quadrilateral are both $\cong$ and $\parallel$, the quadrilateral is a parallelogram |
| 10. $\overline{MN} \parallel \overline{BC}$ | 10. Opposite sides of a $\square$ are $\parallel$ |

> *Technology Exploration*
>
> Use computer software if available.
>
> 1. Construct $\triangle ABC$ (any triangle).
> 2. Where $M$ is the midpoint of $\overline{AB}$ and $N$ is the midpoint of $\overline{AC}$, draw $\overline{MN}$.
> 3. Measure $\angle AMN$ and $\angle B$.
> 4. Show that $m\angle AMN = m\angle B$, which shows that $\overline{MN} \parallel \overline{BC}$.
> 5. Now measure $\overline{MN}$ and $\overline{BC}$.
> 6. Show that $MN = \frac{1}{2}(BC)$. (Measures may not be "perfect.")

Theorem 4.2.5 also asserts that the segment formed by joining the midpoints of two sides of a triangle has a length equal to one-half the length of the third side. This part of the theorem is used in Examples 3 and 4.

EXAMPLE 3

In $\triangle RST$ in Figure 4.17, $M$ and $N$ are the midpoints of $\overline{RS}$ and $\overline{RT}$, respectively.

a) If $ST = 12.7$, find $MN$.
b) If $MN = 15.8$, find $ST$.

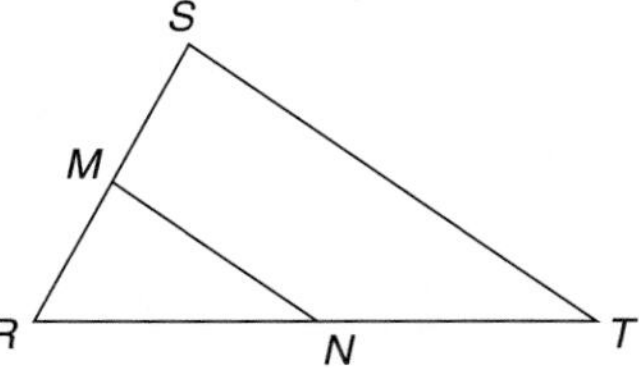


**FIGURE 4.17**

***Solution***

a) $MN = \frac{1}{2}(ST)$, so $MN = \frac{1}{2}(12.7) = 6.35$.
b) $MN = \frac{1}{2}(ST)$, so $15.8 = \frac{1}{2}(ST)$.
Multiplying by 2, $ST = 31.6$.

**Discover!**

Draw a triangle $\triangle ABC$ with midpoint $D$ of $\overline{CA}$ and $E$ of $\overline{CB}$. Cut out $\triangle CDE$ and place it at the base $\overline{AB}$. By sliding $\overline{DE}$ along $\overline{AB}$, what do you find?

ANSWER

$DE = \frac{1}{2}(AB)$ or $AB = 2(DE)$

EXAMPLE 4

*Given:* $\triangle ABC$ in Figure 4.18, with $D$ the midpoint of $\overline{AC}$ and $E$ the midpoint of $\overline{BC}$; $DE = 2x + 1$; $AB = 5x - 1$

*Find:* $x$, $DE$, and $AB$

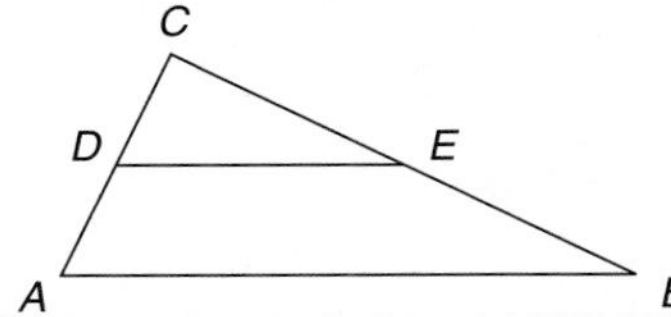


**FIGURE 4.18**

***Solution***

By Theorem 4.2.5,

$$DE = \frac{1}{2}(AB)$$

so

$$2x + 1 = \frac{1}{2}(5x - 1)$$

Multiplying by 2, we have

$$4x + 2 = 5x - 1$$
$$3 = x$$

Therefore, $DE = 2 \cdot 3 + 1 = 7$. Similarly, $AB = 5 \cdot 3 - 1 = 14$.

NOTE: In Example 4, a check shows that $DE = \frac{1}{2}(AB)$.

*Exs. 11–15*

In the final example of this section, we consider the design of a product. Also see related Exercises 17 and 18 of this section.

EXAMPLE 5

In a studio apartment, there is a bed that folds down from the wall. In the vertical position, the design shows drop-down legs of equal length; that is, $AB = CD$ [see Figure 4.19(a)]. Determine the type of quadrilateral $ABDC$, shown in Figure 4.19(b), that is formed when the bed is lowered to a horizontal position.

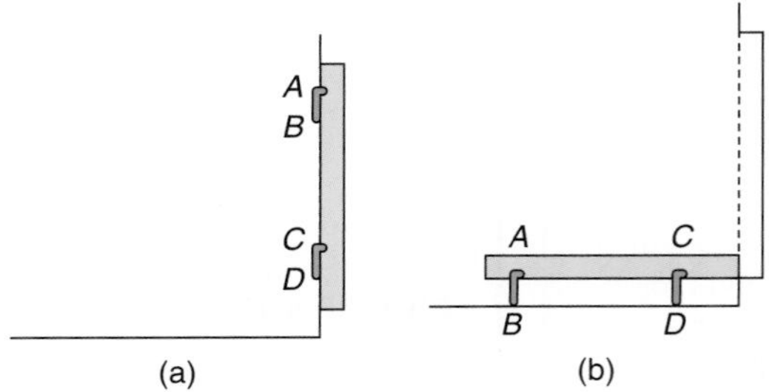


**FIGURE 4.19**

***Solution***

See Figure 4.19(a). Because $AB = CD$, it follows that $AB + BC = BC + CD$; here, $BC$ was added to each side of the equation. But $AB + BC = AC$ and $BC + CD = BD$. Thus $AC = BD$ by substitution.

In Figure 4.19(b), we see that $AB = CD$ and $AC = BD$. Because both pairs of opposite sides of the quadrilateral are congruent, $ABDC$ is a parallelogram.

NOTE: In Section 4.3, we will also show that $ABDC$ of Figure 4.19(b) is a *rectangle* (a special type of parallelogram).

## 4.2 Exercises

1. a) As shown, must quadrilateral $ABCD$ be a parallelogram?
   b) Given the lengths of the sides as shown, is the measure of $\angle A$ unique?

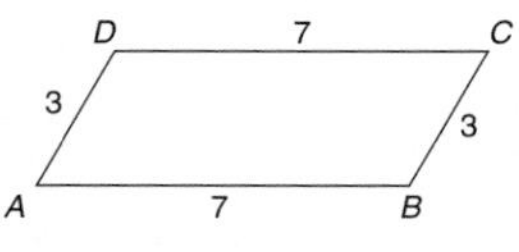


2. a) As shown, must $RSTV$ be a parallelogram?
   b) With measures as indicated, is it necessary that $RS = 8$?

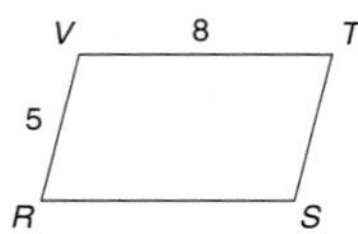


3. In the drawing, suppose that $\overline{WY}$ and $\overline{XZ}$ bisect each other. What type of quadrilateral is $WXYZ$?

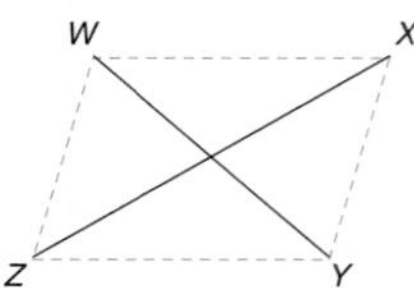


*Exercises 3, 4*

4. In the drawing, suppose that $\overline{ZX}$ is the perpendicular bisector of $\overline{WY}$. What type of quadrilateral is $WXYZ$?

5. A carpenter lays out boards of lengths 8 ft, 8 ft, 4 ft, and 4 ft by placing them end-to-end.

   a) If these are joined at the ends to form a quadrilateral that has the 8-ft pieces connected in order, what type of quadrilateral is formed?
   b) If these are joined at the ends to form a quadrilateral that has the 4-ft and 8-ft pieces alternating, what type of quadrilateral is formed?

6. A carpenter joins four boards of lengths 6 ft, 6 ft, 4 ft, and 4 ft, in that order, to form quadrilateral *ABCD* as shown.

   a) What type of quadrilateral is formed?
   b) How are angles *B* and *D* related?

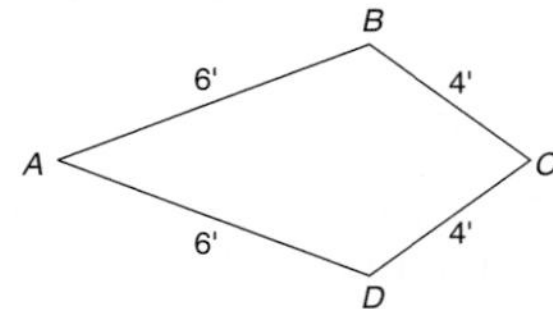


7. In parallelogram *ABCD* (not shown), $AB = 8$, $m\angle B = 110°$, and $BC = 5$. Which diagonal has the greater length?

8. In kite *WXYZ*, the measures of selected angles are shown. Which diagonal of the kite has the greater length?

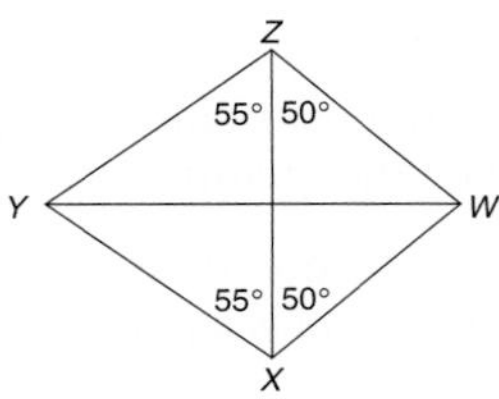


9. In $\triangle ABC$, *M* and *N* are midpoints of $\overline{AC}$ and $\overline{BC}$, respectively. If $AB = 12.36$, how long is $\overline{MN}$?

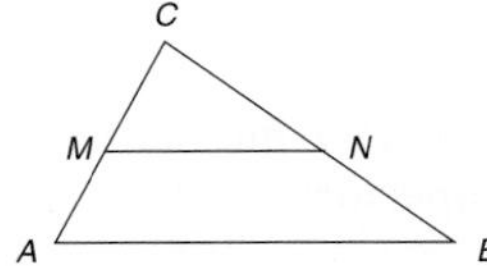


*Exercises 9, 10*

10. In $\triangle ABC$, *M* and *N* are midpoints of $\overline{AC}$ and $\overline{BC}$, respectively. If $MN = 7.65$, how long is $\overline{AB}$?

*In Exercises 11 to 14, assume that X, Y, and Z are midpoints of the sides of $\triangle RST$.*

11. If $RS = 12$, $ST = 14$, and $RT = 16$, find:

    a) $XY$  b) $XZ$  c) $YZ$

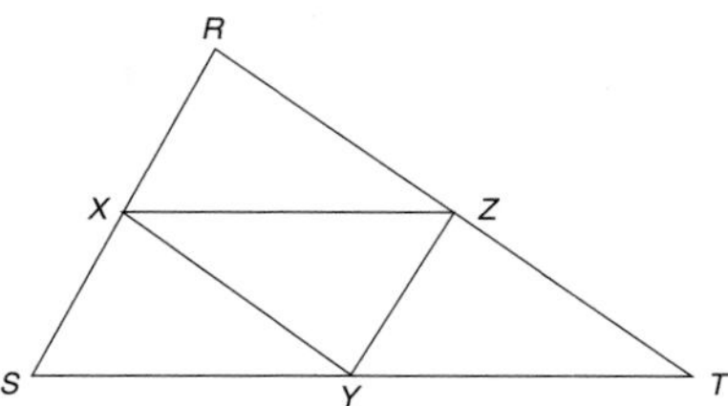


*Exercises 11–14*

12. If $XY = 6$, $YZ = 8$, and $XZ = 10$, find:

    a) $RS$  b) $ST$  c) $RT$

13. If the perimeter (sum of the lengths of all three sides) of $\triangle RST$ is 20, what is the perimeter of $\triangle XYZ$?

14. If the perimeter (sum of the lengths of all three sides) of $\triangle XYZ$ is 12.7, what is the perimeter of $\triangle RST$?

15. Consider any kite.

    a) Does it have line symmetry? If so, describe an axis of symmetry.
    b) Does it have point symmetry? If so, describe the point of symmetry.

16. Consider any parallelogram.

    a) Does it have line symmetry? If so, describe an axis of symmetry.
    b) Does it have point symmetry? If so, describe the point of symmetry.

17. For compactness, the drop-down wheels of a stretcher (or gurney) are folded under it as shown. In order for the board's upper surface to be parallel to the ground when the wheels are dropped, what relationship must exist between $\overline{AB}$ and $\overline{CD}$?

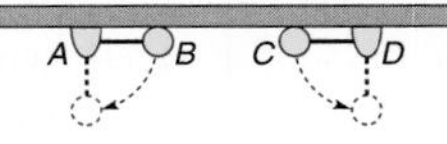


18. For compactness, the drop-down legs of an ironing board fold up under the board. A sliding mechanism at point *A* and the legs being connected at common midpoint *M* cause the board's upper surface to be parallel to the floor. How are $\overline{AB}$ and $\overline{CD}$ related?

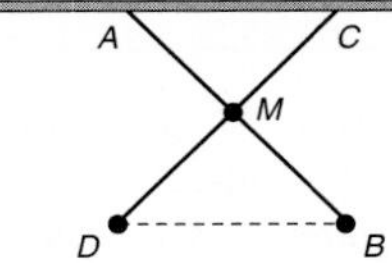

*In Exercises 19 to 24, complete each proof.*

19. *Given:* $\angle 1 \cong \angle 2$ and $\angle 3 \cong \angle 4$
*Prove:* $MNPQ$ is a kite

**PROOF**

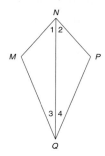

| Statements | Reasons |
|---|---|
| 1. $\angle 1 \cong \angle 2$ and $\angle 3 \cong \angle 4$ | 1. ? |
| 2. $\overline{NQ} \cong \overline{NQ}$ | 2. ? |
| 3. ? | 3. ASA |
| 4. $\overline{MN} \cong \overline{PN}$ and $\overline{MQ} \cong \overline{PQ}$ | 4. ? |
| 5. ? | 5. If a quadrilateral has two pairs of $\cong$ adjacent sides, it is a kite |

20. *Given:* Quadrilateral $ABCD$, with midpoints $E$, $F$, $G$, and $H$ of the sides
*Prove:* $\overline{EF} \parallel \overline{HG}$

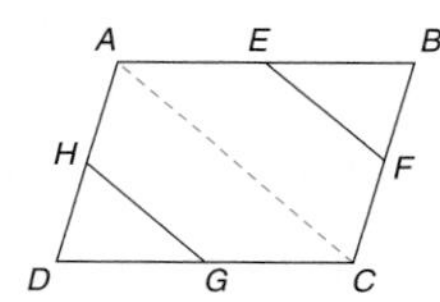


**PROOF**

| Statements | Reasons |
|---|---|
| 1. ? | 1. Given |
| 2. Draw $\overline{AC}$ | 2. Through two points, there is one line |
| 3. In $\triangle ABC$, $\overline{EF} \parallel \overline{AC}$ and in $\triangle ADC$, $\overline{HG} \parallel \overline{AC}$ | 3. ? |
| 4. ? | 4. If two lines are $\parallel$ to the same line, these lines are $\parallel$ to each other |

21. *Given:* $M$-$Q$-$T$ and $P$-$Q$-$R$ such that $MNPQ$ and $QRST$ are $\square$s
*Prove:* $\angle N \cong \angle S$

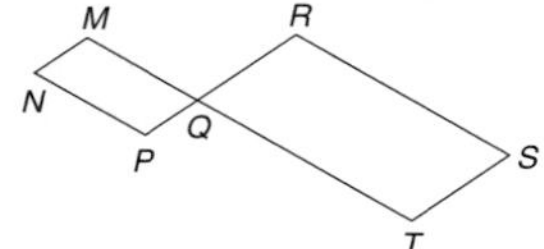


22. *Given:* $\square WXYZ$ with diagonals $\overline{WY}$ and $\overline{XZ}$
*Prove:* $\triangle WMX \cong \triangle YMZ$

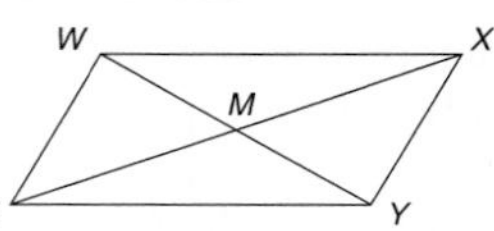


23. *Given:* Kite $HJKL$ with diagonal $\overline{HK}$
*Prove:* $\overrightarrow{HK}$ bisects $\angle LHJ$

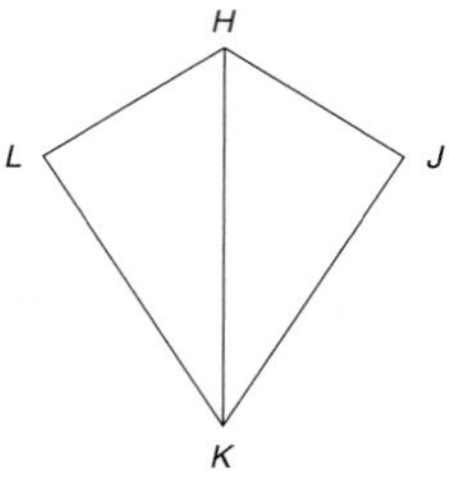


24. *Given:* $\square MNPQ$, with $T$ the midpoint of $\overline{MN}$ and $S$ the midpoint of $\overline{QP}$
*Prove:* $\triangle QMS \cong \triangle NPT$, and $MSPT$ is a $\square$

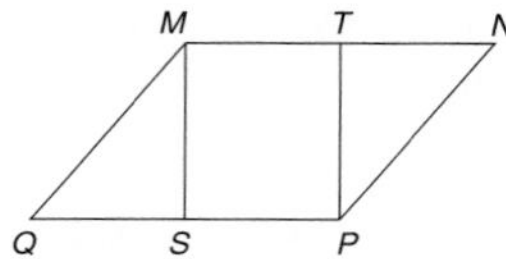


*In Exercises 25 to 28, write a formal proof of each theorem or corollary.*

25. If both pairs of opposite sides of a quadrilateral are congruent, then the quadrilateral is a parallelogram.

26. If the diagonals of a quadrilateral bisect each other, then the quadrilateral is a parallelogram.

27. In a kite, one diagonal is the perpendicular bisector of the other diagonal.

28. One diagonal of a kite bisects two of the angles of the kite.

*In Exercises 29 to 31, $\triangle RST$ has M and N for midpoints of sides $\overline{RS}$ and $\overline{RT}$, respectively.*

29. *Given:* $MN = 2y - 3$
$ST = 3y$
*Find:* $y$, $MN$, and $ST$

30. *Given:* $MN = x^2 + 5$
$ST = x(2x + 5)$
*Find:* $x$, $MN$, and $ST$

31. *Given:* $RM = RN = 2x + 1$
$ST = 5x - 3$
$m\angle R = 60°$
*Find:* $x$, $RM$, and $ST$

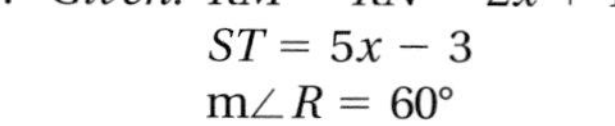

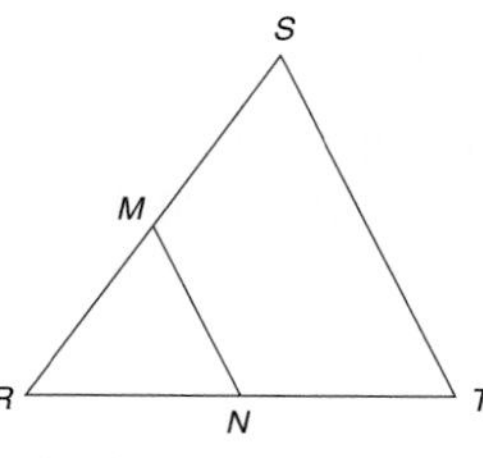

*Exercises 29–31*

32. $RSTV$ is a kite, with $\overline{RS} \perp \overline{ST}$ and $\overline{RV} \perp \overline{VT}$. If $m\angle STV = 40°$, how large is the angle formed by the bisectors of $\angle RST$ and $\angle STV$? By the bisectors of $\angle SRV$ and $\angle RST$?

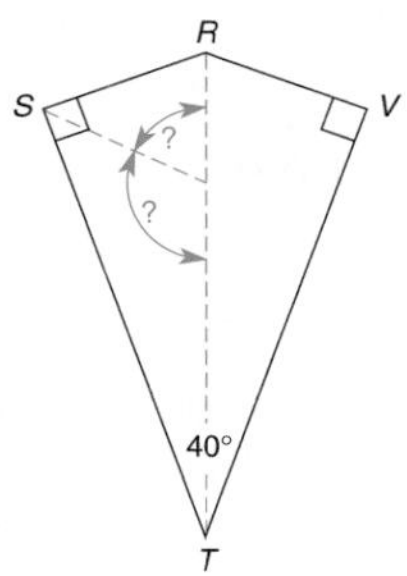


33. In concave kite $ABCD$, there is an interior angle at vertex $B$ that is a reflex angle. Given that $m\angle A = m\angle C = m\angle D = 30°$, find the measure of the indicated reflex angle.

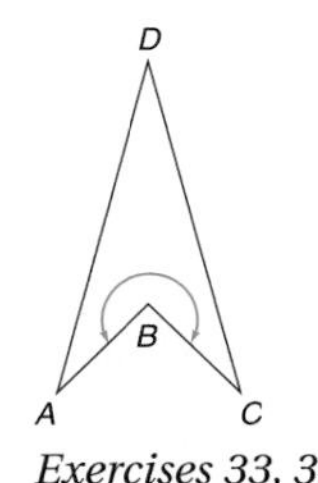

*Exercises 33, 34*

34. If the length of side $\overline{AB}$ (for kite $ABCD$) is 6 in., find the length of $\overline{AC}$ (not shown). Recall that $m\angle A = m\angle C = m\angle D = 30°$

* 35. Prove that the segment that joins the midpoints of two sides of a triangle has a length equal to one-half the length of the third side.

(HINT: In the drawing, $\overline{MN}$ is extended to $D$, a point on $\overline{CD}$. Also, $\overline{CD}$ is parallel to $\overline{AB}$.)

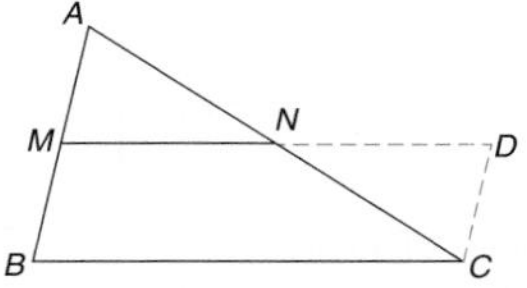


* 36. Prove that when the midpoints of consecutive sides of a quadrilateral are joined in order, the resulting quadrilateral is a parallelogram.

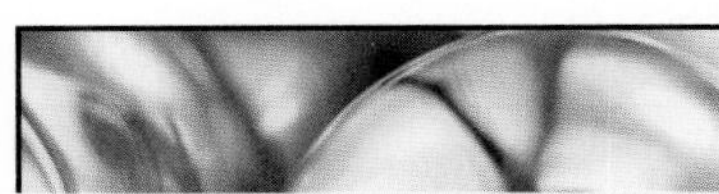

**KEY CONCEPTS**

Rectangle • Square • Rhombus • Pythagorean Theorem

# 4.3 The Rectangle, Square, and Rhombus

## THE RECTANGLE

In this section, we investigate special parallelograms. The first of these is the rectangle (abbreviated "rect."), which is defined as follows:

DEFINITION: A **rectangle** is a parallelogram that has a right angle. (See Figure 4.20.)

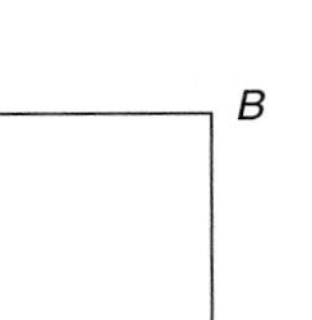

**FIGURE 4.20**

Any reader who is familiar with the rectangle may be confused by the fact that the preceding definition calls for only one right angle. Because a rectangle is a parallelogram by definition, the fact that a rectangle has four right angles is easily proved by applying Corollaries 4.1.3 and 4.1.5. The proof of Corollary 4.3.1 is left to the student.

COROLLARY 4.3.1: All angles of a rectangle are right angles.

**Reminder**

A rectangle is a parallelogram. Thus it has all the properties of a parallelogram, plus some properties of its own.

The following theorem is true for rectangles, but not for parallelograms in general.

THEOREM 4.3.2: The diagonals of a rectangle are congruent.

NOTE: To follow the flow of the proof in Example 1, it may be best to draw triangles *NMQ* and *PQM* of Figure 4.21 separately.

EXAMPLE 1

Complete a proof of Theorem 4.3.2.

*Given:* Rectangle *MNPQ* with diagonals $\overline{MP}$ and $\overline{NQ}$ (See Figure 4.21.)

*Prove:* $\overline{MP} \cong \overline{NQ}$

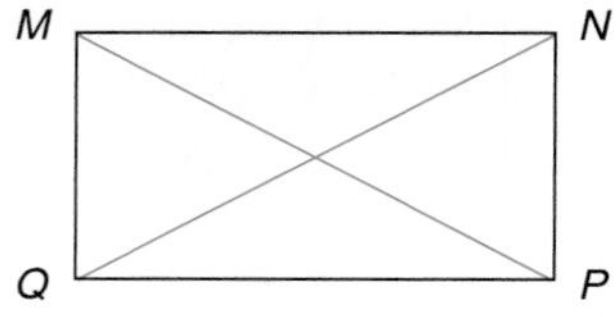


**FIGURE 4.21**

**Discover!**

Given a rectangle *MNPQ* (like a sheet of paper), draw diagonals $\overline{MP}$ and $\overline{NQ}$. From a second sheet, cut out $\triangle MPQ$ (formed by two sides and a diagonal of *MNPQ*). Can you position $\triangle MPQ$ so that it coincides with $\triangle NQP$?

ANSWER

Yes

PROOF

| Statements | Reasons |
|---|---|
| 1. Rectangle *MNPQ* with diagonals $\overline{MP}$ and $\overline{NQ}$ | 1. Given |
| 2. *MNPQ* is a ▱ | 2. By definition, a rectangle is a ▱ with a right angle |
| 3. $\overline{MN} \cong \overline{QP}$ | 3. Opposite sides of a ▱ are ≅ |
| 4. $\overline{MQ} \cong \overline{MQ}$ | 4. Identity |
| 5. $\angle NMQ$ and $\angle PQM$ are right $\angle$s | 5. By Corollary 4.3.1, the four $\angle$s of a rectangle are right $\angle$s |
| 6. $\angle NMQ \cong \angle PQM$ | 6. All right $\angle$s are ≅ |
| 7. $\triangle NMQ \cong \triangle PQM$ | 7. SAS |
| 8. $\overline{MP} \cong \overline{NQ}$ | 8. CPCTC |

*Exs. 1–4*

## THE SQUARE

All rectangles are parallelograms; some parallelograms are rectangles; and some rectangles are *squares.*

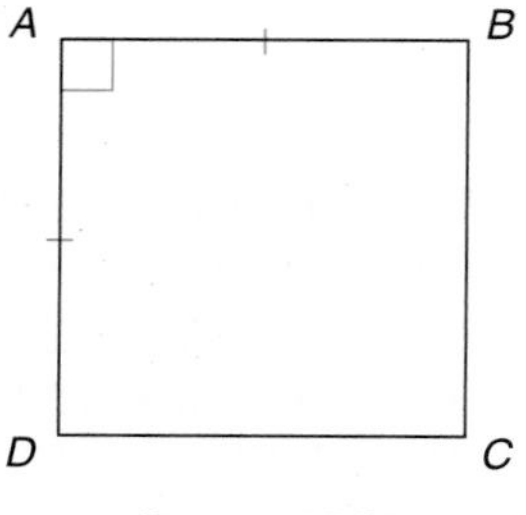


Square *ABCD*

**FIGURE 4.22**

DEFINITION: A **square** is a rectangle that has two congruent adjacent sides. (See Figure 4.22.)

COROLLARY 4.3.3: All sides of a square are congruent.

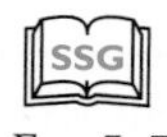

*Exs. 5–7*

Because a square is a type of rectangle, it has four right angles and its diagonals are congruent. Because a square is also a parallelogram, its opposite sides are parallel. For any square, we can show that the diagonals are perpendicular.

In Chapter 7, we measure area in "square units."

## THE RHOMBUS

The next type of quadrilateral we consider is the rhombus. The plural of the word *rhombus* is *rhombi* (pronounced rhŏm-bī).

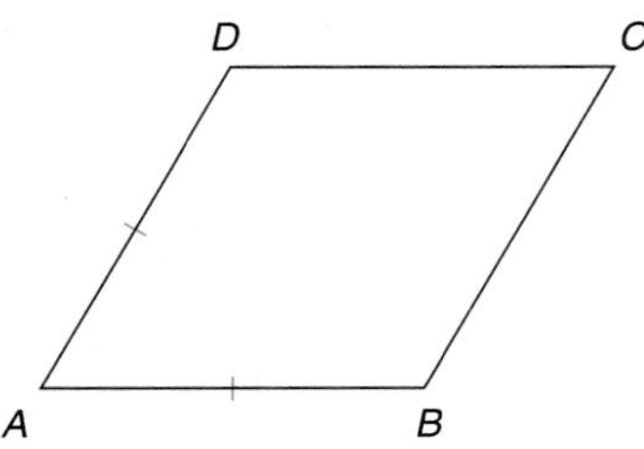


**FIGURE 4.23**

> DEFINITION: A **rhombus** is a parallelogram with two congruent adjacent sides.

In Figure 4.23, the adjacent sides $\overline{AB}$ and $\overline{AD}$ of rhombus *ABCD* are marked congruent. Because a rhombus is a type of parallelogram, it is also necessary that $\overline{AB} \cong \overline{DC}$ and $\overline{AD} \cong \overline{BC}$. Thus we have Corollary 4.3.4.

> COROLLARY 4.3.4: All sides of a rhombus are congruent.

We will use Corollary 4.3.4 in the proof of the following theorem.

> THEOREM 4.3.5: The diagonals of a rhombus are perpendicular.

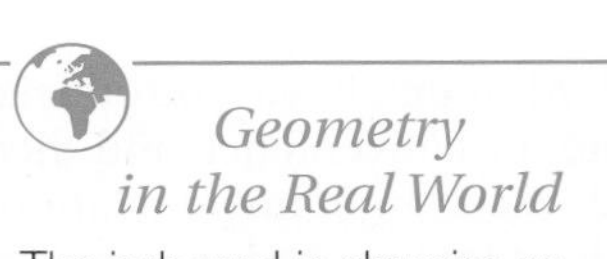


The jack used in changing an automobile tire illustrates the shape of a rhombus.

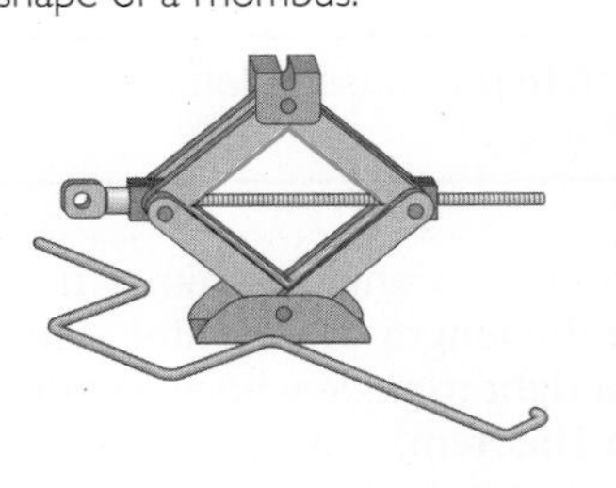

**Discover!**

Sketch regular hexagon *RSTVWX*. Draw diagonals $\overline{RT}$ and $\overline{XV}$. What type of quadrilateral is *RTVX*?

ANSWER

Rectangle

### EXAMPLE 2

Study the picture proof of Theorem 4.3.5. In the proof, pairs of triangles are congruent by the reason SSS.

**PICTURE PROOF OF THEOREM 4.3.5**

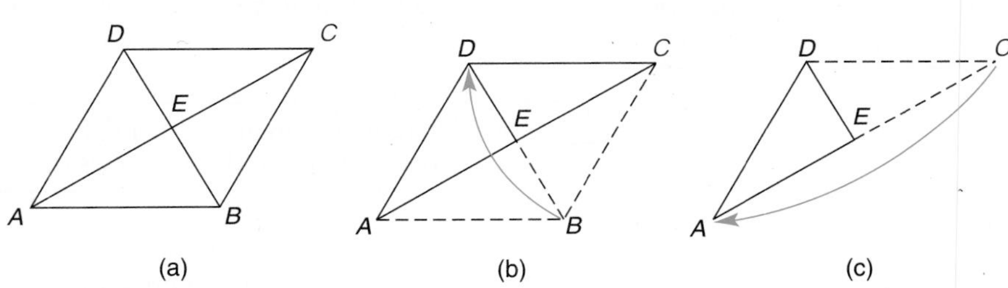


**FIGURE 4.24**

*Given:* Rhombus *ABCD*, with diagonals $\overline{AC}$ and $\overline{DB}$ (See Figure 4.24(a)).

*Prove:* $\overline{AC} \perp \overline{DB}$

*Proof:* Fold $\triangle ABC$ across $\overline{AC}$ to coincide with $\triangle CED$ [see Figure 4.24(b)]. Now fold $\triangle CED$ across half-diagonal $\overline{DE}$ to coincide with $\triangle AED$ [see Figure 4.24(c)]. The four congruent triangles formed in Figure 4.24(c) can be unwrapped to return rhombus *ABCD* of Figure 4.24(a). With four congruent right angles at vertex *E*, we see that $\overline{AC} \perp \overline{DB}$.

SSG *Exs. 8–11*

An alternative definition of *square* is "A square is a rhombus whose adjacent sides form a right angle." Therefore, a further property of a square is that its diagonals are perpendicular.

The Pythagorean Theorem, which deals with right triangles, is also useful in applications involving quadrilaterals that have right angles. In antiquity, the theorem claimed that "the square upon the hypotenuse equals the sum of the squares upon the legs of the right triangle." See Figure 4.25(a). This interpretation involves the area concept, which we study in a later chapter. By counting squares in Figure 4.25(a), one sees that 25 "square units" is the sum of 9 and 16 square units. Our interpretation of the Pythagorean Theorem uses number (length) relationships.

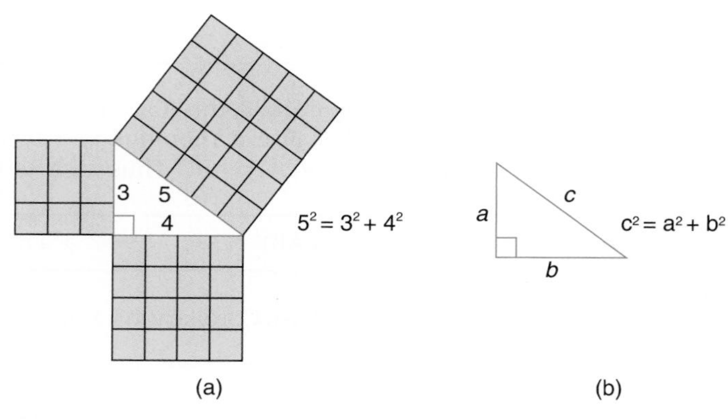


**FIGURE 4.25**

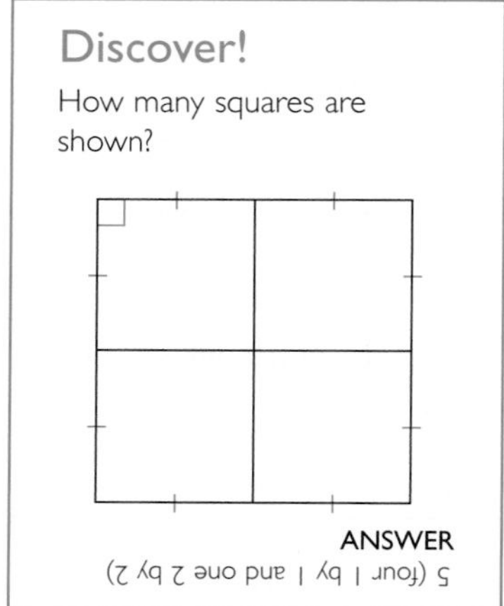


## THE PYTHAGOREAN THEOREM

The Pythagorean Theorem will be proved in Section 5.4. Although it was introduced in Section 3.2, we restate the Pythagorean Theorem here for convenience and then review its application to the *right* triangle in Example 3. When right angle relationships exist in quadrilaterals, we can often apply the "rule of Pythagoras" as well; see Examples 4, 5, and 6.

> **The Pythagorean Theorem** In a right triangle with hypotenuse of length $c$ and legs of lengths $a$ and $b$, it follows that $c^2 = a^2 + b^2$.

Provided that the lengths of two of the sides of a right triangle are known, the Pythagorean Theorem can be applied to determine the length of the third side. In Example 3, we seek the length of the hypotenuse in a right triangle whose lengths of legs are known. When we are using the Pythagorean Theorem, $c$ must represent the length of the hypotenuse; however, either leg can be chosen for length $a$ (or $b$).

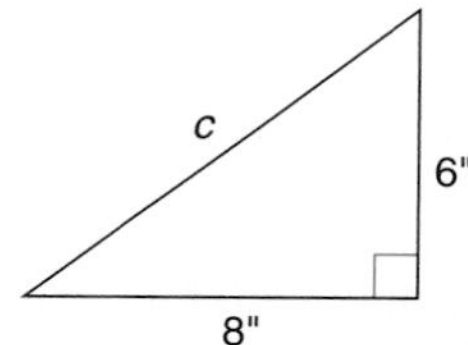


**FIGURE 4.26**

### EXAMPLE 3

What is the length of the hypotenuse of a right triangle whose legs measure 6 in. and 8 in.? (See Figure 4.26.)

***Solution***

$$c^2 = a^2 + b^2$$
$$c^2 = 6^2 + 8^2$$
$$c^2 = 36 + 64 \rightarrow c^2 = 100 \rightarrow c = 10 \text{ in.}$$

In the following example, the diagonal of a rectangle separates it into two right triangles. As shown in Figure 4.27, the diagonal of the rectangle is the hypotenuse of each right triangle formed by the diagonal.

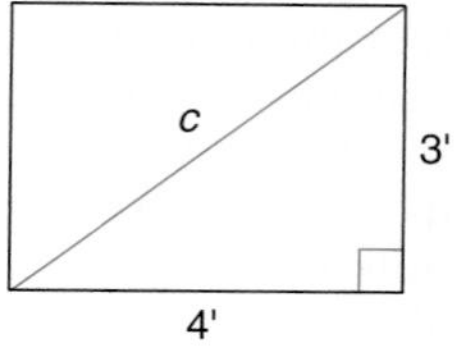


**FIGURE 4.27**

### EXAMPLE 4

What is the length of the diagonal in a rectangle whose sides measure 3 ft and 4 ft?

***Solution***

For each triangle in Figure 4.27, $c^2 = a^2 + b^2$ becomes $c^2 = 3^2 + 4^2$ or $c^2 = 9 + 16$. Then $c^2 = 25$, so $c = 5$. The length of the diagonal is 5 ft.

In Example 5, we use the fact that a rhombus is a parallelogram to justify that its diagonals bisect each other. By Theorem 4.3.5, the diagonals of the rhombus are also perpendicular.

EXAMPLE 5

What is the length of each side of a rhombus whose diagonals measure 10 cm and 24 cm? (See Figure 4.28.)

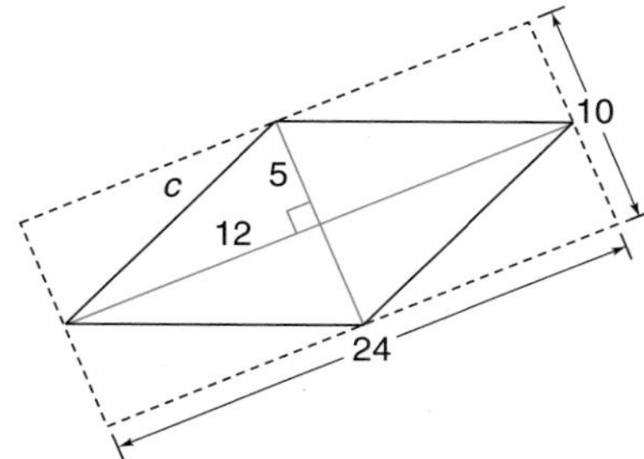


**FIGURE 4.28**

***Solution***

The diagonals of a rhombus are perpendicular bisectors of each other. Thus the diagonals separate the rhombus shown into four congruent right triangles with legs of lengths 5 cm and 12 cm. For each triangle, $c^2 = a^2 + b^2$ becomes $c^2 = 5^2 + 12^2$, or $c^2 = 25 + 144$. Then $c^2 = 169$, so $c = 13$. The length of each side is 13 cm.

EXAMPLE 6

On a softball diamond (actually a square), the distance along the base paths is 60 ft. Using the triangle in Figure 4.29, find the distance from home plate to second base.

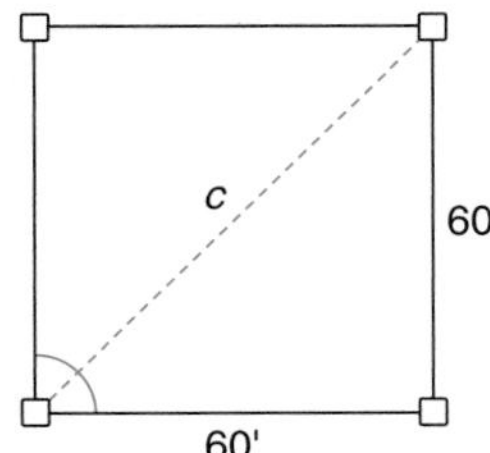


**FIGURE 4.29**

***Solution***

Using $c^2 = a^2 + b^2$, we have

$$c^2 = 60^2 + 60^2$$
$$c^2 = 7200$$

Then $c = \sqrt{7200}$ or $c \approx 84.85$ ft.

*Exs. 12–14*

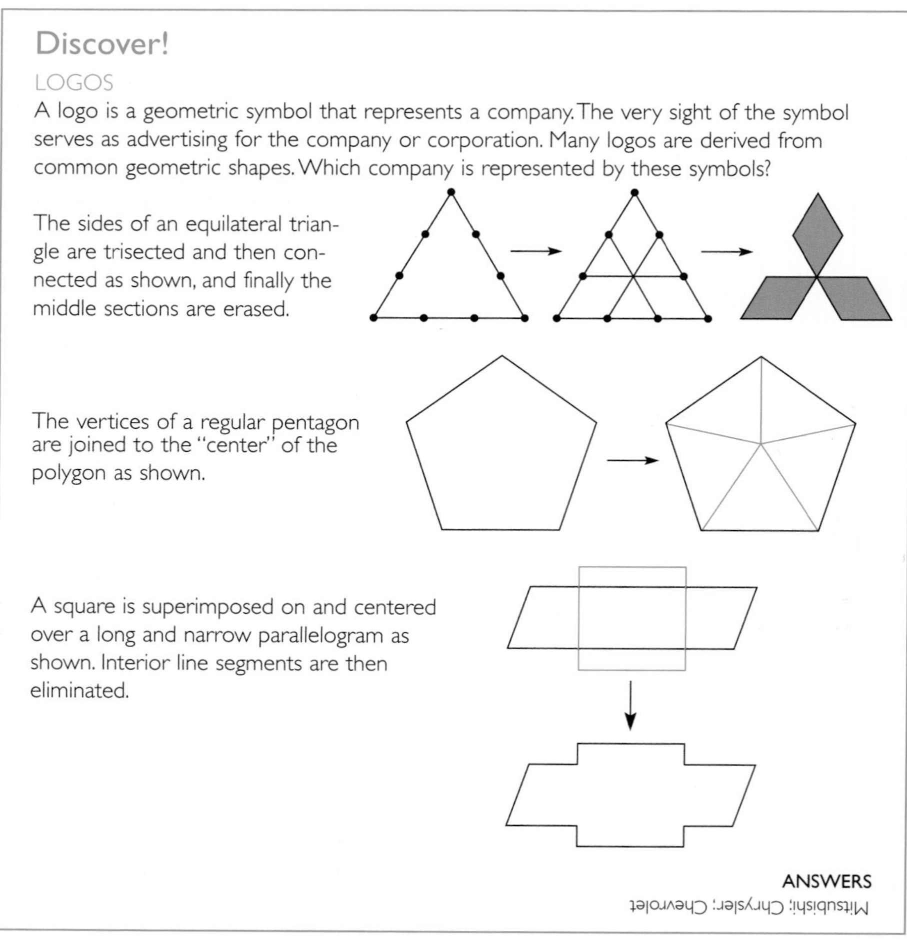

Discover!
LOGOS
A logo is a geometric symbol that represents a company. The very sight of the symbol serves as advertising for the company or corporation. Many logos are derived from common geometric shapes. Which company is represented by these symbols?
The sides of an equilateral triangle are trisected and then connected as shown, and finally the middle sections are erased.
The vertices of a regular pentagon are joined to the "center" of the polygon as shown.
A square is superimposed on and centered over a long and narrow parallelogram as shown. Interior line segments are then eliminated.
ANSWERS
Mitsubishi; Chrysler; Chevrolet


When all vertices of a quadrilateral lie on a circle, the quadrilateral is a *cyclic quadrilateral.* As it happens, all rectangles are cyclic quadrilaterals but no rhombus is a cyclic quadrilateral. The key factor in determining whether a quadrilateral is cyclic lies in the fact that the diagonals must intersect at a point that is equidistant from all four vertices. In Figure 4.30(a), rectangle $ABCD$ is cyclic because $A$, $B$, $C$, and $D$ all lie on the circle. However, rhombus $WXYZ$ in Figure 4.30(b) is *not* cyclic because $X$ and $Z$ cannot lie on the circle when $W$ and $Y$ do lie on the circle.

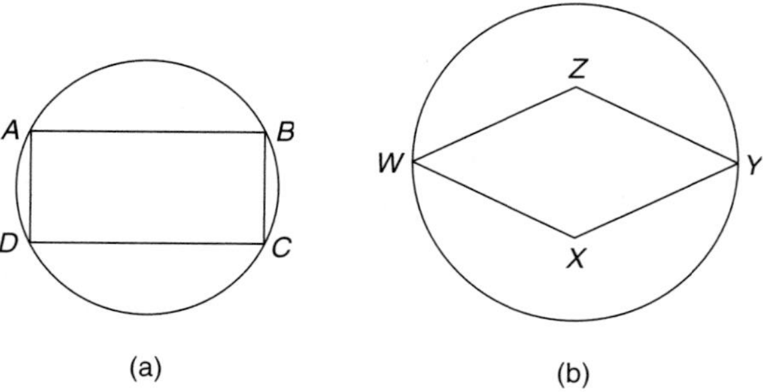


**FIGURE 4.30**

EXAMPLE 7

For cyclic rectangle $ABCD$, $AB = 8$. Diagonal $\overline{DB}$ of the rectangle is also a diameter of the circle and $DB = 10$. Find the perimeter of $ABCD$ shown in Figure 4.31.

***Solution***

$AB = DC = 8$. Let $AD = b$; applying the Pythagorean Theorem with right triangle $ABD$, we find that
$10^2 = 8^2 + b^2$.
Then $100 = 64 + b^2$ and $b^2 = 36$, so $b = \sqrt{36}$ or 6.
In turn, $AD = BC = 6$. The perimeter of $ABCD$ is
$2(8) + 2(6) = 16 + 12 = 28$.

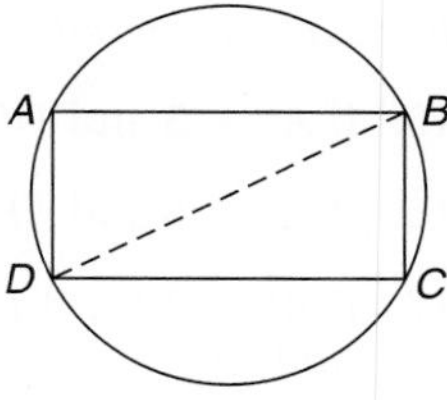

**FIGURE 4.31**

## 4.3 Exercises

1. If diagonal $\overline{DB}$ is congruent to each side of rhombus $ABCD$, what is the measure of $\angle A$? Of $\angle ABC$?

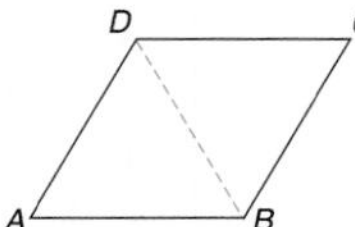


2. If the diagonals of a parallelogram are perpendicular, what can you conclude about the parallelogram? (HINT: Make a number of drawings in which you use only the information suggested.)

3. If the diagonals of a parallelogram are congruent, what can you conclude about the parallelogram?

4. If the diagonals of a parallelogram are perpendicular and congruent, what can you conclude about the parallelogram?

5. If the diagonals of a quadrilateral are perpendicular bisectors of each other (but not congruent), what can you conclude about the quadrilateral?

6. If the diagonals of a rhombus are congruent, what can you conclude about the rhombus?

7. A line segment joins the midpoints of two opposite sides of a rectangle as shown. What can you conclude about $\overline{MN}$ and $MN$?

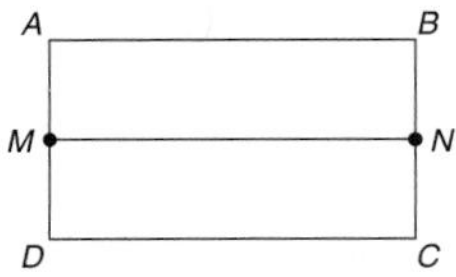


*In Exercises 8 to 10, use the properties of rectangles to solve each problem. Rectangle ABCD is shown in the figure.*

*Exercises 8–10*

8. *Given:* $AB = 5$ and $BC = 12$
   *Find:* $CD$, $AD$, and $AC$ (not shown)

9. *Given:* $AB = 2x + 7$, $BC = 3x + 4$, and $CD = 3x + 2$
   *Find:* $x$ and $DA$

10. *Given:* $AB = x + y$, $BC = x + 2y$, $CD = 2x - y - 1$, and $DA = 3x - 3y + 1$
    *Find:* $x$ and $y$

*In Exercises 11 to 14, consider rectangle MNPQ with diagonals $\overline{MP}$ and $\overline{NQ}$. When the answer is not a whole number, leave a square root answer.*

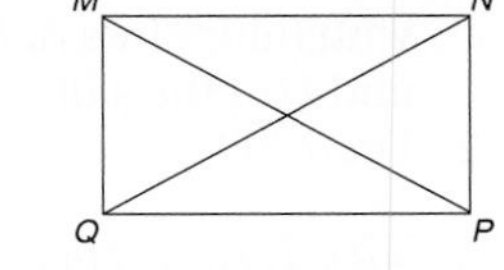


*Exercises 11–14*

11. If $MQ = 6$ and $MN = 8$, find $NQ$ and $MP$.

12. If $QP = 9$ and $NP = 6$, find $NQ$ and $MP$.

13. If $NP = 7$ and $MP = 11$, find $QP$ and $MN$.

14. If $QP = 15$ and $MP = 17$, find $MQ$ and $NP$.

*In Exercises 15 to 18, consider rhombus ABCD with diagonals $\overline{AC}$ and $\overline{DB}$. When the answer is not a whole number, leave a square root answer.*

15. If $AE = 5$ and $DE = 4$, find $AD$.
16. If $AE = 6$ and $EB = 5$, find $AB$.
17. If $AC = 10$ and $DB = 6$, find $AD$.
18. If $AC = 14$ and $DB = 10$, find $BC$.

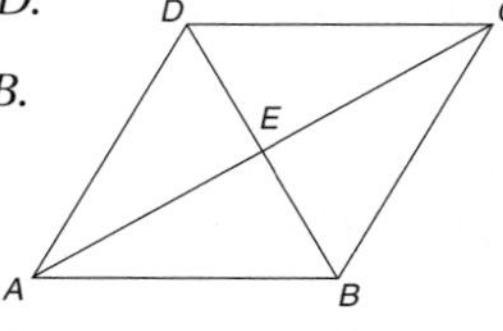

*Exercises 15–18*

19. *Given:* Rectangle $ABCD$ (not shown) with $AB = 8$ and $BC = 6$; $M$ and $N$ are the midpoints of sides $\overline{AB}$ and $\overline{BC}$, respectively
    *Find:* $MN$

20. *Given:* Rhombus $RSTV$ (not shown) with diagonals $\overline{RT}$ and $\overline{SV}$ so that $RT = 8$ and $SV = 6$
    *Find:* $RS$, the length of a side

*For Exercises 21 and 22, let P = {parallelograms}, R = {rectangles}, and H = {rhombi}. Classify as true or false:*

21. $H \subseteq P$ and $R \subseteq P$
22. $R \cup H = P$ and $R \cap H = \varnothing$

*In Exercises 23 and 24, supply the missing statements and reasons.*

23. *Given:* Quadrilateral $PQST$ with midpoints $A$, $B$, $C$, and $D$ of the sides
    *Prove:* $ABCD$ is a ▱

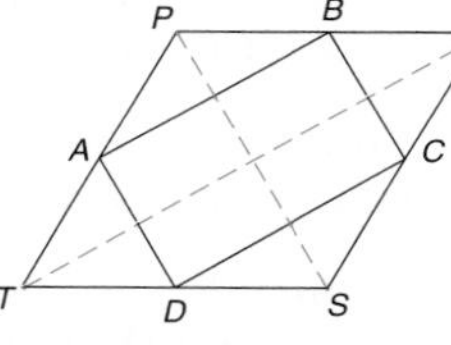


**PROOF**

| Statements | Reasons |
|---|---|
| 1. Quadrilateral $PQST$ with midpoints $A$, $B$, $C$, and $D$ of the sides | 1. ? |
| 2. Draw $\overline{TQ}$ | 2. Through two points, there is one line |
| 3. $\overline{AB} \parallel \overline{TQ}$ in $\triangle TPQ$ | 3. The line joining the midpoints of two sides of a triangle is ∥ to the third side |
| 4. $\overline{DC} \parallel \overline{TQ}$ in $\triangle TSQ$ | 4. ? |
| 5. $\overline{AB} \parallel \overline{DC}$ | 5. ? |
| 6. Draw $\overline{PS}$ | 6. ? |
| 7. $\overline{AD} \parallel \overline{PS}$ in $\triangle TSP$ | 7. ? |
| 8. $\overline{BC} \parallel \overline{PS}$ in $\triangle PSQ$ | 8. ? |
| 9. $\overline{AD} \parallel \overline{BC}$ | 9. ? |
| 10. ? | 10. If both pairs of opposite sides of a quadrilateral are ∥, the quad. is a ▱ |

24. *Given:* Rectangle $WXYZ$ with diagonals $\overline{WY}$ and $\overline{XZ}$
    *Prove:* $\angle 1 \cong \angle 2$

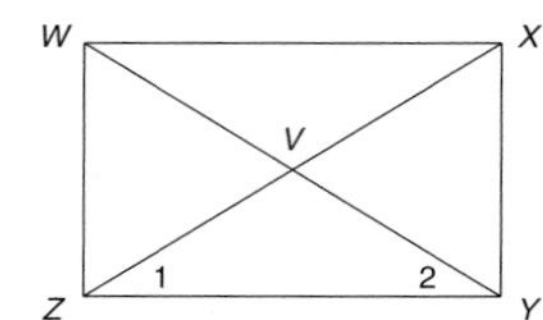


**PROOF**

| Statements | Reasons |
|---|---|
| 1. ? | 1. Given |
| 2. ? | 2. The diagonals of a rectangle are ≅ |
| 3. $\overline{WZ} \cong \overline{XY}$ | 3. The opposite sides of a rectangle (▱) are ≅ |
| 4. $\overline{ZY} \cong \overline{ZY}$ | 4. ? |
| 5. $\triangle XZY \cong \triangle WYZ$ | 5. ? |
| 6. ? | 6. ? |

25. Which type(s) of quadrilateral(s) is(are) necessarily cyclic?
    a) A square b) A parallelogram

26. Which type(s) of quadrilateral(s) is(are) necessarily cyclic?
    a) A kite b) A rectangle

27. Find the perimeter of the cyclic quadrilateral shown.

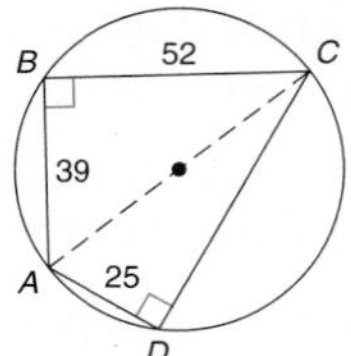


28. Find the perimeter of the square shown.

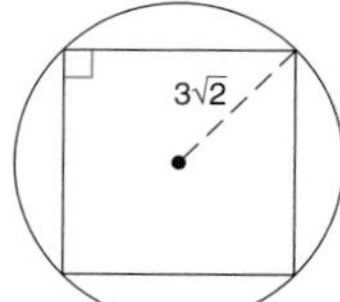

*In Exercises 29 to 31, explain why each statement is true.*

29. All angles of a rectangle are right angles.
30. All sides of a rhombus are congruent.
31. All sides of a square are congruent.

*In Exercises 32 to 37, write a formal proof of each theorem.*

32. The diagonals of a square are perpendicular.
33. A diagonal of a rhombus bisects two angles of the rhombus.
34. If the diagonals of a parallelogram are congruent, the parallelogram is a rectangle.
35. If the diagonals of a parallelogram are perpendicular, the parallelogram is a rhombus.
36. If the diagonals of a parallelogram are congruent and perpendicular, the parallelogram is a square.
37. If the midpoints of the sides of a rectangle are joined in order, the quadrilateral formed is a rhombus.

*In Exercises 38 and 39, you will need to use the square root ($\sqrt{\ }$) function of your calculator.*

38. A wall that is 12 ft long by 8 ft high has a triangular brace along the diagonal. Use a calculator to approximate the length of the brace to the nearest tenth of a foot.

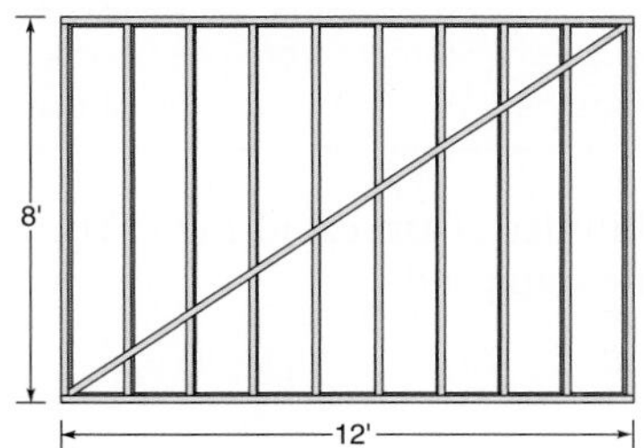


39. A walk-up ramp moves horizontally 20 ft while rising 4 ft. Use a calculator to approximate its length to the nearest tenth of a foot.

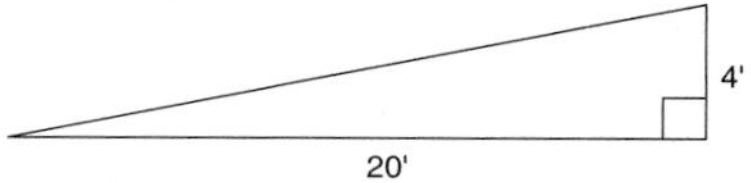


40. a) Argue that the midpoint of the hypotenuse of a right triangle is equidistant from the three vertices of the triangle. Use the fact that the congruent diagonals of a rectangle bisect each other. Be sure to provide a drawing.
    b) Use the relationship from part (a) to find *CM*, the length of the median to the hypotenuse of right $\triangle ABC$, in which $m\angle C = 90°$, $AC = 6$, and $BC = 8$.

41. Two sets of rails (railroad tracks are equally spaced) intersect, but not at right angles. Being as specific as possible, indicate what type of quadrilateral *WXYZ* is formed.

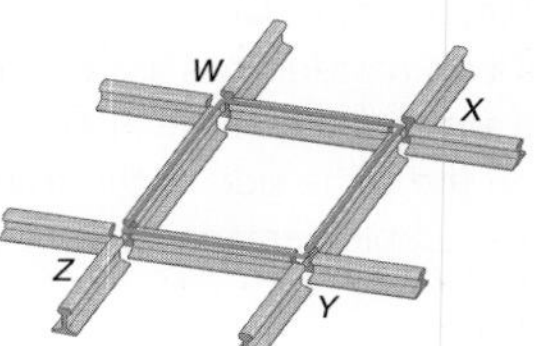


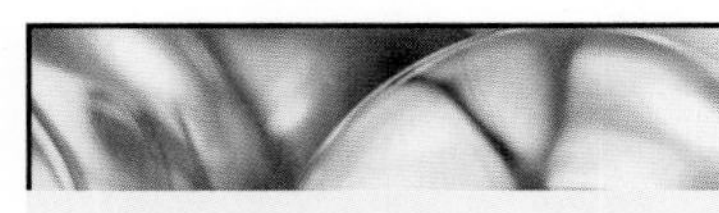

# 4.4 The Trapezoid

**KEY CONCEPTS**

Trapezoid • Bases • Legs • Base Angles • Median • Isosceles Trapezoid

DEFINITION: A **trapezoid** is a quadrilateral with exactly two parallel sides.

Figure 4.32 on page 204 shows trapezoid *HJKL*, in which $\overline{HL} \parallel \overline{JK}$. The parallel sides $\overline{HL}$ and $\overline{JK}$ are **bases,** and the nonparallel sides $\overline{HJ}$ and $\overline{LK}$ are **legs.** Because $\angle J$ and $\angle K$ both have $\overline{JK}$ for a side, they are a pair of **base angles** of the trapezoid; $\angle H$ and $\angle L$ are also a pair of base angles because $\overline{HL}$ is a base.

When the midpoints of the two legs of a trapezoid are joined, the resulting line segment is known as the **median** of the trapezoid. Given that *M* and *N* are the

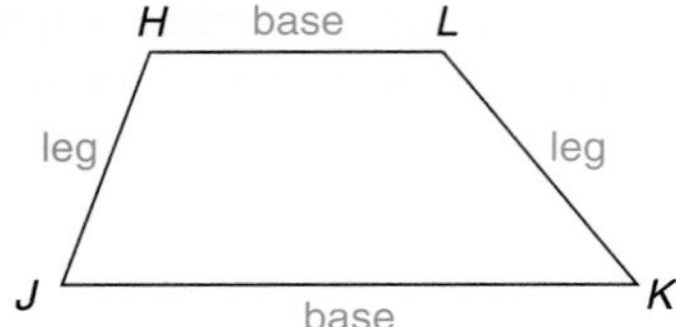


**FIGURE 4.32**

midpoints of the legs $\overline{HJ}$ and $\overline{LK}$ in trapezoid $HJKL$, $\overline{MN}$ is the median of the trapezoid. [See Figure 4.33(a).]

If the two legs of a trapezoid are congruent, the trapezoid is known as an **isosceles trapezoid.** In Figure 4.33(b), $RSTV$ is an **isosceles trapezoid** because $\overline{RV} \cong \overline{ST}$ and $\overline{RS} \parallel \overline{VT}$.

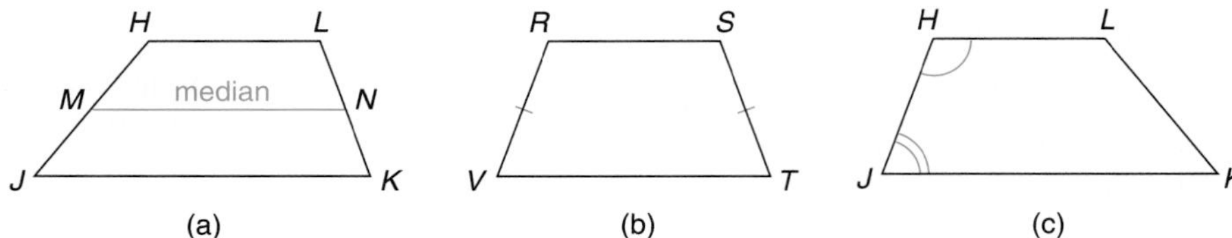


**FIGURE 4.33**

Reminder

If two parallel lines are cut by a transversal, then the interior angles on the same side of the transversal are supplementary.

Every trapezoid contains two pairs of consecutive interior angles that are supplementary. Each of these pairs of angles is formed when parallel lines are cut by a transversal. In Figure 4.33(c), angles $H$ and $J$ are supplementary, as are angles $L$ and $K$. See the "Reminder" at the left.

EXAMPLE 1

In Figure 4.32, suppose that $m\angle H = 107°$ and $m\angle K = 58°$. Find $m\angle J$ and $m\angle L$.

***Solution***

Because $\overline{HL} \parallel \overline{JK}$, $\angle$s $H$ and $J$ are supplementary angles, as are $\angle$s $L$ and $K$. Then $m\angle H + m\angle J = 180$ and $m\angle L + m\angle K = 180$. Substitution leads to $107 + m\angle J = 180$ and $m\angle L + 58 = 180$, so $m\angle J = 73°$ and $m\angle L = 122°$.

DEFINITION: An **altitude** of a trapezoid is a line segment from one vertex of one base of the trapezoid perpendicular to the opposite base (or to an extension of that base).

In Figure 4.34, $\overline{HX}$, $\overline{LY}$, $\overline{JP}$, and $\overline{KQ}$ are altitudes of trapezoid $HJKL$. The length of any altitude of $HJKL$ is called the *height* of the trapezoid.

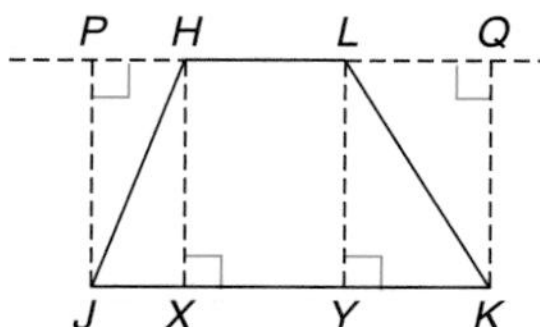


**FIGURE 4.34**

*Exs. 1–6*

Discover!

Using construction paper, cut out two trapezoids that are copies of each other. To accomplish this, hold two pieces of paper together and cut once left and once right. Take the second trapezoid and turn it so that a pair of congruent legs coincide. What type of quadrilateral has been formed?

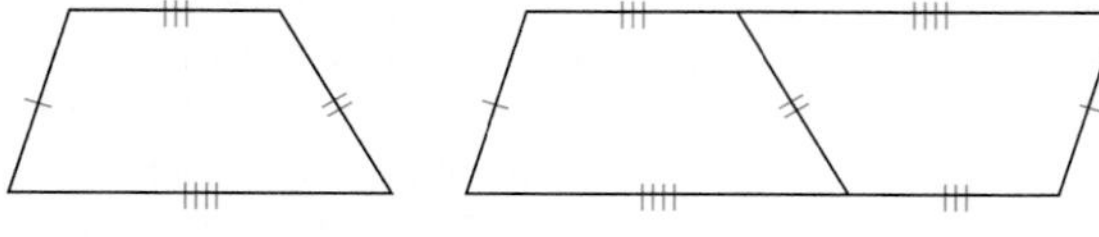

ANSWER

Parallelogram

The preceding activity may provide insight for a number of theorems involving the trapezoid.

> THEOREM 4.4.1: The base angles of an isosceles trapezoid are congruent.

### EXAMPLE 2

Study the picture proof of Theorem 4.4.1.

**PICTURE PROOF OF THEOREM 4.4.1**

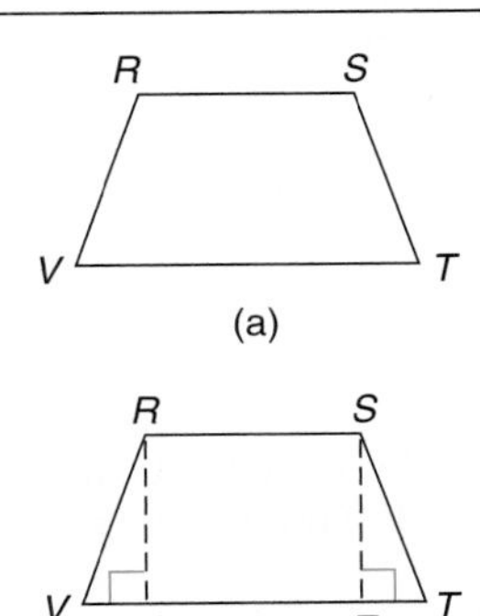


**FIGURE 4.35**

*Given:* Trapezoid $RSTV$ with $\overline{RV} \cong \overline{ST}$ and $\overline{RS} \parallel \overline{VT}$ [See Figure 4.35(a)].

*Prove:* $\angle V \cong \angle T$ and $\angle R \cong \angle S$

*Proof:* By drawing $\overline{RY} \perp \overline{VT}$ and $\overline{SZ} \perp \overline{VT}$, we see that $\overline{RY} \cong \overline{SZ}$ (Theorem 4.1.6). By HL, $\triangle RYV \cong \triangle SZT$ so $\angle V \cong \angle T$ (CPCTC). $\angle R \cong \angle S$ in Figure 4.35(a) because these angles are supplementary to congruent angles ($\angle V$ and $\angle T$).

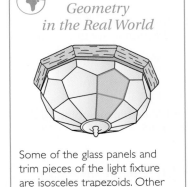

*Geometry in the Real World*

Some of the glass panels and trim pieces of the light fixture are isosceles trapezoids. Other glass panels are pentagons.

The following statement is a corollary of Theorem 4.4.1. Its proof is left to the student.

> COROLLARY 4.4.2: The diagonals of an isosceles trapezoid are congruent.

If diagonals $\overline{AC}$ and $\overline{BD}$ were shown in Figure 4.36 (at the left), they would be congruent.

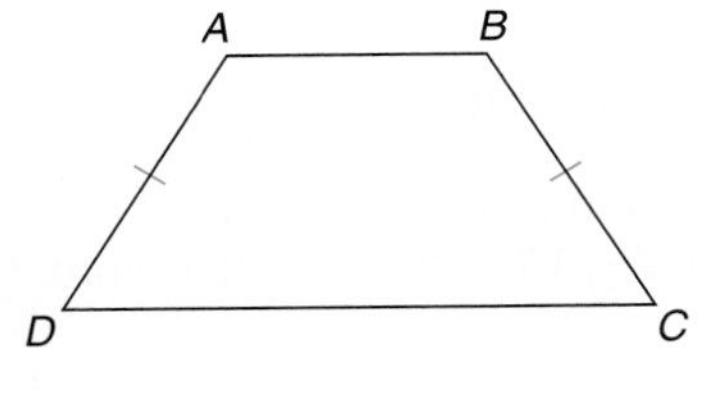


**FIGURE 4.36**

### EXAMPLE 3

Given isosceles trapezoid $ABCD$ with $\overline{AB} \parallel \overline{DC}$ (see Figure 4.36):

a) Find the measures of the angles of $ABCD$ if $m\angle A = 12x + 30$ and $m\angle B = 10x + 46$.
b) Find the length of each diagonal (not shown) if it is known that $AC = 2y - 5$ and $BD = 19 - y$.

***Solution***

a) Because $m\angle A = m\angle B$, $12x + 30 = 10x + 46$, so $2x = 16$ and $x = 8$. Then $m\angle A = 12(8) + 30$ or $126°$, and $m\angle B = 10(8) + 46$ or $126°$. Subtracting $(180 - 126 = 54)$, we determine the supplements of $\angle$s $A$ and $B$. That is, $m\angle C = m\angle D = 54°$.
b) By Corollary 4.4.2, $\overline{AC} \cong \overline{BD}$, so $2y - 5 = 19 - y$. Then $3y = 24$ and $y = 8$. Thus $AC = 2(8) - 5 = 11$. Also $BD = 19 - 8 = 11$.

For completeness, we state two properties of the isosceles trapezoid.

1. An isosceles trapezoid has line symmetry; the axis of symmetry is the perpendicular-bisector of either base.
2. An isosceles trapezoid is cyclic; the center of the circle containing all four vertices of the trapezoid is the point of intersection of the perpendicular bisectors of any two consecutive sides (or of the two legs).

The proof of the following theorem is left as Exercise 33. We apply Theorem 4.4.3 in Examples 4 and 5.

THEOREM 4.4.3: The length of the median of a trapezoid equals one-half the sum of the lengths of the two bases.

NOTE: The length of the median of a trapezoid is the "average" of the lengths of the bases. Where $m$ is the length of the median and $b_1$ and $b_2$ are the lengths of the bases, $m = \frac{1}{2}(b_1 + b_2)$.

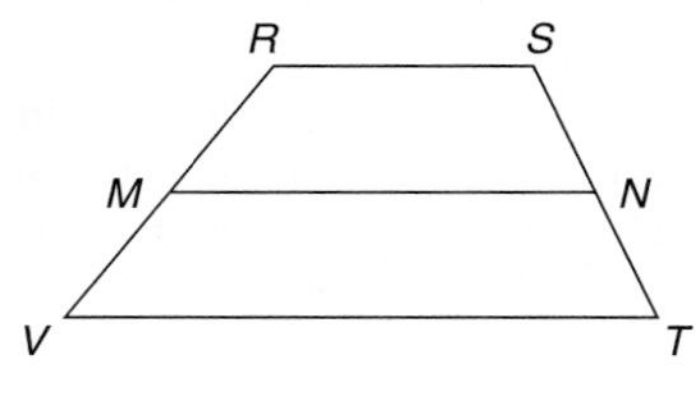

**FIGURE 4.37**

### EXAMPLE 4

In trapezoid $RSTV$ in Figure 4.37, $\overline{RS} \parallel \overline{VT}$ and $M$ and $N$ are the midpoints of $\overline{RV}$ and $\overline{TS}$, respectively. Find the length of median $\overline{MN}$ if $RS = 12$ and $VT = 18$.

***Solution***

Using Theorem 4.4.3, $MN = \frac{1}{2}(RS + VT)$, so $MN = \frac{1}{2}(12 + 18)$, or $MN = \frac{1}{2}(30)$. Thus, $MN = 15$.

### EXAMPLE 5

In trapezoid $RSTV$, $\overline{RS} \parallel \overline{VT}$ and $M$ and $N$ are the midpoints of $\overline{RV}$ and $\overline{TS}$, respectively (see Figure 4.37). Find $MN$, $RS$, and $VT$ if $RS = 2x$, $MN = 3x - 5$, and $VT = 2x + 10$.

***Solution***

Using Theorem 4.4.3, $MN = \frac{1}{2}(RS + VT)$, so

$$3x - 5 = \frac{1}{2}[2x + (2x + 10)] \quad \text{or} \quad 3x - 5 = \frac{1}{2}(4x + 10)$$

Then $3x - 5 = 2x + 5$ and $x = 10$. Now $RS = 2x = 2(10)$, so $RS = 20$. Also, $MN = 3x - 5 = 3(10) - 5$; therefore, $MN = 25$. Finally, $VT = 2x + 10$; therefore, $VT = 2(10) + 10 = 30$.

NOTE: As a check, $MN = \frac{1}{2}(RS + VT)$ leads to the true statement $25 = \frac{1}{2}(20 + 30)$.

THEOREM 4.4.4: The median of a trapezoid is parallel to each base.

The proof of Theorem 4.4.4 is left as Exercise 28. In Figure 4.37, $\overline{MN} \parallel \overline{RS}$ and $\overline{MN} \parallel \overline{VT}$.

*Exs. 7–12*

Theorems 4.4.5 and 4.4.6 enable us to show that a quadrilateral with certain characteristics is an isosceles trapezoid. We state these theorems as follows:

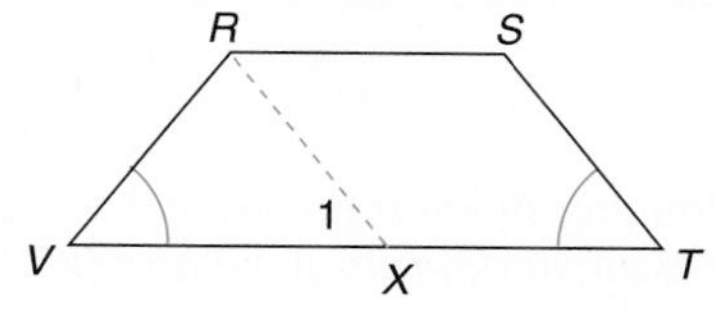

**FIGURE 4.38**

> THEOREM 4.4.5: If two base angles of a trapezoid are congruent, the trapezoid is an isosceles trapezoid.

Consider the following plan for proving Theorem 4.4.5. See Figure 4.38.

*Given:* Trapezoid $RSTV$ with $\overline{RS} \parallel \overline{VT}$ and $\angle V \cong \angle T$

*Prove:* $RSTV$ is an isosceles trapezoid

*Plan:* Draw auxiliary line $\overline{RX}$ parallel to $\overline{ST}$. Now show that $\angle V \cong \angle 1$, so $\overline{RV} \cong \overline{RX}$ in $\triangle RXV$. But $\overline{RX} \cong \overline{ST}$ in parallelogram $RXTS$, so $\overline{RV} \cong \overline{ST}$ and $RSTV$ is isosceles.

> THEOREM 4.4.6: If the diagonals of a trapezoid are congruent, the trapezoid is an isosceles trapezoid.

Theorem 4.4.6 has a lengthy proof, for which we have provided a sketch.

*Given:* Trapezoid $ABCD$ with $\overline{AB} \parallel \overline{DC}$ and $\overline{AC} \cong \overline{DB}$

*Prove:* $ABCD$ is an isosceles trapezoid. See Figure 4.39(a).

*Plan:* Draw $\overline{AF} \perp \overline{DC}$ and $\overline{BE} \perp \overline{DC}$ in Figure 4.39(b). Now we can show that $ABEF$ is a rectangle. Because $\overline{AF} \cong \overline{BE}$, $\triangle AFC \cong \triangle BED$ by HL. Then $\angle ACD \cong \angle BDC$ by CPCTC. With $\overline{DC} \cong \overline{DC}$ by Identity, $\triangle ACD \cong \triangle BDC$ by SAS. Now $\overline{AD} \cong \overline{BC}$ because these are corresponding parts of $\triangle ACD$ and $\triangle BDC$. Then trapezoid $ABCD$ is isosceles.

SSG *Exs. 13–15*

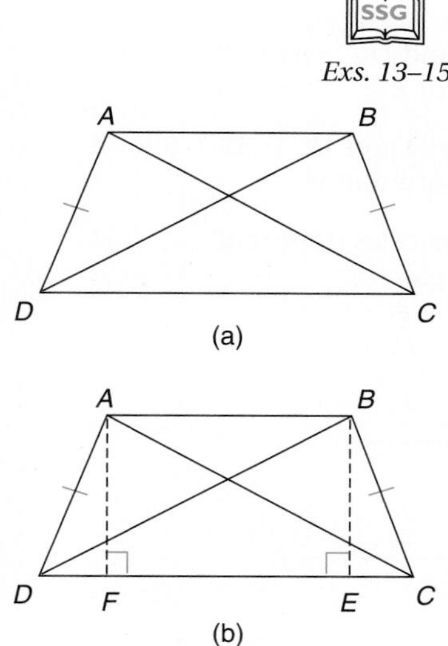

**FIGURE 4.39**

For several reasons, our final theorem is a challenge to prove. Looking at parallel lines $a$, $b$, and $c$ in Figure 4.40, one sees trapezoids such as $ABED$ and $BCFE$. However, the proof (whose "plan" we provide) uses auxiliary lines, parallelograms, and congruent triangles.

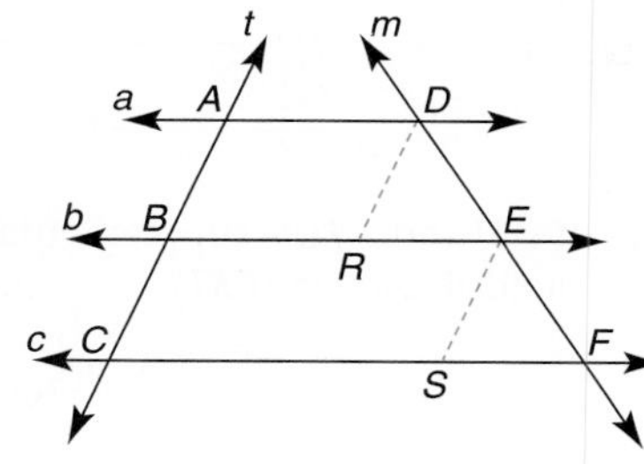

**FIGURE 4.40**

> THEOREM 4.4.7: If three (or more) parallel lines intercept congruent line segments on one transversal, then they intercept congruent line segments on any transversal.

*Given:* Parallel lines $a$, $b$, and $c$ cut by transversal $t$ so that $\overline{AB} \cong \overline{BC}$; also transversal $m$ in Figure 4.40

*Prove:* $\overline{DE} \cong \overline{EF}$

*Plan:* Through $D$ and $E$, draw $\overline{DR} \parallel \overline{AB}$ and $\overline{ES} \parallel \overline{AB}$. In each ▱ formed, $\overline{DR} \cong \overline{AB}$ and $\overline{ES} \cong \overline{BC}$. Given $\overline{AB} \cong \overline{BC}$, it follows that $\overline{DR} \cong \overline{ES}$. By AAS, we can show $\triangle DER \cong \triangle EFS$; then $\overline{DE} \cong \overline{EF}$ by CPCTC.

### EXAMPLE 6

In Figure 4.40, $a \parallel b \parallel c$. If $AB = BC = 7.2$ and $DE = 8.4$, find $EF$.

***Solution***

Using Theorem 4.4.7, we find that $EF = 8.4$.

*Exs. 16, 17*

## 4.4 Exercises

1. Find the measures of the remaining angles of trapezoid *ABCD* (not shown) if $\overline{AB} \parallel \overline{DC}$ and $m\angle A = 58°$ and $m\angle C = 125°$.

2. Find the measures of the remaining angles of trapezoid *ABCD* (not shown) if $\overline{AB} \parallel \overline{DC}$ and $m\angle B = 63°$ and $m\angle D = 118°$.

3. If the diagonals of a trapezoid are congruent, what can you conclude about the trapezoid?

4. If two of the base angles of a trapezoid are congruent, what type of trapezoid is it?

5. What type of quadrilateral is formed when the midpoints of the sides of an isosceles trapezoid are joined in order?

6. In trapezoid *ABCD*, $\overline{MN}$ is the median. Without writing a formal proof, explain why $MN = \frac{1}{2}(AB + DC)$.

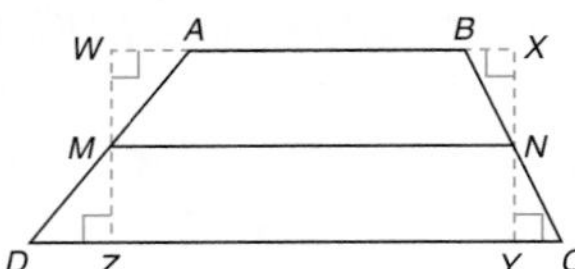


7. If $\angle H$ and $\angle J$ are supplementary, what type of quadrilateral is *HJKL*?

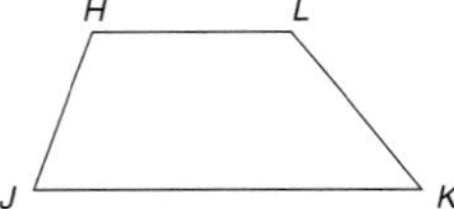


*Exercises 7–8*

8. If $\angle H$ and $\angle J$ are supplementary in *HJKL*, are $\angle K$ and $\angle L$ necessarily supplementary also?

*For Exercises 9 and 10, consider isosceles trapezoid RSTV, the midpoints of the sides being M, N, P, and Q.*

9. Would *RSTV* have symmetry with respect to

   a)  $\overleftrightarrow{MP}$?    b) 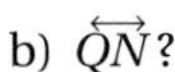 $\overleftrightarrow{QN}$?

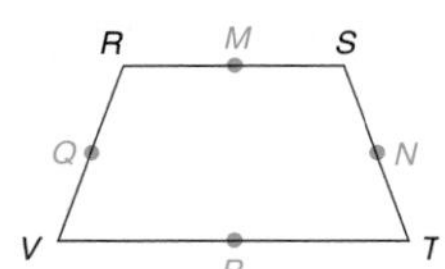


*Exercises 9, 10*

10. a) Does $QN = \frac{1}{2}(RS + VT)$?

    b) Does $MP = \frac{1}{2}(RV + ST)$?

*In Exercises 11 to 16, the drawing shows trapezoid ABCD with $\overline{AB} \parallel \overline{DC}$; also, M and N are midpoints of $\overline{AD}$ and $\overline{BC}$, respectively.*

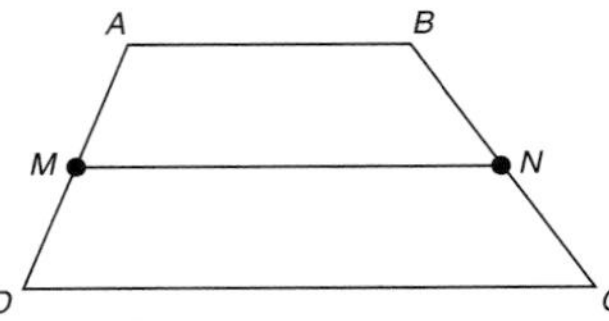


*Exercises 11–16*

11. *Given:* $AB = 7.3$ and $DC = 12.1$
    *Find:* $MN$

12. *Given:* $MN = 6.3$ and $DC = 7.5$
    *Find:* $AB$

13. *Given:* $AB = 8.2$ and $MN = 9.5$
    *Find:* $DC$

14. *Given:* $AB = 7x + 5$, $DC = 4x - 2$, and $MN = 5x + 3$
    *Find:* $x$

15. *Given:* $AB = 6x + 5$ and $DC = 8x - 1$
    *Find:* $MN$, in terms of $x$

16. *Given:* $AB = x + 3y + 4$ and $DC = 3x + 5y - 2$
    *Find:* $MN$, in terms of $x$ and $y$

17. *Given:* *ABCD* is an isosceles trapezoid
    *Prove:* $\triangle ABE$ is isosceles

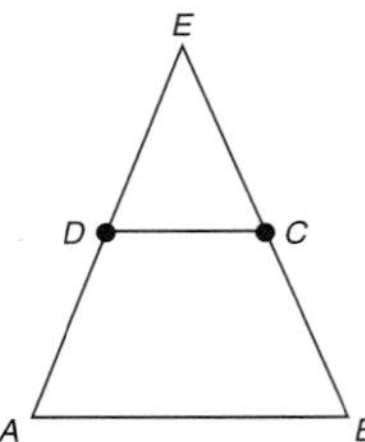


*Exercises 17, 18*

18. *Given:* Isosceles $\triangle ABE$ with $\overline{AE} \cong \overline{BE}$; also, *D* and *C* are midpoints of $\overline{AE}$ and $\overline{BE}$, respectively
    *Prove:* *ABCD* is an isosceles trapezoid

19. In isosceles trapezoid *WXYZ* with bases $\overline{ZY}$ and $\overline{WX}$, $ZY = 8$, $YX = 10$, and $WX = 20$. Find height $h$ (the length of $\overline{ZD}$ or $\overline{YE}$).

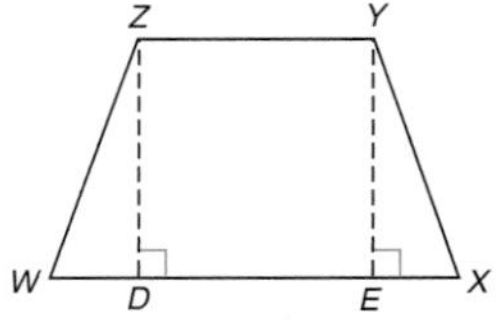


*Exercises 19, 20*

20. In trapezoid $WXYZ$ with bases $\overline{ZY}$ and $\overline{WX}$, $ZY = 12$, $YX = 10$, $WZ = 17$, and $ZD = 8$. Find the length of base $\overline{WX}$. (See figure on page 208.)

21. In isosceles trapezoid $MNPQ$ with $\overline{MN} \parallel \overline{QP}$, diagonal $\overline{MP} \perp \overline{MQ}$. If $PQ = 13$ and $NP = 5$, how long is diagonal $\overline{MP}$?

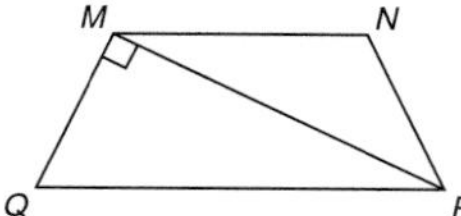


22. In trapezoid $RSTV$, $\overline{RV} \parallel \overline{ST}$, $m\angle SRV = 90°$, and $M$ and $N$ are midpoints of the nonparallel sides. If $ST = 13$, $RV = 17$, and $RS = 16$, how long is $\overline{RN}$?

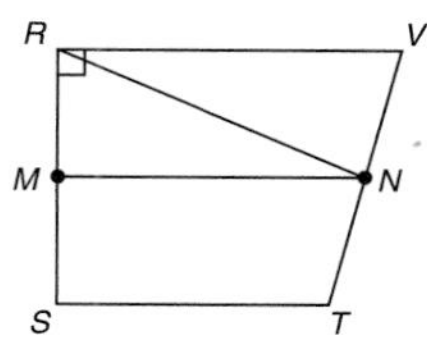


23. Each vertical section of a suspension bridge is in the shape of a trapezoid. For additional support, a vertical cable is placed midway as shown. If the two vertical columns shown have heights of 20 ft and 24 ft and the section is 10 ft wide, what will the height of the cable be?

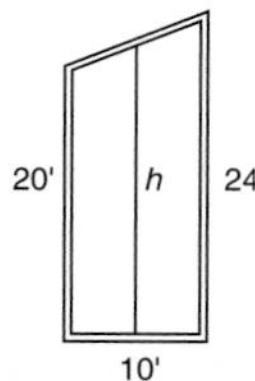


24. The state of Nevada approximates the shape of a trapezoid with these dimensions for boundaries: 340 miles on the north, 515 miles on the east, 435 miles on the south, and 225 miles on the west. If $A$ and $B$ are points located midway across the north and south boundaries, what is the approximate distance from $A$ to $B$?

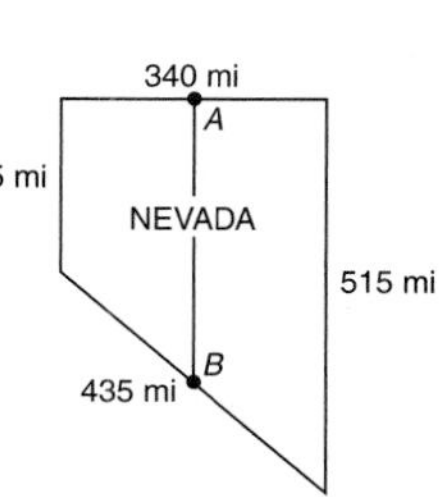


25. In the figure, $a \parallel b \parallel c$ and $B$ is the midpoint of $\overline{AC}$. If $AB = 2x + 3$, $BC = x + 7$, and $DE = 3x + 2$, find the length of $\overline{EF}$.

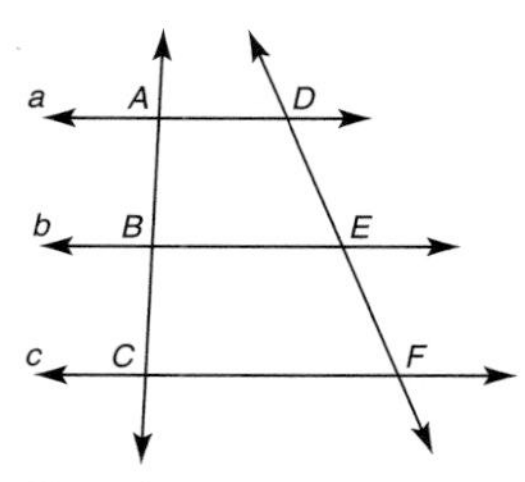

Exercises 25, 26

26. In the figure, $a \parallel b \parallel c$ and $B$ is the midpoint of $\overline{AC}$. If $AB = 2x + 3y$, $BC = x + y + 7$, $DE = 2x + 3y + 3$, and $EF = 5x - y + 2$, find $x$ and $y$.

*In Exercises 27 to 33, complete a formal proof.*

27. The diagonals of an isosceles trapezoid are congruent.

28. The median of a trapezoid is parallel to each base.

29. If two consecutive angles of a quadrilateral are supplementary, the quadrilateral is a trapezoid.

30. If two base angles of a trapezoid are congruent, the trapezoid is an isosceles trapezoid.

31. If three parallel lines intercept congruent segments on one transversal, then they intercept congruent segments on any transversal.

32. If the midpoints of the sides of an isosceles trapezoid are joined in order, then the quadrilateral formed is a rhombus.

33. *Given:* $\overline{EF}$ is the median of trapezoid $ABCD$
*Prove:* $EF = \frac{1}{2}(AB + DC)$
(HINT: Using Theorem 4.4.7, show that $M$ is the midpoint of $\overline{AC}$. For $\triangle ADC$ and $\triangle CBA$, apply Theorem 4.2.5.)

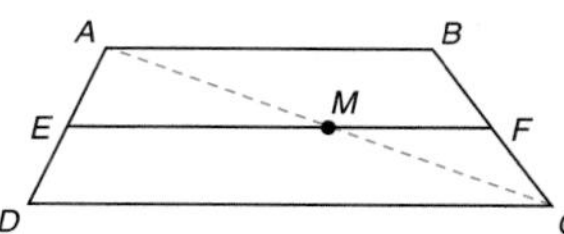

Exercises 33–35

*For Exercises 34 and 35, $\overline{EF}$ is the median of trapezoid ABCD.*

34. In the figure for Exercise 33, suppose that $AB = 12.8$ and $DC = 18.4$. Find:

a) $MF$
b) $EM$
c) $EF$
d) Whether $EF = \frac{1}{2}(AB + DC)$

35. In the figure for Exercise 33, suppose that $EM = 7.1$ and $MF = 3.5$. Find:

a) $AB$
b) $DC$
c) $EF$
d) Whether $EF = \frac{1}{2}(AB + DC)$

36. *Given:* $\overline{AB} \parallel \overline{DC}$
$m\angle A = m\angle B = 56°$
$\overline{CE} \parallel \overline{DA}$ and $\overrightarrow{CF}$ bisects $\angle DCB$
*Find:* $m\angle FCE$

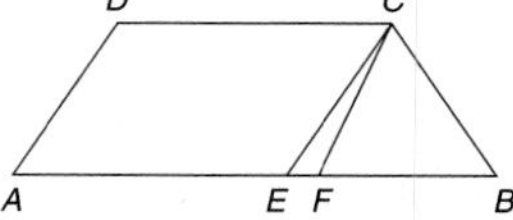

37. In a gambrel style roof, the gable end of a barn has the shape of an isosceles trapezoid surmounted by an isosceles triangle. If $AE = 30$ ft and $BD = 24$ ft, find:

a) $AS$ b) $VD$ c) $CD$ d) $DE$

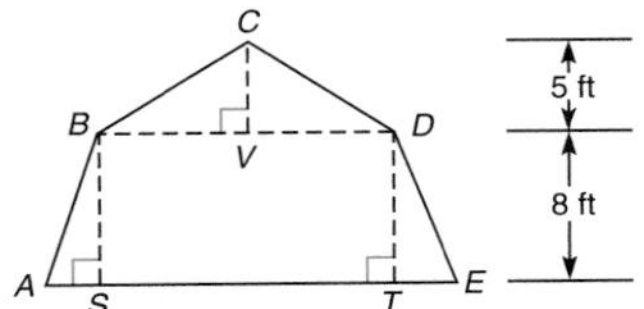


38. Successive steps on a ladder form isosceles trapezoids with the sides. $AH = 2$ ft and $BI = 2.125$ ft.

a) Find $GN$, the width of the bottom step
b) Which step is the median of the trapezoid with bases $\overline{AH}$ and $\overline{GN}$?

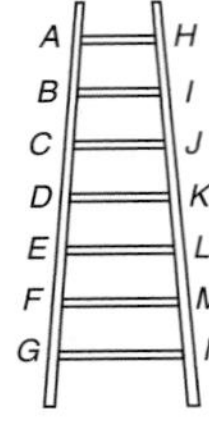


39. The vertical sidewall of an in-ground pool that is 24 ft in length has the shape of a trapezoid. What is the depth of the pool in the middle?

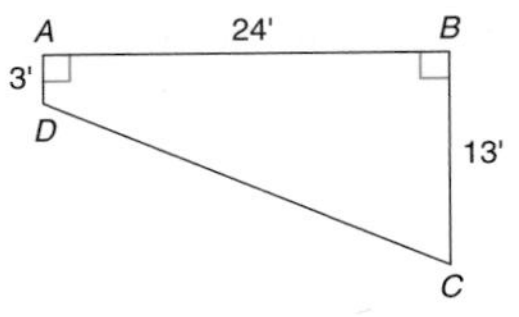


40. For the in-ground pool shown in Exercise 39, find the length of the sloped bottom from point $D$ to point $C$.

# Perspective on History

## SKETCH OF THALES

One of the most significant contributors to the development of geometry was the Greek mathematician Thales of Miletus (625–547 B.C.). Thales is credited with being the "Father of Geometry" because he was the first person to organize geometric thought and utilize the deductive method as a means of verifying propositions (theorems). It is not surprising that Thales made original discoveries in geometry. Just as significant as his discoveries was Thales' persistence in verifying the claims of his predecessors. In this textbook, you will find that propositions such as these are only a portion of those that can be attributed to Thales:

Chapter 1: If two straight lines intersect, the opposite (vertical) angles formed are equal.
Chapter 3: The base angles of an isosceles triangle are equal.
Chapter 5: The sides of similar triangles are proportional.
Chapter 6: An angle inscribed in a semicircle is a right angle.

Thales' knowledge of geometry was matched by the wisdom that he displayed in everyday affairs. For example, he is known to have measured the height of the Great Pyramid of Egypt by comparing the lengths of the shadows cast by the pyramid and by his own staff. Thales also used his insights into geometry to measure the distances from the land to ships at sea.

Perhaps the most interesting story concerning Thales was one related by Aesop (famous for fables). It seems that Thales was on his way to market with his beasts of burden carrying saddlebags filled with salt. Quite by accident, one of the mules discovered that rolling in the stream where he was led to drink greatly reduced this load; of course, this was due to the dissolving of salt in the saddlebags. On subsequent trips, the same mule continued to lighten his load by rolling in the water. Thales soon realized the need to do something (anything!) to modify the mule's behavior. When preparing for the next trip, Thales filled the offensive mule's saddlebags with sponges. When the mule took his usual dive, he found that his load was heavier than ever. Soon the mule realized the need to keep the saddlebags out of the water. In this way, it is said that Thales discouraged the mule from allowing the precious salt to dissolve during later trips to market.

# Perspective on Application

## SQUARE NUMBERS AS SUMS

In algebra, there is a principle that is generally "proved" by a quite sophisticated method known as mathematical induction. However, verification of the principle is much simpler when provided a geometric justification.

In the following paragraphs, we:

1) State the principle
2) Illustrate the principle
3) Provide the geometric justification for the principle

> Where $n$ is a counting number, the sum of the first $n$ positive odd counting numbers is $n^2$.

The principle stated above is illustrated for various choices of $n$.

Where $n = 1$, $1 = 1^2$.
Where $n = 2$, $1 + 3 = 2^2$, or 4.
Where $n = 3$, $1 + 3 + 5 = 3^2$, or 9.
Where $n = 4$, $1 + 3 + 5 + 7 = 4^2$, or 16.

The geometric explanation for this principle utilizes a *wrap-around* effect. Study the diagrams in Figure 4.41.

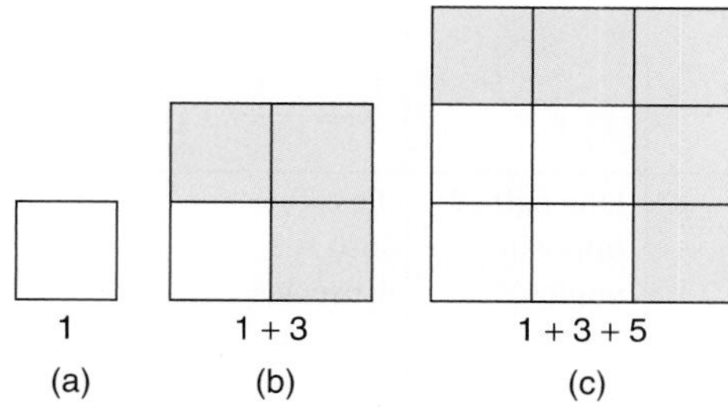


**FIGURE 4.41**

Given a unit square (one with sides of length 1), we build a second square by wrapping 3 unit squares around the first unit square; in Figure 4.41(b), the "wrap-around" is indicated by 3 shaded squares. Now for the second square (sides of length 2), we form the next square by wrapping 5 unit squares around this square; see Figure 4.41(c).

The next figure in the sequence of squares illustrates that

$$1 + 3 + 5 + 7 = 4^2, \text{ or } 16$$

In the "wrap-around," we emphasize that the next number in the sum is an odd number. The "wrap-around" approach adds $2 \times 3 + 1$, or 7 unit squares in Figure 4.42. When building each sequential square, we always add an odd number of unit squares as in Figure 4.42.

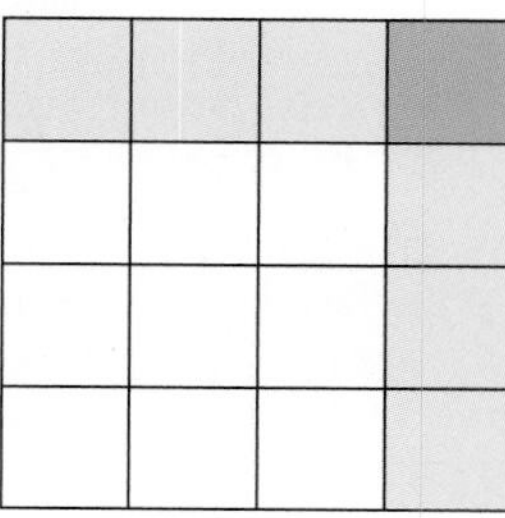

**FIGURE 4.42**

**PROBLEM:**

Use the following principle to answer each question:

Where $n$ is a counting number, the sum of the first $n$ positive odd counting numbers is $n^2$.

a) Find the sum of the first five positive odd integers; that is, find $1 + 3 + 5 + 7 + 9$.
b) Find the sum of the first six positive odd integers.
c) How many positive odd integers were added to obtain the sum 81?

***Solutions***

a) $5^2$, or 25 b) $6^2$, or 36 c) 9, because $9^2 = 81$

# Summary

### A LOOK BACK AT CHAPTER 4

The goal of this chapter has been to develop the properties of quadrilaterals, including special types of quadrilaterals such as the parallelogram, rectangle, and trapezoid. Table 4.1 on page 212 summarizes the properties of quadrilaterals.

### A LOOK AHEAD TO CHAPTER 5

In the next chapter, similarity will be defined for all polygons, with an emphasis on triangles. The Pythagorean Theorem, which we applied in Chapter 4, will be proved in Chapter 5. Special right triangles will be discussed.

### KEY CONCEPTS

**4.1** Quadrilateral • Skew • Parallelogram • Diagonals of a Parallelogram • Altitude

**4.2** Quadrilaterals That Are Parallelograms • Rectangle • Kite

**4.3** Rectangle • Square • Rhombus • Pythagorean Theorem

**4.4** Trapezoid (Bases, Legs, Base Angles, Median) • Isosceles Trapezoid

## TABLE 4.1 AN OVERVIEW OF CHAPTER FOUR

### Properties of Quadrilaterals

| | *Parallelogram* | *Rectangle* | *Rhombus* | *Square* | *Kite* | *Trapezoid* | *Isosceles trapezoid* |
|---|---|---|---|---|---|---|---|
| *Congruent sides* | Both pairs of opposite sides | Both pairs of opposite sides | All four sides | All four sides | Both pairs of adjacent sides | Possible; also see isosceles trapezoid | Pair of legs |
| *Parallel sides* | Both pairs of opposite sides | Both pairs of opposite sides | Both pairs of opposite sides | Both pairs of opposite sides | Generally none | Pair of bases | Pair of bases |
| *Perpendicular sides* | If parallelogram is a rectangle or square | Consecutive pairs | If rhombus is a square | Consecutive pairs | Possible | Possible | Generally none |
| *Congruent angles* | Both pairs of opposite angles | All four angles | Both pairs of opposite angles | All four angles | One pair of opposite angles | Possible; also see isosceles trapezoid | Each pair of base angles |
| *Supplementary angles* | All pairs of consecutive angles | Any two angles | All pairs of consecutive angles | Any two angles | Possibly two pairs | Each pair of leg angles | Each pair of leg angles |
| *Diagonal relationships* | Bisect each other | Congruent; bisect each other | Perpendicular; bisect each other and interior angles | Congruent; perpendicular; bisect each other and interior angles | Perpendicular; one bisects other and two interior angles | Intersect | Congruent |

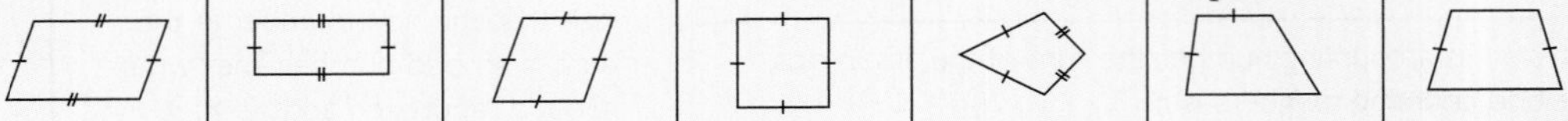

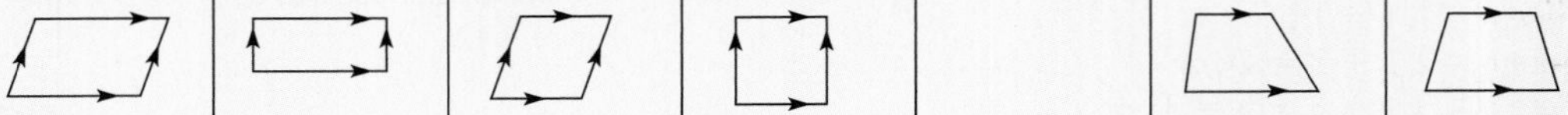

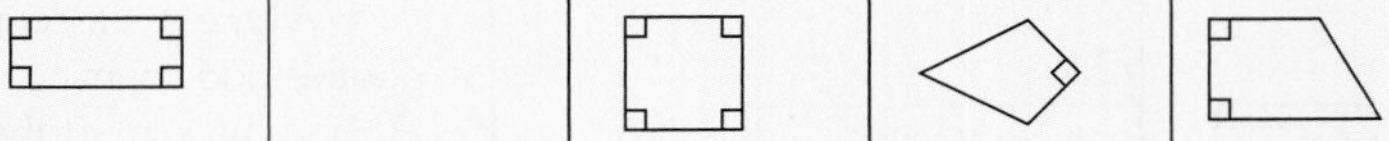

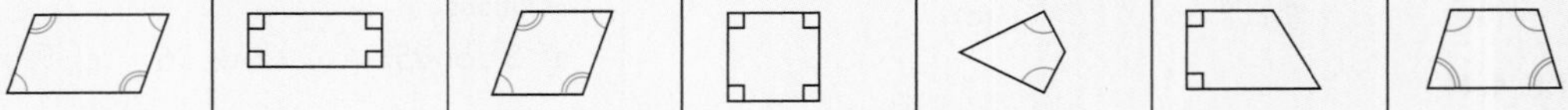

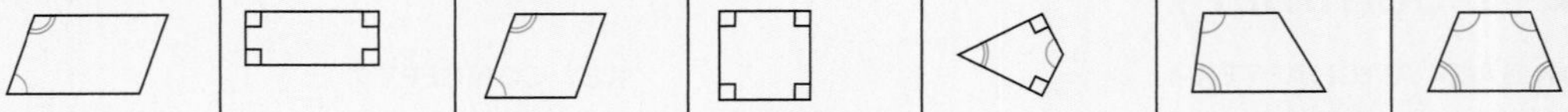

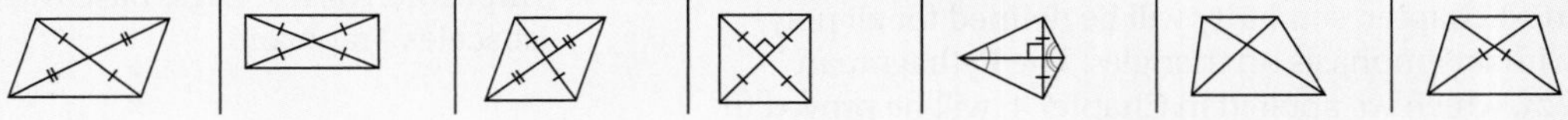

# Chapter 4 Review Exercises

*State whether the statements in Review Exercises 1 to 12 are always true (A), sometimes true (S), or never true (N).*

1. A square is a rectangle.
2. If two of the angles of a trapezoid are congruent, then the trapezoid is isosceles.
3. The diagonals of a trapezoid bisect each other.
4. The diagonals of a parallelogram are perpendicular.
5. A rectangle is a square.
6. The diagonals of a square are perpendicular.
7. Two consecutive angles of a parallelogram are supplementary.
8. Opposite angles of a rhombus are congruent.
9. The diagonals of a rectangle are congruent.
10. The four sides of a kite are congruent.
11. The diagonals of a parallelogram are congruent.
12. The diagonals of a kite are perpendicular bisectors of each other.

13. *Given:* $\square ABCD$
    $CD = 2x + 3$
    $BC = 5x - 4$
    Perimeter of $\square ABCD = 96$ cm
    *Find:* The lengths of the sides of $\square ABCD$

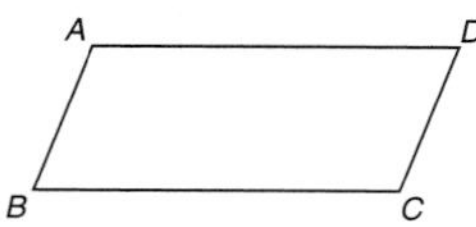


*Exercises 13, 14*

14. *Given:* $\square ABCD$
    $m\angle A = 2x + 6$
    $m\angle B = x + 24$
    *Find:* $m\angle C$

15. The diagonals of $\square ABCD$ (not shown) are perpendicular. If one diagonal has a length of 10 and the other diagonal has a length of 24, find the perimeter of the parallelogram.

16. *Given:* $\square MNOP$
    $m\angle M = 4x$
    $m\angle O = 2x + 50$
    *Find:* $m\angle M$ and $m\angle P$

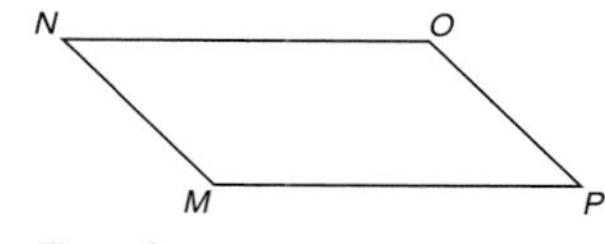


*Exercises 16, 17*

17. Using the information from Exercise 16, determine which diagonal ($\overline{MO}$ or $\overline{PN}$) would be longer.
18. In quadrilateral *ABCD, M* is the midpoint only of $\overline{BD}$ and $\overline{AC} \perp \overline{DB}$ at *M*. What special type of quadrilateral is *ABCD*?
19. In isosceles trapezoid *DEFG*, $\overline{DE} \parallel \overline{GF}$ and $m\angle D = 108°$. Find the measures of the other angles in the trapezoid.
20. One base of a trapezoid has a length of 12.3 cm and the length of the other base is 17.5 cm. Find the length of the median of the trapezoid.
21. In trapezoid *MNOP*, $\overline{MN} \parallel \overline{PO}$ and *R* and *S* are the midpoints of $\overline{MP}$ and $\overline{NO}$, respectively. Find the lengths of the bases if $RS = 15$, $MN = 3x + 2$, and $PO = 2x - 7$.

*In Review Exercises 22 to 24, M and N are the midpoints of* $\overline{FJ}$ *and* $\overline{FH}$*, respectively.*

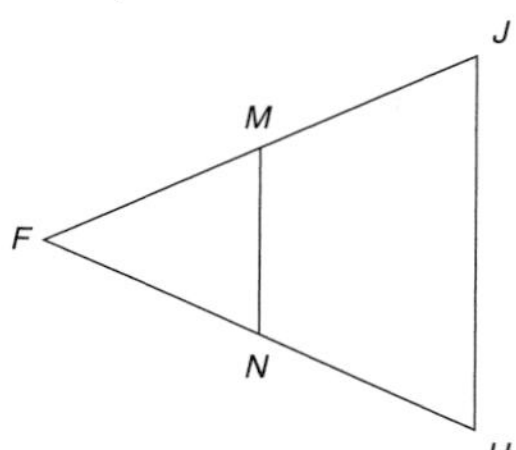


*Exercises 22–24*

22. *Given:* Isosceles $\triangle FJH$ with
    $\overline{FJ} \cong \overline{FH}$
    $FM = 2y + 3$
    $NH = 5y - 9$
    $JH = 2y$
    *Find:* The perimeter of $\triangle FMN$

23. *Given:* $JH = 12$
    $m\angle J = 80°$
    $m\angle F = 60°$
    *Find:* $MN$, $m\angle FMN$, $m\angle FNM$

24. *Given:* $MN = x^2 + 6$
    $JH = 2x(x + 2)$
    *Find:* $x$, $MN$, $JH$

25. *Given:* *ABCD* is a $\square$
    $\overline{AF} \cong \overline{CE}$
    *Prove:* $\overline{DF} \parallel \overline{EB}$

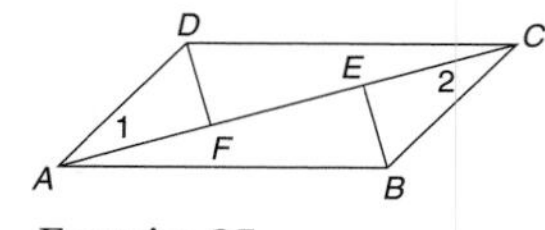


*Exercise 25*

26. *Given:* $ABEF$ is a rectangle
$BCDE$ is a rectangle
$\overline{FE} \cong \overline{ED}$
*Prove:* $\overline{AE} \cong \overline{BD}$ and $\overline{AE} \parallel \overline{BD}$

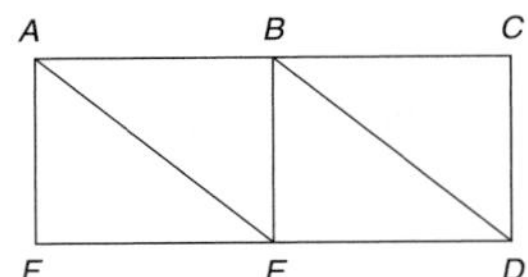


27. *Given:* $\overline{DE}$ is a median of $\triangle ADC$
$\overline{BE} \cong \overline{FD}$
$\overline{EF} \cong \overline{FD}$
*Prove:* $ABCF$ is a ▱

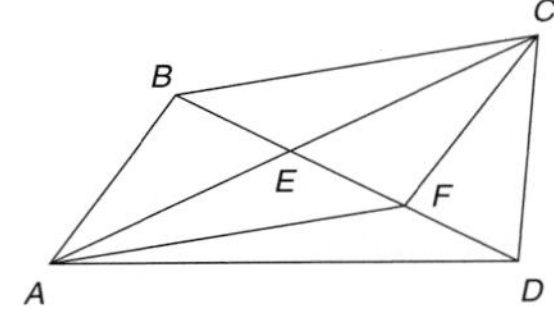


28. *Given:* $\triangle FAB \cong \triangle HCD$
$\triangle EAD \cong \triangle GCB$
*Prove:* $ABCD$ is a ▱

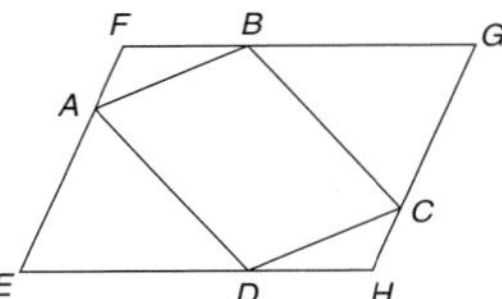


29. *Given:* $ABCD$ is a parallelogram
$\overline{DC} \cong \overline{BN}$
$\angle 3 \cong \angle 4$
*Prove:* $ABCD$ is a rhombus

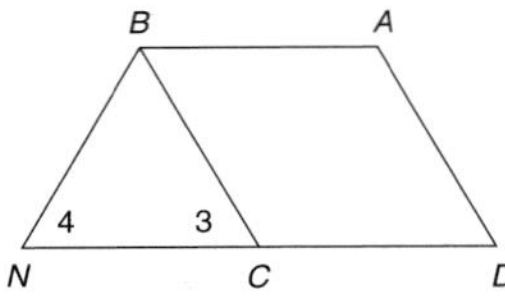


30. *Given:* $\triangle TWX$ is isosceles, with base $\overline{WX}$
$\overline{RY} \parallel \overline{WX}$
*Prove:* $RWXY$ is an isosceles trapezoid

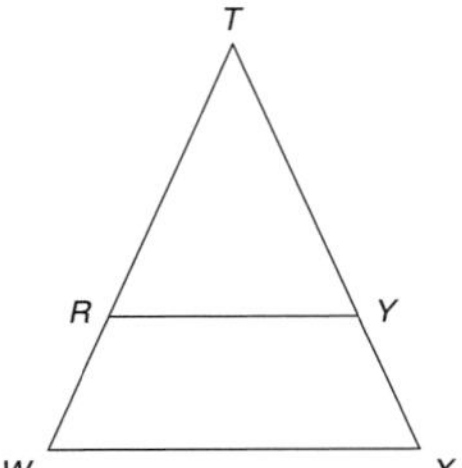


31. Construct a rhombus, given these lengths for the diagonals.

32. Draw rectangle $ABCD$ with $AB = 5$ and $BC = 12$. Include diagonals $\overline{AC}$ and $\overline{BD}$.

   a) How are $\overline{AB}$ and $\overline{BC}$ related?
   b) Find the length of diagonal $\overline{AC}$.

33. Draw rhombus $WXYZ$ with diagonals $\overline{WY}$ and $\overline{XZ}$. Let $\overline{WY}$ name the longer diagonal.

   a) How are diagonals $\overline{WY}$ and $\overline{XZ}$ related?
   b) If $WX = 17$ and $XZ = 16$, find the length of diagonal $\overline{WY}$.

34. Considering parallelograms, kites, rectangles, squares, rhombi, trapezoids, and isosceles trapezoids, which figures have

   a) line symmetry?
   b) point symmetry?

35. What type of quadrilateral is formed when the triangle is reflected across the indicated side?

   a) Isosceles $\triangle ABC$ across $\overline{BC}$
   b) Obtuse $\triangle XYZ$ across $\overline{XY}$

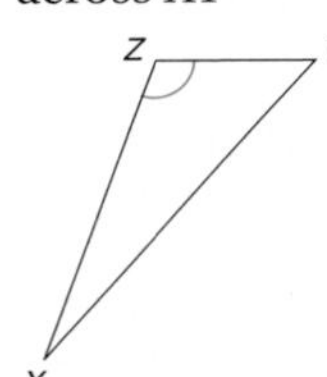

# Chapter 4 Test

1. Consider ▱$ABCD$ as shown.

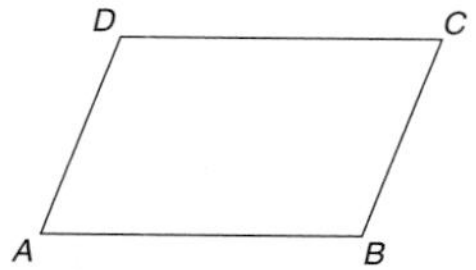


a) How are $\angle A$ and $\angle C$ related? ______________
b) How are $\angle A$ and $\angle B$ related? ______________

2. In ▱$RSTV$ (not shown), $RS = 5.3$ cm and $ST = 4.1$ cm. Find the perimeter of $RSTV$. ______________

3. In ▱$ABCD$, $AD = 5$ and $DC = 9$. If the altitude from vertex $D$ to $\overline{AB}$ has length 4 (that is, $DE = 4$), find the length of $\overline{EB}$. ______________

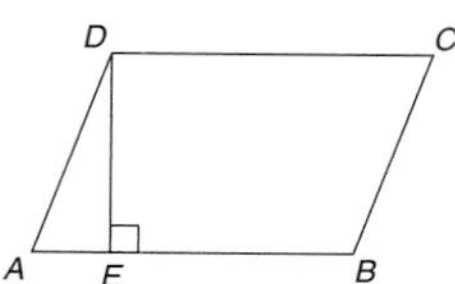


4. In ▱$RSTV$, $m\angle S = 57°$. Which diagonal ($\overline{VS}$ or $\overline{RT}$) would have the greater length? ______________

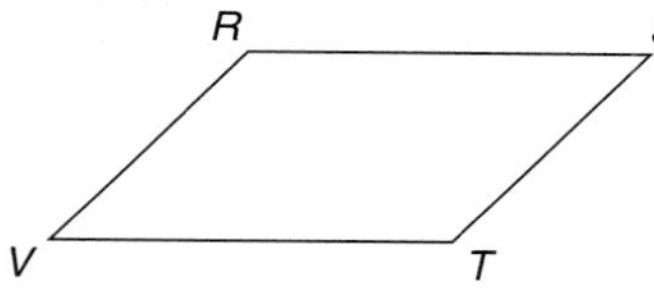


*Exercises 4, 5*

5. In ▱$RSTV$, $VT = 3x - 1$, $TS = 2x + 1$, and $RS = 4(x - 2)$. Find the value of $x$. ______________

6. Complete each statement:

a) If a quadrilateral has two pairs of congruent *adjacent* sides, then the quadrilateral is a(n) ______________.

b) If a quadrilateral has two pairs of congruent *opposite* sides, then the quadrilateral is a(n) ______________.

7. Complete each statement:

a) In ▱$RSTV$, $\overline{RW}$ is the ______________ from vertex $R$ to base $\overline{VT}$.

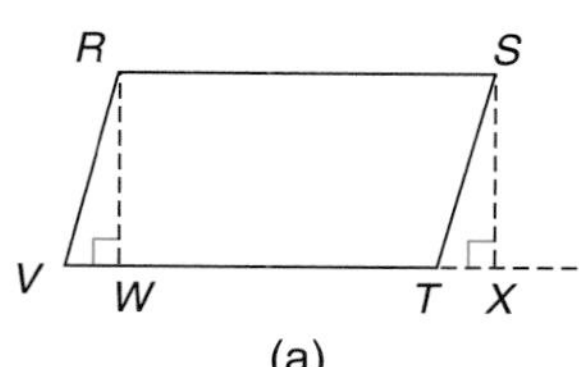


(a)

b) If altitude $\overline{RW}$ of figure (a) is congruent to altitude $\overline{TY}$ of figure (b), then ▱$RSTV$ must also be a(n) ______________.

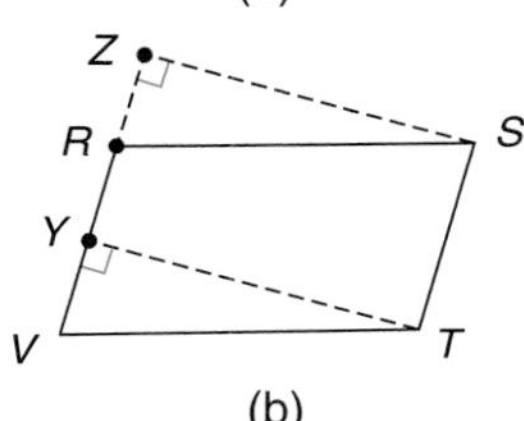


(b)

8. In $\triangle ABC$, $M$ is the midpoint of $\overline{AB}$ and $N$ is the midpoint of $\overline{AC}$.

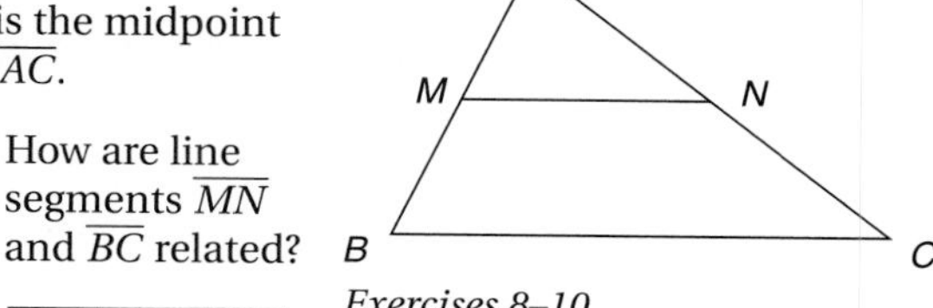


*Exercises 8–10*

a) How are line segments $\overline{MN}$ and $\overline{BC}$ related? ______________

b) Use an equation to state how the lengths $MN$ and $BC$ are related. ______________

9. In $\triangle ABC$, $M$ is the midpoint of $\overline{AB}$ and $N$ is the midpoint of $\overline{AC}$. If $MN = 7.6$ cm, find $BC$. ______________

10. In $\triangle ABC$, $M$ is the midpoint of $\overline{AB}$ and $N$ is the midpoint of $\overline{AC}$. If $MN = 3x - 11$ and $BC = 4x + 24$, find the value of $x$. ______________

11. In rectangle $ABCD$, $AD = 12$ and $DC = 5$. Find the length of diagonal $\overline{AC}$ (not shown). ______________

(HINT: Apply the Pythagorean Theorem.)

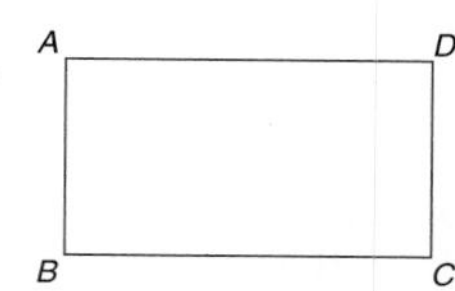


12. In trapezoid $RSTV$, $\overline{RS} \parallel \overline{VT}$.

a) Which sides are the legs of $RSTV$? ______________
b) Name two angles that are supplementary. ______________

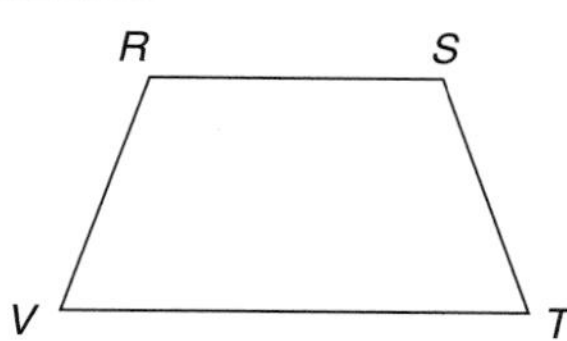


13. In trapezoid $RSTV$, $\overline{RS} \parallel \overline{VT}$ and $\overline{MN}$ is the median. Find the length $MN$ (of median $\overline{MN}$) if $RS = 12.4$ in. and $VT = 16.2$ in. ______________

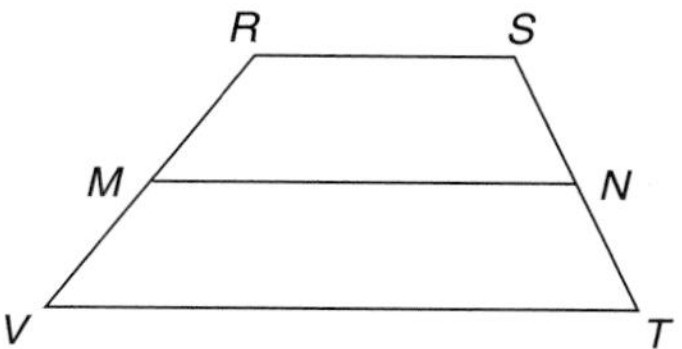


*Exercises 13, 14*

14. In trapezoid $RSTV$ of Exercise 13, $\overline{RS} \parallel \overline{VT}$ and $\overline{MN}$ is the median. Find $x$ if $VT = 2x + 9$, $MN = 6x - 13$, and $RS = 15$. ______

15. This problem consists of completing a proof of the following theorem:
"In a kite, one pair of opposite angles are congruent."

*Given:* Kite $ABCD$; $\overline{AB} \cong \overline{AD}$ and $\overline{BC} \cong \overline{DC}$
*Prove:* $\angle B \cong \angle D$

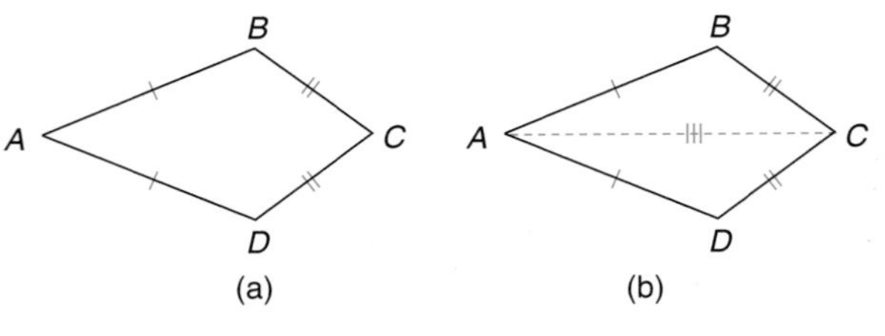


**PROOF**

| Statements | Reasons |
|---|---|
| 1. ______ | 1. ______ |
| 2. Draw $\overline{AC}$. | 2. Through two points, there is exactly one line |
| 3. ______ | 3. Identity |
| 4. $\triangle ACD \cong \triangle ACB$ | 4. ______ |
| 5. ______ | 5. ______ |

16. This problem consists of completing a proof of the following theorem:
"The diagonals of an isosceles trapezoid are congruent."

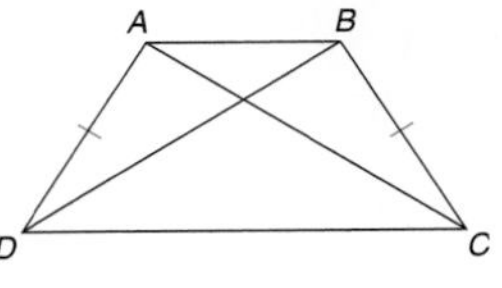


*Given:* Trapezoid $ABCD$ with $\overline{AB} \parallel \overline{DC}$ and $\overline{AD} \cong \overline{BC}$
*Prove:* $\overline{AC} \cong \overline{DB}$

**PROOF**

| Statements | Reasons |
|---|---|
| 1. ______ | 1. ______ |
| 2. $\angle ADC \cong \angle BCD$ | 2. Base $\angle$s of an isosceles trapezoid are ______ |
| 3. $\overline{DC} \cong \overline{DC}$ | 3. ______ |
| 4. $\triangle ADC \cong \triangle BCD$ | 4. ______ |
| 5. ______ | 5. CPCTC |

# Chapter 5

# Similar Triangles

## Chapter Outline



**For online student resources, visit this textbook's website at math.college.hmco.com/students.**

Although of different sizes, the larger and smaller leaves of the water lily have the same shape. Given two squares, one with sides of length 5 inches and the other with sides of length 3 inches, the squares have the same shape. The smaller and larger cylindrical containers found on grocery store shelves also illustrate figures with different sizes but the same shape. In all situations, one figure is simply an enlargement of the other. In geometry, we say that the two objects are *similar* in shape. See Sections 5.2 and 5.3 for more information and further illustrations of similar geometric figures.

The solutions for some applications found in this chapter and later chapters lead to quadratic equations. A review of the methods for solving quadratic equations can be found in Appendix A.4 of this textbook.

KEY CONCEPTS

Ratio • Rate • Proportion • Extremes • Means • Means-Extremes Property • Geometric Mean • Extended Ratio • Extended Proportion

# 5.1 Ratios, Rates, and Proportions

The concepts and techniques discussed in Section 5.1 are often necessary for the geometry applications found throughout this chapter and beyond.

A **ratio** is the quotient $\frac{a}{b}$ (where $b \neq 0$) that provides a comparison between the numbers $a$ and $b$. Because every fraction indicates a division, every fraction represents a ratio. Read "$a$ to $b$," the ratio is sometimes written in the form $a:b$.

It is generally preferable to provide the ratio in simplest form, so the ratio 6 to 8 would be reduced (in fraction form) from $\frac{6}{8}$ to $\frac{3}{4}$. If units of measure are involved, these units must be **commensurable** (convertible to the same unit of measure). When simplifying the ratio of two quantities that are expressed in the same unit, we eliminate the common unit in the process. If two quantities cannot be compared because no common unit of measure is possible, the quantities are **incommensurable.**

**Reminder**

Units are neither needed nor desirable in a simplified ratio.

EXAMPLE 1

Find the best form of each ratio:

a) 12 to 20
b) 12 in. to 24 in.
c) 12 in. to 3 ft (NOTE: 1 ft = 12 in.)
d) 5 lb to 20 oz (NOTE: 1 lb = 16 oz)
e) 5 lb to 2 ft
f) 4 m to 30 cm (NOTE: 1 m = 100 cm)

***Solution***

a) $\frac{12}{20} = \frac{3}{5}$

b) $\frac{12\text{ in.}}{24\text{ in.}} = \frac{12}{24} = \frac{1}{2}$

c) $\frac{12\text{ in.}}{3\text{ ft}} = \frac{12\text{ in.}}{3(12\text{ in.})} = \frac{12\text{ in.}}{36\text{ in.}} = \frac{1}{3}$

d) $\frac{5\text{ lb}}{20\text{ oz}} = \frac{5(16\text{ oz})}{20\text{ oz}} = \frac{80\text{ oz}}{20\text{ oz}} = \frac{4}{1}$

e) $\frac{5\text{ lb}}{2\text{ ft}}$ is incommensurable!

f) $\frac{4\text{ m}}{30\text{ cm}} = \frac{4(100\text{ cm})}{30\text{ cm}} = \frac{400\text{ cm}}{30\text{ cm}} = \frac{40}{3}$

*Geometry in the Real World*

At a grocery store, the cost per unit is a rate that allows the consumer to know which brand is more expensive.

A **rate** is a quotient that compares two quantities that are incommensurable. If an automobile can travel 300 miles along an interstate on a full tank of 10 gallons of gasoline, then its consumption *rate* is $\frac{300\text{ miles}}{10\text{ gallons}}$. In simplified form, the consumption rate is $\frac{30\text{ mi}}{\text{gal}}$, which is read as "30 miles per gallon" and is often abbreviated 30 mpg.

EXAMPLE 2

Simplify each rate. Units are necessary in each answer.

a) $\frac{120\text{ miles}}{5\text{ gallons}}$

b) $\frac{100\text{ meters}}{10\text{ seconds}}$

c) $\frac{12\text{ teaspoons}}{2\text{ quarts}}$

d) $\frac{\$8.45}{5\text{ gallons}}$

***Solution***

a) $\dfrac{120 \text{ mi}}{5 \text{ gal}} = \dfrac{24 \text{ mi}}{\text{gal}}$ (sometimes written 24 mpg)

b) $\dfrac{100 \text{ m}}{10 \text{ s}} = \dfrac{10 \text{ m}}{\text{s}}$

c) $\dfrac{12 \text{ teaspoons}}{2 \text{ quarts}} = \dfrac{6 \text{ teaspoons}}{\text{quart}}$

d) $\dfrac{\$8.45}{5 \text{ gal}} = \dfrac{\$1.69}{\text{gal}}$

*Exs. 1–2*

A **proportion** is a statement that two ratios or two rates are equal. Thus $\frac{a}{b} = \frac{c}{d}$ is a proportion and may be read as "$a$ is to $b$ as $c$ is to $d$." In the order read, $a$ is the *first term* of the proportion, $b$ is the *second term*, $c$ is the *third term*, and $d$ is the *fourth term*. The first and last terms ($a$ and $d$) of the proportion are the **extremes,** whereas the second and third terms ($b$ and $c$) are the **means.**

The following property is convenient for solving many proportions.

PROPERTY 1: **(Means-Extremes Property)**
In a proportion, the product of the means equals the product of the extremes; that is, if $\frac{a}{b} = \frac{c}{d}$ (where $b \neq 0$ and $d \neq 0$), then $a \cdot d = b \cdot c$.

In the false proportion $\frac{9}{12} = \frac{2}{3}$, it is obvious that $9 \cdot 3 \neq 12 \cdot 2$; on the other hand, the truth of the statement $\frac{9}{12} = \frac{3}{4}$ is evident from the fact that $9 \cdot 4 = 12 \cdot 3$. Henceforth, any proportion given in this text is intended to be a true proportion.

EXAMPLE 3

Use the Means-Extremes Property to solve each proportion for $x$.

a) $\dfrac{x}{8} = \dfrac{5}{12}$

b) $\dfrac{x+1}{9} = \dfrac{x-3}{3}$

c) $\dfrac{3}{x} = \dfrac{x}{2}$

d) $\dfrac{x+3}{3} = \dfrac{9}{x-3}$

e) $\dfrac{x+2}{5} = \dfrac{4}{x-1}$

***Solution***

a) $x \cdot 12 = 8 \cdot 5$ (Means-Extremes Property)
$12x = 40$
$x = \dfrac{40}{12} = \dfrac{10}{3}$

b) $3(x + 1) = 9(x - 3)$ (Means-Extremes Property)
$3x + 3 = 9x - 27$
$30 = 6x$
$x = 5$

c) $3 \cdot 2 = x \cdot x$ (Means-Extremes Property)
$x^2 = 6$
$x = \pm\sqrt{6} \approx \pm 2.45$

**Warning** (!)
As you solve a proportion such as $\frac{x}{8} = \frac{5}{12}$, write $12x = 40$ on the next line. Do *not* write $\frac{x}{8} = \frac{5}{12} = 12x = 40$, which would imply that $\frac{5}{12} = 40$.

d) $(x + 3)(x - 3) = 3 \cdot 9$ (Means-Extremes Property)

$$x^2 - 9 = 27$$
$$x^2 - 36 = 0$$
$$(x + 6)(x - 6) = 0 \quad \text{(using factoring)}$$
$$x + 6 = 0 \quad \text{or} \quad x - 6 = 0$$
$$x = -6 \quad \text{or} \quad x = 6$$

e) $(x + 2)(x - 1) = 5 \cdot 4$ (Means-Extremes Property)

$$x^2 + x - 2 = 20$$
$$x^2 + x - 22 = 0$$
$$x = \frac{-b \pm \sqrt{b^2 - 4ac}}{2a} \quad \text{(using Quadratic Formula; see Appendix A.4)}$$
$$= \frac{-1 \pm \sqrt{(1)^2 - 4(1)(-22)}}{2(1)}$$
$$= \frac{-1 \pm \sqrt{1 + 88}}{2}$$
$$= \frac{-1 \pm \sqrt{89}}{2}$$
$$\approx 4.22 \text{ or } -5.22$$

In application problems involving proportions, it is essential to order the related quantities in each ratio or rate. The following example illustrates the care that must be taken in forming the proportion for an application.

EXAMPLE 4

If an automobile can travel 90 mi on 4 gal of gasoline, how far can it travel on 6 gal of gasoline?

***Solution***

By form,

$$\frac{\text{number miles first trip}}{\text{number gallons first trip}} = \frac{\text{number miles second trip}}{\text{number gallons second trip}}$$

Where $x$ represents the number of miles traveled on the second trip, we have

$$\frac{90}{4} = \frac{x}{6}$$
$$4x = 540$$
$$x = 135$$

Thus the car can travel 135 mi on 6 gal of gasoline.

*Geometry in the Real World*

The automobile described in Example 4 has a consumption rate of 22.5 mpg (miles per gallon).

In $\frac{a}{b} = \frac{b}{c}$, where the second and third terms of the proportion are identical, the value of $b$ is known as the **geometric mean** of $a$ and $c$. For example, 6 and $-6$ are the geometric means of 4 and 9 because $\frac{4}{6} = \frac{6}{9}$ and $\frac{4}{-6} = \frac{-6}{9}$. Because applications in geometry generally require positive solutions, we usually seek only the positive geometric mean of $a$ and $c$.

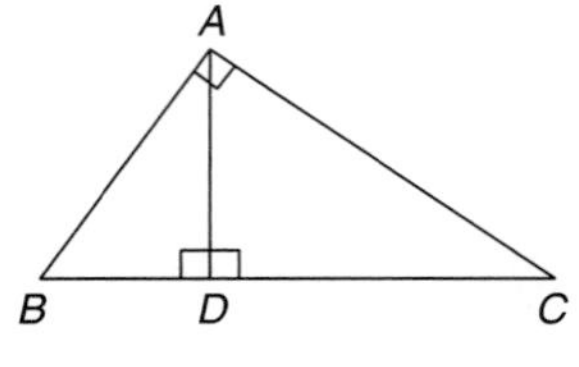


**FIGURE 5.1**

EXAMPLE 5

In Figure 5.1, $AD$ is the geometric mean of $BD$ and $DC$. If $BC = 10$ and $BD = 4$, determine $AD$.

***Solution***

$\frac{BD}{AD} = \frac{AD}{DC}$. Because $DC = BC - BD$, we know that $DC = 10 - 4 = 6$. Therefore,

$$\frac{4}{x} = \frac{x}{6}$$

in which $x$ is the length of $\overline{AD}$. Applying the Means-Extremes Property, we get

$$x^2 = 24$$

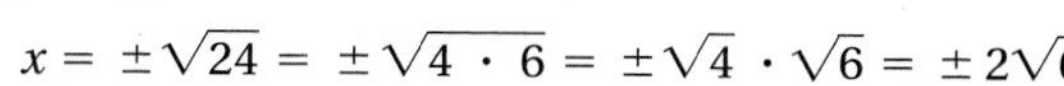

$$x = \pm\sqrt{24} = \pm\sqrt{4 \cdot 6} = \pm\sqrt{4} \cdot \sqrt{6} = \pm 2\sqrt{6}$$

*Exs. 3–6*

To have a permissible length for $\overline{AD}$, the geometric mean is the positive solution. Thus $AD = 2\sqrt{6}$ or $AD \approx 4.90$.

An **extended ratio** compares more than two quantities and must be expressed in a form such as *a:b:c* or *d:e:f:g*. If you know that the angles of a triangle are 90°, 60°, and 30°, then the ratio that compares these measures is 90:60:30, or 3:2:1 (because 90, 60, and 30 have the greatest common factor of 30). Unknown quantities in the ratio *a:b:c:d:* . . . are generally represented by variable expressions such as *ax, bx, cx, dx,* . . . .

### EXAMPLE 6

Suppose that the perimeter of a quadrilateral is 70 and the lengths of the sides are in the ratio 2:3:4:5. Find the measure of each side.

***Solution***

Let the lengths of the sides be represented by $2x$, $3x$, $4x$, and $5x$. Then

$$\begin{aligned} 2x + 3x + 4x + 5x &= 70 \\ 14x &= 70 \\ x &= 5 \end{aligned}$$

Because $2x = 10$, $3x = 15$, $4x = 20$, and $5x = 25$, the lengths of the sides are 10, 15, 20, and 25.

It is possible to solve certain problems in more ways than one, as is illustrated in the next example. However, the solution is unique and is not altered by the method chosen.

### EXAMPLE 7

The measures of two complementary angles are in the ratio 2 to 3. Find the measure of each angle.

***Solution***

Let the first of the complementary angles have measure $x$; then the second has measure $90 - x$. Thus we have

$$\frac{x}{90 - x} = \frac{2}{3}$$

Using the Means-Extremes Property, we have

$$3x = 2(90 - x)$$
$$3x = 180 - 2x$$
$$5x = 180$$
$$x = 36$$
$$90 - x = 54$$

The angles have measures of 36° and 54°.

***Alternative Solution***

Because the measures of the angles are in the ratio 2:3, let their measures be $2x$ and $3x$. Because the angles are complementary,

$$2x + 3x = 90$$
$$5x = 90$$
$$x = 18$$

*Exs. 7–9*

Now $2x = 36$ and $3x = 54$, so the measures of the two angles are 36° and 54°.

Some additional properties of proportions follow. Because they are not cited as often as the Means-Extremes Property, they are not given titles.

PROPERTY 2: In a proportion, the means or the extremes (or both the means and the extremes) may be interchanged; that is, if $\frac{a}{b} = \frac{c}{d}$ (where $a$, $b$, $c$, and $d$ are nonzero), then $\frac{a}{c} = \frac{b}{d}$, $\frac{d}{b} = \frac{c}{a}$, and $\frac{d}{c} = \frac{b}{a}$.

NOTE: The last proportion is the inverted form of the given proportion. In any of the equivalent proportions, $a \cdot d = b \cdot c$ (product of means = product of extremes).

When we are given, say, the proportion $\frac{2}{3} = \frac{8}{12}$, Property 2 enables us to draw conclusions such as

1. $\frac{2}{8} = \frac{3}{12}$ (means interchanged)
2. $\frac{12}{3} = \frac{8}{2}$ (extremes interchanged))
3. $\frac{3}{2} = \frac{12}{8}$ (both sides inverted)

PROPERTY 3: If $\frac{a}{b} = \frac{c}{d}$ (where $b \neq 0$ and $d \neq 0$), then $\frac{a + b}{b} = \frac{c + d}{d}$ and $\frac{a - b}{b} = \frac{c - d}{d}$.

NOTE: If $\frac{a}{b} = \frac{c}{d}$, then $\frac{a}{b} + 1 = \frac{c}{d} + 1$. In turn, $\frac{a}{b} + \frac{b}{b} = \frac{c}{d} + \frac{d}{d}$.

Given the proportion $\frac{2}{3} = \frac{8}{12}$, Property 3 enables us to draw conclusions such as

1. $\frac{2 + 3}{3} = \frac{8 + 12}{12}$ $\left(\text{each side simplifies to } \frac{5}{3}\right)$
2. $\frac{2 - 3}{3} = \frac{8 - 12}{12}$ $\left(\text{each side equals } -\frac{1}{3}\right)$

SSG *Exs. 10, 11*

Just as there are extended ratios, there are also **extended proportions,** such as

$$\frac{a}{b} = \frac{c}{d} = \frac{e}{f} = \cdots$$

Suggested by different numbers of servings of a particular recipe, the statement below is an extended proportion comparing numbers of eggs to numbers of cups of milk:

$$\frac{2 \text{ eggs}}{3 \text{ cups}} = \frac{4 \text{ eggs}}{6 \text{ cups}} = \frac{6 \text{ eggs}}{9 \text{ cups}}$$

EXAMPLE 8

In the triangles shown in Figure 5.2, $\frac{AB}{DE} = \frac{AC}{DF} = \frac{BC}{EF}$. Find the lengths of $\overline{DF}$ and $\overline{EF}$.

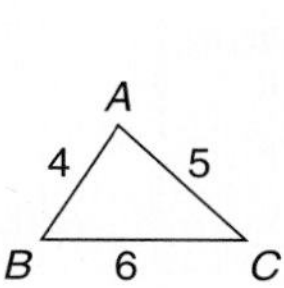


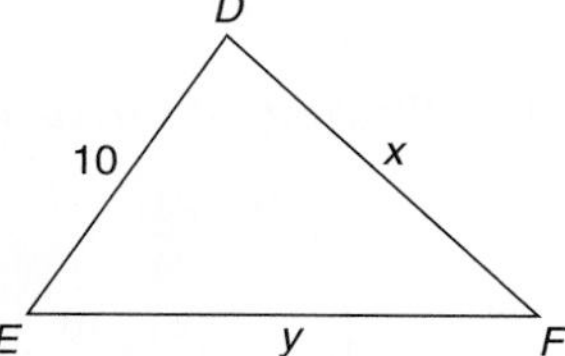


**FIGURE 5.2**

***Solution***

Substituting into the proportion $\frac{AB}{DE} = \frac{AC}{DF} = \frac{BC}{EF}$, we have

$$\frac{4}{10} = \frac{5}{x} = \frac{6}{y}$$

From the equation

$$\frac{4}{10} = \frac{5}{x}$$

it follows that $4x = 50$ and that $x = DF = 12.5$. Using the equation

$$\frac{4}{10} = \frac{6}{y}$$

we find that $4y = 60$, so $y = EF = 15$.

*Exs. 12, 13*

## Discover!

THE GOLDEN RATIO

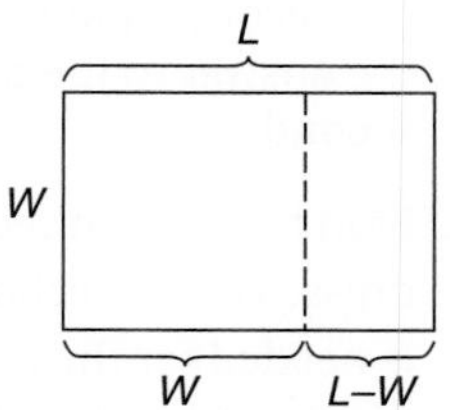


It is believed that the "ideal" rectangle is most attractive when a square can be removed in such a way as to leave a smaller rectangle with the same shape as the original rectangle. As we shall find, the rectangles are known as *similar* in shape. Upon removal of the square, the similarity in the shapes of the rectangles requires that $\frac{W}{L} = \frac{L-W}{W}$. To discover the relationship between $L$ and $W$, we choose $W = 1$ and solve the equation $\frac{1}{L} = \frac{L-1}{1}$ for $L$. The solution is $L = \frac{1+\sqrt{5}}{2}$. The ratio comparing length to width is known as the *golden ratio.* Because $L = \frac{1+\sqrt{5}}{2}$ when $W = 1$ and $\frac{1+\sqrt{5}}{2} \approx 1.62$, the *ideal* rectangle has a length that is approximately 1.62 times its width; that is, $L \approx 1.62W$.

## 5.1 Exercises

*In Exercises 1 to 4, give the ratios in simplified form.*

1. a) 12 to 15 b) 12 in. to 15 in. c) 1 ft to 18 in. d) 1 ft to 18 oz
2. a) 20 to 36 b) 24 oz to 52 oz c) 20 oz to 2 lb (1 lb = 16 oz) d) 2 lb to 20 oz
3. a) 15:24 b) 2 ft:2 yd (1 yd = 3 ft) c) 2 m:150 cm (1 m = 100 cm) d) 2 m:1 lb
4. a) 24:32 b) 12 in.:2 yd c) 150 cm:2 m d) 1 gal:24 mi

*In Exercises 5 to 14, find the value of x in each proportion.*

5. a) $\frac{x}{4} = \frac{9}{12}$ b) $\frac{7}{x} = \frac{21}{24}$
6. a) $\frac{x-1}{10} = \frac{3}{5}$ b) $\frac{x+1}{6} = \frac{10}{12}$
7. a) $\frac{x-3}{8} = \frac{x+3}{24}$ b) $\frac{x+1}{6} = \frac{4x-1}{18}$
8. a) $\frac{9}{x} = \frac{x}{16}$ b) $\frac{32}{x} = \frac{x}{2}$
9. a) $\frac{x}{4} = \frac{7}{x}$ b) $\frac{x}{6} = \frac{3}{x}$
10. a) $\frac{x+1}{3} = \frac{10}{x+2}$ b) $\frac{x-2}{5} = \frac{12}{x+2}$
11. a) $\frac{x+1}{x} = \frac{10}{2x}$ b) $\frac{2x+1}{x+1} = \frac{14}{3x-1}$
12. a) $\frac{x+1}{2} = \frac{7}{x-1}$ b) $\frac{x+1}{3} = \frac{5}{x-2}$
13. a) $\frac{x+1}{x} = \frac{2x}{3}$ b) $\frac{x+1}{x-1} = \frac{2x}{5}$
14. a) $\frac{x+1}{x} = \frac{x}{x-1}$ b) $\frac{x+2}{x} = \frac{2x}{x-2}$
15. Sarah ran the 300-m hurdles in 47.7 sec. In meters per second, find the rate at which Sarah ran. Give the answer to the nearest tenth of a meter per second.
16. Fran has been hired to sew the dance troupe's dresses for the school musical. If $13\frac{1}{3}$ yd of material is needed for the four dresses, find the rate that describes the amount of material needed for each dress.

*In Exercises 17 to 22, use proportions to solve each problem.*

17. A recipe calls for 4 eggs and 3 cups of milk. To prepare for a larger number of guests, a cook uses 14 eggs. How many cups of milk are needed?
18. If a school secretary copies 168 worksheets for a class of 28 students, how many must be prepared for a class of 32 students?
19. An electrician installs 20 electrical outlets in a new six-room house. Assuming proportionality, how many outlets should be installed in a new construction having seven rooms? (Round up to an integer.)
20. The secretarial pool (15 secretaries in all) on one floor of a corporate complex has access to four copy machines. If there are 23 secretaries on a different floor, approximately what number of copy machines should be available? (Assume a proportionality.)
21. Assume that *AD* is the geometric mean of *BD* and *DC* in $\triangle ABC$ shown in the accompanying drawing.
    a) Find *AD* if *BD* = 6 and *DC* = 8.
    b) Find *BD* if *AD* = 6 and *DC* = 8.

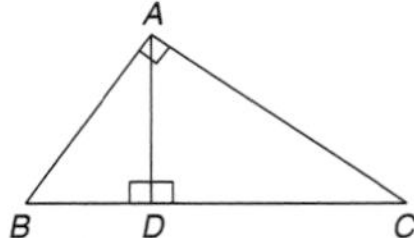


*Exercises 21, 22*

22. In the drawing, assume that *AB* is the geometric mean of *BD* and *BC*.
    a) Find *AB* if *BD* = 6 and *DC* = 10.
    b) Find *DC* if *AB* = 10 and *BC* = 15.
23. The salaries of a secretary, a salesperson, and a vice president for a retail sales company are in the ratio 2:3:5. If their combined annual salaries amount to $124,500, what is the annual salary of each?
24. If the measures of the angles of a quadrilateral are in the ratio of 2:3:4:6, find the measure of each angle.
25. The measures of two complementary angles are in the ratio 4:5. Find the measure of each angle, using the two methods shown in Example 7.

26. The measures of two supplementary angles are in the ratio of 2:7. Find the measure of each angle, using the two methods of Example 7.

27. If 1 in. equals 2.54 cm, use a proportion to convert 12 in. to centimeters.
(HINT: $\frac{2.54\text{ cm}}{1\text{ in.}} = \frac{x\text{ cm}}{12\text{ in.}}$)

28. If 1 kilogram equals 2.2 pounds, use a proportion to convert 12 pounds to kilograms.

29. For the quadrilaterals shown, $\frac{MN}{WX} = \frac{NP}{XY} = \frac{PQ}{YZ} = \frac{MQ}{WZ}$. If $MN = 7$, $WX = 3$, and $PQ = 6$, find $YZ$.

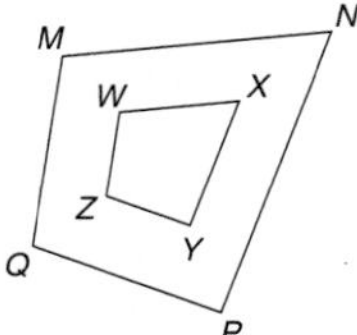


*Exercises 29, 30*

30. For this exercise, use the drawing and extended ratio of Exercise 29. If $NP = 2 \cdot XY$ and $WZ = 3\frac{1}{2}$, find $MQ$.

31. Two numbers $a$ and $b$ are in the ratio 3:4. If the first number is decreased by 2 and the second is decreased by 1, they are in the ratio 2:3. Find $a$ and $b$.

32. If the ratio of the measure of the complement of an angle to the measure of its supplement is 1:4, what is the measure of the angle?

33. On a blueprint, a 1-in. scale corresponds to 3 ft. To show a room with actual dimensions 12 ft wide by 14 ft long, what dimensions should be shown on the blueprint?

34. To find the golden ratio (see Discover! on page 223), solve the equation $\frac{1}{L} = \frac{L-1}{1}$ for $L$.
(HINT: You will need the Quadratic Formula.)

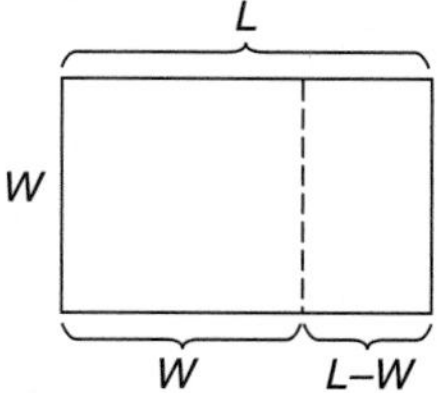


35. Find:

a) The exact length of an ideal rectangle with width $W = 5$ by solving $\frac{5}{L} = \frac{L-5}{5}$

b) The approximate length of an ideal rectangle with width $W = 5$ by using $L \approx 1.62W$

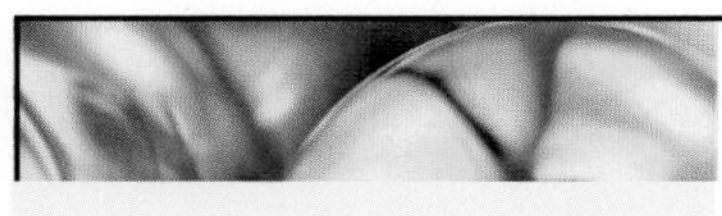

**KEY CONCEPTS**

Similar Polygons • Congruent Polygons • Corresponding Vertices, Angles, and Sides

# 5.2 Similar Polygons

When two geometric figures have exactly the same shape, they are **similar;** the symbol for "is similar to" is $\sim$. When two figures have the same shape ($\sim$) and all corresponding parts have equal ($=$) measures, the two figures are **congruent** ($\cong$). Note that the symbol for congruence combines the symbols for similarity and equality. In fact, we include the following property for emphasis.

> Two congruent polygons are also similar polygons.

Two-dimensional figures can be similar, as are $\triangle ABC$ and $\triangle DEF$ in Figure 5.3, but it is also possible for three-dimensional figures to be similar. Similar orange juice containers are shown in Figures 5.4(a) and (b). Informally, two figures are "similar" if one is an enlargement of the other. Thus a tuna fish can and an orange juice can are *not* similar, even if both are right-circular cylinders [see Figures 5.4(b) and (c)]. We will consider cylinders in greater detail in Chapter 8.

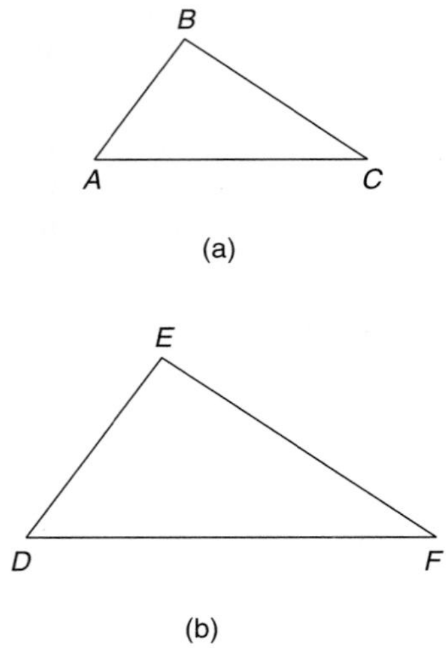

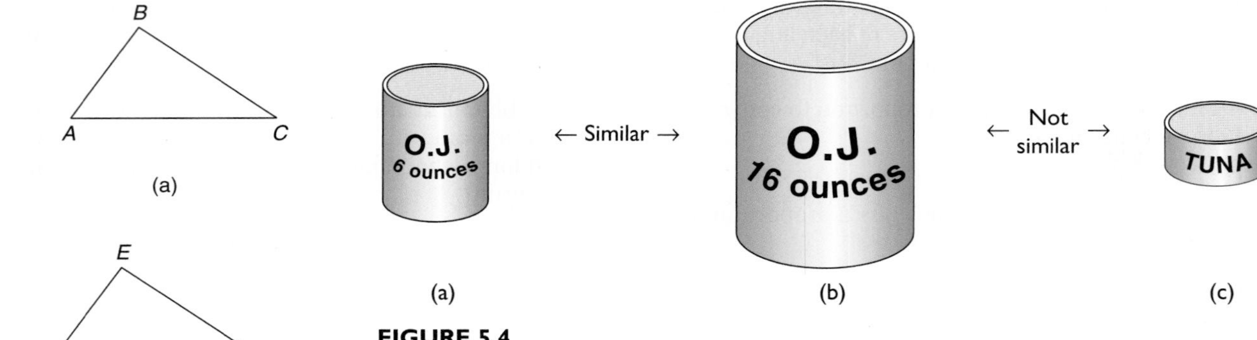


(a)

E
D
F

(b)

**FIGURE 5.3**

(a) (b) (c)

**FIGURE 5.4**

Our discussion of similarity will generally be limited to plane figures.

For two polygons to be similar, one requirement is that each angle of one polygon must be congruent to the corresponding angle of the other. Although this congruence of angles is necessary for similarity of polygons, it alone is not sufficient to establish similarity. The vertices of the congruent angles are **corresponding vertices** of the similar polygons. If $\angle A$ in one polygon is congruent to $\angle M$ in the second polygon, then vertex $A$ corresponds to vertex $M$, and this is symbolized $A \leftrightarrow M$; we can indicate that $\angle A$ corresponds to $\angle M$ by writing $\angle A \leftrightarrow \angle M$. A pair of angles like $\angle A$ and $\angle M$ are **corresponding angles,** and the sides determined by consecutive and corresponding vertices are **corresponding sides** of the similar polygons. For instance, if $A \leftrightarrow M$ and $B \leftrightarrow N$, then $\overline{AB}$ corresponds to $\overline{MN}$.

> **Discover!**
>
> When a transparency is projected onto a screen, the image created is similar to the projected figure.

EXAMPLE 1

Given that quadrilaterals $ABCD$ and $HJKL$ are similar, with congruent angles indicated in Figure 5.5, name the vertices, angles, and sides that correspond to each other.

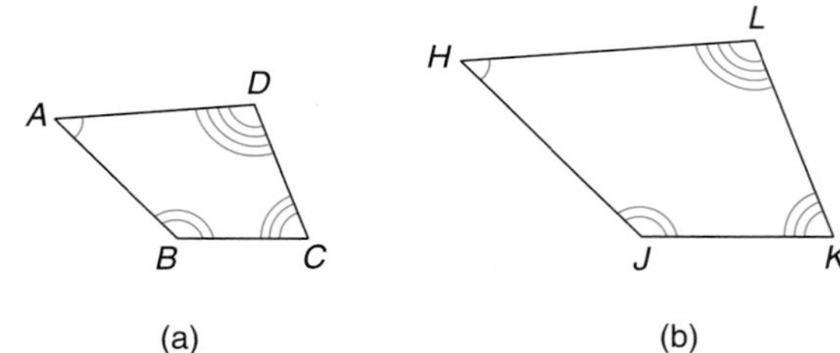


(a) (b)

**FIGURE 5.5**

***Solution***

Because $\angle A \cong \angle H$, it follows that

$$A \leftrightarrow H \quad \text{and} \quad \angle A \leftrightarrow \angle H$$

Similarly,

$$B \leftrightarrow J \quad \text{and} \quad \angle B \leftrightarrow \angle J$$
$$C \leftrightarrow K \quad \text{and} \quad \angle C \leftrightarrow \angle K$$
$$D \leftrightarrow L \quad \text{and} \quad \angle D \leftrightarrow \angle L$$

When pairs of consecutive and corresponding vertices are associated, the corresponding sides are included between the corresponding angles (or vertices).

$$\overline{AB} \leftrightarrow \overline{HJ}, \quad \overline{BC} \leftrightarrow \overline{JK}, \quad \overline{CD} \leftrightarrow \overline{KL}, \quad \text{and} \quad \overline{AD} \leftrightarrow \overline{HL}$$

Geometry in Nature

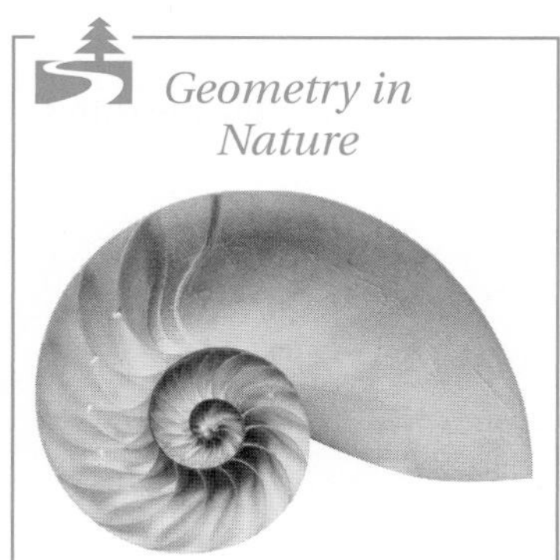

The segments of the chambered nautilus are similar (not congruent) in shape.

**DEFINITION:** Two polygons are **similar** if and only if two conditions are satisfied:

1. All pairs of corresponding angles are congruent.
2. All pairs of corresponding sides are proportional.

The second condition for similarity requires that the following extended proportion exists for the sides of the similar quadrilaterals of Example 1.

$$\frac{AB}{HJ} = \frac{BC}{JK} = \frac{CD}{KL} = \frac{AD}{HL}$$

Note that *both* conditions for similarity are necessary! Although condition 1 is satisfied for square *EFGH* and rectangle *RSTU* [see Figures 5.6(a) and (b)], the figures are not similar—that is, one is not an enlargement of the other—because the extended proportion is not true. On the other hand, condition 2 is satisfied for square *EFGH* and rhombus *WXYZ* [see Figures 5.6(a) and (c)], but the figures are not similar because the pairs of corresponding angles are not congruent.

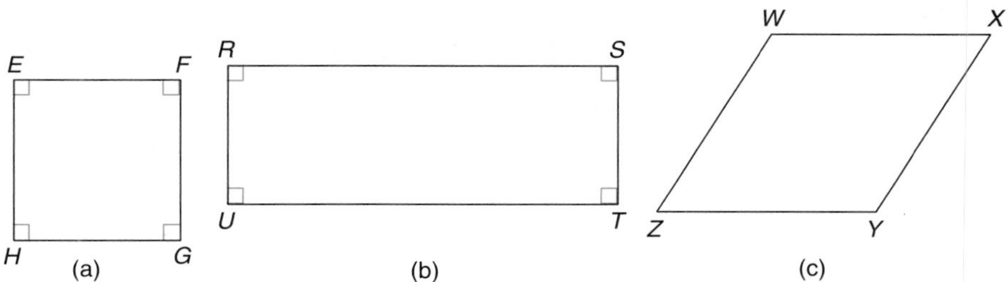


**FIGURE 5.6**

EXAMPLE 2

Which figures must be similar?

a) Any two isosceles triangles
b) Any two regular pentagons
c) Any two rectangles
d) Any two squares

***Solution***

a) No; $\angle$ pairs need not be $\cong$, nor do the pairs of sides need to be proportional.
b) Yes; all angles are congruent (measure 108° each), and all pairs of sides are proportional.
c) No; all angles measure 90°, but the pairs of sides are not necessarily proportional.
d) Yes; all angles measure 90°, and all pairs of sides are proportional.

*Exs. 1–4*

The practice of naming corresponding vertices in consecutive order for the two polygons is most convenient! For instance, if pentagon *ABCDE* is similar to pentagon *MNPQR*, then we know that $A \leftrightarrow M$, $B \leftrightarrow N$, $C \leftrightarrow P$, $D \leftrightarrow Q$, $E \leftrightarrow R$, $\angle A \cong \angle M$, $\angle B \cong \angle N$, $\angle C \cong \angle P$, $\angle D \cong \angle Q$, and $\angle E \cong \angle R$. Because of the indicated correspondence of vertices, we also know that

$$\frac{AB}{MN} = \frac{BC}{NP} = \frac{CD}{PQ} = \frac{DE}{QR} = \frac{EA}{RM}$$

EXAMPLE 3

If $\triangle ABC \sim \triangle DEF$ in Figure 5.7, use the indicated measures to find the measures of the remaining parts of each of the triangles.

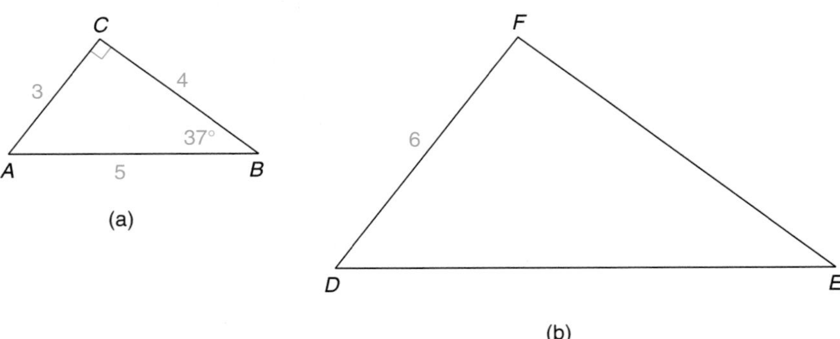


**FIGURE 5.7**

***Solution***

Because the sum of the measures of the angles of a triangle is 180°,

$$m\angle A = 180 - (90 + 37) = 53°$$

And because of the similarity and the corresponding vertices,

$$m\angle D = 53°, \quad m\angle E = 37°, \quad \text{and} \quad m\angle F = 90°$$

The proportion that relates the lengths of the sides is

$$\frac{AC}{DF} = \frac{CB}{FE} = \frac{AB}{DE} \quad \text{so} \quad \frac{3}{6} = \frac{4}{FE} = \frac{5}{DE}$$

From $\frac{3}{6} = \frac{4}{FE}$, we see that

$$3 \cdot FE = 6 \cdot 4 = 24$$
$$FE = 8$$

From $\frac{3}{6} = \frac{5}{DE}$, we see that

$$3 \cdot DE = 6 \cdot 5 = 30$$
$$DE = 10$$

In a proportion, the ratios can all be inverted; thus Example 3 could have been solved by using the proportion

$$\frac{DF}{AC} = \frac{FE}{CB} = \frac{DE}{AB}$$

In an extended proportion, the ratios must all be equal to the same constant value. By designating this number (which is often called the "constant of proportionality") by $k$, we see that

$$\frac{DF}{AC} = k, \quad \frac{FE}{CB} = k, \quad \text{and} \quad \frac{DE}{AB} = k$$

*Exs. 5–10*

It follows that $DF = k \cdot AC$, $FE = k \cdot CB$, and $DE = k \cdot AB$. In Example 3, this constant of proportionality had the value $k = 2$, which means that the length of each side of the larger triangle was twice the length of the corresponding side of the smaller triangle.

If $k > 1$, the similarity leads to an enlargement, or *stretch*. If $0 < k < 1$, the similarity results in a *shrink*.

The constant of the previous discussion is also used to *scale* a map, a diagram, or a blueprint. As a consequence, scaling problems can be solved by using proportions.

EXAMPLE 4

On a map, a length of 1 inch represents a distance of 30 miles. On the map, how far apart should two cities appear if they are actually 140 miles apart along a straight line?

***Solution***

Where $x =$ the map distance desired (in inches),

$$\frac{1}{30} = \frac{x}{140}$$

Then $30x = 140$ and $x = 4\frac{2}{3}$ inches.

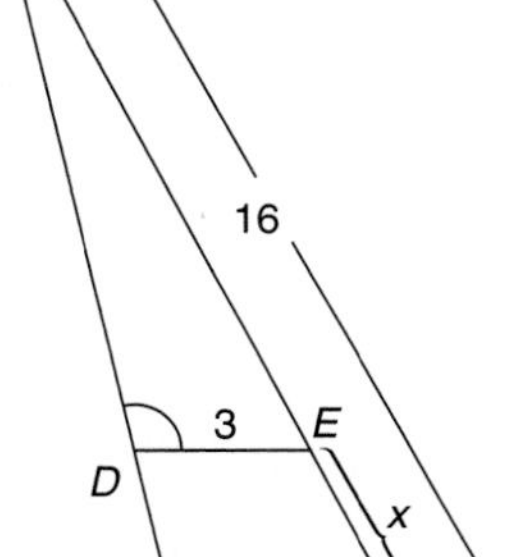

**FIGURE 5.8**

EXAMPLE 5

In Figure 5.8, $\triangle ABC \sim \triangle ADE$ so that $\angle ADE \cong \angle B$. If $DE = 3$, $AC = 16$, and $EC = BC$, find the length $BC$.

***Solution***

From the similar triangles, we have $\frac{DE}{BC} = \frac{AE}{AC}$. With $AC = AE + EC$ and representing the lengths of the congruent segments ($\overline{EC}$ and $\overline{BC}$) by $x$, we have

$$16 = AE + x \qquad \text{so} \qquad AE = 16 - x$$

Substituting into the proportion, we have

$$\frac{3}{x} = \frac{16 - x}{16}$$

It follows that

$$\begin{aligned} x(16 - x) &= 3 \cdot 16 \\ 16x - x^2 &= 48 \\ x^2 - 16x + 48 &= 0 \\ (x - 4)(x - 12) &= 0 \end{aligned}$$

Now $x$ (or $BC$) equals 4 or 12. Each length is acceptable, but the scaled drawings differ, as illustrated in Figure 5.9.

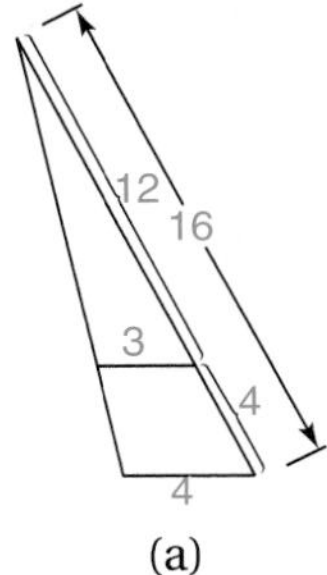

(a)

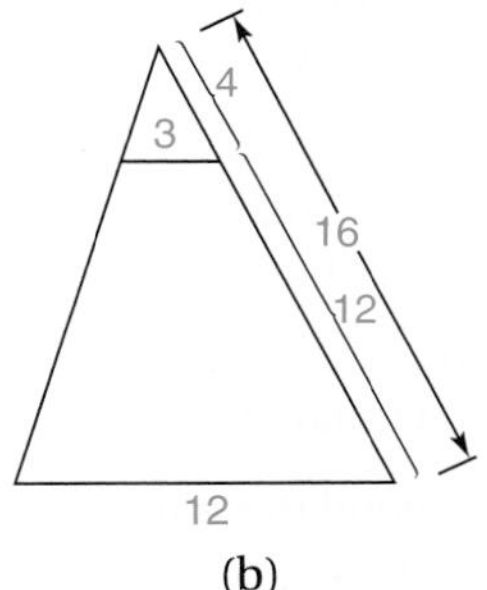

(b)

**FIGURE 5.9**

The following example uses a method called *shadow reckoning.* Not a new technique, this method of calculating a length dates back more than 2500 years. It was used by Thales to estimate the height of the pyramids in Egypt. In application, the method assumes (correctly) that $\triangle ABC \sim \triangle DEF$. Note that $\angle A \cong \angle D$ and $\angle C \cong \angle F$.

EXAMPLE 6

Darnell is curious about the height of a flagpole that stands in front of his school. Darnell, who is 6 ft tall, casts a shadow that he paces off at 9 ft. He walks the length of the shadow of the flagpole, a distance of 30 ft. How tall is the flagpole?

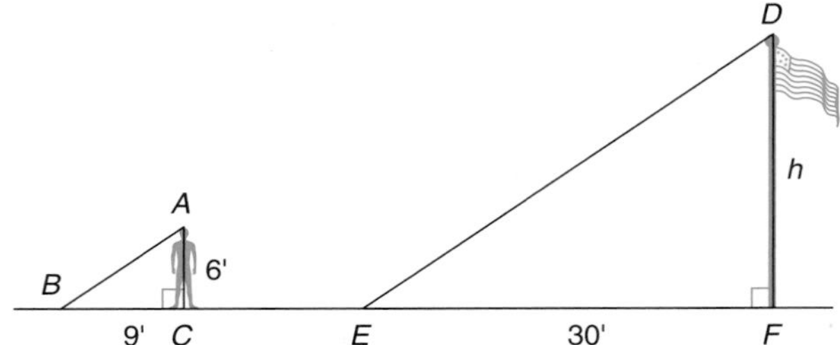


**FIGURE 5.10**

***Solution***

In Figure 5.10, $\triangle ABC \sim \triangle DEF$. From similar triangles, we know that $\frac{AC}{DF} = \frac{BC}{EF}$ or $\frac{AC}{BC} = \frac{DF}{EF}$ by interchanging the means.

Where $h$ is the height of the flagpole, substitution into the second proportion leads to

$$\frac{6}{9} = \frac{h}{30} \rightarrow 9h = 180 \rightarrow h = 20$$

The height of the flagpole is 20 ft.

*Exs. 11–13*

## 5.2 Exercises

1. a) What is true of any pair of corresponding angles of two similar polygons?
   b) What is true of any pairs of corresponding sides of two similar polygons?
2. a) Are any two quadrilaterals similar?
   b) Are any two squares similar?
3. a) Are any two regular pentagons similar?
   b) Are any two equiangular pentagons similar?
4. a) Are any two equilateral hexagons similar?
   b) Are any two regular hexagons similar?

*In Exercises 5 and 6, refer to the drawing.*

5. a) Given that $A \leftrightarrow X$, $B \leftrightarrow T$, and $C \leftrightarrow N$, write a statement claiming that the triangles shown are similar.
   b) Given that $A \leftrightarrow N$, $C \leftrightarrow X$, and $B \leftrightarrow T$, write a statement claiming that the triangles shown are similar.

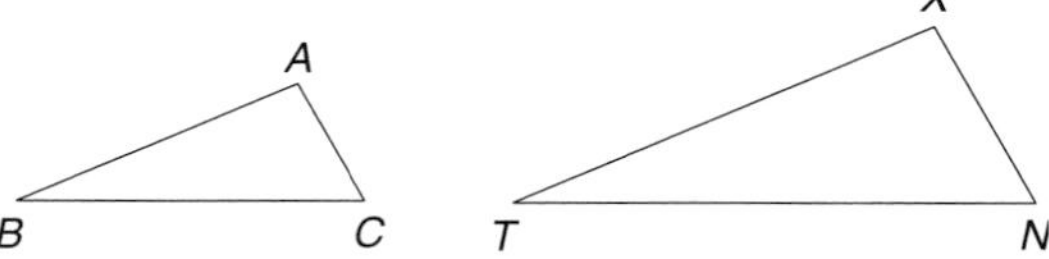


*Exercises 5, 6*

6. a) If $\triangle ABC \sim \triangle XTN$, which angle of $\triangle ABC$ corresponds to $\angle N$ of $\triangle XTN$?
   b) If $\triangle ABC \sim \triangle XTN$, which side of $\triangle XTN$ corresponds to side $\overline{AC}$ of $\triangle ABC$?

7. A **sphere** is the three-dimensional surface that contains all points in space lying at a fixed distance from a point known as the center of the sphere. Consider the two spheres shown. Are these two spheres similar? Are any two spheres similar? Explain.

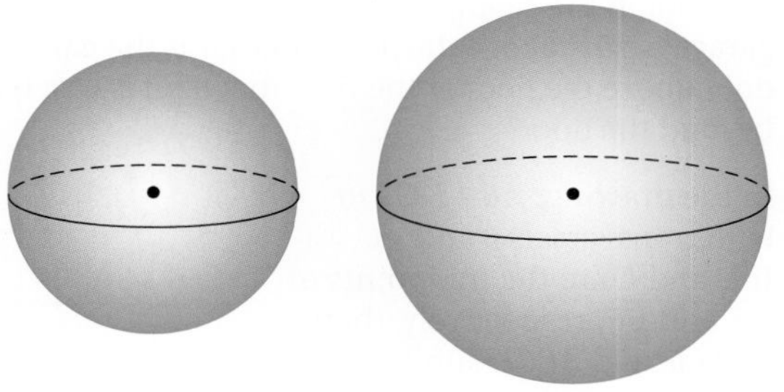

8. Given that rectangle $ABCE$ is similar to rectangle $MNPR$ and that $\triangle CDE \sim \triangle PQR$, what can you conclude regarding pentagon $ABCDE$ and pentagon $MNPQR$?

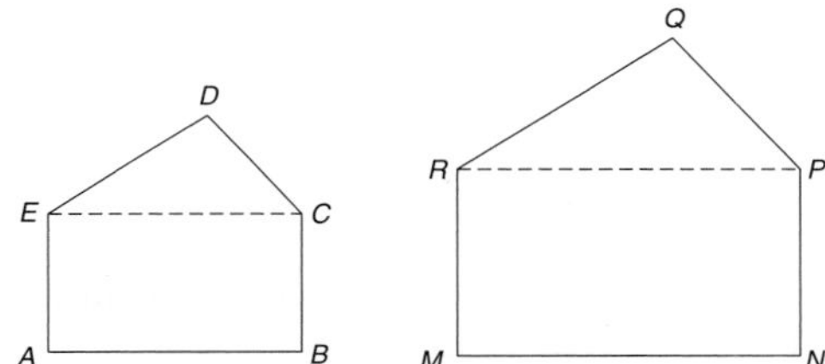


9. *Given:* $\triangle MNP \sim \triangle QRS$, $m\angle M = 56°$, $m\angle R = 82°$, $MN = 9$, $QR = 6$, $RS = 7$, $MP = 12$
   *Find:* a) $m\angle N$ c) $NP$
   b) $m\angle P$ d) $QS$

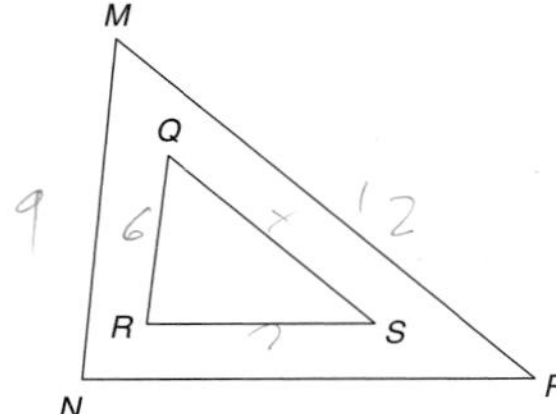


10. *Given:* $\triangle ABC \sim \triangle PRC$, $m\angle A = 67°$, $PC = 5$, $CR = 12$, $PR = 13$, $AB = 26$
    *Find:* a) $m\angle B$
    b) $m\angle RPC$
    c) $AC$
    d) $CB$

11. a) Does the similarity relationship have a **reflexive** property for triangles (and polygons in general)?
    b) Is there a **symmetric** property for the similarity of triangles (and polygons)?
    c) Is there a **transitive** property for the similarity of triangles (and polygons)?

12. Using the names of properties from Exercise 11, identify the property illustrated by each statement:
    a) If $\triangle 1 \sim \triangle 2$, then $\triangle 2 \sim \triangle 1$.
    b) If $\triangle 1 \sim \triangle 2$, $\triangle 2 \sim \triangle 3$, and $\triangle 3 \sim \triangle 4$, then $\triangle 1 \sim \triangle 4$.
    c) $\triangle 1 \sim \triangle 1$

13. In the drawing, $\triangle HJK \sim \triangle FGK$. If $HK = 6$, $KF = 8$, and $HJ = 4$, find $FG$.

14. In the drawing, $\triangle HJK \sim \triangle FGK$. If $HK = 6$, $KF = 8$, and $FG = 5$, find $HJ$.

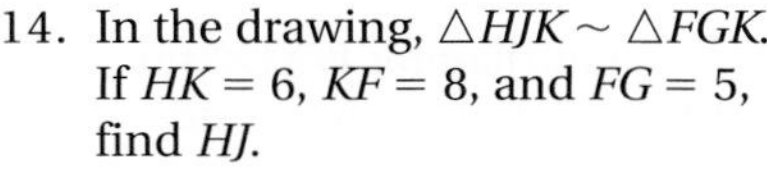

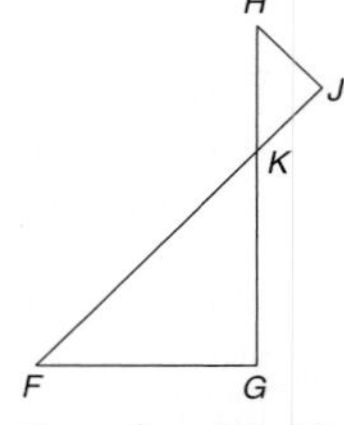

*Exercises 13, 14*

15. Quadrilateral $ABCD \sim$ quadrilateral $HJKL$. If $m\angle A = 55°$, $m\angle J = 128°$, and $m\angle D = 98°$, find $m\angle K$.

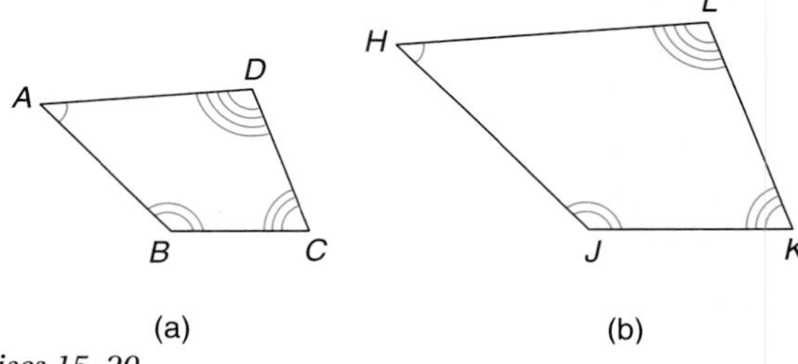

*Exercises 15–20*

16. Quadrilateral $ABCD \sim$ quadrilateral $HJKL$. If $m\angle A = x$, $m\angle J = x + 50$, $m\angle D = x + 35$, and $m\angle K = 2x - 45$, find $x$.

17. Quadrilateral $ABCD \sim$ quadrilateral $HJKL$. If $AB = 5$, $BC = n$, $HJ = 10$, and $JK = n + 3$, find $n$.

18. Quadrilateral $ABCD \sim$ quadrilateral $HJKL$. If $m\angle D = 90°$, $AD = 8$, $DC = 6$, and $HL = 12$, find the length of diagonal $\overline{HK}$ (not shown).

19. Quadrilateral $ABCD \sim$ quadrilateral $HJKL$. If $m\angle A = 2x + 4$, $m\angle H = 68°$, and $m\angle D = 3x - 6$, find $m\angle L$.

20. Quadrilateral $ABCD \sim$ quadrilateral $HJKL$. If $m\angle A = m\angle K = 70°$, and $m\angle B = 110°$, what types of quadrilaterals are $ABCD$ and $HJKL$?

*In Exercises 21 to 24, $\triangle ADE \sim \triangle ABC$.*

21. *Given:* $DE = 4$, $AE = 6$, $EC = BC$
*Find:* $BC$

22. *Given:* $DE = 5$, $AD = 8$, $DB = BC$
*Find:* $AB$
(HINT: Find $DB$ first.)

23. *Given:* $DE = 4$, $AC = 20$,
$EC = BC$
*Find:* $BC$

24. *Given:* $AD = 4$, $AC = 18$, $DB = AE$
*Find:* $AE$

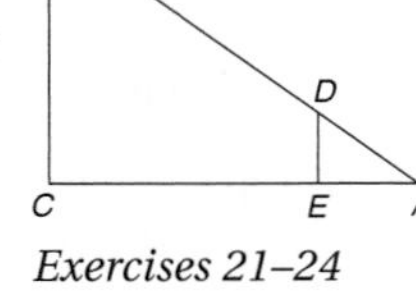

*Exercises 21–24*

25. Pentagon $ABCDE \sim$ pentagon $GHJKL$ (not shown), $AB = 6$, and $GH = 9$. If the perimeter of $ABCDE$ is 50, find the perimeter of $GHJKL$.

26. Quadrilateral $MNPQ \sim$ quadrilateral $WXYZ$ (not shown), $PQ = 5$, and $YZ = 7$. If the longest side of $MNPQ$ is of length 8, find the length of the longest side of $WXYZ$.

27. A blueprint represents the 72-ft length of a building by a line segment of length 6 in. What length on the blueprint would be used to represent the height of this 30-ft-tall building?

28. A technical drawing shows the $3\frac{1}{2}$-ft lengths of the legs of a baby's swing by line segments 3 in. long. If the diagram should indicate the legs are $2\frac{1}{2}$ ft apart at the base, what length represents this distance on the diagram?

*In Exercises 29 to 32, use the fact that triangles are similar.*

29. A person who is walking away from a 10-ft lamppost casts a shadow 6 ft long. If the person is at a distance of 10 ft from the lamppost at that moment, what is the person's height?

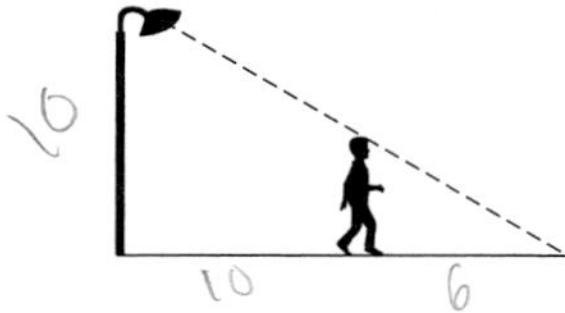

30. With 100 ft of string out, a kite is 64 ft above ground level. When the girl flying the kite pulls in 40 ft of string, the angle formed by the string and the ground does not change. What is the height of the kite above the ground after the 40 ft of string have been taken in?

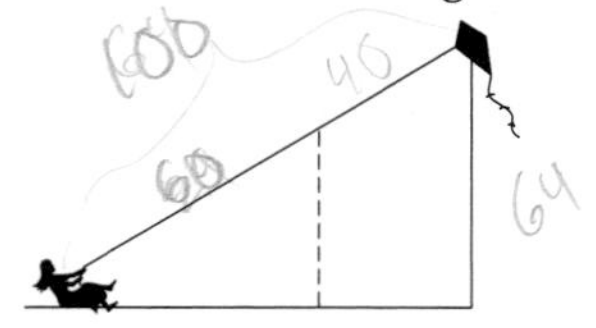

31. While admiring a rather tall tree, Fred notes that the shadow of his 6-ft frame has a length of 3 paces. On the level ground, he walks off the complete shadow of the tree in 37 paces. How tall is the tree?

32. As a garage door closes, light is cast 6 ft beyond the base of the door (as shown in the accompanying drawing) by a light fixture that is set in the garage ceiling 10 ft back from the door. If the ceiling of the garage is 10 ft above the floor, how far is the garage door above the floor at the time that light is cast 6 ft beyond the door?

33. In the drawing, $\overleftrightarrow{AB} \parallel \overleftrightarrow{DC} \parallel \overleftrightarrow{EF}$ with transversals $\ell$ and $m$. If $D$ and $C$ are the midpoints of $\overline{AE}$ and $\overline{BF}$, respectively, then is trapezoid $ABCD$ similar to trapezoid $DCFE$?

34. In the drawing, $\overleftrightarrow{AB} \parallel \overleftrightarrow{DC} \parallel \overleftrightarrow{EF}$. Suppose that transversals $\ell$ and $m$ are also parallel. $D$ and $C$ are the midpoints of $\overline{AE}$ and $\overline{BF}$, respectively. Is parallelogram $ABCD$ similar to parallelogram $DCFE$?

*Exercises 33, 34*

35. Given $\triangle ABC$, a second triangle ($\triangle XTN$) is constructed so that $\angle X \cong \angle A$ and $\angle N \cong \angle C$.

a) Is $\angle T$ congruent to $\angle B$?
b) Using intuition (appearance), does it seem that $\triangle XTN$ is similar to $\triangle ABC$?

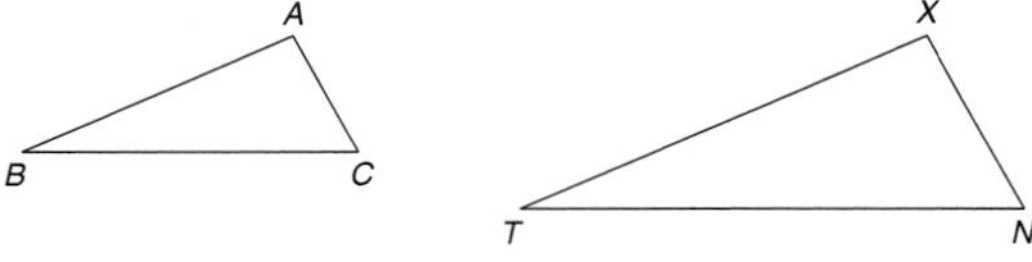


36. Given $\triangle RST$, a second triangle ($\triangle UVW$) is constructed so that $UV = 2(RS)$, $VW = 2(ST)$, and $WU = 2(RT)$.

a) What is the constant value of the ratios $\frac{UV}{RS}$, $\frac{VW}{ST}$, and $\frac{WU}{RT}$?
b) Using intuition (appearance), does it seem that $\triangle UVW$ is similar to $\triangle RST$?

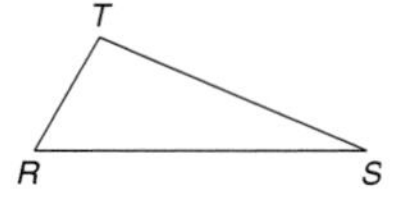


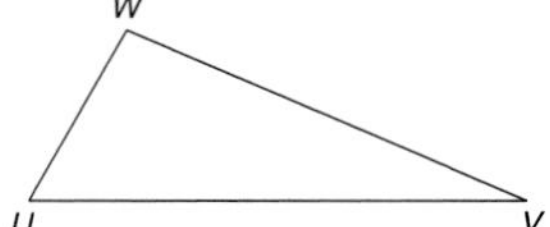

# 5.3 Proving Triangles Similar

**KEY CONCEPTS**

AAA • AA • CSSTP • CASTC • SAS~ • SSS~

Because of the difficulty of establishing proportional sides, our definition of similar polygons (and therefore of similar triangles) is almost impossible to use as a method of proof. Fortunately, some easier methods are available for proving triangles similar. If two triangles are carefully sketched or constructed so that their angles are congruent, they will appear to be similar, as shown in Figure 5.11.

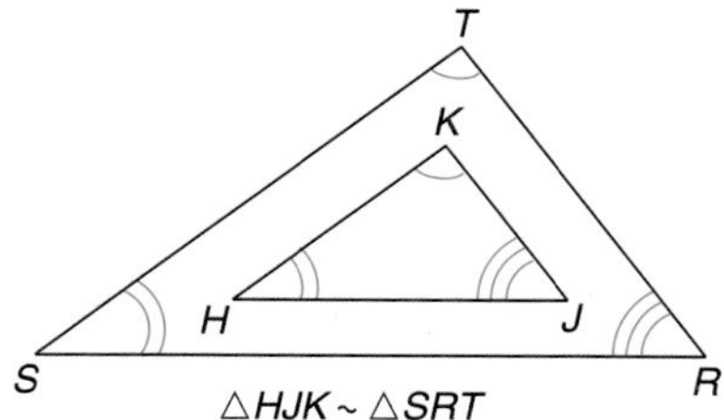


**FIGURE 5.11**

*Technology Exploration*

Use a calculator if available. On a sheet of paper, draw two similar triangles, $\triangle ABC$ and $\triangle DEF$. To accomplish this, use your protractor to form three pairs of congruent corresponding angles. Using a ruler, measure $\overline{AB}$, $\overline{BC}$, $\overline{AC}$, $\overline{DE}$, $\overline{EF}$, and $\overline{DF}$. Show that $\frac{AB}{DE} = \frac{BC}{EF} = \frac{AC}{DF}$.

NOTE: Answers are not "perfect."

POSTULATE 15:
If the three angles of one triangle are congruent to the three angles of a second triangle, then the triangles are similar (AAA).

Corollary 5.3.1 of Postulate 15 follows from knowing that if two angles of one triangle are congruent to two angles of another triangle, then the third angles *must* also be congruent. See Corollary 2.4.4.

COROLLARY 5.3.1: If two angles of one triangle are congruent to two angles of another triangle, then the triangles are similar (AA).

Rather than use AAA to prove triangles similar, we use AA because it requires fewer steps.

EXAMPLE 1

Provide a two-column proof of the following problem.

*Given:* $\overline{AB} \parallel \overline{DE}$ in Figure 5.12

*Prove:* $\triangle ABC \sim \triangle EDC$

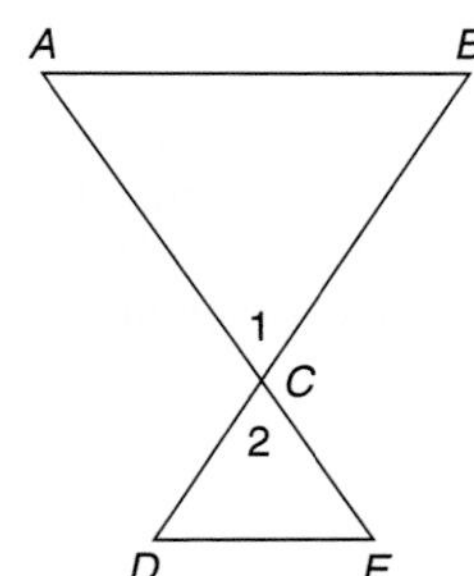


**FIGURE 5.12**

**PROOF**

| Statements | Reasons |
|---|---|
| 1. $\overline{AB} \parallel \overline{DE}$ | 1. Given |
| 2. $\angle A \cong \angle E$ | 2. If two $\parallel$ lines are cut by a transversal, the alternate interior angles are $\cong$ |
| 3. $\angle 1 \cong \angle 2$ | 3. Vertical angles are $\cong$ |
| 4. $\triangle ABC \sim \triangle EDC$ | 4. AA |

In some instances, we wish to prove some relationship beyond the similarity of triangles. This is possible through the definition of similarity. The following consequences of the definition of similarity are often cited as reasons in a proof. The first fact, abbreviated CSSTP, is used in Example 2. Although the statement involves triangles, we realize that the corresponding sides of any two similar polygons are proportional. That is, the ratio of any pair of corresponding sides equals the ratio of another pair of corresponding sides. The second fact, abbreviated CASTC, is used in Example 4.

CSSTP: Corresponding sides of similar triangles are proportional.

*Exs. 1–4*

CASTC: Corresponding angles of similar triangles are congruent.

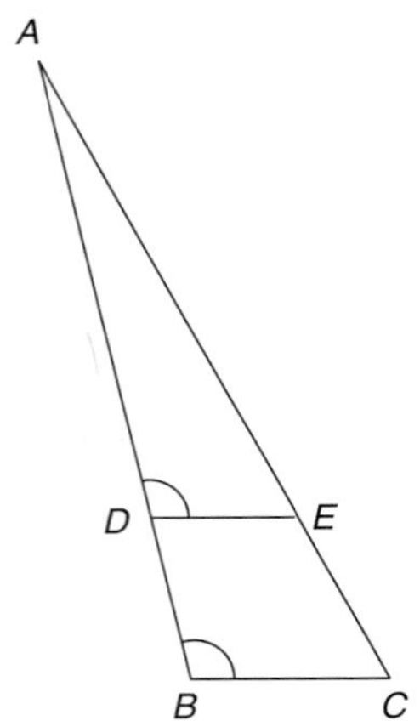


**FIGURE 5.13**

EXAMPLE 2

Complete the following two-column proof.

*Given:* $\angle ADE \cong \angle B$ in Figure 5.13

*Prove:* $\frac{DE}{BC} = \frac{AE}{AC}$

**PROOF**

| Statements | Reasons |
|---|---|
| 1. $\angle ADE \cong \angle B$ | 1. Given |
| 2. $\angle A \cong \angle A$ | 2. Identity |
| 3. $\triangle ADE \sim \triangle ABC$ | 3. AA |
| 4. $\frac{DE}{BC} = \frac{AE}{AC}$ | 4. CSSTP |

NOTE: In this proof, $DE$ appears above $BC$ because the sides with these names lie opposite $\angle A$ in the two similar triangles. $AE$ and $AC$ are the lengths of the sides opposite the congruent and corresponding angles $\angle ADE$ and $\angle B$. That is, corresponding sides of similar triangles always lie opposite corresponding angles.

THEOREM 5.3.2: The lengths of the corresponding altitudes of similar triangles have the same ratio as the lengths of any pair of corresponding sides.

The proof of this theorem is left to the student; see Exercise 33. Note that this proof also requires the use of CSSTP.

In Example 3, you are asked to prove that the product of lengths of two segments equals the product of lengths of two other segments. Here is a plan for establishing such a relationship:

1. Use AA to show that two triangles are similar, such as $\triangle ABC \sim \triangle DEF$.
2. Use CSSTP to form a proportion by choosing two ratios from

$$\frac{AB}{DE} = \frac{BC}{EF} = \frac{AC}{DF}$$

3. Use the Means-Extremes Property to obtain equal products, such as

$$AB \cdot EF = DE \cdot BC$$

when $\frac{AB}{DE} = \frac{BC}{EF}$ is the proportion chosen in step 2.

The paragraph style proof is generally used in upper-level mathematics classes. These paragraph proofs are no more than modified two-column proofs. Compare the following two-column proof, based upon Figure 5.14, to the paragraph proof found in Example 3.

*Given:* $\angle M \cong \angle Q$ in Figure 5.14

*Prove:* $NP \cdot QR = RP \cdot MN$

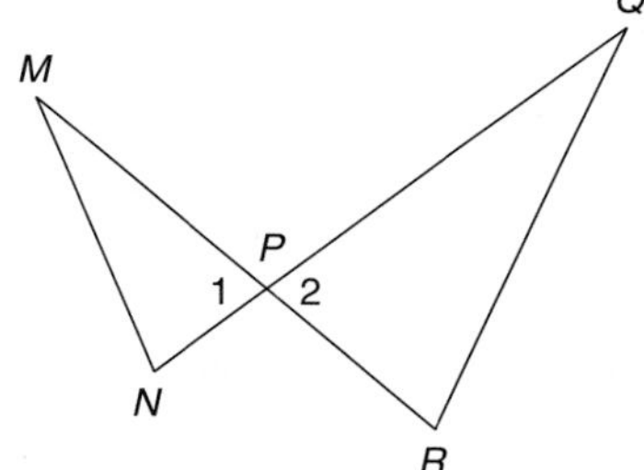


**FIGURE 5.14**

**PROOF**

| Statements | Reasons |
|---|---|
| 1. $\angle M \cong \angle Q$ | 1. Given (hypothesis) |
| 2. $\angle 1 \cong \angle 2$ | 2. Vertical angles are $\cong$ |
| 3. $\triangle MPN \sim \triangle QPR$ | 3. AA |
| 4. $\frac{NP}{RP} = \frac{MN}{QR}$ | 4. CSSTP |
| 5. $NP \cdot QR = RP \cdot MN$ | 5. Means-Extremes Property |

EXAMPLE 3

Use a paragraph proof to complete this problem.

*Given:* $\angle M \cong \angle Q$ in Figure 5.14

*Prove:* $NP \cdot QR = RP \cdot MN$

*Proof:* By hypothesis, $\angle M \cong \angle Q$. Also, $\angle 1 \cong \angle 2$ by the fact that vertical angles are congruent. Now $\triangle MPN \sim \triangle QPR$ by AA. Using CSSTP, $\frac{NP}{RP} = \frac{MN}{QR}$. Then $NP \cdot QR = RP \cdot MN$ by the Means-Extremes Property.

NOTE: In the proof, the sides selected for the proportion were carefully chosen. The statement to be proved suggested that we include *NP*, *QR*, *RP*, and *MN* in the proportion.

*Exs. 5–7*

In addition to AA, there are other methods that can be used to establish similar triangles. To distinguish the following techniques for showing triangles similar from methods for proving triangles congruent, we use SAS~ and SSS~ to identify these theorems. We prove SAS~ in Example 6 and prove SSS~ at our website.

> THEOREM 5.3.3 (SAS~): If an angle of one triangle is congruent to an angle of a second triangle and the pairs of sides including the angles are proportional, then the triangles are similar.

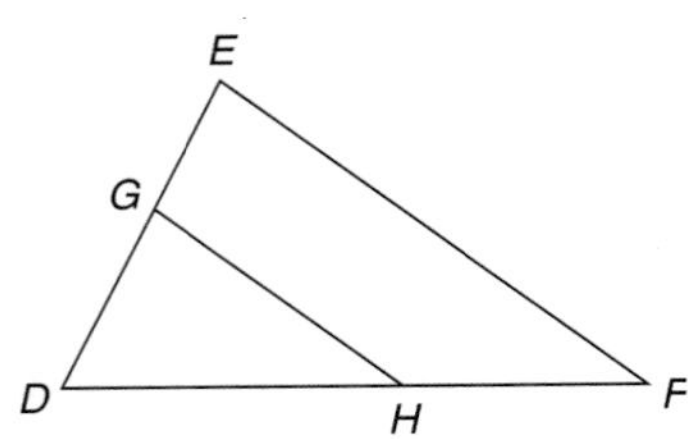


**FIGURE 5.15**

EXAMPLE 4

In Figure 5.15, $\frac{DG}{DE} = \frac{DH}{DF}$. Also, $m\angle E = x$, $m\angle D = x + 22$, and $m\angle DHG = x - 10$.

Find the value of $x$ and the measure of each angle.

***Solution***

With $\angle D \cong \angle D$ (Identity) and $\frac{DG}{DE} = \frac{DH}{DF}$ (Given), $\triangle DGH \sim \triangle DEF$ by SAS~. By CASTC, $\angle F \cong \angle DHG$, so $m\angle F = x - 10$. Thus, the sum of angles in $\triangle DEF$ is $x + x + 22 + x - 10 = 180$, or $3x + 12 = 180$. Then $3x = 168$ and $x = 56$. In turn, $m\angle E = \angle DGH = 56°$, $m\angle F = m\angle DHG = 46°$, and $m\angle D = 78°$.

Warning (!)
SSS and SAS prove that triangles are congruent. SSS~ and SAS~ prove that triangles are similar.

> THEOREM 5.3.4 (SSS~): If the three sides of one triangle are proportional to the three corresponding sides of a second triangle, then the triangles are similar.

EXAMPLE 5

Which method (AA, SAS~, or SSS~) establishes that $\triangle ABC \sim \triangle XTN$? See Figure 5.16.

a) $\angle A \cong \angle X$, $AC = 6$, $XN = 9$, $AB = 8$, and $XT = 12$
b) $AB = 6$, $AC = 4$, $BC = 8$, $XT = 9$, $XN = 6$, and $TN = 12$

***Solution***

a) SAS~; $\frac{AC}{XN} = \frac{AB}{XT}$ b) SSS~; $\frac{AB}{XT} = \frac{AC}{XN} = \frac{BC}{TN}$

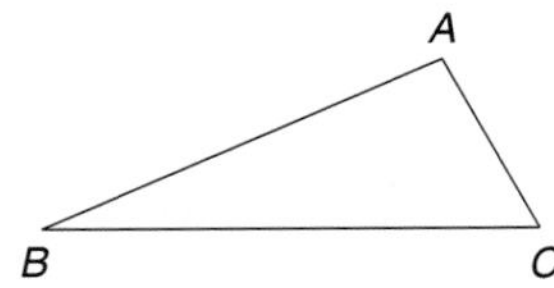


X
T
N

**FIGURE 5.16**

*Exs. 8–10*

We close this section by proving Theorem 5.3.3 (SAS~). In order to achieve this goal, we prove a helping theorem by the indirect method. In Figure 5.17, we say that sides $\overline{CA}$ and $\overline{CB}$ are divided proportionally by $\overline{DE}$ if $\frac{DA}{CD} = \frac{EB}{CE}$.

> LEMMA 5.3.5: If a line segment divides two sides of a triangle proportionally, then this line segment is parallel to the third side of the triangle.

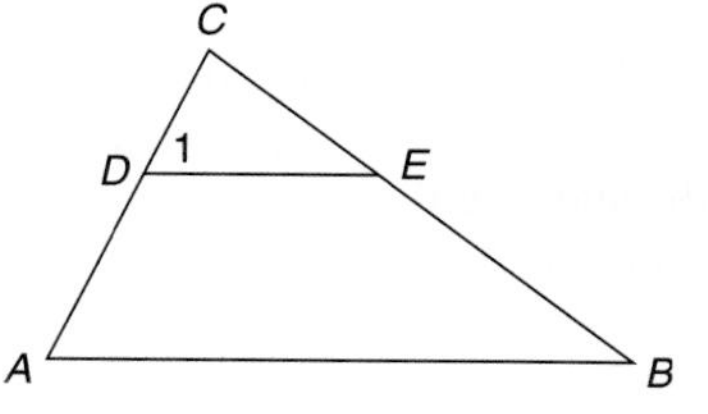


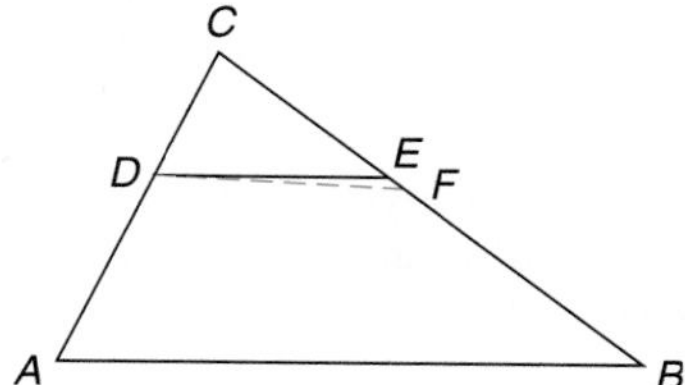


**FIGURE 5.17**

*Given:* $\triangle ABC$ with $\frac{DA}{CD} = \frac{EB}{CE}$

*Prove:* $\overline{DE} \parallel \overline{AB}$

*Proof:* $\frac{DA}{CD} = \frac{EB}{CE}$ in $\triangle ABC$. Applying Property 3 of Section 5.1, we have $\frac{CD + DA}{CD} = \frac{CE + EB}{CE}$, so $\frac{CA}{CD} = \frac{CB}{CE}(*)$.

Now suppose that $\overline{DE}$ is not parallel to $\overline{AB}$. Through $D$, we draw $\overline{DF} \parallel \overline{AB}$. It follows that $\angle CDF \cong \angle A$. With $\angle C \cong \angle C$, it follows that $\triangle CDF \sim \triangle CAB$ by the reason AA. By CSSTP, we have $\frac{CA}{CD} = \frac{CB}{CF}(**)$. Using the starred statements and substitution, we see that $\frac{CB}{CE} = \frac{CB}{CF}$ $\left(\text{both ratios are equal to } \frac{CA}{CD}\right)$.

By the Means-Extremes Property, $CB \cdot CF = CB \cdot CE$. Dividing each side of the last equation by $CB$, we find that $CF = CE$. That is, $F$ must coincide with $E$; it follows that $\overline{DE} \parallel \overline{AB}$.

In Example 6, we use Lemma 5.3.5 to prove the SAS~ theorem.

### EXAMPLE 6

*Given:* $\triangle ABC$ and $\triangle DEC$; $\frac{CA}{CD} = \frac{CB}{CE}$

*Prove:* $\triangle ABC \sim \triangle DEC$

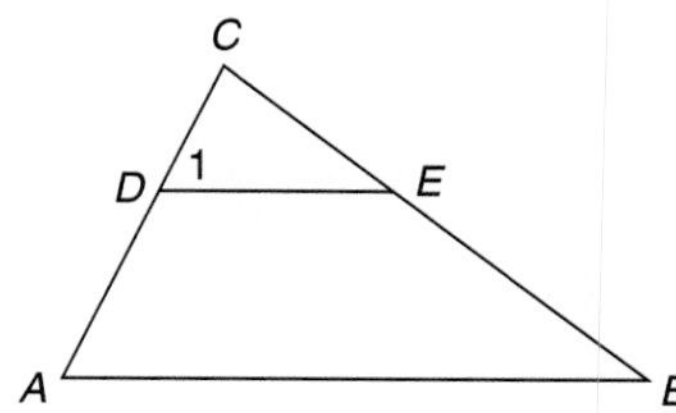


| Statements | Reasons |
|---|---|
| 1. $\triangle ABC$ and $\triangle DEC$; $\frac{CA}{CD} = \frac{CB}{CE}$ | 1. Given |
| 2. $\frac{CA - CD}{CD} = \frac{CB - CE}{CE}$ | 2. Property 3 of Section 5.1 |
| 3. $\frac{DA}{CD} = \frac{EB}{CE}$ | 3. Substitution |
| 4. $\therefore \overline{DE} \parallel \overline{AB}$ | 4. Lemma 5.3.5 |
| 5. $\angle 1 \cong \angle A$ | 5. If 2 ∥ lines are cut by a trans., corr. ∠s are ≅ |
| 6. $\angle C \cong \angle C$ | 6. Identity |
| 7. $\triangle ABC \sim \triangle DEC$ | 7. AA |

*Exs. 11, 12*

## 5.3 Exercises

1. What is the acronym used to represent the statement "Corresponding angles of similar triangles are congruent?"

2. What is the acronym used to represent the statement "Corresponding sides of similar triangles are proportional?"

3. Classify as true or false:

   a) If the vertex angles of two isosceles triangles are congruent, the triangles are similar.
   b) Any two equilateral triangles are similar.

4. Classify as true or false:

   a) If the midpoints of two sides of a triangle are joined, the triangle formed is similar to the original triangle.
   b) Any two isosceles triangles are similar.

*In Exercises 5 to 8, name the method (AA, SSS~, or SAS~) that is used to show that the triangles are similar.*

5. $WU = \frac{3}{2} \cdot TR$, $WV = \frac{3}{2} \cdot TS$, and $UV = \frac{3}{2} \cdot RS$

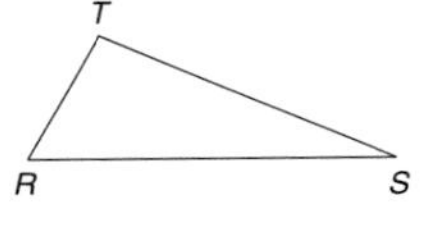


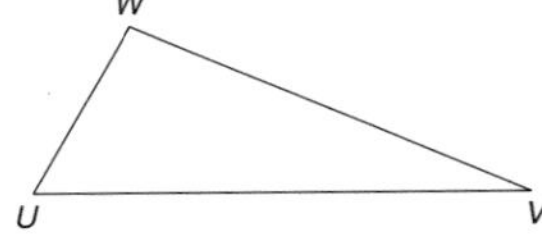


*Exercises 5–8*

6. $\angle T \cong \angle W$ and $\angle R \cong \angle U$

7. $\angle T \cong \angle W$ and $\frac{TR}{WU} = \frac{TS}{WV}$

8. $\frac{TR}{WU} = \frac{TS}{WV} = \frac{RS}{UV}$

*In Exercises 9 and 10, name the method that explains why $\triangle DGH \sim \triangle DEF$.*

9. $\frac{DG}{DE} = \frac{DH}{DF}$

10. $DE = 3 \cdot DG$ and $DF = 3 \cdot DH$

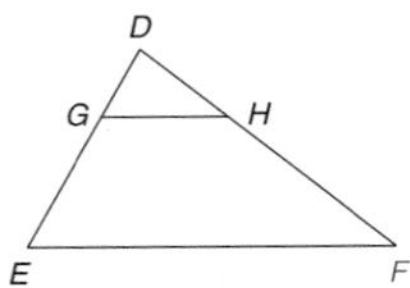


*Exercises 9, 10*

*In Exercises 11 to 14, provide the missing reasons.*

11. *Given:* $\square RSTV$; $\overline{VW} \perp \overline{RS}$; $\overline{VX} \perp \overline{TS}$
    *Prove:* $\triangle VWR \sim \triangle VXT$

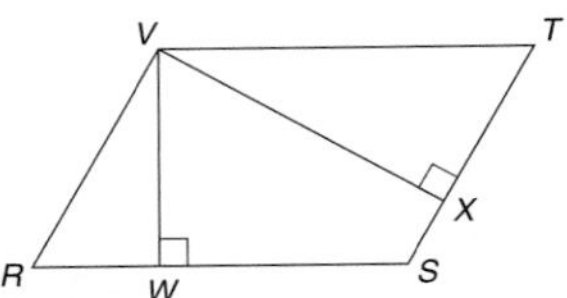


**PROOF**

| Statements | Reasons |
|---|---|
| 1. $\square RSTV$; $\overline{VW} \perp \overline{RS}$; $\overline{VX} \perp \overline{TS}$ | 1. ? |
| 2. $\angle VWR$ and $\angle VXT$ are rt. $\angle$s | 2. ? |
| 3. $\angle VWR \cong \angle VXT$ | 3. ? |
| 4. $\angle R \cong \angle T$ | 4. ? |
| 5. $\triangle VWR \sim \triangle VXT$ | 5. ? |

12. *Given:* $\triangle DET$ and $\square ABCD$
    *Prove:* $\triangle ABE \sim \triangle CTB$

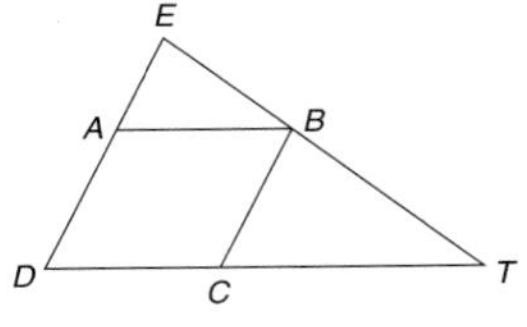


**PROOF**

| Statements | Reasons |
|---|---|
| 1. $\triangle DET$ and $\square ABCD$ | 1. ? |
| 2. $\overline{AB} \parallel \overline{DT}$ | 2. Opposite sides of a $\square$ are $\parallel$ |
| 3. $\angle EBA \cong \angle T$ | 3. ? |
| 4. $\overline{ED} \parallel \overline{CB}$ | 4. ? |
| 5. $\angle E \cong \angle CBT$ | 5. ? |
| 6. $\triangle ABE \sim \triangle CTB$ | 6. ? |

13. *Given:* $\triangle ABC$; $M$ and $N$ are midpoints of $\overline{AB}$ and $\overline{AC}$, respectively
*Prove:* $\triangle AMN \sim \triangle ABC$

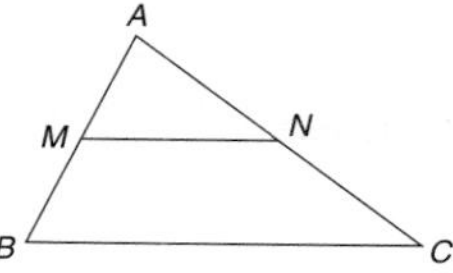


**PROOF**

| Statements | Reasons |
|---|---|
| 1. $\triangle ABC$; $M$ and $N$ are the midpoints of $\overline{AB}$ and $\overline{AC}$, respectively | 1. ? |
| 2. $AM = \frac{1}{2}(AB)$ and $AN = \frac{1}{2}(AC)$ | 2. ? |
| 3. $MN = \frac{1}{2}(BC)$ | 3. ? |
| 4. $\frac{AM}{AB} = \frac{1}{2}$, $\frac{AN}{AC} = \frac{1}{2}$, and $\frac{MN}{BC} = \frac{1}{2}$ | 4. ? |
| 5. $\frac{AM}{AB} = \frac{AN}{AC} = \frac{MN}{BC}$ | 5. ? |
| 6. $\triangle AMN \sim \triangle ABC$ | 6. ? |

14. *Given:* $\triangle XYZ$ with $\overline{XY}$ trisected at $P$ and $Q$ and $\overline{YZ}$ trisected at $R$ and $S$
*Prove:* $\triangle XYZ \sim \triangle PYR$

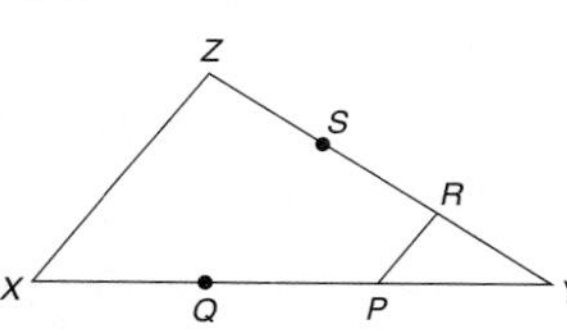


**PROOF**

| Statements | Reasons |
|---|---|
| 1. $\triangle XYZ$; $\overline{XY}$ trisected at $P$ and $Q$; $\overline{YZ}$ trisected at $R$ and $S$ | 1. ? |
| 2. $\frac{YR}{YZ} = \frac{1}{3}$ and $\frac{YP}{YX} = \frac{1}{3}$ | 2. Definition of trisect |
| 3. $\frac{YR}{YZ} = \frac{YP}{YX}$ | 3. ? |
| 4. $\angle Y \cong \angle Y$ | 4. ? |
| 5. $\triangle XYZ \sim \triangle PYR$ | 5. ? |

*In Exercises 15 to 22, complete each proof.*

15. *Given:* $\overline{MN} \perp \overline{NP}$, $\overline{QR} \perp \overline{RP}$
*Prove:* $\triangle MNP \sim \triangle QRP$

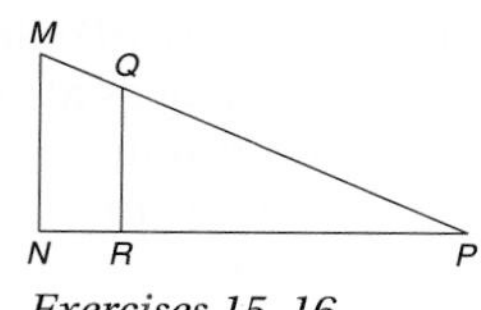

*Exercises 15, 16*

**PROOF**

| Statements | Reasons |
|---|---|
| 1. ? | 1. Given |
| 2. $\angle$s $N$ and $QRP$ are right $\angle$s | 2. ? |
| 3. ? | 3. All right $\angle$s are $\cong$ |
| 4. $\angle P \cong \angle P$ | 4. ? |
| 5. ? | 5. ? |

16. *Given:* $\overline{MN} \parallel \overline{QR}$ (See figure for Exercise 15.)
*Prove:* $\triangle MNP \sim \triangle QRP$

**PROOF**

| Statements | Reasons |
|---|---|
| 1. ? | 1. Given |
| 2. $\angle M \cong \angle RQP$ | 2. ? |
| 3. ? | 3. If two $\parallel$ lines are cut by a transversal, the corresponding $\angle$s are $\cong$ |
| 4. ? | 4. ? |

17. *Given:* $\angle H \cong \angle F$
*Prove:* $\triangle HJK \sim \triangle FGK$

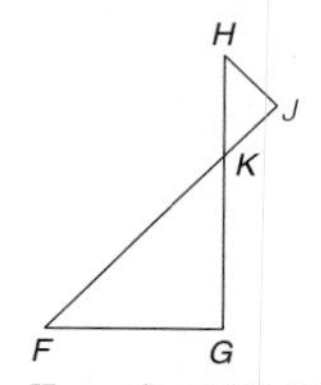

*Exercises 17, 18*

**PROOF**

| Statements | Reasons |
|---|---|
| 1. ? | 1. Given |
| 2. $\angle HKJ \cong \angle FKG$ | 2. ? |
| 3. ? | 3. ? |

18. *Given:* $\overline{HJ} \perp \overline{JF}$, $\overline{HG} \perp \overline{FG}$ (See figure for Exercise 17.)
*Prove:* $\triangle HJK \sim \triangle FGK$

**PROOF**

| Statements | Reasons |
|---|---|
| 1. ? | 1. Given |
| 2. $\angle$s $G$ and $J$ are right $\angle$s | 2. ? |
| 3. $\angle G \cong \angle J$ | 3. ? |
| 4. $\angle HKJ \cong \angle GKF$ | 4. ? |
| 5. ? | 5. ? |

19. *Given:* $\frac{RQ}{NM} = \frac{RS}{NP} = \frac{QS}{MP}$
*Prove:* $\angle N \cong \angle R$

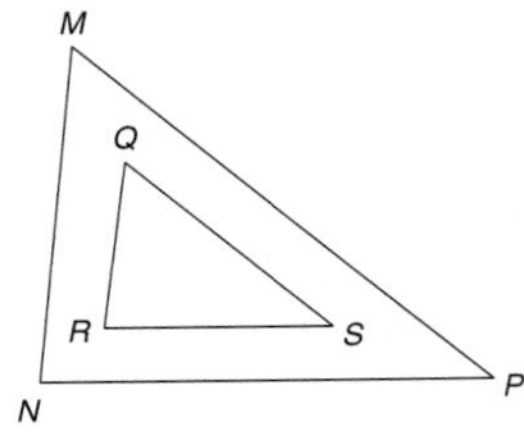


**PROOF**

| Statements | Reasons |
|---|---|
| 1. ? | 1. Given |
| 2. ? | 2. SSS~ |
| 3. ? | 3. CASTC |

20. *Given:* $\frac{DG}{DE} = \frac{DH}{DF}$
*Prove:* $\angle DGH \cong \angle E$

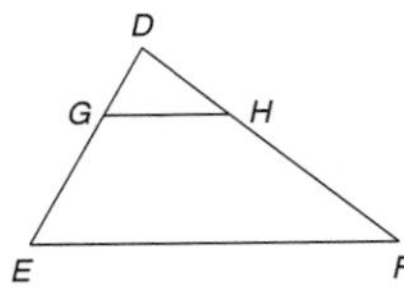


**PROOF**

| Statements | Reasons |
|---|---|
| 1. ? | 1. ? |
| 2. $\angle D \cong \angle D$ | 2. ? |
| 3. $\triangle DGH \sim \triangle DEF$ | 3. ? |
| 4. ? | 4. ? |

21. *Given:* $\overline{RS} \parallel \overline{UV}$
*Prove:* $\frac{RT}{VT} = \frac{RS}{VU}$

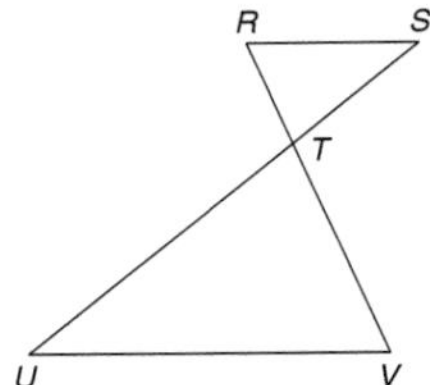


**PROOF**

| Statements | Reasons |
|---|---|
| 1. ? | 1. ? |
| 2. $\angle R \cong \angle V$ and $\angle S \cong \angle U$ | 2. ? |
| 3. ? | 3. AA |
| 4. ? | 4. ? |

22. *Given:* $\overline{AB} \parallel \overline{DC}$, $\overline{AC} \parallel \overline{DE}$
*Prove:* $\frac{AB}{DC} = \frac{BC}{CE}$

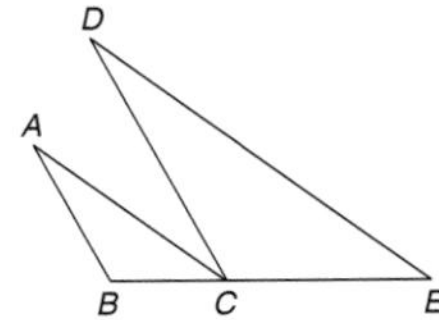


**PROOF**

| Statements | Reasons |
|---|---|
| 1. $\overline{AB} \parallel \overline{DC}$ | 1. ? |
| 2. ? | 2. If 2 ∥ lines are cut by a trans. corr. ∠s are ≅ |
| 3. ? | 3. Given |
| 4. $\angle ACB \cong \angle E$ | 4. ? |
| 5. $\triangle ACB \sim \triangle DEC$ | 5. ? |
| 6. ? | 6. ? |

*In Exercises 23 to 26, $\triangle ABC \sim \triangle DBE$.*

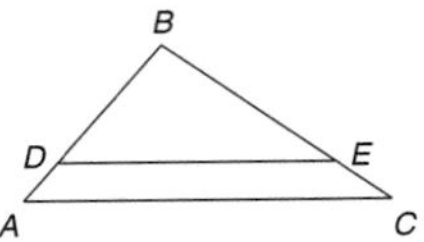


*Exercises 23–26*

23. *Given:* $AC = 8$, $DE = 6$, $CB = 6$
*Find:* $EB$
(HINT: Let $EB = x$, and solve an equation.)

24. *Given:* $AC = 10$, $CB = 12$
$E$ is the midpoint of $\overline{CB}$
*Find:* $DE$

25. *Given:* $AC = 10$, $DE = 8$, $AD = 4$
*Find:* $DB$

26. *Given:* $CB = 12$, $CE = 4$, $AD = 5$
*Find:* $DB$

27. $\triangle CDE \sim \triangle CBA$ with $\angle CDE \cong \angle B$. If $CD = 10$, $DA = 8$, and $CE = 6$, find $EB$.

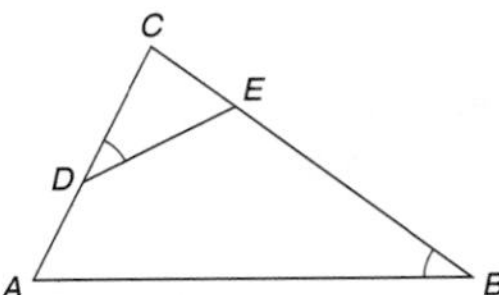


*Exercises 27, 28*

28. $\triangle CDE \sim \triangle CBA$ with $\angle CDE \cong \angle B$. If $CD = 10$, $CA = 16$, and $EB = 12$, find $CE$.

29. $\triangle ABF \sim \triangle CBD$ with obtuse angles at vertices $D$ and $F$ as indicated. If $m\angle B = 45°$, $m\angle C = x$ and $m\angle AFB = 4x$, find $x$.

30. $\triangle ABF \sim \triangle CBD$ with obtuse angles at vertices $D$ and $F$. If $m\angle B = 44°$ and $m\angle A : m\angle CDB = 1{:}3$, find $m\angle A$.

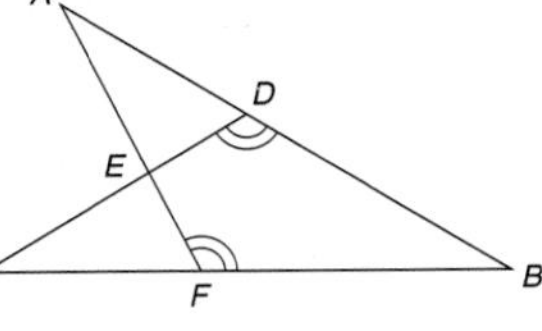


*Exercises 29, 30*

*In Exercise 31, provide a two-column proof.*

31. *Given:* $\overline{AB} \parallel \overline{DF}$, $\overline{BD} \parallel \overline{FG}$
*Prove:* $\triangle ABC \sim \triangle EFG$

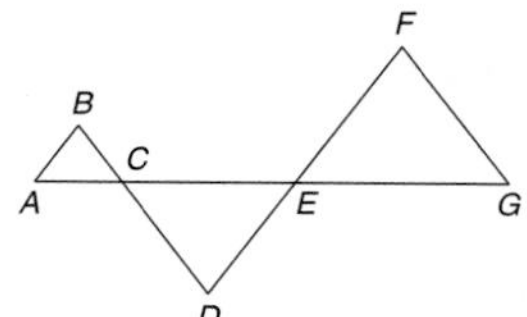


*In Exercise 32, provide a paragraph proof.*

32. *Given:* $\overline{RS} \perp \overline{AB}$, $\overline{CB} \perp \overline{AC}$
*Prove:* $\triangle BSR \sim \triangle BCA$

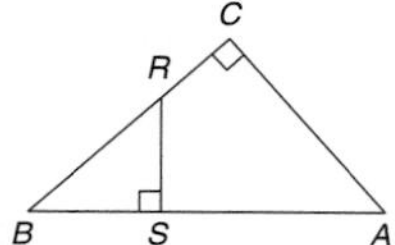


33. Use a two-column proof to prove the following theorem: "The lengths of the corresponding altitudes of similar triangles have the same ratio as the lengths of any pair of corresponding sides."
*Given:* $\triangle DEF \sim \triangle MNP$; $\overline{DG}$ and $\overline{MQ}$ are altitudes
*Prove:* $\dfrac{DG}{MQ} = \dfrac{DE}{MN}$

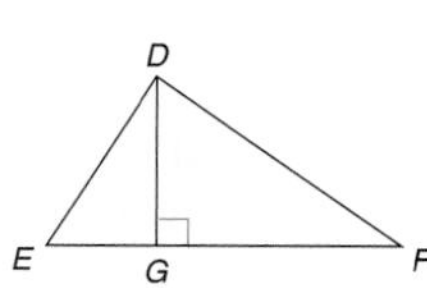


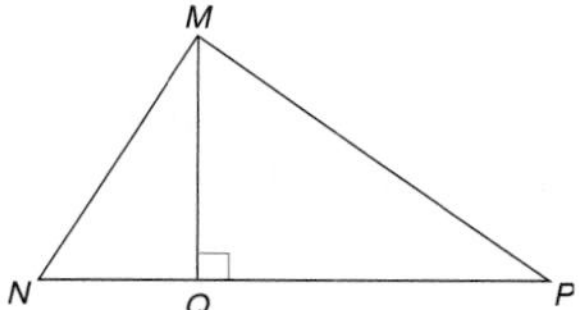


34. Provide a paragraph proof for the following problem.
*Given:* $\overline{RS} \parallel \overline{YZ}$, $\overline{RU} \parallel \overline{XZ}$
*Prove:* $RS \cdot ZX = ZY \cdot RT$

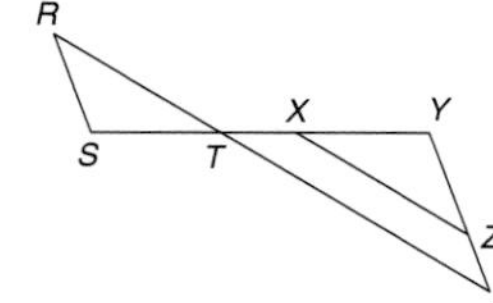


35. Use the result of Exercise 11 to do the following problem. In ▱$MNPQ$, $QP = 12$ and $QM = 9$. The length of altitude $\overline{QR}$ (to side $\overline{MN}$) is 6. Find the length of altitude $\overline{QS}$ from $Q$ to $\overline{PN}$.

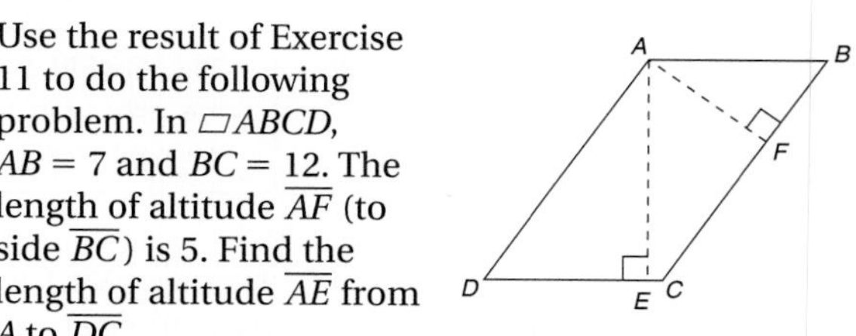


36. Use the result of Exercise 11 to do the following problem. In ▱$ABCD$, $AB = 7$ and $BC = 12$. The length of altitude $\overline{AF}$ (to side $\overline{BC}$) is 5. Find the length of altitude $\overline{AE}$ from $A$ to $\overline{DC}$.

37. The distance across a pond is to be measured indirectly by using similar triangles. If $XY = 160$ ft, $YW = 40$ ft, $TY = 120$ ft, and $WZ = 50$ ft, find $XT$.

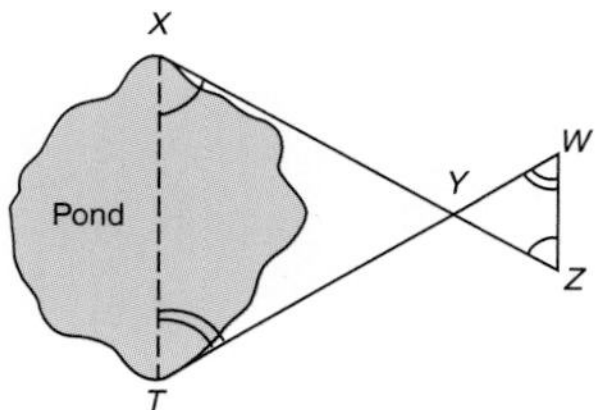


38. In the figure, $\angle ABC \cong \angle ADB$. Find $AB$ if $AD = 2$ and $DC = 6$.

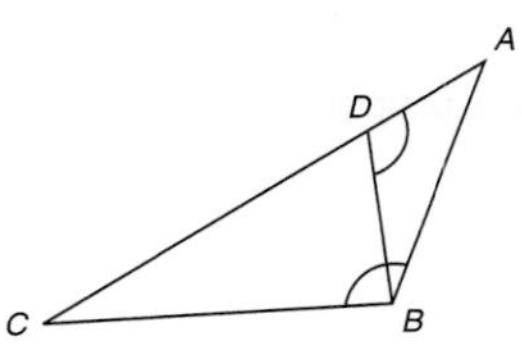


39. Prove that the altitude drawn to the hypotenuse of a right triangle separates the right triangle into two right triangles that are similar to each other and to the original right triangle.

40. Prove that the line segment joining the midpoints of two sides of a triangle determines a triangle that is similar to the original triangle.

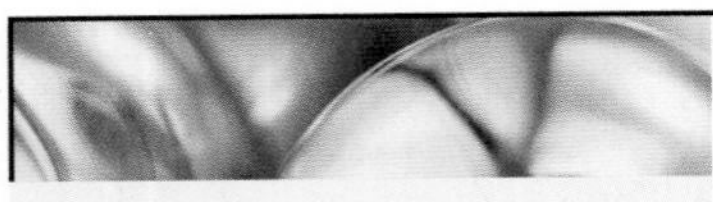

# 5.4 The Pythagorean Theorem

**KEY CONCEPTS**

Pythagorean Theorem • Converse of Pythagorean Theorem • Pythagorean Triple

The following theorem, which was proved in Exercise 39 of Section 5.3, will enable us to prove the well-known Pythagorean Theorem.

> THEOREM 5.4.1: The altitude drawn to the hypotenuse of a right triangle separates the right triangle into two right triangles that are similar to each other and to the original right triangle.

Theorem 5.4.1 is illustrated by Figure 5.18, in which the right triangle $\triangle ABC$ has its right angle at vertex $C$ so that $\overline{CD}$ is the altitude to hypotenuse $\overline{AB}$. The smaller triangles are shown in Figures 5.18(b) and (c), and the original triangle is shown in Figure 5.18(d). Note the matched arcs indicating congruent angles.

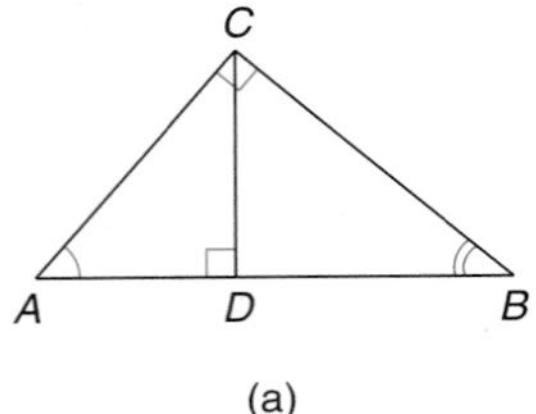


(a)

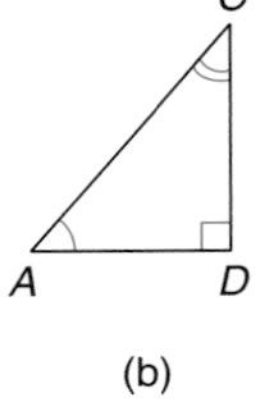


(b)

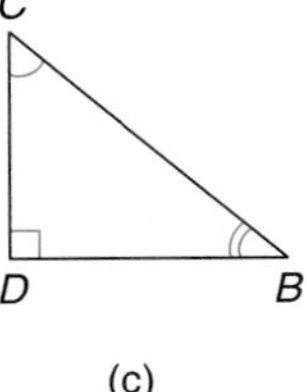


(c)

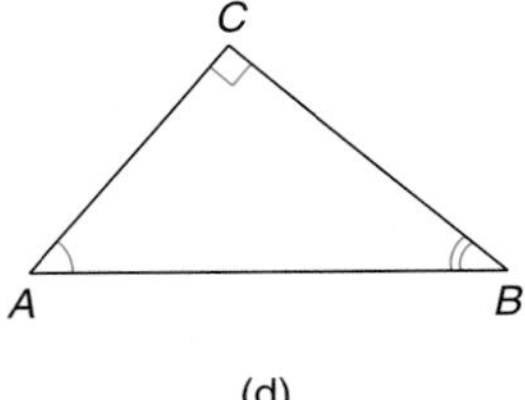


(d)

**FIGURE 5.18**

Reminder

CSSTP means "corresponding sides of similar triangles are proportional."

In Figure 5.18(a), $\overline{AD}$ and $\overline{DB}$ are known as *segments* (parts) of the hypotenuse $\overline{AB}$. Furthermore, $\overline{AD}$ is the segment of the hypotenuse *adjacent* to (next to) leg $\overline{AC}$, and $\overline{BD}$ is the segment of the hypotenuse *adjacent* to leg $\overline{BC}$. Proof of the following theorem is left as an exercise.

> THEOREM 5.4.2: The length of the altitude to the hypotenuse of a right triangle is the geometric mean of the lengths of the segments of the hypotenuse.

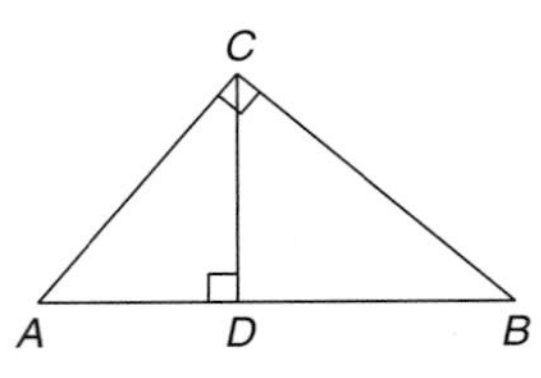


**FIGURE 5.19**

*Given:* $\triangle ABC$ in Figure 5.19, with right $\angle ACB$, $\overline{CD} \perp \overline{AB}$

*Prove:* $\frac{AD}{CD} = \frac{CD}{DB}$

*Plan for Proof:* Show that $\triangle ADC \sim \triangle CDB$. Then use CSSTP.

In the proportion $\frac{AD}{CD} = \frac{CD}{DB}$, recall that $CD$ is a geometric mean because the second and the third terms are identical.

The proof of the following lemma is left as an exercise.

C
A
D
B

**FIGURE 5.20**

> LEMMA 5.4.3: The length of each leg of a right triangle is the geometric mean of the length of the hypotenuse and the length of the segment of the hypotenuse adjacent to that leg.

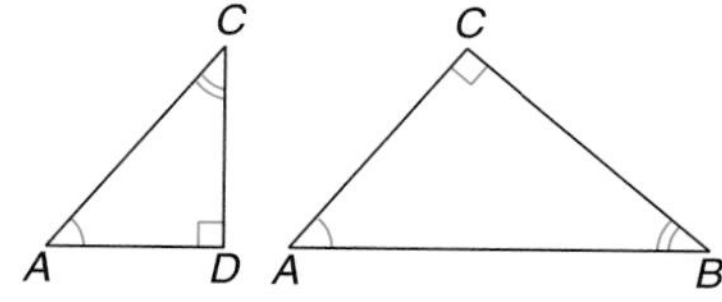


**FIGURE 5.21**

*Given:* $\triangle ABC$ with right $\angle ACB$; $\overline{CD} \perp \overline{AB}$ (See Figure 5.20.)

*Prove:* $\dfrac{AB}{AC} = \dfrac{AC}{AD}$

*Plan:* Show that $\triangle ADC \sim \triangle ACB$. See Figure 5.21. Then use CSSTP.

NOTE: Although $\overline{AD}$ and $\overline{DB}$ are both segments of the hypotenuse, $\overline{AD}$ is the segment adjacent to $\overline{AC}$.

*Exs. 1, 2*

Lemma 5.4.3 opens the doors to a proof of the famous Pythagorean Theorem, one of the most frequently applied relationships in geometry. Although the theorem's title gives credit to the Greek geometer Pythagoras, many other proofs are known, and the ancient Chinese were aware of the relationship before the time of Pythagoras.

THEOREM 5.4.4: **(Pythagorean Theorem)** The square of the length of the hypotenuse of a right triangle is equal to the sum of the squares of the lengths of the legs.

Thus, where $c$ is the length of the hypotenuse and $a$ and $b$ are the lengths of the legs, $c^2 = a^2 + b^2$.

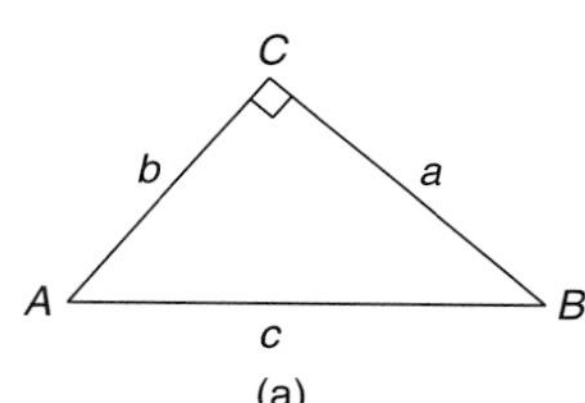


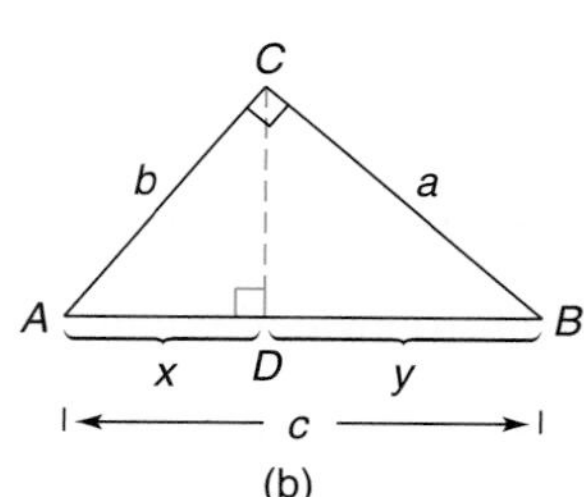


**FIGURE 5.22**

*Given:* In Figure 5.22(a), $\triangle ABC$ with right $\angle C$

*Prove:* $c^2 = a^2 + b^2$

*Proof:* Draw $\overline{CD} \perp \overline{AB}$, as shown in Figure 5.22(b). Denote $AD = x$ and $DB = y$. By Lemma 5.4.3,

$$\frac{c}{b} = \frac{b}{x} \quad \text{and} \quad \frac{c}{a} = \frac{a}{y}$$

Therefore, $\quad b^2 = cx \quad$ and $\quad a^2 = cy$

Using the Addition Property of Equality, we have

$$a^2 + b^2 = cy + cx = c(y + x)$$

But $y + x = x + y = AD + DB = AB = c$. Thus

$$a^2 + b^2 = c(c) = c^2$$

EXAMPLE 1

Given $\triangle RST$ with right $\angle S$ in Figure 5.23, find:

a) $RT$ if $RS = 3$ and $ST = 4$
b) $RT$ if $RS = 4$ and $ST = 6$
c) $RS$ if $RT = 13$ and $ST = 12$
d) $ST$ if $RS = 6$ and $RT = 9$

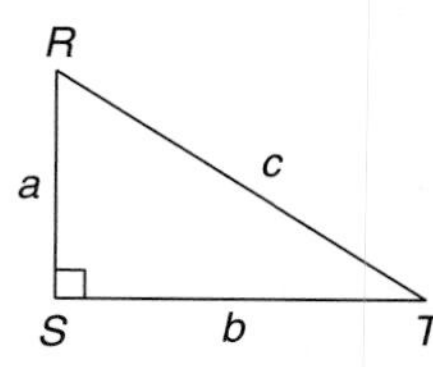


**FIGURE 5.23**

***Solution***

With right $\angle S$, the hypotenuse is $\overline{RT}$. Then $RT = c$, $RS = a$, and $ST = b$.

a) $3^2 + 4^2 = c^2 \rightarrow 9 + 16 = c^2$
$c^2 = 25$
$c = 5$; $RT = 5$

b) $4^2 + 6^2 = c^2 \rightarrow 16 + 36 = c^2$
$c^2 = 52$
$c = \sqrt{52} = \sqrt{4 \cdot 13} = \sqrt{4} \cdot \sqrt{13} = 2\sqrt{13}$
$RT = 2\sqrt{13} \approx 7.21$

c) $a^2 + 12^2 = 13^2 \rightarrow a^2 + 144 = 169$
$a^2 = 25$
$a = 5$; $RS = 5$

**Discover!**

A video titled "The Rule of Pythagoras" is available through Project Mathematics at Cal Tech University in Pasadena, CA. It is well worth watching!

d) $6^2 + b^2 = 9^2 \rightarrow 36 + b^2 = 81$

$$b^2 = 45$$
$$b = \sqrt{45} = \sqrt{9 \cdot 5} = \sqrt{9} \cdot \sqrt{5} = 3\sqrt{5}$$
$$ST = 3\sqrt{5} \approx 6.71$$

*Exs. 3, 4*

The converse of the Pythagorean Theorem is also true.

> THEOREM 5.4.5: **(Converse of Pythagorean Theorem)** If $a$, $b$, and $c$ are the lengths of the three sides of a triangle, with $c$ the length of the longest side, and if $c^2 = a^2 + b^2$, then the triangle is a right triangle with the right angle opposite the side of length $c$.

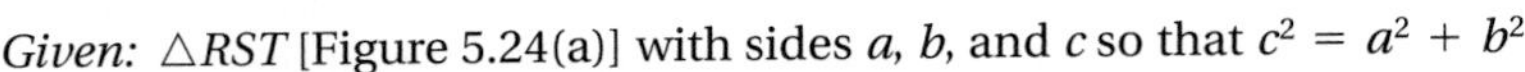

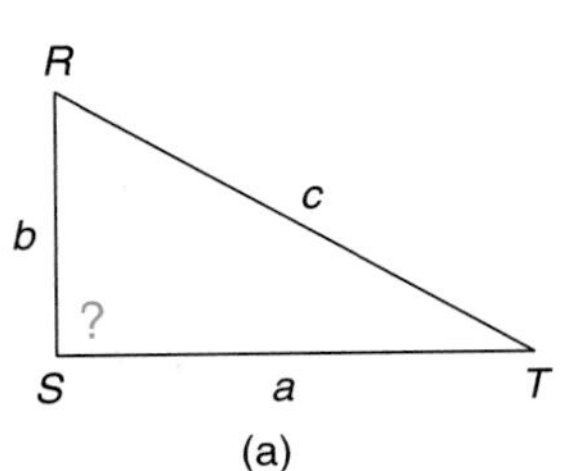


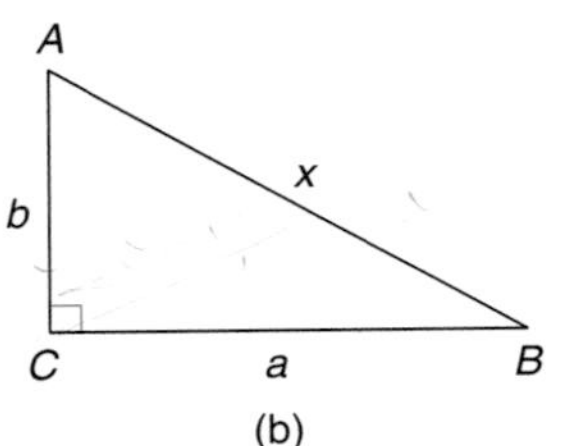


**FIGURE 5.24**

*Given:* $\triangle RST$ [Figure 5.24(a)] with sides $a$, $b$, and $c$ so that $c^2 = a^2 + b^2$

*Prove:* $\triangle RST$ is a right triangle.

*Proof:* We are given $\triangle RST$ for which $c^2 = a^2 + b^2$. Construct the right $\triangle ABC$, which has legs of lengths $a$ and $b$ and a hypotenuse of length $x$. [See Figure 5.24(b).] By the Pythagorean Theorem, $x^2 = a^2 + b^2$. By substitution, $x^2 = c^2$ and $x = c$. Thus $\triangle RTS \cong \triangle ABC$ by SSS. Then $\angle S$, opposite the side of length $c$, must be $\cong$ to $\angle C$, the right $\angle$ of $\triangle ABC$. Then $\angle S$ is a right $\angle$, and $\triangle RST$ is a right triangle.

## EXAMPLE 2

Which of the following can be the lengths of the sides of a right triangle?

a) $a = 5$, $b = 12$, $c = 13$
b) $a = 15$, $b = 8$, $c = 17$
c) $a = 7$, $b = 9$, $c = 10$
d) $a = \sqrt{2}$, $b = \sqrt{3}$, $c = \sqrt{5}$

***Solution***

a) Because $5^2 + 12^2 = 13^2$ (that is, $25 + 144 = 169$), this triangle is a right triangle.
b) Because $15^2 + 8^2 = 17^2$ (that is, $225 + 64 = 289$), this triangle is a right triangle.
c) $7^2 + 9^2 = 49 + 81 = 130$, which is not $10^2$ (that is, 100), so this triangle is not a right triangle.
d) Because $(\sqrt{2})^2 + (\sqrt{3})^2 = (\sqrt{5})^2$ leads to $2 + 3 = 5$, this triangle is a right triangle.

*Exs. 5, 6*

## EXAMPLE 3

A ladder 12 ft long is leaning against a wall so that its base is 4 ft from the wall at ground level (see Figure 5.25). How far up the wall does the ladder reach?

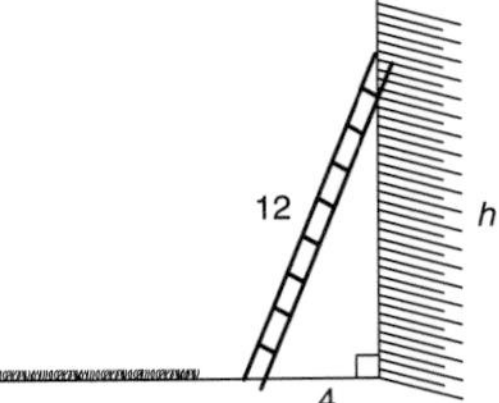


**FIGURE 5.25**

> **Discover!**
>
> Construct a triangle with sides of lengths 3 inches, 4 inches, and 5 inches. Measure the angles of the triangle. Is there a right angle?
>
> ANSWER
>
> Yes, opposite the 5 inch side.

***Solution***

The desired height is represented by $h$, so we have

$$\begin{aligned} 4^2 + h^2 &= 12^2 \\ 16 + h^2 &= 144 \\ h^2 &= 128 \\ h &= \sqrt{128} = \sqrt{64 \cdot 2} = \sqrt{64} \cdot \sqrt{2} = 8\sqrt{2} \end{aligned}$$

The height is represented exactly by $h = 8\sqrt{2}$, which is approximately 11.31 ft.

Reminder

The diagonals of a rhombus are perpendicular bisectors of each other.

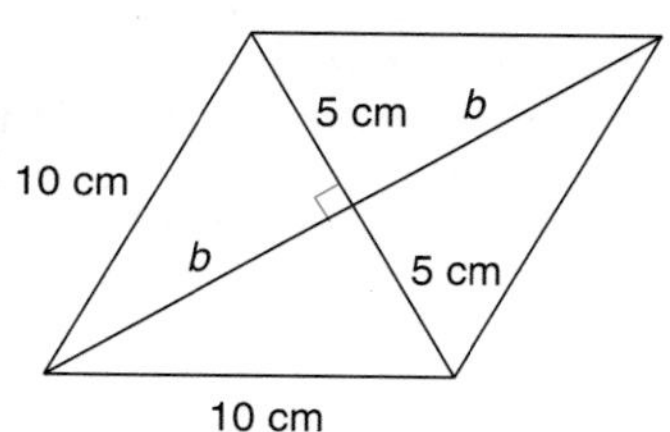


**FIGURE 5.26**

EXAMPLE 4

One diagonal of a rhombus has the same length, 10 cm, as each side (see Figure 5.26). How long is the other diagonal?

***Solution***

Because the diagonals are perpendicular bisectors of each other, four right $\triangle$s are formed. For each right $\triangle$, a side of the rhombus is the hypotenuse. Half of the length of each diagonal is the length of a leg of each right triangle. Therefore,

$$\begin{aligned} 5^2 + b^2 &= 10^2 \\ 25 + b^2 &= 100 \\ b^2 &= 75 \\ b &= \sqrt{75} = \sqrt{25 \cdot 3} = \sqrt{25} \cdot \sqrt{3} = 5\sqrt{3} \end{aligned}$$

Thus the length of the whole diagonal is $10\sqrt{3}$ cm $\approx$ 17.32 cm.

Example 5 also uses the Pythagorean Theorem, but it is considerably more complicated than Example 4. Indeed, it is one of those situations that may require some insight to solve. Note that the triangle described in Example 5 is *not* a right triangle, because $4^2 + 5^2 \neq 6^2$.

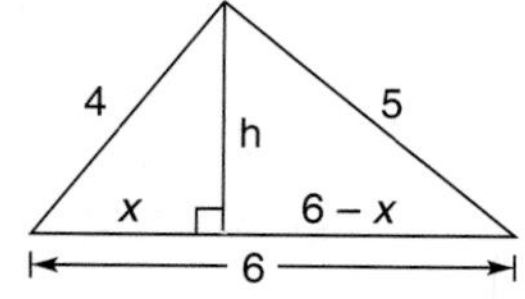


**FIGURE 5.27**

EXAMPLE 5

A triangle has sides of lengths 4, 5, and 6, as shown in Figure 5.27. Find the length of the altitude to the side of length 6.

***Solution***

The altitude to the side of length 6 separates that side into two parts whose lengths are given by $x$ and $6 - x$. Using the two right triangles formed, we have, by the Pythagorean Theorem,

$$x^2 + h^2 = 4^2 \qquad \text{and} \qquad (6 - x)^2 + h^2 = 5^2$$

Subtracting the first equation from the second, we can calculate $x$.

$$\begin{aligned} 36 - 12x + x^2 + h^2 &= 25 \\ x^2 + h^2 &= 16 \\ \hline 36 - 12x \qquad &= 9 \qquad \text{(subtraction)} \\ -12x &= -27 \\ x &= \frac{27}{12} = \frac{9}{4} \end{aligned}$$

Now we use $x = \frac{9}{4}$ to find $h$.

$$x^2 + h^2 = 4^2$$

$$\left(\frac{9}{4}\right)^2 + h^2 = 4^2$$

$$\frac{81}{16} + h^2 = 16$$

$$\frac{81}{16} + h^2 = \frac{256}{16}$$

$$h^2 = \frac{175}{16}$$

$$h = \frac{\sqrt{175}}{4} = \frac{\sqrt{25 \cdot 7}}{4} = \frac{\sqrt{25} \cdot \sqrt{7}}{4} = \frac{5\sqrt{7}}{4} \approx 3.31$$

It is now possible to prove the HL method for the congruence of triangles, a method that was introduced in Section 3.2.

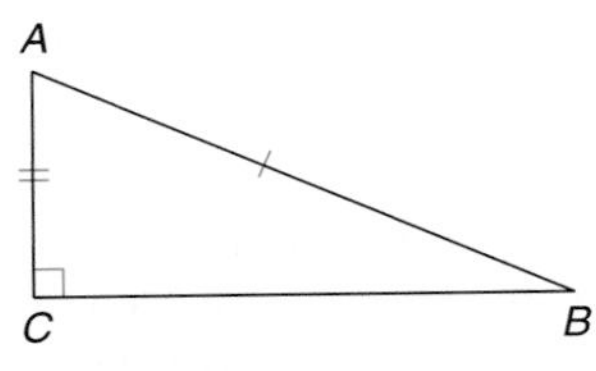


> THEOREM 5.4.6: If the hypotenuse and a leg of one right triangle are congruent to the hypotenuse and a leg of a second right triangle, then the triangles are congruent (HL).

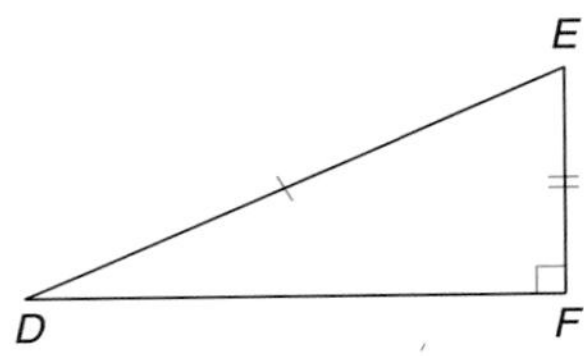


**FIGURE 5.28**

*Given:* Right $\triangle ABC$ with right $\angle C$ and right $\triangle DEF$ with right $\angle F$ (see Figure 5.28); $\overline{AB} \cong \overline{DE}$ and $\overline{AC} \cong \overline{EF}$

*Prove:* $\triangle ABC \cong \triangle EDF$

*Proof:* With right $\angle C$, the hypotenuse of $\triangle ABC$ is $\overline{AB}$; similarly, $\overline{DE}$ is the hypotenuse of right $\triangle EDF$. Because $\overline{AB} \cong \overline{DE}$, we denote the common length by $c$; that is, $AB = DE = c$. Because $\overline{AC} \cong \overline{EF}$, we also have $AC = EF = a$. Then

$$a^2 + (BC)^2 = c^2 \quad \text{and} \quad a^2 + (DF)^2 = c^2 \text{ which leads to}$$

$$BC = \sqrt{c^2 - a^2} \quad \text{and} \quad DF = \sqrt{c^2 - a^2}$$

Then $BC = DF$ so that $\overline{BC} \cong \overline{DF}$. Hence, $\triangle ABC \cong \triangle EDF$ by SSS.

*Exs. 7, 8*

Our work with the Pythagorean Theorem would be incomplete if we did not address two concerns. The first, Pythagorean triples, involves natural (or counting) numbers as possible choices of $a$, $b$, and $c$. The second leads to a classification of triangles according to the lengths of their sides; see Theorem 5.4.7.

## PYTHAGOREAN TRIPLES

> DEFINITION: A **Pythagorean triple** is a set of three natural numbers $(a, b, c)$ for which $a^2 + b^2 = c^2$.

Three sets of Pythagorean triples encountered in this section are (3, 4, 5), (5, 12, 13), and (8, 15, 17). These numbers will always fit the sides of a right triangle.

Natural-number multiples of any of these triples will also constitute Pythagorean triples. For example, doubling (3, 4, 5) yields (6, 8, 10), which is also a Pythagorean triple. In Figure 5.29, the triangles are similar by SSS~.

The Pythagorean triple (3, 4, 5) also leads to (9, 12, 15), (12, 16, 20), and (15, 20, 25). The Pythagorean triple (5, 12, 13) leads to triples such as (10, 24, 26) and (15, 36, 39). Basic Pythagorean triples that are used less frequently include (7, 24, 25), (9, 40, 41), and (20, 21, 29).

**FIGURE 5.29**

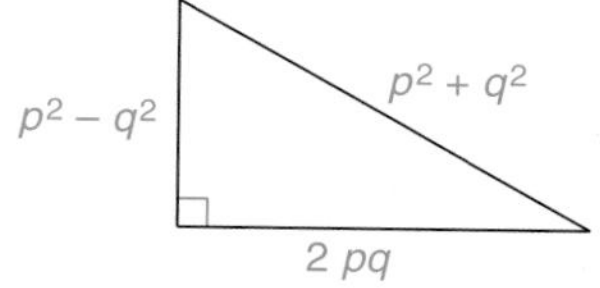


**FIGURE 5.30**

Pythagorean triples can be generated by using any of several formulas. One formula uses $2pq$ for one leg, $p^2 - q^2$ for the other leg, and $p^2 + q^2$ for the hypotenuse, where $p$ and $q$ are natural numbers and $p > q$. (See Figure 5.30.)

Table 5.1 lists some Pythagorean triples corresponding to choices for $p$ and $q$. The triples printed in boldface type are basic triples, also known as primitive triples. In application, knowledge of these triples and their multiples will save you considerable time and effort. In the final column, the resulting triple is provided, in order, from $a$ (small) to $c$ (large).

**TABLE 5.1**

Pythagorean Triples

| $p$ | $q$ | $a$ (or $b$) $p^2 - q^2$ | $b$ (or $a$) $2pq$ | $c$ $p^2 + q^2$ | $(a, b, c)$ |
|---|---|---|---|---|---|
| 2 | 1 | 3 | 4 | 5 | **(3, 4, 5)** |
| 3 | 1 | 8 | 6 | 10 | (6, 8, 10) |
| 3 | 2 | 5 | 12 | 13 | **(5, 12, 13)** |
| 4 | 1 | 15 | 8 | 17 | **(8, 15, 17)** |
| 4 | 3 | 7 | 24 | 25 | **(7, 24, 25)** |
| 5 | 1 | 24 | 10 | 26 | (10, 24, 26) |
| 5 | 2 | 21 | 20 | 29 | **(20, 21, 29)** |
| 5 | 3 | 16 | 30 | 34 | (16, 30, 34) |
| 5 | 4 | 9 | 40 | 41 | **(9, 40, 41)** |

*Exs. 9–11*

## THE CONVERSE OF THE PYTHAGOREAN THEOREM

The Converse of the Pythagorean Theorem allows us to recognize a right triangle by knowing the lengths of its sides. A variation on the Converse allows us to determine whether a triangle is acute or obtuse. This theorem is stated without proof.

THEOREM 5.4.7: Let $a$, $b$, and $c$ represent the lengths of the three sides of a triangle, with $c$ the length of the longest side.

1. If $c^2 > a^2 + b^2$, then the triangle is obtuse and the obtuse angle lies opposite the side of length $c$.
2. If $c^2 < a^2 + b^2$, then the triangle is acute.

EXAMPLE 6

Determine the type of triangle represented if the lengths of its sides are as follows:

a) 4, 5, 7 b) 6, 7, 8 c) 9, 12, 15 d) 3, 4, 9

***Solution***

a) Because $c$ is the longest side, $c = 7$, and we have $7^2 > 4^2 + 5^2$, or $49 > 16 + 25$; the triangle is obtuse.
b) Choosing $c = 8$, we have $8^2 < 6^2 + 7^2$, or $64 < 36 + 49$; the triangle is acute.
c) Choosing $c = 15$, we have $15^2 = 9^2 + 12^2$, or $225 = 81 + 144$; the triangle is a right triangle.
d) Because $9 > 3 + 4$, no triangle is possible. (Remember that the sum of the lengths of two sides of a triangle must be greater than the length of the third side.)

*Exs. 12, 13*

## 5.4 Exercises

1. By naming the vertices in order, state three different triangles that are similar to each other.

2. Use Theorem 5.4.2 to form a proportion in which $SV$ is a geometric mean. (HINT: $\triangle SVT \sim \triangle RVS$)

3. Use Lemma 5.4.3 to form a proportion in which $RS$ is a geometric mean. (HINT: $\triangle RVS \sim \triangle RST$)

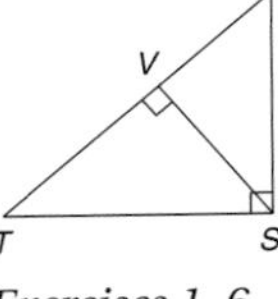


Exercises 1–6

4. Use Lemma 5.4.3 to form a proportion in which $TS$ is a geometric mean. (HINT: $\triangle TVS \sim \triangle TSR$)

5. Use Theorem 5.4.2 to find $RV$ if $SV = 6$ and $VT = 8$.

6. Use Lemma 5.4.3 to find $RT$ if $RS = 6$ and $VR = 4$.

7. Find the length of $\overline{DF}$ if:

   a) $DE = 8$ and $EF = 6$
   b) $DE = 5$ and $EF = 3$

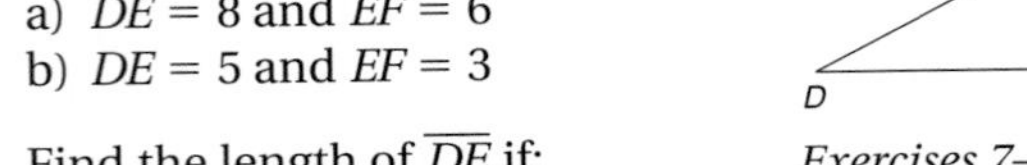


Exercises 7–10

8. Find the length of $\overline{DE}$ if:

   a) $DF = 13$ and $EF = 5$
   b) $DF = 12$ and $EF = 6\sqrt{3}$

9. Find $EF$ if:

   a) $DF = 17$ and $DE = 15$
   b) $DF = 12$ and $DE = 8\sqrt{2}$

10. Find $DF$ if:

    a) $DE = 12$ and $EF = 5$
    b) $DE = 12$ and $EF = 6$

11. Determine whether each triple $(a, b, c)$ is a Pythagorean triple.

    a) $(3, 4, 5)$
    b) $(4, 5, 6)$
    c) $(5, 12, 13)$
    d) $(6, 13, 15)$

12. Determine whether each triple $(a, b, c)$ is a Pythagorean triple.

    a) $(8, 15, 17)$
    b) $(10, 13, 19)$
    c) $(6, 8, 10)$
    d) $(11, 17, 20)$

13. Determine the type of triangle represented if the lengths of its sides are:

    a) $a = 4$, $b = 3$, and $c = 5$
    b) $a = 4$, $b = 5$, and $c = 6$
    c) $a = 2$, $b = \sqrt{3}$, and $c = \sqrt{7}$
    d) $a = 3$, $b = 8$, and $c = 15$

14. Determine the type of triangle represented if the lengths of its sides are:

    a) $a = 1.5$, $b = 2$, and $c = 2.5$
    b) $a = 20$, $b = 21$, and $c = 29$
    c) $a = 10$, $b = 12$, and $c = 16$
    d) $a = 5$, $b = 7$, and $c = 9$

15. A guy wire 25 ft long supports an antenna at a point that is 20 ft above the base of the antenna. How far from the base of the antenna is the guy wire secured?

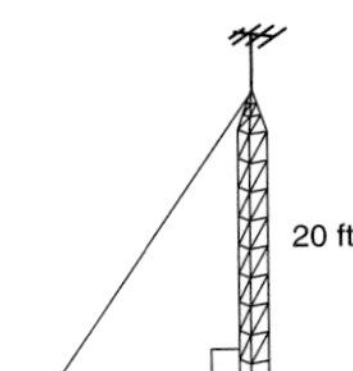


16. A strong wind holds a kite 30 ft above the earth in a position 40 ft across the ground. How much string does the girl have out (to the kite)?

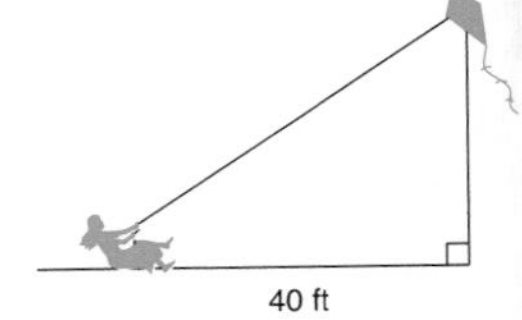


17. A boat is 6 m below the level of a pier and 12 m from the pier as measured across the water. How much rope is needed to reach the boat?

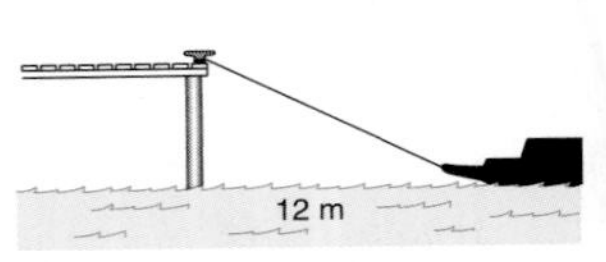


18. A hot-air balloon is held in place by the ground crew at a point that is 21 ft from a point directly beneath the balloon. If the rope is of length 29 ft, how far above ground level is the balloon?

19. A drawbridge that is 104 ft in length is raised at its midpoint so that the uppermost points are 8 ft apart. How far has each of the midsections been raised?

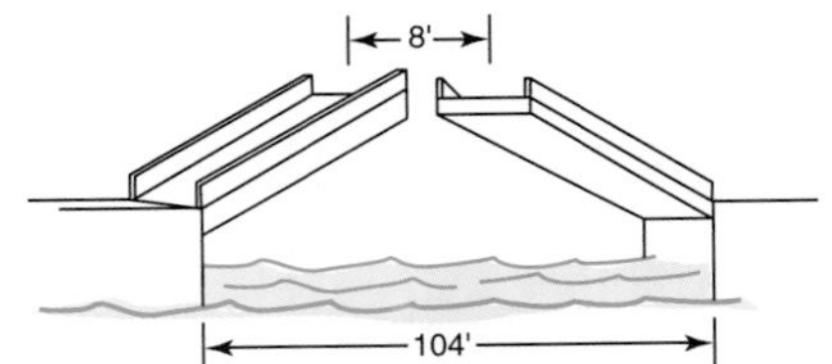

20. A drawbridge that is 136 ft in length is raised at its midpoint so that the uppermost points are 16 ft apart. How far has each of the midsections been raised?
(HINT: Consider the drawing for Exercise 19.)

21. A rectangle has a width of 16 cm and a diagonal of length 20 cm. How long is the rectangle?

22. A right triangle has legs of lengths $x$ and $2x + 2$ and a hypotenuse of length $2x + 3$. What are the lengths of its sides?

23. A rectangle has base $x + 3$, altitude $x + 1$, and diagonals $2x$ each. What are the lengths of its base, altitude, and diagonals?

24. The diagonals of a rhombus measure 6 m and 8 m. How long are each of the congruent sides?

25. Each side of a rhombus measures 12 in. If one diagonal is 18 in. long, how long is the other diagonal?

26. An isosceles right triangle has a hypotenuse of length 10 cm. How long is each leg?

27. Each leg of an isosceles right triangle has a length of $6\sqrt{2}$ in. What is the length of the hypotenuse?

28. In right $\triangle ABC$ with right $\angle C$, $AB = 10$ and $BC = 8$. Find the length of $\overline{MB}$ if $M$ is the midpoint of $\overline{AC}$.

29. In right $\triangle ABC$ with right $\angle C$, $AB = 17$ and $BC = 15$. Find the length of $\overline{MN}$ if $M$ and $N$ are the midpoints of $\overline{AB}$ and $\overline{BC}$, respectively.

30. Find the length of the altitude to the 10-in. side of a triangle whose sides are 6, 8, and 10 in. in length.

31. Find the length of the altitude to the 26-in. side of a triangle whose sides are 10, 24, and 26 in. in length.

32. In quadrilateral $ABCD$, $\overline{BC} \perp \overline{AB}$ and $\overline{DC} \perp$ diagonal $\overline{AC}$. If $AB = 4$, $BC = 3$, and $DC = 12$, determine $DA$.

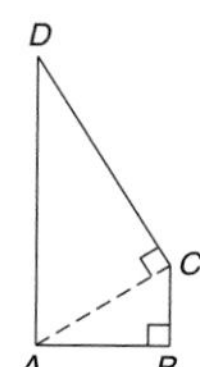


33. In quadrilateral $RSTU$, $\overline{RS} \perp \overline{ST}$ and $\overline{UT} \perp$ diagonal $\overline{RT}$. If $RS = 6$, $ST = 8$, and $RU = 15$, find $UT$.

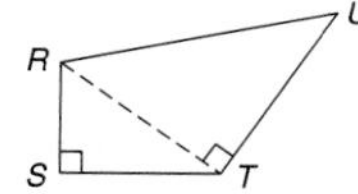


34. *Given:* $\triangle ABC$ is not a right $\triangle$
*Prove:* $a^2 + b^2 \neq c^2$

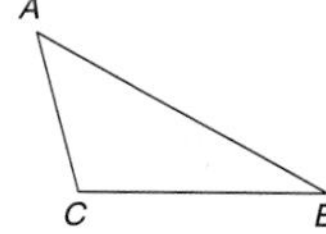


* 35. If $a = p^2 - q^2$, $b = 2pq$, and $c = p^2 + q^2$, show that $c^2 = a^2 + b^2$.

36. Given that the line segment shown has length 1, construct a line segment whose length is $\sqrt{2}$.

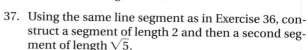

*Exercises 36, 37*

37. Using the same line segment as in Exercise 36, construct a segment of length 2 and then a second segment of length $\sqrt{5}$.

38. When the rectangle in the accompanying drawing (whose dimensions are 16 by 9) is cut into pieces and rearranged, a square can be formed. What is the perimeter of this square?

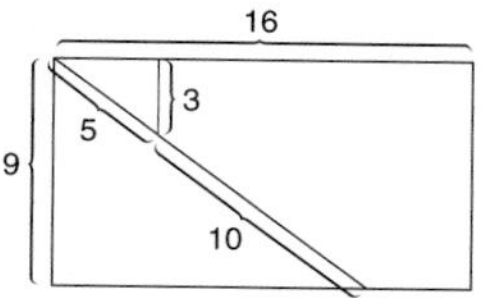


39. $A$, $C$, and $F$ are three of the vertices of the cube shown in the accompanying figure. Given that each face of the cube is a square, what is the measure of angle $ACF$?

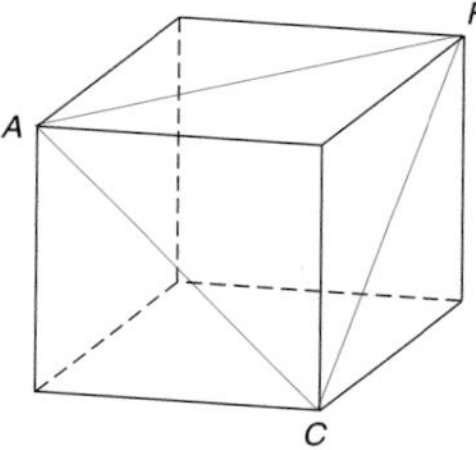


* 40. Find the length of the altitude to the 8-in. side of a triangle whose sides are 4, 6, and 8 in. long.
(HINT: See Example 5.)

41. In the figure, square $RSTV$ has its vertices on the sides of square $WXYZ$ as shown. If $ZT = 5$ and $TY = 12$, find $TS$. Also find $RT$.

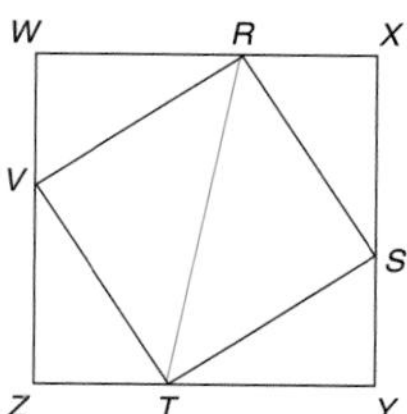


42. Prove that if $(a, b, c)$ is a Pythagorean triple and $n$ is a natural number, then $(na, nb, nc)$ is also a Pythagorean triple.

43. Use Figure 5.19 to prove Theorem 5.4.2.

44. Use Figure 5.21 to prove Lemma 5.4.3.

# 5.5 Special Right Triangles

**KEY CONCEPTS**
The 45°-45°-90° Triangle • The 30°-60°-90° Triangle

Many of the calculations that we do in this section involve square root radicals. To understand some of these calculations better, it may be necessary to review the Properties of Square Roots in Appendix A.4.

Certain right triangles occur so often that they deserve more attention than others. The special right triangles we will consider have angle measures of 45°, 45°, and 90° or of 30°, 60°, and 90°.

## THE 45°-45°-90° RIGHT TRIANGLE

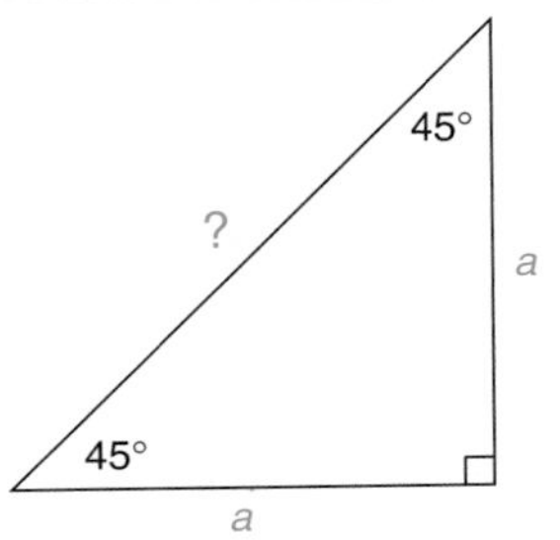


**FIGURE 5.31**

In the 45-45-90 triangle, the legs are opposite the congruent angles and are also congruent. Rather than using $a$ and $b$ to represent the lengths of the legs, we use $a$ for both lengths, as shown in Figure 5.31. It then follows, by the Pythagorean Theorem, that

$$\begin{aligned} c^2 &= a^2 + a^2 \\ c^2 &= 2a^2 \\ c &= \sqrt{2a^2} \\ c &= \sqrt{2} \cdot \sqrt{a^2} \\ c &= a\sqrt{2} \end{aligned}$$

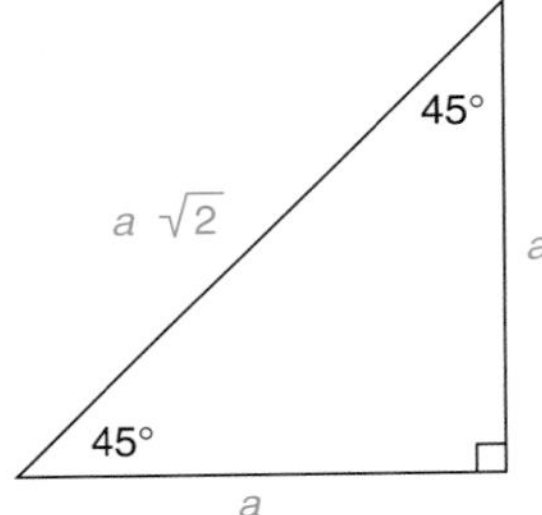


**FIGURE 5.32**

THEOREM 5.5.1: **(45-45-90 Theorem)** In a triangle whose angles measure 45°, 45°, and 90°, the hypotenuse has a length equal to the product of $\sqrt{2}$ and the length of either leg.

*Exs. 1–3*

It is better to memorize the sketch in Figure 5.32 than to repeat the steps preceding the 45-45-90 Theorem.

**Reminder**
If two angles of a triangle are congruent, then the sides opposite these angles are congruent.

EXAMPLE 1

Find the lengths of the missing sides in each triangle in Figure 5.33.

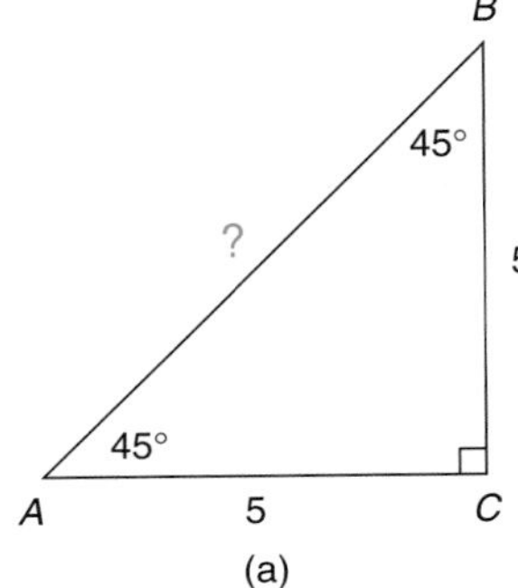


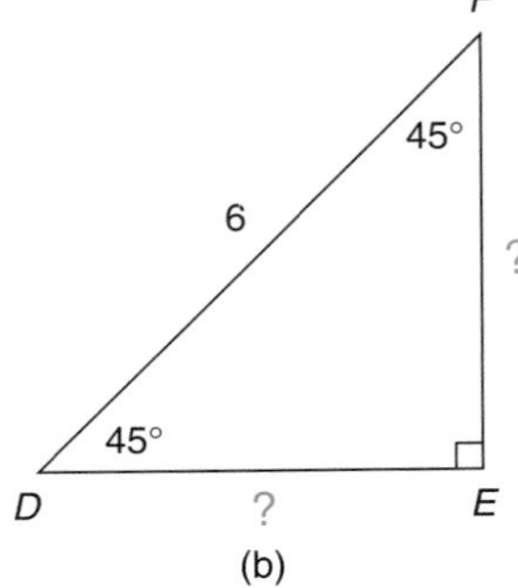


**FIGURE 5.33**

***Solution***

a) The length of hypotenuse $\overline{AB}$ is $5\sqrt{2}$, the product of $\sqrt{2}$ and the length of either of the equal legs.

b) Let $a$ denote the length of $\overline{DE}$ and of $\overline{EF}$. The length of hypotenuse $\overline{DF}$ is $a\sqrt{2}$.

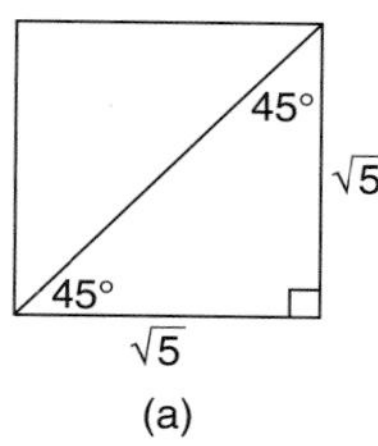


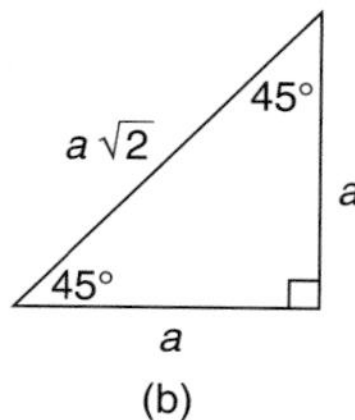


**FIGURE 5.34**

Then $a\sqrt{2} = 6$, so $a = \frac{6}{\sqrt{2}}$. Simplifying yields

$$\begin{aligned} a &= \frac{6}{\sqrt{2}} \cdot \frac{\sqrt{2}}{\sqrt{2}} \\ &= \frac{6\sqrt{2}}{2} \\ &= 3\sqrt{2} \end{aligned}$$

Therefore, $DE = EF = 3\sqrt{2} \approx 4.24$.

NOTE: If we use the Pythagorean Theorem to solve Example 1, the solution in part (a) can be found by solving the equation $5^2 + 5^2 = c^2$ and the solution in part (b) can be found by solving $a^2 + a^2 = 6^2$.

*Exs. 4–7*

EXAMPLE 2

Each side of a square has a length of $\sqrt{5}$. Find the length of a diagonal.

***Solution***

The square shown in Figure 5.34(a) is separated into two 45°-45°-90° triangles. With each of the congruent legs represented by $a$ in Figure 5.34(b), we see that $a = \sqrt{5}$ and the diagonal (hypotenuse length) is $a \cdot \sqrt{2} = \sqrt{5} \cdot \sqrt{2}$, so $a = \sqrt{10} \approx 3.16$.

## THE 30°-60°-90° RIGHT TRIANGLE

The second special triangle is the 30°-60°-90° triangle.

THEOREM 5.5.2: **(30-60-90 Theorem)** In a triangle whose angles measure 30°, 60°, and 90°, the hypotenuse has a length equal to twice the length of the shorter leg, and the length of the longer leg is the product of $\sqrt{3}$ and the length of the shorter leg.

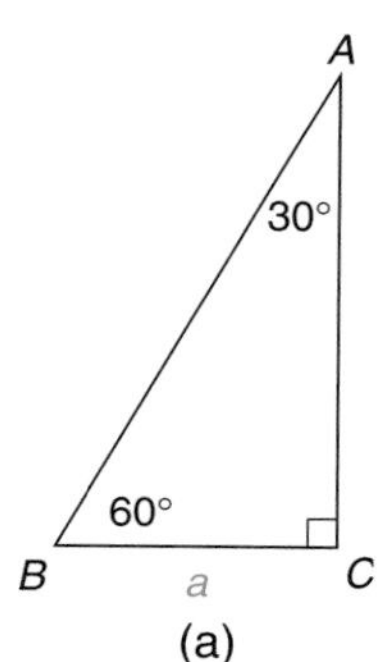


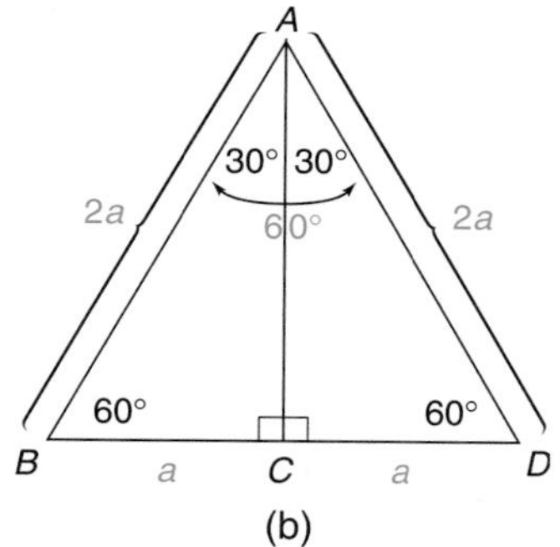


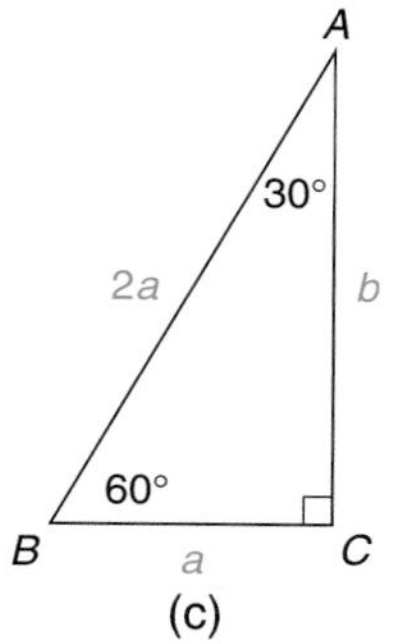


**FIGURE 5.35**

EXAMPLE 3

Study the picture proof of Theorem 5.5.2. See Figure 5.35(a).

**PICTURE PROOF OF THEOREM 5.5.2**

*Given:* $\triangle ABC$ with $m\angle A = 30°$, $m\angle B = 60°$, $m\angle C = 90°$, and $BC = a$

*Prove:* $AB = 2a$ and $AC = a\sqrt{3}$

*Proof:* We reflect $\triangle ABC$ across $\overline{AC}$ to form an equiangular and therefore equilateral $\triangle ABD$. As shown in Figures 5.35(b) and (c), we have $AB = 2a$. To find $b$ in Figure 5.35(c), we apply the Pythagorean Theorem.

$$\begin{aligned} c^2 &= a^2 + b^2 \\ (2a)^2 &= a^2 + b^2 \\ 4a^2 &= a^2 + b^2 \\ 3a^2 &= b^2, \end{aligned}$$

So

$$\begin{aligned} b^2 &= 3a^2 \\ b &= \sqrt{3a^2} \\ b &= \sqrt{3} \cdot \sqrt{a^2} \\ b &= a\sqrt{3} \end{aligned}$$

That is, $AC = a\sqrt{3}$.

It would be best to memorize the sketch in Figure 5.36. So that you will more easily recall which expression is used for each side, remember that the lengths of the sides follow the same order as the angles opposite them. Thus,

*Exs. 8–10*

> Opposite the $30° \angle$ (smallest angle) is $a$ (shortest side).
> Opposite the $60° \angle$ (middle angle) is $a\sqrt{3}$ (middle side).
> Opposite the $90° \angle$ (largest angle) is $2a$ (longest side).

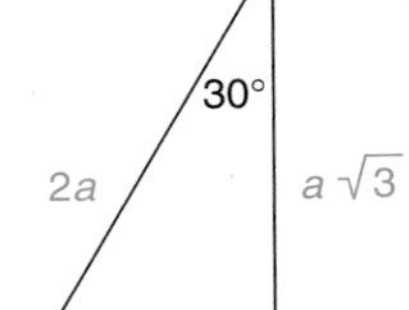


**FIGURE 5.36**

EXAMPLE 4

Find the lengths of the missing sides of each triangle in Figure 5.37.

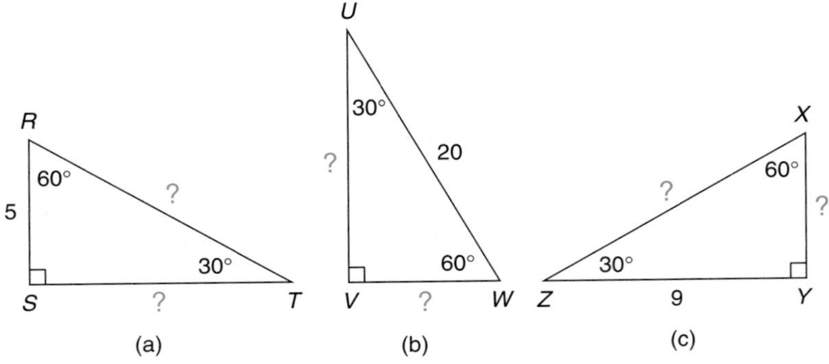


**FIGURE 5.37**

***Solution***

a) $RT = 2 \cdot RS = 2 \cdot 5 = 10$
$ST = RS\sqrt{3} = 5\sqrt{3} \approx 8.66$

b) $UW = 2 \cdot VW \rightarrow 20 = 2 \cdot VW \rightarrow VW = 10$
$UV = VW\sqrt{3} = 10\sqrt{3} \approx 17.32$

c) $ZY = XY\sqrt{3} \rightarrow 9 = XY \cdot \sqrt{3} \rightarrow XY = \frac{9}{\sqrt{3}} = \frac{9}{\sqrt{3}} \cdot \frac{\sqrt{3}}{\sqrt{3}}$
$= \frac{9\sqrt{3}}{3} = 3\sqrt{3} \approx 5.20$

$XZ = 2 \cdot XY = 2 \cdot 3\sqrt{3} = 6\sqrt{3} \approx 10.39$

EXAMPLE 5

Each side of an equilateral triangle measures 6 in. Find the length of an altitude of the triangle.

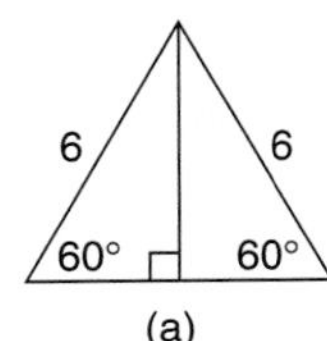


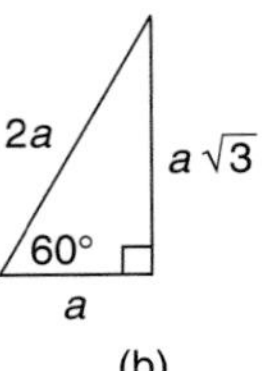


**FIGURE 5.38**

**Solution**

The equilateral triangle shown in Figure 5.38(a) is separated into two 30°-60°-90° triangles by the altitude. In the 30°-60°-90° triangle in Figure 5.38(b), the side of the equilateral triangle becomes the hypotenuse, so $2a = 6$ and $a = 3$. The altitude lies opposite the 60° angle of the 30°-60°-90° triangle, so its length is $a\sqrt{3}$ or $3\sqrt{3}$ in. $\approx$ 5.20 in.

The converse of Theorem 5.5.1 is true and is described in the following theorem.

*Exs. 11–13*

THEOREM 5.5.3: If the length of the hypotenuse of a right triangle equals the product of $\sqrt{2}$ and the length of either leg, then the angles of the triangle measure 45°, 45°, and 90°.

***Proof***

In Figure 5.39, we represent the length of the hypotenuse by $a\sqrt{2}$, where $a$ is the length of either leg. In a right triangle, the angles that lie opposite the congruent legs are also congruent. In a right triangle, the acute angles are complementary, so each of the congruent acute angles measures 45°.

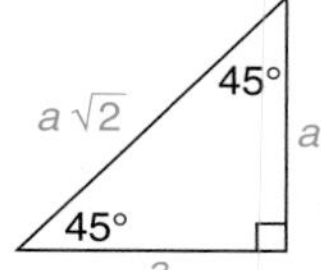


**FIGURE 5.39**

EXAMPLE 6

In right $\triangle RST$, $RS = ST$. (See Figure 5.40.) What are the measures of the angles of the triangle? If $RT = 12\sqrt{2}$, what is the length of $\overline{RS}$ (or $\overline{ST}$)?

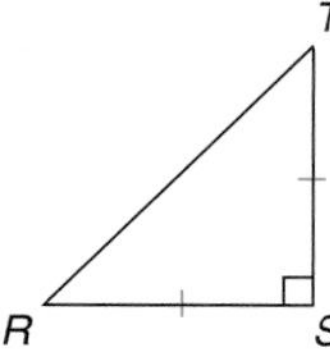


**FIGURE 5.40**

**Solution**

The longest side is the hypotenuse $\overline{RT}$, so the right angle is $\angle S$ and $m\angle S = 90°$. Because $\overline{RS} \cong \overline{ST}$, the congruent acute angles are $\angle$s $R$ and $T$ and $m\angle R = m\angle T = 45°$. Because $RT = 12\sqrt{2}$, $RS = ST = 12$.

The converse of Theorem 5.5.2 is also true and can be proved by the indirect method. Rather than construct the proof, we state and apply this theorem. See Figure 5.41.

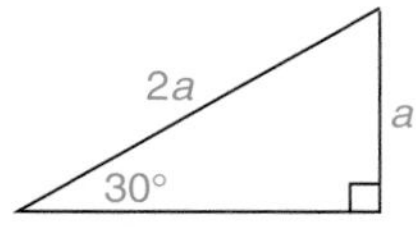


**FIGURE 5.41**

THEOREM 5.5.4: If the length of the hypotenuse of a right triangle is twice the length of one leg of the triangle, then the angle of the triangle opposite that leg measures 30°.

An equivalent form of this theorem is stated as follows:

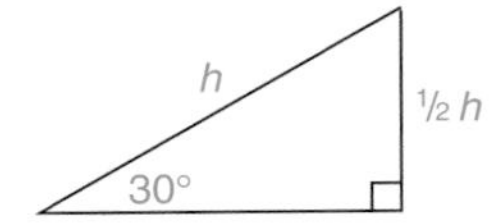

FIGURE 5.42

If one leg of a right triangle has a length equal to one-half the length of the hypotenuse, then the angle of the triangle opposite that leg measures 30° (see Figure 5.42).

EXAMPLE 7

In right $\triangle ABC$ with right $\angle C$, $AB = 24.6$ and $BC = 12.3$ (see Figure 5.43). What are the measures of the angles of the triangle? Also, what is the length of $\overline{AC}$?

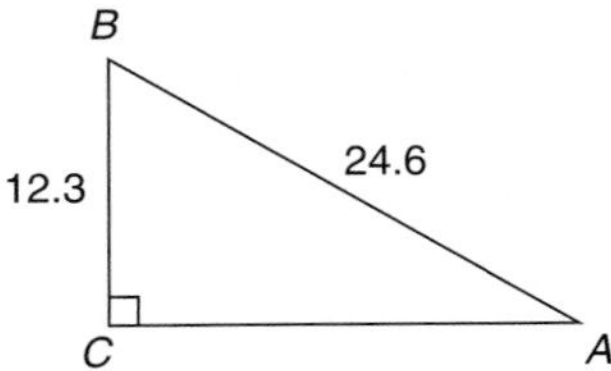

FIGURE 5.43

***Solution***

Because $\angle C$ is a right angle, $m\angle C = 90°$ and $\overline{AB}$ is the hypotenuse. Because $BC = \frac{1}{2}(AB)$, the angle opposite $\overline{BC}$ measures 30°. Thus $m\angle A = 30°$ and $m\angle B = 60°$.

Because $\overline{AC}$ lies opposite the 60° angle, $AC = (12.3)\sqrt{3} \approx 21.3$.

## 5.5 Exercises

1. For the 45°-45°-90° triangle shown, suppose that $AC = a$. Find:

   a) $BC$ b) $AB$

2. For the 45°-45°-90° triangle shown, suppose that $AB = a\sqrt{2}$. Find:

   a) $AC$ b) $BC$

*Exercises 1, 2*

3. For the 30°-60°-90° triangle shown, suppose that $XZ = a$. Find:

   a) $YZ$ b) $XY$

4. For the 30°-60°-90° triangle shown, suppose that $XY = 2a$. Find:

   a) $XZ$ b) $YZ$

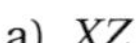

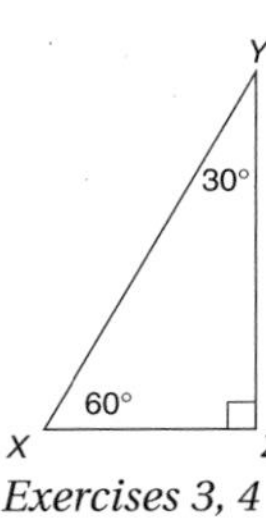

*Exercises 3, 4*

*In Exercises 5 to 22, find the missing lengths. Give your answers in both simplest radical form and as approximations to two decimal places.*

5. *Given:* Right $\triangle XYZ$ with $m\angle X = 45°$ and $XZ = 8$
   *Find:* $YZ$ and $XY$

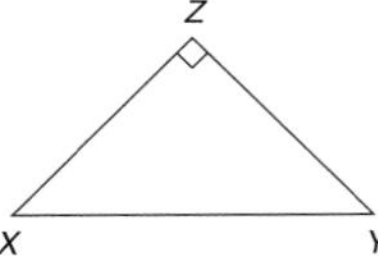

*Exercises 5–8*

6. *Given:* Right $\triangle XYZ$ with $\overline{XZ} \cong \overline{YZ}$ and $XY = 10$
   *Find:* $XZ$ and $YZ$

7. *Given:* Right $\triangle XYZ$ with $\overline{XZ} \cong \overline{YZ}$ and $XY = 10\sqrt{2}$
   *Find:* $XZ$ and $YZ$

8. *Given:* Right $\triangle XYZ$ with $m\angle X = 45°$ and $XY = 12\sqrt{2}$
   *Find:* $XZ$ and $YZ$

9. *Given:* Right $\triangle DEF$ with $m\angle E = 60°$ and $DE = 5$
*Find:* $DF$ and $FE$

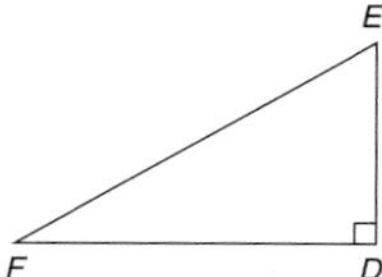


*Exercises 9–12*

10. *Given:* Right $\triangle DEF$ with $m\angle F = 30°$ and $FE = 12$
*Find:* $DF$ and $DE$

11. *Given:* Right $\triangle DEF$ with $m\angle E = 60°$ and $FD = 12\sqrt{3}$
*Find:* $DE$ and $FE$

12. *Given:* Right $\triangle DEF$ with $m\angle E = 2 \cdot m\angle F$ and $EF = 12\sqrt{3}$
*Find:* $DE$ and $DF$

13. *Given:* Rectangle $HJKL$ with diagonals $\overline{HK}$ and $\overline{JL}$
$m\angle HKL = 30°$
*Find:* $HL$, $HK$, and $MK$

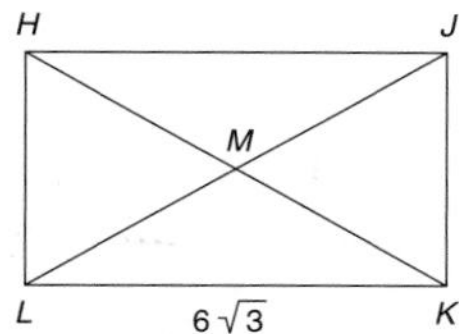


14. *Given:* Right $\triangle RST$ with
$RT = 6\sqrt{2}$ and $m\angle STV = 150°$
*Find:* $RS$ and $ST$

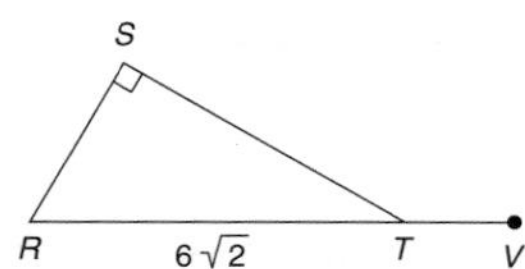


*In Exercises 15–19, create drawings as needed.*

15. *Given:* $\triangle ABC$ with $m\angle A = m\angle B = 45°$ and $BC = 6$
*Find:* $AC$ and $AB$

16. *Given:* Right $\triangle MNP$ with $MP = PN$ and $MN = 10\sqrt{2}$
*Find:* $PM$ and $PN$

17. *Given:* $\triangle RST$ with $m\angle T = 30°$, $m\angle S = 60°$, and $ST = 12$
*Find:* $RS$ and $RT$

18. *Given:* $\triangle XYZ$ with $\overline{XY} \cong \overline{XZ} \cong \overline{YZ}$
$\overline{ZW} \perp \overline{XY}$
$YZ = 6$
*Find:* $ZW$

19. *Given:* Square $ABCD$ with diagonals $\overline{DB}$ and $\overline{AC}$ intersecting at $E$
$DC = 5\sqrt{3}$
*Find:* $DB$

20. *Given:* $\triangle NQM$ with angles as shown in the drawing
$\overline{MP} \perp \overline{NQ}$
*Find:* $NM$, $MP$, $MQ$, $PQ$, and $NQ$

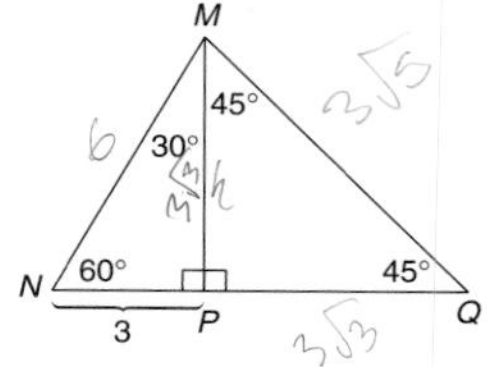


21. *Given:* $\triangle XYZ$ with angles as shown in the drawing
*Find:* $XY$
(HINT: Compare this drawing to the one for Exercise 20.)

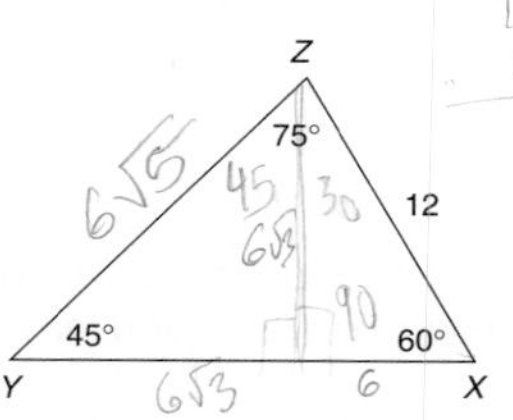


22. *Given:* Rhombus $ABCD$ in which diagonals $\overline{AC}$ and $\overline{DB}$ intersect at point $E$; $DB = AB = 8$
*Find:* $AC$

23. A carpenter is working with a board that is $3\frac{3}{4}$ in. wide. After marking off a point down the side of length $3\frac{3}{4}$ in., the carpenter makes a cut along $\overline{BC}$ with a saw. What is the measure of the angle ($\angle ACB$) that is formed?

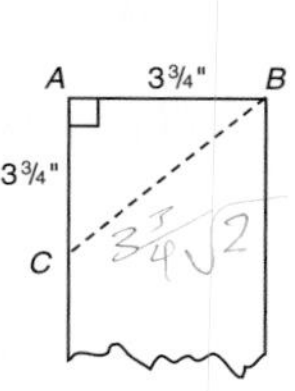


24. To unload groceries from a delivery truck at the Piggly Wiggly Market, an 8-ft ramp that rises 4 ft to the door of the trailer is used. What is the measure of the indicated angle ($\angle D$)?

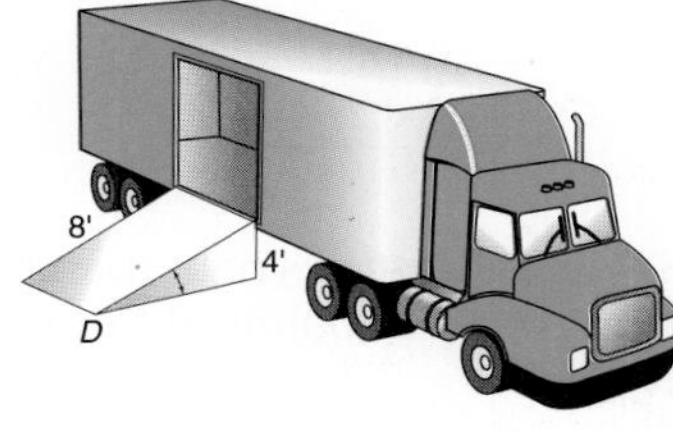


25. A jogger runs along two sides of an open rectangular lot. If the first side of the lot is 200 ft long and the diagonal distance across the lot is 400 ft, what is the measure of the angle formed by the 200-ft and 400-ft dimensions? To the nearest foot, how much farther does the jogger run by traveling the two sides of the block rather than the diagonal distance across the lot?

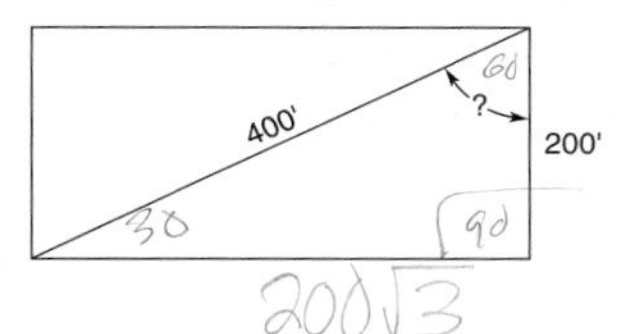

26. Thelma's boat leaves the dock at the same time that Gina's boat leaves the dock. Thelma's boat travels due east at 12 mph. Gina's boat travels at 24 mph in the direction N 30° E. To the nearest tenth of a mile, how far apart will the boats be in half an hour?

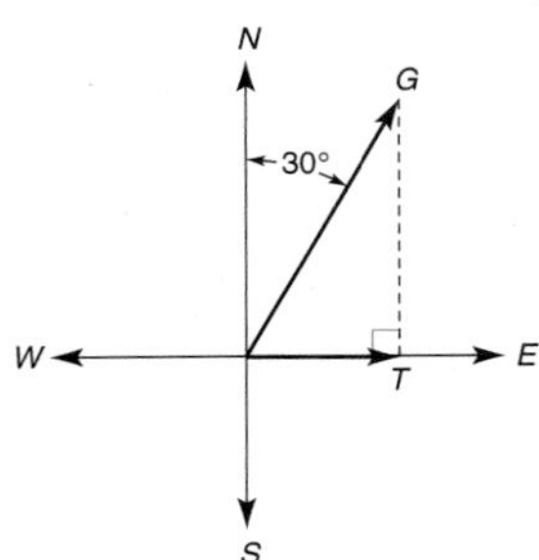


*In Exercises 27 to 33, give both exact solutions and approximate solutions to two decimal places.*

27. *Given:* In $\triangle ABC$, $\overrightarrow{AD}$ bisects $\angle BAC$
$m\angle B = 30°$ and $AB = 12$
*Find:* $DC$ and $DB$

28. *Given:* In $\triangle ABC$, $\overrightarrow{AD}$ bisects $\angle BAC$
$AB = 20$ and $AC = 10$
*Find:* $DC$ and $DB$

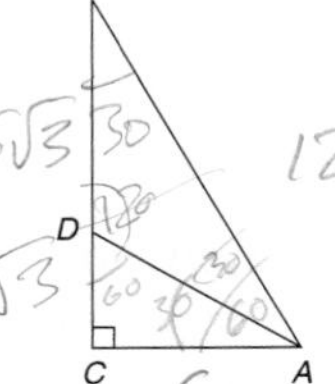


*Exercises 27, 28*

29. *Given:* $\triangle MNQ$ is equiangular and $NR = 6$
$\overrightarrow{NR}$ bisects $\angle MNQ$
$\overrightarrow{QR}$ bisects $\angle MQN$
*Find:* $NQ$

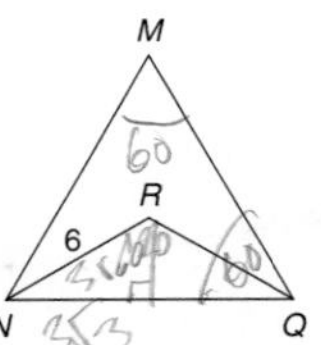


30. *Given:* $\triangle STV$ is an isosceles right triangle
$M$ and $N$ are midpoints of $\overline{ST}$ and $\overline{SV}$
*Find:* $MN$

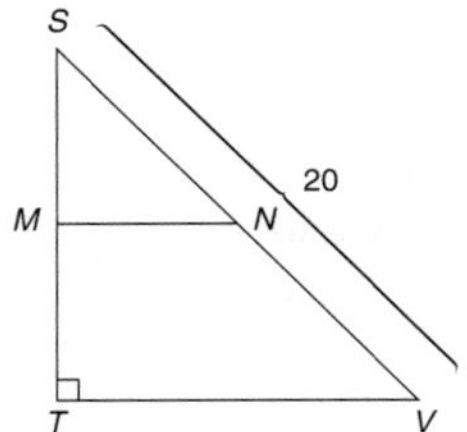


31. *Given:* Right $\triangle ABC$ with $m\angle C = 90°$ and $m\angle BAC = 60°$; point $D$ on $\overline{BC}$; $\overrightarrow{AD}$ bisects $\angle BAC$ and $AB = 12$
*Find:* $BD$

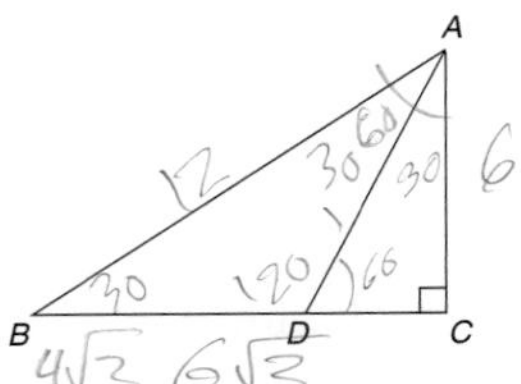


*Exercises 31, 32*

32. *Given:* Right $\triangle ABC$ with $m\angle C = 90°$ and $m\angle BAC = 60°$; point $D$ on $\overline{BC}$; $\overrightarrow{AD}$ bisects $\angle BAC$ and $AC = 2\sqrt{3}$
*Find:* $BD$

33. *Given:* $\triangle ABC$ with $m\angle A = 45°$, $m\angle B = 30°$, and $BC = 12$
*Find:* $AB$
(HINT: Use altitude $\overline{CD}$ from $C$ to $\overline{AB}$ as an auxiliary line.)

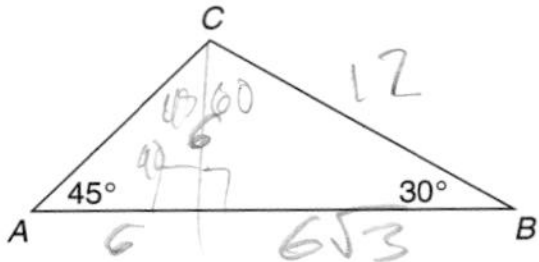


*34. *Given:* Isosceles trapezoid $MNPQ$ with $QP = 12$ and $m\angle M = 120°$; the bisectors of $\angle$s $MQP$ and $NPQ$ meet at point $T$ on $\overline{MN}$
*Find:* The perimeter of $MNPQ$

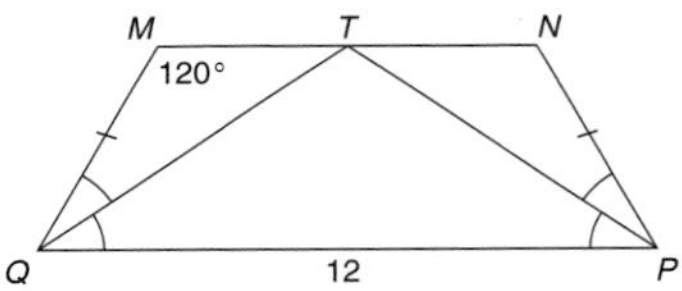


35. In regular hexagon $ABCDEF$, $AB = 6$ inches. Find the exact length of:
a) Diagonal $\overline{BF}$
b) Diagonal $\overline{CF}$

36. In regular hexagon $ABCDEF$, the length of $AB$ is $x$ centimeters. In terms of $x$, find the length of:
a) Diagonal $\overline{BF}$
b) Diagonal $\overline{CF}$

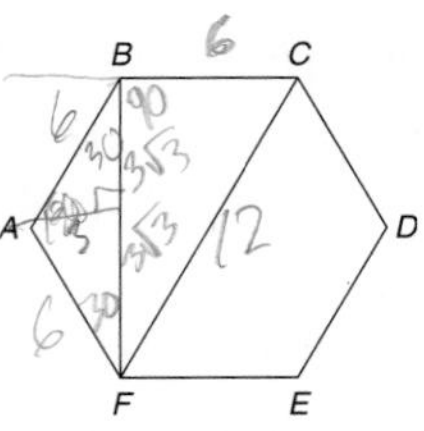


*Exercises 35, 36*

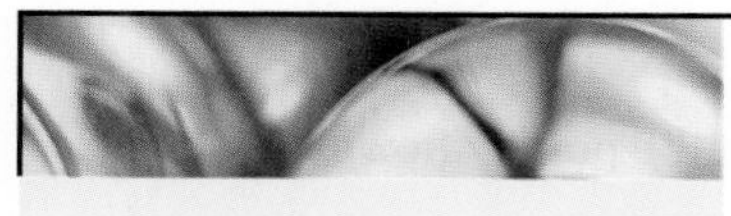

**KEY CONCEPTS**

Segments Divided Proportionally • The Angle-Bisector Theorem • Ceva's Theorem

# 5.6 Segments Divi

In this section, we begin with an inform
*tionally.* Suppose that three children hav
by their parents. Equal monthly deposi
child since birth. If the ages of the childr
for simplicity) and the total in the accou
should receive can be found by solving t

$$2x + 4x \cdot$$

Solving this equation leads to the solution \$1200 for the 2-year-old, \$2400 for the 4-year-old, and \$3600 for the 6-year-old. We say that the amount has been divided proportionally. Expressed as a proportion, this is

$$\frac{1200}{2} = \frac{2400}{4} = \frac{3600}{6}$$

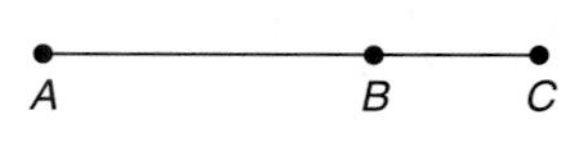

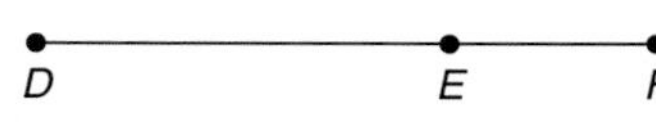

**FIGURE 5.44**

In Figure 5.44, $\overline{AC}$ and $\overline{DF}$ are divided proportionally at points $B$ and $E$ if

$$\frac{AB}{DE} = \frac{BC}{EF} \quad \text{or} \quad \frac{AB}{BC} = \frac{DE}{EF}$$

Of course, a pair of segments may be divided proportionally by several points, as shown in Figure 5.45. In this case, $\overline{RW}$ and $\overline{HM}$ are divided proportionally when

$$\frac{RS}{HJ} = \frac{ST}{JK} = \frac{TV}{KL} = \frac{VW}{LM}$$

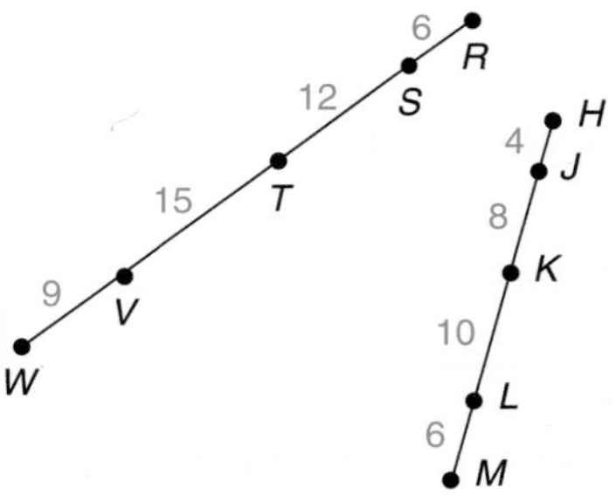

**FIGURE 5.45**

EXAMPLE 1

In Figure 5.46, points $D$ and $E$ divide $\overline{AB}$ and $\overline{AC}$ proportionally. If $AD = 4$, $DB = 7$, and $EC = 6$, find $AE$.

***Solution***

$\frac{AD}{AE} = \frac{DB}{EC}$, so $\frac{4}{x} = \frac{7}{6}$, where $x = AE$. Then $7x = 24$, so $x = AE = \frac{24}{7} = 3\frac{3}{7}$.

A property that will be proved in Exercise 29 of this section is

$$\text{If } \frac{a}{b} = \frac{c}{d}, \text{ then } \frac{a+c}{b+d} = \frac{a}{b} = \frac{c}{d}$$

In words, we may restate this property as follows:

> *The fraction whose numerator and denominator are determined, respectively, by adding numerators and denominators of equal fractions is equal to each of those equal fractions.*

Here is a numerical example of this claim:

$$\text{If } \frac{2}{3} = \frac{4}{6}, \text{ then } \frac{2+4}{3+6} = \frac{2}{3} = \frac{4}{6}$$

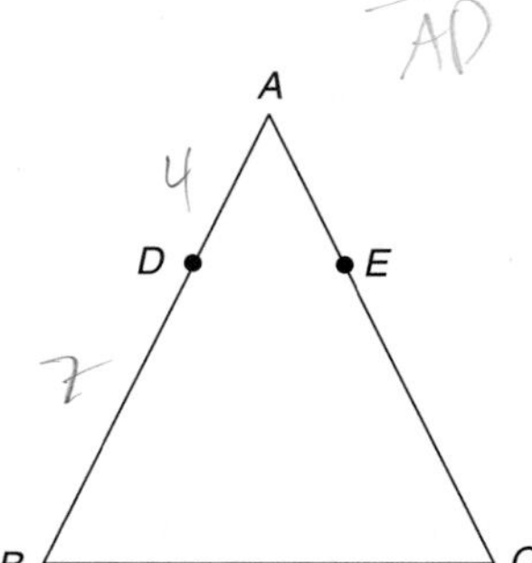

**FIGURE 5.46**

In Example 2, the preceding property is necessary as a reason.

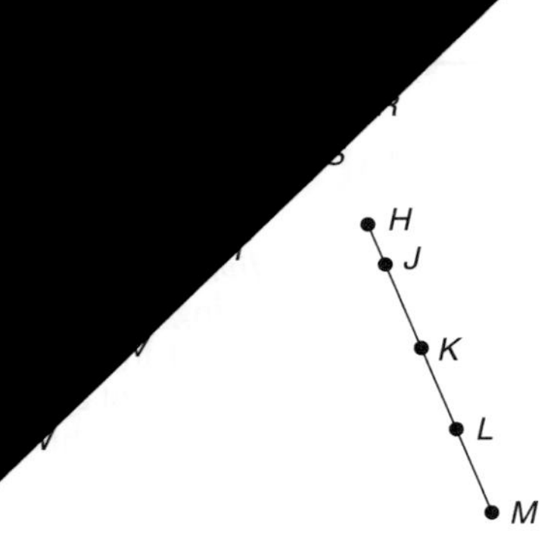

**FIGURE 5.47**

SSG
*Exs. 1, 2*

EXAMPLE 2

*Given:* $\overline{RW}$ and $\overline{HM}$ are divided proportionally at the points shown in Figure 5.47.

*Prove:* $\frac{RT}{HK} = \frac{TW}{KM}$

*Proof:* $\overline{RW}$ and $\overline{HM}$ are divided proportionally so that

$$\frac{RS}{HJ} = \frac{ST}{JK} = \frac{TV}{KL} = \frac{VW}{LM}$$

Using the property that if $\frac{a}{b} = \frac{c}{d}$, then $\frac{a+c}{b+d} = \frac{a}{b} = \frac{c}{d}$, we have

$$\frac{RS}{HJ} = \frac{RS + ST}{HJ + JK} = \frac{TV + VW}{KL + LM} = \frac{TV}{KL}$$

Because $RS + ST = RT$, $HJ + JK = HK$, $TV + VW = TW$, and $KL + LM = KM$,

$$\frac{RT}{HK} = \frac{TW}{KM}$$

Two properties that were introduced earlier (Property 3 of Section 5.1) are now recalled.

$$\text{If } \frac{a}{b} = \frac{c}{d}, \textit{ then } \frac{a \pm b}{b} = \frac{c \pm d}{d}$$

The subtraction part of the property is needed for the proof of Theorem 5.6.1.

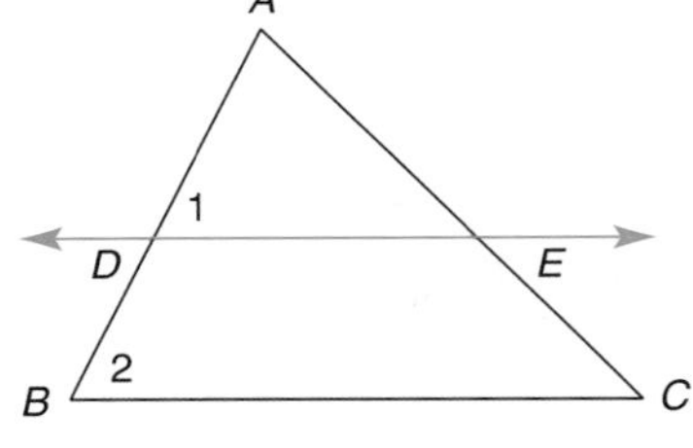


**FIGURE 5.48**

THEOREM 5.6.1: If a line is parallel to one side of a triangle and intersects the other two sides, then it divides these sides proportionally.

*Given:* In Figure 5.48, $\triangle ABC$ with $\overleftrightarrow{DE} \parallel \overline{BC}$ and with $\overleftrightarrow{DE}$ intersecting $\overline{AB}$ at $D$ and $\overline{AC}$ at $E$

*Prove:* $\frac{AD}{DB} = \frac{AE}{EC}$

*Proof:* Because $\overleftrightarrow{DE} \parallel \overline{BC}$, $\angle 1 \cong \angle 2$. With $\angle A$ as a common angle for $\triangle ADE$ and $\triangle ABC$, it follows by AA that these triangles are similar. Now

$$\frac{AB}{AD} = \frac{AC}{AE} \quad \text{by CSSTP}$$

By Property 3 of Section 5.1,

$$\frac{AB - AD}{AD} = \frac{AC - AE}{AE}$$

Because $AB - AD = DB$ and $AC - AE = EC$, the proportion becomes

$$\frac{DB}{AD} = \frac{EC}{AE}$$

Exs. 3–6

Inverting both fractions gives the desired conclusion:

$$\frac{AD}{DB} = \frac{AE}{EC}$$

---

COROLLARY 5.6.2: When three (or more) parallel lines are cut by a pair of transversals, the transversals are divided proportionally by the parallel lines.

*Given:* $p_1 \parallel p_2 \parallel p_3$ in Figure 5.49

*Prove:* $\frac{AB}{BC} = \frac{DE}{EF}$

**PICTURE PROOF OF COROLLARY 5.6.2**

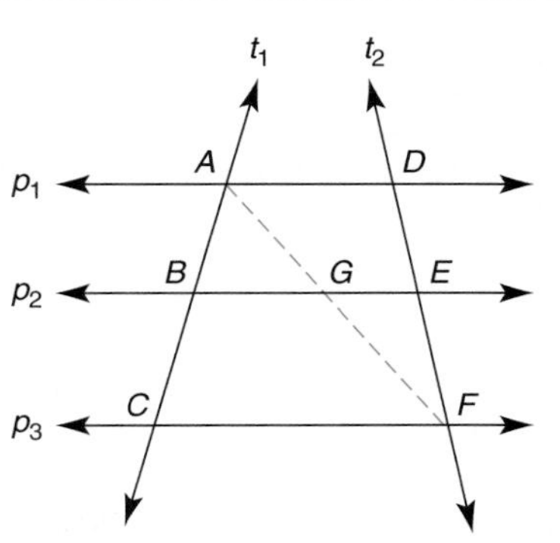


**FIGURE 5.49**

In Figure 5.49, draw $\overline{AF}$ as an auxiliary line segment.

Based upon Theorem 5.6.1, we see that $\frac{AB}{BC} = \frac{AG}{GF}$ in $\triangle ACF$ and that $\frac{AG}{GF} = \frac{DE}{EF}$ in $\triangle ADF$.

By the Transitive Property of Equality, $\frac{AB}{BC} = \frac{DE}{EF}$.

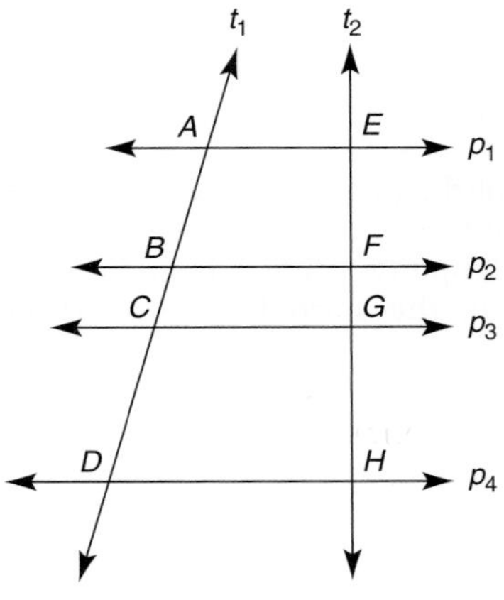


**FIGURE 5.50**

NOTE: By interchanging the means, we can write this proportion in the form $\frac{AB}{DE} = \frac{BC}{EF}$.

EXAMPLE 3

Given parallel lines $p_1$, $p_2$, $p_3$, and $p_4$ cut by $t_1$ and $t_2$ so that $AB = 4$, $EF = 3$, $BC = 2$, and $GH = 5$, find FG and CD. (See Figure 5.50.)

***Solution***

Because the transversals are divided proportionally,

$$\frac{AB}{EF} = \frac{BC}{FG} = \frac{CD}{GH}$$

so

$$\frac{4}{3} = \frac{2}{FG} = \frac{CD}{5}$$

Then $\quad 4 \cdot FG = 6 \quad$ and $\quad 3 \cdot CD = 20$

$$FG = \frac{3}{2} = 1\tfrac{1}{2} \quad \text{and} \quad CD = \frac{20}{3} = 6\tfrac{2}{3}$$

Exs. 7, 8

The following activity leads us to the relationship described in Theorem 5.6.3.

## Discover!

On a piece of paper, draw or construct $\triangle ABC$ whose sides measure $AB = 4$, $BC = 6$, and $AC = 5$. Then construct the angle bisector $\overrightarrow{BD}$ of $\angle B$. How does $\frac{AB}{AD}$ compare to $\frac{BC}{DC}$?

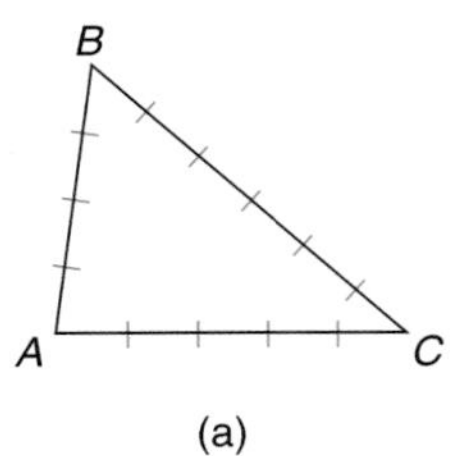

(a)

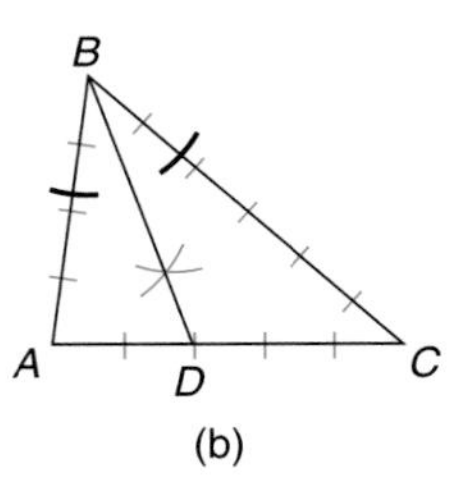

(b)

ANSWER

Though not by chance, it may come as a surprise that $\frac{AB}{AD} = \frac{BC}{DC}$ (that is, $\frac{4}{2} = \frac{6}{3}$) and $\frac{AB}{BC} = \frac{AD}{DC}$ $\left(\frac{4}{6} = \frac{2}{3}\right)$. It seems that the bisector of an angle included by two sides of a triangle separates the third side into segments whose lengths are proportional to the lengths of the two sides forming the angle.

The proof of Theorem 5.6.3 requires the use of Theorem 5.6.1.

> THEOREM 5.6.3: **(The Angle-Bisector Theorem)** If a ray bisects one angle of a triangle, then it divides the opposite side into segments whose lengths are proportional to the lengths of the two sides that form the bisected angle.

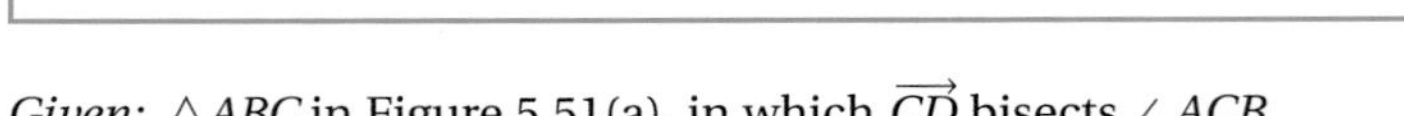

*Given:* $\triangle ABC$ in Figure 5.51(a), in which $\overrightarrow{CD}$ bisects $\angle ACB$

*Prove:* $\frac{AD}{AC} = \frac{DB}{CB}$

*Proof:* We begin by extending $\overline{BC}$ beyond $C$ (there is only one line through $B$ and $C$) to meet the line through $A$ drawn parallel to $\overline{DC}$. [See Figure 5.51(b).] Let $E$ be the point of intersection (these lines must intersect; otherwise $\overline{AE}$ would have two parallels, $\overline{BC}$ and $\overline{CD}$, through point $C$).

Because $\overline{CD} \parallel \overline{EA}$, we have

$$\frac{EC}{AD} = \frac{CB}{DB} \quad (*)$$

by Theorem 5.6.1. Now $\angle 1 \cong \angle 2$ (because of the angle bisector), $\angle 1 \cong \angle 3$ (corresponding angles for parallel lines), and $\angle 2 \cong \angle 4$ (alternate interior angles for parallel lines). By the Transitive Property, $\angle 3 \cong \angle 4$, so $\triangle ACE$ is isosceles with $\overline{EC} \cong \overline{AC}$. Using substitution, the starred (*) proportion becomes

$$\frac{AC}{AD} = \frac{CB}{DB} \quad \text{or} \quad \frac{AD}{AC} = \frac{DB}{CB} \quad \text{by inversion}$$

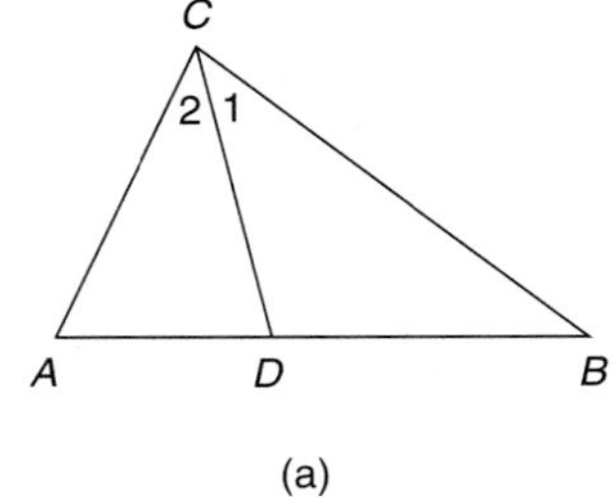

(a)

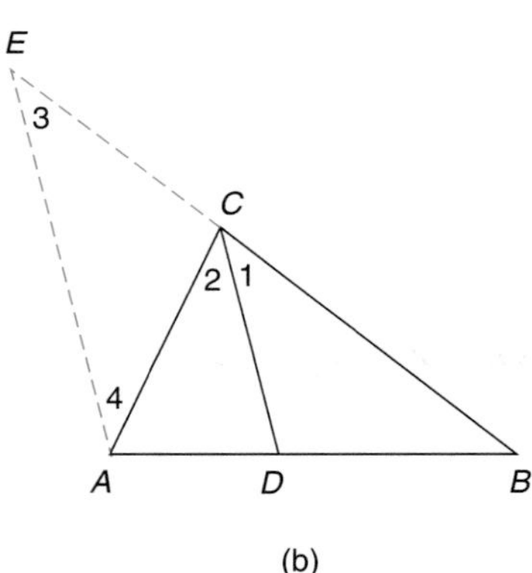

(b)

**FIGURE 5.51**

The Prove statement of the preceding theorem indicates that one form of the proportionality described is given by

$$\frac{\text{segment at left}}{\text{side at left}} = \frac{\text{segment at right}}{\text{side at right}}$$

Equivalently, the proportion could state

$$\frac{\text{segment at left}}{\text{segment at right}} = \frac{\text{side at left}}{\text{side at right}}$$

Other forms of the proportion are also possible!

EXAMPLE 4

For $\triangle XYZ$ in Figure 5.52, $XY = 3$ and $YZ = 5$. If $\overrightarrow{YW}$ bisects $\angle XYZ$ and $XW = 2$, find $XZ$.

***Solution***

Let $WZ = x$. We know that $\frac{YX}{XW} = \frac{YZ}{WZ}$, so $\frac{3}{2} = \frac{5}{x}$.

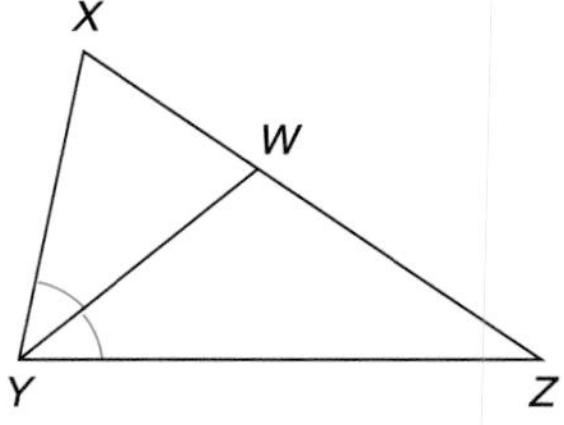

**FIGURE 5.52**

Therefore,

$$3x = 10$$
$$x = \frac{10}{3} = 3\tfrac{1}{3}$$

Then $WZ = 3\frac{1}{3}$.

*Exs. 9–13*

Because $XZ = XW + WZ$, we have $XZ = 2 + 3\frac{1}{3} = 5\frac{1}{3}$.

EXAMPLE 5

In Figure 5.52, $\triangle XYZ$ has sides of lengths $XY = 3$, $YZ = 4$, and $XZ = 5$. If $\overrightarrow{YW}$ bisects $\angle XYZ$, find $XW$ and $WZ$.

***Solution***

Let $XW = y$; then $WZ = 5 - y$, and $\frac{XY}{YZ} = \frac{XW}{WZ}$ becomes $\frac{3}{4} = \frac{y}{5 - y}$. From this proportion, we can find $y$.

$$3(5 - y) = 4y$$
$$15 - 3y = 4y$$
$$15 = 7y$$
$$y = \frac{15}{7}$$

Then $XW = \frac{15}{7} = 2\frac{1}{7}$ and $WZ = 5 - 2\frac{1}{7} = 2\frac{6}{7}$.

In the following example, we provide an alternative solution to a problem of the type found in Example 5.

EXAMPLE 6

In Figure 5.52, which is repeated here, $\triangle XYZ$ is isosceles with $\overline{XZ} \cong \overline{YZ}$. If $XY = 3$ and $YZ = 6$, find $XW$ and $WZ$.

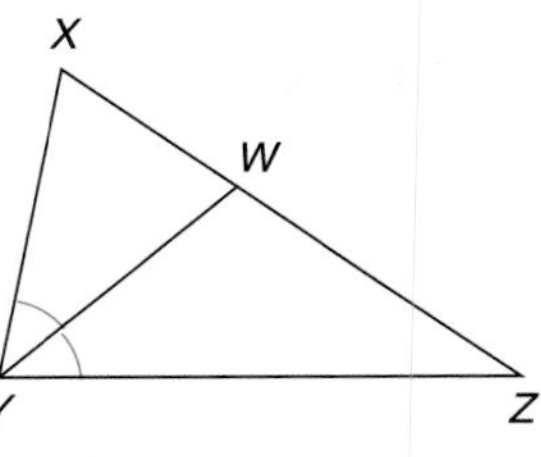


***Solution***

Because the ratio $XY{:}YZ$ is 3:6, or 1:2, the ratio $XW{:}WZ$ is also 1:2. Thus, we can represent these lengths by

$$XW = a \qquad \text{and} \qquad WZ = 2a$$

With $XZ = 6$ in the isosceles triangle, the statement $XW + WZ = XZ$ becomes $a + 2a = 6$, so $3a = 6$, and $a = 2$. Now $XW = 2$ and $WZ = 4$.

You will find the proof of the following theorem in the Perspective on History section at the end of this chapter. In Ceva's Theorem, point $D$ is *any* point in the interior of the triangle. See Figure 5.53(a). The auxiliary lines needed to complete the proof of Ceva's Theorem are shown in Figure 5.53(b). In the figure, line $\ell$ is drawn through vertex $C$ so that it is parallel to $\overline{AB}$. $\overline{BE}$ and $\overline{AF}$ are extended to meet $\ell$ at $R$ and $S$, respectively.

> THEOREM 5.6.4: **(Ceva's Theorem)** Let point $D$ be any point in the interior of $\triangle ABC$, and let $\overline{BE}$, $\overline{AF}$, and $\overline{CG}$ be the line segments determined by $D$ and vertices of $\triangle ABC$. Then the product of the ratios of the lengths of the segments of each of the three sides (taken in order from a given vertex of the triangle) equals 1; that is,
>
> $$\frac{AG}{GB}\cdot\frac{BF}{FC}\cdot\frac{CE}{EA}=1$$

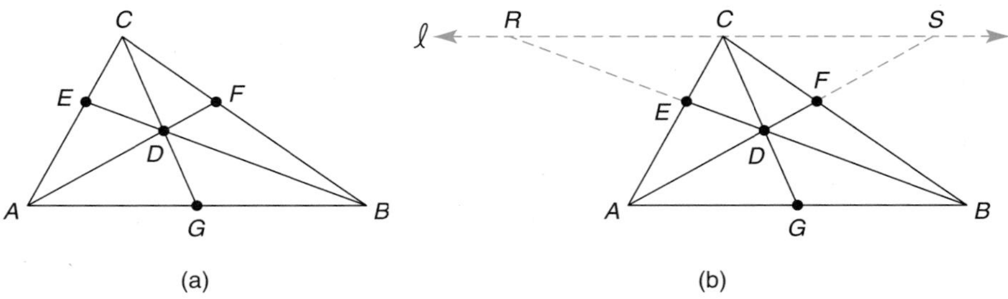


**FIGURE 5.53**

We will apply Ceva's Theorem in Example 7.

EXAMPLE 7

In $\triangle RST$ with interior point $D$, $RG = 6$, $GS = 4$, $SH = 4$, $HT = 3$, and $KR = 5$. Find $TK$. See Figure 5.54.

***Solution***

Let $TK = a$. Applying Ceva's Theorem and following a counterclockwise path beginning at vertex $R$, we have $\frac{RG}{GS}\cdot\frac{SH}{HT}\cdot\frac{TK}{KR} = 1$. Then $\frac{6}{4}\cdot\frac{4}{3}\cdot\frac{a}{5} = 1$ and so $\frac{\cancel{6}^2}{\cancel{4}_1}\cdot\frac{\cancel{4}^1}{\cancel{3}_1}\cdot\frac{a}{5} = 1$ becomes $\frac{2a}{5} = 1$. Then $2a = 5$ and $a = 2.5$; thus, $TK = 2.5$.

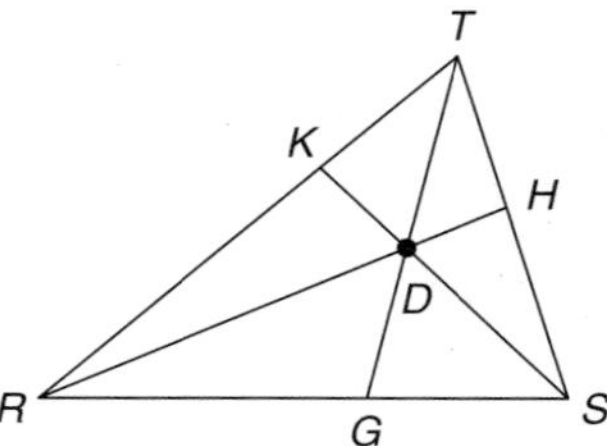


**FIGURE 5.54**

SSG

*Ex. 14*

## 5.6 Exercises

1. In preparing a certain recipe, a chef uses 5 oz of ingredient A, 4 oz of ingredient B, and 6 oz of ingredient C. If 90 oz of this dish are needed, how many ounces of each ingredient should be used?

2. In a chemical mixture, 2 g of chemical A are used for each gram of chemical B, and 3 g of chemical C are needed for each gram of B. If 72 g of the mixture are prepared, what amount (in grams) of each chemical is needed?

3. Given that $\frac{AB}{EF} = \frac{BC}{FG} = \frac{CD}{GH}$, do the following proportions hold?

   a) $\frac{AC}{EG} = \frac{CD}{GH}$

   b) $\frac{AB}{EF} = \frac{BD}{FH}$

4. Given that $\overleftrightarrow{XY} \parallel \overline{TS}$, do the following proportions hold?

   a) $\frac{TX}{XR} = \frac{RY}{YS}$

   b) $\frac{TR}{XR} = \frac{SR}{YR}$

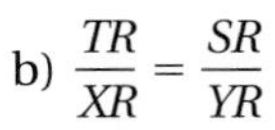

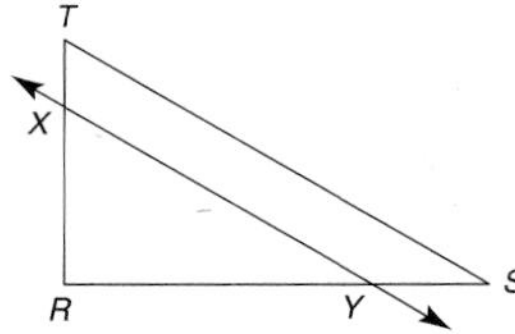

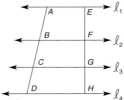

5. *Given:* $\ell_1 \parallel \ell_2 \parallel \ell_3 \parallel \ell_4$, $AB = 5$, $BC = 4$, $CD = 3$, $EH = 10$
   *Find:* $EF$, $FG$, $GH$

6. *Given:* $\ell_1 \parallel \ell_2 \parallel \ell_3 \parallel \ell_4$, $AB = 7$, $BC = 5$, $CD = 4$, $EF = 6$
   *Find:* $FG$, $GH$, $EH$

*Exercises 5, 6*

7. *Given:* $\ell_1 \parallel \ell_2 \parallel \ell_3$, $AB = 4$, $BC = 5$, $DE = x$, $EF = 12 - x$
   *Find:* $x$, $DE$, $EF$

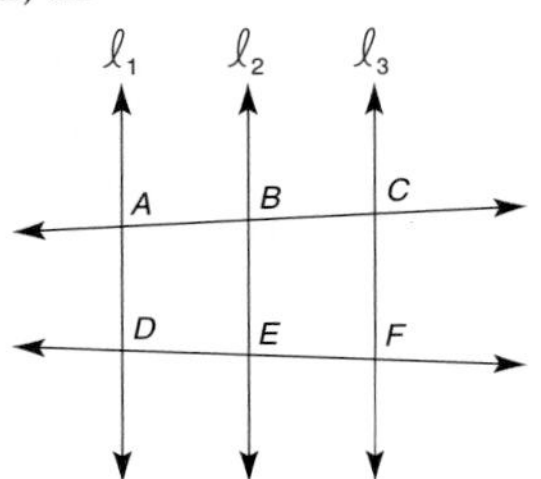


*Exercises 7, 8*

8. *Given:* $\ell_1 \parallel \ell_2 \parallel \ell_3$, $AB = 5$, $BC = x$, $DE = x - 2$, $EF = 7$
   *Find:* $x$, $BC$, $DE$ (See figure for Exercise 7.)

9. *Given:* $\overleftrightarrow{DE} \parallel \overline{BC}$, $AD = 5$, $DB = 12$, $AE = 7$
   *Find:* $EC$

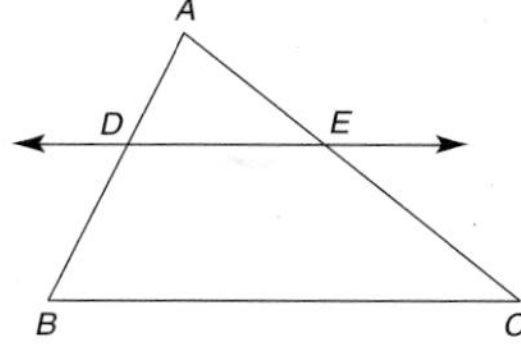


*Exercises 9–12*

10. *Given:* $\overleftrightarrow{DE} \parallel \overline{BC}$, $AD = 6$, $DB = 10$, $AC = 20$
    *Find:* $EC$

11. *Given:* $\overleftrightarrow{DE} \parallel \overline{BC}$, $AD = a - 1$, $DB = 2a + 2$, $AE = a$, $EC = 4a - 5$
    *Find:* $a$ and $AD$

12. *Given:* $\overleftrightarrow{DE} \parallel \overline{BC}$, $AD = 5$, $DB = a + 3$, $AE = a + 1$, $EC = 3(a - 1)$
    *Find:* $a$ and $EC$

13. *Given:* $\overrightarrow{RW}$ bisects $\angle SRT$
    Do the following equalities hold?

    a) $SW = WT$

    b) $\frac{RS}{RT} = \frac{SW}{WT}$

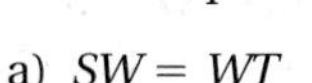

*Exercises 13, 14*

14. *Given:* $\overrightarrow{RW}$ bisects $\angle SRT$
    Do the following equalities hold?

    a) $\frac{RS}{SW} = \frac{RT}{WT}$

    b) $m\angle S = m\angle T$

15. *Given:* $\overrightarrow{UT}$ bisects $\angle WUV$, $WU = 8$, $UV = 12$, $WT = 6$
    *Find:* $TV$

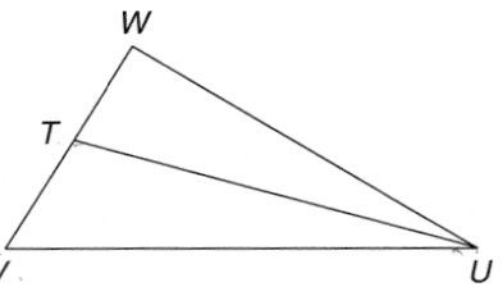


*Exercises 15, 16*

16. *Given:* $\overrightarrow{UT}$ bisects $\angle WUV$, $WU = 9$, $UV = 12$, $WV = 9$
    *Find:* $WT$

17. *Given:* $\overrightarrow{NQ}$ bisects $\angle MNP$, $NP = MQ$, $QP = 8$, $MN = 12$
*Find:* $NP$

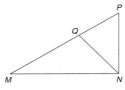

18. *Given:* In $\triangle ABC$, $\overrightarrow{AD}$ bisects $\angle BAC$
$AB = 20$ and $AC = 16$
*Find:* $DC$ and $DB$

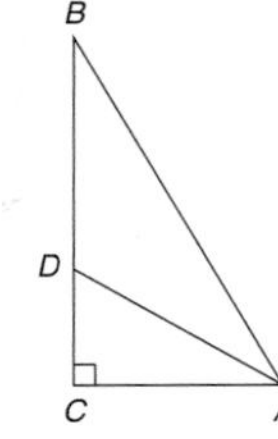


19. In $\triangle ABC$, $\angle ACB$ is trisected by $\overrightarrow{CD}$ and $\overrightarrow{CE}$ so that $\angle 1 \cong \angle 2 \cong \angle 3$. Write two different proportions that follow from this information.

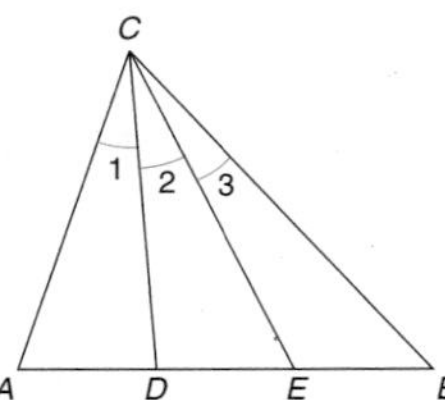


20. In $\triangle ABC$, $m\angle CAB = 80°$, $m\angle ACB = 60°$, and $m\angle ABC = 40°$. With the angle bisectors as shown, which line segment is longer?

a) $\overline{AE}$ or $\overline{EC}$? b) $\overline{CD}$ or $\overline{DB}$? c) $\overline{AF}$ or $\overline{FB}$?

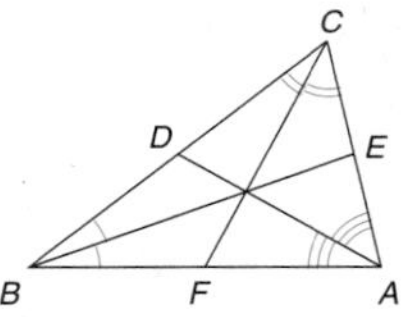


21. In right $\triangle RST$ (not shown) with right $\angle S$, $\overrightarrow{RV}$ bisects $\angle SRT$ so that $V$ lies on side $\overline{ST}$. If $RS = 6$, $ST = 6\sqrt{3}$, and $RT = 12$, find $SV$ and $VT$.

22. *Given:* $AC$ is the geometric mean between $AD$ and $AB$.
$AD = 4$, and $DB = 6$
*Find:* $AC$

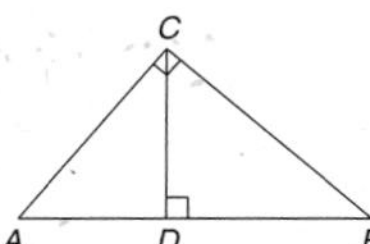


23. *Given:* $\overrightarrow{RV}$ bisects $\angle SRT$,
$RS = x - 6$, $SV = 3$,
$RT = 2 - x$, and
$VT = x + 2$
*Find:* $x$
(HINT: You will need to apply the Quadratic Formula.)

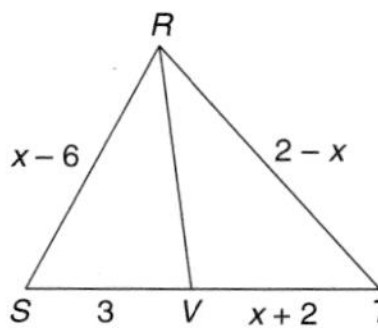


24. *Given:* $\overrightarrow{MR}$ bisects $\angle NMP$, $MN = 2x$, $NR = x$, $RP = x + 1$, and $MP = 3x - 1$
*Find:* $x$

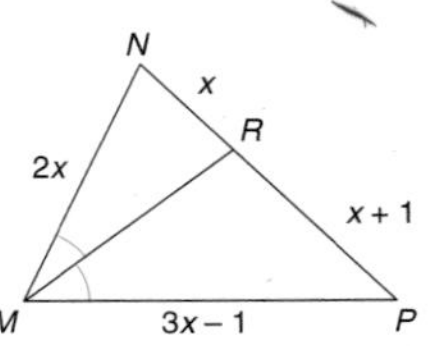


25. Given point $D$ in the interior of $\triangle RST$, which statement(s) is(are) true?

a) $\frac{RK}{KT} \cdot \frac{TH}{HS} \cdot \frac{GS}{RG} = 1$

b) $\frac{TK}{KR} \cdot \frac{RG}{GS} \cdot \frac{SH}{HT} = 1$

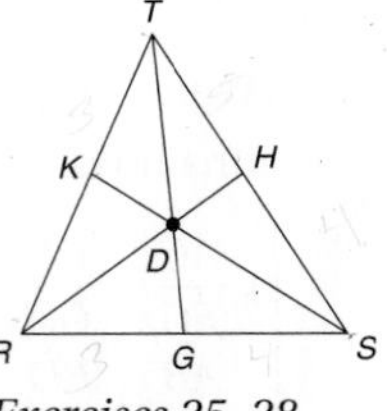

*Exercises 25–28*

26. In $\triangle RST$ shown in Exercise 25, suppose that $\overline{RH}$, $\overline{TG}$, and $\overline{SK}$ are medians. Find the value of:

a) $\frac{RK}{KT}$ b) $\frac{TH}{HS}$

27. Given point $D$ in the interior of $\triangle RST$, suppose that $RG = 3$, $GS = 4$, $SH = 4$, $HT = 5$, and $KT = 3$. Find $RK$.

28. Given point $D$ in the interior of $\triangle RST$, suppose that $RG = 2$, $GS = 3$, $SH = 3$, and $HT = 4$. Find $\frac{KT}{KR}$.

29. Complete the proof of this property:

If $\frac{a}{b} = \frac{c}{d}$, then $\frac{a+c}{b+d} = \frac{a}{b}$ and $\frac{a+c}{b+d} = \frac{c}{d}$

**PROOF**

| Statements | Reasons |
|---|---|
| 1. $\frac{a}{b} = \frac{c}{d}$ | 1. ? |
| 2. $b \cdot c = a \cdot d$ | 2. ? |
| 3. $ab + bc = ab + ad$ | 3. ? |
| 4. $b(a + c) = a(b + d)$ | 4. ? |
| 5. $\frac{a+c}{b+d} = \frac{a}{b}$ | 5. Means-Extremes Property (symmetric form) |
| 6. $\frac{a+c}{b+d} = \frac{c}{d}$ | 6. ? |

30. *Given:* $\triangle RST$, with $\overleftrightarrow{XY} \parallel \overline{RT}$, $\overleftrightarrow{YZ} \parallel \overline{RS}$
*Prove:* $\frac{RX}{XS} = \frac{ZT}{RZ}$

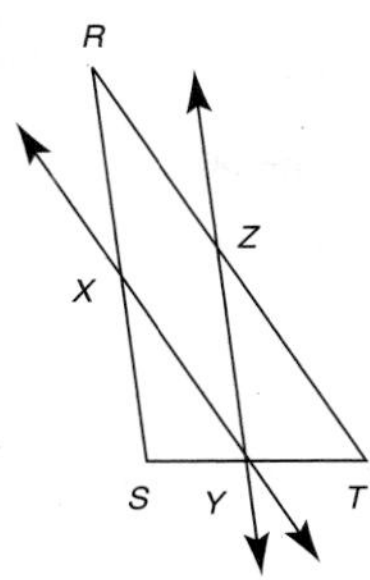


31. Use Theorem 5.6.1 and the drawing to complete the proof of this theorem: "If a line is parallel to one side of a triangle and passes through the midpoint of a second side, then it will pass through the midpoint of the third side."

*Given:* $\triangle RST$ with $M$ the midpoint of $\overline{RS}$; $\overleftrightarrow{MN} \parallel \overline{ST}$
*Prove:* $N$ is the midpoint of $\overline{RT}$

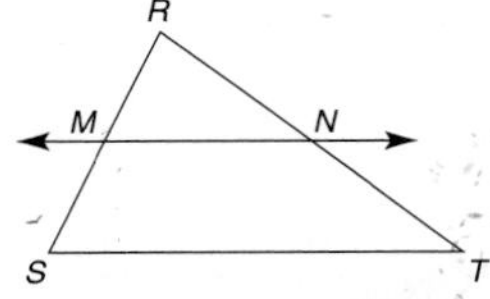


32. Use Exercise 31 and the following drawing to complete the proof of this theorem: "The length of the median of a trapezoid is one-half the sum of the lengths of the two bases."

*Given:* Trapezoid $ABCD$ with median $\overline{MN}$
*Prove:* $MN = \frac{1}{2}(AB + CD)$

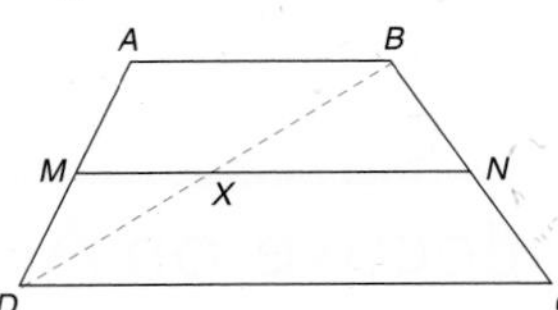


33. Use Theorem 5.6.3 to complete the proof of this theorem: "If the bisector of an angle of a triangle also bisects the opposite side, then the triangle is an isosceles triangle."

*Given:* $\triangle XYZ$; $\overrightarrow{YW}$ bisects $\angle XYZ$; $\overline{WX} \cong \overline{WZ}$
*Prove:* $\triangle XYZ$ is isosceles
(HINT: Use a proportion to show that $YX = YZ$.)

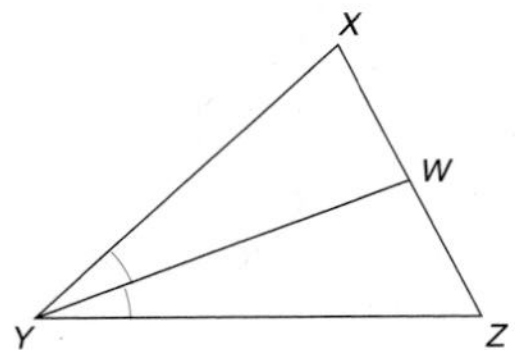


*34. In right $\triangle ABC$ (not shown) with right $\angle C$, $\overrightarrow{AD}$ bisects $\angle BAC$ so that $D$ lies on side $\overline{CB}$. If $AC = 6$ and $DC = 3$, find $BD$ and $AB$.
(HINT: Let $BD = x$ and $AB = 2x$. Then use the Pythagorean Theorem.)

*35. *Given:* $\triangle ABC$ (not shown) is isosceles with $m\angle ABC = m\angle C = 72°$; $\overrightarrow{BD}$ bisects $\angle ABC$ and $AB = 1$
*Find:* $BC$

*36. *Given:* $\triangle RST$ with right $\angle RST$; $m\angle R = 30°$ and $ST = 6$; $\angle RST$ is trisected by $\overrightarrow{SM}$ and $\overrightarrow{SN}$
*Find:* $TN$, $NM$, and $MR$

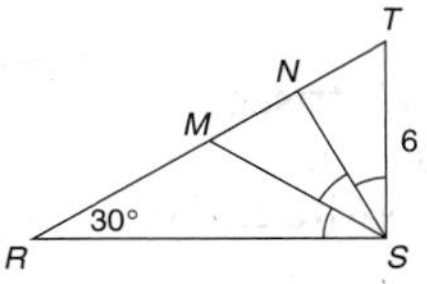


*37. In the figure, the angle bisectors of $\triangle ABC$ intersect at a point in the interior of the triangle. If $BC = 5$, $BA = 6$, and $CA = 4$, find:

a) $CD$ and $DB$ (HINT: Use Theorem 5.6.3.)
b) $CE$ and $EA$
c) $BF$ and $FA$
d) Use results from parts (a), (b), and (c) to show that $\frac{BD}{DC} \cdot \frac{CE}{EA} \cdot \frac{AF}{FB} = 1$.

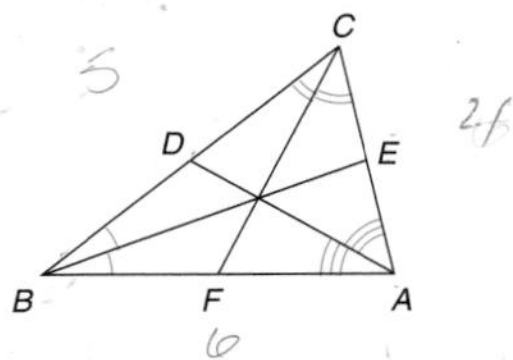


*38. In $\triangle RST$, the altitudes of the triangle intersect at a point in the interior of the triangle. The lengths of the sides of $\triangle RST$ are $RS = 14$, $ST = 15$, and $TR = 13$.

a) If $TX = 12$, find $RX$ and $XS$.
(HINT: Use the Pythagorean Theorem)
b) If $RY = \frac{168}{15}$, find $TY$ and $YS$.
c) If $SZ = \frac{168}{13}$, find $ZR$ and $TZ$.
d) Use results from parts (a), (b), and (c) to show that $\frac{RX}{XS} \cdot \frac{SY}{YT} \cdot \frac{TZ}{ZR} = 1$.

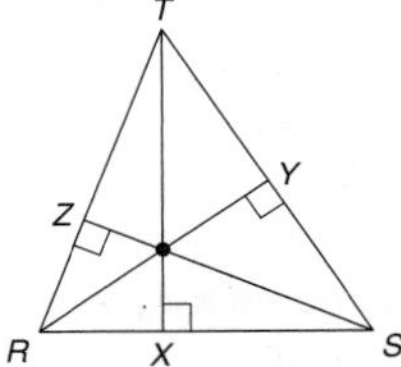

# Perspective on History

## CEVA'S PROOF

Giovanni Ceva (1647–1736) was the Italian mathematician for whom Ceva's Theorem is named. Although his theorem is difficult to believe, its proof is not lengthy. The proof follows.

THEOREM 5.6.4: **(Ceva's Theorem)** Let point $D$ be any point in the interior of $\triangle ABC$, and let $\overline{BE}$, $\overline{AF}$, and $\overline{CG}$ be the line segments determined by $D$ and vertices of $\triangle ABC$. Then the product of the ratios of the segments of each of the three sides (taken in order from a given vertex of the triangle) equals 1; that is, $\frac{AG}{GB} \cdot \frac{BF}{FC} \cdot \frac{CE}{EA} = 1$.

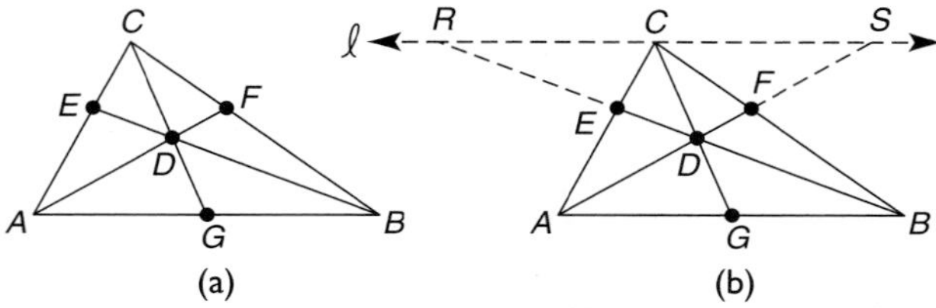


**FIGURE 5.55**

### Proof

Given $\triangle ABC$ with interior point $D$ [see Figure 5.55(a)], draw a line $\ell$ through point $C$ that is parallel to $\overline{AB}$. Now extend $\overline{BE}$ to meet $\ell$ at point $R$. Likewise, extend $\overline{AF}$ to meet $\ell$ at point $S$. See Figure 5.55(b). With similar triangles, we will be able to substitute desired ratios into the obvious statement $\frac{CS}{CR} \cdot \frac{AB}{CS} \cdot \frac{CR}{AB} = 1$ (*), in which each numerator has a matching denominator. Because $\triangle AGD \sim \triangle SCD$ by AA, we have $\frac{AG}{CS} = \frac{GD}{CD}$. Also with $\triangle DGB \sim \triangle DCR$, we have $\frac{GD}{CD} = \frac{GB}{CR}$. By the Transitive Property of Equality, $\frac{AG}{CS} = \frac{GB}{CR}$, and by interchanging the means, we see that $\frac{AG}{BG} = \frac{CS}{CR}$. [The first ratio, $\frac{AG}{BG}$, of this proportion will replace the ratio $\frac{CS}{CR}$ in the starred (*) statement.]

From the fact that $\triangle CSF \sim \triangle BAF$, $\frac{AB}{SC} = \frac{BF}{FC}$. [The second ratio, $\frac{BF}{FC}$, of this proportion will replace the ratio $\frac{AB}{CS}$ in the starred (*) statement.]

With $\triangle RCE \sim \triangle BAE$, $\frac{CE}{EA} = \frac{CR}{AB}$. [The first ratio, $\frac{CE}{EA}$, of this proportion replaces $\frac{CR}{AB}$ in the starred (*) statement.] Making the indicated substitutions into the starred statement, we have

$$\frac{AG}{GB} \cdot \frac{BF}{FC} \cdot \frac{CE}{EA} = 1$$

# Perspective on Application

## AN UNUSUAL APPLICATION OF SIMILAR TRIANGLES

The following problem is one that can be solved in many ways. If methods of calculus are applied, the solution is found through many complicated and tedious calculations. The simplest solution, which follows, utilizes geometry and similar triangles.

*Problem:* A hiker is at a location 450 ft downstream from his campsite. He is 200 ft away from the straight stream, and his tent is 100 ft away, as shown in Figure 5.56(a). Across the flat field, he sees that a spark from his campfire has ignited the tent. Taking the empty bucket he is carrying, he runs to the river to get water and then on to the tent. To what point on the river should he run to minimize the distance he travels?

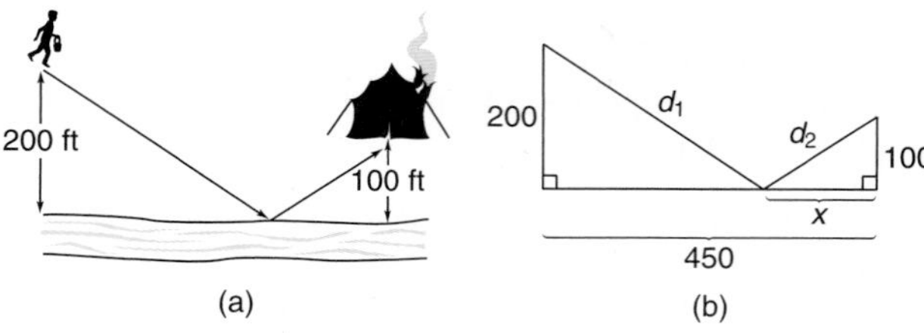


**FIGURE 5.56**

We wish to determine $x$ in Figure 5.56(b) so that the total distance $D = d_1 + d_2$ is as small as possible. Consider three possible choices of this point on the river. These are suggested by dashed, dotted, and solid lines in Figure 5.57(a). Also consider the reflections of the triangles across the river. [See Figure 5.57(b).]

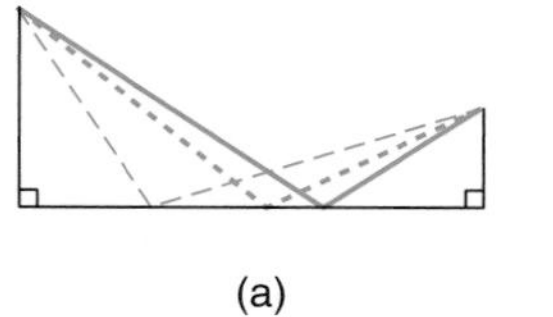

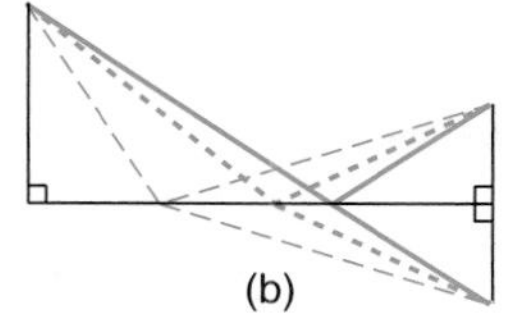

(a) (b)

**FIGURE 5.57**

The minimum distance $D$ occurs where the segments of lengths $d_1$ and $d_2$ form a straight line. That is, the configuration with the solid line segments minimizes the distance. In that case, the triangle at left and the reflected triangle at right are similar. (See Figure 5.58.)

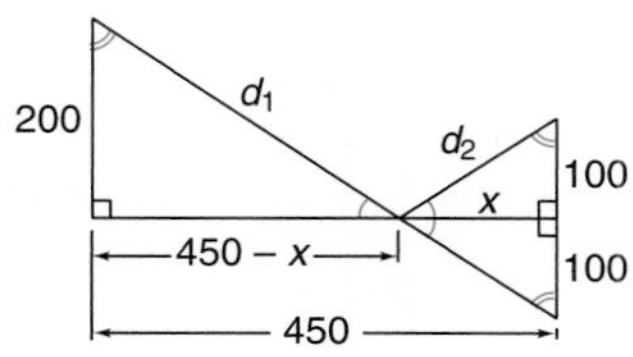


**FIGURE 5.58**

Thus
$$\frac{200}{100} = \frac{450 - x}{x}$$
$$200x = 100(450 - x)$$
$$200x = 45{,}000 - 100x$$
$$300x = 45{,}000$$
$$x = 150$$

Accordingly, the desired point on the river is 300 ft (determined by $450 - x$) upstream from the hiker's location.

# Summary

### A LOOK BACK AT CHAPTER 5

One goal of this chapter has been to define similarity for two polygons. We postulated a method for proving triangles similar and showed that proportions are a consequence of similar triangles, a line parallel to one side of a triangle, and a ray bisecting one angle of a triangle. The Pythagorean Theorem and its converse were proved. We discussed the 30°-60°-90° triangle, the 45°-45°-90° triangle, and other special right triangles with sides forming Pythagorean triples. The final section developed the concept segments divided proportionally.

### A LOOK AHEAD TO CHAPTER 6

In the next chapter, we will begin our work with the circle. Segments and lines of the circle will be defined, as will special angles in a circle. Several theorems dealing with the measurements of these angles and line segments will be proved. Our work with constructions will enable us to deal with the locus of points and the concurrence of lines.

### KEY CONCEPTS

5.1 Ratio • Rate • Proportion • Extremes • Means • Means-Extremes Property • Geometric Mean • Extended Ratio • Extended Proportion

5.2 Similar Polygons • Congruent Polygons • Corresponding Vertices, Angles, and Sides

5.3 AAA • AA • CSSTP • CASTC • SAS~ and SSS~

5.4 Pythagorean Theorem • Converse of Pythagorean Theorem • Pythagorean Triple

5.5 The 45-45-90 Triangle • The 30-60-90 Triangle

5.6 Segments Divided Proportionally • The Angle-Bisector Theorem • Ceva's Theorem

## TABLE 5.2 AN OVERVIEW OF CHAPTER FIVE

### Methods of Proving Triangles Similar ($\triangle ABC \sim \triangle DEF$)

| *Figure (Note marks.)* | *Method* | *Steps Needed in Proof* |
|---|---|---|
| 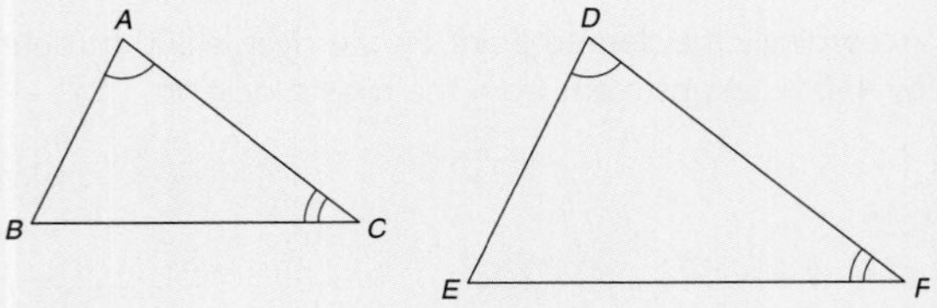 A, B, C; D, E, F | AA | $\angle A \cong \angle D$; $\angle C \cong \angle F$ |
| 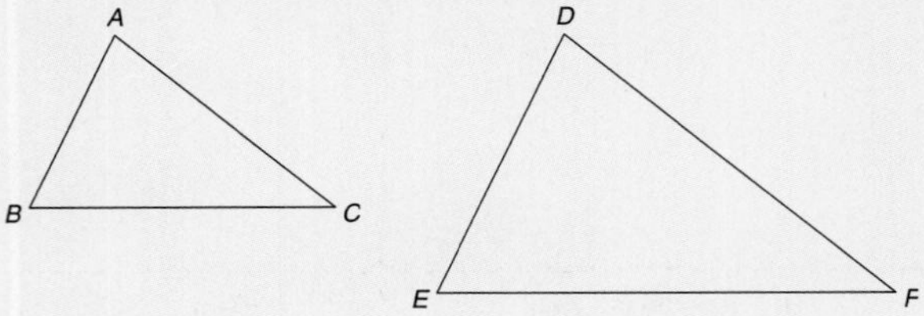 A, B, C; D, E, F | SSS ~ | $\frac{AB}{DE} = \frac{AC}{DF} = \frac{BC}{EF} = k$ <br> ($k$ is a constant.) |
| 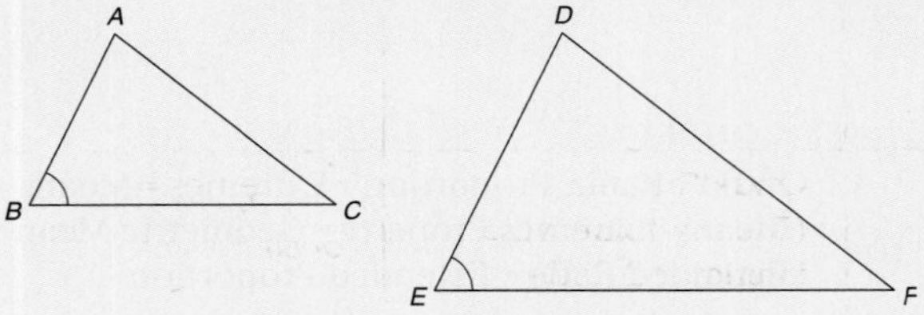 A, B, C; D, E, F | SAS ~ | $\frac{AB}{DE} = \frac{BC}{EF} = k$ <br> $\angle B \cong \angle E$ |

### Special Relationships

| *Figure* | *Relationship* | *Conclusion(s)* |
|---|---|---|
| 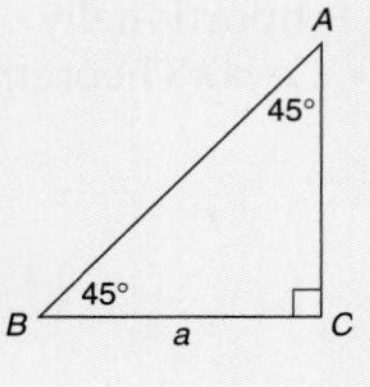 A, 45°, 45°, B, a, C | 45°-45°-90° $\triangle$ <br> NOTE: $BC = a$ | $AC = a$ <br> $AB = a\sqrt{2}$ |
| 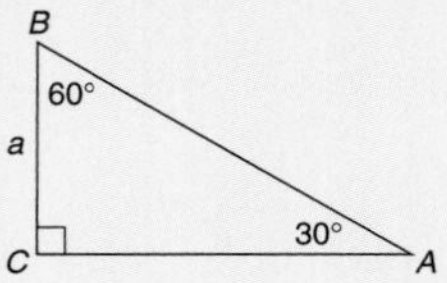 B, 60°, a, C, 30°, A | 30°-60°-90° $\triangle$ <br> NOTE: $BC = a$ | $AC = a\sqrt{3}$ <br> $AB = 2a$ |

## TABLE 5.2 CONT.

### Segments Divided Proportionally

| *Figure* | *Relationship* | *Conclusion* |
|---|---|---|
| 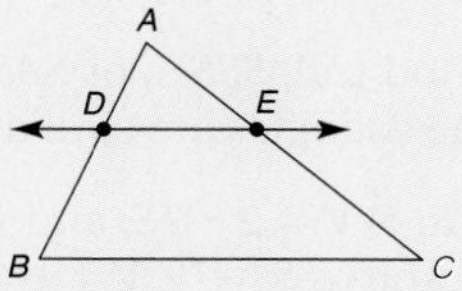 | $\overleftrightarrow{DE} \parallel \overline{BC}$ | $\frac{AD}{DB} = \frac{AE}{EC}$ or $\frac{AD}{AE} = \frac{DB}{EC}$ |
| 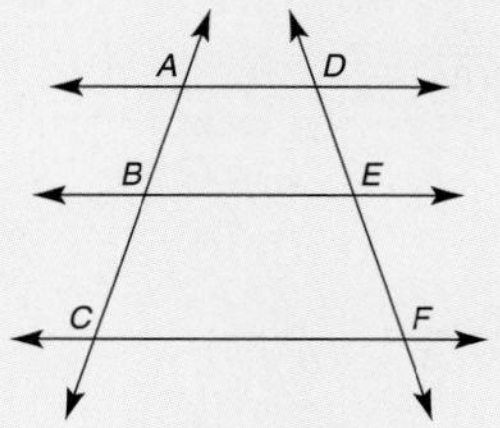 | $\overleftrightarrow{AD} \parallel \overleftrightarrow{BE} \parallel \overleftrightarrow{CF}$ | $\frac{AB}{BC} = \frac{DE}{EF}$ or $\frac{AB}{DE} = \frac{BC}{EF}$ |
| 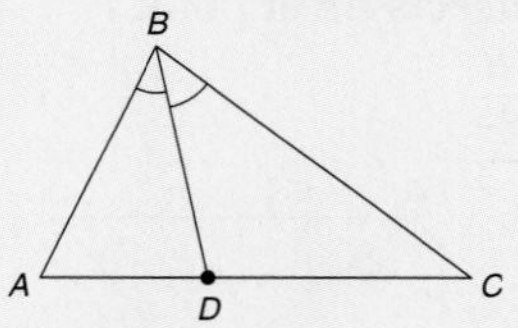 | $\overrightarrow{BD}$ bisects $\angle ABC$ | $\frac{AB}{BC} = \frac{AD}{DC}$ or $\frac{AB}{AD} = \frac{BC}{DC}$ |
| 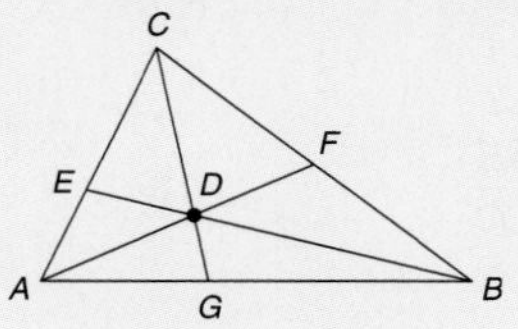 | Ceva's Theorem ($D$ is any point in the interior of $\triangle ABC$.) | $\frac{AG}{GB} \cdot \frac{BF}{FC} \cdot \frac{CE}{EA} = 1$ or equivalent |

# Chapter 5 Review Exercises

*Answer true or false for Review Exercises 1 to 7.*

1. The ratio of 12 hr to 1 day is 2 to 1.
2. If the numerator and the denominator of a ratio are multiplied by 4, the new ratio equals the given ratio.
3. The value of a ratio must be less than 1.
4. The three numbers 6, 14, and 22 are in a ratio of 3:7:11.
5. To express a ratio correctly, the terms must have the same unit of measure.
6. The ratio 3:4 is the same as the ratio 4:3.
7. If the second and third terms of a proportion are equal, then either is the geometric mean of the first and fourth terms.
8. Find the value(s) of $x$ in each proportion:

   a) $\frac{x}{6} = \frac{3}{x}$ e) $\frac{x-2}{x-5} = \frac{2x+1}{x-1}$

   b) $\frac{x-5}{3} = \frac{2x-3}{7}$ f) $\frac{x(x+5)}{4x+4} = \frac{9}{5}$

   c) $\frac{6}{x+4} = \frac{2}{x+2}$ g) $\frac{x-1}{x+2} = \frac{10}{3x-2}$

   d) $\frac{x+3}{5} = \frac{x+5}{7}$ h) $\frac{x+7}{2} = \frac{x+2}{x-2}$

*Use proportions to solve Review Exercises 9 to 11.*

9. Four containers of fruit juice cost \$2.52. How much do six containers cost?
10. Two packages of M&Ms cost 69¢. How many packages can you buy for \$2.25?
11. A rug measuring 20 square meters costs \$132. How much would a 12 square-meter rug of the same material cost?
12. The ratio of the measures of the sides of a quadrilateral is 2:3:5:7. If the perimeter is 68, find the length of each side.
13. The length and width of a rectangle are 18 and 12, respectively. A similar rectangle has length 27. What is its width?
14. The sides of a triangle are 6, 8, and 9. The shortest side of a similar triangle is 15. How long are its other sides?
15. The ratio of the measure of the supplement of an angle to that of the complement of the angle is 5:2. Find the measure of the supplement.
16. Name the method (*AA*, *SSS~*, or *SAS~*) that is used to show that the triangles are similar.

   a) $WU = 2 \cdot TR$, $WV = 2 \cdot TS$, and $UV = 2 \cdot RS$
   b) $\angle T \cong \angle W$ and $\angle S \cong \angle V$
   c) $\angle T \cong \angle W$ and $\frac{TR}{WU} = \frac{TS}{WV}$
   d) $\frac{TR}{WU} = \frac{TS}{WV} = \frac{RS}{UV}$

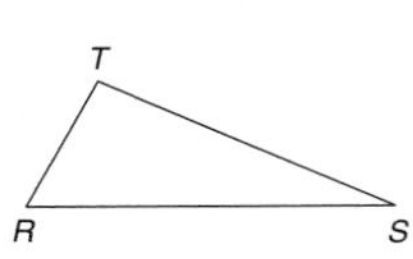


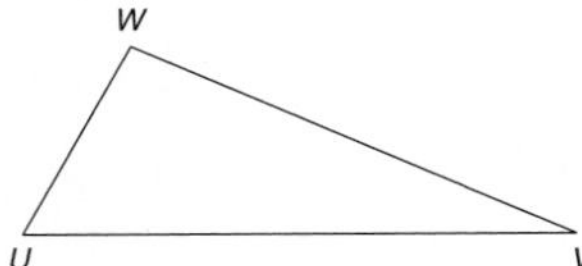


17. *Given:* $ABCD$ is a parallelogram
   $\overline{DB}$ intersects $\overline{AE}$ at point $F$

   *Prove:* $\frac{AF}{EF} = \frac{AB}{DE}$

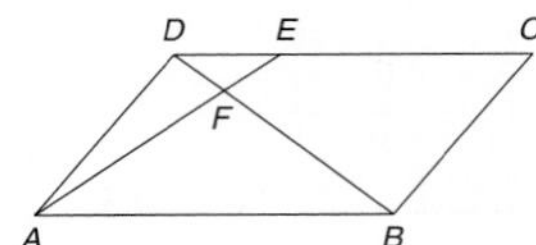


18. *Given:* $\angle 1 \cong \angle 2$

   *Prove:* $\frac{AB}{AC} = \frac{BE}{CD}$

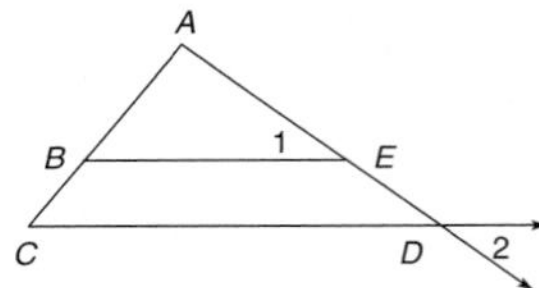


19. *Given:* $\triangle ABC \sim \triangle DEF$ (not shown)
   $m\angle A = 50°$, $m\angle E = 33°$
   $m\angle D = 2x + 40$

   *Find:* $x$, $m\angle F$

20. *Given:* In $\triangle ABC$ and $\triangle DEF$ (not shown)
   $\angle B \cong \angle F$ and $\angle C \cong \angle E$
   $AC = 9$, $DE = 3$, $DF = 2$, $FE = 4$

   *Find:* $AB$, $BC$

*For Review Exercises 21 to 23,* $\overline{DE} \parallel \overline{AC}$.

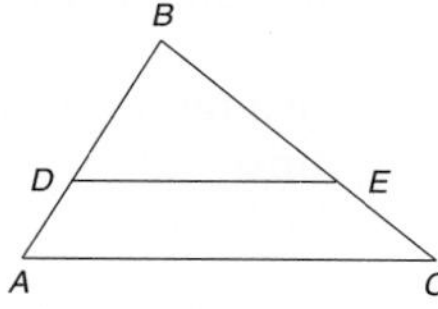


*Exercises 21–23*

21. $BD = 6$, $BE = 8$, $EC = 4$, $AD = ?$

22. $AD = 4$, $BD = 8$, $DE = 3$, $AC = ?$

23. $AD = 2$, $AB = 10$, $BE = 5$, $BC = ?$

*For Review Exercises 24 to 26,* $\overrightarrow{GJ}$ *bisects* $\angle FGH$.

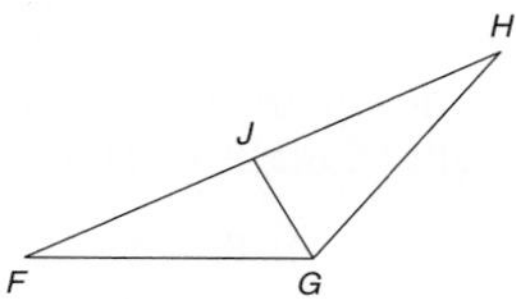


*Exercises 24–26*

24. *Given:* $FG = 10$, $GH = 8$, $FJ = 7$
 *Find:* $JH$

25. *Given:* $GF{:}GH = 1{:}2$, $FJ = 5$
 *Find:* $JH$

26. *Given:* $FG = 8$, $HG = 12$, $FH = 15$
 *Find:* $FJ$

27. *Given:* $\overleftrightarrow{EF} \parallel \overleftrightarrow{GO} \parallel \overleftrightarrow{HM} \parallel \overleftrightarrow{JK}$, with transversals $\overline{FJ}$ and $\overline{EK}$
 $FG = 2$, $GH = 8$, $HJ = 5$, $EM = 6$
 *Find:* $EO$, $EK$

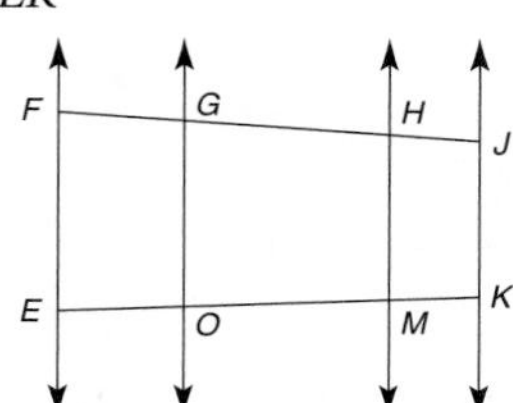


28. Prove that if a line bisects one side of a triangle and is parallel to a second side, then it bisects the third side.

29. Prove that the diagonals of a trapezoid divide themselves proportionally.

30. *Given:* $\triangle ABC$ with right $\angle BAC$
 $\overline{AD} \perp \overline{BC}$

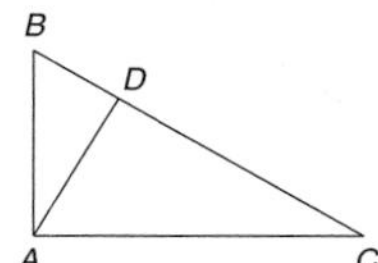


 a) $BD = 3$, $AD = 5$, $DC = ?$
 b) $AC = 10$, $DC = 4$, $BD = ?$
 c) $BD = 2$, $BC = 6$, $BA = ?$
 d) $BD = 3$, $AC = 3\sqrt{2}$, $DC = ?$

31. *Given:* $\triangle ABC$ with right $\angle ABC$
 $\overline{BD} \perp \overline{AC}$

 a) $BD = 12$, $AD = 9$, $DC = ?$
 b) $DC = 5$, $BC = 15$, $AD = ?$
 c) $AD = 2$, $DC = 8$, $AB = ?$
 d) $AB = 2\sqrt{6}$, $DC = 2$, $AD = ?$

32. In the drawings shown, find $x$.

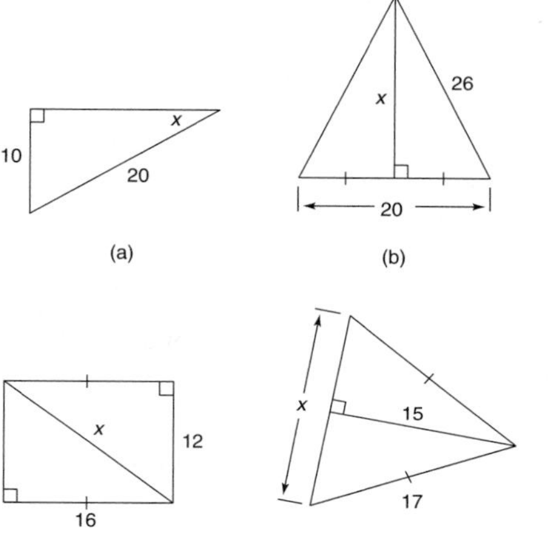


33. *Given:* $ABCD$ is a rectangle
 $E$ is the midpoint of $\overline{BC}$
 $AB = 16$, $CF = 9$, $AD = 24$
 *Find:* $AE$, $EF$, $AF$

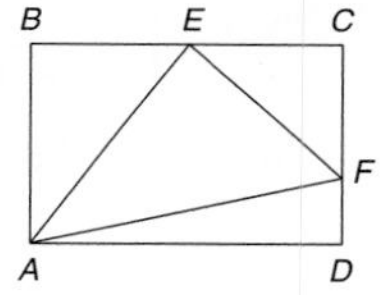


34. Find the length of a diagonal of a square whose side is 4 in. long.

35. Find the length of a side of a square whose diagonal is 6 cm long.

36. Find the length of a side of a rhombus whose diagonals are 48 cm and 14 cm long.

37. Find the length of an altitude of an equilateral triangle whose side is 10 in. long.

38. Find the length of a side of an equilateral triangle if an altitude is 6 in. long.

39. The lengths of three sides of a triangle are 13 cm, 14 cm, and 15 cm. Find the length of the altitude to the 14-cm side.

40. In the drawings, find $x$ and $y$.

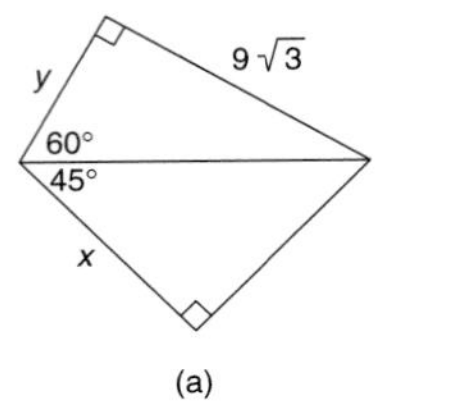

(a)

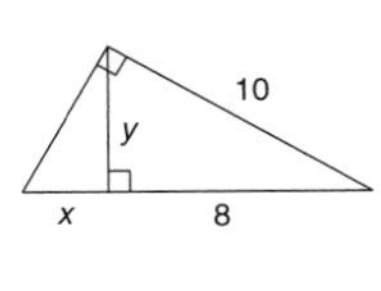

(b)

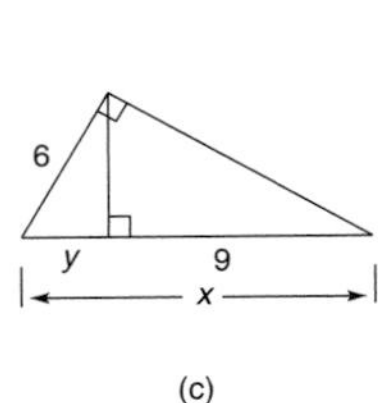

(c)

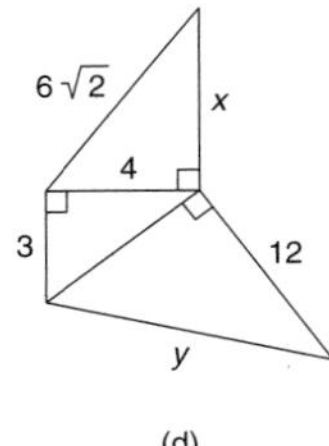

(d)

41. An observation aircraft flying at a height of 12 km has detected a Brazilian ship at a distance of 20 km from the aircraft and in line with an American ship that is 13 km from the aircraft. How far apart are the U.S. and Brazilian ships?

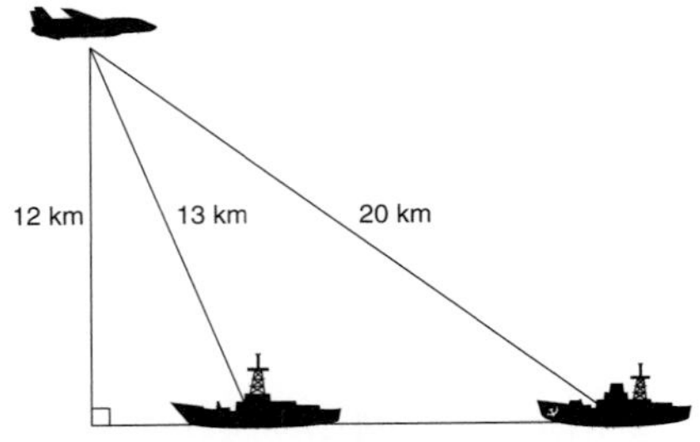


42. Tell whether each set of numbers represents the lengths of the sides of an acute triangle, of an obtuse triangle, of a right triangle, or of no triangle:

a) 12, 13, 14
b) 11, 5, 18
c) 9, 15, 18
d) 6, 8, 10
e) 8, 7, 16
f) 8, 7, 6
g) 9, 13, 8
h) 4, 2, 3

# Chapter 5 Test

1. Reduce to its simplest form:

   a) The ratio 12:20 ______
   b) The rate $\frac{200 \text{ miles}}{8 \text{ gallons}}$ ______

2. Solve each proportion for $x$. Show your work!

   a) $\frac{x}{5} = \frac{8}{13}$ ______ b) $\frac{x+1}{5} = \frac{16}{x-1}$ ______

3. The measures of two complementary angles are in the ratio 1:5. Find the measure of each angle. Smaller: ______; Larger: ______

4. $\triangle RTS \sim \triangle UWV$.

   a) Find $m\angle W$ if $m\angle R = 67°$ and $m\angle S = 21°$.

   ______

   b) Find $WV$ if $RT = 4$, $UW = 6$, and $TS = 8$.

   ______

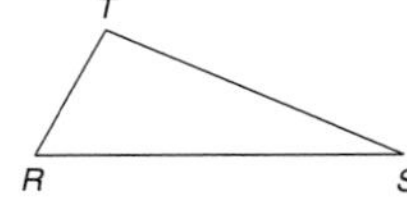


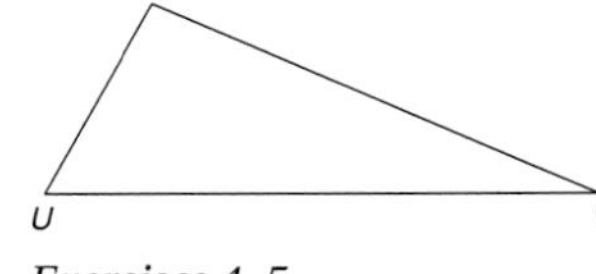

*Exercises 4, 5*

5. Give the reason (AA, SAS~, or SSS~) why $\triangle RTS \sim \triangle UTW$.

   a) $\angle R \cong \angle U$ and $\frac{TR}{WU} = \frac{RS}{UV}$ ______
   b) $\angle S \cong \angle V$; $\angle T$ and $\angle W$ are right angles ______

6. In right triangle $ABC$, $\overline{CD}$ is the altitude from $C$ to hypotenuse $\overline{AB}$. Name three triangles that are similar to each other.

   ______

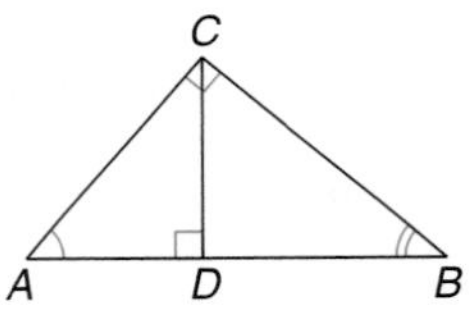


7. In $\triangle ABC$, $m\angle C = 90°$. Use a square root radical to represent:

   a) $c$, if $a = 5$ and $b = 4$

   ______

   b) $a$, if $b = 6$ and $c = 8$

   ______

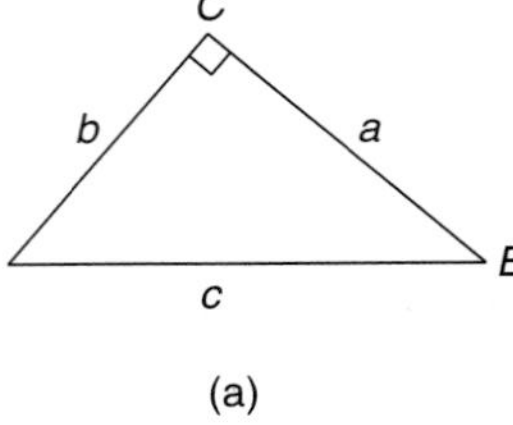

(a)

8. Given its lengths of sides, is $\triangle RST$ a right triangle?

   a) $a = 15$, $b = 8$, and $c = 17$ ______ (Yes or No)
   b) $a = 11$, $b = 8$, and $c = 15$ ______ (Yes or No)

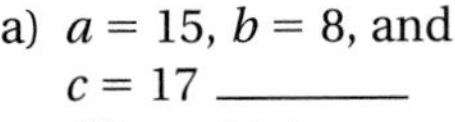

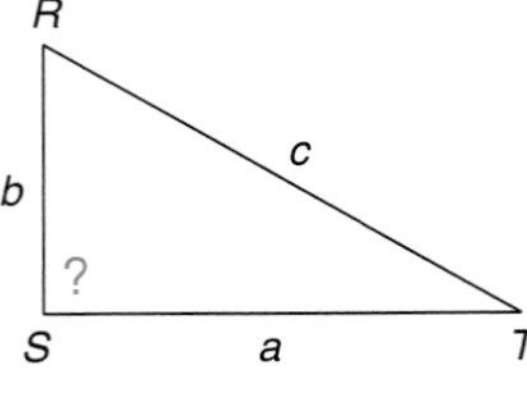

9. Given quadrilateral $ABCD$ with diagonal $\overline{AC}$. If $\overline{BC} \perp \overline{AB}$ and $\overline{AC} \perp \overline{DC}$, find $DA$ if $AB = 4$, $BC = 3$, and $DC = 8$. Express the answer as a square root radical. __________

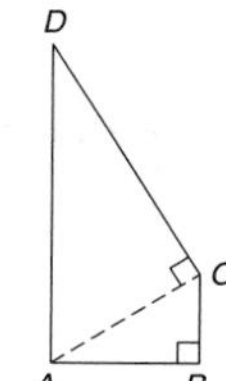


10. In $\triangle XYZ$, $\overline{XZ} \cong \overline{YZ}$ and $\angle Z$ is a right angle.

 a) Find $XY$ if $XZ = 10$ in.

 ________

 b) Find $XZ$ if $XY = 8\sqrt{2}$ cm.

 ________

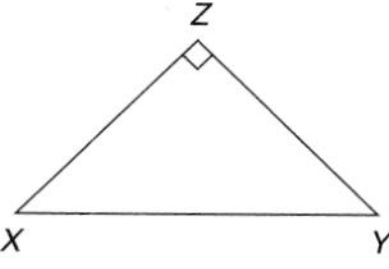


11. In $\triangle DEF$, $\angle D$ is a right angle and $m\angle F = 30°$.

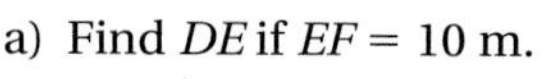

 a) Find $DE$ if $EF = 10$ m.

 ________

 b) Find $EF$ if $DF = 6\sqrt{3}$ ft.

 ________

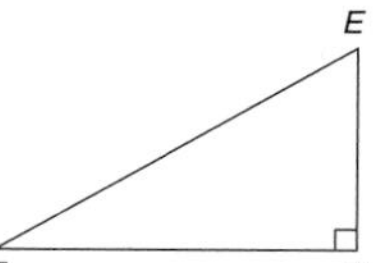


12. In $\triangle ABC$, $\overleftrightarrow{DE} \parallel \overline{BC}$. If $AD = 6$, $DB = 8$, and $AE = 9$, find $EC$. ________

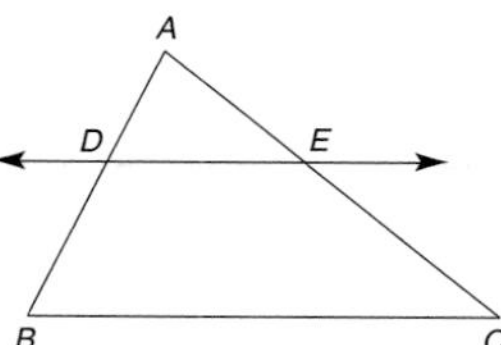


13. In $\triangle MNP$, $\overrightarrow{NQ}$ bisects $\angle MNP$. If $PN = 6$, $MN = 9$, and $MP = 10$, find $PQ$ and $QM$.

 $PQ =$ ______; $QM =$ ______

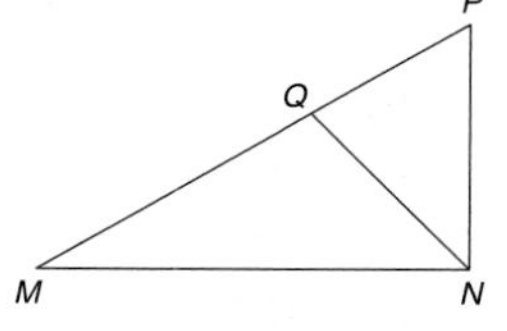


14. For $\triangle ABC$, the three angle bisectors are shown. Find the product $\frac{AE}{EC} \cdot \frac{CD}{DB} \cdot \frac{BF}{FA}$. __________

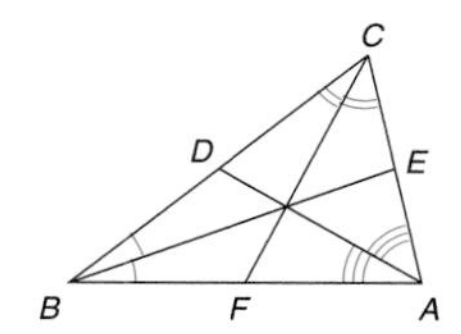


*In Exercises 15 and 16, complete the statements and reasons in each proof.*

15. *Given:* $\overline{MN} \parallel \overline{QR}$
 *Prove:* $\triangle MNP \sim \triangle QRP$

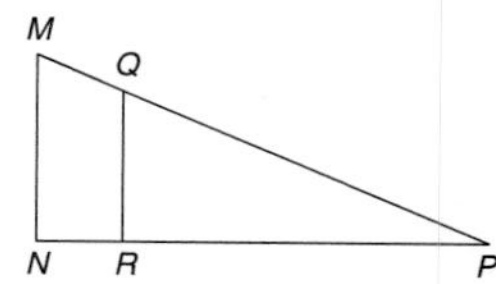


| Statements | Reasons |
|---|---|
| 1. __________ | 1. __________ |
| 2. $\angle N \cong \angle QRP$ | 2. If 2 ∥ lines are cut by a trans., ________ |
| 3. __________ | 3. Identity |
| 4. $\triangle MNP \sim \triangle QRP$ | 4. __________ |

16. *Given:* In $\triangle ABC$, $P$ is the midpoint of $\overline{AC}$, and $R$ is the midpoint of $\overline{CB}$.
 *Prove:* $\angle PRC \cong \angle B$

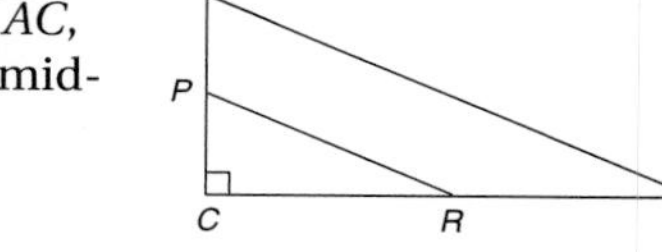


| Statements | Reasons |
|---|---|
| 1. $\triangle ABC$ | 1. __________ |
| 2. $\angle C \cong \angle C$ | 2. __________ |
| 3. $P$ is the midpoint of $\overline{AC}$, and $R$ is the midpoint of $\overline{CB}$ | 3. __________ |
| 4. $\frac{PC}{AC} = \frac{1}{2}$ and $\frac{CR}{CB} = \frac{1}{2}$ | 4. Definition of midpoint |
| 5. $\frac{PC}{AC} = \frac{CR}{CB}$ | 5. __________ |
| 6. $\triangle CPR \sim \triangle CAB$ | 6. __________ |
| 7. __________ | 7. CASTC |

# Chapter 6

# Circles

When we consider something as simple as a pancake, as functional as a gear or pulley, or as essential as a wire-spoke wheel, we generally think of a circle. In this chapter, we deal with circles and develop their properties, which are logical consequences of the principles developed in previous chapters. The use of circular gears in mechanical applications, as suggested by the bicycle, has taken place over many hundreds of years. See related Exercises 40 and 41 of Section 6.3.

## Chapter Outline



**For online student resources, visit this textbook's website at math.college.hmco.com/students.**

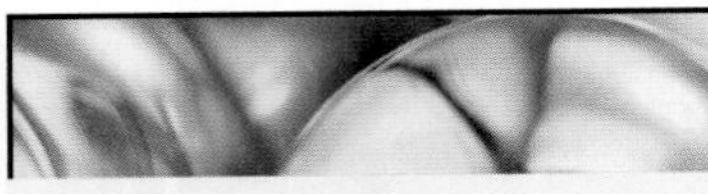

KEY CONCEPTS

Circle • Congruent Circles • Concentric Circles • Center • Radius • Diameter • Chord • Semicircle • Arc • Major Arc • Minor Arc • Intercepted Arc • Congruent Arcs • Central Angle • Inscribed Angle

# 6.1 Circles and Related Segments and Angles

In this chapter, we will introduce terminology related to the circle, some methods of measurement, and many properties of the circle.

> DEFINITION: A **circle** is the set of all points in a plane that are at a fixed distance from a given point known as the *center* of the circle.

A circle is named by its center point. In Figure 6.1, point $P$ is the center of the circle. The symbol for circle is $\odot$, so the circle in Figure 6.1 is $\odot P$. Points $A$, $B$, $C$, and $D$ are points *of* (or *on*) the circle. Points $P$ (the center) and $R$ are in the *interior* of circle $P$; points $G$ and $H$ are in the *exterior* of the circle.

In $\odot Q$ of Figure 6.2, $\overline{SQ}$ is a radius of the circle. A **radius** is a segment that joins the center of the circle to a point on the circle. $\overline{SQ}$, $\overline{TQ}$, $\overline{VQ}$, and $\overline{WQ}$ are **radii** (plural of *radius*) of $\odot Q$. By definition, $SQ = TQ = VQ = WQ$.

The following statement is a consequence of the definition of a circle.

**Warning** If the phrase "in a plane" is omitted from the definition of a circle, the result is the definition of a sphere.

> All radii of a circle are congruent.

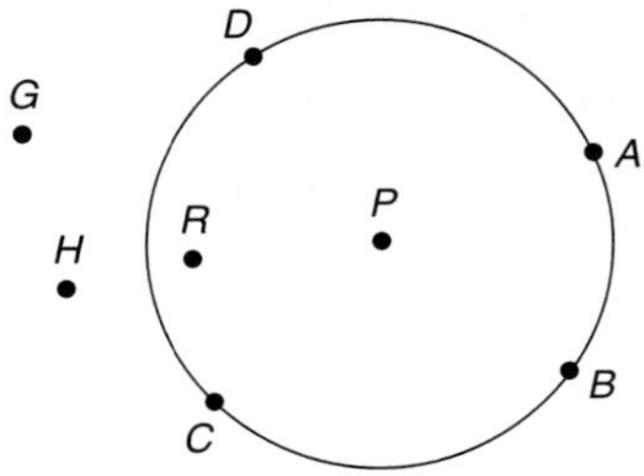

FIGURE 6.1

A line segment that joins two points of a circle (such as $\overline{SW}$ in Figure 6.2) is a **chord** of the circle. A **diameter** of a circle is a chord that contains the center of the circle; in Figure 6.2, $\overline{TW}$ is a diameter of $\odot Q$.

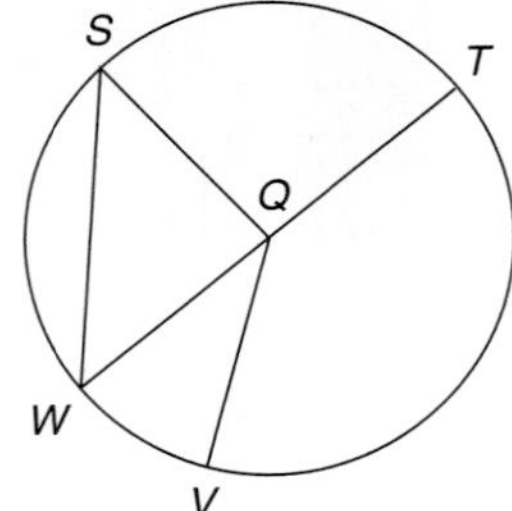

FIGURE 6.2

> DEFINITION: **Congruent circles** are two or more circles that have congruent radii.

In Figure 6.3, circles $P$ and $Q$ are congruent because their radii have equal lengths. We can slide $\odot P$ to the right to coincide with $\odot Q$.

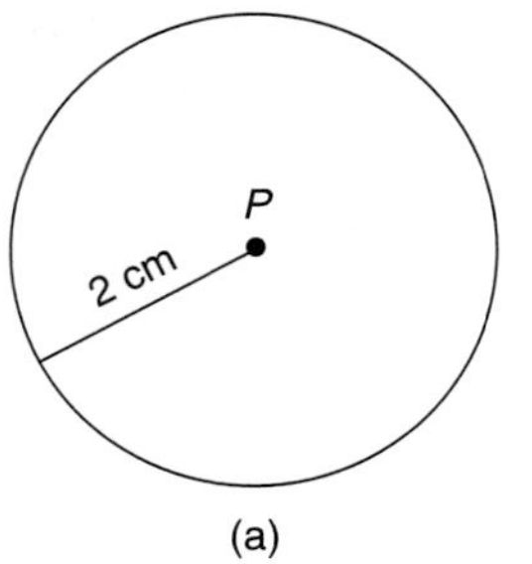


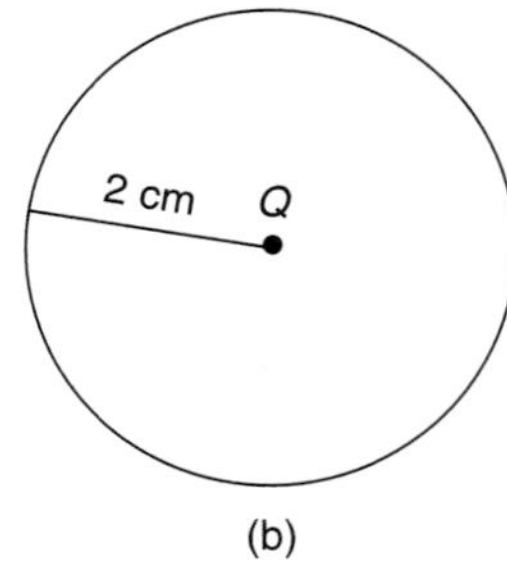


FIGURE 6.3

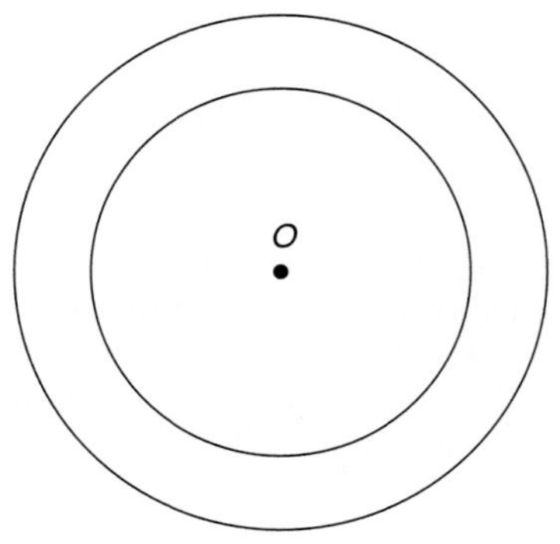


FIGURE 6.4

> DEFINITION: **Concentric circles** are coplanar circles that have a common center.

The concentric circles in Figure 6.4 have the common center $O$.

In $\odot P$ of Figure 6.5, the part of the circle shown from point $A$ to point $B$ is **arc** $AB$, symbolized by $\overset{\frown}{AB}$. If $\overline{AC}$ is a diameter, then $\overset{\frown}{ABC}$ (three letters are used for clarity) is a **semicircle.** In Figure 6.5, a **minor arc** like $\overset{\frown}{AB}$ is part of a semicircle; a **major arc** such as $\overset{\frown}{ABCD}$ (also denoted by $\overset{\frown}{ABD}$ or $\overset{\frown}{ACD}$) is more than a semicircle but less than the entire circle.

> DEFINITION: A **central angle** of a circle is an angle whose vertex is the center of the circle and whose sides are radii of the circle.

*Exs. 1–3*

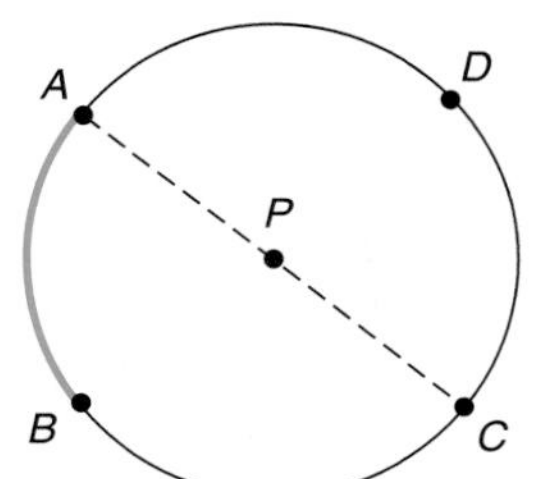


FIGURE 6.5

In Figure 6.6, $\angle NOP$ is a central angle of $\odot O$. The **intercepted arc** of $\angle NOP$ is $\overset{\frown}{NP}$. The intercepted arc of an angle is determined by the two points of intersection of the angle with the circle and all points of the arc in the interior of the angle.

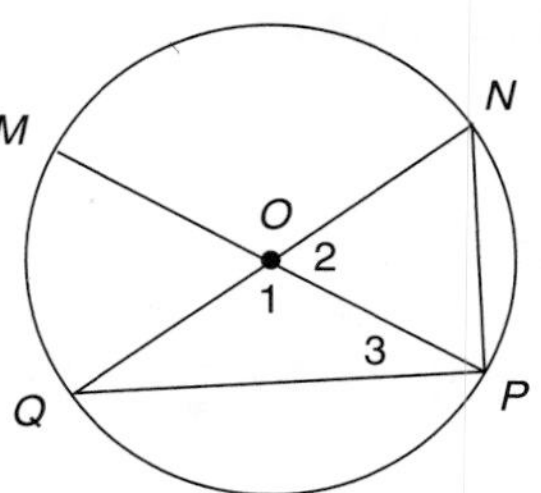


FIGURE 6.6

In Example 1, we "check" the terminology just introduced.

### EXAMPLE 1

In Figure 6.6, $\overline{MP}$ and $\overline{NQ}$ intersect at $O$, the center of the circle. Name:

a) All four radii (shown)
b) Both diameters (shown)
c) All four chords (shown)
d) One central angle
e) One minor arc
f) One semicircle
g) One major arc
h) Intercepted arc of $\angle MON$
i) Central angle that intercepts $\overset{\frown}{NP}$

***Solution***

a) $\overline{OM}$, $\overline{OQ}$, $\overline{OP}$, and $\overline{ON}$
b) $\overline{MP}$ and $\overline{QN}$
c) $\overline{MP}$, $\overline{QN}$, $\overline{QP}$, and $\overline{NP}$
d) $\angle QOP$ (other answers are possible)
e) $\overset{\frown}{NP}$ (other answers are possible)
f) $\overset{\frown}{MQP}$ (other answers are possible)
g) $\overset{\frown}{MQN}$ (can be named $\overset{\frown}{MQPN}$; other answers are possible)
h) $\overset{\frown}{MN}$ (lies in the interior of $\angle MON$)
i) $\angle NOP$ (also called $\angle 2$)

The following statement is a consequence of the Segment-Addition Postulate.

> In a circle, the length of a diameter is twice that of a radius.

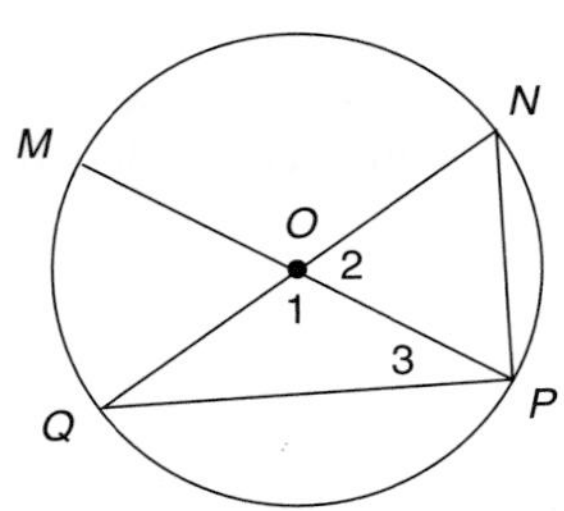

**FIGURE 6.6**

EXAMPLE 2

$\overline{QN}$ is a diameter of $\odot O$ in Figure 6.6 and $PN = ON = 12$. Find the length of chord $\overline{QP}$.

***Solution***

Because $PN = ON$ and $ON = OP$, $\triangle NOP$ is equilateral. Then $m\angle 2 = m\angle N = m\angle NPO = 60°$. Also, $OP = OQ$, so $\triangle POQ$ is isosceles with $m\angle 1 = 120°$, because this angle is supplementary to $\angle 2$. Now $m\angle Q = m\angle 3 = 30°$ because the sum of the measures of the angles of $\triangle POQ$ is 180°. If $m\angle N = 60°$ and $m\angle Q = 30°$, then $\triangle NPQ$ is a right $\triangle$ whose angle measures are 30°, 60°, and 90°. It follows that $QP = PN \cdot \sqrt{3} = 12\sqrt{3}$.

THEOREM 6.1.1: A radius that is perpendicular to a chord bisects the chord.

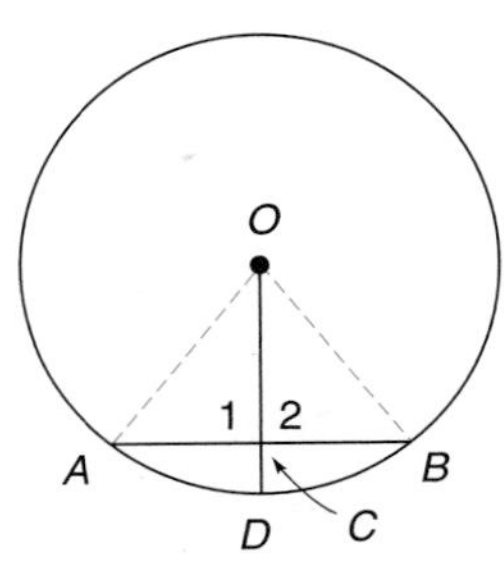

**FIGURE 6.7**

*Given:* $\overline{OD} \perp \overline{AB}$ in $\odot O$ (See Figure 6.7.)

*Prove:* $\overline{OD}$ bisects $\overline{AB}$

*Proof:* $\overline{OD} \perp \overline{AB}$ in $\odot O$. Draw radii $\overline{OA}$ and $\overline{OB}$. Now $\overline{OA} \cong \overline{OB}$ because all radii of a circle are $\cong$. Because $\angle 1$ and $\angle 2$ are right $\angle$s and $\overline{OC} \cong \overline{OC}$, we see that $\triangle OCA \cong \triangle OCB$ by HL. Then $\overline{AC} \cong \overline{CB}$ by CPCTC, so $\overline{OD}$ bisects $\overline{AB}$.

## ANGLE AND ARC RELATIONSHIPS IN THE CIRCLE

In Figure 6.8, the sum of the measures of the angles about point $O$ (angles determined by perpendicular diameters $\overline{AC}$ and $\overline{BD}$) is 360°. Similarly, the circle can be separated into 360 equal arcs, *each of which measures 1° of arc measure;* that is, each arc would be intercepted by a central angle measuring 1°. Our description of arc measure leads to the following postulate.

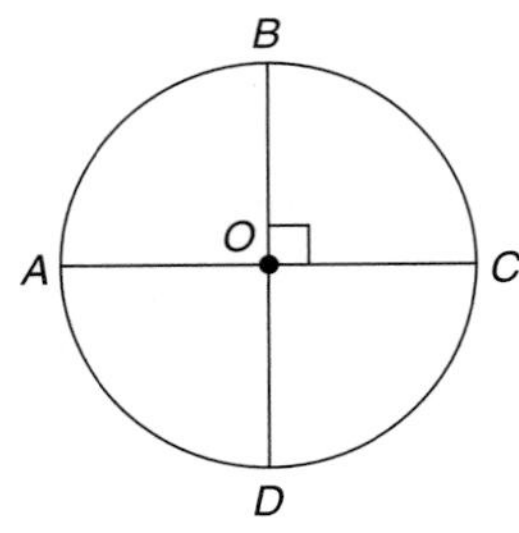

**FIGURE 6.8**

POSTULATE 16: **(Central Angle Postulate)**
In a circle, the degree measure of a central angle is equal to the degree measure of its intercepted arc.

If $m\widehat{AB} = 90°$ in Figure 6.8, then $m\angle AOB = 90°$. The related reflex angle composed of three right angles measures 270°.

In Figure 6.8, $m\widehat{AB} = 90°$, $m\widehat{BCD} = 180°$, and $m\widehat{AD} = 90°$. It follows that $m\widehat{AB} + m\widehat{BCD} + m\widehat{AD} = 360°$. Consequently, we have the following generalization.

The sum of the measures of the consecutive arcs that form a circle is exactly 360°.

In $\odot Y$ [Figure 6.9(a)], if $m\angle XYZ = 76°$, then $m\widehat{XZ} = 76°$ by the Central Angle Postulate. If two arcs have equal degree measures [Figures 6.9(b) and (c)] but are parts of two circles with unequal radii, then these arcs will not coincide. This observation leads to the following definition.

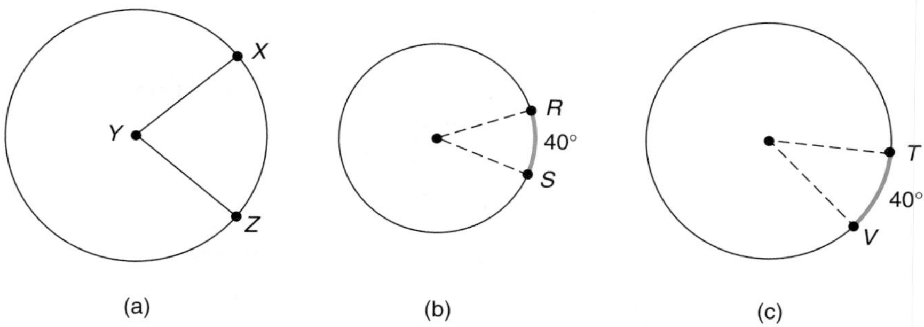


**FIGURE 6.9**

> DEFINITION: In a circle or congruent circles, **congruent arcs** are arcs with equal measures.

*Exs. 4–10*

To clarify the definition of congruent arcs, consider the concentric circles (having the same center) in Figure 6.10. Here the degree measure of $\angle AOB$ of the smaller circle is the same as the degree measure of $\angle COD$ of the larger circle. Even though $m\overset{\frown}{AB} = m\overset{\frown}{CD}$, $\overset{\frown}{AB} \not\cong \overset{\frown}{CD}$ because the arcs would not coincide.

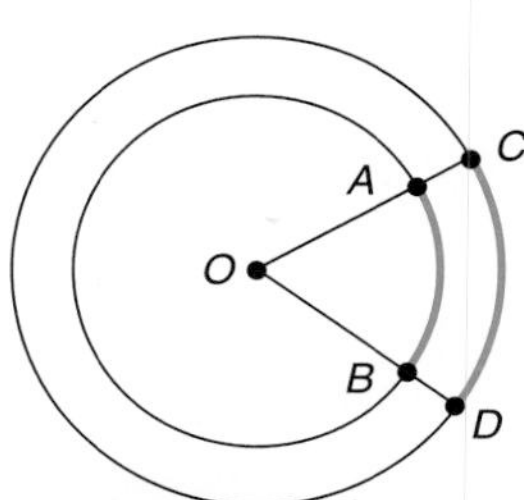


**FIGURE 6.10**

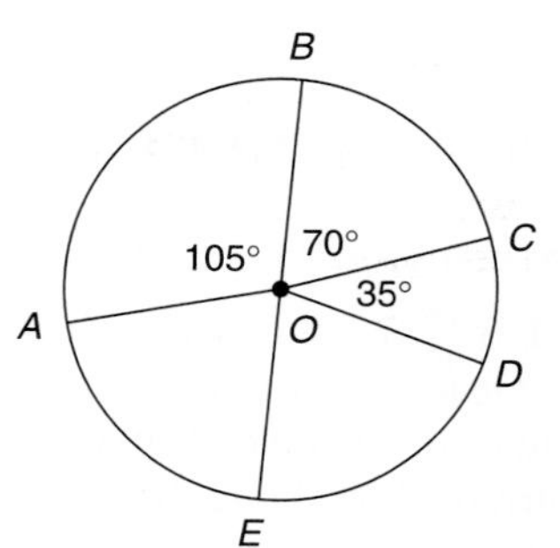


**FIGURE 6.11**

EXAMPLE 3

In $\odot O$ in Figure 6.11, $\overrightarrow{OE}$ bisects $\angle AOD$. Using the measures indicated, find:

a) $m\overset{\frown}{AB}$ b) $m\overset{\frown}{BC}$ c) $m\overset{\frown}{BD}$ d) $m\angle AOD$
e) $m\overset{\frown}{AE}$ f) $m\overset{\frown}{ACE}$ g) whether $\overset{\frown}{AE} \cong \overset{\frown}{ED}$
h) Measure of the reflex angle that intercepts $\overset{\frown}{ABCD}$

***Solution***

a) 105° b) 70° c) 105° d) 150°, from 360 − (105 + 70 + 35) e) 75° because the corresponding central angle ($\angle AOE$) is the result of bisecting $\angle AOD$, which was found to be 150° f) 285° (from 360 − 75, the measure of $\overset{\frown}{AE}$) g) The arcs are congruent because both measure 75° and both are found in the same circle. h) 210° (from 105° + 70° + 35°)

In Example 3(c), note that $m\overset{\frown}{BD} = m\overset{\frown}{BC} + m\overset{\frown}{CD}$. Because $D$ lies between $B$ and $A$ so that the union of $\overset{\frown}{BD}$ and $\overset{\frown}{DA}$ is a major arc, $m\overset{\frown}{BDA} = m\overset{\frown}{BD} + m\overset{\frown}{DA}$. With this understanding, we have the following postulate.

> POSTULATE 17: **(Arc-Addition Postulate)**
> If $B$ lies between $A$ and $C$ on a circle, then $m\overset{\frown}{AB} + m\overset{\frown}{BC} = m\overset{\frown}{ABC}$.

The drawing in Figure 6.12(a) further supports the claim in Postulate 17.

Given points $A$, $B$, and $C$ on $\odot O$ as shown in Figure 6.12(a), suppose that radii $\overline{OA}$, $\overline{OB}$, and $\overline{OC}$ are drawn. Because

$$m\angle AOB + m\angle BOC = m\angle AOC$$

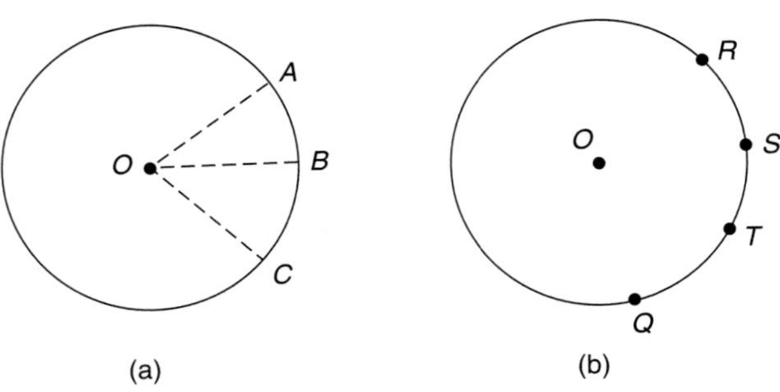


**FIGURE 6.12**

by the Angle-Addition Postulate, it follows that

$$m\widehat{AB} + m\widehat{BC} = m\widehat{ABC}$$

The reason for writing $\widehat{ABC}$, rather than $\widehat{AC}$, in stating the Arc-Addition Postulate is to avoid confusing minor arc $\widehat{AC}$ with the major arc with endpoints at $A$ and $C$. It is easy to show that $m\widehat{ABC} - m\widehat{BC} = m\widehat{AB}$.

The Arc-Addition Postulate can easily be extended to include more than two arcs. In Figure 6.12(b), $m\widehat{RS} + m\widehat{ST} + m\widehat{TQ} = m\widehat{RSTQ}$.

If $m\widehat{RS} = m\widehat{ST}$ in Figure 6.12(b), then point $S$ is the **midpoint** of $\widehat{RT}$, and $\widehat{RT}$ is **bisected** at point $S$.

In Example 4, we use the fact that the entire circle measures 360°.

**FIGURE 6.13**

EXAMPLE 4

Determine the measure of the angle formed by the hands of a clock at 3:12 P.M. (See Figure 6.13.)

***Solution***

The minute hand moves through 12 minutes, which is $\frac{12}{60}$ or $\frac{1}{5}$ of an hour. Thus the minute hand points in a direction whose angle measure from the vertical is $\frac{1}{5}(360°)$ or 72°. At exactly 3 P.M., the hour hand would form an angle of 90° with the vertical. However, gears inside the clock also turn the hour hand through $\frac{1}{5}$ of the 30° arc from the 3 toward the 4; that is, the hour hand moves another $\frac{1}{5}(30°)$ or 6° to form an angle of 96° with the vertical. The angle between the hands must measure 96° − 72° or 24°.

As we have seen, the measure of an arc can be used to measure the corresponding central angle. The measure of an arc can also be used to measure other types of angles related to the circle, including the inscribed angle.

DEFINITION: An **inscribed angle** of a circle is an angle whose vertex is a point on the circle and whose sides are chords of the circle.

The word *inscribed* is often linked to the word *inside.*

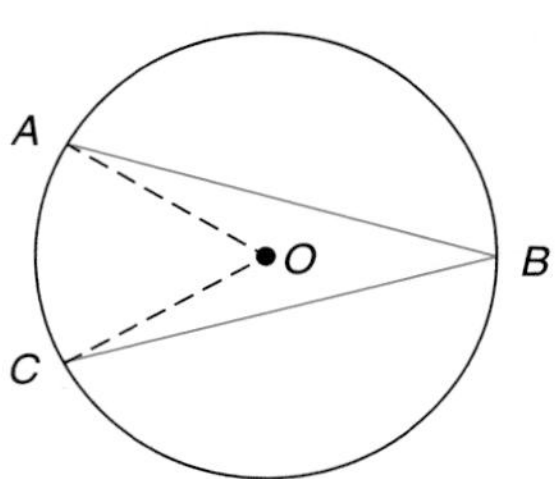


FIGURE 6.14

## Discover!

In Figure 6.14, $\angle B$ is the inscribed angle whose sides are chords $\overline{BA}$ and $\overline{BC}$.
a) Use a protractor to find the measure of central $\angle AOC$. b) Also find the measure of $\overset{\frown}{AC}$.
c) Finally, measure inscribed $\angle B$. d) How is the measure of inscribed $\angle B$ related to the measure of its intercepted arc $\overset{\frown}{AC}$?

ANSWERS

a) 56° b) 56° c) 28° d) $m\angle B = \frac{1}{2}m\overset{\frown}{AC}$

As suggested by the preceding Discover! exercise, the relationship between the measure of an inscribed angle and its intercepted arc is true in general.

**Reminder**

The measure of an exterior angle of a triangle equals the sum of the measures of the two remote interior angles.

THEOREM 6.1.2: The measure of an inscribed angle of a circle is one-half the measure of its intercepted arc.

The proof of Theorem 6.1.2 must be divided into three cases:

*CASE 1.* One side of the inscribed angle is a diameter.
*CASE 2.* The diameter to the vertex of the inscribed angle lies in the interior of the angle.
*CASE 3.* The diameter to the vertex of the inscribed angle lies in the exterior of the angle.

*Technology Exploration*

Use computer software if available:

1) Create circle $O$ with inscribed angle $RST$.
2) Include radius $\overline{OR}$ in the figure. See Figure 6.16.
3) Measure $\overset{\frown}{RT}$, $\angle ROT$, and $\angle RST$.
4) Show that: $m\angle ROT = m\overset{\frown}{RT}$ and $m\angle RST = \frac{1}{2}m\overset{\frown}{RT}$

The proof of Case 1 follows, but proofs of the other cases are left as exercises for the student. Drawings for Cases 2 and 3 are found in Figure 6.15.

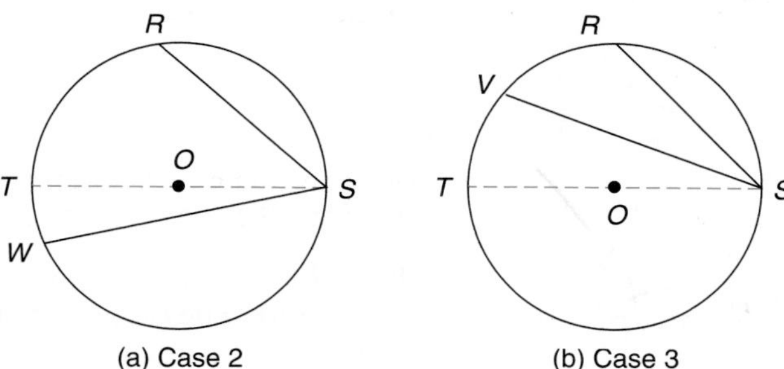


FIGURE 6.15

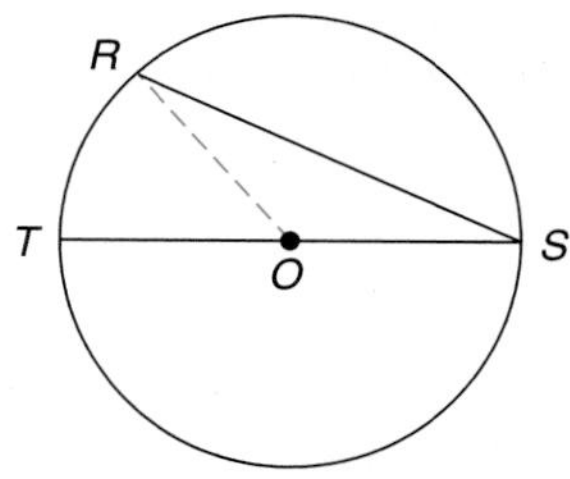


FIGURE 6.16

*Given:* $\odot O$ with inscribed $\angle RST$ and diameter $\overline{ST}$ (Figure 6.16)

*Prove:* $m\angle S = \frac{1}{2}m\overset{\frown}{RT}$

*Proof of Case 1:* We begin by constructing radius $\overline{RO}$. Then $m\angle ROT = m\overset{\frown}{RT}$ because the central angle has a measure equal to the measure of its intercepted arc. With $\overline{OR} \cong \overline{OS}$, $\triangle ROS$ is isosceles and $m\angle R = m\angle S$. Now the exterior angle of the triangle is $\angle ROT$, so

$$m\angle ROT = m\angle R + m\angle S$$

Because $m\angle R = m\angle S$, $m\angle ROT = 2(m\angle S)$. Then $m\angle S = \frac{1}{2}m\angle ROT$. With $m\angle ROT = m\overset{\frown}{RT}$, we have $m\angle S = \frac{1}{2}m\overset{\frown}{RT}$ by substitution.

*Exs. 11–15*

Although proofs in this chapter generally take the less formal paragraph form, it remains necessary to justify each statement of the proof.

THEOREM 6.1.3: In a circle (or in congruent circles), congruent minor arcs have congruent central angles. (See Figure 6.17.)

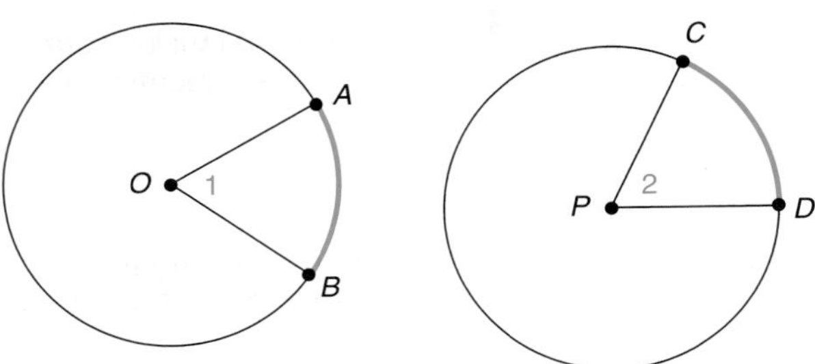


If $\widehat{AB} \cong \widehat{CD}$ in congruent circles $O$ and $P$, then $\angle 1 \cong \angle 2$ by Theorem 6.1.3.

**FIGURE 6.17**

We suggest that the student make a drawing to illustrate each of the next three theorems. Some of the proofs depend on auxiliary radii.

THEOREM 6.1.4: In a circle (or in congruent circles), congruent central angles have congruent arcs.

THEOREM 6.1.5: In a circle (or in congruent circles), congruent chords have congruent minor (major) arcs.

THEOREM 6.1.6: In a circle (or in congruent circles), congruent arcs have congruent chords.

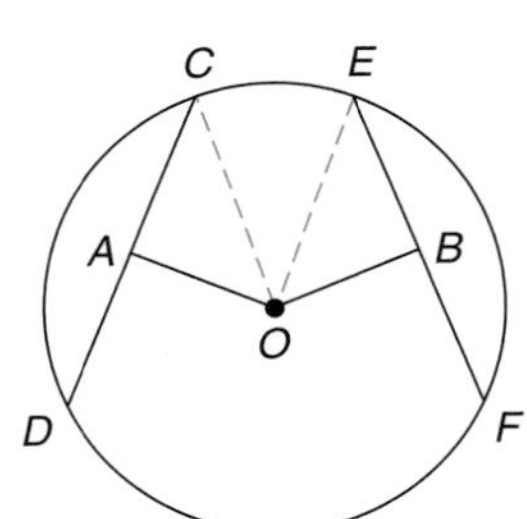


**FIGURE 6.18**

On the basis of an earlier definition, we define the distance from the center of a circle to a chord to be the length of the perpendicular segment joining the center to that chord.

Congruent triangles are used to prove the next two theorems.

THEOREM 6.1.7: Chords that are at the same distance from the center of a circle are congruent.

*Given:* $\overline{OA} \perp \overline{CD}$ and $\overline{OB} \perp \overline{EF}$ in $\odot O$ (See Figure 6.18.)
$\overline{OA} \cong \overline{OB}$

*Prove:* $\overline{CD} \cong \overline{EF}$

*Proof:* Draw radii $\overline{OC}$ and $\overline{OE}$. With $\overline{OA} \perp \overline{CD}$ and $\overline{OB} \perp \overline{EF}$, $\angle OAC$ and $\angle OBE$ are right $\angle$s. $\overline{OA} \cong \overline{OB}$ is given, and $\overline{OC} \cong \overline{OE}$ because all radii of a circle are congruent. $\triangle OAC$ and $\triangle OBE$ are right triangles. Thus $\triangle OAC \cong \triangle OBE$ by HL.

By CPCTC, $\overline{CA} \cong \overline{BE}$ so $CA = BE$. Then $2(CA) = 2(BE)$. But $2(CA) = CD$ because $A$ is the midpoint of chord $\overline{CD}$. ($\overline{OA}$ bisects chord $\overline{CD}$ because $\overline{OA}$ is part of a radius. See Theorem 6.1.1). Likewise, $2(BE) = EF$, and it follows that

$$CD = EF \quad \text{and} \quad \overline{CD} \cong \overline{EF}$$

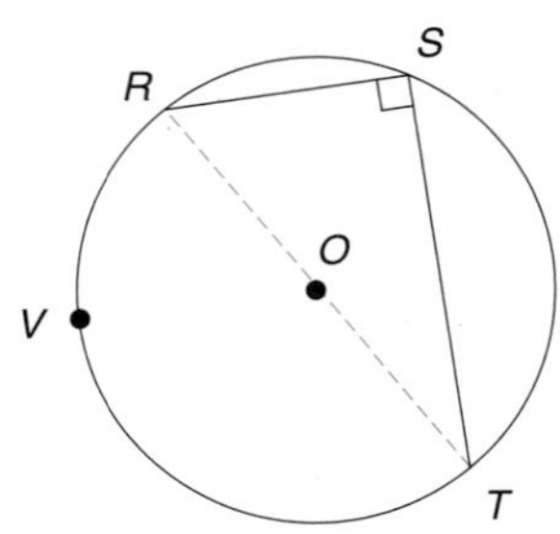


**FIGURE 6.19**

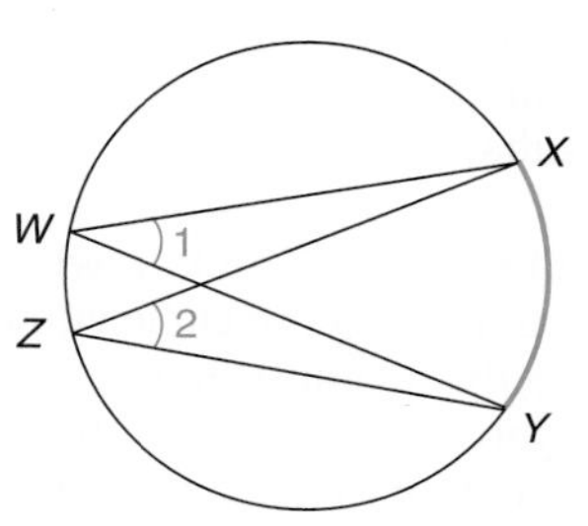


**FIGURE 6.20**

> THEOREM 6.1.8: Congruent chords are located at the same distance from the center of a circle.

The student should make a drawing to illustrate Theorem 6.1.8.

Proofs of the remaining theorems are left as exercises.

> THEOREM 6.1.9: An angle inscribed in a semicircle is a right angle.

Theorem 6.1.9 is illustrated in Figure 6.19, on page 282, where $\angle S$ is inscribed in the semicircle $\widehat{RST}$. Note that $\angle S$ also intercepts semicircle $\widehat{RVT}$.

> THEOREM 6.1.10: If two inscribed angles intercept the same arc, then these angles are congruent.

*Exs. 16, 17*

Theorem 6.1.10 is illustrated in Figure 6.20. Note that $\angle 1$ and $\angle 2$ both intercept $\widehat{XY}$. Because $m\angle 1 = \frac{1}{2}m\widehat{XY}$ and $m\angle 2 = \frac{1}{2}m\widehat{XY}$, $\angle 1 \cong \angle 2$.

## 6.1 Exercises

*For Exercises 1 to 8, use the figure provided.*

1. If $m\widehat{AC} = 58°$, find $m\angle B$.

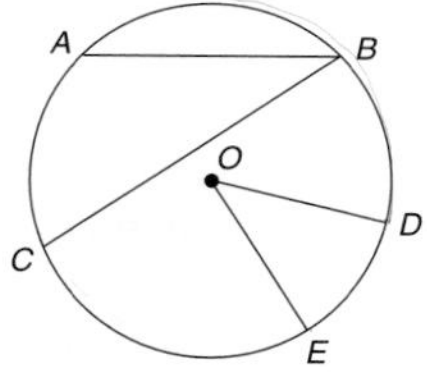


*Exercises 1–8*

2. If $m\widehat{DE} = 46°$, find $m\angle O$.
3. If $m\widehat{DE} = 47.6°$, find $m\angle O$.
4. If $m\widehat{AC} = 56.4°$, find $m\angle B$.
5. If $m\angle B = 28.3°$, find $m\widehat{AC}$.
6. If $m\angle O = 48.3°$, find $m\widehat{DE}$.
7. If $m\widehat{DE} = 47°$, find the measure of the reflex angle that intercepts $\widehat{DBACE}$.
8. If $m\widehat{ECABD} = 312°$, find $m\angle DOE$.
9. *Given:* $\overline{AO} \perp \overline{OB}$ and $\overline{OC}$ bisects $\widehat{ACB}$ in $\odot O$

   *Find:* a) $m\widehat{AB}$
   b) $m\widehat{ACB}$
   c) $m\widehat{BC}$
   d) $m\angle AOC$

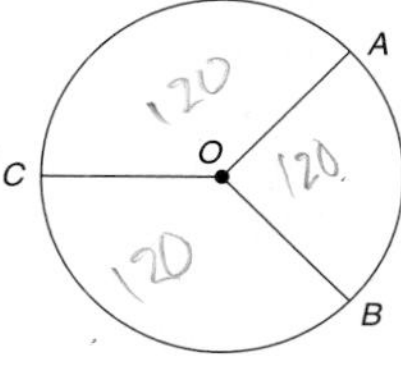


10. *Given:* $ST = \frac{1}{2}(SR)$ in $\odot Q$
    $\overline{SR}$ is a diameter

    *Find:* a) $m\widehat{ST}$
    b) $m\widehat{TR}$
    c) $m\widehat{STR}$
    d) $m\angle S$

    (HINT: Draw $\overline{QT}$.)

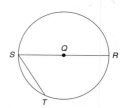

11. *Given:* $\odot Q$ in which $m\widehat{AB}:m\widehat{BC}:m\widehat{CA} = 2:3:4$

    *Find:* a) $m\widehat{AB}$
    b) $m\widehat{BC}$
    c) $m\widehat{CA}$
    d) $m\angle 1$ ($\angle AQB$)
    e) $m\angle 2$ ($\angle CQB$)
    f) $m\angle 3$ ($\angle CQA$)
    g) $m\angle 4$ ($\angle CAQ$)
    h) $m\angle 5$ ($\angle QAB$)
    i) $m\angle 6$ ($\angle QBC$)

    (HINT: Let $m\widehat{AB} = 2x$, $m\widehat{BC} = 3x$, and $m\widehat{CA} = 4x$.)

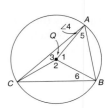

12. *Given:* $m\angle DOE = 76°$ and $m\angle EOG = 82°$ in $\odot O$
    $\overline{EF}$ is a diameter

    *Find:* a) $m\widehat{DE}$
    b) $m\widehat{DF}$
    c) $m\angle F$
    d) $m\angle DGE$
    e) $m\angle EHG$
    f) Whether $m\angle EHG = \frac{1}{2}(m\widehat{EG} + m\widehat{DF})$

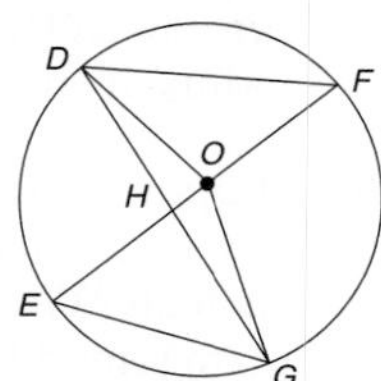


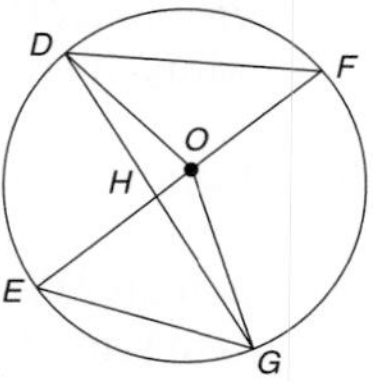

13. *Given:* $\odot O$ with $\overline{AB} \cong \overline{AC}$ and $m\angle BOC = 72°$
*Find:* a) $m\widehat{BC}$
b) $m\widehat{AB}$
c) $m\angle A$
d) $m\angle ABC$
e) $m\angle ABO$

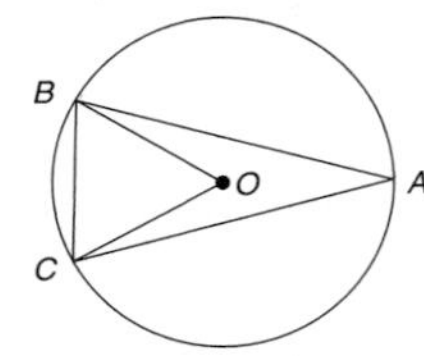


14. In $\odot O$ (not shown), $\overline{OA}$ is a radius, $\overline{AB}$ is a diameter, and $\overline{AC}$ is a chord.

a) How does $OA$ compare to $AB$?
b) How does $AC$ compare to $AB$?
c) How does $AC$ compare to $OA$?

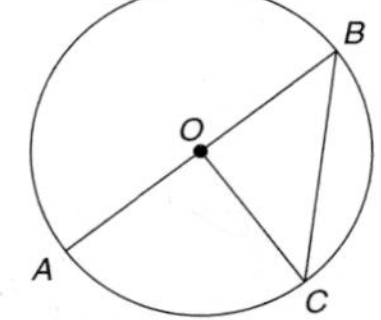

*Exercise 15*

15. *Given:* In $\odot O$, $\overline{OC} \perp \overline{AB}$ and $OC = 6$
*Find:* a) $AB$
b) $BC$

16. *Given:* Concentric circles with center $Q$
$SR = 3$ and $RQ = 4$
$\overline{QS} \perp \overline{TV}$ at $R$
*Find:* a) $RV$
b) $TV$

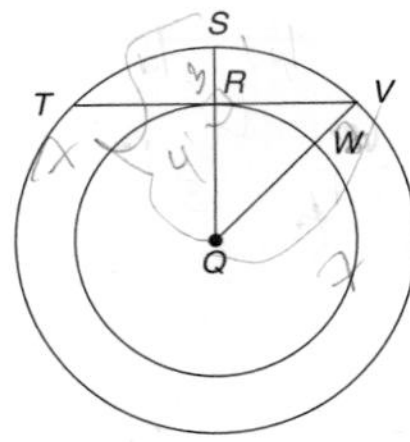

*Exercises 16, 17*

17. *Given:* Concentric circles with center $Q$
$TV = 8$ and $VW = 2$
$\overline{RQ} \perp \overline{TV}$
*Find:* $RQ$ (HINT: Let $RQ = x$.)

18. $\overline{AB}$ is the **common chord** of $\odot O$ and $\odot Q$. If $AB = 12$ and each circle has a radius of length 10, how long is $\overline{OQ}$?

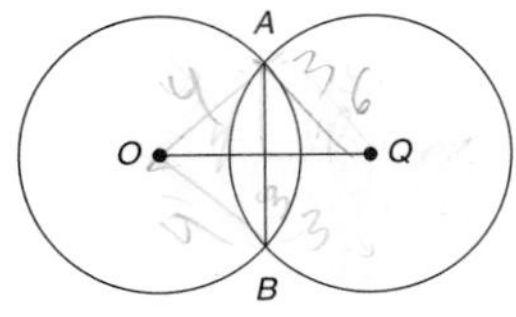

*Exercises 18, 19*

19. Circles $O$ and $Q$ have the common chord $\overline{AB}$. If $AB = 6$, $\odot O$ has a radius of length 4, and $\odot Q$ has a radius of length 6, how long is $\overline{OQ}$?

20. Suppose that a circle is divided into three congruent arcs by points $A$, $B$, and $C$. What is the measure of each arc? What type of figure results when $A$, $B$, and $C$ are joined by segments?

21. Suppose that a circle is divided by points $A$, $B$, $C$, and $D$ into four congruent arcs. What is the measure of each arc? If these points are joined in order, what type of quadrilateral results?

22. Following the pattern of Exercises 20 and 21, what type of figure results from dividing the circle equally by five points and joining those points in order? What type of polygon is formed by joining consecutively the $n$ points that separate the circle into $n$ congruent arcs?

23. Consider a circle or congruent circles, and explain why each statement is true:

a) Congruent arcs have congruent central angles.
b) Congruent central angles have congruent arcs.
c) Congruent chords have congruent arcs.
d) Congruent arcs have congruent chords.
e) Congruent central angles have congruent chords.
f) Congruent chords have congruent central angles.

24. State the measure of the angle formed by the minute hand and the hour hand of a clock when the time is

a) 1:30 P.M. b) 2:20 A.M.

25. State the measure of the angle formed by the hands of the clock at

a) 6:30 P.M. b) 5:40 A.M.

26. Five points are equally spaced on a circle. A five-pointed star (pentagram) is formed by joining nonconsecutive points two at a time. What is the degree measure of an arc determined by two consecutive points?

27. A ceiling fan has five equally spaced blades. What is the measure of the angle formed by two consecutive blades?

28. Repeat Exercise 27, but with the ceiling fan having six equally spaced blades.

29. An amusement park ride (the "Octopus") has eight support arms that are equally spaced about a circle. What is the measure of the central angle formed by two consecutive arms?

*In Exercises 30 and 31, complete each proof.*

30. *Given:* Diameters $\overline{AB}$ and $\overline{CD}$ intersecting at $E$ in $\odot E$
*Prove:* $\widehat{AC} \cong \widehat{DB}$

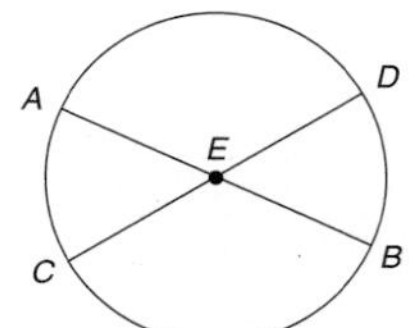


**PROOF**

| Statements | Reasons |
|---|---|
| 1. ? | 1. Given |
| 2. $\angle AEC \cong \angle DEB$ | 2. ? |
| 3. $m\angle AEC = m\angle DEB$ | 3. ? |
| 4. $m\angle AEC = m\widehat{AC}$ and $m\angle DEB = m\widehat{DB}$ | 4. ? |
| 5. $m\widehat{AC} = m\widehat{DB}$ | 5. ? |
| 6. ? | 6. If two arcs of a circle have the same measure, they are ≅ |

31. *Given:* $\overline{MN} \parallel \overline{OP}$ in $\odot O$
*Prove:* $m\widehat{MQ} = 2(m\widehat{NP})$

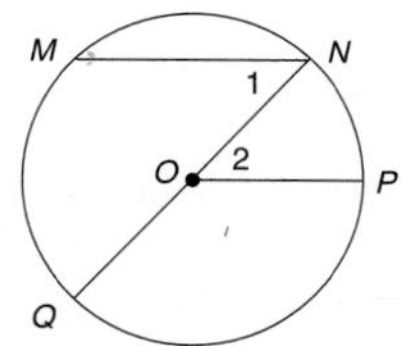


**PROOF**

| Statements | Reasons |
|---|---|
| 1. ? | 1. Given |
| 2. $\angle 1 \cong \angle 2$ | 2. ? |
| 3. $m\angle 1 = m\angle 2$ | 3. ? |
| 4. $m\angle 1 = \frac{1}{2}(m\widehat{MQ})$ | 4. ? |
| 5. $m\angle 2 = m\widehat{NP}$ | 5. ? |
| 6. $\frac{1}{2}(m\widehat{MQ}) = m\widehat{NP}$ | 6. ? |
| 7. $m\widehat{MQ} = 2(m\widehat{NP})$ | 7. Multiplication Prop. of Equality |

*In Exercises 32 to 37, write a paragraph proof.*

32. *Given:* $\overline{RS}$ and $\overline{TV}$ are diameters of $\odot W$
*Prove:* $\triangle RST \cong \triangle VTS$

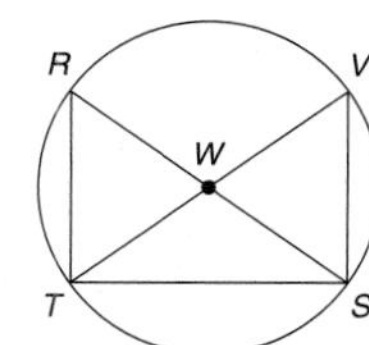


33. *Given:* Chords $\overline{AB}$, $\overline{BC}$, $\overline{CD}$, and $\overline{AD}$ in $\odot O$
*Prove:* $\triangle ABE \sim \triangle CDE$

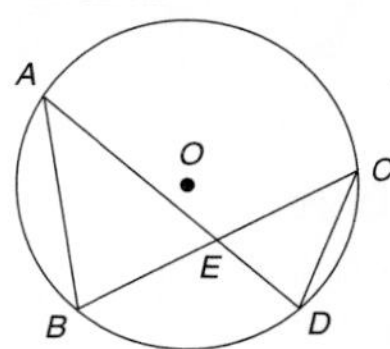


34. Congruent chords are at the same distance from the center of a circle.

35. A radius perpendicular to a chord bisects the arc of that chord.

36. An angle inscribed in a semicircle is a right angle.

37. If two inscribed angles intercept the same arc, then these angles are congruent.

38. If $\overleftrightarrow{MN} \parallel \overleftrightarrow{PQ}$ in $\odot O$, explain why $MNPQ$ is an isosceles trapezoid. (HINT: Draw a diagonal.)

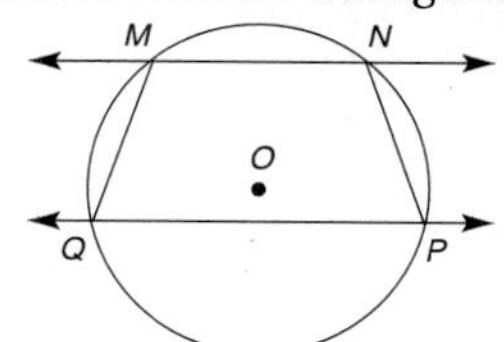


39. If $\widehat{ST} \cong \widehat{TV}$, explain why $\triangle STV$ is an isosceles triangle.

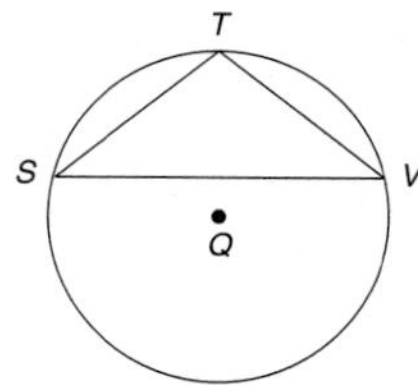


*40. Use a paragraph proof to complete this exercise.

*Given:* $\odot O$ with chords $\overline{AB}$ and $\overline{BC}$, radii $\overline{AO}$ and $\overline{OC}$
*Prove:* $m\angle ABC < m\angle AOC$

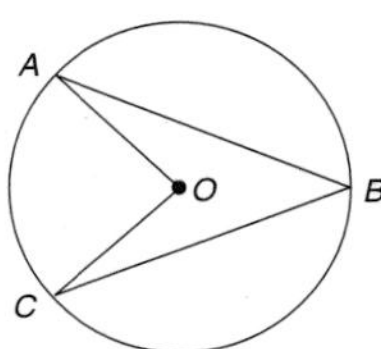


41. Prove Case 2 of Theorem 6.1.2.

42. Prove Case 3 of Theorem 6.1.2.

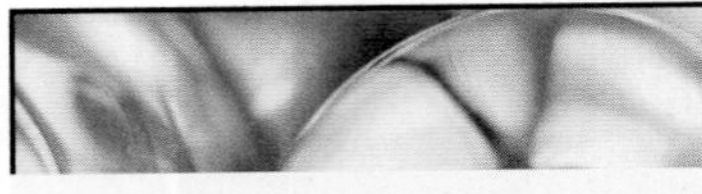

# 6.2 More Angle Measures in the Circle

**KEY CONCEPTS**

Tangent • Point of Tangency • Secant • Polygon Inscribed in a Circle • Circumscribed Circle • Polygon Circumscribed about a Circle • Inscribed Circle • Interior and Exterior of a Circle

We begin this section by considering lines, rays, and segments that are related to the circle. We assume the lines and circles are coplanar.

> DEFINITION: A **tangent** is a line that intersects a circle at exactly one point; the point of intersection is the **point of contact**, or **point of tangency.**

The term *tangent* also applies to a segment or ray that is part of a tangent line to a circle. In each case, the tangent touches the circle at one point.

> DEFINITION: A **secant** is a line (or segment or ray) that intersects a circle at exactly two points.

In Figure 6.21(a), line $s$ is a secant to $\odot O$; also, line $t$ is a tangent to $\odot O$ and point $C$ is its point of contact. In Figure 6.21(b), $\overline{AB}$ is a tangent to $\odot Q$ and point $T$ is its point of tangency; $\overleftrightarrow{CD}$ is a secant with points of intersection at $E$ and $F$.

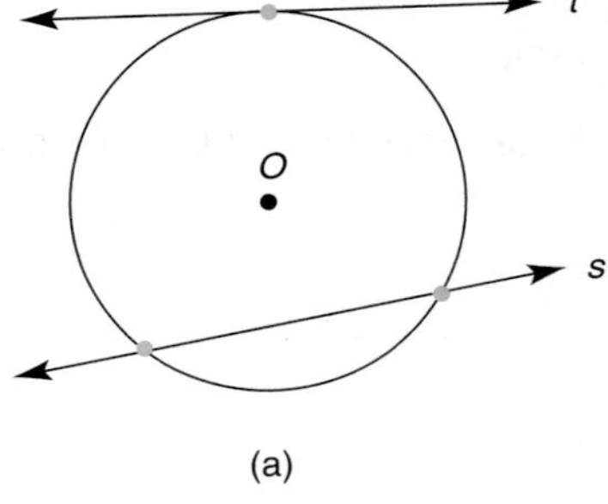


(a)

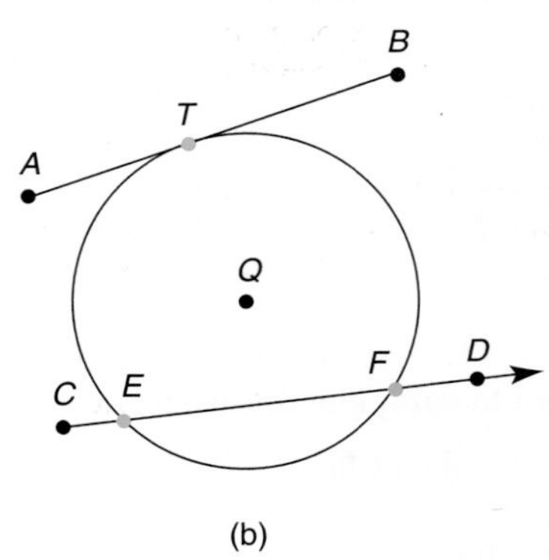


(b)

**FIGURE 6.21**

> DEFINITION: A polygon is **inscribed in a circle** if its vertices are points on the circle and its sides are chords of the circle. Equivalently, the circle is said to be **circumscribed about the polygon.** The polygon inscribed in a circle is further described as a **cyclic polygon.**

In Figure 6.22, $\triangle ABC$ is inscribed in $\odot O$ and quadrilateral $RSTV$ is inscribed in $\odot Q$. Conversely, $\odot O$ is circumscribed about $\triangle ABC$ and $\odot Q$ is circumscribed about quadrilateral $RSTV$. Note that $\overline{AB}$, $\overline{BC}$, and $\overline{AC}$ are chords of $\odot O$ and that $\overline{RS}$, $\overline{ST}$, $\overline{TV}$, and $\overline{RV}$ are chords of $\odot Q$. Quadrilateral $RSTV$ and $\triangle ABC$ are cyclic polygons.

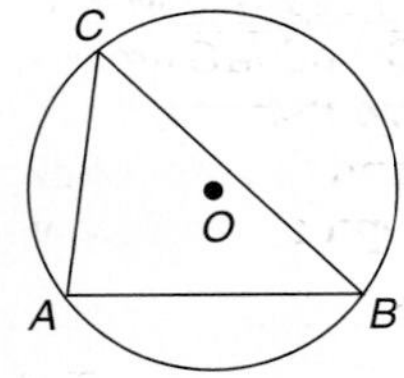


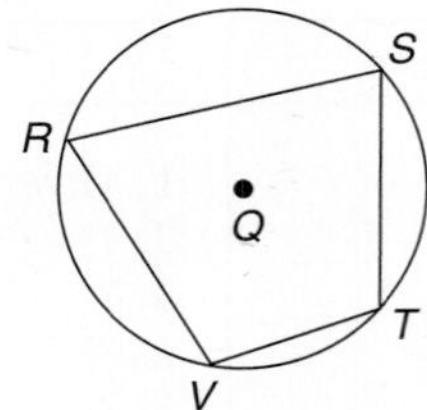


**FIGURE 6.22**

## Discover!

Draw any circle and call it $\odot O$. Now choose four points on $\odot O$ (in order, call these points $A$, $B$, $C$, and $D$). Join these points to form quadrilateral $ABCD$ inscribed in $\odot O$. Measure each of the inscribed angles ($\angle A$, $\angle B$, $\angle C$, and $\angle D$).

a) Find the sum $m\angle A + m\angle C$.
b) How are $\angle$s $A$ and $C$ related?
c) Find the sum $m\angle B + m\angle D$.
d) How are $\angle$s $B$ and $D$ related?

ANSWERS

a) 180° b) Supplementary c) 180° d) Supplementary

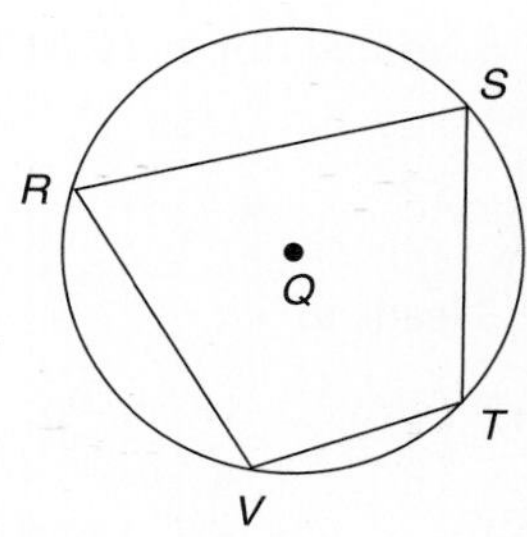


**FIGURE 6.23**

Reminder

A quadrilateral is said to be cyclic if its vertices lie on a circle.

The preceding Discover! activity prepares the way for the next theorem.

> THEOREM 6.2.1: If a quadrilateral is inscribed in a circle, the opposite angles are supplementary.
>
> *Alternative Form:* The opposite angles of a cyclic quadrilateral are supplementary.

The proof of Theorem 6.2.1 follows.

*Given:* $RSTV$ is inscribed in $\odot Q$ (See Figure 6.23.)

*Prove:* $\angle R$ and $\angle T$ are supplementary

*Proof:* From Section 6.1, an inscribed angle is equal in measure to one-half the measure of its intercepted arc. Because $m\angle R = \frac{1}{2}m\widehat{STV}$ and $m\angle T = \frac{1}{2}m\widehat{SRV}$, it follows that

$$m\angle R + m\angle T = \frac{1}{2}m\widehat{STV} + \frac{1}{2}m\widehat{SRV}$$

$$= \frac{1}{2}(m\widehat{STV} + m\widehat{SRV})$$

Because $\widehat{STV}$ and $\widehat{SRV}$ form the entire circle, $m\widehat{STV} + m\widehat{SRV} = 360°$. By substitution,

$$m\angle R + m\angle T = \frac{1}{2}(360°) = 180°$$

By definition, $\angle R$ and $\angle T$ are supplementary.

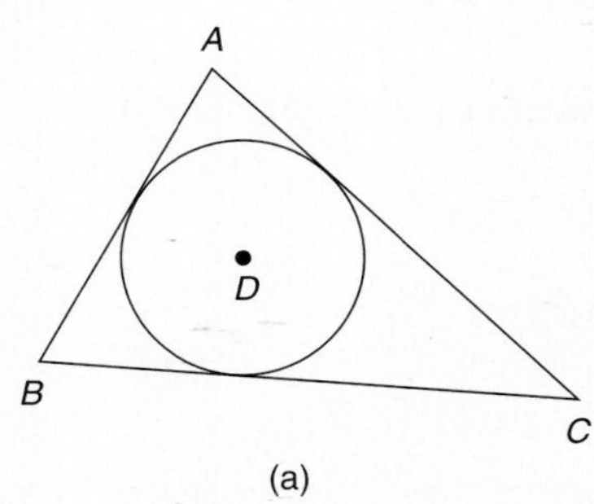


(a)

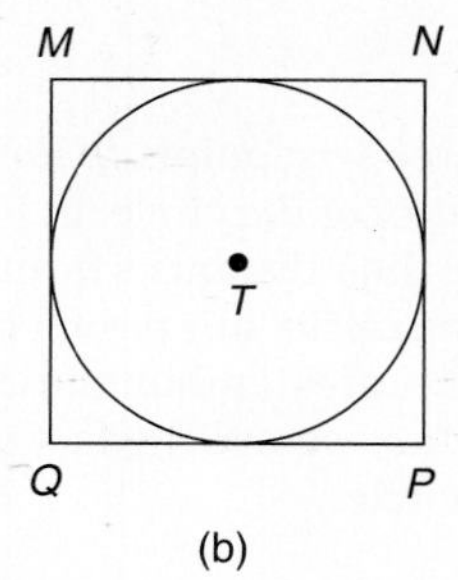


(b)

**FIGURE 6.24**

> DEFINITION: A polygon is **circumscribed about a circle** if all sides of the polygon are line segments tangent to the circle; also, the circle is said to be **inscribed in the polygon.**

In Figure 6.24(a), $\triangle ABC$ is circumscribed about $\odot D$. In Figure 6.24(b), square $MNPQ$ is circumscribed about $\odot T$. Furthermore, $\odot D$ is inscribed in $\triangle ABC$, and $\odot T$ is inscribed in square $MNPQ$. Note that $\overline{AB}$, $\overline{AC}$, and $\overline{BC}$ are tangents to $\odot D$ and that $\overline{MN}$, $\overline{NP}$, $\overline{PQ}$, and $\overline{MQ}$ are tangents to $\odot T$.

We know that a central angle has a measure equal to the measure of its intercepted arc and that an inscribed angle has a measure equal to one-half the measure of its intercepted arc. Now we consider another type of angle in the circle.

> THEOREM 6.2.2: The measure of an angle formed by two chords that intersect within a circle is one-half the sum of the measures of the arcs intercepted by the angle and its vertical angle.

*Exs. 1–6*

In Figure 6.25(a) on page 288, $\angle 1$ intercepts $\widehat{DB}$ and $\angle AEC$ intercepts $\widehat{AC}$. According to Theorem 6.2.2,

$$m\angle 1 = \frac{1}{2}(m\widehat{AC} + m\widehat{DB})$$

To prove Theorem 6.2.2, we draw auxiliary line segment $\overline{CB}$ [see $\odot$Figure 6.25(b)].

*Given:* Chords $\overline{AB}$ and $\overline{CD}$ intersect at point $E$ in $\odot O$

*Prove:* $m\angle 1 = \frac{1}{2}(m\widehat{AC} + m\widehat{DB})$

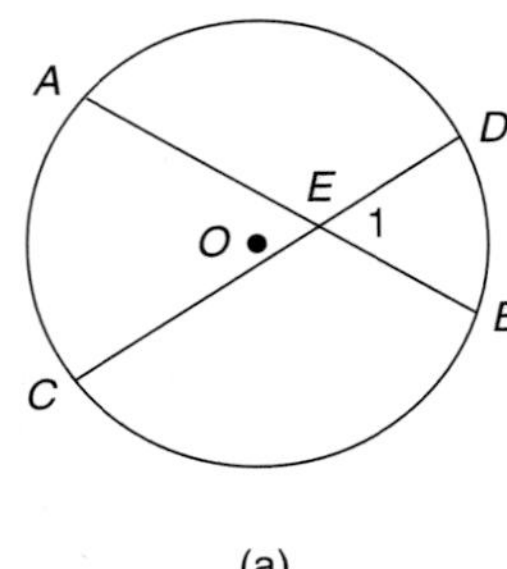


(a)

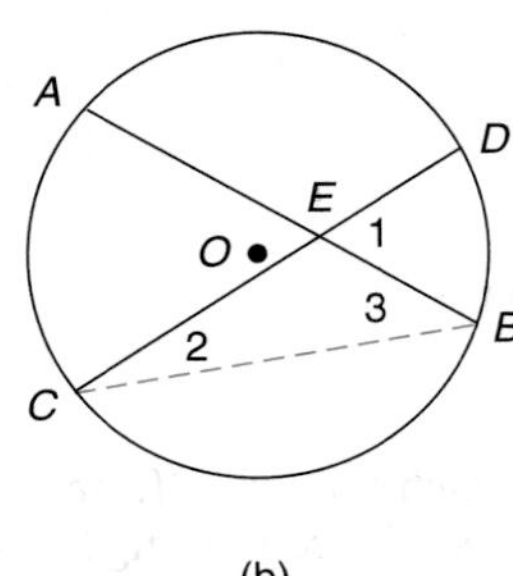


(b)

**FIGURE 6.25**

*Proof:* Draw $\overline{CB}$. Now $m\angle 1 = m\angle 2 + m\angle 3$ because $\angle 1$ is an exterior $\angle$ of $\triangle CBE$. Because $\angle 2$ and $\angle 3$ are inscribed angles of $\odot O$,

$$m\angle 2 = \frac{1}{2}m\widehat{DB} \text{ and } m\angle 3 = \frac{1}{2}m\widehat{AC}$$

Substitution into the equation $m\angle 1 = m\angle 2 + m\angle 3$ leads to

$$m\angle 1 = \frac{1}{2}m\widehat{DB} + \frac{1}{2}m\widehat{AC}$$
$$= \frac{1}{2}(m\widehat{DB} + m\widehat{AC})$$

Equivalently,

$$m\angle 1 = \frac{1}{2}(m\widehat{AC} + m\widehat{DB})$$

Next, we apply Theorem 6.2.2.

EXAMPLE 1

In Figure 6.25(a), $m\widehat{AC} = 84°$ and $m\widehat{DB} = 62°$. Find $m\angle 1$.

***Solution***

By Theorem 6.2.2,

$$m\angle 1 = \frac{1}{2}(m\widehat{AC} + m\widehat{DB})$$
$$= \frac{1}{2}(84° + 62°)$$
$$= \frac{1}{2}(146°) = 73°$$

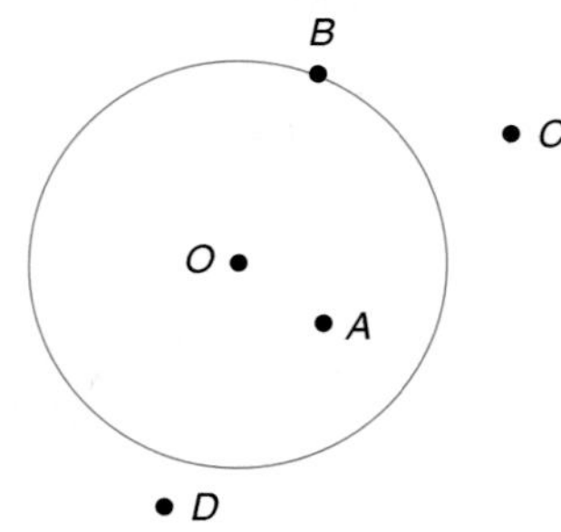


**FIGURE 6.26**

Recall that a circle separates points in the plane into three sets: points *in the interior* of the circle, points *on* the circle, and points *in the exterior* of the circle. In Figure 6.26, point $A$ and center $O$ are in the **interior** of $\odot O$ because their distances from center $O$ are less than the length of the radius. Point $B$ is on the circle, but points $C$ and $D$ are in the **exterior** of $\odot O$ because their distances from $O$ are greater than the length of the radius. (See Exercise 44.) In the proof of Theorem 6.2.3, we use the fact that a tangent to a circle cannot contain an interior point of the circle.

THEOREM 6.2.3: The radius (or any other line through the center of a circle) drawn to a tangent at the point of tangency is perpendicular to the tangent at that point.

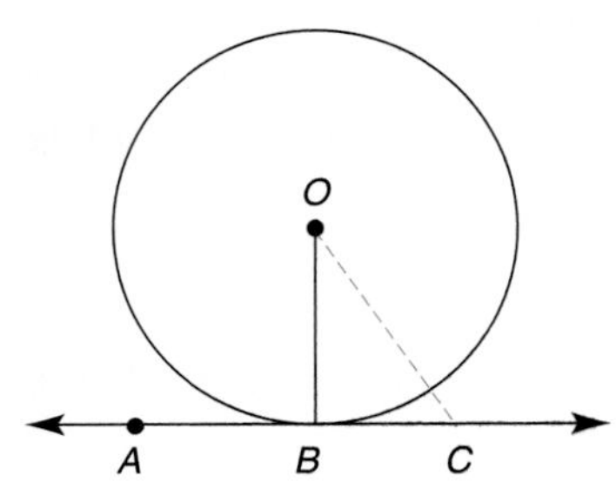


**FIGURE 6.27**

*Given:* $\odot O$ with tangent $\overleftrightarrow{AB}$; point $B$ is the point of tangency (See Figure 6.27.)

*Prove:* $\overline{OB} \perp \overleftrightarrow{AB}$

*Proof:* $\odot O$ has tangent $\overleftrightarrow{AB}$ and radius $\overline{OB}$. Let $C$ name any point on $\overleftrightarrow{AB}$ except $B$. Now $OC > OB$ because $C$ lies in the exterior of the circle. It follows that $\overline{OB} \perp \overleftrightarrow{AB}$ because the shortest distance from a point to a line is determined by the perpendicular segment from that point to the line.

The following example illustrates an application of Theorem 6.2.3.

EXAMPLE 2

A shuttle going to the moon has reached a position that is 5 mi above its surface. If the radius of the moon is 1080 mi, how far to the horizon can the NASA crew members see? (See Figure 6.28.)

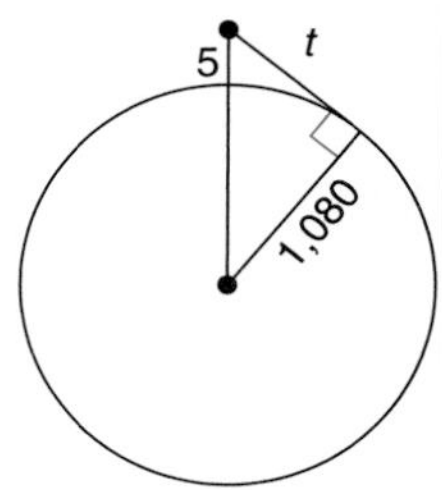


**FIGURE 6.28**

***Solution***

According to Theorem 6.2.3, the tangent determining the line of sight and the radius of the moon form a right angle. In the right triangle determined, let $t$ represent the desired distance. Using the Pythagorean Theorem,

$$1085^2 = t^2 + 1080^2$$
$$1{,}177{,}225 = t^2 + 1{,}166{,}400$$
$$t^2 = 10{,}825 \rightarrow t = \sqrt{10{,}825} \approx 104 \text{ mi}$$

*Exs. 7–10*

A consequence of Theorem 6.2.3 is Corollary 6.2.4, which follows. Of the three possible cases indicated in Figure 6.29, only the first is proved; the remaining two are left as exercises for the student. See Exercises 42 and 43.

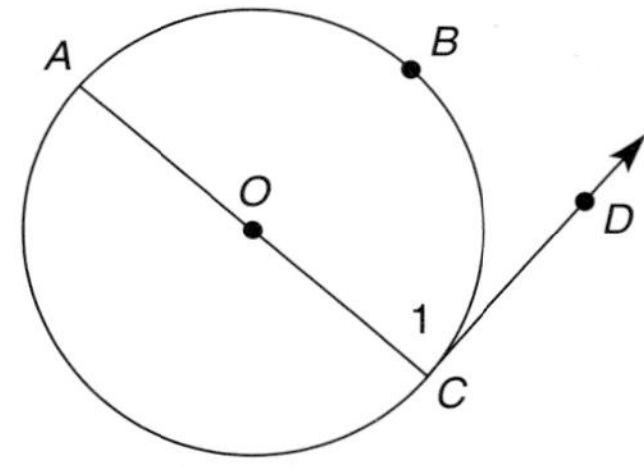


(a) Case 1
The chord is a diameter.

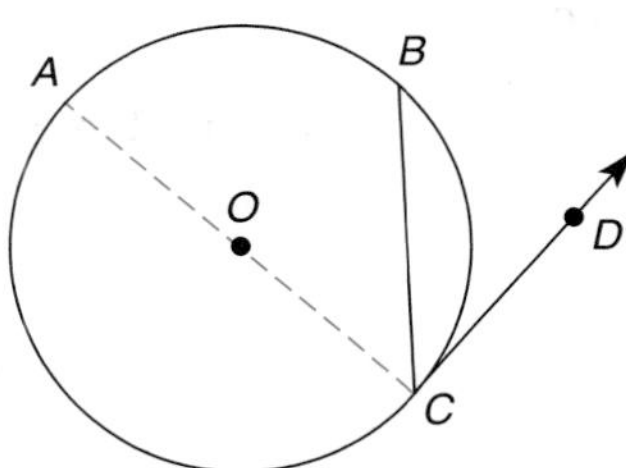


(b) Case 2
The diameter is in the exterior of the angle.

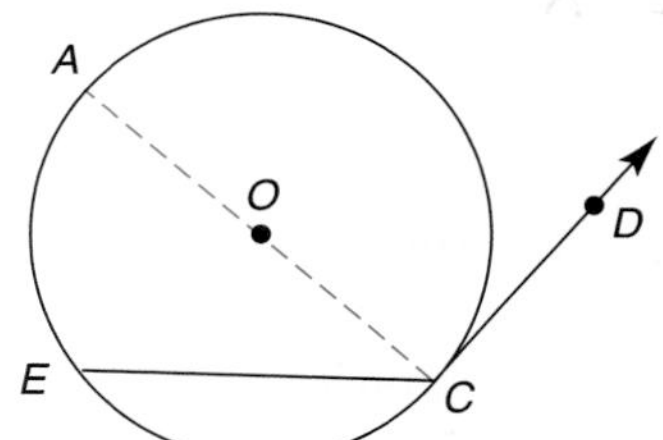


(c) Case 3
The diameter lies in the interior of the angle.

**FIGURE 6.29**

COROLLARY 6.2.4: The measure of an angle formed by a tangent and a chord drawn to the point of tangency is one-half the measure of the intercepted arc. (See Figure 6.29.)

*Given:* Chord $\overline{CA}$ (which is a diameter) and tangent $\overrightarrow{CD}$ [See Figure 6.29(a).]

*Prove:* $m\angle 1 = \frac{1}{2}m\widehat{ABC}$.

*Proof:* By Theorem 6.2.3, $\overline{AC} \perp \overrightarrow{CD}$. Then $\angle 1$ is a right angle and $m\angle 1 = 90°$. Because the intercepted arc $\widehat{ABC}$ is a semicircle, $m\widehat{ABC} = 180°$. Thus it follows that $m\angle 1 = \frac{1}{2}m\widehat{ABC}$.

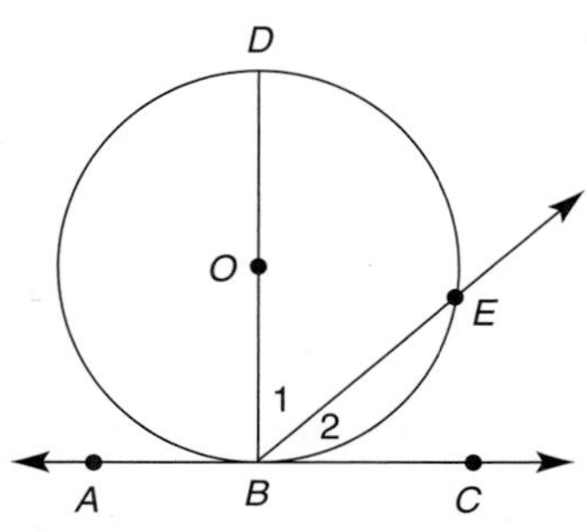


**FIGURE 6.30**

EXAMPLE 3

*Given:* In Figure 6.30, $\odot O$ with diameter $\overline{DB}$ and $m\widehat{DE} = 84°$

*Find:*
a) $m\angle 1$
b) $m\angle 2$
c) $m\angle ABD$
d) $m\angle ABE$

***Solution***

See Figure 6.30 on page 289.

a) $\angle 1$ is an inscribed angle; $m\angle 1 = \frac{1}{2}m\widehat{DE} = 42°$

b) With $m\widehat{DE} = 84°$ and $\widehat{DEB}$ a semicircle, $m\widehat{BE} = 180° - 84° = 96°$.
By Corollary 6.2.4, $m\angle 2 = \frac{1}{2}m\widehat{BE} = \frac{1}{2}(96°) = 48°$.

c) Because $\overline{DB}$ is perpendicular to $\overleftrightarrow{AB}$, $m\angle ABD = 90°$.

d) $m\angle ABE = m\angle ABD + m\angle 1 = 90° + 42° = 132°$

*Exs. 11, 12*

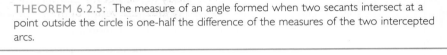

THEOREM 6.2.5: The measure of an angle formed when two secants intersect at a point outside the circle is one-half the difference of the measures of the two intercepted arcs.

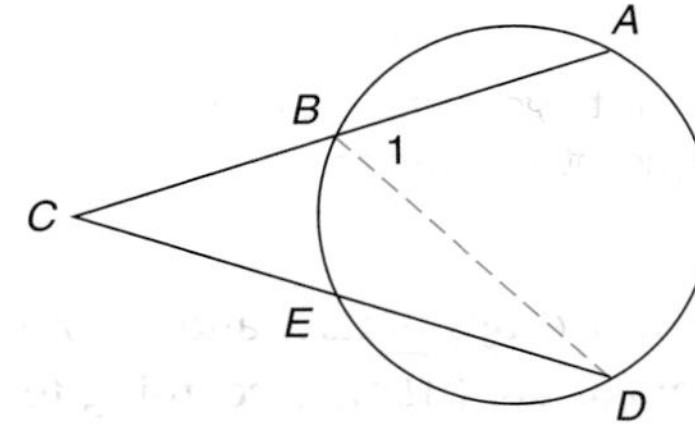



**FIGURE 6.31**

*Given:* Secants $\overline{AC}$ and $\overline{DC}$ as shown in Figure 6.31

*Prove:* $m\angle C = \frac{1}{2}(m\widehat{AD} - m\widehat{BE})$

*Proof:* Draw $\overline{BD}$ to form $\triangle BCD$. Then the measure of the exterior angle of $\triangle BCD$ is given by

$$m\angle 1 = m\angle C + m\angle D$$

so
$$m\angle C = m\angle 1 - m\angle D$$

$\angle 1$ and $\angle D$ are inscribed angles, so $m\angle 1 = \frac{1}{2}m\widehat{AD}$ and $m\angle D = \frac{1}{2}m\widehat{BE}$. Then

$$m\angle C = \frac{1}{2}m\widehat{AD} - \frac{1}{2}m\widehat{BE}$$

so
$$m\angle C = \frac{1}{2}(m\widehat{AD} - m\widehat{BE})$$

*Technology Exploration*

Use computer software if available.

1) Form a circle containing points $A$ and $D$.
2) From external point $C$, draw secants $\overline{CA}$ and $\overline{CD}$. Designate points of intersection as $B$ and $E$. See Figure 6.31.
3) Measure $\widehat{AD}$, $\widehat{BE}$, and $\angle C$.
4) Show that $m\angle C = \frac{1}{2}(m\widehat{AD} - m\widehat{BE})$.

NOTE: In an application of Theorem 6.2.5, one subtracts the smaller arc measure from the larger arc measure.

EXAMPLE 4

*Given:* In $\odot O$ of Figure 6.32, $m\angle AOB = 136°$ and $m\angle DOC = 46°$

*Find:* $m\angle E$

***Solution***

If $m\angle AOB = 136°$, then $m\widehat{AB} = 136°$. If $m\angle DOC = 46°$, then $m\widehat{DC} = 46°$. By Theorem 6.2.5,

$$\begin{aligned} m\angle E &= \frac{1}{2}(m\widehat{AB} - m\widehat{DC}) \\ &= \frac{1}{2}(136° - 46°) \\ &= \frac{1}{2}(90°) = 45° \end{aligned}$$

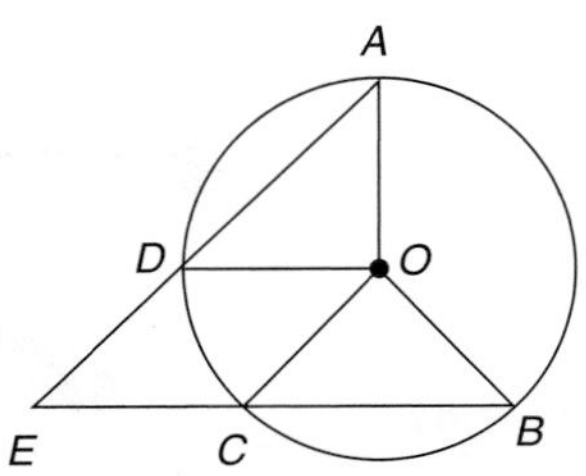



**FIGURE 6.32**

Theorems 6.2.5–6.2.7 will show that any angle formed by two lines that intersect outside a circle has a measure equal to one-half of the difference of the measures of the two intercepted arcs. The next two theorems are not proved, but the auxiliary lines shown will help complete the proofs.

> THEOREM 6.2.6: If an angle is formed by a secant and a tangent that intersect in the exterior of a circle, then the measure of the angle is one-half the difference of the measures of its intercepted arcs.

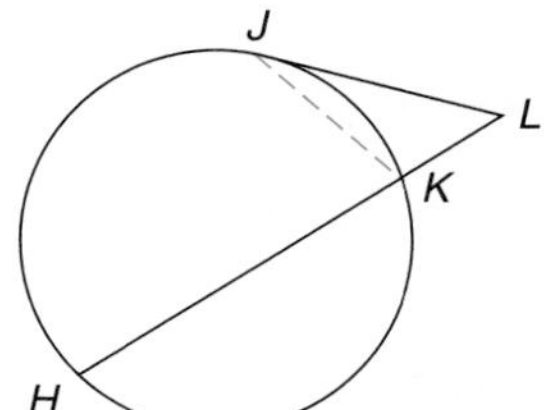


**FIGURE 6.33**

According to Theorem 6.2.6,

$$m\angle L = \frac{1}{2}(m\widehat{HJ} - m\widehat{JK})$$

in Figure 6.33.

A quick study of the figures that illustrate Theorems 6.2.5–6.2.7 shows that the smaller arc is "nearer" the vertex of the angle and that the larger arc is "farther from" the vertex.

> THEOREM 6.2.7: If an angle is formed by two intersecting tangents, then the measure of the angle is one-half the difference of the measures of the intercepted arcs.

In Figure 6.34(a), $\angle ABC$ intercepts the two arcs determined by points $A$ and $C$. The small arc is a minor arc ($\widehat{AC}$) and the large arc is a major arc ($\widehat{ADC}$). According to Theorem 6.2.7,

$$m\angle ABC = \frac{1}{2}(m\widehat{ADC} - m\widehat{AC})$$

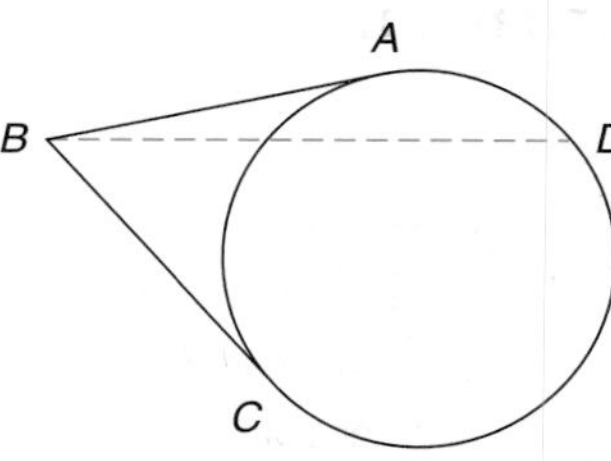


(a)

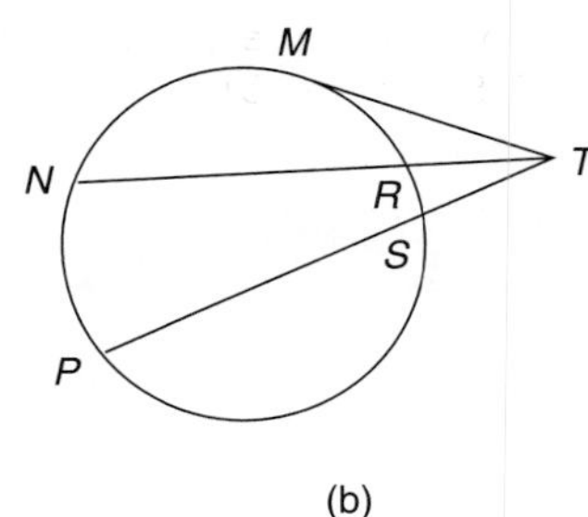


(b)

**FIGURE 6.34**

EXAMPLE 5

*Given:* In Figure 6.34(b), $m\widehat{MN} = 70°$, $m\widehat{NP} = 88°$, $m\widehat{MR} = 46°$, and $m\widehat{RS} = 26°$

*Find:* a) $m\angle MTN$
b) $m\angle NTP$
c) $m\angle MTP$

***Solution***

a) $m\angle MTN = \frac{1}{2}(m\widehat{MN} - m\widehat{MR})$
$= \frac{1}{2}(70° - 46°)$
$= \frac{1}{2}(24°) = 12°$

b) $m\angle NTP = \frac{1}{2}(m\widehat{NP} - m\widehat{RS})$
$= \frac{1}{2}(88° - 26°)$
$= \frac{1}{2}(62°) = 31°$

c) $m\angle MTP = m\angle MTN + m\angle NTP$
Using results from (a) and (b),
$m\angle MTP = 12° + 31° = 43°$

Before considering the final example of this section, let's review the methods used to measure different types of angles related to a circle. These are summarized in Table 6.1.

**TABLE 6.1**

**Methods for Measuring Angles Related to a Circle**

| *Location of the Vertex of the Angle* | *Rule for Measuring the Angle* |
|---|---|
| *Center* of the circle | The *measure* of the intercepted arc |
| In the *interior* of the circle | *One-half the sum* of the measures of the intercepted arcs |
| *On* the circle | *One-half the measure* of the intercepted arc |
| In the *exterior* of the circle | *One-half the difference* of the measures of the two intercepted arcs |

*Exs. 13–18*

EXAMPLE 6

Given that $m\angle 1 = 46°$ in Figure 6.35, find the measures of $\widehat{AB}$ and $\widehat{ACB}$.

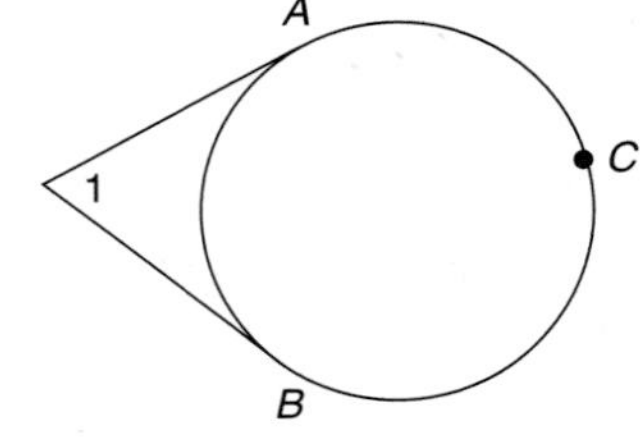


**FIGURE 6.35**

***Solution***

Let $m\widehat{AB} = x$ and $m\widehat{ACB} = y$. Now

$$m\angle 1 = \frac{1}{2}(m\widehat{ACB} - m\widehat{AB})$$

so

$$46 = \frac{1}{2}(y - x)$$

Multiplying by 2, we have $92 = y - x$.

Also, $y + x = 360$ because these two arcs form the entire circle. We add these equations as shown.

$$\begin{array}{r} y + x = 360 \\ \underline{y - x = \phantom{0}92} \\ 2y \phantom{+ x} = 452 \\ y = 226 \end{array}$$

Because $x + y = 360$, we know that $x + 226 = 360$ and $x = 134$. Then $m\widehat{AB} = 134°$ and $m\widehat{ACB} = 226°$.

**FIGURE 6.36**

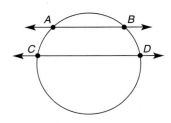

THEOREM 6.2.8: If two parallel lines intersect a circle, the intercepted arcs between these lines are congruent. (See Figure 6.36.)

Where $\overleftrightarrow{AB} \parallel \overleftrightarrow{CD}$ in Figure 6.36, it follows that $\widehat{AC} \cong \widehat{BD}$. Equivalently, $m\widehat{AC} = m\widehat{BD}$. The proof of Theorem 6.2.8 is left as an exercise.

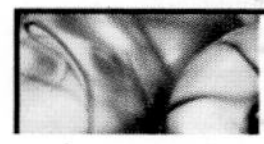

## 6.2 Exercises

1. *Given:* $m\widehat{AB} = 92°$
   $m\widehat{DA} = 114°$
   $m\widehat{BC} = 138°$
   *Find:* a) $m\angle 1$ ($\angle DAC$)
   b) $m\angle 2$ ($\angle ADB$)
   c) $m\angle 3$ ($\angle AFB$)
   d) $m\angle 4$ ($\angle DEC$)
   e) $m\angle 5$ ($\angle CEB$)

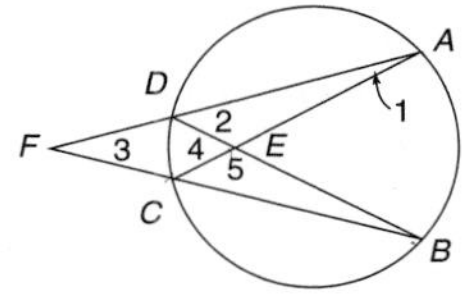


*Exercises 1, 2*

2. *Given:* $m\widehat{DC} = 30°$ and $\widehat{DABC}$ is trisected at points $A$ and $B$
   *Find:* a) $m\angle 1$
   b) $m\angle 2$
   c) $m\angle 3$
   d) $m\angle 4$
   e) $m\angle 5$

3. *Given:* Circle $O$ with diameter $\overline{RS}$, tangent $\overrightarrow{SW}$, chord $\overline{TS}$, and $m\widehat{RT} = 26°$.
   *Find:* a) $m\angle WSR$
   b) $m\angle RST$
   c) $m\angle WST$

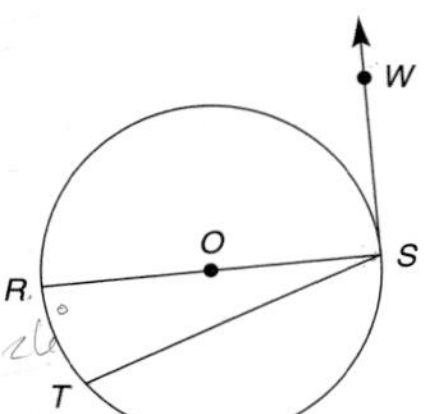


*Exercises 3–5*

4. Find $m\widehat{RT}$ if $m\angle RST{:}m\angle RSW = 1{:}5$.

5. Find $m\angle RST$ if $m\widehat{RT}{:}m\widehat{TS} = 1{:}4$.

6. Is it possible for
   a) an inscribed rectangle in a circle to have a diameter for a side? Explain.
   b) a circumscribed rectangle about a circle to be a square? Explain.

7. *Given:* In $\odot Q$, $\overline{PR}$ contains $Q$, $\overline{MR}$ is a tangent, $m\widehat{MP} = 112°$, $m\widehat{MN} = 60°$, and $m\widehat{MT} = 46°$
   *Find:* a) $m\angle MRP$
   b) $m\angle 1$
   c) $m\angle 2$

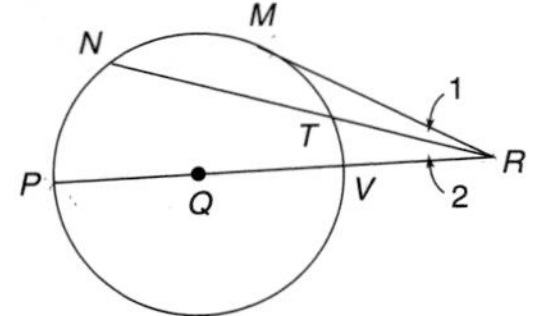


8. *Given:* $\overrightarrow{AB}$ and $\overrightarrow{AC}$ are tangent to $\odot O$, $m\widehat{BC} = 126°$
   *Find:* a) $m\angle A$
   b) $m\angle ABC$
   c) $m\angle ACB$

9. *Given:* Tangents $\overrightarrow{AB}$ and $\overrightarrow{AC}$ to $\odot O$
   $m\angle ACB = 68°$
   *Find:* a) $m\widehat{BC}$
   b) $m\widehat{BDC}$
   c) $m\angle ABC$
   d) $m\angle A$

*Exercises 8, 9*

10. *Given:* $m\angle 1 = 72°$, $m\widehat{DC} = 34°$
    *Find:* a) $m\widehat{AB}$
    b) $m\angle 2$

11. *Given:* $m\angle 2 = 36°$
    $m\widehat{AB} = 4 \cdot m\widehat{DC}$
    *Find:* a) $m\widehat{AB}$
    b) $m\angle 1$
    (HINT: Let $m\widehat{DC} = x$ and $m\widehat{AB} = 4x$.)

*Exercises 10, 11*

*In Exercises 12 and 13, R and T are points of tangency.*

12. *Given:* $m\angle 3 = 42°$
    *Find:* a) $m\widehat{RT}$
    b) $m\widehat{RST}$

13. *Given:* $\widehat{RS} \cong \widehat{ST} \cong \widehat{RT}$
    *Find:* a) $m\widehat{RT}$
    b) $m\widehat{RST}$
    c) $m\angle 3$

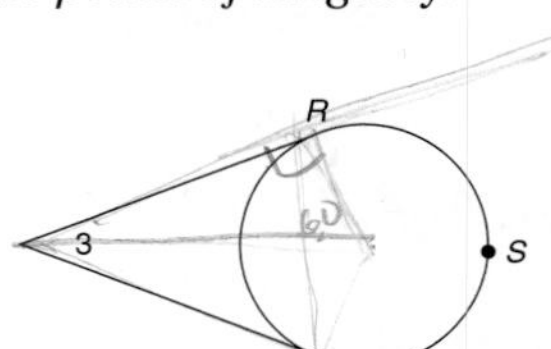


*Exercises 12, 13*

14. *Given:* $m\angle 1 = 63°$
    $m\widehat{RS} = 3x + 6$
    $m\widehat{VT} = x$
    *Find:* $m\widehat{RS}$

15. *Given:* $m\angle 2 = 124°$
    $m\widehat{TV} = x + 1$
    $m\widehat{SR} = 3(x + 1)$
    *Find:* $m\widehat{TV}$

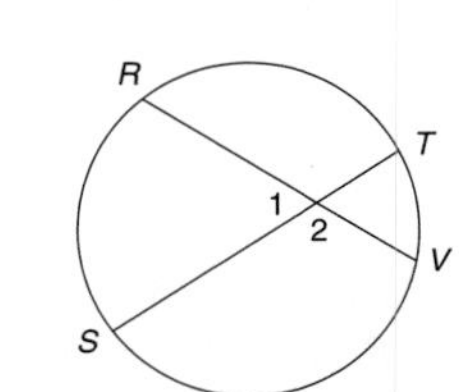


*Exercises 14, 15*

16. *Given:* $m\angle 1 = 71°$
    $m\angle 2 = 33°$
    *Find:* $m\widehat{CE}$ and $m\widehat{BD}$

17. *Given:* $m\angle 1 = 62°$
    $m\angle 2 = 26°$
    *Find:* $m\widehat{CE}$ and $m\widehat{BD}$

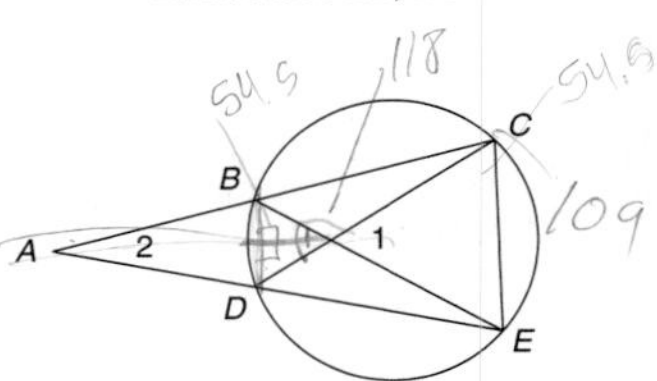


*Exercises 16, 17*

18. a) How are $\angle R$ and $\angle T$ related?
    b) Find m$\angle R$ if m$\angle T = 112°$.

19. a) How are $\angle S$ and $\angle V$ related?
    b) Find m$\angle V$ if m$\angle S = 73°$.

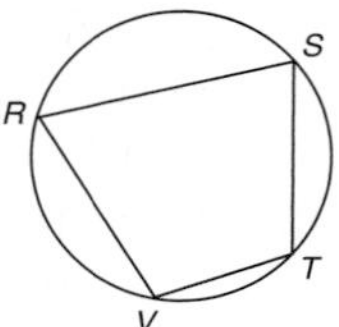

*Exercises 18, 19*

20. A quadrilateral *RSTV* is circumscribed about a circle so that its tangent sides are at the endpoints of two intersecting diameters.

    a) What type of quadrilateral is *RSTV*?
    b) If the diameters are also perpendicular, what type of quadrilateral is *RSTV*?

*In Exercises 21 and 22, complete each proof.*

21. *Given:* $\overline{AB}$ and $\overline{AC}$ are tangents to $\odot O$ from point *A*
    *Prove:* $\triangle ABC$ is isosceles

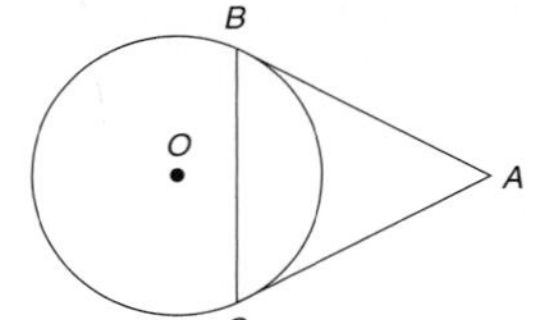


**PROOF**

| Statements | Reasons |
|---|---|
| 1. ? | 1. Given |
| 2. $m\angle B = \frac{1}{2}(m\widehat{BC})$ and $m\angle C = \frac{1}{2}(m\widehat{BC})$ | 2. ? |
| 3. $m\angle B = m\angle C$ | 3. ? |
| 4. $\angle B \cong \angle C$ | 4. ? |
| 5. ? | 5. If two $\angle$s of a $\triangle$ are $\cong$, the sides opposite the $\angle$s are $\cong$ |
| 6. ? | 6. If two sides of a $\triangle$ are $\cong$, the $\triangle$ is isosceles |

22. *Given:* $\overline{RS} \parallel \overline{TQ}$
    *Prove:* $\widehat{RT} \cong \widehat{SQ}$

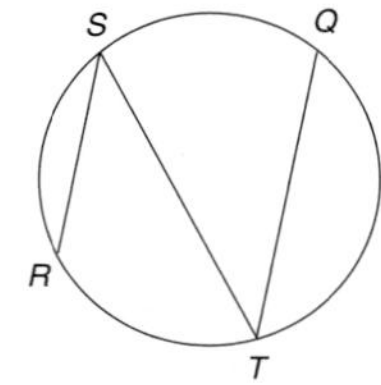


**PROOF**

| Statements | Reasons |
|---|---|
| 1. $\overline{RS} \parallel \overline{TQ}$ | 1. ? |
| 2. $\angle S \cong \angle T$ | 2. ? |
| 3. ? | 3. If two $\angle$s are $\cong$, the $\angle$s are $=$ in measure |
| 4. $m\angle S = \frac{1}{2}(m\widehat{RT})$ | 4. ? |
| 5. $m\angle T = \frac{1}{2}(m\widehat{SQ})$ | 5. ? |
| 6. $\frac{1}{2}(m\widehat{RT}) = \frac{1}{2}(m\widehat{SQ})$ | 6. ? |
| 7. $m\widehat{RT} = m\widehat{SQ}$ | 7. Multiplication Property of Equality |
| 8. ? | 8. If two arcs of a $\odot$ are $=$ in measure, the arcs are $\cong$ |

*In Exercises 23 to 25, complete a paragraph proof.*

23. *Given:* Tangent $\overline{AB}$ to $\odot O$ at point *B*
    $m\angle A = m\angle B$
    *Prove:* $m\widehat{BD} = 2 \cdot m\widehat{BC}$

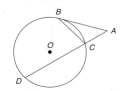

24. *Given:* Diameter $\overline{AB} \perp \overline{CE}$ at *D*
    *Prove:* *CD* is the geometric mean of *AD* and *DB*

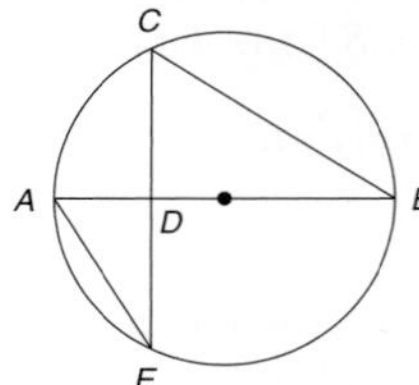


*In Exercises 25 and 26, $\overline{CA}$ and $\overline{CB}$ are tangents.*

25. *Given:* $m\widehat{AB} = x$
    *Prove:* $m\angle 1 = 180° - x$

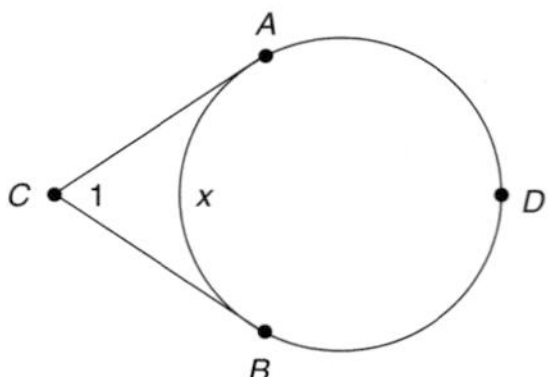

*Exercises 25, 26*

26. Use the result from Exercise 25 to find $m\angle 1$ if $m\widehat{AB} = 104°$.

27. An airplane reaches an altitude of 3 mi above the earth. Assuming a clear day and that a passenger has binoculars, how far can the passenger see? (HINT: The radius of the earth is approximately 4000 mi.)

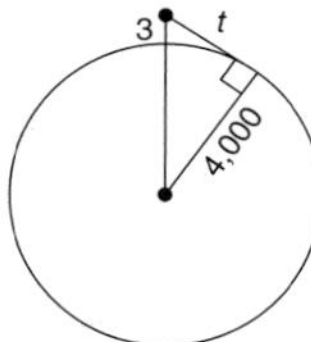


28. From the veranda of a beachfront hotel, Manny is searching the seascape through his binoculars. A ship suddenly appears on the horizon. If Manny is 80 ft above the earth, how far is the ship out at sea? (HINT: See Exercise 27 and note that 1 mi = 5280 ft.)

29. For the five-pointed star (pentagram), find the measures of ∠1 and ∠2.

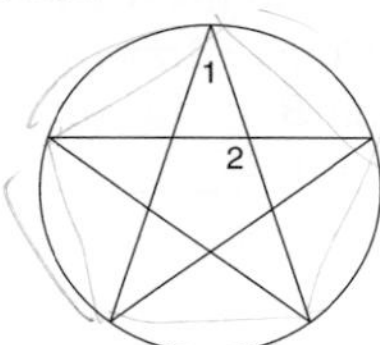


30. A six-pointed star (hexagram) is inscribed in a circle (see Exercise 29). Find m∠1, an angle of the hexagram with vertex on the circle. Then find m∠2, an interior angle of the regular hexagon that lies in the interior of the circle.

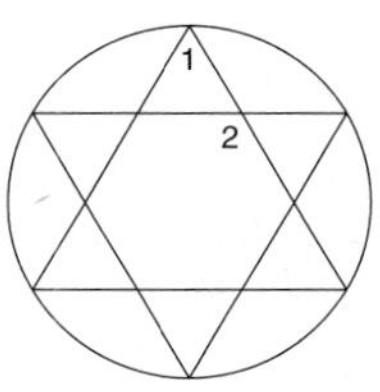


31. A satellite dish in the shape of a regular dodecagon (12 sides) is nearly "circular." Find:

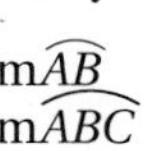

a) $m\widehat{AB}$
b) $m\widehat{ABC}$
c) m∠ABC (inscribed angle)

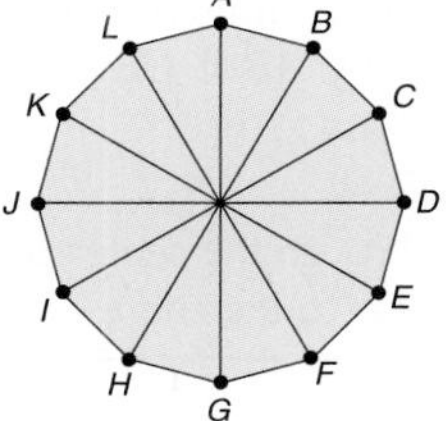


32. In the figure shown, △RST ~ △WVT by the reason AA. Name two pairs of congruent angles in these similar triangles.

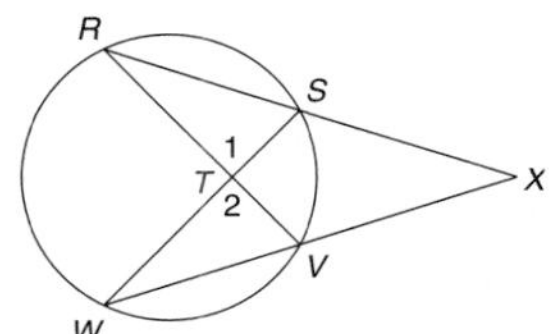


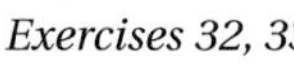
*Exercises 32, 33*

33. In the figure shown, △RXV ~ △WXS by the reason AA. Name two pairs of congruent angles in these similar triangles.

$*$ 34. On a fitting for a hex wrench, the distance from the center O to a vertex is 5 mm. The length of radius $\overline{OB}$ of the circle is 10 mm. If $\overline{OC} \perp \overline{DE}$ at F, how long is $\overline{FC}$?

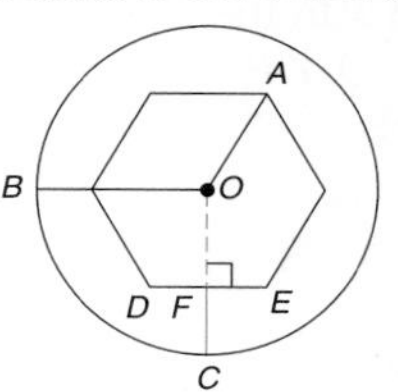


$*$ 35. *Given:* $\overline{AB}$ is a diameter in ⊙O
M is the midpoint of chord $\overline{AC}$
N is the midpoint of chord $\overline{CB}$
$MB = \sqrt{73}$, $AN = 2\sqrt{13}$
*Find:* The length of diameter $\overline{AB}$

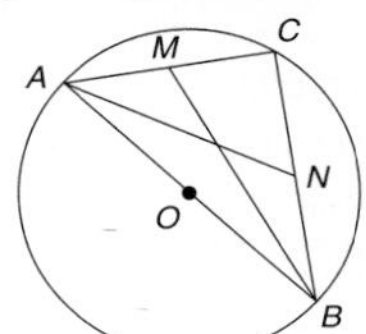


36. A surveyor sees a circular planetarium through a 60° angle. If the surveyor is 45 ft from the door, what is the diameter of the planetarium?

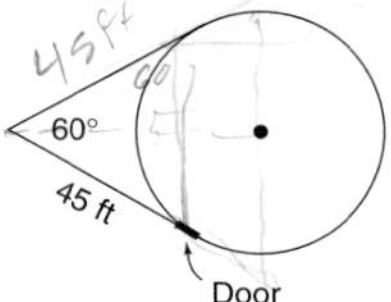


*In Exercises 37 to 45, provide a paragraph proof. Be sure to provide a drawing, Given, and Prove where needed.*

37. If two parallel lines intersect a circle, then the intercepted arcs between these lines are congruent. (HINT: See Figure 6.36. Draw chord $\overline{AD}$.)

38. The line joining the centers of two circles that intersect at two points is the perpendicular bisector of the common chord.

39. If a trapezoid is inscribed in a circle, then it is an isosceles trapezoid.

40. If a parallelogram is inscribed in a circle, then it is a rectangle.

41. If one side of an inscribed triangle is a diameter, then the triangle is a right triangle.

42. Prove Case 2 of Corollary 6.2.4: The measure of an angle formed by a tangent and a chord drawn to the point of tangency is one-half the measure of the intercepted arc. (See Figure 6.29.)

43. Prove Case 3 of Corollary 6.2.4. (See Figure 6.29.)

44. *Given:* $\odot O$ with $P$ in its exterior; $O$-$Y$-$P$
*Prove:* $OP > OY$

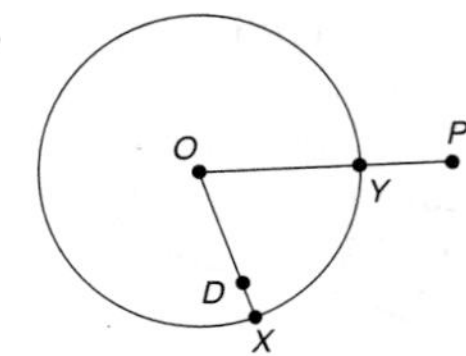


45. *Given:* Quadrilateral $RSTV$ inscribed in $\odot Q$
*Prove:* $m\angle R + m\angle T = m\angle V + m\angle S$

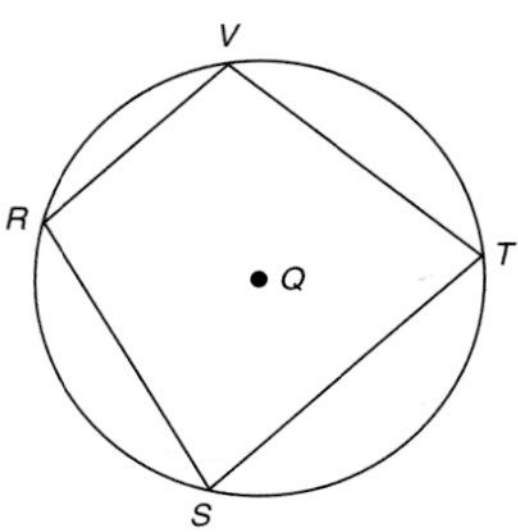


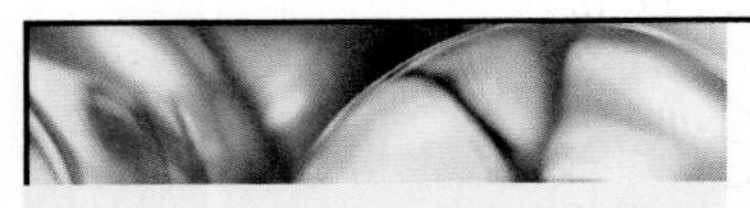

**KEY CONCEPTS**

Tangent Circles • Internally Tangent Circles • Externally Tangent Circles • Line of Centers • Common Tangent • Common External Tangents • Common Internal Tangents

# 6.3 Line and Segment Relationships in the Circle

In this section, we turn our attention to line and line segment relationships in the circle. Because some of these statements (such as Theorems 6.3.1–6.3.3) are so similar in wording, the student is strongly encouraged to make drawings and then compare the information that is given in each theorem to the conclusion of that theorem.

> THEOREM 6.3.1: If a line is drawn through the center of a circle perpendicular to a chord, then it bisects the chord and its arc.

NOTE: Note that the term *arc* generally refers to the minor arc, even though the major arc is also bisected.

*Given:* $\overleftrightarrow{AB} \perp$ chord $\overline{CD}$ in circle $A$ (See Figure 6.37.)

*Prove:* $\overline{CB} \cong \overline{BD}$ and $\overset{\frown}{CE} \cong \overset{\frown}{ED}$

The proof is left as an exercise for the student.

HINT: Draw $\overline{AC}$ and $\overline{AD}$.

Even though the Prove statement does not match the conclusion of Theorem 6.3.1, we know that $\overline{CD}$ is bisected by $\overleftrightarrow{AB}$ if $\overline{CB} \cong \overline{BD}$ and that $\overset{\frown}{CD}$ is bisected by $\overleftrightarrow{AE}$ if $\overset{\frown}{CE} \cong \overset{\frown}{ED}$.

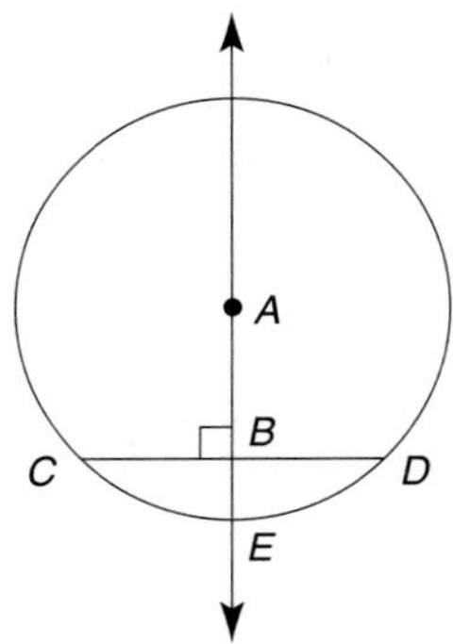


**FIGURE 6.37**

> THEOREM 6.3.2: If a line through the center of a circle bisects a chord other than a diameter, then it is perpendicular to the chord.

*Given:* Circle $O$; $\overleftrightarrow{OM}$ is the bisector of chord $\overline{RS}$ (See Figure 6.38.)

*Prove:* $\overleftrightarrow{OM} \perp \overline{RS}$

The proof is left as an exercise for the student.

HINT: Draw radii $\overline{OR}$ and $\overline{OS}$.

Figure 6.39(a) illustrates the following theorem. However, Figure 6.39(b) is used in the proof.

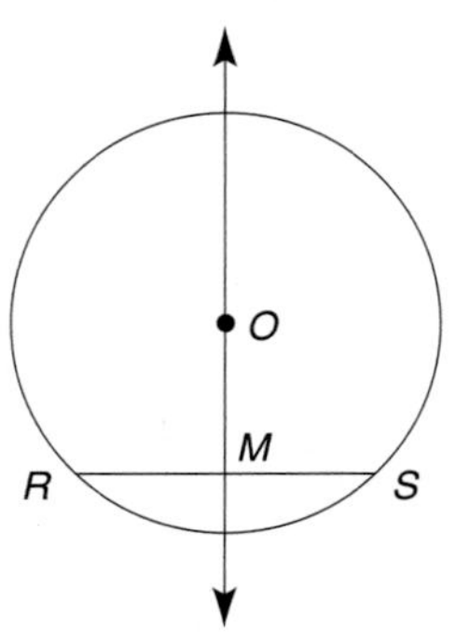


**FIGURE 6.38**

> THEOREM 6.3.3: The perpendicular bisector of a chord contains the center of the circle.

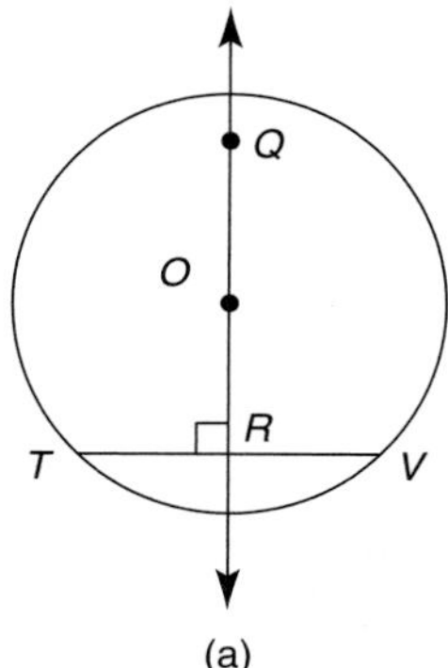


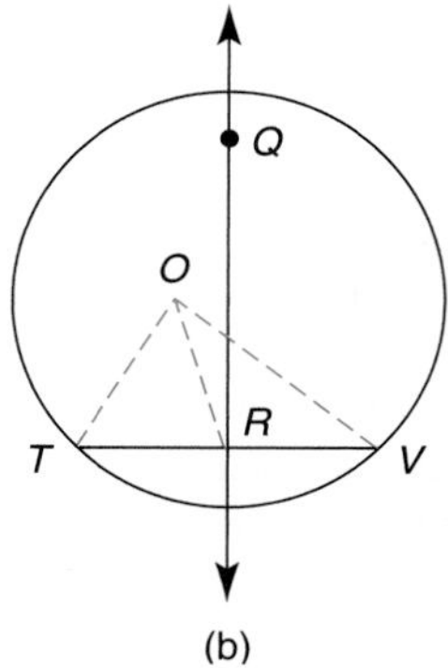


**FIGURE 6.39**

*Given:* In Figure 6.39(a), $\overleftrightarrow{QR}$ is the perpendicular bisector of chord $\overline{TV}$ in $\odot O$

*Prove:* $\overleftrightarrow{QR}$ contains point $O$

*Proof (by indirect method):* Suppose that $O$ is not on $\overleftrightarrow{QR}$. Draw $\overline{OR}$ and radii $\overline{OT}$ and $\overline{OV}$. [See Figure 6.39(b).] Because $\overleftrightarrow{QR}$ is the perpendicular bisector of $\overline{TV}$, $R$ must be the midpoint of $\overline{TV}$; then $\overline{TR} \cong \overline{RV}$. Also, $\overline{OT} \cong \overline{OV}$ (all radii of a $\odot$ are $\cong$). With $\overline{OR} \cong \overline{OR}$ by identity, we have $\triangle ORT \cong \triangle ORV$ by SSS.

Now $\angle ORT \cong \angle ORV$ by CPCTC. It follows that $\overline{OR} \perp \overline{TV}$ because these lines (segments) meet to form congruent adjacent angles.

Then $\overline{OR}$ is the perpendicular bisector of $\overline{TV}$. But $\overleftrightarrow{QR}$ is also the perpendicular bisector of $\overline{TV}$, which contradicts the uniqueness of the perpendicular bisector of a segment.

Then the supposition must be false, and it follows that center $O$ is on $\overleftrightarrow{QR}$, the perpendicular bisector of chord $\overline{TV}$.

EXAMPLE 1

*Given:* In Figure 6.40, $\odot O$ has a radius of length 5
$\overline{OE} \perp \overline{CD}$ at $B$ and $OB = 3$

*Find:* $CD$

***Solution***

Draw radius $\overline{OC}$. By the Pythagorean Theorem,

$$\begin{aligned} (OC)^2 &= (OB)^2 + (BC)^2 \\ 5^2 &= 3^2 + (BC)^2 \\ 25 &= 9 + (BC)^2 \\ (BC)^2 &= 16 \\ BC &= 4 \end{aligned}$$

According to Theorem 6.3.1, we know that $CD = 2 \cdot BC$; then it follows that $CD = 2 \cdot 4 = 8$.

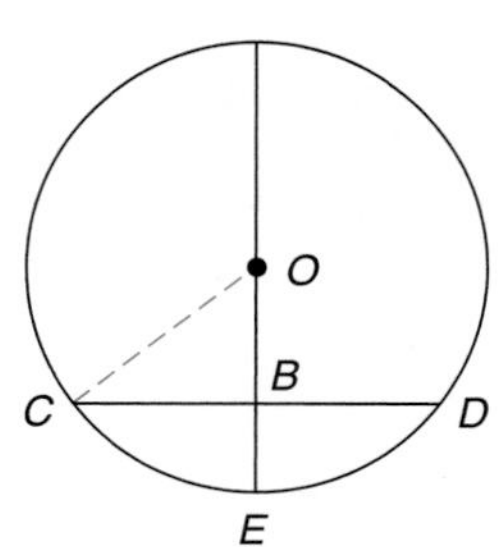


**FIGURE 6.40**

## CIRCLES THAT ARE TANGENT

*Exs. 1–4*

In this section, we assume that two circles are coplanar. Although concentric circles do not intersect, they do share a common center. For the concentric circles shown in Figure 6.41, the tangent of the smaller circle is a chord of the larger circle.

If two circles touch at one point, they are **tangent circles.** In Figure 6.42, circles $P$ and $Q$ are **internally tangent,** whereas circles $O$ and $R$ are **externally tangent.**

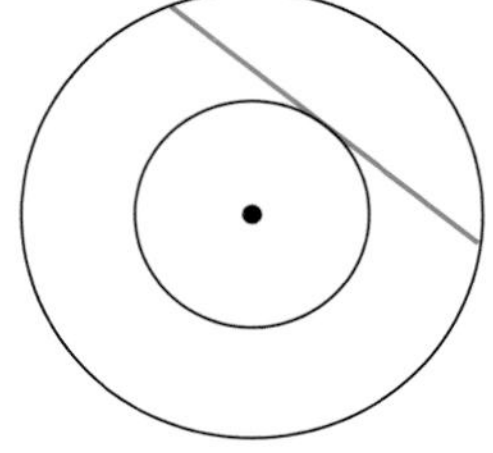

**FIGURE 6.41**

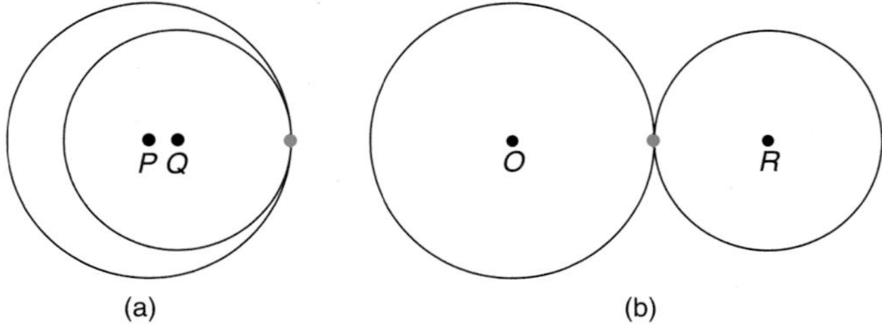


**FIGURE 6.42**

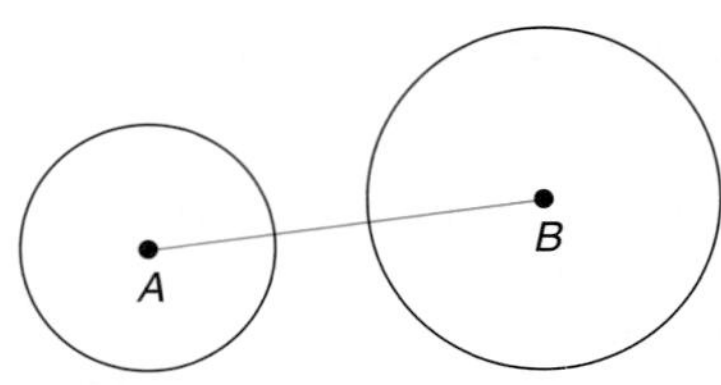


**FIGURE 6.43**

DEFINITION: For two circles with different centers, the **line of centers** is the line (or line segment) containing the centers of both circles.

As the definition suggests, the line segment joining the centers of two circles is also commonly called the line of centers of the two circles. In Figure 6.43, $\overleftrightarrow{AB}$ or $\overline{AB}$ is the line of centers for circles $A$ and $B$.

Geometry in the Real World

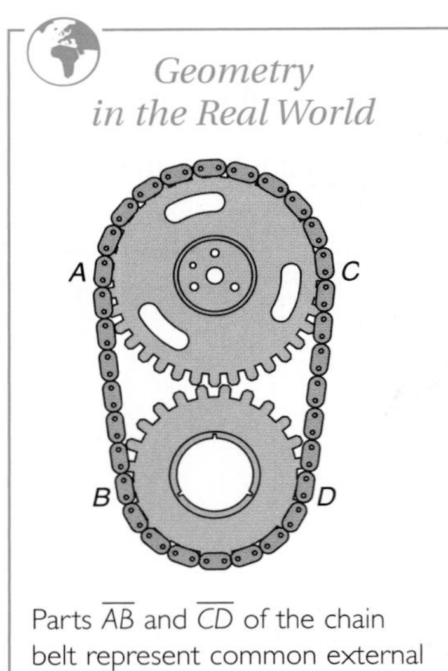


Parts $\overline{AB}$ and $\overline{CD}$ of the chain belt represent common external tangents to the circular gears.

## COMMON TANGENT LINES TO CIRCLES

A line segment that is tangent to each of two circles is a **common tangent** for these circles. If the common tangent *does not* intersect the line of centers, it is a **common external tangent.** In Figure 6.44, circles $P$ and $Q$ have one common external tangent, $\overleftrightarrow{ST}$; circles $A$ and $B$ have two common external tangents, $\overleftrightarrow{WX}$ and $\overleftrightarrow{YZ}$.

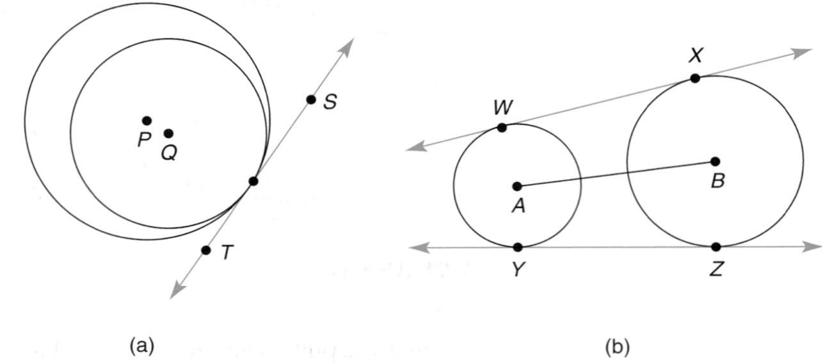


**FIGURE 6.44**

*Exs. 5–6*

If the common tangent *does* intersect the line of centers for two circles, it is a **common internal tangent** for the two circles. In Figure 6.45, $\overleftrightarrow{DE}$ is a common internal tangent for externally tangent circles $O$ and $R$; $\overleftrightarrow{AB}$ and $\overleftrightarrow{CD}$ are common internal tangents for $\odot M$ and $\odot N$.

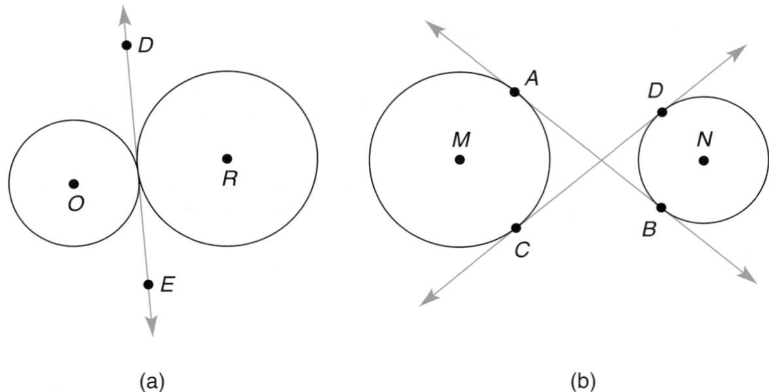


**FIGURE 6.45**

Discover!

Measure the lengths of tangent segments $\overline{AB}$ and $\overline{AC}$ of Figure 6.46. How do $AB$ and $AC$ compare?

ANSWER

They are equal.

THEOREM 6.3.4: The tangent segments to a circle from an external point are congruent.

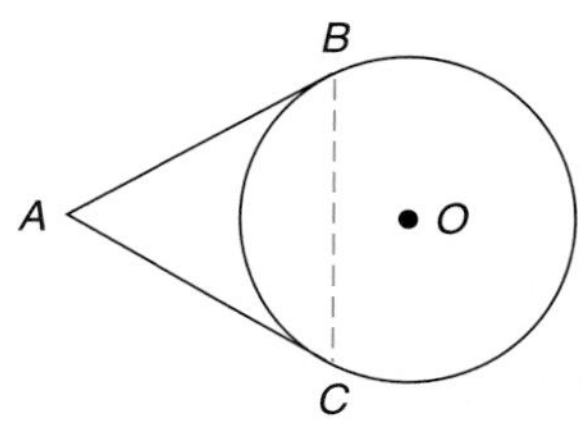


**FIGURE 6.46**

*Given:* In Figure 6.46, $\overline{AB}$ and $\overline{AC}$ are tangents to $\odot O$ from point $A$

*Prove:* $\overline{AB} \cong \overline{AC}$

*Proof:* Draw $\overline{BC}$. Now $m\angle B = \frac{1}{2}m\widehat{BC}$ and $m\angle C = \frac{1}{2}m\widehat{BC}$. Then $\angle B \cong \angle C$ because these angles have equal measures. In turn, the sides opposite $\angle B$ and $\angle C$ of $\triangle ABC$ are congruent. That is, $\overline{AB} \cong \overline{AC}$.

We apply Theorem 6.3.4 in Examples 2 and 3.

EXAMPLE 2

A belt used in an automobile engine wraps around two pulleys with different lengths of radii. Explain why the straight pieces named $\overline{AB}$ and $\overline{CD}$ have the same length. (See Figure 6.47.)

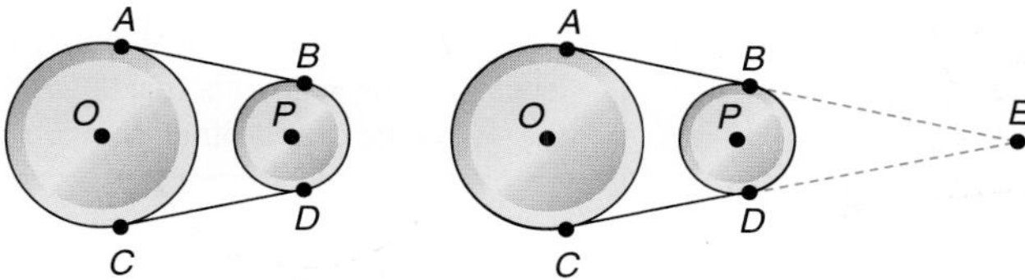


**FIGURE 6.47**

***Solution***

Because the pulley centered at $O$ has the larger radius length, we extend $\overline{AB}$ and $\overline{CD}$ to meet at point $E$. Because $E$ is an external point to both $\odot O$ and $\odot P$, we know that $EB = ED$ and $EA = EC$ by Theorem 6.3.4. By subtracting equals from equals, $EA - EB = EC - ED$. Because $EA - EB = AB$ and $EC - ED = CD$, it follows that $AB = CD$.

> **Discover!**
>
> Place three coins of the same size together so that they all touch each other. What type of triangle is formed by joining their centers?
>
> ANSWER
>
> Equilateral or Equiangular

EXAMPLE 3

The circle shown in Figure 6.48 is inscribed in $\triangle ABC$; $AB = 9$, $BC = 8$, and $AC = 7$. Find the lengths $AM$, $MB$, and $NC$.

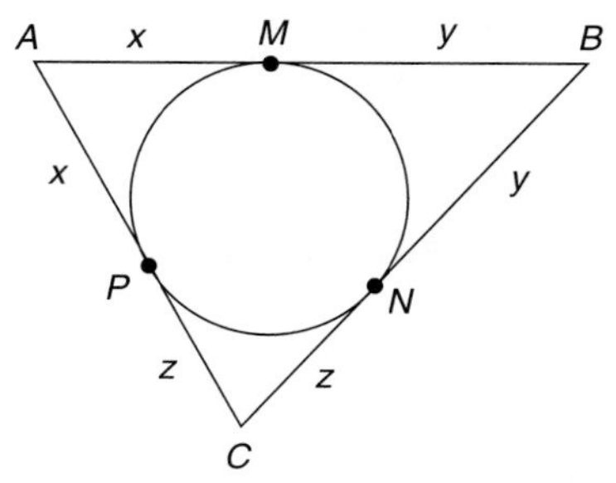


**FIGURE 6.48**

***Solution***

Because the tangent segments from an external point are $\cong$, we can let

$$AM = AP = x$$
$$BM = BN = y$$
$$NC = CP = z$$

Now

$$x + y = 9 \quad (\text{from } AB = 9)$$
$$y + z = 8 \quad (\text{from } BC = 8)$$
$$x + z = 7 \quad (\text{from } AC = 7)$$

Subtracting the second equation from the first, we have

$$\begin{array}{rcl} x + y \phantom{+ z} & = & 9 \\ y + z & = & 8 \\ \hline x \phantom{+ y} - z & = & 1 \end{array}$$

Now we use this new equation along with the third equation on the previous page and add:

$$\begin{array}{r} x - z = 1 \\ \underline{x + z = 7} \\ 2x = 8 \rightarrow x = 4 \rightarrow AM = 4 \end{array}$$

Because $x = 4$ and $x + y = 9$, $y = 5$. Then $BM = 5$. Because $x = 4$ and $x + z = 7$, $z = 3$, so $NC = 3$. Summarizing, $AM = 4$, $BM = 5$, and $NC = 3$.

*Exs. 7–10*

## LENGTHS OF SEGMENTS IN A CIRCLE

To complete this section, we consider three relationships involving the lengths of chords, secants, or tangents. The first theorem is proved, but the proofs of the remaining theorems are left as exercises for the student.

AA is the method used to prove triangles similar in this section.

THEOREM 6.3.5: If two chords intersect within a circle, then the product of the lengths of the segments (parts) of one chord is equal to the product of the lengths of the segments of the other chord.

*Given:* Circle $O$ with chords $\overline{RS}$ and $\overline{TQ}$ intersecting at point $V$ (See Figure 6.49.)

*Prove:* $RV \cdot VS = TV \cdot VQ$

*Proof:* Draw $\overline{RT}$ and $\overline{QS}$. In $\triangle RTV$ and $\triangle QSV$, we have $\angle 1 \cong \angle 2$ (vertical $\angle$s). Also, $\angle R$ and $\angle Q$ are inscribed angles that intercept the same arc (namely $\overset{\frown}{TS}$), so $\angle R \cong \angle Q$. By $AA$, $\triangle RTV \sim \triangle QSV$. Using CSSTP, $\frac{RV}{VQ} = \frac{TV}{VS}$ and so $RV \cdot VS = TV \cdot VQ$.

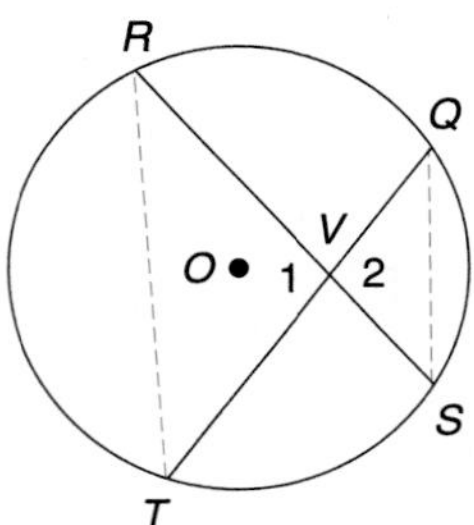


**FIGURE 6.49**

Use computer software if available.

1) Draw a circle with chords $\overline{HJ}$ and $\overline{LM}$ intersecting at point $P$. (See Figure 6.50.)
2) Measure $\overline{MP}$, $\overline{PL}$, $\overline{HP}$, and $\overline{PJ}$.
3) Show that $MP \cdot PL = HP \cdot PJ$. (Answers are not "perfect.")

EXAMPLE 4

In Figure 6.50, $HP = 4$, $PJ = 5$, and $LP = 8$. Find $PM$.

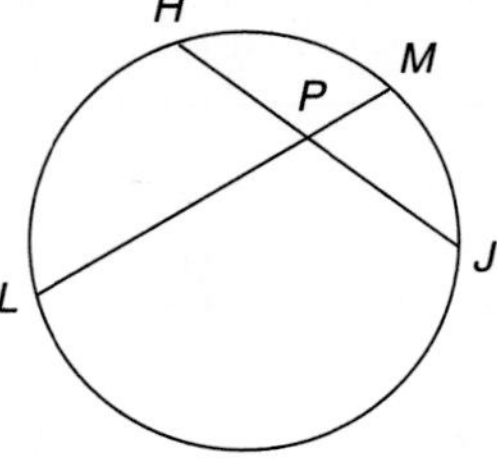


**FIGURE 6.50**

***Solution***

Applying Theorem 6.3.5, $HP \cdot PJ = LP \cdot PM$. Then

$$\begin{aligned} 4 \cdot 5 &= 8 \cdot PM \\ 8 \cdot PM &= 20 \\ PM &= 2.5 \end{aligned}$$

EXAMPLE 5

In Figure 6.50 on page 300, $HP = 6$, $PJ = 4$, and $LM = 11$. Find $LP$ and $PM$.

**Solution**

Because $LP + PM = LM$, it follows that $PM = LM - LP$. If $LM = 11$ and $LP = x$, then $PM = 11 - x$. Now $HP \cdot PJ = LP \cdot PM$ becomes

$$\begin{aligned} 6 \cdot 4 &= x(11 - x) \\ 24 &= 11x - x^2 \\ x^2 - 11x + 24 &= 0 \\ (x - 3)(x - 8) &= 0, \text{ so } x - 3 = 0 \text{ or } x - 8 = 0 \\ x = 3 \quad &\text{or} \quad x = 8 \end{aligned}$$

Therefore, $LP = 3$ or $LP = 8$

If $LP = 3$, then $PM = 8$; conversely, if $LP = 8$, then $PM = 3$. That is, the segments of chord $\overline{LM}$ have lengths of 3 and 8.

SSG *Exs. 11–13*

In Figure 6.51, we say that secant $\overline{AB}$ has internal segment (part) $\overline{RB}$ and external segment (part) $\overline{AR}$.

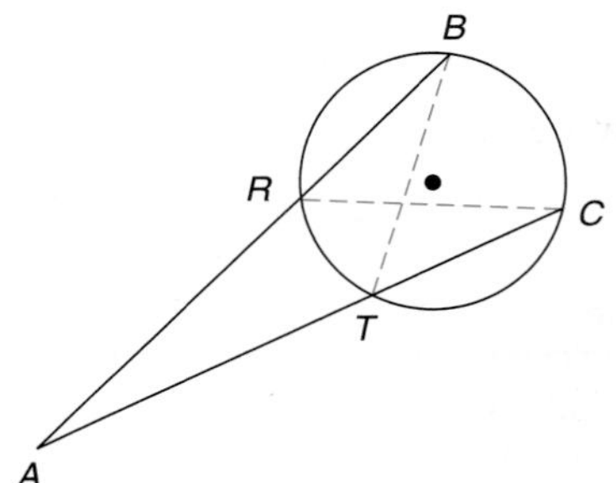


**FIGURE 6.51**

THEOREM 6.3.6: If two secant segments are drawn to a circle from an external point, then the products of the lengths of each secant with its external segment are equal.

*Given:* Secants $\overline{AB}$ and $\overline{AC}$ for the circle in Figure 6.51

*Prove:* $AB \cdot RA = AC \cdot TA$

The proof is left as an exercise for the student.

HINT: First use the auxiliary lines shown to prove that $\triangle ABT \sim \triangle ACR$.

EXAMPLE 6

*Given:* In Figure 6.51, $AB = 14$, $BR = 5$, and $TC = 5$

*Find:* $AC$ and $TA$

**Solution**

Let $AC = x$. Because $AT + TC = AC$, we have $AT + 5 = x$, so $TA = x - 5$. If $AB = 14$ and $BR = 5$, then $AR = 9$. The statement $AB \cdot RA = AC \cdot TA$ becomes

$$\begin{aligned} 14 \cdot 9 &= x(x - 5) \\ 126 &= x^2 - 5x \\ x^2 - 5x - 126 &= 0 \\ (x - 14)(x + 9) &= 0, \text{ so } x - 14 = 0 \text{ or } x + 9 = 0 \\ x = 14 \text{ or } x &= -9 \end{aligned}$$

($x = -9$ is discarded because the length of $\overline{AC}$ cannot be negative.)

Thus $AC = 14$, so $TA = 9$.

THEOREM 6.3.7: If a tangent segment and a secant segment are drawn to a circle from an external point, then the square of the length of the tangent equals the product of the length of the secant with the length of its external segment.

*Given:* Tangent $\overline{TV}$ and secant $\overline{TW}$ in Figure 6.52

*Prove:* $(TV)^2 = TW \cdot TX$

The proof is left as an exercise for the student.

HINT: Use the auxiliary lines shown to prove that $\triangle TVW \sim \triangle TXV$.

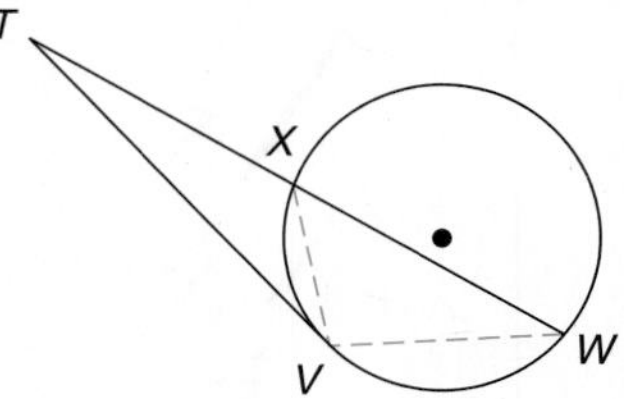


**FIGURE 6.52**

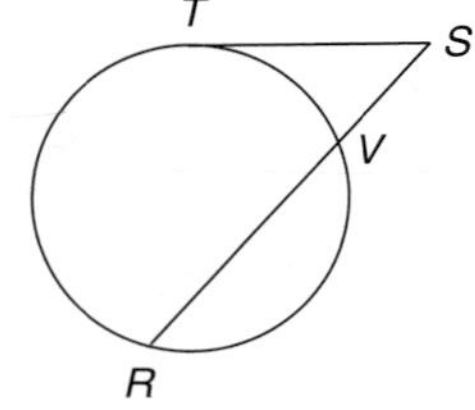


**FIGURE 6.53**

EXAMPLE 7

*Given:* In Figure 6.53, $SV = 3$ and $VR = 9$

*Find:* $ST$

***Solution***

If $SV = 3$ and $VR = 9$, then $SR = 12$. Using Theorem 6.3.7, we find that

$$(ST)^2 = SR \cdot SV$$
$$(ST)^2 = 12 \cdot 3$$
$$(ST)^2 = 36$$
$$ST = 6 \text{ or } -6$$

*Exs. 14–17*

Because $ST$ cannot be negative, $ST = 6$.

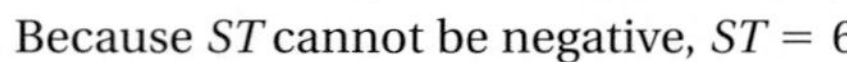

## 6.3 Exercises

1. *Given:* $\odot O$ with $\overline{OE} \perp \overline{CD}$
   $CD = OC$
   *Find:* $m\widehat{CF}$

2. *Given:* $OC = 8$ and $OE = 6$
   $\overline{OE} \perp \overline{CD}$ in $\odot O$
   *Find:* $CD$

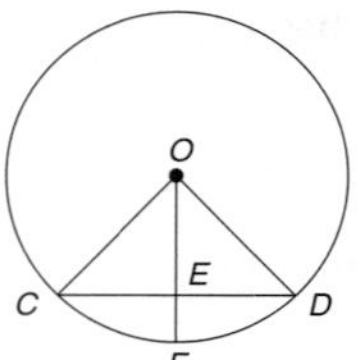


*Exercises 1, 2*

3. *Given:* $\overline{OV} \perp \overline{RS}$ in $\odot O$
   $OV = 9$ and $OT = 6$
   *Find:* $RS$

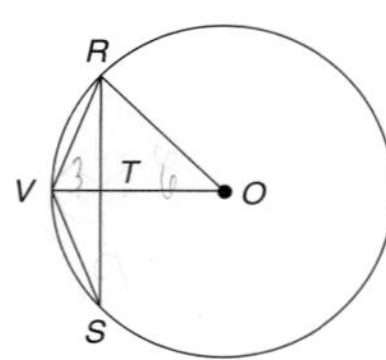


*Exercises 3, 4*

4. *Given:* $V$ is the midpoint of $\widehat{RS}$ in $\odot O$
   $m\angle S = 15°$ and $OT = 6$
   *Find:* $OR$

5. Sketch two circles that have:
   a) No common tangents
   b) Exactly one common tangent
   c) Exactly two common tangents
   d) Exactly three common tangents
   e) Exactly four common tangents

6. Two congruent intersecting circles $B$ and $D$ (not shown) have a line (segment) of centers $\overline{BD}$ and a common chord $\overline{AC}$ that are congruent. Explain why quadrilateral $ABCD$ is a square.

*In the figure for Exercises 7 to 14, O is the center of the circle. See Theorem 6.3.5.*

7. *Given:* $AE = 6$, $EB = 4$, $DE = 8$
   *Find:* $EC$

8. *Given:* $DE = 12$, $EC = 5$, $AE = 8$
   *Find:* $EB$

9. *Given:* $AE = 8$, $EB = 6$, $DC = 16$
   *Find:* $DE$ and $EC$

10. *Given:* $AE = 7$, $EB = 5$, $DC = 12$
    *Find:* $DE$ and $EC$

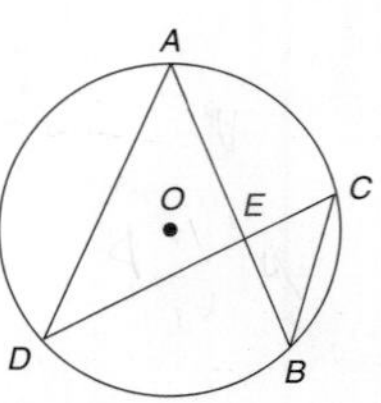


*Exercises 7–14*

11. *Given:* $AE = 6$, $EC = 3$, $AD = 8$
*Find:* $CB$

12. *Given:* $AD = 10$, $BC = 4$, $AE = 7$
*Find:* $EC$

13. *Given:* $AE = 9$ and $EB = 8$; $DE:EC = 2:1$
*Find:* $DE$ and $EC$

14. *Given:* $AE = 6$ and $EB = 4$; $DE:EC = 3:1$
*Find:* $DE$ and $EC$

*For Exercises 15–18, see Theorem 6.3.6.*

15. *Given:* $AB = 6$, $BC = 8$, $AE = 15$
*Find:* $DE$

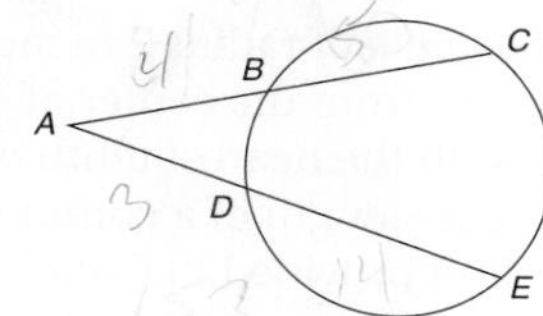


*Exercises 15–18*

16. *Given:* $AC = 12$, $AB = 6$, $AE = 14$
*Find:* $AD$

17. *Given:* $AB = 4$, $BC = 5$, $AD = 3$
*Find:* $DE$

18. *Given:* $AB = 5$, $BC = 6$, $AD = 6$
*Find:* $AE$

*In the figure for Exercises 19 to 22, $\overline{RS}$ is tangent to the circle at S. See Theorem 6.3.7.*

19. *Given:* $RS = 8$ and $RV = 12$
*Find:* $RT$

20. *Given:* $RT = 4$ and $TV = 6$
*Find:* $RS$

21. *Given:* $\overline{RS} \cong \overline{TV}$ and $RT = 6$
*Find:* $RS$
(HINT: Use the Quadratic Formula.)

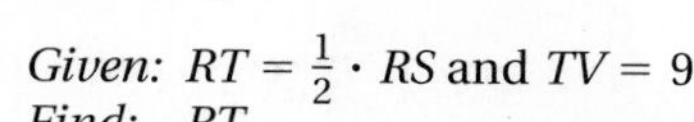


*Exercises 19–22*

22. *Given:* $RT = \frac{1}{2} \cdot RS$ and $TV = 9$
*Find:* $RT$

23. For the two circles in Figures (a), (b), and (c), find the total number of common tangents (internal and external).

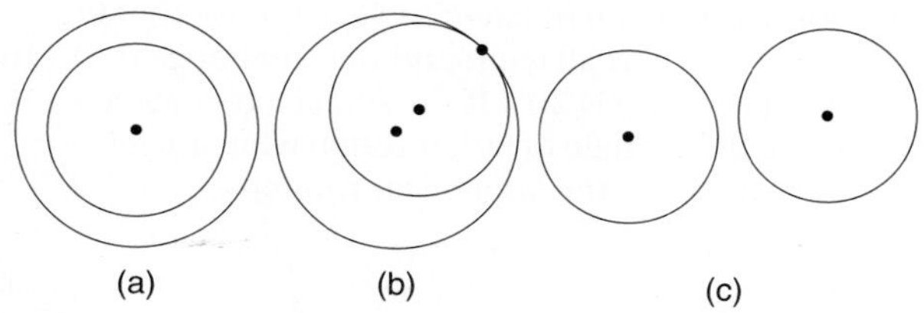


24. For the two circles in Figures (a), (b), and (c), find the total number of common tangents (internal and external).

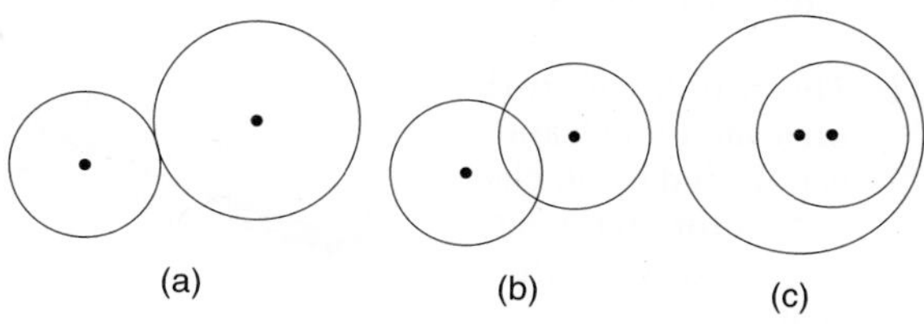


*In Exercises 25 to 28, provide a paragraph proof.*

25. *Given:* $\odot O$ and $\odot Q$ are tangent at point $F$
Secant $\overline{AC}$ to $\odot O$
Secant $\overline{AE}$ to $\odot Q$
Common internal tangent $\overline{AF}$
*Prove:* $AC \cdot AB = AE \cdot AD$

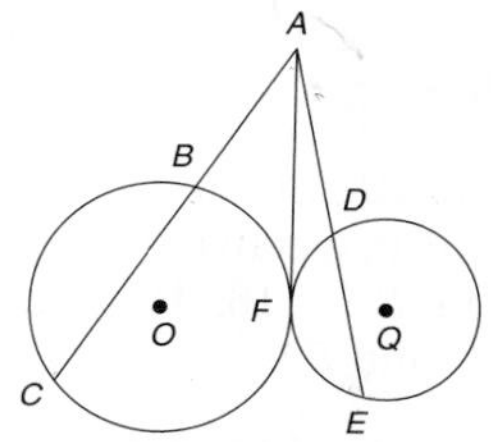


26. *Given:* $\odot O$ with $\overline{OM} \perp \overline{AB}$ and $\overline{ON} \perp \overline{BC}$
$\overline{OM} \cong \overline{ON}$
*Prove:* $\triangle ABC$ is isosceles

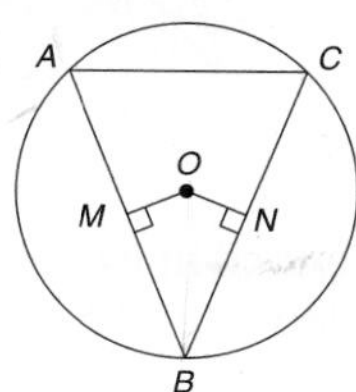


27. *Given:* Quadrilateral $ABCD$ is circumscribed about $\odot O$
*Prove:* $AB + CD = DA + BC$

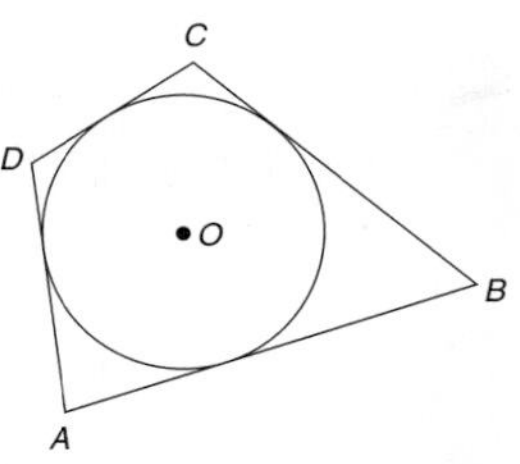


28. *Given:* $\overline{AB} \cong \overline{DC}$ in $\odot P$
*Prove:* $\triangle ABD \cong \triangle CDB$

29. Does it follow from Exercise 28 that $\triangle ADE$ is also congruent to $\triangle CBE$? What can you conclude about $\overline{AE}$ and $\overline{CE}$ in the drawing? What can you conclude about $\overline{DE}$ and $\overline{EB}$?

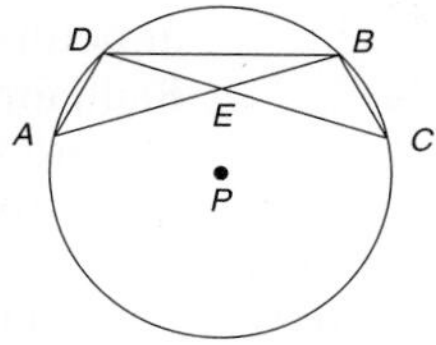


*Exercises 28, 29*

30. In $\odot O$ (not shown), $\overline{RS}$ is a diameter and $T$ is the midpoint of semicircle $\widehat{RTS}$. What is the value of the ratio $\frac{RT}{RS}$? The ratio $\frac{RT}{RO}$?

31. The cylindrical brush on a vacuum cleaner is powered by an electric motor. In the figure, the drive shaft is at point $D$. If $m\widehat{AC} = 160°$, find the measure of the angle formed by the drive belt at point $D$; that is, find $m\angle D$.

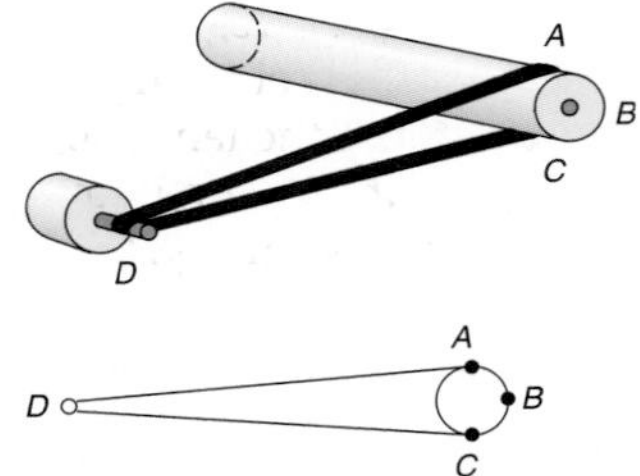


Exercises 31, 32

32. The drive mechanism on a treadmill is powered by an electric motor. In the figure, find $m\angle D$ if $m\widehat{ABC}$ is 36° larger than $m\widehat{AC}$.

*33. *Given:* Tangents $\overline{AB}$, $\overline{BC}$, and $\overline{AC}$ to $\odot O$ at points $M$, $N$, and $P$, respectively
$AB = 14$, $BC = 16$, $AC = 12$
*Find:* $AM$, $PC$, and $BN$

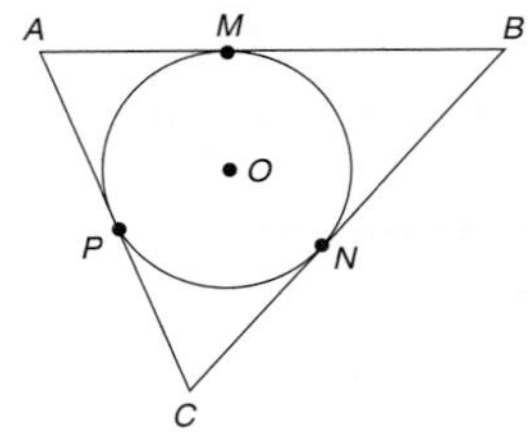


*34. *Given:* $\odot Q$ is inscribed in isosceles right $\triangle RST$
The perimeter of $\triangle RST$ is $8 + 4\sqrt{2}$
*Find:* $TM$

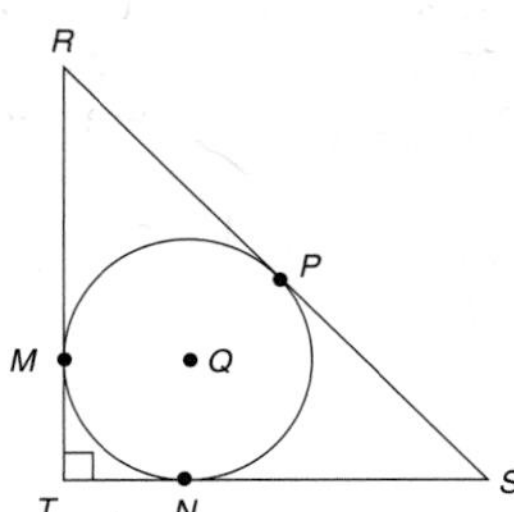


*35. *Given:* $\overline{AB}$ is an external tangent to $\odot O$ and $\odot Q$ at points $A$ and $B$; Radii for $\odot O$ and $\odot Q$ are 4 and 9, respectively
*Find:* $AB$
(HINT: The line of centers $\overline{OQ}$ contains point $C$, the point at which $\odot O$ and $\odot Q$ are tangent.)

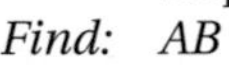

36. The center of a circle of radius 3 inches is at a distance of 20 inches from the center of a circle of radius 9 inches. What is the exact length of common internal tangent $\overline{AB}$?
(HINT: Use similar triangles to find $OD$ and $DP$. Then apply the Pythagorean Theorem twice.)

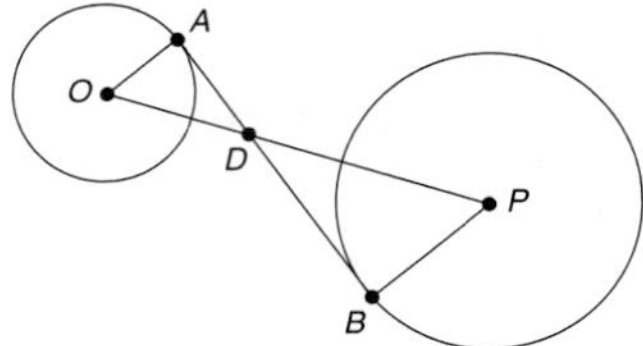


Exercises 36, 37

37. The center of a circle of radius 2 inches is at a distance of 10 inches from the center of a circle of radius 3 inches. To the nearest tenth of an inch, what is the approximate length of a common internal tangent? Use the hint provided in Exercise 36.

38. Circles $O$, $P$, and $Q$ are tangent (as shown) at points $X$, $Y$, and $Z$. Being as specific as possible, explain what type of triangle $\triangle PQO$ is if:

a) $OX = 2$, $PY = 3$, $QZ = 1$
b) $OX = 2$, $PY = 3$, $QZ = 2$

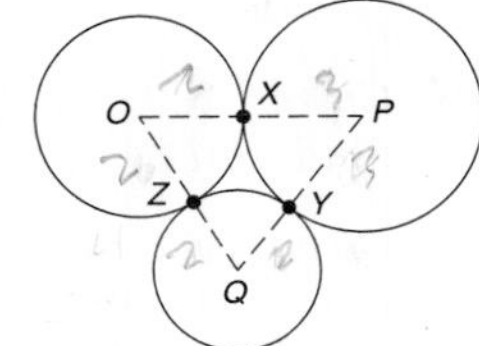


Exercises 38, 39

39. Circles $O$, $P$, and $Q$ are tangent (as shown) at points $X$, $Y$, and $Z$. Being as specific as possible, explain what type of triangle $\triangle PQO$ is if:

a) $OX = 3$, $PY = 4$, $QZ = 1$
b) $OX = 2$, $PY = 2$, $QZ = 2$

*40. If the larger gear has 30 teeth and the smaller gear has 18, then the gear ratio (larger to smaller) is 5:3. When the larger gear rotates through an angle of 60°, through what angle measure does the smaller gear rotate?

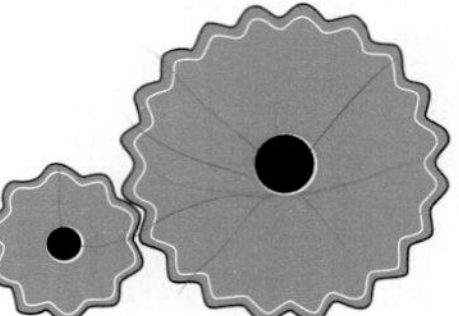

Exercises 40, 41

*41. For the drawing in Exercise 40, suppose that the larger gear has 20 teeth and the smaller gear has 10 (the gear ratio is 2:1). If the smaller gear rotates through an angle of 90°, through what angle measure does the larger gear rotate?

*In Exercises 42 to 45, prove the stated theorem.*

42. If a line is drawn through the center of a circle perpendicular to a chord, then it bisects the chord and its minor arc. See Figure 6.37.
(NOTE: The major arc is also bisected by the line.)

43. If a line is drawn through the center of a circle to the midpoint of a chord other than a diameter, then it is perpendicular to the chord. See Figure 6.38.

44. If two secant segments are drawn to a circle from an external point, then the products of the lengths of each secant with its external segment are equal. See Figure 6.51.

45. If a tangent segment and a secant segment are drawn to a circle from an external point, then the square of the length of the tangent equals the product of the length of the secant with the length of its external segment. See Figure 6.52.

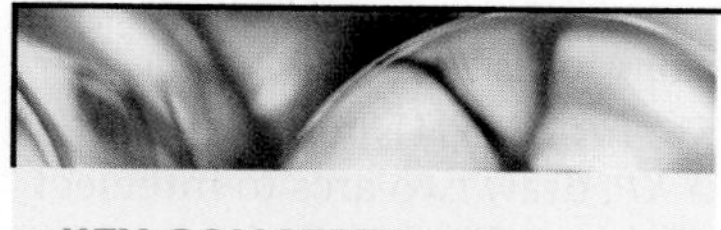

**KEY CONCEPTS**

Construction of Tangents to a Circle • Inequalities in the Circle

# 6.4 Some Constructions and Inequalities for the Circle

In Section 6.3, we proved that the radius drawn to a tangent at the point of contact is perpendicular to the tangent at that point. We now show, by using an indirect proof, that the converse of that theorem is also true. Recall that there is only one line perpendicular to a given line at a point on that line.

THEOREM 6.4.1: The line that is perpendicular to the radius of a circle at its endpoint on the circle is a tangent to the circle.

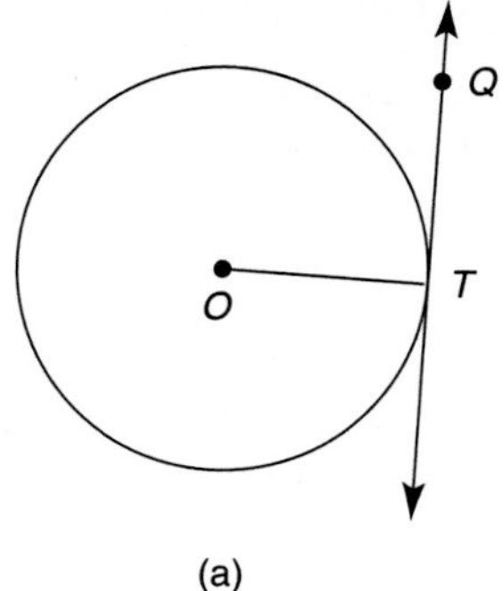


(a)

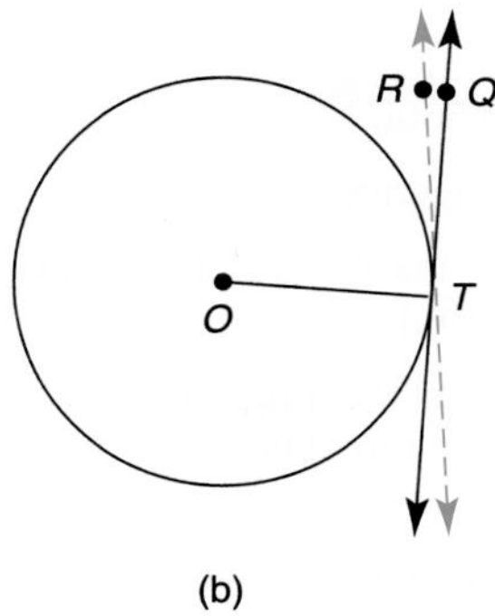


(b)

**FIGURE 6.54**

*Given:* In Figure 6.54(a), $\odot O$ with radius $\overline{OT}$
$\overleftrightarrow{QT} \perp \overline{OT}$

*Prove:* $\overleftrightarrow{QT}$ is a tangent to $\odot O$ at point $T$

*Proof:* Suppose that $\overleftrightarrow{QT}$ is not a tangent to $\odot O$ at $T$. Then the tangent (call it $\overleftrightarrow{RT}$) can be drawn at $T$, the point of tangency. [See Figure 6.54(b).]

Now $\overline{OT}$ is the radius to tangent $\overleftrightarrow{RT}$ at $T$, and because a radius drawn to a tangent at the point of contact of the tangent is perpendicular to the tangent, $\overline{OT} \perp \overleftrightarrow{RT}$. But $\overline{OT} \perp \overleftrightarrow{QT}$ by hypothesis. Thus two lines are perpendicular to $\overline{OT}$ at point $T$, contradicting the fact that there is only one line perpendicular to a line at a point on the line. Therefore, $\overleftrightarrow{QT}$ must be the tangent to $\odot O$ at point $T$.

## CONSTRUCTIONS OF TANGENTS TO CIRCLES

CONSTRUCTION 8:
To construct a tangent to a circle at a point on the circle.

The strategy used in Construction 8 is based on Theorem 6.4.1. We will draw a radius (extended beyond the circle). At the point on the circle (point $X$ in Figure 6.55), we construct the line perpendicular to $\overline{PX}$. The constructed line ($\overleftrightarrow{WX}$ in Figure 6.55) is tangent to circle $P$ at point $X$.

*Given:* $\odot P$ with point $X$ on the circle [See Figure 6.55(a) on page 306.]

*Construct:* A tangent $\overleftrightarrow{XW}$ to $\odot P$ at point $X$

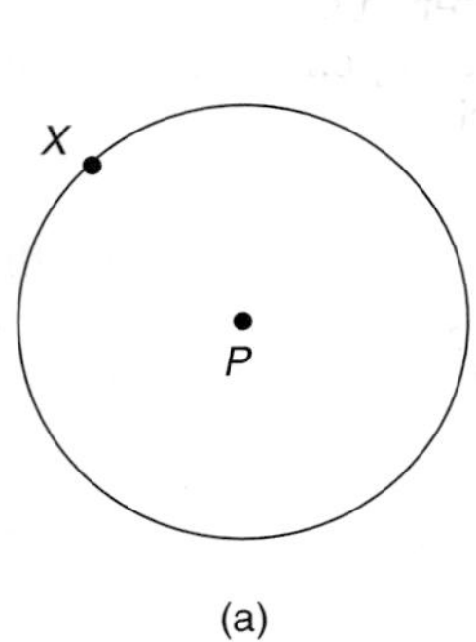


(a)

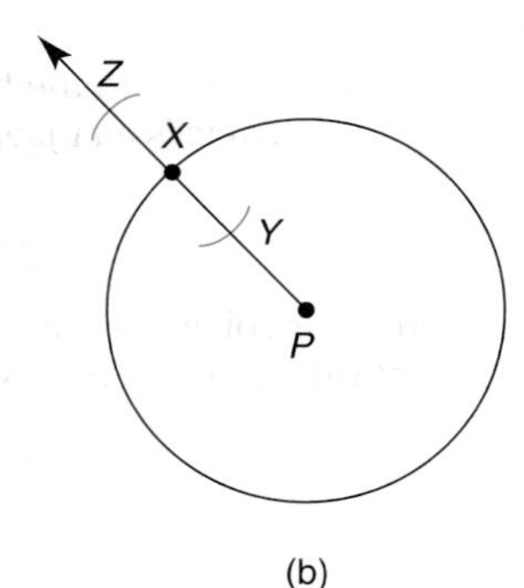


(b)

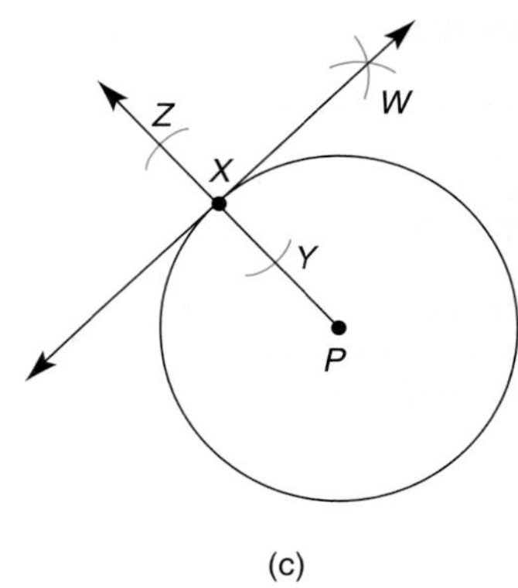


(c)

**FIGURE 6.55**

*Construction:* First draw radius $\overline{PX}$ and extend it to form $\overrightarrow{PX}$. Using $X$ as the center and any radius length less than $XP$, draw two arcs to intersect $\overrightarrow{PX}$ at points $Y$ and $Z$, as shown in Figure 6.55(b).

Now complete the construction of the perpendicular to $\overrightarrow{PX}$ at point $P$. From $Y$ and $Z$, mark arcs with equal radii of length greater than $XY$. Calling the point of intersection $W$ [see Figure 6.55(c)], draw $\overleftrightarrow{XW}$, the desired tangent to $\odot P$ at point $X$.

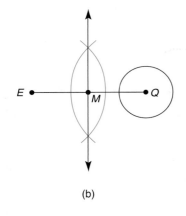

EXAMPLE 1

Make a drawing so that points $A$, $B$, $C$, and $D$ are on $\odot O$ in that order. If tangents are constructed at points $A$, $B$, $C$, and $D$, what type of quadrilateral will be formed by the tangent segments if

a) $m\widehat{AB} = m\widehat{CD}$ and $m\widehat{BC} = m\widehat{AD}$?
b) all arcs $\widehat{AB}$, $\widehat{BC}$, $\widehat{CD}$, and $\widehat{DA}$ are congruent?

***Solution***

a) A rhombus (all sides are congruent)
b) A square (all four $\angle$s are right $\angle$s; all sides $\cong$)

We now consider a more difficult construction.

CONSTRUCTION 9:

To construct a tangent to a circle from an external point.

*Given:* $\odot Q$ and external point E [See Figure 6.56(a).]

*Construct:* A tangent $\overline{ET}$, with T as the point of tangency

*Construction:* Draw $\overline{EQ}$. Construct the perpendicular bisector of $\overline{EQ}$, to intersect $\overline{EQ}$ at its midpoint $M$. [See Figure 6.56(b).]

With $M$ as center and $MQ$ (or $ME$) as the length of radius, construct a circle. The points of intersection of circle $M$ with circle $Q$ are designated by $T$ and $V$.

Draw $\overline{ET}$, the desired tangent. [See Figure 6.56(c).] (Note that $\overline{EV}$, if drawn, would also be a tangent to $\odot Q$).

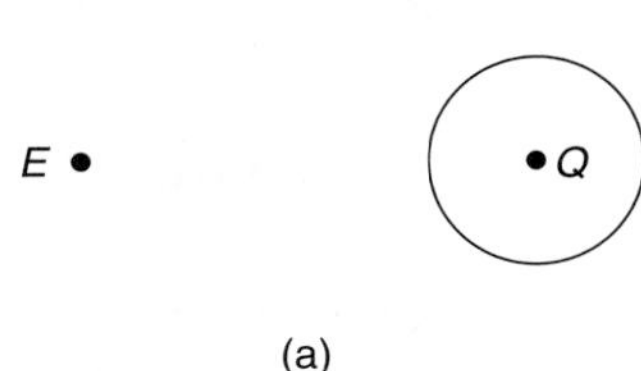


(a)

(b)

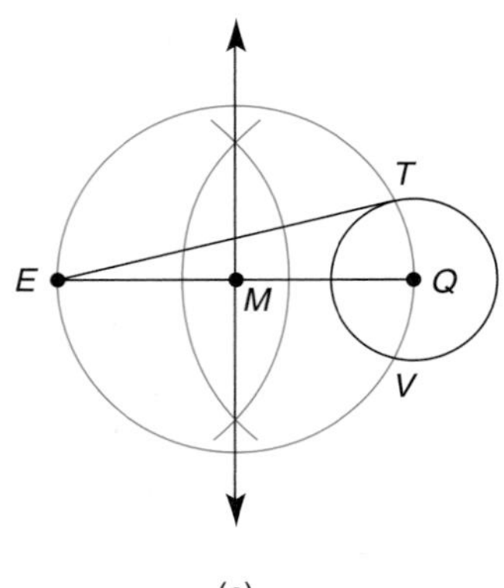


(c)

**FIGURE 6.56**

*Exs. 1–3*

In the preceding construction, $\overline{QT}$ (not shown) is a radius of the smaller circle $Q$. In the larger circle $M$, $\angle ETQ$ is an inscribed angle that intercepts a semicircle. Thus $\angle ETQ$ is a right angle and $\overline{ET} \perp \overline{TQ}$. Because the line drawn perpendicular to the radius of a circle at its endpoint on the circle is a tangent to the circle, $\overline{ET}$ is a tangent to circle $Q$.

## INEQUALITIES IN THE CIRCLE

The remaining theorems involve inequalities in the circle. For the theorems that are stated, the exercise set that follows emphasizes applications of these theorems.

> THEOREM 6.4.2: In a circle (or in congruent circles) containing two unequal central angles, the larger angle corresponds to the larger intercepted arc.

*Given:* $\odot O$ with central angles $\angle 1$ and $\angle 2$ in Figure 6.57; $m\angle 1 > m\angle 2$

*Prove:* $m\widehat{AB} > m\widehat{CD}$

*Proof:* In $\odot O$, $m\angle 1 > m\angle 2$. By the Central Angle Postulate, $m\angle 1 = m\widehat{AB}$ and $m\angle 2 = m\widehat{CD}$. By substitution, $m\widehat{AB} > m\widehat{CD}$.

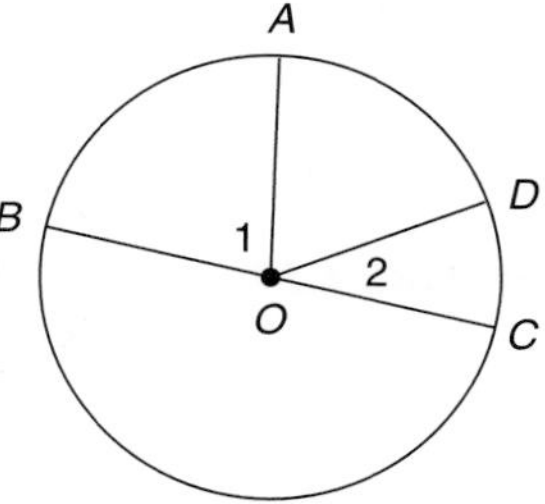


**FIGURE 6.57**

The converse of Theorem 6.4.2 follows and it is also readily proved.

> THEOREM 6.4.3: In a circle (or in congruent circles) containing two unequal arcs, the larger arc corresponds to the larger central angle.

*Given:* In Figure 6.57, $\odot O$ with $\widehat{AB}$ and $\widehat{CD}$
$m\widehat{AB} > m\widehat{CD}$

*Prove:* $m\angle 1 > m\angle 2$

The proof is left as an exercise for the student.

EXAMPLE 2

*Given:* In Figure 6.58, $\odot Q$ with $m\widehat{RS} > m\widehat{TV}$.

a) Using Theorem 6.4.3, what can you conclude regarding the measures of $\angle RQS$ and $\angle TQV$?

b) What does intuition suggest regarding $RS$ and $TV$?

***Solution***

a) $m\angle RQS > m\angle TQV$

b) $RS > TV$

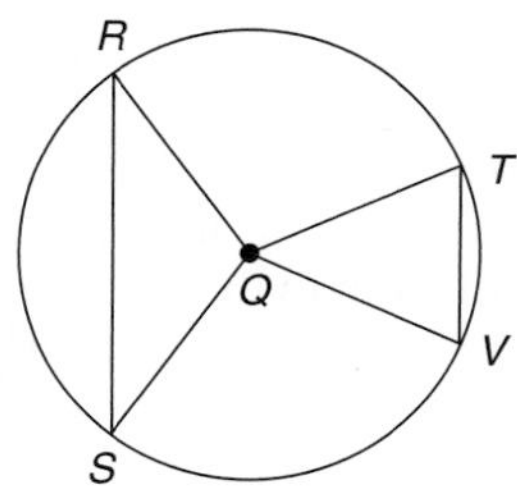


**FIGURE 6.58**

Before we apply Theorem 6.4.4 and prove Theorem 6.4.5, we consider the following activity. It is worth mentioning that the proof of Theorem 6.4.4, which is not provided, is similar to that of Theorem 6.4.5.

### Discover!

In Figure 6.59, $\overline{PT}$ measures the distance from center $P$ to chord $\overline{EF}$. Likewise, $\overline{PR}$ measures the distance from $P$ to chord $\overline{AB}$. Using a ruler, show that $PR > PT$. How do the lengths of chords $\overline{AB}$ and $\overline{EF}$ compare?

ANSWER
$AB < EF$

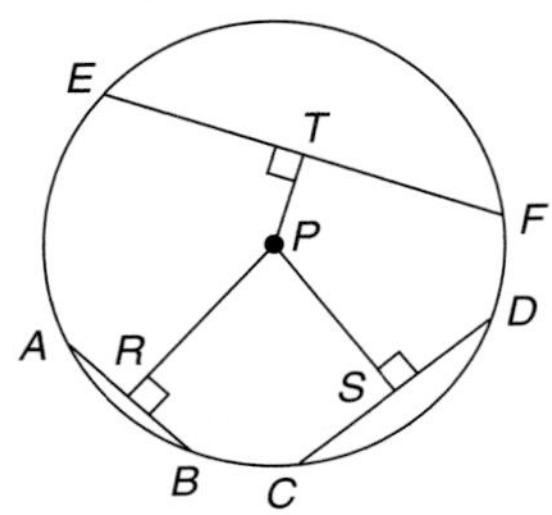


**FIGURE 6.59**

THEOREM 6.4.4: In a circle (or in congruent circles) containing two unequal chords, the shorter chord is at the greater distance from the center of the circle.

EXAMPLE 3

In circle $P$ of Figure 6.59, any radius has a length of 6 cm, and the chords have lengths $AB = 4$ cm, $DC = 6$ cm, and $EF = 10$ cm. Let $\overline{PR}$, $\overline{PS}$, and $\overline{PT}$ name perpendicular segments to these chords from center $P$.

a) Of $\overline{PR}$, $\overline{PS}$, and $\overline{PT}$, which is longest?
b) Of $\overline{PR}$, $\overline{PS}$, and $\overline{PT}$, which is shortest?

***Solution***

a) $\overline{PR}$ is longest, according to Theorem 6.4.4.
b) $\overline{PT}$ is shortest.

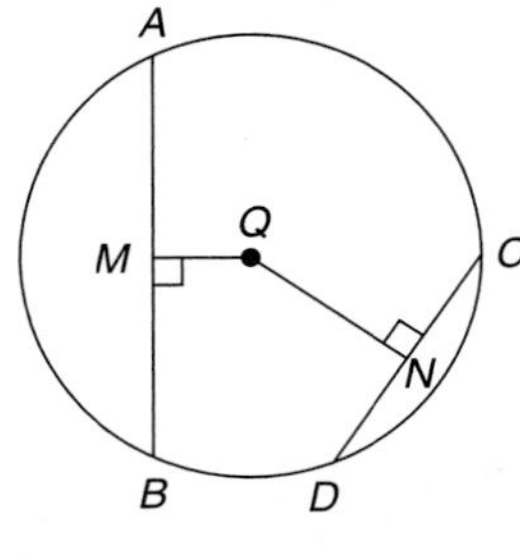


(a)

In the proof of Theorem 6.4.5, $a$ and $b$ represent the lengths of line segments. Then $a$ and $b$ are positive. If $a < b$, then $a^2 < b^2$; the converse is also true.

THEOREM 6.4.5: In a circle (or in congruent circles) containing two unequal chords, the chord nearer the center of the circle has the greater length.

*Given:* In Figure 6.60(a), $\odot Q$ with chords $\overline{AB}$ and $\overline{CD}$
$\overline{QM} \perp \overline{AB}$ and $\overline{QN} \perp \overline{CD}$
$QM < QN$

*Prove:* $AB > CD$

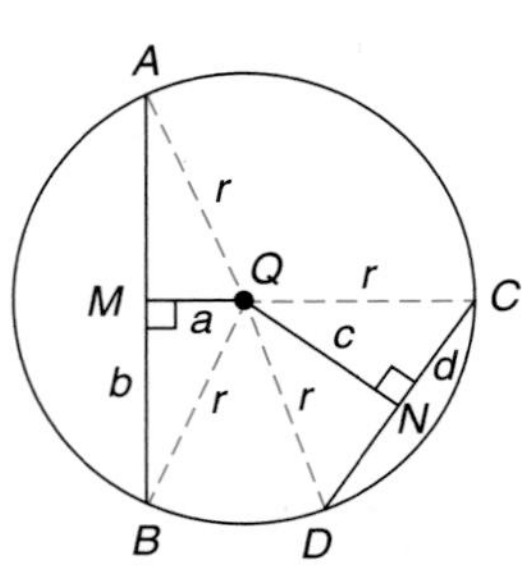


(b)

**FIGURE 6.60**

*Proof:* In Figure 6.60(b), we represent the lengths of $\overline{QM}$ and $\overline{QN}$ by $a$ and $c$, respectively. Draw radii $\overline{QA}$, $\overline{QB}$, $\overline{QC}$, and $\overline{QD}$, and denote all lengths by $r$. $\overline{QM}$ is the perpendicular bisector of $\overline{AB}$, and $\overline{QN}$ is the perpendicular bisector of $\overline{CD}$, because a radius perpendicular to a chord bisects the chord and its arc. Let $MB = b$ and $NC = d$.

With right angles at $M$ and $N$, we see that $\triangle QMB$ and $\triangle QNC$ are right triangles.

According to the Pythagorean Theorem, $r^2 = a^2 + b^2$ and $r^2 = c^2 + d^2$, so $b^2 = r^2 - a^2$ and $d^2 = r^2 - c^2$. If $QM < QN$, then $a < c$ and $a^2 < c^2$. Multiplication by $-1$ reverses the order of this inequality; therefore, $-a^2 > -c^2$. Adding $r^2$, we have $r^2 - a^2 > r^2 - c^2$ or $b^2 > d^2$, which implies that $b > d$. If $b > d$, then $2b > 2d$. But $AB = 2b$ whereas $CD = 2d$. Therefore, $AB > CD$.

*Exs. 4–9*

It is important that the phrase *minor arc* be used in our final theorems. The proof of Theorem 6.4.6 is left to the student. For the second theorem, the proof is provided because it is more involved. In each theorem, the chord and related minor arc share common endpoints.

> THEOREM 6.4.6: In a circle (or in congruent circles) containing two unequal chords, the longer chord corresponds to the greater minor arc.

If $AB > CD$ in Figure 6.61, then $m\widehat{AB} > m\widehat{CD}$.

> THEOREM 6.4.7: In a circle (or in congruent circles) containing two unequal minor arcs, the greater minor arc corresponds to the longer of the chords related to these arcs.

*Given:* In Figure 6.61(a), $\odot O$ with $m\widehat{AB} > m\widehat{CD}$ and chords $\overline{AB}$ and $\overline{CD}$

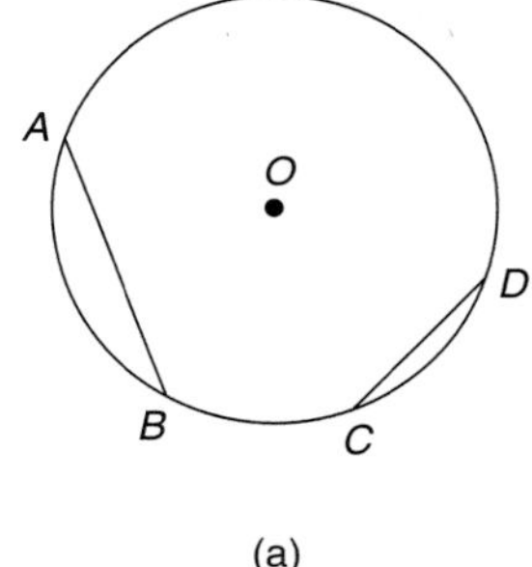


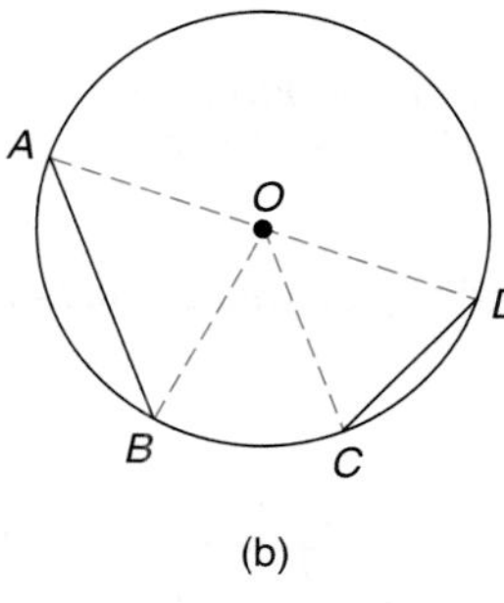


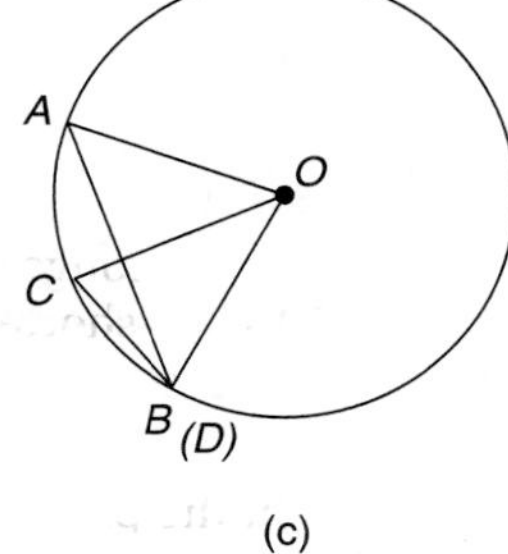


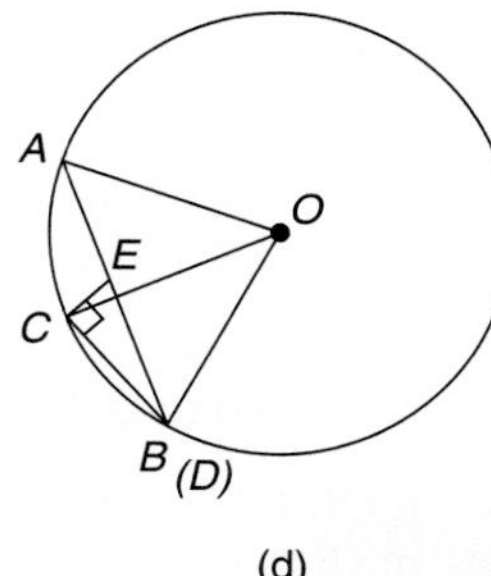


**FIGURE 6.61**

*Prove:* $AB > CD$

*Proof:* In circle $O$, draw radii $\overline{OA}$, $\overline{OB}$, $\overline{OC}$, and $\overline{OD}$. Because $m\widehat{AB} > m\widehat{CD}$, it follows that $m\angle AOB > m\angle COD$ because the larger arc in a circle corresponds to a larger central angle. [See Figure 6.61(b).]

We now rotate $\triangle COD$ to the position on the circle for which $D$ coincides with $B$, as shown in Figure 6.61(c). Because radii $\overline{OC}$ and $\overline{OB}$ are congruent, $\triangle COD$ is isosceles; also, $m\angle C = m\angle ODC$.

In $\triangle COD$, $m\angle COD + m\angle C + m\angle CDO = 180°$. Because $m\angle COD$ is positive, we have $m\angle C + m\angle CDO < 180°$ and $2 \cdot m\angle C < 180°$ by substitution. Therefore, $m\angle C < 90°$.

*Exs. 10–16*

Now construct the perpendicular segment to $\overline{CD}$ at point $C$, as shown in Figure 6.61(d). Denote the intersection of the perpendicular segment and $\overline{AB}$ by point $E$. Because $\triangle DCE$ is a right $\triangle$ with hypotenuse $\overline{EB}$, $EB > CD$ (*). Because $AB = AE + EB$ and $AE > 0$, we have $AB > EB$ (*). By the Transitive Property, the starred (*) statements reveal that $AB > CD$.

NOTE: In the preceding proof, $\overline{CE}$ must intersect $\overline{AB}$ at some point between $A$ and $B$. If it were to intersect at $A$, the measure of inscribed $\angle BCA$ would have to be more than 90°; this follows from the facts that $\widehat{AB}$ is a minor arc and that the intercepted arc for $\angle BCA$ would have to be a major arc.

## 6.4 Exercises

*In Exercises 1 to 8, use the figure provided.*

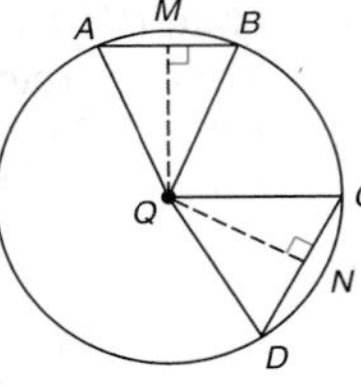

Exercises 1–8

1. If $m\widehat{CD} < m\widehat{AB}$, write an inequality that compares $m\angle CQD$ and $m\angle AQB$.

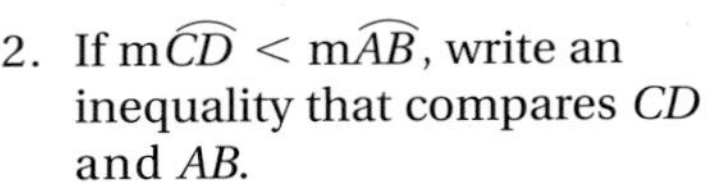

2. If $m\widehat{CD} < m\widehat{AB}$, write an inequality that compares $CD$ and $AB$.
3. If $m\widehat{CD} < m\widehat{AB}$, write an inequality that compares $QM$ and $QN$.
4. If $m\widehat{CD} < m\widehat{AB}$, write an inequality that compares $m\angle A$ and $m\angle C$.
5. If $m\angle CQD < m\angle AQB$, write an inequality that compares $CD$ to $AB$.
6. If $m\angle CQD < m\angle AQB$, write an inequality that compares $QM$ to $QN$.
7. If $m\widehat{CD}:m\widehat{AB} = 3:2$, write an inequality that compares $QM$ to $QN$.
8. If $QN:QM = 5:6$, write an inequality that compares $m\widehat{AB}$ to $m\widehat{CD}$.
9. Construct a circle $O$ and choose some point $D$ on the circle. Now construct the tangent to circle $O$ at point $D$.
10. Construct a circle $P$ and choose three points $R$, $S$, and $T$ on the circle. Construct the triangle that has its sides tangent to the circle at $R$, $S$, and $T$.
11. $X$, $Y$, and $Z$ are on circle $O$ such that $m\widehat{XY} = 120°$, $m\widehat{YZ} = 130°$, and $m\widehat{XZ} = 110°$. Suppose that triangle $XYZ$ is drawn and that the triangle $ABC$ is constructed with its sides tangent to circle $O$ at $X$, $Y$, and $Z$. Are $\triangle XYZ$ and $\triangle ABC$ similar triangles?

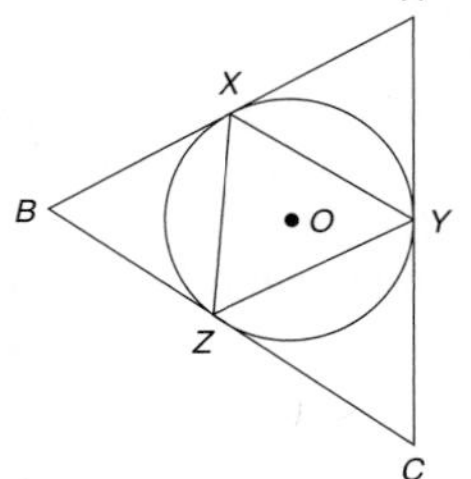


12. Construct the two tangent segments to circle $P$ (not shown) from external point $E$.
13. Point $V$ is in the exterior of circle $Q$ (not shown) such that $\overline{VQ}$ is equal in length to the diameter of circle $Q$. Construct the two tangents to circle $Q$ from point $V$. Then determine the measure of the angle that has vertex $V$ and has the tangents as sides.
14. Given circle $P$ and points $R$-$P$-$T$ such that $R$ and $T$ are in the exterior of circle $P$, suppose that tangents are constructed from $R$ and $T$ to form a quadrilateral (as shown). Identify the type of quadrilateral formed
    a) when $RP > PT$.
    b) when $RP = PT$.

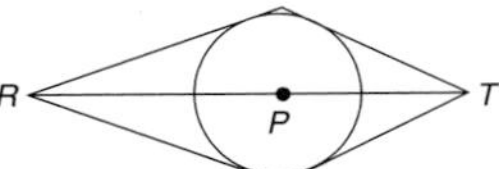


15. Given parallel chords $\overline{AB}$, $\overline{CD}$, $\overline{EF}$, and $\overline{GH}$ in circle $O$, which chord has the greatest length? Which has the least length? Why?

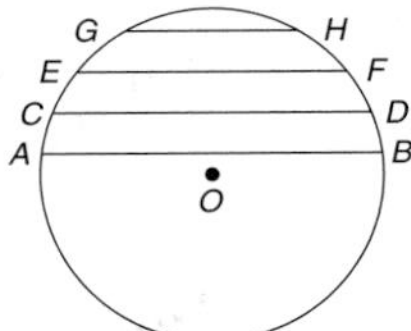


16. Given chords $\overline{MN}$, $\overline{RS}$, and $\overline{TV}$ in $\odot Q$ such that $QZ > QY > QX$, which chord has the greatest length? Which has the least length? Why?

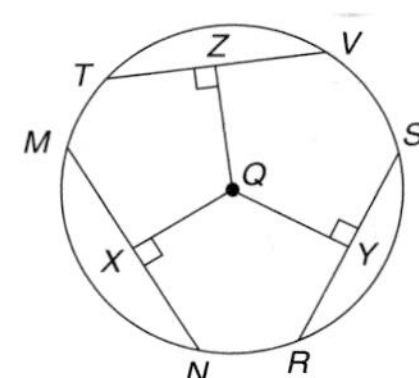


17. Given circle $O$ with radius $\overline{OT}$, tangent $\overleftrightarrow{AD}$, and line segments $\overline{OA}$, $\overline{OB}$, $\overline{OC}$, and $\overline{OD}$:
    a) Which line segment drawn from $O$ has the smallest length?
    b) If $m\angle 1 = 40°$, $m\angle 2 = 50°$, $m\angle 3 = 45°$, and $m\angle 4 = 30°$, which line segment from point $O$ has the greatest length?

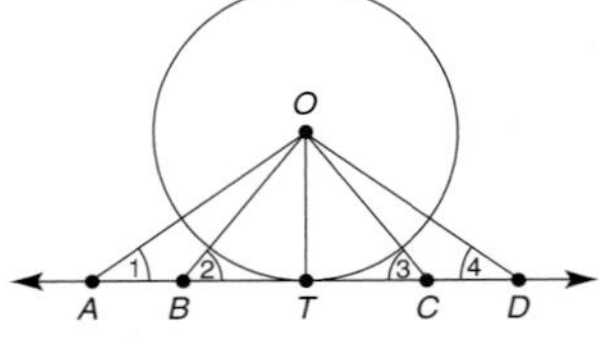


18. a) If $m\widehat{RS} > m\widehat{TV}$, write an inequality to compare $m\angle 1$ with $m\angle 2$.
    b) If $m\angle 1 > m\angle 2$, write an inequality to compare $m\widehat{RS}$ with $m\widehat{TV}$.

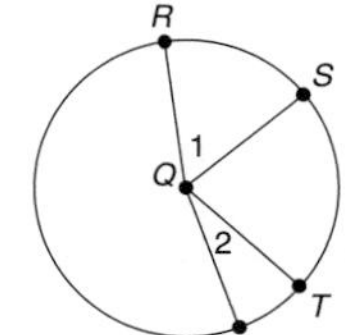


19. a) If $MN > PQ$, write an inequality to compare the measures of minor arcs $\widehat{MN}$ and $\widehat{PQ}$.
    b) If $MN > PQ$, write an inequality to compare the measures of major arcs $\widehat{MPN}$ and $\widehat{PMQ}$.

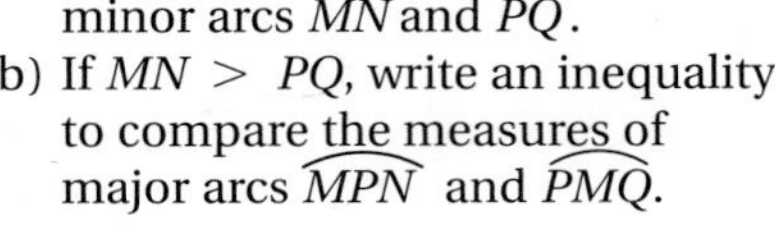

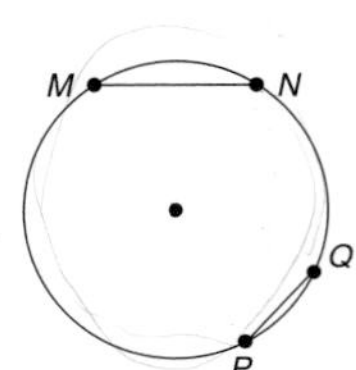

20. a) If $m\widehat{XY} > m\widehat{YZ}$, write an inequality to compare the measures of inscribed angles 1 and 2.
b) If $m\angle 1 < m\angle 2$, write an inequality to compare the measures of $\widehat{XY}$ and $\widehat{YZ}$.

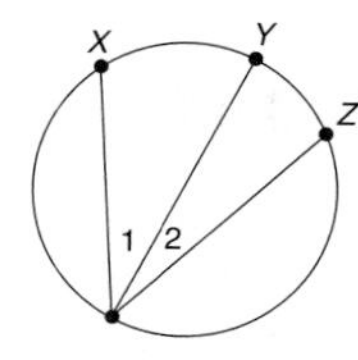


21. Quadrilateral $ABCD$ is inscribed in circle $P$ (not shown). If $\angle A$ is an acute angle, what type of angle is $\angle C$?

22. Quadrilateral $RSTV$ is inscribed in circle $Q$ (not shown). If arcs $\widehat{RS}$, $\widehat{ST}$, and $\widehat{TV}$ are all congruent, what type of quadrilateral is $RSTV$?

23. In circle $O$, points $A$, $B$, and $C$ are on the circle such that $m\widehat{AB} = 60°$ and $m\widehat{BC} = 40°$.

a) How are $m\angle AOB$ and $m\angle BOC$ related?
b) How are $AB$ and $BC$ related?

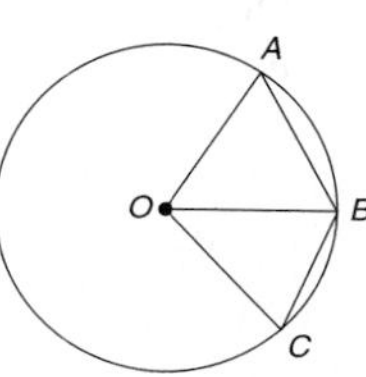


*Exercises 23–25*

24. In $\odot O$, $AB = 6$ cm and $BC = 4$ cm.

a) How are $m\angle AOB$ and $m\angle BOC$ related?
b) How are $m\widehat{AB}$ and $m\widehat{BC}$ related?

25. In $\odot O$, $m\angle AOB = 70°$ and $m\angle BOC = 30°$. See the figure above.

a) How are $m\widehat{AB}$ and $m\widehat{BC}$ related?
b) How are $AB$ and $BC$ related?

26. Triangle $ABC$ is inscribed in circle $O$; $AB = 5$, $BC = 6$, and $AC = 7$.

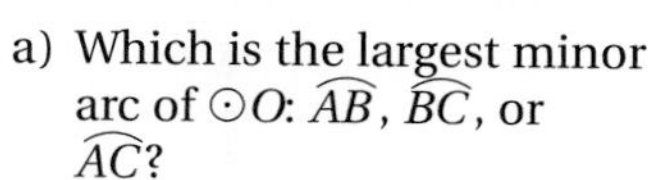

a) Which is the largest minor arc of $\odot O$: $\widehat{AB}$, $\widehat{BC}$, or $\widehat{AC}$?

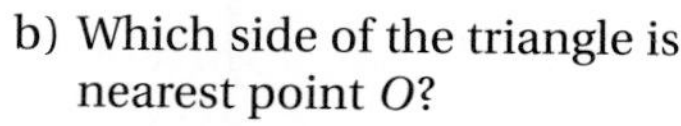

b) Which side of the triangle is nearest point $O$?

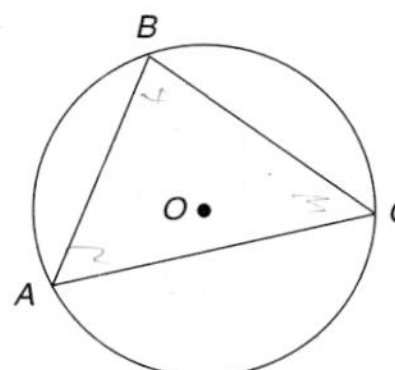


*Exercises 26–29*

27. Given circle $O$ with $m\widehat{BC} = 120°$ and $m\widehat{AC} = 130°$:

a) Which angle of triangle $ABC$ is smallest?
b) Which side of triangle $ABC$ is nearest point $O$?

28. Given that $m\widehat{AC}:m\widehat{BC}:m\widehat{AB} = 4:3:2$ in circle $O$:

a) Which arc is largest?
b) Which chord is longest?

29. Given that $m\angle A:m\angle B:m\angle C = 2:4:3$ in circle $O$:

a) Which angle is largest?
b) Which chord is longest?

30. Circle $O$ has a diameter of length 20 cm. Chord $\overline{AB}$ has length 12 cm, and chord $\overline{CD}$ has length 10 cm. How much closer is $\overline{AB}$ than $\overline{CD}$ to point $O$?

31. Circle $P$ has a radius of length 8 in. Points $A$, $B$, $C$, and $D$ lie on circle $P$ in such a way that $m\angle APB = 90°$ and $m\angle CPD = 60°$. How much closer to point $P$ is chord $\overline{AB}$ than $\overline{CD}$?

32. A tangent $\overline{ET}$ is constructed to circle $Q$ from external point $E$. Which angle and which side of triangle $QTE$ are largest? Which angle and which side are smallest?

33. Two congruent circles, $\odot O$ and $\odot P$, do not intersect. Construct a common external tangent for $\odot O$ and $\odot P$.

34. Explain why the following statement is incorrect:

"In a circle (or in congruent circles) containing two unequal chords, the longer chord corresponds to the greater major arc."

35. Prove: In a circle containing two unequal arcs, the larger arc corresponds to the larger central angle.

36. Prove: In a circle containing two unequal chords, the longer chord corresponds to the larger central angle. (HINT: You may use any theorems stated in this section.)

*37. In $\odot O$, chord $\overline{AB} \parallel$ chord $\overline{CD}$. Radius $\overline{OE}$ is perpendicular to $\overline{AB}$ and $\overline{CD}$ at points $M$ and $N$, respectively. If $OE = 13$, $AB = 24$, and $CD = 10$, then the distance from $O$ to $\overline{CD}$ is greater than the distance from $O$ to $\overline{AB}$. Determine how much farther chord $\overline{CD}$ is from center $O$ than chord $\overline{AB}$ is from center $O$; that is, find $MN$.

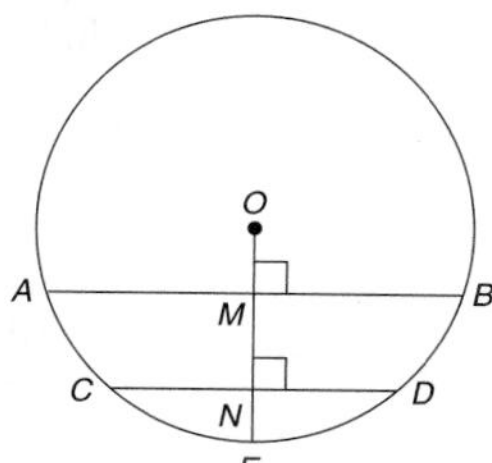


*38. In $\odot P$, whose radius has length 8 in., $m\widehat{AB} = m\widehat{BC} = 60°$. Because $m\widehat{AC} = 120°$, chord $\overline{AC}$ is longer than either of the congruent chords $\overline{AB}$ and $\overline{BC}$. Determine how much longer $\overline{AC}$ is than $\overline{AB}$; that is, find the exact value and the approximate value of $AC - AB$.

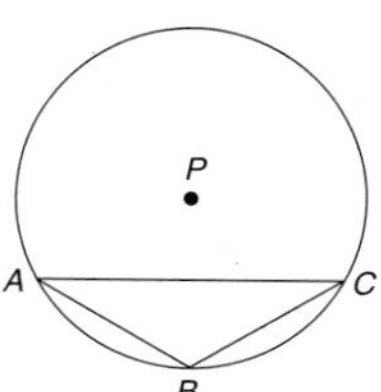

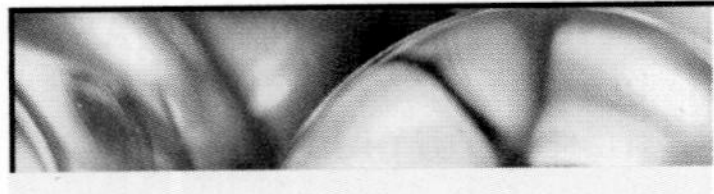

KEY CONCEPTS

Locus of Points in a Plane
• Locus of Points in Space

# 6.5 Locus of Points

At times we need to describe a set of points that satisfy a given condition or set of conditions. The term used to describe the resulting geometric figure is *locus,* the plural of which is *loci* (pronounced lō-sī). The English word *location* is derived from the Latin word *locus.*

DEFINITION: A **locus** is the set of all points and only those points that satisfy a given condition (or set of conditions).

In this definition, the phrase "all points and only those points" has a dual meaning:

1. All points of the locus satisfy the given condition.
2. All points satisfying the given condition are included in the locus.

The set of points satisfying a given locus can be a well-known geometric figure such as a line or a circle. In Examples 1, 2, and 3, several points are located and then connected in order to form the locus.

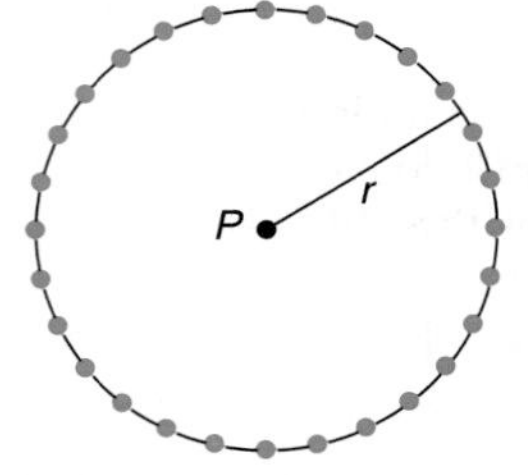


**FIGURE 6.62**

EXAMPLE 1

Describe the locus of points in a plane that are at a fixed distance ($r$) from a given point ($P$).

***Solution***

The locus is the circle with center $P$ and radius $r$. (See Figure 6.62.)

EXAMPLE 2

Describe the locus of points in a plane that are equidistant from two fixed points ($P$ and $Q$).

***Solution***

The locus is a line that is the perpendicular bisector of $\overline{PQ}$. (See Figure 6.63, in which $PX = QX$ for any point $X$ on line $t$.)

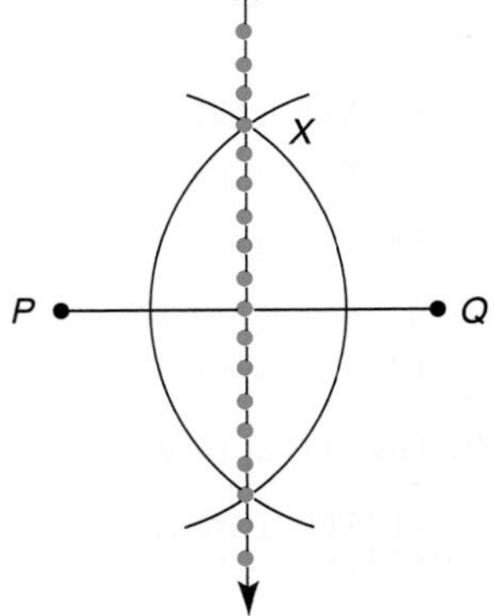


**FIGURE 6.63**

EXAMPLE 3

Describe the locus of points in a plane that are equidistant from the sides of an angle ($\angle ABC$) in that plane.

***Solution***

The locus is the ray that bisects $\angle ABC$. (See Figure 6.64.)

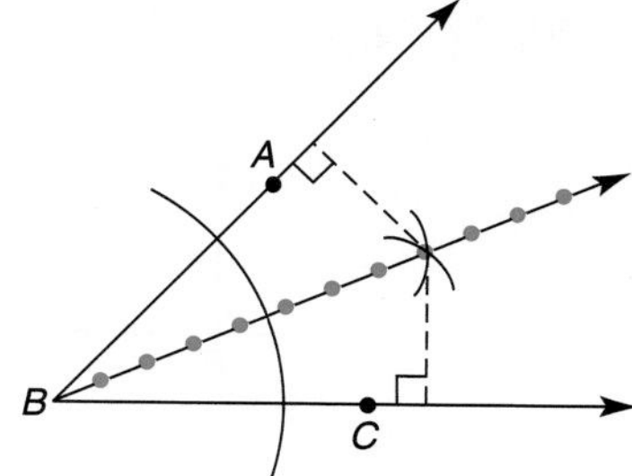


**FIGURE 6.64**

Some definitions are given in a locus format; for example, the following is an alternative definition of the term **circle.**

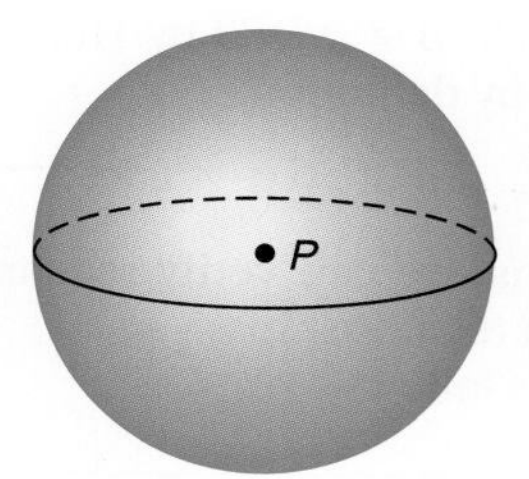


**FIGURE 6.65**

DEFINITION: A **circle** is the locus of points in a plane that are at a fixed distance from a given point.

Each of the preceding examples includes the phrase "in a plane." If that phrase is omitted, the locus is found "in space." For instance, the locus of points that are at a fixed distance from a given point is actually a *sphere* (the three-dimensional object in Figure 6.65); the sphere has the fixed point as center, and the fixed distance determines the length of the radius. Unless otherwise stated, we will consider the locus to be restricted to a plane.

*Exs. 1–4*

EXAMPLE 4

Describe the locus of points *in space* that are equidistant from two parallel planes ($P$ and $Q$).

***Solution***

The locus is the plane parallel to each of the given planes and midway between them. (See Figure 6.66.)

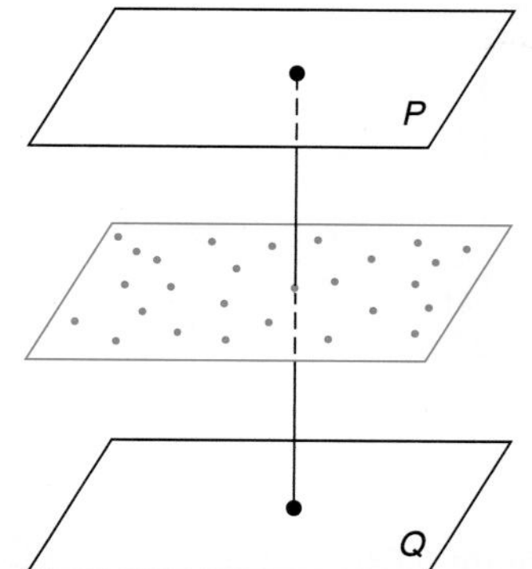


**FIGURE 6.66**

There are two very important theorems involving the locus concept. The results of these two theorems will be used in Section 6.6. When we verify the locus theorems, we *must* establish two results:

1. If a point is in the locus, then it satisfies the condition.
2. If a point satisfies the condition, then it is a point of the locus.

THEOREM 6.5.1: The locus of points in a plane and equidistant from the sides of an angle is the angle bisector.

***Proof***

(Note that *both* parts i and ii are necessary.)

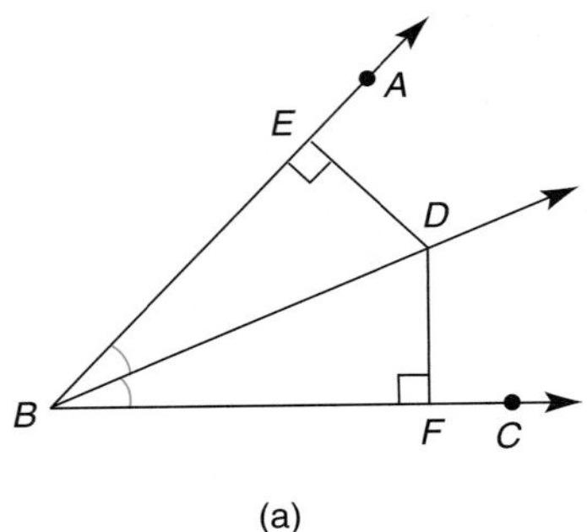

(a)

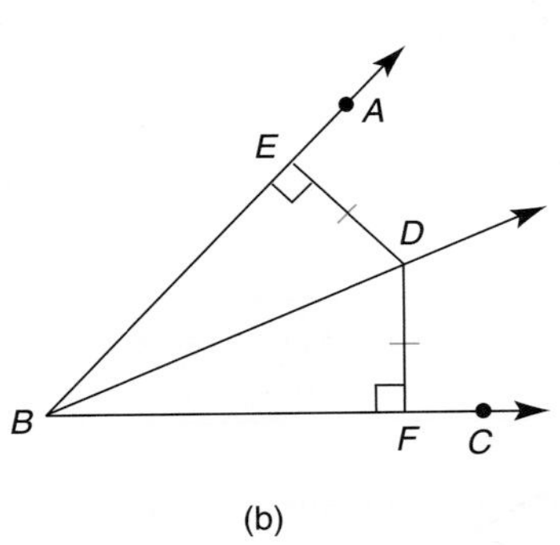

(b)

**FIGURE 6.67**

i) If a point is on the angle bisector, then it is equidistant from the sides of the angle.

*Given:* $\overrightarrow{BD}$ bisects $\angle ABC$
$\overline{DE} \perp \overrightarrow{BA}$ and $\overline{DF} \perp \overrightarrow{BC}$

*Prove:* $\overline{DE} \cong \overline{DF}$

*Proof:* In Figure 6.67(a), $\overrightarrow{BD}$ bisects $\angle ABC$; thus $\angle ABD \cong \angle CBD$. $\overline{DE} \perp \overrightarrow{BA}$ and $\overline{DF} \perp \overrightarrow{BC}$, so $\angle DEB$ and $\angle DFB$ are $\cong$ right $\angle$s. By Identity, $\overline{BD} \cong \overline{BD}$. By AAS, $\triangle DEB \cong \triangle DFB$. Then $\overline{DE} \cong \overline{DF}$ by CPCTC.

ii) If a point is equidistant from the sides of an angle, then it is on the angle bisector.

*Given:* $\angle ABC$ such that $\overline{DE} \perp \overrightarrow{BA}$ and $\overline{DF} \perp \overrightarrow{BC}$
$\overline{DE} \cong \overline{DF}$

*Prove:* $\overrightarrow{BD}$ bisects $\angle ABC$; that is, $D$ is on the bisector of $\angle ABC$

*Proof:* In Figure 6.67(b), $\overline{DE} \perp \overrightarrow{BA}$ and $\overline{DF} \perp \overrightarrow{BC}$, so $\angle DEB$ and $\angle DFB$ are right triangles. $\overline{DE} \cong \overline{DF}$ by hypothesis. Also, $\overline{BD} \cong \overline{BD}$. $\triangle DEB \cong \triangle DFB$ by HL. Then $\angle ABD \cong \angle CBD$ by CPCTC, so $\overrightarrow{BD}$ bisects $\angle ABC$ by definition.

In both construction problems and locus problems, we can verify results. In locus problems, we must remember to demonstrate two relationships!

A second important theorem about a locus of points follows.

> THEOREM 6.5.2: The locus of points in a plane that are equidistant from the endpoints of a line segment is the perpendicular bisector of that line segment.

*Exs. 5, 6*

***Proof***

i) If a point is equidistant from the endpoints of a line segment, then it lies on the perpendicular bisector of the line segment.

*Given:* $\overline{AB}$ and point $X$ not on $\overline{AB}$, so that $AX = BX$ [See Figure 6.68(a).]

*Prove:* $X$ lies on the perpendicular bisector of $\overline{AB}$

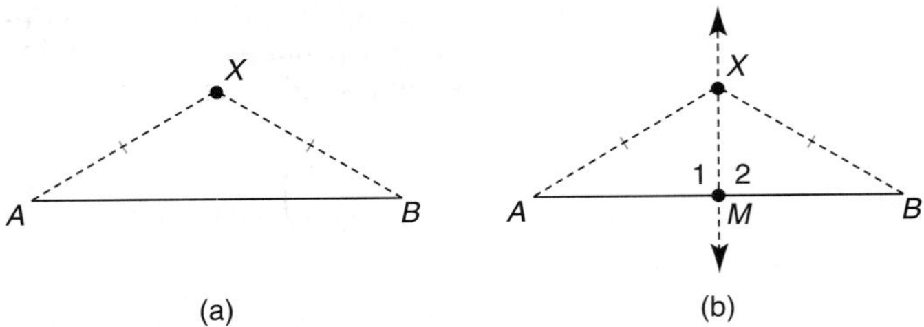


**FIGURE 6.68**

*Proof:* Let $M$ represent the midpoint of $\overline{AB}$. [See Figure 6.68(b).] Then $\overline{AM} \cong \overline{MB}$. Because $AX = BX$, we know that $\overline{AX} \cong \overline{BX}$. Because $\overline{XM} \cong \overline{XM}$, $\triangle AMX \cong \triangle BMX$ by SSS. By CPCTC, $\angle$s 1 and 2 are congruent and $\overleftrightarrow{MX} \perp \overline{AB}$. By definition, $\overleftrightarrow{MX}$ is the perpendicular bisector of $\overline{AB}$, so $X$ lies on the perpendicular bisector of $\overline{AB}$.

ii) If a point is on the perpendicular bisector of a line segment, then the point is equidistant from the endpoints of the line segment.

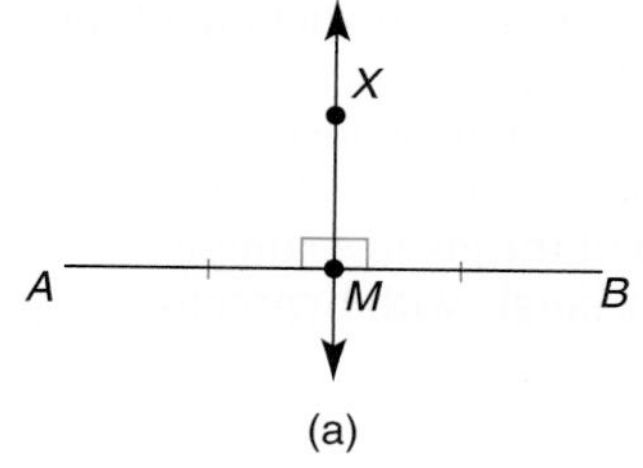

(a)

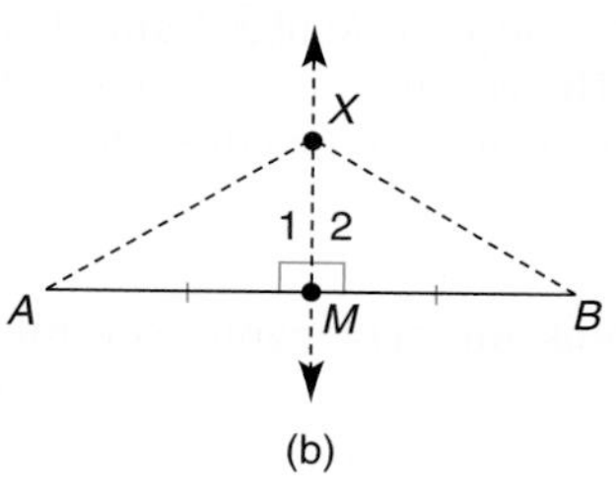

(b)

**FIGURE 6.69**

*Given:* Point $X$ lies on $\overleftrightarrow{MX}$, the perpendicular bisector of $\overline{AB}$ [See Figure 6.69(a).]

*Prove:* $X$ is equidistant from $A$ and $B$ ($AX = XB$) [See Figure 6.69(b).]

*Proof:* $X$ is on the perpendicular bisector of $\overline{AB}$, so $\angle$s 1 and 2 are congruent right angles and $\overline{AM} \cong \overline{MB}$. With $\overline{XM} \cong \overline{XM}$, $\triangle$s $AMX$ and $BMX$ are congruent by SAS; in turn, $\overline{XA} \cong \overline{XB}$ by CPCTC. Then $XA = XB$ and $X$ is equidistant from $A$ and $B$.

We now return to further considerations of a locus in a plane.

Suppose that a given line segment in a fixed location is to be used as the hypotenuse of a right triangle. How might you locate possible positions for the vertex of the right angle? One method might be to draw 30° and 60° angles at the endpoints so that the remaining angle formed must measure 90° [see Figure 6.70(a)]. This is only one possibility, but because of symmetry, it actually provides four permissible points, which are indicated in Figure 6.70(b). This problem is completed in Example 5.

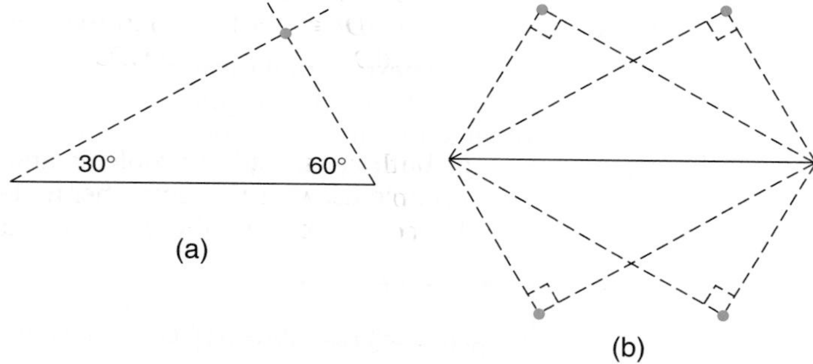

(a) (b)

*Exs. 7, 8*

**FIGURE 6.70**

## EXAMPLE 5

Find the locus of the vertex of the right angle of a right triangle if the hypotenuse is $\overline{AB}$ in Figure 6.71(a).

**Reminder**
An angle inscribed in a semicircle is a right angle.

***Solution***

Rather than use the "hit or miss" approach for locating the possible vertices (as suggested in the paragraph preceding this example), recall that an angle inscribed in a semicircle is a right angle.

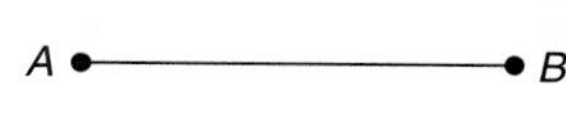

(a)

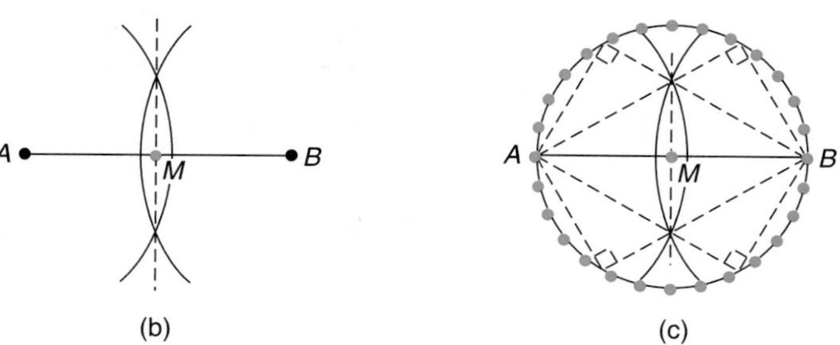

(b) (c)

**FIGURE 6.71**

Thus we construct the circle whose center is the midpoint $M$ of the hypotenuse and whose radius equals one-half the length of the hypotenuse. First, the midpoint $M$ of the hypotenuse $\overline{AB}$ is located. [See Figure 6.71(b)].

With the length of the radius of the circle equal to one-half the length of the hypotenuse, the circle with center $M$ is drawn in Figure 6.71(c).

The locus of the vertex of the right angle of a right triangle whose hypotenuse is given is the circle whose center is at the midpoint of the given segment and whose radius is equal in length to half the length of the given segment. Every point (except $A$ and $B$) on $\odot M$ is the vertex of a right triangle with hypotenuse $\overline{AB}$; see Theorem 6.1.9.

In Example 5, the construction involves locating the midpoint $M$ of $\overline{AB}$, and this is found by the method for the perpendicular bisector. The compass is then opened to a radius whose length is $MA$ or $MB$, and the circle is drawn. When a construction is performed, it falls into one of two categories:

1. A basic construction method
2. A construction problem that may require several steps and may involve several basic construction methods (like Example 5)

The next example also falls into category 2.

Recall that the diagonals of a rhombus are perpendicular and also bisect each other. With this information, we can locate the vertices of the rhombus whose diagonals (lengths) are known.

EXAMPLE 6

Construct rhombus $ABCD$ given its diagonals $\overline{AC}$ and $\overline{BD}$. (See Figure 6.72.)

***Solution***

To begin, we construct the perpendicular bisector of $\overline{AC}$; we know that the remaining vertices $B$ and $D$ must lie on this line. See Figure 6.72(a), in which $M$ is the midpoint of $\overline{AC}$.

To locate the midpoint of $\overline{BD}$, we construct its perpendicular bisector as well. See Figure 6.72(b), in which the midpoint is also called $M$.

Using an arc length equal to one-half the length of $\overline{BD}$ [like $MB$ in Figure 6.72(c)], we mark off this distance both above and below $\overline{AC}$ on the perpendicular bisector determined in Figure 6.72(a). See the result in Figure 6.72(c).

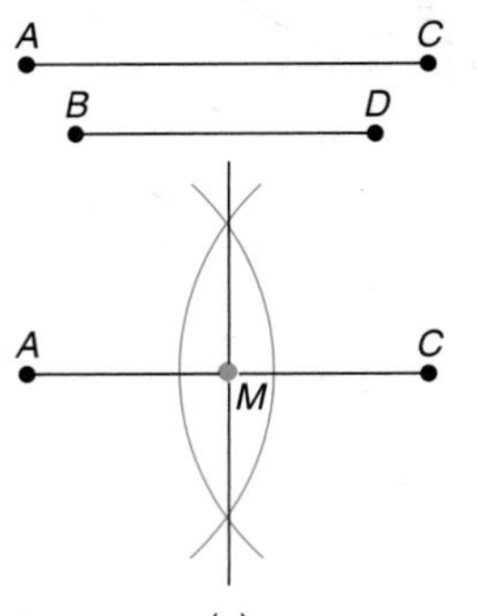


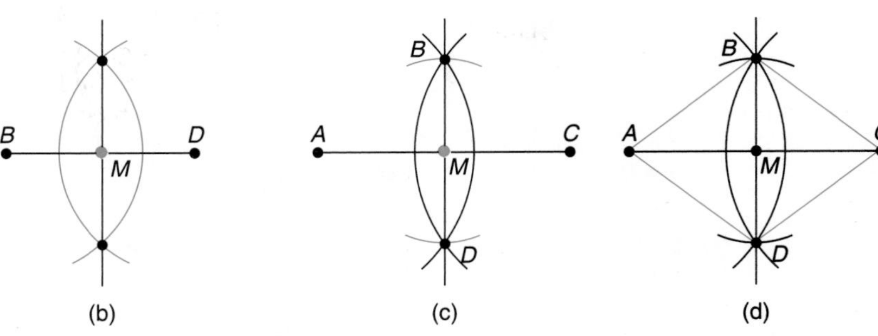


**FIGURE 6.72**

Using the marked arcs to locate (determine) points $B$ and $D$, we join $A$ to $B$, $B$ to $C$, $C$ to $D$, and $D$ to $A$. The completed rhombus $ABCD$ is shown in Figure 6.72(d).

*Exs. 9, 10*

## 6.5 Exercises

1. In the figure, which of the points $A$, $B$, $C$, $D$, and $E$ belong to "the locus of points in the plane that are at distance $r$ from point $P$"?

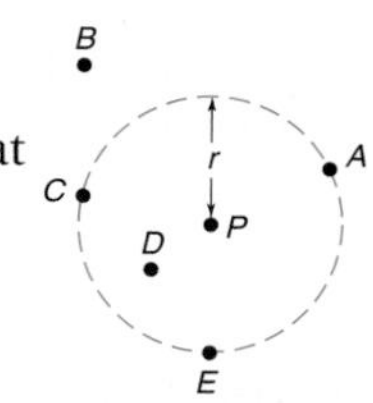


2. In the figure, which of the points $F$, $G$, $H$, $J$, and $K$ belong to "the locus of points in the plane that are at distance $r$ from line $\ell$"?

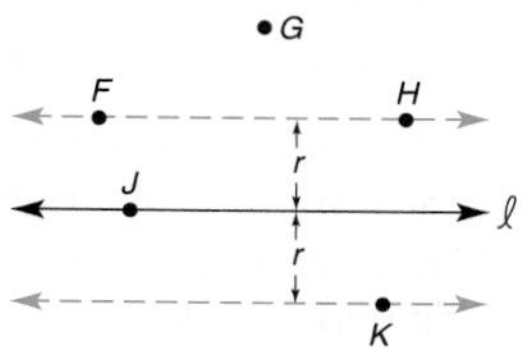


*In Exercises 3 to 8, use the drawing provided.*

3. *Given:* Obtuse $\triangle ABC$
*Construct:* The bisector of $\angle ABC$

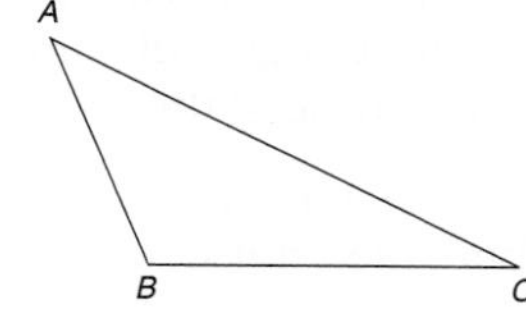


*Exercises 3–8*

4. *Given:* Obtuse $\triangle ABC$
*Construct:* The bisector of $\angle BAC$

5. *Given:* Obtuse $\triangle ABC$
*Construct:* The perpendicular bisector of $\overline{AB}$

6. *Given:* Obtuse $\triangle ABC$
*Construct:* The perpendicular bisector of $\overline{AC}$

7. *Given:* Obtuse $\triangle ABC$
*Construct:* The altitude from $A$ to $\overline{BC}$
(HINT: Extend $\overline{BC}$.)

8. *Given:* Obtuse $\triangle ABC$
*Construct:* The altitude from $B$ to $\overline{AC}$

9. *Given:* Right $\triangle RST$
*Construct:* The median from $S$ to $\overline{RT}$

10. *Given:* Right $\triangle RST$
*Construct:* The median from $R$ to $\overline{ST}$

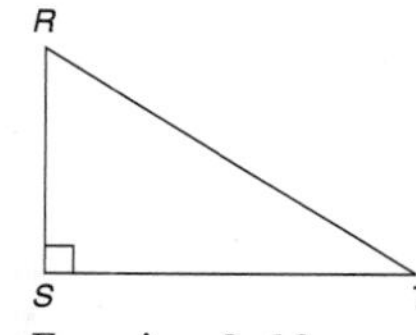


*Exercises 9–10*

*In Exercises 11 to 22, sketch and describe each locus in the plane.*

11. Find the locus of points that are at a given distance from a fixed line.

12. Find the locus of points that are equidistant from two given parallel lines.

13. Find the locus of points that are at a distance of 3 in. from a fixed point $O$.

14. Find the locus of points that are equidistant from two fixed points $A$ and $B$.

15. Find the locus of points that are equidistant from three noncollinear points $D$, $E$, and $F$.

16. Find the locus of the midpoints of the radii of a circle $O$ that has a radius of length 8 cm.

17. Find the locus of the midpoints of all chords of circle $Q$ that are parallel to diameter $\overline{PR}$.

18. Find the locus of points in the interior of a right triangle with sides of 6 in., 8 in., and 10 in. and at a distance of 1 in. from the triangle.

19. Find the locus of points that are equidistant from two given intersecting lines.

*20. Find the locus of points that are equidistant from a fixed line and a point not on that line.
(NOTE: This figure is known as a *parabola.*)

21. Given that lines $p$ and $q$ intersect, find the locus of points that are at a distance of 1 cm from line $p$ and also at a distance of 2 cm from line $q$.

22. Given that congruent circles $O$ and $P$ have radii of length 4 in. and that the line of centers has length 6 in., find the locus of points that are 1 in. from each circle.

*In Exercises 23 to 30, sketch and describe the locus of points in space.*

23. Find the locus of points that are at a given distance from a fixed line.

24. Find the locus of points that are equidistant from two fixed points.

25. Find the locus of points that are at a distance of 2 cm from a sphere whose radius is 5 cm.

26. Find the locus of points that are at a given distance from a given plane.

27. Find the locus of points that are the midpoints of the radii of a sphere whose center is point $O$ and whose radius has a length of 5 m.

*28. Find the locus of points that are equidistant from three noncollinear points $D$, $E$, and $F$.

29. In a room, find the locus of points that are equidistant from the parallel ceiling and floor, which are 8 ft apart.

30. Find the locus of points that are equidistant from all points on the surface of a sphere with center point $Q$.

*In Exercises 31 and 32, use the method of proof of Theorem 6.5.1 to justify each construction method.*

31. The perpendicular bisector method.

32. The construction of a perpendicular to a line from a point outside the line.

*In Exercises 33 to 36, refer to the line segments shown.*

33. Construct an isosceles right triangle that has hypotenuse $\overline{AB}$.

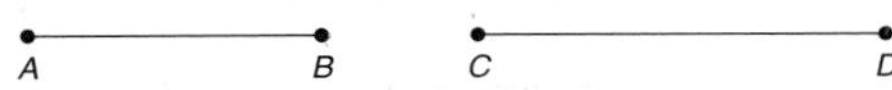


*Exercises 33–36*

34. Construct a rhombus whose sides are equal in length to $AB$, and so that one diagonal of the rhombus has length $CD$.

35. Construct an isosceles triangle in which each leg has length $CD$ and the altitude to the base has length $AB$.

36. Construct an equilateral triangle in which the altitude to any side has length $AB$.

37. Construct the three angle bisectors and then the inscribed circle for obtuse $\triangle RST$.

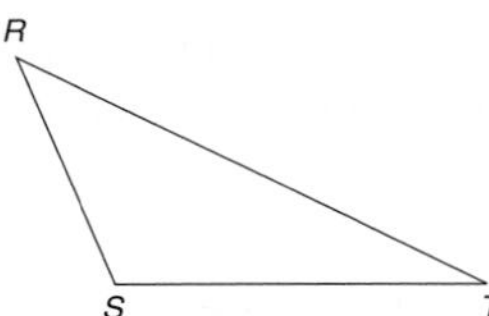


*Exercises 37, 38*

38. Construct the three perpendicular bisectors of sides and then the circumscribed circle for obtuse $\triangle RST$.

39. Use the following theorem to locate the center of the circle of which $\widehat{RT}$ is a part.

*Theorem:* The perpendicular bisector of a chord passes through the center of a circle.

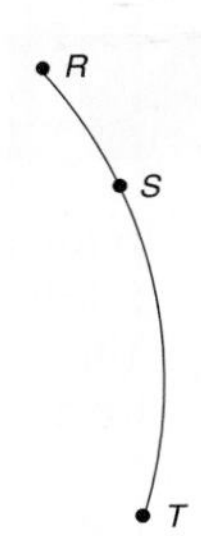


*40. Use the following theorem to construct the geometric mean of the numerical lengths of the segments $\overline{WX}$ and $\overline{YZ}$.

*Theorem:* The length of the altitude to the hypotenuse of a right triangle is the geometric mean between the lengths of the segments of the hypotenuse.

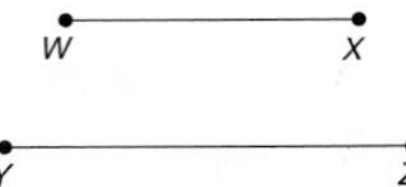


41. Use the following theorem to construct a triangle similar to the given triangle but with sides that are twice the length of those of the given triangle.

*Theorem:* If the three pairs of sides for two triangles are in proportion, then those triangles are similar (SSS~).

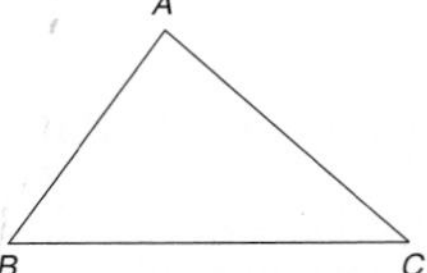


*42. Verify this locus theorem:

The locus of points equidistant from two fixed points is the perpendicular bisector of the segment joining those points.

# 6.6 Concurrence of Lines

**KEY CONCEPTS**

Concurrent Lines • Incenter • Incircle • Circumcenter • Circumcircle • Orthocenter • Centroid

In this section, we consider lines that share a common point.

> DEFINITION: A number of lines are **concurrent** if they have exactly one point in common.

**Discover!**

A computer software program can be useful in demonstrating the concurrence of the lines described in each theorem in this section.

The three lines in Figure 6.73 are concurrent at point $A$. The three lines in Figure 6.74 are not concurrent even though any pair of lines (like $r$ and $s$) do intersect.

Parts of lines (rays or segments) are concurrent if they are parts of concurrent lines and the parts share a common point.

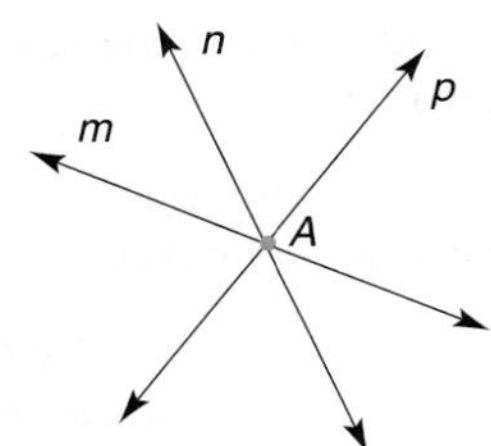


*m*, *n*, and *p are* concurrent

**FIGURE 6.73**

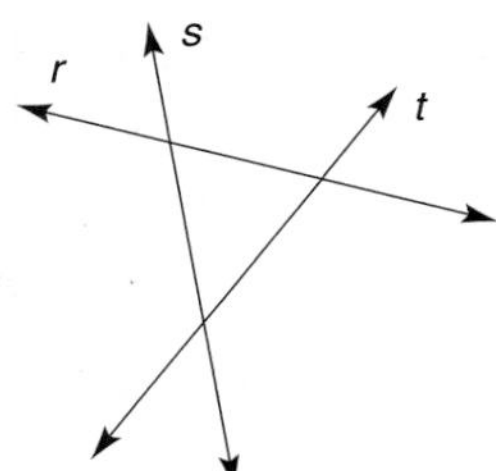


*r*, *s*, and *t* are *not* concurrent

**FIGURE 6.74**

*Exs. 1, 2*

> THEOREM 6.6.1: The three angle bisectors of the angles of a triangle are concurrent.

For the informal proofs of this section, no Given or Prove is stated. In more advanced courses, these parts of the proof are understood.

**EXAMPLE 1**

Give an informal proof of Theorem 6.6.1.

***Proof***

In Figure 6.75(a), the bisectors of $\angle BAC$ and $\angle ABC$ intersect at point $E$.

Because the angle bisector of $\angle BAC$ is the locus of points equidistant from the sides of $\angle BAC$, we know that $\overline{EM} \cong \overline{EN}$ in Figure 6.75(b). Similarly, we know that $\overline{EM} \cong \overline{EP}$ because $E$ is on the angle bisector of $\angle ABC$.

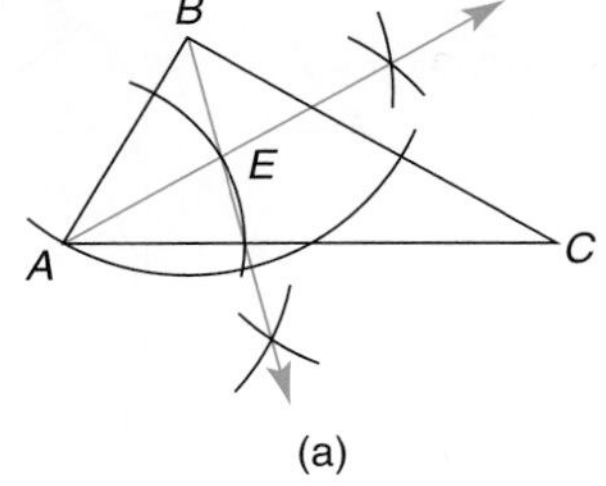


(a)

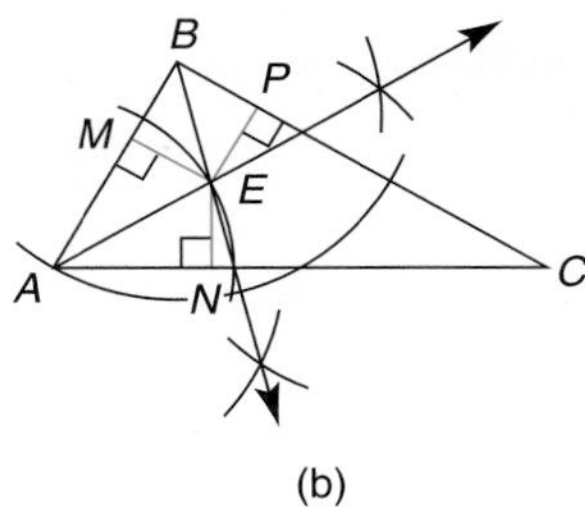


(b)

**FIGURE 6.75**

**Reminder**

A point on the bisector of an angle is equidistant from the sides of the angle.

By the Transitive Property of Congruence, it follows that $\overline{EP} \cong \overline{EN}$.

Because the angle bisector is the locus of points equidistant from the sides of the angle, we know that $E$ is also on the bisector of the third angle, $\angle ACB$. Thus, the angle bisectors are concurrent.

The point $E$ at which the angle bisectors meet in Example 1 is the **incenter** of the triangle. As the following example shows, the term *incenter* is well deserved because this point is the *center* of the *in*scribed circle of the triangle.

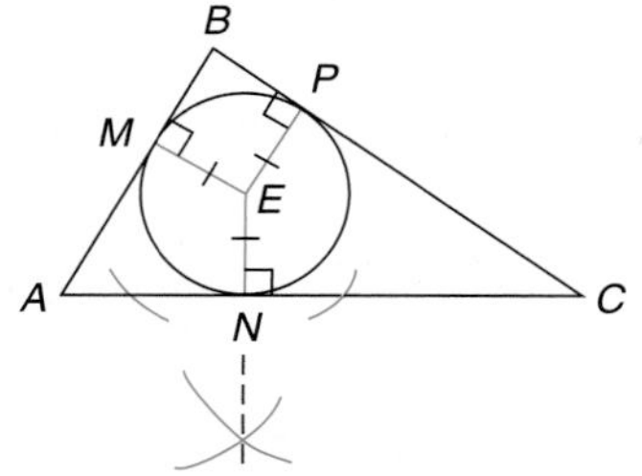


**FIGURE 6.76**

EXAMPLE 2

Complete the construction of the inscribed circle for $\triangle ABC$ in Figure 6.75(b).

***Solution***

Having found the incenter $E$, we need the length of the radius. Because $\overline{EN} \perp \overline{AC}$ (as shown in Figure 6.76), the length of $\overline{EN}$ is the desired radius; thus, the circle is completed.

NOTE: The sides of the triangle are tangents for the inscribed circle, which is called the **incircle** of the triangle.

It is also possible to circumscribe a circle about a given triangle. The construction depends on the following theorem, the proof of which is sketched in Example 3.

*Exs. 3–7*

THEOREM 6.6.2: The three perpendicular bisectors of the sides of a triangle are concurrent.

EXAMPLE 3

Give an informal proof of Theorem 6.6.2. See $\triangle ABC$ in Figure 6.77.

***Proof***

Let $\overline{FS}$ and $\overline{FR}$ name the perpendicular bisectors of sides $\overline{BC}$ and $\overline{AC}$, respectively. See Figure 6.77(a). Using Theorem 6.5.2, the point of concurrency $F$ is equidistant from the endpoints of $\overline{BC}$; thus $\overline{BF} \cong \overline{FC}$. In the same manner, $\overline{AF} \cong \overline{FC}$. By the Transitive Property, it follows that $\overline{AF} \cong \overline{BF}$; again citing Theorem 6.5.2, $F$ must be on the perpendicular bisector of $\overline{AB}$, because this point is equidistant from the endpoints of $\overline{AB}$. Thus, $F$ is the point of concurrence.

Reminder

A point on the perpendicular bisector of a line segment is equidistant from the endpoints of the line segment.

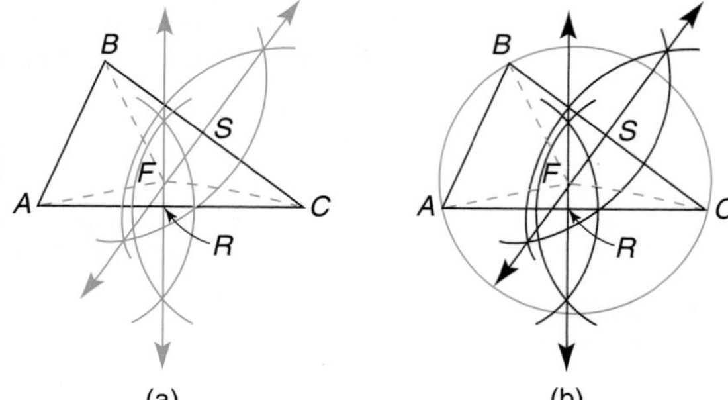


**FIGURE 6.77**

The point at which the perpendicular bisectors of the sides of a triangle meet is the **circumcenter** of the triangle. This is easily remembered as the *center* of the *circumscribed* circle.

EXAMPLE 4

Complete the construction of the circumscribed circle for $\triangle ABC$ that was given in Figure 6.77(a).

***Solution***

We have already identified the center of the circle as point $F$. To complete the construction, we use $F$ as the center and a radius of length equal to the distance from $F$ to any one of the vertices $A$, $B$, or $C$. The circumscribed circle is shown in Figure 6.77(b).

NOTE: The sides of the inscribed triangle are chords of the circumscribed circle, which is called the **circumcircle** of the triangle.

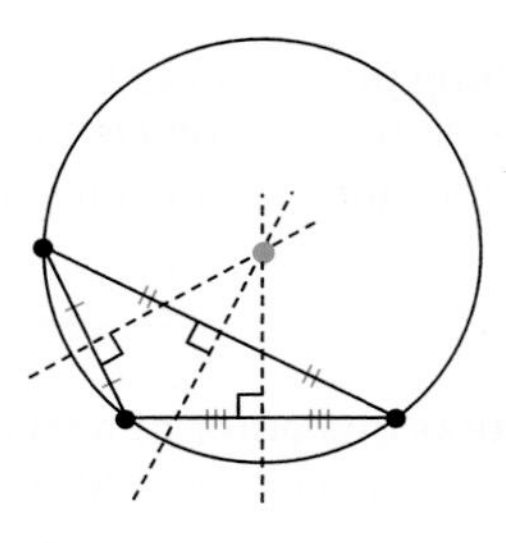

FIGURE 6.78

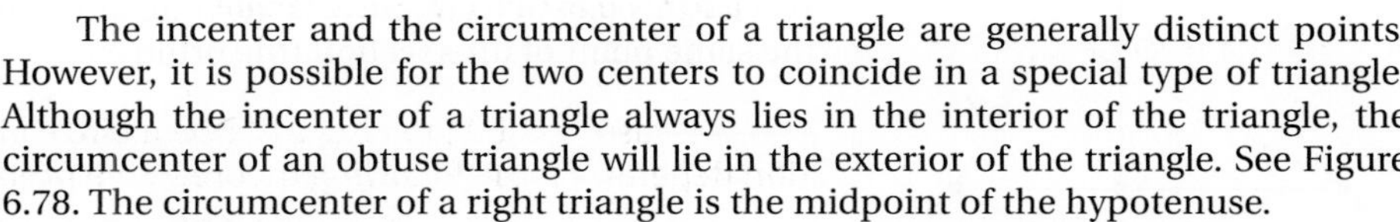

The incenter and the circumcenter of a triangle are generally distinct points. However, it is possible for the two centers to coincide in a special type of triangle. Although the incenter of a triangle always lies in the interior of the triangle, the circumcenter of an obtuse triangle will lie in the exterior of the triangle. See Figure 6.78. The circumcenter of a right triangle is the midpoint of the hypotenuse.

*Exs. 8–12*

To complete the discussion of concurrence, we include theorems involving the altitudes of a triangle and the medians of a triangle.

THEOREM 6.6.3: The three altitudes of a triangle are concurrent.

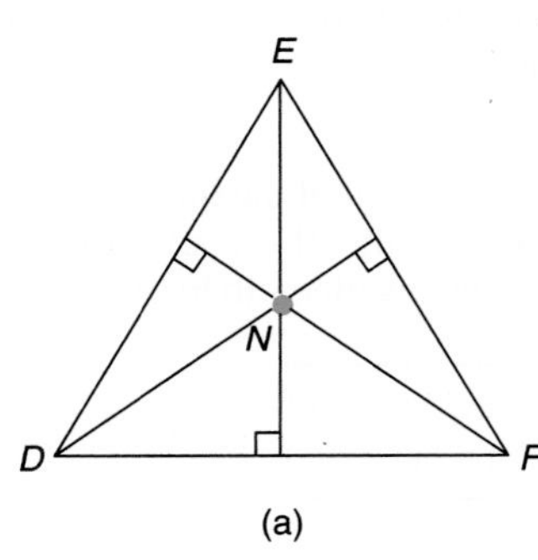


(a)

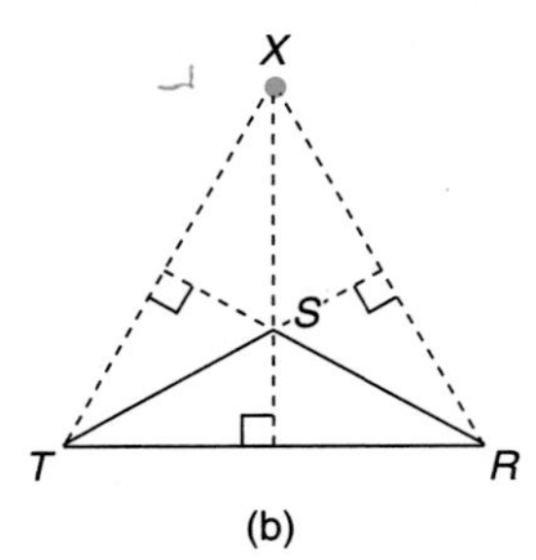


(b)

FIGURE 6.79

The point of concurrence for the three altitudes of a triangle is the **orthocenter** of the triangle. In Figure 6.79(a), point $N$ is the orthocenter of $\triangle DEF$. For the obtuse triangle $RST$ shown in Figure 6.79(b), we see that orthocenter $X$ lies in the exterior.

Rather than prove Theorem 6.6.3, we sketch a part of that proof. Consider $\triangle MNP$, shown on page 322 with its altitudes in Figure 6.80(a). To prove that the altitudes are concurrent requires

1. that we draw auxiliary lines through $N$ parallel to $\overline{MP}$, through $M$ parallel to $\overline{NP}$, and through $P$ parallel to $\overline{NM}$. [See Figure 6.80(b) on page 322.]
2. that we show that the altitudes of $\triangle MNP$ are perpendicular bisectors of the sides of the newly formed $\triangle RST$; thus altitudes $\overline{PX}$, $\overline{MY}$, and $\overline{NZ}$ are concurrent (a consequence of Theorem 6.6.2).

*Sketch of proof that $\overline{PX}$ is the $\perp$ bisector of $\overline{RS}$:*
Because $\overline{PX}$ is an altitude of $\triangle MNP$, $\overline{PX} \perp \overline{MN}$. But $\overline{RS} \parallel \overline{MN}$ by construction. Because a line perpendicular to one of two parallel lines must be perpendicular to the other, we have $\overline{PX} \perp \overline{RS}$. Now we need to show that $\overline{PX}$ bisects $\overline{RS}$. By construction, $\overline{MR} \parallel \overline{NP}$ and $\overline{RP} \parallel \overline{MN}$, so $MRPN$ is a parallelogram. Then $\overline{MN} \cong \overline{RP}$ because the opposite sides of a parallelogram are congruent. By construction, $MPSN$ is also a parallelogram and $\overline{MN} \cong \overline{PS}$. Then $\overline{RP} \cong \overline{PS}$ because $\overline{MN}$ is congruent to each segment. Thus, $\overline{RS}$ is bisected at point $P$, and $\overline{PX}$ is the $\perp$ bisector of $\overline{RS}$.

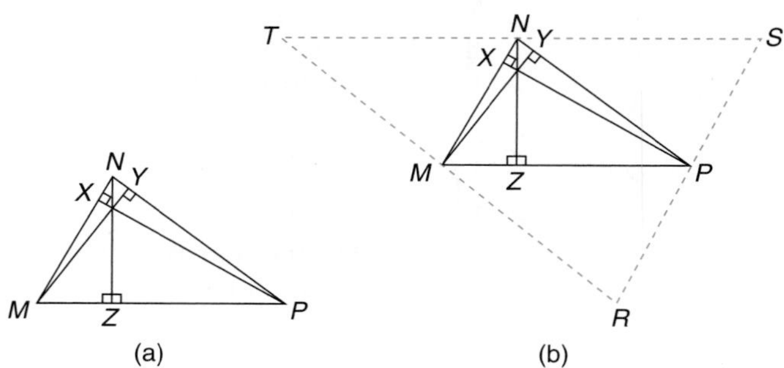


**FIGURE 6.80**

In Figure 6.80(b), similar arguments (leading to one long proof) could be used to show that $\overline{NZ}$ is the $\perp$ bisector of $\overline{TS}$ and that $\overline{MY}$ is the $\perp$ bisector of $\overline{TR}$. Because the concurrent perpendicular bisectors of the sides of $\triangle RST$ are also the altitudes of $\triangle MNP$, these altitudes must be concurrent.

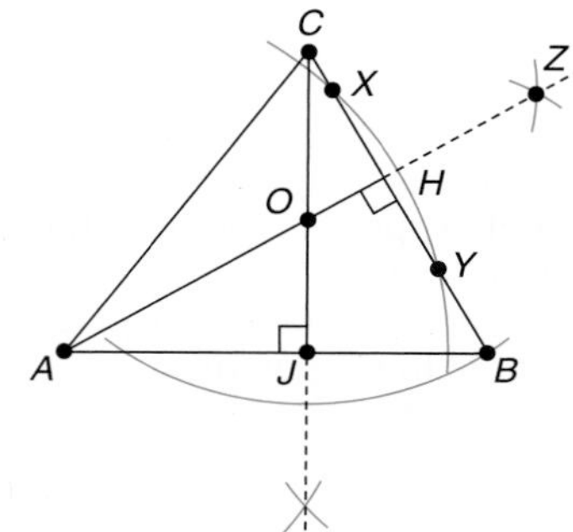


**FIGURE 6.81**

The intersection of any two altitudes determines the orthocenter of a triangle. We use this fact in Example 5. If the third altitude were constructed, it would contain the same point of intersection (the orthocenter).

EXAMPLE 5

Construct the orthocenter of $\triangle ABC$ in Figure 6.81.

***Solution***

First construct the altitude from $A$ to $\overline{BC}$; here, we draw an arc from $A$ to intersect $\overline{BC}$ at $X$ and $Y$. Now draw equal arcs from $X$ and $Y$ to intersect at $Z$. $\overline{AH}$ is the desired altitude. Repeat the process to construct altitude $\overline{CJ}$ from vertex $C$ to side $\overline{AB}$. The point of intersection $O$ is the orthocenter of $\triangle ABC$.

*Exs. 13–16*

Recall that a median of a triangle joins a vertex to the midpoint of the opposite side of the triangle. Through construction, we can show that the three medians of a triangle are concurrent. We will discuss the proof of the following theorem in Chapter 9.

THEOREM 6.6.4: The three medians of a triangle are concurrent at a point that is two-thirds the distance from any vertex to the midpoint of the opposite side.

The point of concurrence for the three medians is the **centroid** of the triangle. In Figure 6.82, point $C$ is the centroid of $\triangle RST$. According to Theorem 6.6.4, $RC = \frac{2}{3}(RM)$, $SC = \frac{2}{3}(SN)$, and $TC = \frac{2}{3}(TP)$.

**FIGURE 6.82**

Discover!

On a piece of paper draw a triangle and its medians. Label the figure like Figure 6.82.

a) Find the value of $\frac{RC}{RM}$.

b) Find the value of $\frac{SC}{CN}$.

ANSWERS

(a) $\frac{2}{3}$ (b) 2

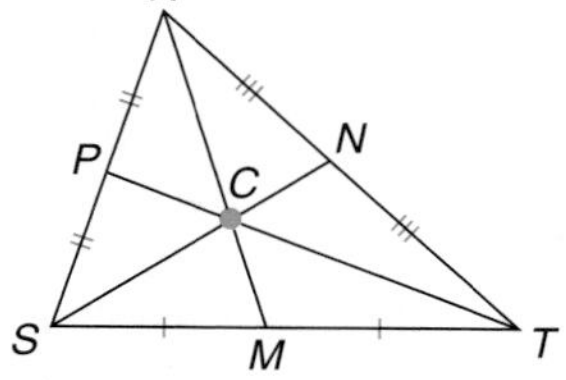

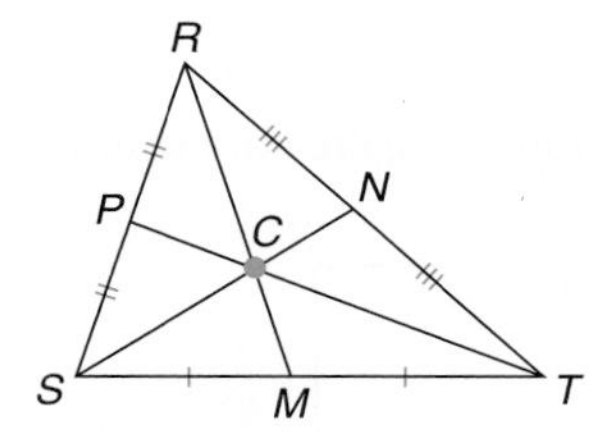


**FIGURE 6.82**

EXAMPLE 6

Suppose that the medians of $\triangle RST$ in Figure 6.82 have the lengths $RM = 12$, $SN = 15$, and $TP = 18$. If the centroid of $\triangle RST$ is point $C$, find the length of:

a) $RC$ b) $CM$ c) $SC$

***Solution***

a) $RC = \frac{2}{3}(RM)$, so $RC = \frac{2}{3}(12) = 8$.
b) $CM = RM - RC$, so $CM = 12 - 8 = 4$.
c) $SC = \frac{2}{3}(SN)$, so $SC = \frac{2}{3}(15) = 10$.

EXAMPLE 7

*Given:* In Figure 6.83(a), isosceles $\triangle RST$ with $RS = RT = 15$, and $ST = 18$; medians $\overline{RZ}$, $\overline{TX}$, and $\overline{SY}$ meet at centroid $Q$.

*Find:* $RQ$

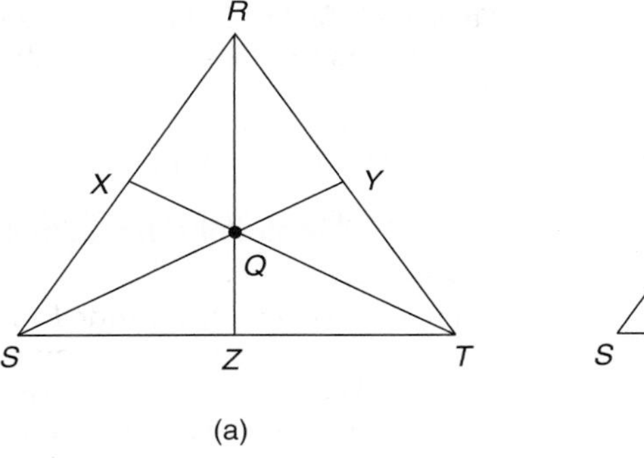


(a)

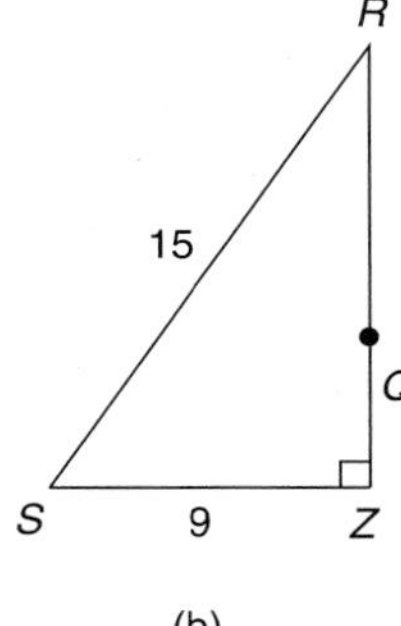


(b)

**FIGURE 6.83**

***Solution***

Median $\overline{RZ}$ separates $\triangle RST$ into two congruent right triangles, $\triangle RZS$ and $\triangle RZT$; this follows from SSS. With $Z$ the midpoint of $\overline{ST}$, $SZ = 9$.

Using the Pythagorean Theorem with $\triangle RZS$ in Figure 6.83(b), we have

$$\begin{aligned} (RS)^2 &= (RZ)^2 + (SZ)^2 \\ 15^2 &= (RZ)^2 + 9^2 \\ 225 &= (RZ)^2 + 81 \\ (RZ)^2 &= 144 \\ RZ &= 12 \end{aligned}$$

By Theorem 6.6.4,

$$RQ = \frac{2}{3}(RZ) = \frac{2}{3}(12) = 8$$

SSG *Exs. 17–22*

It is *possible* for the angle bisectors of certain quadrilaterals to be concurrent. Likewise, the perpendicular bisectors of the sides of a quadrilateral *can* be concurrent. Of course, there are four angle bisectors and four perpendicular bisectors of sides to consider. In Example 8, we explore this situation.

EXAMPLE 8

Use intuition and Figure 6.84 to decide which of the following are concurrent.

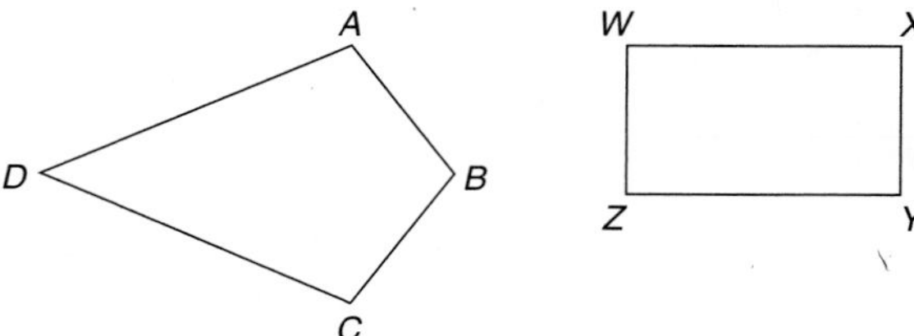


**FIGURE 6.84**

a) The angle bisectors of a kite
b) The perpendicular bisectors of the sides of a kite
c) The angle bisectors of a rectangle
d) The perpendicular bisectors of the sides of a rectangle

***Solution***

a) The angle bisectors of the kite are concurrent.
b) The $\perp$ bisectors of the sides of the kite are not concurrent (unless $\angle A$ and $\angle C$ are both right angles).
c) The angle bisectors of the rectangle are not concurrent (unless the rectangle is a square).
d) The $\perp$ bisectors of the sides of the rectangle are concurrent (the circumcenter is also the point of intersection of diagonals).

NOTE: The student should make drawings to verify the results in Example 8.

The centroid of a triangular region is sometimes called its *center of mass* or *center of gravity*. This is because the region of uniform thickness "balances" upon the point known as its centroid. Consider the following activity.

## Discover!

Take a piece of cardboard or heavy poster paper. Draw a triangle on the paper and cut out the triangular shape. Now use a ruler to mark the midpoints of each side and draw the medians to locate the centroid. Place the triangle on the point of a pen or pencil at the centroid and see how well you can balance the triangular region.

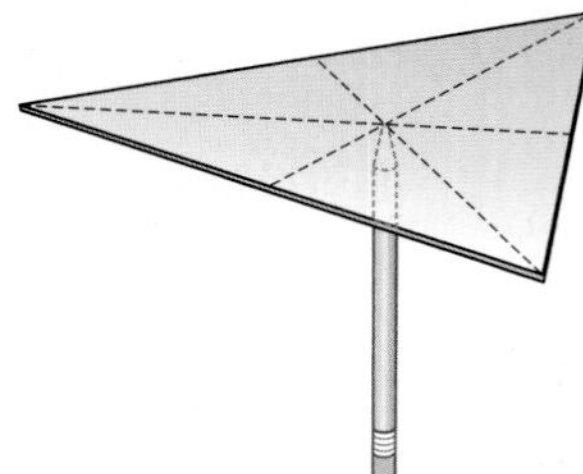

## 6.6 Exercises

1. In the figure, are lines $m$, $n$, and $p$ concurrent?

2. If one exists, name the point of concurrence for lines $m$, $n$, and $p$.

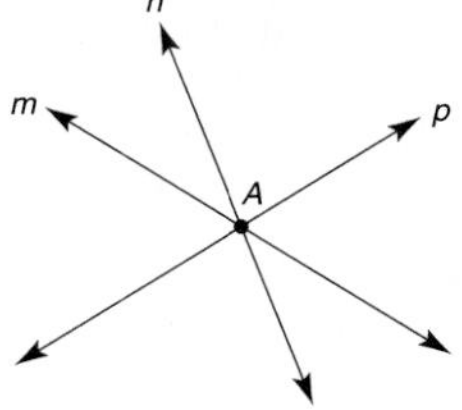


*Exercises 1, 2*

3. What is the general name of the point of concurrence for the three angle bisectors of a triangle?

4. What is the general name of the point of concurrence for the three altitudes of a triangle?

5. What is the general name of the point of concurrence for the three perpendicular bisectors of sides of a triangle?

6. What is the general name of the point of concurrence for the three medians of a triangle?

7. Which lines or line segments or rays must be drawn or constructed in a triangle to locate its

   a) incenter?
   b) circumcenter?
   c) orthocenter?
   d) centroid?

8. Is it really necessary to construct all three angle bisectors of the angles of a triangle to locate its incenter?

9. Is it really necessary to construct all three perpendicular bisectors of the sides of a triangle to locate its circumcenter?

10. To locate the orthocenter, is it necessary to construct all three altitudes of a right triangle?

11. For what type of triangle are the angle bisectors, the medians, the perpendicular bisectors of sides, and the altitudes all the same?

12. What point on a right triangle is the orthocenter of the right triangle?

13. What point on a right triangle is the circumcenter of the right triangle?

14. Must the centroid of an isosceles triangle lie on the altitude to the base?

15. Draw a triangle and, by construction, find its incenter.

16. Draw an acute triangle and, by construction, find its circumcenter.

17. Draw an obtuse triangle and, by construction, find its circumcenter.

18. Draw an acute triangle and, by construction, find its orthocenter.

19. Draw an obtuse triangle and, by construction, find its orthocenter. (HINT: You will have to extend the sides opposite the acute angles.)

20. Draw an acute triangle and, by construction, find the centroid of the triangle. (HINT: Begin by constructing the perpendicular bisectors of the sides.)

21. Draw an obtuse triangle and, by construction, find the centroid of the triangle. (HINT: Begin by constructing the perpendicular bisectors of the sides.)

22. Is the incenter always located in the interior of the triangle?

23. Is the circumcenter always located in the interior of the triangle?

24. Find the length of the radius of the inscribed circle for a right triangle whose legs measure 6 and 8.

25. Find the distance from the circumcenter to each vertex of an equilateral triangle whose sides have the length 10.

26. A triangle has angles measuring 30°, 30°, and 120°. If the congruent sides measure 6 units each, find the length of the radius of the circumscribed circle.

27. *Given:* Isosceles $\triangle RST$
    $RS = RT = 17$ and $ST = 16$
    Medians $\overline{RZ}$, $\overline{TX}$, and $\overline{SY}$ meet at centroid $Q$
    *Find:* $RQ$ and $SQ$

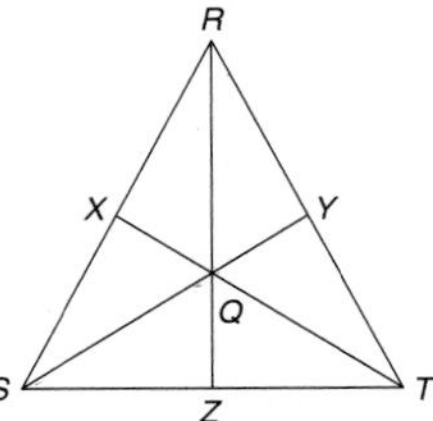


*Exercises 27, 28*

28. *Given:* Isosceles $\triangle RST$
    $RS = RT = 10$ and $ST = 16$
    Medians $\overline{RZ}$, $\overline{TX}$, and $\overline{SY}$ meet at $Q$
    *Find:* $RQ$ and $QT$

29. In $\triangle MNP$, medians $\overline{MB}$, $\overline{NA}$, and $\overline{PC}$ intersect at centroid $Q$.

a) If $MQ = 8$, find $QB$.
b) If $QC = 3$, find $PQ$.
c) If $AQ = 3.5$, find $AN$.

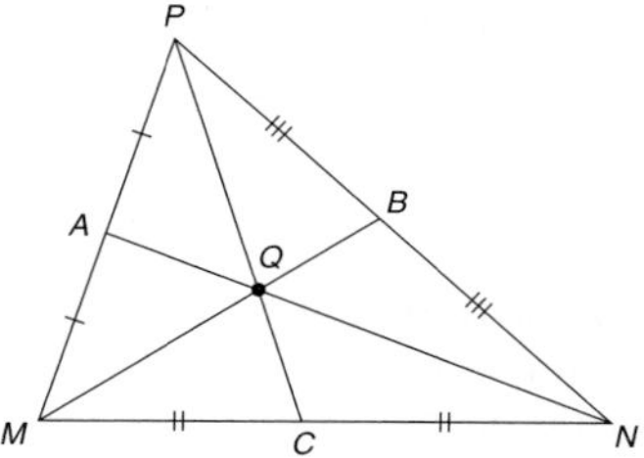


*Exercises 29, 30*

30. In $\triangle MNP$, medians $\overline{MB}$, $\overline{NA}$, and $\overline{PC}$ intersect at centroid $Q$.

a) Find $QB$ if $MQ = 8.2$.
b) Find $PQ$ if $QC = \frac{7}{2}$.
c) Find $AN$ if $AQ = 4.6$.

31. Draw a triangle. Construct its inscribed circle.

32. Draw a triangle. Construct its circumscribed circle.

33. For what type of triangle will the incenter and the circumcenter be the same?

34. Does a rectangle have (a) an incenter? (b) a circumcenter?

35. Does a square have (a) an incenter? (b) a circumcenter?

36. Does a regular pentagon have (a) an incenter? (b) a circumcenter?

37. Does a rhombus have (a) an incenter? (b) a circumcenter?

38. Does an isosceles trapezoid have (a) an incenter? (b) a circumcenter?

39. A distributing company plans an Illinois location that would be the same distance from each of its principal delivery sites at Chicago, St. Louis, and Indianapolis. Use a construction method to locate the approximate position of the distributing company.
(NOTE: Trace the outline of the two states on your own paper.)

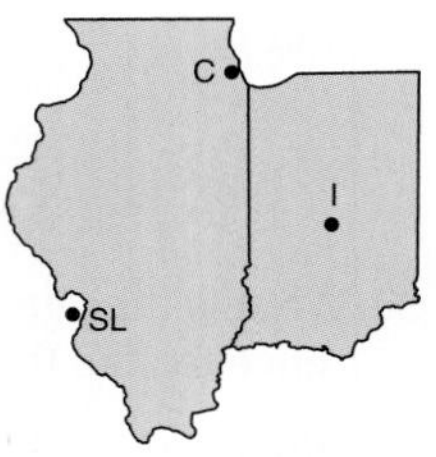


40. There are plans to locate a disaster response agency in an area that is prone to tornadic activity. The agency is to be located at equal distances from Wichita, Tulsa, and Oklahoma City. Use a construction method to locate the approximate position of the agency.
(NOTE: Trace the outline of the two states on your own paper.)

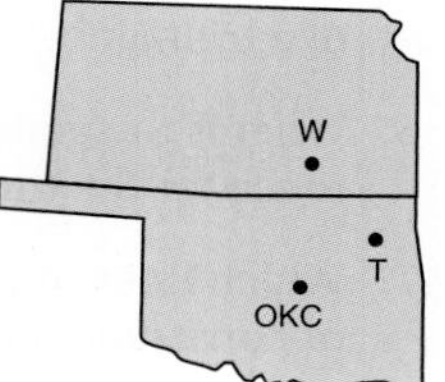


# Perspective on History

## THE VALUE OF $\pi$

In geometry, any two figures that have the same shape are described as similar. Because all circles have the same shape, we say that all circles are similar to each other. Just as a proportionality exists among the corresponding sides of similar triangles, we can demonstrate a proportionality among the circumferences (distances around) and diameters (distances across) of circles. By representing the circumferences of the circles in Figure 6.85 by $C_1$, $C_2$, and $C_3$ and their corresponding lengths of diameters by $d_1$, $d_2$, and $d_3$, we are claiming that

$$\frac{C_1}{d_1} = \frac{C_2}{d_2} = \frac{C_3}{d_3} = k$$

for some constant of proportionality $k$.

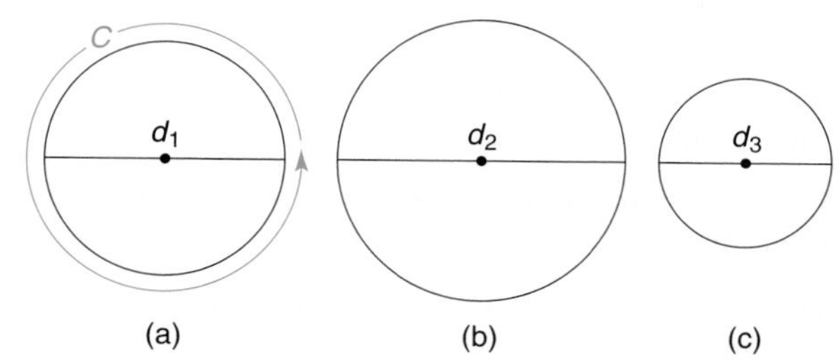


**FIGURE 6.85**

We denote the constant $k$ described above by the Greek letter $\pi$. Thus $\pi = \frac{C}{d}$ in any circle. It follows that $C = \pi d$ or $C = 2\pi r$ (because $d = 2r$ in any circle). In applying these formulas for the circumference of a circle, we often leave $\pi$ in the answer so that the result is exact. When an approximation for the circumference (and later for the area) of a circle is needed, several common substitutions are used for $\pi$. Among these are $\pi \approx \frac{22}{7}$ and $\pi \approx 3.14$. A calculator may display the value $\pi \approx 3.1415926535$.

Because $\pi$ is needed in many applications involving the circumference or area of a circle, its approximation is often necessary; but finding an accurate approximation of $\pi$ was not quickly or easily done. The formula for circumference can be expressed as $C = 2\pi r$, but the formula for the area of the circle is $A = \pi r^2$. This and other area formulas will be given more attention in Chapter 7.

Several references to the value of $\pi$ are made in literature. One of the earliest comes from the Bible; the passage from I Kings, Chapter 7, verse 23, describes the distance around a vat as three times the distance across the vat (which suggests that $\pi$ equals 3, a very rough approximation). Perhaps no greater accuracy was needed in applications at that time.

In the content of the Rhind papyrus (a document over 3000 years old), the Egyptian scribe Ahmes gives the formula for the area of a circle as $\left(d - \frac{1}{9}d\right)^2$. To determine the Egyptian approximation of $\pi$, we need to expand this expression as follows:

$$\left(d - \frac{1}{9}d\right)^2 = \left(\frac{8}{9}d\right)^2 = \left(\frac{8}{9} \cdot 2r\right)^2 = \left(\frac{16}{9}r\right)^2 = \frac{256}{81}r^2$$

In the formula for the area of the circle, the value of $\pi$ is the multiplier (coefficient) of $r^2$. Because this coefficient is $\frac{256}{81}$ (which has the decimal equivalent of 3.1604), the Egyptians had a better approximation of $\pi$ than was given in the book of I Kings.

Archimedes, the brilliant Greek geometer, knew that the formula for the area of a circle was $A = \frac{1}{2}Cr$ (with $C$ the circumference and $r$ the length of radius). His formula was equivalent to the one we use today and is developed as follows:

$$A = \frac{1}{2}Cr = \frac{1}{2}(2\pi r)r = \pi r^2$$

The second proposition of Archimedes' work *Measure of the Circle* develops a relationship between the area of a circle and the area of the square in which it is inscribed. (See Figure 6.86.) Specifically, Archimedes claimed that the ratio of the area of the circle to that of the square was 11:14. This leads to the following set of equations and to an approximation of the value of $\pi$.

$$\frac{\pi r^2}{(2r)^2} \approx \frac{11}{14}$$

$$\frac{\pi r^2}{4r^2} \approx \frac{11}{14}$$

$$\frac{\pi}{4} \approx \frac{11}{14}$$

$$\pi \approx 4 \cdot \frac{11}{14} \approx \frac{22}{7}$$

**FIGURE 6.86**

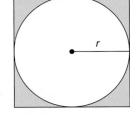

Archimedes later improved his approximation of $\pi$ by showing that

$$3\frac{10}{71} < \pi < 3\frac{1}{7}$$

Today's calculators provide excellent approximations for the irrational number $\pi$. We should recall, however, that $\pi$ is an irrational number that can be expressed exactly only by the unique symbol $\pi$.

# Perspective on Application

## THE NINE-POINT CIRCLE

In the study of geometry, there is a curiosity known as the Nine-Point Circle—a curiosity because its practical value consists of the reasoning needed to verify its plausibility.

In $\triangle ABC$, we locate these points:

> $M$, $N$, and $P$, the midpoints of the sides of $\triangle ABC$, $D$, $E$, and $F$, points on $\triangle ABC$ determined by its altitudes, and $X$, $Y$, and $Z$, the midpoints of the line segments determined by orthocenter $O$ and the vertices of $\triangle ABC$.

Through these nine points, it is possible to draw or construct the circle shown. See Figure 6.87.

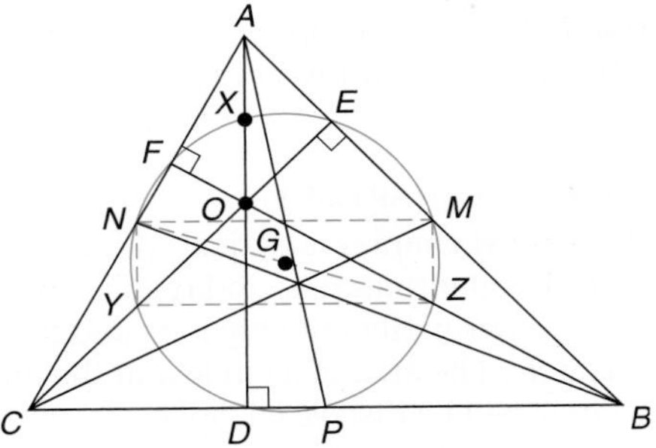

**FIGURE 6.87**

Consider the following drawing, in which we reason from Figure 6.87. To understand why the nine-point circle can be drawn, we show that the quadrilateral $NMZY$ is both a parallelogram and a rectangle. Because $\overline{NM}$ joints the midpoints of $\overline{AC}$ and $\overline{AB}$, we know that $\overline{NM} \parallel \overline{CB}$ and $NM = \frac{1}{2}(CB)$. Likewise, $Y$ and $Z$ are midpoints of the sides of $\triangle OBC$, so $\overline{YZ} \parallel \overline{CB}$ and $YZ = \frac{1}{2}(CB)$. By Theorem 4.2.1, $NMZY$ is a parallelogram. Then $\overline{NY}$ must be parallel to $\overline{MZ}$. With $\overline{CB} \perp \overline{AD}$, it follows that $\overline{NM}$ must be perpendicular to $\overline{AD}$ as well. In turn, $\overline{MZ} \perp \overline{NM}$ and $NMZY$ is a rectangle. See Figure 6.88. It is possible to circumscribe a circle about any rectangle; in fact, the length of the radius of the circumscribed circle is one-half the length of a diagonal of the rectangle, so we choose $r = \frac{1}{2}(NZ) = NG$. This circle certainly contains the points $N$, $M$, $Z$, and $Y$.

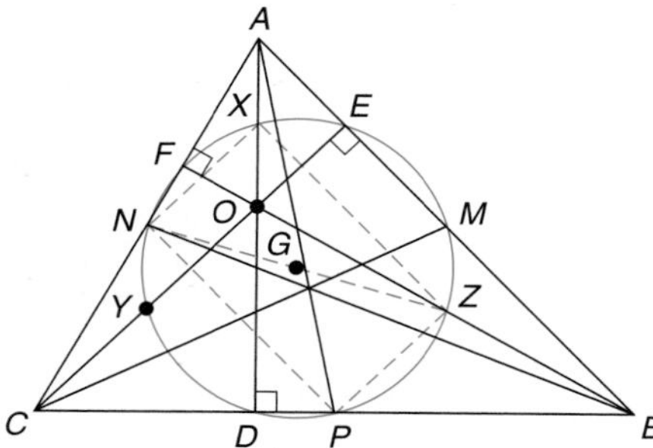


**FIGURE 6.88**

Although we do not provide the details, it can also be shown that quadrilateral $XZPN$ is a rectangle as well. Further, $\overline{NZ}$ is also a diagonal of rectangle $XZPN$. Then we can choose the radius of the circumscribed circle for rectangle $XZPN$ to have the length $r = \frac{1}{2}(NZ) = NG$. See Figure 6.88. Having the same center $G$ and the same length of radius $r$ as the circle that was circumscribed about rectangle $NMZY$, we see that the same circle must contain points $N$, $X$, $M$, $Z$, $P$, and $Y$.

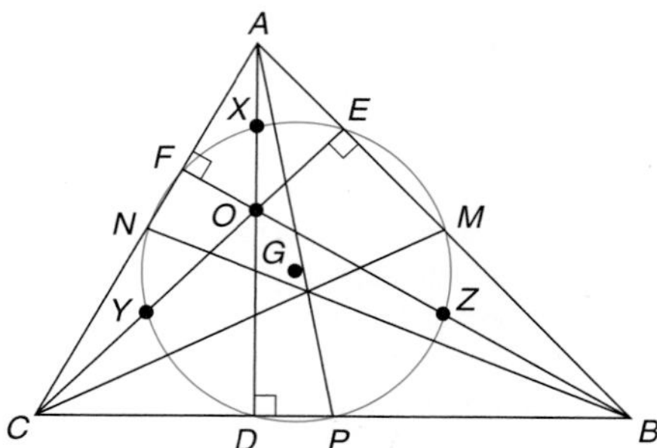


**FIGURE 6.89**

Finally, we need to show that the circle in Figure 6.89 with center $G$ and radius $r = \frac{1}{2}(NZ)$ will contain the points $D$, $E$, and $F$. This can be done by an indirect argument. If we suppose that these points do *not* lie on the circle, then we contradict the fact that an angle inscribed in a semicircle must be a right angle. Of course, $\overline{AD}$, $\overline{BF}$, and $\overline{CE}$ were altitudes of $\triangle ABC$, so inscribed angles at $D$, $E$, and $F$ must measure 90°; in turn, these angles must lie inside semicircles, so $D$, $E$, and $F$ are on the same circle that has center $G$ and radius $r$. Thus, the circle described in the preceding paragraphs is the anticipated nine-point circle!

# Summary

**A LOOK BACK AT CHAPTER 6**

One goal in this chapter has been to classify angles inside, on, and outside the circle. Formulas for finding the measures of these angles were developed. Line and line segments related to a circle were defined, and some ways of finding the measures of these segments were described. Theorems involving inequalities in a circle were proved. Using the concept of locus, we justified several of the basic constructions as well as the concurrence of lines.

**A LOOK AHEAD TO CHAPTER 7**

Our goal in the next chapter is to deal with the areas of triangles, certain quadrilaterals, and regular polygons. We will consider perimeters of polygons and the circumference of a circle. The area of a circle and the area of a sector of a circle will be discussed. Special right triangles will play an important role in determining the areas of these plane figures.

**KEY CONCEPTS**

**6.1** Circle • Congruent Circles • Concentric Circles • Center of the Circle • Radius • Diameter • Chord • Semicircle • Arc • Major Arc • Minor Arc • Intercepted Arc • Congruent Arcs • Central Angle • Inscribed Angle

**6.2** Tangent • Point of Tangency • Secant • Polygon Inscribed in a Circle • Circumscribed Circle • Polygon Circumscribed about a Circle • Inscribed Circle • Interior and Exterior of a Circle

**6.3** Tangent Circles • Internally Tangent Circles • Externally Tangent Circles • Line of Centers • Common Tangent • Common External Tangents • Common Internal Tangents

**6.4** Constructions of Tangents to a Circle • Inequalities in the Circle

**6.5** Locus of Points in a Plane • Locus of Points in Space

**6.6** Concurrent Lines • Incenter • Incircle • Circumcenter • Circumcircle • Orthocenter • Centroid

## TABLE 6.2 AN OVERVIEW OF CHAPTER SIX

### Selected Properties of Circles

| *Figure* | *Angle Measure* | *Segment Relationships* |
|---|---|---|
| Central angle 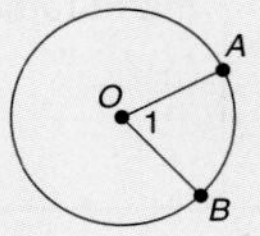<br>O, 1, A, B | $m\angle 1 = m\widehat{AB}$ | $OA = OB$ |
| Inscribed angle 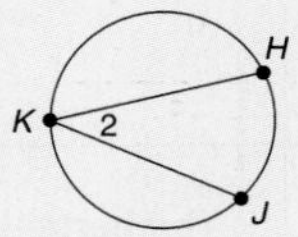<br>K, 2, H, J | $m\angle 2 = \frac{1}{2}m\widehat{HJ}$ | Generally, $HK \neq KJ$ |
| Angle formed by intersecting chords 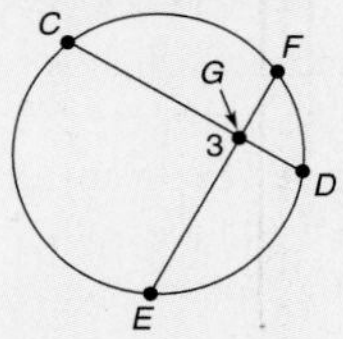<br>C, F, G, 3, D, E | $m\angle 3 = \frac{1}{2}(m\widehat{CE} + m\widehat{FD})$ | $CG \cdot GD = EG \cdot GF$ |
| Angle formed by intersecting secants 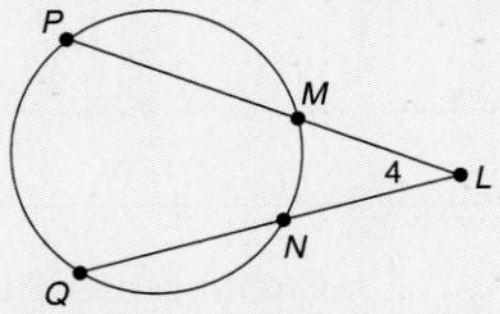<br>P, M, 4, L, N, Q | $m\angle 4 = \frac{1}{2}(m\widehat{PQ} - m\widehat{MN})$ | $PL \cdot LM = QL \cdot LN$ |
| Angle formed by intersecting tangents 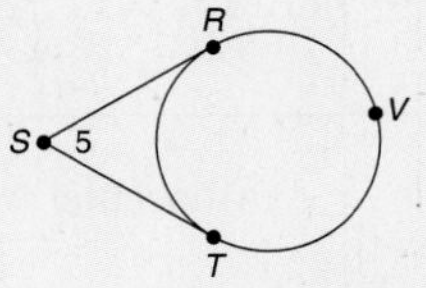<br>R, S, 5, V, T | $m\angle 5 = \frac{1}{2}(m\widehat{RVT} - m\widehat{RT})$ | $SR = ST$ |
| Angle formed by radius drawn to tangent 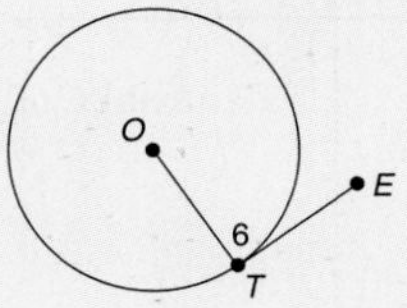<br>O, E, 6, T | $m\angle 6 = 90°$ | $\overline{OT} \perp \overline{TE}$ |

*continued*

**TABLE 6.2 CONT.**

## Selected Locus Problems (in a plane)

| Locus | Figure | Description |
|---|---|---|
| Locus of points that are at a fixed distance $r$ from fixed point $P$ | 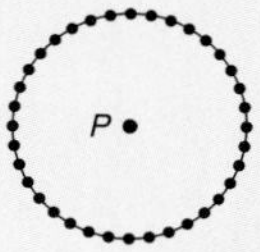 | The circle with center $P$ and radius $r$ |
| Locus of points that are equidistant from the sides of an angle | 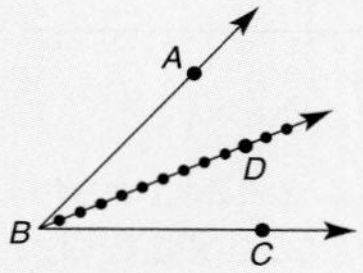 | The angle bisector $\overrightarrow{BD}$ of $\angle ABC$ |
| Locus of points that are equidistant from the endpoints of a line segment | 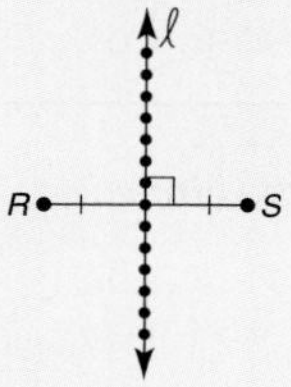 | The perpendicular bisector $\ell$ of $\overline{RS}$ |

## Concurrence of Lines (in a triangle)

| Type of Lines | Figure | Point of Concurrence |
|---|---|---|
| Angle bisectors | 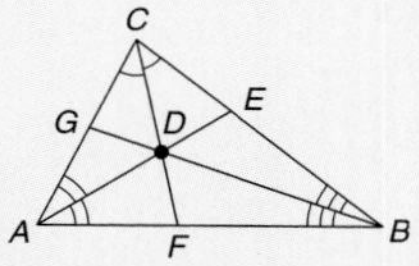 | Incenter $D$ of $\triangle ABC$ |
| Perpendicular bisectors of the sides | 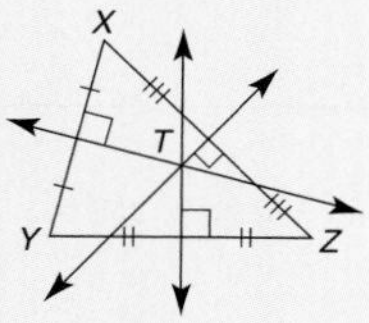 | Circumcenter $T$ of $\triangle XYZ$ |
| Altitudes | 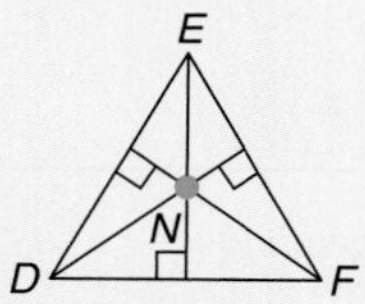 | Orthocenter $N$ of $\triangle DEF$ |
| Medians | 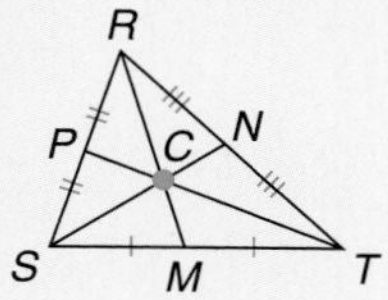 | Centroid $C$ of $\triangle RST$ |

# Chapter 6 Review Exercises

1. The radius of a circle is 15 mm. The length of a chord is 24 mm. Find the distance from the center of the circle to the chord.
2. Find the length of a chord that is 8 cm from the center of a circle that has a radius of 17 cm.
3. Two circles intersect and have a common chord 10 in. long. The radius of one circle is 13 in. and the centers of the circles are 16 in. apart. Find the radius of the other circle.
4. Two circles intersect and they have a common chord 12 cm long. The measure of the angles formed by the common chord and a radius of each circle to the points of intersection of the circles is 45°. Find the radius of each circle.

*In Review Exercises 5 to 10, $\overrightarrow{BA}$ is tangent to the circle in the figure shown.*

5. $m\angle B = 25°$, $m\widehat{AD} = 140°$, $m\widehat{DC} = ?$

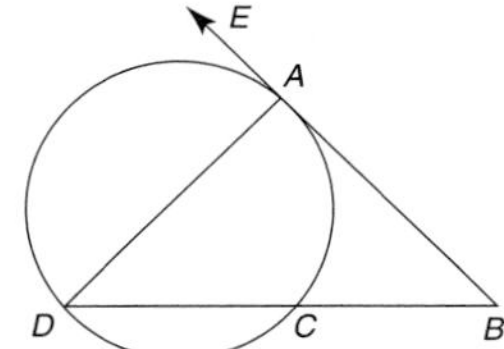


*Exercises 5–10*

6. $m\widehat{ADC} = 310°$, $m\widehat{AD} = 120°$, $m\angle B = ?$
7. $m\angle EAD = 70°$, $m\angle B = 30°$, $m\widehat{AC} = ?$
8. $m\angle D = 40°$, $m\widehat{DC} = 130°$, $m\angle B = ?$
9. *Given:* $C$ is the midpoint of $\widehat{ACD}$ and $m\angle B = 40°$
   *Find:* $m\widehat{AD}$, $m\widehat{AC}$, $m\widehat{DC}$
10. *Given:* $m\angle B = 35°$ and $m\widehat{DC} = 70°$
    *Find:* $m\widehat{AD}$, $m\widehat{AC}$
11. *Given:* $\odot O$ with tangent $\ell$ and $m\angle 1 = 46°$
    *Find:* $m\angle 2$, $m\angle 3$, $m\angle 4$, $m\angle 5$

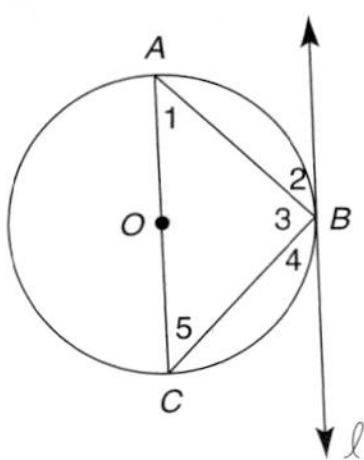


*Exercises 11, 12*

12. *Given:* $\odot O$ with tangent $\ell$ and $m\angle 5 = 40°$ (See the figure for Exercise 11.)
    *Find:* $m\angle 1$, $m\angle 2$, $m\angle 3$, $m\angle 4$
13. Two circles are concentric. A chord of the larger circle is also tangent to the smaller circle. The radius of one circle is 20, and the radius of the other is 16. Find the length of the chord.
14. Two parallel chords of a circle each have length 16. The distance between these chords is 12. Find the radius of the circle.

*In Review Exercises 15 to 22, state whether the statements are always true (A), sometimes true (S), or never true (N).*

15. In a circle, congruent chords are equidistant from the center.
16. If a triangle is inscribed in a circle and one of its sides is a diameter, then the triangle is an isosceles triangle.
17. If a central angle and an inscribed angle of a circle intercept the same arc, then they are congruent.
18. A trapezoid can be inscribed in a circle.
19. If a parallelogram is inscribed in a circle, then each of its diagonals must be a diameter.
20. If two chords of a circle are not congruent, then the shorter chord is nearer the center of the circle.
21. Tangents to a circle at the endpoints of a diameter are parallel.
22. Two concentric circles have at least one point in common.
23. a) $m\widehat{AB} = 80°$, $m\angle AEB = 75°$, $m\widehat{CD} = ?$
    b) $m\widehat{AC} = 62°$, $m\angle DEB = 45°$, $m\widehat{BD} = ?$
    c) $m\widehat{AB} = 88°$, $m\angle P = 24°$, $m\angle CED = ?$
    d) $m\angle CED = 41°$, $m\widehat{CD} = 20°$, $m\angle P = ?$
    e) $m\angle AEB = 65°$, $m\angle P = 25°$, $m\widehat{AB} = ?$, $m\widehat{CD} = ?$
    f) $m\angle CED = 50°$, $m\widehat{AC} + m\widehat{BD} = ?$

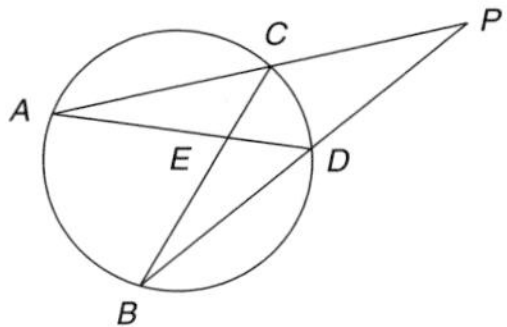

24. Given that $\overline{CF}$ is a tangent to the circle shown:

a) $CF = 6$, $AC = 12$, $BC = ?$
b) $AG = 3$, $BE = 10$, $BG = 4$, $DG = ?$
c) $AC = 12$, $BC = 4$, $DC = 3$, $CE = ?$
d) $AG = 8$, $GD = 5$, $BG = 10$, $GE = ?$
e) $CF = 6$, $AB = 5$, $BC = ?$
f) $EG = 4$, $GB = 2$, $AD = 9$, $GD = ?$
g) $AC = 30$, $BC = 3$, $CD = ED$, $ED = ?$
h) $AC = 9$, $BC = 5$, $ED = 12$, $CD = ?$
i) $ED = 8$, $DC = 4$, $FC = ?$
j) $FC = 6$, $ED = 9$, $CD = ?$

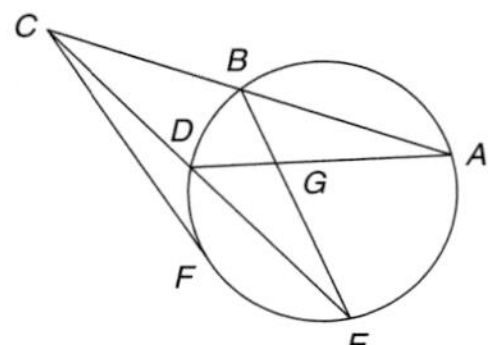


25. *Given:* $\overline{DF} \cong \overline{AC}$ in $\odot O$
$OE = 5x + 4$
$OB = 2x + 19$
*Find:* $OE$

26. *Given:* $\overline{OE} \cong \overline{OB}$ in $\odot O$
$DF = x(x - 2)$
$AC = x + 28$
*Find:* $DE$ and $AC$

*Exercises 25, 26*

*In Review Exercises 27 to 29, give a proof for each statement.*

27. *Given:* $\overline{DC}$ is tangent to circles $B$ and $A$ at points $D$ and $C$, respectively
*Prove:* $AC \cdot ED = CE \cdot BD$

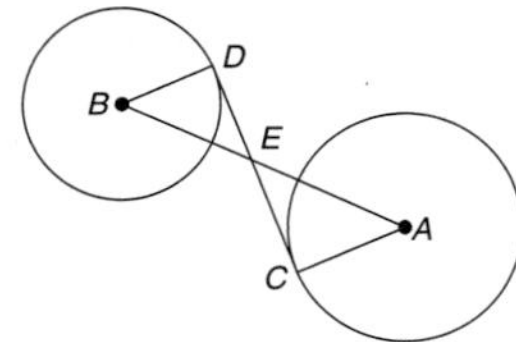


28. *Given:* $\odot O$ with $\overline{EO} \perp \overline{BC}$, $\overline{DO} \perp \overline{BA}$, $\overline{EO} \cong \overline{OD}$
*Prove:* $\widehat{BC} \cong \widehat{BA}$

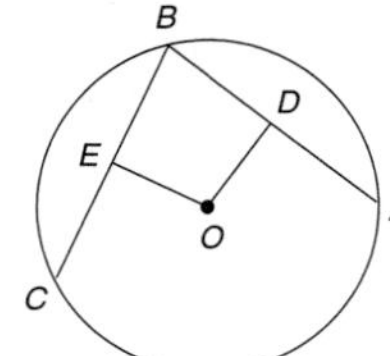


29. *Given:* $\overline{AP}$ and $\overline{BP}$ are tangent to $\odot Q$ at $A$ and $B$
$C$ is the midpoint of $\widehat{AB}$
*Prove:* $\overrightarrow{PC}$ bisects $\angle APB$

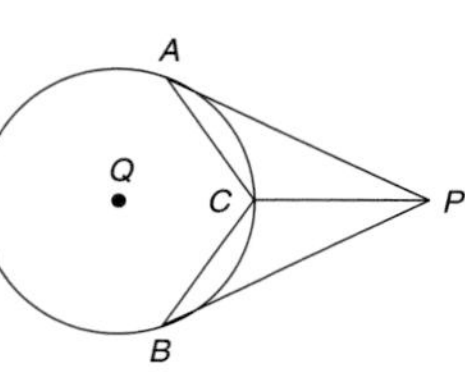


30. *Given:* $\odot O$ with diameter $\overline{AC}$ and tangent $\overleftrightarrow{DE}$
$m\widehat{AD} = 136°$ and $m\widehat{BC} = 50°$
*Find:* The measures of the angles, $\angle 1$ through $\angle 10$

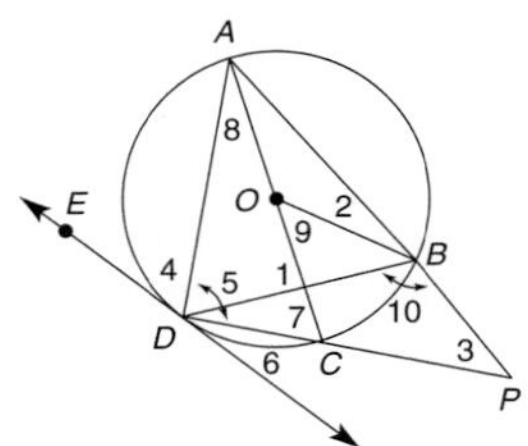


31. A square is inscribed in a circle with a radius of 6 cm. Find the perimeter of the square.

32. A 30°-60°-90° triangle is inscribed in a circle with a radius of 5 cm. Find the perimeter of the triangle.

33. A circle is inscribed in a right triangle. The radius of the circle is 6 cm, and the hypotenuse is 29 cm. Find the lengths of the two segments of the hypotenuse that are determined by the point of tangency.

34. *Given:* $\odot O$ is inscribed in $\triangle ABC$
$AB = 9$, $BC = 13$, $AC = 10$
*Find:* $AD$, $BE$, $FC$

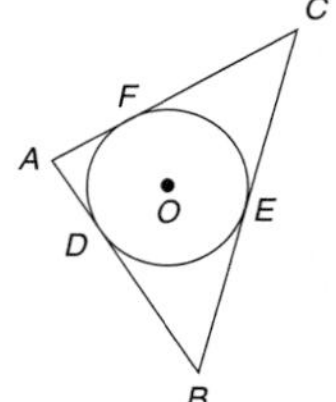


35. In $\odot Q$ with $\triangle ABQ$ and $\triangle CDQ$, $m\widehat{AB} > m\widehat{CD}$. Also, $\overline{QP} \perp \overline{AB}$ and $\overline{QR} \perp \overline{CD}$.

a) How are $AB$ and $CD$ related?
b) How are $QP$ and $QR$ related?
c) How are $\angle A$ and $\angle C$ related?

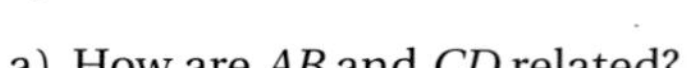

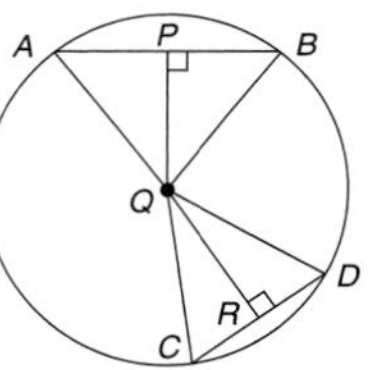


36. In $\odot O$ (not shown), secant $\overleftrightarrow{AB}$ intersects the circle at $A$ and $B$; $C$ is a point on $\overleftrightarrow{AB}$ in the exterior of the circle.

a) Construct the tangent to $\odot O$ at point $B$.
b) Construct the tangents to $\odot O$ from point $C$.

*In Review Exercises 37 and 38, use the figure shown.*

37. Construct a right triangle so that one leg has length $AB$ and the other has length twice $AB$.

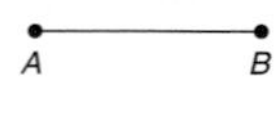


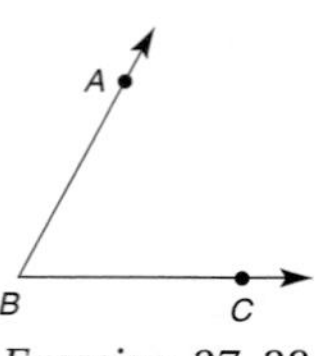


*Exercises 37, 38*

38. Construct a rhombus with side $\overline{AB}$ and $\angle ABC$.

*In Review Exercises 39 to 41, sketch and describe the locus in a plane.*

39. Find the locus of the midpoints of the radii of a circle.

40. Find the locus of the centers of all circles passing through two given points.

41. What is the locus of the center of a penny that rolls around and remains tangent to a half-dollar?

*In Exercises 42 and 43, sketch and describe the locus in space.*

42. Find the locus of points less than 3 units from a given point.

43. Find the locus of points equidistant from two parallel planes.

*In Review Exercises 44 to 49, use construction methods with the accompanying figure.*

44. *Given:* $\triangle ABC$
    *Find:* The incenter

45. *Given:* $\triangle ABC$
    *Find:* The circumcenter

46. *Given:* $\triangle ABC$
    *Find:* The orthocenter

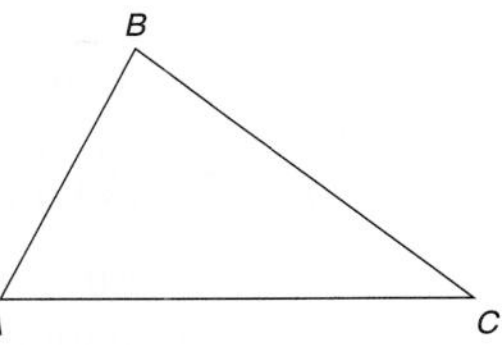

*Exercises 44–49*

47. *Given:* $\triangle ABC$
    *Find:* The centroid

48. Use the result from Exercise 44 to inscribe a circle in $\triangle ABC$.

49. Use the result from Exercise 45 to circumscribe a circle about $\triangle ABC$.

50. *Given:* $\triangle ABC$ with medians $\overline{AE}$, $\overline{DC}$, $\overline{BF}$
    *Find:* a) $BG$ if $BF = 18$
    b) $GE$ if $AG = 4$
    c) $DG$ if $CG = 4\sqrt{3}$

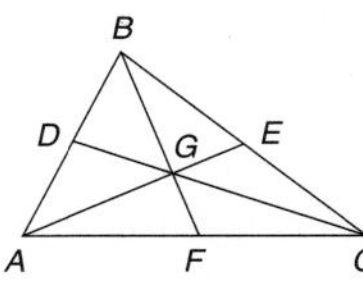

*Exercises 50, 51*

51. *Given:* $\triangle ABC$ with medians $\overline{AE}$, $\overline{DC}$, $\overline{BF}$
    $AG = 2x + 2y$, $GE = 2x - y$
    $BG = 3y + 1$, $GF = x$
    *Find:* $BF$ and $AE$

## Chapter 6 Test

1. a) If $m\widehat{AB} = 88°$, then $m\widehat{ACB} =$ ________.
   b) If $m\widehat{AB} = 92°$ and $C$ is the midpoint of major arc $ACB$, then $m\widehat{AC} =$ ________.

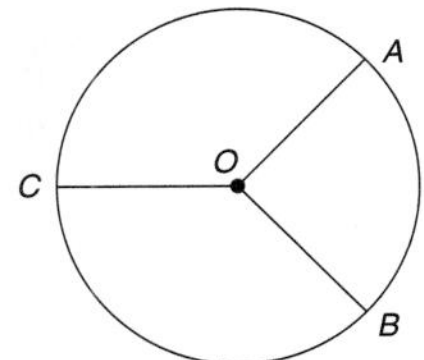


2. a) If $m\widehat{BC} = 69°$, then $m\angle BOC =$ ________.
   b) If $m\widehat{BC} = 64°$, then $m\angle BAC =$ ________.

3. a) If $m\angle BAC = 24°$, then $m\widehat{BC} =$ ________.
   b) If $\widehat{AB} \cong \widehat{AC}$, then $\triangle ABC$ is a(n) ____________ triangle.

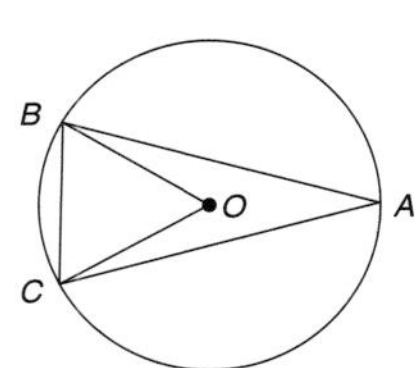

*Exercises 2, 3*

4. Complete each theorem:
   a) An angle inscribed in a semicircle is a(n) ____________ angle.
   b) The two tangent segments drawn to a circle from an external point are ____________.

5. Given that $m\widehat{AB} = 106°$ and $m\widehat{DC} = 32°$, find:
   a) $m\angle 1$ ________
   b) $m\angle 2$ ________

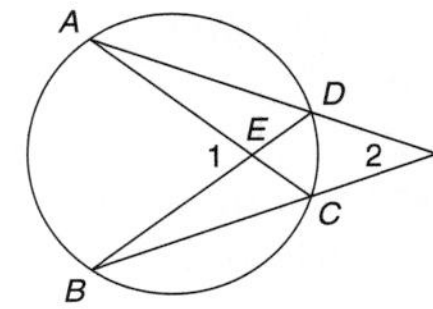

6. Given that m$\widehat{RT}$ = 146°, find:

   a) m$\widehat{RST}$ ________
   b) m∠3 ________

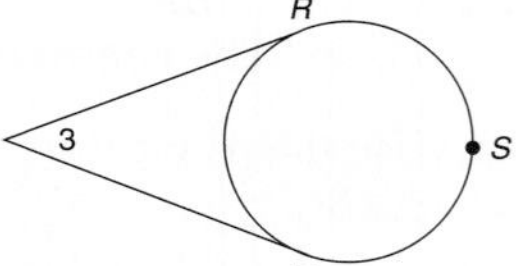

7. a) Because point $Q$ is their common center, these circles are known as ____________ circles.

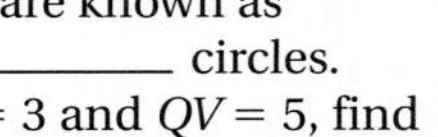

   b) If $RQ = 3$ and $QV = 5$, find the length of chord $\overline{TV}$. ________

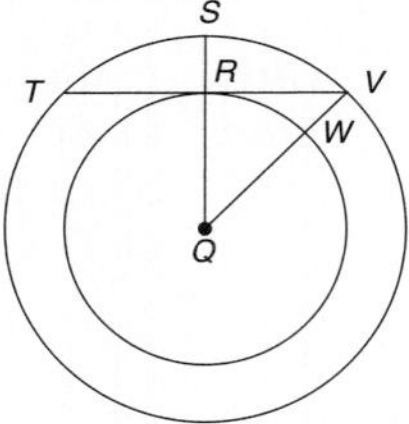

8. For the circles described and shown, how many common tangents do they possess?

   a) Internally tangent circles ________

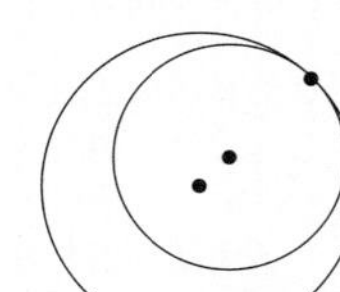

   b) Circles that intersect in two points ________

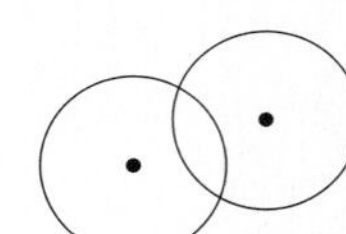

9. a) If $HP = 4$, $PJ = 5$, and $PM = 2$, find $LP$. ________
   b) If $HP = x + 1$, $PJ = x - 1$, $LP = 8$, and $PM = 3$, find $x$.
   ________

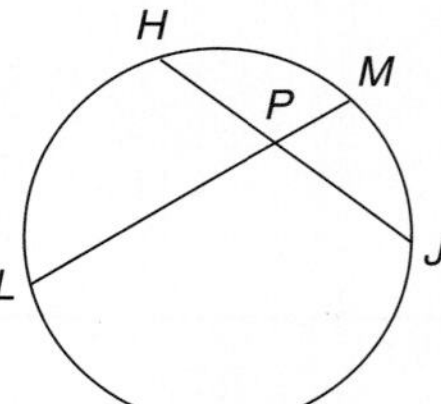

10. Construct the tangent line to ⊙$P$ at point $X$.

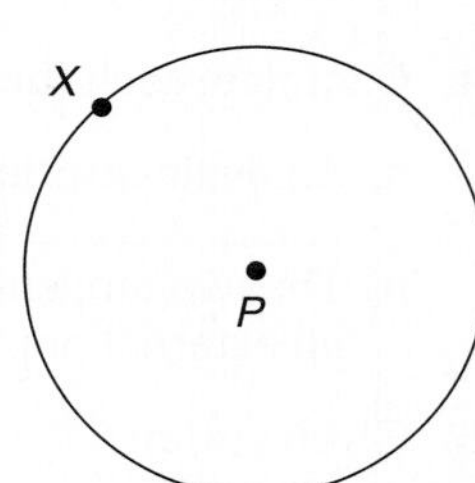

11. a) If m$\widehat{AB}$ > m$\widehat{CD}$, write an inequality that compares m∠$AQB$ and m∠$CQD$.
    ____________
    b) If $QR > QP$, write an inequality that compares $AB$ and $CD$.
    ____________

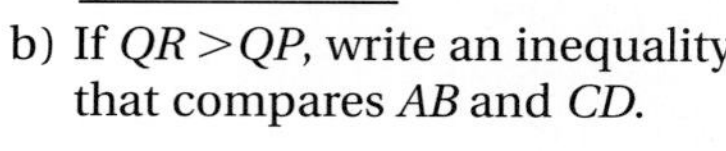

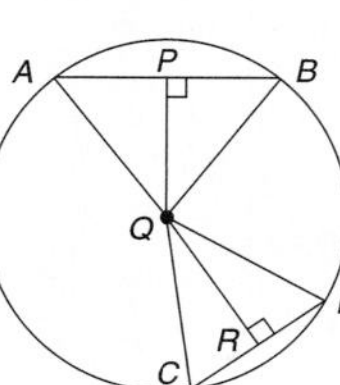

12. a) Describe the locus of points in a plane that are located at a distance of 3 centimeters from point $P$.

    ________________________________________

    ________________________________________

    b) Describe the locus of points in space that are located at a distance of 3 centimeters from point $P$.

    ________________________________________

    ________________________________________

13. a) Describe the locus of points in a plane that are equidistant from points $A$ and $B$.

    ________________________________________

    ________________________________________

    b) Describe the locus of points in space that are equidistant from points $A$ and $B$.

    ________________________________________

    ________________________________________

14. For a given triangle (such as △$ABC$), what is the name of the point of concurrency for

    a) the three angle bisectors? ____________
    b) the three medians? ____________

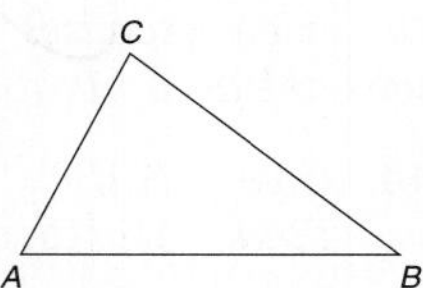

15. For △$MNP$, the three medians are concurrent at point $Q$.

    a) If $MQ = 12$, find $MB$. ________
    b) If $PQ = x + 4$ and $QC = x$, find $x$. ________

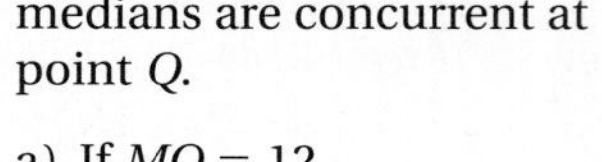

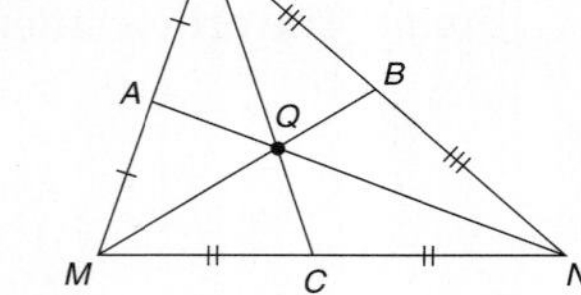

16. Complete the missing statements and reasons in the following proof.

    *Given:* In ⊙$O$, chords $\overline{AD}$ and $\overline{BC}$ intersect at $E$.
    *Prove:* $\frac{AE}{CE} = \frac{BE}{DE}$

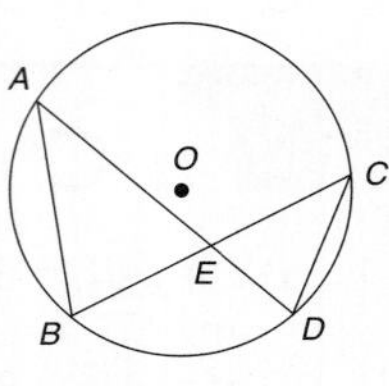

**PROOF**

| Statements | Reasons |
|---|---|
| 1. ____________ | 1. ____________ |
| 2. ∠$AEB$ ≅ ∠$DEC$ | 2. ____________ |
| 3. ∠$B$ ≅ ∠$D$ | 3. If two inscribed angles intercept the same arc, these angles are congruent |
| 4. △$ABE$ ~ △$CDE$ | 4. ____________ |
| 5. ____________ | 5. CSSTP |

# Chapter 7

# Areas of Polygons and Circles

This aerial photograph provides a panoramic view of the Chateau de Vaux-le-Vicomte and its gardens in France, which date back to the year 1661. The rectangular and square plots shown have a measurable size, known as area. In this chapter, we calculate the area of a region in units such as square feet or square yards. For example, see Exercise 24 in Section 7.1 and Exercise 35 in Section 7.2. We will explore many area applications in Chapter 7.

## Chapter Outline



**For online student resources, visit this textbook's website at math.college.hmco.com/students.**

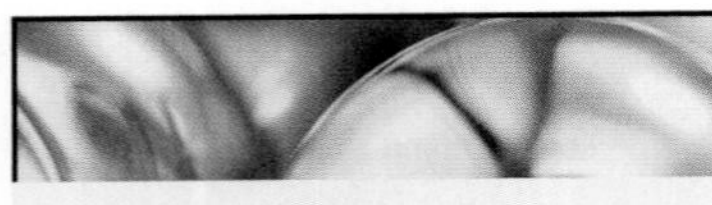

# 7.1 Area and Initial Postulates

**KEY CONCEPTS**

Plane Region • Square Unit • Area Postulates • Area of Rectangle, Parallelogram, and Triangle • Altitude and Base of a Parallelogram or Triangle

A line segment is measured in linear units such as inches, centimeters, or yards. When a line segment measures 5 centimeters, we write $AB = 5$ cm or simply $AB = 5$ (if the units are apparent or are not stated). The instrument of measure is the ruler.

Lines are *one-dimensional*; that is, we speak only of the dimension "length" when measuring a line segment. On the other hand, a plane is an infinite *two-dimensional* surface. A closed or bounded portion of the plane is called a **region.**

When a region such as $R$ in plane $M$ [see Figure 7.1(a)] is measured, we call this measure the "area of the plane region." The unit used to measure area is called a **square unit** because it is a square with each side of length 1. The measure of the area of region $R$ is the number of non-overlapping square units that can be placed in (fit into) the region.

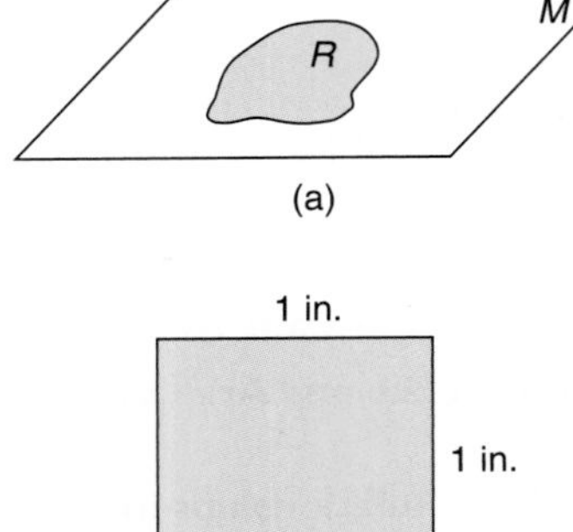


**FIGURE 7.1**

> Square units (not linear units) are used to measure area. Using an exponent, we write square inches as $in^2$. The unit represented by Figure 7.1(b) is 1 square inch or 1 $in^2$.

One application of area involves measuring the floor area to be covered by carpeting, which is often measured in square yards ($yd^2$). Another application involves calculating the number of squares of shingles needed to cover a roof. (Here a "square" is the number of shingles needed to cover a 100-$ft^2$ section of the roof.)

In Figure 7.2, the regions have measurable areas and are bounded by figures encountered in earlier chapters. A region is **bounded** if we can distinguish between its interior and its exterior; in calculating area, we measure the amount of interior region.

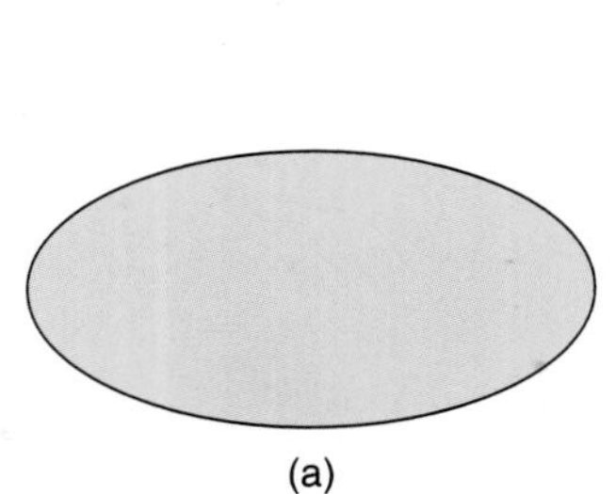
(a)

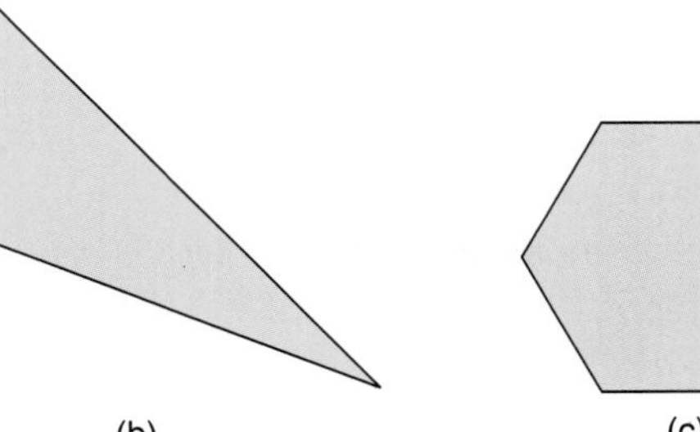
(b)

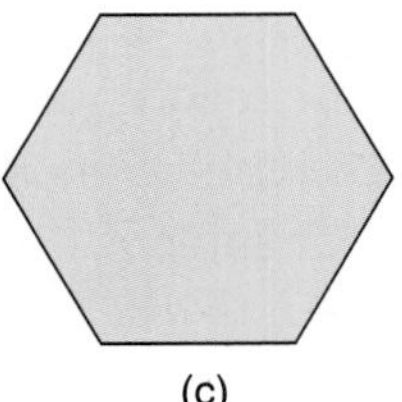
(c)

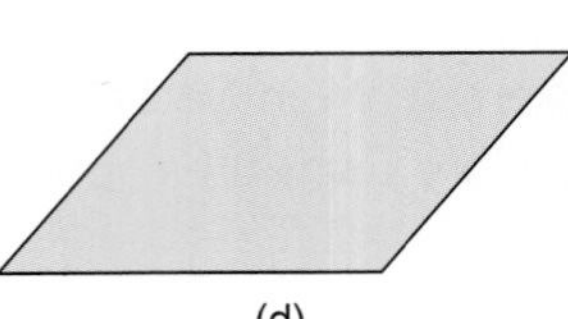
(d)

**FIGURE 7.2**

> We can measure the area of the region within a triangle [see Figure 7.2(b)]. However, we cannot actually measure the area of the triangle itself (three line segments do not have area). Nonetheless, the area of the region within a triangle is commonly referred to as the *area of the triangle.*

The preceding discussion does not formally define a region or its area. These are accepted as the undefined terms in the following postulate.

**POSTULATE 18: (Area Postulate)**
Corresponding to every bounded region is a unique positive number $A$, known as the area of that region.

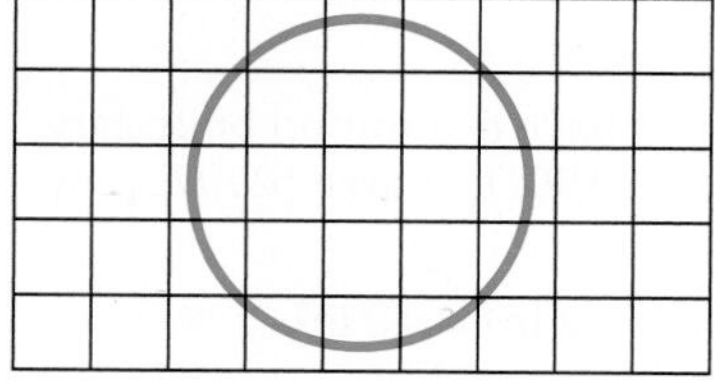

**FIGURE 7.3**

One way to estimate the area of a region is to place it in a grid, as shown in Figure 7.3. Counting only the number of whole squares inside the region gives an approximation that is less than the actual area. On the other hand, counting squares that are inside or partially inside provides an approximation that is greater than the actual area. A fair estimate of the area of a region is often given by the average of the smaller and larger approximations just described. If the area of the circle shown in Figure 7.3 is between 9 and 21 square units, we might estimate its area to be $\frac{9 + 21}{2}$ or 15 square units.

To develop another property of area, we consider $\triangle ABC$ and $\triangle DEF$ (which are congruent) in Figure 7.4. One triangle can be placed over the other so that they coincide. How are the areas of the two triangles related? The answer is found in the following postulate.

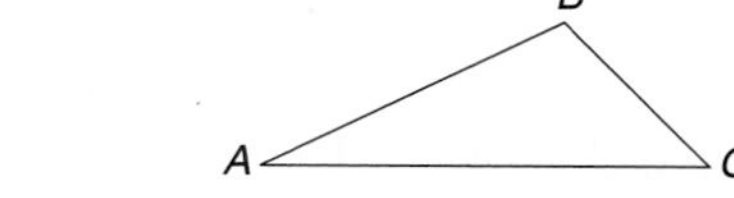


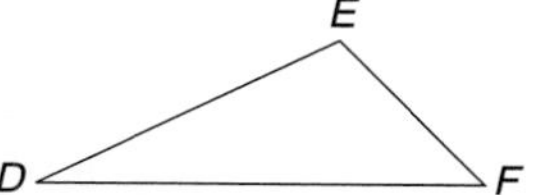


**FIGURE 7.4**

**Discover!**

Complete this analogy: An inch is to the length of a line segment as a __?__ __?__ is to the area of a plane region.

ANSWER

Square inch

**POSTULATE 19:**
If two closed plane figures are congruent, then their areas are equal.

**EXAMPLE 1**

In Figure 7.5, points $B$ and $C$ trisect $\overline{AD}$; $\overline{EC} \perp \overline{AD}$. Name two triangles with equal areas.

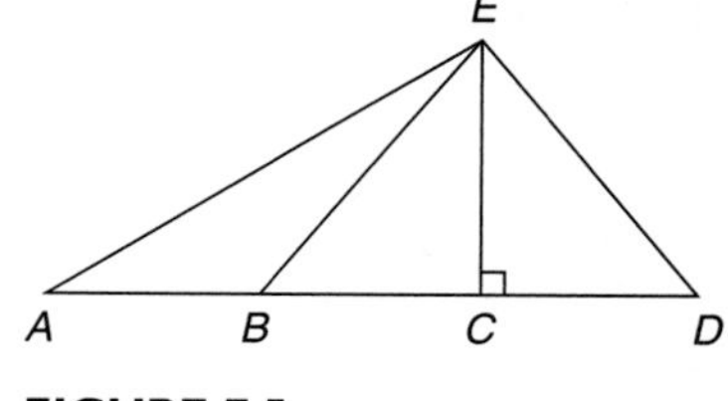


**FIGURE 7.5**

***Solution***

$\triangle ECB \cong \triangle ECD$ by SAS. Then $\triangle ECB$ and $\triangle ECD$ have equal areas according to Postulate 19.

NOTE: $\triangle EBA$ is also equal in area to $\triangle ECB$ and $\triangle ECD$, but this relationship cannot be established until we consider Theorem 7.1.3.

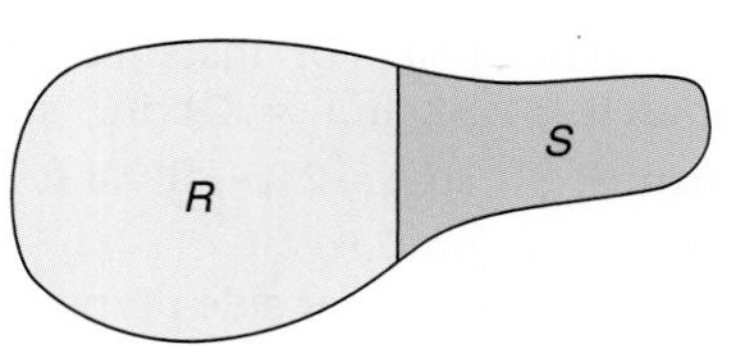


**FIGURE 7.6**

Consider Figure 7.6. The entire region is bounded by a curve and then subdivided by a line segment into smaller regions $R$ and $S$. These regions have a common boundary and do not overlap. Because a numerical area can be associated with each region $R$ and $S$, the area of $R \cup S$ (read as "$R$ union $S$" and meaning region $R$ joined to region $S$) is equal to the sum of the areas of $R$ and $S$. This leads to Postulate 20, in which $A_R$ represents the "area of region $R$," $A_S$ represents the "area of region $S$," and $A_{R \cup S}$ represents the "area of region $R \cup S$."

**POSTULATE 20: (Area-Addition Postulate)**
Let $R$ and $S$ be two enclosed regions that do not overlap. Then

$$A_{R \cup S} = A_R + A_S$$

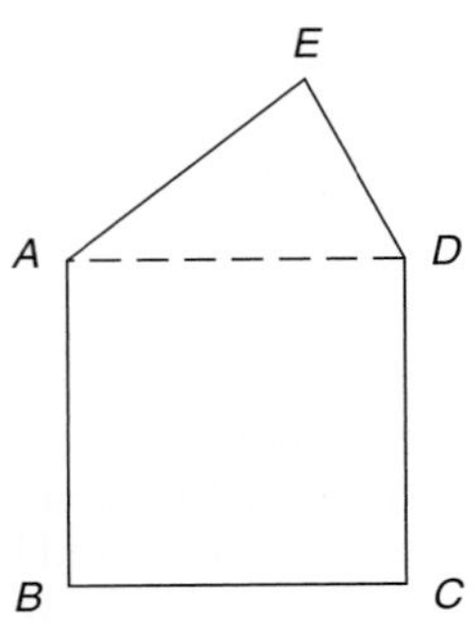


**FIGURE 7.7**

EXAMPLE 2

In Figure 7.7, the pentagon *ABCDE* is formed by square *ABCD* and $\triangle ADE$. If the area of the square is 36 in$^2$ and that of $\triangle ADE$ is 12 in$^2$, find the area of pentagon *ABCDE*.

***Solution***

Because square *ABCD* and $\triangle ADE$ do not overlap and have a common boundary $\overline{AD}$, we have Area (pentagon *ABCDE*) = area (square *ABCD*) + area ($\triangle ADE$). By the Area-Addition Postulate,

$$\text{Area (pentagon } ABCDE) = 36 \text{ in}^2 + 12 \text{ in}^2 = 48 \text{ in}^2$$

It is convenient to provide a subscript for *A* (area) that names the figure whose area is indicated. The principle used in Example 2 is conveniently and compactly stated in the form

$$A_{ABCDE} = A_{ABCD} + A_{ADE}$$

*Exs. 1–5*

## AREA OF A RECTANGLE

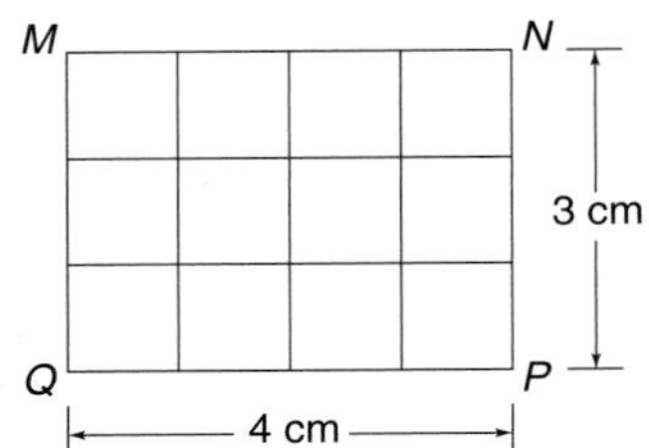


**FIGURE 7.8**

Discover!

Study rectangle *MNPQ* in Figure 7.8, and note that it has dimensions of 3 cm and 4 cm. The number of squares, 1 cm on a side, in the rectangle is 12. Rather than count the number of squares in the figure, how can you calculate the area?

ANSWER

Multiply 3 × 4 = 12.

In the preceding Discover!, the unit of area is cm$^2$. Multiplication of dimensions is handled like algebraic multiplication. Compare

$$3x \cdot 4x = 12x^2 \qquad \text{and} \qquad 3 \text{ cm} \cdot 4 \text{ cm} = 12 \text{ cm}^2$$

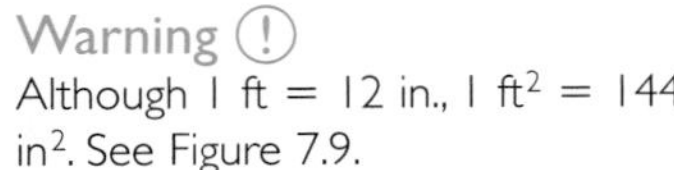

Warning (!)
Although 1 ft = 12 in., 1 ft$^2$ = 144 in$^2$. See Figure 7.9.

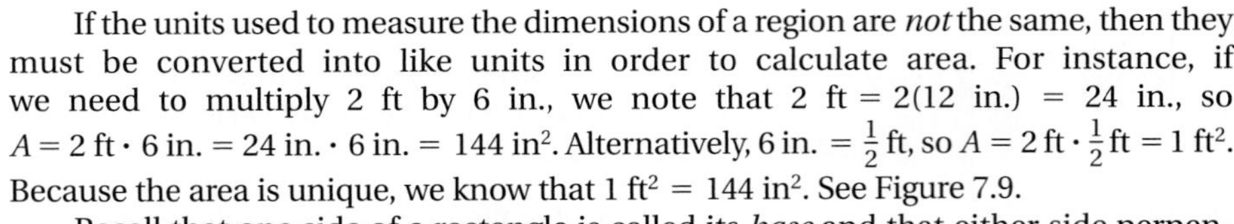

If the units used to measure the dimensions of a region are *not* the same, then they must be converted into like units in order to calculate area. For instance, if we need to multiply 2 ft by 6 in., we note that 2 ft = 2(12 in.) = 24 in., so $A = 2 \text{ ft} \cdot 6 \text{ in.} = 24 \text{ in.} \cdot 6 \text{ in.} = 144 \text{ in}^2$. Alternatively, 6 in. $= \frac{1}{2}$ ft, so $A = 2 \text{ ft} \cdot \frac{1}{2} \text{ ft} = 1 \text{ ft}^2$. Because the area is unique, we know that 1 ft$^2$ = 144 in$^2$. See Figure 7.9.

Recall that one side of a rectangle is called its *base* and that either side perpendicular to the base is called the *altitude* of the rectangle.

POSTULATE 21:
The area *A* of a rectangle whose base has length *b* and whose altitude has length *h* is given by $A = bh$.

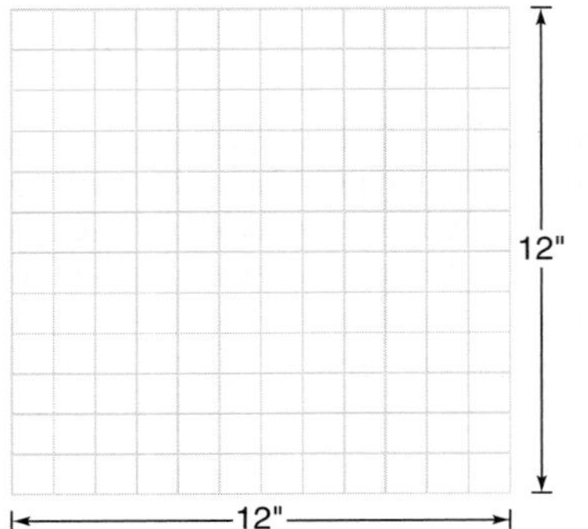


**FIGURE 7.9**

It is also common to describe the dimensions of a rectangle as its length $\ell$ and its width *w*. The area of the rectangle is then written $A = \ell w$.

**FIGURE 7.10**

EXAMPLE 3

Find the area of rectangle $ABCD$ in Figure 7.10 if $AB = 12$ cm and $AD = 7$ cm.

***Solution***

Because it makes little difference which dimension is chosen as base $b$ and which as altitude $h$, we arbitrarily choose $AB = b = 12$ cm and $AD = h = 7$ cm. Then

$$\begin{aligned} A &= bh \\ &= 12 \text{ cm} \cdot 7 \text{ cm} \\ &= 84 \text{ cm}^2 \end{aligned}$$

If units are not provided for the dimensions of a region, we assume that they are alike. In such a case, we simply give the area as a number of square units.

THEOREM 7.1.1: The area $A$ of a square whose sides are each of length $s$ is given by $A = s^2$.

SSG *Exs. 6–10*

No proof is given for Theorem 7.1.1, which follows immediately from Postulate 21.

## AREA OF A PARALLELOGRAM

A rectangle's altitude is one of its sides, but that is not true of a parallelogram. An **altitude** of a parallelogram is a perpendicular segment from one side to the opposite side, known as the **base.** A side may have to be extended in order to show this altitude-base relationship in a drawing. In Figure 7.11(a), if $\overline{RS}$ is designated as the base, then any of the segments $\overline{ZR}$, $\overline{VX}$, or $\overline{YS}$ is an altitude corresponding to that base (or, for that matter, to base $\overline{VT}$).

> **Discover!**
>
> Because congruent squares "cover" a plane region, it is common to measure area in "square units." It is also possible to cover the region with congruent equilateral triangles; however, area is generally *not* measured in "triangular units." Is it possible to cover a plane region with
> (a) congruent regular pentagons?
> (b) congruent regular hexagons?
>
> ANSWERS
> (a) No (b) Yes

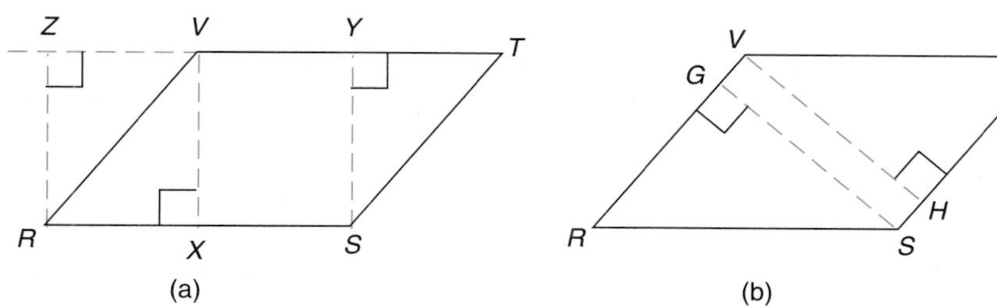


**FIGURE 7.11**

Another look at $\square RSTV$ [in Figure 7.11(b)] shows that $\overline{ST}$ (or $\overline{VR}$) could just as well have been chosen as the base. Possible choices for the corresponding altitude in this case include $\overline{VH}$ and $\overline{GS}$. In the theorem that follows, it is necessary to select a base and an altitude drawn to that base!

THEOREM 7.1.2: The area $A$ of a parallelogram with a base of length $b$ and with corresponding altitude of length $h$ is given by

$$A = bh$$

The proof of Theorem 7.1.2 follows on p. 340.

**Discover!**

On a geoboard (or pegboard), a parallelogram is formed by a rubber band. With base $b = 6$ and altitude $h = 4$, count wholes and halves to find the area of the parallelogram.

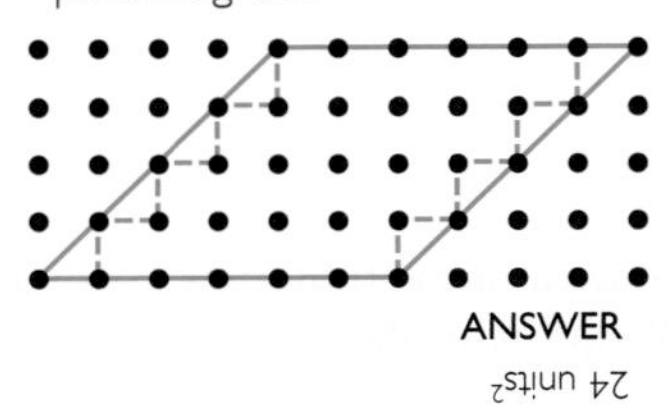

ANSWER

24 units²

*Given:* In Figure 7.12(a), ▱$RSTV$ with $\overline{VX} \perp \overline{RS}$
$RS = b$ and $VX = h$

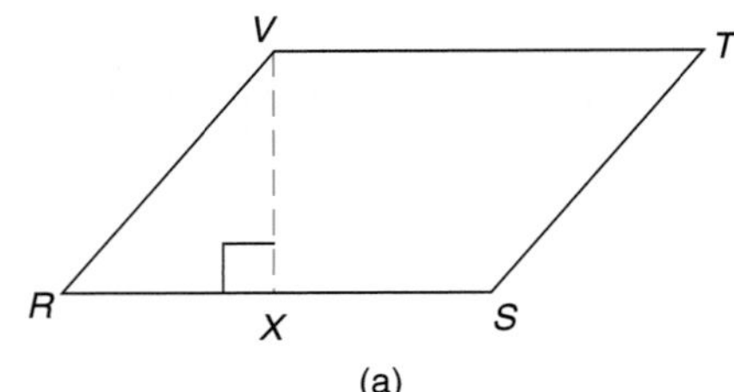


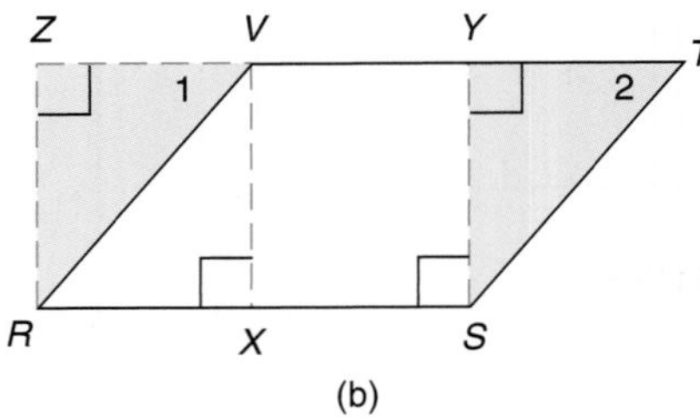


**FIGURE 7.12**

*Prove:* $A_{RSTV} = bh$

*Proof:* Construct $\overline{YS} \perp \overline{VT}$ and $\overline{RZ} \perp \overline{VT}$, in which $Z$ lies on an extension of $\overline{VT}$, as shown in Figure 7.12(b). Right $\angle Z$ and right $\angle SYT$ are $\cong$. Also, $\overline{ZR} \cong \overline{SY}$ because parallel lines are everywhere equidistant.

Because $\angle 1$ and $\angle 2$ are $\cong$ corresponding angles for parallel segments $\overline{VR}$ and $\overline{TS}$, $\triangle RZV \cong \triangle SYT$ by AAS. Then $A_{RZV} = A_{SYT}$ because congruent $\triangle$s have equal areas.

Because $A_{RSTV} = A_{RSYV} + A_{SYT}$, it follows that $A_{RSTV} = A_{RSYV} + A_{RZV}$. But $RSYV \cup RZV$ is rectangle RSYZ, which has the area $bh$.
Therefore, $A_{RSTV} = A_{RSYZ} = bh$.

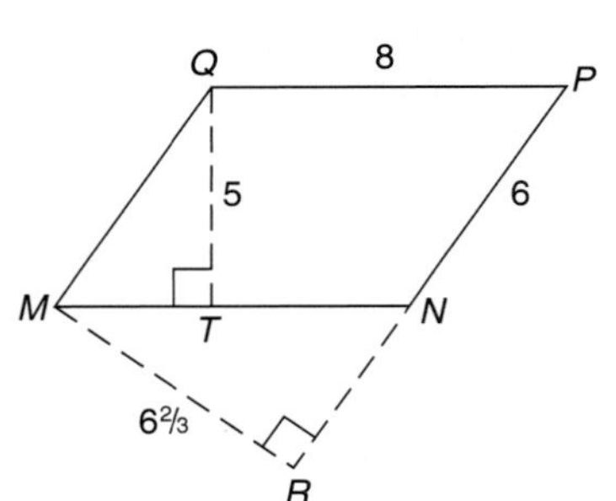


**FIGURE 7.13**

## EXAMPLE 4

Given that all dimensions in Figure 7.13 are in inches, find the area of ▱$MNPQ$ by using base

a) $MN$. b) $PN$.

***Solution***

a) $MN = QP = b = 8$, and the corresponding altitude is of length $QT = h = 5$. Then

$$A = 8 \text{ in.} \cdot 5 \text{ in.}$$
$$= 40 \text{ in}^2$$

b) $PN = b = 6$, so the corresponding altitude length is $MR = h = 6\frac{2}{3}$. Then

$$A = 6 \cdot 6\tfrac{2}{3}$$
$$= 6 \cdot \tfrac{20}{3}$$
$$= 40 \text{ in}^2$$

NOTE: The area of the ▱ is unchanged when a different base and its corresponding altitude are used to calculate its area. See Postulate 18.

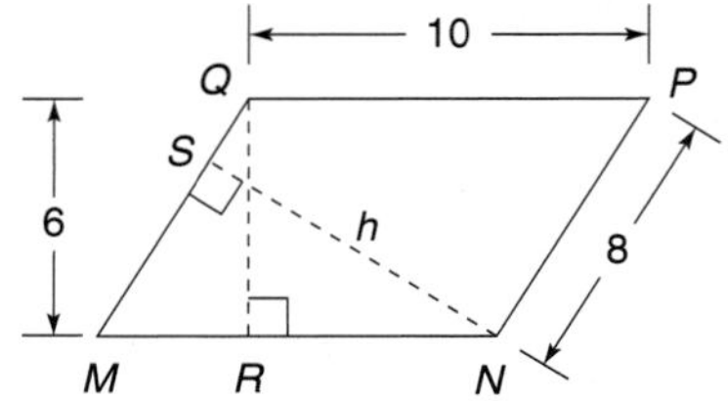


**FIGURE 7.14**

## EXAMPLE 5

*Given:* In Figure 7.14, ▱$MNPQ$ with $PN = 8$ and $QP = 10$
Altitude $\overline{QR}$ to base $\overline{MN}$ has length $QR = 6$

*Find:* $SN$, the length of the altitude between $\overline{QM}$ and $\overline{PN}$

***Solution***

Choosing $MN = b = 10$ and $QR = h = 6$, we see that

$$A = bh = 10 \cdot 6 = 60$$

Now we choose $PN = b = 8$ and $SN = h$, so $A = 8h$. Because the area of the parallelogram is unique, it follows that

$$8h = 60$$
$$h = \frac{60}{8} = 7.5$$

*Exs. 11–14*

that is, $SN = 7.5$

## AREA OF A TRIANGLE

The next formula we consider is used to calculate the area of a triangle. It follows easily from the formula for the area of a parallelogram. In the formula, any side of the triangle can be chosen as its base.

THEOREM 7.1.3: The area $A$ of a triangle whose base has length $b$ and whose corresponding altitude has length $h$ is given by

$$A = \frac{1}{2}bh$$

Following is a picture proof of Theorem 7.1.3.

### PICTURE PROOF OF THEOREM 7.1.3

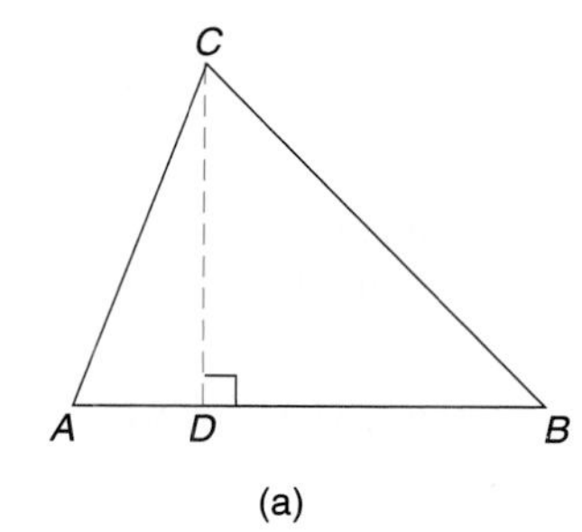


(a)

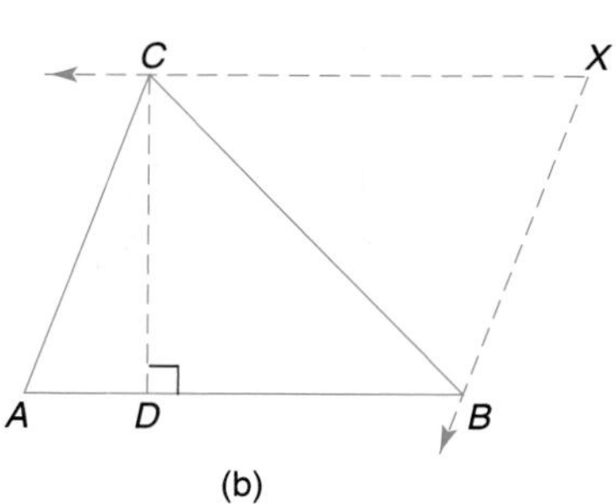


(b)

**FIGURE 7.15**

*Given:* In Figure 7.15(a), $\triangle ABC$ with $\overline{CD} \perp \overline{AB}$; $AB = b$ and $CD = h$

*Prove:* $A = \frac{1}{2}bh$

*Proof:* Let lines through $C$ parallel to $\overline{AB}$ and through $B$ parallel to $\overline{AC}$ meet at point $X$ [see Figure 7.15(b)]. With $\square ABXC$ and congruent triangles $ABC$ and $XCB$, we see that $A_{ABC} = \frac{1}{2} \cdot A_{ABXC} = \frac{1}{2}bh$.

EXAMPLE 6

In the figure, find the area of $\triangle ABC$ if $AB = 10$ cm and $CD = 7$ cm.

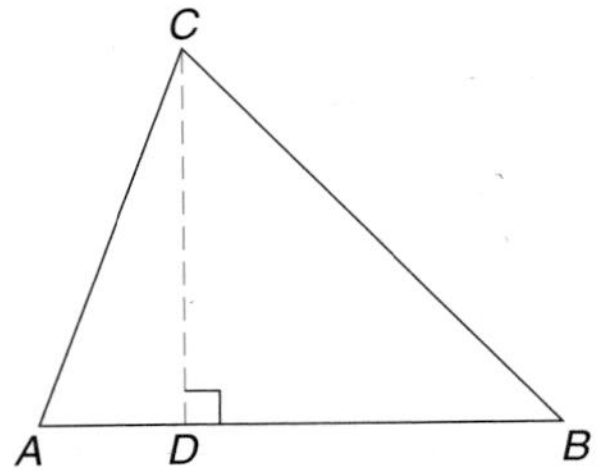


***Solution***

With $\overline{AB}$ as base, $b = 10$ cm.
The corresponding altitude for base $\overline{AB}$ is $\overline{CD}$, so $h = 7$ cm. Now

$$A = \frac{1}{2}bh$$

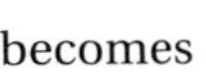

becomes $$A = \frac{1}{2} \cdot 10 \text{ cm} \cdot 7 \text{ cm}$$
$$A = 35 \text{ cm}^2$$

The following theorem is a corollary of Theorem 7.1.3.

**Warning** (!)
The phrase *area of a polygon* really means the area of the region enclosed by the polygon.

COROLLARY 7.1.4: The area of a right triangle with legs of lengths $a$ and $b$ is given by $A = \frac{1}{2}ab$.

In the proof of Corollary 7.1.4, the length of either leg can be chosen as the base. In turn, the length of the altitude to the base chosen is the length of the remaining leg. This follows from the fact, "The legs of a right triangle are perpendicular."

EXAMPLE 7

*Given:* In Figure 7.16, right $\triangle MPN$ with $PN = 8$ and $MN = 17$

*Find:* $A_{MNP}$

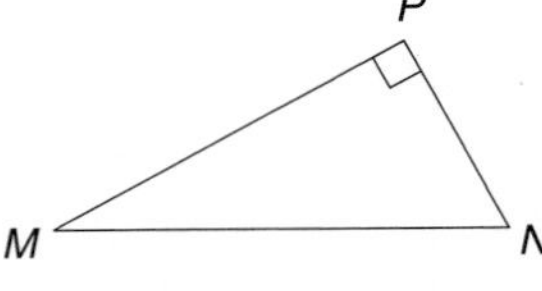


**FIGURE 7.16**

***Solution***

With $\overline{PN}$ as one leg of $\triangle MPN$, we need the length of the second leg $\overline{PM}$. By the Pythagorean Theorem,

$$17^2 = (PM)^2 + 8^2$$
$$289 = (PM)^2 + 64$$

Then $(PM)^2 = 225$, so $PM = 15$.
With $PN = a = 8$ and $PM = b = 15$,

$$A = \frac{1}{2}ab$$

becomes $$A = \frac{1}{2} \cdot 8 \cdot 15 = 60 \text{ units}^2$$

*Exs. 15–20*

## 7.1 Exercises

1. Suppose that two triangles have equal areas. Are the triangles congruent? Why or why not? Are two squares with equal areas necessarily congruent? Why or why not?

2. The area of the square is 12, and the area of the circle is 30. Does the area of the entire shaded region equal 42? Why or why not?

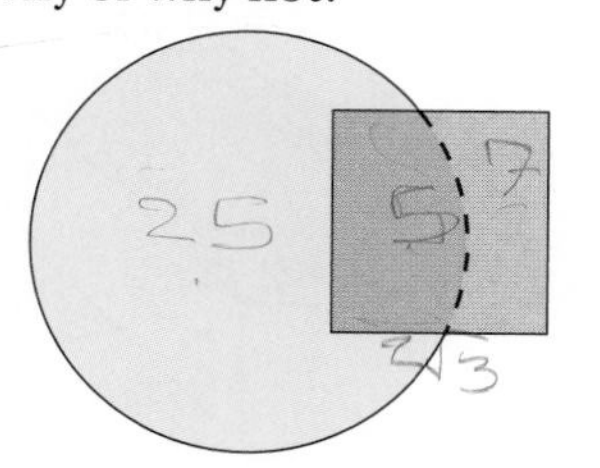

*Exercises 2, 3*

3. Consider the information in Exercise 2, but suppose you know that the area of the region defined by the intersection of the square and the circle measures 5. What is the area of the entire shaded region?

4. If *MNPQ* is a rhombus, which formula from this section should be used to calculate its area?

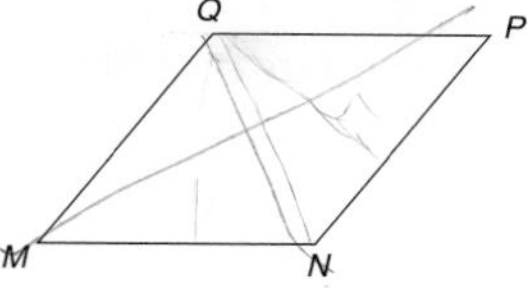


*Exercises 4–6*

5. In rhombus *MNPQ*, how does the length of the altitude from *Q* to $\overline{PN}$ compare to the length of the altitude from *Q* to $\overline{MN}$? Explain.

6. When the diagonals of rhombus *MNPQ* are drawn, how do the areas of the four resulting smaller triangles compare to each other and to the area of the given rhombus?

7. $\triangle ABC$ is an obtuse triangle with obtuse angle *A*. $\triangle DEF$ is an acute triangle. How do the areas of $\triangle ABC$ and $\triangle DEF$ compare?

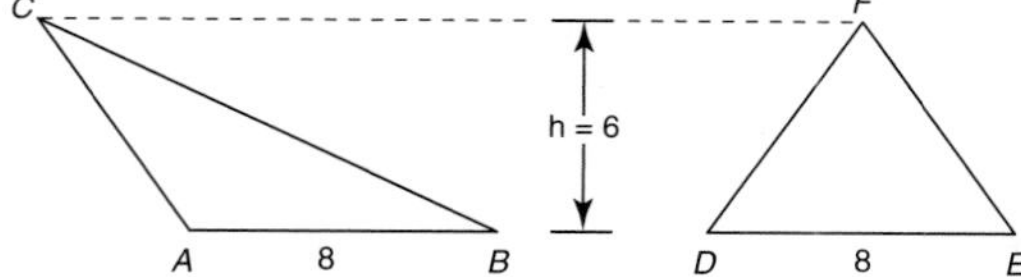


*Exercises 7, 8*

8. Are $\triangle ABC$ and $\triangle DEF$ congruent?

*In Exercises 9 to 18, find the areas of the figures shown or described.*

9. A rectangle's length is 6 cm, and its width is 9 cm.

10. A right triangle has one leg measuring 20 in. and a hypotenuse measuring 29 in.

11. A 45-45-90 triangle has a leg measuring 6 m.

12. A triangle's altitude to the 15-in. side measures 8 in.

13. 14.

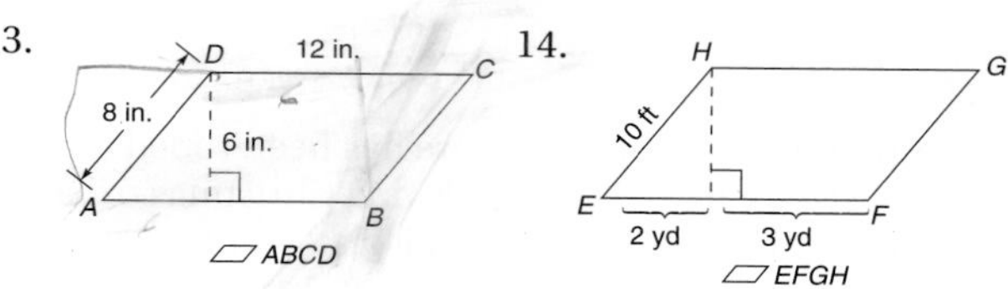


15. 16.

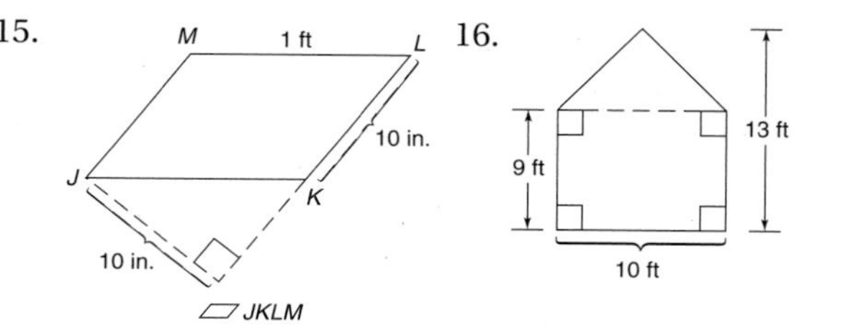


17.

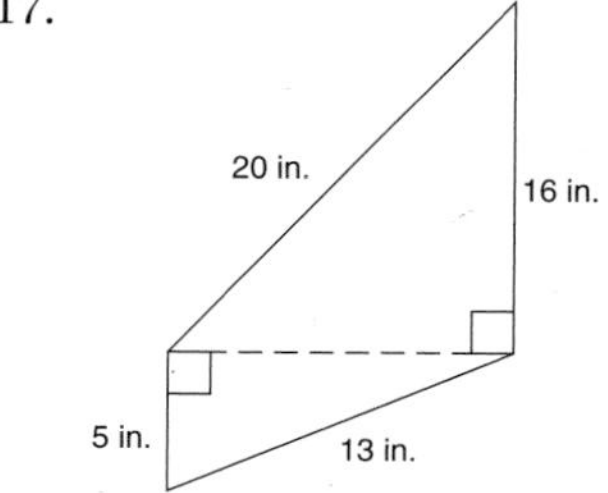


18.

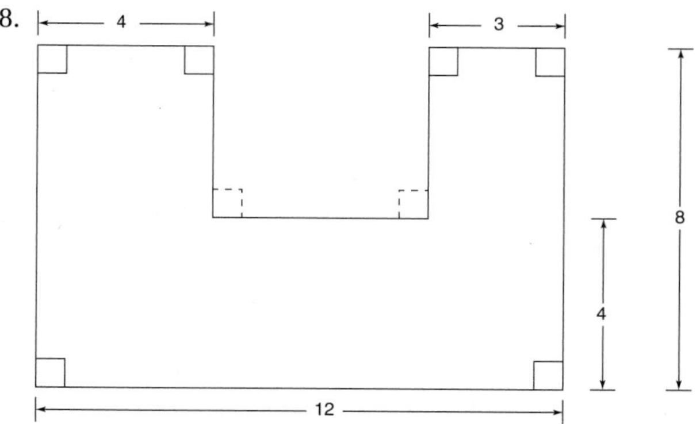

*In Exercises 19 to 22, find the area of the shaded region.*

19.

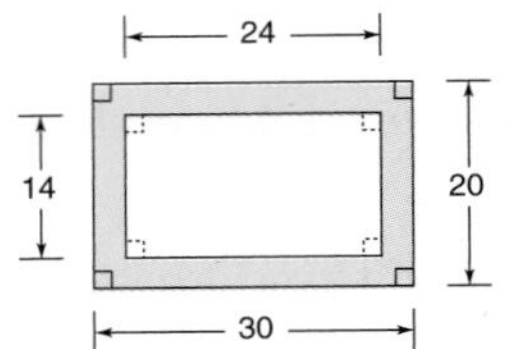


20.

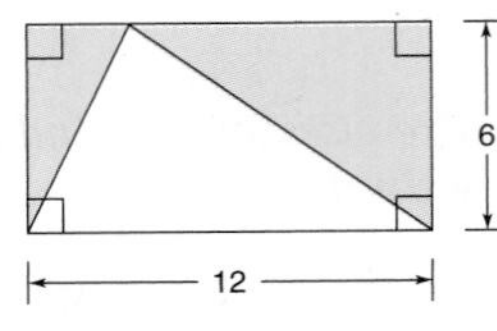


21.

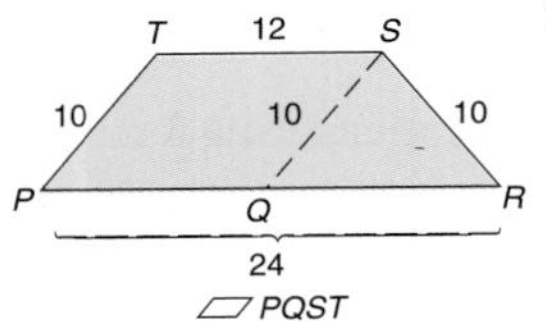


22.

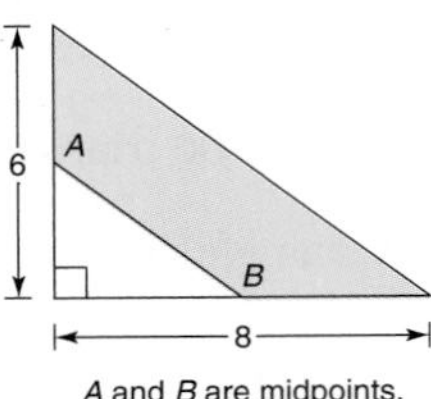


23. A triangular corner of a store has been roped off to be used as an area for displaying Christmas ornaments. Find the area of the display section.

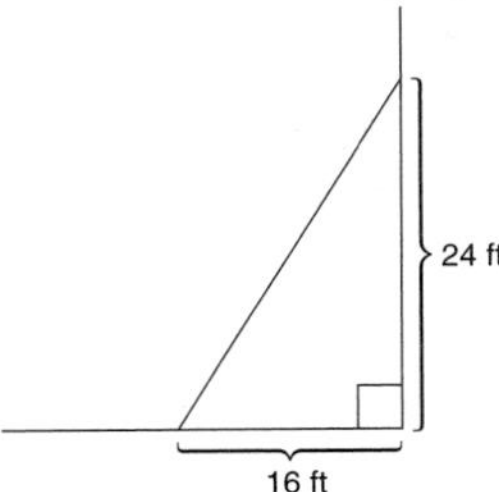


24. Carpeting is to be purchased for the family room and hallway shown. What is the area to be covered?

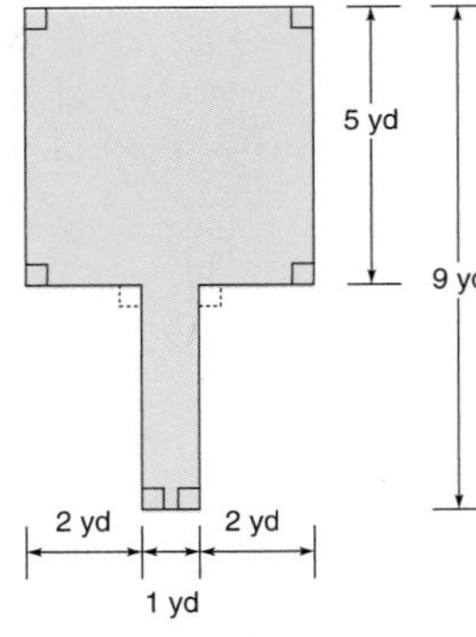


25. The exterior wall (the gabled end of the house shown) remains to be painted.

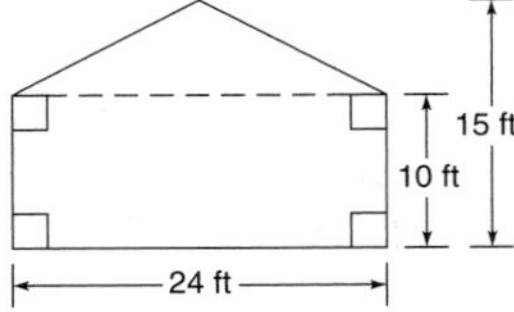


a) What is the area of the outside wall?
b) If each gallon of paint covers approximately 105 ft², how many gallons of paint must be purchased?
c) If each gallon of paint is on sale for $15.50, what is the total cost of the paint?

26. The roof of the house shown needs to be shingled.

a) Considering that both the front and back sections of the roof have equal areas, find the total area to be shingled.
b) If roofing is sold in squares (each covering 100 ft²), how many squares are needed to complete the work?
c) If each square costs $22.50 and an extra square is allowed for trimming around vents, what is the total cost of the shingles?

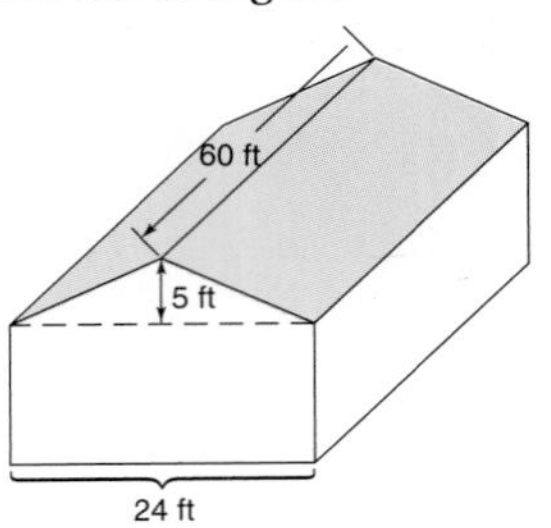


27. A beach tent is designed so that one side is open. Find the number of square feet of canvas needed to make the tent.

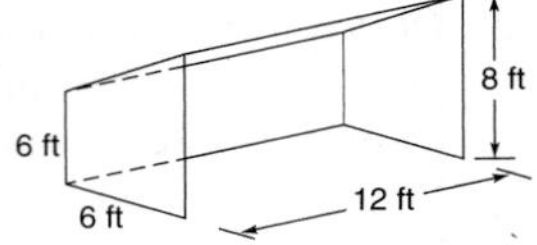


28. Gary and Carolyn plan to build the deck shown.

a) Find the total floor space (area) of the deck.
b) Find the approximate cost of building the deck if the estimated cost is $3.20 per ft².

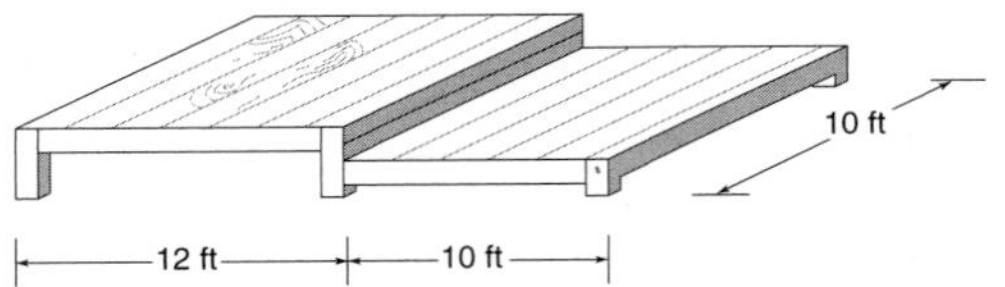


29. A *square yard* is a square with sides 1 yard in length.

a) How many *square feet* are in 1 square yard?
b) How many *square inches* are in 1 square yard?

30. The following problem is based on this theorem: "A median of a triangle separates it into two triangles of equal area."

a) Given $\triangle RST$ with median $\overline{RV}$, explain why $A_{RSV} = A_{RVT}$.
b) If $A_{RST} = 40.8$ cm², find $A_{RSV}$.

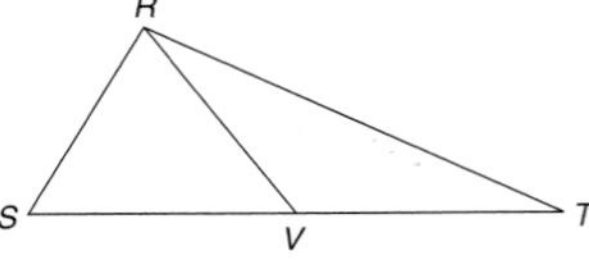

*For Exercises 31 and 32, X is the midpoint of $\overline{VT}$ and Y is the midpoint of $\overline{TS}$.*

31. If $A_{RSTV} = 48 \text{ cm}^2$, find $A_{RYTX}$.

32. If $A_{RYTX} = 13.5 \text{ in}^2$, find $A_{RSTV}$.

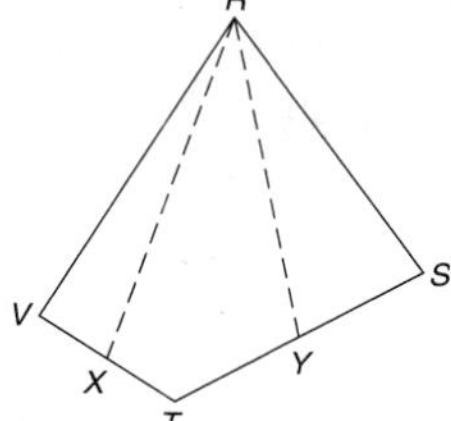

*Exercises 31, 32*

33. Given $\triangle ABC$ with midpoints $M$, $N$, and $P$ of the sides, explain why $A_{ABC} = 4 \cdot A_{MNP}$.

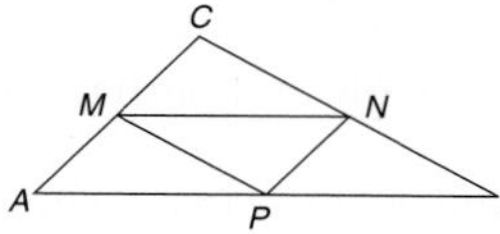


*In Exercises 34 to 36, provide paragraph proofs.*

34. *Given:* Right $\triangle ABC$
*Prove:* $h = \frac{ab}{c}$

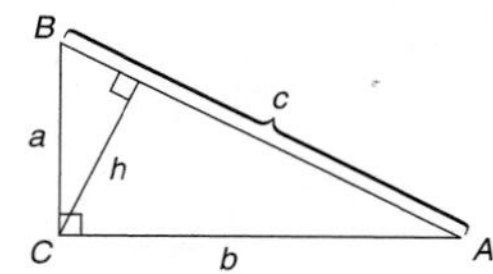


35. *Given:* Square $HJKL$ with $LJ = d$
*Prove:* $A_{HJKL} = \frac{d^2}{2}$

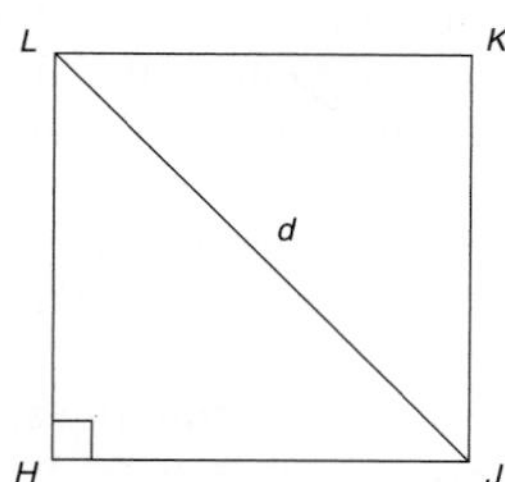


36. *Given:* $\square RSTV$ with $\overline{VW} \cong \overline{VT}$
*Prove:* $A_{RSTV} = (RS)^2$

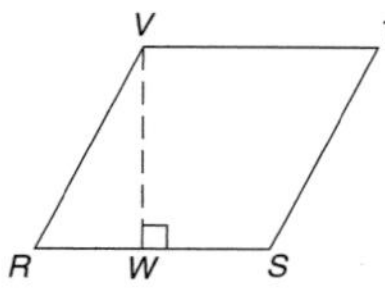


37. *Given:* The area of right $\triangle ABC$ (not shown) is $40 \text{ in}^2$.
$m\angle C = 90°$
$AC = x$
$BC = x + 2$
*Find:* $x$

38. The lengths of the legs of a right triangle are consecutive even integers. The numerical value of the area is three times that of the longer leg. Find the lengths of the legs of the triangle.

* 39. *Given:* $\triangle ABC$, whose sides are 13 in., 14 in., and 15 in.
*Find:* a) $BD$, the length of the altitude to the 14-in. side (HINT: Use the Pythagorean Theorem twice.)
b) The area of $\triangle ABC$, using the result from part (a)

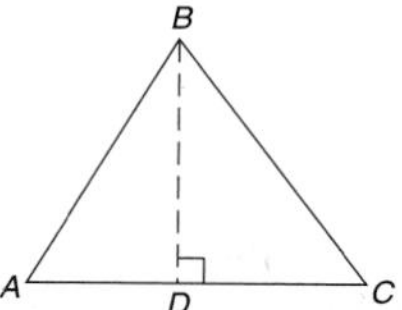

*Exercises 39, 40*

* 40. *Given:* $\triangle ABC$, whose sides are 10 cm, 17 cm, and 21 cm
*Find:* a) $BD$, the length of the altitude to the 21-cm side
b) The area of $\triangle ABC$, using the result from part (a)

41. If the base of a rectangle is increased by 20 percent and the altitude is increased by 30 percent, by what percentage is the area increased?

42. If the base of a rectangle is increased by 20 percent but the altitude is decreased by 30 percent, by what percentage is the area changed? Is this an increase or a decrease in area?

43. Given region $R \cup S$, explain why $A_{R \cup S} > A_R$.

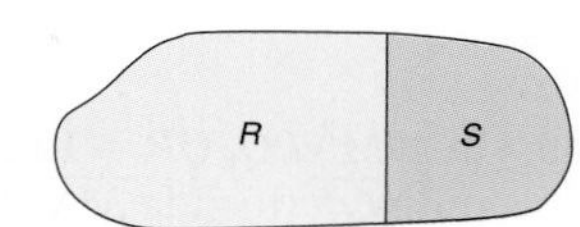


44. Given region $R \cup S \cup T$, explain why $A_{R \cup S \cup T} = A_R + A_S + A_T$.

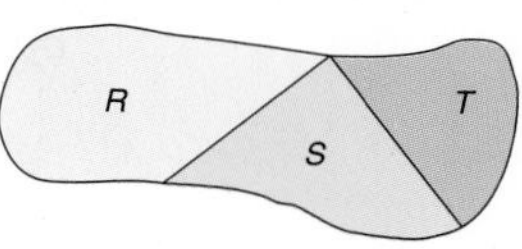


45. The algebra method of FOIL multiplication is illustrated geometrically in the drawing. Use the drawing with rectangular regions to complete the following rule: $(a + b)(c + d) =$ ________________

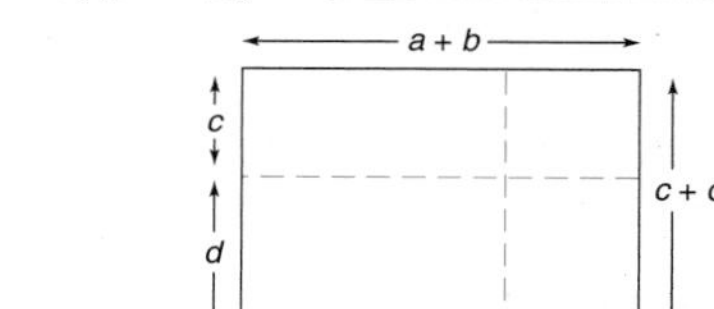

46. Use the square configuration to complete the following algebra rule:
$(a + b)^2 =$ ________________________
(NOTE: Simplify where possible.)

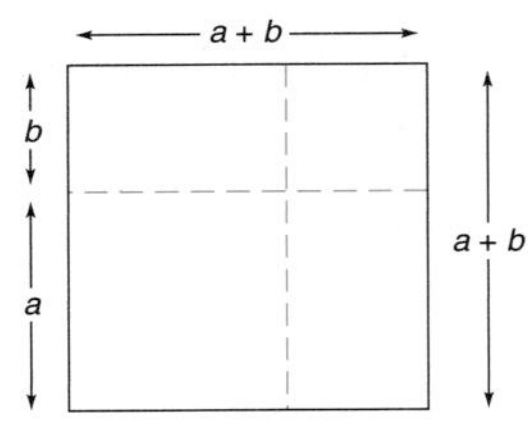


*In Exercises 47 to 50, use the fact that the area of the polygon is unique.*

47. In the right triangle, find the length of the altitude drawn to the hypotenuse.

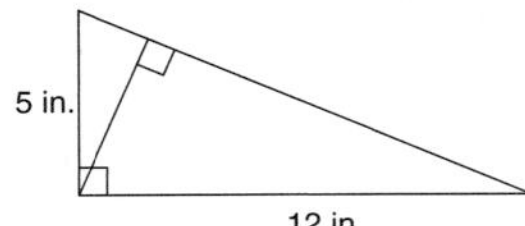


48. In the triangle whose sides are 13, 20, and 21 cm long, the length of the altitude drawn to the 21-cm side is 12 cm. Find the lengths of the remaining altitudes of the triangle.

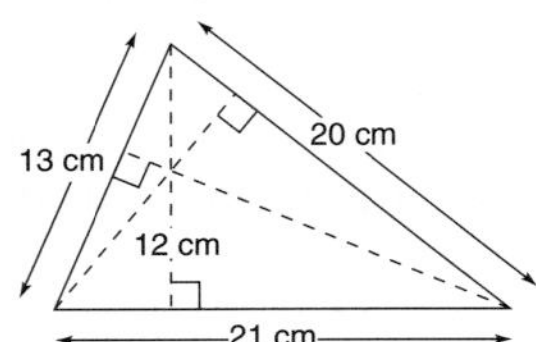


49. In ▱$MNPQ$, $QP = 12$ and $QM = 9$. The length of altitude $\overline{QR}$ (to side $\overline{MN}$) is 6. Find the length of altitude $\overline{QS}$ from $Q$ to $\overline{PN}$.

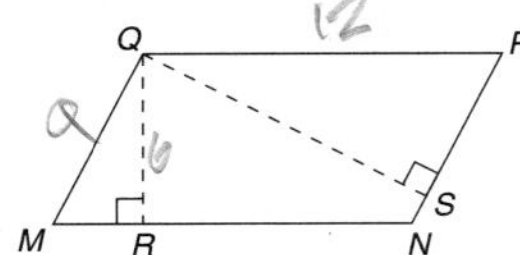


50. In ▱$ABCD$, $AB = 7$ and $BC = 12$. The length of altitude $\overline{AF}$ (to side $\overline{BC}$) is 5. Find the length of altitude $\overline{AE}$ from $A$ to $\overline{DC}$.

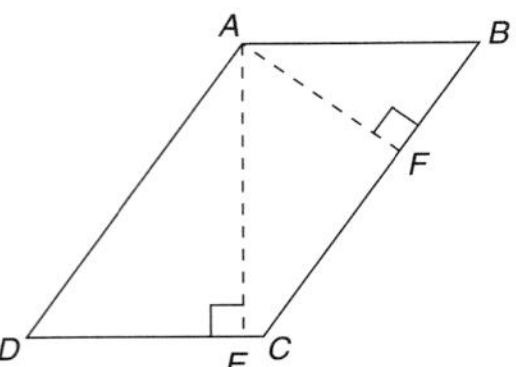


51. a) Find a lower estimate of the area of the figure by counting whole squares within the figure.
b) Find an upper estimate of the area of the figure by counting whole and partial squares within the figure.
c) Use the average of the results in parts (a) and (b) to provide a better estimate of the area of the figure.
d) Does intuition suggest that the area estimate of part (c) is the exact answer?

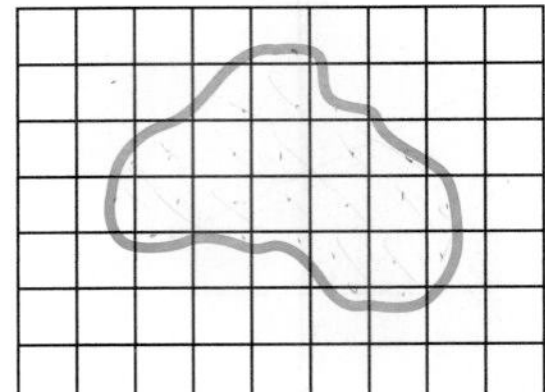

52. a) Find a lower estimate of the area of the figure by counting whole squares within the figure.
b) Find an upper estimate of the area of the figure by counting whole and partial squares within the figure.
c) Use the average of the results in parts (a) and (b) to provide a better estimate of the area of the figure.
d) Does intuition suggest that the area estimate of part (c) is the exact answer?

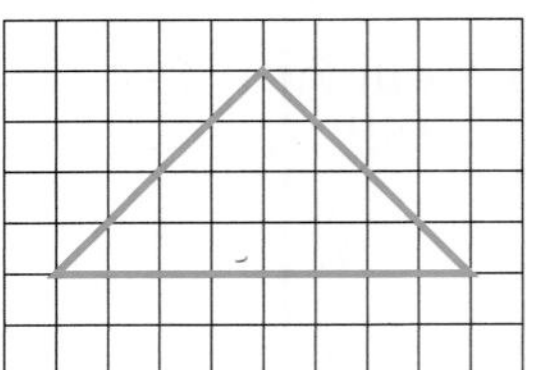

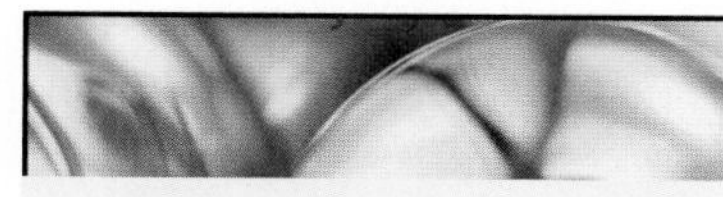

# 7.2 Perimeter and Area of Polygons

**KEY CONCEPTS**

Perimeter of a Polygon • Semiperimeter of a Triangle • Heron's Formula • Brahmagupta's Formula • Area of Trapezoid, Rhombus, and Kite • Areas of Similar Polygons

We begin this section with a reminder of the meaning of perimeter.

> DEFINITION: The **perimeter** of a polygon is the sum of the lengths of all sides of the polygon.

Table 7.1 summarizes perimeter formulas for types of triangles, and Table 7.2 summarizes formulas for the perimeters of selected types of quadrilaterals. However, it is more important to understand the concept of perimeter than to memorize formulas. See if you can explain each formula.

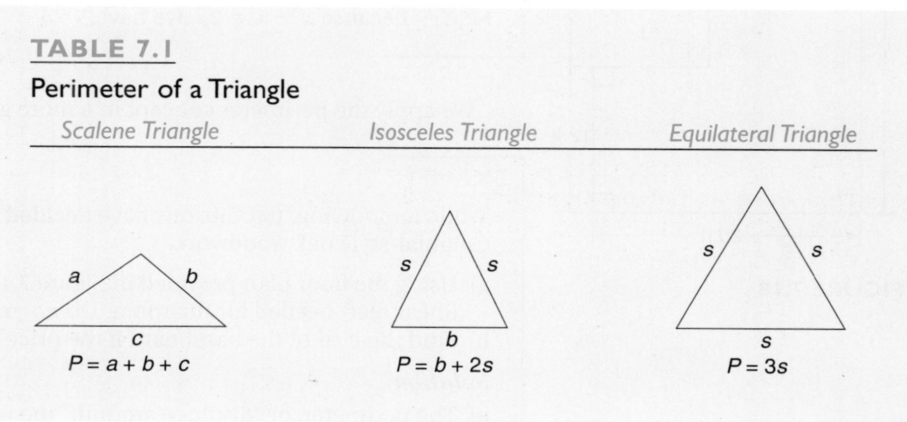

**TABLE 7.1**

**Perimeter of a Triangle**

| *Scalene Triangle* | *Isosceles Triangle* | *Equilateral Triangle* |
|---|---|---|
| $P = a + b + c$ | $P = b + 2s$ | $P = 3s$ |

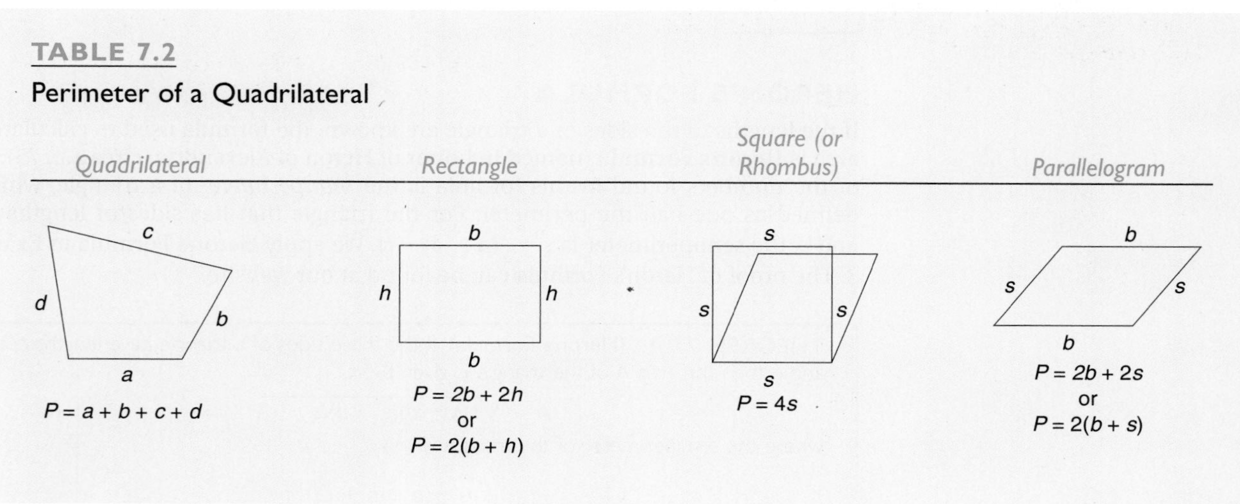

**TABLE 7.2**

**Perimeter of a Quadrilateral**

| *Quadrilateral* | *Rectangle* | *Square (or Rhombus)* | *Parallelogram* |
|---|---|---|---|
| $P = a + b + c + d$ | $P = 2b + 2h$ or $P = 2(b + h)$ | $P = 4s$ | $P = 2b + 2s$ or $P = 2(b + s)$ |

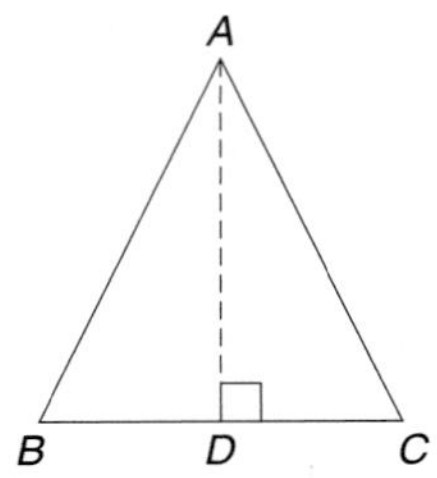


**FIGURE 7.17**

EXAMPLE 1

Find the perimeter of $\triangle ABC$ shown in Figure 7.17 if:

a) $AB = 5$ in., $AC = 6$ in., and $BC = 7$ in.
b) Altitude $AD = 8$ cm, $BC = 6$ cm, and $\overline{AB} \cong \overline{AC}$

***Solution***

a) $P_{ABC} = AB + AC + BC$
$= 5 + 6 + 7$
$= 18$ in.

b) With $\overline{AB} \cong \overline{AC}$, $\triangle ABC$ is isosceles. Then $\overline{AD}$ is the $\perp$ bisector of $\overline{BC}$. If $BC = 6$, it follows that $DC = 3$. Using the Pythagorean Theorem, we have

$$(AD)^2 + (DC)^2 = (AC)^2$$
$$8^2 + 3^2 = (AC)^2$$
$$64 + 9 = (AC)^2$$
$$AC = \sqrt{73}$$

Now $P_{ABC} = 6 + \sqrt{73} + \sqrt{73} = 6 + 2\sqrt{73} \approx 23.09$ cm.

NOTE: Because $x + x = 2x$, we have $\sqrt{73} + \sqrt{73} = 2\sqrt{73}$.

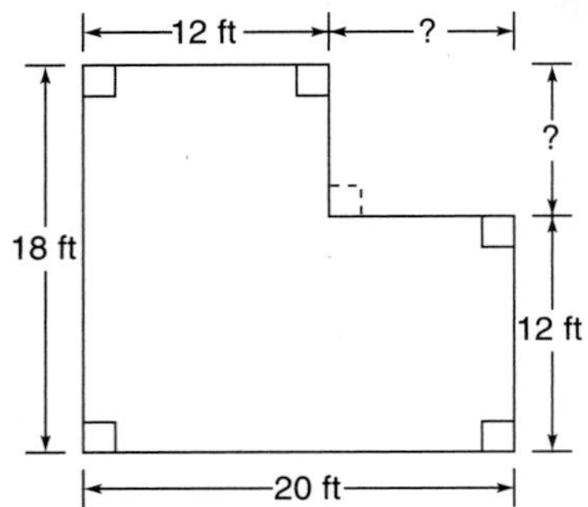


**FIGURE 7.18**

We apply the perimeter concept in a more general manner in Example 2.

EXAMPLE 2

While remodeling, the Gibsons have decided to replace the old woodwork with Colonial-style oak woodwork.

a) Using the floor plan provided in Figure 7.18, find the amount of baseboard (in linear feet) needed for the room. Do *not* make any allowances for doors!
b) Find the cost of the baseboard if the price is \$1.32 per linear foot.

***Solution***

a) The perimeter, or "distance around," the room is

$$12 + 6 + 8 + 12 + 20 + 18 = 76 \text{ linear feet}$$

b) The cost is $76 \cdot \$1.32 = \$100.32$.

*Exs. 1–4*

## HERON'S FORMULA

If the lengths of the sides of a triangle are known, the formula used to calculate the area is **Heron's Formula** (named in honor of Heron of Alexandria, circa A.D. 75). One of the numbers found in this formula is the *semiperimeter* of a triangle, which is defined as one-half the perimeter. For the triangle that has sides of lengths $a$, $b$, and $c$, the semiperimeter is $s = \frac{1}{2}(a + b + c)$. We apply Heron's Formula in Example 3. The proof of Heron's Formula can be found at our website.

THEOREM 7.2.1: **(Heron's Formula)** If the three sides of a triangle have lengths $a$, $b$, and $c$, then the area $A$ of the triangle is given by

$$A = \sqrt{s(s - a)(s - b)(s - c)},$$

where the semiperimeter of the triangle is

$$s = \frac{1}{2}(a + b + c)$$

EXAMPLE 3

Find the area of a triangle which has sides of lengths 4, 13, and 15. (See Figure 7.19.)

***Solution***

If we designate the sides as $a = 4$, $b = 13$, and $c = 15$, the semiperimeter of the triangle is given by $s = \frac{1}{2}(4 + 13 + 15) = \frac{1}{2}(32) = 16$. Therefore,

$$\begin{aligned} A &= \sqrt{s(s-a)(s-b)(s-c)} \\ &= \sqrt{16(16-4)(16-13)(16-15)} \\ &= \sqrt{16(12)(3)(1)} = \sqrt{576} = 24 \text{ units}^2 \end{aligned}$$

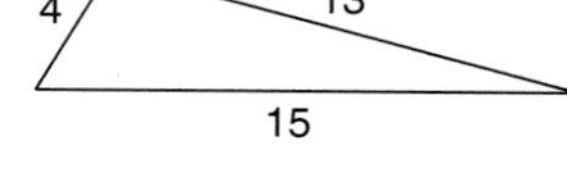


**FIGURE 7.19**

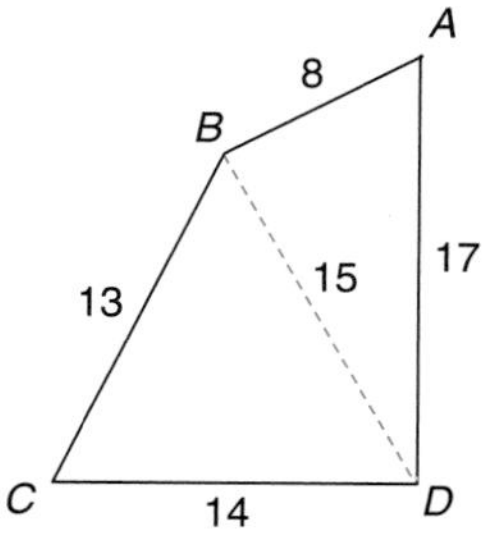

**FIGURE 7.20**

If the lengths of the sides of a quadrilateral are known, we can apply Heron's Formula to find the area when the length of a diagonal is also known. In quadrilateral *ABCD* in Figure 7.20, Heron's Formula can be used to show that the area of $\triangle ABD$ is 60 and the area of $\triangle BCD$ is 84. Thus the area of quadrilateral *ABCD* is 144 units$^2$.

The following theorem is named in honor of Brahmagupta, a Hindu mathematician born in A.D. 598. We include the theorem without its rather lengthy proof. As it happens, Heron's Formula for the area of any triangle is actually a special case of Brahmagupta's Formula, which is used to determine the area of a cyclic quadrilateral. In Brahmagupta's Formula, as in Heron's Formula, the letter *s* represents the numerical value of the semiperimeter. The formula is applied in essentially the same manner as Heron's Formula. See Exercises 11, 12, 41, and 42 of this section.

THEOREM 7.2.2: **(Brahmagupta's Formula)** For a cyclic quadrilateral with sides of lengths *a*, *b*, *c*, and *d*, the area is given by

$$A = \sqrt{(s-a)(s-b)(s-c)(s-d)},$$

where $s = \frac{1}{2}(a + b + c + d)$

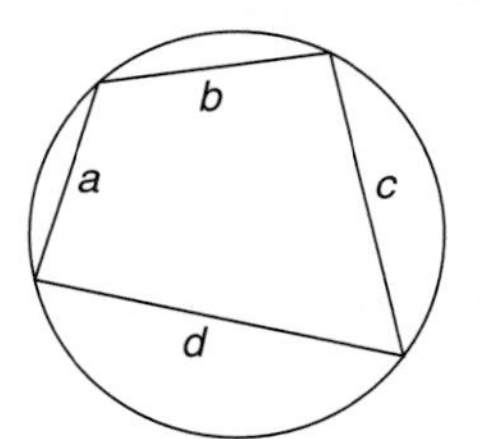


Brahmagupta's Formula becomes Heron's Formula when the length *d* of the fourth side shrinks (the length *d* approaches 0) so that the quadrilateral becomes a triangle with sides of lengths *a*, *b*, and *c*.

The remaining theorems of this section contain *numerical subscripts.* In practice, subscripts enable us to compare like quantities. For instance, the lengths of the two unequal bases of a trapezoid are written $b_1$ (read "b sub 1") and $b_2$. In particular, $b_1$ represents the numerical length of the first base, and $b_2$ represents the length of the second base. The following chart illustrates the use of numerical subscripts.

*Exs. 5–8*

| *Theorem* | *Subscripted Symbol* | *Meaning* |
|---|---|---|
| Theorem 7.2.3 | $b_1$ | Length of *first* base of trapezoid |
| Corollary 7.2.5 | $d_2$ | Length of *second* diagonal of rhombus |
| Theorem 7.2.7 | $A_1$ | Area of *first* triangle |

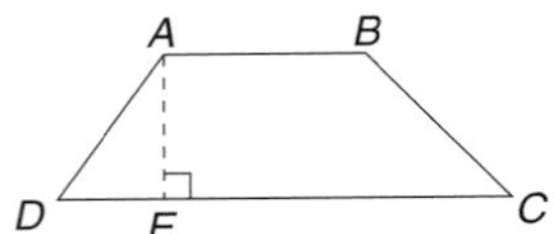


**FIGURE 7.21**

## AREA OF A TRAPEZOID

Recall that the two parallel sides of a trapezoid are its *bases*. The *altitude* is any line segment that is drawn perpendicular from one base to the other. In Figure 7.21, $\overline{AB}$ and $\overline{DC}$ are bases and $\overline{AE}$ is an altitude for the trapezoid.

We use the more common formula for the area of a triangle (namely, $A = \frac{1}{2}bh$) to develop our remaining theorems. In Theorem 7.2.3, $b_1$ and $b_2$ represent the lengths of the bases of the trapezoid. (In some textbooks, $b$ represents the length of the *shorter* base and $B$ represents the length of the *longer* base.)

THEOREM 7.2.3: The area $A$ of a trapezoid whose bases have lengths $b_1$ and $b_2$ and whose altitude has length $h$ is given by

$$A = \frac{1}{2}h(b_1 + b_2)$$

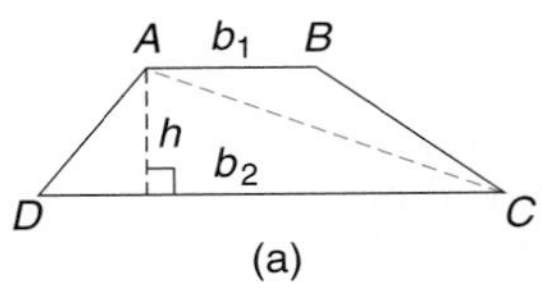


(a)

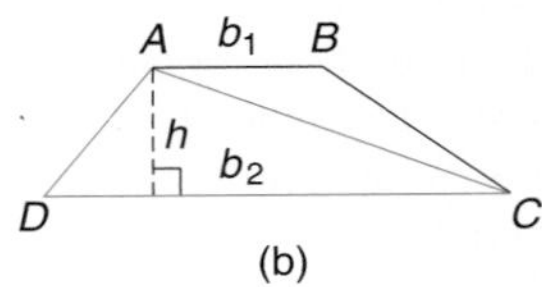


(b)

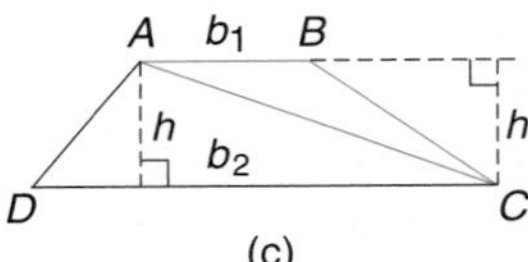


(c)

**FIGURE 7.22**

*Given:* Trapezoid $ABCD$ with $\overline{AB} \parallel \overline{DC}$

*Prove:* $A_{ABCD} = \frac{1}{2}h(b_1 + b_2)$

*Proof:* Draw $\overline{AC}$ as shown in Figure 7.22(a). Now $\triangle ADC$ has an altitude of length $h$ and a base of length $b_2$. As shown in Figure 7.22(b),

$$A_{ADC} = \frac{1}{2}hb_2$$

Also, $\triangle ABC$ has an altitude of length $h$ and a base of length $b_1$. [See Figure 7.22(c).] Then

$$A_{ABC} = \frac{1}{2}hb_1$$

Thus

$$\begin{aligned} A_{ABCD} &= A_{ABC} + A_{ADC} \\ &= \frac{1}{2}hb_1 + \frac{1}{2}hb_2 \\ &= \frac{1}{2}h(b_1 + b_2) \end{aligned}$$

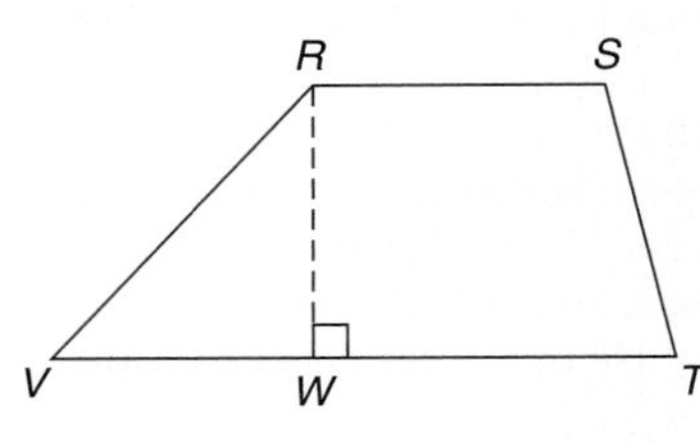


**FIGURE 7.23**

EXAMPLE 4

Find the area of the trapezoid in Figure 7.23 if $RS = 5$, $TV = 13$, and $RW = 6$.

***Solution***

Let $RS = 5 = b_1$ and $TV = 13 = b_2$. Also, $RW = h = 6$. Now,

$$A = \frac{1}{2}h(b_1 + b_2)$$

becomes

$$\begin{aligned} A &= \frac{1}{2} \cdot 6(5 + 13) \\ &= \frac{1}{2} \cdot 6 \cdot 18 \\ &= 3 \cdot 18 = 54 \text{ units}^2 \end{aligned}$$

The following activity reinforces the formula for the area of a trapezoid.

## Discover!

Cut out two trapezoids that are copies of each other and place one next to the other to form a parallelogram.

a) How long is the base of the parallelogram? b) What is the area of the parallelogram?
c) What is the area of the trapezoid?

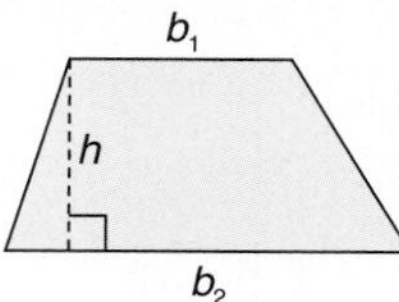


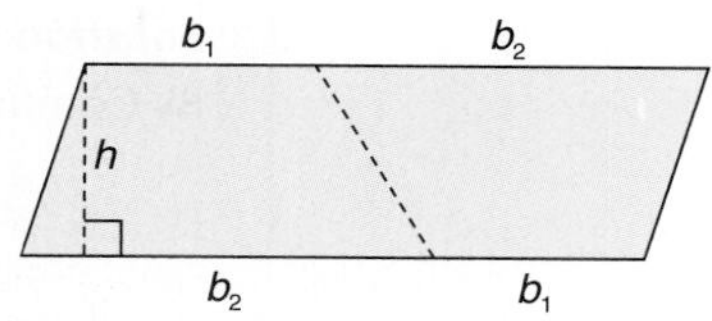


ANSWERS

a) $b_1 + b_2$ b) $h(b_1 + b_2)$ c) $\frac{1}{2}h(b_1 + b_2)$

Exs. 9–12

## QUADRILATERALS WITH PERPENDICULAR DIAGONALS

The following theorem leads to Corollaries 7.2.5 and 7.2.6, where the formula is also used to find the area of a rhombus and kite.

THEOREM 7.2.4: The area of any quadrilateral with perpendicular diagonals of lengths $d_1$ and $d_2$ is given by

$$A = \frac{1}{2}d_1d_2$$

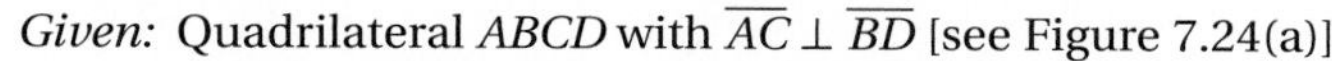

*Given:* Quadrilateral $ABCD$ with $\overline{AC} \perp \overline{BD}$ [see Figure 7.24(a)].

*Prove:* $A_{ABCD} = \frac{1}{2}d_1d_2$

*Proof:* Through points $A$ and $C$, draw lines parallel to $\overline{DB}$. Likewise, draw lines parallel to $\overline{AC}$ through points $B$ and $D$. Let the points of intersection of these lines be $R$, $S$, $T$, and $V$, as shown in Figure 7.24(b). Because each of the quadrilaterals $ARDE$, $ASBE$, $BECT$, and $CEDV$ is a parallelogram containing a right angle, each is a rectangle. Furthermore, $A_{\triangle ADE} = \frac{1}{2} \cdot A_{ARDE}$, $A_{\triangle ABE} = \frac{1}{2} \cdot A_{ASBE}$, $A_{\triangle BEC} = \frac{1}{2} \cdot A_{BECT}$, and $A_{\triangle DEC} = \frac{1}{2} \cdot A_{CEDV}$.

Then $A_{ABCD} = \frac{1}{2} \cdot A_{RSTV}$. But $RSTV$ is a rectangle, because it is a parallelogram containing a right angle. Because $RSTV$ has dimensions $d_1$ and $d_2$ [see Figure 7.24(b)], its area is $d_1d_2$. By substitution, $A_{ABCD} = \frac{1}{2}d_1d_2$.

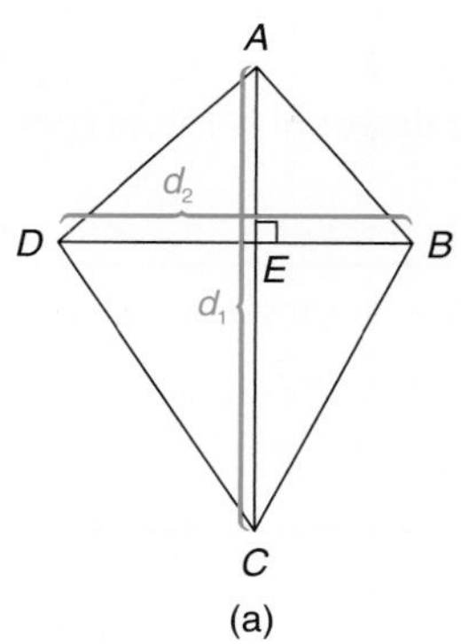

(a)

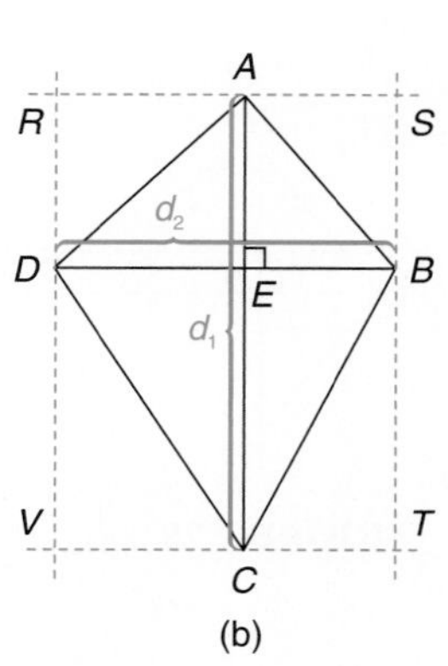

(b)

**FIGURE 7.24**

## AREA OF A RHOMBUS

Recall that a rhombus is a parallelogram with two congruent adjacent sides; in turn, we proved that all four sides were congruent. Because the diagonals of a rhombus are perpendicular, we have the following corollary of Theorem 7.2.4.

COROLLARY 7.2.5: The area $A$ of a rhombus whose diagonals have lengths $d_1$ and $d_2$ is given by

$$A = \frac{1}{2}d_1d_2$$

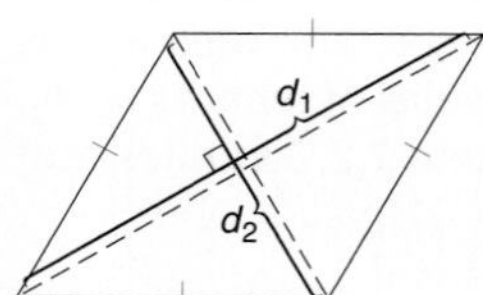


**FIGURE 7.25**

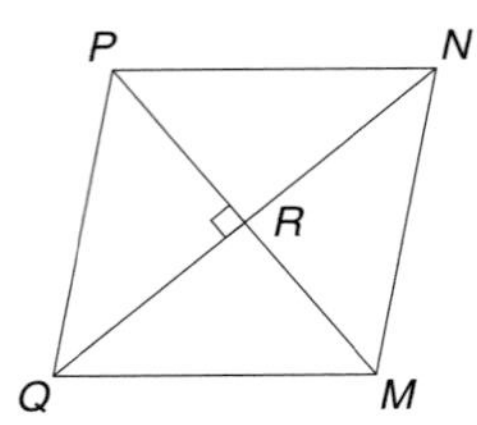

**FIGURE 7.26**

Corollary 7.2.5 and Corollary 7.2.6 are immediate consequences of Theorem 7.2.4. Example 5 illustrates Corollary 7.2.5.

EXAMPLE 5

Find the area of the rhombus *MNPQ* in Figure 7.26 if $MP = 12$ and $NQ = 16$.

***Solution***

By Corollary 7.2.5,

$$A_{MNPQ} = \frac{1}{2}d_1d_2 = \frac{1}{2} \cdot 12 \cdot 16 = 96 \text{ units}^2$$

In problems involving the rhombus, we often use the fact that diagonals are perpendicular to find needed measures. If the length of a side and the length of either diagonal are known, the length of the other diagonal can be found by applying the Pythagorean Theorem.

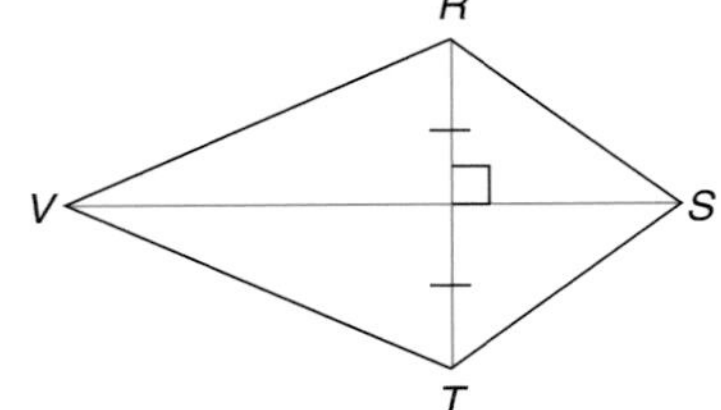


**FIGURE 7.27**

## AREA OF A KITE

For a kite, we proved in Exercise 27 of Section 4.2 that one diagonal is the perpendicular bisector of the other. (See Figure 7.27.)

COROLLARY 7.2.6: The area $A$ of a kite whose diagonals have lengths $d_1$ and $d_2$ is given by

$$A = \frac{1}{2}d_1d_2$$

We apply Corollary 7.2.6 in Example 6.

EXAMPLE 6

Find the length of $\overline{RT}$ in Figure 7.28 if the area of the kite *RSTV* is 360 in.$^2$ and $SV = 30$ in.

***Solution***

$A = \frac{1}{2}d_1d_2$ becomes $360 = \frac{1}{2}(30)d$, in which $d$ is the length of the remaining diagonal $\overline{RT}$. Then $360 = 15d$, which means that $d = 24$. Then $RT = 24$ in.

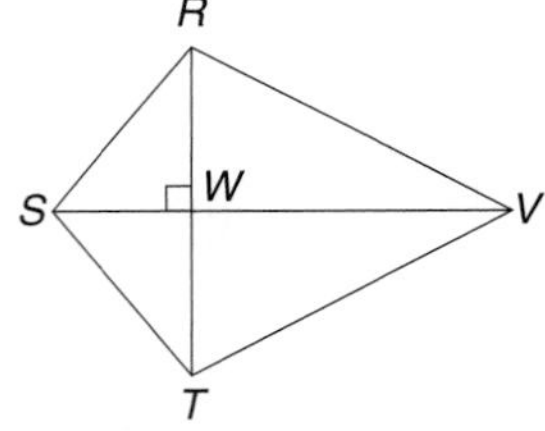


**FIGURE 7.28**

*Exs. 13–17*

## AREAS OF SIMILAR POLYGONS

The following theorem compares the areas of similar triangles. In Figure 7.29, we refer to the areas of the similar triangles as $A_1$ and $A_2$. The triangle with area $A_1$ has sides of lengths $a_1$, $b_1$, and $c_1$, and the triangle with area $A_2$ has sides of lengths $a_2$, $b_2$, and $c_2$. Where $a_1$ corresponds to $a_2$, $b_1$ to $b_2$, and $c_1$ to $c_2$, Theorem 7.2.7 implies that

$$\frac{A_1}{A_2} = \left(\frac{a_1}{a_2}\right)^2 \quad \text{or} \quad \frac{A_1}{A_2} = \left(\frac{b_1}{b_2}\right)^2 \quad \text{or} \quad \frac{A_1}{A_2} = \left(\frac{c_1}{c_2}\right)^2$$

Reminder 

Corresponding altitudes of similar triangles have the same ratio as any pair of corresponding sides.

We prove only the first relationship; the other proofs are analogous.

**THEOREM 7.2.7:** The ratio of the areas of two similar triangles equals the square of the ratio of the lengths of any two corresponding sides; that is,

$$\frac{A_1}{A_2} = \left(\frac{a_1}{a_2}\right)^2$$

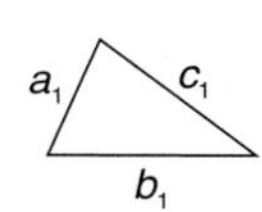


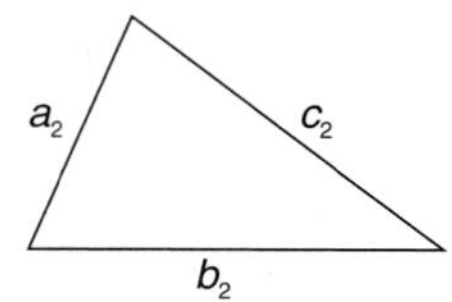


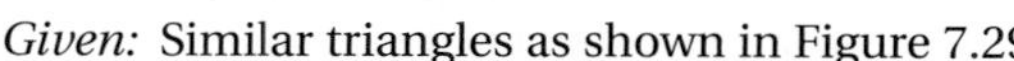

*Given:* Similar triangles as shown in Figure 7.29

*Prove:* $\frac{A_1}{A_2} = \left(\frac{a_1}{a_2}\right)^2$

*Proof:* For the similar triangles, $h_1$ and $h_2$ are the respective lengths of altitudes to the corresponding sides of lengths $b_1$ and $b_2$. Now $A_1 = \frac{1}{2}b_1h_1$ and $A_2 = \frac{1}{2}b_2h_2$, so

$$\frac{A_1}{A_2} = \frac{\frac{1}{2}b_1h_1}{\frac{1}{2}b_2h_2} \quad \text{or} \quad \frac{A_1}{A_2} = \frac{\frac{1}{2}}{\frac{1}{2}} \cdot \frac{b_1}{b_2} \cdot \frac{h_1}{h_2}$$

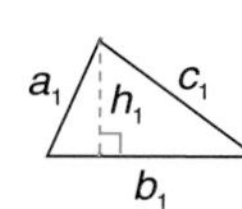


Simplifying, we have

$$\frac{A_1}{A_2} = \frac{b_1}{b_2} \cdot \frac{h_1}{h_2}$$

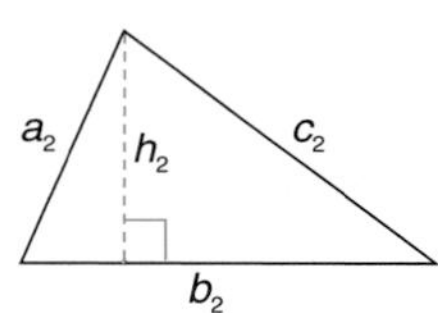


**FIGURE 7.29**

Because the triangles are similar, we know that $\frac{b_1}{b_2} = \frac{a_1}{a_2}$. Because corresponding altitudes of similar triangles have the same ratio as a pair of corresponding sides (Theorem 5.3.2), we also know that $\frac{h_1}{h_2} = \frac{a_1}{a_2}$. Through substitution, $\frac{A_1}{A_2} = \frac{b_1}{b_2} \cdot \frac{h_1}{h_2}$ becomes $\frac{A_1}{A_2} = \frac{a_1}{a_2} \cdot \frac{a_1}{a_2}$. Then $\frac{A_1}{A_2} = \left(\frac{a_1}{a_2}\right)^2$.

Because Theorem 7.2.7 can be extended to any pair of similar polygons, we could also prove that the ratio of the areas of two squares equals the square of the ratio of the lengths of any two sides. We apply this relationship in Example 7.

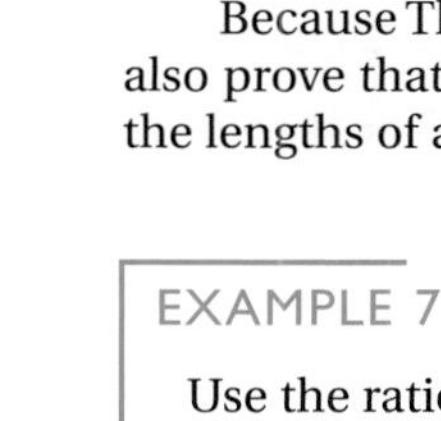

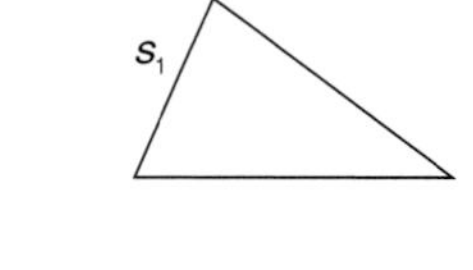


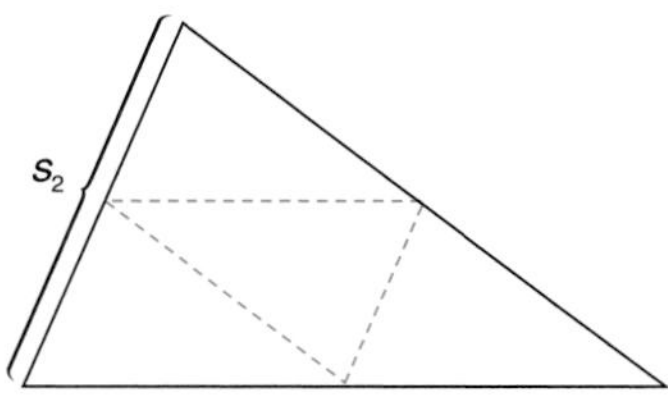


**FIGURE 7.30**

EXAMPLE 7

Use the ratio $\frac{A_1}{A_2}$ to compare the areas of:

a) Two similar triangles in which the sides of the first triangle are $\frac{1}{2}$ as long as the sides of the second triangle
b) Two squares in which each side of the first square is 3 times as long as each side of the second square

***Solution***

a) $s_1 = \frac{1}{2}s_2$, so $\frac{s_1}{s_2} = \frac{1}{2}$. (See Figure 7.30.)

Now $\frac{A_1}{A_2} = \left(\frac{s_1}{s_2}\right)^2$, so that $\frac{A_1}{A_2} = \left(\frac{1}{2}\right)^2$ or

$\frac{A_1}{A_2} = \frac{1}{4}$. That is, the area of the first

triangle is $\frac{1}{4}$ the area of the second triangle.

b) $s_1 = 3s_2$, so $\frac{s_1}{s_2} = 3$. (See Figure 7.31.)

$\frac{A_1}{A_2} = \left(\frac{s_1}{s_2}\right)^2$, so that $\frac{A_1}{A_2} = (3)^2$ or $\frac{A_1}{A_2} = 9$. That is, the area of the first square is 9 times the area of the second square.

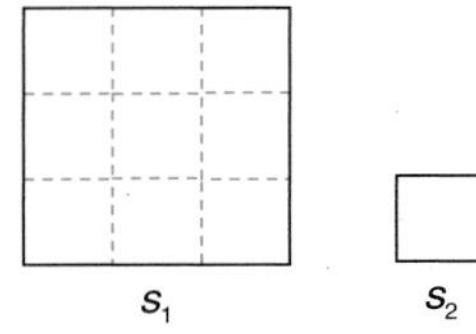


**FIGURE 7.31**

*Exs. 18–21*

NOTE: For Example 7, Figures 7.30 and 7.31 provide visual insight into the relationship described in Theorem 7.2.7.

## 7.2 Exercises

*In Exercises 1 to 8, find the perimeter of each figure.*

1.
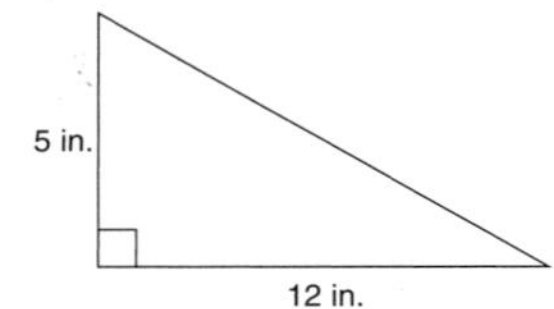


2.
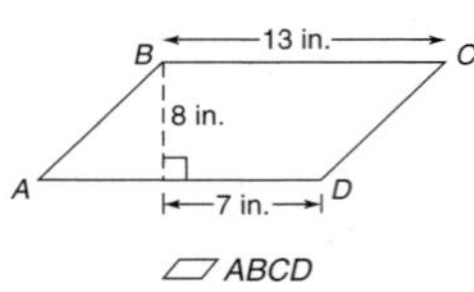


▱ABCD

3.
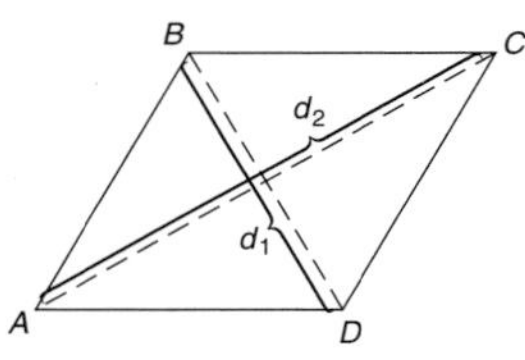


▱ABCD with $\overline{AB} \cong \overline{BC}$

$d_1$ = 4 m

$d_2$ = 10 m

4.
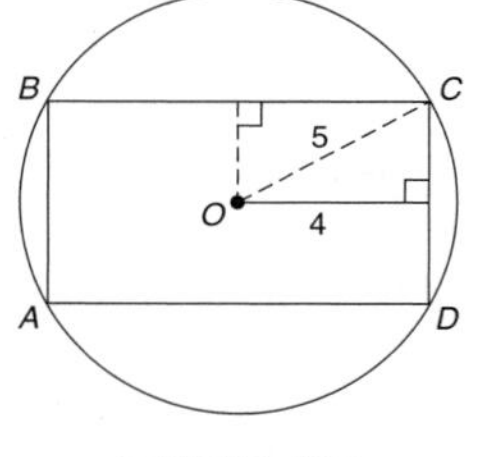


▱ABCD in ⊙O

5.
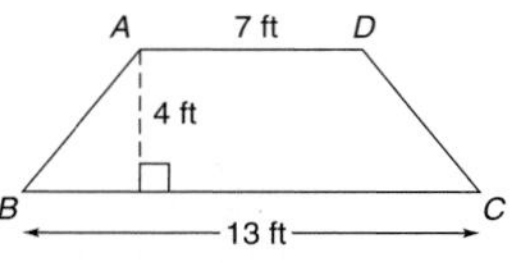


Trapezoid *ABCD* with $\overline{AB} \cong \overline{DC}$

6.
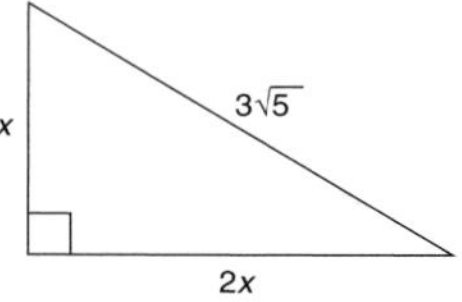


7.
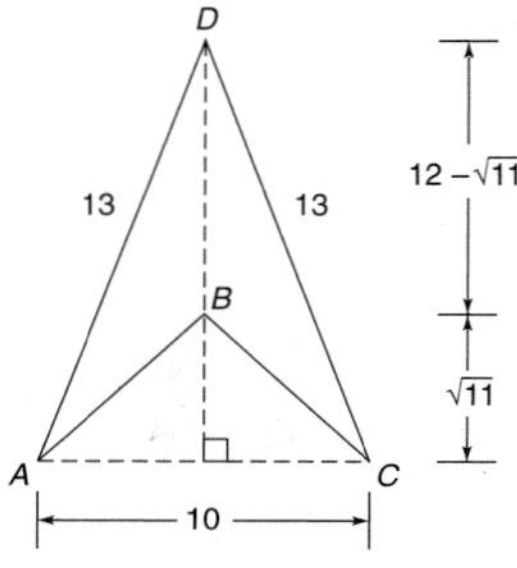


$\overline{AB} \cong \overline{BC}$ in concave quadrilateral *ABCD*

8.
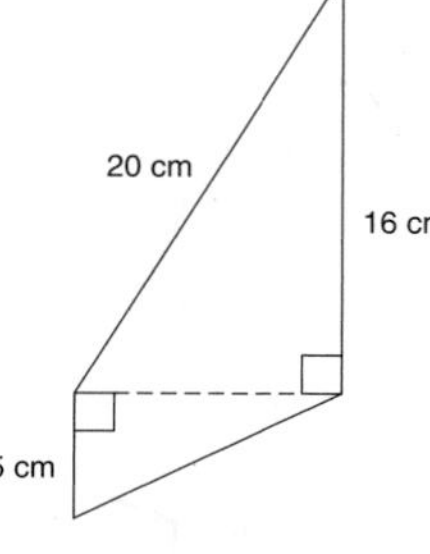


*In Exercises 9 and 10, use Heron's Formula.*

9. Find the area of a triangle whose sides measure 13 in., 14 in., and 15 in.

10. Find the area of a triangle whose sides measure 10 cm, 17 cm, and 21 cm.

*For Exercises 11 and 12, use Brahmagupta's Formula.*

11. For cyclic quadrilateral *ABCD*, find the area if *AB* = 39 mm, *BC* = 52 mm, *CD* = 25 mm, and *DA* = 60 mm.

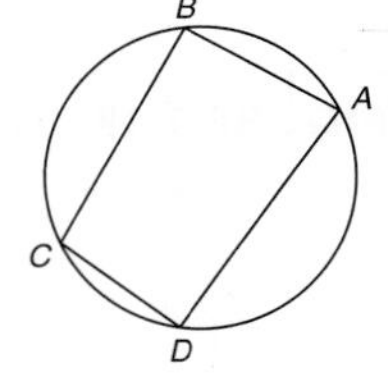


12. For cyclic quadrilateral *ABCD*, find the area if *AB* = 6 cm, *BC* = 7 cm, *CD* = 2 cm, and *DA* = 9 cm.

*In Exercises 13 to 18, find the area of the given polygon.*

13.
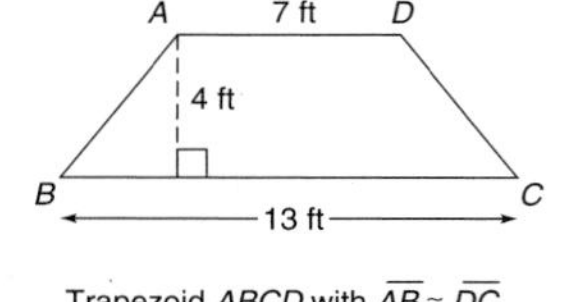


Trapezoid *ABCD* with $\overline{AB} \cong \overline{DC}$

14.
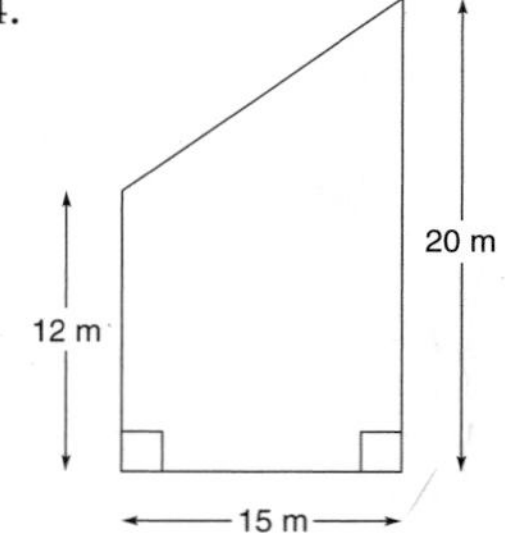


15.
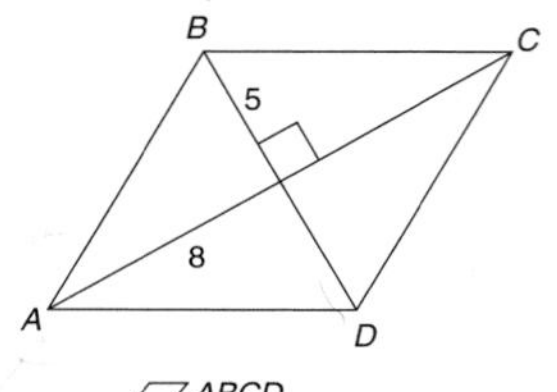


▱ABCD

16.
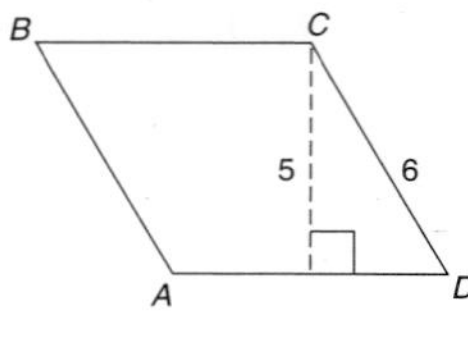


▱ABCD with $\overline{BC} \cong \overline{CD}$

17. 18.

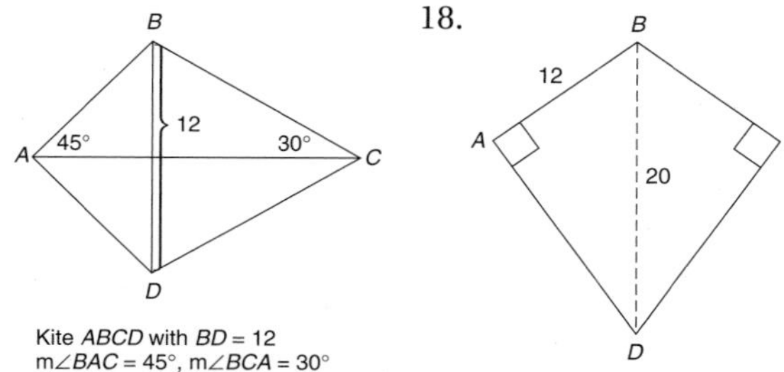


Kite $ABCD$ with $BD = 12$
$m\angle BAC = 45°$, $m\angle BCA = 30°$

Kite $ABCD$

19. In a triangle of perimeter 76 in., the length of the first side is twice the length of the second side, and the length of the third side is 12 in. more than the length of the second side. Find the lengths of the three sides.

20. In a triangle whose area is 72 in$^2$, the base has a length of 8 in. Find the length of the corresponding altitude.

21. A trapezoid has an area of 96 cm$^2$. If the altitude has a length of 8 cm and one base has a length of 9 cm, find the length of the other base.

22. The numerical difference between the area of a square and the perimeter of that square is 32. Find the length of a side of the square.

23. Find the ratio $\frac{A_1}{A_2}$ of the areas of two similar triangles if:

    a) The ratio of corresponding sides is $\frac{s_1}{s_2} = \frac{3}{2}$.
    b) The lengths of the sides of the first triangle are 6, 8, and 10 in., and those of the second triangle are 3, 4, and 5 in.

24. Find the ratio $\frac{A_1}{A_2}$ of the areas of two similar rectangles if:

    a) The ratio of corresponding sides is $\frac{s_1}{s_2} = \frac{2}{5}$
    b) The length of the first rectangle is 6 m, and the length of the second rectangle is 4 m

*In Exercises 25 and 26, give a paragraph form of proof. Provide drawings as needed.*

25. *Given:* Equilateral $\triangle ABC$ with each side of length $s$
    *Prove:* $A_{ABC} = \frac{s^2}{4}\sqrt{3}$ (HINT: Use Heron's Formula.)

26. *Given:* Isosceles $\triangle MNQ$ with $QM = QN = s$ and $MN = 2a$
    *Prove:* $A_{MNQ} = a\sqrt{s^2 - a^2}$ (NOTE: $s > a$)

*In Exercises 27 to 30, find the area of the figure shown.*

27. *Given:* In $\odot O$, $OA = 5$, $BC = 6$, and $CD = 4$
    *Find:* $A_{ABCD}$

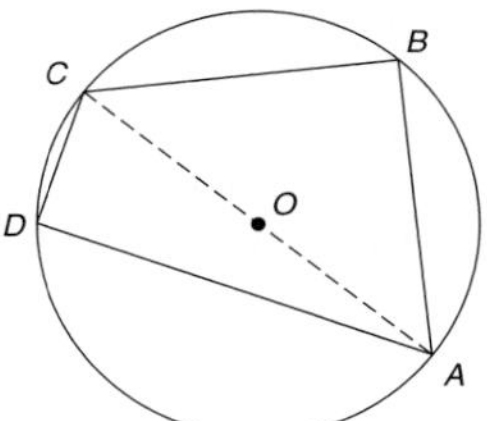


28. *Given:* Hexagon $RSTVWX$ with $\overline{WV} \parallel \overline{XT} \parallel \overline{RS}$
    $RS = 10$
    $ST = 8$
    $TV = 5$
    $WV = 16$
    $\overline{WX} \cong \overline{VT}$
    *Find:* $A_{RSTVWX}$

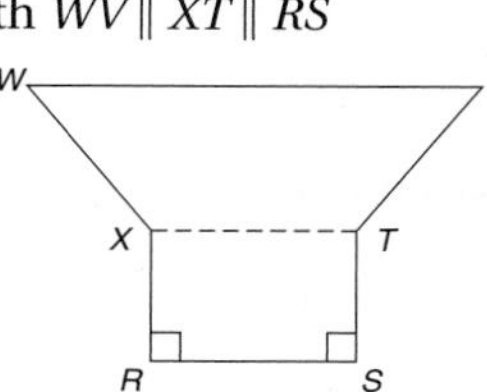


29. *Given:* Pentagon $ABCDE$ with $\overline{DC} \cong \overline{DE}$
    $AE = AB = 5$
    $BC = 12$
    *Find:* $A_{ABCDE}$

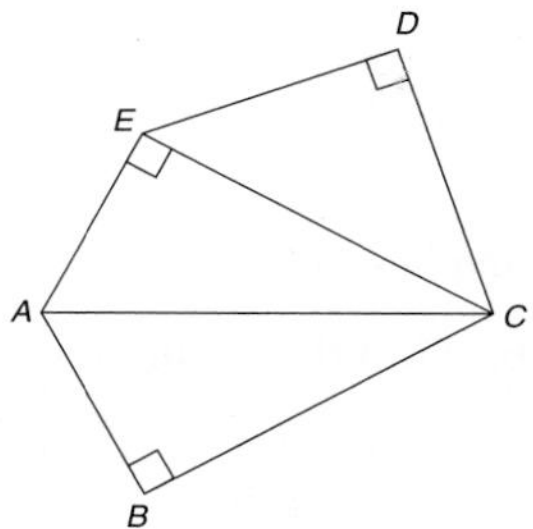


30. *Given:* Pentagon $RSTVW$ with $m\angle VRS = m\angle VSR = 60°$, $RS = 8\sqrt{2}$, and $\overline{RW} \cong \overline{WV} \cong \overline{VT} \cong \overline{TS}$
    *Find:* $A_{RSTVW}$

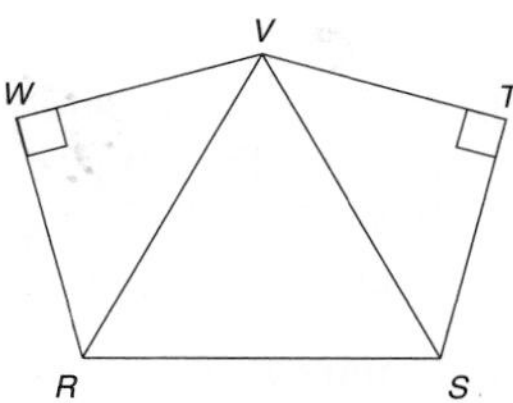


31. Mary Frances has a rectangular garden plot that encloses an area of 48 yd$^2$. If 28 yd of fencing is purchased to enclose the garden, what are the dimensions of the rectangular plot?

32. The perimeter of a right triangle is 12 m. If the hypotenuse has a length of 5 m, find the lengths of the two legs.

33. Farmer Watson wishes to fence a rectangular plot of ground measuring 245 ft by 140 ft.

    a) What amount of fencing is needed?
    b) What is the total cost of the fencing if it costs \$0.59 per foot?

34. The farmer in Exercise 33 has decided to take the fencing purchased and use it to enclose the subdivided plots shown.

    a) What are the overall dimensions of the rectangular enclosure shown?
    b) What is the total area of the enclosures shown?

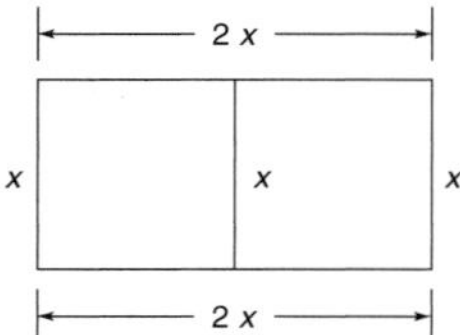


35. Find the area of the room whose floor plan is shown.

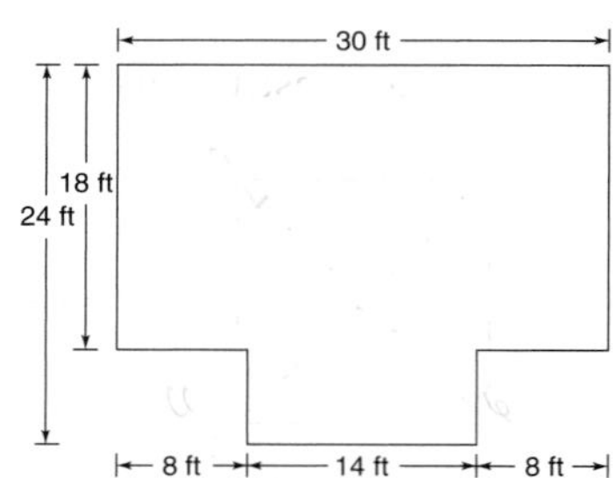


*Exercises 35, 36*

36. Find the perimeter of the room in Exercise 35.

37. Examine several rectangles, each with a perimeter of 40 in., and find the dimensions of the rectangle that has the largest area. What type of figure has the largest area?

38. Examine several rectangles, each with an area of 36 $\text{in}^2$, and find the dimensions of the rectangle that has the smallest perimeter. What type of figure has the smallest perimeter?

39. Square *RSTV* is inscribed in square *WXYZ* as shown. If $ZT = 5$ and $TY = 12$, find:

    a) The perimeter of *RSTV*
    b) The area of *RSTV*

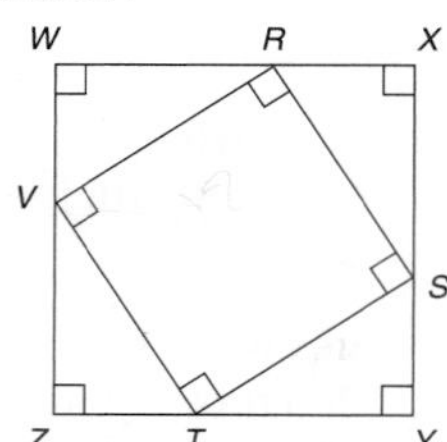


*Exercises 39, 40*

40. Square *RSTV* is inscribed in square *WXYZ* as shown. If $ZT = 8$ and $TY = 15$, find:

    a) The perimeter of *RSTV*
    b) The area of *RSTV*

41. Although not all kites are cyclic, one with sides of lengths 5 in., 1 ft, 1 ft, and 5 in. would be cyclic. Find the area of this kite. Give the resulting area in *square inches.*

42. Although not all trapezoids are cyclic, one with bases of lengths 12 cm and 28 cm and both legs of length 10 cm would be cyclic. Find the area of this isosceles trapezoid.

*For Exercises 43 and 44, use this information: Let a, b, and c be the integer lengths of the sides of a triangle. If the area of the triangle is also an integer, then (a, b, c) is known as a Heron triple.*

43. Which of these are Heron triples?

    a) (5, 6, 7) b) (13, 14, 15)

44. Which of these are Heron triples?

    a) (9, 10, 17) b) (8, 10, 12)

45. Prove that the area of a trapezoid whose altitude has length $h$ and whose median has length $m$ is $A = hm$.

*For Exercises 46 and 47, use the formula found in Exercise 45.*

46. Find the area of a trapezoid with an altitude of length 4.2 m and a median of length 6.5 m.

47. Find the area of a trapezoid with an altitude of length $5\frac{1}{3}$ ft and a median of length $2\frac{1}{4}$ ft.

48. Prove that the area of a square whose diagonal length is $d$ is $A = \frac{1}{2}d^2$.

*For Exercises 49 and 50, use the formula found in Exercise 48.*

49. Find the area of a square whose diagonal has length $\sqrt{10}$ in.

50. Find the area of a square whose diagonal has length 14.5 cm.

*51. The shaded region is that of a trapezoid. Determine the height of the trapezoid.

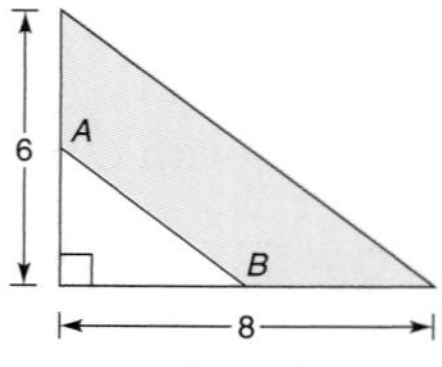


A and B are midpoints.

**KEY CONCEPTS**

Regular Polygon • Center and Central Angle of a Regular Polygon • Radius and Apothem of a Regular Polygon • Area of a Regular Polygon

# 7.3 Regular Polygons and Area

In this section, our main goal is to develop a formula for the area of any **regular polygon.** However, several interesting properties of regular polygons are developed as we move toward this goal. For instance, every regular polygon has both an inscribed circle and a circumscribed circle; furthermore, these two circles are concentric. In Example 1, we use angle bisectors of the angles of a square to locate the center of the inscribed circle. The center, which is found by using the bisectors of any two consecutive angles, is equidistant from the sides of the square.

**Reminder**

A regular polygon is both equilateral and equiangular.

EXAMPLE 1

Given square *ABCD* in Figure 7.32(a), construct inscribed ⊙*O*.

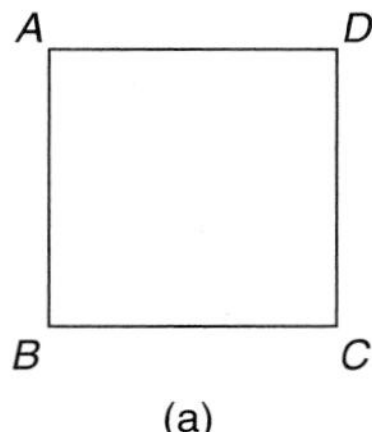


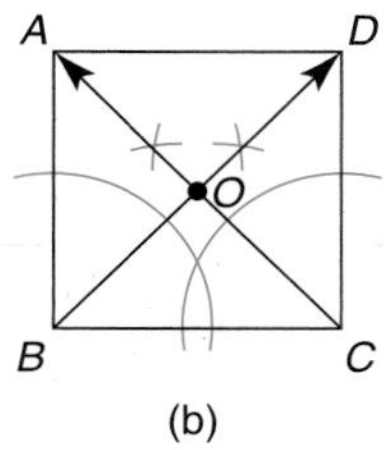


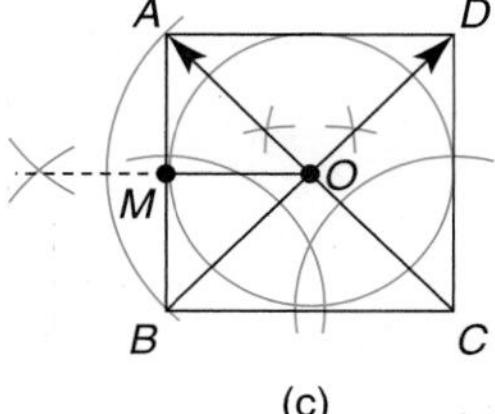


**FIGURE 7.32**

***Solution***

The center of an inscribed circle must lie at the same distance from each side. Center *O* is the point of concurrency of the angle bisectors of the square. In Figure 7.32(b), we construct the angle bisectors of $\angle B$ and $\angle C$ to identify point *O*. Constructing $\overline{OM} \perp \overline{AB}$, *OM* is the distance from *O* to $\overline{AB}$ and the length of the radius of the inscribed circle. Finally we construct inscribed ⊙*O* with radius $\overline{OM}$ in Figure 7.32(c).

In Example 2, we use the perpendicular bisectors of two consecutive sides of a regular hexagon to locate the center of the circumscribed circle. The center determines a point that is equidistant from the vertices of the hexagon.

EXAMPLE 2

Given regular hexagon *MNPQRS* in Figure 7.33(a), construct circumscribed ⊙*X*.

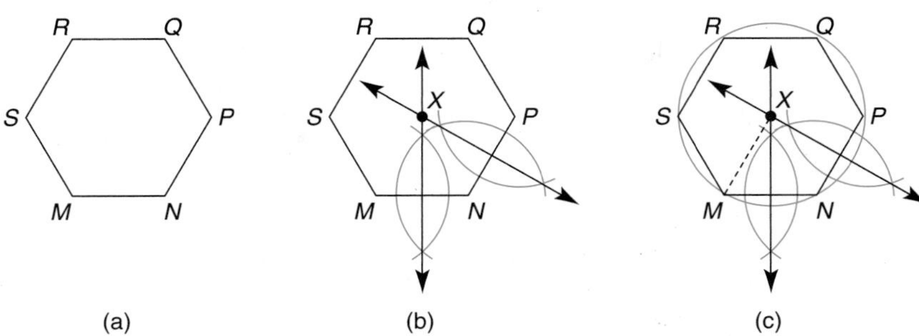


**FIGURE 7.33**

***Solution***

The center of a circumscribed circle must lie at the same distance from each vertex of the hexagon. Center $X$ is the point of concurrency of the perpendicular bisectors of two consecutive sides of the hexagon. In Figure 7.33(b) on page 357, we construct the perpendicular bisectors of $\overline{MN}$ and $\overline{NP}$ to locate point $X$. Where $XM$ is the distance from $X$ to vertex $M$, we use radius $\overline{XM}$ to construct circumscribed $\odot X$ in Figure 7.33(c).

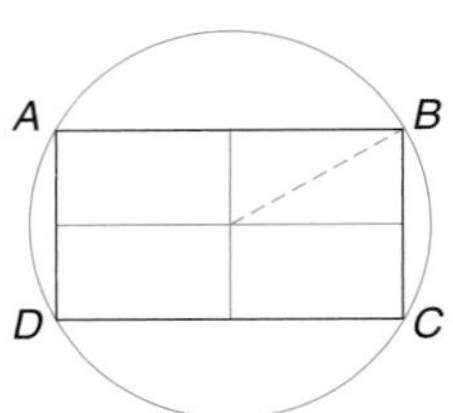


**FIGURE 7.34**

For a rectangle, which is not a regular polygon, we can only circumscribe the circle (see Figure 7.34). Why? For a rhombus (also not a regular polygon), we can only inscribe the circle (see Figure 7.35). Why?

As we shall see, we can construct the inscribed and circumscribed circles for regular polygons because they are both equilateral and equiangular. A few of the regular polygons are shown in Figure 7.36.

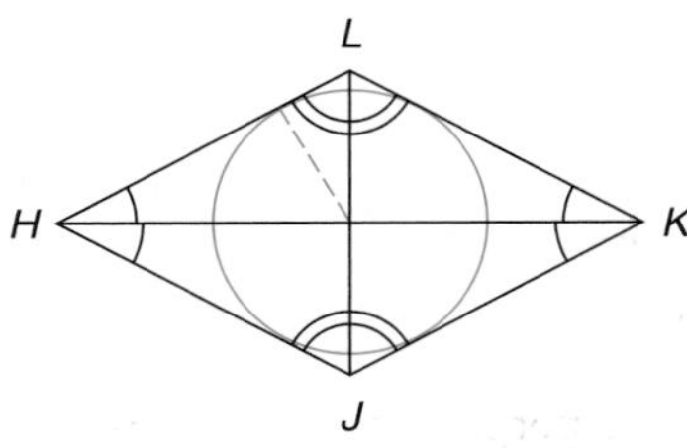


**FIGURE 7.35**

*Exs. 1–6*

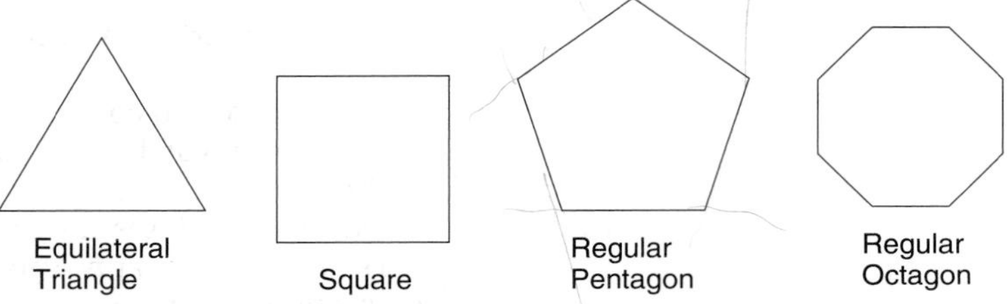


**FIGURE 7.36**

We saw earlier that the sum of the measures of the interior angles of any polygon is given by the formula

$$S = (n - 2)180$$

where $n$ is the number of sides of the polygon. The sum of the measures of the exterior angles is always 360°.

EXAMPLE 3

a) Find the measure of each interior angle of a regular polygon with 15 sides.
b) Find the number of sides of a regular polygon if each interior angle measures 144°.

***Solution***

a) Because all of the $n$ angles have equal measures, the formula for the measure of each interior angle,

$$I = \frac{(n - 2)180}{n}$$

becomes

$$I = \frac{(15 - 2)180}{15}$$

which simplifies to 156°.

b) Because $I = 144°$, we can determine the number of sides by solving the equation

$$\frac{(n - 2)180}{n} = 144$$

Then
$$(n - 2)180 = 144n$$
$$180n - 360 = 144n$$
$$36n = 360$$
$$n = 10$$

NOTE: In Example 3(a), we could have found the measure of each exterior angle and then used the fact that the interior angle is its supplement. In Example 3(b), the supplement of the interior angle is the exterior angle; then we could have used the formula $E = \frac{360}{n}$ to find $n$. Because $I = 144°$, the value of $E$ is 36°.

Exs. 7, 8

Regular polygons allow us to inscribe and to circumscribe a circle. The proof of the following theorem will establish the following relationships:

1. The centers of the inscribed and circumscribed circles of a regular polygon are the same.
2. The angle bisectors of two consecutive angles or the perpendicular bisectors of two consecutive sides can be used to locate the common center of the inscribed circle and the circumscribed circle.
3. The inscribed circle's radius is any line segment from the center drawn perpendicular to a side of the regular polygon; the radius of the circumscribed circle joins the center to any vertex of the regular polygon.

THEOREM 7.3.1: A circle can be circumscribed about (or inscribed in) any regular polygon.

*Given:* Regular polygon *ABCDEF* [See Figure 7.37(a).]

*Prove:* A circle *O* can be circumscribed about *ABCDEF* and a circle with center *O* can be inscribed in *ABCDEF*.

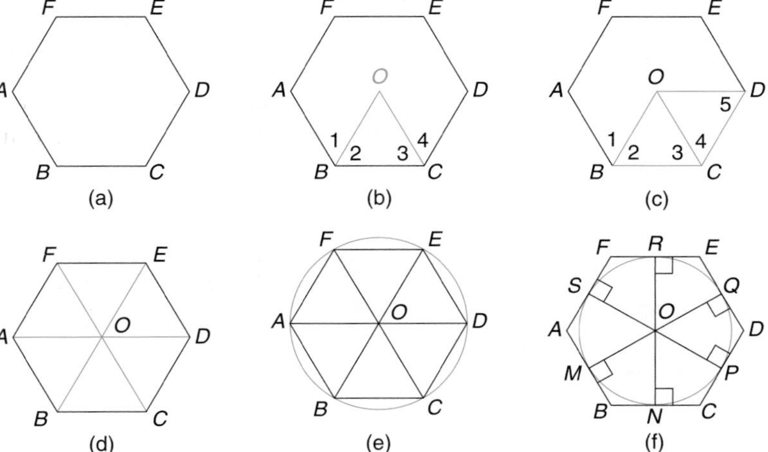


**FIGURE 7.37**

*Proof:* Let point $O$ be the point at which the angle bisectors for $\angle ABC$ and $\angle BCD$ meet. [See Figure 7.37(b) on page 359.] Then $\angle 1 \cong \angle 2$ and $\angle 3 \cong \angle 4$.

Because $\angle ABC \cong \angle BCD$ (by the definition of a regular polygon), it follows that

$$\frac{1}{2}\text{m}\angle ABC = \frac{1}{2}\text{m}\angle BCD$$

In turn, $\text{m}\angle 2 = \text{m}\angle 3$, so $\angle 2 \cong \angle 3$. Then $\overline{OB} \cong \overline{OC}$ (sides opposite $\cong$ $\angle$s of a $\triangle$).

From the facts that $\angle 3 \cong \angle 4$, $\overline{OC} \cong \overline{OC}$, and $\overline{BC} \cong \overline{CD}$, it follows that $\triangle OCB \cong \triangle OCD$ by SAS. [See Figure 7.37(c).] In turn, $\overline{OC} \cong \overline{OD}$ by CPCTC, so $\angle 4 \cong \angle 5$ because these lie opposite $\overline{OC}$ and $\overline{OD}$.

Because $\angle 5 \cong \angle 4$ and $\text{m}\angle 4 = \frac{1}{2}\text{m}\angle BCD$, it follows that $\text{m}\angle 5 = \frac{1}{2}\text{m}\angle BCD$. But $\angle BCD \cong \angle CDE$ because these are angles of a regular polygon. Thus $\text{m}\angle 5 = \frac{1}{2}\text{m}\angle CDE$, and $\overline{OD}$ bisects $\angle CDE$.

By continuing this procedure, we can show that $\overline{OE}$ bisects $\angle DEF$, $\overline{OF}$ bisects $\angle EFA$, and $\overline{OA}$ bisects $\angle FAB$. Therefore, the resulting $\triangle AOB$, $\triangle BOC$, $\triangle COD$, $\triangle DOE$, $\triangle EOF$, and $\triangle FOA$ are congruent by ASA. [See Figure 7.37(d).] By CPCTC, $\overline{OA} \cong \overline{OB} \cong \overline{OC} \cong \overline{OD} \cong \overline{OE} \cong \overline{OF}$. With $O$ as center and $\overline{OA}$ as radius, circle $O$ can be circumscribed about $ABCDEF$, as shown in Figure 7.37(e).

Because corresponding altitudes of $\cong$ $\triangle$s are also congruent, we see that $\overline{OM} \cong \overline{ON} \cong \overline{OP} \cong \overline{OQ} \cong \overline{OR} \cong \overline{OS}$, where these are the altitudes to the bases of the triangles.

Again with $O$ as center, but now with a radius equal in length to $OM$, we complete the inscribed circle in $ABCDEF$. [See Figure 7.37(f).]

In the proof of Theorem 7.3.1 a regular hexagon was drawn. The method of proof would not change, regardless of the number of sides of the polygon chosen. In the proof, point $O$ was the common center of the circumscribed and inscribed circles for $ABCDEF$.

Because any regular polygon can be inscribed in a circle, any regular polygon is cyclic.

> DEFINITION: The **center of a regular polygon** is the common center for the inscribed and circumscribed circles of the polygon.

NOTE: The preceding definition does not tell us how to locate the center of a regular polygon. The center is the intersection of the angle bisectors of two consecutive angles; alternatively, the intersection of the perpendicular bisectors of two consecutive sides can be used to locate the center of the regular polygon. Note that a regular polygon has a center, whether or not either of the related circles is shown.

In Figure 7.38, point $O$ is the center of the regular pentagon $RSTVW$. In this figure, $\overline{OR}$ is called a "radius" of the regular pentagon.

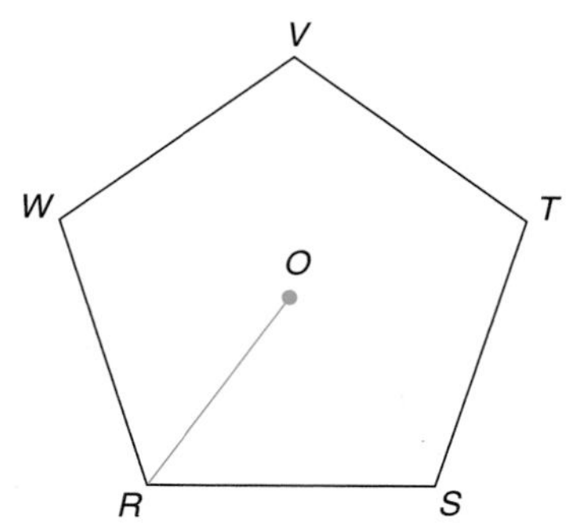

**FIGURE 7.38**

> DEFINITION: A **radius of a regular polygon** is any line segment that joins the center of the regular polygon to one of its vertices.

In the proof of Theorem 7.3.1, we saw that "All radii of a regular polygon are congruent."

> DEFINITION: An **apothem** of a regular polygon is any line segment drawn from the center of that polygon perpendicular to one of the sides.

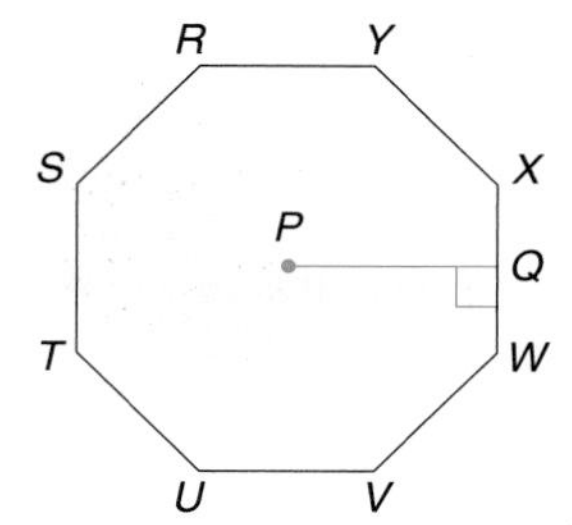


**FIGURE 7.39**

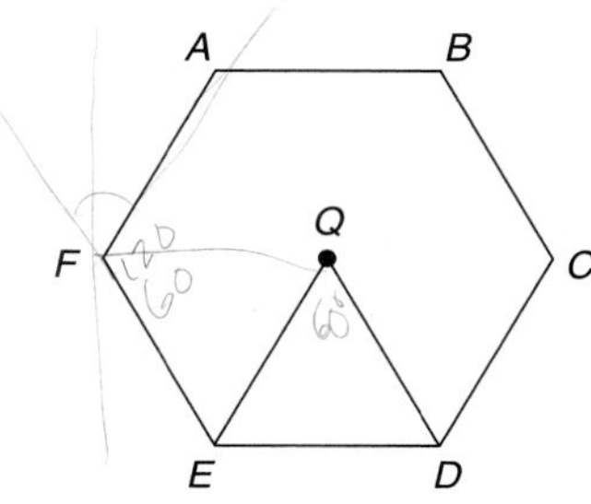


**FIGURE 7.40**

In regular octagon *RSTUVWXY* with center *P* (see Figure 7.39), the segment $\overline{PQ}$ is an apothem. Any regular polygon of *n* sides has *n* apothems and *n* radii. The proof of Theorem 7.3.1 establishes that "All apothems of a regular polygon are congruent."

DEFINITION: A **central angle of a regular polygon** is an angle formed by two consecutive radii of the regular polygon.

In regular hexagon *ABCDEF* with center *Q* (see Figure 7.40), angle *EQD* is a central angle. Because of the congruences of the triangles in the proof of Theorem 7.3.1, "All central angles of a regular polygon are congruent."

THEOREM 7.3.2: The measure of the central angle of a regular polygon of *n* sides is given by $c = \frac{360}{n}$.

We apply Theorem 7.3.2 in Example 4.

EXAMPLE 4

a) Find the measure of the central angle of a regular polygon of 9 sides.
b) Find the number of sides of a regular polygon whose central angle measures 72°.

***Solution***

a) $c = \frac{360}{9} = 40°$
b) $72 = \frac{360}{n} \rightarrow 72n = 360 \rightarrow n = 5$ sides

The next two theorems are stated without proof.

THEOREM 7.3.3: Any radius of a regular polygon bisects the angle at the vertex to which it is drawn.

THEOREM 7.3.4: Any apothem of a regular polygon bisects the side of the polygon to which it is drawn.

EXAMPLE 5

Given that each side of regular hexagon *ABCDEF* has the length 4 in., find the length of:

a) Radius $\overline{QE}$
b) Apothem $\overline{QG}$

***Solution***

a) By Theorem 7.3.2, the measure of $\angle EQD$ is $\frac{360°}{6}$, or 60°. With $\overline{QE} \cong \overline{QD}$, $\triangle QED$ is equiangular and equilateral. Then $QE = 4$ in.
b) With apothem $\overline{QG}$ as shown, $\triangle QEG$ is a 30°-60°-90° triangle. By Theorem 7.3.4, $EG = 2$ in.
With $\overline{QG}$ opposite the 60° angle of $\triangle QEG$, it follows that $QG = 2\sqrt{3}$ in.

*Exs. 9–18*

## AREA OF A REGULAR POLYGON

We have laid the groundwork for determining the area of a regular polygon. In the proof of Theorem 7.3.5, the figure chosen is a regular pentagon; however, the proof applies to regular polygons of any number of sides.

It is also worth noting that the perimeter $P$ of a regular polygon is the sum of its equal sides. If there are $n$ sides and each has length $s$, the perimeter of the regular polygon is $P = ns$.

> THEOREM 7.3.5: The area $A$ of a regular polygon whose apothem has length $a$ and whose perimeter $P$ is given by
>
> $$A = \frac{1}{2}aP$$

*Given:* Regular polygon $ABCDE$ in Figure 7.41(a) such that $OF = a$ and the perimeter of $ABCDE$ is $P$

*Prove:* $A_{ABCDE} = \frac{1}{2}aP$

*Proof:* From center $O$, draw radii $\overline{OA}$, $\overline{OB}$, $\overline{OC}$, $\overline{OD}$, and $\overline{OE}$. [See Figure 7.41(b).] Now $\triangle AOB$, $\triangle BOC$, $\triangle COD$, $\triangle DOE$, and $\triangle EOA$ are all $\cong$ by SSS. Where $s$ represents the length of each of the congruent sides of the regular polygon and $a$ is the length of an apothem, the area of each $\triangle$ is $\frac{1}{2}sa$ (from $A = \frac{1}{2}bh$). Therefore, the area of the pentagon is

$$A_{ABCDE} = \left(\frac{1}{2}sa\right) + \left(\frac{1}{2}sa\right) + \left(\frac{1}{2}sa\right) + \left(\frac{1}{2}sa\right) + \left(\frac{1}{2}sa\right)$$
$$= \frac{1}{2}a(s + s + s + s + s)$$

Because the sum $s + s + s + s + s$ represents the perimeter of the polygon, we have

$$A_{ABCDE} = \frac{1}{2}aP$$

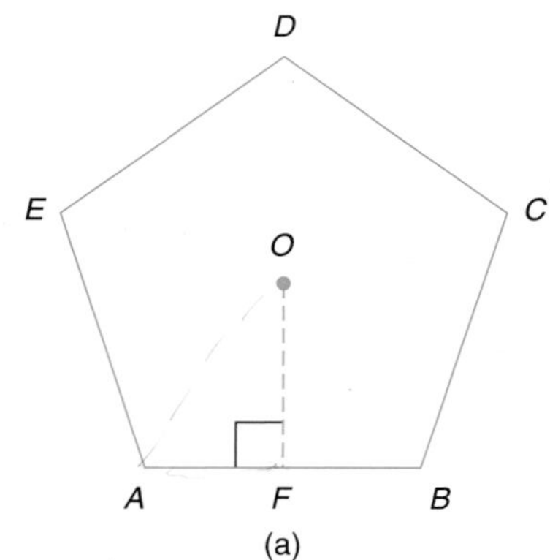

(a)

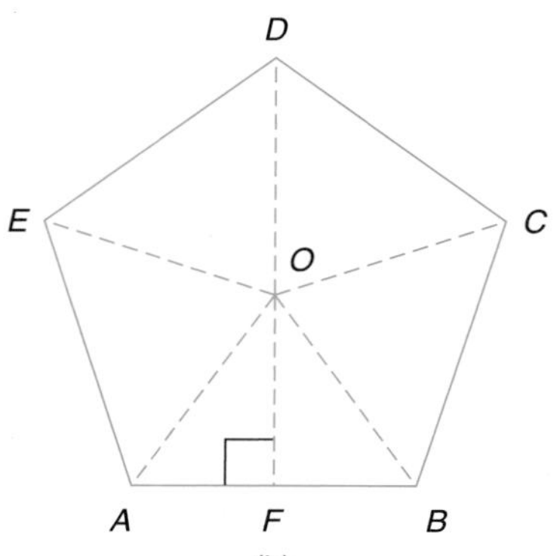

(b)

**FIGURE 7.41**

### EXAMPLE 6

In Figure 7.41(a), find the area of the regular pentagon $ABCDE$ with center $O$ if $OF = 4$ and $AB = 5.9$.

***Solution***

$OF = a = 4$ and $AB = 5.9$. Therefore, $P = 5(5.9)$ or $P = 29.5$. Consequently,

$$A_{ABCDE} = \frac{1}{2} \cdot 4(29.5)$$
$$= 59 \text{ units}^2$$

### EXAMPLE 7

Find the exact area of equilateral triangle $ABC$ in Figure 7.42 if each side measures 12 in.

***Solution***

In $\triangle ABC$, the perimeter is $P = 3 \cdot 12$ or 36 in.

To find the length $a$ of an apothem, we draw the radius $\overline{OA}$ from center $O$ to point $A$ and the apothem $\overline{OM}$ from $O$ to side $\overline{AB}$. Because the radius bisects $\angle BAC$, $m\angle OAB = 30°$. Because apothem $\overline{OM} \perp \overline{AB}$, $m\angle OMA = 90°$. Using the 30°-60°-90° relationship in $\triangle OMA$, we see that $a\sqrt{3} = 6$. Thus

$$a = \frac{6}{\sqrt{3}} = \frac{6}{\sqrt{3}} \cdot \frac{\sqrt{3}}{\sqrt{3}} = \frac{6\sqrt{3}}{3} = 2\sqrt{3}$$

Now $A = \frac{1}{2}aP$ becomes $A = \frac{1}{2} \cdot 2\sqrt{3} \cdot 36 = 36\sqrt{3}$ in$^2$.

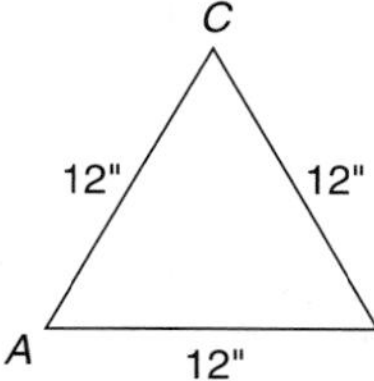


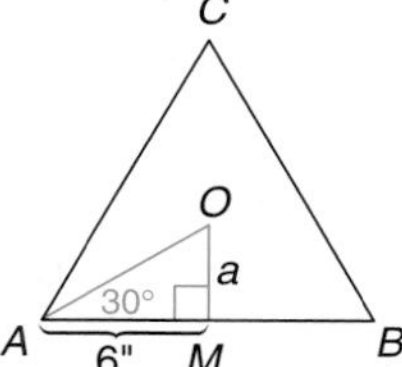


**FIGURE 7.42**

NOTE: Using the calculator's value for $\sqrt{3}$ leads to an approximation of the area rather than to an exact area.

*Exs. 19–22*

## Discover!

TESSELLATIONS

Tessellations are patterns composed strictly of interlocking and non-overlapping regular polygons. All of the regular polygons of a given number of sides will be congruent. Tessellations are commonly used in design, but especially in flooring (tiles and vinyl sheets). A *pure tessellation* is one formed by using only one regular polygon in the pattern. An *impure tessellation* is one formed by using two different regular polygons.

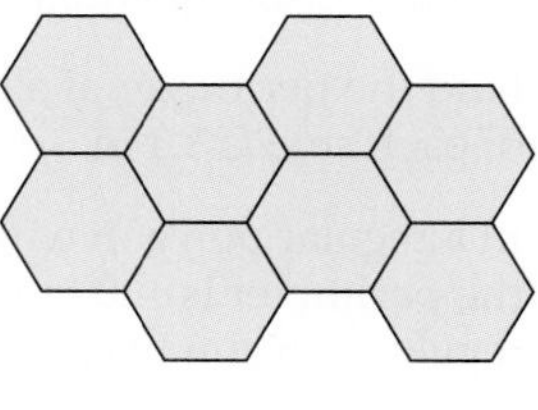

In the accompanying pure tessellation, only the regular hexagon appears. In nature, the beehive has compartments that are regular hexagons. Note that the adjacent angles' measures must sum to 360°; in this case, 120° + 120° + 120°= 360°. It would also be possible to form a pure tessellation of congruent squares because the sum of the adjacent angles' measures would be 90° + 90° + 90° + 90° = 360°.

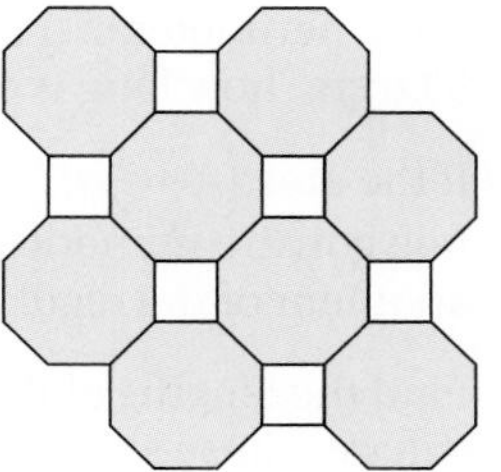

In the impure tessellation shown, the regular octagon and the square are used. In Champaign-Urbana, sidewalks found on the University of Illinois campus use this tessellation pattern. Again it is necessary that the sum of the adjacent angles' measures be 360°; for this impure tessellation, 135° + 135° + 90° = 360°.

a) Can congruent equilateral triangles be used to form a pure tessellation?

b) Can two regular hexagons and a square be used to build an impure tessellation?

ANSWERS

a) Yes, because 6 × 60° = 360° b) No, because 120° + 120° + 90° ≠ 360°

## 7.3 Exercises

1. Describe, if possible, how you would inscribe a circle within kite $ABCD$.

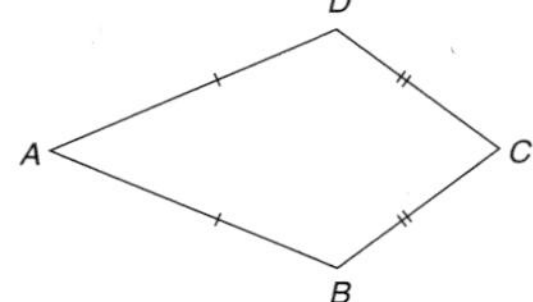


*Exercises 1, 2*

2. What condition must be satisfied for it to be possible to circumscribe a circle about kite $ABCD$?

3. Describe, if possible, how you would inscribe a circle in rhombus $JKLM$.

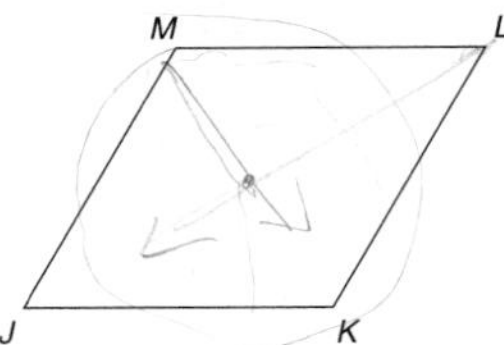


4. What condition must be satisfied for it to be possible to circumscribe a circle about trapezoid $RSTV$?

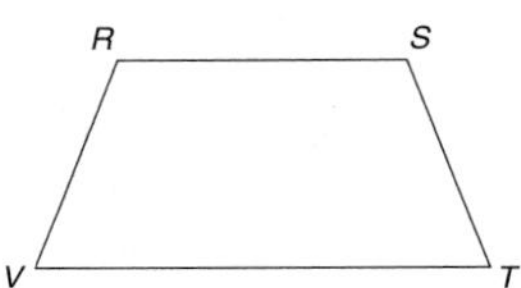


*In Exercises 5 to 8, perform constructions.*

5. Inscribe a regular octagon within a circle.

6. Inscribe an equilateral triangle within a circle.

7. Circumscribe a square about a circle.

8. Circumscribe an equilateral triangle about a circle.

9. Find the perimeter of a regular octagon if the length of each side is 3.4 in.

10. In a regular polygon with each side of length 6.5 cm, the perimeter is 130 cm. How many sides does the regular polygon have?

11. If the perimeter of a regular dodecagon (12 sides) is 99.6 cm, how long is each side?

12. If the area $\left(A = \frac{1}{2}aP\right)$ and the perimeter of a regular polygon are numerically equal, find the length of the apothem of the regular polygon.

13. Find the lengths of the apothem and the radius of a square whose sides have length 10 in.

14. Find the lengths of the apothem and the radius of a regular hexagon whose sides have length 6 cm.

15. Find the lengths of the side and the radius of an equilateral triangle whose apothem's length is 8 ft.

16. Find the lengths of the side and the radius of a regular hexagon whose apothem's length is 10 m.

17. Find the measure of the central angle of a regular polygon of

    a) three sides.
    b) four sides.
    c) five sides.
    d) six sides.

18. Find the measure of the central angle of a regular polygon of

    a) 8 sides.
    b) 10 sides.
    c) 9 sides.
    d) 12 sides.

19. Find the number of sides of a regular polygon that has a central angle measuring

    a) 90°.
    b) 45°.
    c) 60°.
    d) 24°.

20. Find the number of sides of a regular polygon that has a central angle measuring

    a) 30°.
    b) 72°.
    c) 36°.
    d) 20°.

*In Exercises 21 to 30, use the formula $A = \frac{1}{2}aP$ to find the area of the regular polygon described.*

21. Find the area of a regular pentagon with an apothem of length $a = 5.2$ cm and each side of length $s = 7.5$ cm.

22. Find the area of a regular pentagon with an apothem of length $a = 6.5$ in. and each side of length $s = 9.4$ in.

23. Find the area of a regular octagon with an apothem of length $a = 9.8$ in. and each side of length $s = 8.1$ in.

24. Find the area of a regular octagon with an apothem of length $a = 7.9$ ft and each side of length $s = 6.5$ ft.

25. Find the area of a regular hexagon whose sides have length 6 cm.

26. Find the area of a square whose apothem measures 5 cm.

27. Find the area of an equilateral triangle whose radius measures 10 in.

28. Find the approximate area of a regular pentagon whose apothem measures 6 in. and each of whose sides measures approximately 8.9 in.

29. In a regular polygon of 12 sides, the measure of each side is 2 in., and the measure of an apothem is $(2 + \sqrt{3})$ in. Find the area of this regular polygon.

30. In a regular octagon, the measure of each apothem is 4 cm, and each side measures $8(\sqrt{2} - 1)$ cm. Find the area of this regular polygon.

31. Find the ratio of the area of a square circumscribed about a circle to the area of a square inscribed in the circle.

32. Given regular hexagon *ABCDEF* with each side of length 6, find the length of diagonal $\overline{AC}$.
(HINT: With *G* on $\overline{AC}$, draw $\overline{BG} \perp \overline{AC}$.)

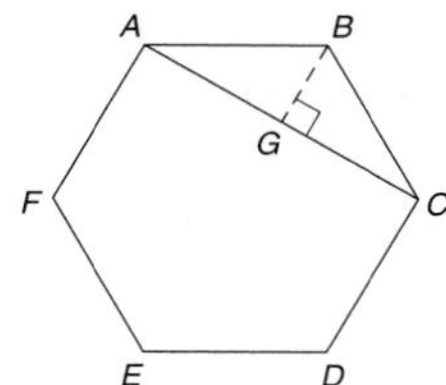


33. Given regular octagon *RSTUVWXY* with each side of length 4, find the length of diagonal *RU*.
(HINT: Extended sides, as shown, form a square.)

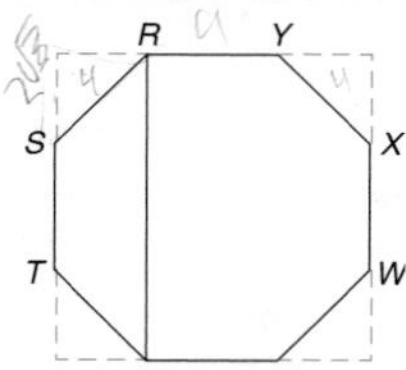


34. Regular octagon *ABCDEFGH* is inscribed in a circle whose radius is $\frac{7}{2}\sqrt{2}$ cm. Considering that the area of the octagon is less than the area of the circle and greater than the area of the square *ACEG*, find the two integers between which the area of the octagon must lie.
(NOTE: For the circle, use $A = \pi r^2$ with $\pi \approx \frac{22}{7}$.)

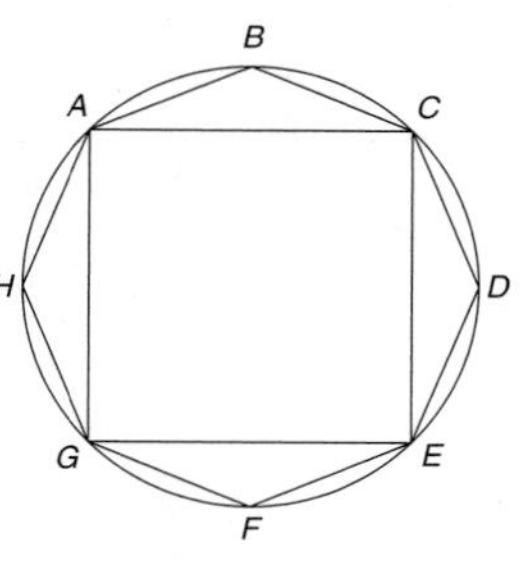


35. *Given:* Regular pentagon *RSTVQ* with equilateral $\triangle PQR$
*Find:* $m\angle VPS$

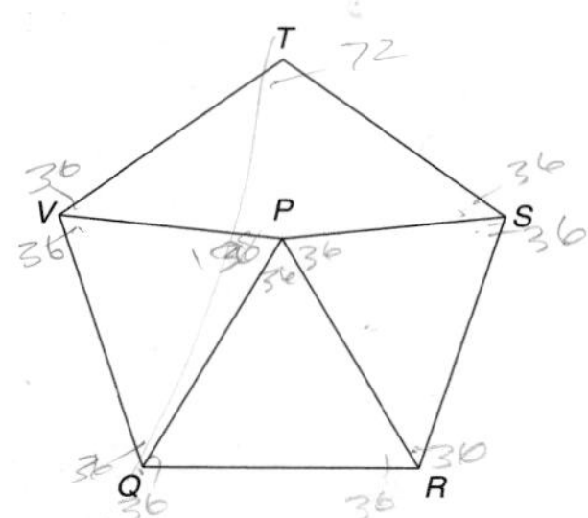


36. *Given:* Regular pentagon *JKLMN* (not shown) with diagonals $\overline{LN}$ and $\overline{KN}$
*Find:* $m\angle LNK$

*37. *Prove:* If a circle is divided into *n* congruent arcs ($n \geq 3$), the chords determined by joining consecutive endpoints of these arcs form a regular polygon.

*38. *Prove:* If a circle is divided into *n* congruent arcs ($n \geq 3$), the tangents drawn at the endpoints of these arcs form a regular polygon.

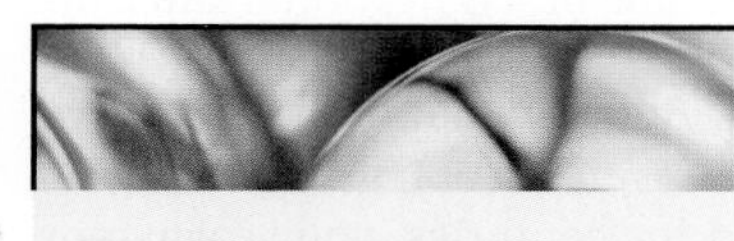

**KEY CONCEPTS**

Circumference of a Circle • $\pi$ (Pi) • Length of an Arc • Limit • Area of a Circle

# 7.4 Circumference and Area of a Circle

In geometry, any two figures that have the same shape are described as similar. For this reason, we say that all circles are similar to each other. Just as a proportionality exists among the sides of similar triangles, experimentation shows that there is a proportionality among the circumferences (distances around) and diameters (distances across) of circles; see the Discover! activity on page 366. Representing the

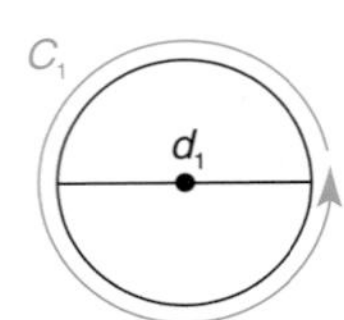


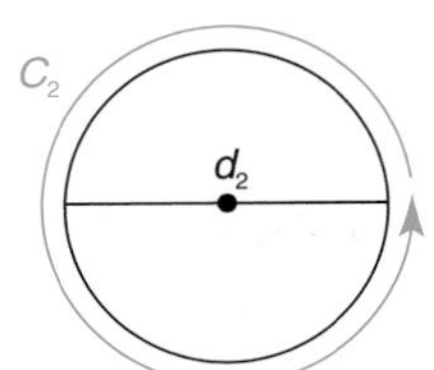


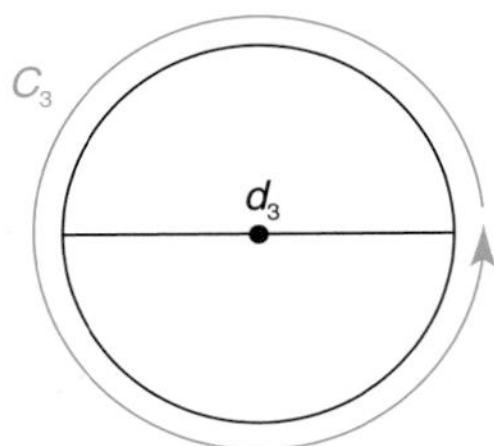


**FIGURE 7.43**

circumferences of the circles in Figure 7.43 by $C_1$, $C_2$, and $C_3$ and the diameters by $d_1$, $d_2$, and $d_3$, we claim that

$$\frac{C_1}{d_1} = \frac{C_2}{d_2} = \frac{C_3}{d_3} = k$$

where $k$ is the constant of proportionality.

POSTULATE 22:
The ratio of the circumference of a circle to the length of its diameter is a unique positive constant.

### Discover!

Find an object of circular shape, such as the lid of a jar. Using a flexible tape measure (such as a seamstress or carpenter might use), measure both the distance around (circumference) and the distance across (length of diameter) the circle. Now divide the circumference $C$ by the diameter length $d$. What is your result?

ANSWER
The ratio should be slightly larger than 3.

The constant of proportionality $k$ described in the opening paragraph of this section, in Postulate 22, and in the Discover! activity is represented by the Greek letter $\pi$ (pi).

### Technology Exploration

Use computer software if available.

1) Draw a circle with center $O$.
2) Through $O$, draw diameter $\overline{AB}$.
3) Measure the circumference $C$ and length $d$ of diameter $\overline{AB}$.
4) Show that $\frac{C}{d} \approx 3.14$

DEFINITION: $\pi$ is the ratio between the circumference $C$ and the diameter length $d$ of any circle; thus $\pi = \frac{C}{d}$ in any circle.

THEOREM 7.4.1: The circumference of a circle is given by the formula

$$C = \pi d \quad \text{or} \quad C = 2\pi r$$

*Given:* Circle $O$ with radius $r$ (See Figure 7.44.)

*Prove:* $C = 2\pi r$

*Proof:* By Postulate 22, $\pi = \frac{C}{d}$. Multiplying each side of the equation by $d$, we have $C = \pi d$. Because $d = 2r$ (the diameter's length is twice that of the radius), the formula for the circumference can be written $C = \pi(2r)$, or $C = 2\pi r$.

*Exs. 1–2*

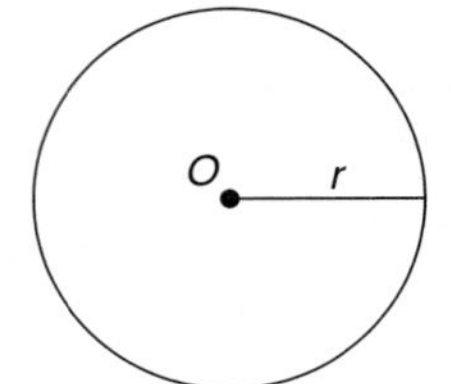


**FIGURE 7.44**

## VALUE OF $\pi$

In calculating the circumference of a circle, we generally leave the symbol $\pi$ in the answer in order to state an *exact* result. However, the value of $\pi$ is irrational and cannot be represented exactly by a common fraction or by a terminating decimal. When an approximation is needed for $\pi$, we use a calculator. Approximations of $\pi$ that have been commonly used throughout history include $\pi \approx \frac{22}{7}$, $\pi \approx 3.14$, and $\pi \approx 3.1416$. Although these approximate values have been used for centuries, your calculator provides greater accuracy. A calculator may show that $\pi \approx 3.141592654$.

EXAMPLE 1

In $\odot O$ in Figure 7.45, $OA = 7$ cm. Using $\pi \approx \frac{22}{7}$,

a) find the approximate circumference C of $\odot O$.
b) find the approximate length of the minor arc $\overset{\frown}{AB}$.

Reminder
More background regarding the value of $\pi$ can be found in the Perspective on History in Chapter 6.

***Solution***

a) $C = 2\pi r$
$= 2 \cdot \frac{22}{7} \cdot 7$
$= 44$ cm

b) Because the degree of measure of $\widehat{AB}$ is 90°, we have $\frac{90}{360}$ or $\frac{1}{4}$ of the circumference for the arc length. Then

$$\text{length of } \widehat{AB} = \frac{90}{360} \cdot 44 = \frac{1}{4} \cdot 44 = 11 \text{ cm}$$

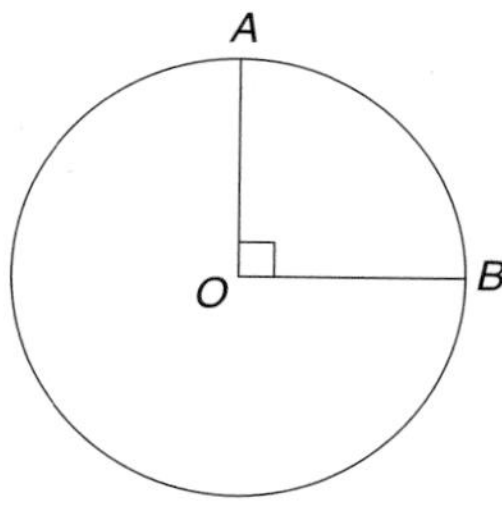

**FIGURE 7.45**

EXAMPLE 2

The exact circumference of a circle is $17\pi$ in.

a) Find the length of the radius.
b) Find the length of the diameter.

***Solution***

a) $C = 2\pi r$
$17\pi = 2\pi r$
$\frac{17\pi}{2\pi} = \frac{2\pi r}{2\pi}$
$r = \frac{17}{2} = 8.5$ in.

b) Because $d = 2r$, $d = 2(8.5)$ or $d = 17$ in.

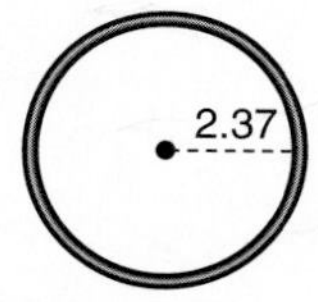

**FIGURE 7.46**

EXAMPLE 3

A thin circular rubber gasket is used as a seal to prevent oil from leaking from a tank (see Figure 7.46). If the gasket has a radius of 2.37 in., use the value of $\pi$ provided by your calculator to find the circumference of the gasket to the nearest hundredth of an inch.

***Solution***

Using the calculator with $C = 2\pi r$, we have $C = 2 \cdot \pi \cdot 2.37$ or $C \approx 14.89114918$. Rounding to the nearest hundredth of an inch, $C \approx 14.89$ in.

*Exs. 3–5*

## LENGTH OF AN ARC

In Example 1(b), we used the phrase *length of arc* without a definition. Informally, the length of an arc is the distance between the endpoints of the arc as though it were measured along a straight line. If we measured one-third of the circumference of the rubber gasket (a 120° arc) in Example 3, we would expect the length to be slightly less than 5 in. This measurement could be accomplished by holding that part of the gasket taut in a straight line, but not so tightly that it would be stretched.

Two further observations can be made with regard to the measurement of arc length.

1. The ratio of the degree measure $m$ of the arc to 360 (the degree measure of the entire circle) is the same as the ratio of the length $\ell$ of the arc to the circumference; that is, $\frac{m}{360} = \frac{\ell}{C}$.

2. Just as $m\widehat{AB}$ denotes the degree measure of an arc, $\ell\widehat{AB}$ denotes the length of the arc. Whereas $m\widehat{AB}$ is measured in degrees, $\ell\widehat{AB}$ is measured in linear units such as inches, feet, or centimeters.

THEOREM 7.4.2: In a circle whose circumference is $C$, the length $\ell$ of an arc whose degree measure is $m$ is given by

$$\ell = \frac{m}{360} \cdot C$$

NOTE: For arc $AB$, $\ell\widehat{AB} = \frac{m\widehat{AB}}{360} \cdot C$.

*Exs. 6–8*

EXAMPLE 4

Find the approximate length of major arc $ABC$ in a circle of radius 7 in. if $m\widehat{AC} = 45°$. (See Figure 7.47.) Use $\pi \approx \frac{22}{7}$.

***Solution***

$m\widehat{ABC} = 360° - 45° = 315°$. Theorem 7.4.2 tells us that $\ell\widehat{ABC} = \frac{m\widehat{ABC}}{360} \cdot C$, or $\ell\widehat{ABC} = \frac{315}{360} \cdot 2 \cdot \frac{22}{7} \cdot 7$, which can be simplified to $\ell\widehat{ABC} = 38\frac{1}{2}$ in.

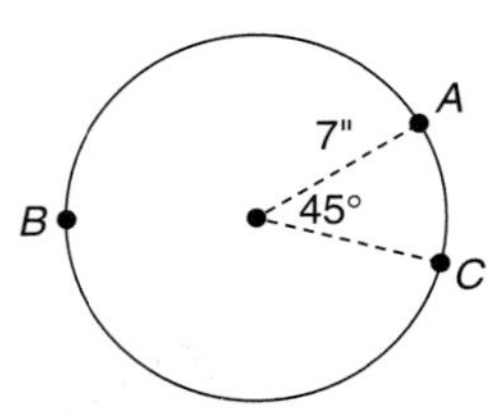


**FIGURE 7.47**

## LIMITS

In the discussion that follows, we use the undefined term *limit;* in practice, a limit represents a numerical measure. In some situations, we seek an upper limit, a lower limit, or both. The following example illustrates this notion.

EXAMPLE 5

Find the upper limit (largest possible number) for the length of a chord in a circle whose length of radius is 5 cm.

***Solution***

By considering several chords in the circle in Figure 7.48, we see that the greatest possible length of a chord is that of a diameter. Thus the limit of the length of a chord is 10 cm.

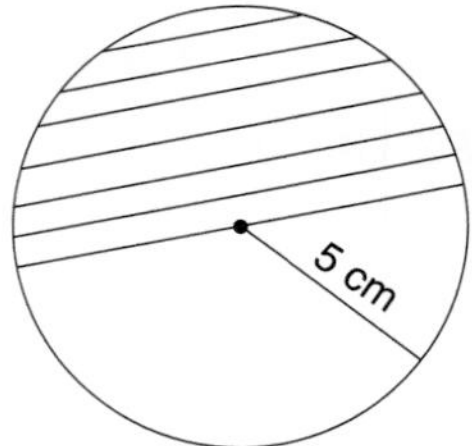


**FIGURE 7.48**

*Exs. 9–11*

NOTE: In Example 5, the lower limit is 0.

## AREA OF A CIRCLE

Now consider the problem of finding the area of a circle. To do so, let a regular polygon of $n$ sides be inscribed in the circle. As we allow $n$ to grow larger (often written as $n \rightarrow \infty$ and read as "$n$ approaches infinity"), two observations can be made:

1. The length of an apothem of the regular polygon approaches the length of a radius of the circle as its limit ($a \rightarrow r$).
2. The perimeter of the regular polygon approaches the circumference of the circle as its limit ($P \rightarrow C$).

Geometry in the Real World

A measuring wheel can be used by a police officer to find the length of skid marks or by a cross-country coach to determine the length of a running course.

In Figure 7.49, the area of an inscribed regular polygon with $n$ sides approaches the area of the circle as its limit as $n$ increases. Using observations 1 and 2, we make the following claim. Because the formula for the area of a regular polygon is

$$A = \frac{1}{2}aP$$

the area of the circumscribed circle is given by the limit

$$A = \frac{1}{2}rC$$

Because $C = 2\pi r$, this formula becomes

$$A = \frac{1}{\not{2}}r(\not{2}\pi r) \text{ or } A = \pi r^2$$

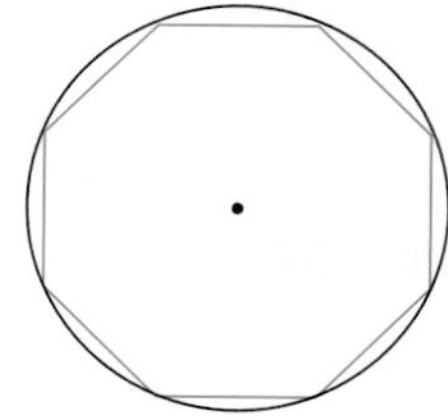

**FIGURE 7.49**

> THEOREM 7.4.3: The area $A$ of a circle whose radius has length $r$ is given by $A = \pi r^2$.

## Discover!

AREA OF A CIRCLE

Use a protractor to divide a circle into several congruent "sectors." For instance, 15° central angles will divide the circle into $\frac{360}{15} = 24$ sectors. If these sectors are alternated as shown, the resulting figure approximates a parallelogram. With the parallelogram having a base of length $\pi r$ (half the circumference of the circle) and an altitude of length $r$ (radius of the circle), the area of the parallelogram (and of the circle) can be seen to be $A = (\pi r)r$, or $A = \pi r^2$.

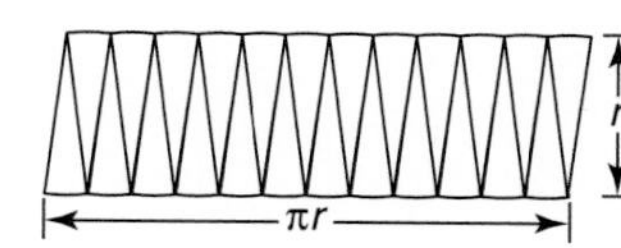


### EXAMPLE 6

Find the approximate area of a circle whose radius has a length of 10 in. (Use $\pi \approx 3.14$.)

***Solution***

$A = \pi r^2$ becomes $A = 3.14(10)^2$. Then

$$A = 3.14(100) = 314 \text{ in}^2$$

### EXAMPLE 7

The approximate area of a circle is 38.5 cm². Find the length of the radius of the circle. $\left(\text{Use } \pi \approx \frac{22}{7}\right)$.

***Solution***

$A = \pi r^2$ becomes $38.5 = \frac{22}{7} \cdot r^2$, or $\frac{77}{2} = \frac{22}{7} \cdot r^2$. Multiplying each side of the equation by $\frac{7}{22}$, we have

$$\frac{7}{\underset{2}{\not{22}}} \cdot \frac{\overset{7}{\not{77}}}{2} = \frac{\not{7}}{\not{22}} \cdot \frac{\not{22}}{\not{7}} \cdot r^2$$

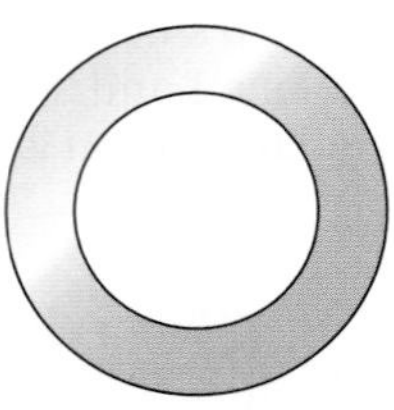

FIGURE 7.50

or $$r^2 = \frac{49}{4}$$

Taking the positive square root for the approximate length of radius,

$$r = \sqrt{\frac{49}{4}} = \frac{\sqrt{49}}{\sqrt{4}} = \frac{7}{2} = 3.5 \text{ cm}$$

*Exs. 12–17*

A plane figure bounded by concentric circles is known as a *ring* or *annulus* (see Figure 7.50). The piece of hardware known as a *washer* has the shape of an annulus.

EXAMPLE 8

A machine cuts washers from a flat piece of metal. The radius of the inside circular boundary of the washer is 0.3 in., and the radius of the outer circular boundary is 0.5 in. What is the area of the annulus? Give both an exact answer and an approximate answer rounded to tenths of a square inch. Using the approximate answer, determine the number of square inches of material used to produce 1000 washers. Figure 7.50 illustrates the shape of a washer.

***Solution***

Where $R$ is the larger radius and $r$ is the smaller radius, $A = \pi R^2 - \pi r^2$. Then $A = \pi(0.5)^2 - \pi(0.3)^2$, or $A = 0.16\pi$. The exact number of square inches used in producing a washer is $0.16\pi \text{ in}^2$, or approximately $0.5 \text{ in}^2$. When 1000 washers are produced, approximately $500 \text{ in}^2$ of metal is used.

NOTE: Many people have a difficult time remembering which expression ($2\pi r$ or $\pi r^2$) is used in the formula for the circumference or area of a circle. This is understandable, because each expression contains a 2, a radius $r$, and the factor $\pi$. To remember that $C = 2\pi r$ gives the circumference and $A = \pi r^2$ gives the area, *think about the units involved.* Considering a circle of radius 3 in., $C = 2\pi r$ becomes $C = 2 \times 3.14 \times 3$ in., or Circumference equals 18.84 *inches.* (We measure the *distance around* a circle in *linear* units such as inches.) For the circle of radius 3 in., $A = \pi r^2$ becomes $A = 3.14 \times 3 \text{ in.} \times 3 \text{ in.}$ or Area equals $28.26 \text{ in}^2$. (We measure the area of a circular region in *square* units.)

## 7.4 Exercises

1. Find the exact circumference and area of a circle whose radius has length 8 cm.

2. Find the exact circumference and area of a circle whose diameter has length 10 in.

3. Find the approximate circumference and area of a circle whose radius has length $10\frac{1}{2}$ in. Use $\pi \approx \frac{22}{7}$.

4. Find the approximate circumference and area of a circle whose diameter is 20 cm. Use $\pi \approx 3.14$.

5. Find the exact lengths of a radius and a diameter of a circle whose circumference is:

   a) $44\pi$ in. b) $60\pi$ ft

6. Find the approximate lengths of a radius and a diameter of a circle whose circumference is:

   a) 88 in. $\left(\text{Use } \pi \approx \frac{22}{7}.\right)$ b) 157 m (Use $\pi \approx 3.14$.)

7. Find the exact lengths of a radius and a diameter of a circle whose area is:

   a) $25\pi$ in$^2$ b) $2.25\pi$ cm$^2$

8. Find the exact length of a radius and the exact circumference of a circle whose area is:

   a) $36\pi$ m$^2$ b) $6.25\pi$ ft$^2$

9. Find the exact length $\ell\widehat{AB}$, where $\widehat{AB}$ refers to the minor arc of the circle.

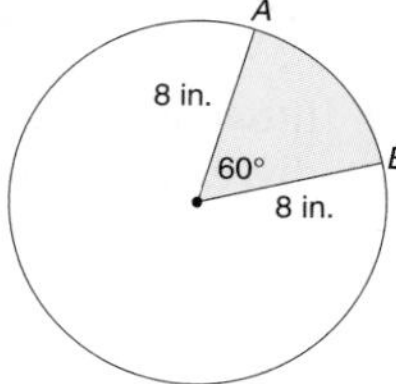


10. Find the exact length $\ell\widehat{CD}$ of the minor arc shown.

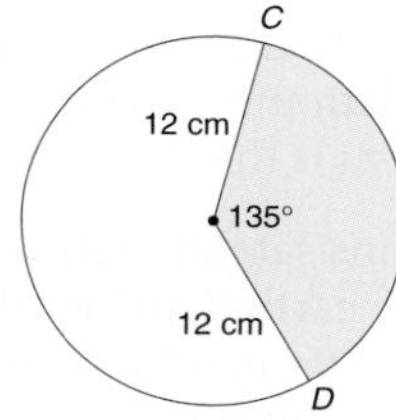


11. Use your calculator value of $\pi$ to find the approximate circumference of a circle with radius 12.38 in.

12. Use your calculator value of $\pi$ to find the approximate area of a circle with radius 12.38 in.

13. A metal circular disk whose area is 143 cm$^2$ is used as a knockout on an electrical service in a factory. Use your calculator value of $\pi$ to find the radius of the disk to the nearest tenth of a centimeter.

14. A circular lock washer whose outside circumference measures 5.48 cm is used in an electric box to hold an electrical cable in place. Use your calculator value of $\pi$ to find the outside radius to the nearest tenth of a centimeter.

15. The central angle corresponding to a circular brake shoe measures 60°. Approximately how long is the curved surface of the brake shoe if the circle has a radius of 7 in.?

16. Use your calculator to find the approximate radius and diameter of a circle with area 56.35 in$^2$.

17. A rectangle has a perimeter of 16 in. What is the limit (largest possible value) of the area of the rectangle?

18. Two sides of a triangle measure 5 in. and 7 in. What is the limit of the length of the third side?

19. Let $N$ be any point on side $\overline{BC}$ of the right triangle $ABC$. Find the upper and lower limits for the length of $\overline{AN}$.

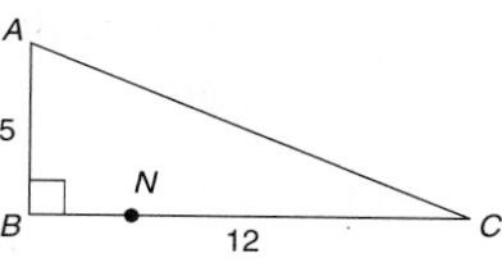


20. What is the limit of m$\angle RTS$ if $T$ lies in the interior of the shaded region?

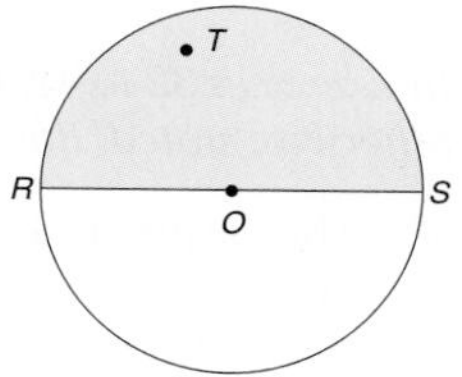


*In Exercises 21 to 24, find the exact areas of the shaded regions.*

21.

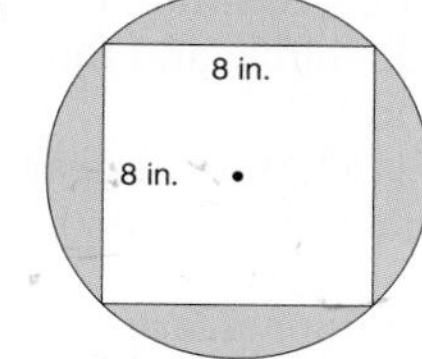


Square inscribed in a circle

22.

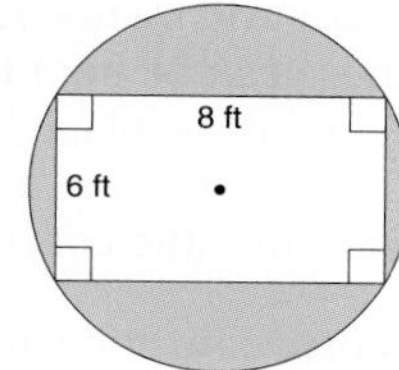


23.

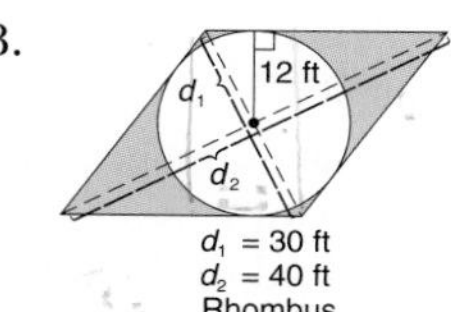


24.

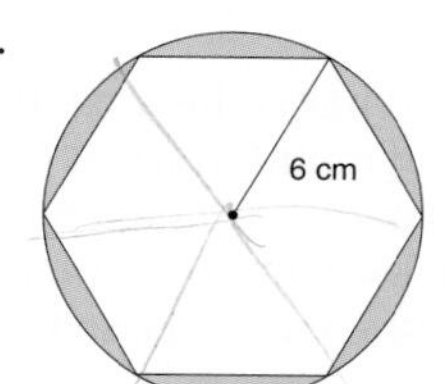


Regular hexagon inscribed in a circle

*In Exercises 25 and 26, use your calculator value of $\pi$ to solve each problem. Round answers to the nearest integer.*

25. Find the length of the radius of a circle whose area is 154 cm$^2$.

26. Find the length of the diameter of a circle whose circumference is 157 in.

27. Assuming that a 90° arc has an exact length of $4\pi$ in., find the length of the radius of the circle.

28. The ratio of the circumferences of two circles is 2:1. What is the ratio of their areas?

29. Given concentric circles with radii of lengths $R$ and $r$, where $R > r$, explain why $Area_{\text{ring}} = \pi(R + r)(R - r)$.

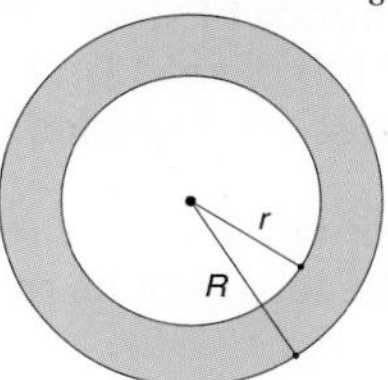

30. Given a circle with diameter of length $d$, explain why $A_{\text{circle}} = \frac{1}{4}\pi d^2$.

31. The radii of two concentric circles differ in length by exactly 1 in. If their areas differ by exactly $7\pi$ in$^2$, find the lengths of the radii of the two circles.

*In Exercises 32 to 42, use your calculator value of $\pi$ unless otherwise stated. Round answers to two decimal places.*

32. The carpet in the circular entryway of a church needs to be replaced. The diameter of the circular region to be carpeted is 18 ft.

   a) What length (in feet) of a metal protective strip is needed to bind the circumference of the carpet?
   b) If the metal strips are sold in lengths of 6 ft, how many will be needed? (NOTE: Assume that these can be bent to follow the circle and that they can be placed end to end.)
   c) If the cost of the metal strip is \$1.59 per linear foot, find the cost of the metal strips needed.

33. At center court on a gymnasium floor, a large circular emblem is to be painted. The circular design has a radius of 8 ft.

   a) What is the area to be painted?
   b) If a pint of paint covers 70 ft$^2$, how many pints of paint are needed to complete the job?
   c) If each pint of paint costs \$2.95, find the cost of the paint needed.

34. A track is to be constructed around the football field at a junior high school. If the straightaways are 100 yd in length, what length of radius is needed for each of the semicircles shown if the total length around the track is to be 440 yd?

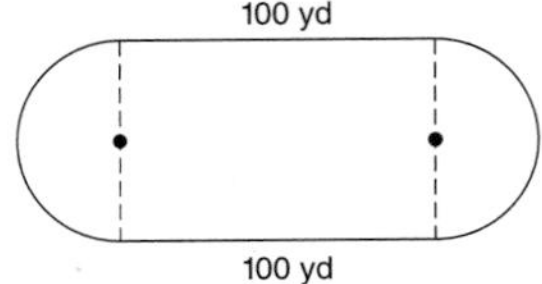


35. A circular grass courtyard at a shopping mall has a 40-ft diameter. This area needs to be reseeded.

   a) What is the total area to be reseeded? (Use $\pi \approx 3.14$.)
   b) If 1 lb of seed is to be used to cover a 60-ft$^2$ region, how many pounds of seed will be needed?
   c) If the cost of 1 lb of seed is \$1.65, what is the total cost of the grass seed needed?

36. Find the approximate area of a regular polygon that has 20 sides if the length of its radius is 7 cm.

37. Find the approximate perimeter of a regular polygon that has 20 sides if the length of its radius is 7 cm.

38. In a two-pulley system, the centers of the pulleys are 20 in. apart. If the radius of each pulley measures 6 in., how long is the belt used in the pulley system?

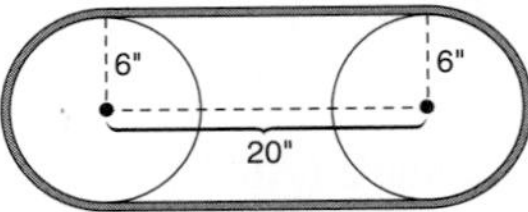


39. If two gears, each of radius 4 in., are used in a chain drive system with a chain of length 54 in., what is the distance between the centers of the gears?

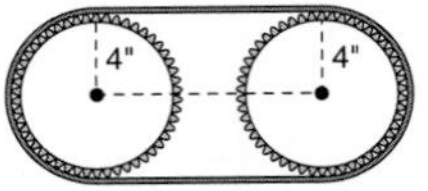


40. A pizza with a 12-in. diameter costs \$6.95. A 16-in. diameter pizza with the same ingredients costs \$9.95. Which pizza is the better buy?

41. A communications satellite forms a circular orbit 375 mi above the earth. If the earth's radius is approximately 4000 mi, what distance is traveled by the satellite in one complete orbit?

42. The radius of the Ferris wheel's circular path is 40 ft. If a "ride" of 12 revolutions is made in 3 minutes, at what rate in *feet per second* is the passenger in a cart moving during the ride?

* 43. Given that the length of each side of a rhombus is 8 cm and that an interior angle (shown) measures 60°, find the area of the inscribed circle.

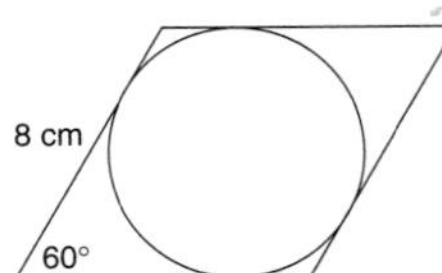

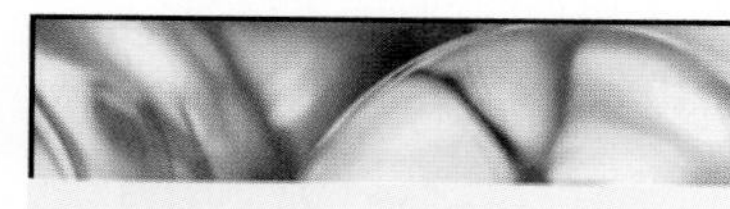

**KEY CONCEPTS**

Sector • Area and Perimeter of Sector • Segment of a Circle • Area and Perimeter of Segment • Area of Triangle with Inscribed Circle

# 7.5 More Area Relationships in the Circle

DEFINITION: A **sector** of a circle is a region bounded by two radii of the circle and an arc intercepted by those radii. (See Figure 7.51.)

A sector will generally be shaded to avoid confusion about whether the arc is intended to be a major or a minor arc. In simple terms, the sector of a circle generally has the shape of a piece of pie.

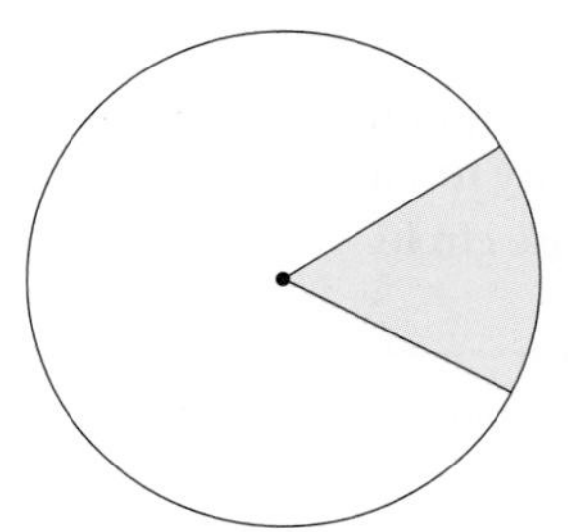

**FIGURE 7.51**

## AREA OF A SECTOR

Just as the length of an arc is part of the circle's circumference, the area of a sector is part of the area of this circle. When fractions are illustrated by using circles, $\frac{1}{4}$ is represented by shading a 90° sector, and $\frac{1}{3}$ is represented by shading a 120° sector (see Figure 7.52). Thus we make the following assumption about the measure of the area of a sector.

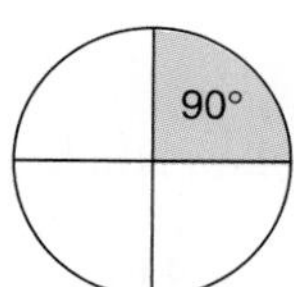


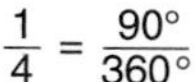


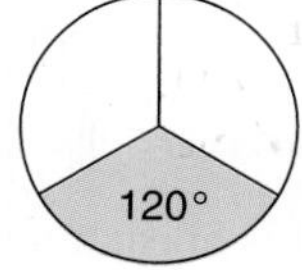


$\frac{1}{3} = \frac{120°}{360°}$

**FIGURE 7.52**

POSTULATE 23:
The ratio of the degree measure $m$ of the arc (or central angle) of a sector to 360° is the same as the ratio of the area of the sector to the area of the circle; that is,

$$\frac{\text{area of sector}}{\text{area or circle}} = \frac{m}{360}.$$

THEOREM 7.5.1: In a circle of radius $r$, the area $A$ of a sector whose arc has degree measure $m$ is given by

$$A = \frac{m}{360}\pi r^2$$

Theorem 7.5.1 follows directly from Postulate 23.

EXAMPLE 1

If m∠$O$ = 100°, find the area of the 100° sector shown in Figure 7.53. Use your calculator and round the answer to the nearest hundredth.

***Solution***

$$A = \frac{m}{360}\pi r^2$$

becomes $$A = \frac{100}{360} \cdot \pi \cdot 10^2 \approx 87.27 \text{ in}^2$$

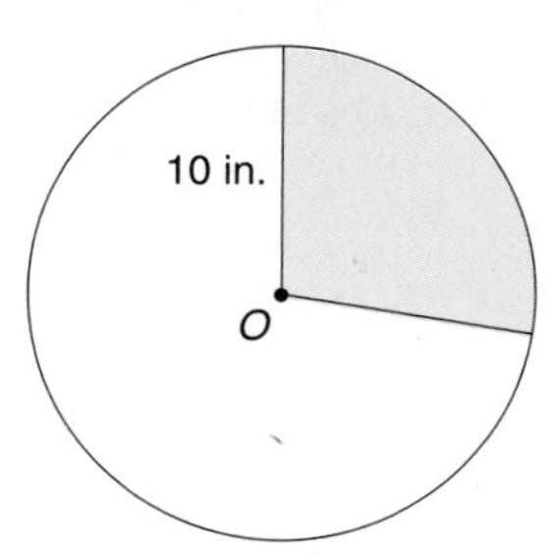


**FIGURE 7.53**

In applications with circles, we sometimes need exact answers for circumference and area; in such cases, we simply leave $\pi$ in the result. For instance, in a circle of radius length 5 in., the exact circumference is $10\pi$ in. and the exact area is expressed as $25\pi$ in$^2$.

Because a sector is bounded by two radii and an arc, the perimeter of a sector is the sum of the lengths of the two radii and the length of its arc. In Example 2, we apply this formula, $P_{\text{sector}} = 2r + \ell\widehat{AB}$.

**EXAMPLE 2**

Find the perimeter of the sector shown in Figure 7.53 on page 373. Use the calculator value of $\pi$ and round your answer to the nearest hundredth of an inch.

***Solution***

Because $r = 10$ and $m\angle O = 100°$, $\ell\widehat{AB} = \frac{100}{360} \cdot 2 \cdot \pi \cdot 10 \approx 17.45$ in. Now $P_{\text{sector}} = 2r + \ell\widehat{AB}$ becomes $P_{\text{sector}} = 2(10) + 17.45 \approx 37.45$ in.

Because a semicircle is one-half of a circle, a semicircular region corresponds to a central angle of 180°. As stated in the following corollary to Theorem 7.5.1, the area of the semicircle is $\frac{180}{360}$ (or one-half) the area of the entire circle.

*Exs. 1–6*

**COROLLARY 7.5.2:** The area of a semicircular region of radius $r$ is $A = \frac{1}{2}\pi r^2$.

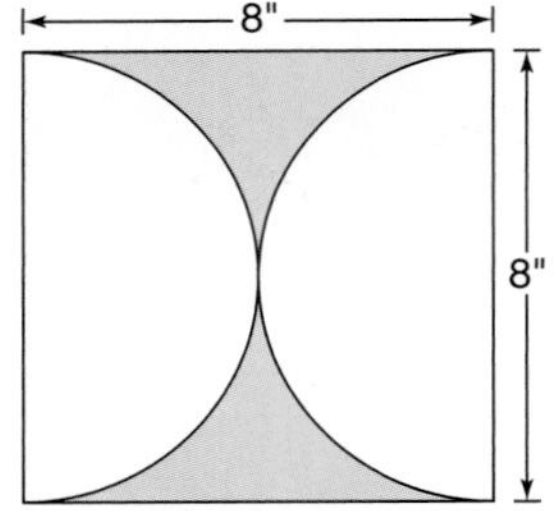


**FIGURE 7.54**

**EXAMPLE 3**

In Figure 7.54, a square of side 8 in. is shown with semicircles cut away. Find the exact shaded area by leaving $\pi$ in the answer.

***Solution***

To find the shaded area $A$, we see that $A + 2 \cdot A_{\text{semicircle}} = A_{\text{square}}$. It follows that $A = A_{\text{square}} - 2 \cdot A_{\text{semicircle}}$.

If the side of the square is 8 in., then the radius of each semicircle is 4 in. Now $A = 8^2 - 2\left(\frac{1}{2}\pi \cdot 4^2\right)$, or $A = 64 - 2(8\pi)$, so $A = (64 - 16\pi)$ in$^2$.

**Discover!**

In statistics, a pie chart can be used to represent the breakdown of a budget. In the pie chart shown, a 90° sector (one-fourth the area of the circle) is shaded to show that 25% of a person's income (one-fourth of the income) is devoted to rent payment. What degree measure of sector must be shaded if a sector indicates that 20% of the person's income is used for a car payment?

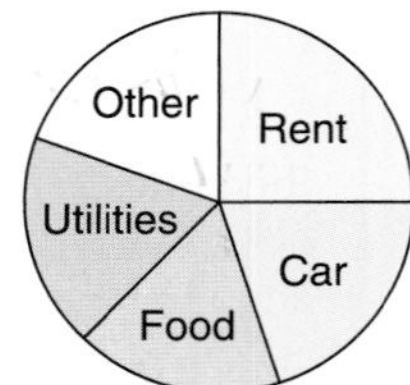


ANSWER

72° (from 20% of 360°)

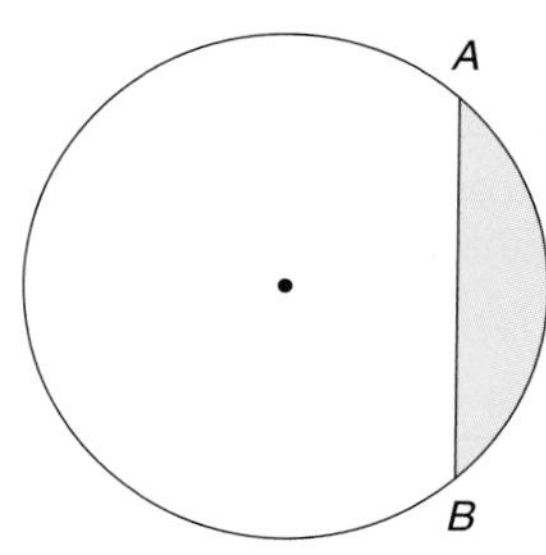


FIGURE 7.55

## AREA OF A SEGMENT

DEFINITION: A **segment** of a circle is a region bounded by a chord and its minor (or major) arc.

In Figure 7.55, the segment is bounded by chord $\overline{AB}$ and its minor arc $\overset{\frown}{AB}$. Again, we avoid confusion by shading the segment whose area or perimeter we seek.

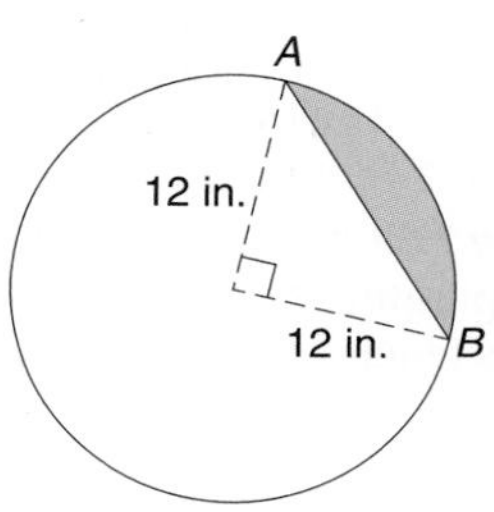


FIGURE 7.56

EXAMPLE 4

Find the exact area of the segment bounded by a chord and an arc whose measure is 90°. The radius has length 12 in., as shown in Figure 7.56.

***Solution***

Let $A_\triangle$ represent the area of the triangle shown. Because $A_\triangle + A_{\text{segment}} = A_{\text{sector}}$,

$$\begin{aligned} A_{\text{segment}} &= A_{\text{sector}} - A_\triangle \\ &= \frac{90}{360} \cdot \pi \cdot 12^2 - \frac{1}{2} \cdot 12 \cdot 12 \\ &= (36\pi - 72)\text{ in}^2. \end{aligned}$$

In Example 4, the boundaries of the segment shown are chord $\overline{AB}$ and minor arc $\overset{\frown}{AB}$. Therefore, the perimeter of the segment is given by $P_{\text{segment}} = AB + \ell\overset{\frown}{AB}$. We use this formula in Example 5.

EXAMPLE 5

Find the exact perimeter of the segment described in Example 4 (see Figure 7.57). Then use your calculator to approximate this answer to the nearest hundredth of an inch.

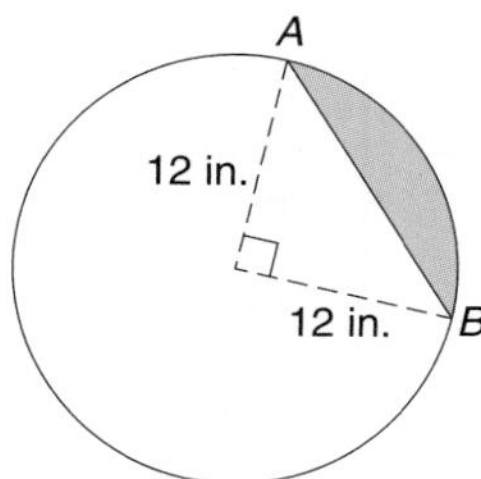


FIGURE 7.57

***Solution***

Because $\ell\overset{\frown}{AB} = \frac{90}{360} \cdot 2 \cdot \pi \cdot r$, we have $\ell\overset{\frown}{AB} = \frac{1}{4} \cdot 2 \cdot \pi \cdot 12 = 6\pi$ in.

Using either the Pythagorean Theorem or the 45°-45°-90° relationship, $AB = 12\sqrt{2}$.

Now $P_{\text{segment}} = AB + \ell\overset{\frown}{AB}$ becomes $P_{\text{segment}} = (12\sqrt{2} + 6\pi)$ in. Using a calculator, we find that the approximate perimeter is 35.82 in.

*Exs. 7–11*

## AREA OF A TRIANGLE WITH AN INSCRIBED CIRCLE

THEOREM 7.5.3: Where $P$ represents the perimeter of a triangle and $r$ represents the length of the radius of its inscribed circle, the area of the triangle is given by

$$A = \frac{1}{2}rP$$

### PICTURE PROOF OF THEOREM 7.5.3

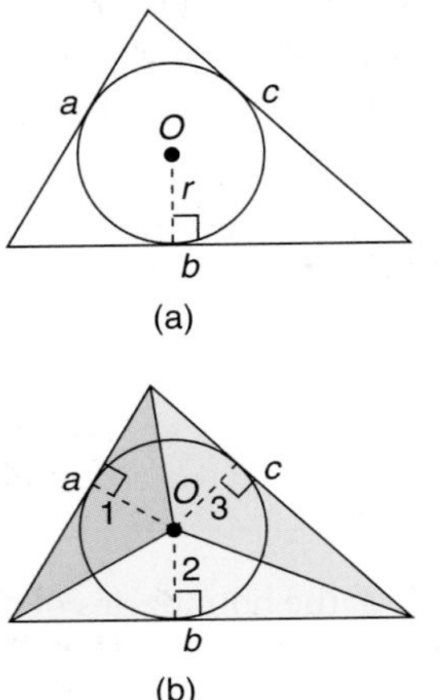


**FIGURE 7.58**

*Given:* A triangle with perimeter $P$, whose sides measure $a$, $b$, and $c$; the radius of the inscribed circle measures $r$. See Figure 7.58(a).

*Prove:* $A = \frac{1}{2}rP$

*Proof:* In Figure 7.58(b), the triangle has been separated into three smaller triangles (each with altitude $r$). Hence

$$A = A_1 + A_2 + A_3$$
$$A = \frac{1}{2}r \cdot a + \frac{1}{2}r \cdot b + \frac{1}{2}r \cdot c$$
$$A = \frac{1}{2}r(a + b + c)$$
$$A = \frac{1}{2}rP$$

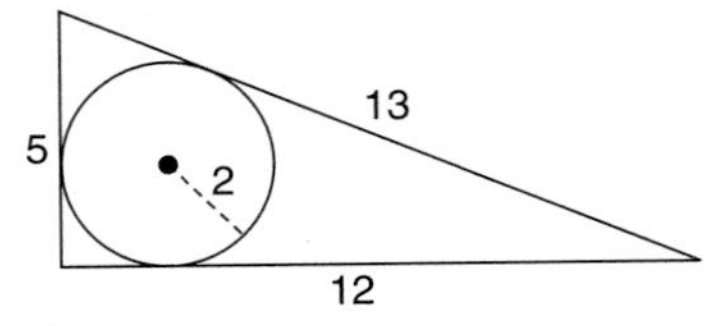


**FIGURE 7.59**

EXAMPLE 6

Find the area of a triangle whose sides measure 5 cm, 12 cm, and 13 cm if the radius of the inscribed circle is 2 cm. (See Figure 7.59.)

***Solution***

Using the given lengths of sides, the perimeter of the triangle is $P = 5 + 12 + 13 = 30$ cm. Using $A = \frac{1}{2}rP$, we have $A = \frac{1}{2} \cdot 2 \cdot 30$, or $A = 30 \text{ cm}^2$.

Because the triangle shown in Example 6 is a right triangle ($5^2 + 12^2 = 13^2$), the area of the triangle could have been determined by using either $A = \frac{1}{2}ab$ or $A = \sqrt{s(s - a)(s - b)(s - c)}$. The advantage provided by Theorem 7.5.3 lies in applications where we need to determine the length of the radius of the inscribed circle of a triangle.

EXAMPLE 7

In an attic, wooden braces supporting the roof form a triangle whose sides measure 4 ft, 6 ft, and 6 ft; see Figure 7.60 on page 377. To the nearest inch, find the radius of the largest circular cold-air duct that can be run through the opening formed by the braces.

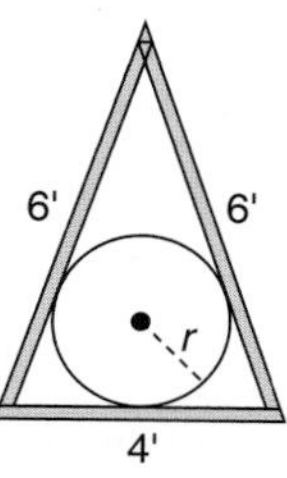


**FIGURE 7.60**

*Exs. 11–15*

***Solution***

Where $s$ is the semiperimeter of the triangle, Heron's Formula states that $A = \sqrt{s(s-a)(s-b)(s-c)}$. Because $s = \frac{1}{2}(a+b+c) = \frac{1}{2}(4+6+6) = 8$, we have $A = \sqrt{8(8-4)(8-6)(8-6)} = \sqrt{8(4)(2)(2)} = \sqrt{128}$. We can simplify the area expression to $\sqrt{64} \cdot \sqrt{2}$, so $A = 8\sqrt{2}$ ft$^2$.

Recalling Theorem 7.5.3, we know that $A = \frac{1}{2}rP$. Substitution leads to $8\sqrt{2} = \frac{1}{2}r(4+6+6)$, or $8\sqrt{2} = 8r$. Then $r = \sqrt{2}$. Where $r \approx 1.414$ ft, it follows that $r \approx 1.414(12$ in.$)$, or $r \approx 16.97$ in. $\approx 17$ in.

NOTE: If the ductwork is a flexible plastic tubing, the duct having radius 17 in. can probably be used. If the ductwork were a rigid metal or heavy plastic, the radius might need to be restricted to perhaps 16 in.

## 7.5 Exercises

1. In the circle, the radius length is 10 in. and the length of $\widehat{AB}$ is 14 in. What is the perimeter of the shaded sector?

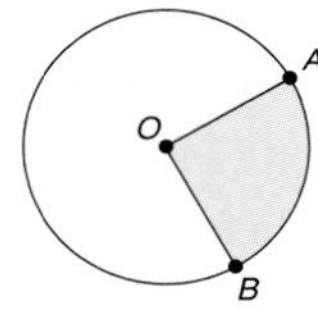


*Exercises 1, 2*

2. If the area of the circle is 360 in$^2$, what is the area of the sector if its central angle measures 90°?

3. If the area of the 120° sector is 50 cm$^2$, what is the area of the entire circle?

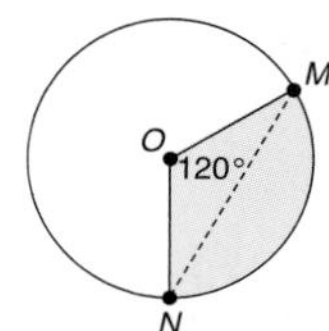


*Exercises 3, 4*

4. If the area of the 120° sector is 40 cm$^2$ and the area of $\triangle MON$ is 16 cm$^2$, what is the area of the segment bounded by chord $\overline{MN}$ and $\widehat{MN}$?

5. Suppose that a circle of radius $r$ is inscribed in an equilateral triangle whose sides have length $s$. Find an expression for the area of the triangle in terms of $r$ and $s$.
(HINT: Use Theorem 7.5.3.)

6. Suppose that a circle of radius $r$ is inscribed in a rhombus each of whose sides has length $s$. Find an expression for the area of the rhombus in terms of $r$ and $s$.

7. Find the perimeter of a segment of a circle whose boundaries are a chord measuring 24 mm (millimeters) and an arc of length 30 mm.

8. A sector with perimeter 30 in. has a bounding arc of length 12 in. Find the length of the radius of the circle.

9. A circle is inscribed in a triangle having sides of lengths 6 in., 8 in., and 10 in. If the radius of the inscribed circle is 2 in., use $A = \frac{1}{2}rP$ to find the area of the triangle.

10. A circle is inscribed in a triangle having sides of lengths 5 in., 12 in., and 13 in. If the radius of the inscribed circle is 2 in., use $A = \frac{1}{2}rP$ to find the area of the triangle.

11. A triangle with sides of lengths 3 in., 4 in., and 5 in. has an area of 6 in$^2$. What is the length of the radius of the inscribed circle?

12. The approximate area of a triangle with sides of lengths 3 in., 5 in., and 6 in. is 7.48 in$^2$. What is the length of the radius of the inscribed circle?

13. Find the exact perimeter and area of the sector shown.

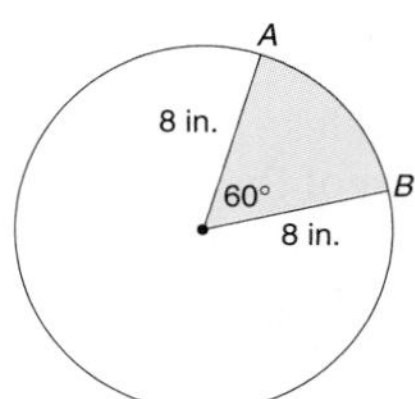

14. Find the exact perimeter and area of the sector shown.

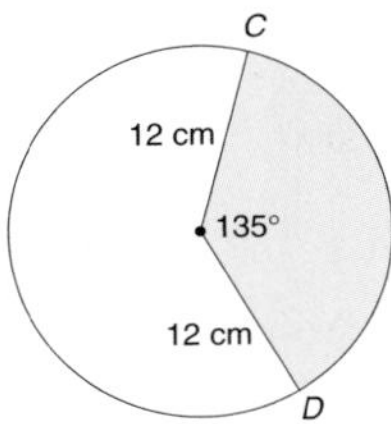


15. Find the approximate perimeter of the sector shown. Answer to the nearest hundredth.

16. Find the approximate area of the sector shown. Answer to the nearest hundredth.

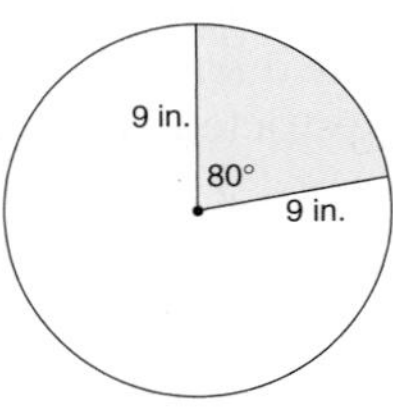


*Exercises 15, 16*

17. Find the exact perimeter and area of the segment shown, given that m$\angle O = 60°$ and $OA = 12$ in.

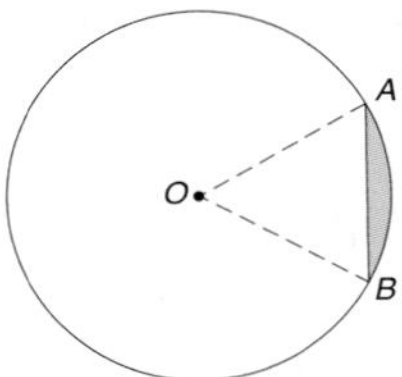


*Exercises 17, 18*

18. Find the exact perimeter and area of the segment shown, given that m$\angle O = 120°$ and $AB = 10$ in.

*In Exercises 19 and 20, find the exact areas of the shaded regions.*

19.

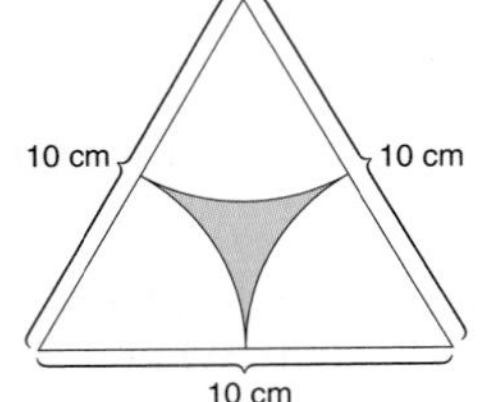


20.

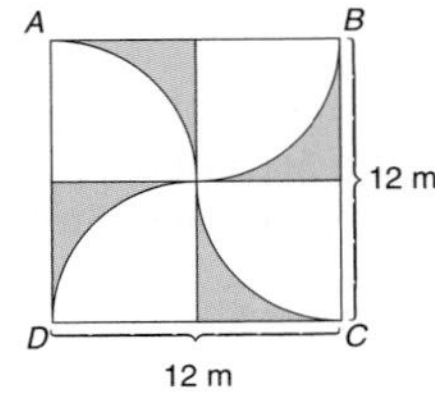


Square $ABCD$

21. Assuming that the exact area of a sector determined by a 40° arc is $\frac{9}{4}\pi \text{cm}^2$, find the length of the radius of the circle.

22. For concentric circles with radii of lengths 3 in. and 6 in., find the area of the smaller segment determined by a chord of the larger circle that is also a tangent of the smaller circle.

*23. A circle can be inscribed in the trapezoid shown. Find the area of that circle.

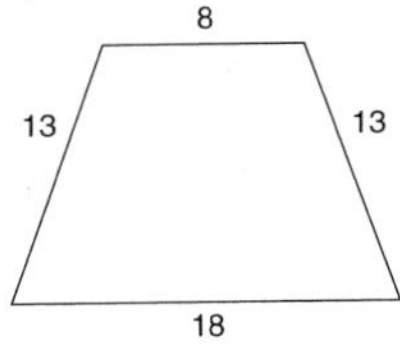


*24. A circle can be inscribed in an equilateral triangle each of whose sides has length 10 cm. Find the area of that circle.

25. In a circle whose radius has length 12 m, the length of an arc is $6\pi$ m. What is the degree measure of that arc?

26. At the Pizza Dude restaurant, a 12-in. pizza costs \$3.40 to make, and the manager wants to make at least \$2.20 from the sale of each pizza. If the pizza will be sold by the slice and each pizza is cut into 6 pieces, what is the minimum charge per slice?

27. At the Pizza Dude restaurant, pizza is sold by the slice. If the pizza is cut into 6 pieces, then the selling price is \$1.25 per slice. If the pizza is cut into 8 pieces, then each slice is sold for \$0.95. In which way will the Pizza Dude restaurant clear more money from sales?

28. Determine a formula for the area of the shaded region determined by the square and its inscribed circle.

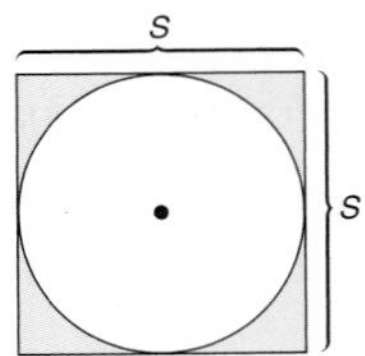


29. Determine a formula for the area of the shaded region determined by the circle and its inscribed square.

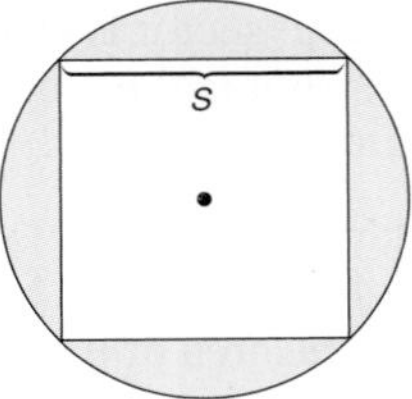


30. Find a formula for the area of the shaded region, which represents one-fourth of an annulus (ring).

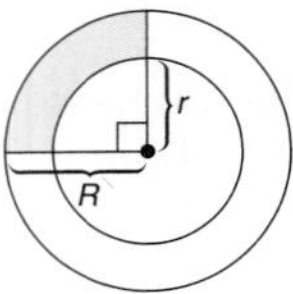

31. A company logo on the side of a building shows an isosceles triangle with an inscribed circle. If the sides of the triangle are 10 ft, 13 ft, and 13 ft, find the radius of the inscribed circle.

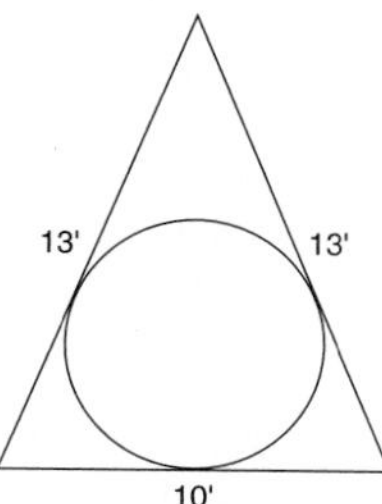


32. In a right triangle with sides of lengths $a$, $b$, and $c$ (where $c$ is the hypotenuse), show that the length of the radius of the inscribed circle is $r = \frac{ab}{a + b + c}$.

33. In a triangle with sides of lengths $a$, $b$, and $c$ and semiperimeter $s$, show that the length of the radius of the inscribed circle is

$$r = \frac{2\sqrt{s(s - a)(s - b)(s - c)}}{a + b + c}$$

34. Use the results from Exercises 32 and 33 to find the length of the radius of the inscribed circle for a triangle with sides of lengths

a) 8, 15, and 17. b) 7, 9, and 12.

35. Use the results from Exercises 32 and 33 to find the length of the radius of the inscribed circle for a triangle with sides of lengths

a) 7, 24, and 25. b) 9, 10, and 17.

36. Three pipes, each of radius 4 in., are stacked as shown. What is the height of the stack?

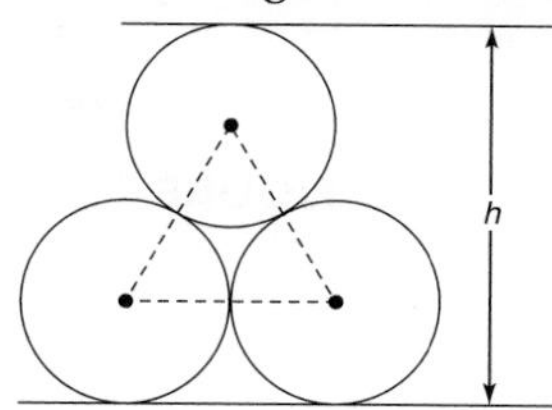


37. A windshield wiper rotates through a 120° angle as it cleans a windshield. From the point of rotation, the wiper blade begins at a distance of 4 in. and ends at a distance of 18 in. (The wiper blade is 14 inches in length.) Find the area cleaned by the wiper blade.

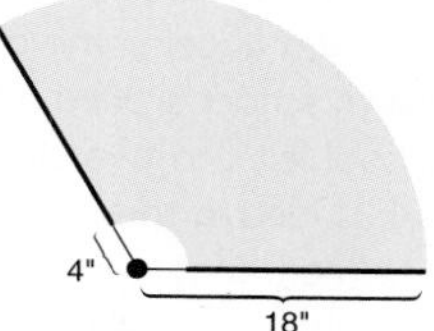


38. A goat is tethered to a barn by a 12-ft chain. If the chain is connected to the barn at a point 6 ft from one end of the barn, what is the area of the pasture that the goat is able to graze?

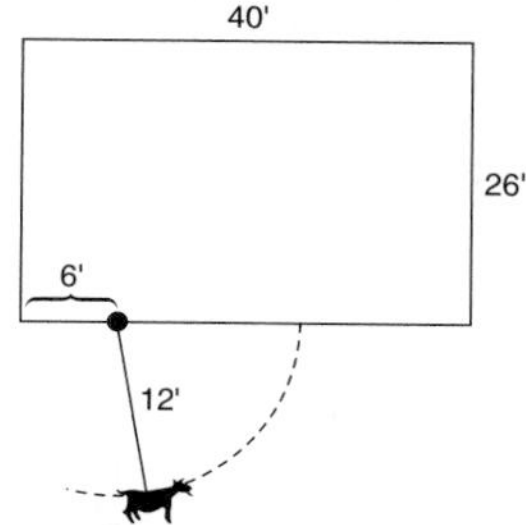


39. An exit ramp from one freeway onto another freeway forms a 90° arc of a circle. The ramp is scheduled for resurfacing. As shown, its inside radius is 370 ft, and its outside radius is 380 ft. What is the area of the ramp?

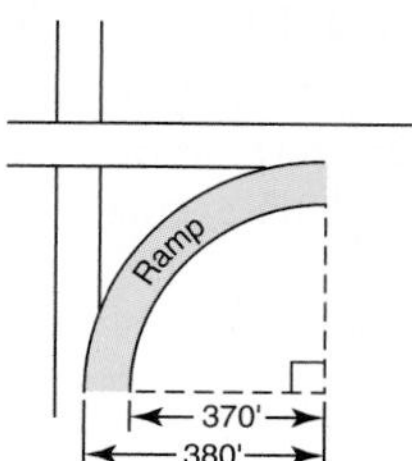

# Perspective on History

## SKETCH OF PYTHAGORAS

Pythagoras (circa 580–500 B.C.) was a Greek philosopher and mathematician. Having studied under some of the great minds of the day, he formed his own school around 529 B.C. in Crotona, Italy.

Students of his school fell into two classes, the listeners and the elite Pythagoreans. Included in the Pythagoreans were brilliant students, including 28 women, and all were faithful followers of Pythagoras. The Pythagoreans, who adhered to a rigid set of beliefs, were guided by the principle "Knowledge is the greatest purification."

The apparent areas of study for the Pythagoreans included arithmetic, music, geometry, and astronomy, but underlying principles that led to a cult-like existence included self-discipline, temperance, purity, and obedience. The Pythagoreans recognized fellow members by using the pentagram (five-pointed star) as their symbol. With their focus on virtue, politics, and religion, the members of the group saw themselves as above others. Because of their belief in *transmigration* (movement of the soul after death to another human or animal), the Pythagoreans refused to eat meat or fish. On one occasion, it is said that Pythagoras came upon a person beating a dog. Approaching that person, Pythagoras said, "Stop beating the dog, for in this dog lives the soul of my friend; I recognize him by his voice."

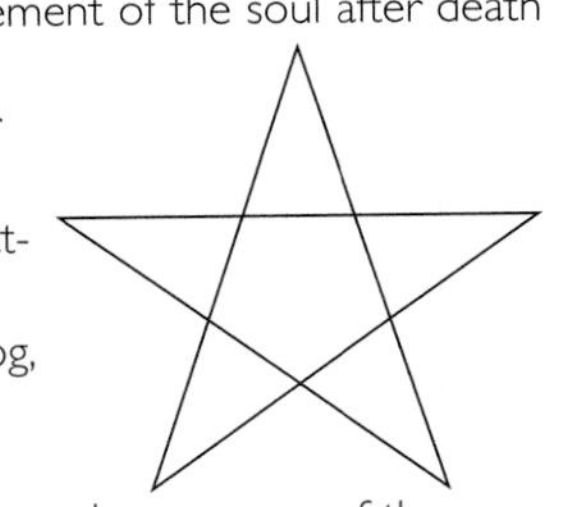

In time, the secrecy, clannishness, and supremacy of the Pythagoreans led to suspicion and fear on the part of other factions of society. Around 500 B.C., the revolution against the Pythagoreans led to the burning of their primary meeting house. Although many of the Pythagoreans died in the ensuing inferno, it is unclear whether Pythagoras himself died or escaped.

# Perspective on Application

## ANOTHER LOOK AT THE PYTHAGOREAN THEOREM

Some of the many proofs of the Pythagorean Theorem depend on area relationships. One such proof was devised by President James A. Garfield (1831–1881), twentieth president of the United States.

In his proof, the right triangle with legs $a$ and $b$ and hypotenuse $c$ is introduced into a trapezoid, as shown in Figure 7.61(b).

In Figure 7.61(b), the points $A$, $B$, and $C$ are collinear. With $\angle 1$ and $\angle 2$ being complementary and the sum of the angles' measures about point $B$ being 180°, it follows that $\angle 3$ is a right angle.

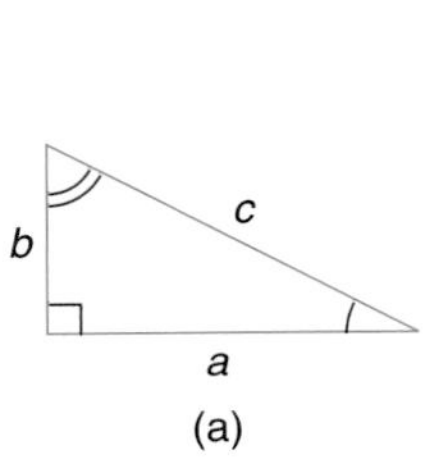


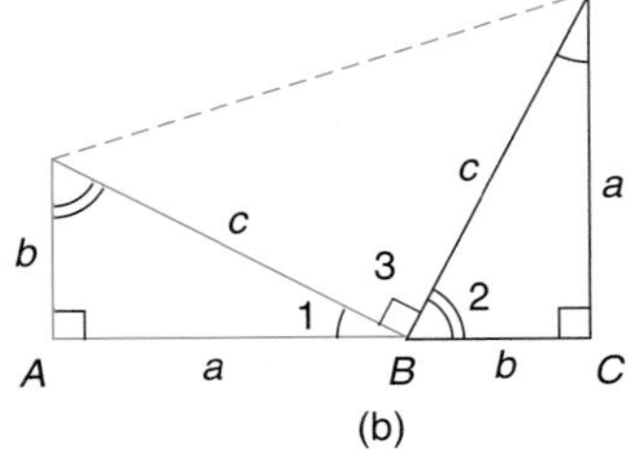


**FIGURE 7.61**

If the drawing is perceived as a trapezoid (as shown in Figure 7.62), the area is given by

$$\begin{aligned} A &= \frac{1}{2}h(b_1 + b_2) \\ &= \frac{1}{2}(a + b)(a + b) \\ &= \frac{1}{2}(a + b)^2 \\ &= \frac{1}{2}(a^2 + 2ab + b^2) \\ &= \frac{1}{2}a^2 + ab + \frac{1}{2}b^2 \end{aligned}$$



**FIGURE 7.62**

Now we treat the trapezoid as a composite of three triangles in Figure 7.63.

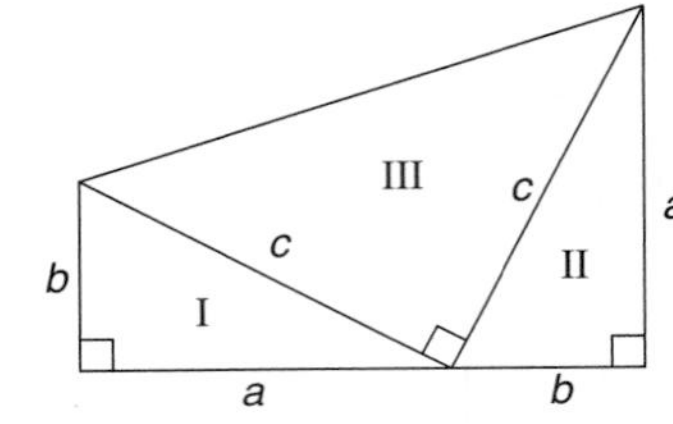


**FIGURE 7.63**

The total area of regions (triangles) I, II, and III is given by

$$\begin{aligned} A &= A_I + A_{II} + A_{III} \\ &= \frac{1}{2}ab + \frac{1}{2}ab + \left(\frac{1}{2}c \cdot c\right) \\ &= ab + \frac{1}{2}c^2 \end{aligned}$$

Equating the areas of the trapezoid in Figure 7.62 and the composite in Figure 7.63, we find that

$$\begin{aligned} \frac{1}{2}a^2 + ab + \frac{1}{2}b^2 &= ab + \frac{1}{2}c^2 \\ \frac{1}{2}a^2 + \frac{1}{2}b^2 &= \frac{1}{2}c^2 \end{aligned}$$

Multiplying by 2, we have

$$a^2 + b^2 = c^2$$

The earlier proof (over 2000 years earlier!) of this theorem by the Greek mathematician Pythagoras is found in many historical works on geometry. It is not difficult to see the relationship between the two proofs.

In the proof credited to Pythagoras, a right triangle with legs of lengths $a$ and $b$ and hypotenuse of length $c$ is reproduced several times to form a square. Again, points $A$, $B$, $C$ (and $C$, $D$, $E$; and so on) must be collinear. [See Figure 7.64(c).]

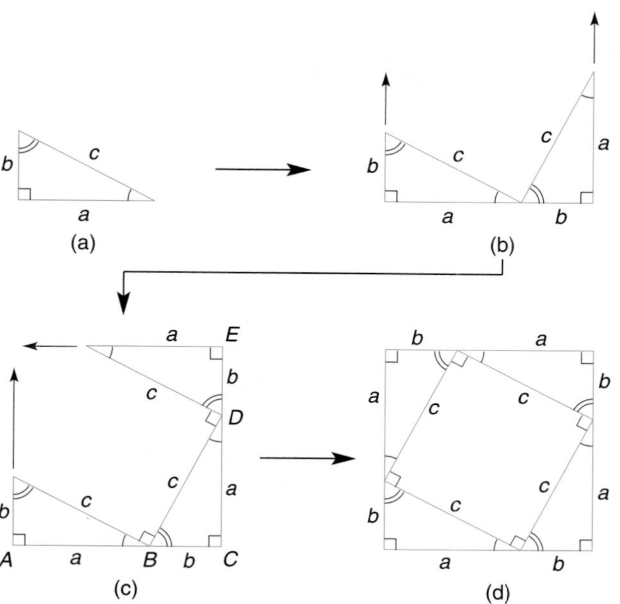


**FIGURE 7.64**

The area of the large square in Figure 7.65(a) is given by

$$\begin{aligned} A &= (a + b)^2 \\ &= a^2 + 2ab + b^2 \end{aligned}$$

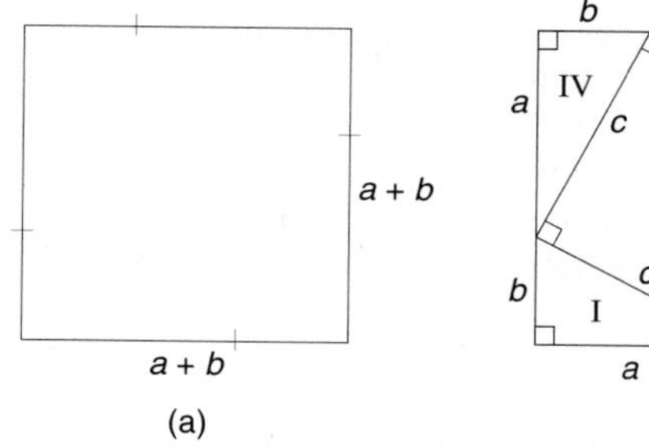


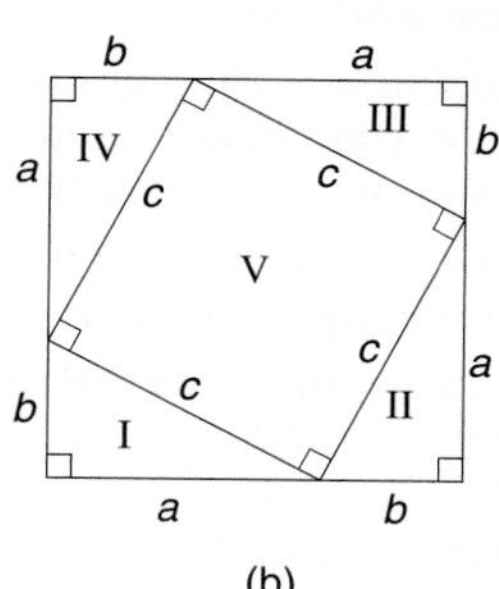


**FIGURE 7.65**

Considering the composite in Figure 7.65(b), we find that

$$\begin{aligned} A &= A_I + A_{II} + A_{III} + A_{IV} + A_V \\ &= 4 \cdot A_I + A_V \end{aligned}$$

because the four right triangles are congruent. Then

$$\begin{aligned} A &= 4\left(\frac{1}{2}ab\right) + c^2 \\ &= 2ab + c^2 \end{aligned}$$

Again, because of the uniqueness of area, the results (area of square and area of composite) must be equal. Then

$$\begin{aligned} a^2 + 2ab + b^2 &= 2ab + c^2 \\ a^2 + b^2 &= c^2 \end{aligned}$$

Another look at the proofs by President Garfield and by Pythagoras makes it clear that the results must be consistent. In Figure 7.66, observe that Garfield's trapezoid must have one-half the area of Pythagoras' square, while maintaining the relationship that

$$c^2 = a^2 + b^2$$

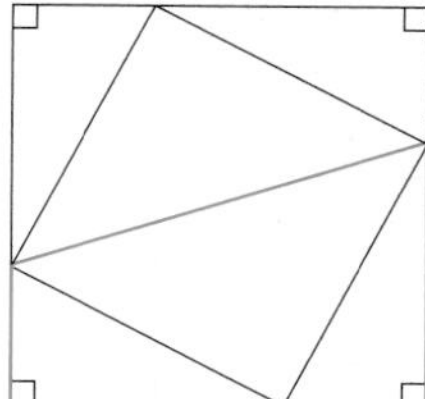

**FIGURE 7.66**

# Summary

### A LOOK BACK AT CHAPTER 7

Our goal in this chapter was to determine the areas of triangles, certain quadrilaterals, and regular polygons. We also explored the circumference and area of a circle and the area of a sector of a circle. The area of a circle is sometimes approximated by using $\pi \approx 3.14$ or $\pi \approx \frac{22}{7}$. At other times, the exact area is given by leaving $\pi$ in the answer.

### A LOOK AHEAD TO CHAPTER 8

Our goal in the next chapter is to deal with a type of geometry known as solid geometry. We will find the surface areas of solids with polygonal or circular bases. We will also find the volumes of these solid figures. Select polyhedra will be discussed.

### KEY CONCEPTS

7.1 Plane Region • Square Unit • Area Postulates • Area of a Rectangle, Parallelogram, and Triangle • Altitude and Base of a Parallelogram or Triangle

7.2 Perimeter of a Polygon • Semiperimeter of a Triangle • Heron's Formula • Brahmagupta's Formula • Area of a Trapezoid, Rhombus, and Kite • Areas of Similar Polygons

7.3 Regular Polygon • Center and Central Angle of a Regular Polygon • Radius and Apothem of a Regular Polygon • Area of a Regular Polygon

7.4 Circumference of a Circle • $\pi$ (Pi) • Length of an Arc • Limit • Area of a Circle

7.5 Sector • Area and Perimeter of a Sector • Segment of a Circle • Area and Perimeter of a Segment • Area of a Triangle with Inscribed Circle

### TABLE 7.3 AN OVERVIEW OF CHAPTER SEVEN

**Area and Perimeter Relationships**

| Figure | Drawing | Area | Perimeter or Circumference |
|---|---|---|---|
| Rectangle | w, ℓ | $A = \ell w$ (or $A = bh$) | $P = 2\ell + 2w$ (or $P = 2b + 2h$) |
| Square | s | $A = s^2$ | $P = 4s$ |
| Parallelogram | s, h, b | $A = bh$ | $P = 2b + 2s$ |
| Triangle | a, h, c, b | $A = \frac{1}{2}bh$<br>$A = \sqrt{s(s-a)(s-b)(s-c)}$,<br>where $s = \frac{1}{2}(a+b+c)$ | $P = a + b + c$ |
| Right triangle | c, a, b | $A = \frac{1}{2}ab$ | $P = a + b + c$ |

**TABLE 7.3 CONT.**

## Area and Perimeter Relationships

| Figure | Drawing | Area | Perimeter or Circumference |
|---|---|---|---|
| Trapezoid | $b_1$, $s_1$, $h$, $s_2$, $b_2$ | $A = \frac{1}{2}h(b_1 + b_2)$ | $P = s_1 + s_2 + b_1 + b_2$ |
| Rhombus (diagonals of lengths $d_1$ and $d_2$) | $s$, $d_1$, $d_2$ | $A = \frac{1}{2}d_1d_2$ | $P = 4s$ |
| Kite (diagonals of lengths $d_1$ and $d_2$) | $b$, $s$, $d_1$, $d_2$, $b$, $s$ | $A = \frac{1}{2}d_1d_2$ | $P = 2b + 2s$ |
| Regular polygon ($n$ sides; $s$ is length of side; $a$ is length of apothem) | $a$, $s$ | $A = \frac{1}{2}aP$ ($P$ = perimeter) | $P = ns$ |
| Circle | $r$ | $A = \pi r^2$ | $C = 2\pi r$ |
| Sector ($m\widehat{AB}$ is degree measure of $\widehat{AB}$ and of central angle $AOB$) | $A$, $O$, $B$ | $A = \frac{m\widehat{AB}}{360°}\pi r^2$ | $P = 2r + \ell\widehat{AB}$, where $\ell\widehat{AB} = \frac{m\widehat{AB}}{360°} \cdot 2\pi r$ |
| Triangle with inscribed circle of radius $r$ | $a$, $c$, $r$, $b$ | $A = \frac{1}{2}rP$ ($P$ = perimeter) | $P = a + b + c$ |

# Chapter 7 Review Exercises

*In Review Exercises 1 to 3, draw a figure that enables you to solve each problem.*

1. *Given:* $\square ABCD$ with $BD = 34$ and $BC = 30$
   $m\angle C = 90°$
   *Find:* $A_{ABCD}$

2. *Given:* $\square ABCD$ with $AB = 8$ and $AD = 10$
   *Find:* $A_{ABCD}$ if:
   a) $m\angle A = 30°$
   b) $m\angle A = 60°$
   c) $m\angle A = 45°$

3. *Given:* $\square ABCD$ with $\overline{AB} \cong \overline{BD}$ and $AD = 10$
   $\overline{BD} \perp \overline{DC}$
   *Find:* $A_{ABCD}$

*In Review Exercises 4 and 5, draw $\triangle ABC$, if necessary, to solve each problem.*

4. *Given:* $AB = 26$, $BC = 25$, and $AC = 17$
   *Find:* $A_{ABC}$

5. *Given:* $AB = 30$, $BC = 26$, and $AC = 28$
   *Find:* $A_{ABC}$

6. *Given:* Trapezoid $ABCD$, with $\overline{AB} \cong \overline{CD}$, $BC = 6$, $AD = 12$, and $AB = 5$
   *Find:* $A_{ABCD}$

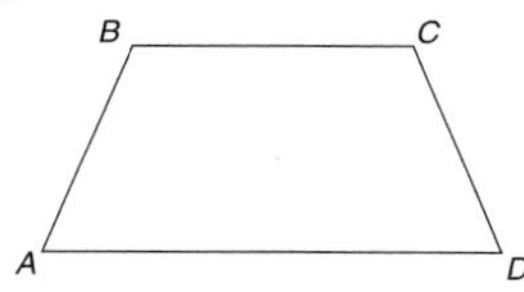


*Exercises 6, 7*

7. *Given:* Trapezoid $ABCD$, with $AB = 6$ and $BC = 8$, $\overline{AB} \cong \overline{CD}$
   *Find:* $A_{ABCD}$ if:
   a) $m\angle A = 45°$
   b) $m\angle A = 30°$
   c) $m\angle A = 60°$

8. Find the area and the perimeter of a rhombus whose diagonals have lengths 18 in. and 24 in.

9. Tom Morrow wants to buy some fertilizer for his yard. The lot size is 140 ft by 160 ft. The outside measurements of his house are 80 ft by 35 ft. The driveway measures 30 ft by 20 ft. All shapes are rectangular.

   a) What is the square footage of his yard that needs to be fertilized?
   b) If each bag of fertilizer covers 5000 ft², how many bags should Tom buy?
   c) If the fertilizer costs $18 per bag, what is his total cost?

10. Alice's mother wants to wallpaper two adjacent walls in Alice's bedroom. She also wants to put a border along the top of all four walls. The bedroom is 9 ft by 12 ft by 8 ft high.

    a) If each double roll covers approximately 60 ft² and the wallpaper is sold in double rolls only, how many double rolls are needed?
    b) If the border is sold in rolls of 5 yd each, how many rolls of the border are needed?

11. *Given:* Isosceles trapezoid $ABCD$
    Equilateral $\triangle FBC$
    Right $\triangle AED$
    $BC = 12$, $AB = 5$, and $ED = 16$
    *Find:* a) $A_{EAFD}$
    b) Perimeter of $EAFD$

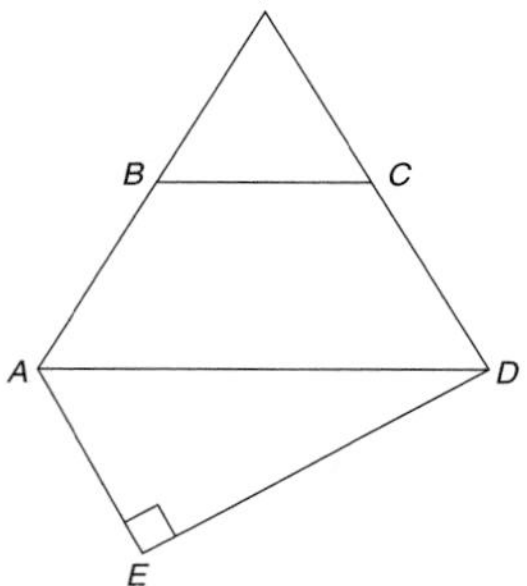


12. *Given:* Kite $ABCD$ with $AB = 10$, $BC = 17$, and $BD = 16$
    *Find:* $A_{ABCD}$

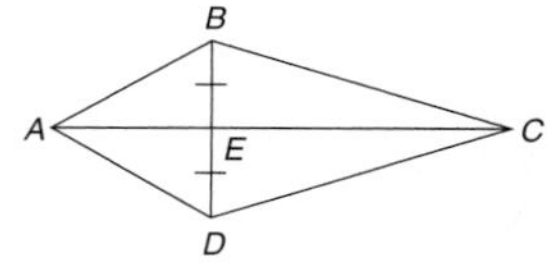


13. One side of a rectangle is 2 cm longer than a second side. If the area is 35 cm², find the dimensions of the rectangle.

14. One side of a triangle is 10 cm longer than a second side, and the third side is 5 cm longer than the second side. The perimeter of the triangle is 60 cm.

    a) Find the lengths of the three sides.
    b) Find the area of the triangle.

15. Find the area of $\triangle ABD$ as shown.

16. Find the area of an equilateral triangle if each of its sides has length 12 cm.

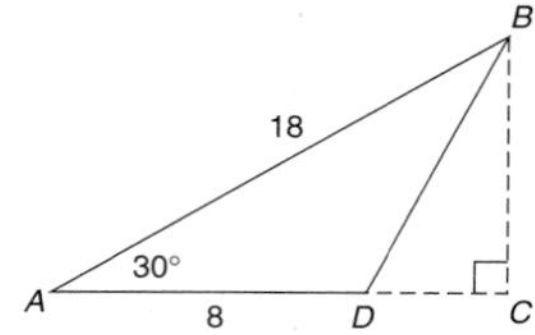

17. If $\overline{AC}$ is a diameter of $\odot O$, find the area of the shaded triangle.

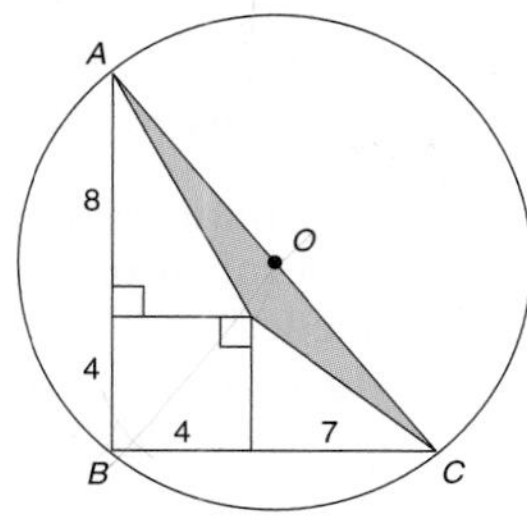

Exercise 17

18. For a regular pentagon, find the measure of each:
    a) Central angle
    b) Interior angle
    c) Exterior angle

19. Find the area of a regular hexagon each of whose sides has length 8 ft.

20. The area of an equilateral triangle is $108\sqrt{3}$ in$^2$. If the length of each side of the triangle is $12\sqrt{3}$ in, find the length of an apothem of the triangle.

21. Find the area of a regular hexagon whose apothem has length 9 in.

22. A regular polygon has each central angle equal to 45°.
    a) How many sides does the regular polygon have?
    b) If each side is 5 cm and each apothem is approximately 6 cm, what is the approximate area of the polygon?

23. Can a circle be circumscribed about each of the following figures? Why or why not?
    a) Parallelogram
    b) Rhombus
    c) Rectangle
    d) Square

24. Can a circle be inscribed in each of the following figures? Why or why not?
    a) Parallelogram
    b) Rhombus
    c) Rectangle
    d) Square

25. The radius of a circle inscribed in an equilateral triangle is 7 in. Find the area of the triangle.

26. The Turners want to carpet the cement around their rectangular pool. The dimensions for the rectangular area formed by the pool and its cement walkway are 20 ft by 30 ft. The pool is 12 ft by 24 ft.
    a) How many square feet need to be covered?
    b) Carpet is sold only by the square yard. Approximately how many square yards does the area in part (a) represent?
    c) If the carpet costs $9.97 per square yard, what will be the total cost of the carpet?

*Find the exact areas of the shaded regions in Exercises 27 to 31.*

27.

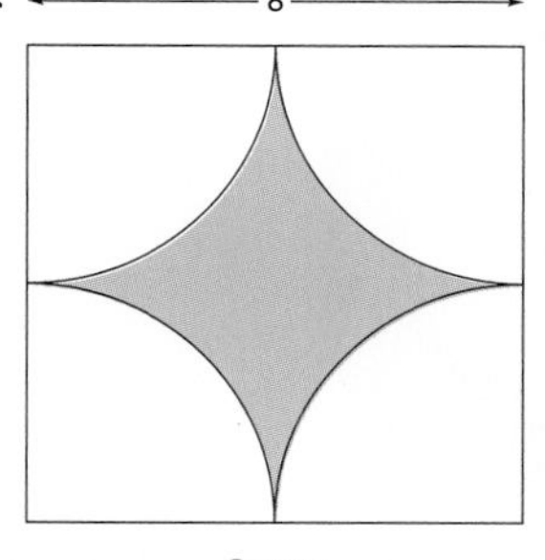

Square

28.

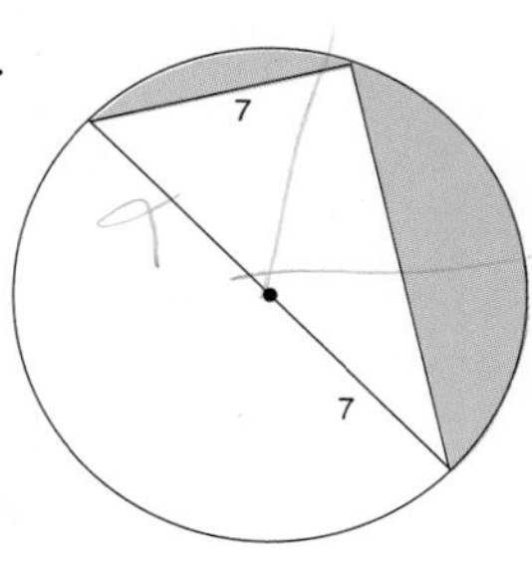


29.

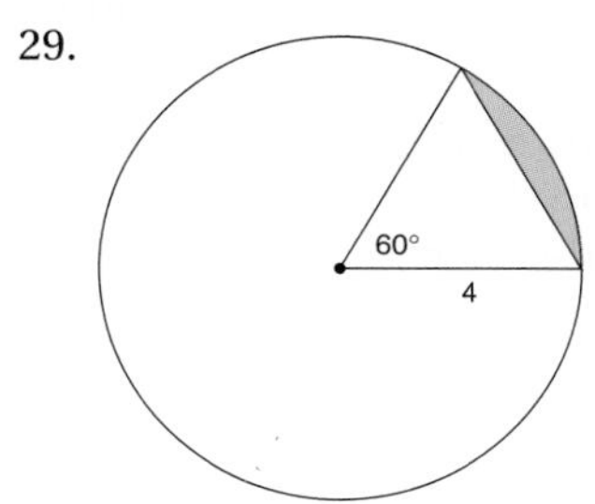


30.

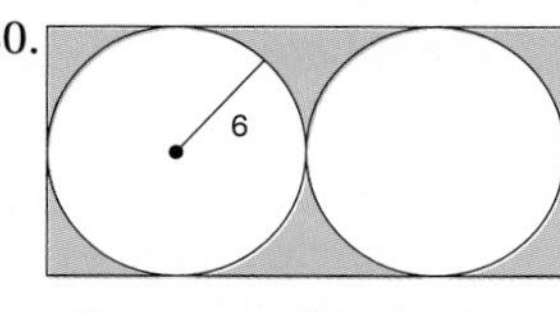

Two ≅ tangent circles, inscribed in a rectangle

31.


Equilateral triangle

32. The arc of a sector measures 40°. Find the exact length of the arc and the exact area of the sector if the radius measures $3\sqrt{5}$ cm.

33. The circumference of a circle is 66 ft.
    a) Find the diameter of the circle, using $\pi \approx \frac{22}{7}$.
    b) Find the area of the circle, using $\pi \approx \frac{22}{7}$.

34. A circle has an exact area of $27\pi$ ft$^2$.
    a) What is the area of a sector of this circle if the arc of the sector measures 80°?
    b) What is the exact perimeter of the sector in part (a)?

35. An isosceles right triangle is inscribed in a circle that has a diameter of 12 in. Find the exact area between one of the legs of the triangle and its corresponding arc.

36. *Given:* Concentric circles with radii of lengths $R$ and $r$, with $R > r$
*Prove:* $Area_{ring} = \pi(BC)^2$

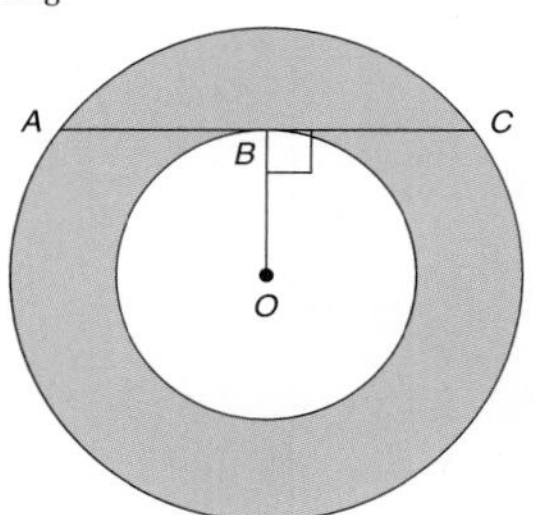


37. Prove that the area of a circle circumscribed about a square is twice the area of the circle inscribed within the square.

38. Prove that if semicircles are constructed on each of the sides of a right triangle, then the area of the semicircle on the hypotenuse is equal to the sum of the areas of the semicircles on the two legs.

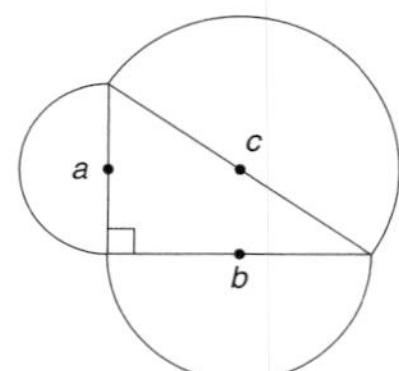


39. Jeff and Helen want to carpet their family room, except for the entranceway and the semicircle in front of the fireplace, both of which they want to tile.

a) How many square yards of carpeting are needed?
b) How many square feet are to be tiled?

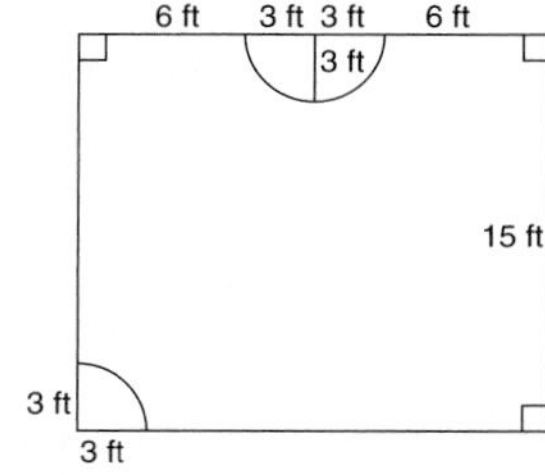


40. Sue and Dave's semicircular driveway is to be resealed, and then flowers are to be planted on either side.

a) What is the number of square feet to be resealed?
b) If the cost of resealing is \$0.18 per square foot, what is the total cost?
c) If individual flowers are to be planted 1 foot from the edge of the driveway at intervals of approximately 1 foot on both sides of the driveway, how many flowers are needed?

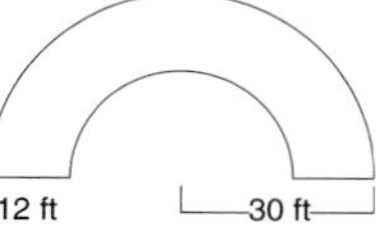

## Chapter 7 Test

1. Complete each statement.

a) Given that the length and the width of a rectangle are measured in inches, its area is measured in ________.
b) If two closed plane figures are congruent, then their areas are ________.

2. Give each formula.

a) The formula for the area of a square whose sides are of length $s$ is ________.
b) The formula for the circumference of a circle with radius length $r$ is ________.

3. Determine whether the statement is True or False.

a) The area of a circle with radius length $r$ is given by $A = \pi r^2$. ________
b) With corresponding sides of similar polygons having the ratio $\frac{s_1}{s_2} = \frac{1}{2}$, the ratio of their areas is $\frac{A_1}{A_2} = \frac{1}{2}$. ________

4. If the area of rectangle $ABCD$ is 46 $\text{cm}^2$, find the area of $\triangle ABE$.
________

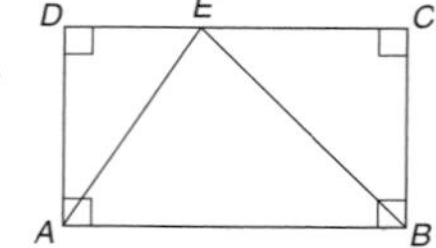

5. In *square feet*, find the area of ▱*EFGH*. ________

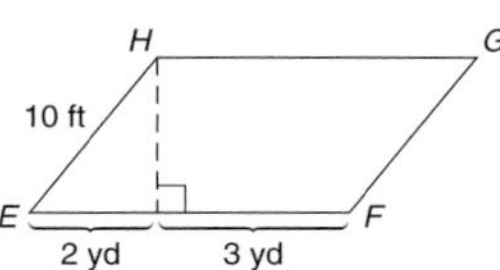


6. Find the area of rhombus *MNPQ* given that $QN = 8$ ft and $PM = 6$ ft. ________

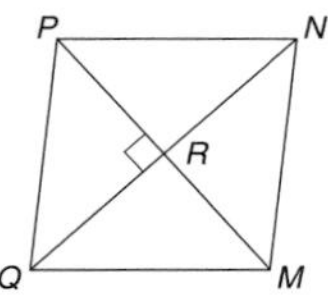


7. Use Heron's Formula, $A = \sqrt{s(s-a)(s-b)(s-c)}$, to find the exact area of a triangle that has lengths of sides 4 cm, 13 cm, and 15 cm. ________

8. In trapezoid *ABCD*, $AB = 7$ ft and $DC = 13$ ft. If the area of trapezoid *ABCD* is 60 ft$^2$, find the length of altitude $\overline{AE}$. ________

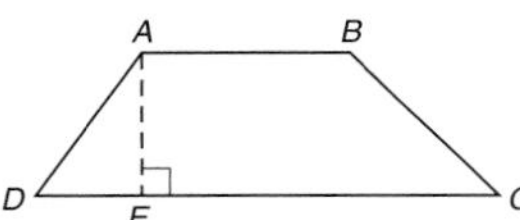


9. A regular pentagon has an apothem of length 4.0 in. and each side is of length $s = 5.8$ in. For the regular pentagon, find its:

a) Perimeter ________
b) Area ________

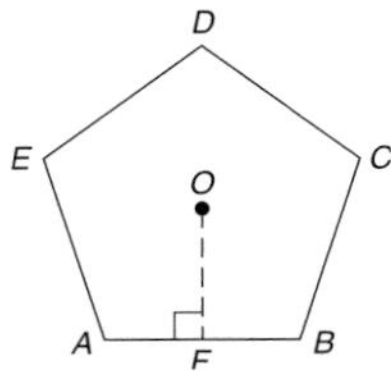


10. For the circle shown below, the length of the radius is $r = 5$ in. Find the exact:

a) Circumference ________
b) Area ________
(HINT: Leave $\pi$ in the answer in order to achieve exactness.)

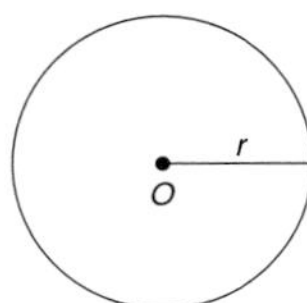


11. Where $\pi \approx \frac{22}{7}$, find the approximate length of $\widehat{AC}$.
$\ell\widehat{AC} \approx$ ________

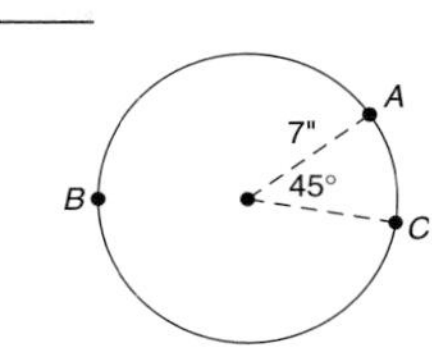


12. Where $\pi \approx 3.14$, find the approximate area of a circle (not shown) whose *diameter* measures 20 cm.

________

13. In the figure, a square is inscribed in a circle. If each side of the square measures $4\sqrt{2}$ in., find an expression for the exact area of the shaded region.

________

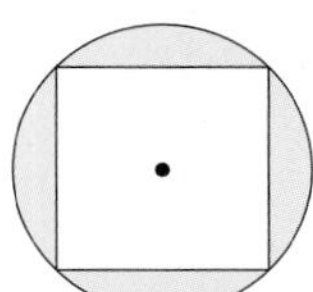

Square inscribed in a circle

14. Find the exact area of the 135° sector shown.

________

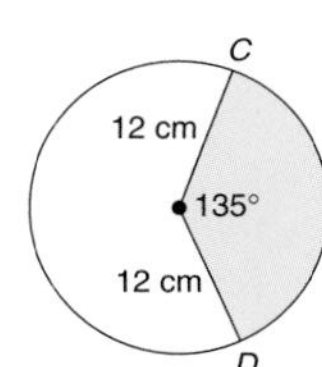


15. Find the exact area of the shaded segment. ________

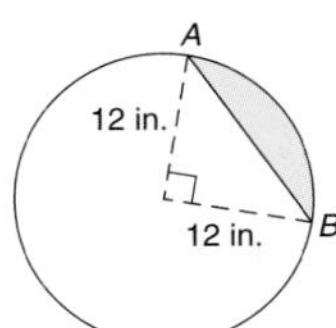


16. The area of a right triangle whose sides have lengths 5 in., 12 in., and 13 in. is exactly 30 in$^2$. Use the formula $A = \frac{1}{2}rP$ to find the length of the radius of the circle that can be inscribed in this triangle. ________

# Chapter 8

# Surfaces and Solids

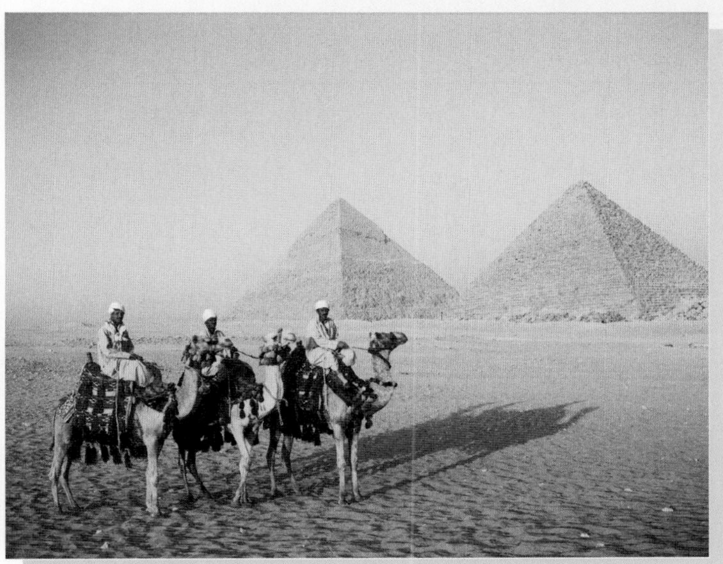

Located near Cairo, Egypt, the Great Pyramids illustrate one of the types of solids that we study in Chapter 8. The architectural designs of buildings illustrate other solid shapes that we encounter in this chapter. The real world is three-dimensional; that is, solids have length, width, and depth. Each solid determines a bounded region of space that has a measure known as volume. The units used to measure volume include the cubic foot and the cubic meter. In Section 8.2, the same technique used to determine the volume of the pyramid in Example 5 could be used to find the volumes of the Great Pyramids as well.

## Chapter Outline



**For online student resources, visit this textbook's website at math.college.hmco.com/students.**

**KEY CONCEPTS**

Prisms (Right and Oblique) • Bases • Altitude • Vertices • Edges • Faces • Lateral Area • Total (Surface) Area • Volume • Regular Prism • Cube • Cubic Unit

# 8.1 Prisms, Area, and Volume

## PRISMS

Suppose that two congruent polygons lie in parallel planes in such a way that their corresponding sides are parallel. If the corresponding vertices of these polygons [such as A and $A'$ in Figure 8.1(a)] are joined by line segments, then the "solid" that results is a **prism.** The congruent figures that lie in the parallel planes are the **bases** of the prism. The parallel planes need not be shown in the drawings of prisms. In practice, the "solid" includes the interior of the actual prism; in reality, the prism encloses a surface in space that does not include interior points.

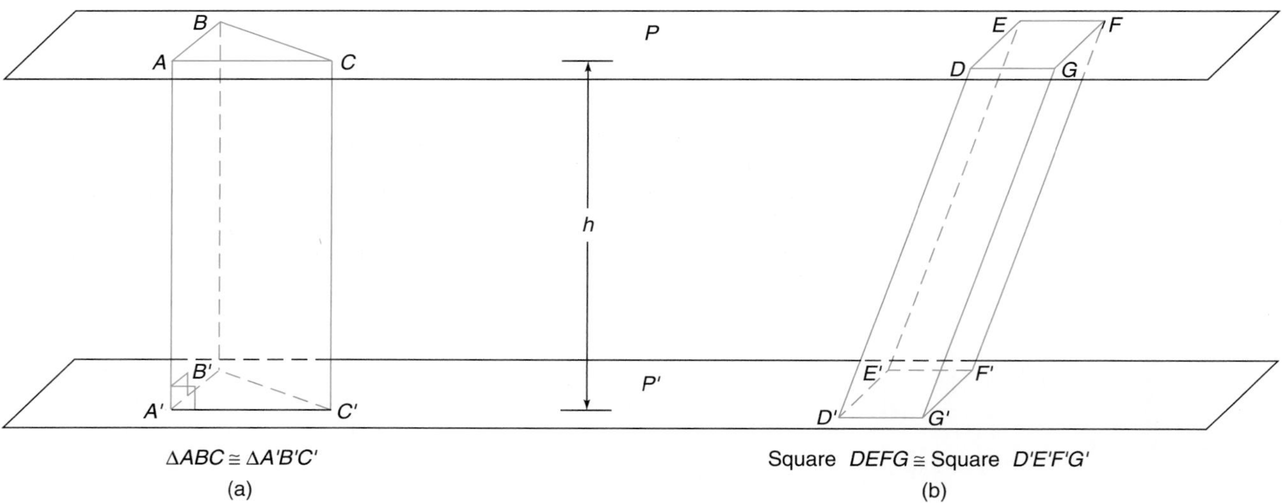


$\triangle ABC \cong \triangle A'B'C'$
(a)

Square $DEFG \cong$ Square $D'E'F'G'$
(b)

**FIGURE 8.1**

In Figure 8.1(a), $\overline{AB}$, $\overline{AC}$, $\overline{BC}$, $\overline{A'B'}$, $\overline{A'C'}$, and $\overline{B'C'}$ are **base edges** and $\overline{AA'}$, $\overline{BB'}$, and $\overline{CC'}$ are **lateral edges** of the prism. Because the lateral edges of this prism are perpendicular to its base edges, the **lateral faces** (like quadrilateral $ACC'A'$) are rectangles. Points $A$, $B$, $C$, $A'$, $B'$, and $C'$ are the **vertices** of the prism.

In Figure 8.1(b), the lateral edges of the prism are not perpendicular to its base edges. This relationship between the lateral edge and the base edge is often described as **oblique** (slanted). For the oblique prism, the lateral faces are parallelograms. Considering the prisms in Figure 8.1, we are led to the following definitions.

> DEFINITION: A **right prism** is a prism in which the lateral edges are perpendicular to the base edges at their points of intersection. An **oblique prism** is a prism in which the parallel lateral edges are oblique to the base edges at their points of intersection.

Part of the description used to classify a prism depends on its base. For instance, the prism in Figure 8.1(a) is a *right triangular prism*; in this case, the word *right* describes the prism, whereas the word *triangular* refers to the triangular base. Similarly, the prism in Figure 8.1(b) is an *oblique square prism*, assuming that the bases are squares. Both prisms in Figure 8.1 have an **altitude** (a perpendicular segment joining the planes that contain the bases) of length $h$.

EXAMPLE 1

Name each type of prism in Figure 8.2.

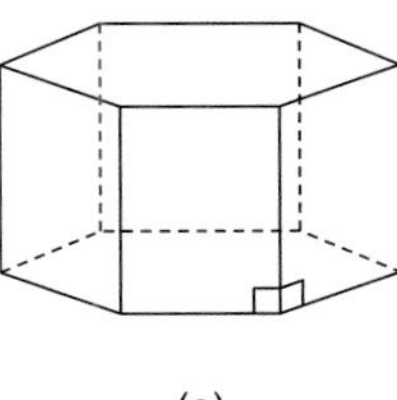

(a)

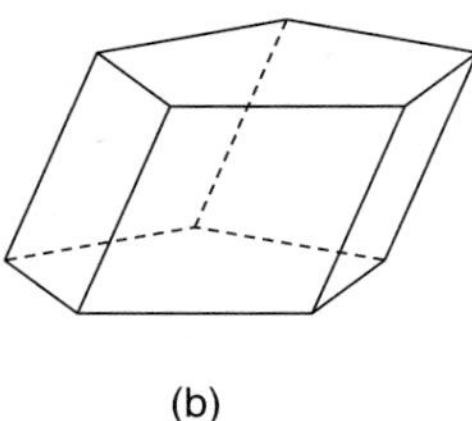

(b)

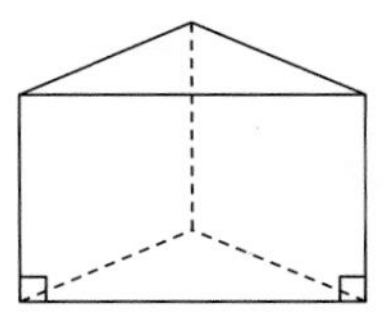


(c)

**FIGURE 8.2**

***Solution***

a) The lateral edges are perpendicular to the base edges of the hexagonal base. The prism is a *right hexagonal prism.*
b) The lateral edges are oblique to the base edges of the pentagonal base. The prism is an *oblique pentagonal prism.*
c) The lateral edges are perpendicular to the base edges of the triangular base. Because the base is equilateral, the prism is a *right equilateral triangular prism.*

*Exs. 1, 2*

## AREA OF A PRISM

DEFINITION: The **lateral area** $L$ of a prism is the sum of the areas of all lateral faces.

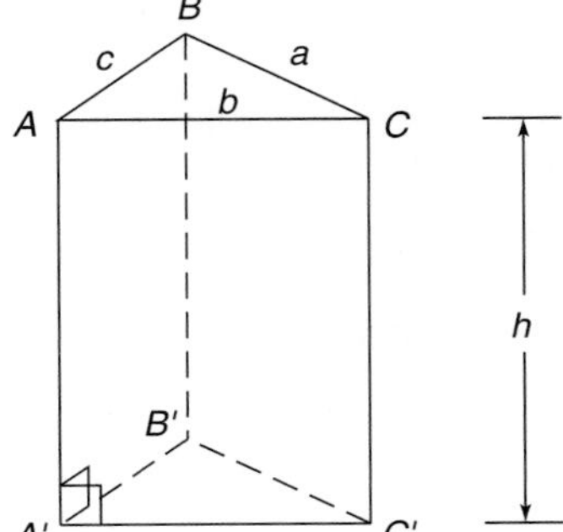


**FIGURE 8.3**

In the right triangular prism of Figure 8.3, $a$, $b$, and $c$ are the lengths of the sides of either base. These dimensions are used along with the length of the altitude (denoted by $h$) to calculate the lateral area, the sum of the areas of rectangles $ACC'A'$, $ABB'A'$, and $BCC'B'$. The lateral area $L$ of the right triangular prism can be found as follows:

$$\begin{aligned} L &= ah + bh + ch \\ &= h(a + b + c) \\ &= hP \end{aligned}$$

where $P$ is the perimeter of a base of the prism. This formula, $L = hP$, is valid for finding the lateral area of any *right* prism. Although lateral faces of an oblique prism are parallelograms, the formula $L = hP$ is also used to find its lateral area.

THEOREM 8.1.1: The lateral area $L$ of any prism whose altitude has measure $h$ and whose base has perimeter $P$ is given by $L = hP$.

Many students (and teachers) find it easier to calculate the lateral area of a prism without using the formula $L = hP$. We illustrate this in the following example.

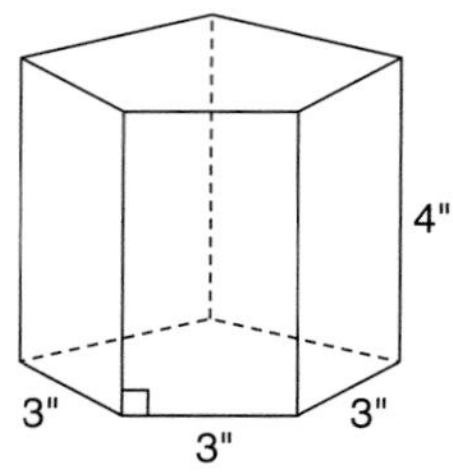


**FIGURE 8.4**

EXAMPLE 2

The bases of the right prism shown in Figure 8.4 are equilateral pentagons with sides of length 3 in. each. If the altitude of the prism is 4 in., find the lateral area of the prism.

***Solution***

Each lateral face is a rectangle with dimensions 3 in. by 4 in. The area of each rectangular face is 3 in. $\times$ 4 in. $=$ 12 in$^2$. Because there are five congruent lateral faces, the lateral area of the pentagonal prism is $5 \times 12$ in.$^2 = 60$ in$^2$.

NOTE: When applied in Example 2, the formula $L = hP$ leads to $L = 4$ in. $\times$ 15 in. $= 60$ in$^2$.

DEFINITION: For any prism, the **total area** $T$ is the sum of the lateral area and the areas of the bases.

NOTE: The total area of the prism is also known as its surface area.

Both bases and lateral faces are known as *faces* of a prism. Thus the total area $T$ of the prism is the sum of the areas of all its faces.

Recalling Heron's Formula, we know that the base area $B$ of the right triangular prism in Figure 8.3 can be found by the formula

$$B = \sqrt{s(s - a)(s - b)(s - c)}$$

in which $s$ is the semiperimeter of the triangular base. We use Heron's Formula in Example 3.

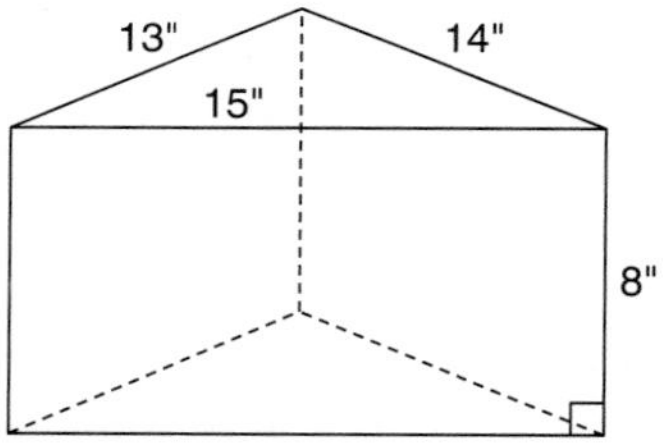

**FIGURE 8.5**

EXAMPLE 3

Find the total area of the right triangular prism with an altitude of length 8 in. if the sides of the triangular bases have lengths of 13 in., 14 in., and 15 in. (See Figure 8.5.)

***Solution***

The lateral area is found by adding the areas of the three rectangular lateral faces. That is,

$$\begin{aligned} L &= 8 \text{ in.} \cdot 13 \text{ in.} + 8 \text{ in.} \cdot 14 \text{ in.} + 8 \text{ in.} \cdot 15 \text{ in.} \\ &= 104 \text{ in}^2 + 112 \text{ in}^2 + 120 \text{ in}^2 = 336 \text{ in}^2 \end{aligned}$$

We use Heron's Formula to find the area of each base. With $s = \frac{1}{2}(13 + 14 + 15)$, or $s = 21$, $B = \sqrt{21(21 - 13)(21 - 14)(21 - 15)} = \sqrt{21(8)(7)(6)} = \sqrt{7056} = 84$.

Calculating the total area (or surface area) of the triangular prism,

$$T = 336 + 2(84) \qquad \text{or} \qquad T = 504 \text{ in}^2$$

A more general formula for the total area of a prism follows.

THEOREM 8.1.2: The total area $T$ of any prism with lateral area $L$ and base area $B$ is given by $T = L + 2B$.

**PICTURE PROOF OF THEOREM 8.1.2**

*Given:* The pentagonal prism of Figure 8.6(a)

*Prove:* $T = L + 2B$

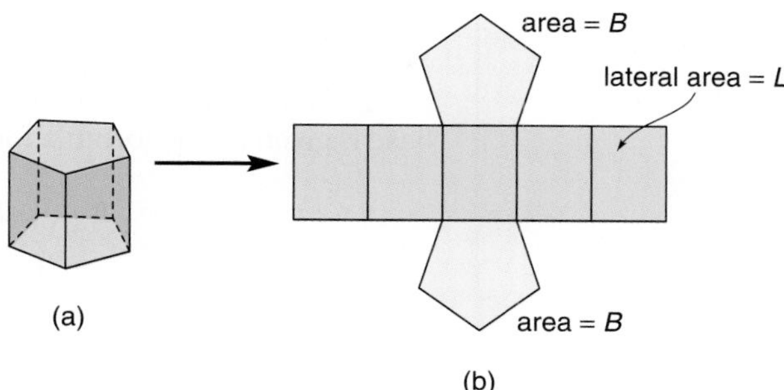


**FIGURE 8.6**

*Proof:* When the prism is "taken apart" and laid flat, we see that the total area depends upon the lateral area (shaded darker) and the areas of the two bases; that is,

$$T = L + 2B$$

DEFINITION: A **regular prism** is a right prism whose bases are regular polygons.

Henceforth, Figure 8.2(c) on page 391 will be called a regular triangular prism.

In the following example, each base of the prism is a regular hexagon. Because the prism is a right prism, the lateral faces are congruent rectangles.

**EXAMPLE 4**

Find the lateral area $L$ and the surface area $T$ of the regular hexagonal prism in Figure 8.7(a).

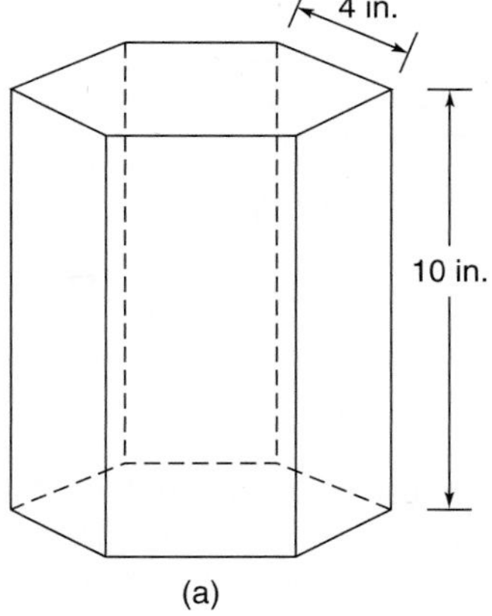


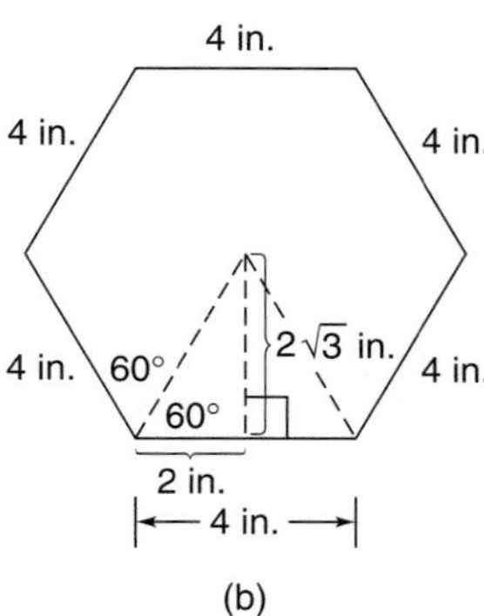


**FIGURE 8.7**

***Solution***

In Figure 8.7 on page 393, there are six congruent faces, each rectangular and with dimensions of 4 in. by 10 in. Then

$$\begin{aligned} L &= 6(4 \cdot 10) \\ &= 240 \text{ in}^2 \end{aligned}$$

For the regular hexagonal base [see Figure 8.7(b)], the apothem measures $a = 2\sqrt{3}$ in., and the perimeter is $P = 6 \cdot 4 = 24$ in. Then the area $B$ of each base is given by the formula for the area of a regular polygon.

$$\begin{aligned} B &= \frac{1}{2}aP \\ &= \frac{1}{2} \cdot 2\sqrt{3} \cdot 24 \\ &= 24\sqrt{3} \text{ in}^2 \approx 41.57 \text{ in}^2 \end{aligned}$$

Now

$$\begin{aligned} T &= L + 2B \\ &= (240 + 48\sqrt{3}) \text{ in}^2 \approx 323.14 \text{ in}^2 \end{aligned}$$

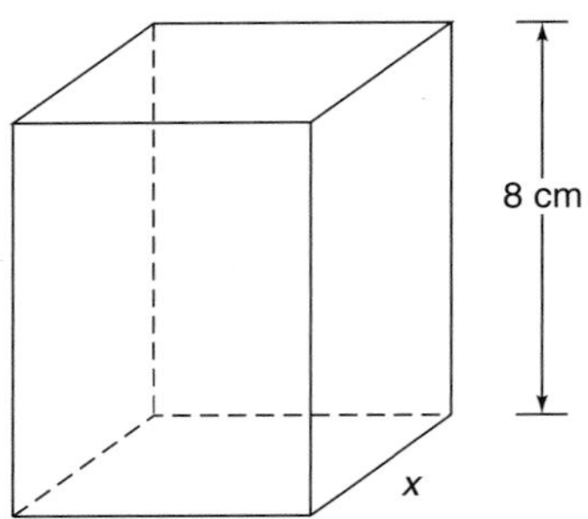


**FIGURE 8.8**

EXAMPLE 5

The total area of the right square prism in Figure 8.8 is 210 $\text{cm}^2$. Find the length of a side of the square base if the altitude of the prism is 8 cm.

***Solution***

Let $x$ be the length of a side of the square. Then the area of the base is $B = x^2$ and the area of each of the four lateral faces is $8x$. Therefore,

$$\underbrace{2(x^2)}_{\text{2 bases}} + \underbrace{4(8x)}_{\text{4 lateral faces}} = 210$$

$$\begin{aligned} 2x^2 + 32x &= 210 \\ 2x^2 + 32x - 210 &= 0 \\ x^2 + 16x - 105 &= 0 && \text{dividing by 2} \\ (x + 21)(x - 5) &= 0 && \text{factoring} \end{aligned}$$

$$x + 21 = 0 \quad \text{or} \quad x - 5 = 0$$

$$x = -21 \quad \text{or} \quad x = 5 \qquad \text{reject } -21 \text{ as a solution}$$

Then each side of the square base measures 5 cm.

*Exs. 3–7*

DEFINITION: A **cube** is a right square prism whose edges are congruent.

The cube is very important in determining the volume of a solid. See Figure 8.9.

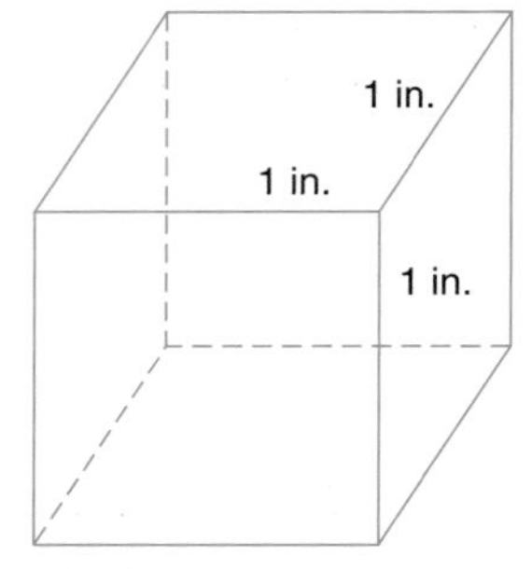


**FIGURE 8.9**

## VOLUME OF A PRISM

To introduce the notion of *volume*, we realize that a prism encloses a portion of space. Without a formal definition, we say that **volume** is a number that measures the amount of enclosed space. To begin, we need a unit for measuring volume. Just as the meter can be used to measure length and the square yard can be used to measure area, a **cubic unit** is used to measure the amount of space enclosed within a bounded region of space. One such unit is described next.

The volume enclosed by the cube shown in Figure 8.9 is 1 cubic inch or 1 $\text{in}^3$. The volume of a solid is the number of cubic units within the solid. Thus we assume that the volume of any solid is a positive number of cubic units.

**POSTULATE 24: (Volume Postulate)**
Corresponding to every solid is a unique positive number $V$ known as the volume of that solid.

The simplest figure for which we can determine volume is the **right rectangular prism.** Such a solid might be described as a **parallelpiped** or as a "box." Because boxes are used as containers for storage and shipping (like a boxcar), it is important to calculate volume as a measure of capacity. A right rectangular prism is shown in Figure 8.10; its dimensions are length $\ell$, width $w$, and height (or altitude) $h$.

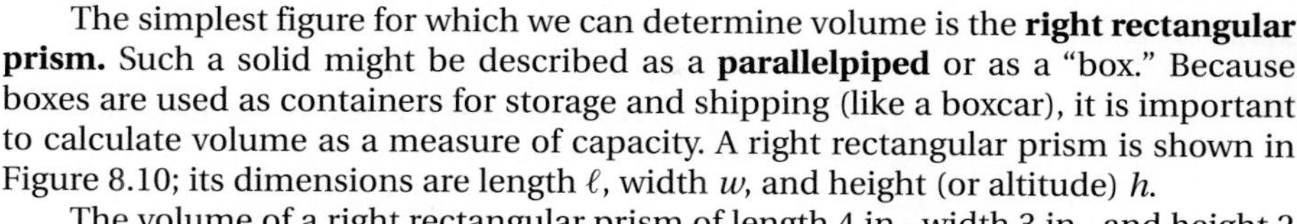

**FIGURE 8.10**

The volume of a right rectangular prism of length 4 in., width 3 in., and height 2 in. is easily shown to be 24 in$^3$. The volume is the product of the three dimensions of the given solid. We see not only that $4 \cdot 3 \cdot 2 = 24$ but also that the units of volume are in. $\cdot$ in. $\cdot$ in. $=$ in$^3$. Figures 8.11(a) and (b) illustrate that the 4 by 3 by 2 box must have the volume 24 cubic units. We see that there are four layers of blocks, each of which is a 2 by 3 configuration of 6 units$^3$. Figure 8.11 also provides some insight into our next postulate.

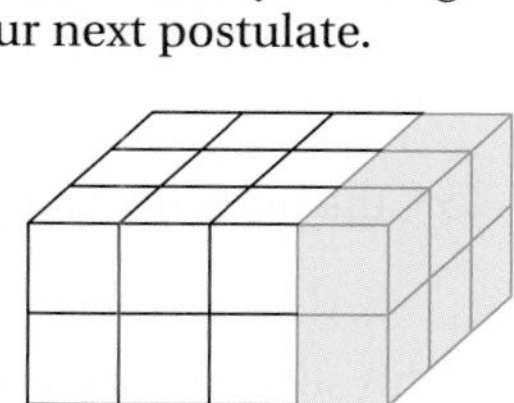
(a)

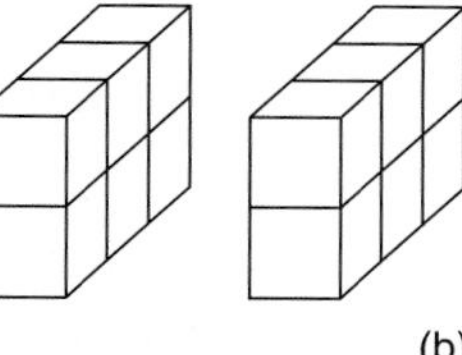
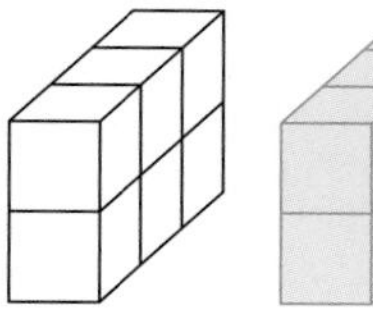
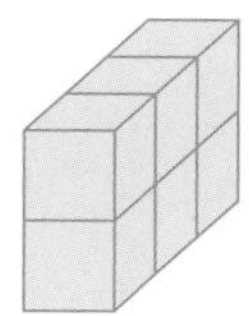
(b)

**FIGURE 8.11**

*Geometry in the Real World*

The frozen solids found in ice cube trays usually approximate the shapes of cubes.

**POSTULATE 25:**
The volume of a right rectangular prism is given by

$$V = \ell wh$$

where $\ell$ measures the length, $w$ the width, and $h$ the altitude of the prism.

To apply this formula, the units used for dimensions $\ell$, $w$, and $h$ must also be alike.

**EXAMPLE 6**

Find the volume of a box whose dimensions are 1 ft, 8 in., and 10 in. (See Figure 8.12.)

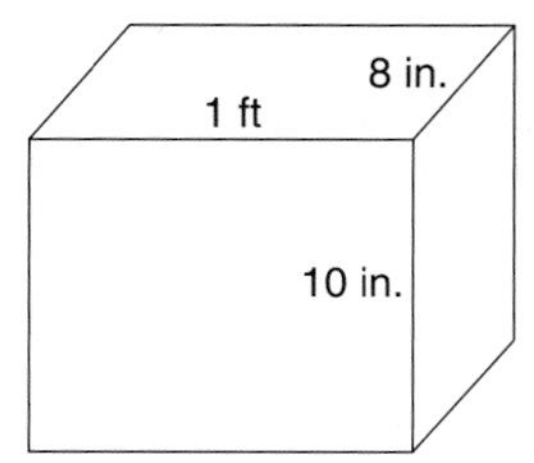

**FIGURE 8.12**

***Solution***

Although it makes no difference which dimension is chosen for $\ell$ or $w$ or $h$, it is most important that the units of measure be the same. Thus 1 ft is replaced by 12 in. in the formula for volume:

$$\begin{aligned} V &= \ell wh \\ &= 12 \text{ in.} \cdot 8 \text{ in.} \cdot 10 \text{ in.} \\ &= 960 \text{ in}^3 \end{aligned}$$

Note that the formula for the volume of the right rectangular prism, $V = \ell wh$, could be replaced by the formula $V = Bh$, where $B$ is the area of the base of the prism; that is, $B = \ell w$. As stated in the next postulate, this volume relationship is true for right prisms in general.

**Warning** ⓘ
The uppercase $B$ found in formulas in this chapter represents the area of the base of a solid; because the base is a plane region, $B$ is measured in square units.

**POSTULATE 26:**
The volume of a right prism is given by

$$V = Bh$$

where $B$ is the area of a base and $h$ is the altitude of the prism

*Exs. 9–13*

In real-world applications, the formula $V = Bh$ is valid for oblique prisms as well as right prisms.

**EXAMPLE 7**

Find the volume of the right hexagonal prism in Figure 8.7 on page 393.

***Solution***
We found that the area of the hexagonal base was $24\sqrt{3}$ in$^2$. Because the altitude of the hexagonal prism is 10 in., the volume is $V = Bh$, or $V = (24\sqrt{3}\text{ in.}^2)(10\text{ in.})$. Then $V = 240\sqrt{3}\text{ in.}^3 \approx 415.69\text{ in}^3$.

NOTE: Just as $x^2 \cdot x = x^3$, the units in Example 7 are in.$^2 \cdot$ in. $=$ in.$^3$

In the final example of this section, we use the fact that 1 yd$^3$ = 27 ft$^3$. In the cube shown in Figure 8.13, each dimension measures 1 yd, or 3 ft. The cube's volume is given by 1 yd · 1 yd · 1 yd = 1 yd$^3$ or 3 ft · 3 ft · 3 ft = 27 ft$^3$.

***Technology Exploration***
On your calculator, determine the method of "cubing." That is, find a value such as $2.1^3$. On many calculators, we enter 2.1, a caret ∧, and 3.

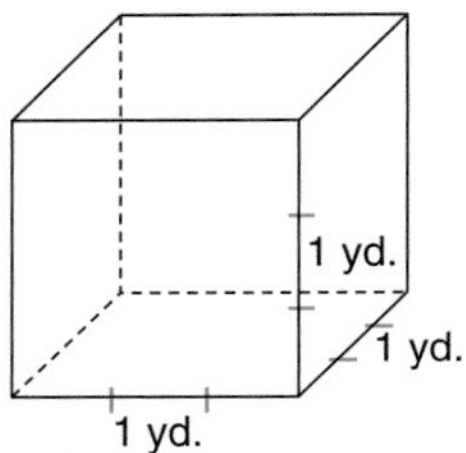


**FIGURE 8.13**

**EXAMPLE 8**

The Tarantinos are having a concrete driveway poured at their house. The section to be poured is rectangular, measuring 12 ft by 40 ft, and is 4 in. deep. How many cubic yards of concrete are needed?

***Solution***
Using $V = \ell wh$, we must be consistent with units. Thus $\ell = 12$ ft, $w = 40$ ft, and $h = \frac{1}{3}$ ft (from 4 in.). Now

$$V = 12\text{ ft} \cdot 40\text{ ft} \cdot \frac{1}{3}\text{ ft}$$
$$V = 160\text{ ft}^3$$

To change 160 ft$^3$ to cubic yards, we divide by 27 to obtain $5\frac{25}{27}$ yd$^3$.

NOTE: The Tarantinos will be charged for 6 yd$^3$ of concrete, the result of rounding upward.

*Exs. 14, 15*

## 8.1 Exercises

1. Consider the solid shown.
   a) Does it appear to be a prism?
   b) Is it right or oblique?
   c) What type of base(s) does the solid have?
   d) Name the type of solid.
   e) What type of figure is each lateral face?

2. Consider the solid shown.
   a) Does it appear to be a prism?
   b) Is it right or oblique?
   c) What type of base(s) does the solid have?
   d) Name the type of solid.
   e) What type of figure is each lateral face?

3. Consider the hexagonal prism shown in Exercise 1.
   a) How many vertices does it have?
   b) How many edges (lateral edges plus base edges) does it have?
   c) How many faces (lateral faces plus bases) does it have?

4. Consider the triangular prism shown in Exercise 2.
   a) How many vertices does it have?
   b) How many edges (lateral edges plus base edges) does it have?
   c) How many faces (lateral faces plus bases) does it have?

5. If each edge of the hexagonal prism in Exercise 1 is measured in centimeters, what unit is used to measure its (a) surface area? (b) volume?

6. If each edge of the triangular prism in Exercise 2 is measured in inches, what unit is used to measure its (a) lateral area? (b) volume?

7. Suppose that each of the bases of the hexagonal prism in Exercise 1 has an area of 12 $cm^2$ and that each lateral face has an area of 18 $cm^2$. Find the total (surface) area of the prism.

8. Suppose that each of the bases of the triangular prism in Exercise 2 has an area of 3.4 $in^2$ and that each lateral face has an area of 4.6 $in^2$. Find the total (surface) area of the prism.

9. Suppose that each of the bases of the hexagonal prism in Exercise 1 has an area of 12 $cm^2$ and that the altitude of the prism measures 10 cm. Find the volume of the prism.

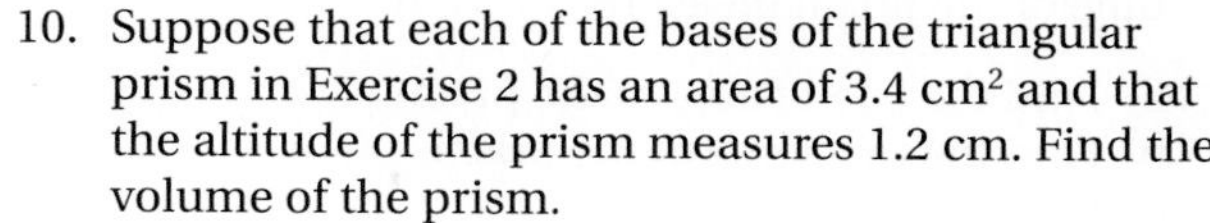

10. Suppose that each of the bases of the triangular prism in Exercise 2 has an area of 3.4 $cm^2$ and that the altitude of the prism measures 1.2 cm. Find the volume of the prism.

11. A solid is an octagonal prism.
    a) How many vertices does it have?
    b) How many lateral edges does it have?
    c) How many base edges are there in all?

12. A solid is a pentagonal prism.
    a) How many vertices does it have?
    b) How many lateral edges does it have?
    c) How many base edges are there in all?

13. Generalize the results found in Exercises 11 and 12 by answering each of the following questions. Assume that the number of sides in each base of the prism is $n$. For the prism, what is the
    a) number of vertices?
    b) number of lateral edges?
    c) number of base edges?
    d) total number of edges?
    e) number of lateral faces?
    f) number of bases?
    g) total number of faces?
    (NOTE: Upper and lower faces = bases)

14. In the accompanying regular pentagonal prism, suppose that each base edge measures 6 in. and that the apothem of the base measures 4.1 in. The altitude of the prism measures 10 in.
    a) Find the lateral area of the prism.
    b) Find the total area of the prism.
    c) Find the volume of the prism.

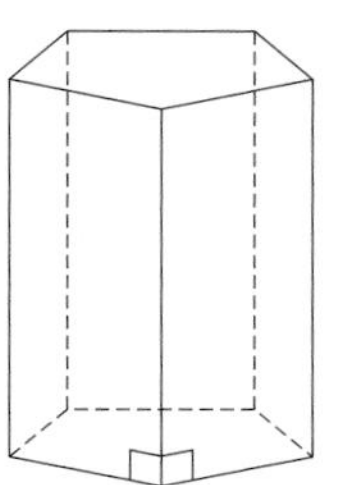

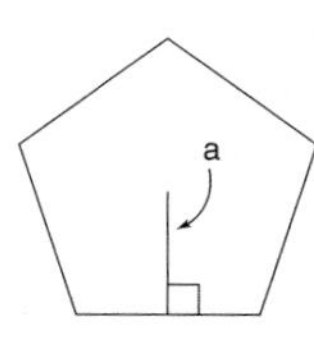


*Exercises 14, 15*

15. In the regular pentagonal prism on page 397, suppose that each base edge measures 9.2 cm and that the apothem of the base measures 6.3 cm. The altitude of the prism measures 14.6 cm.

    a) Find the lateral area of the prism.
    b) Find the total area of the prism.
    c) Find the volume of the prism.

16. For a right triangular prism, suppose that the sides of the triangular base measure 4 m, 5 m, and 6 m. The altitude is 7 m.

    a) Find the lateral area of the prism.
    b) Find the total area of the prism.
    c) Find the volume of the prism.

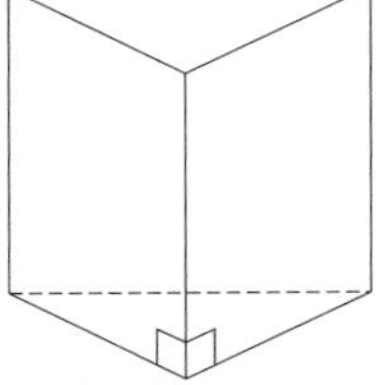

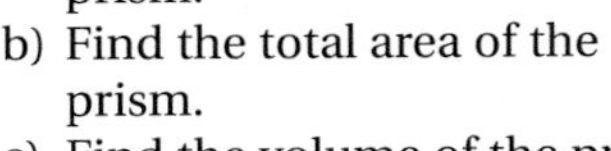

*Exercises 16, 17*

17. For a right triangular prism, suppose that the sides of the triangular base measure 3 ft, 4 ft, and 5 ft. The altitude is 6 ft.

    a) Find the lateral area of the prism.
    b) Find the total area of the prism.
    c) Find the volume of the prism.

18. Given that 100 cm = 1 m, find the number of cubic centimeters in 1 cubic meter.

19. Given that 12 in. = 1 ft, find the number of cubic inches in 1 cubic foot.

20. A cereal box measures 2 in. by 8 in. by 10 in. What is the volume of the box? How many square inches of cardboard make up its surface? (Disregard any hidden flaps.)

21. The measures of the sides of the square base of a box are twice the measure of the height of the box. If the volume of the box is 108 in$^3$, find the dimensions of the box.

22. For a given box, the height measures 4 m. If the length of the rectangular base is 2 m greater than the width of the base and the lateral area $L$ is 96 m$^2$, find the dimensions of the box.

23. For the box shown, the total area is 94 cm$^2$. Determine the value of $x$.

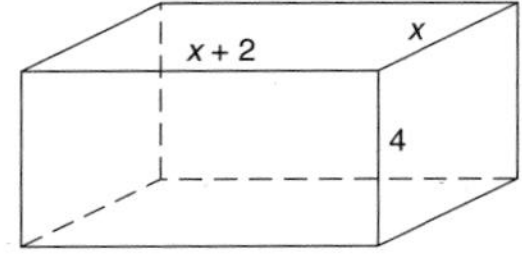


*Exercises 23, 24*

24. If the volume of the box is 252 in$^3$, find the value of $x$. (See the figure for Exercise 23.)

25. The box with dimensions indicated is to be constructed of materials that cost 1 cent per square inch for the lateral surface and 2 cents per square inch for the bases. What is the total cost of constructing the box?

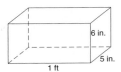

26. A hollow steel door is 32 in. wide by 80 in. tall by $1\frac{3}{8}$ in. thick. How many cubic inches of foam insulation are needed to fill the door?

27. A storage shed is in the shape of a pentagonal prism. The front represents one of its two bases. If the shed is 10 ft deep, what is the storage capacity (volume) of its interior?

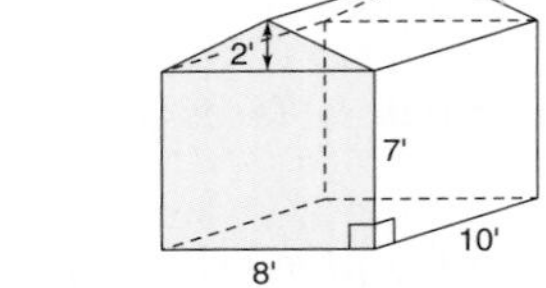


28. A storage shed is in the shape of a trapezoidal prism. Each trapezoid represents one of its bases. With dimensions as shown, what is the storage capacity (volume) of its interior?

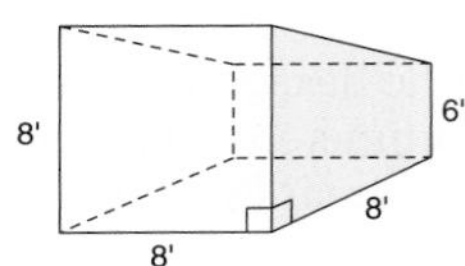


29. A cube is a right square prism in which all edges have the same length. For the cube with edge $e$,

    a) show that the total area is $T = 6e^2$.
    b) find the total area if $e = 4$ cm.
    c) show that the volume is $V = e^3$.
    d) find the volume if $e = 4$ cm.

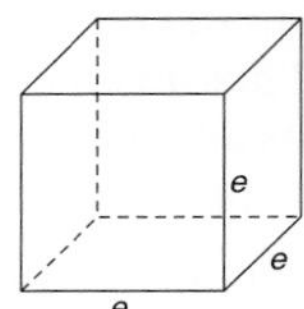


*Exercises 29–31*

30. Use the formulas and drawing in Exercise 29 to find (a) the total area $T$ and (b) the volume $V$ of a cube with edges of length 5.3 ft each.

*31. A diagonal of a cube joins two vertices so that the remaining points on the diagonal lie in the interior of the cube. Show that the diagonal of the cube having edges of length $e$ is $e\sqrt{3}$ units long.

32. A concrete pad 4 in. thick is to have a length of 36 ft and a width of 30 ft. How many cubic yards of concrete must be poured? (HINT: $1 \text{ yd}^3 = 27 \text{ ft}^3$)

33. A raised flower bed is 2 ft high by 12 ft wide by 15 ft long. The mulch, soil, and peat mixture used to fill the raised bed costs $9.60 per cubic yard. What is the total cost of the ingredients used to fill the raised garden?

34. In excavating for a new house, a contractor digs a hole in the shape of a right rectangular prism. The dimensions of the hole are 54 ft long by 36 ft wide by 9 ft deep. How many cubic yards of dirt were removed?

35. An open box is formed by cutting congruent squares from the four corners of a square piece of cardboard that has a length of 24 in. per side. If the congruent squares that are removed have sides that measure 6 in. each, what is the volume of the box formed by folding and sealing the "flaps"?

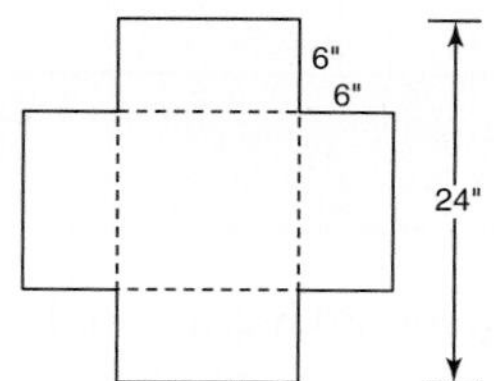


36. Repeat Exercise 35, but with the four congruent squares with sides of length 6 in. being cut from the corners of a rectangular piece of poster board that is 20 in. wide by 30 in. long.

37. An aquarium is "box-shaped" with dimensions of 2 ft by 1 ft by 8 in. If $1 \text{ ft}^3$ corresponds to 7.5 gal of water, what is the water capacity of the aquarium in *gallons?*

38. The gasoline tank on an automobile is "box-shaped" with dimensions of 24 in. by 20 in. by 9 in. If $1 \text{ ft}^3$ corresponds to 7.5 gal of gasoline, what is the capacity of the automobile's fuel tank in *gallons?*

*For Exercises 39–41, consider the oblique regular pentagonal prism shown. Each side of the base measures 12 cm, and the altitude measures 12 cm.*

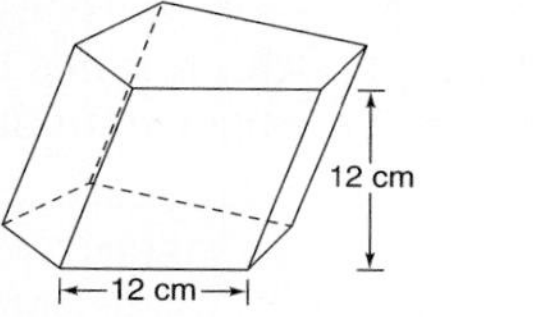


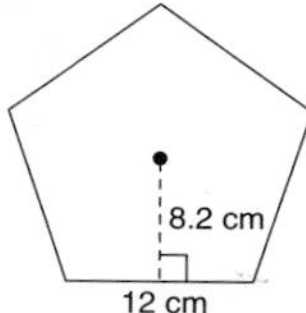


*Exercises 39–41*

39. Find the lateral area of the prism. (HINT: Each lateral face is a parallelogram.)

40. Find the total area of the prism.

41. Find the volume of the prism.

42. It can be shown that the length of a diagonal of a right rectangular prism with dimensions $\ell$, $w$, and $h$ is given by $d = \sqrt{\ell^2 + w^2 + h^2}$. Use this formula to find the length of the diagonal when $\ell = 12$ in., $w = 4$ in., and $h = 3$ in.

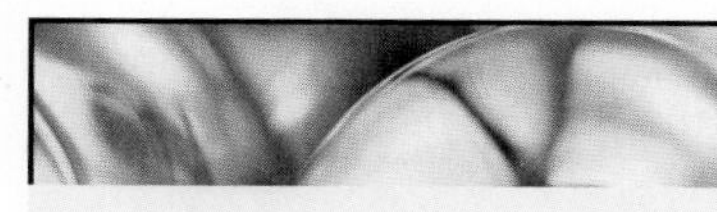

## 8.2 Pyramids, Area, and Volume

**KEY CONCEPTS**

Pyramid • Base • Altitude • Vertices • Edges • Faces • Vertex of a Pyramid • Regular Pyramid • Slant Height of a Regular Pyramid • Lateral Area • Total (Surface) Area • Volume

The solids shown in Figure 8.14 on the following page are **pyramids.** In Figure 8.14(a), point $A$ is non-coplanar with square $BCDE$. In Figure 8.14(b), $F$ is non-coplanar with $\triangle GHJ$. In these pyramids, the non-coplanar point has been joined (by drawing line segments) to each vertex of the square and to each vertex of the triangle, respectively. Every pyramid has exactly one base. Square $BCDE$ is the base of the first pyramid, and $\triangle GHJ$ is the base of the second pyramid. Point $A$ is known as the **vertex** of the **square pyramid;** likewise, point $F$ is the vertex of the **triangular pyramid.**

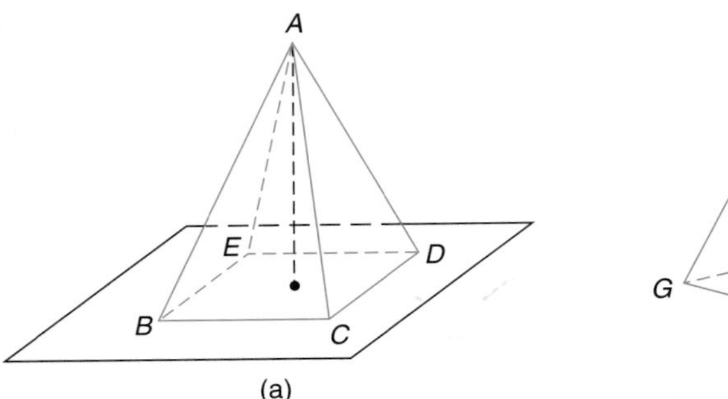


**FIGURE 8.14**

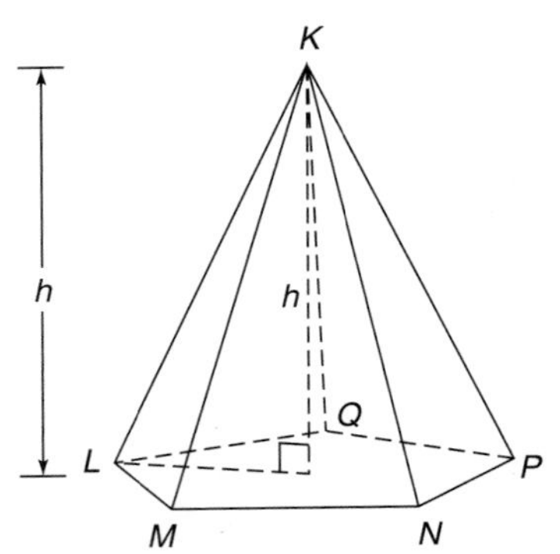


**FIGURE 8.15**

The pyramid in Figure 8.15 is a **pentagonal pyramid.** It has vertex $K$, pentagon $LMNPQ$ for its **base,** and **lateral edges** $\overline{KL}$, $\overline{KM}$, $\overline{KN}$, $\overline{KP}$, and $\overline{KQ}$. Although $K$ is called *the* vertex of the pyramid, there are actually six vertices: $K$, $L$, $M$, $N$, $P$, and $Q$. The sides of the base $\overline{LM}$, $\overline{MN}$, $\overline{NP}$, $\overline{PQ}$, and $\overline{QL}$ are **base edges.** All **lateral faces** of a pyramid are triangles; $\triangle KLM$ is one of the five lateral faces of the pentagonal pyramid. Including base $LMNPQ$, this pyramid has a total of six faces. The **altitude** of the pyramid, of length $h$, is the line segment from the vertex $K$ perpendicular to the plane of the base.

DEFINITION: A **regular pyramid** is a pyramid whose base is a regular polygon and whose lateral edges are all congruent.

Suppose that the pyramid in Figure 8.15 is a regular pentagonal pyramid. Then the lateral faces are necessarily congruent to each other; in Figure 8.15, $\triangle KLM \cong \triangle KMN \cong \triangle KNP \cong \triangle KPQ \cong \triangle KQL$ by SSS. Each lateral face is an isosceles triangle. In a regular pyramid, the altitude joins the vertex of the pyramid to the center of the regular polygon that is the base.

*Exs. 1, 2*

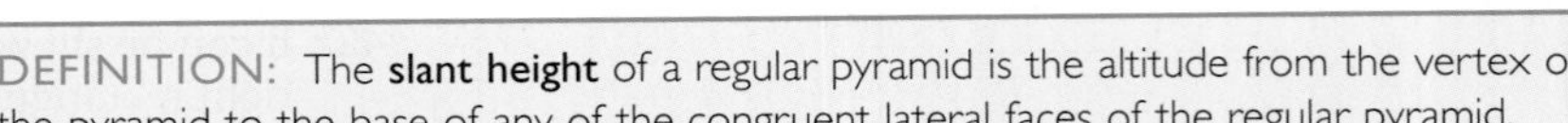
DEFINITION: The **slant height** of a regular pyramid is the altitude from the vertex of the pyramid to the base of any of the congruent lateral faces of the regular pyramid.

NOTE: Among pyramids, only a regular pyramid has a slant height.

In our formulas and explanations, we use $\ell$ to represent the length of the slant height of a regular pyramid.

EXAMPLE 1

For a regular square pyramid with altitude 4 in. and base edges of length 6 in. each, find the length of the slant height $\ell$. (See Figure 8.16 on page 401.)

***Solution***

In Figure 8.16, it can be shown that the apothem to any side has length 3 in. (one-half the length of the side of the square base). Also, the slant height is the hypotenuse of a right triangle with legs equal to the lengths of the altitude and the apothem. By the Pythagorean Theorem, we have

$$\ell^2 = a^2 + h^2$$
$$\ell^2 = 3^2 + 4^2$$
$$\ell^2 = 9 + 16$$
$$\ell^2 = 25$$
$$\ell = 5 \text{ in.}$$

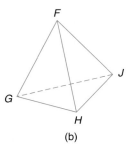

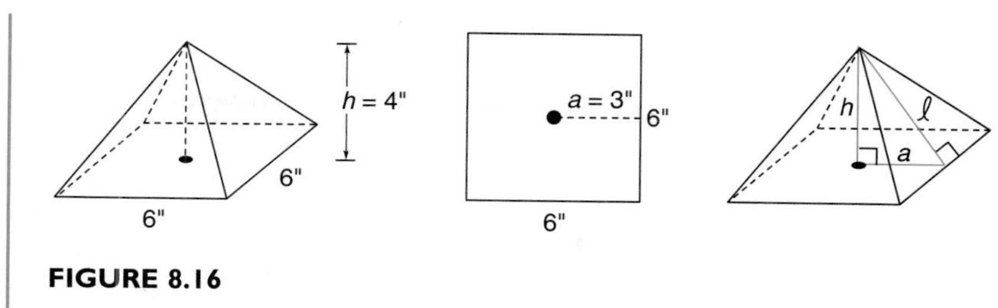

**FIGURE 8.16**

The following fact was used in the solution of Example 1; see the pyramid above and at the right. We accept Theorem 8.2.1 without further proof than Figure 8.16 provides.

*Exs. 3, 4*

THEOREM 8.2.1: In a regular pyramid, the length $a$ of the apothem of the base, the altitude $h$, and the slant height $\ell$ satisfy the Pythagorean Theorem; that is, $\ell^2 = a^2 + h^2$ in every regular pyramid.

## SURFACE AREA OF A PYRAMID

To lay the groundwork for the next theorem, we justify the result by "taking apart" one of the regular pyramids and laying it out flat. We will use a regular hexagonal pyramid for this purpose, but the argument is similar if the base is any regular polygon.

When the lateral faces of the regular pyramid are folded down into the plane, as shown in Figure 8.17, the shaded lateral area is the sum of the areas of the triangular lateral faces. Using $A = \frac{1}{2}bh$, we find that the area of each face is $\frac{1}{2} \cdot s \cdot \ell$ (each side of the base of the pyramid has length $s$, and the slant height is $\ell$). The combined areas of the triangles give the lateral area. Because there are $n$ triangles,

$$L = n \cdot \frac{1}{2} \cdot s \cdot \ell$$
$$= \frac{1}{2} \cdot \ell(n \cdot s)$$
$$= \frac{1}{2}\ell P$$

where $P$ is the perimeter of the base.

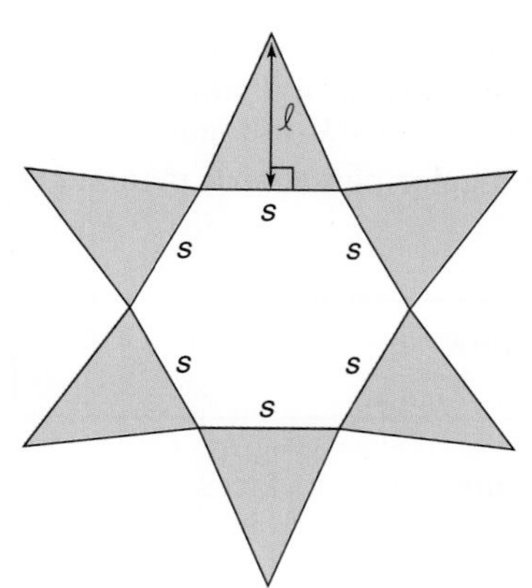

**FIGURE 8.17**

THEOREM 8.2.2: The lateral area $L$ of a regular pyramid with slant height of length $\ell$ and perimeter $P$ of the base is given by

$$L = \frac{1}{2}\ell P$$

We will illustrate the use of Theorem 8.2.2 in Example 2.

EXAMPLE 2

Find the lateral area of a regular pentagonal pyramid if the sides of the base measure 8 cm and the lateral edges measure 10 cm each [see Figure 8.18(a)].

***Solution***

For the triangular lateral face [see Figure 8.18(b)], we find the length of the slant height by applying the Pythagorean Theorem:

$$4^2 + \ell^2 = 10^2, \text{ so } 16 + \ell^2 = 100$$
$$\ell^2 = 84$$
$$\ell = \sqrt{84} = \sqrt{4 \cdot 21} = \sqrt{4} \cdot \sqrt{21} = 2\sqrt{21}$$

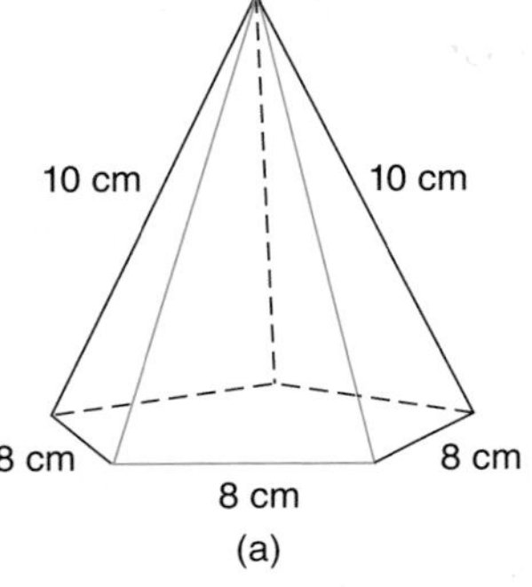


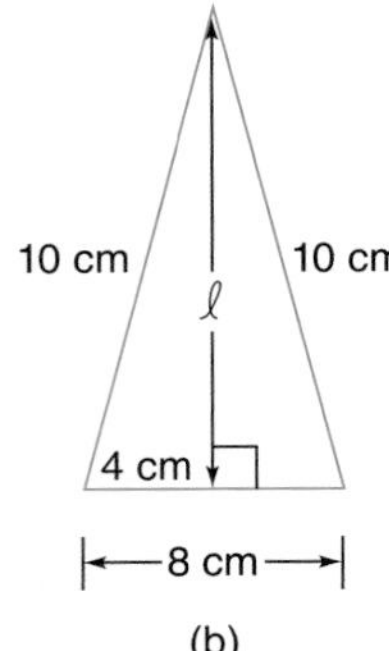


**FIGURE 8.18**

Now $L = \frac{1}{2}\ell P$ becomes $L = \frac{1}{2} \cdot 2\sqrt{21} \cdot (5 \cdot 8) = \frac{1}{2} \cdot 2\sqrt{21} \cdot 40 = 40\sqrt{21}\text{ cm}^2 \approx 183.30\text{ cm}^2$.

It may be easier to find the lateral area of a regular pyramid without using the formula of Theorem 8.2.2; simply find the area of one lateral face and multiply by the number of faces. In Example 2, for instance, the area of each triangular face is $\frac{1}{2} \cdot 8 \cdot 2\sqrt{21}$ or $8\sqrt{21}$; thus the lateral area of the regular pentagonal pyramid is $5 \cdot 8\sqrt{21} = 40\sqrt{21}\text{ cm}^2$.

THEOREM 8.2.3: The total area (surface area) $T$ of a pyramid with lateral area $L$ and base area $B$ is given by $T = L + B$.

The total area $T$ of the pyramid is the sum of the areas of all its faces.

EXAMPLE 3

Find the total area of a regular square pyramid that has base edges of length 4 ft and lateral edges of length 6 ft. [See Figure 8.19(a).]

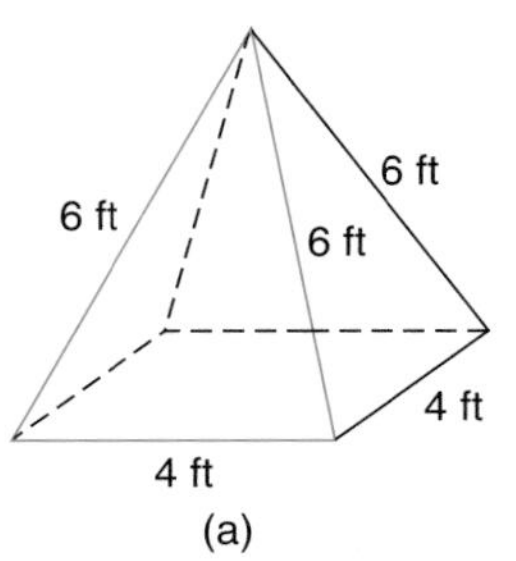


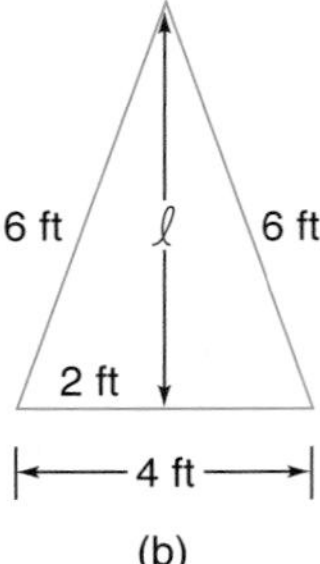


**FIGURE 8.19**

***Solution***

To determine the lateral area, we need the length of the slant height. [See Figure 8.19(b).]

$$\ell^2 + 2^2 = 6^2$$
$$\ell^2 + 4 = 36$$
$$\ell^2 = 32$$
$$\ell = \sqrt{32} = \sqrt{16 \cdot 2} = \sqrt{16} \cdot \sqrt{2} = 4\sqrt{2}$$

The lateral area is $L = \frac{1}{2}\ell P$. Therefore,

$$L = \frac{1}{2} \cdot 4\sqrt{2}(16) = 32\sqrt{2}\ \text{ft}^2$$

Because the area of the square base is 16 ft$^2$, the total area is

$$T = 16 + 32\sqrt{2} \approx 61.25\ \text{ft}^2$$

SSG *Exs. 5–7*

The pyramid in Figure 8.20(a) is a regular square pyramid rather than just a square pyramid. The pyramid shown in Figure 8.20(b) is oblique. It does not have congruent lateral edges or congruent faces.

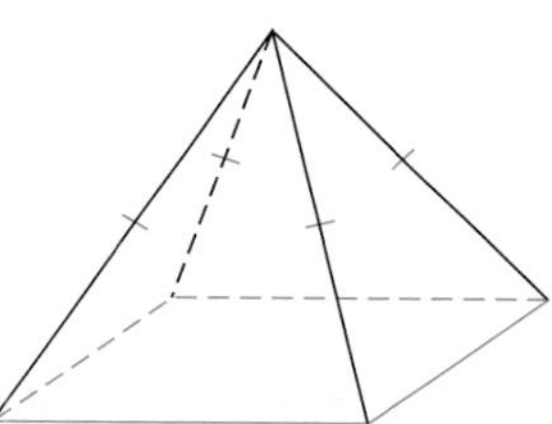

Regular square pyramid
(a)

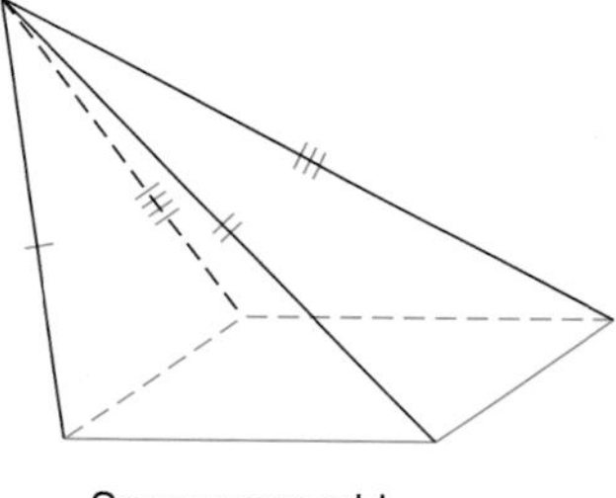

Square pyramid
(b)

**FIGURE 8.20**

## VOLUME OF A PYRAMID

The final theorem in this section is presented without any attempt to construct the proof. In an advanced course such as calculus, the statement can be proved. The factor "one-third" in the formula for the volume of a pyramid provides exact results. This formula can be applied to any pyramid, even one that is not regular. See Figure 8.20(b), in which the length of altitude $h$ is the perpendicular distance from the vertex to the plane of the square base. Read the Discover! box in the margin at left before moving on to Theorem 8.2.4 and its applications.

### Discover!

There are kits that contain a hollow pyramid and a hollow prism that have congruent bases and the same altitude. Fill the pyramid with water and then empty the water into the prism.

a) How many times did you have to empty the pyramid in order to fill the prism?

b) As a fraction, the volume of the pyramid is what part of the volume of the prism?

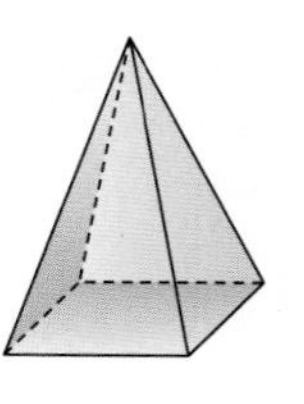

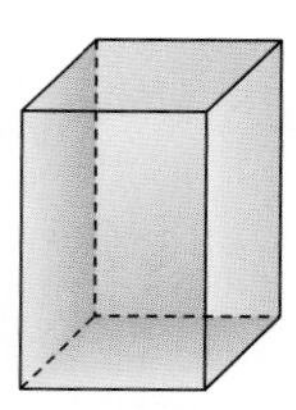

ANSWERS

(a) Three times (b) $\frac{1}{3}$

THEOREM 8.2.4: The volume $V$ of a pyramid having a base area $B$ and an altitude of length $h$ is given by

$$V = \frac{1}{3}Bh$$

### EXAMPLE 4

Find the volume of the regular square pyramid with altitude $h = 4$ in. and base edges of length $s = 6$ in. (This was the pyramid of Example 1.)

***Solution***

The area of the square base is $B = (6 \text{ in.})^2$ or $36 \text{ in}^2$. Because $h = 4$ in., the formula $V = \frac{1}{3}Bh$ becomes

$$V = \frac{1}{3}(36 \text{ in}^2)(4 \text{ in.}) = 48 \text{ in}^3$$

To find the volume of a pyramid, we use the formula $V = \frac{1}{3}Bh$. In many applications, it is necessary to determine $B$ or $h$ from other information that has been provided. In Example 5, calculating the length of the altitude $h$ is a challenge! In Example 6, the difficulty lies in finding the area of the base. Before we consider either problem, Table 8.1 will remind us of the types of units necessary in different types of measure.

**TABLE 8.1**

| *Type of Measure* | *Geometric Measure* | *Type of Unit* |
|---|---|---|
| Linear | Length of segment, such as length of slant height | in., cm, etc. |
| Area | Amount of plane region enclosed, such as area of lateral face | $\text{in}^2$, $\text{cm}^2$, etc. |
| Volume | Amount of space enclosed, such as volume of a pyramid | $\text{in}^3$, $\text{cm}^3$, etc. |

In Example 5, we apply the following theorem. This application of the Pythagorean Theorem relates the lengths of the lateral edge, the radius of the base, and the altitude of a *regular* pyramid. Figure 8.21(c) provides a visual interpretation of the theorem.

THEOREM 8.2.5: In a regular pyramid, the lengths of altitude $h$, radius $r$ of the base, and lateral edge $e$ satisfy the Pythagorean Theorem; that is, $e^2 = h^2 + r^2$.

EXAMPLE 5

Find the volume of the regular square pyramid in Figure 8.21(a).

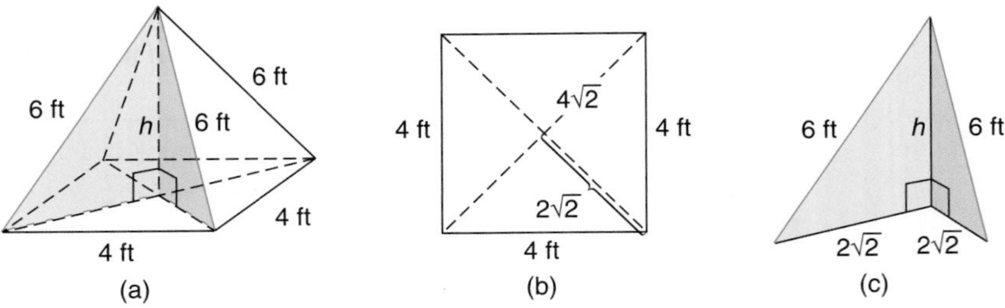


**FIGURE 8.21**

***Solution***

The length of the altitude (of the pyramid) is represented by $h$, which is determined as follows.

First we see that this altitude meets the diagonals of the square base at their common midpoint [see Figure 8.21(b)]. Because each diagonal has the length $4\sqrt{2}$ ft by the 45°-45°-90° relationship, we have a right triangle whose legs are of lengths $2\sqrt{2}$ ft and $h$, and the hypotenuse has length 6 ft (the length of the lateral edge). See Figure 8.21(c) in which $r = 2\sqrt{2}$ and $e = 6$.

Applying Theorem 8.25 in Figure 8.21(c), we have

$$h^2 + (2\sqrt{2})^2 = 6^2$$
$$h^2 + 8 = 36$$
$$h^2 = 28$$
$$h = \sqrt{28} = \sqrt{4 \cdot 7} = \sqrt{4} \cdot \sqrt{7} = 2\sqrt{7}$$

The area of the square base is $B = 4^2$, or $B = 16\text{ ft}^2$. Now we have

$$\begin{aligned} V &= \frac{1}{3}Bh \\ &= \frac{1}{3}(16)(2\sqrt{7}) \\ &= \frac{32}{3}\sqrt{7}\text{ ft}^3 \approx 28.22\text{ ft}^3 \end{aligned}$$

EXAMPLE 6

Find the volume of a regular hexagonal pyramid whose base edges have length 4 in. and whose altitude measures 12 in. [See Figure 8.22(a).]

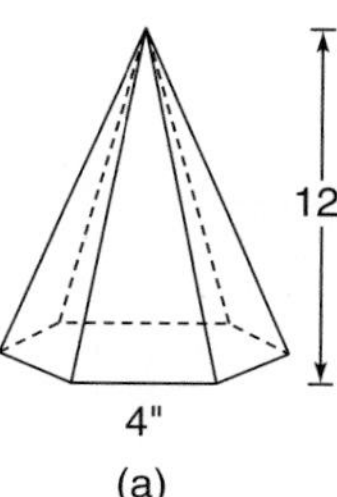

(a)

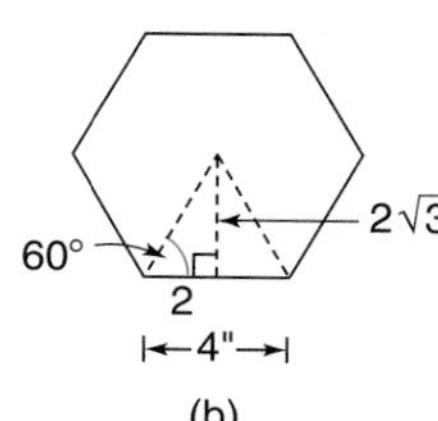

(b)

**FIGURE 8.22**

***Solution***

In the formula $V = \frac{1}{3}Bh$, the altitude is $h = 12$. To find the area of the base, we use the formula $B = \frac{1}{2}aP$ (this was written $A = \frac{1}{2}aP$ in Chapter 7). In the 30°-60°-90° triangle formed by the apothem, radius, and side of the regular hexagon, we see that

$$a = 2\sqrt{3}\text{ in.} \qquad \text{[See Figure 8.22(b)]}$$

Now $B = \frac{1}{2} \cdot 2\sqrt{3} \cdot (6 \cdot 4)$, or $B = 24\sqrt{3}\text{ in}^2$.

In turn, $V = \frac{1}{3}Bh$ becomes $V = \frac{1}{3}(24\sqrt{3})(12)$, so $V = 96\sqrt{3}\text{ in}^3$.

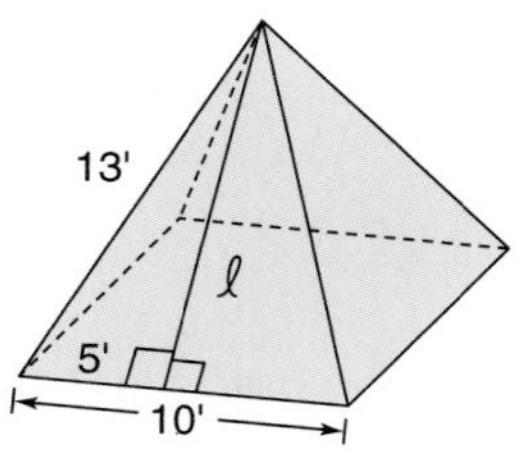


**FIGURE 8.23**

EXAMPLE 7

A church steeple has the shape of a regular square pyramid. Measurements taken show that the base edges measure 10 ft and that the length of a lateral edge is 13 ft. To determine the amount of roof needing to be reshingled, find the lateral area of the pyramid. (See Figure 8.23.)

Reminder

It is sometimes easier to find the lateral area without memorizing and using another new formula.

Exs. 8–11

***Solution***

The slant height $\ell$ of each triangular face is determined by solving the equation

$$5^2 + \ell^2 = 13^2$$
$$25 + \ell^2 = 169$$
$$\ell^2 = 144$$
$$\ell = 12$$

The formula $A = \frac{1}{2}bh$ becomes $A = \frac{1}{2} \cdot 10 \cdot 12 = 60 \text{ ft}^2$. Considering the four lateral faces, the area to be reshingled measures

$$L = 4 \cdot 60 \text{ ft}^2 \qquad \text{or} \qquad L = 240 \text{ ft}^2$$

Plane and solid figures may have line symmetry and point symmetry. However, solid figures may also have **plane symmetry.** Thus, a plane can be drawn for which a point of the space figure has a corresponding point on the opposite side of the plane at the same distance.

Each solid in Figure 8.24 has more than one plane of symmetry. In Figure 8.24(a), the plane of symmetry shown is determined by the midpoints of the indicated edges of the "box." In Figure 8.24(b), the plane determined by the vertex and the midpoints of opposite sides of the square base leads to plane symmetry for the pyramid.

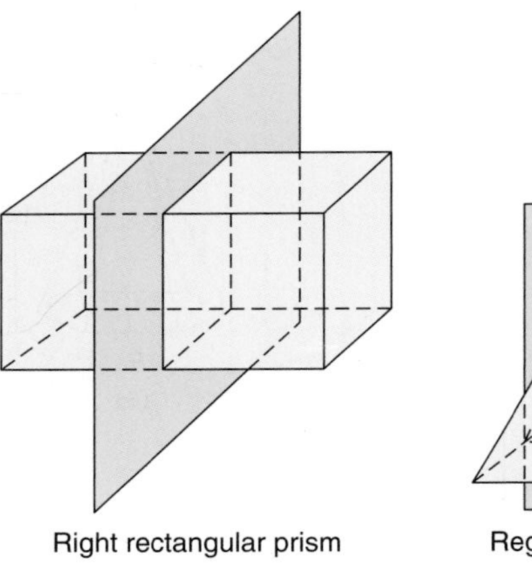

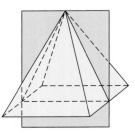

Right rectangular prism

(a)

Regular square pyramid

(b)

**FIGURE 8.24**

## 8.2 Exercises

*In Exercises 1 to 4, name the solid that is shown. Answers are based on Sections 8.1 and 8.2.*

1. a)

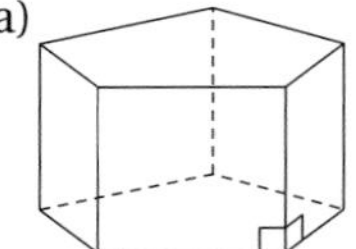

Bases are not regular.

b)

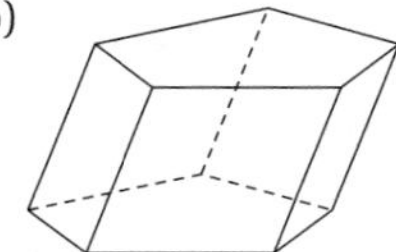

Bases are not regular.

2. a)

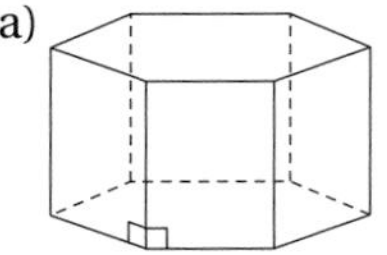

Bases are regular.

b)

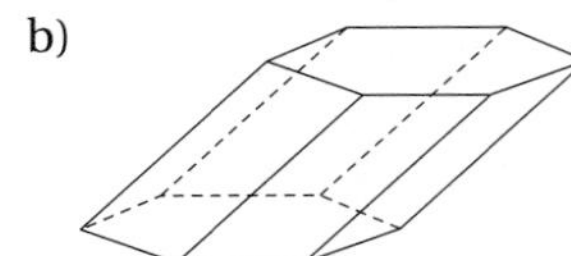

Bases are not regular.

3. a)

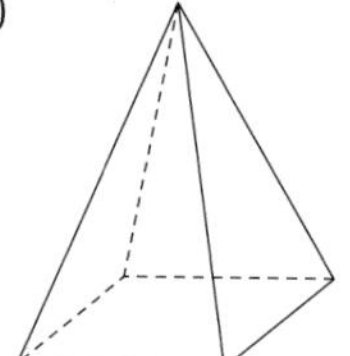

Lateral faces are congruent; base is a square.

b)

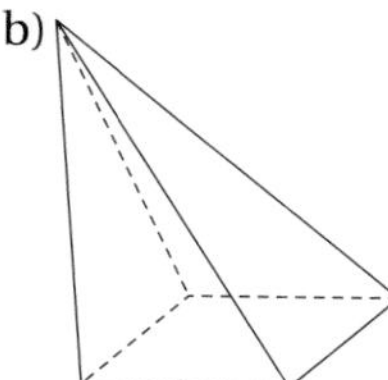

Base is a square.

4. a)

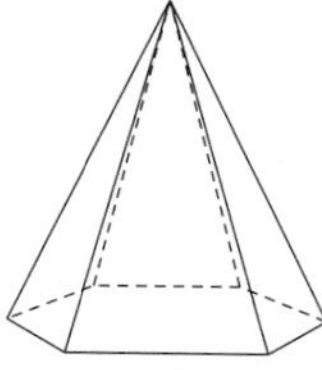

Lateral faces are congruent; base is a regular polygon.

b)

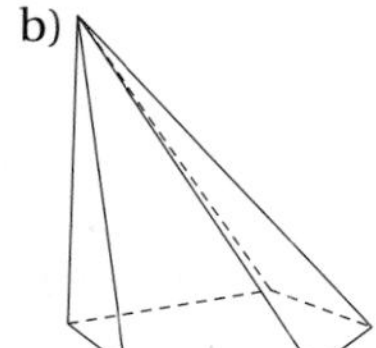

Lateral faces are not congruent.

5. In the solid shown, base *ABCD* is a square.

   a) Is the solid a prism or a pyramid?
   b) Name the vertex of the pyramid.
   c) Name the lateral edges.
   d) Name the lateral faces.
   e) Is the solid a regular square pyramid?

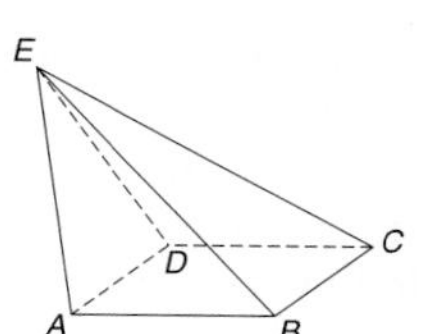


*Exercises 5, 7, 9, 11*

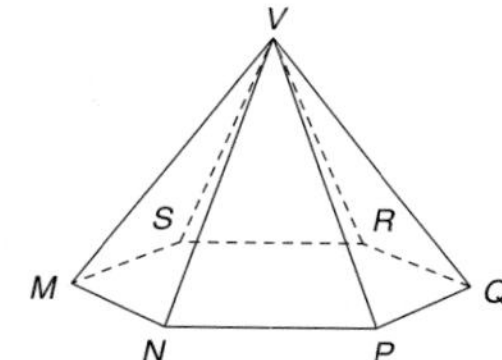


*Exercises 6, 8, 10, 12*

6. In the solid shown, the base is a regular hexagon.

   a) Name the vertex of the pyramid.
   b) Name the base edges of the pyramid.
   c) Assuming that lateral edges are congruent, are the lateral faces also congruent?
   d) Assuming that lateral edges are congruent, is the solid a regular hexagonal pyramid?

7. Consider the square pyramid in Exercise 5.

   a) How many vertices does it have?
   b) How many edges (lateral edges plus base edges) does it have?
   c) How many faces (lateral faces plus bases) does it have?

8. Consider the hexagonal pyramid in Exercise 6.

   a) How many vertices does it have?
   b) How many edges (lateral edges plus base edges) does it have?
   c) How many faces (lateral faces plus bases) does it have?

9. Suppose that the lateral faces of the pyramid in Exercise 5 have areas $A_{ABE} = 12 \text{ in}^2$, $A_{BCE} = 16 \text{ in}^2$, $A_{CED} = 12 \text{ in}^2$, and $A_{ADE} = 10 \text{ in}^2$. If each side of the square base measures 4 in., find the total surface area of the pyramid.

10. Suppose that the base of the hexagonal pyramid in Exercise 6 has an area of 41.6 $\text{cm}^2$ and that each lateral face has an area of 20 $\text{cm}^2$. Find the total (surface) area of the pyramid.

11. Suppose that the base of the square pyramid in Exercise 5 has an area of 16 $\text{cm}^2$ and that the altitude of the pyramid measures 6 cm. Find the volume of the square pyramid.

12. Suppose that the base of the hexagonal pyramid in Exercise 6 has an area of 41.6 $\text{cm}^2$ and that the altitude of the pyramid measures 3.7 cm. Find the volume of the hexagonal pyramid.

13. Assume that the number of sides in the base of a pyramid is $n$. Generalize the results found in earlier exercises by answering each of the following questions.

    a) What is the number of vertices?
    b) What is the number of lateral edges?
    c) What is the number of base edges?
    d) What is the total number of edges?
    e) What is the number of lateral faces?
    f) What is the total number of faces?
    (NOTE: Lateral faces and base = faces)

14. Refer to the prisms of Exercises 1 and 2. Which of these have symmetry with respect to one (or more) plane(s)?

15. Refer to the pyramids of Exercises 3 and 4. Which of these have symmetry with respect to one (or more) plane(s)?

16. Consider any regular pyramid. Indicate which line segment has the greater length:

    a) Slant height or altitude?
    b) Lateral edge or radius of the base?

17. Consider any regular pyramid. Indicate which line segment has the greater length:

    a) Slant height or apothem of base
    b) Lateral edge or slant height

*In Exercises 18 and 19, use Theorem 8.2.1; that is, the apothem a, the altitude h, and the slant height ℓ of a regular pyramid are related by the equation* $\ell^2 = a^2 + h^2$.

18. In a regular square pyramid whose base edges measure 8 in., the apothem of the base measures 4 in. If the altitude of the pyramid is 8 in., find the length of its slant height.

19. In a regular hexagonal pyramid whose base edges measure $2\sqrt{3}$ in., the apothem of the base measures 3 in. If the slant height of the pyramid is 5 in., find the length of its altitude.

20. In the regular pentagonal pyramid, each lateral edge measures 8 in., and each base edge measures 6 in. The apothem of the base measures 4.1 in.

 a) Find the lateral area of the pyramid.
 b) Find the total area of the pyramid.

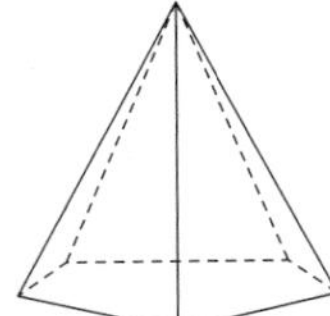

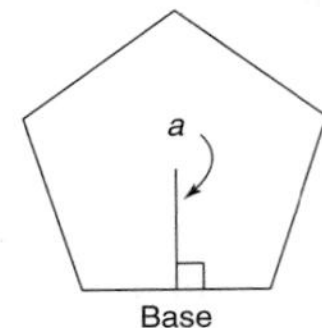


*Exercises 20, 21*

21. In the pentagonal pyramid, suppose that each base edge measures 9.2 cm and that the apothem of the base measures 6.3 cm. The altitude of the pyramid measures 14.6 cm.

 a) Find the base area of the pyramid.
 b) Find the volume of the pyramid.

22. For the regular square pyramid shown, suppose that the sides of the square base measure 10 m each and that the lateral edges measure 13 m each.

 a) Find the lateral area of the pyramid.
 b) Find the total area of the pyramid.
 c) Find the volume of the pyramid.

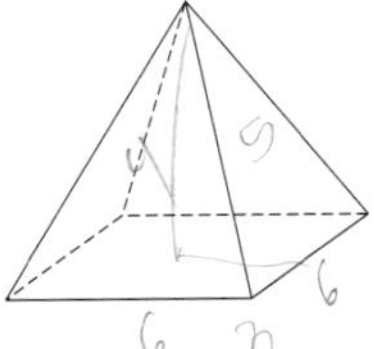

*Exercises 22, 23*

23. For the regular square pyramid shown, suppose that the sides of the square base measure 6 ft each and that the altitude is 4 ft.

 a) Find the lateral area of the pyramid.
 b) Find the total area of the pyramid.
 c) Find the volume of the pyramid.

24. a) Find the lateral area $L$ of the regular hexagonal pyramid.
 b) Find the total area $T$ of the pyramid.
 c) Find the volume $V$ of the pyramid.

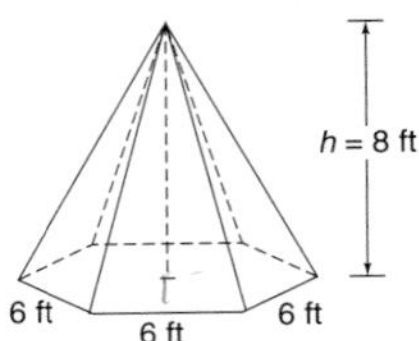


25. For a regular square pyramid, suppose that the altitude has a measure equal to that of the edges of the base. If the volume of the pyramid is 72 in³, find the total area of the pyramid.

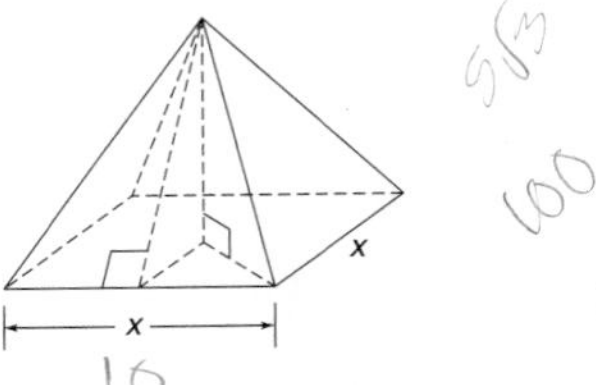


*Exercises 25, 26*

26. For a regular square pyramid, the slant height of each lateral face has a measure equal to that of each edge of the base. If the lateral area is 200 in², find the volume of the pyramid.

27. A church steeple in the shape of a regular square pyramid needs to be reshingled. The part to be covered corresponds to the lateral area of the square pyramid. If each lateral edge measures 17 ft and each base edge measures 16 ft, how many square feet of shingles need to be replaced?

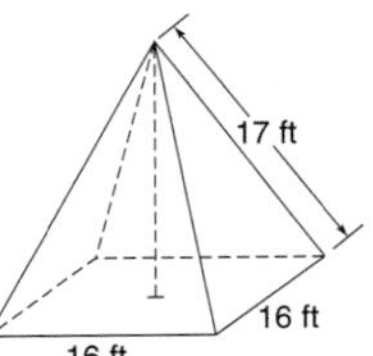


*Exercises 27, 28*

28. Before the shingles of the steeple (see Exercise 27) are replaced, an exhaust fan is to be installed in the steeple. To determine what size exhaust fan should be installed, it is necessary to know the volume of air in the attic (steeple). Find the volume of the regular square pyramid described in Exercise 27.

29. A teepee is constructed by using 12 poles. The construction leads to a regular pyramid with a dodecagon (12 sides) for the base. With the base as shown, and knowing that the altitude of the teepee is 15 ft, find its volume.

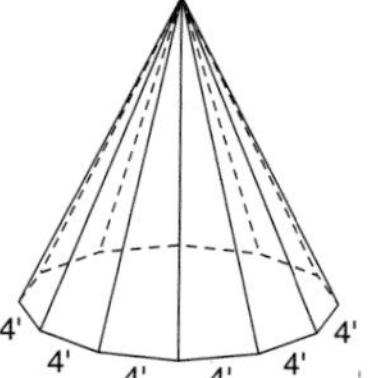

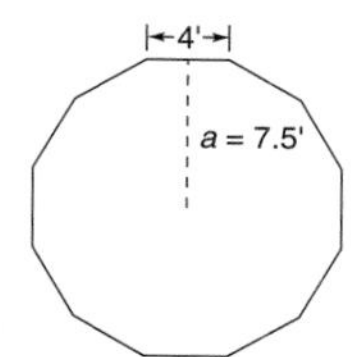


*Exercises 29, 30*

30. For its occupants to be protected from the elements, it was necessary that the teepee in Exercise 29 be enclosed. Find the amount of area to be covered; that is, determine the lateral area of the regular dodecagonal pyramid. Recall that its altitude measures 15 ft.

31. The street department's storage building, which is used to store the rock, gravel, and salt used on the city's roadways, is built in the shape of a regular hexagonal pyramid. The altitude of the pyramid has the same length as any side of the base. If the volume of the interior is 11,972 ft$^3$, find, to the nearest foot, the length of the altitude and of each side of the base.

32. The foyer planned as an addition to an existing church is designed as a regular octagonal pyramid. Each side of the octagonal floor has a length of 10 ft, and its apothem measures 12 ft. If 800 ft$^2$ of plywood is needed to cover the exterior of the foyer (that is, the lateral area of the pyramid is 800 ft$^2$), what is the altitude of the foyer?

33. The exhaust chute on a wood chipper has a shape like the part of a pyramid known as the *frustrum of the pyramid.* With dimensions as indicated, find the volume (capacity) of the chipper's exhaust chute.

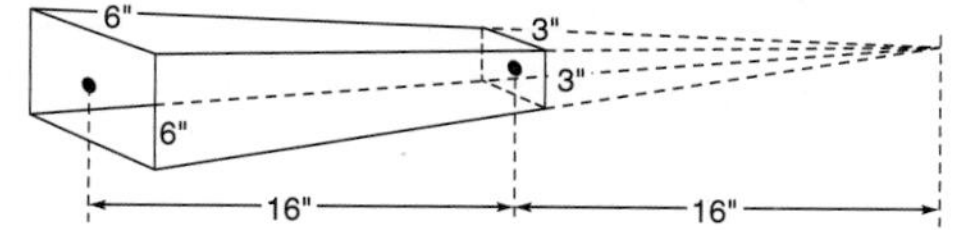


34. A popcorn container at a movie theater has the shape of a frustrum of a pyramid (see Exercise 33). With dimensions as indicated, find the volume (capacity) of the container.

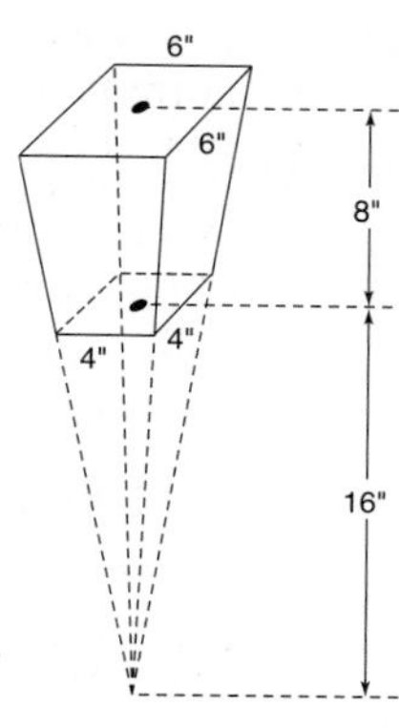


35. A regular tetrahedron is a regular triangular pyramid in which all faces (lateral faces and base) are congruent. If each edge has length $e$,

a) show that the area of each face is $A = \frac{e^2\sqrt{3}}{4}$.
b) show that the total area of the tetrahedron is $T = e^2\sqrt{3}$.
c) find the total area if each side measures $e = 4$ in.

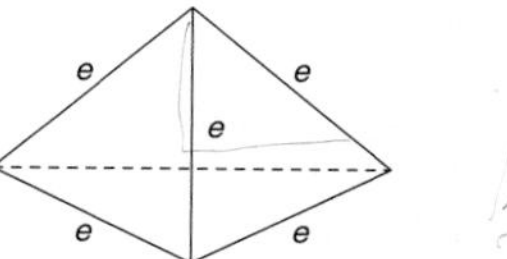

*Exercises 35, 36*

$\ast$ 36. Each edge of a regular tetrahedron (see Exercise 35) has length $e$.

a) Show that the altitude of the tetrahedron measures $h = \frac{\sqrt{2}}{\sqrt{3}}e$.
b) Show that the volume of the tetrahedron is $V = \frac{\sqrt{2}}{12}e^3$.
c) Find the volume of the tetrahedron if each side measures $e = 4$ in.

37. Consider the accompanying figure. When the four congruent isosceles triangles are folded upward, a regular square pyramid is formed. What is the surface area (total area) of the pyramid?

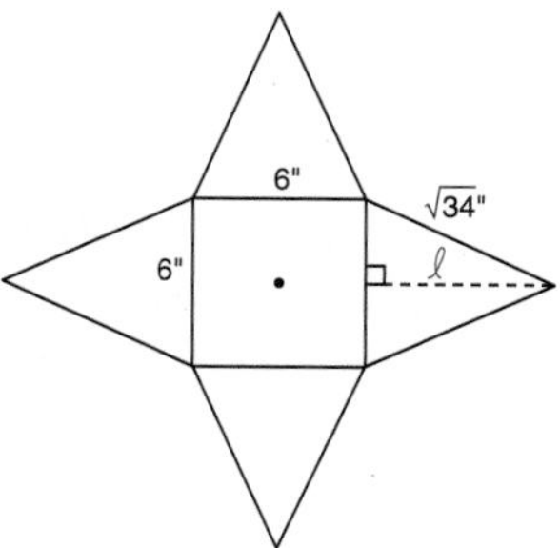


*Exercises 37, 38*

38. Find the volume of the regular square pyramid that was formed in Exercise 37.

39. Where $e_1$ and $e_2$ are the lengths of two corresponding edges (or altitudes) of similar prisms or pyramids, the ratio of their volumes is $\frac{V_1}{V_2} = \left(\frac{e_1}{e_2}\right)^3$. Write a ratio to compare volumes for two similar regular square pyramids in which $e_1 = 4$ in. and $e_2 = 2$ in.

40. Use the information from Exercise 39 to find the ratio of volumes $\frac{V_1}{V_2}$ for two cubes in which $e_1 = 2$ cm and $e_2 = 6$ cm.
(NOTE: $\frac{V_1}{V_2}$ can be found by determining the actual volumes of the cubes.)

41. A hexagonal pyramid (not regular) with base *ABCDEF* has plane symmetry with respect to a plane determined by vertex *G* and vertices *A* and *D* of its base. If the volume of the pyramid with vertex *G* and base *ABCD* is 19.7 in$^3$, find the volume of the given hexagonal pyramid.

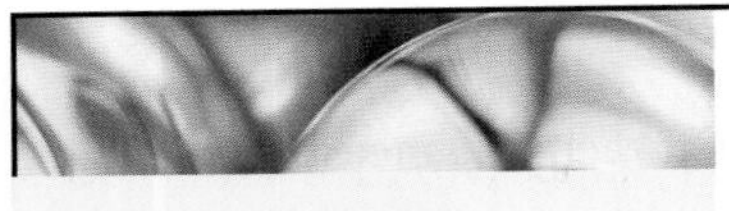

## 8.3 Cylinders and Cones

**KEY CONCEPTS**

Cylinders (Right and Oblique) • Bases and Altitude of a Cylinder • Axis of a Cylinder • Cones (Right and Oblique) • Base and Altitude of a Cone • Vertex and Slant Height of a Cone • Axis of a Cone • Lateral Area • Total Area • Volume • Solid of Revolution • Axis of a Solid of Revolution

### CYLINDERS

Consider the solids in Figure 8.25, in which congruent circles lie in parallel planes. For the circles on the left, suppose that centers $O$ and $O'$ are joined by $\overline{OO'}$; similarly, suppose that $\overline{QQ'}$ joins the centers of the circles on the right. Let segments such as $\overline{XX'}$ join two points of the circles on the left, so that $\overline{XX'} \parallel \overline{OO'}$. If all such segments (like $\overline{XX'}$, $\overline{YY'}$, and $\overline{ZZ'}$) are parallel to each other, then a **cylinder** is generated. Because $\overline{OO'}$ is not perpendicular to planes $P$ and $P'$, the solid on the left is an **oblique circular cylinder.** With $\overline{QQ'}$ perpendicular to planes $P$ and $P'$, the solid on the right is a **right circular cylinder.** For both cylinders, the distance $h$ between the planes $P$ and $P'$ is the length of the **altitude** of the cylinder. The congruent circles are known as the **bases** of each cylinder.

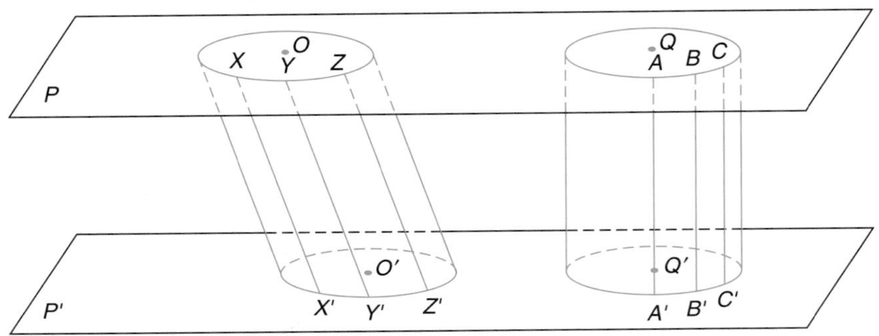


**FIGURE 8.25**

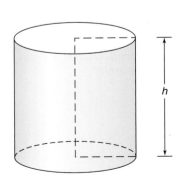

**FIGURE 8.26**

A right circular cylinder is shown in Figure 8.26; however, the parallel planes (like $P$ and $P'$ in Figure 8.25) are generally not pictured. The segment joining the centers of the two circular bases is known as the **axis** of the cylinder. For a right circular cylinder, it is necessary that the axis be perpendicular to the planes of the circular bases; in such a case, the length of the altitude $h$ is the length of the axis.

## SURFACE AREA OF A CYLINDER

### Discover!

Think of the aluminum can pictured in Figure (a) as a right circular cylinder. The cylinder's circular bases are the lid and bottom of the can, and the lateral surface is the "label" of the can. If the label were sliced downward by a perpendicular line between the planes, removed, and rolled out flat, it would be rectangular in shape. As shown in Figure (b), that rectangle would have a length equal to the circumference of the circular base and a width equal to the height of the cylinder. Thus the lateral area is given by $A = bh$, which becomes $L = Ch$, or $L = 2\pi rh$.

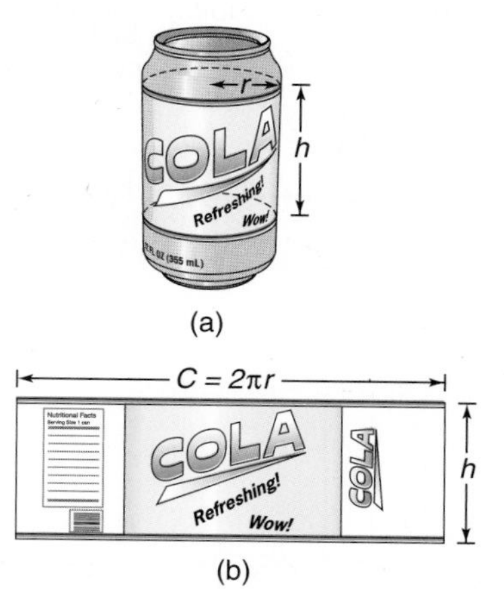

*Exs. 1, 2*

The formula for the lateral area of a right circular cylinder (found in the following theorem) should be compared to the formula $L = hP$, the lateral area of a right prism whose base has perimeter $P$.

THEOREM 8.3.1: The lateral area $L$ of a right circular cylinder with altitude of length $h$ and circumference $C$ of the base is given by $L = hC$.

*Alternative Form:* The lateral area of a right circular cylinder can be expressed in the form $L = 2\pi rh$, where $r$ is the length of the radius of the circular base.

Rather than construct a formal proof of Theorem 8.3.1, consider the activity in the Discover! box above.

THEOREM 8.3.2: The total area $T$ of a right circular cylinder with base area $B$ and lateral area $L$ is given by $T = L + 2B$.

*Alternative Form:* Where $r$ is the length of the radius of the base and $h$ is the length of the altitude of the cylinder, the total area can be expressed in the form $T = 2\pi rh + 2\pi r^2$.

### EXAMPLE 1

For the right circular cylinder shown in Figure 8.27, find the

a) exact lateral area $L$.
b) exact surface area $T$.

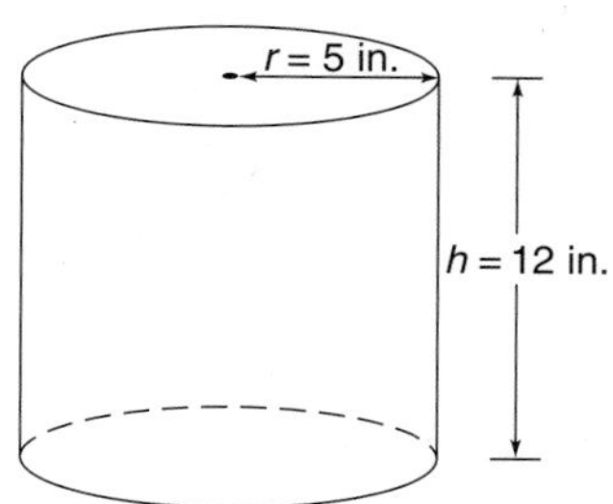


**FIGURE 8.27**

***Solution***

a) $L = 2\pi rh$
$= 2 \cdot \pi \cdot 5 \cdot 12$
$= 120\pi \text{ in}^2$

b) $T = L + 2B$
$= 2\pi rh + 2\pi r^2$
$= 2 \cdot \pi \cdot 5 \cdot 12 + 2 \cdot \pi \cdot 5^2$
$= 120\pi + 50\pi$
$= 170\pi \text{ in}^2$

SSG *Exs. 3–5*

## VOLUME OF A CYLINDER

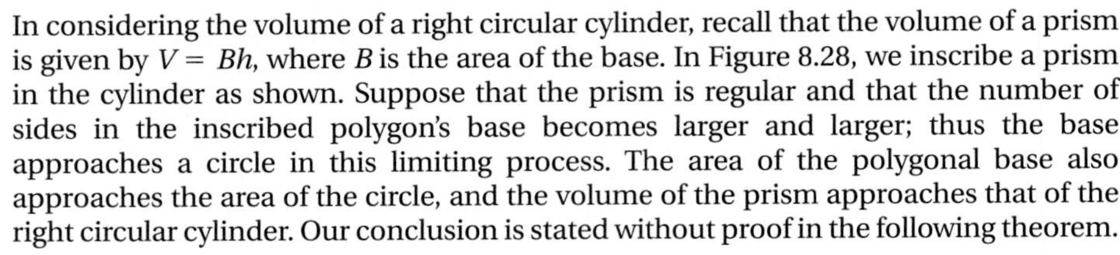

In considering the volume of a right circular cylinder, recall that the volume of a prism is given by $V = Bh$, where $B$ is the area of the base. In Figure 8.28, we inscribe a prism in the cylinder as shown. Suppose that the prism is regular and that the number of sides in the inscribed polygon's base becomes larger and larger; thus the base approaches a circle in this limiting process. The area of the polygonal base also approaches the area of the circle, and the volume of the prism approaches that of the right circular cylinder. Our conclusion is stated without proof in the following theorem.

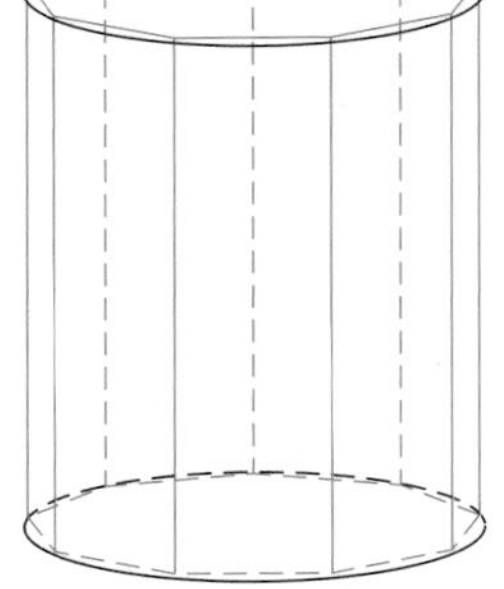

**FIGURE 8.28**

> THEOREM 8.3.3: The volume $V$ of a right circular cylinder with base area $B$ and altitude of length $h$ is given by $V = Bh$.
>
> *Alternative Form:* Where $r$ is the length of the radius of the base, the volume for the cylinder can be written $V = \pi r^2 h$.

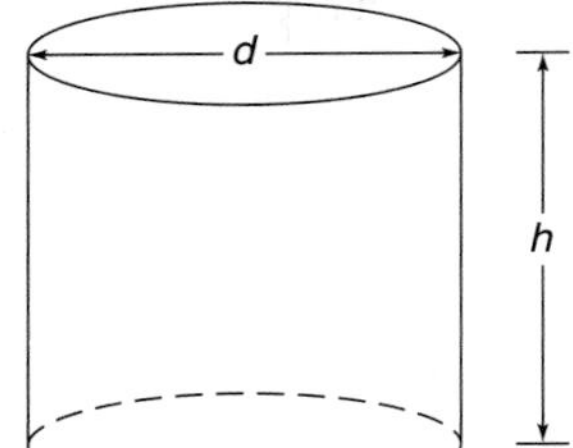


**FIGURE 8.29**

EXAMPLE 2

If $d = 4$ cm and $h = 3.5$ cm, use a calculator to find the approximate volume of the right circular cylinder shown in Figure 8.29. Give the answer correct to two decimal places.

***Solution***

$d = 4$, so $r = 2$. Thus $V = Bh$ or $V = \pi r^2 h$ becomes

$$V = \pi \cdot 2^2(3.5)$$
$$= \pi \cdot 4(3.5) = 14\pi \approx 43.98 \text{ cm}^3$$

EXAMPLE 3

In the right circular cylinder shown in Figure 8.29, suppose that the height equals the diameter of the circular base. If the exact volume is $128\pi \text{ in}^3$, find the exact lateral area $L$ of the cylinder.

***Solution***

$$h = 2r$$

so $V = \pi r^2 h$

becomes $V = \pi r^2(2r)$
$= 2\pi r^3$

Then $2\pi r^3 = 128\pi$, and dividing by $2\pi$,

$$r^3 = 64$$
$$r = 4$$
$$h = 8 \quad \text{(from } h = 2r\text{)}$$

Now $L = 2\pi rh$
$= 2 \cdot \pi \cdot 4 \cdot 8$
$= 64\pi \text{ in}^2$

Table 8.2 should help us to recall and compare the area and volume formulas found in Sections 8.1 and 8.3.

**TABLE 8.2**

| | *Lateral Area* | *Total Area* | *Volume* |
|---|---|---|---|
| *Prism* | $L = hP$ | $T = L + 2B$ | $V = Bh$ |
| *Cylinder* | $L = hC$ | $T = L + 2B$ | $V = Bh$ |

*Exs. 6, 7*

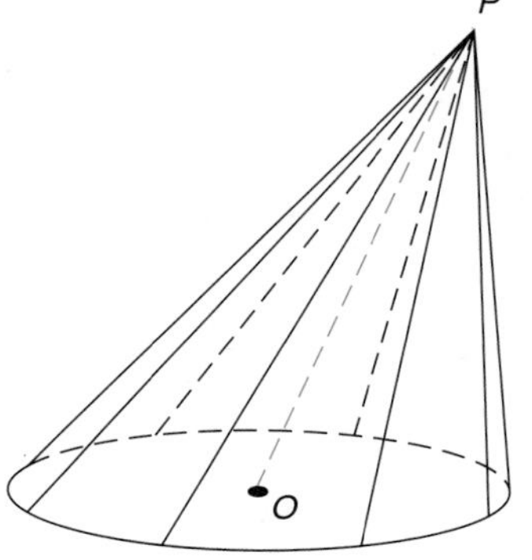


**FIGURE 8.30**

## CONES

In Figure 8.30, consider point $P$, which lies outside the plane containing circle $O$. A surface known as a **cone** results when line segments are drawn from $P$ to points on the circle. However, if $P$ is joined to all possible points on the circle as well as to points in the interior of the circle, a solid is formed. If $\overline{PO}$ is not perpendicular to the plane of circle $O$ in Figure 8.30, the cone is an **oblique circular cone.**

In Figures 8.30 and 8.31, point $P$ is the **vertex** of the cone, and circle $O$ is the **base.** The segment $\overline{PO}$, which joins the vertex to the center of the circular base, is the **axis** of the cone. If the axis is perpendicular to the plane containing the base, as in Figure 8.31, the cone is a **right circular cone.** In any cone, the perpendicular segment from the vertex to the plane of the base is the **altitude** of the cone. In a right circular cone, the length $h$ of the altitude equals the length of the axis. For a right circular cone, and only for this type of cone, any line segment that joins the vertex to a point on the circle is a **slant height** of the cone; we will denote the length of the slant height by $\ell$ as shown in Figure 8.31.

*Exs. 8, 9*

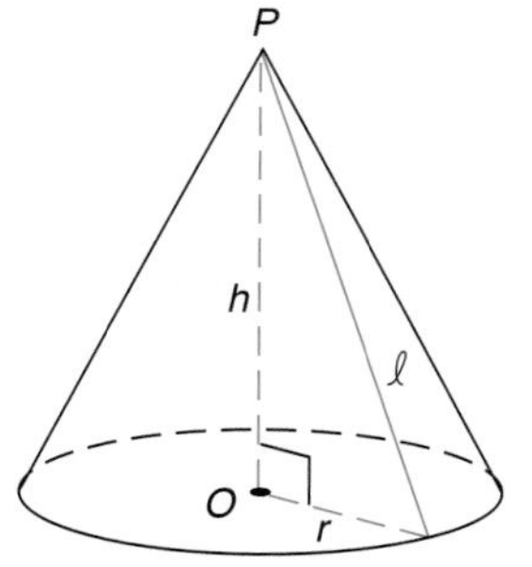


**FIGURE 8.31**

## SURFACE AREA OF A CONE

Recall now that the lateral area for a regular pyramid is given by $L = \frac{1}{2}\ell P$. For a right circular cone, consider an inscribed regular pyramid as in Figure 8.32. As the number of sides of the inscribed polygon's base grows larger, the perimeter of the inscribed polygon approaches the circumference of the circle as a limit. In addition, the slant height of the congruent triangular faces approaches that of the slant height of the cone. Thus the lateral area of the right circular cone can be compared to $L = \frac{1}{2}\ell P$; for the cone we have

$$L = \frac{1}{2}\ell C$$

in which $C$ is the circumference of the base. The fact that $C = 2\pi r$ leads to

$$L = \frac{1}{2}\ell(2\pi r)$$

so

$$L = \pi r\ell$$

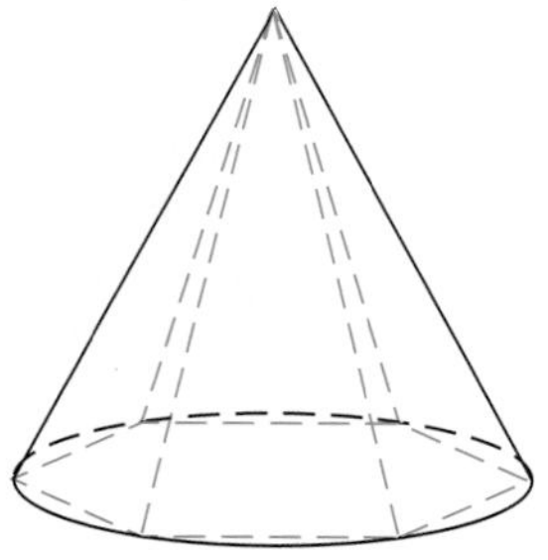

**FIGURE 8.32**

> THEOREM 8.3.4: The lateral area $L$ of a right circular cone with slant height of length $\ell$ and circumference $C$ of the base is given by $L = \frac{1}{2}\ell C$.
>
> *Alternative Form:* Where $r$ is the length of the radius of the base, $L = \pi r\ell$.

The following theorem follows easily and is given without proof.

THEOREM 8.3.5: The total area $T$ of a right circular cone with base area $B$ and lateral area $L$ is given by $T = B + L$.

*Alternative Form:* Where $r$ is the length of the radius of the base and $\ell$ is the length of the slant height, $T = \pi r^2 + \pi r\ell$

EXAMPLE 4

For the right circular cone in which $r = 3$ cm and $h = 6$ cm (see Figure 8.33), find the

a) exact and approximate lateral area $L$.
b) exact and approximate total area $T$.

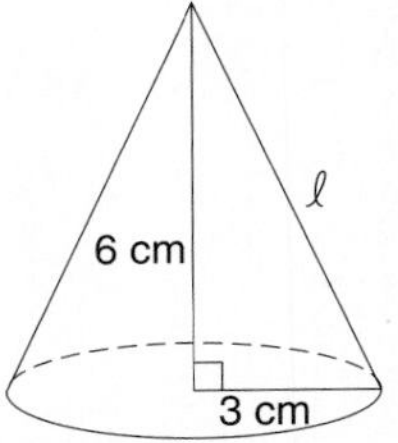

**FIGURE 8.33**

***Solution***

a) We need the length of the slant height $\ell$ for each problem part, so we apply the Pythagorean Theorem:

$$\begin{aligned} \ell^2 &= r^2 + h^2 \\ &= 3^2 + 6^2 \\ &= 9 + 36 = 45 \\ \ell &= \sqrt{45} = \sqrt{9 \cdot 5} \\ &= \sqrt{9} \cdot \sqrt{5} = 3\sqrt{5} \end{aligned}$$

Using the alternative form of $L = \frac{1}{2}\ell C$, namely $L = \pi r\ell$, we have

$$\begin{aligned} L &= \pi \cdot 3 \cdot 3\sqrt{5} \\ &= 9\pi\sqrt{5}\text{ cm}^2 \approx 63.22\text{ cm}^2 \end{aligned}$$

b) We also have

$$\begin{aligned} T &= B + L \\ &= \pi r^2 + \pi r\ell \\ &= \pi \cdot 3^2 + \pi \cdot 3 \cdot 3\sqrt{5} \\ &= (9\pi + 9\pi\sqrt{5})\text{ cm}^2 \approx 91.50\text{ cm}^2 \end{aligned}$$

The following theorem was demonstrated in the solution of Example 4.

THEOREM 8.3.6: In a right circular cone, the lengths of the radius $r$ (of the base), the altitude $h$, and the slant height $\ell$ satisfy the Pythagorean Theorem; that is, $\ell^2 = r^2 + h^2$ in every right circular cone.

*Exs. 10, 11*

## VOLUME OF A CONE

Recall that the volume of a pyramid is given by the formula $V = \frac{1}{3}Bh$. Consider a regular pyramid inscribed in a right circular cone. If its number of sides increases indefinitely, the volume of the pyramid approaches that of the right circular cone (see Figure 8.34). Then the volume of the right circular cone is $V = \frac{1}{3}Bh$. Because the area of the base of the cone is $B = \pi r^2$, an alternative formula for the volume of the cone is

$$V = \frac{1}{3}\pi r^2 h$$

**FIGURE 8.34**

We state this result as a theorem.

**Discover!**

Complete this analogy:
Prism is to Cylinder as
Pyramid is to ___.

ANSWER

Cone

THEOREM 8.3.7: The volume $V$ of a right circular cone with base area $B$ and altitude of length $h$ is given by $V = \frac{1}{3}Bh$.

*Alternative Form:* Where $r$ is the length of the radius of the base, the formula for the volume of the cone is usually written $V = \frac{1}{3}\pi r^2 h$.

Table 8.3 should help us to recall and compare the area and volume formulas found in Sections 8.2 and 8.3.

**TABLE 8.3**

| | *Lateral Area* | *Total Area* | *Volume* | *Slant Height* |
|---|---|---|---|---|
| *Pyramid* | $L = \frac{1}{2}\ell P$ | $T = B + L$ | $V = \frac{1}{3}Bh$ | $\ell^2 = a^2 + h^2$ |
| *Cone* | $L = \frac{1}{2}\ell C$ | $T = B + L$ | $V = \frac{1}{3}Bh$ | $\ell^2 = r^2 + h^2$ |

NOTE: The formulas that contain the slant height are used only with the regular pyramid and the right circular cone.

*Exs. 12, 13*

## SOLIDS OF REVOLUTION

Suppose that part of the boundary for a plane region is a line segment. When the plane region is revolved about this line segment, the locus of points generated in space is called a **solid of revolution.** The complete 360° rotation moves the region about the edge until the region returns to its original position. The side (edge) used is called the **axis** of the resulting solid of revolution. Consider Example 5.

EXAMPLE 5

Describe the solid of revolution that results when

a) a rectangular region of dimensions 2 ft by 5 ft is revolved about the 5-ft side [see Figure 8.35(a)].
b) a semicircular region of radius 3 cm is revolved about the diameter shown in Figure 8.35(b).

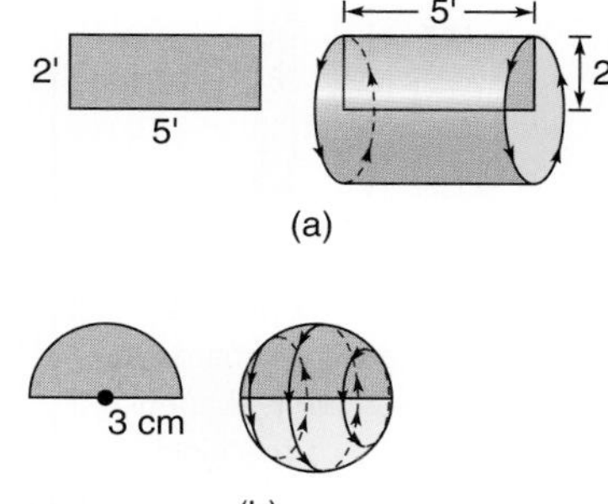


**FIGURE 8.35**

***Solution***

a) In Figure 8.35(a), the rectangle on the left is revolved about the 5-ft side to form the solid on the right. The solid of revolution generated is a right circular cylinder that has a base radius of 2 ft and an altitude of 5 ft.
b) In Figure 8.35(b), the semicircle on the left is revolved about its diameter to form the solid on the right. The solid of revolution generated is a *sphere* with a radius of length 3 cm.

NOTE: We will study the sphere in greater detail in Section 8.4.

EXAMPLE 6

Determine the exact volume of the solid of revolution formed when the region bounded by a right triangle with legs of lengths 4 in. and 6 in. is revolved about the 6-in. side. The triangular region is shown in Figure 8.36(a).

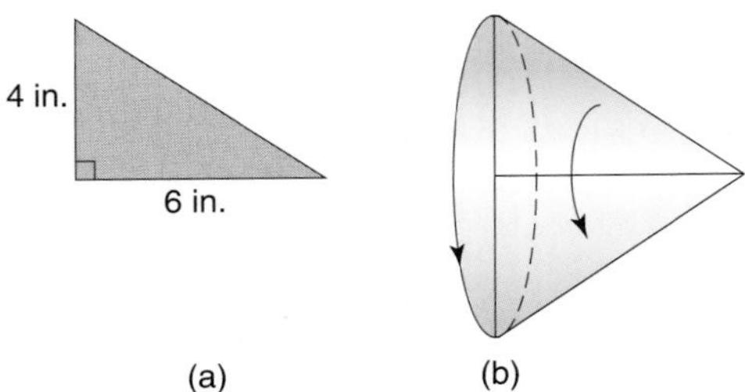


**FIGURE 8.36**

***Solution***

As shown in Figure 8.36(b), the resulting solid is a cone whose altitude measures 6 in. and whose radius of the base measures 4 in.

Using $V = \frac{1}{3}Bh$, we have

$$V = \frac{1}{3}\pi r^2 h$$
$$= \frac{1}{3} \cdot \pi \cdot 4^2 \cdot 6 = 32\pi \text{ in}^3$$

*Geometry in the Real World*

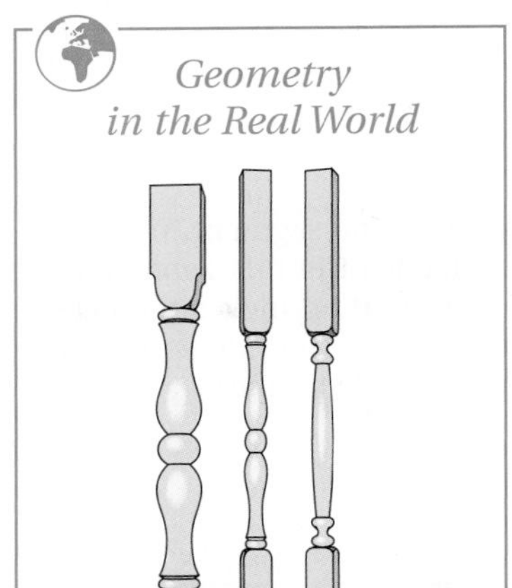

Spindles are examples of solids of revolution. As the piece of wood is rotated, the ornamental part of each spindle is shaped and smoothed by a machine (wood lathe).

It may come as a surprise that the formulas used to calculate the volumes of an oblique circular cylinder and a right circular cylinder are identical. To see why the formula $V = Bh$ or $V = \pi r^2 h$ can be used to calculate the volume of an oblique circular cylinder, consider the stacks of pancakes shown in Figures 8.37(a) and 8.37(b). With each stack $h$ units high, the volume is the same regardless of whether the stack is vertical or oblique.

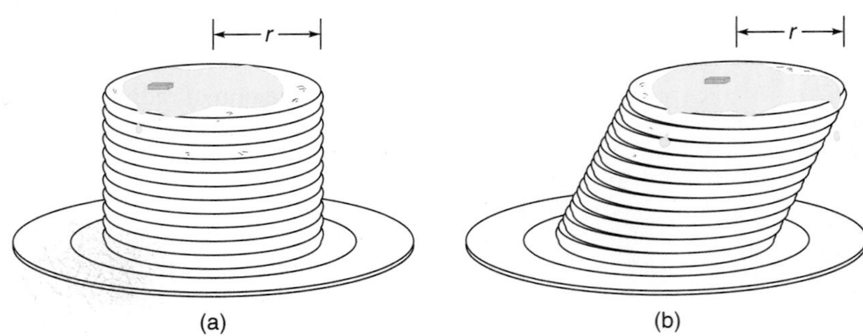


**FIGURE 8.37**

*Exs. 14, 15*

It is also true that the formula for the volume of an oblique circular cone is $V = \frac{1}{3}Bh$ or $V = \frac{1}{3}\pi r^2 h$. In fact, the motivating argument preceding Theorem 8.3.7 would be repeated, with the exception that the inscribed pyramid is oblique.

## 8.3 Exercises

1. Does a right circular cylinder such as an aluminum can have

   a) symmetry with respect to at least one plane?
   b) symmetry with respect to at least one line?
   c) symmetry with respect to a point?

2. Does a right circular cone such as a wizard's cap have

   a) symmetry with respect to at least one plane?
   b) symmetry with respect to at least one line?
   c) symmetry with respect to a point?

3. For the right circular cylinder, suppose that $r = 5$ in. and $h = 6$ in. Find the exact and approximate:

   a) Lateral area
   b) Total area
   c) Volume

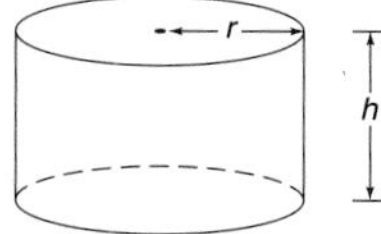


*Exercises 3, 4*

4. Suppose that $r = 12$ cm and $h = 15$ cm in the right circular cylinder. Find the exact and approximate:

   a) Lateral area
   b) Total area
   c) Volume

5. The tin can shown at the right has the indicated dimensions. Estimate the number of square inches of tin required for its construction. (HINT: Include the lid and the base in the result.)

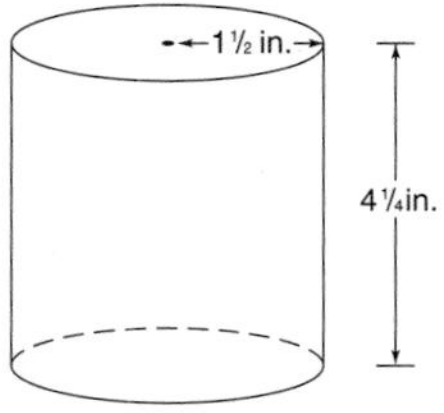


*Exercises 5, 6*

6. What is the volume of the tin can? If it contains 16 oz of green beans, what is the volume of the can used for 20 oz of green beans? Assume a proportionality between weight and volume.

7. If the exact volume of a right circular cylinder is $200\pi$ cm$^3$ and its altitude measures 8 cm, what is the measure of the radius of the circular base?

8. Suppose that the volume of an aluminum can is to be $9\pi$ in$^3$. Find the dimensions of the can if the diameter of the base is three-fourths the length of the altitude.

9. For an aluminum can, the lateral surface area is $12\pi$ in$^2$. If the length of the altitude is 1 in. greater than the length of the radius of the circular base, find the dimensions of the can.

10. Find the altitude of a storage tank in the shape of a right circular cylinder that has a circumference measuring $6\pi$ m and a volume measuring $81\pi$ m$^3$.

11. Find the volume of the oblique circular cylinder. The axis meets the plane of the base to form a 45° angle.

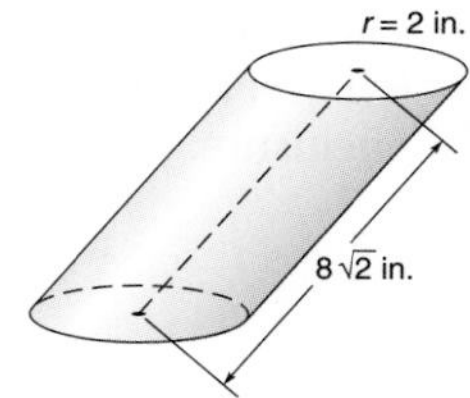


12. A cylindrical orange juice container has metal bases of radius 1 in. and a cardboard lateral surface 3 in. high. If the cost of the metal used is 0.5 cent per square inch and the cost of the cardboard is 0.2 cent per square inch, what is the approximate cost of constructing one container? Let $\pi \approx 3.14$.

*In Exercises 13 to 18, use the fact that $r^2 + h^2 = \ell^2$ in a right circular cone (Theorem 8.3.6).*

13. Find the slant height $\ell$ of a right circular cone with $r = 4$ cm and $h = 6$ cm.

14. Find the slant height $\ell$ of a right circular cone with $r = 5.2$ ft and $h = 3.9$ ft.

15. Find the altitude $h$ of a right circular cone in which the diameter of the base measures $d = 9.6$ m and $\ell = 5.2$ m.

16. Find the radius of base $r$ of a right circular cone in which $h = 6$ yd and $\ell = 8$ yd.

17. Find the slant height $\ell$ of a right circular cone with $r = 6$ in., length of altitude $h$, and $\ell = 2h$ in.

18. Find the radius $r$ of a right circular cone with $\ell = 12$ in. and $h = 3r$ in.

19. The oblique circular cone has altitude and a diameter of base that are each of length 6 cm. The line segment joining the vertex to the center of the base is the axis of the cone. What is the length of the axis?

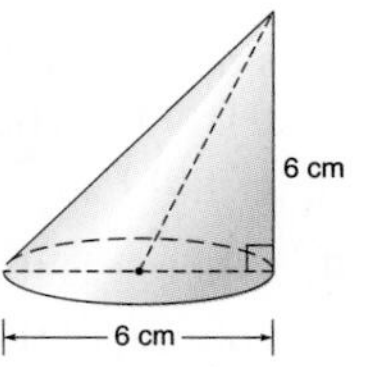


20. For the accompanying right circular cone, $h = 6$ m and $r = 4$ m. Find the exact and approximate:

    a) Lateral area
    b) Total area
    c) Volume

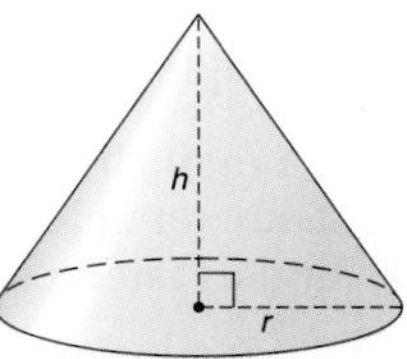


*Exercises 20, 21*

21. For the right circular cone on the previous page, suppose that $h = 7$ in. and $r = 6$ in. Find the exact and approximate:

 a) Lateral area
 b) Total area
 c) Volume

22. The teepee has a circular floor with a radius equal to 6 ft and a height of 15 ft. Find the volume of the enclosure.

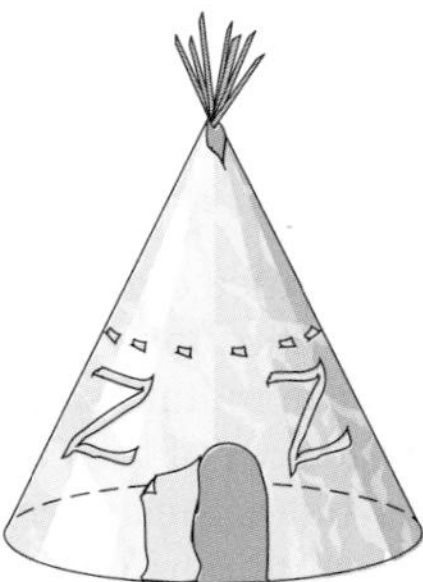

23. A rectangle has dimensions of 6 in. by 3 in. Find the exact volume of the solid of revolution formed when the rectangle is rotated about its 6-in. side.

24. A rectangle has dimensions of 6 in. by 3 in. Find the exact volume of the solid of revolution formed when the rectangle is rotated about its 3-in. side.

25. A triangle has sides that measure 15 cm, 20 cm, and 25 cm. Find the exact volume of the solid of revolution formed when the triangle is revolved about the side of length 15 cm.

26. A triangle has sides that measure 15 cm, 20 cm, and 25 cm. Find the exact volume of the solid of revolution formed when the triangle is revolved about the side of length 20 cm.

27. A triangle has sides that measure 15 cm, 20 cm, and 25 cm. Find the exact volume of the solid of revolution formed when the triangle is revolved about the side of length 25 cm.
 (HINT: The altitude to the 25-cm side has length 12 cm.)

28. Where $r$ is the length of the radius of a sphere, the volume of the sphere is given by $V = \frac{4}{3}\pi r^3$. Find the exact volume of the sphere that was formed in Example 5(b).

29. If a right circular cone has a circular base with a diameter of length 10 cm and a volume of $100\pi$ cm$^3$, find its lateral area.

30. A right circular cone has a slant height of 12 ft and a lateral area of $96\pi$ ft$^2$. Find its volume.

31. A solid is formed by cutting a conical section away from a right circular cylinder. If the radius measures 6 in. and the altitude measures 8 in., what is the volume of the resulting solid?

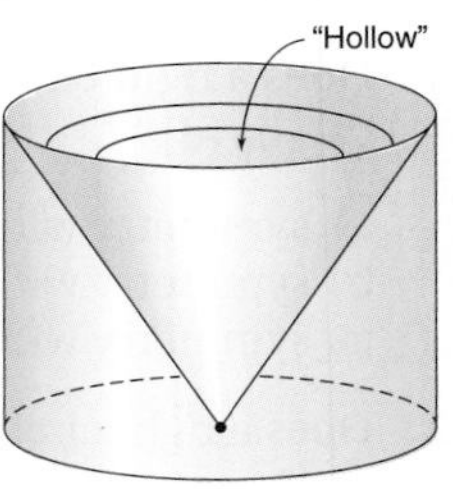


*In Exercises 32 and 33, give a paragraph proof for each claim.*

32. The total area $T$ of a right circular cylinder whose altitude is of length $h$ and whose circular base has a radius of length $r$ is given by $T = 2\pi r(r + h)$.

33. The volume $V$ of a washer that has an inside radius of length $r$, an outside radius of length $R$, and an altitude of measure $h$ is given by $V = \pi h(R + r)(R - r)$.

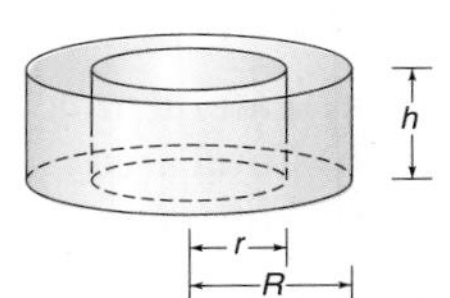


34. For a right circular cone, the slant height has a measure equal to twice that of the radius of the base. If the total area of the cone is $48\pi$ in$^2$, what are the dimensions of the cone?

35. For a right circular cone, the ratio of the slant height to the radius is 5:3. If the volume of the cone is $96\pi$ in$^3$, find the lateral area of the cone.

36. If the radius and height of a right circular cylinder are both doubled to form a larger cylinder, what is the ratio of the volume of the larger cylinder to the volume of the smaller cylinder?
 (NOTE: The two cylinders are said to be "similar.")

37. For the two similar cylinders in Exercise 36, what is the ratio of the lateral area of the larger cylinder to that of the smaller cylinder?

38. For a right circular cone, the dimensions are $r = 6$ cm and $h = 8$ cm. If the radius is doubled while the height is made half as large in forming a new cone, will the volumes of the two cones be equal?

39. A cylindrical storage tank has a depth of 5 ft and a radius measuring 2 ft. If each cubic foot can hold 7.5 gal of gasoline, what is the total storage capacity of the tank (measured in gallons)?

40. If the tank in Exercise 39 needs to be painted and 1 pt of paint covers 50 ft$^2$, how many pints are needed to paint the exterior of the storage tank?

41. A frustrum of a cone is the portion of the cone bounded between the circular base and a plane parallel to the base. With dimensions as indicated, show that the volume of the frustrum of the cone is $V = \frac{1}{3}\pi R^2 H - \frac{1}{3}\pi r^2 h$.

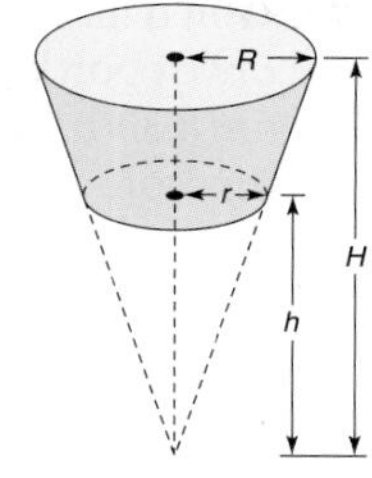


*In Exercises 42 and 43, use the formula from Exercise 41. Similar triangles were used to find h and H.*

42. A margarine tub has the shape of the frustrum of a cone. With the lower base having diameter 11 cm and the upper base having diameter 14 cm, the volume of such a container $6\frac{2}{3}$ cm tall can be determined by using $R = 7$ cm, $r = 5.5$ cm, $H = 32\frac{2}{3}$ cm, and $h = 26$ cm. Find its volume.

43. A container of yogurt has the shape of the frustrum of a cone. With the lower base having diameter 6 cm and the upper base having diameter 8 cm, the volume of such a container 7.5 cm tall can be determined by using $R = 4$ cm, $r = 3$ cm, $H = 30$ cm, and $h = 22.5$ cm. Find its volume.

44. An oil refinery has storage tanks in the shape of right circular cylinders. Each tank has a height of 16 ft and a radius of 10 ft for its circular base. If 1 $\text{ft}^3$ of volume contains 7.5 gal of oil, what is the capacity of the fuel tank in *gallons?* Round the result to the nearest hundred (of gallons).

45. A farmer has a fuel tank in the shape of a right circular cylinder. The tank has a height of 6 ft and a radius of 1.5 ft for its circular base. If 1 $\text{ft}^3$ of volume contains 7.5 gal of gasoline, what is the capacity of the fuel tank in *gallons?*

46. When radii $\overline{OA}$ and $\overline{OB}$ are placed so that they coincide, a 240° sector of a circle is sealed to form a right circular cone. If the radius of the circle is 6.4 cm, what is the approximate lateral area of the cone that is formed? Use a calculator and round the answer to the nearest *tenth* of a square inch.

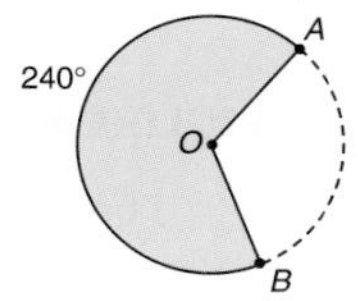


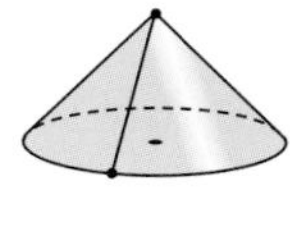

47. A lawn roller in the shape of a right circular cylinder has a radius of 18 in. and a length (height) of 4 ft. Find the area rolled during one complete revolution of the roller. Use the calculator value of $\pi$, and give the answer to the nearest square foot.

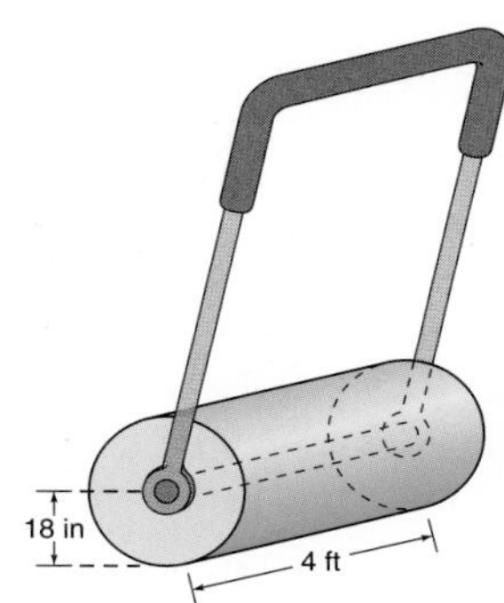


**KEY CONCEPTS**

Dihedral Angle • Polyhedron (Convex and Concave) • Vertices • Edges and Faces • Euler's Equation • Regular Polyhedrons (Tetrahedron, Hexahedron, Octahedron, Dodecahedron, Icosahedron) • Sphere (Center, Radius, Diameter, Great Circle, Hemisphere) • Surface Area and Volume of a Sphere

# 8.4 Polyhedrons and Spheres

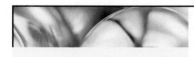

## POLYHEDRONS

When two planes intersect, the angle formed by two half-planes with a common edge (the line of intersection) is a **dihedral angle.** The angle shown in Figure 8.38 is such an angle. In Figure 8.38, the measure of the dihedral angle is the same as that of the angle determined by two rays that

1. have a vertex on the edge.
2. lie in the planes so that they are perpendicular to the edge.

A **polyhedron** (plural *polyhedrons or polyhedra*) is a solid bounded by plane regions. Polygons form the **faces** of the solid, and the segments common to these polygons are the **edges** of the polyhedron. Endpoints of the edges are the **vertices** of the polyhedron. When a polyhedron is **convex,**

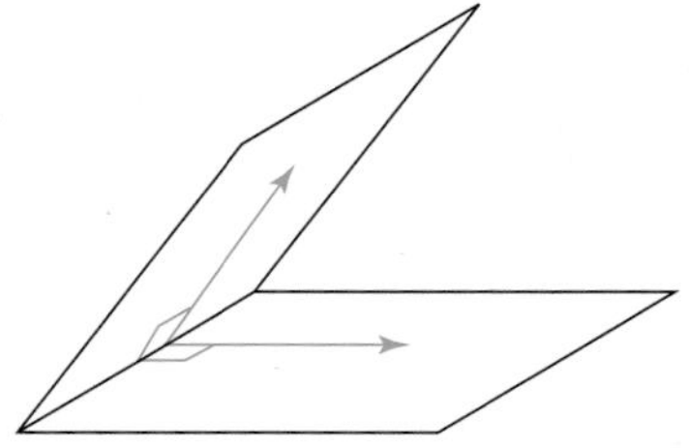

**FIGURE 8.38**

each face determines a plane for which all remaining faces lie on the same side of that plane. Figure 8.39(a) illustrates a convex polyhedron, and Figure 8.39(b) illustrates a **concave** polyhedron; as shown in Figure 8.39(b), a line segment containing two vertices lies in the exterior of the concave polyhedron.

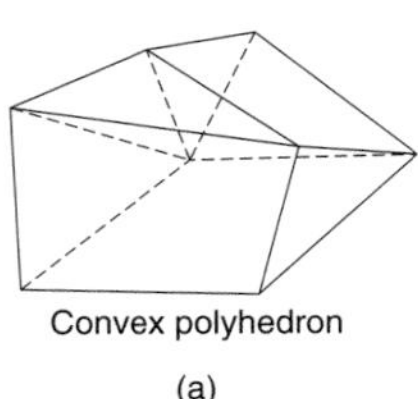
Convex polyhedron

(a)

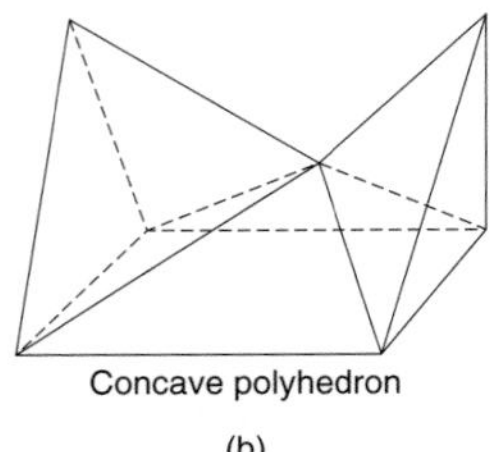
Concave polyhedron

(b)

**FIGURE 8.39**

The prisms and pyramids discussed in Sections 8.1 and 8.2 were special types of polyhedrons. For instance, a pentagonal pyramid can be described as a hexahedron because it has six faces. Because some of their surfaces do not lie in planes, the cylinders and cones of Section 8.3 are not polyhedrons.

Leonhard Euler (Swiss, 1707–1763) found that the number of vertices, edges, and faces of any polyhedron are related by **Euler's equation.** This equation is given in the following theorem, which is stated without proof.

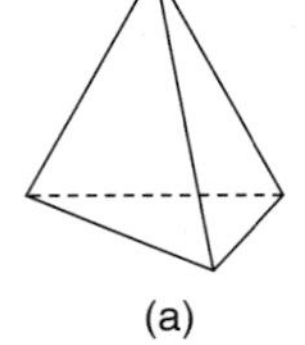
(a)

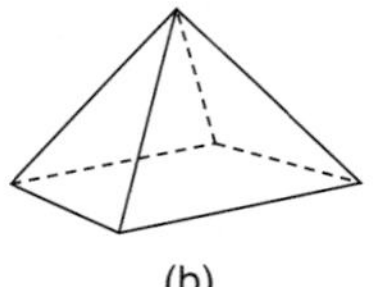
(b)

**FIGURE 8.40**

THEOREM 8.4.1: **(Euler's equation)** The number of vertices $V$, the number of edges $E$, and the number of faces $F$ of a polyhedron are related by the equation

$$V + F = E + 2$$

EXAMPLE 1

Verify Euler's equation for the (a) tetrahedron and (b) square pyramid shown in Figure 8.40.

***Solution***

a) The tetrahedron has four vertices ($V = 4$), six edges ($E = 6$), and four faces ($F = 4$), so the equation becomes $4 + 4 = 6 + 2$, which is true.
b) The pyramid has five vertices ("vertex" + 4 vertices from the base), eight edges (4 base edges + 4 lateral edges), and five faces (4 triangular faces + 1 square base). Now $V + F = E + 2$ becomes $5 + 5 = 8 + 2$, which is also true.

*Exs. 1–5*

## REGULAR POLYHEDRONS

DEFINITION: A **regular polyhedron** is a convex polyhedron whose faces are congruent regular polygons arranged in such a way that adjacent faces form congruent dihedral angles.

The five regular polyhedrons are as follows:

1. Regular **tetrahedron,** which has 4 faces (congruent equilateral triangles)
2. Regular **hexahedron** (or **cube**), which has 6 faces (congruent squares)
3. Regular **octahedron,** which has 8 faces (congruent equilateral triangles)
4. Regular **dodecahedron,** which has 12 faces (congruent regular pentagons)
5. Regular **icosahedron,** which has 20 faces (congruent equilateral triangles)

Four of the regular polyhedrons are shown in Figure 8.41.

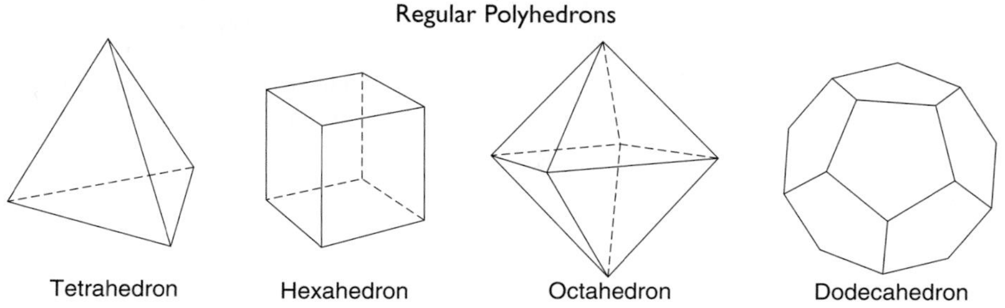

Regular Polyhedrons

Tetrahedron Hexahedron Octahedron Dodecahedron

**FIGURE 8.41**

Because each regular polyhedron has a central point, each solid is said to have a center. Except for the tetrahedron, these polyhedrons have point symmetry. Each solid also has line symmetry and plane symmetry.

*Geometry in the Real World*

Polyhedra dice are used in numerous games.

EXAMPLE 2

Consider a die that is a regular tetrahedron. Assuming that each face has an equal chance of being rolled, what is the likelihood (probability) that one roll produces (a) a "1"? (b) a result larger than "1"?

***Solution***

a) With four equally likely results (1, 2, 3, and 4), the probability of a "1" is $\frac{1}{4}$.

b) With four equally likely results (1, 2, 3, and 4) and three "favorable" outcomes (2, 3, and 4), the probability of rolling a number larger than a "1" is $\frac{3}{4}$.

SSG *Exs. 6, 7*

## SPHERES

Another type of solid with which you are familiar is the sphere. Although the surface of a basketball correctly depicts the sphere, we often use the term *sphere* to refer to a solid like a baseball as well.

Reminder

The sphere was defined as a locus of points in Chapter 6.

In space, the sphere is characterized in three ways:

1. A **sphere** is the set of all points at a fixed distance $r$ from a given point $O$. Point $O$ is known as the **center** of the sphere, even though it is not a part of the spherical surface.
2. A **sphere** is the surface determined when a circle (or semicircle) is rotated about any of its diameters.
3. A **sphere** is the surface that represents the theoretical limit of an "inscribed" regular polyhedron whose number of faces increases without limit.

NOTE: In characterization 3, suppose that the number of faces of the regular polyhedron could grow without limit. In theory, the resulting regular polyhedra would appear more "spherical" as the number of faces increases without limit. In reality, a regular polyhedron can have no more than 20 faces (the regular icosahedron). It will be convenient and necessary to use this third characterization of the sphere when we determine the formula for its volume.

Each characterization of the sphere has its advantages.

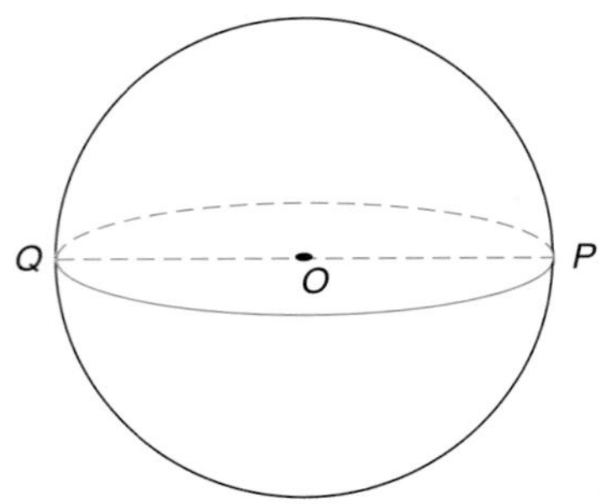


**FIGURE 8.42**

*Characterization 1:* In Figure 8.42, a sphere was generated as the locus of points in space at a distance $r$ from point $O$. The line segment $\overline{OP}$ is a **radius** of sphere $O$, and $\overline{QP}$ is a **diameter** of the sphere. The intersection of a sphere and a plane that contains its center is a **great circle** of the sphere. For the earth, the equator is a great circle that separates the earth into two **hemispheres.**

## Discover!

Suppose that you use scissors to cut out each pattern. (You may want to copy and enlarge this page.) Then glue or tape the indicated tabs (shaded) to form regular polyhedra. Which regular polyhedron is formed in each pattern?

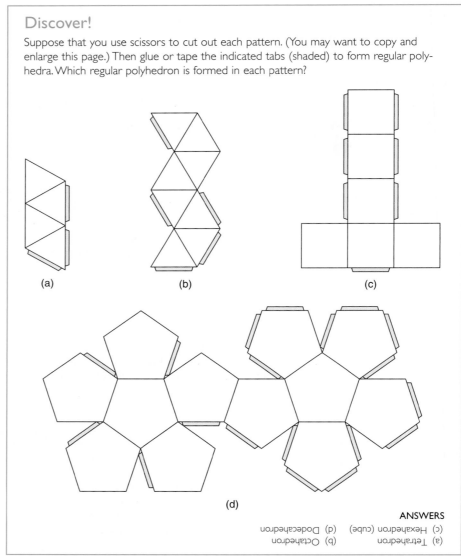

(a) (b) (c)

(d)

ANSWERS

(a) Tetrahedron (b) Octahedron
(c) Hexahedron (cube) (d) Dodecahedron

## SURFACE AREA OF A SPHERE

*Characterization 2:* The following theorem claims that the surface area of a sphere equals four times the area of a great circle of that sphere. This theorem, which is proved in calculus, treats the sphere as a surface of revolution.

THEOREM 8.4.2: The surface area $S$ of a sphere whose radius has length $r$ is given by $S = 4\pi r^2$.

Geometry in the Real World

Fruits such as oranges have the shape of a sphere.

EXAMPLE 3

Find the surface area of a sphere whose radius is $r = 7$ in. Use your calculator to approximate the result.

***Solution***

$$S = 4\pi r^2 \rightarrow S = 4\pi \cdot 7^2 = 196\pi \text{ in}^2$$

Then $S \approx 615.75 \text{ in}^2$.

*Exs. 8–10*

Although half of a circle is called a semicircle, half of a sphere is generally called a hemisphere.

## VOLUME OF A SPHERE

*Characterization 3:* The third description of the sphere enables us to find its volume. To accomplish this, we treat the sphere as the theoretical limit of an inscribed regular polyhedron whose number of faces $n$ increases without limit. The polyhedron can be separated into $n$ pyramids; the center of the sphere is the vertex of each pyramid. As $n$ increases, the altitude of each pyramid approaches the radius of the sphere in length. Next we find the sum of the volumes of these pyramids, the limit of which is the volume of the sphere.

In Figure 8.43, one of the pyramids described in the preceding paragraph is shown. We designate the height of each and every pyramid by $h$. Where the areas of the bases of the pyramids are written $B_1$, $B_2$, $B_3$, and so on, the sum of the volumes of the $n$ pyramids forming the polyhedron is

$$\frac{1}{3}B_1 h + \frac{1}{3}B_2 h + \frac{1}{3}B_3 h + \cdots + \frac{1}{3}B_n h$$

Next we write the volume of the polyhedron in the form

$$\frac{1}{3}h(B_1 + B_2 + B_3 + \cdots + B_n)$$

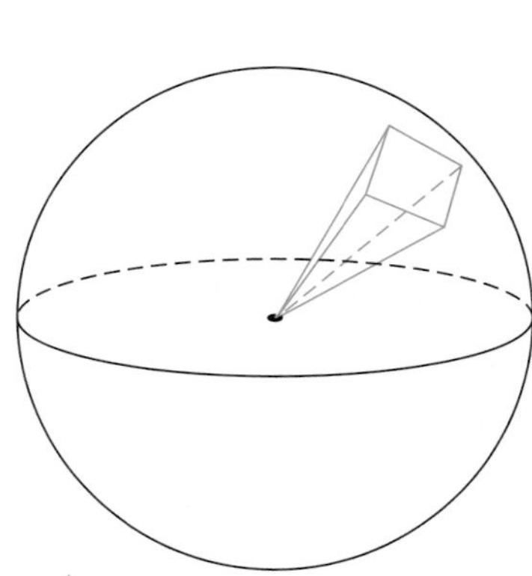

**FIGURE 8.43**

As $n$ increases, $h \rightarrow r$ and $B_1 + B_2 + B_3 + \cdots + B_n \rightarrow S$, the surface area of the sphere. Because the surface area of the sphere is $S = 4\pi r^2$, the sum approaches the following limit as the volume of the sphere:

$$\frac{1}{3}h(B_1 + B_2 + B_3 + \cdots + B_n) \rightarrow \frac{1}{3}rS \quad \text{or} \quad \frac{1}{3}r \cdot 4\pi r^2 = \frac{4}{3}\pi r^3$$

This leads us to the following theorem.

THEOREM 8.4.3: The volume $V$ of a sphere with a radius of length $r$ is given by $V = \frac{4}{3}\pi r^3$.

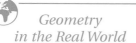

## Discover!

A farmer's silo is a composite shape. That is, it is actually composed of two solids. What are they?

ANSWER

Cylinder and hemisphere

### Technology Exploration

Determine the method of calculating "cube roots" on your calculator. Then show that $\sqrt[3]{27} = 3$.

EXAMPLE 4

Find the exact volume of a sphere whose length of radius is 1.5 in.

***Solution***

This calculation can be done more easily if we replace 1.5 by $\frac{3}{2}$.

$$\begin{aligned} V &= \frac{4}{3}\pi r^3 \\ &= \frac{4}{3} \cdot \pi \cdot \frac{3}{2} \cdot \frac{3}{2} \cdot \frac{3}{2} \\ &= \frac{9\pi}{2} \text{ in}^3 \end{aligned}$$

EXAMPLE 5

A spherical propane gas storage tank has a volume of $\frac{792}{7}$ ft$^3$. Using $\pi \approx \frac{22}{7}$, find the radius of the sphere.

***Solution***

$V = \frac{4}{3}\pi r^3$, which becomes $\frac{792}{7} = \frac{4}{3} \cdot \frac{22}{7} \cdot r^3$. Then $\frac{88}{21}r^3 = \frac{792}{7}$. In turn,

$$\frac{21}{88} \cdot \frac{88}{21}r^3 = \frac{21}{88} \cdot \frac{792}{7} \rightarrow r^3 = 27 \rightarrow r = \sqrt[3]{27} \rightarrow r = 3$$

The radius of the tank is 3 ft.

Just as two concentric circles have the same center but different lengths of radii, two spheres also can be concentric. This fact is the basis for the following example.

EXAMPLE 6

A child's hollow plastic ball has an inside diameter of 10 in. and is approximately $\frac{1}{8}$ in. thick (see the cross-section of the ball in Figure 8.44). Approximately how many cubic inches of plastic were needed to construct the ball?

***Solution***

The volume of plastic used is the difference between the outside volume and the inside volume. Where $R$ denotes the length of the outside radius and $r$ denotes the length of the inside radius, $R \approx 5.125$ and $r = 5$.

$$V = \frac{4}{3}\pi R^3 - \frac{4}{3}\pi r^3, \quad \text{so} \quad V = \frac{4}{3}\pi(5.125)^3 - \frac{4}{3}\pi \cdot 5^3$$

Then $V \approx 563.86 - 523.60 \approx 40.26$

The volume of plastic used was approximately 40.26 in$^3$.

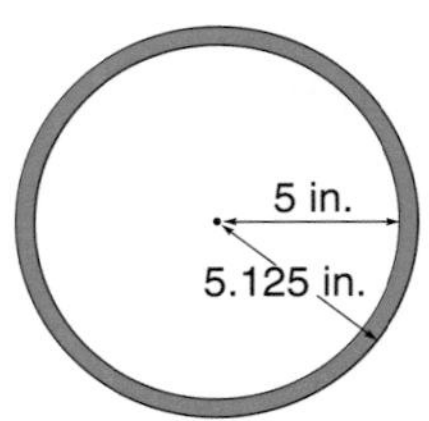


**FIGURE 8.44**

Like circles, spheres may have tangent lines; however, spheres also have tangent planes. As shown in Figure 8.45, it is also possible for spheres to be tangent to each other.

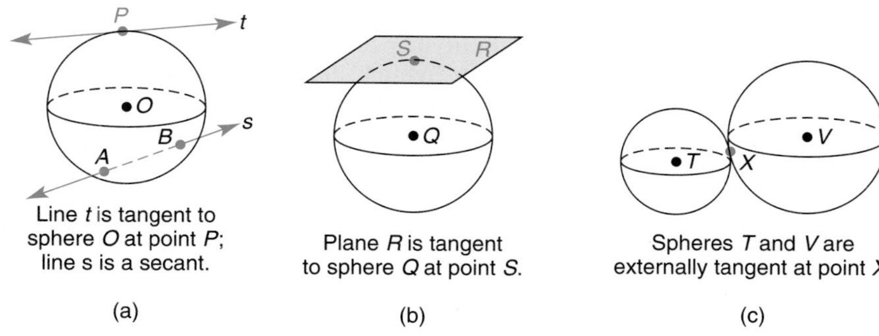


Line $t$ is tangent to sphere $O$ at point $P$; line s is a secant.

(a)

Plane $R$ is tangent to sphere $Q$ at point $S$.

(b)

Spheres $T$ and $V$ are externally tangent at point $X$.

(c)

**FIGURE 8.45**

SSG *Exs. 11–13*

## MORE SOLIDS OF REVOLUTION

In Section 8.3, each solid of revolution was generated by revolving a plane region about a horizontal line segment. It is also possible to form a solid of revolution by rotating a region about a vertical or oblique line segment. Every solid determined in this manner will have symmetry about the line. See Examples 7 and 8.

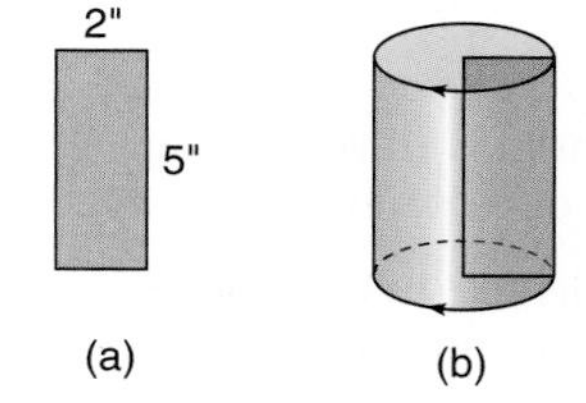


(a) (b)

**FIGURE 8.46**

### EXAMPLE 7

Describe the solid of revolution formed when the rectangular region having dimensions of 2 in. by 5 in. [see Figure 8.46(a)] is rotated about a vertical side of length 5 in. Then find the exact volume of the solid formed [see Figure 8.46(b)].

***Solution***

The solid formed is a right circular cylinder with radius of base $r = 2$ and altitude $h = 5$. The formula we use to find the volume, $V = \pi r^2 h$, becomes

$$V = \pi \cdot 2^2 \cdot 5, \text{ so } V = 20\pi \text{ in}^3$$

### EXAMPLE 8

Describe the solid of revolution formed when a semicircular region having a vertical diameter of length 12 cm [see Figure 8.47(a)] is revolved about that diameter. Then find the exact volume of the solid formed [see Figure 8.47(b)].

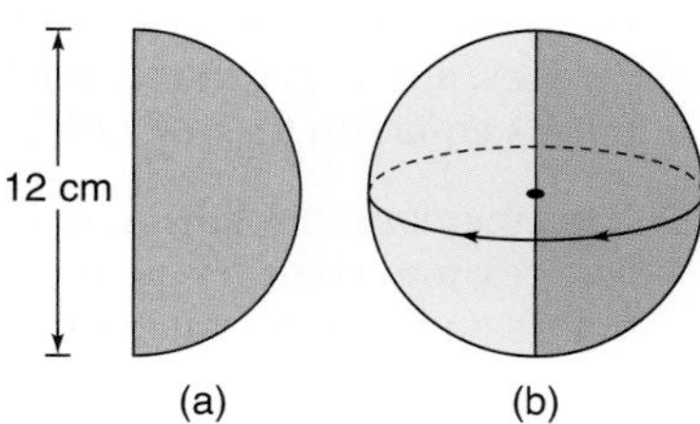


(a) (b)

**FIGURE 8.47**

***Solution***

The solid formed is a sphere with length of radius $r = 6$ cm. The formula we use to find the volume, $V = \frac{4}{3}\pi r^3$, becomes $V = \frac{4}{3}\pi \cdot 6^3$, which simplifies to $V = 288\pi \text{ cm}^3$.

When a circular region is revolved about a line in the circle's exterior, a doughnut-shaped solid results. The formal name of the resulting solid of revolution, shown in Figure 8.48, is the *torus*. Calculus is necessary to calculate both the surface area and the volume of the torus.

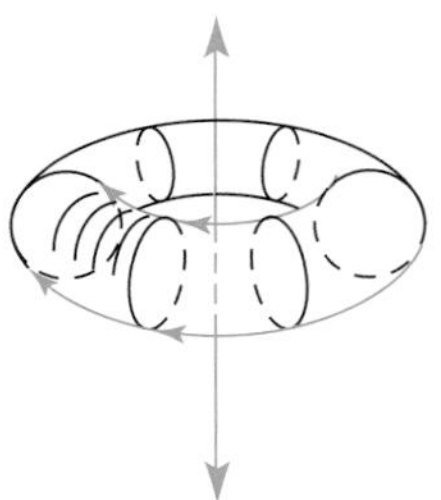

*Exs. 14–16* **FIGURE 8.48**

## 8.4 Exercises

1. Which of these two polyhedrons is concave? Note that the interior dihedral angle formed by the planes containing $\triangle EJF$ and $\triangle KJF$ is larger than 180°.

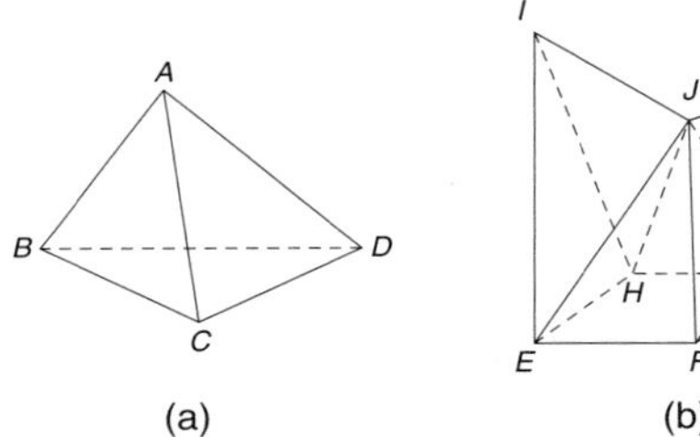


(a) (b)

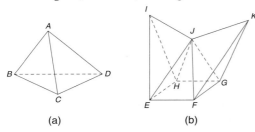

2. For Figure (a) of Exercise 1, find the number of faces, vertices, and edges in the polyhedron. Then verify Euler's equation for that polyhedron.

3. For Figure (b) of Exercise 1, find the number of faces, vertices, and edges in the polyhedron. Then verify Euler's equation for that polyhedron.

4. For a regular tetrahedron, find the number of faces, vertices, and edges in the polyhedron. Then verify Euler's equation for that polyhedron.

5. For a regular hexahedron, find the number of faces, vertices, and edges in the polyhedron. Then verify Euler's equation for that polyhedron.

6. A regular polyhedron has 12 edges and 8 vertices.
   a) Use Euler's equation to find the number of faces.
   b) Use the result from part (a) to name the regular polyhedron.

7. A regular polyhedron has 12 edges and 6 vertices.
   a) Use Euler's equation to find the number of faces.
   b) Use the result from part (a) to name the regular polyhedron.

8. A polyhedron (not regular) has 10 vertices and 7 faces. How many edges does it have?

9. A polyhedron (not regular) has 14 vertices and 21 edges. How many faces must it have?

*In Exercises 10 to 12, the probability is the ratio $\frac{\textit{number of favorable outcomes}}{\textit{number of possible outcomes}}$. Use Example 2 of this section as a guide.*

10. Assume that a die of the most common shape, a hexahedron, is rolled. What is the likelihood that
    a) a "2" results?
    b) an even number results?
    c) the result is larger than 2?

11. Assume that a die in the shape of a dodecahedron is rolled. What is the probability that
    a) an even number results?
    b) a prime number (2, 3, 5, 7, or 11) results?
    c) the result is larger than 2?

12. Assume that a die in the shape of an icosahedron is rolled. What is the likelihood that
    a) an odd number results?
    b) a prime number (2, 3, 5, 7, 11, 13, 17, or 19) results?
    c) the result is larger than 2?

13. In sphere $O$, the length of radius $\overline{OP}$ is 6 in. Find the length of the chord:

a) $\overline{QR}$ if $m\angle QOR = 90°$
b) $\overline{QS}$ if $m\angle SOP = 60°$

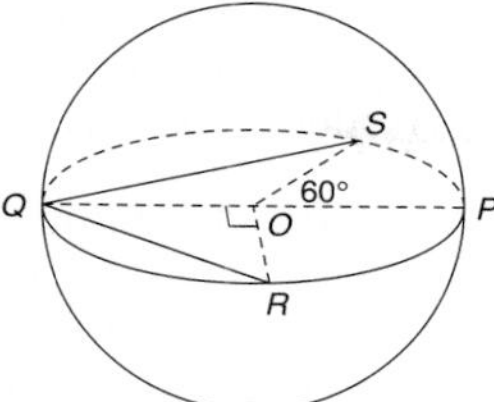


*Exercises 13, 14*

14. Find the approximate surface area and volume of the sphere if $OP = 6$ in. Use your calculator.

15. Find the total area (surface area) of a regular octahedron if the area of each face is 5.5 in$^2$.

16. Find the total area (surface area) of a regular dodecahedron (12 faces) if the area of each face is 6.4 cm$^2$.

17. Find the total area (surface area) of a regular hexahedron if each edge has a length of 4.2 cm.

18. Find the total area (surface area) of a regular tetrahedron if each edge has a length of 6 in.

19. The total area (surface area) of a regular hexahedron is 105.84 m$^2$. Find the

a) area of each face.
b) length of each edge.

20. The total area (surface area) of a regular octahedron is $32\sqrt{3}$ ft$^2$. Find the

a) area of each face.
b) length of each edge.

21. The surface of a soccer ball is composed of 12 regular pentagons and 20 regular hexagons. With each side of each regular polygon measuring 4.5 cm, the area of each regular pentagon is 34.9 cm$^2$ and the area of each regular hexagon is 52.5 cm$^2$.

a) What is the surface area of the soccer ball?
b) If the material used to construct the ball costs 0.6 cent per square centimeter, what is the cost of the materials used in construction?

22. A calendar is determined by using each of the 12 faces of a regular dodecahedron for one month of the year. With each side of the regular pentagonal face measuring 4 cm, the area of each face is approximately 27.5 cm$^2$.

a) What is the total surface area of the calendar?
b) If the material used to construct the calendar costs 0.8 cent per square centimeter, what is the cost of the materials used in construction?

23. A sphere is inscribed within a right circular cylinder whose altitude and diameter have equal measures.

a) Find the ratio of the surface area of the cylinder to that of the sphere.
b) Find the ratio of the volume of the cylinder to that of the sphere.

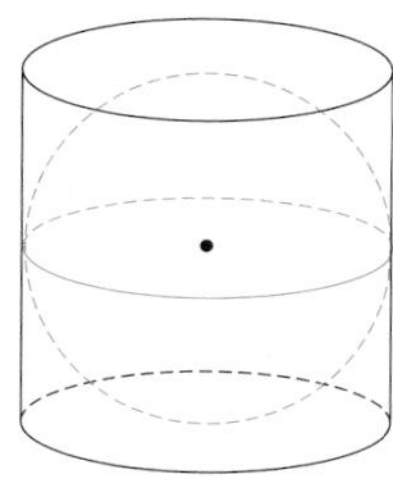

24. Given that a right circular cylinder is inscribed within a sphere, what is the least possible volume of the cylinder? (HINT: Consider various lengths for radius and altitude.)

25. In calculus, it can be shown that the largest possible volume for the inscribed right circular cylinder in Exercise 24 occurs when its altitude has a length equal to the diameter of the circular base. Find the length of the radius and the altitude of the cylinder of greatest volume if the radius of the sphere is 6 in.

26. Given that a *regular* polyhedron of $n$ faces is inscribed in a sphere of radius 6 in., find the maximum (largest) possible volume for the polyhedron.

27. A right circular cone is inscribed in a sphere. If the slant height of the cone has a length equal to that of its diameter, find the length of the

a) radius of the base of the cone.
b) altitude of the cone.

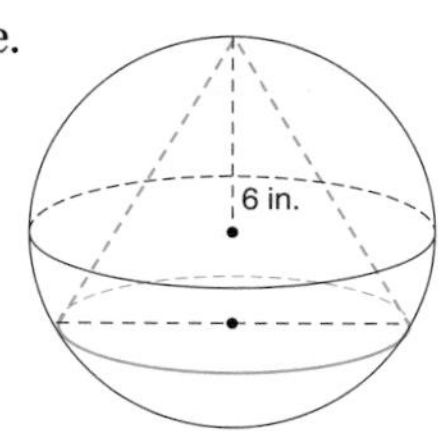


The radius of the sphere has a length of 6 in.

28. A sphere is inscribed in a right circular cone whose slant height has a length equal to the diameter of its base. What is the length of the radius of the sphere if the slant height and the diameter of the cone both measure 12 cm?

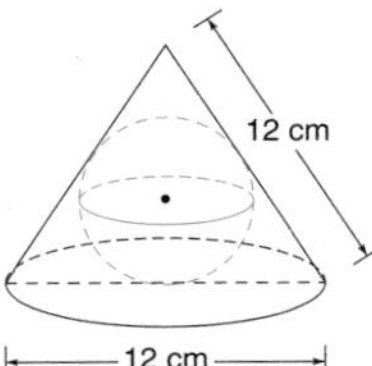


*In Exercises 29 and 30, use the calculator value of $\pi$.*

29. For a sphere whose radius has length 3 m, find the approximate
   a) surface area.
   b) volume.

30. For a sphere whose radius has length 7 cm, find the approximate
   a) surface area.
   b) volume.

31. A sphere has a volume equal to $\frac{99}{7}$ in$^3$. Determine the length of the radius of the sphere. $\left(\text{Let } \pi \approx \frac{22}{7}.\right)$

32. A sphere has a surface area equal to 154 in$^2$. Determine the length of the radius of the sphere. $\left(\text{Let } \pi \approx \frac{22}{7}.\right)$

33. The spherical storage tank described in Example 5 had a length of radius of 3 ft. Because the tank needs to be painted, we need to find its surface area. Also determine the number of pints of rust-proofing paint needed to paint the tank if 1 pt covers approximately 40 ft$^2$. Use your calculator.

34. An observatory has the shape of a right circular cylinder surmounted by a hemisphere. If the radius of the cylinder is 14 ft and its altitude measures 30 ft, what is the surface area of the observatory? If 1 gal of paint covers 300 ft$^2$, how many gallons are needed to paint the surface if it requires two coats? Use your calculator.

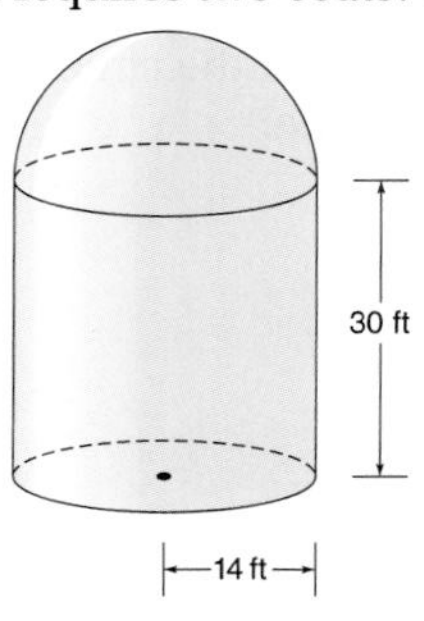


35. A leather soccer ball has an inside diameter of 8.5 in. and a thickness of 0.1 in. Find the volume of leather needed for its construction. Use your calculator.

36. An ice cream cone is filled with ice cream as shown. What is the volume of the ice cream? Use your calculator.

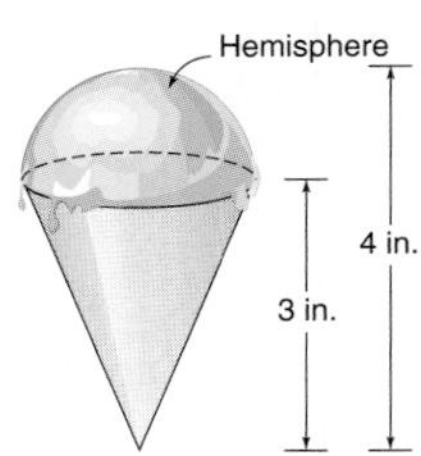


*For Exercises 37 to 42, make drawings as needed.*

37. Can two spheres
   a) be internally tangent?
   b) have no points in common?

38. If two spheres intersect at more than one point, what type of geometric figure is determined by their intersection?

39. Two planes are tangent to a sphere at the endpoints of a diameter. How are the planes related?

40. Plane $R$ is tangent to sphere $O$ at point $T$. How are radius $\overline{OT}$ and plane $R$ related?

41. Two tangent segments are drawn to sphere $Q$ from external point $E$. Where $A$ and $B$ are the points of tangency on sphere $Q$, how are $\overline{EA}$ and $\overline{EB}$ related?

42. How many common tangent planes do two externally tangent spheres have?

43. Suppose that a semicircular region with a vertical diameter of length 6 is rotated about that diameter. Determine the exact surface area and the exact volume of the resulting solid of revolution.

44. Suppose that a semicircular region with a vertical diameter of length 4 is rotated about that diameter. Determine the exact surface area and the exact volume of the resulting solid of revolution.

45. Sketch the torus that results when the given circle of radius 1 is revolved about the horizontal line that lies 4 units below the center of that circle.

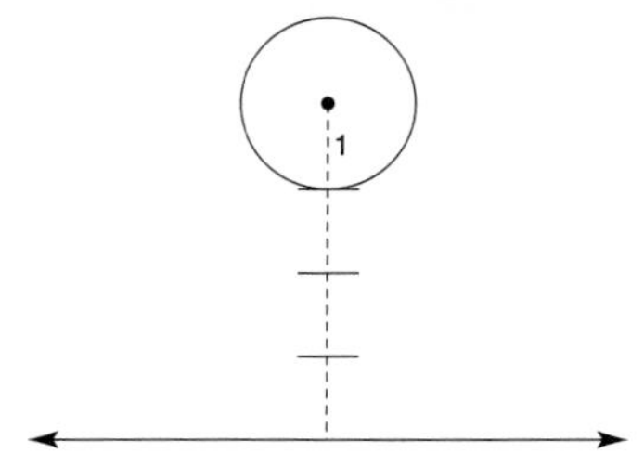

46. Sketch the solid that results when the given circle of radius 1 is revolved about the horizontal line that lies 1 unit below the center of that circle.

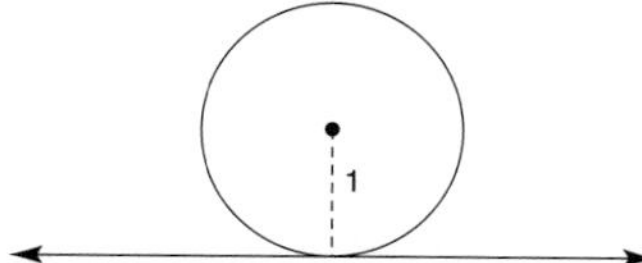


47. Explain how the following formula used in Example 6 was obtained:

$$V = \frac{4}{3}\pi R^3 - \frac{4}{3}\pi r^3$$

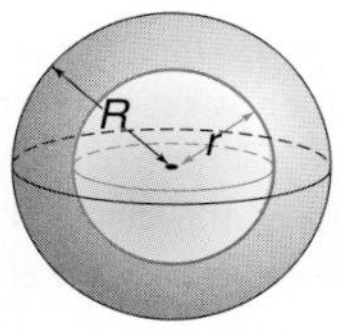


48. Derive a formula for the total surface area of the hollow-core sphere. (NOTE: Include both interior and exterior surface areas.)

# Perspective on History

## SKETCH OF RENÉ DESCARTES

René Descartes was born in Tours, France, on March 31, 1596, and died in Stockholm, Sweden, on February 11, 1650. He was a contemporary of Galileo, the Italian scientist responsible for many discoveries in the science of dynamics. Descartes was also a friend of the French mathematicians Marin Mersenne (Mersenne Numbers) and Blaise Pascal (Pascal's Triangle).

As a small child, René Descartes was in poor health much of the time. Because he spent so much time reading in bed during his illnesses, he became a very well educated young man. When Descartes was older and stronger, he joined the French army. It was during his time as a soldier that Descartes had three dreams that vastly influenced his future. The dreams, dated to November 10, 1619, shaped his philosophy and laid the framework for his discoveries in mathematics.

Descartes resigned his commission with the army in 1621 so that he could devote his life to studying philosophy, science, and mathematics. In the ensuing years, Descartes came to be highly regarded as a philosopher and mathematician and was invited to the learning centers of France, Holland, and Sweden.

Descartes's work in mathematics, in which he used an oblique coordinate system as a means of representing points, led to the birth of analytical geometry. His convention for locating points was eventually replaced by a coordinate system with perpendicular axes. In this system, algebraic equations could be represented by geometric figures; subsequently, many conjectured

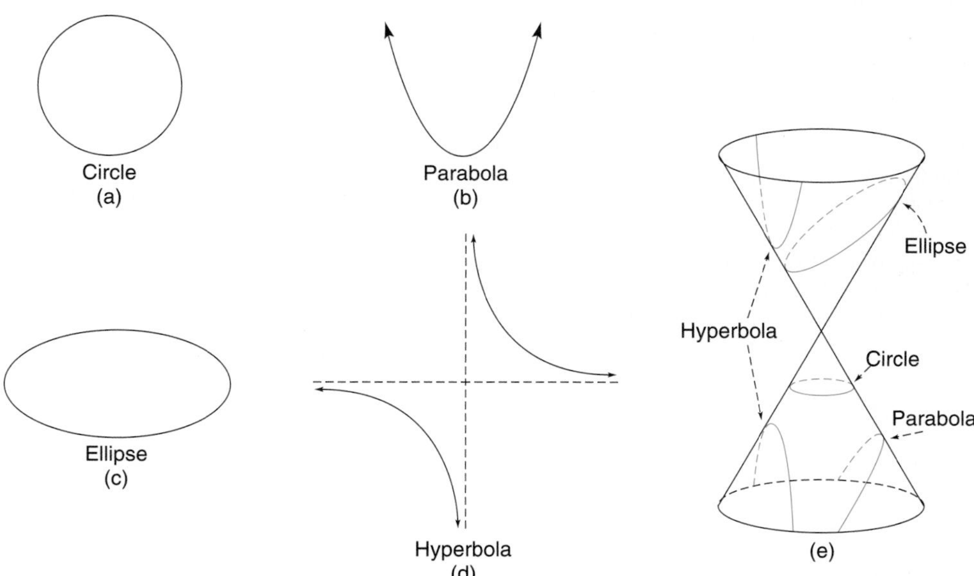


**FIGURE 8.49**

properties of these figures could be established through algebraic (analytic) proof. The rectangular coordinate system (which is called the Cartesian system in honor of Descartes) can also be used to locate the points of intersection of geometric figures such as lines and circles. Much of the material in Chapter 9 depends on his work.

Generally, the phrase *conic sections* refers to four geometric figures: the **circle,** the **parabola,** the **ellipse,** and the **hyperbola.** These figures are shown in Figure 8.49 on page 429 individually and also in relation to the upper and lower nappes of a cone. The conic sections are formed when a plane intersects the nappes of a cone.

Other mathematical works of Descartes were devoted to the study of tangent lines to curves. The notion of a tangent to a curve is illustrated in Figure 8.50; this concept is the basis for the branch of mathematics known as **differential calculus.**

Descartes's final contributions to mathematics involved his standardizing the use of many symbols. To mention a few of these, Descartes used (1) $a^2$ rather than $aa$ and $a^3$ rather than $aaa$; (2) $ab$ to indicate multiplication; and (3) $a$, $b$, and $c$ as constants and $x$, $y$, and $z$ as variables.

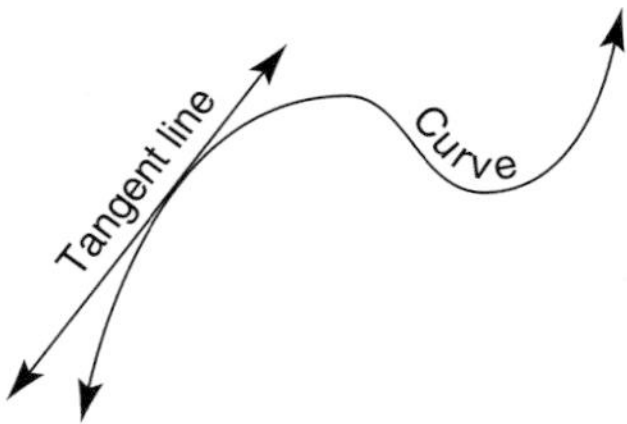


**FIGURE 8.50**

# Perspective on Application

## BIRDS IN FLIGHT

The following application of geometry is not so much practical as it is classical.

Two birds have been attracted to bird feeders that rest atop vertical poles. The bases of these poles where their perches are located are 20 ft apart. The poles are themselves 10 ft and 16 ft tall. See Figure 8.51. Each bird eyes birdseed that has fallen on the ground at the base of the other pole. Leaving their perches, the birds fly in a straight-line path toward their goal. Avoiding a collision in flight, the birds take paths that have them pass at a common point X. How far is the point above ground level?

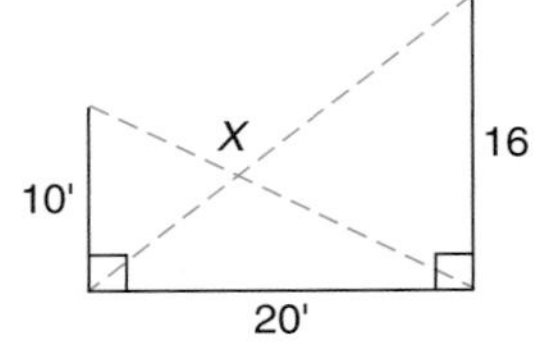


**FIGURE 8.51**

The solution to the problem follows. However, we redraw the figure to indicate that the 20-ft distance between poles is separated along the ground into line segments with lengths of $a$ and $20 - a$ as shown. See Figures 8.52(a), (b), and (c).

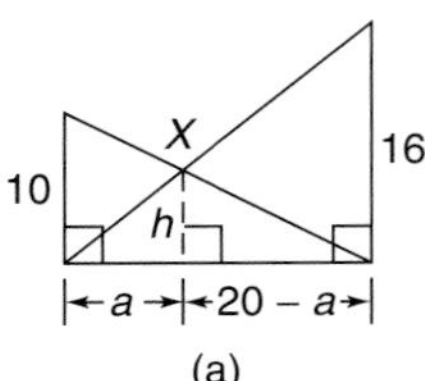


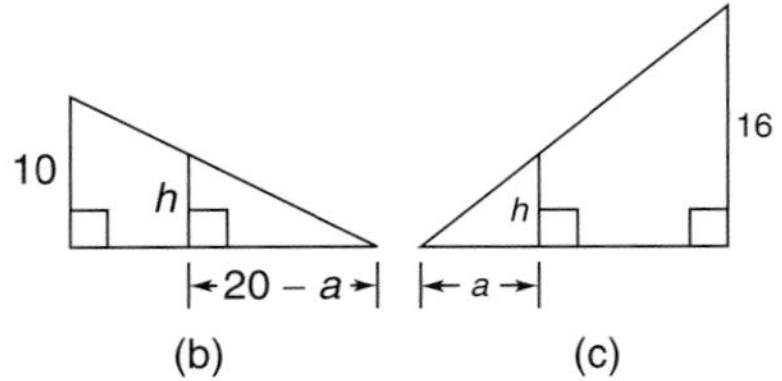


**FIGURE 8.52**

From Figures 8.52(b) and (c), we form the following equations based upon similarity of the right triangles.

$$\frac{10}{20} = \frac{h}{20 - a} \quad \text{and} \quad \frac{16}{20} = \frac{h}{a}$$

By the Means-Extremes Property of Proportions, $10(20 - a) = 20h$ and $16a = 20h$.
By substitution,

$$\begin{aligned} 10(20 - a) &= 16a \\ 200 - 10a &= 16a \\ 26a &= 200 \end{aligned}$$

$$a = \frac{200}{26} \quad \text{or} \quad a = \frac{100}{13}$$

From the fact that $16a = 20h$, we see that

$$16 \cdot \frac{100}{13} = 20h$$

so

$$h = \frac{1}{\cancel{20}_{1}} \cdot 16 \cdot \frac{\cancel{100}^{5}}{13} = \frac{80}{13} = 6\frac{2}{13}\text{ ft}$$

The point at which the birds flew past each other was $6\frac{2}{13}$ feet above the ground.

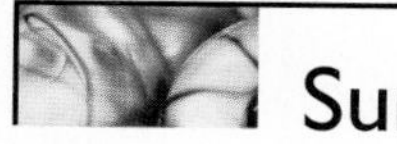

# Summary

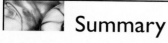

### A LOOK BACK AT CHAPTER 8

Our goal in this chapter was to deal with a type of geometry known as solid geometry. We found formulas for the lateral area, the total area (surface area), and the volume of prisms, pyramids, cylinders, cones, and spheres. Some of the formulas used in this chapter were developed using the concept of "limit."

### A LOOK AHEAD TO CHAPTER 9

Our focus in the next chapter is analytic (or coordinate) geometry. This type of geometry relates algebra and geometry. Formulas for the midpoint of a line segment, the length of a line segment, and the slope of a line will be developed. We will not only graph the equations of lines but also determine equations for given lines. We will see that proofs of many theorems can be completed by using analytic geometry.

### KEY CONCEPTS

8.1 Prisms (Right and Oblique) • Bases • Altitude • Vertices • Edges • Faces • Lateral Area • Total (Surface) Area • Volume • Regular Prism • Cube • Cubic Unit

8.2 Pyramid • Base • Altitude • Vertices • Edges • Faces • Vertex of a Pyramid • Regular Pyramid • Slant Height of a Regular Pyramid • Lateral Area • Total (Surface) Area • Volume

8.3 Cylinders (Right and Oblique) • Bases and Altitude of a Cylinder • Axis of a Cylinder • Cones (Right and Oblique) • Base and Altitude of a Cone • Vertex and Slant Height of a Cone • Axis of a Cone • Lateral Area • Total Area • Volume • Solid of Revolution • Axis of a Solid of Revolution

8.4 Dihedral Angle • Polyhedron (Convex and Concave) • Vertices • Edges and Faces • Euler's Equation • Regular Polyhedrons (Tetrahedron, Hexahedron, Octahedron, Dodecahedron, and Icosahedron) • Sphere (Center, Radius, Diameter, Great Circle, Hemisphere) • Surface Area and Volume of a Sphere

### TABLE 8.4 AN OVERVIEW OF CHAPTER EIGHT

**Volume and Area Relationships for Solids**

| *Solid* | *Figure* | *Volume* | *Area* |
|---|---|---|---|
| Rectangular prism (box) | | $V = \ell wh$ | $T = 2\ell w + 2\ell h + 2wh$ |
| Cube | | $V = e^3$ | $T = 6e^2$ |
| Prism (right prism shown) | | $V = Bh$ ($B$ = area of base) | $L = hP$ ($P$ = perimeter of base); $T = L + 2B$ |
| Regular pyramid (with slant height $\ell$) | | $V = \frac{1}{3}Bh$ ($B$ = area of base) | $L = \frac{1}{2}\ell P$ ($P$ = perimeter of base); $T = L + B$ NOTE: $\ell^2 = a^2 + h^2$ |

*continued*

**TABLE 8.4 CONT.**

**Volume and Area Relationships for Solids**

| Solid | Figure | Volume | Area |
|---|---|---|---|
| Right circular cylinder | | $V = Bh$ or<br>$V = \pi r^2 h$ | $L = 2\pi rh$;<br>$T = L + 2B$ or<br>$T = 2\pi rh + 2\pi r^2$ |
| Right circular cone (with slant height $\ell$) | | $V = \frac{1}{3}Bh$ or<br>$V = \frac{1}{3}\pi r^2 h$ | $L = \pi r\ell$;<br>$T = L + B$ or<br>$T = \pi r\ell + \pi r^2$<br>NOTE: $\ell^2 = r^2 + h^2$ |
| Sphere | | $V = \frac{4}{3}\pi r^3$ | $S = 4\pi r^2$ |

## Chapter 8 Review Exercises

1. Each side of the base of a right octagonal prism is 7 in. long. The altitude of the prism measures 12 in. Find the lateral area.

2. The base of a right prism is a triangle whose sides measure 7 cm, 8 cm, and 12 cm. The altitude of the prism measures 11 cm. Calculate the lateral area of the right prism.

3. The height of a square box is 2 in. more than three times the length of a side of the base. If the lateral area is 480 $\text{in}^2$, find the dimensions of the box and the volume of the box.

4. The base of a right prism is a rectangle whose length is 3 cm more than its width. If the altitude of the prism is 12 cm and the lateral area is 360 $\text{cm}^2$, find the total area and the volume of the prism.

5. The base of a right prism is a triangle whose sides have lengths of 9 in., 15 in., and 12 in. The height of the prism is 10 in. Find the
   a) lateral area.
   b) total area.
   c) volume.

6. The base of a right prism is a regular hexagon whose sides are 8 cm in length. The altitude of the prism is 13 cm. Find the
   a) lateral area. b) total area. c) volume.

7. A regular square pyramid has a base whose sides are of length 10 cm each. The altitude of the pyramid measures 8 cm. Find the length of the slant height.

8. A regular hexagonal pyramid has a base whose sides are of length $6\sqrt{3}$ in. each. If the slant height is 12 in., find the length of the altitude of the pyramid.

9. The radius of the base of a right circular cone measures 5 in. If the altitude of the cone measures 7 in., what is the length of the slant height?

10. The diameter of the base of a right circular cone is equal in length to the slant height. If the altitude of the cone is 6 cm, find the length of the radius of the base.

11. The slant height of a regular square pyramid measures 15 in. One side of the base measures 18 in. Find the
    a) lateral area. b) total area. c) volume.

12. The base of a regular pyramid is an equilateral triangle each of whose sides is 12 cm. The altitude of the pyramid is 8 cm. Find the exact and approximate
    a) lateral area.
    b) total area.
    c) volume.

13. The radius of the base of a right circular cylinder is 6 in. The height of the cylinder is 10 in. Find the exact
    a) lateral area.
    b) total area.
    c) volume.

14. a) For the trough in the shape of a half-cylinder, find the volume of water it will hold. (Use $\pi \approx 3.14$ and disregard the thickness.)

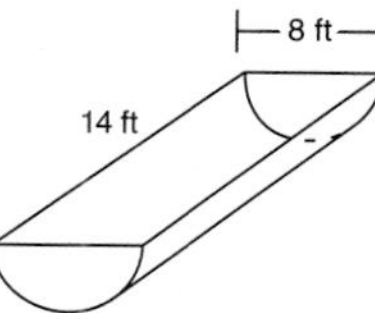


    b) If the trough is to be painted inside and out, find the number of square feet to be painted. (Use $\pi \approx 3.14$.)

15. The slant height of a right circular cone is 12 cm. The angle formed by the slant height and the altitude is 30°. Find the exact and approximate
    a) lateral area.
    b) total area.
    c) volume.

16. The volume of a right circular cone is $96\pi$ in$^3$. If the radius of the base is 6 in., find the length of the slant height.

17. Find the surface area of a sphere if the radius has the length 7 in. Use $\pi \approx \frac{22}{7}$.

18. Find the volume of a sphere if the diameter has the length 12 cm. Use $\pi \approx 3.14$.

19. The solid shown consists of a hemisphere (half of a sphere), a cylinder, and a cone. Find the exact volume of the solid.

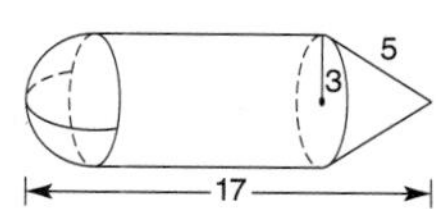


20. If the radius of one sphere is three times as long as the radius of another sphere, how do the surface areas of the spheres compare? How do the volumes compare?

21. Find the volume of the solid of revolution that results when a right triangle with legs of lengths 5 in. and 7 in. is rotated about the 7-in. leg. Use $\pi \approx \frac{22}{7}$.

22. Find the exact volume of the solid of revolution that results when a rectangular region with dimensions of 6 cm and 8 cm is rotated about a side of length 8 cm.

23. Find the exact volume of the solid of revolution that results when a semicircular region with diameter of length 4 in. is rotated about that diameter.

24. A plastic pipe is 3 ft long and has an inside radius of 4 in. and an outside radius of 5 in. How many cubic inches of plastic are in the pipe? (Use $\pi \approx 3.14$.)

25. A sphere with a diameter of 14 in. is inscribed in a hexahedron. Find the exact volume of the space inside the hexahedron but outside the sphere.

26. a) An octahedron has ________ faces that are ________.
    b) A tetrahedron has ________ faces that are ________.
    c) A dodecahedron has ________ faces that are ________.

27. A drug manufacturing company wants to manufacture a capsule that contains a spherical pill inside. The diameter of the pill is 4 mm, and the capsule is cylindrical with hemispheres on either end. The length of the capsule between the two hemispheres is 10 mm. What is the exact volume that the capsule will hold, excluding the volume of the pill?

28. For each of the following solids, verify Euler's equation by determining $V$, the number of vertices; $E$, the number of edges; and $F$, the number of faces.
    a) Right octagonal prism
    b) Tetrahedron
    c) Octahedron

29. Find the volume of cement used in the block shown.

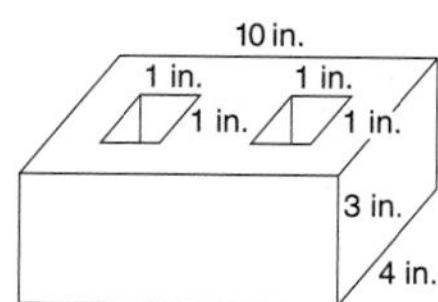


30. Given a die in the shape of a regular octahedron, find the probability that one roll produces
    a) an even-number result.
    b) a result of 4 or more.

31. Find the total surface area of
    a) a regular dodecahedron with each face having an area of 6.5 in.$^2$
    b) a regular tetrahedron with each edge measuring 4 cm.

32. Three spheres are tangent to each other in pairs. They have radii of 1 in., 2 in., and 3 in., respectively. What type of triangle is formed by the lines of center?

# Chapter 8 Test

1. For the regular pentagonal prism shown below, find the total number of:

   a) Edges ________ b) Faces ________

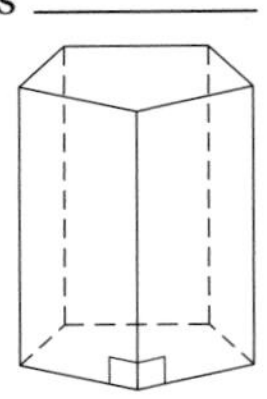

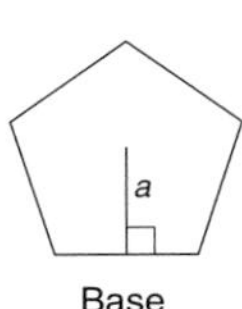


*Exercises 1, 2*

2. For the regular pentagonal base, each edge measures 3.2 cm and the apothem measures 2 cm.

   a) Find the area of the base $\left(\text{use } A = \frac{1}{2}aP\right)$. ________
   b) Find the total area of the regular pentagonal prism if its altitude measures 5 cm. ________
   c) Find the volume of the prism. ________

3. For the regular square pyramid shown, find the total number of:

   a) Vertices ________ b) Lateral faces ________

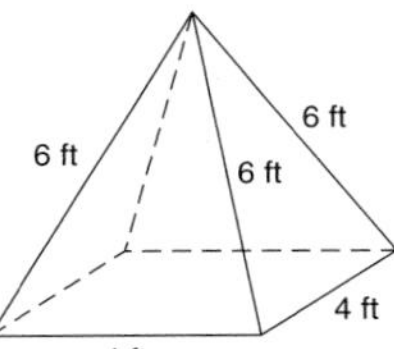


*Exercises 3, 4*

4. For the regular square pyramid shown above, find:

   a) The lateral area ________
   b) The total area ________

5. For the regular square pyramid shown, find the slant height.

   ________

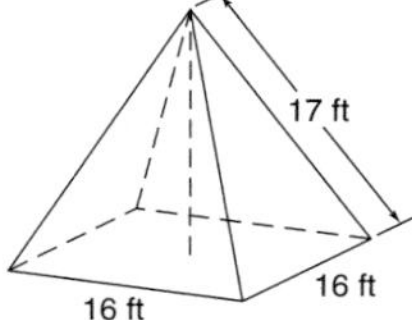


6. Find the altitude of a regular square pyramid (not shown) if each edge of the base measures 8 in. and the slant height of the pyramid is 5 in. ________

7. Find the volume of the regular square pyramid shown if each edge of the base measures 5 ft and the altitude measures 6 ft.

   ________

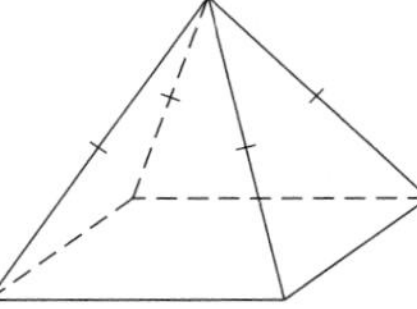

8. Determine whether the statement is True or False.

   a) A right circular cone has exactly two bases.

   ________

   b) The lateral area $L$ of a right circular cylinder with radius of base $r$ and altitude $h$ is given by $L = 2\pi rh$. ________

9. Determine whether the statement is True or False.

   a) The volume of a right circular cone is given by $V = \frac{1}{3}Bh$, which can also be expressed in the form $V = \frac{1}{3}\pi r^2 h$. ________
   b) A regular dodecahedron has exactly 12 faces.

   ________

10. Recall Euler's Formula, $V + F = E + 2$. For a certain polyhedron, there are eight faces and six vertices. How many edges does it have? ________

11. Find the slant height of the right circular cone at the right. Leave the answer in simplified radical form.

    ________

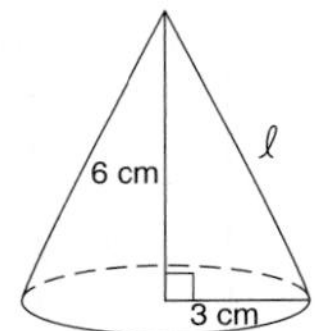


12. For the right circular cylinder shown, $r = 4$ cm and $h = 6$ cm. Find the exact:

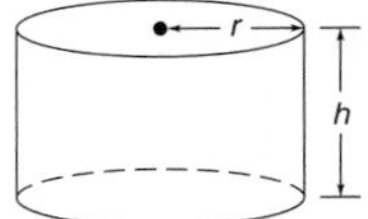


    a) Lateral area ________
    b) Volume ________

13. The exact volume of a right circular cone (not shown) is $32\pi$ in³. If the length of the base radius is 4 in., find the length of the altitude of the cone.

    ________

14. Assume that a die used for gaming is in the shape of a regular octahedron. The faces are numbered 1, 2, 3, 4, . . . , and 8. When this die is rolled once, what is the probability that the roll produces

    a) an even number result? ________
    b) a result greater than or equal to 6? ________

15. A spherical storage tank has a radius of 10 ft. Use the calculator's stored value of $\pi$ to find to nearest tenth of unit the approximate:

    a) Surface area of the sphere ________
    b) Volume of the sphere ________

# Chapter 9

# Analytic Geometry

The French mathematician René Descartes is considered the father of analytic geometry. His inspiration relating algebra and geometry, the Cartesian coordinate system, was a major breakthrough in the development of much of mathematics. The photograph illustrates the use of a GPS (global positioning system). The system allows one to pinpoint locations such as that of the vehicle being driven or that of a destination. On the map, locations identified by the latitude and longitude are comparable to points whose $x$ and $y$ coordinates locate a position in the Cartesian coordinate system.

## Chapter Outline



**For online student resources, visit this textbook's website at math.college.hmco.com/students.**

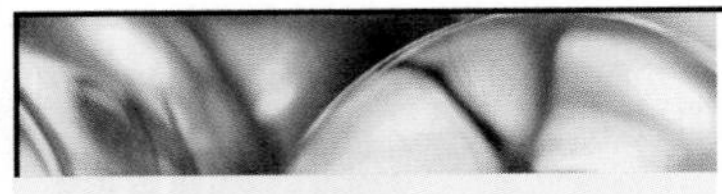

**KEY CONCEPTS**
Analytic Geometry • Cartesian Coordinate System • $x$ Axis • $y$ Axis • Quadrants • Origin • $x$ Coordinate • $y$ Coordinate • Ordered Pair • Distance Formula • Linear Equation • Midpoint Formula

# 9.1 The Rectangular Coordinate System

Graphing the solution sets for $3x - 2 = 7$ and $3x - 2 > 7$ required a single number line to indicate the value of $x$. (See Appendices A.2 and A.3 for further information.) In this chapter, we deal with equations containing two variables; to relate such algebraic statements to plane geometry, we will need two number lines.

The study of the relationships between number pairs and points is usually referred to as **analytic geometry.** The **Cartesian coordinate system** or **rectangular coordinate system** is the plane that results when two number lines intersect perpendicularly at the origin (the point corresponding to the number 0) of each line. The horizontal number line is known as the ***x* axis,** and its numerical coordinates increase from left to right. On the vertical number line, the ***y* axis,** values increase from bottom to top; see Figure 9.1. The two axes separate the plane into four sections known as **quadrants;** the quadrants are numbered I, II, III, and IV, as shown. The point that marks the common origin of the two number lines is the **origin** of the rectangular coordinate system. It is convenient to identify the origin as (0, 0); this notation indicates that the ***x* coordinate** (listed first) is 0 and also that the ***y* coordinate** (listed second) is 0.

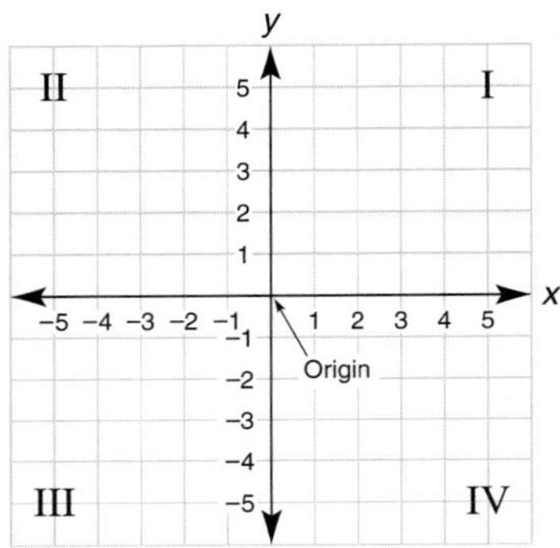


**FIGURE 9.1**

In the coordinate system in Figure 9.2, the point $(3, -2)$ is shown. For each point, we have the order $(x, y)$; these pairs are referred to as **ordered pairs** because $x$ must precede $y$. To plot (or locate) this point, we see that $x = 3$ and that $y = -2$. Thus the point is located by moving 3 units to the right of the origin and then 2 units down from the $x$ axis. The dashed lines shown are used to emphasize the reason why it is called the rectangular coordinate system. Note that this point $(3, -2)$ could also have been located by first moving down 2 units and then moving 3 units to the right of the $y$ axis. This point is located in Quadrant IV. In Figure 9.2, ordered pairs of plus and minus signs characterize the signs of the coordinates of a point in each quadrant.

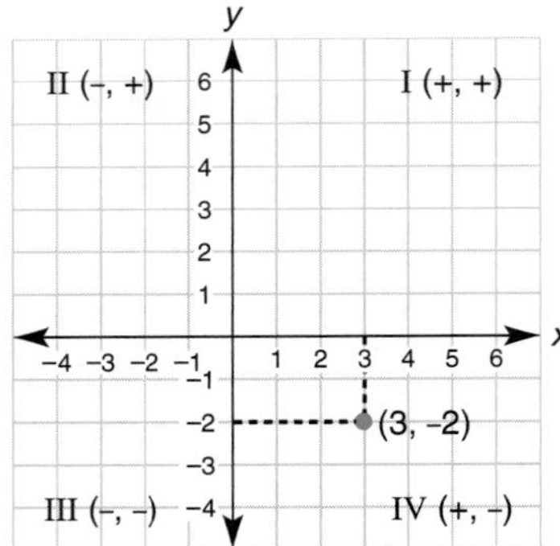


**FIGURE 9.2**

### EXAMPLE 1

Plot points $A(-3, 4)$ and $B(2, 4)$, and find the distance between them.

***Solution***

Point $A$ is located by moving 3 units to the left of the origin and then 4 units up from the $x$ axis. Point $B$ is located by moving 2 units to the right of the origin and then 4 units up from the $x$ axis. In Figure 9.3, $\overline{AB}$ is a horizontal segment.

In the rectangular coordinate system, $ABCD$ is a rectangle in which $DC = 5$; $\overline{DC}$ is easily measured because it lies on the $x$ axis. Because the opposite sides of a rectangle are congruent, it follows that $AB = 5$.

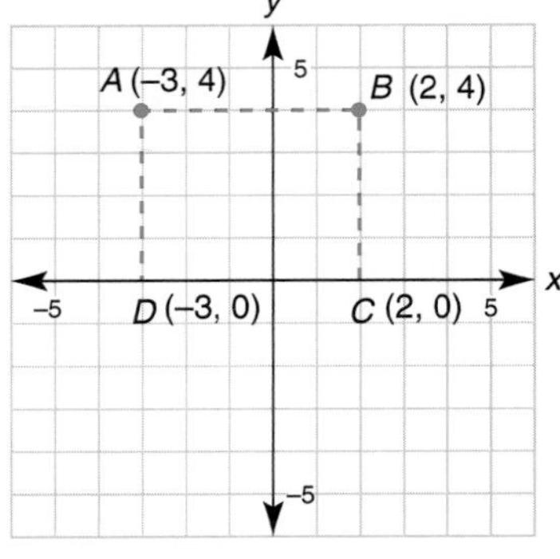


**FIGURE 9.3**

*Exs. 1–4*

In Example 1, the points $(-3, 4)$ and $(2, 4)$ have the same $y$ coordinates. In this case, the distance between the points on a horizontal line is merely the positive difference in the $x$ coordinates; thus the distance is $2 - (-3)$, or 5. It is also easy to find the distance between two points on a vertical line. When the $x$ coordinates are the same, the distance between points is the positive difference in the $y$ coordinates. In Figure 9.3, where $C$ is (2, 0) and $B$ is (2, 4), the distance between the points is $4 - 0$ or 4.

**DEFINITION:** Given points $A(x_1, y_1)$ and $B(x_2, y_1)$ on a horizontal line segment $\overline{AB}$, the **distance** between these points is

$$AB = x_2 - x_1 \text{ if } x_2 > x_1 \quad \text{or} \quad AB = x_1 - x_2 \text{ if } x_1 > x_2$$

In the preceding definition, repeated $y$ coordinates characterize a horizontal line segment. In the following definition, repeated $x$ coordinates determine a vertical line segment. In each definition, the distance is found by subtracting the smaller from the larger of the two unequal coordinates.

**DEFINITION:** Given points $C(x_1, y_1)$ and $D(x_1, y_2)$ on a vertical line segment $\overline{CD}$, the **distance** between these points is

$$CD = y_2 - y_1 \text{ if } y_2 > y_1 \quad \text{or} \quad CD = y_1 - y_2 \text{ if } y_1 > y_2$$

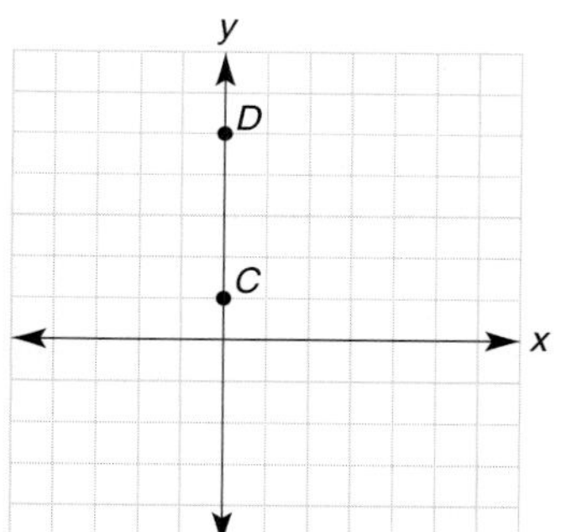


**FIGURE 9.4**

EXAMPLE 2

In Figure 9.4, name the coordinates of points $C$ and $D$, and find the distance between them. The $x$ coordinates of $C$ and $D$ are identical.

***Solution***

$C$ is the point (0, 1) because $C$ is 1 unit above the origin; similarly, $D$ is the point (0, 5). We designate the coordinates of point $C$ by $x_1 = 0$ and $y_1 = 1$ and the coordinates of point $D$ by $x_1 = 0$ and $y_2 = 5$. Using the preceding definition,

$$CD = y_2 - y_1 = 5 - 1 = 4$$

We now turn our attention to the more general problem of finding the distance between any two points. For the proof, see Figure 9.5 on page 438.

## THE DISTANCE FORMULA

The following formula enables us to find the distance between two points that lie on a "slanted" line.

**THEOREM 9.1.1:** **(Distance Formula)** The distance between two points $(x_1, y_1)$ and $(x_2, y_2)$ is given by the formula

$$d = \sqrt{(x_2 - x_1)^2 + (y_2 - y_1)^2}$$

**Discover!**

Plot the points $A(0, 0)$ and $B(4, 3)$. Now find $AB$ by using the Pythagorean Theorem. To accomplish this, you will need to form a path from $A$ to $B$ along horizontal and vertical line segments.

ANSWER
5

***Proof***

In the coordinate system in Figure 9.5 are points $P_1\ (x_1, y_1)$ and $P_2\ (x_2, y_2)$. In addition to drawing the segment joining these points, we draw an auxiliary horizontal segment through $P_1$ and an auxiliary vertical segment through $P_2$; these meet at point $C$ $(x_2, y_1)$ in Figure 9.5(a). Using Figure 9.5(b) and the definitions for lengths of horizontal and vertical segments,

$$P_1C = x_2 - x_1 \quad \text{and} \quad P_2C = y_2 - y_1$$

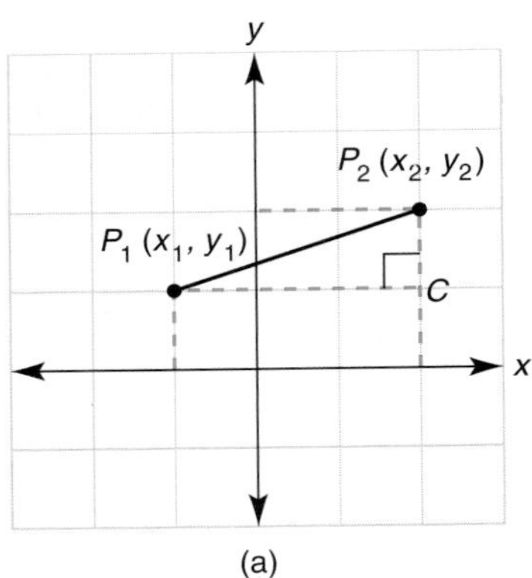

(a)

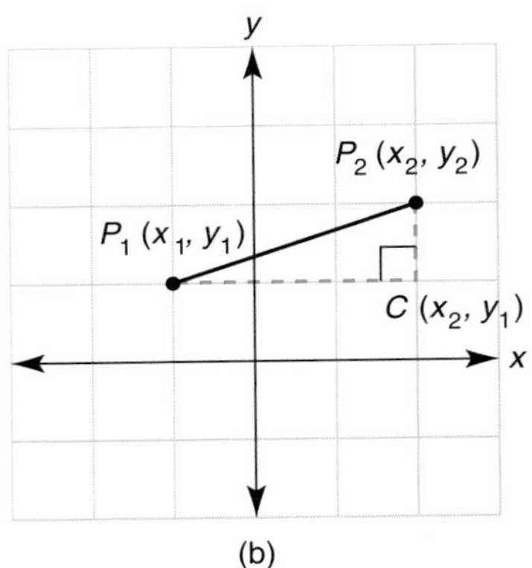

(b)

**FIGURE 9.5**

In right triangle $P_1P_2C$ in Figure 9.5(b), let $d = P_1P_2$. By the Pythagorean Theorem,

$$d^2 = (x_2 - x_1)^2 + (y_2 - y_1)^2$$

Taking the positive square root for length yields

$$d = \sqrt{(x_2 - x_1)^2 + (y_2 - y_1)^2}$$

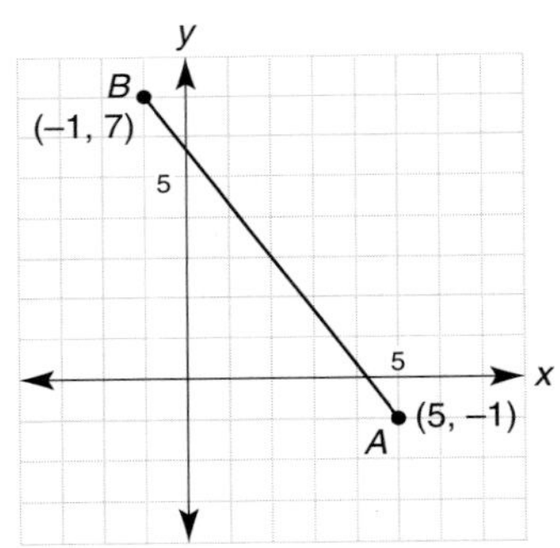


**FIGURE 9.6**

EXAMPLE 3

In Figure 9.6, find the distance between points $A\ (5, -1)$ and $B\ (-1, 7)$.

***Solution***

Using the Distance Formula and choosing $x_1 = 5$ and $y_1 = -1$ (from point $A$) and $x_2 = -1$ and $y_2 = 7$ (from point $B$), we obtain

$$\begin{aligned} d &= \sqrt{(-1 - 5)^2 + [7 - (-1)]^2} \\ &= \sqrt{(-6)^2 + (8)^2} = \sqrt{36 + 64} \\ &= \sqrt{100} = 10 \end{aligned}$$

NOTE: If the coordinates of point $A$ were designated as $x_2 = 5$ and $y_2 = -1$ and those of point $B$ were designated as $x_1 = -1$ and $y_1 = 7$, the distance would remain the same.

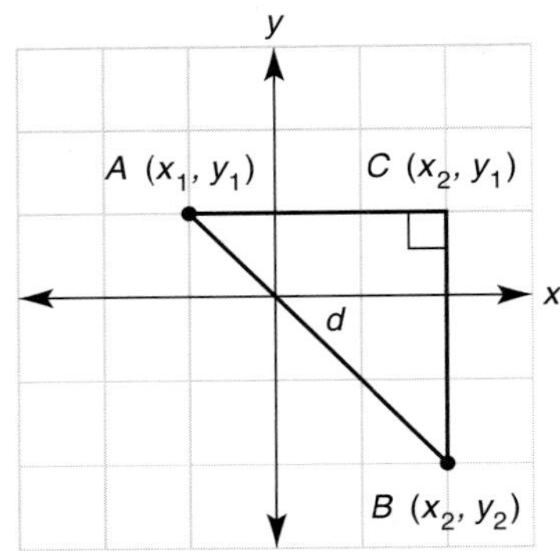


**FIGURE 9.7**

Looking back at the proof of the Distance Formula, Figure 9.5 shows only one of several possible placements of points. If the placement had been as shown in Figure 9.7, then we would have had $AC = x_2 - x_1$ because $x_2 > x_1$, and $BC = y_1 - y_2$ because $y_1 > y_2$. The Pythagorean Theorem leads to what looks like a different result:

$$d^2 = (x_2 - x_1)^2 + (y_1 - y_2)^2$$

But this can be converted to the earlier formula by using the fact that

$$(y_1 - y_2)^2 = (y_2 - y_1)^2$$

This follows from the fact that $(-a)^2 = a^2$ for any real number $a$.

The following example reminds us of the form of a **linear equation,** an equation whose graph is a straight line. In general, this form is $Ax + By = C$ for constants $A$, $B$, and $C$ (where $A$ and $B$ do not both equal 0). We will consider the graphing of linear equations in Section 9.2.

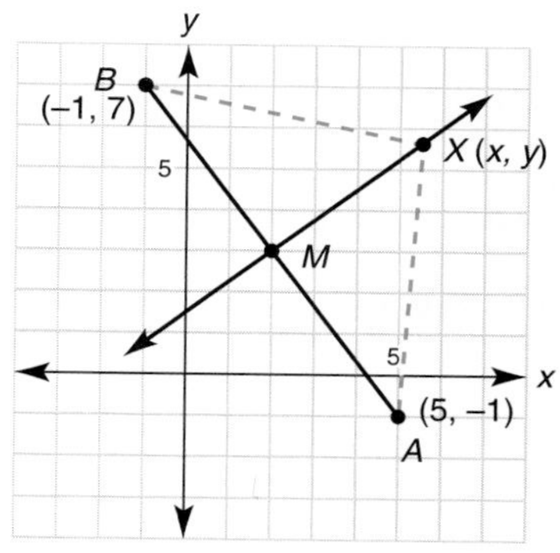


**FIGURE 9.8**

### EXAMPLE 4

Find the equation that describes all points $(x, y)$ that are equidistant from $A\,(5, -1)$ and $B\,(-1, 7)$.

***Solution***

In Chapter 6, we saw that the locus of points equidistant from two fixed points is a line. This line, $\overleftrightarrow{MX}$ in Figure 9.8, is the perpendicular bisector of $\overline{AB}$.

If $X$ is on the locus, then $AX = BX$. By the Distance Formula, we have

$$\sqrt{(x-5)^2 + [y-(-1)]^2} = \sqrt{[x-(-1)]^2 + (y-7)^2}$$

or

$$(x-5)^2 + (y+1)^2 = (x+1)^2 + (y-7)^2$$

after simplifying and squaring. Then

$$x^2 - 10x + 25 + y^2 + 2y + 1 = x^2 + 2x + 1 + y^2 - 14y + 49$$

Eliminating $x^2$ and $y^2$ terms by subtraction leads to the equation

$$-12x + 16y = 24$$

When we divide by 4, the equation of the line becomes

$$-3x + 4y = 6$$

If we divide the equation $-12x + 16y = 24$ by $-4$, an equivalent solution is

$$3x - 4y = -6$$

*Exs. 5–8*

NOTE: The equations $-3x + 4y = 6$ and $3x - 4y = -6$ are said to be **equivalent** because their solutions are the same. For instance, $(-2, 0)$, $(2, 3)$, and $(6, 6)$ are all solutions for *both* equations.

## THE MIDPOINT FORMULA

In Figure 9.8, point $M$ was the midpoint of $\overline{AB}$. It will be shown in Example 5(a) that $M$ is the point $(2, 3)$.

A generalized midpoint formula is given in Theorem 9.1.2. The result shows that the coordinates of the midpoint $M$ of a line segment are the averages of the coordinates of the endpoints. See the Discover! activity at the left.

To prove the Midpoint Formula, we need to establish two things:

1. $BM + MA = BA$, which establishes that the three points $A$, $M$, and $B$ are collinear
2. $BM = MA$, which establishes that point $M$ is midway between $A$ and $B$

> **Discover!**
>
> On a number line, $x_2$ lies to the right of $x_1$. Then $x_2 > x_1$ and the distance between points is $(x_2 - x_1)$. To find the number $a$ that is midway between $x_1$ and $x_2$, we add one-half the distance $(x_2 - x_1)$ to $x_1$. That is,
>
> $$a = x_1 + \frac{1}{2}(x_2 - x_1).$$
>
> Complete the simplification of $a$.
>
> ANSWER
>
> $a = \frac{1}{2}(x_1 + x_2)$ or $a = \frac{x_1 + x_2}{2}$

THEOREM 9.1.2: **(Midpoint Formula)** The midpoint $M$ of the line segment joining $A(x_1, y_1)$ and $B(x_2, y_2)$ has coordinates $x_M$ and $y_M$, where

$$(x_M, y_M) = \left(\frac{x_1 + x_2}{2}, \frac{y_1 + y_2}{2}\right)$$

that is,

$$M = \left(\frac{x_1 + x_2}{2}, \frac{y_1 + y_2}{2}\right)$$

EXAMPLE 5

Use the Midpoint Formula to find the midpoint of the line segment joining:

a) $(5, -1)$ and $(-1, 7)$

b) $(a, b)$ and $(c, d)$

***Solution***

a) Using the Midpoint Formula and setting $x_1 = 5$, $y_1 = -1$, $x_2 = -1$, and $y_2 = 7$, we have

$$M = \left(\frac{5 + (-1)}{2}, \frac{-1 + 7}{2}\right) = \left(\frac{4}{2}, \frac{6}{2}\right), \text{ so } M = (2, 3)$$

b) Using the Midpoint Formula and setting $x_1 = a$, $y_1 = b$, $x_2 = c$, and $y_2 = d$, we have

$$M = \left(\frac{a + c}{2}, \frac{b + d}{2}\right)$$

In part (a) of Example 5, it may be helpful to make a sketch of the segment; this will enable you to test whether your solution appears to be reasonable! In part (b), we are generalizing the coordinates in preparation for the analytic geometry proofs that appear later in the chapter. In those sections, we will choose the $x$ and $y$ values of each point in such a way as to be as general as possible. When the Midpoint Formula is used in a proof, it is a good idea to select coordinates for a point such as $(2a, 2b)$ so that division by 2 will not introduce fractions.

*Exs. 9–12*

## PROOF OF THE MIDPOINT FORMULA (OPTIONAL)

For the segment joining $P_1$ and $P_2$, we designate the midpoint by $M$, as shown in Figure 9.9. Let the coordinates of $M$ be designated by $(x_M, y_M)$. Now construct horizontal segments through $P_1$ and $M$ and vertical segments through $M$ and $P_2$ to intersect at points $A$ and $B$, as shown in Figure 9.9(b). Because $\angle A$ and $\angle B$ are right angles, $\angle A \cong \angle B$.

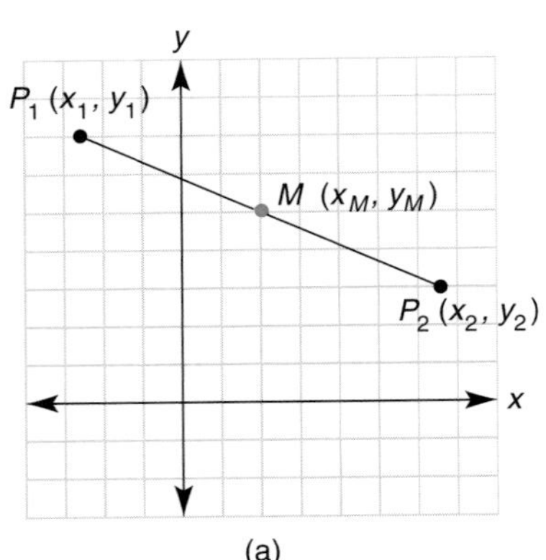


(a)

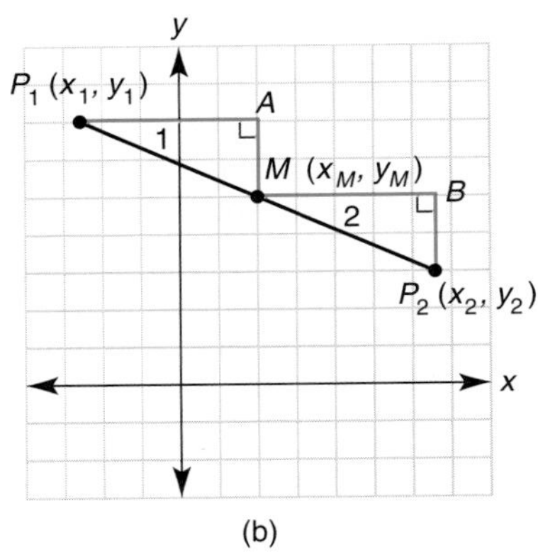


(b)

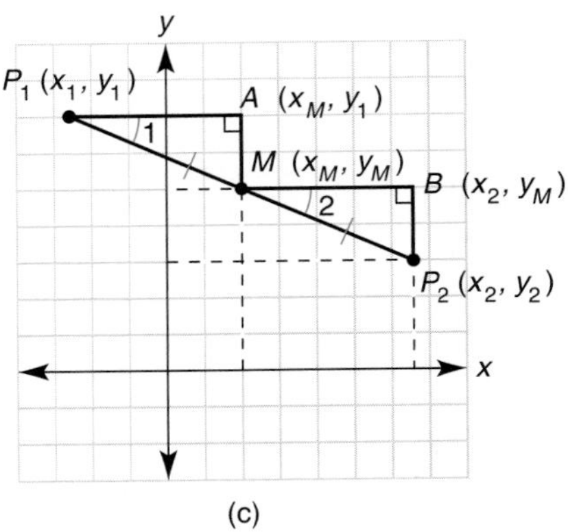


(c)

**FIGURE 9.9**

Because $\overline{P_1A}$ and $\overline{MB}$ are both horizontal, these segments are parallel. Then $\angle 1 \cong \angle 2$. With $\overline{P_1M} \cong \overline{MP_2}$ by the definition of a midpoint, we have $\triangle P_1AM \cong \triangle MBP_2$ by AAS. Because $A$ is the point $(x_M, y_1)$, we have $P_1A = x_M - x_1$. Likewise, the coordinates of $B$ are $(x_2, y_M)$, so $MB = x_2 - x_M$. Because $\overline{P_1A} \cong \overline{MB}$ by

**Discover!**

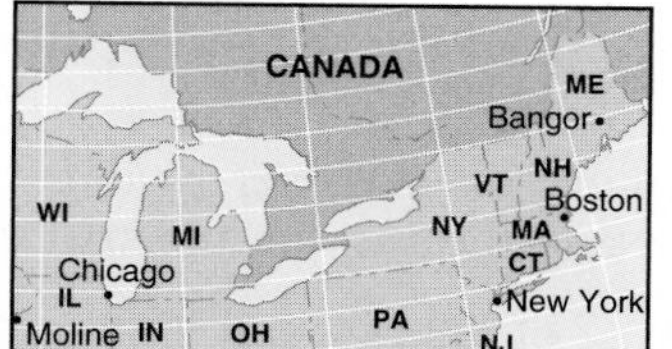


On a map, the approximate coordinates (latitude and longitude) of Bangor, Maine, are 45°N, 70°W and of Moline, Illinois, are 41°N, 90°W. If Niagara Falls has coordinates that are "mid-way" between those of Bangor and Moline, express its location in coordinates of latitude and longitude.

ANSWER

43°N, 80°W

CPCTC, we represent the common length of the segments $\overline{P_1A}$ and $\overline{MB}$ by $a$. From the first equation, $x_M - x_1 = a$, so $x_M = x_1 + a$. From the second equation, $x_2 - x_M = a$, so $x_2 = x_M + a$. Substituting $x_1 + a$ for $x_M$ in the second equation, we have

$$(x_1 + a) + a = x_2$$
$$x_1 + 2a = x_2$$

Then

$$2a = x_2 - x_1$$

so

$$a = \frac{x_2 - x_1}{2}$$

It follows that

$$\begin{aligned} x_M &= x_1 + a \\ &= x_1 + \frac{x_2 - x_1}{2} \\ &= \frac{2x_1}{2} + \frac{x_2 - x_1}{2} \\ &= \frac{x_1 + x_2}{2} \end{aligned}$$

The $y$ coordinate of the midpoint can be determined in a similar manner, to show that $y_M = \frac{y_1 + y_2}{2}$. Then

$$M = \left(\frac{x_1 + x_2}{2}, \frac{y_1 + y_2}{2}\right)$$

The following example is based upon the definitions of symmetry with respect to a line and symmetry with respect to a point. Review Section 2.6 if necessary.

EXAMPLE 6

Draw a coordinate system and plot the point $A(2, -3)$. Find point $B$ if points $A$ and $B$ have symmetry with respect to:

a) The $y$ axis b) The $x$ axis c) The origin

***Solution***

a) $(-2, -3)$ b) $(2, 3)$ c) $(-2, 3)$

*Exs. 13–15*

## 9.1 Exercises

1. Plot and then label the points $A(0, -3)$, $B(3, -4)$, $C(5, 6)$, $D(-2, -5)$, and $E(-3, 5)$.

2. Give the coordinates of each point $A$, $B$, $C$, $D$, and $E$. Also name the quadrant in which each point lies.

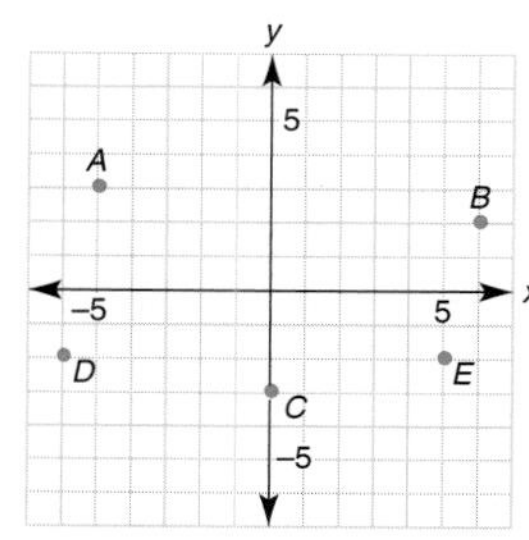


3. Find the distance between each pair of points:

   a) $(5, -3)$ and $(5, 1)$
   b) $(-3, 4)$ and $(5, 4)$
   c) $(0, 2)$ and $(0, -3)$
   d) $(-2, 0)$ and $(7, 0)$

4. If the distance between $(-2, 3)$ and $(-2, a)$ is 5 units, find all possible values of $a$.

5. If the distance between $(b, 3)$ and $(7, 3)$ is 3.5 units, find all possible values of $b$.

6. Find an expression for the distance between $(a, b)$ and $(a, c)$ if $b > c$.

7. Find the distance between each pair of points:

a) $(0, -3)$ and $(4, 0)$
b) $(-2, 5)$ and $(4, -3)$
c) $(3, 2)$ and $(5, -2)$
d) $(a, 0)$ and $(0, b)$

8. Find the distance between each pair of points:

a) $(-3, -7)$ and $(2, 5)$
b) $(0, 0)$ and $(-2, 6)$
c) $(-a, -b)$ and $(a, b)$
d) $(2a, 2b)$ and $(2c, 2d)$

9. Find the midpoint of the line segment that joins each pair of points:

a) $(0, -3)$ and $(4, 0)$
b) $(-2, 5)$ and $(4, -3)$
c) $(3, 2)$ and $(5, -2)$
d) $(a, 0)$ and $(0, b)$

10. Find the midpoint of the line segment that joins each pair of points:

a) $(-3, -7)$ and $(2, 5)$
b) $(0, 0)$ and $(-2, 6)$
c) $(-a, -b)$ and $(a, b)$
d) $(2a, 2b)$ and $(2c, 2d)$

11. Points $A$ and $B$ have *symmetry with respect to the origin O*. Find the coordinates of $B$ if $A$ is the point:

a) $(3, -4)$
b) $(0, 2)$
c) $(a, 0)$
d) $(b, c)$

12. Points $A$ and $B$ have *symmetry with respect to point* $C(2, 3)$. Find the coordinates of $B$ if $A$ is the point:

a) $(3, -4)$
b) $(0, 2)$
c) $(5, 0)$
d) $(a, b)$

13. Points $A$ and $B$ have *symmetry with respect to point* $C$. Find the coordinates of $C$ given the points:

a) $A(3, -4)$ and $B(5, -1)$
b) $A(0, 2)$ and $B(0, 6)$
c) $A(5, -3)$ and $B(2, 1)$
d) $A(2a, 0)$ and $B(0, 2b)$

14. Points $A$ and $B$ have *symmetry with respect to* the $x$ axis. Find the coordinates of $B$ if $A$ is the point:

a) $(3, -4)$
b) $(0, 2)$
c) $(0, a)$
d) $(b, c)$

15. Points $A$ and $B$ have *symmetry with respect to* the $x$ axis. Find the coordinates of $A$ if $B$ is the point:

a) $(5, 1)$
b) $(0, 5)$
c) $(2, a)$
d) $(b, c)$

16. Points $A$ and $B$ have *symmetry with respect to the vertical line where* $x = 2$. Find the coordinates of $A$ if $B$ is the point:

a) $(5, 1)$
b) $(0, 5)$
c) $(-6, a)$
d) $(b, c)$

17. Points $A$ and $B$ have *symmetry with respect to* the $y$ axis. Find the coordinates of $A$ if $B$ is the point:

a) $(3, -4)$
b) $(2, 0)$
c) $(a, 0)$
d) $(b, c)$

18. Points $A$ and $B$ have *symmetry with respect to* either the $x$ axis or the $y$ axis. Name the axis of symmetry for:

a) $A(3, -4)$ and $B(3, 4)$
b) $A(2, 0)$ and $B(-2, 0)$
c) $A(3, -4)$ and $B(-3, -4)$
d) $A(a, b)$ and $B(a, -\text{b})$

19. Points $A$ and $B$ have *symmetry with respect to* a vertical line ($x = a$) or a horizontal line ($y = b$). Give an equation like $x = 3$ for the axis of symmetry for:

a) $A(3, -4)$ and $B(5, -4)$
b) $A(a, 0)$ and $B(a, 2\text{b})$
c) $A(7, -4)$ and $B(-3, -4)$
d) $A(\text{a}, 7)$ and $B(a, -1)$

*In Exercises 20 to 22, apply the Midpoint Formula.*

20. $M(3, -4)$ is the midpoint of $\overline{AB}$, in which $A$ is the point $(-5, 7)$. Find the coordinates of $B$.

21. $M(2.1, -5.7)$ is the midpoint of $\overline{AB}$, in which $A$ is the point $(1.7, 2.3)$. Find the coordinates of $B$.

22. A circle has its center at the point $(-2, 3)$. If one endpoint of a diameter is at $(3, -5)$, find the other endpoint of the diameter.

23. A rectangle $ABCD$ has three of its vertices at $A(2, -1)$, $B(6, -1)$, and $C(6, 3)$. Find the fourth vertex $D$ and the area of rectangle $ABCD$.

24. A rectangle $MNPQ$ has three of its vertices at $M(0, 0)$, $N(a, 0)$, and $Q(0, b)$. Find the fourth vertex $P$ and the area of the rectangle $MNPQ$.

25. Use the Distance Formula to determine the type of triangle that has these vertices:

a) $A(0, 0)$, $B(4, 0)$, and $C(2, 5)$
b) $D(0, 0)$, $E(4, 0)$, and $F(2, 2\sqrt{3})$
c) $G(-5, 2)$, $H(-2, 6)$, and $K(2, 3)$

26. Use the method of Example 4 to find the equation of the line that describes all points equidistant from the points $(-3, 4)$ and $(3, 2)$.

27. Use the method of Example 4 to find the equation of the line that describes all points equidistant from the points $(1, 2)$ and $(4, 5)$.

28. For coplanar points $A$, $B$, and $C$, suppose that you have used the Distance Formula to show that $AB = 5$, $BC = 10$, and $AC = 15$. What can you conclude regarding points $A$, $B$, and $C$?

29. If two vertices of an equilateral triangle are at (0, 0) and $(2a, 0)$, what point is the third vertex?

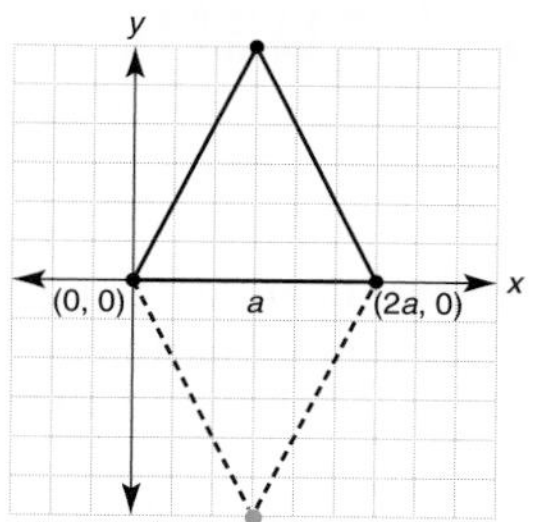


30. The rectangle whose vertices are $A(0, 0)$, $B(a, 0)$, $C(a, b)$, and $D(0, b)$ is shown. Use the Distance Formula to draw a conclusion concerning the lengths of the diagonals $\overline{AC}$ and $\overline{BD}$.

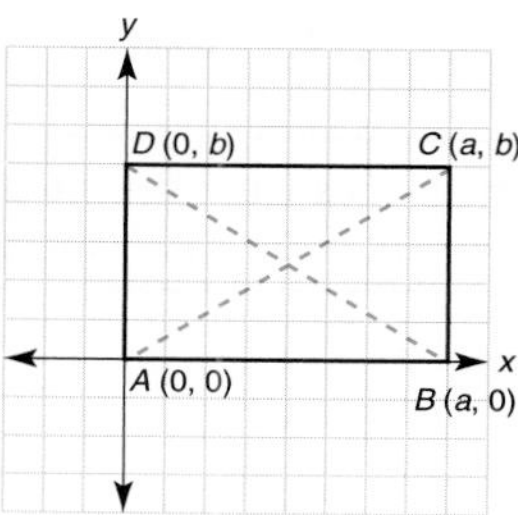


*31. There are two points on the $y$ axis that are located a distance of 6 units from the point (3, 1). Determine the coordinates of each point.

*32. There are two points on the $x$ axis that are located a distance of 6 units from the point (3, 1). Determine the coordinates of each point.

33. The triangle that has vertices at $M(-4, 0)$, $N(3, -1)$, and $Q(2, 4)$ has been boxed in as shown. Find the area of $\triangle MNQ$.

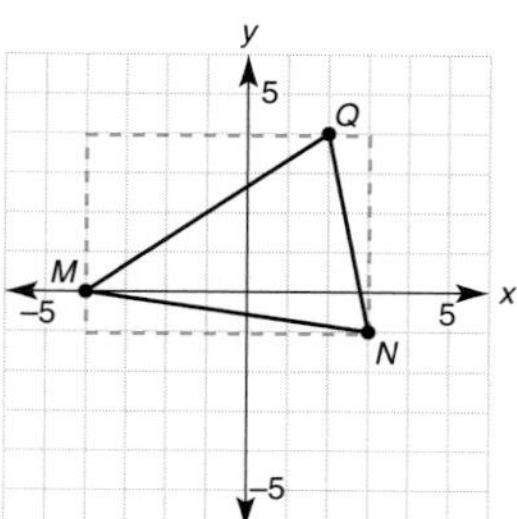


*Exercises 33, 34*

34. Use the method suggested in Exercise 33 to find the area of $\triangle RST$, with $R(-2, 4)$, $S(-1, -2)$, and $T(6, 5)$.

35. Determine the area of $\triangle ABC$ if $A = (2, 1)$, $B = (5, 3)$, and $C$ is the reflection of $B$ across the $x$ axis.

36. Find the area of $\triangle ABC$ in Exercise 35, but assume that $C$ is the reflection of $B$ across the $y$ axis.

*For Exercises 37 to 42, refer to Examples 5 and 6 of Section 8.3.*

37. Find the exact volume of the right circular cone that results when the triangular region with vertices at (0, 0), (5, 0), and (0, 9) is rotated about the

a) $x$ axis. b) $y$ axis.

38. Find the exact volume of the solid that results when the triangular region with vertices at (0, 0), (6, 0), and (6, 4) is rotated about the

a) $x$ axis. b) $y$ axis.

39. Find the exact volume of the solid formed when the rectangular region with vertices at (0, 0), (6, 0), (6, 4), and (0, 4) is revolved about the

a) $x$ axis. b) $y$ axis.

40. Find the exact volume of the solid formed when the region bounded in Quadrant I by the axes and the lines $x = 9$ and $y = 5$ is revolved about the

a) $x$ axis. b) $y$ axis.

41. Find the exact lateral area of each solid in Exercise 40.

*42. Find the volume of the solid formed when the triangular region having vertices at (2, 0), (4, 0), and (2, 4) is rotated about the $y$ axis.

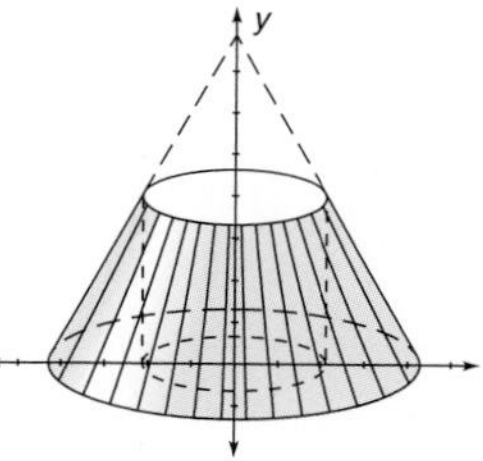


43. Following a 90° counterclockwise rotation about the origin, the image of $A(3, 1)$ is point $B(-1, 3)$. What is the image of point $A$ following a counterclockwise rotation of

a) 180° about the origin?
b) 270° about the origin?
c) 360° about the origin?

44. Consider the point $C(a, b)$. What is the image of $C$ after a counterclockwise rotation of

a) 90° about the origin?
b) 180° about the origin?
c) 360° about the origin?

45. Given the point $D(3, 2)$, find the image of $D$ after a counterclockwise rotation of

a) 90° about the point $E(3, 4)$.
b) 180° about the point $F(4, 5)$.
c) 360° about the point $G(a, b)$.

**KEY CONCEPTS**

Graphs of Equations • $x$ Intercept • $y$ Intercept • Slope • Slope Formula • Negative Reciprocal

# 9.2 Graphs of Linear Equations and Slope

In Section 9.1, we were reminded that the general form of the equation of a line is $Ax + By = C$ (where $A$ and $B$ do not both equal 0). Some examples of linear equations are $2x + 3y = 12$, $3x - 4y = 12$, and $3x = -6$; as we shall see, the graph of each of these equations is a line.

## THE GRAPH OF AN EQUATION

DEFINITION: The **graph of an equation** is the set of all points $(x, y)$ in the rectangular coordinate system whose ordered pairs satisfy the equation.

### EXAMPLE 1

Draw the graph of the equation $2x + 3y = 12$.

***Solution***

We begin by completing a table. It is convenient to use one point for which $x = 0$, a second point for which $y = 0$, and a third point as a check for collinearity.

$$x = 0 \rightarrow 2(0) + 3y = 12 \rightarrow y = 4$$
$$y = 0 \rightarrow 2x + 3(0) = 12 \rightarrow x = 6$$
$$x = 3 \rightarrow 2(3) + 3y = 12 \rightarrow y = 2$$

| $x$ | $y$ | $(x, y)$ |
|---|---|---|
| 0 | 4 | (0, 4) |
| 6 | 0 | (6, 0) |
| 3 | 2 | (3, 2) |

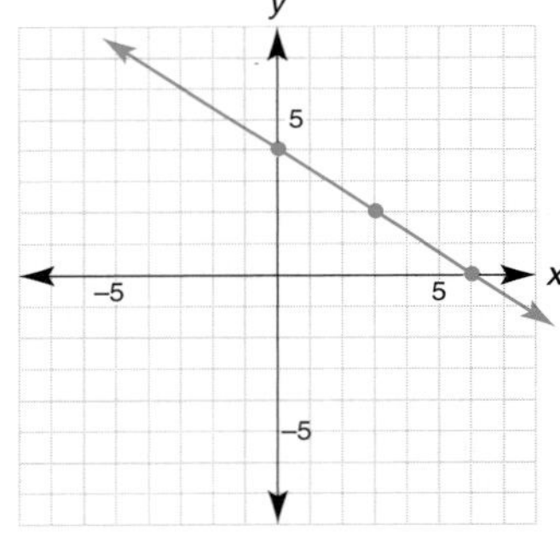


**FIGURE 9.10**

*Exs. 1–3*

Upon plotting the third point, we see that the three points are collinear; the graph of a linear equation must be a straight line, as shown in Figure 9.10.

*Technology Exploration*

Use a graphing calculator if one is available.

1) To graph $2x + 3y = 12$, solve for $y$.
2) Enter your result from part (1) as the value of $Y_1$. $[Y_1 = -(\frac{2}{3})x + 4]$
3) Now GRAPH to see the line of Figure 9.10.

Because the graph in Example 1 is a locus, every point on the line must also satisfy the given equation. It is easy to see that the point $(-3, 6)$ lies on the line shown in Figure 9.10. Notice that this ordered pair also satisfies the equation $2x + 3y = 12$; that is, $2(-3) + 3(6) = 12$ or $-6 + 18 = 12$.

For the equation in Example 1, the number 6 is known as the ***x* intercept** because (6, 0) is the point at which the graph crosses the $x$ axis; similarly, the number 4 is known as the ***y* intercept.** Most linear equations have two intercepts; these are generally represented by $a$ (the $x$ intercept) and $b$ (the $y$ intercept).

For the equation $Ax + By = C$, we determine the
a) $x$ intercept by choosing $y = 0$. Solve for $x$.
b) $y$ intercept by choosing $x = 0$. Solve for $y$.

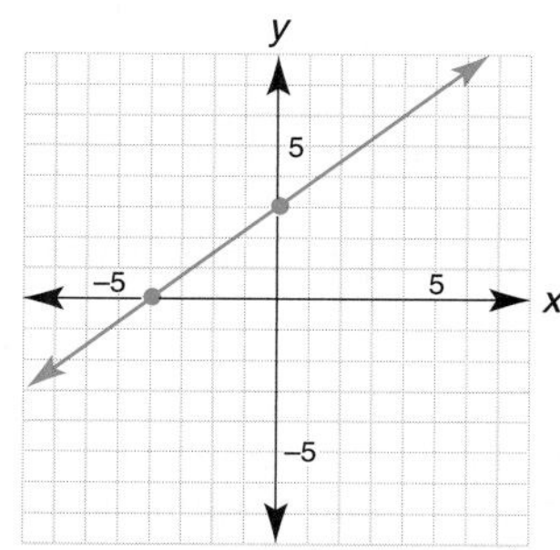


FIGURE 9.11

EXAMPLE 2

Find the $x$ and $y$ intercepts of the equation $3x - 4y = -12$, and use them to graph the equation.

***Solution***

The $x$ intercept is found when $y = 0$; $3x - 4(0) = -12$, so $x = -4$. The $x$ intercept is $a = -4$, so $(-4, 0)$ is on the graph. The $y$ intercept results when $x = 0$; $3(0) - 4y = -12$, so $y = 3$. The $y$ intercept is $b = 3$, so $(0, 3)$ is on the graph. Once the points $(-4, 0)$ and $(0, 3)$ are plotted, the graph can be completed by drawing the line through these points (see Figure 9.11).

As we shall see in Example 3, a linear equation may have only one intercept. Is it possible for a linear equation to have no intercepts at all?

EXAMPLE 3

Draw the graphs of the following equations:

a) $x = -2$  b) $y = 3$

***Solution***

First note that each equation is a linear equation:

$$x = -2 \text{ is equivalent to } (1 \cdot x) + (0 \cdot y) = -2$$
$$y = 3 \text{ is equivalent to } (0 \cdot x) + (1 \cdot y) = 3$$

a) The equation $x = -2$ claims that the value of $x$ is $-2$ regardless of the value of $y$; this leads to the following table:

| $x$ | $y$ | $\rightarrow$ | $(x, y)$ |
|---|---|---|---|
| $-2$ | $-2$ | $\rightarrow$ | $(-2, -2)$ |
| $-2$ | $0$ | $\rightarrow$ | $(-2, 0)$ |
| $-2$ | $5$ | $\rightarrow$ | $(-2, 5)$ |

b) The equation $y = 3$ claims that the value of $y$ is 3 regardless of the value of $x$; this leads to the following table:

| $x$ | $y$ | $\rightarrow$ | $(x, y)$ |
|---|---|---|---|
| $-4$ | $3$ | $\rightarrow$ | $(-4, 3)$ |
| $0$ | $3$ | $\rightarrow$ | $(0, 3)$ |
| $5$ | $3$ | $\rightarrow$ | $(5, 3)$ |

The graphs of the equations are shown in Figure 9.12.

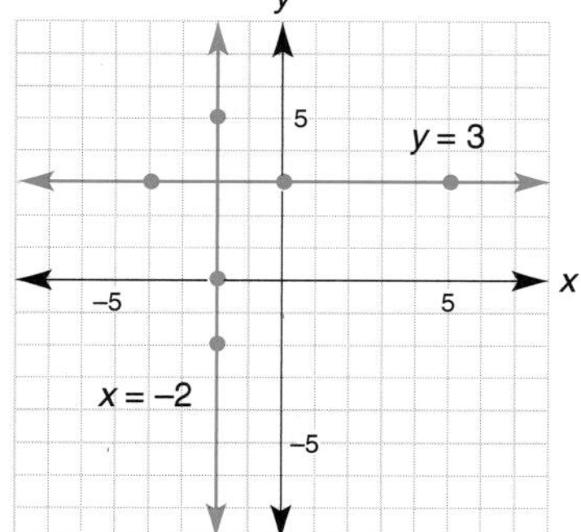


FIGURE 9.12

*Exs. 4–8*

NOTE: When an equation can be written in the form $x = a$ (for constant $a$), its graph is the vertical line containing the point $(a, 0)$. When an equation can be written in the form $y = b$ (for constant $b$), its graph is the horizontal line containing the point $(0, b)$.

## THE SLOPE OF A LINE

Most lines are oblique—that is, the line is neither horizontal nor vertical. Especially for oblique lines, it is convenient to describe the amount of "slant" by a number called the *slope* of the line.

> **DEFINITION: (Slope Formula)** The slope of the line that contains the points $(x_1, y_1)$ and $(x_2, y_2)$ is given by
>
> $$m = \frac{y_2 - y_1}{x_2 - x_1} \quad \text{for } x_1 \neq x_2$$

NOTE: When $x_1 = x_2$, the denominator of the Slope Formula becomes 0 and we say that the slope of the line is undefined.

Whereas the uppercase italic $M$ means midpoint, we use the lowercase italic $m$ for slope. Other terms used to describe the slope of a line include *pitch* and *grade*. A carpenter may say that a roofline has a $\frac{5}{12}$ pitch. [See Figure 9.13(a).] In constructing a stretch of roadway, an engineer may say that this part of the roadway has a grade of $\frac{3}{100}$, or 3 percent. [See Figure 9.13(b).]

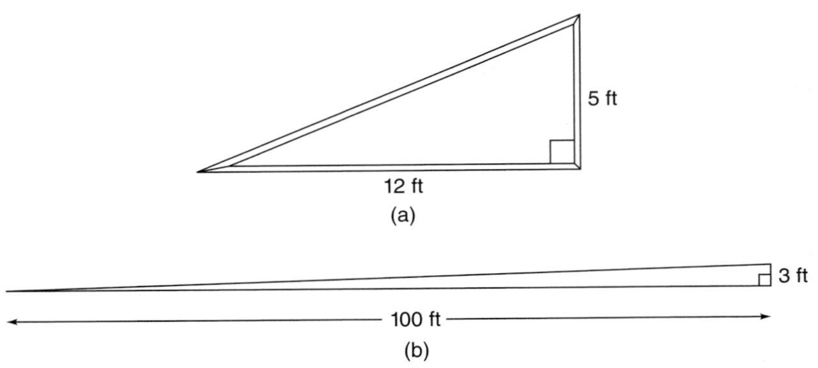


**FIGURE 9.13**

Whether in geometry, carpentry, or engineering, the **slope** of a line is a number. The slope is the ratio of the change along the vertical to the change along the horizontal, for any two points on the line in question. A line that "rises" from left to right has a *positive* slope, and a line that "falls" from left to right has a *negative* slope. (See Figure 9.14.)

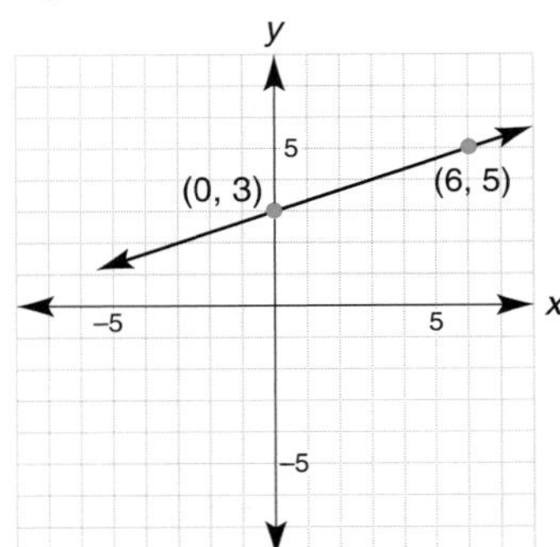


$$m = \frac{y_2 - y_1}{x_2 - x_1}$$

$$m = \frac{5 - 3}{6 - 0} = \frac{2}{6} = \frac{1}{3}$$

(a)

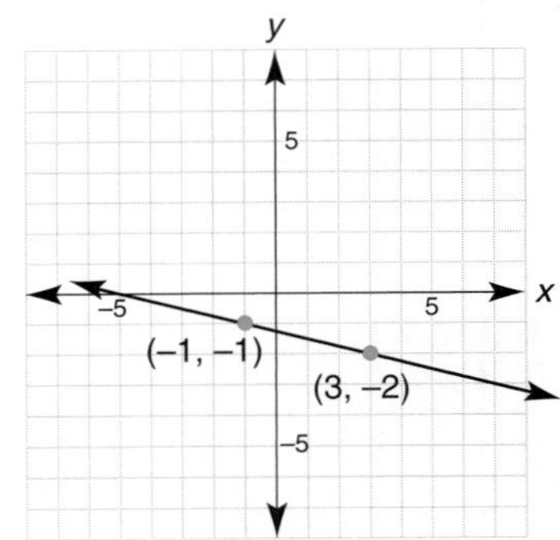


$$m = \frac{y_2 - y_1}{x_2 - x_1}$$

$$m = \frac{-2 - (-1)}{3 - (-1)} = -\frac{1}{4}$$

(b)

**FIGURE 9.14**

Any horizontal line has slope 0; any vertical line has an undefined slope. Figure 9.15 shows an example of each of these types of lines.

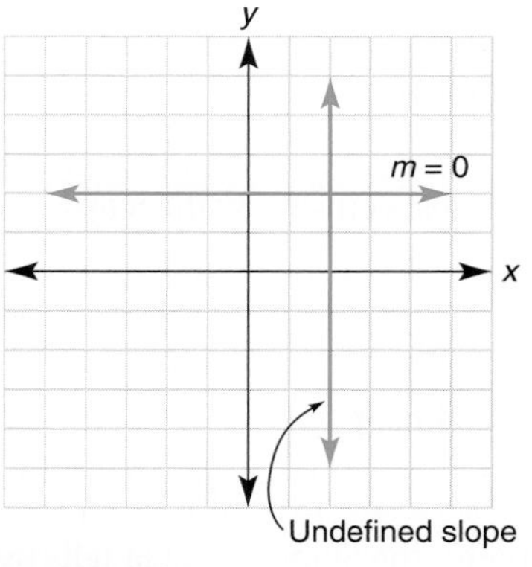


**FIGURE 9.15**

EXAMPLE 4

Without graphing, find the slope of the line that contains:

a) (2, 2) and (5, 3)    b) (1, −1) and (1, 3)

***Solution***

a) Using the Slope Formula and choosing $x_1 = 2$, $y_1 = 2$, $x_2 = 5$, and $y_2 = 3$, we have

$$m = \frac{3-2}{5-2} = \frac{1}{3}$$

NOTE: If drawn, this line will slant upward from left to right.

b) Let $x_1 = 1$, $y_1 = -1$, $x_2 = 1$, and $y_2 = 3$. Then we calculate

$$m = \frac{3-(-1)}{1-1} = \frac{4}{0}$$

which is undefined.

NOTE: If drawn, the line in part (b) will be vertical because the $x$ coordinates are the same.

Reminder

In the Slope Formula, be sure to write the difference in values of $y$ in the numerator; that is,

$$m = \frac{y_2 - y_1}{x_2 - x_1}$$

*Exs. 9–12*

The slope of a line is unique; that is, the slope does not change when:

1. The order of the two points is reversed in the Slope Formula.
2. Different points on the line are selected.

The first situation is true due to the fact that $\frac{-a}{-b} = \frac{a}{b}$. The second situation is more difficult to explain, but depends on similar triangles.

For an explanation of point 2, consider Figure 9.16, in which points $P_1$, $P_2$, and $P_3$ are collinear. What we wish to show is that the slope of line $\ell$ is the same whether $P_1$ and $P_2$, or $P_2$ and $P_3$, are used in the Slope Formula. If horizontal and vertical segments are drawn as shown in Figure 9.16, we can show that triangles $P_1P_2A$ and $P_2P_3B$ are similar.

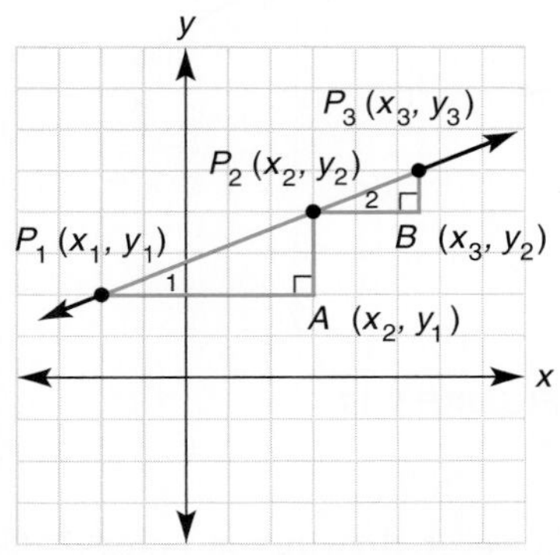


**FIGURE 9.16**

The similarity follows from the facts that $\angle 1 \cong \angle 2$ (because $\overline{P_1A} \parallel \overline{P_2B}$) and that $\angle A$ and $\angle B$ are right angles. Then $\frac{P_2A}{P_3B} = \frac{P_1A}{P_2B}$ because these are corresponding sides of similar triangles. By interchanging the means, we have $\frac{P_2A}{P_1A} = \frac{P_3B}{P_2B}$. But

$$\frac{P_2A}{P_1A} = \frac{y_2 - y_1}{x_2 - x_1} \quad \text{and} \quad \frac{P_3B}{P_2B} = \frac{y_3 - y_2}{x_3 - x_2}$$

so

$$\frac{y_2 - y_1}{x_2 - x_1} = \frac{y_3 - y_2}{x_3 - x_2}$$

Thus the slope is not changed by our having used either pair of points. For that matter, a fourth point could have been shown, in which case the slope of the line would not be changed if $P_1$ and $P_2$, or $P_3$ and $P_4$, were used; in every case, the slopes agree because of similar triangles.

In summary, if points $P_1$, $P_2$, and $P_3$ are collinear, then the slopes of $\overline{P_1P_2}$, $\overline{P_1P_3}$, and $\overline{P_2P_3}$ are the same. The converse of this statement, which is also true, is applied in Example 5.

EXAMPLE 5

Are the points $A(2, -3)$, $B(5, 1)$, and $C(-4, -11)$ collinear?

***Solution***

Let $m_{\overline{AB}}$ and $m_{\overline{BC}}$ represent the slopes of $\overline{AB}$ and $\overline{BC}$, respectively. By the Slope Formula, we have

$$m_{\overline{AB}} = \frac{1 - (-3)}{5 - 2} = \frac{4}{3} \quad \text{and} \quad m_{\overline{BC}} = \frac{-11 - 1}{-4 - 5} = \frac{-12}{-9} = \frac{4}{3}$$

Because $m_{\overline{AB}} = m_{\overline{BC}}$, it follows that $A$, $B$, and $C$ are collinear.

As we trace a line from one point to a second point, the Slope Formula tells us that

$$m = \frac{\text{change in } y}{\text{change in } x} \quad \text{or} \quad m = \frac{\text{vertical change}}{\text{horizontal change}}$$

This interpretation is used in Example 6.

EXAMPLE 6

Draw the line through $(-1, 5)$ with slope $m = -\frac{2}{3}$.

***Solution***

First we plot the point $(-1, 5)$. The slope can be written as $m = \frac{-2}{3}$. Thus we let the change in $y$ from the first to the second point be $-2$ while the change in $x$ is 3. From the first point $(-1, 5)$, we locate the second point by moving 2 units down and 3 units to the right. The line is then drawn as shown in Figure 9.17.

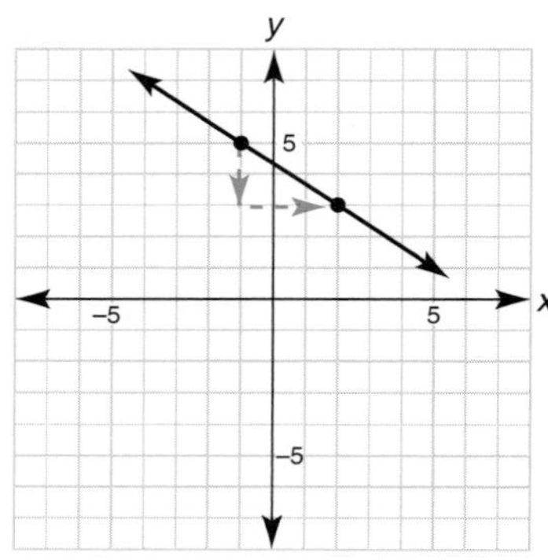


**FIGURE 9.17**

*Exs. 13–15*

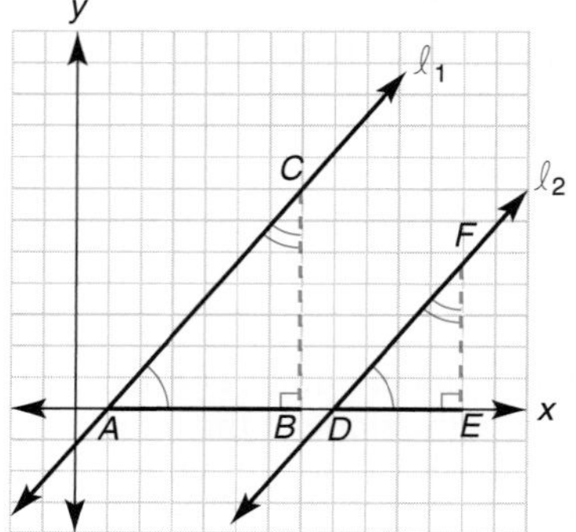


**FIGURE 9.18**

Two theorems are now stated without proof. However, drawings are provided, and these are to be used in the following exercises. Each proof depends on the fact that similar triangles are created through the use of the auxiliary segments included in the drawings.

THEOREM 9.2.1: If two nonvertical lines are parallel, then their slopes are equal.

*Alternative Form:* If $\ell_1 \parallel \ell_2$, then $m_1 = m_2$.

In Figure 9.18, note that $\overline{AC} \parallel \overline{DF}$. Also, $\overline{AB}$ and $\overline{DE}$ are horizontal, and $\overline{BC}$ and $\overline{EF}$ are auxiliary vertical segments. In the proof of Theorem 9.2.1, the goal is to show that $m_{\overline{AC}} = m_{\overline{DF}}$. The converse of Theorem 9.2.1 is also true; that is, if $m_1 = m_2$, then $\ell_1 \parallel \ell_2$.

> THEOREM 9.2.2: If two lines (neither horizontal nor vertical) are perpendicular, then the product of their slopes is $-1$.
>
> *Alternative Form:* If $\ell_1 \perp \ell_2$, then $m_1 \cdot m_2 = -1$.

In Figure 9.19, auxiliary segments have been included. To prove Theorem 9.2.2, we need to show that $m_{\overline{AC}} \cdot m_{\overline{CE}} = -1$. When the product of the slopes is $-1$, we say that the slopes are **negative reciprocals.** The converse of Theorem 9.2.2 is also true; if $m_1 \cdot m_2 = -1$, then $\ell_1 \perp \ell_2$.

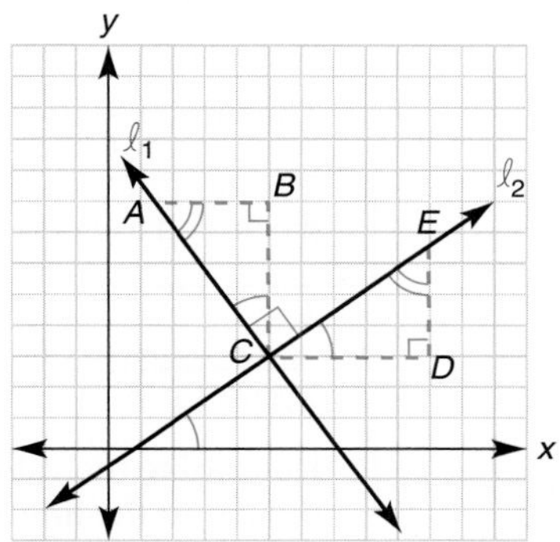


**FIGURE 9.19**

EXAMPLE 7

Given the points $A(-2, 3)$, $B(2, 1)$, $C(-1, 8)$, and $D(7, 3)$, are $\overline{AB}$ and $\overline{CD}$ parallel, perpendicular, or neither? (See Figure 9.20.)

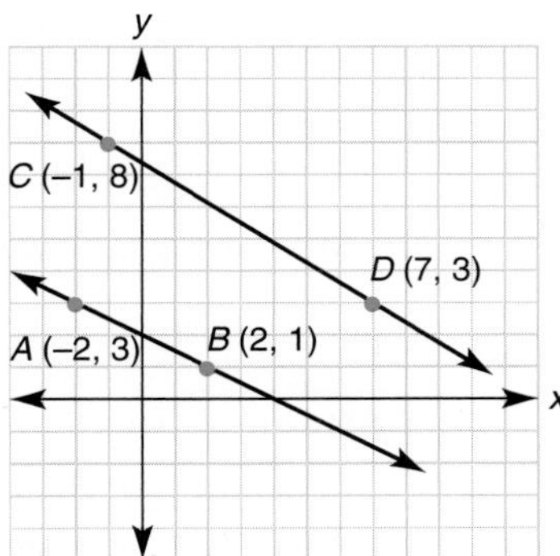


**FIGURE 9.20**

***Solution***

$$m_{\overline{AB}} = \frac{1 - 3}{2 - (-2)} = \frac{-2}{4} = -\frac{1}{2}$$

$$m_{\overline{CD}} = \frac{3 - 8}{7 - (-1)} = \frac{-5}{8} \text{ or } -\frac{5}{8}$$

Because $m_{\overline{AB}} \neq m_{\overline{CD}}$, $\overline{AB} \nparallel \overline{CD}$. The slopes are not negative reciprocals, so $\overline{AB}$ is not perpendicular to $\overline{CD}$. Neither relationship holds for $\overline{AB}$ and $\overline{CD}$.

*Exs. 16–18*

In Example 7, it was worthwhile to sketch the lines described. It is apparent from Figure 9.20 that no special relationship exists between the lines. A sketch may help show that a relationship does *not* exist, but sketching is not a dependable method for showing that lines are parallel or perpendicular.

### EXAMPLE 8

Are the lines that are the graphs of $2x + 3y = 6$ and $3x - 2y = 12$ parallel, perpendicular, or neither?

***Solution***

Because $2x + 3y = 6$ contains the points (3, 0) and (0, 2), its slope is $\frac{2 - 0}{0 - 3} = -\frac{2}{3}$. The line $3x - 2y = 12$ contains (0, −6) and (4, 0); thus its slope is equal to $\frac{0 - (-6)}{4 - 0} = \frac{6}{4}$ or $\frac{3}{2}$. Because the product of the slopes is $-\frac{2}{3} \cdot \frac{3}{2}$, or $-1$, the lines described are perpendicular.

### EXAMPLE 9

Determine the value of $a$ for which the line through (2, −3) and (5, $a$) is perpendicular to the line $3x + 4y = 12$.

***Solution***

The line $3x + 4y = 12$ contains the points (4, 0) and (0, 3); this line has the slope

$$m = \frac{3 - 0}{0 - 4} = -\frac{3}{4}$$

For the two lines to be perpendicular, the second line must have slope $\frac{4}{3}$. Using the Slope Formula, we find that the second line has the slope

$$\frac{a - (-3)}{5 - 2}$$

so

$$\frac{a + 3}{3} = \frac{4}{3}$$

Multiplying by 3, we obtain $a + 3 = 4$, so $a = 1$.

### EXAMPLE 10

In Figure 9.21, show that the quadrilateral $ABCD$, with vertices $A(0, 0)$, $B(a, 0)$, $C(a, b)$, and $D(0, b)$, is a rectangle.

***Solution***

By applying the Slope Formula, we see that

$$m_{\overline{AB}} = \frac{0 - 0}{a - 0} = 0$$

and

$$m_{\overline{DC}} = \frac{b - b}{a - 0} = 0$$

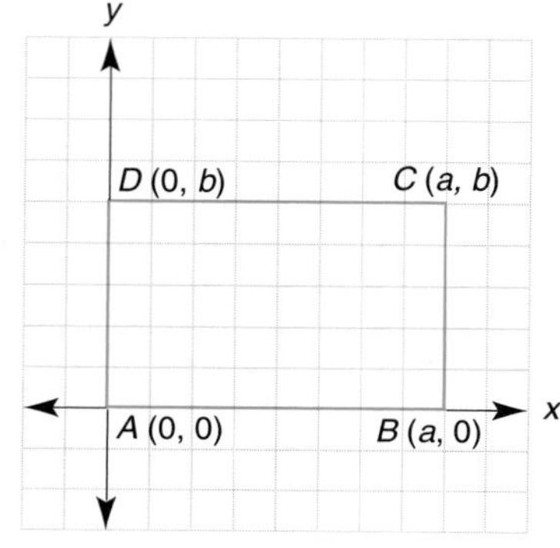


**FIGURE 9.21**

Then $\overline{AB}$ and $\overline{DC}$ are horizontal and therefore are parallel to each other.

For $\overline{DA}$ and $\overline{CB}$, the slopes are undefined because the denominators in the Slope Formula both equal 0. Then $\overline{DA}$ and $\overline{CB}$ are both vertical and therefore parallel to each other.

Thus $ABCD$ is a parallelogram. With $\overline{AB}$ being horizontal and $\overline{DA}$ vertical, it follows that $\overline{DA} \perp \overline{AB}$. Therefore, $ABCD$ is a rectangle by definition.

*Ex. 19*

## 9.2 Exercises

*In Exercises 1 to 8, draw the graph of each equation. Name any intercepts.*

1. $3x + 4y = 12$
2. $3x + 5y = 15$
3. $x - 2y = 5$
4. $x - 3y = 4$
5. $2x + 6 = 0$
6. $3y - 9 = 0$
7. $\frac{1}{2}x + y = 3$
8. $\frac{2}{3}x - y = 1$
9. Find the slopes of the lines containing:
   a) (2, −3) and (4, 5)
   b) (3, −2) and (3, 7)
   c) (1, −1) and (2, −2)
   d) (−2.7, 5) and (−1.3, 5)
   e) $(a, b)$ and $(c, d)$
   f) $(a, 0)$ and $(0, b)$
10. Find the slopes of the lines containing:
    a) (3, −5) and (−1, 2)
    b) (−2, −3) and (−5, −7)
    c) $(2\sqrt{2}, -3\sqrt{6})$ and $(3\sqrt{2}, 5\sqrt{6})$
    d) $(\sqrt{2}, \sqrt{7})$ and $(\sqrt{2}, \sqrt{3})$
    e) $(a, 0)$ and $(a + b, c)$
    f) $(a, b)$ and $(-b, -a)$
11. Find $x$ so that $\overline{AB}$ has slope $m$, where:
    a) $A$ is (2, −3), $B$ is $(x, 5)$, and $m = 1$
    b) $A$ is $(x, -1)$, $B$ is (3, 5), and $m = -0.5$
12. Find $y$ so that $\overline{CD}$ has slope $m$, where:
    a) $C$ is (2, −3), $D$ is $(4, y)$, and $m = \frac{3}{2}$
    b) $C$ is (−1, −4), $D$ is $(3, y)$, and $m = -\frac{2}{3}$
13. Are these points collinear?
    a) $A(-2, 5)$, $B(0, 2)$, and $C(4, -4)$
    b) $D(-1, -1)$, $E(2, -2)$, and $F(5, -5)$
14. Are these points collinear?
    a) $A(-1, -2)$, $B(3, 2)$, and $C(5, 5)$
    b) $D(a, c - d)$, $E(b, c)$, and $F(2b - a, c + d)$
15. Parallel lines $\ell_1$ and $\ell_2$ have slopes $m_1$ and $m_2$, respectively. Find $m_2$ if $m_1$ equals:
    a) $\frac{3}{4}$ b) $-\frac{5}{3}$ c) $-2$ d) $\frac{a-b}{c}$
16. Parallel lines $\ell_1$ and $\ell_2$ have slopes $m_1$ and $m_2$, respectively. Find $m_2$ if $m_1$ equals:
    a) $\frac{4}{5}$ b) $-\frac{1}{5}$ c) 3 d) $\frac{f+g}{h+j}$
17. Perpendicular lines $\ell_1$ and $\ell_2$ have slopes $m_1$ and $m_2$, respectively. Find $m_2$ if $m_1$ equals:
    a) $-\frac{1}{2}$ b) $\frac{3}{4}$ c) 3 d) $\frac{f+g}{h+j}$
18. Perpendicular lines $\ell_1$ and $\ell_2$ have slopes $m_1$ and $m_2$, respectively. Find $m_2$ if $m_1$ equals:
    a) 5 b) $-\frac{5}{3}$ c) $-\frac{1}{2}$ d) $\frac{a-b}{c}$

*In Exercises 19 to 22, state whether the lines are parallel, perpendicular, the same, or none of these.*

19. $2x + 3y = 6$ and $2x - 3y = 12$
20. $2x + 3y = 6$ and $4x + 6y = -12$
21. $2x + 3y = 6$ and $3x - 2y = 12$
22. $2x + 3y = 6$ and $4x + 6y = 12$
23. Find $x$ such that the points $A(x, 5)$, $B(2, 3)$, and $C(4, -5)$ are collinear.
24. Find $a$ such that the points $A(1, 3)$, $B(4, 5)$, and $C(a, a)$ are collinear.
25. Find $x$ such that the line through (2, −3) and (3, 2) is perpendicular to the line through (−2, 4) and $(x, -1)$.

26. Find $x$ such that the line through (2, −3) and (3, 2) is parallel to the line through (−2, 4) and ($x$, −1).

*In Exercises 27 to 32, draw the line described.*

27. Through (3, −2) and with $m = 2$

28. Through (−2, −5) and with $m = \frac{5}{7}$

29. With $y$ intercept 5 and with $m = -\frac{3}{4}$

30. With $x$ intercept −3 and with $m = 0.25$

31. Through (−2, 1) and parallel to the line $2x - y = 6$

32. Through (−2, 1) and perpendicular to the line that has intercepts $a = -2$ and $b = 3$

33. Use slopes to decide whether the triangle with vertices at (6, 5), (−3, 0), and (4, −2) is a right triangle.

34. If $A(2, 2)$, $B(7, 3)$, and $C(4, x)$ are the vertices of a right triangle with right angle $C$, find the value of $x$.

* 35. If (2, 3), (5, −2), and (7, 2) are three vertices (not necessarily consecutive) of a parallelogram, find the possible locations of the fourth vertex.

36. Three vertices of rectangle $ABCD$ are $A(-5, 1)$, $B(-2, -3)$, and $C(6, y)$. Find the value of $y$ and also the fourth vertex.

37. Show that quadrilateral $RSTV$ is an isosceles trapezoid.

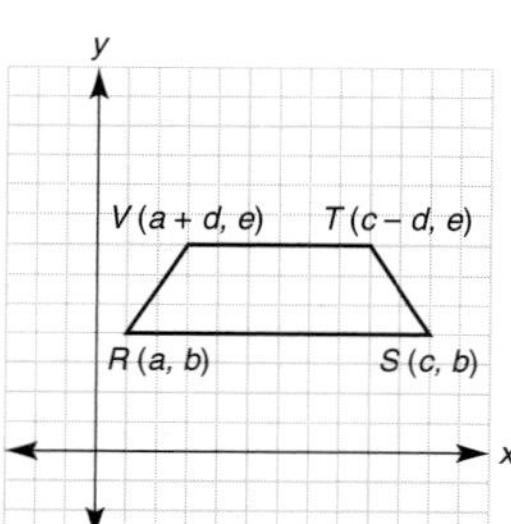


38. Show that quadrilateral $ABCD$ is a parallelogram.

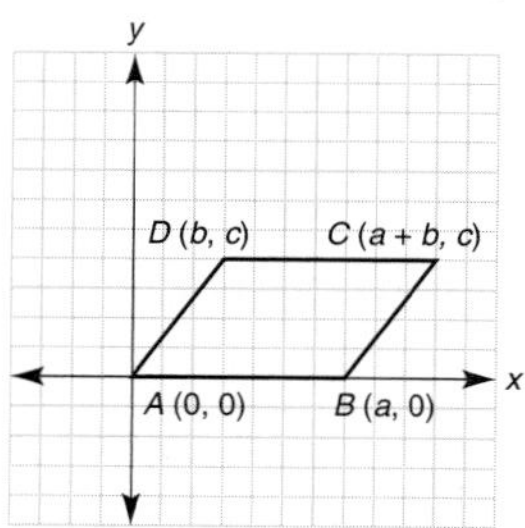


39. Quadrilateral $EFGH$ has the vertices $E(0,0)$, $F(a,0)$, $G(a + b, c)$, and $H(2b, 2c)$. Verify that $EFGH$ is a trapezoid by showing that the slopes of two sides are equal.

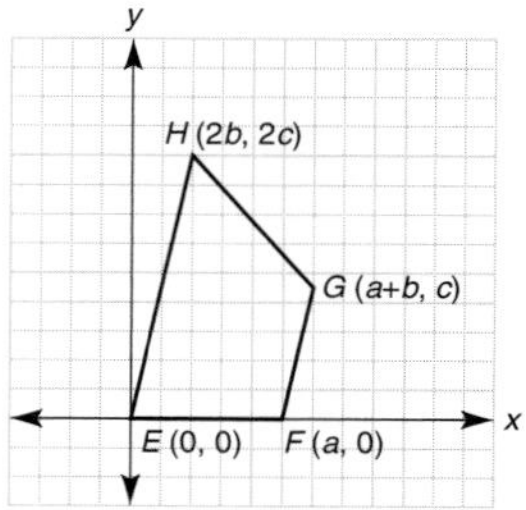


40. Find an equation involving $a, b, c, d$, and $e$ if $\overleftrightarrow{AC} \perp \overleftrightarrow{BC}$. (HINT: Use slopes.)

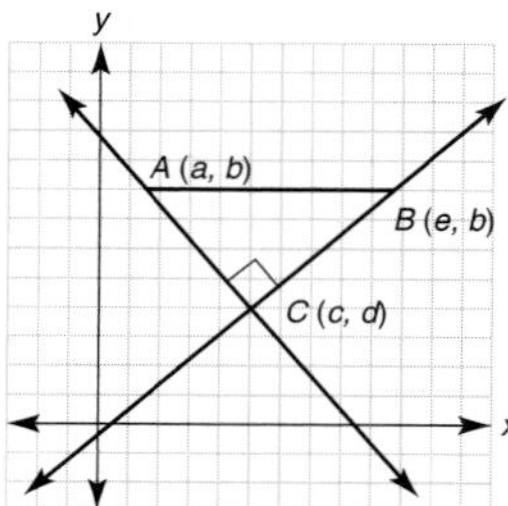


41. Prove that if two nonvertical lines are parallel, then their slopes are equal. (HINT: See Figure 9.18.)

* 42. Prove that if two lines (neither horizontal nor vertical) are perpendicular, then the product of their slopes is −1. (HINT: See Figure 9.19. You need to show and use the fact that $\triangle ABC \sim \triangle EDC$.)

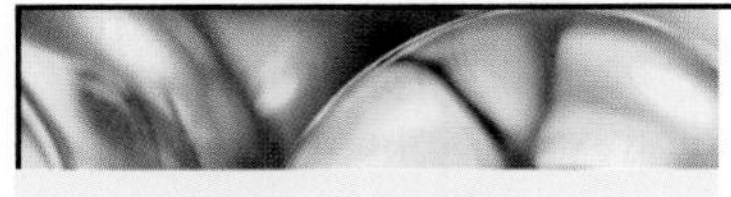

# 9.3 Preparing to Do Analytic Proofs

**KEY CONCEPTS**
Formulas and Relationships • Placement of Figure

In this section, our goal is to lay the groundwork for analytic proofs of geometric theorems. An analytic proof requires the use of the coordinate system and the application of the formulas found in earlier sections of this chapter. Because of the need for these formulas, a summary follows. Be sure that you have these formulas memorized and know when and how to use them.

Exs. 1–6

FORMULAS OF ANALYTIC GEOMETRY

| | |
|---|---|
| Distance | $d = \sqrt{(x_2 - x_1)^2 + (y_2 - y_1)^2}$ |
| Midpoint | $M = \left(\frac{x_1 + x_2}{2}, \frac{y_1 + y_2}{2}\right)$ |
| Slope | $m = \frac{y_2 - y_1}{x_2 - x_1}$ |
| Special relationships for lines | $\ell_1 \parallel \ell_2 \leftrightarrow m_1 = m_2$<br>$\ell_1 \perp \ell_2 \leftrightarrow m_1 \cdot m_2 = -1$ |

NOTE: Neither $\ell_1$ nor $\ell_2$ is a vertical line in the preceding claims.

To see how the preceding list might be used in this and the next section, consider the following examples.

EXAMPLE 1

Suppose that you are to prove the following relationships:

a) Two lines are parallel.
b) Two lines are perpendicular.
c) Two line segments are congruent.

Which formula(s) would you need to use? How would you complete your proof?

***Solution***

a) Use the Slope Formula first, to find the slope of each line. Then show that the slopes are equal.
b) Use the Slope Formula first, to find the slope of each line. Then show that $m_1 \cdot m_2 = -1$.
c) Use the Distance Formula first, to find the length of each line segment. Then show that the resulting lengths are equal.

The following example has a proof that is subtle. A drawing is provided to help you understand the concept.

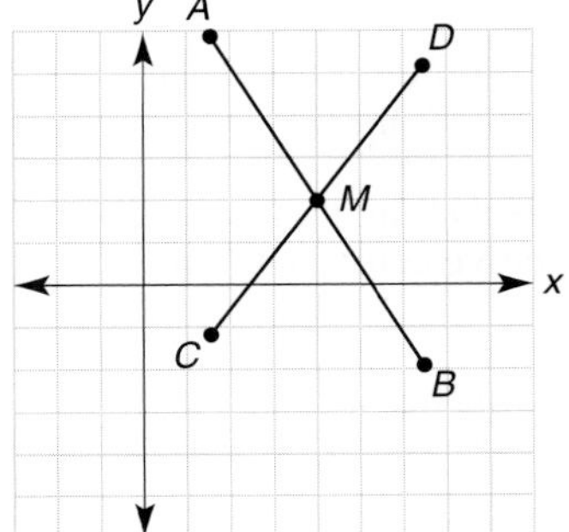


**FIGURE 9.22**

EXAMPLE 2

How can the Midpoint Formula be used to verify that the two line segments shown in Figure 9.22 bisect each other?

***Solution***

If $\overline{AB}$ bisects $\overline{CD}$, and conversely, then $M$ is the common midpoint of the two segments. (See Figure 9.22.) The Midpoint Formula is used to find the midpoint of each line segment, and the results are then shown to be the same point. This establishes that each segment has been bisected by a point that is on the other segment.

**EXAMPLE 3**

Suppose that line $\ell_1$ has slope $\frac{c}{d}$. Use this fact to identify the slopes of the following lines:

a) $\ell_2$ if $\ell_1 \parallel \ell_2$  b) $\ell_3$ if $\ell_1 \perp \ell_3$

***Solution***

a) $m_2 = \frac{c}{d}$ because $m_1 = m_2$ when $\ell_1 \parallel \ell_2$.

b) $m_3 = -\frac{d}{c}$ because $m_1 \cdot m_3 = -1$ when $\ell_1 \perp \ell_3$.

**EXAMPLE 4**

What can you conclude if you know that the point $(p, q)$ lies on the line $y = mx + b$?

***Solution***

Because $(p, q)$ is on the line, it is also a solution for the equation $y = mx + b$. Therefore, $q = mp + b$.

*Exs. 7–9*

To construct proofs of geometric theorems by analytic methods, we must use the hypothesis to determine the drawing. Unlike the drawings in Chapters 1–8, the figure must be placed in the coordinate system. Making the drawing requires careful placement of the figure and proper naming of the vertices, using coordinates of the rectangular system. The following guidelines should prove helpful in positioning the figure and in naming its vertices.

**MAKING THE DRAWING FOR AN ANALYTIC PROOF**

Some considerations for preparing the drawing:

1. Coordinates of the vertices must be general; for instance, you may use $(a, b)$ as a vertex, but do *not* use (2, 3).
2. Make the drawing satisfy the hypothesis without providing any additional qualities; if the theorem describes a rectangle, draw and label a rectangle but *not* a square.
3. For simplicity in your calculations, drop the figure into the rectangular coordinate system in such a manner that
   a) as many 0 coordinates are used as possible.
   b) the remaining coordinates represent positive numbers because of your positioning of the remaining vertices in Quadrant I.

NOTE: In some cases, it is convenient to place a figure so that it has symmetry with respect to the *y* axis, in which case some negative coordinates are present.

4. When possible, use horizontal and vertical segments because you know their parallel and perpendicular relationships.
5. Use as few variable names in the coordinates as possible.

**Discover!**

The geoboard (pegboard) creates a coordinate system of its own. Even though the coordinates of vertices are not named, describe the type of triangle represented by:
a) $\triangle ABC$  b) $\triangle DEF$

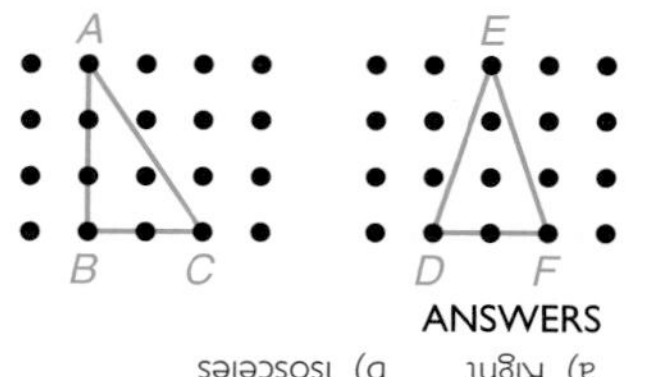


ANSWERS

a) Right b) Isosceles

Now we consider Example 5, which clarifies the preceding list of suggestions. As you observe the drawing in each part of the example, imagine that $\triangle ABC$ has been cut out of a piece of cardboard and dropped into the coordinate system in the position indicated. Because we have freedom of placement, we want to choose the positioning that allows us the simplest possible solution for a proof or problem.

EXAMPLE 5

Suppose that you are asked to make a drawing for the following theorem, which is to be proved analytically: "The midpoint of the hypotenuse of a right triangle is equidistant from the three vertices of the triangle." Explain why the placement of right $\triangle ABC$ in each part of Figure 9.23 is poorly done.

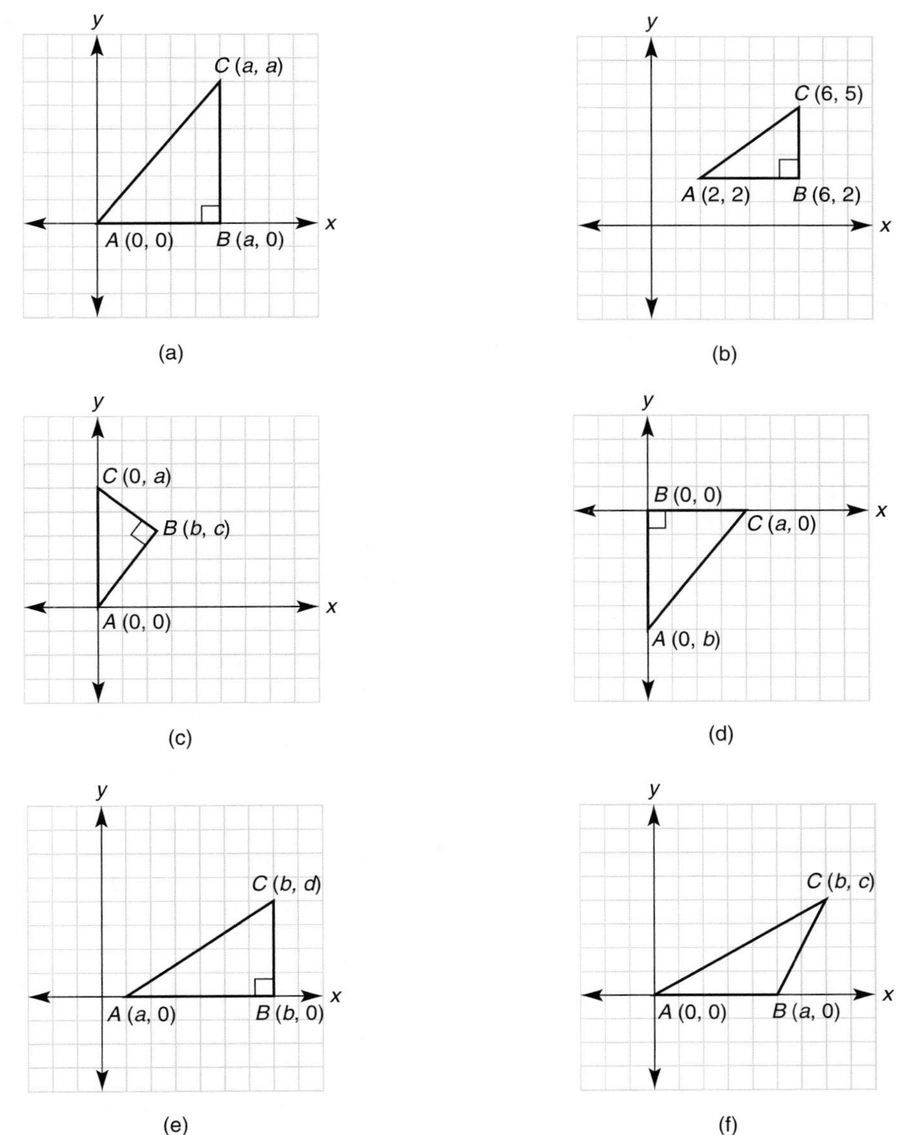


**FIGURE 9.23**

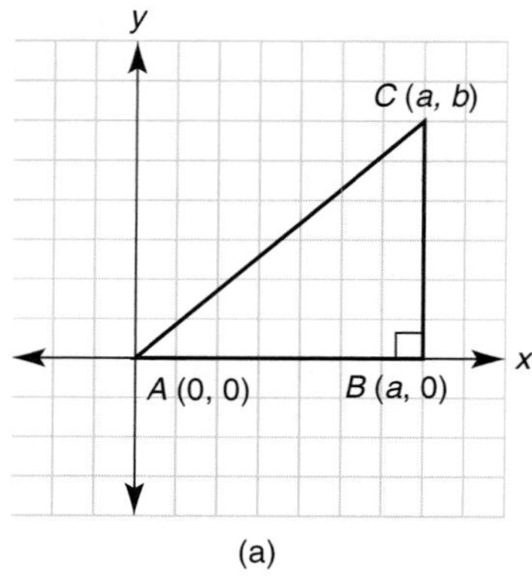

(a)

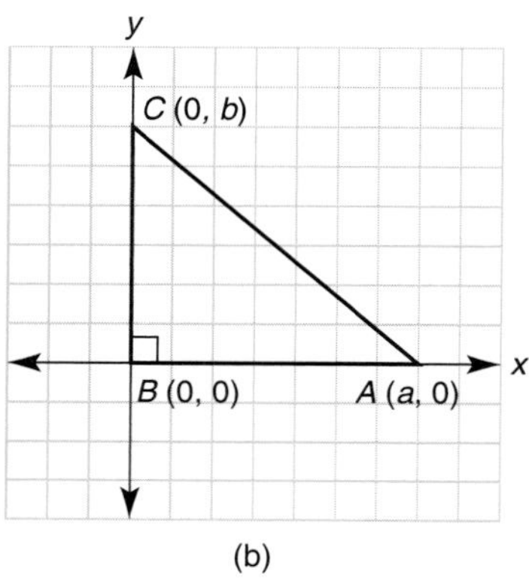

(b)

**FIGURE 9.24**

***Solution***

Refer to Figure 9.23 on page 455.

a) The choice of vertices causes $AB = BC$, so the triangle is also an isosceles triangle. This contradicts point 2 of the list of suggestions.
b) The coordinates are too specific! This contradicts point 1 of the list. A proof with these coordinates would *not* establish the general case.
c) The drawing does not make use of horizontal and vertical lines to obtain the right angle. This violates point 4 of the list.
d) This placement fails point 3 of the list, because $b$ is a negative number. The length of $\overline{AB}$ would be $-b$, which could be confusing.
e) This placement fails point 3 because we have not used as many 0 coordinates as we could have used. As we shall see, it also fails point 5.
f) This placement fails point 2. The triangle is not a right triangle unless $a = b$.

In Example 5, we wanted to place $\triangle ABC$ so that we met as many of the conditions previously listed as possible. Two convenient placements are given in Figure 9.24. The triangle in Figure 9.24(b) is slightly better than the one in 9.24(a) in that it uses four 0 coordinates rather than three. Another advantage of Figure 9.24(b) is that the placement forces angle $B$ to be a right angle, because the $x$ and $y$ axes are perpendicular.

We now turn our attention to the conclusion of the theorem. A second list examines some considerations for proving statements analytically.

*Exs. 10–15*

USING THE CONCLUSION TO DO AN ANALYTIC PROOF

Three considerations for using the conclusion as a guide:

1. If the conclusion is a conjunction "$P$ and $Q$," be sure to verify both parts of the conclusion.
2. The following pairings indicate how to prove statements of the type shown in the left column.

| *To prove the conclusion:* | *Use the:* |
|---|---|
| a) Segments are congruent or have equal lengths (like $AB = CD$). | Distance Formula |
| b) Segments are parallel (like $\overline{AB} \parallel \overline{CD}$). | Slope Formula (need $m_{\overline{AB}} = m_{\overline{CD}}$) |
| c) Segments are perpendicular (like $\overline{AB} \perp \overline{CD}$). | Slope Formula (need $m_{\overline{AB}} \cdot m_{\overline{CD}} = -1$) |
| d) A segment is bisected. | Distance Formula |
| e) Segments bisect each other. | Midpoint Formula |

3. Anticipate the proof by thinking of the steps of the proof in reverse order; that is, reason "back" from the conclusion.

EXAMPLE 6

a) Provide an ideal drawing for the following theorem: The midpoint of the hypotenuse of a right triangle is equidistant from the three vertices of the triangle.

b) By studying the theorem, name at least two of the formulas that will be used to complete the proof.

***Solution***

a) We improve Figure 9.24(b) by giving the value $2a$ to the $x$ coordinate of $A$ and the value $2b$ to the $y$ coordinate of $C$. (A factor of 2 makes it easier to calculate and represent the midpoint $M$ of $\overline{AC}$. See Figure 9.25.)

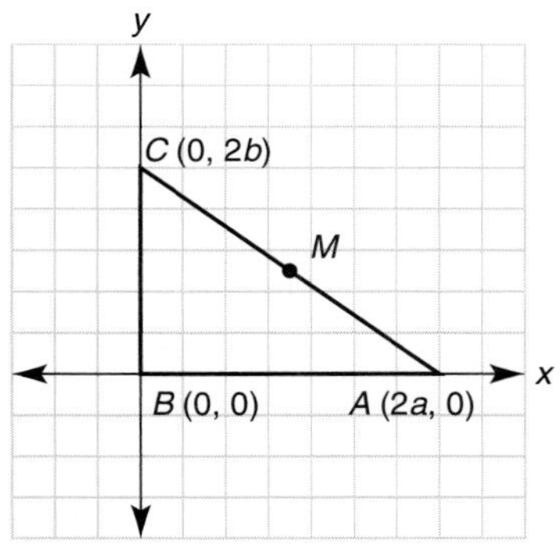


**FIGURE 9.25**

b) The Midpoint Formula is applied to describe the midpoint of $\overline{AC}$. Using the formula, we find that

$$M = \left(\frac{x_1 + x_2}{2}, \frac{y_1 + y_2}{2}\right) = \left(\frac{2a + 0}{2}, \frac{0 + 2b}{2}\right) = \left(\frac{2a}{2}, \frac{2b}{2}\right)$$

so the midpoint is $(a, b)$. The Distance Formula will also be needed because the theorem states that the distances from $M$ to $A$, from $M$ to $B$, and from $M$ to $C$ should all be equal.

Discover!

The geoboard (pegboard) creates a coordinate system of its own. Describe the type of quadrilateral represented by *ABCD*.

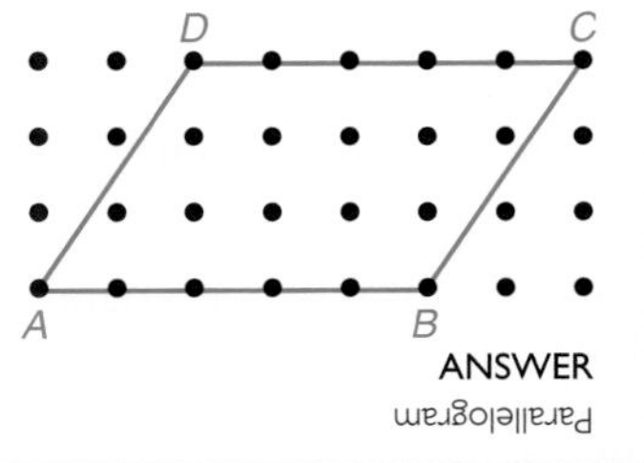


ANSWER

Parallelogram

The purpose of our next example is to demonstrate efficiency in the labeling of vertices. Our goal is to use fewer variables in characterizing points found in Figure 9.26.

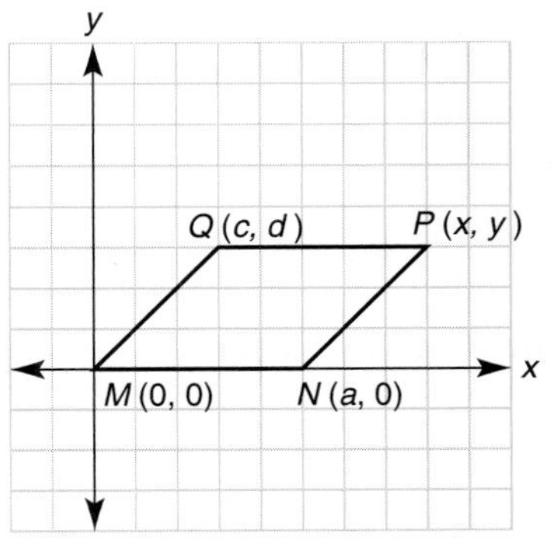


**FIGURE 9.26**

EXAMPLE 7

If $MNPQ$ is a parallelogram in Figure 9.26, find the coordinates of point $P$ in terms of $a$, $c$, and $d$.

***Solution***

Consider $\square MNPQ$ in Figure 9.26. For the moment, we refer to point $P$ as $(x, y)$.

Because $\overline{MN} \parallel \overline{QP}$, we have $m_{\overline{MN}} = m_{\overline{QP}}$. But $m_{\overline{MN}} = \frac{0 - 0}{a - 0} = 0$ and $m_{\overline{QP}} = \frac{y - d}{x - c}$, so we are led to the equation

$$\frac{y - d}{x - c} = 0 \rightarrow y - d = 0 \rightarrow y = d$$

Now $P$ is described by $(x, d)$. Because $\overline{MQ} \parallel \overline{NP}$, we are also led to the equation equating slopes of these segments. But

$$m_{\overline{MQ}} = \frac{d-0}{c-0} = \frac{d}{c} \qquad \text{and} \qquad m_{\overline{NP}} = \frac{d-0}{x-a} = \frac{d}{x-a}$$

so

$$\frac{d}{c} = \frac{d}{x-a}$$

By using the Means-Extremes Property, we have

$$\begin{aligned} d(x-a) &= d \cdot c && \text{with } d \neq 0 \\ x - a &= c && \text{dividing by } d \\ x &= a + c && \text{adding } a \end{aligned}$$

Therefore, $P$ is the point $(a + c, d)$.

In retrospect, Example 7 shows that $\square MNPQ$ is characterized by vertices $M(0, 0)$, $N(a, 0)$, $P(a + c, d)$, and $Q(c, d)$. Because $\overline{MN}$ and $\overline{QP}$ are horizontal segments, it is obvious that $\overline{MN} \parallel \overline{QP}$. Both $\overline{MQ}$ (starting at $M$) and $\overline{NP}$ (starting at $N$) trace paths that move along each segment *c units to the right* and *d units upward.* Thus, the slopes of $\overline{MQ}$ and $\overline{NP}$ are both $\frac{d}{c}$, and it follows that $\overline{MQ} \parallel \overline{NP}$. See the improved Figure 9.27.

In Example 7, we named the coordinates of the vertices of a parallelogram with the fewest possible letters. We now extend our result in Example 7 to allow for a rhombus—a parallelogram with two congruent adjacent sides.

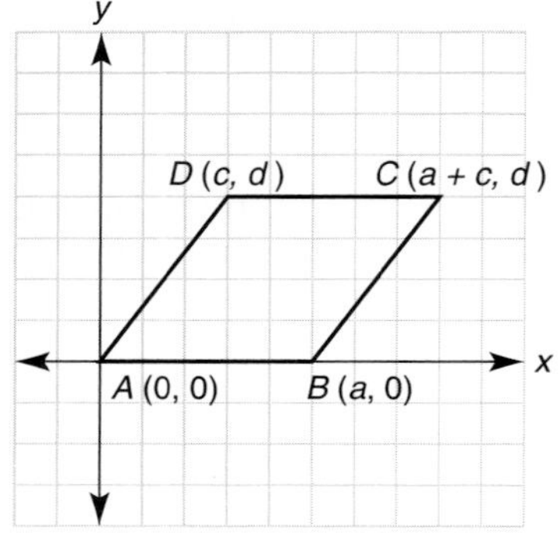


**FIGURE 9.27**

EXAMPLE 8

In Figure 9.27, find an equation that relates $a$, $c$, and $d$ if $\square ABCD$ is a rhombus.

***Solution***

As we saw in Example 7, the coordinates of the vertices of $ABCD$ define a parallelogram. For emphasis, we note that $\overline{AB} \parallel \overline{DC}$ and $\overline{AD} \parallel \overline{BC}$ because

$$m_{\overline{AB}} = m_{\overline{DC}} = 0 \qquad \text{and} \qquad m_{\overline{AD}} = m_{\overline{BC}} = \frac{d}{c}$$

For Figure 9.27 to represent a rhombus, it is necessary that $AB = AD$. Now $AB = a - 0 = a$ because $\overline{AB}$ is a horizontal segment. To find an expression for the length of $\overline{AD}$, we need to use the Distance Formula.

$$\begin{aligned} AD &= \sqrt{(x_2 - x_1)^2 + (y_2 - y_1)^2} \\ &= \sqrt{(c-0)^2 + (d-0)^2} \\ &= \sqrt{c^2 + d^2} \end{aligned}$$

Because $AB = AD$, we are led to $a = \sqrt{c^2 + d^2}$. Squaring, we have the desired relationship, $a^2 = c^2 + d^2$.

SSG *Exs. 16–18*

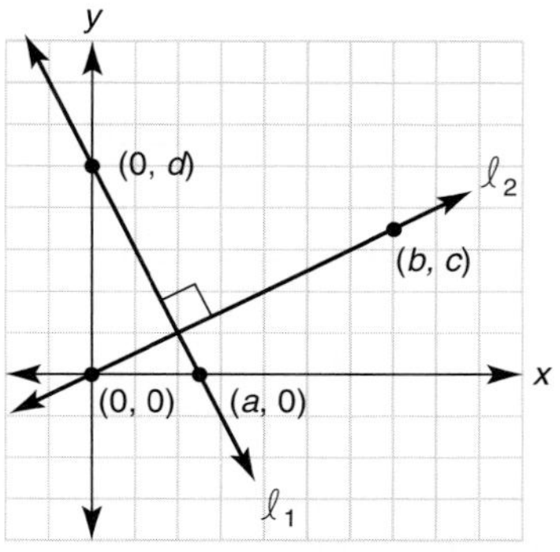


**FIGURE 9.28**

EXAMPLE 9

If $\ell_1 \perp \ell_2$ in Figure 9.28, find a relationship among the variables *a, b, c,* and *d.*

***Solution***

First we find the slopes of lines $\ell_1$ and $\ell_2$. For $\ell_1$, we have

$$m_1 = \frac{0 - d}{a - 0} = -\frac{d}{a}$$

For $\ell_2$, we have

$$m_2 = \frac{c - 0}{b - 0} = \frac{c}{b}$$

With $\ell_1 \perp \ell_2$, it follows that $m_1 \cdot m_2 = -1$. Substituting the slopes found above into the equation $m_1 \cdot m_2 = -1$, we have

$$-\frac{d}{a} \cdot \frac{c}{b} = -1 \qquad \text{so} \qquad -\frac{dc}{ab} = -1$$

Equivalently, $\frac{dc}{ab} = 1$ and $dc = ab$.

## 9.3 Exercises

1. Find an expression for:

   a) The distance between $(a, 0)$ and $(0, a)$
   b) The slope of the segment joining $(a, b)$ and $(c, d)$

2. Find the coordinates of the midpoint of the segment that joins the points

   a) $(a, 0)$ and $(0, b)$  b) $(2a, 0)$ and $(0, 2b)$

3. Find the slope of the line containing the points

   a) $(a, 0)$ and $(0, a)$  b) $(a, 0)$ and $(0, b)$

4. Find the slope of the line that is:

   a) Parallel to the line containing $(a, 0)$ and $(0, b)$
   b) Perpendicular to the line through $(a, 0)$ and $(0, b)$

*In Exercises 5 to 10, the real numbers a, b, c, and d are positive.*

5. Consider the triangle with vertices at $A(0, 0)$, $B(a, 0)$, and $C(a, b)$. Explain why $\triangle ABC$ is a right triangle.

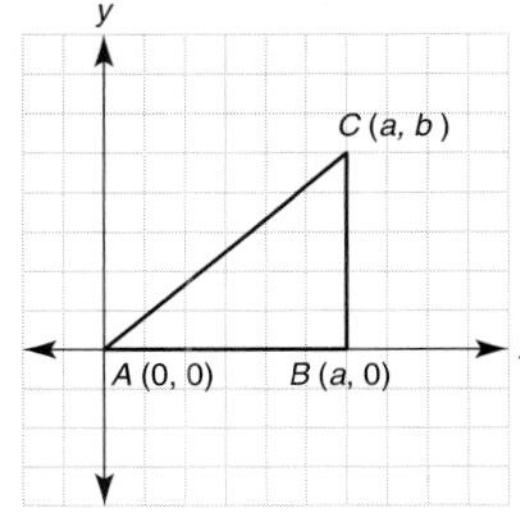


6. Consider the triangle with vertices at $R(-a, 0)$, $S(a, 0)$, and $T(0, b)$. Explain why $\triangle RST$ is an isosceles triangle.

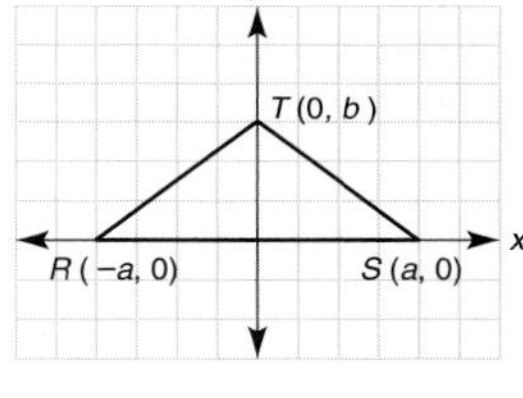


7. Consider the quadrilateral with vertices at $M(0, 0)$, $N(a, 0)$, $P(a + b, c)$, and $Q(b, c)$. Explain why $MNPQ$ is a parallelogram.

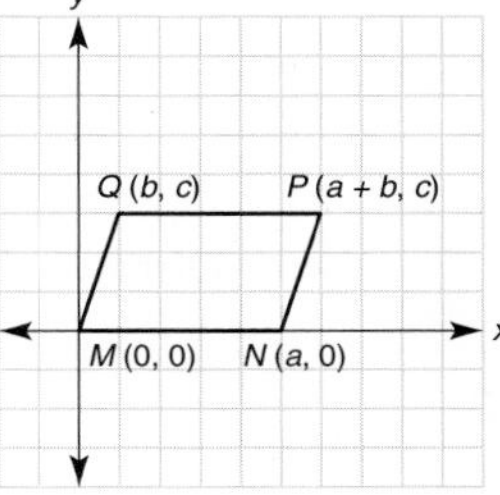


8. Consider the quadrilateral with vertices at $A(0, 0)$, $B(a, 0)$, $C(b, c)$, and $D(d, c)$. Explain why $ABCD$ is a trapezoid.

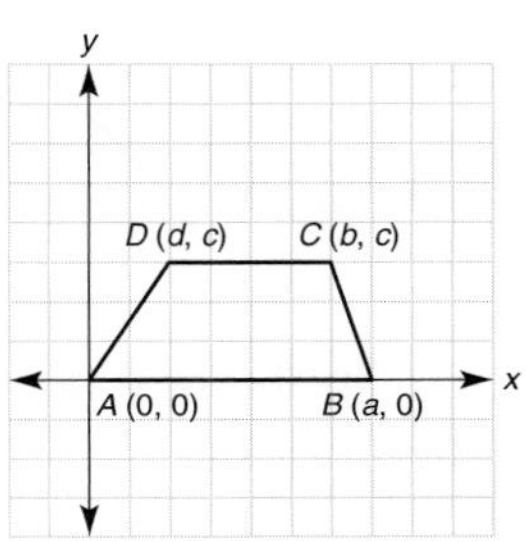

9. Consider the quadrilateral with vertices at $M(0, 0)$, $N(a, 0)$, $P(a, b)$, and $Q(0, b)$. Explain why $MNPQ$ is a rectangle.

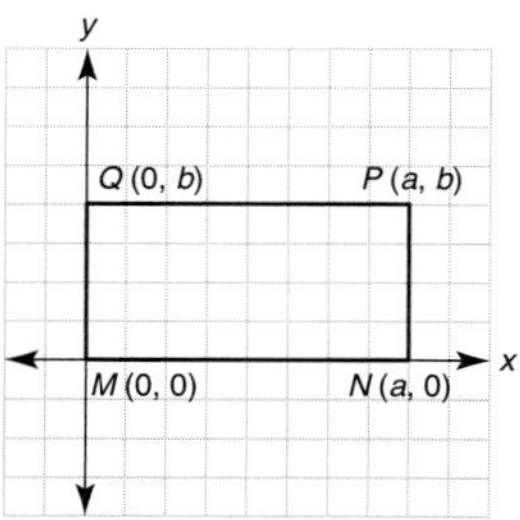


10. Consider the quadrilateral with vertices at $R(0, 0)$, $S(a, 0)$, $T(a, a)$, and $V(0, a)$. Explain why $RSTV$ is a square.

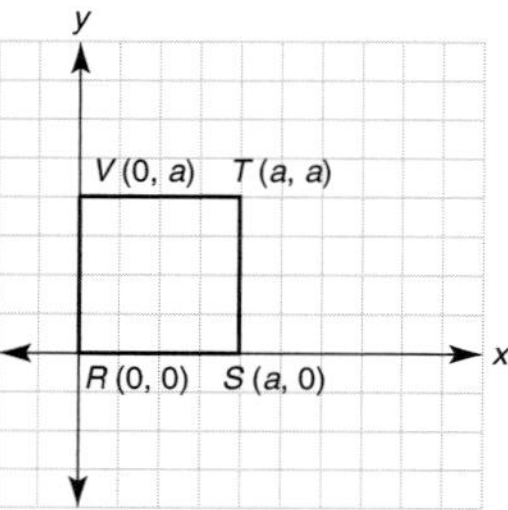


*In Exercises 11 to 16, supply the missing coordinates for the vertices, using as few variables as possible.*

11.

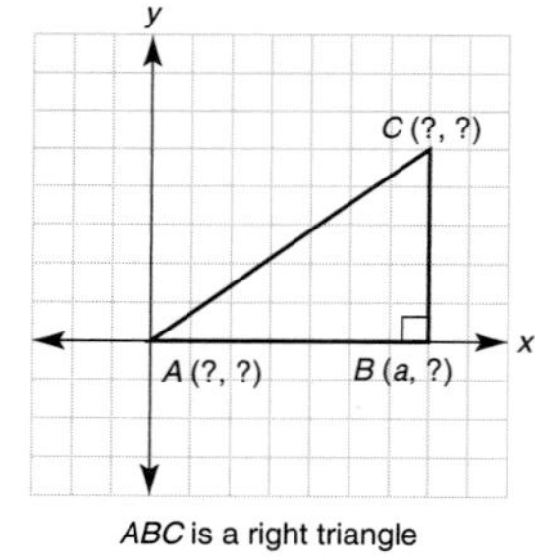


*ABC* is a right triangle

12.

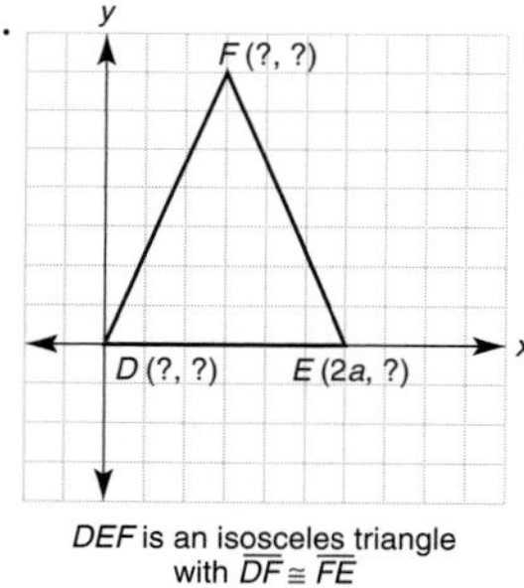


*DEF* is an isosceles triangle with $\overline{DF} \cong \overline{FE}$

13.

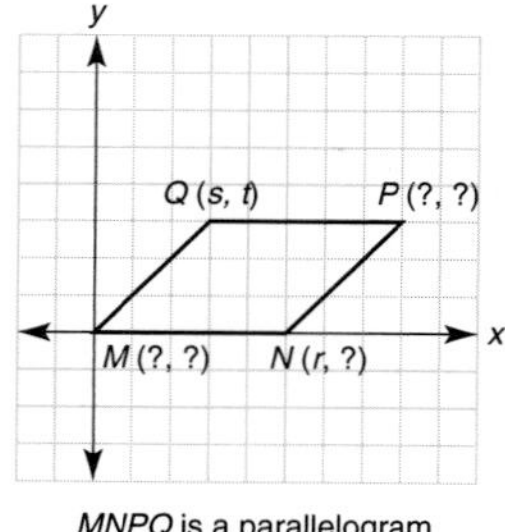


*MNPQ* is a parallelogram

14.

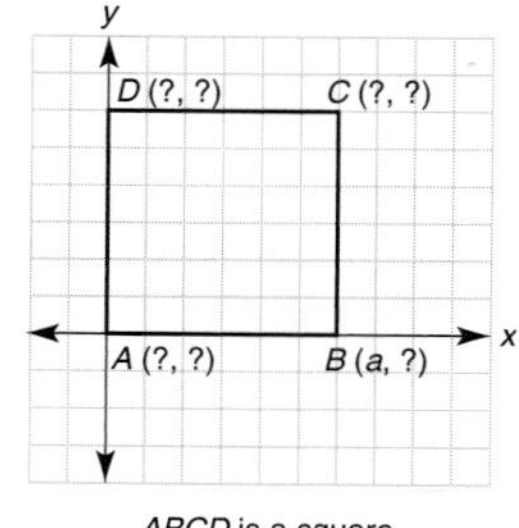


*ABCD* is a square

15.

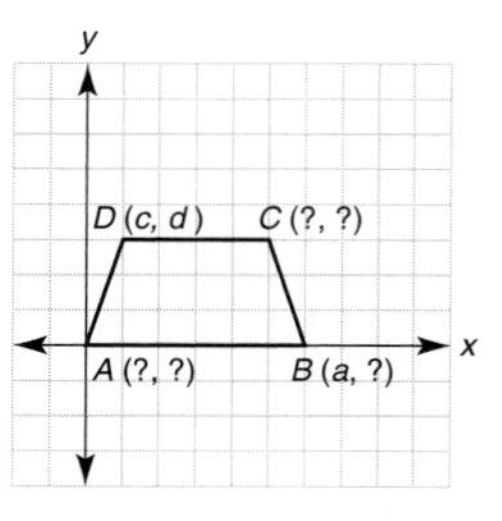


*ABCD* is an isosceles trapezoid; $\overline{AB} \parallel \overline{DC}$ and $\overline{AD} \cong \overline{BC}$

16.

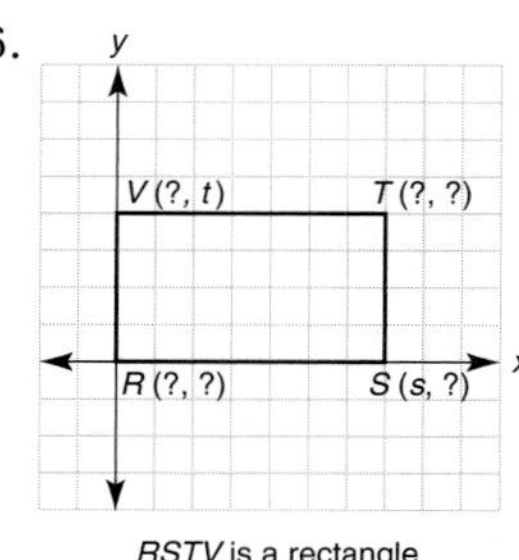


*RSTV* is a rectangle

*In Exercises 17 to 22, draw an ideally placed figure in the coordinate system; then name the coordinates of each vertex of the figure.*

17. a) A square
 b) A square (midpoints of sides are needed)

18. a) A rectangle
 b) A rectangle (midpoints of sides are needed)

19. a) A parallelogram
 b) A parallelogram (midpoints of sides are needed)

20. a) A triangle
 b) A triangle (midpoints of sides are needed)

21. a) An isosceles triangle
 b) An isosceles triangle (midpoints of sides are needed)

22. a) A trapezoid
 b) A trapezoid (midpoints of sides are needed)

*In Exercises 23 to 28, find the equation (relationship) requested. Then eliminate fractions and square root radicals from the equation.*

23. If ▱$MNPQ$ is a rhombus, state an equation that relates $r$, $s$, and $t$.

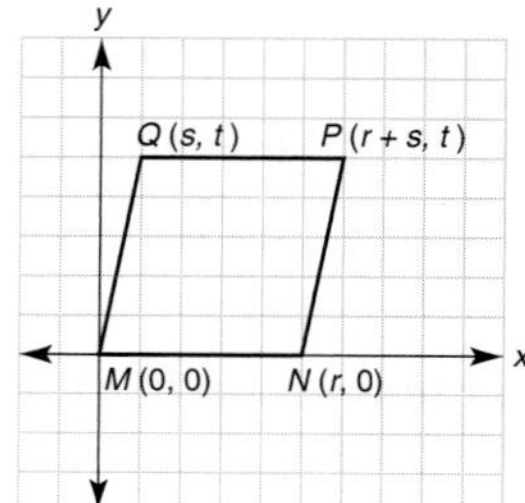


24. For ▱$RSTV$, suppose that $RT = VS$. State an equation that relates $s$, $t$, and $v$.

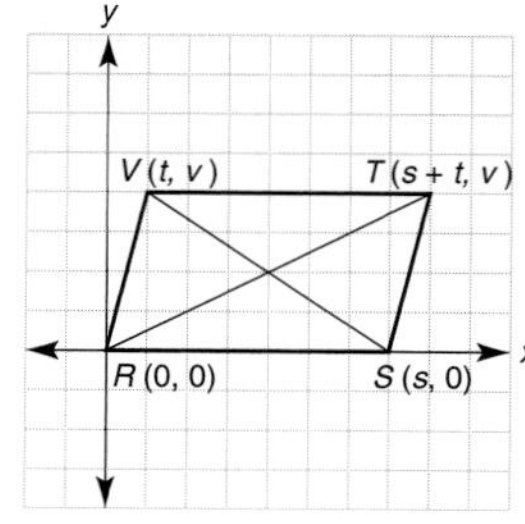

25. For ▱$ABCD$, suppose that diagonals $\overline{AC}$ and $\overline{DB}$ are perpendicular. State an equation that relates $a$, $b$, and $c$.

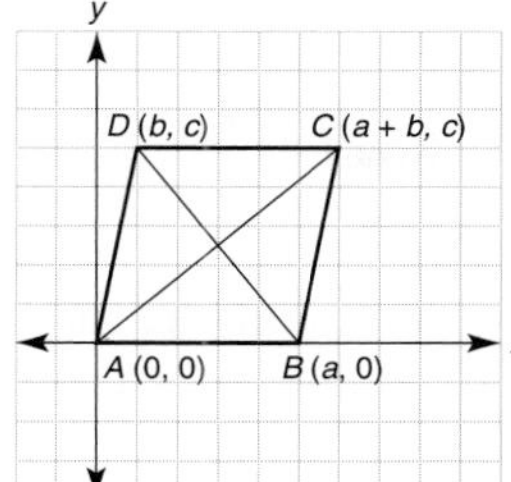


26. For quadrilateral $RSTV$, suppose that $\overline{RV} \parallel \overline{ST}$. State an equation that relates $m$, $n$, $p$, $q$, and $r$.

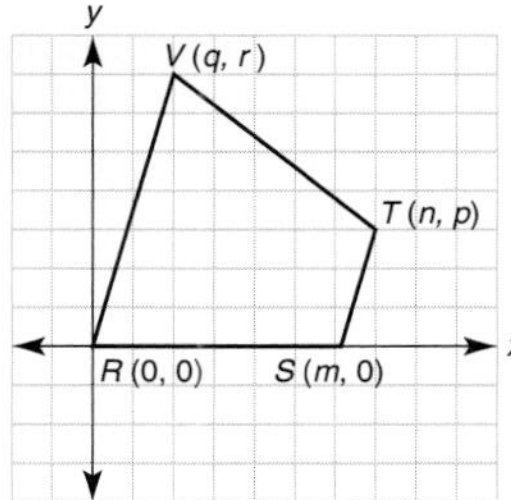


27. Suppose that $\triangle ABC$ is an equilateral triangle. State an equation that relates variables $a$ and $b$.

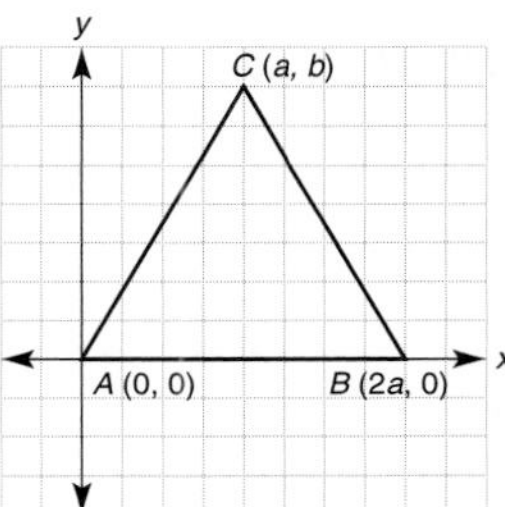


28. Suppose that $\triangle RST$ is an isosceles triangle, with $\overline{RS} \cong \overline{RT}$. State an equation that relates $s$, $t$, and $v$.

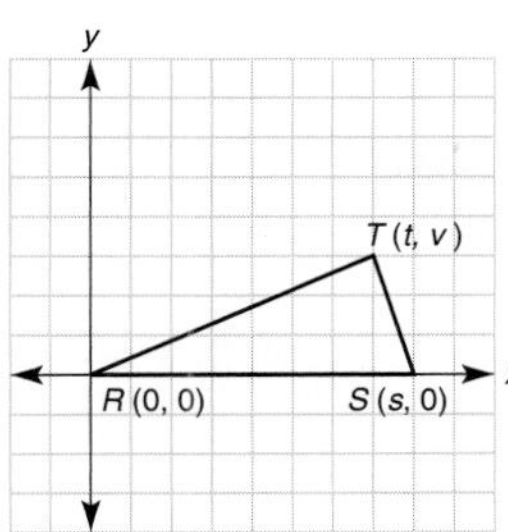


29. The drawing shows isosceles $\triangle ABC$ with $\overline{AC} \cong \overline{BC}$.

a) What type of number is $a$?
b) What type of number is $-a$?
c) Find an expression for the length of $\overline{AB}$.

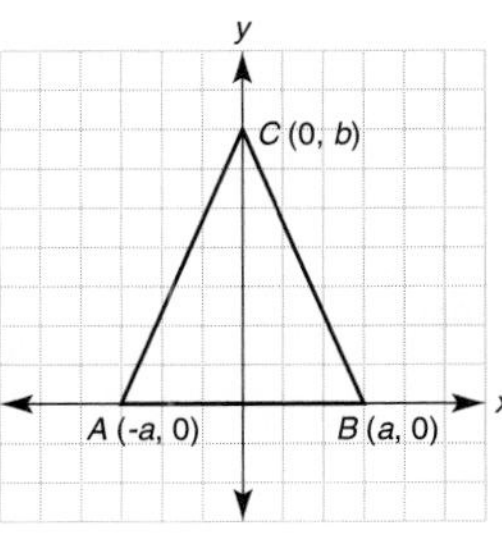


30. The drawing shows parallelogram $RSTV$.

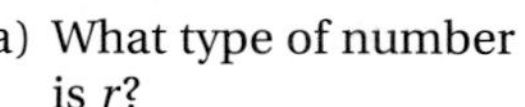

a) What type of number is $r$?
b) Find an expression for $RS$.
c) Describe the coordinate $t$ in terms of the other variables.

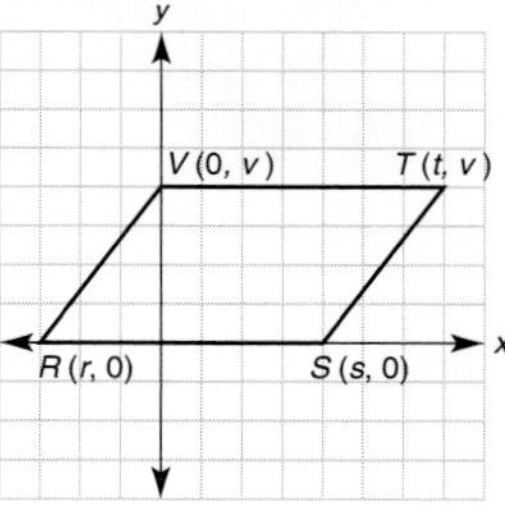


31. Which formula would you use to establish each of the following claims?

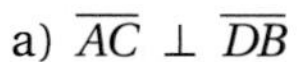

a) $\overline{AC} \perp \overline{DB}$
b) $AC = DB$
c) $\overline{DB}$ and $\overline{AC}$ bisect each other
d) $\overline{AD} \parallel \overline{BC}$

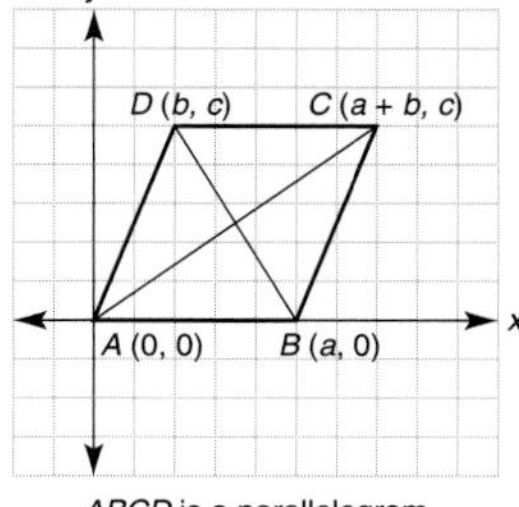

*ABCD* is a parallelogram

32. Which formula would you use to establish each of the following claims?

a) The coordinates of $X$ are $(d, c)$.
b) $m_{\overline{VT}} = 0$
c) $\overline{VT} \parallel \overline{RS}$
d) The length of $\overline{RV}$ is $2\sqrt{d^2 + c^2}$.

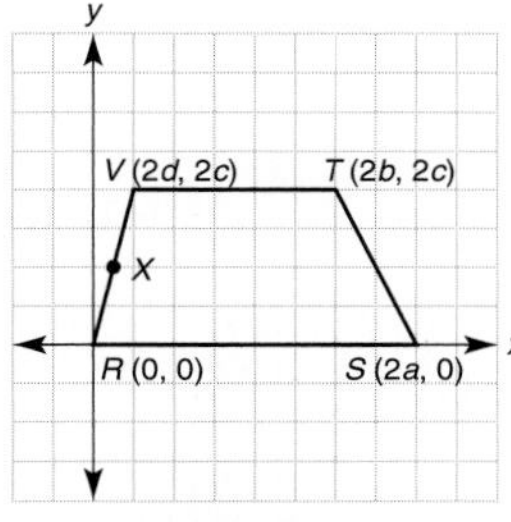

Trapezoid *RSTV*; *X* is the midpoint of $\overline{RV}$

*In Exercises 33 to 36, draw and label a well-placed figure in the coordinate system for each theorem. Do not attempt to prove the theorem!*

33. The line segment joining the midpoints of the two nonparallel sides of a trapezoid is parallel to each base of the trapezoid.

34. If the midpoints of the sides of a quadrilateral are joined in order, the resulting quadrilateral is a parallelogram.

35. The diagonals of a rectangle are equal in length.

36. The diagonals of a rhombus are perpendicular to each other.

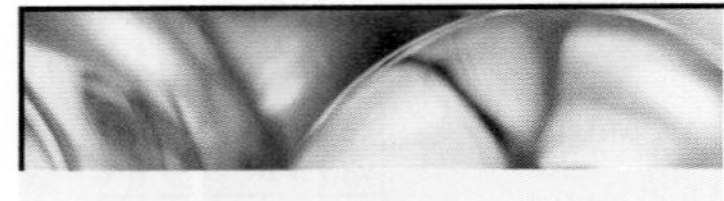

**KEY CONCEPTS**

Analytic Proof • Synthetic Proof

# 9.4 Analytic Proofs

When we use algebra along with the rectangular coordinate system to prove a geometric theorem, the proof is termed **analytic.** The analytic (algebraic) approach relies heavily on the placement of the figure in the coordinate system and on the application of the Distance Formula, the Midpoint Formula, or the Slope Formula (at the appropriate time). In order to contrast analytic proof with **synthetic** proof (the two-column or paragraph proofs used in earlier chapters), we repeat in this section some of our earlier theorems.

In Section 9.3, we saw how to place triangles having special qualities in the coordinate system. We review this information in Table 9.1, and then, in Example 1, we will consider the proof of a theorem involving triangles. In Table 9.1, you will find that the figure determined by any positive numerical choices of *a, b,* and *c* matches the type of triangle described.

**TABLE 9.1**

**Analytic Proof: Suggestions for Placement of the Triangle**

y, x, $C\ (b, c)$, $A\ (0, 0)$, $B\ (a, 0)$

General Triangle

y, x, $C\ (2b, 2c)$, $A\ (0, 0)$, $B\ (2a, 0)$

General Triangle (Midpoints)

y, x, $C\ (a, b)$, $A\ (0, 0)$, $B\ (2a, 0)$

Isosceles Triangle

y, x, $C\ (2a, 2b)$, $A\ (0, 0)$, $B\ (4a, 0)$

Isosceles Triangle (Midpoints)

y, x, $C\ (0, b)$, $A\ (0, 0)$, $B\ (a, 0)$

Right Triangle

y, x, $C\ (0, b)$, $A\ (-a, 0)$, $B\ (a, 0)$

Equilateral Triangle (where $2a = \sqrt{a^2 + b^2}$, so $3a^2 = b^2$)

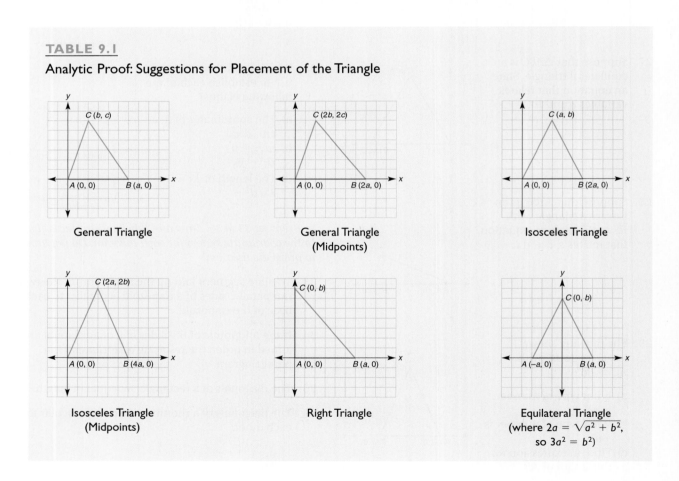

In Table 9.2, we review convenient placements for types of quadrilaterals. As placed, the variables named represent positive numbers.

**TABLE 9.2**

**Analytic Proof: Suggestions for Placement of the Quadrilateral**

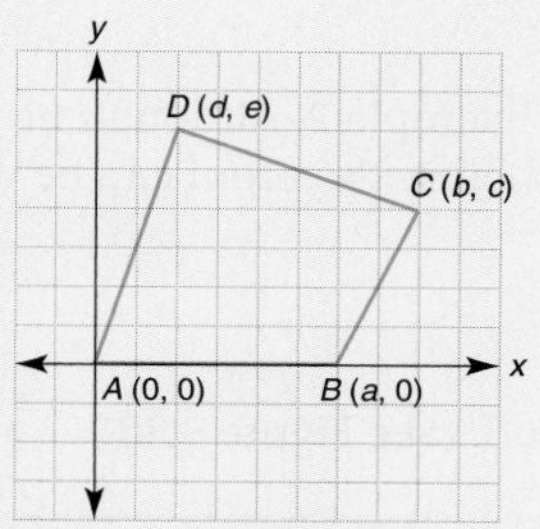


General Quadrilateral

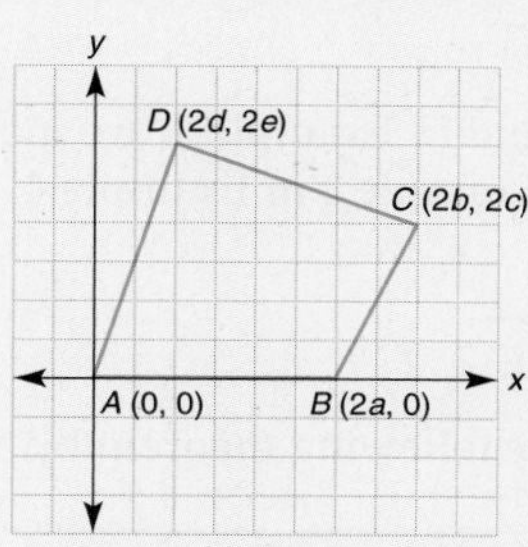


General Quadrilateral (Midpoints)

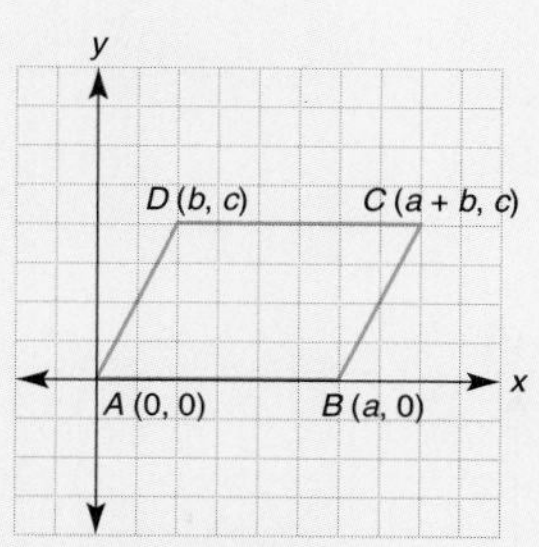


Parallelogram

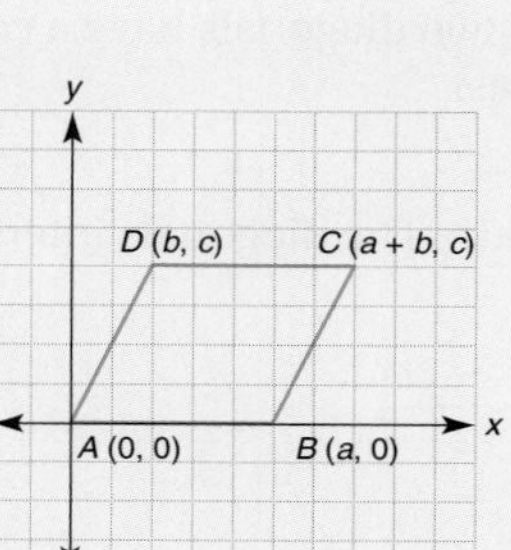


Rhombus (where $a = \sqrt{b^2 + c^2}$ so $a^2 = b^2 + c^2$)

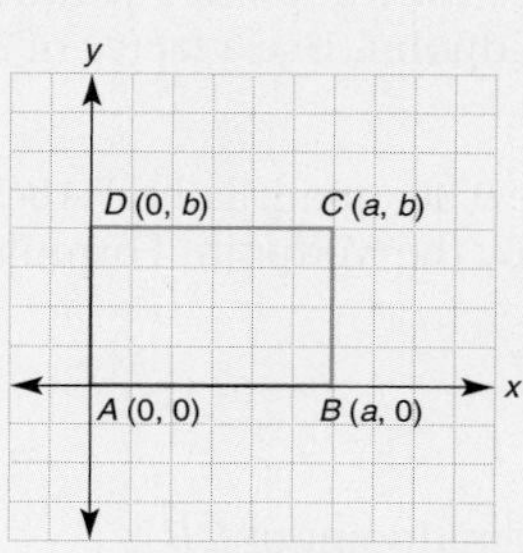


Rectangle

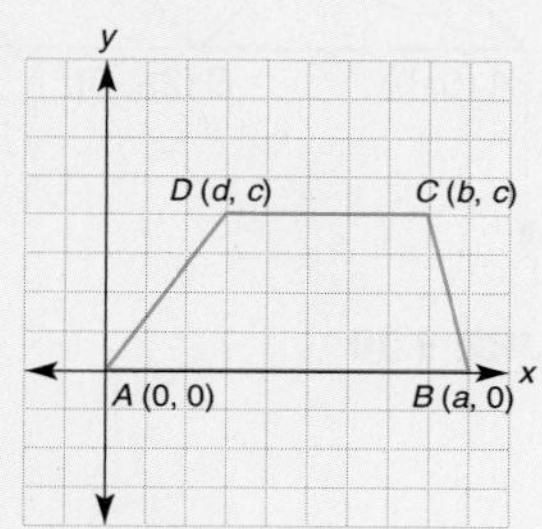


Trapezoid

*Exs. 1–4*

EXAMPLE 1

Prove the following theorem by the analytic method (see Figure 9.29).

> THEOREM 9.4.1: The line segment determined by the midpoints of two sides of a triangle is parallel to the third side.

*Plan:* Use the Slope Formula; if $m_{\overline{MN}} = m_{\overline{AC}}$, then $\overline{MN} \parallel \overline{AC}$.

*Proof:*

As shown in Figure 9.29, $\triangle ABC$ has vertices at $A(0, 0)$, $B(2a, 0)$, and $C(2b, 2c)$. With $M$ the midpoint of $\overline{BC}$,

$$M = \left(\frac{2a + 2b}{2}, \frac{0 + 2c}{2}\right) \text{ which simplifies to } (a + b, c)$$

With $N$ the midpoint of $\overline{AB}$,

$$N = \left(\frac{0 + 2a}{2}, \frac{0 + 0}{2}\right) \text{ which simplifies to } (a, 0)$$

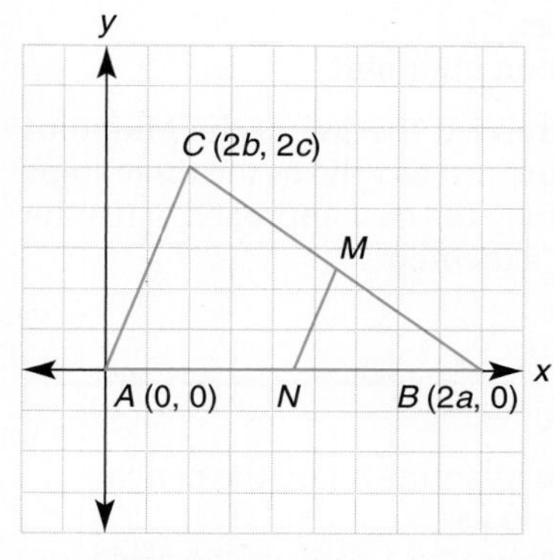


**FIGURE 9.29**

Next we apply the Slope Formula in order to determine $m_{\overline{MN}}$ and $m_{\overline{AC}}$. Now $m_{\overline{MN}} = \frac{c - 0}{(a + b) - a} = \frac{c}{b}$; also, $m_{\overline{AC}} = \frac{2c - 0}{2b - 0} = \frac{2c}{2b} = \frac{c}{b}$. Because $m_{\overline{MN}} = m_{\overline{AC}}$, we see that $\overline{MN} \parallel \overline{AC}$.

As we did in Example 1, we include a "plan" for Example 2. Although no plan is shown for Example 3 or Example 4, one is necessary before the proof can be written.

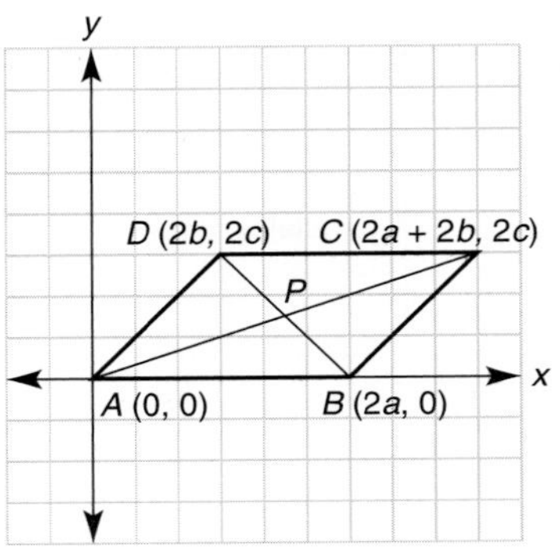


**FIGURE 9.30**

## EXAMPLE 2

Prove the following theorem by the analytic method. (See Figure 9.30.)

> THEOREM 9.4.2: The diagonals of a parallelogram bisect each other.

*Plan:* Use the Midpoint Formula to show that the two diagonals have a common midpoint. Use a factor of 2 in the coordinates.

*Proof:*

In Figure 9.30, quadrilateral *ABCD* is a parallelogram. The diagonals intersect at point *P*. By the Midpoint Formula, we have

$$M_{\overline{AC}} = \left(\frac{0 + (2a + 2b)}{2}, \frac{0 + 2c}{2}\right)$$
$$= (a + b, c)$$

Also, the midpoint of $\overline{DB}$ is

$$M_{\overline{DB}} = \left(\frac{2a + 2b}{2}, \frac{0 + 2c}{2}\right)$$
$$= (a + b, c)$$

Thus $(a + b, c)$ is the common midpoint of the two diagonals and must be the point of intersection of $\overline{AC}$ and $\overline{DB}$. Then $\overline{AC}$ and $\overline{DB}$ must bisect each other at point *P*.

*Exs. 5–9*

The proof of Theorem 9.4.2 is not unique! In Section 9.5, we could prove Theorem 9.4.2 by using a three-step proof:

1. Find the equations of the two lines.
2. Determine the point of intersection of these lines.
3. Show that this point of intersection is the common midpoint.

But the phrase *bisect each other* in Theorem 9.4.2 implied the use of the Midpoint Formula. Our approach to Example 2 was far easier and just as valid as the three steps described. The use of the Midpoint Formula is generally the best approach when the phrase *bisect each other* appears in the statement of a theorem.

We now outline the method of analytic proof.

### COMPLETING AN ANALYTIC PROOF

1. Read the theorem carefully to distinguish the hypothesis and the conclusion. The hypothesis characterizes the figure to use.
2. Use the hypothesis (and nothing more) to determine a convenient placement of the figure in the rectangular coordinate system. Then label the figure. See Tables 9.1 and 9.2.

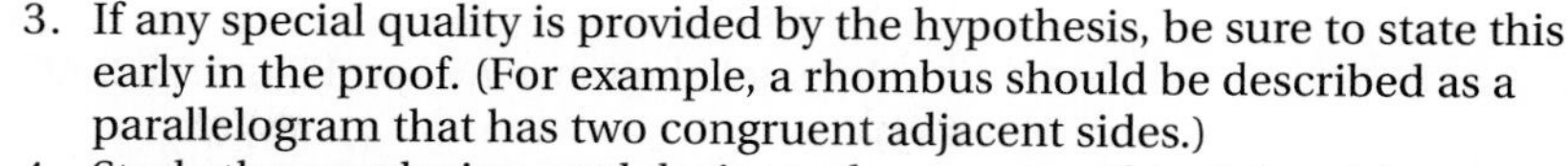

3. If any special quality is provided by the hypothesis, be sure to state this early in the proof. (For example, a rhombus should be described as a parallelogram that has two congruent adjacent sides.)
4. Study the conclusion, and devise a plan to prove this claim; this may involve reasoning back from the conclusion step by step until the hypothesis is reached.
5. Write the proof, being careful to order the statements properly and to justify each statement.

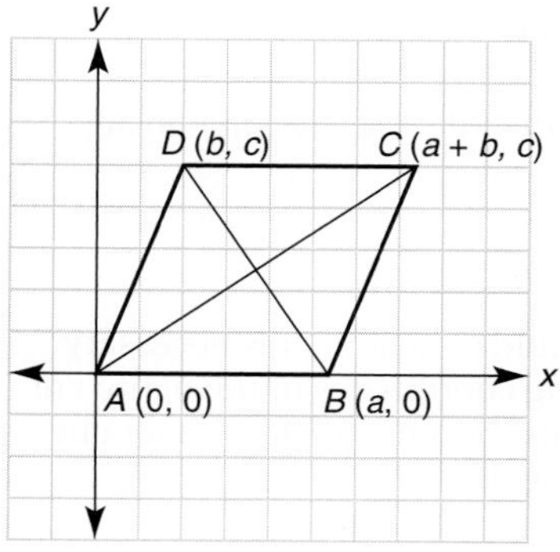


**FIGURE 9.31**

EXAMPLE 3

Prove Theorem 9.4.3 by the analytic method. (See Figure 9.31.)

THEOREM 9.4.3: The diagonals of a rhombus are perpendicular.

***Solution***

In Figure 9.31, *ABCD* has the coordinates of a parallelogram. Because $\square ABCD$ is a rhombus, $AB = AD$. Then $a = \sqrt{b^2 + c^2}$ by the Distance Formula, and squaring gives $a^2 = b^2 + c^2$. The Slope Formula leads to

$$m_{\overline{AC}} = \frac{c - 0}{(a + b) - 0} \quad \text{and} \quad m_{\overline{DB}} = \frac{0 - c}{a - b}$$

so

$$m_{\overline{AC}} = \frac{c}{a + b} \quad \text{and} \quad m_{\overline{DB}} = \frac{-c}{a - b}$$

Then the product of the slopes of the diagonals is

$$\begin{aligned} m_{\overline{AC}} \cdot m_{\overline{DB}} &= \frac{c}{a + b} \cdot \frac{-c}{a - b} \\ &= \frac{-c^2}{a^2 - b^2} \\ &= \frac{-c^2}{(b^2 + c^2) - b^2} \qquad \text{(replaced } a^2 \text{ by } b^2 + c^2\text{)} \\ &= \frac{-c^2}{c^2} = -1 \end{aligned}$$

Then $\overline{AC} \perp \overline{DB}$ because the product of their slopes equals $-1$.

Reminder

We prove that lines are perpendicular by showing that the product of their slopes is $-1$.

In Example 3, we had to use the condition that two adjacent sides of the rhombus were congruent in order to complete the proof. Had that condition been omitted, the product of slopes could not have been shown to equal $-1$. In general, the diagonals of a parallelogram are not perpendicular.

In our next example, we consider the proof of the converse of an earlier theorem. Although it is easy to complete an analytic proof of the statement "The diagonals of a rectangle are equal in length," the proof of the converse is not as straightforward.

*Exs. 10, 11*

EXAMPLE 4

Prove Theorem 9.4.4 by the analytic method. (See Figure 9.32.)

THEOREM 9.4.4: If the diagonals of a parallelogram are equal in length, then the parallelogram is a rectangle.

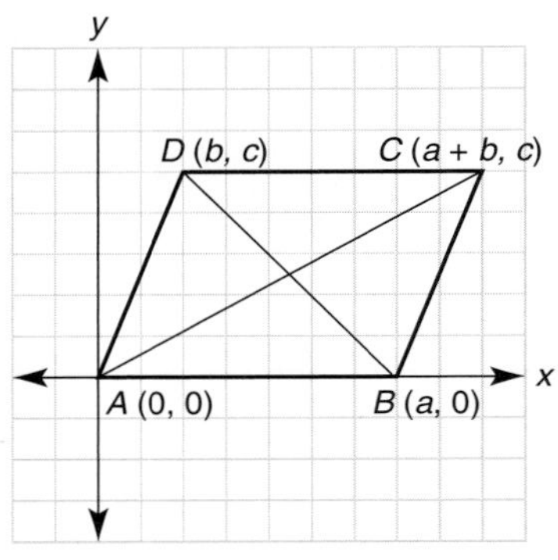


**FIGURE 9.32**

Exs. 12, 13

***Solution***

In parallelogram $ABCD$ in Figure 9.32, $AC = DB$. Applying the Distance Formula,

$$AC = \sqrt{[(a+b)-0]^2 + (c-0)^2}$$

and

$$DB = \sqrt{(a-b)^2 + (0-c)^2}$$

Because the diagonals have the same length,

$$\sqrt{(a+b)^2 + c^2} = \sqrt{(a-b)^2 + (-c)^2}$$

$$(a+b)^2 + c^2 = (a-b)^2 + (-c)^2 \quad \text{squaring}$$

$$a^2 + 2ab + b^2 + c^2 = a^2 - 2ab + b^2 + c^2 \quad \text{simplifying}$$

$$4ab = 0$$

$$a \cdot b = 0 \quad \text{dividing by 4}$$

Thus $\quad a = 0 \quad$ or $\quad b = 0$

Because $a \neq 0$ (otherwise points $A$ and $B$ would coincide), it is necessary that $b = 0$, so point $D$ is on the $y$ axis. The resulting coordinates of the figure are $A(0, 0)$, $B(a, 0)$, $C(a, c)$, and $D(0, c)$. Then $ABCD$ must be a rectangle with a right angle at $A$ because $\overline{AB}$ is horizontal and $\overline{AD}$ is vertical.

## 9.4 Exercises

*In Exercises 1 to 17, complete an analytic proof for each theorem.*

1. The diagonals of a rectangle are equal in length.
2. The opposite sides of a parallelogram are equal in length.
3. The diagonals of a square are perpendicular bisectors of each other.
4. The diagonals of an isosceles trapezoid are equal in length.
5. The median from the vertex of an isosceles triangle to the base is perpendicular to the base.
6. The medians to the congruent sides of an isosceles triangle are equal in length.
7. The segments that join the midpoints of the consecutive sides of a quadrilateral form a parallelogram.
8. The segments that join the midpoints of the opposite sides of a quadrilateral bisect each other.
9. The segments that join the midpoints of the consecutive sides of a rectangle form a rhombus.
10. The segments that join the midpoints of the consecutive sides of a rhombus form a rectangle.
11. The midpoint of the hypotenuse of a right triangle is equidistant from the three vertices of the triangle.
12. The median of a trapezoid is parallel to the bases of the trapezoid and has a length equal to one-half the sum of the lengths of the two bases.
13. The segment that joins the midpoints of two sides of a triangle is parallel to the third side and has a length equal to one-half the length of the third side.
14. The perpendicular bisector of the base of an isosceles triangle contains the vertex of the triangle.
15. If the midpoint of one side of a rectangle is joined to the endpoints of the opposite side, then an isosceles triangle is formed.

* 16. If the median to one side of a triangle is also an altitude of the triangle, then the triangle is isosceles.

* 17. If the diagonals of a parallelogram are perpendicular, then the parallelogram is a rhombus.

18. Use the analytic method to decide what type of quadrilateral is formed when the midpoints of the consecutive sides of a parallelogram are joined by line segments.
19. Use the analytic method to decide what type of triangle is formed when the midpoints of the sides of an isosceles triangle are joined by line segments.

20. Use slopes to verify that the graphs of the equations

$$Ax + By = C \qquad \text{and} \qquad Ax + By = D$$

are parallel. (NOTE: $A \neq 0$, $B \neq 0$, and $C \neq D$.)

21. Use slopes to verify that the graphs of the equations

$$Ax + By = C \qquad \text{and} \qquad Bx - Ay = D$$

are perpendicular. (NOTE: $A \neq 0$ and $B \neq 0$)

22. Use the result in Exercise 20 to find the equation of the line that contains (4, 5) and is parallel to the graph of $2x + 3y = 6$.

23. Use the result in Exercise 21 to find the equation of the line that contains (4, 5) and is perpendicular to the graph of $2x + 3y = 6$.

24. Use the Distance Formula to show that the circle with center (0, 0) and radius length $r$ has the equation $x^2 + y^2 = r^2$.

25. Use the result in Exercise 24 to find the equation of the circle with center (0, 0) and radius length $r = 3$.

26. Use the result in Exercise 24 to find the equation of the circle that has center (0, 0) and contains the point (3, 4).

27. Would the theorem of Exercise 7 remain true for a concave quadrilateral like the one shown?

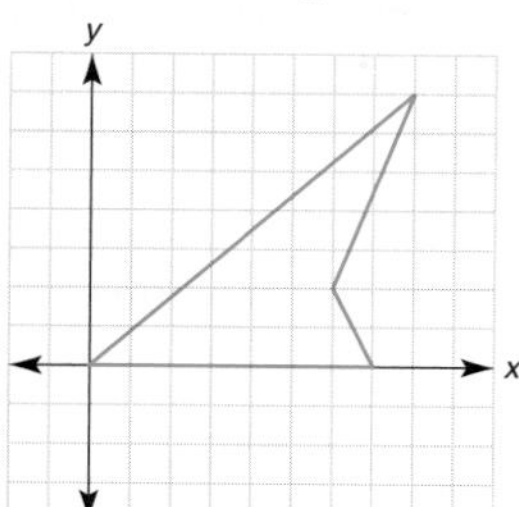


Exercise 27

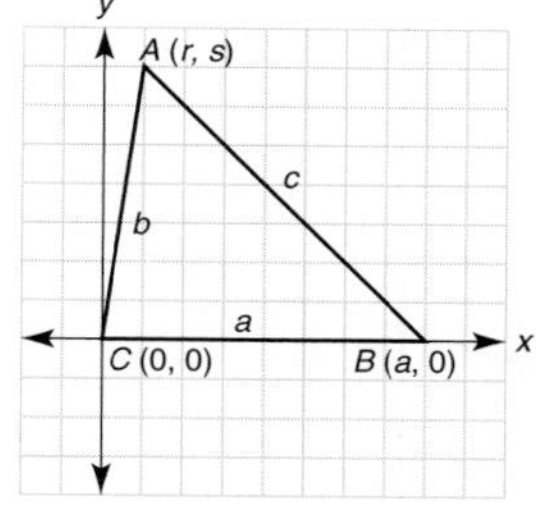


Exercise 28

*28. Complete an analytic proof of the following theorem: In a triangle that has sides of lengths $a$, $b$, and $c$, if $c^2 = a^2 + b^2$, then the triangle is a right triangle.

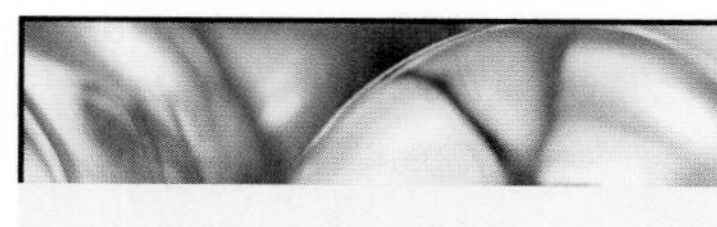

**KEY CONCEPTS**

Slope-Intercept Form of a Line • Point-Slope Form of a Line • Systems of Equations

# 9.5 Equations of Lines

In Section 9.2, we saw that equations such as $2x + 3y = 6$ and $4x - 12y = 60$ have graphs that are lines. To graph an equation of the general form $Ax + By = C$, that equation is often replaced with an equivalent equation of the form $y = mx + b$. For instance, $2x + 3y = 6$ can be transformed into $y = -\frac{2}{3}x + 2$; such equations are known as *equivalent* because their ordered-pair solutions (and graphs) are identical. In particular, we must express a linear equation in the form $y = mx + b$ in order to plot it on a graphing calculator.

EXAMPLE 1

Write the equation $4x - 12y = 60$ in the form $y = mx + b$.

***Solution***

Given $4x - 12y = 60$, we subtract $4x$ from each side of the equation to obtain $-12y = -4x + 60$. Dividing by $-12$ yields

$$\frac{-12y}{-12} = \frac{-4x}{-12} + \frac{60}{-12}$$

Then $y = \frac{1}{3}x - 5$.

*Exs. 1–3*

## SLOPE-INTERCEPT FORM OF A LINE

We now turn our attention to a method for finding the equation of a line. In the following technique, the equation can be found if the slope and the $y$ intercept of the line are known. The form $y = mx + b$ is known as the Slope-Intercept Form of a line.

> THEOREM 9.5.1: **(Slope-Intercept Form of a Line)** The line whose slope is $m$ and whose $y$ intercept is $b$ has the equation $y = mx + b$.

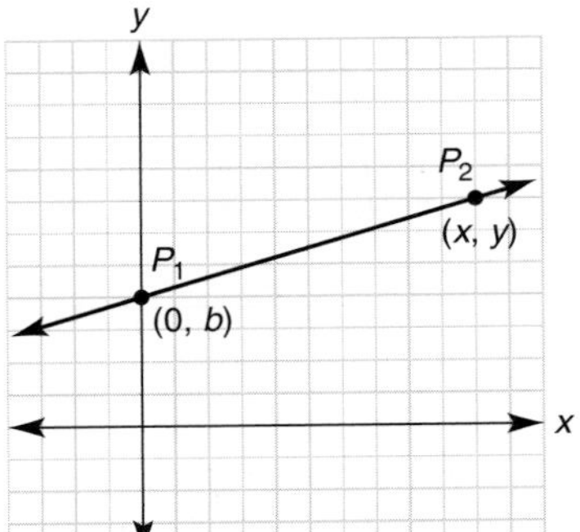


**FIGURE 9.33**

*Proof*

Consider the line whose slope is $m$ (see Figure 9.33). Using the Slope Formula

$$m = \frac{y_2 - y_1}{x_2 - x_1}$$

we designate $(x, y)$ as $P_2$ and $(0, b)$ as $P_1$. Then

$$m = \frac{y - b}{x - 0} \quad \text{or} \quad m = \frac{y - b}{x}$$

Multiplying by $x$, we have $mx = y - b$. Then $mx + b = y$, so $y = mx + b$.

> **Discover!**
>
> Use a graphing calculator to graph $Y_1 = x$, $Y_2 = x^2$, and $Y_3 = x^3$. Which of these is (are) a line(s)?
>
> ANSWER
>
> $Y_1 = x$

### EXAMPLE 2

Find the general equation $Ax + By = C$ for the line with slope $m = -\frac{2}{3}$ and $y$ intercept $-2$.

***Solution***

With $y = mx + b$, we have

$$y = -\frac{2}{3}x - 2$$

Multiplying by 3, we obtain

$$3y = -2x - 6 \qquad \text{so} \qquad 2x + 3y = -6$$

NOTE: An equivalent and correct solution is $-2x - 3y = 6$.

It is often easier to graph an equation if it is in the form $y = mx + b$. When an equation can be changed to this form, we know that its graph is a line that has slope $m$ and contains $(0, b)$.

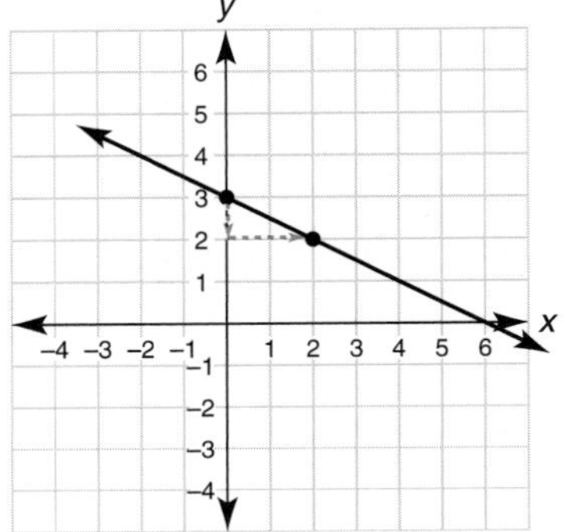


**FIGURE 9.34**

### EXAMPLE 3

Draw the graph of $\frac{1}{2}x + y = 3$.

***Solution***

Solving for $y$, we have $y = -\frac{1}{2}x + 3$. Then $m = -\frac{1}{2}$ and the $y$ intercept is 3. We first plot the point (0, 3). Because $m = -\frac{1}{2}$ or $\frac{-1}{2}$, the vertical change $-1$ corresponds to a horizontal change of $+2$. Thus the second point is located 1 unit down from and 2 units to the right of the first point. The line is drawn in Figure 9.34.

Exs. 4–8

## POINT-SLOPE FORM OF A LINE

If slope $m$ and a point other than the $y$ intercept of a line are known, we do not use the Slope-Intercept Form to find the equation of the line. Instead, the Point-Slope Form of the equation of a line is used. This form is also used when the coordinates of two points of the line are known; in that case, the value of $m$ is found by the Slope Formula. The form $y - y_1 = m(x - x_1)$ is known as the Point-Slope Form of a line.

> THEOREM 9.5.2: **(Point-Slope Form of a Line)** The line that has slope $m$ and contains the point $(x_1, y_1)$ has the equation
>
> $$y - y_1 = m(x - x_1)$$

***Proof***

Let $P_1$ be the given point $(x_1, y_1)$ on the line, and let $P_2$ be $(x, y)$, which represents any other point on the line. (See Figure 9.35.) Using the Slope Formula, we have

$$m = \frac{y - y_1}{x - x_1}$$

Multiplying the equation by $(x - x_1)$ yields

$$m(x - x_1) = y - y_1$$

It follows that

$$y - y_1 = m(x - x_1)$$

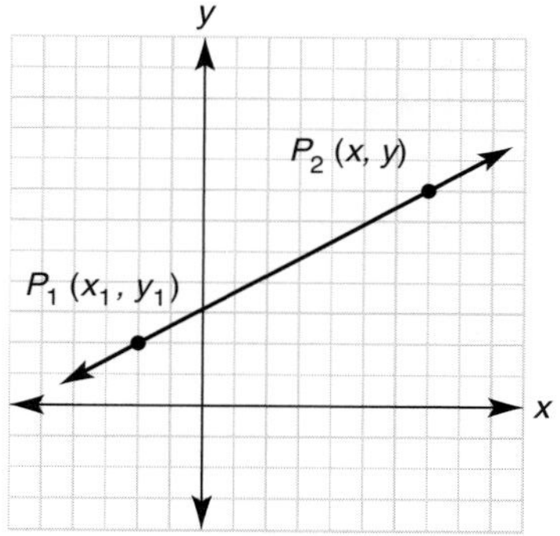


**FIGURE 9.35**

EXAMPLE 4

Find the general equation, $Ax + By = C$, for the line that has the slope $m = 2$ and contains the point $(-1, 3)$.

***Solution***

We have $m = 2$, $x_1 = -1$, and $y_1 = 3$. Applying the Point-Slope Form, we find that the line in Figure 9.36 has the equation

$$y - 3 = 2[x - (-1)]$$
$$y - 3 = 2(x + 1)$$
$$y - 3 = 2x + 2$$
$$-2x + y = 5$$

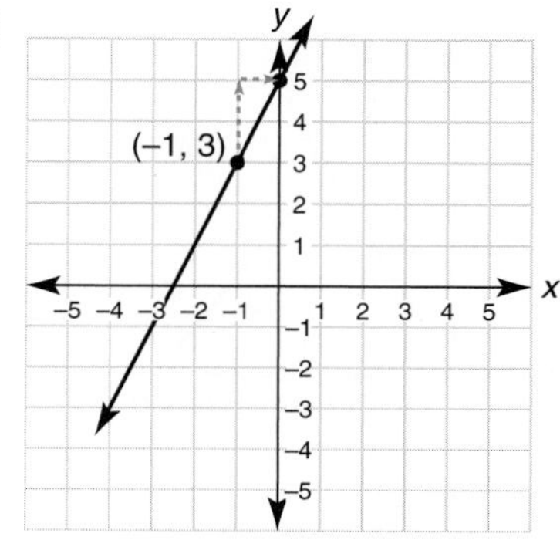


**FIGURE 9.36**

An equivalent answer for Example 4 is the equation $2x - y = -5$. The form $y = 2x + 5$ emphasizes that the slope is $m = 2$ and the $y$ intercept is $(0, 5)$. With $m = 2$, or $\frac{2}{1}$, the vertical change of 2 corresponds to a horizontal change of 1. See Figure 9.36.

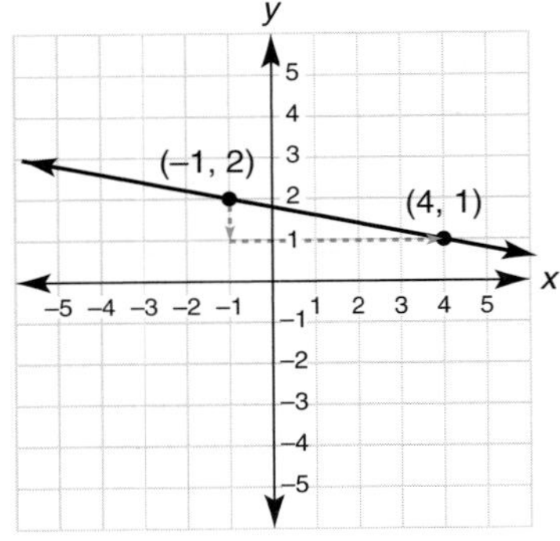


**FIGURE 9.37**

### EXAMPLE 5

Find an equation for the line containing the points $(-1, 2)$ and $(4, 1)$.

***Solution***

To use the Point-Slope Form, we need to know the slope of the line (see Figure 9.37). When we choose $P_1\ (-1, 2)$ and $P_2\ (4, 1)$, the Slope Formula reads

$$m = \frac{1 - 2}{4 - (-1)} = \frac{-1}{5} = -\frac{1}{5}$$

Therefore, $$y - 2 = -\frac{1}{5}[x - (-1)]$$

Then $$y - 2 = -\frac{1}{5}[x + 1]$$

and $$y - 2 = -\frac{1}{5}x - \frac{1}{5}$$

Multiplying the equation by 5, we obtain

$$5y - 10 = -1x - 1 \qquad \text{so} \qquad x + 5y = 9$$

NOTE: Other forms of the answer are $-x - 5y = -9$ and $y = -\frac{1}{5}x + \frac{9}{5}$. In any correct form, the given points must satisfy the equation.

In Example 6, we use the Point-Slope Form to find an equation for a median of a triangle.

*Exs. 9–12*

### EXAMPLE 6

For $\triangle ABC$, the vertices are $A(0, 0)$, $B(2a, 0)$, and $C(2b, 2c)$. Find the equation of median $\overline{CM}$ in the form $y = mx + b$. See Figure 9.38.

***Solution***

For $\overline{CM}$ to be a median of $\triangle ABC$, $M$ must be the midpoint of $\overline{AB}$. Then

$$M = \left(\frac{0 + 2a}{2}, \frac{0 + 0}{2}\right) = (a, 0)$$

To determine an equation for $\overline{CM}$, we also need to know its slope. With $M(a, 0)$ and $C(2b, 2c)$ on $\overline{CM}$, the slope is $m_{\overline{CM}} = \frac{2c - 0}{2b - a}$ or $\frac{2c}{2b - a}$. With $M = (a, 0)$ as the point on the line, $y - y_1 = m(x - x_1)$ becomes

$$y - 0 = \frac{2c}{2b - a}(x - a) \qquad \text{or} \qquad y = \frac{2c}{2b - a}x - \frac{2ac}{2b - a}$$

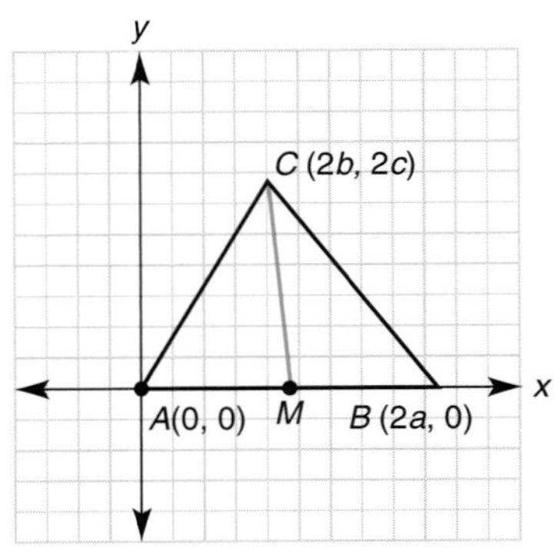


**FIGURE 9.38**

## SOLVING SYSTEMS OF EQUATIONS

In earlier chapters, we solved systems of equations such as

$$\begin{aligned} x + 2y &= 6 \\ 2x - \phantom{2}y &= 7 \end{aligned}$$

by using the Addition Property or the Subtraction Property of Equality. We review the method in Example 7. The solution for the system is an ordered pair; in fact, the solution is the point of intersection of the graphs of the given equations.

*Technology Exploration*

Use a graphing calculator if one is available.

1) Solve each equation of Example 7 for $y$.
2) Graph $Y_1 = -\left(\frac{1}{2}\right)x + 3$ and $Y_2 = 2x - 7$.
3) Use the [intersect] feature to show that the solution for the system is (4, 1).

EXAMPLE 7

Solve the following system by using algebra:

$$\begin{cases} x + 2y = 6 \\ 2x - \phantom{2}y = 7 \end{cases}$$

***Solution***

When we multiply the second equation by 2, the system becomes

$$\begin{cases} x + 2y = 6 \\ 4x - 2y = 14 \end{cases}$$

Adding these equations yields $5x = 20$ so that $x = 4$. Substituting $x = 4$ into the first equation, we get $4 + 2y = 6$, so $2y = 2$. Then $y = 1$. The solution is the ordered pair (4, 1).

Another method for solving a system of equations is geometric and requires graphing. Solving by graphing amounts to finding the point of intersection of the linear graphs. That point is the ordered pair that is the common solution (when one exists) for the two equations. Notice that Example 8 repeats the system of Example 7.

EXAMPLE 8

Solve the following system by graphing:

$$\begin{cases} x + 2y = 6 \\ 2x - \phantom{2}y = 7 \end{cases}$$

***Solution***

Each equation is changed to the form $y = mx + b$ so that the slope and the $y$ intercept are used in graphing:

$$x + 2y = 6 \rightarrow 2y = -1x + 6 \rightarrow y = -\frac{1}{2}x + 3$$
$$2x - y = 7 \rightarrow -y = -2x + 7 \rightarrow y = 2x - 7$$

The graph of $y = -\frac{1}{2}x + 3$ is a line with $y$ intercept 3 and slope $m = -\frac{1}{2}$. The graph of $y = 2x - 7$ is a line with $y$ intercept $-7$ and slope $m = 2$.

The graphs are drawn in the same coordinate system. See Figure 9.39. The point of intersection (4, 1) is the common solution for each of the given equations and thus is the solution of the system.

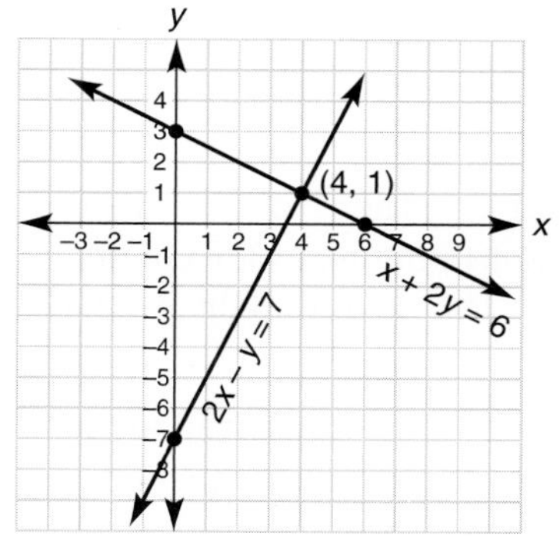


**FIGURE 9.39**

NOTE: To check the result of Examples 7 and 8, we show that (4, 1) satisfies both of the given equations:

$$x + 2y = 6 \rightarrow 4 + 2(1) = 6 \text{ is } \textit{true.}$$
$$2x - y = 7 \rightarrow 2(4) - 1 = 7 \text{ is } \textit{true.}$$

For the solution to be verified, both statements resulting from substitution into the given equations must be true. If either or both statements are false, we do not have the solution of the system.

Advantages of the method of solving a system of equations by graphing include the following:

1. It is easy to understand why a system such as

$$\begin{cases} x + 2y = 6 \\ 2x - y = 7 \end{cases} \quad \text{can be replaced by} \quad \begin{cases} x + 2y = 6 \\ 4x - 2y = 14 \end{cases}$$

when we are solving by addition or subtraction. We know that the graphs of $2x - y = 7$ and $4x - 2y = 14$ are the same line because each can be changed to the form $y = 2x - 7$.

2. It is easy to understand why a system such as

$$\begin{cases} x + 2y = 6 \\ 2x + 4y = -4 \end{cases}$$

has no solution. In Figure 9.40, the graphs are parallel lines.

The first equation is equivalent to $y = -\frac{1}{2}x + 3$, and the second equation can be changed to $y = -\frac{1}{2}x - 1$. Both lines have slope $m = -\frac{1}{2}$ and therefore are parallel.

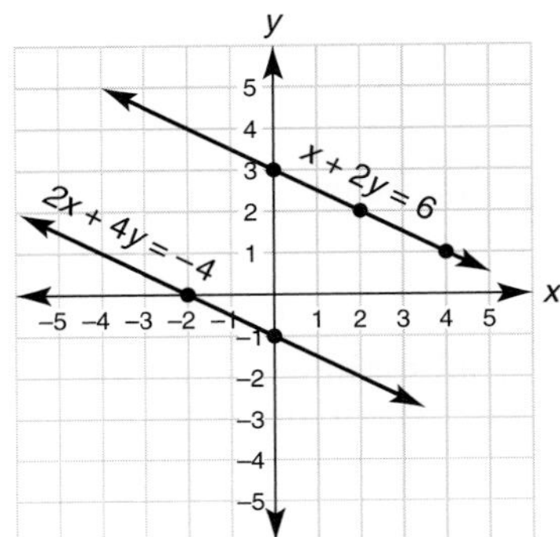


**FIGURE 9.40**

Algebraic substitution can also be used to solve a system of equations. In our approach, we write each equation in the form $y = mx + b$ and then equate the expressions for $y$. Once the $x$ coordinate of the solution is known, we substitute this value of $x$ into either equation to find the value of $y$.

EXAMPLE 9

Use substitution to solve $\begin{cases} x + 2y = 6 \\ 2x - y = 7 \end{cases}$

***Solution***

Solving for $y$, we have

$$x + 2y = 6 \rightarrow 2y = -1x + 6 \rightarrow y = -\frac{1}{2}x + 3$$
$$2x - y = 7 \rightarrow -1y = -2x + 7 \rightarrow y = 2x - 7$$

Equating the expressions for $y$ yields $-\frac{1}{2}x + 3 = 2x - 7$. Then $-2\frac{1}{2}x = -10$, or $-2.5x = -10$. Dividing by $-2.5$, we obtain $x = 4$. Substitution of 4 for $x$ in the equation $y = 2x - 7$ leads to $y = 2(4) - 7$, so $y = 1$. The solution is the ordered pair (4, 1).

NOTE: Substitution of $x = 4$ into the equation $y = -\frac{1}{2}x + 3$ would lead to the same value of $y$, namely $y = 1$. Thus one can substitute into either equation.

*Exs. 13–16*

The method illustrated in Example 9 is also used in our final example. In the proof of Theorem 9.5.3, we use equations of lines to determine the centroid of a triangle.

EXAMPLE 10

Formulate a plan to complete the proof of Theorem 9.5.3. See Figure 9.41.

> THEOREM 9.5.3: The three medians of a triangle are concurrent at a point that is two-thirds the distance from any vertex to the midpoint of the opposite side.

***Solution***

The proof can be completed as follows:

1. Find the coordinates of the two midpoints $X$ and $Y$. See Figures 9.41(a) and 9.41(b). Note that
$$X = (a + b, c) \qquad \text{and} \qquad Y = (b, c)$$
2. Find the equations of the lines containing $\overline{AX}$ and $\overline{BY}$. The equations for $\overline{AX}$ and $\overline{BY}$ are $y = \frac{c}{a+b}x$ and $y = \frac{-c}{2a-b}x + \frac{2ac}{2a-b}$, respectively.
3. Find the point of intersection $Z$ of $\overline{AX}$ and $\overline{BY}$, as shown in Figure 9.41(b). Solving the system provides the solution
$$Z = \left(\frac{2}{3}(a + b), \frac{2}{3}c\right)$$
4. It can now be shown that $AZ = \frac{2}{3} \cdot AX$ and $BZ = \frac{2}{3} \cdot BY$. See Figure 9.41(b), in which we can show that
$$AZ = \frac{2}{3}\sqrt{(a + b)^2 + c^2} \qquad \text{and} \qquad AX = \sqrt{(a + b)^2 + c^2}$$
5. It can also be shown that point $Z$ lies on the third median $\overline{CW}$, whose equation is $y = \frac{2c}{2b - a}(x - a)$. See Figure 9.41(c).
6. We can also show that $CZ = \frac{2}{3} \cdot CW$, which completes the proof.

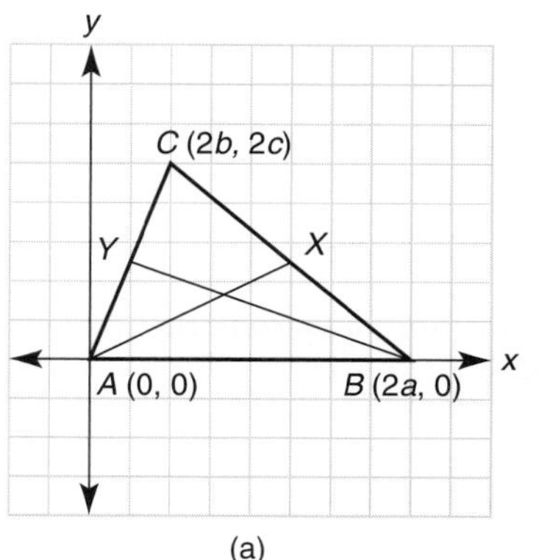

(a)

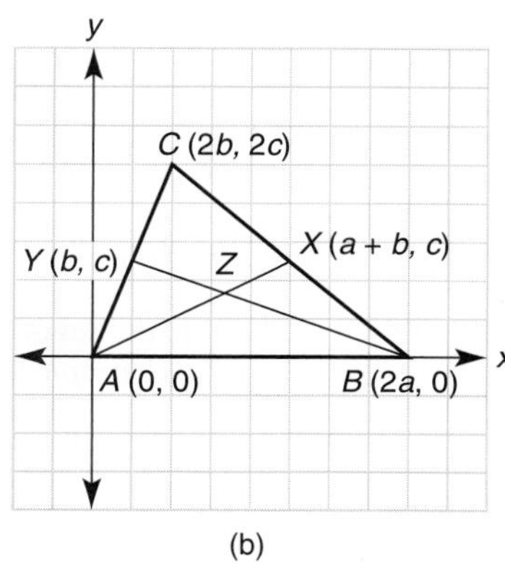

(b)

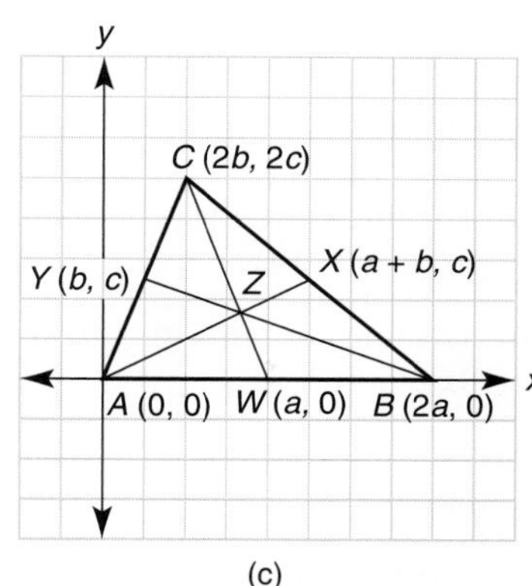

(c)

**FIGURE 9.41**

## 9.5 Exercises

*In Exercises 1 to 4, use division to write an equation of the form $Ax + By = C$ that is equivalent to the one provided. Then write the given equation in the form $y = mx + b$.*

1. $8x + 16y = 48$
2. $15x - 35y = 105$
3. $-6x + 18y = -240$
4. $27x - 36y = 108$

*In Exercises 5 to 8, draw the graph of each equation by using the method of Example 3.*

5. $y = 2x - 3$
6. $y = -2x + 5$
7. $\frac{2}{5}x + y = 6$
8. $3x - 2y = 12$

*In Exercises 9 to 24, find the equation of the line described. Leave the solution in the form $Ax + By = C$.*

9. The line has slope $m = -\frac{2}{3}$ and contains $(0, 5)$.
10. The line has slope $m = -3$ and contains $(0, -2)$.
11. The line contains $(2, 4)$ and $(0, 6)$.
12. The line contains $(-2, 5)$ and $(2, -1)$.
13. The line contains $(0, -1)$ and $(3, 1)$.
14. The line contains $(-2, 0)$ and $(4, 3)$.
15. The line contains $(0, b)$ and $(a, 0)$.
16. The line contains $(b, c)$ and has slope $d$.
17. The line has intercepts $a = 2$ and $b = -2$.
18. The line has intercepts $a = -3$ and $b = 5$.
19. The line contains $(-1, 5)$ and is parallel to the line $5x + 2y = 10$.
20. The line contains $(0, 3)$ and is parallel to the line $3x + y = 7$.
21. The line contains $(0, -4)$ and is perpendicular to the line $y = \frac{3}{4}x - 5$.
22. The line contains $(2, -3)$ and is perpendicular to the line $2x - 3y = 6$.
23. The line is the perpendicular bisector of the line segment that joins $(3, 5)$ and $(5, -1)$.
24. The line is the perpendicular bisector of the line segment that joins $(-4, 5)$ and $(1, 1)$.

*In Exercises 25 and 26, find the equation of the line in the form $y = mx + b$.*

25. The line contains $(g, h)$ and is *perpendicular* to the line $y = \frac{a}{b}x + c$.
26. The line contains $(g, h)$ and is *parallel* to the line $y = \frac{a}{b}x + c$.

*In Exercises 27 to 32, use graphing to find the point of intersection of the two lines. Use Example 8 as a guide.*

27. $y = \frac{1}{2}x - 3$ and $y = \frac{1}{3}x - 2$
28. $y = 2x + 3$ and $y = 3x$
29. $2x + y = 6$ and $3x - y = 19$
30. $\frac{1}{2}x + y = -3$ and $\frac{3}{4}x - y = 8$
31. $4x + 3y = 18$ and $x - 2y = 10$
32. $2x + 3y = 3$ and $3x - 2y = 24$

*In Exercises 33 to 38, use algebra to find the point of intersection of the two lines whose equations are provided. Use Example 7 as a guide.*

33. $2x + y = 8$ and $3x - y = 7$
34. $2x + 3y = 7$ and $x + 3y = 2$
35. $2x + y = 11$ and $3x + 2y = 16$
36. $x + y = 1$ and $4x - 2y = 1$
37. $2x + 3y = 4$ and $3x - 4y = 23$
38. $5x - 2y = -13$ and $3x + 5y = 17$

*In Exercises 39 to 42, use substitution to solve the system. Use Example 9 as a guide.*

39. $y = \frac{1}{2}x - 3$ and $y = \frac{1}{3}x - 2$
40. $y = 2x + 3$ and $y = 3x$
41. $y = a$ and $y = bx + c$
42. $x = d$ and $y = fx + g$
43. For $\triangle ABC$, the vertices are $A(0, 0)$, $B(a, 0)$, and $C(b, c)$. In terms of $a$, $b$, and $c$, find the coordinates of the orthocenter of $\triangle ABC$. (The orthocenter is the point of concurrence for the altitudes of a triangle.)

44. For isosceles $\triangle PNQ$, the vertices are $P(-2a, 0)$, $N(2a, 0)$, and $Q(0, 2b)$. In terms of $a$ and $b$, find the coordinates of the circumcenter of $\triangle PNQ$. (The circumcenter is the point of concurrence for the perpendicular bisectors of the sides of a triangle.)

*In Exercises 45 and 46, complete an analytic proof for each theorem.*

45. The altitudes of a triangle are concurrent.

46. The perpendicular bisectors of the sides of a triangle are concurrent.

47. Describe the steps of the procedure that enables us to find the distance from a point $P(a, b)$ to the line $Ax + By = C$.

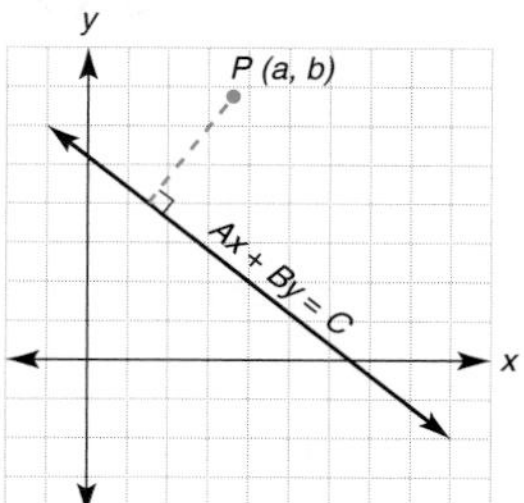


## Perspective on History

### THE BANACH-TARSKI PARADOX

In the 1920s, two Polish mathematicians proposed a mathematical dilemma to their colleagues. Known as the Banach-Tarski paradox, their proposal has puzzled students of geometry for decades. What was most baffling was that the proposal indicated that matter could be created through rearrangement of the pieces of a figure! The following steps outline the Banach-Tarski paradox.

First consider the square whose sides are each of length 8. [See Figure 9.42(a).] By counting squares or by applying a formula, it is clear that the 8-by-8 square must have an area of 64 square units. We now subdivide the square (as shown) to form two right triangles and two trapezoids. Note the dimensions indicated on each piece of the square in Figure 9.42(b).

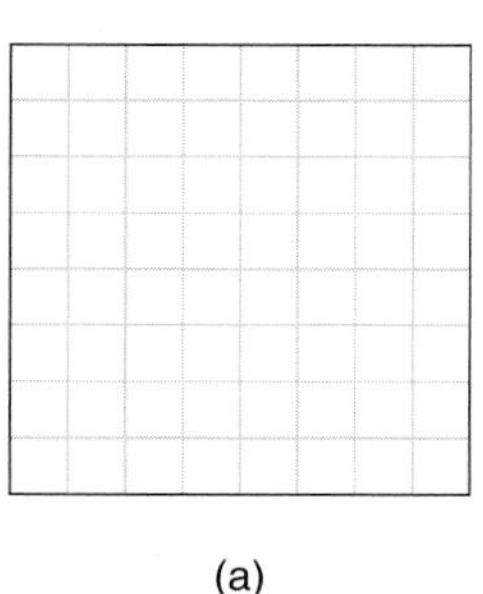
(a)

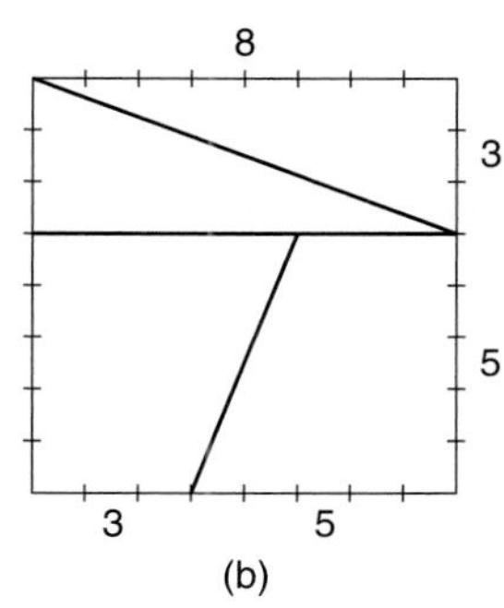

(b)

**FIGURE 9.42**

The parts of the square are now rearranged to form a rectangle (see Figure 9.43) whose dimensions are 13 and 5. This rectangle clearly has an area that measures 65 square units, 1 square unit more than the given square! How is it possible that the second figure has an area greater than the first?

The puzzle is real, but you may also sense that something is wrong. This paradox can be explained by considering the slopes of lines. The triangles, which have legs of lengths 3 and 8, determine a hypotenuse whose slope is $-\frac{3}{8}$. Although the side of the trapezoid appears to be collinear with the hypotenuse, it actually has a slope of $-\frac{2}{5}$. It was easy to accept that the segments were collinear because the slopes are nearly equal; in fact, $-\frac{3}{8} = -0.375$ and $-\frac{2}{5} = -0.400$. In Figure 9.44 (which is somewhat exaggerated), a very thin parallelogram appears in the space between the original segments of the cut-up square. One may quickly conclude that the area of that parallelogram is 1 square unit, and the paradox has been resolved once more!

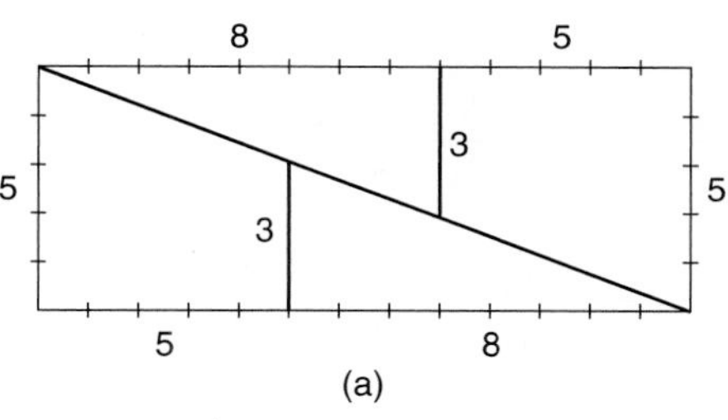

(a)

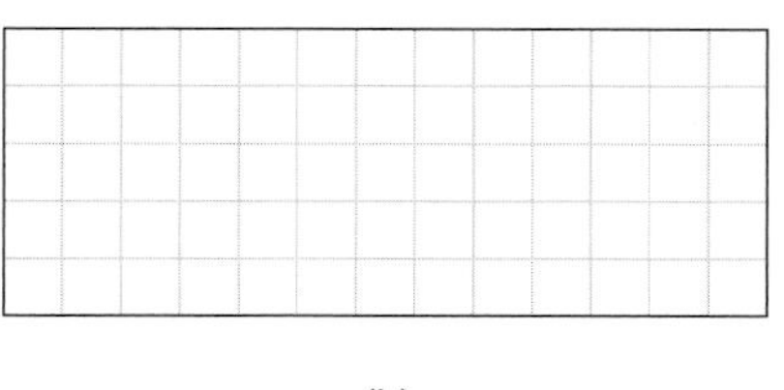
(b)

**FIGURE 9.43**

**FIGURE 9.44**

# Perspective on Application

## THE POINT-OF-DIVISION FORMULAS

The subject of this feature is a generalization of the formulas that led to the Midpoint Formula. Recall that the midpoint of the line segment that joins $A(x_1, y_1)$ to $B(x_2, y_2)$ is given by $M = \left(\frac{x_1 + x_2}{2}, \frac{y_1 + y_2}{2}\right)$, which is derived from the formulas $x = x_1 + \frac{1}{2}(x_2 - x_1)$ and $y = y_1 + \frac{1}{2}(y_2 - y_1)$. The formulas for a more general location of point between $A$ and $B$ follow; to better understand how these formulas can be applied, we note that $r$ represents the fractional part of the distance from point $A$ to point $B$ on $\overline{AB}$; in the Midpoint Formula, $r = \frac{1}{2}$.

> Point-of-Division Formulas: Let $A(x_1, y_1)$ and $B(x_2, y_2)$ represent the endpoints of $\overline{AB}$. Where $r$ represents a common fraction $(0 < r < 1)$, the coordinates of the point $P$ that lies this part $r$ of the distance from $A$ to $B$ is given by
>
> $x = x_1 + r(x_2 - x_1)$ and $y = y_1 + r(y_2 - y_1)$

The following table clarifies the use of the formulas above.

**TABLE 9.3**

| Value of $r$ | Location of Point $P$ on $\overline{AB}$ |
|---|---|
| $\frac{1}{3}$ | Point $P$ lies $\frac{1}{3}$ of the distance from $A$ to $B$ |
| $\frac{3}{4}$ | Point $P$ lies $\frac{3}{4}$ of the distance from $A$ to $B$ |

### EXAMPLE 1

Find the point $P$ on $\overline{AB}$ that is one-third of the distance from $A(-1, 2)$ to $B(8, 5)$.

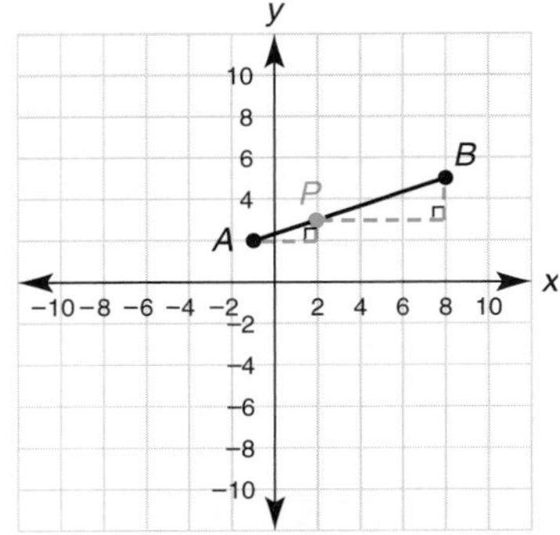


***Solution***

$x = -1 + \frac{1}{3}(8 - [-1])$ and $y = 2 + \frac{1}{3}(5 - 2)$

Then $x = -1 + \frac{1}{3}(9)$, so $x = -1 + 3$ or $2$

also, $y = 2 + \frac{1}{3}(3)$, so $y = 2 + 1$ or $3$

The desired point is $P(2, 3)$.

NOTE: See the figure above in which similar triangles can be used to explain why point $P$ is the desired point.

In some higher-level courses, the value of $r$ is not restricted to values between 0 and 1. For instance, we could choose $r = 2$ or $r = -1$. For such values of $r$, the point $P$ produced by the Point-of-Division Formulas remains collinear with $A$ and $B$. However, the point $P$ that is produced does not lie between $A$ and $B$.

# Summary

### A LOOK BACK AT CHAPTER 9

Our goal in this chapter was to relate algebra and geometry. This relationship is called analytic geometry or coordinate geometry. Formulas for the length of a line segment, the midpoint of a line segment, and the slope of a line were developed. We found the equation for a line and used it for graphing. Analytic proofs were provided for a number of theorems of geometry.

### A LOOK AHEAD TO CHAPTER 10

In the next chapter, we will again deal with the right triangle. Three trigonometric ratios (sine, cosine, and tangent) will be defined for an acute angle of the right triangle in terms of its sides. An area formula for triangles will be derived using the sine ratio. We will also prove the Law of Sines and the Law of Cosines for acute triangles.

### KEY CONCEPTS

9.1 Analytic Geometry • Cartesian Coordinate System • Rectangular Coordinate System • *x* Axis • *y* Axis • Quadrants • Origin • *x* Coordinate • *y* Coordinate • Ordered Pair • Distance Formula • Linear Equation • Midpoint Formula

9.2 Graphs of Equations • *x* Intercept, *y* Intercept • Slope • Slope Formula • Negative Reciprocal

9.3 Formulas and Relationships • Placement of Figure

9.4 Analytic Proof • Synthetic Proof

9.5 Slope-Intercept Form of a Line • Point-Slope Form of a Line • Systems of Equations

# Chapter 9 Review Exercises

1. Find the distance between each pair of points:
   a) $(6, 4)$ and $(6, -3)$
   b) $(1, 4)$ and $(-5, 4)$
   c) $(-5, 2)$ and $(7, -3)$
   d) $(x - 3, y + 2)$ and $(x, y - 2)$
2. Find the distance between each pair of points:
   a) $(2, -3)$ and $(2, 5)$
   b) $(3, -2)$ and $(-7, -2)$
   c) $(-4, 1)$ and $(4, 5)$
   d) $(x - 2, y - 3)$ and $(x + 4, y + 5)$
3. Find the midpoint of the line segment that joins each pair of points in Exercise 1.
4. Find the midpoint of the line segment that joins each pair of points in Exercise 2.
5. Find the slope of the line joining each pair of points in Exercise 1.
6. Find the slope of the line joining each pair of points in Exercise 2.
7. $(2, 1)$ is the midpoint of $\overline{AB}$, in which $A$ has coordinates $(8, 10)$. Find the coordinates of $B$.
8. The $y$ axis is the perpendicular bisector of $\overline{RS}$. Find the coordinates of $R$ if $S$ is the point $(-3, 7)$.
9. If $A$ has coordinates $(2, 1)$ and $B$ has coordinates $(x, 3)$, find $x$ such that the slope of $\overleftrightarrow{AB}$ is $-3$.
10. If $R$ has coordinates $(-5, 2)$ and $S$ has coordinates $(2, y)$, find $y$ such that the slope of $\overleftrightarrow{RS}$ is $\frac{-6}{7}$.
11. Without graphing, determine whether the pairs of lines are parallel, perpendicular, the same, or none of these:
    a) $x + 3y = 6$ and $3x - y = -7$
    b) $2x - y = -3$ and $y = 2x - 14$
    c) $y + 2 = -3(x - 5)$ and $2y = 6x + 11$
    d) $0.5x + y = 0$ and $2x - y = 10$
12. Determine whether the points $(-6, 5)$, $(1, 7)$, and $(16, 10)$ are collinear.
13. Find $x$ such that $(-2, 3)$, $(x, 6)$, and $(8, 8)$ are collinear.
14. Draw the graph of $3x + 7y = 21$, and name the $x$ intercept $a$ and the $y$ intercept $b$.
15. Draw the graph of $4x - 3y = 9$ by changing the equation to Slope-Intercept Form.
16. Draw the graph of $y + 2 = \frac{-2}{3}(x - 1)$.

17. Write the equation for:

a) The line through (2, 3) and (−3, 6)
b) The line through (−2, −1) and parallel to the line through (6, −3) and (8, −9)
c) The line through (3, −2) and perpendicular to the line $x + 2y = 4$
d) The line through (−3, 5) and parallel to the $x$ axis

18. Show that the triangle whose vertices are $A(-2, -3)$, $B(4, 5)$, and $C(-4, 1)$ is a right triangle.

19. Show that the triangle whose vertices are $A(3, 6)$, $B(-6, 4)$, and $C(1, -2)$ is an isosceles triangle.

20. Show that the quadrilateral whose vertices are $R(-5, -3)$, $S(1, -11)$, $T(7, -6)$, and $V(1, 2)$ is a parallelogram.

*In Exercises 21 and 22, find the intersection of the graphs of the two equations by graphing.*

21. $4x - 3y = -3$
$x + 2y = 13$

22. $y = x + 3$
$y = 4x$

*In Exercises 23 and 24, solve the systems of equations in Exercises 21 and 22 by using algebraic methods.*

23. Refer to Exercise 21.

24. Refer to Exercise 22.

25. Three of the four vertices of a parallelogram are (0, −2), (6, 8), and (10, 1). Find the possibilities for the coordinates of the remaining vertex.

26. $A(3, 1)$, $B(5, 9)$, and $C(11, 3)$ are the vertices of $\triangle ABC$.

a) Find the length of the median from $B$ to $\overline{AC}$.
b) Find the slope of the altitude from $B$ to $\overline{AC}$.
c) Find the slope of a line through $B$ parallel to $\overline{AC}$.

*In Exercises 27 to 30, supply the missing coordinates for the vertices, using as few variables as possible.*

27.

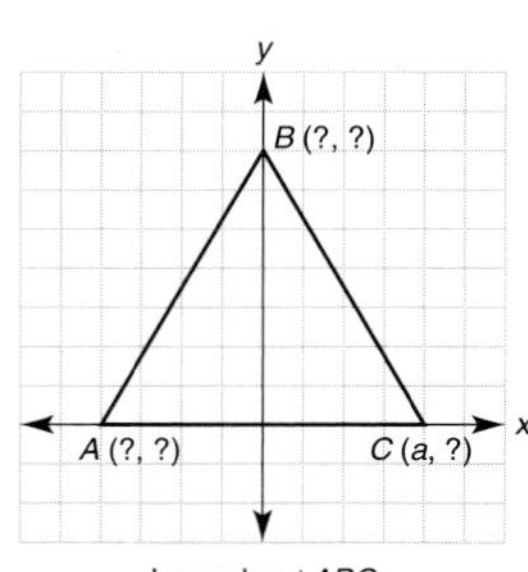


Isosceles $\triangle ABC$ with base $\overline{AC}$

28.

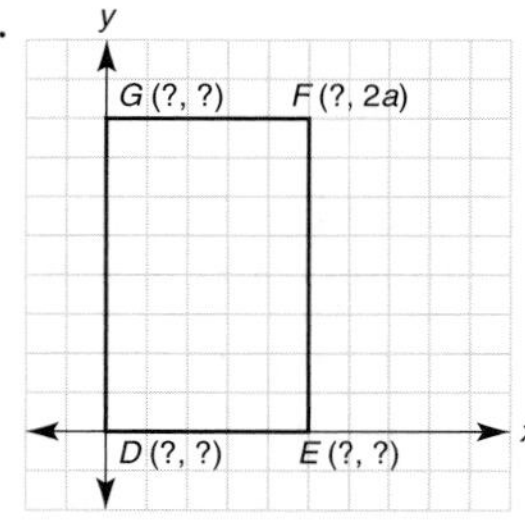


Rectangle $DEFG$ with $DG = 2 \cdot DE$

29. 

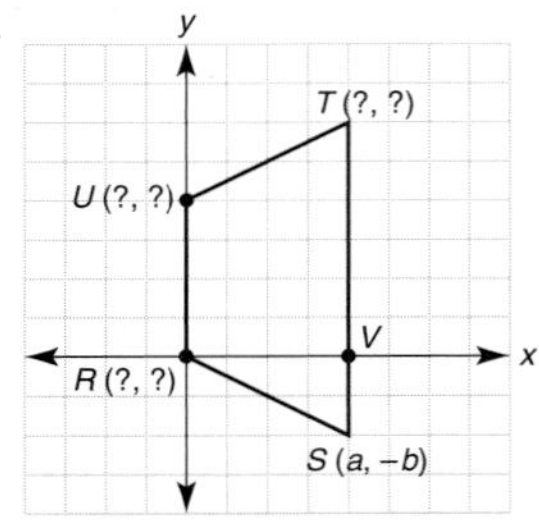


Isosceles trapezoid $RSTU$ with $\overline{RV} \cong \overline{RU}$

30.

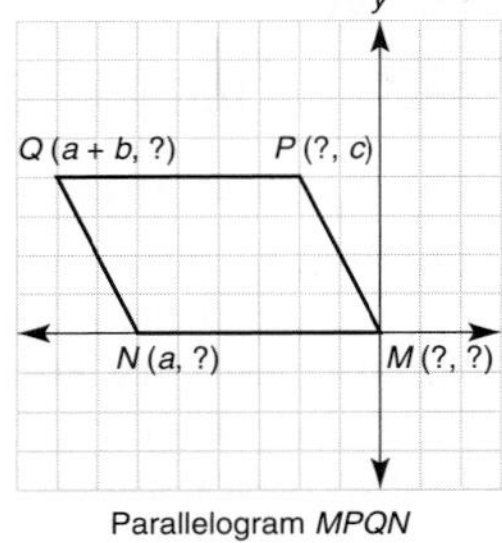


Parallelogram $MPQN$

31. $A(2a, 2b)$, $B(2c, 2d)$, and $C(0, 2e)$ are the vertices of $\triangle ABC$.

a) Find the length of the median from $C$ to $\overline{AB}$.
b) Find the slope of the altitude from $B$ to $\overline{AC}$.
c) Find the equation of the altitude from $B$ to $\overline{AC}$.

*Prove the statements in Exercises 32 to 36 by using analytic geometry.*

32. The segments that join the midpoints of consecutive sides of a parallelogram form another parallelogram.

33. If the diagonals of a rectangle are perpendicular, then the rectangle is a square.

34. If the diagonals of a trapezoid are equal in length, then the trapezoid is an isosceles trapezoid.

35. If two medians of a triangle are equal in length, then the triangle is isosceles.

36. The segments joining the midpoints of consecutive sides of an isosceles trapezoid form a rhombus.

# Chapter 9 Test

1. In the coordinate system provided, give the coordinates of:

    a) Point $A$ in the form $(x, y)$ ________
    b) Point $B$ in the form $(x, y)$ ________

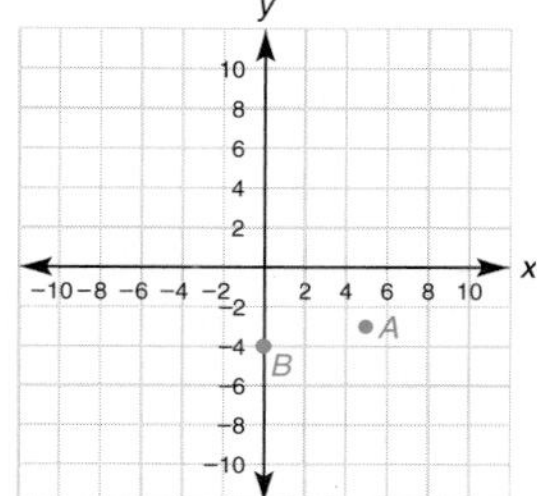

*Exercises 1–4*

2. In the coordinate system for Exercise 1, plot and label each point:

    $C(-6, 1)$ and $D(0, 9)$

3. Use $d = \sqrt{(x_2 - x_1)^2 + (y_2 - y_1)^2}$ to find the length of $\overline{CD}$. See Exercise 2. ________

4. In the form $(x, y)$, determine the midpoint of $\overline{CD}$ as described in Exercise 2. ________

5. Complete the following table of $x$ and $y$ coordinates of points on the graph of the equation $2x + 3y = 12$.

| $x$ | 0 | 3 | | 9 |
|---|---|---|---|---|
| $y$ | | | 4 | |

6. Using the table from Exercise 5, sketch the graph of $2x + 3y = 12$.

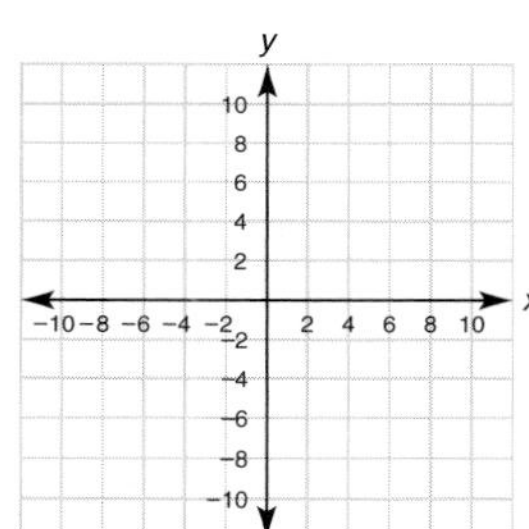

*Exercises 5–6*

7. Find the slope $m$ of a line containing these points:

    a) $(-1, 3)$ and $(2, -6)$ ________
    b) $(a, b)$ and $(c, d)$ ________

8. Line $\ell$ has slope $m = \frac{2}{3}$. Find the slope of any line that is:

    a) Parallel to $\ell$ ________
    b) Perpendicular to $\ell$ ________

9. What type of quadrilateral $ABCD$ is represented if its vertices are $A(0, 0)$, $B(a, 0)$, $C(a + b, c)$, and $D(b, c)$?

    ________

10. For quadrilateral $ABCD$ of Exercise 9 to be a rhombus, it would be necessary that $AB = AD$. Using $a$, $b$, and $c$ (as in Exercise 9), write the equation stating that $AB = AD$. ________

11. Being as specific as possible, describe the polygon shown in each figure.

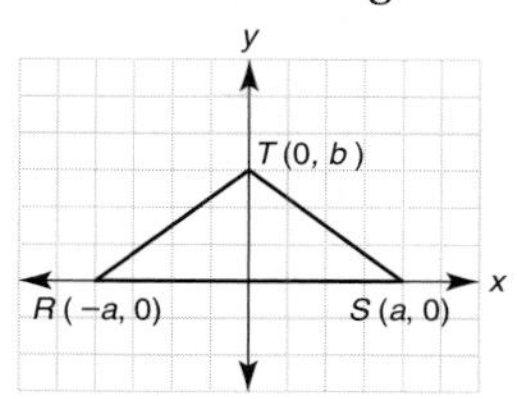


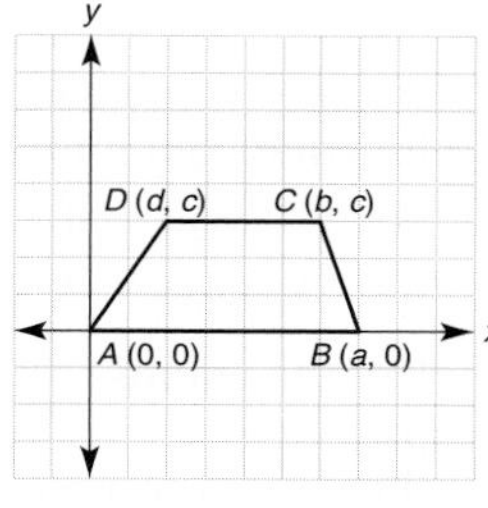


a) ________________ b) ________________

12. What formula (by name) is used to establish that

    a) two lines are parallel? ________
    b) two line segments are congruent? ________

13. Using as few variables as possible, state the coordinates of each point if $\triangle DEF$ is isosceles with $\overline{DF} \cong \overline{FE}$.

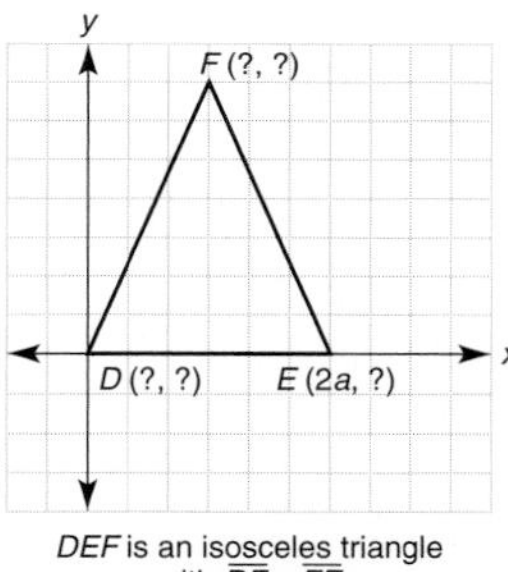

$DEF$ is an isosceles triangle with $\overline{DF} \cong \overline{FE}$

$D$ ( , ), $E$ (2$a$, ), $F$ ( , )

14. For proving the theorem "The midpoint of the hypotenuse of a right triangle is equidistant from all three vertices," which drawing is best? ________

a)

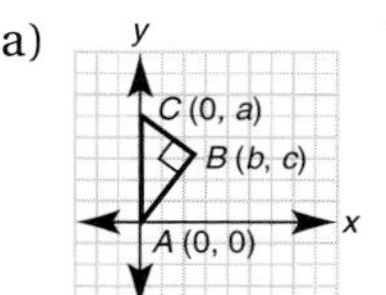


b)

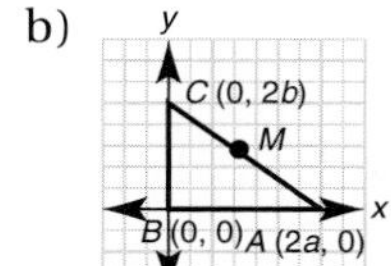


c)

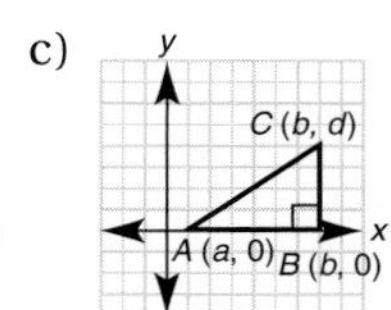


15. In the figure, we see that $m_{\overline{RS}} = m_{\overline{VT}} = 0$. Find the equation that relates $r$, $s$, and $t$ if it is known that $RSTV$ is a parallelogram.

    ________

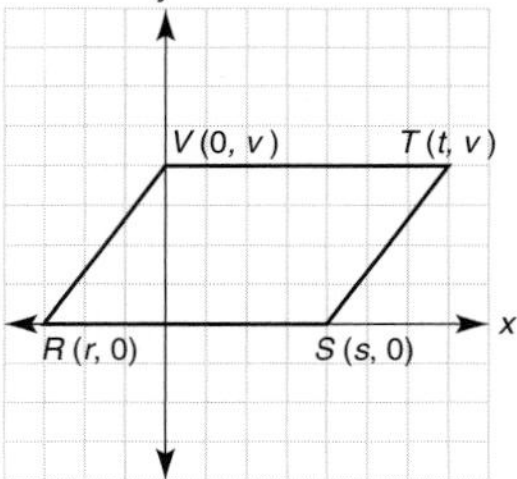

16. In the form $y = mx + b$, find the equation of the line that:

a) Contains the points (0, 4) and (2, 6) ________

b) Contains (0, −3) and is parallel to the line $y = \frac{3}{4}x - 5$ ________

17. Use $y - y_1 = m(x - x_1)$ to find the equation of the line that contains $(a, b)$ and is perpendicular to the line $y = -\frac{1}{c}x + d$. Leave the answer (equation) in the form $y = mx + b$. ________

18. Use the graphs provided to solve the system consisting of the equations $x + 2y = 6$ and $2x - y = 7$.

________

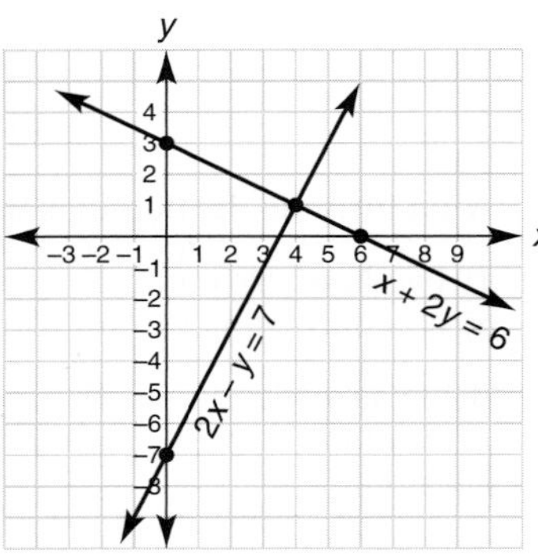


19. Use algebra to solve the system consisting of the equations $5x - 2y = -13$ and $3x + 5y = 17$.

________

20. Use the drawing provided to complete the proof of the theorem "The line segment that joins the midpoints of two sides of a triangle is parallel to the third side of the triangle."

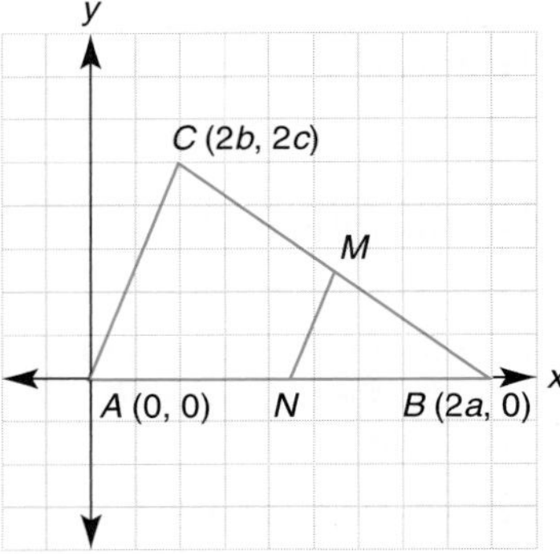


*Proof:* Given $\triangle ABC$ with vertices as shown, let $M$ and $N$ name the midpoints of sides $\overline{CB}$ and $\overline{AB}$, respectively. Then ________________________________

________________________________

________________________________

________________________________

# Chapter 10

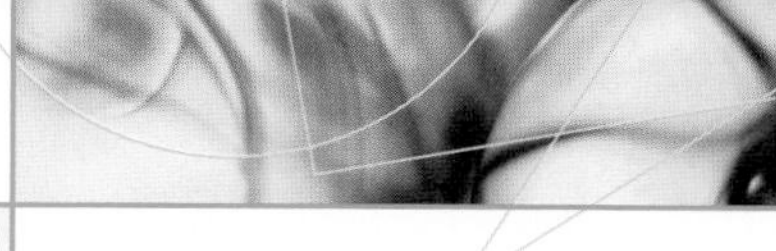

# Introduction to Trigonometry

The lighthouse beacon sends both a "Welcome" message as well as a "Caution" as the boat nears the rocky shore. Methods of trigonometry enable the boat captain to determine the distance from his or her boat to the lighthouse.

The word trigonometry refers to the measurement of triangles. In this chapter, you will discover methods for measuring the angles as well as the lengths of sides of right triangles when the measures of other parts of the triangle are known. These techniques can be extended to other types of triangles.

For this chapter, you will need to use a scientific or graphing calculator.

## Chapter Outline



**For online student resources, visit this textbook's website at math.college.hmco.com/students.**

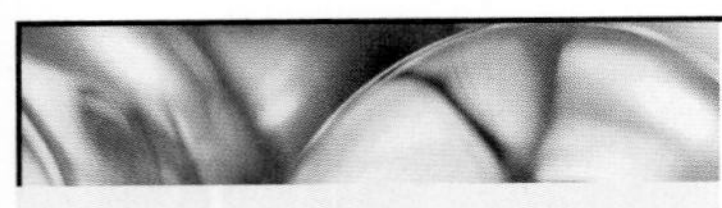

# 10.1 The Sine Ratio and Applications

**KEY CONCEPTS**

Greek Letters: $\alpha$, $\beta$, $\gamma$, $\theta$ • Opposite Side (Leg) • Hypotenuse • Sine Ratio: $\sin\theta = \frac{\text{opposite}}{\text{hypotenuse}}$ • Angle of Elevation • Angle of Depression

In this section, we will deal strictly with similar right triangles. In Figure 10.1, $\triangle ABC \sim \triangle DEF$ and $\angle C$ and $\angle F$ are right angles. Consider corresponding angles $A$ and $D$. If we compare the length of the side opposite each angle to the length of the hypotenuse of each triangle, we see that

$$\frac{BC}{AB} = \frac{EF}{DE} \quad \text{or} \quad \frac{3}{5} = \frac{6}{10}$$

In the two similar right triangles, the ratio of this pair of corresponding sides depends on the measure of acute $\angle A$ (or $\angle D$ because $m\angle A = m\angle D$); for this angle, the numerical value of the ratio

$$\frac{\text{length of side opposite the acute angle}}{\text{length of hypotenuse}}$$

is unique. This ratio becomes smaller for smaller measures of $\angle A$ and larger for larger measures of $\angle A$. This ratio is unique for each measure of an acute angle even though the lengths of the sides of the two similar right triangles containing the angle are different.

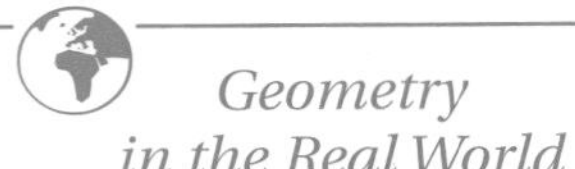

*Geometry in the Real World*

A surveyor uses trigonometry to find both angle measurements and distances.

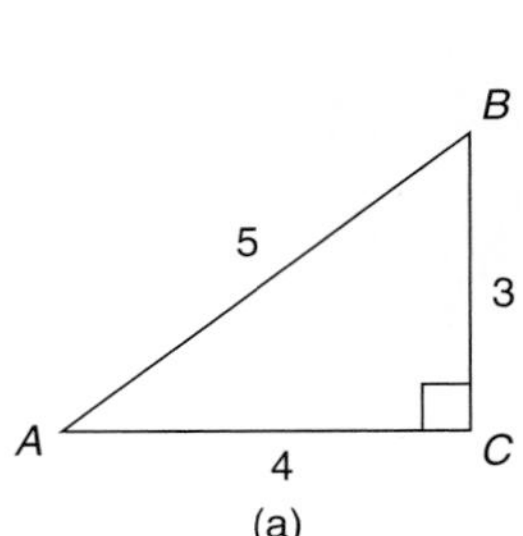


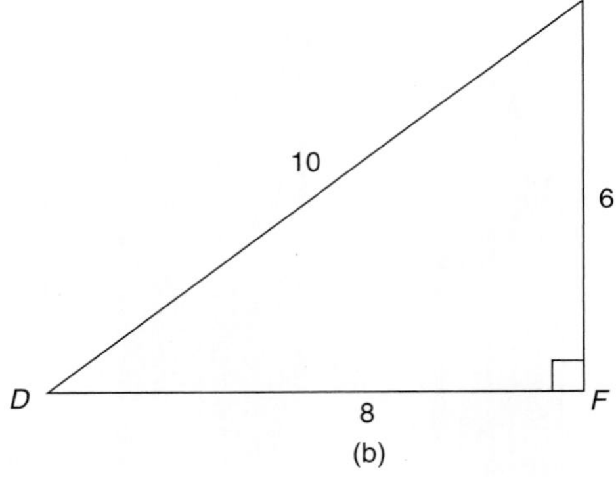


**FIGURE 10.1**

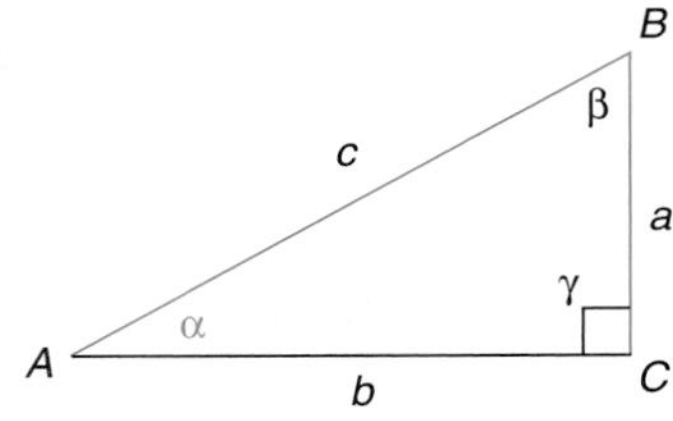


**FIGURE 10.2**

In Figure 10.2, we name the measures of the angles of the right triangle by the Greek letters $\alpha$ (alpha) at vertex $A$, $\beta$ (beta) at vertex $B$, and $\gamma$ (gamma) at vertex $C$. The lengths of the sides opposite vertices $A$, $B$, and $C$ are $a$, $b$, and $c$, respectively. Relative to the acute angle, the lengths of the sides of the right triangle in the following definition are described as "opposite" and "hypotenuse." The word **opposite** is used to mean the length of the side opposite the angle named; the word **hypotenuse** is used to mean the length of the hypotenuse.

DEFINITION: In a right triangle, the **sine ratio** for an acute angle is the ratio $\frac{\text{opposite}}{\text{hypotenuse}}$.

NOTE: In right $\triangle ABC$ in Figure 10.2, we say that $\sin\alpha = \frac{a}{c}$ and $\sin\beta = \frac{b}{c}$, where "sin" is an abbreviation of the word *sine* (pronounced like *sign*).

EXAMPLE 1

In Figure 10.3, find sin $\alpha$ and sin $\beta$ for right $\triangle ABC$.

***Solution***

$a = 3$, $b = 4$, and $c = 5$. Therefore,

$$\sin \alpha = \frac{a}{c} = \frac{3}{5}$$

and

$$\sin \beta = \frac{b}{c} = \frac{4}{5}$$

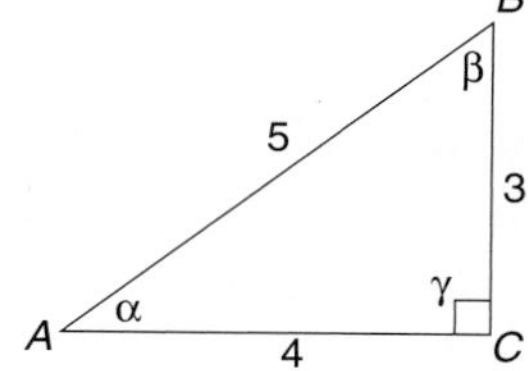


**FIGURE 10.3**

NOTE: Some textbooks use sin $\alpha$ and sin $A$ interchangeably. In Example 1, it is possible to state that $\sin A = \frac{3}{5}$ and $\sin B = \frac{4}{5}$.

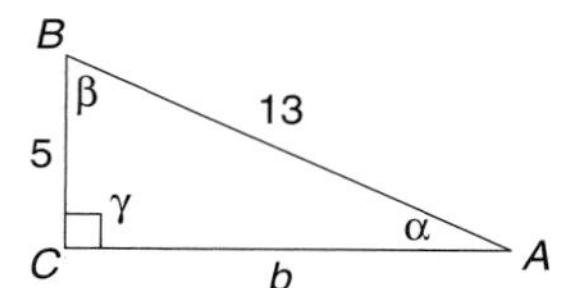


**FIGURE 10.4**

EXAMPLE 2

In Figure 10.4, find sin $\alpha$ and sin $\beta$ for right $\triangle ABC$.

***Solution***

Where $a = 5$ and $c = 13$, we know that $b = 12$ because (5, 12, 13) is a Pythagorean triple. We verify this result using the Pythagorean Theorem.

$$c^2 = a^2 + b^2$$
$$13^2 = 5^2 + b^2$$
$$169 = 25 + b^2$$
$$b^2 = 144$$
$$b = 12$$

Therefore, $\sin \alpha = \frac{a}{c} = \frac{5}{13}$ and $\sin \beta = \frac{b}{c} = \frac{12}{13}$

The following activity is designed to give you a better understanding of the meaning of expressions such as sin 53°.

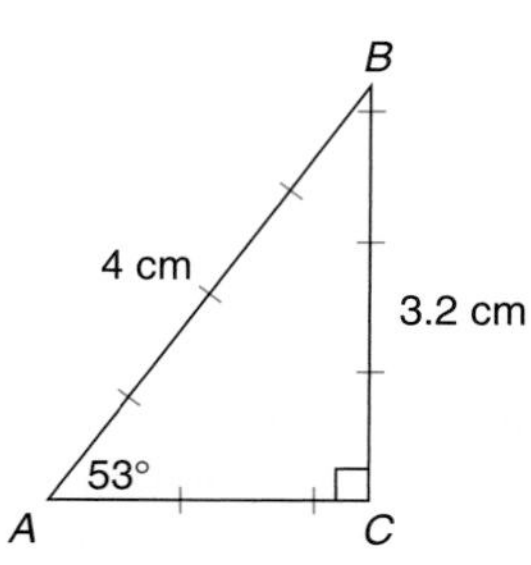


## Discover!

Given that an acute angle of a right triangle measures 53°, find the approximate value of sin 53°. We can estimate the value of sin 53° as follows (refer to the triangle at the left).

1. Draw right $\triangle ABC$ so that $\alpha = 53°$ and $\gamma = 90°$.
2. For convenience, mark off the length of the hypotenuse as 4 cm.
3. Using a ruler, measure the length of the leg opposite the angle measuring 53°. It is approximately 3.2 cm long.
4. Now divide $\frac{\text{opposite}}{\text{hypotenuse}}$ or $\frac{3.2}{4}$ to find that $\sin 53° \approx 0.8$.

NOTE: A calculator provides greater accuracy than this geometric approach by giving the result $\sin 53° \approx 0.7986$.

Repeat the procedure in the preceding Discover! activity and use it to find an approximation for sin 37°. You will need to use the Pythagorean Theorem to find $AC$. You should find that $\sin 37° \approx 0.6$.

*Exs. 1–5*

Although the sine ratios for angle measures are readily available on a calculator, we can justify several of the calculator's results by using special triangles. For certain angles, we can find *exact* results whereas the calculator provides approximations.

Recall the 30-60-90 relationship, in which the side opposite the 30° angle has a length equal to one-half that of the hypotenuse; the remaining leg has a length equal to the product of the length of the shorter leg and $\sqrt{3}$. In this case, we see that $\sin 30° = \frac{x}{2x} = \frac{1}{2}$. See Figure 10.5.

It is also true that $\sin 60° = \frac{x\sqrt{3}}{2x} = \frac{\sqrt{3}}{2}$. Although the exact value of sin 30° is 0.5 and the exact value of sin 60° is $\frac{\sqrt{3}}{2}$, a calculator would give an approximate value for sin 60° such as 0.8660254. If we round the ratio for sin 60° to four decimal places, then $\sin 60° \approx 0.8660$. Use your calculator to show that $\frac{\sqrt{3}}{2} \approx 0.8660$.

**Warning** ⓘ

Be sure to write $\sin \alpha = \frac{5}{13}$ or $\sin 54° \approx 0.8090$. It is incorrect to write "sin" in a claim without naming the angle or its measure; for example, $\sin = \frac{5}{13}$ and $\sin \approx 0.8090$ are both absolutely meaningless.

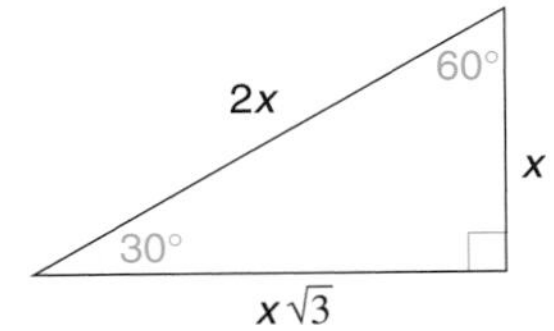


**FIGURE 10.5**

### EXAMPLE 3

Find exact and approximate values for sin 45°.

***Solution***

Using the 45°-45°-90° triangle in Figure 10.6, we see that $\sin 45° = \frac{x}{x\sqrt{2}} = \frac{1}{\sqrt{2}}$. Equivalently, $\sin 45° = \frac{\sqrt{2}}{2}$.

A calculator approximation is $\sin 45° \approx 0.7071$.

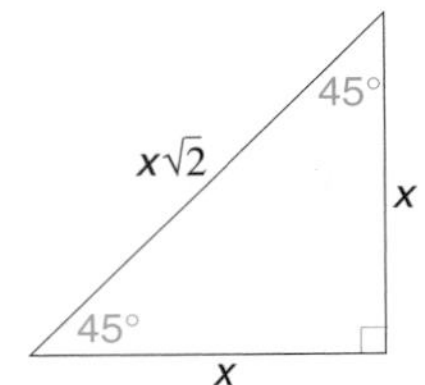


**FIGURE 10.6**

We will now use an earlier result (from Section 5.6) to determine the sine ratios for angles that measure 15° and 75°. Recall that an angle bisector of one angle of a triangle divides the opposite side into two segments that are proportional to the sides forming the bisected angle. Using this fact in the 30°-60°-90° triangle in Figure 10.7, we are led to the proportion

$$\frac{x}{1 - x} = \frac{\sqrt{3}}{2}$$

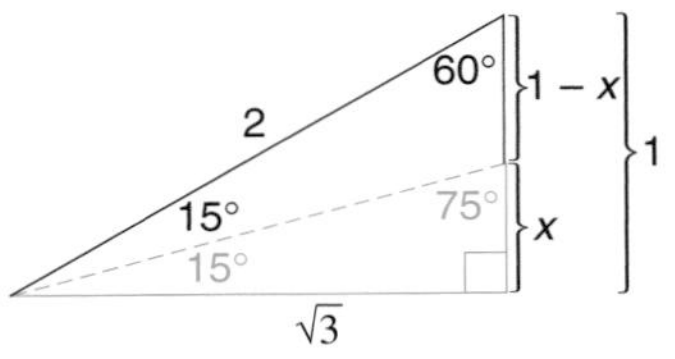

**FIGURE 10.7**

Applying the Means-Extremes Property,

$$\begin{aligned} 2x &= \sqrt{3} - x\sqrt{3} \\ 2x + x\sqrt{3} &= \sqrt{3} \\ (2 + \sqrt{3})x &= \sqrt{3} \\ x &= \frac{\sqrt{3}}{2 + \sqrt{3}} \approx 0.4641 \end{aligned}$$

The number 0.4641 is the length of the side that is opposite the 15° angle of the 15°-75°-90° triangle (see Figure 10.8). Using the Pythagorean Theorem, we can show that the length of the hypotenuse is approximately 1.79315. In turn, $\sin 15° = \frac{0.46410}{1.79315} \approx 0.2588$. Using the same triangle, $\sin 75° = \frac{1.73205}{1.79315} \approx 0.9659$.

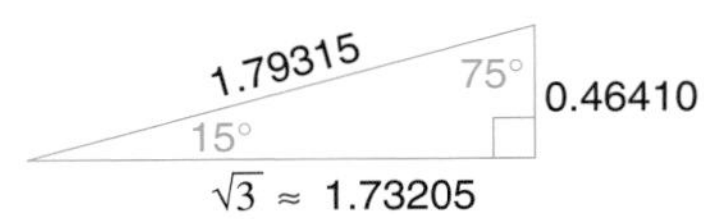


**FIGURE 10.8**

We now begin to formulate a small table of values of sine ratios. In Table 10.1, the Greek letter $\theta$ (theta) designates the angle measure in degrees. The second column has the heading sin $\theta$ and provides the ratio for the corresponding angle; this ratio is generally given to four decimal places of accuracy. Note that the values of sin $\theta$ increase as $\theta$ increases in measure.

**Warning** (!)

Note that $\sin\left(\frac{1}{2}\theta\right) \neq \frac{1}{2}\sin\theta$ in Table 10.1. If $\theta = 60°$, $\sin 30° \neq \frac{1}{2}\sin 60°$ because $0.5000 \neq \frac{1}{2}(0.8660)$.

**TABLE 10.1**

**Sine Ratios**

| $\theta$ | $\sin\theta$ |
|---|---|
| 15° | 0.2588 |
| 30° | 0.5000 |
| 45° | 0.7071 |
| 60° | 0.8660 |
| 75° | 0.9659 |

NOTE: Most values provided in tables or given by a calculator are approximations. Although we use the equality symbol (=) when reading values from a table (or calculator), the solutions to the problems that follow are generally approximations.

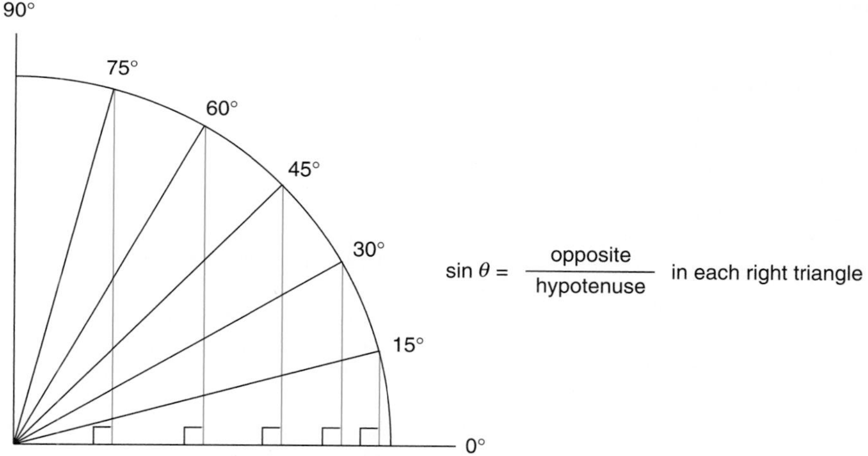


**FIGURE 10.9**

In Figure 10.9, let $\angle\theta$ be the acute angle whose measure increases as shown. In the figure, note that the length of the hypotenuse is constant—it is always equal to the length of the radius of the circle. However, the side opposite $\angle\theta$ gets larger as $\theta$ increases. In fact, as $\theta$ approaches 90° ($\theta \to 90°$), the length of the leg opposite $\angle\theta$ approaches the length of the hypotenuse. As $\theta \to 90°$, $\sin\theta \to 1$. As $\theta$ decreases, $\sin\theta$ also decreases. As $\theta$ decreases ($\theta \to 0°$), the length of the side opposite $\angle\theta$ approaches 0. As $\theta \to 0°$, $\sin\theta \to 0$. These observations lead to the following definition.

DEFINITION: $\sin 0° = 0$ and $\sin 90° = 1$

*Exs. 6–10*

NOTE: A calculator will verify these results.

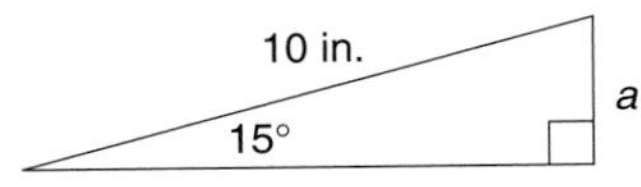


**FIGURE 10.10**

EXAMPLE 4

Using Table 10.1, find the length of $a$ in Figure 10.10 to the nearest tenth of an inch.

***Solution***

$$\sin 15° = \frac{\text{opposite}}{\text{hypotenuse}} = \frac{a}{10}$$

From the table, we have $\sin 15° = 0.2588$.

$$\frac{a}{10} = 0.2588 \quad \text{by substitution}$$

$$a = 2.588$$

Therefore, $a \approx 2.6$ in. when rounded to tenths.

In an application problem, the sine ratio can be used to find the measure of either a side or an angle of a triangle. To find the sine ratio of the angle involved, you may use a table of ratios or a calculator. Table 10.2 provides ratios for many more angle measures than does Table 10.1. As with calculators, the sine ratios found in tables are only approximations.

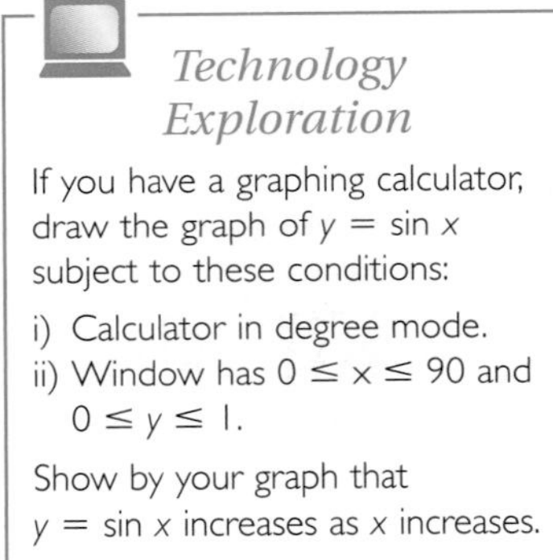

Technology Exploration
If you have a graphing calculator, draw the graph of y = sin x subject to these conditions:
i) Calculator in degree mode.
ii) Window has 0 ≤ x ≤ 90 and 0 ≤ y ≤ 1.
Show by your graph that y = sin x increases as x increases.


**TABLE 10.2**

Sine Ratios

| $\theta$ | $\sin \theta$ | $\theta$ | $\sin \theta$ | $\theta$ | $\sin \theta$ | $\theta$ | $\sin \theta$ |
|---|---|---|---|---|---|---|---|
| 0° | 0.0000 | 23° | 0.3907 | 46° | 0.7193 | 69° | 0.9336 |
| 1° | 0.0175 | 24° | 0.4067 | 47° | 0.7314 | 70° | 0.9397 |
| 2° | 0.0349 | 25° | 0.4226 | 48° | 0.7431 | 71° | 0.9455 |
| 3° | 0.0523 | 26° | 0.4384 | 49° | 0.7547 | 72° | 0.9511 |
| 4° | 0.0698 | 27° | 0.4540 | 50° | 0.7660 | 73° | 0.9563 |
| 5° | 0.0872 | 28° | 0.4695 | 51° | 0.7771 | 74° | 0.9613 |
| 6° | 0.1045 | 29° | 0.4848 | 52° | 0.7880 | 75° | 0.9659 |
| 7° | 0.1219 | 30° | 0.5000 | 53° | 0.7986 | 76° | 0.9703 |
| 8° | 0.1392 | 31° | 0.5150 | 54° | 0.8090 | 77° | 0.9744 |
| 9° | 0.1564 | 32° | 0.5299 | 55° | 0.8192 | 78° | 0.9781 |
| 10° | 0.1736 | 33° | 0.5446 | 56° | 0.8290 | 79° | 0.9816 |
| 11° | 0.1908 | 34° | 0.5592 | 57° | 0.8387 | 80° | 0.9848 |
| 12° | 0.2079 | 35° | 0.5736 | 58° | 0.8480 | 81° | 0.9877 |
| 13° | 0.2250 | 36° | 0.5878 | 59° | 0.8572 | 82° | 0.9903 |
| 14° | 0.2419 | 37° | 0.6018 | 60° | 0.8660 | 83° | 0.9925 |
| 15° | 0.2588 | 38° | 0.6157 | 61° | 0.8746 | 84° | 0.9945 |
| 16° | 0.2756 | 39° | 0.6293 | 62° | 0.8829 | 85° | 0.9962 |
| 17° | 0.2924 | 40° | 0.6428 | 63° | 0.8910 | 86° | 0.9976 |
| 18° | 0.3090 | 41° | 0.6561 | 64° | 0.8988 | 87° | 0.9986 |
| 19° | 0.3256 | 42° | 0.6691 | 65° | 0.9063 | 88° | 0.9994 |
| 20° | 0.3420 | 43° | 0.6820 | 66° | 0.9135 | 89° | 0.9998 |
| 21° | 0.3584 | 44° | 0.6947 | 67° | 0.9205 | 90° | 1.0000 |
| 22° | 0.3746 | 45° | 0.7071 | 68° | 0.9272 | | |

NOTE: In later sections, we will use the calculator (rather than tables) to find values of trigonometric ratios such as sin 36°.

EXAMPLE 5

Find sin 36°, using

a) Table 10.2. b) a scientific or graphing calculator.

***Solution***

a) Find 36° under the heading $\theta$. Now read the number under the sin $\theta$ heading: sin 36° = 0.5878

b) On a scientific calculator that is *in degree mode,* use the following key sequence:

3 → 6 → sin → **0.5878**

The result is sin 36° = 0.5878, correct to four decimal places.

NOTE 1: The boldfaced number in the box represents the final answer.

NOTE 2: The key sequence for a graphing calculator follows. Here, the calculator is in degree mode and the answer is rounded to four decimal places.

sin → 3 → 6 → ENTER → **0.5878**

The entry may appear in the form sin (36).

The table or a calculator can also be used to find the measure of an angle. This is possible when the sine of the angle is known.

EXAMPLE 6

If sin $\theta = 0.7986$, find $\theta$ to the nearest degree by using

a) Table 10.2. b) a calculator.

***Solution***

a) Find 0.7986 under the heading sin $\theta$. Now look to the left to find the degree measure of the angle in the $\theta$ column:

$$\sin \theta = 0.7986 \rightarrow \theta = 53°$$

b) On some scientific calculators, you can use the following key sequence (while in degree mode) to find $\theta$:

. → 7 → 9 → 8 → 6 → inv → sin → **53**

The combination "inv" and "sin" yields the angle whose sine ratio is known, so $\theta = 53°$.

NOTE: On a graphing calculator that is in degree mode, use this sequence:

$\sin^{-1}$ → . → 7 → 9 → 8 → 6 → ENTER → **53**

This entry may appear in the form $\sin^{-1}(.7986)$. The expression $\sin^{-1}(.7986)$ means "the angle whose sine is 0.7986." The calculator function $\boxed{\sin^{-1}}$ is found by pressing 2nd followed by sin.

*Exs. 11–15*

In most application problems, a drawing provides a good deal of information and affords some insight into the method of solution. For some drawings and applications, the phrases *angle of elevation* and *angle of depression* are used. These angles are measured from the horizontal as illustrated in Figures 10.11(a) and (b). In Figure

10.11(a), the angle $\alpha$ measured upward from the horizontal ray is the **angle of elevation.** In Figure 10.11(b), the angle $\beta$ measured downward from the horizontal ray is the **angle of depression.**

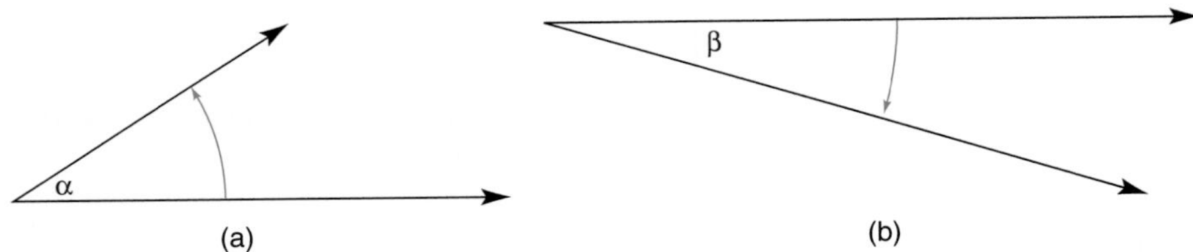


**FIGURE 10.11**

EXAMPLE 7

The tower for a radio station stands 200 ft tall. A guy wire 250 ft long supports the antenna, as shown in Figure 10.12. Find the measure of the angle of elevation $\alpha$ to the nearest degree.

***Solution***

$$\sin \alpha = \frac{\text{opposite}}{\text{hypotenuse}} = \frac{200}{250} = 0.8$$

From Table 10.2 (or from a calculator), we find that the angle whose sine ratio is 0.8 is $\alpha \approx 53°$.

SSG *Exs. 16, 17*

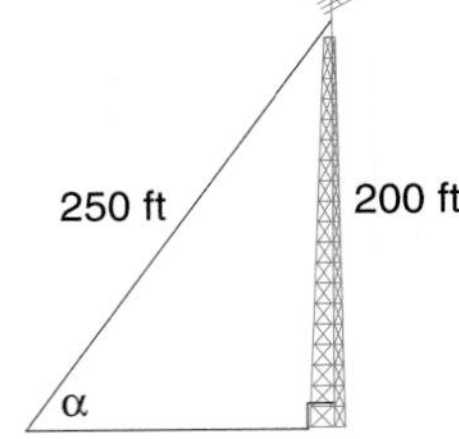


**FIGURE 10.12**

## 10.1 Exercises

*In Exercises 1 to 6, find* $\sin \alpha$ *and* $\sin \beta$ *for the triangle shown.*

1. 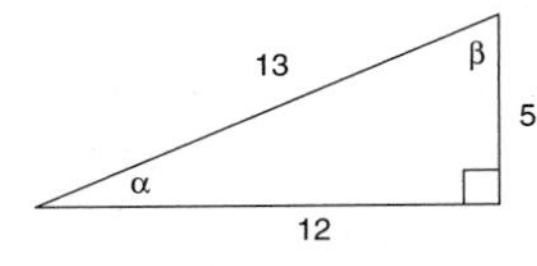


2. 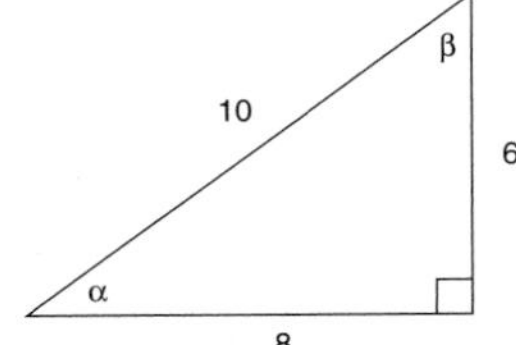


3. 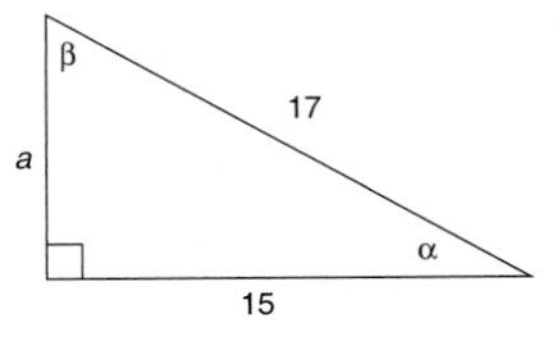


4. 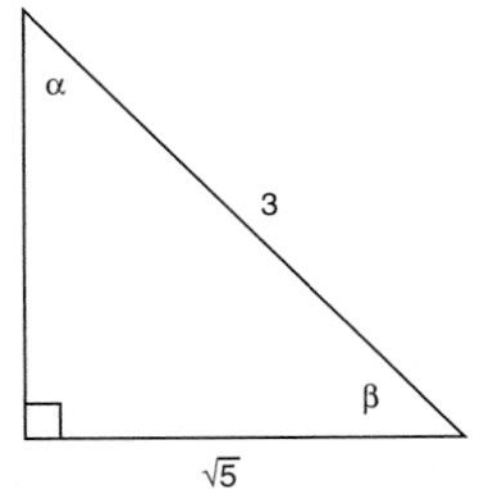


5. 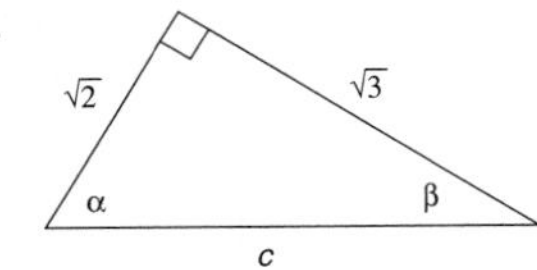


6. 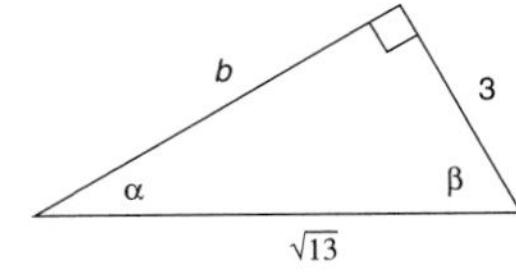


*In Exercises 7 to 14, use either Table 10.2 or a calculator to find the sine of the indicated angle to four decimal places.*

7. sin 90°
8. sin 0°
9. sin 17°
10. sin 23°
11. sin 82°
12. sin 46°
13. sin 72°
14. sin 57°

*In Exercises 15 to 20, find the lengths of the sides indicated by the variables. Use either Table 10.2 or a calculator and round answers to the nearest tenth of a unit.*

15.

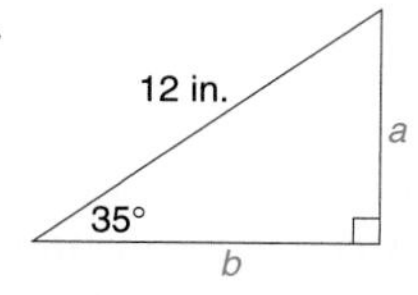


16.

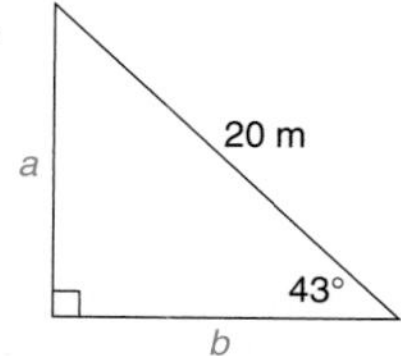


17.

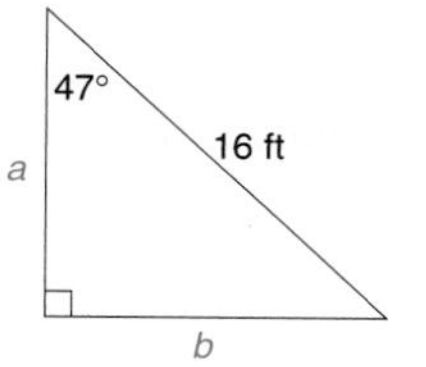


18.

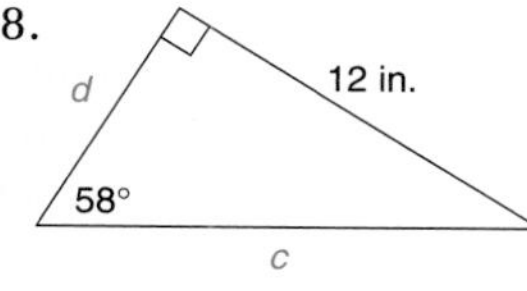


19.

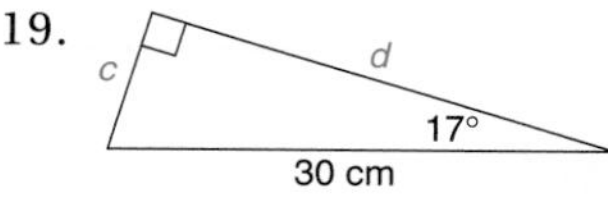


20.

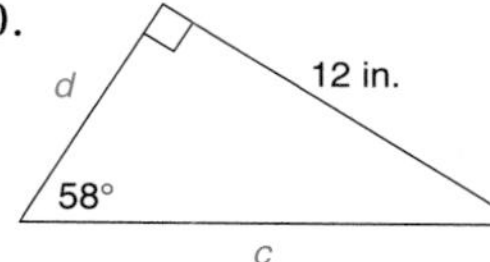


*In Exercises 21 to 26, find the measures of the angles named to the nearest degree.*

21.

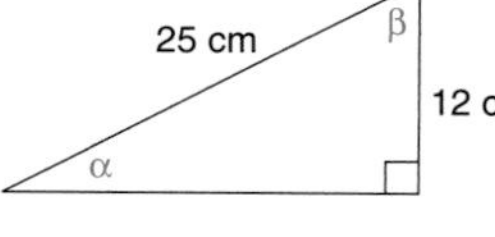


22.

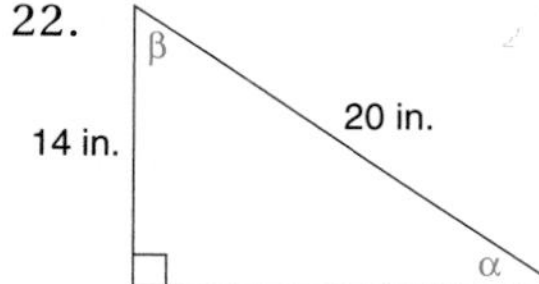


23.

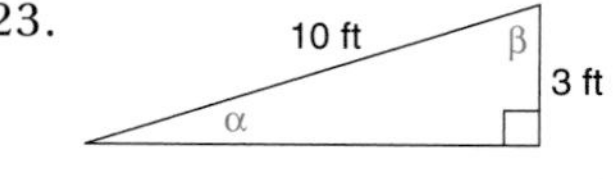


24.

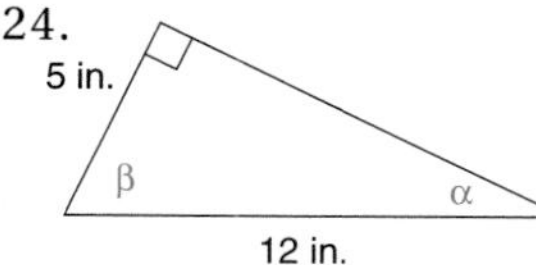


25.

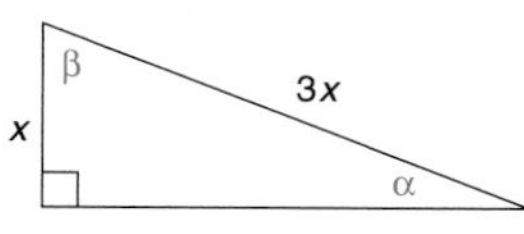


26.

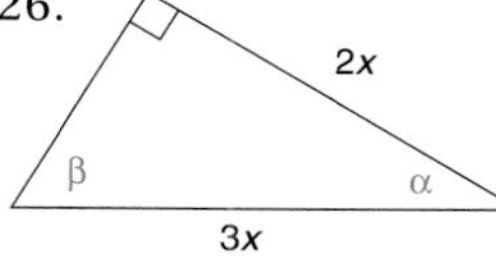


*In Exercises 27 to 34, use the drawings where provided to solve each problem. Angle measures should be given to the nearest degree; distances should be given to the nearest tenth of a unit.*

27. The pitch or slope of a roofline is 5 to 12. Find the measure of angle $\alpha$.

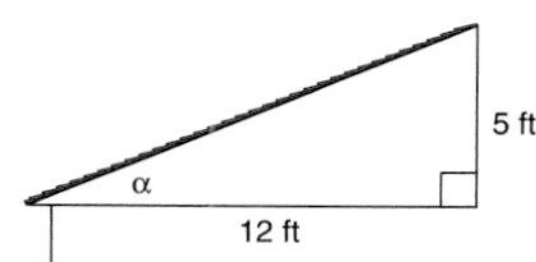


28. A kite is flying at an angle of elevation of 67° from a point on the ground. If 100 ft of kite string is out, how far is the kite above the ground?

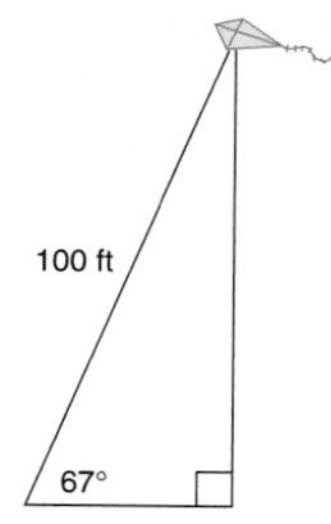


29. Danny sees a balloon that is 100 ft above the ground. If the angle of elevation from Danny to the balloon is 75°, how far from Danny is the balloon?

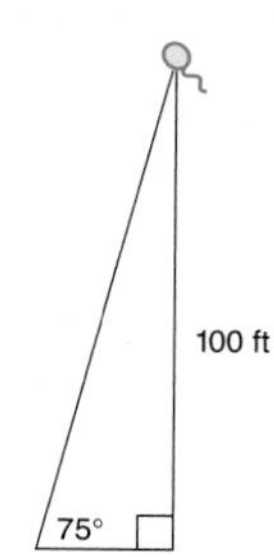


30. Over a 2000-ft span of highway through a hillside, there is a 100-ft rise in the roadway. What is the measure of the angle formed by the road and the horizontal?

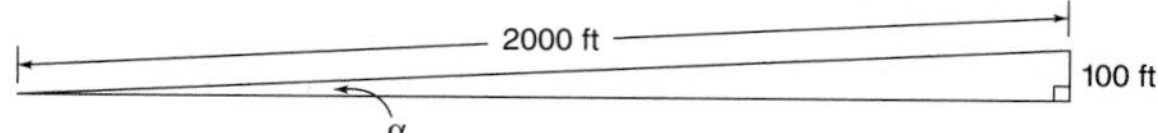


31. From a cliff, a person observes an automobile through an angle of depression of 23°. If the cliff is 50 ft high, how far is the automobile from the person?

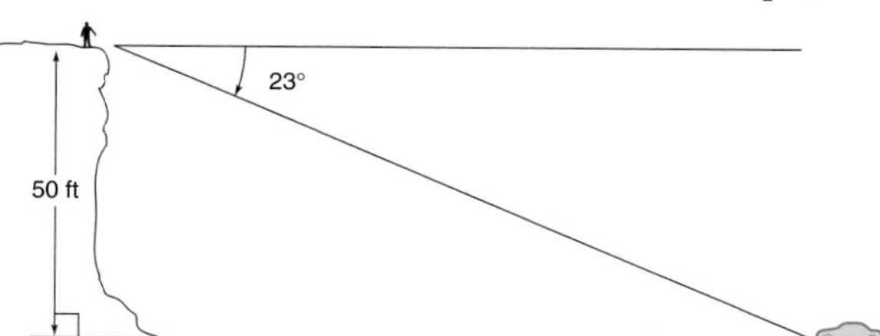


32. A 12-ft rope secures a rowboat to a pier that is 4 ft above the water. Assume that the lower end of the rope is at "water level." What is the angle formed by the rope and the water?

33. A 10-ft ladder is leaning against a vertical wall so that the bottom of the ladder is 4 ft away from the base of the wall. How large is the angle formed by the ladder and the wall?

34. An airplane flying at the rate of 350 feet per second begins to climb at an angle of 10°. What is the increase in altitude over the next 15 seconds?

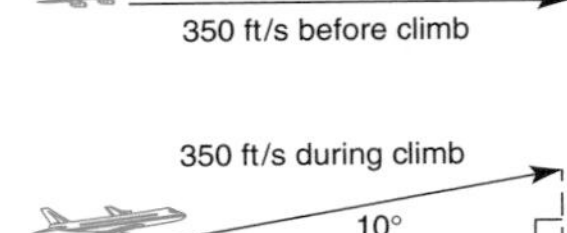

*For Exercises 35 to 38, make drawings as needed.*

35. In parallelogram $ABCD$, $AB = 6$ ft and $AD = 10$ ft. If $m\angle A = 65°$ and $\overline{BE}$ is the altitude to $\overline{AD}$, find:

    a) $BE$ correct to tenths
    b) The area of $\square ABCD$

36. In right $\triangle ABC$, $\gamma = 90°$ and $\beta = 55°$. If $AB = 20$ in., find:

    a) $a$ (the length of $\overline{BC}$) correct to tenths
    b) $b$ (the length of $\overline{AC}$) correct to tenths
    c) The area of right $\triangle ABC$

37. In a right circular cone, the slant height is 13 cm and the altitude is 10 cm. To the nearest degree, find the measure of the angle $\theta$ that is formed by the radius and slant height.

38. In a right circular cone, the slant height is 13 cm. Where $\theta$ is the angle formed by the radius and the slant height, $\theta = 48°$. Find the length of the altitude of the cone, correct to tenths.

**KEY CONCEPTS**

Adjacent Side (Leg) • Cosine Ratio: $\cos\theta = \frac{\text{adjacent}}{\text{hypotenuse}}$ • Identity: $\sin^2\theta + \cos^2\theta = 1$

# 10.2 The Cosine Ratio and Applications

Again we deal strictly with similar right triangles, as shown in Figure 10.13. $\overline{BC}$ is the leg opposite angle $A$, and we say that $\overline{AC}$ is the leg **adjacent** to angle $A$. In the two triangles, the ratios of the form

$$\frac{\text{length of adjacent leg}}{\text{length of hypotenuse}}$$

are equal; that is,

$$\frac{AC}{AB} = \frac{DF}{DE} \quad \text{or} \quad \frac{4}{5} = \frac{8}{10}$$

This relationship follows from the fact that corresponding sides of similar triangles are proportional.

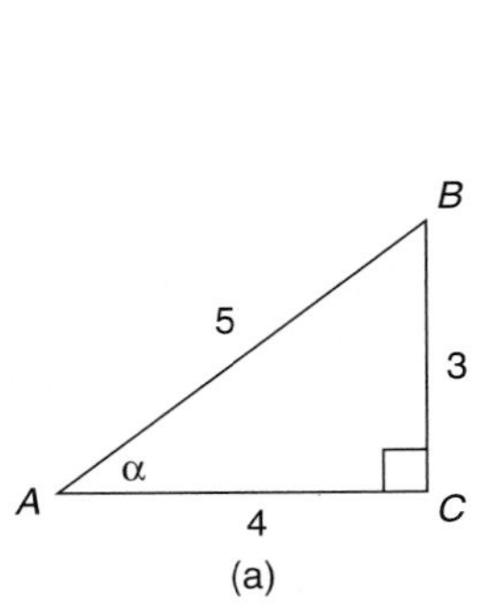


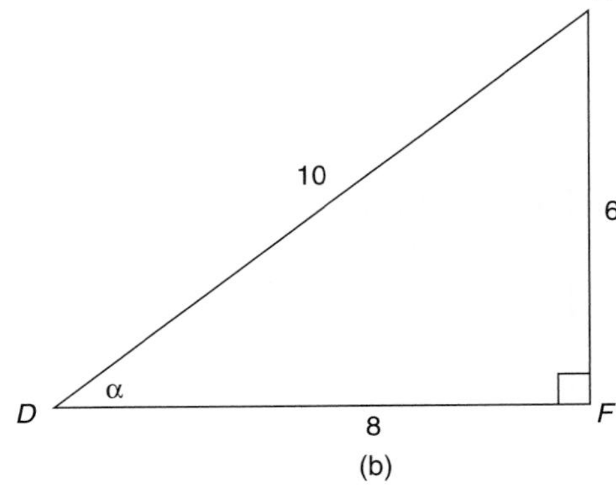


**FIGURE 10.13**

As with the sine ratio, the *cosine ratio* depends on the measure of acute angle $A$ (or $D$) in Figure 10.13. In the following definition, the term *adjacent* refers to the length of the leg that is adjacent to the angle named.

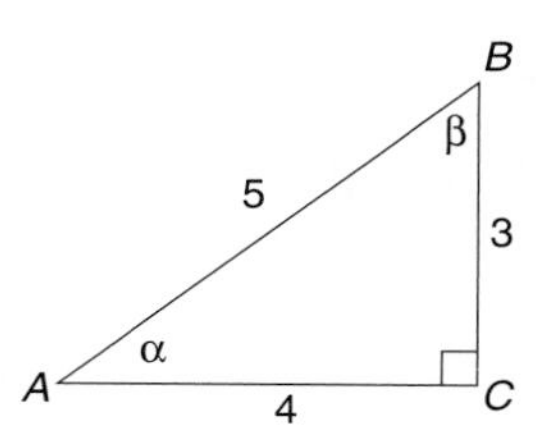


**FIGURE 10.14**

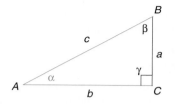

> DEFINITION: In a right triangle, the **cosine ratio** for an acute angle is the ratio $\frac{\text{adjacent}}{\text{hypotenuse}}$.

NOTE: In right $\triangle ABC$ in Figure 10.14, we have $\cos\alpha = \frac{b}{c}$ and $\cos\beta = \frac{a}{c}$, in which "cos" is an abbreviated form of the word *cosine*.

**FIGURE 10.15** (triangle with $A$, $B$, $C$; sides 5, 3, 4; angles α, β)

### EXAMPLE 1

Find $\cos\alpha$ and $\cos\beta$ for right $\triangle ABC$ in Figure 10.15.

***Solution***

$a = 3$, $b = 4$, and $c = 5$ for the triangle shown in Figure 10.15. Because $b$ is the length of the leg adjacent to $\alpha$ and $a$ is the length of the leg adjacent to $\beta$,

$$\cos\alpha = \frac{b}{c} = \frac{4}{5} \qquad \text{and} \qquad \cos\beta = \frac{a}{c} = \frac{3}{5}$$

### EXAMPLE 2

Find $\cos\alpha$ and $\cos\beta$ for right $\triangle ABC$ in Figure 10.16.

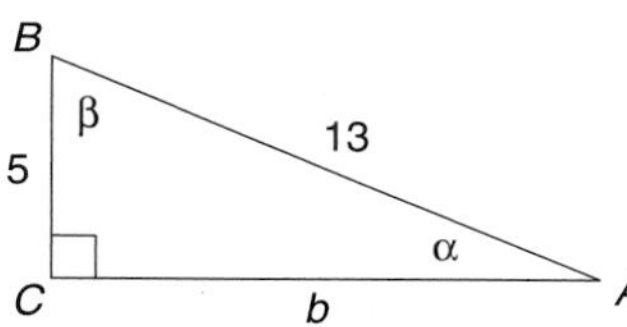


**FIGURE 10.16**

*Exs. 1–5*

***Solution***

$a = 5$ and $c = 13$. Then $b = 12$ from the Pythagorean triple (5, 12, 13). Consequently,

$$\cos\alpha = \frac{b}{c} = \frac{12}{13} \qquad \text{and} \qquad \cos\beta = \frac{a}{c} = \frac{5}{13}$$

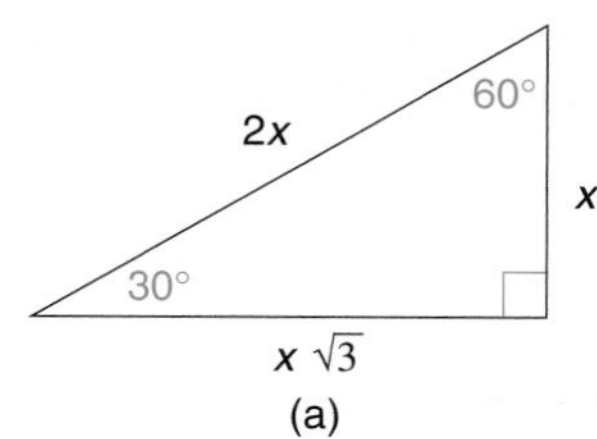


(a)

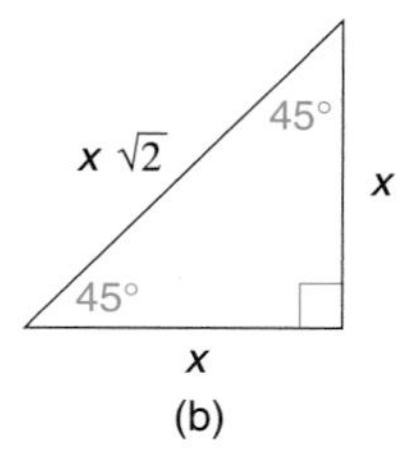


(b)

**FIGURE 10.17**

Just as the sine ratio of any angle is unique, the cosine ratio of any angle is also unique. Using the 30°-60°-90° and 45°-45°-90° triangles of Figure 10.17, we see that

$$\cos 30° = \frac{x\sqrt{3}}{2x} = \frac{\sqrt{3}}{2} \approx 0.8660$$

$$\cos 45° = \frac{x}{x\sqrt{2}} = \frac{1}{\sqrt{2}} = \frac{\sqrt{2}}{2} \approx 0.7071$$

$$\cos 60° = \frac{x}{2x} = \frac{1}{2} = 0.5$$

Now we use the 15°-75°-90° triangle shown in Figure 10.18 on page 492 to find cos 75° and cos 15°. From Section 10.1, $\sin 15° = \frac{a}{c}$ and $\sin 15° = 0.2588$. But

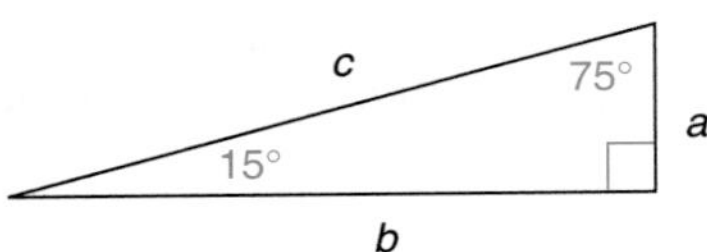


**FIGURE 10.18**

$\cos 75° = \frac{a}{c}$, so $\cos 75° = 0.2588$. Similarly, because $\sin 75° = \frac{b}{c} = 0.9659$, we see that $\cos 15° = \frac{b}{c} = 0.9659$.

In Figure 10.19, we see that the cosine ratios become larger as $\theta$ decreases and become smaller as $\theta$ increases. To understand why, consider the definition

$$\cos \theta = \frac{\text{length of adjacent leg}}{\text{length of hypotenuse}}$$

and Figure 10.19. Recall that the symbol $\rightarrow$ is read "approaches." As $\theta \rightarrow 0°$, length of adjacent leg $\rightarrow$ length of hypotenuse, and therefore $\cos 0° \rightarrow 1$. Similarly, $\cos 90° \rightarrow 0$ because the adjacent leg grows smaller as $\theta \rightarrow 90°$. Consequently, we have the following definition.

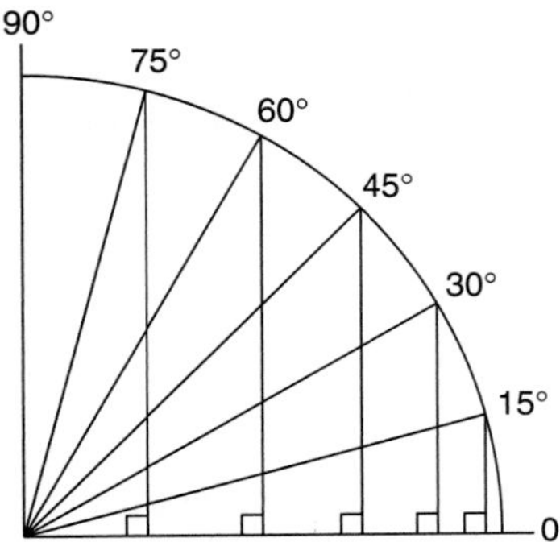


**FIGURE 10.19**

DEFINITION: $\cos 0° = 1$ and $\cos 90° = 0$

NOTE: A calculator will verify the results found in this definition.

We summarize cosine ratios in Table 10.3.

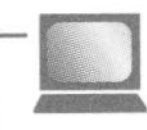

*Technology Exploration*

If you have a graphing calculator, draw the graph of $y = \cos x$ subject to these conditions:

i) Calculator in degree mode.

ii) Window has $0 \leq x \leq 90$ and $0 \leq y \leq 1$.

Show by your graph that $y = \cos x$ decreases as $x$ increases.

**TABLE 10.3**

**Cosine Ratios**

| $\theta$ | $\cos \theta$ |
|---|---|
| 0° | 1.0000 |
| 15° | 0.9659 |
| 30° | 0.8660 |
| 45° | 0.7071 |
| 60° | 0.5000 |
| 75° | 0.2588 |
| 90° | 0.0000 |

*Exs. 6–10*

Some textbooks provide an expanded table of cosine ratios comparable to Table 10.2 for sine ratios. Although this text does not provide an expanded table of cosine ratios, we illustrate the application of such a table in Example 3.

EXAMPLE 3

Using Table 10.3, find the length of $b$ in Figure 10.20 correct to the nearest tenth.

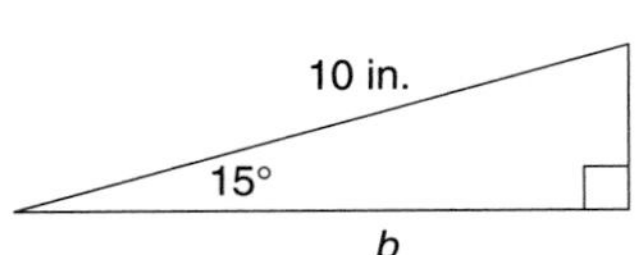


**FIGURE 10.20**

***Solution***

$\cos 15° = \frac{\text{adjacent}}{\text{hypotenuse}} = \frac{b}{10}$ from the triangle. Also, $\cos 15° = 0.9659$ from the table. Then

$$\frac{b}{10} = 0.9659 \quad \text{(because both equal cos 15°)}$$

$$b = 9.659$$

Therefore, $b \approx 9.7$ in.

when rounded to the nearest tenth of an inch.

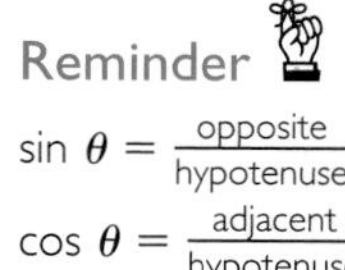


$\sin \theta = \frac{\text{opposite}}{\text{hypotenuse}}$

$\cos \theta = \frac{\text{adjacent}}{\text{hypotenuse}}$

In a right triangle, the cosine ratio can often be used to find an unknown length or an unknown angle measure. Whereas the sine ratio requires that we use *opposite* and *hypotenuse,* the cosine ratio requires that we use *adjacent* and *hypotenuse.*

An equation of the form $\sin \alpha = \frac{a}{c}$ or $\cos \alpha = \frac{b}{c}$ contains three variables: for the equation $\cos \alpha = \frac{b}{c}$, the variables are $\alpha$, $b$, and $c$. When the values of two of the variables are known, the value of the third variable can be determined. However, we must decide which trigonometric ratio is needed to solve the problem.

EXAMPLE 4

In Figure 10.21, which trigonometric ratio would you use to find

a) $\alpha$, if $a$ and $c$ are known?
b) $b$, if $\alpha$ and $c$ are known?
c) $c$, if $a$ and $\alpha$ are known?
d) $\beta$, if $a$ and $c$ are known?

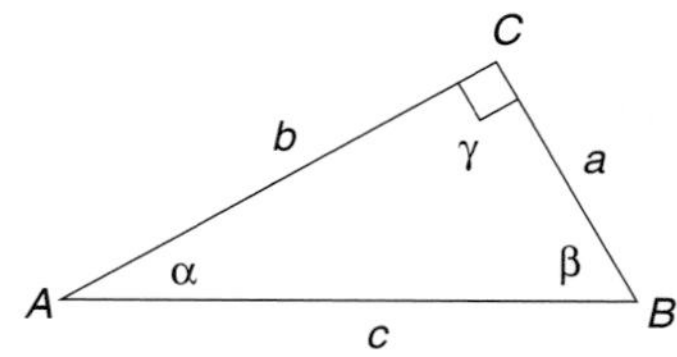


**FIGURE 10.21**

***Solution***

a) sine, because $\sin \alpha = \frac{a}{c}$ and $a$ and $c$ are known
b) cosine, because $\cos \alpha = \frac{b}{c}$ and $\alpha$ and $c$ are known
c) sine, because $\sin \alpha = \frac{a}{c}$ and $a$ and $\alpha$ are known
d) cosine, because $\cos \beta = \frac{a}{c}$ and $a$ and $c$ are known

To solve application problems, you generally use a calculator.

EXAMPLE 5

Find cos 67° correct to four decimal places by using a scientific calculator.

***Solution***

On a scientific calculator that is in degree mode, use the following key sequence:

[6] → [7] → [cos] → **0.3907**

Using a graphing calculator (in degree mode), follow this key sequence:

[cos] → [6] → [7] → [ENTER] → **0.3907**

That is, $\cos 67° \approx 0.3907$.

EXAMPLE 6

Use a calculator to find the measure of angle $\theta$ to the nearest degree if $\cos \theta = 0.5878$.

***Solution***

Using a scientific calculator (in degree mode), follow this key sequence:

[.] → [5] → [8] → [7] → [8] → [inv] → [cos] → **54**

Using a graphing calculator (in degree mode), follow this key sequence:

[$\cos^{-1}$] → [.] → [5] → [8] → [7] → [8] → [ENTER] → **54**

Thus $\theta = 54°$.

NOTE: By pressing [2nd] and [cos] on a graphing calculator, you obtain [$\cos^{-1}$].

*Exs. 11–15*

### EXAMPLE 7

For a regular pentagon, the length of the apothem is 12 in. Find the length of the pentagon's radius to the nearest tenth of an inch.

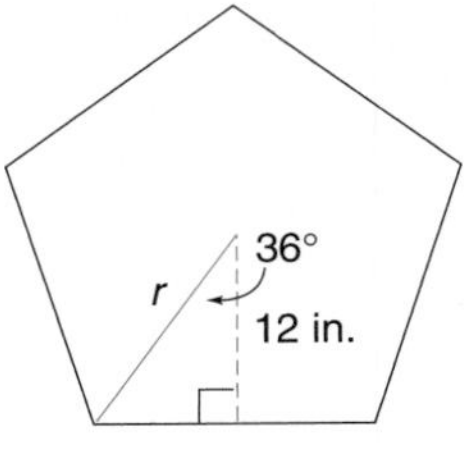

**FIGURE 10.22**

***Solution***

The central angle of the regular pentagon measures $\frac{360}{5}$, or 72°. An apothem bisects this angle, so the angle formed by the apothem and the radius measures 36°.

In Figure 10.22,

$$\cos 36° = \frac{\text{adjacent}}{\text{hypotenuse}} = \frac{12}{r}$$

Using a calculator, $\cos 36° = 0.8090$. Then $\frac{12}{r} = 0.8090$ and $0.8090r = 12$. By division, $r \approx 14.8$ in.

NOTE: The solution in Example 7 can be calculated as $r = \frac{12}{\cos 36°}$.

We now consider the proof of a statement that is called an **identity** because it is true for all angles; we refer to this statement as a theorem. As you will see, the statement is based entirely on the Pythagorean Theorem.

THEOREM 10.2.1: In any right triangle in which $\alpha$ is the measure of an acute angle,

$$\sin^2 \alpha + \cos^2 \alpha = 1$$

NOTE: $\sin^2 \alpha$ means $(\sin \alpha)^2$ and $\cos^2 \alpha$ means $(\cos \alpha)^2$.

***Proof***

In Figure 10.23, $\sin \alpha = \frac{a}{c}$ and $\cos \alpha = \frac{b}{c}$. Then

$$\sin^2 \alpha + \cos^2 \alpha = \left(\frac{a}{c}\right)^2 + \left(\frac{b}{c}\right)^2 = \frac{a^2}{c^2} + \frac{b^2}{c^2} = \frac{a^2 + b^2}{c^2}$$

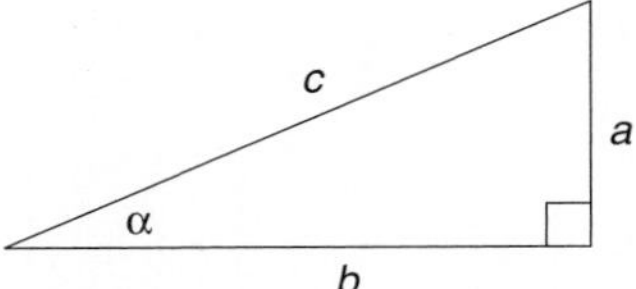

**FIGURE 10.23**

In the right triangle in Figure 10.23, $a^2 + b^2 = c^2$ by the Pythagorean Theorem. Substituting $c^2$ for $a^2 + b^2$, we have

$$\sin^2 \alpha + \cos^2 \alpha = \frac{c^2}{c^2} = 1$$

It follows that $\sin^2 \alpha + \cos^2 \alpha = 1$ for any angle $\alpha$.

NOTE: Use your calculator to show that $(\sin 67°)^2 + (\cos 67°)^2 = 1$. Theorem 10.2.1 is also true for $\alpha = 0°$ and $\alpha = 90°$.

EXAMPLE 8

In right triangle $ABC$ (not shown), $\sin \alpha = \frac{2}{3}$. Find $\cos \alpha$.

***Solution***

$$\sin^2 \alpha + \cos^2 \alpha = 1$$
$$\left(\frac{2}{3}\right)^2 + \cos^2 \alpha = 1$$
$$\frac{4}{9} + \cos^2 \alpha = 1$$
$$\cos^2 \alpha = \frac{5}{9}$$

*Exs. 16, 17*

Therefore, $\cos \alpha = \sqrt{\frac{5}{9}} = \frac{\sqrt{5}}{\sqrt{9}} = \frac{\sqrt{5}}{3}$.

NOTE: Because $\cos \alpha > 0$, $\cos \alpha = \frac{\sqrt{5}}{3}$ rather than $-\frac{\sqrt{5}}{3}$.

In Section 10.3, we will encounter additional trigonometric identities embedded in Exercises 33–36.

## 10.2 Exercises

*In Exercises 1 to 6, find cos α and cos β.*

1.

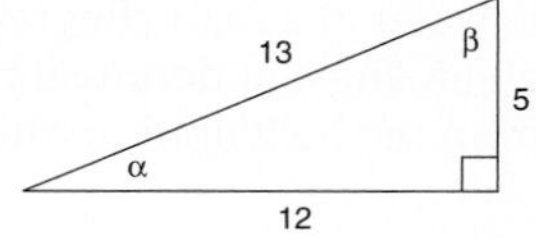


2.

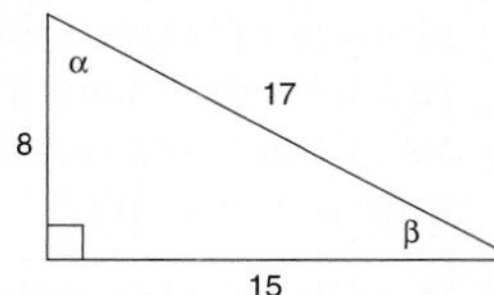


3.

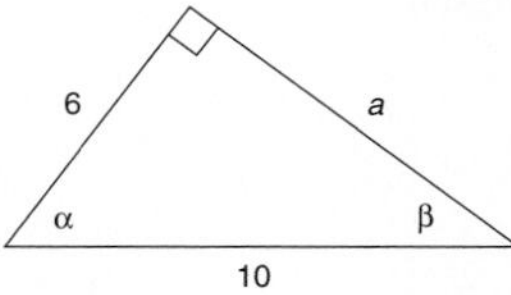


4.

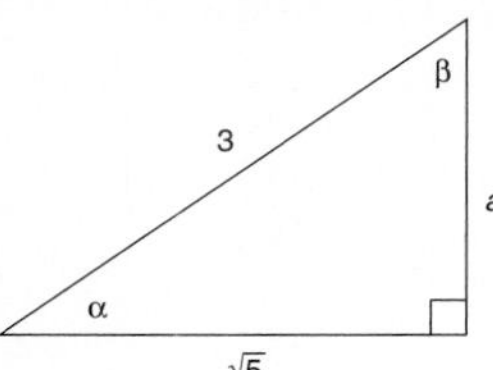


5.

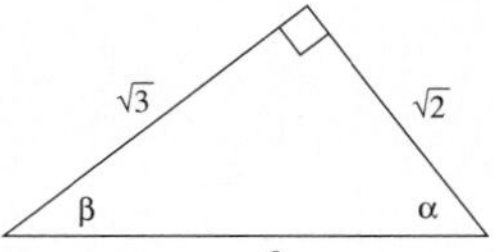


6.

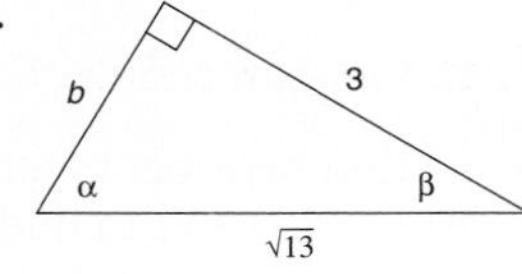


7. In Exercises 1 to 6:

   a) Why does $\sin \alpha = \cos \beta$?
   b) Why does $\cos \alpha = \sin \beta$?

8. Using the right triangle from Exercise 1, show that $\sin^2 \alpha + \cos^2 \alpha = 1$.

*In Exercises 9 to 16, use a scientific calculator to find the indicated cosine ratio to four decimal places.*

9. cos 23°
10. cos 0°
11. cos 17°
12. cos 73°
13. cos 90°
14. cos 42°
15. cos 82°
16. cos 7°

*In Exercises 17 to 22, use either the sine ratio or the cosine ratio to find the lengths of the indicated sides of the triangle, correct to the nearest tenth of a unit.*

17.

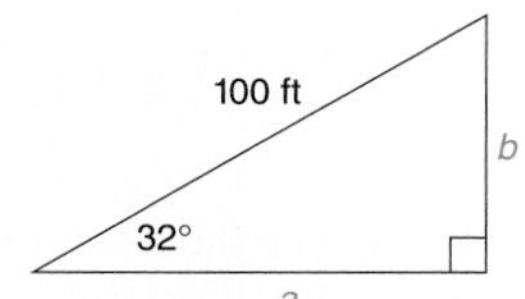


18.

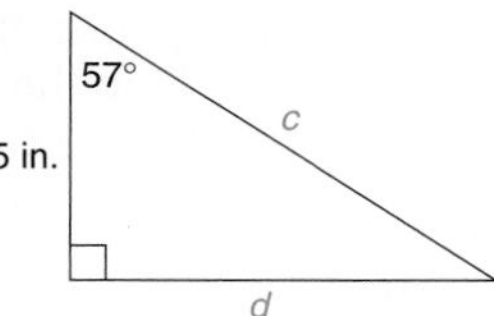


19.

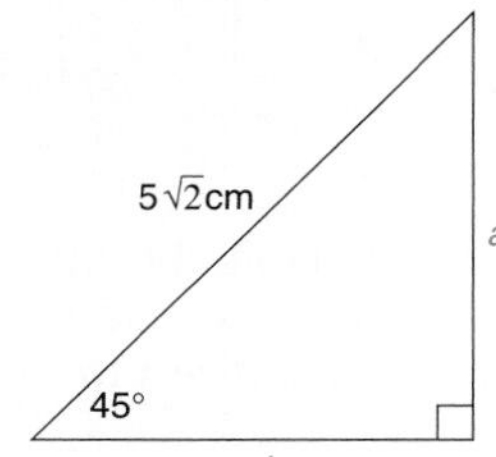


20.

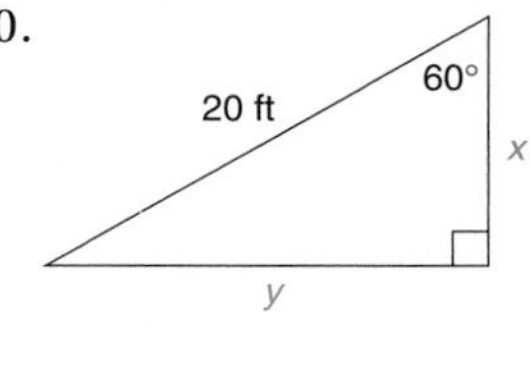

21.

22.

*In Exercises 23 to 28, use the sine ratio or the cosine ratio as needed to find the measure of each indicated angle to the nearest degree.*

23.
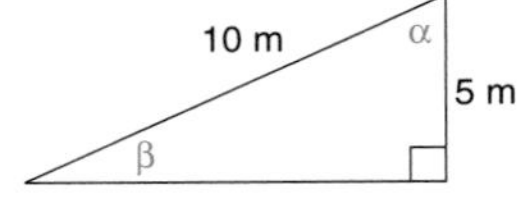


24.
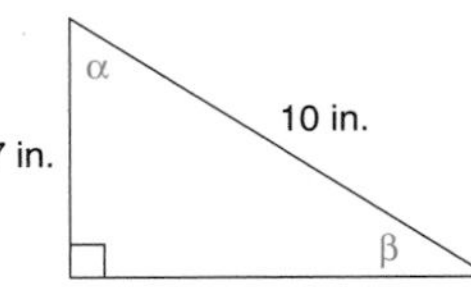


25.
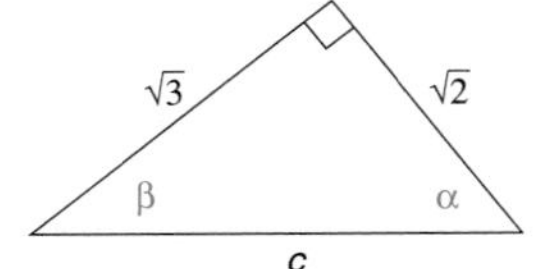


26.
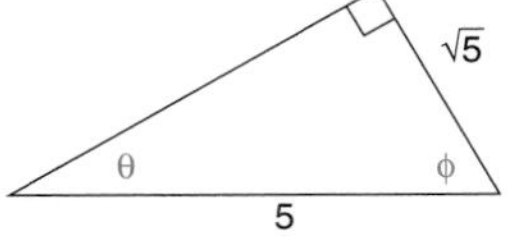


27.
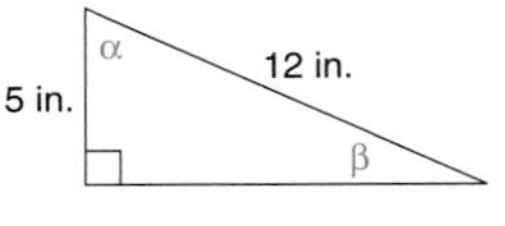


28.
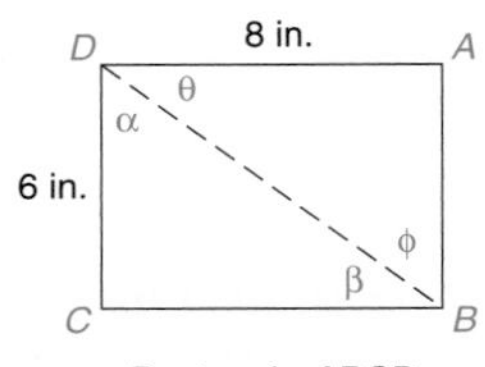

Rectangle *ABCD*

*In Exercises 29 to 37, angle measures should be given to the nearest degree; distances should be given to the nearest tenth of a unit.*

29. In building a garage onto his house, Gene wants to use a sloped 12-ft roof to cover an expanse that is 10 ft wide. Find the measure of angle $\theta$.

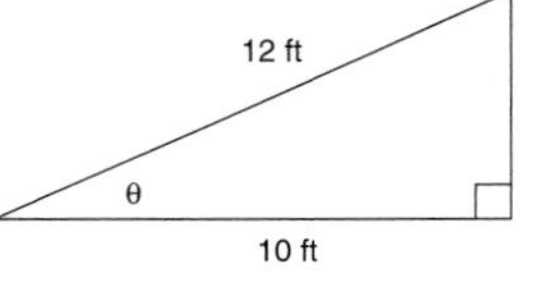


30. Gene redesigned the garage from Exercise 29 so that the 12-ft roof would rise 2 ft as shown. Find the measure of angle $\theta$.

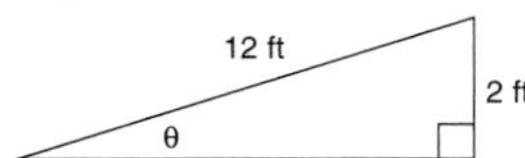


31. When an airplane is descending to land, the angle of depression is 5°. When the plane has a reading of 100 ft on the altimeter, what is its distance $x$ from touchdown?

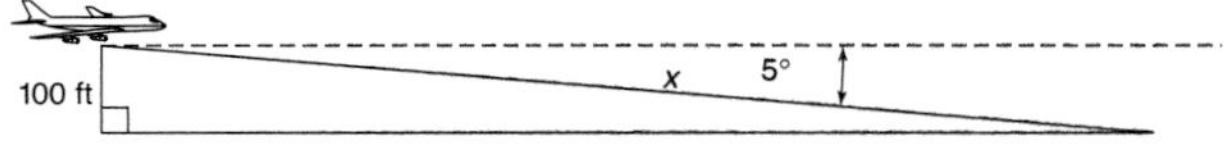


32. At a point 200 ft from the base of a cliff, the top of the cliff is seen through an angle of elevation of 37°. How tall is the cliff?

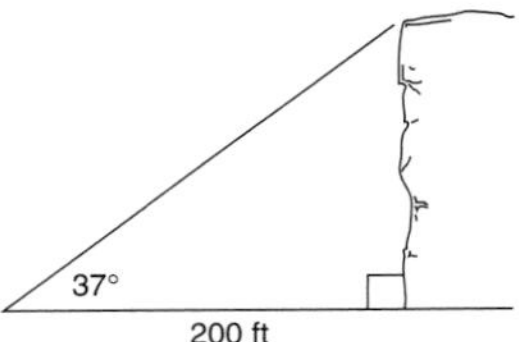


33. Find the length of each apothem in a regular pentagon whose radii measure 10 in. each.

34. Dale looks up to see his friend Lisa waving from her apartment window 30 ft from him. If Dale is standing 10 ft from the building, what is the angle of elevation as Dale looks up at Lisa?

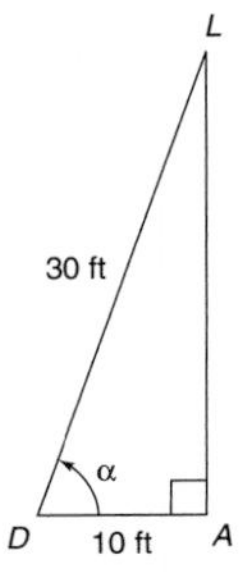


35. Find the length of the radius in a regular decagon for which each apothem has a length of 12.5 cm.

36. In searching for survivors of a boating accident, a helicopter moves horizontally across the ocean at an altitude of 200 ft above the water. If a man clinging to a life raft is seen through an angle of depression of 12°, what is the distance from the helicopter to the man in the water?

* 37. What is the size of the angle $\alpha$ formed by a diagonal of a cube and one of its edges?

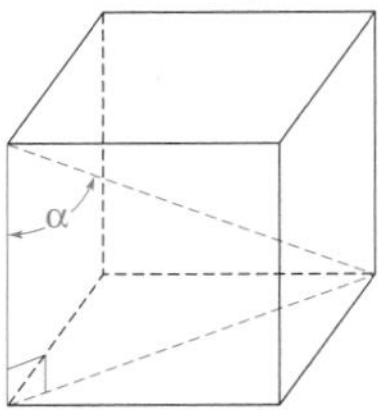


38. In the right circular cone,

a) find $r$ correct to tenths.
b) use $L = \pi r \ell$ to find the lateral area of the cone.

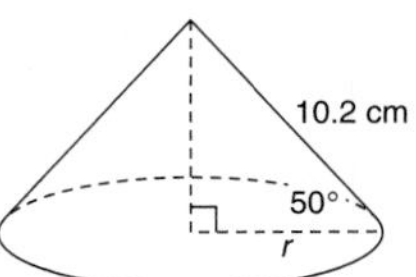


39. In parallelogram *ABCD*, find, to the nearest degree:

a) m$\angle A$   b) m$\angle B$

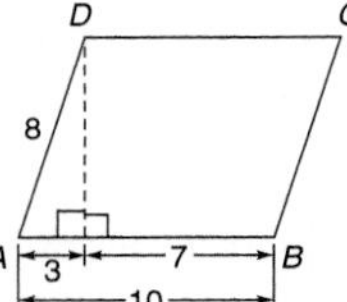

40. A ladder is carried horizontally through an L-shaped turn in a hallway. Show that the ladder has the length $L = \frac{6}{\sin\theta} + \frac{6}{\cos\theta}$.

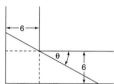

41. Use the drawing provided to show that the area of the isosceles triangle is $A = s^2 \sin\theta\cos\theta$.

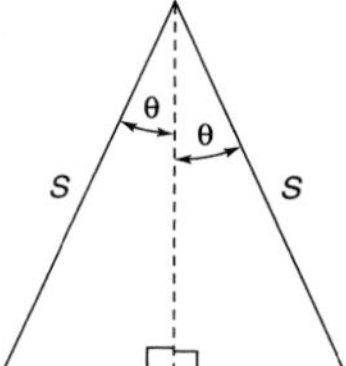


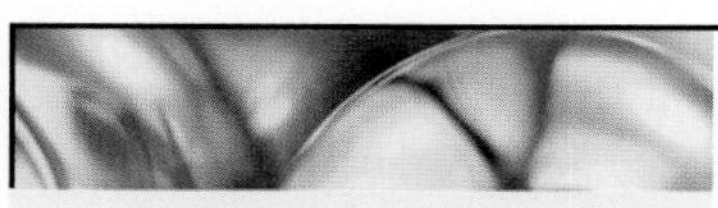

# 10.3 The Tangent Ratio and Other Ratios

**KEY CONCEPTS**

Tangent Ratio: $\tan\theta = \frac{\text{opposite}}{\text{adjacent}}$ • Cotangent • Secant • Cosecant

As in Sections 10.1 and 10.2, we deal strictly with right triangles in Section 10.3. The third trigonometric ratio is the **tangent** ratio, which is defined for an acute angle of the right triangle by

$$\frac{\text{length of leg } \textit{opposite} \text{ acute angle}}{\text{length of leg } \textit{adjacent} \text{ to acute angle}}$$

Like the sine ratio, the tangent ratio increases as the measure of the acute angle increases. Unlike the sine and cosine ratios, whose values range from 0 to 1, the value of the tangent ratio is from 0 upward; that is, there is no greatest value for the tangent.

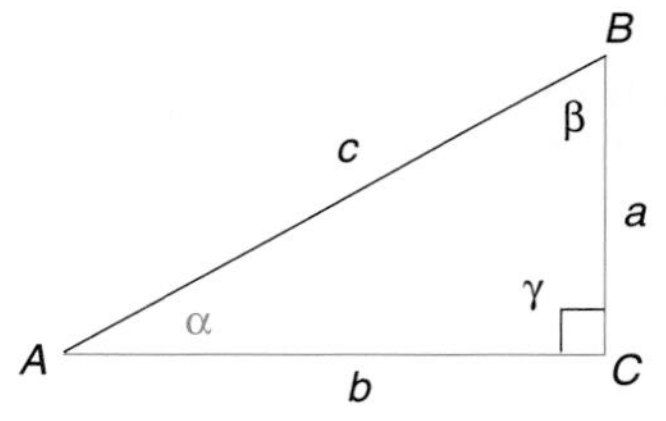


**FIGURE 10.24**

DEFINITION: In a right triangle, the **tangent ratio** for an acute angle is the ratio $\frac{\text{opposite}}{\text{adjacent}}$.

NOTE: In right $\triangle ABC$ in Figure 10.24, $\tan\alpha = \frac{a}{b}$ and $\tan\beta = \frac{b}{a}$, in which "tan" is an abbreviated form of the word tangent.

*Technology Exploration*

If you have a graphing calculator, draw the graph of $y = \tan x$ subject to these conditions:

i) Calculator in degree mode.
ii) Window has $0 \le x \le 90$ and $0 \le y \le 4$

Show that $y = \tan x$ increases as $x$ increases.

EXAMPLE 1

Find the values of $\tan\alpha$ and $\tan\beta$ for the triangle in Figure 10.25.

***Solution***

Using the fact that the tangent ratio is $\frac{\text{opposite}}{\text{adjacent}}$, we find that

$$\tan\alpha = \frac{a}{b} = \frac{8}{15}$$

and

$$\tan\beta = \frac{b}{a} = \frac{15}{8}$$

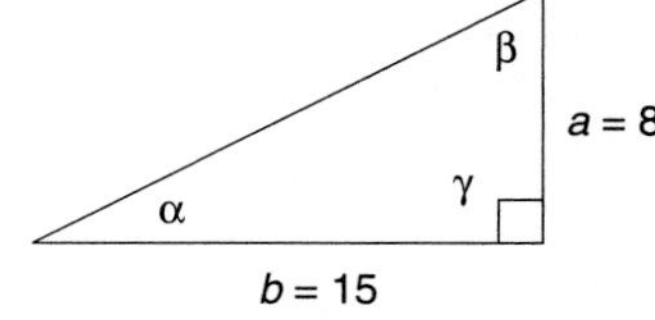


**FIGURE 10.25**

*Exs. 1–4*

The value of $\tan\theta$ changes from 0 for a 0° angle to an immeasurably large value as the measure of the acute angle approaches 90°. That the tangent ratio $\frac{\text{opposite}}{\text{adjacent}}$

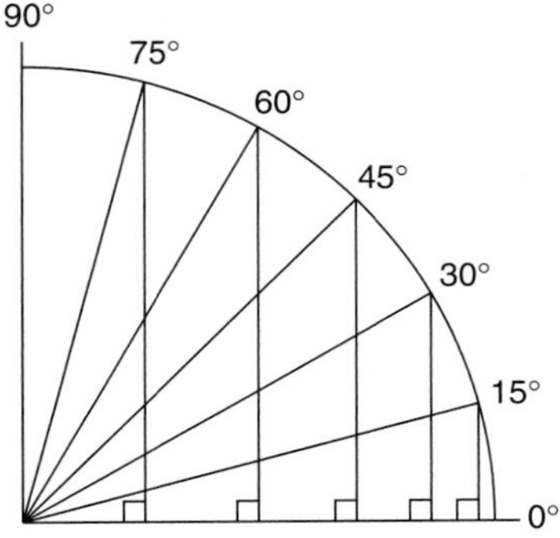


FIGURE 10.26

becomes infinitely large as $\theta \to 90°$ follows from the fact that the denominator becomes smaller and approaches 0 as the numerator increases.

Study Figure 10.26 to see why the value of the tangent of an angle grows immeasurably large as the angle approaches 90° in size. We often express this relationship by writing: As $\theta \to 90°$, $\tan\theta \to \infty$. The symbol $\infty$ is read "infinity" and implies that tan 90° is not measurable; thus tan 90° is *undefined.*

> DEFINITION: $\tan 0° = 0$ and tan 90° is undefined.

NOTE: Use your calculator to verify that $\tan 0° = 0$. What happens when you use your calculator to find tan 90°?

Certain tangent ratios are found through special triangles. By observing the triangles in Figure 10.27 and using the fact that $\tan\theta = \frac{\text{opposite}}{\text{adjacent}}$, we have

$$\tan 30° = \frac{x}{x\sqrt{3}} = \frac{1}{\sqrt{3}} = \frac{\sqrt{3}}{3} \approx 0.5774$$

$$\tan 45° = \frac{x}{x} = 1$$

$$\tan 60° = \frac{x\sqrt{3}}{x} = \sqrt{3} \approx 1.7321$$

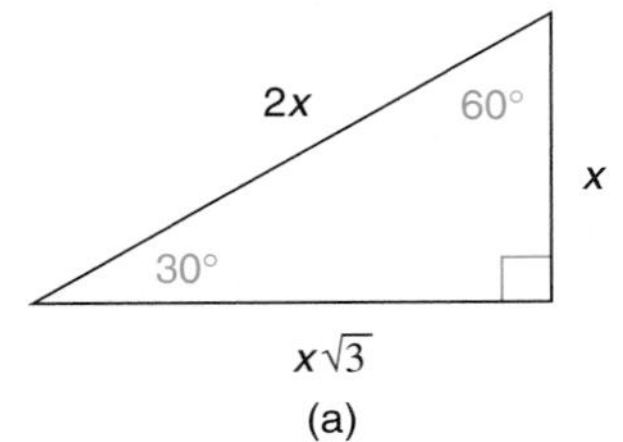


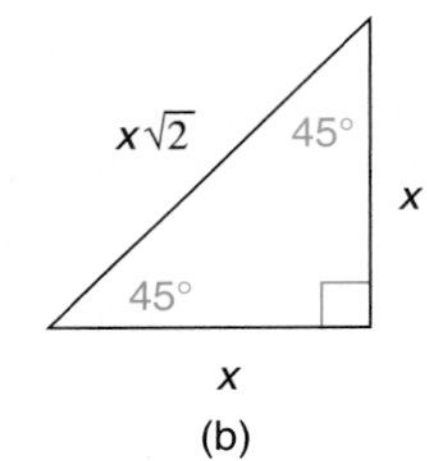


FIGURE 10.27

*Exs. 5–8*

We apply the tangent ratio in Example 2.

EXAMPLE 2

A ski lift moves each chair through an angle of 25°, as shown in Figure 10.28. What vertical change (rise) accompanies a horizontal change (run) of 845 ft?

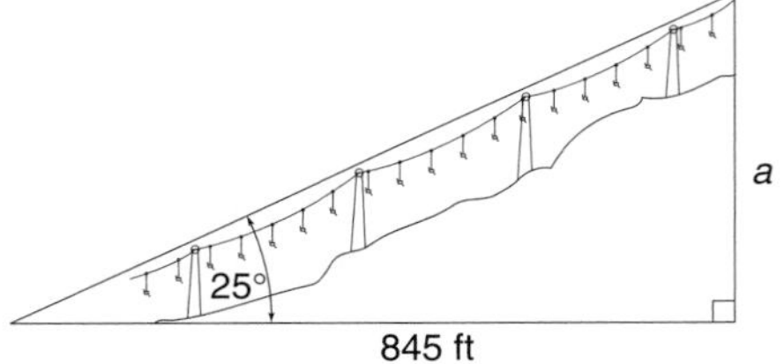


FIGURE 10.28

***Solution***

In the triangle, $\tan 25° = \frac{\text{opposite}}{\text{adjacent}} = \frac{a}{845}$. From $\tan 25° = \frac{a}{845}$, we multiply by 845 to obtain $a = 845 \cdot \tan 25°$. Using a calculator, we find that $a \approx 394$ ft.

The tangent ratio can also be used to find the measure of an angle if the lengths of the legs of a right triangle are known. This is illustrated in Example 3.

EXAMPLE 3

An airplane is seen flying just over Mission Rock, which is 1 mi away. If Mission Rock is known to be 135 ft high and the airplane is 50 ft above it, then what is the angle of elevation through which the plane is seen?

### Geometry in the Real World

In the coordinate system shown, we see that the slope of the line is $m = \tan \alpha$.

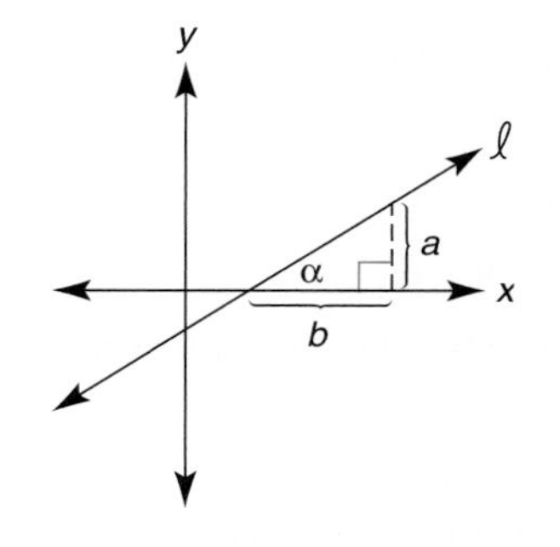


***Solution***

From Figure 10.29 and the fact that 1 mi = 5280 ft,

$$\tan \theta = \frac{\text{opposite}}{\text{adjacent}} = \frac{185}{5280}$$

Then $\tan \theta \approx 0.0350$, so $\theta = 2°$ to the nearest degree.

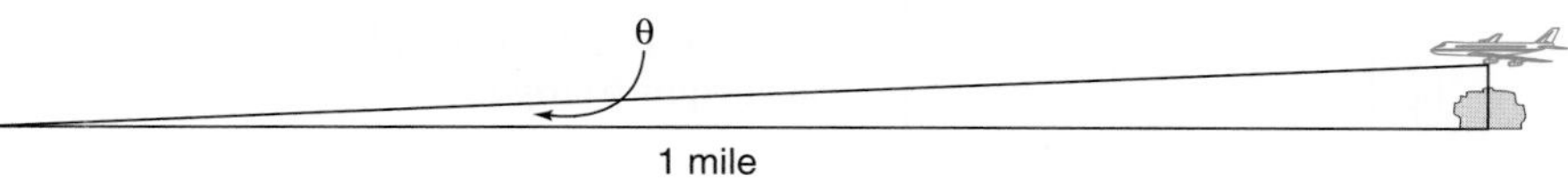


**FIGURE 10.29**

NOTE: The solution for Example 3 required the use of a calculator. When we use a scientific calculator in degree mode, the typical key sequence is

0 → . → 0 → 3 → 5 → inv → tan → **2**

When we use a graphing calculator in degree mode, the typical key sequence is

$\tan^{-1}$ → . → 0 → 3 → 5 → 0 → ENTER → **2**

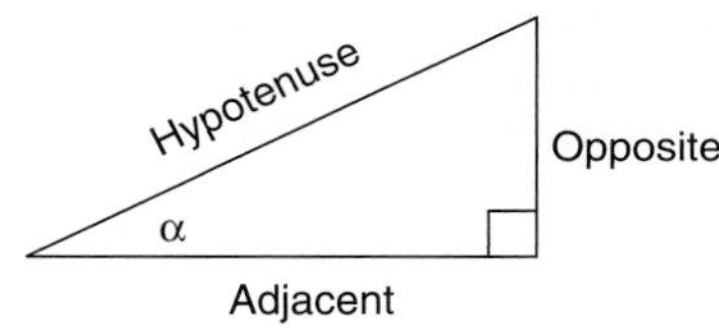


**FIGURE 10.30**

For the right triangle in Figure 10.30, we now have three ratios that can be used in problem solving. These are summarized as follows:

$$\sin \alpha = \frac{\text{opposite}}{\text{hypotenuse}}$$
$$\cos \alpha = \frac{\text{adjacent}}{\text{hypotenuse}}$$
$$\tan \alpha = \frac{\text{opposite}}{\text{adjacent}}$$

*Exs. 9–11*

The equation $\tan \alpha = \frac{a}{b}$ contains three variables: $\alpha$, $a$, and $b$. If the values of two of the variables are known, the value of the third variable can be found.

EXAMPLE 4

In Figure 10.31, name the ratio that should be used to find:

a) $a$, if $\alpha$ and $c$ are known
b) $\alpha$, if $a$ and $b$ are known
c) $\beta$, if $a$ and $c$ are known
d) $b$, if $a$ and $\beta$ are known

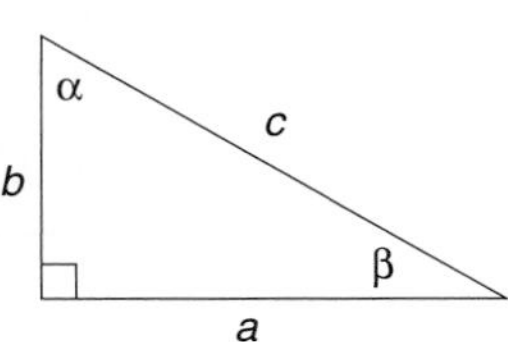


**FIGURE 10.31**

***Solution***

a) sine, because $\sin \alpha = \frac{a}{c}$
b) tangent, because $\tan \alpha = \frac{a}{b}$
c) cosine, because $\cos \beta = \frac{a}{c}$
d) tangent, because $\tan \beta = \frac{b}{a}$

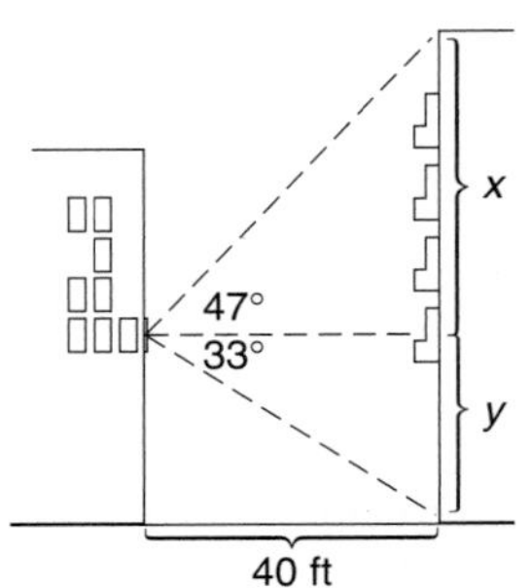


**FIGURE 10.32**

EXAMPLE 5

Two apartment buildings are 40 feet apart. From a window in her apartment, Izzi can see the top of the other apartment building through an angle of elevation of 47°. She can also see the base of the other building through an angle of depression of 33°. Approximately how tall is the other building?

***Solution***

In Figure 10.32, the height of the building is the sum $x + y$. Using the upper and lower right triangles,

$$\tan 47° = \frac{x}{40} \quad \text{and} \quad \tan 33° = \frac{y}{40}$$

Now $\quad x = 40 \cdot \tan 47° \quad \text{and} \quad y = 40 \cdot \tan 33°$

Then $x \approx 43$ and $y \approx 26$, so $x + y \approx 43 + 26 = 69$. The building is approximately 69 ft tall.

NOTE: In Example 5, you can determine the height of the building $(x + y)$ by entering the expression $40 \cdot \tan 47° + 40 \cdot \tan 33°$ on your calculator.

There are a total of six trigonometric ratios. We define the remaining ratios for completeness; however, we will be able to solve all application problems in this chapter by using only the sine, cosine, and tangent ratios. The remaining ratios are the **cotangent** (abbreviated "cot"), **secant** (abbreviated "sec"), and **cosecant** (abbreviated "csc"). These are defined in terms of the right triangle shown in Figure 10.33.

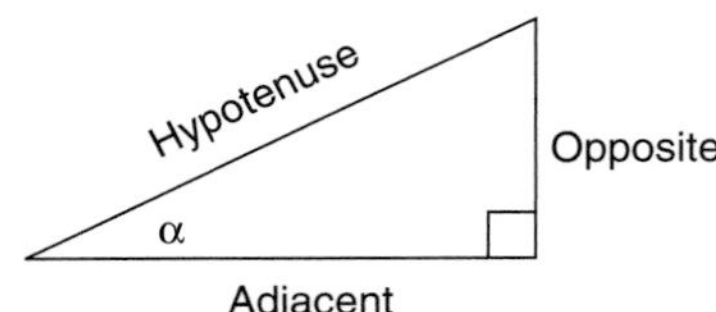


**FIGURE 10.33**

$$\cot \alpha = \frac{\text{adjacent}}{\text{opposite}}$$
$$\sec \alpha = \frac{\text{hypotenuse}}{\text{adjacent}}$$
$$\csc \alpha = \frac{\text{hypotenuse}}{\text{opposite}}$$

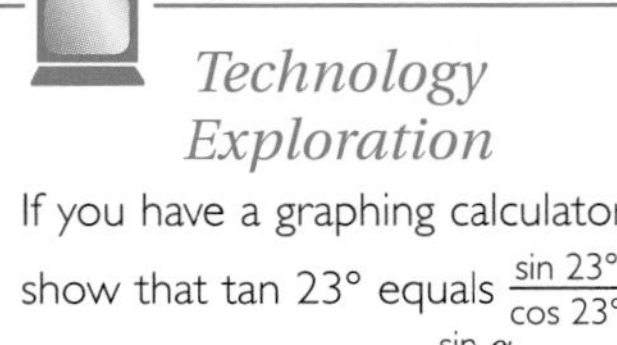

*Technology Exploration*

If you have a graphing calculator, show that $\tan 23°$ equals $\frac{\sin 23°}{\cos 23°}$. The identity $\tan \alpha = \frac{\sin \alpha}{\cos \alpha}$ is true as long as $\cos \alpha \neq 0$.

For the fraction $\frac{a}{b}$ (where $b \neq 0$), the reciprocal is $\frac{b}{a}$ $(a \neq 0)$. It is easy to see that $\cot \alpha$ is the reciprocal of $\tan \alpha$; $\sec \alpha$ is the reciprocal of $\cos \alpha$; and $\csc \alpha$ is the reciprocal of $\sin \alpha$. In the following chart, we invert the trigonometric ratio on the left to obtain the reciprocal ratio named on the right.

| TRIGONOMETRIC RATIO | RECIPROCAL RATIO |
|---|---|
| sine $\alpha = \frac{\text{opposite}}{\text{hypotenuse}}$ | cosecant $\alpha = \frac{\text{hypotenuse}}{\text{opposite}}$ |
| cosine $\alpha = \frac{\text{adjacent}}{\text{hypotenuse}}$ | secant $\alpha = \frac{\text{hypotenuse}}{\text{adjacent}}$ |
| tangent $\alpha = \frac{\text{opposite}}{\text{adjacent}}$ | cotangent $\alpha = \frac{\text{adjacent}}{\text{opposite}}$ |

Calculators display only the sine, cosine, and tangent ratios. By using the reciprocal key, [1/x] or [$x^{-1}$], you can obtain the values for the remaining ratios. See Example 6 for details.

EXAMPLE 6

Use a calculator to evaluate

a) csc 37° b) cot 51° c) sec 84°

***Solution***

a) First we use the calculator to find sin 37° ≈ 0.6081. Now use the [1/x] or [$x^{-1}$] key to show that csc 37° ≈ 1.6616.

b) First we use the calculator to find tan 51° ≈ 1.2349. Now use the [1/x] or [$x^{-1}$] key to show that cot 51° ≈ 0.8098.

c) First we use the calculator to find cos 84° ≈ 0.1045. Now use the [1/x] or [$x^{-1}$] key to show that sec 84° ≈ 9.5668

*Exs. 12–16*

NOTE: In part (a), the value of csc 37° can be determined by using the following display on a graphing calculator: $(\sin 37)^{-1}$. Similar displays can be used in parts (b) and (c).

In Example 7, we are reminded that a calculator is not necessary for all calculations.

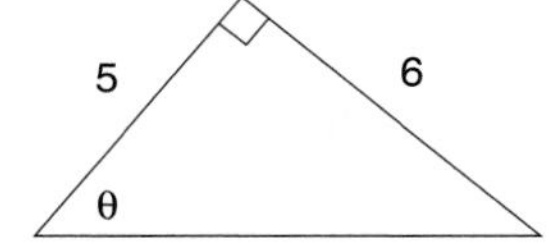

**FIGURE 10.34**

EXAMPLE 7

For the triangle in Figure 10.34, find the exact values of all six trigonometric ratios for angle $\theta$.

***Solution***

We will need the length of the hypotenuse, which we find by the Pythagorean Theorem. With $c$ the length of the hypotenuse,

$$\begin{aligned} c^2 &= 5^2 + 6^2 \\ c^2 &= 25 + 36 \\ c^2 &= 61 \\ c &= \sqrt{61} \end{aligned}$$

Therefore,

$$\sin\theta = \frac{\text{opposite}}{\text{hypotenuse}} = \frac{6}{\sqrt{61}} = \frac{6}{\sqrt{61}} \cdot \frac{\sqrt{61}}{\sqrt{61}} = \frac{6\sqrt{61}}{61}$$

$$\cos\theta = \frac{\text{adjacent}}{\text{hypotenuse}} = \frac{5}{\sqrt{61}} = \frac{5}{\sqrt{61}} \cdot \frac{\sqrt{61}}{\sqrt{61}} = \frac{5\sqrt{61}}{61}$$

$$\tan\theta = \frac{\text{opposite}}{\text{adjacent}} = \frac{6}{5}$$

$$\cot\theta = \frac{\text{adjacent}}{\text{opposite}} = \frac{5}{6}$$

$$\sec\theta = \frac{\text{hypotenuse}}{\text{adjacent}} = \frac{\sqrt{61}}{5}$$

$$\csc\theta = \frac{\text{hypotenuse}}{\text{opposite}} = \frac{\sqrt{61}}{6}$$

NOTE: The arrows in Example 7 remind us which ratios are reciprocals of each other.

EXAMPLE 8

Evaluate the ratio named by using the given ratio:

a) $\tan\theta$, if $\cot\theta = \frac{2}{3}$
b) $\sin\alpha$, if $\csc\alpha = 1.25$
c) $\sec\beta$, if $\cos\beta = \frac{\sqrt{3}}{2}$
d) $\csc\gamma$, if $\sin\gamma = 1$

***Solution***

a) If $\cot\theta = \frac{2}{3}$, then $\tan\theta = \frac{3}{2}$ (the reciprocal of $\cot\theta$).
b) If $\csc\alpha = 1.25$ or $\frac{5}{4}$, then $\sin\alpha = \frac{4}{5}$ (the reciprocal of $\csc\alpha$).
c) If $\cos\beta = \frac{\sqrt{3}}{2}$, then $\sec\beta = \frac{2}{\sqrt{3}}$ or $\frac{2\sqrt{3}}{3}$ (the reciprocal of $\cos\beta$).
d) If $\sin\gamma = 1$, then $\csc\gamma = 1$ (the reciprocal of $\sin\gamma$).

EXAMPLE 9

To the nearest degree, how large is $\theta$ in the triangle in Figure 10.35?

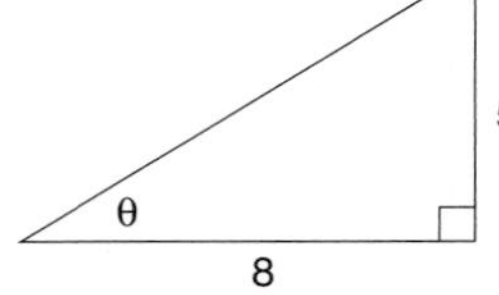

**FIGURE 10.35**

***Solution***

Because the lengths "opposite" and "adjacent" are known, we can use the tangent ratio to find $\theta$.

$$\tan\theta = \frac{5}{8}$$

With a scientific calculator, we determine $\theta$ by using the key sequence

5 → ÷ → 8 → = → inv → tan → **32**.

When we use a graphing calculator, the key sequence is

tan⁻¹ → ( → 5 → ÷ → 8 → ) → ENTER → **32**. Thus $\theta \approx 32°$.

In the application exercises that follow this section, you will have to decide which trigonometric ratio enables you to solve the problem. The Pythagorean Theorem can be used as well.

EXAMPLE 10

As his fishing vessel moves into the bay, the captain notes that the angle of elevation to the top of the lighthouse is 11°. If the lighthouse is 200 ft tall, how far is the vessel from the lighthouse? See Figure 10.36 on page 503.

***Solution***

Again we use the tangent ratio; in Figure 10.36,

$$\tan 11° = \frac{200}{x}$$
$$x \cdot \tan 11° = 200$$
$$x = \frac{200}{\tan 11°} \approx 1028.91$$

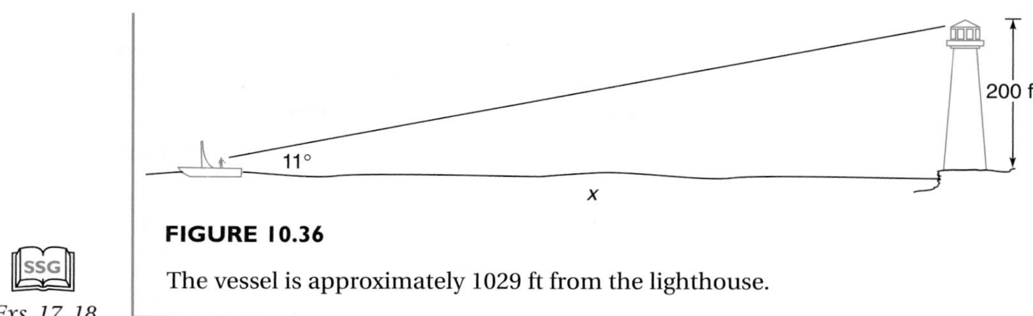


**FIGURE 10.36**

The vessel is approximately 1029 ft from the lighthouse.

*Exs. 17, 18*

## 10.3 Exercises

*In Exercises 1 to 4, find tan α and tan β for each triangle.*

1.
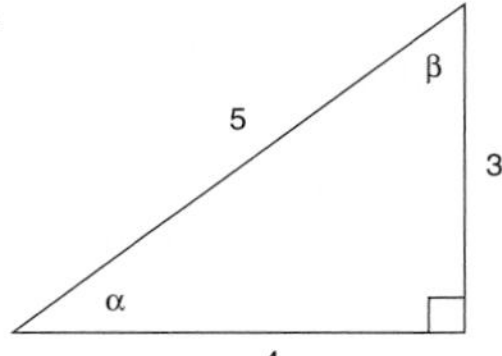


2.
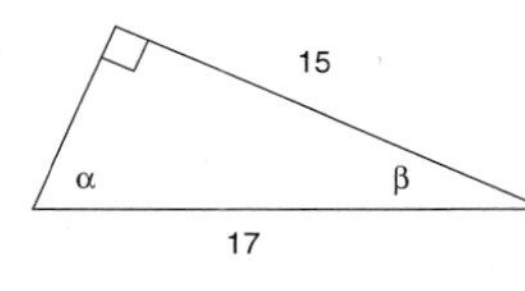


3.
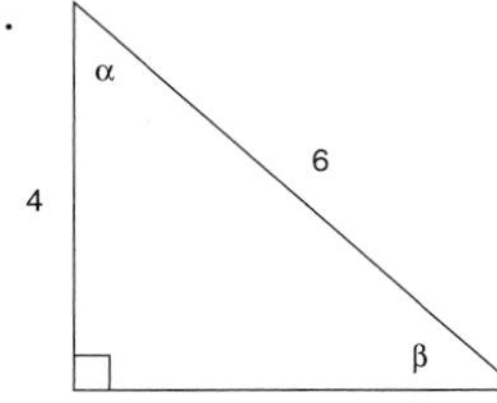


4.
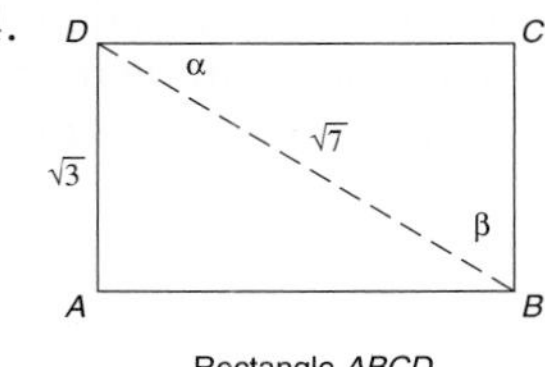


*In Exercises 5 to 10, find the value (or expression) for each of the six trigonometric ratios of angle α. Use the Pythagorean Theorem as needed.*

5.
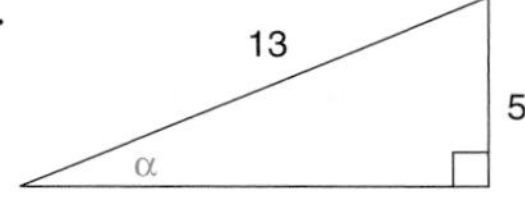


6.
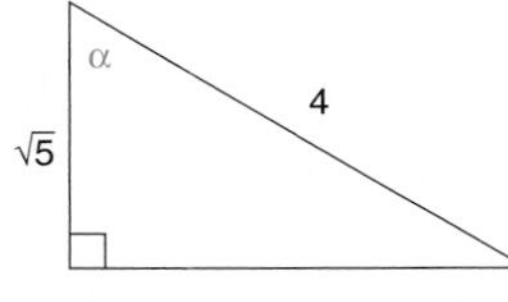


7.
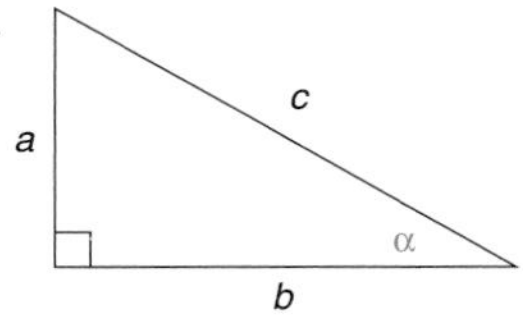


8.
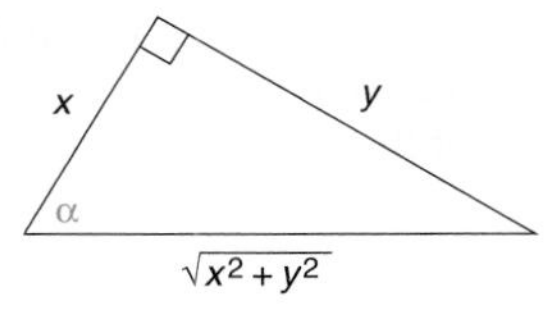


9.
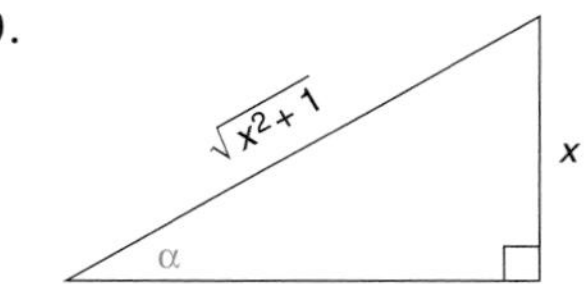


10.
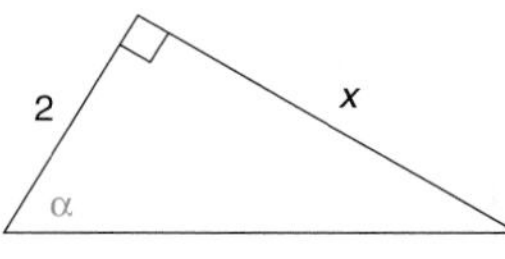


*In Exercises 11 to 14, use a calculator to find the indicated tangent ratio correct to four decimal places.*

11. tan 15°
12. tan 45°
13. tan 57°
14. tan 78°

*In Exercises 15 to 20, use the sine, cosine, or tangent ratio to find the lengths of the indicated sides to the nearest tenth of a unit.*

15.
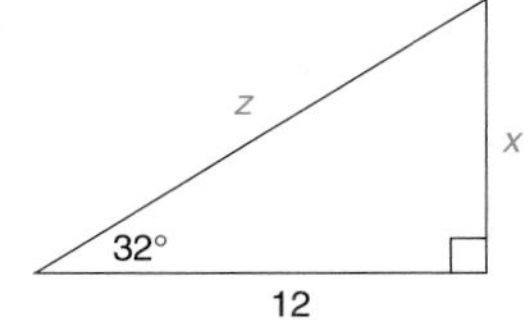


16.
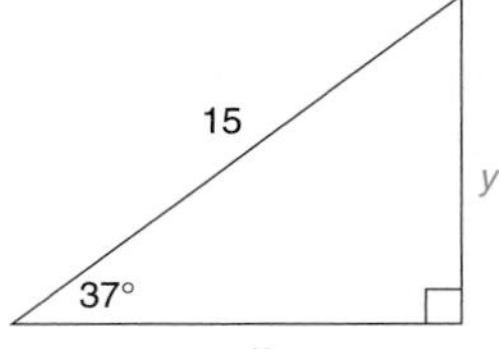


17.
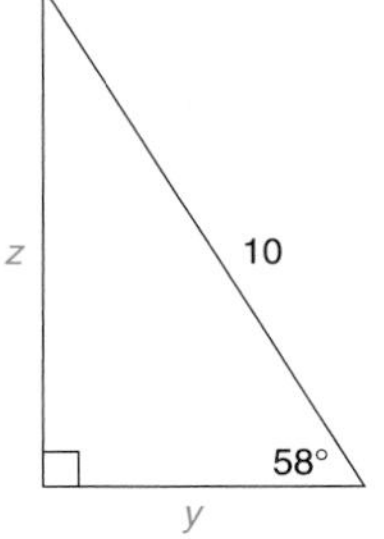

18.

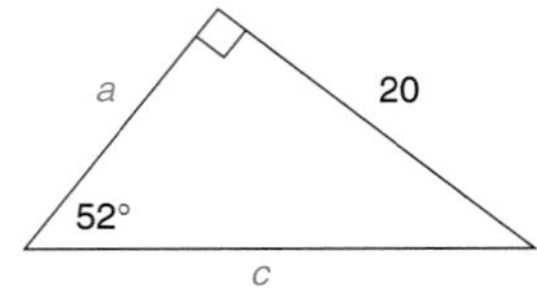


19.

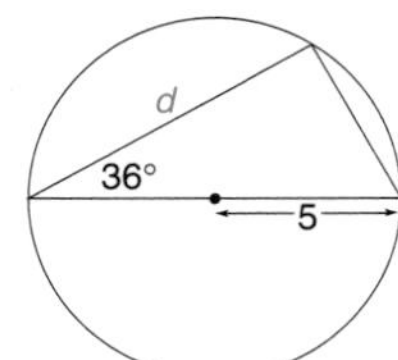


20.

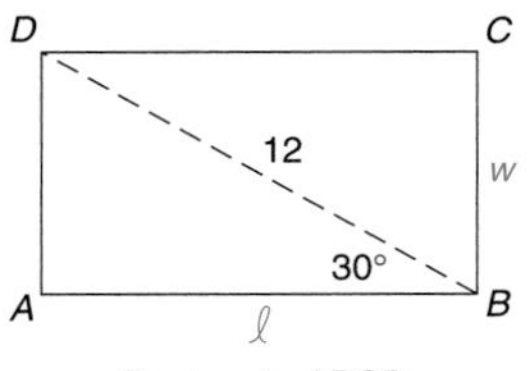


Rectangle *ABCD*

*In Exercises 21 to 26, use the sine, cosine, or tangent ratio to find the indicated angle measures to the nearest degree.*

21.

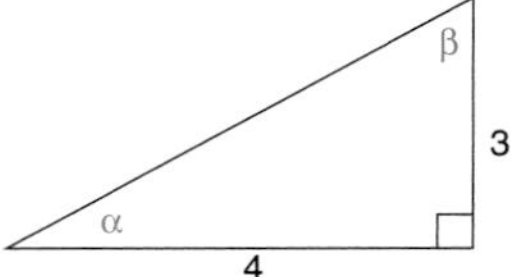


22.

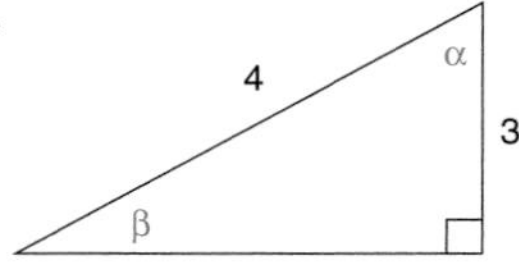


23.

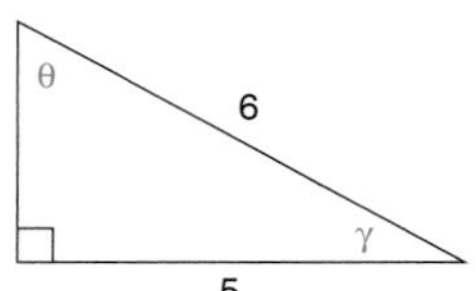


24.

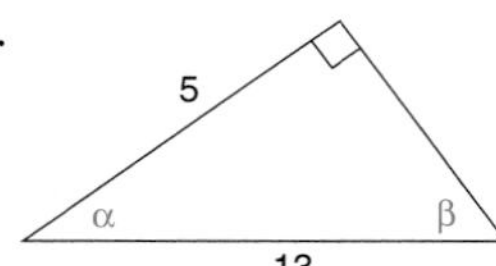


25.

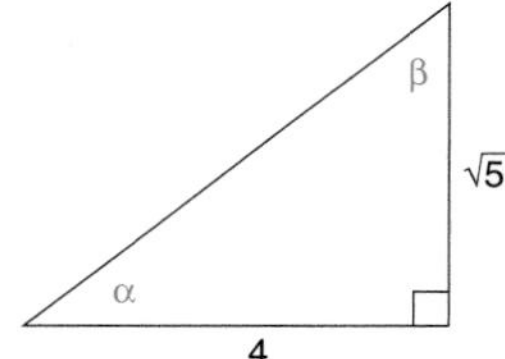


26.

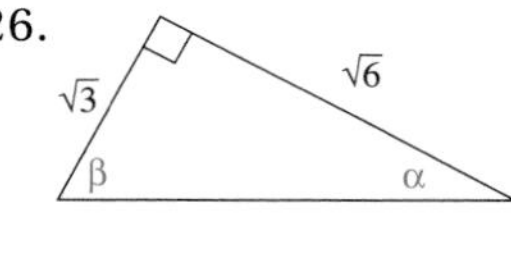


*In Exercises 27 to 32, use a calculator and reciprocal relationships to find each ratio correct to four decimal places.*

27. cot 34° 28. sec 15°

29. csc 30° 30. cot 67°

31. sec 42° 32. csc 72°

*In Exercises 33 to 36, we expand the list of trigonometric identities. As you may recall (see page 494), an identity is a statement that is true for all permissible choices of the variable.*

33. a) For $\alpha \neq 90°$, prove the identity

$$\tan \alpha = \frac{\sin \alpha}{\cos \alpha}$$

(HINT: $\sin \alpha = \frac{a}{c}$ and $\cos \alpha = \frac{b}{c}$.)

b) Use your calculator to show that tan 23° and $\frac{\sin 23°}{\cos 23°}$ are equivalent.

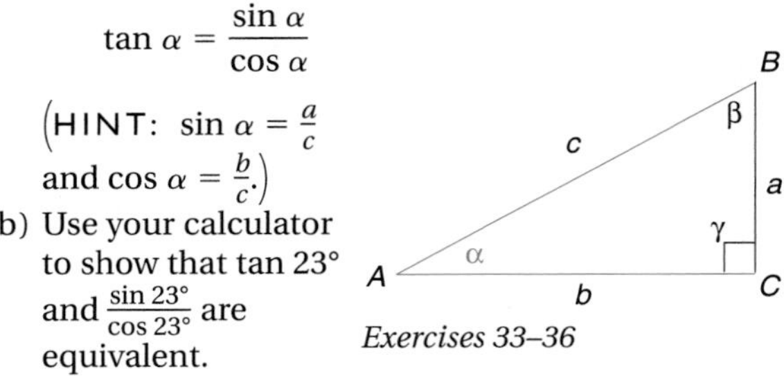


*Exercises 33–36*

34. a) For $\alpha \neq 0°$, prove the identity $\cot \alpha = \frac{\cos \alpha}{\sin \alpha}$.

b) Use your calculator to determine cot 57° by dividing cos 57° by sin 57°.

35. a) For $\alpha \neq 90°$, prove the identity $\sec \alpha = \frac{1}{\cos \alpha}$.

b) Use your calculator to determine sec 82°.

36. a) For $\alpha \neq 0°$, prove the identity $\csc \alpha = \frac{1}{\sin \alpha}$.

b) Use your calculator to determine csc 12.3°.

*In Exercises 37 to 43, angle measures should be given to the nearest degree; distances should be given to the nearest tenth of a unit.*

37. When her airplane is descending to land, the pilot notes an angle of depression of 5°. If the altimeter shows an altitude reading of 120 ft, what is the distance *x* of the plane from touchdown?

38. The top of a lookout tower is seen from a point 270 ft from its base. If the angle of elevation is 37°, how tall is the tower?

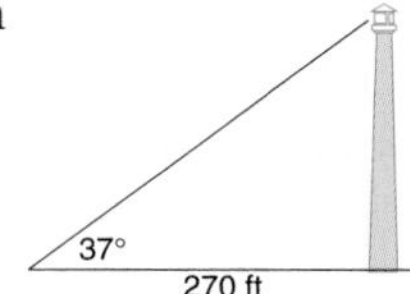


39. Find the length of the apothem to each of the 6-in. sides of a regular pentagon.

*40. What is the measure of the angle between the diagonal of a cube and the diagonal of the face of the cube?

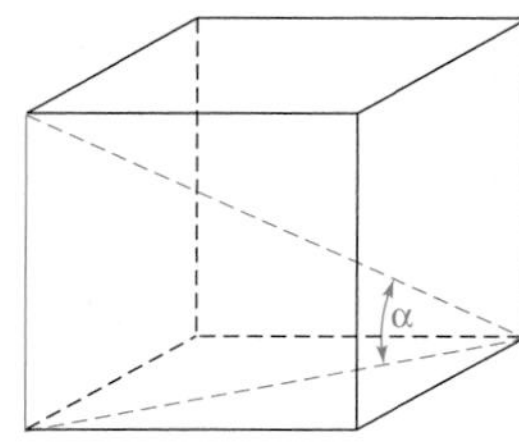

41. Upon approaching a house, Liz hears Lynette shout to her. Liz, who is standing 10 ft from the house, looks up to see Lynette in the third-story window approximately 32 ft away. What is the measure of the angle of elevation as Liz looks up at Lynette?

* 42. While a helicopter hovers 1000 ft above the water, its pilot spies a man in a lifeboat through an angle of depression of 28°. Along a straight line, a rescue boat can also be seen through an angle of depression of 14°. How far is the rescue boat from the lifeboat?

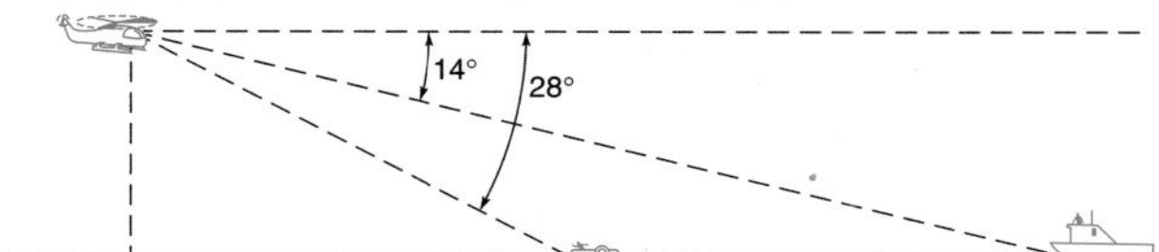


* 43. From atop a 200-ft lookout tower, a fire is spotted due north through an angle of depression of 12°. Firefighters located 1000 ft due east of the tower must work their way through heavy foliage to the fire. By their compasses, through what angle (measured from the north toward the west) must the firefighters travel?

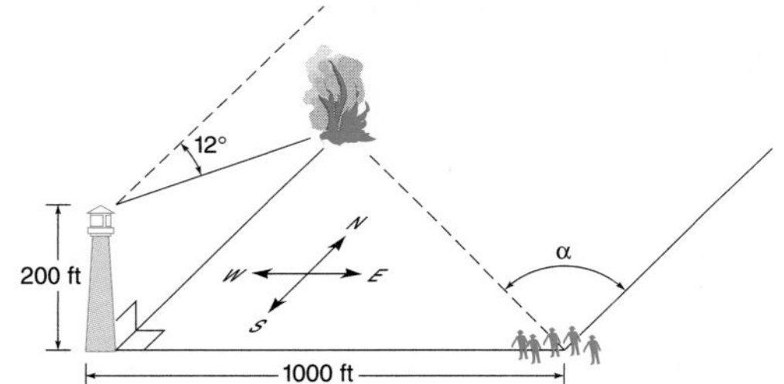

44. In the triangle shown, find each measure to the nearest tenth of a unit.

a) $x$ b) $y$

c) $A$, the area of the triangle

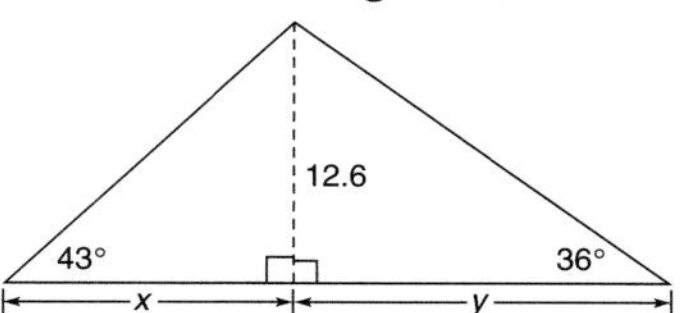


45. At an altitude of 12,000 ft, a pilot sees two towns through angles of depression of 37° and 48° as shown. To the nearest ten feet, how far apart are the towns?

46. Consider the regular square pyramid shown.

a) Find the slant height $\ell$ correct to tenths. (HINT: Apothem $a = 3$)

b) Use $\ell$ from part (a) to find the lateral area $L$ of the pyramid.

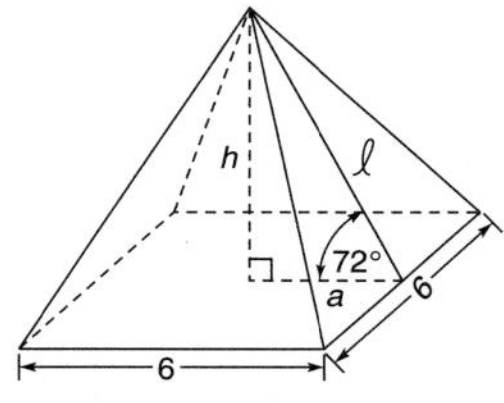

*Exercises 46, 47*

47. Consider the regular square pyramid shown.

a) Find the altitude $h$ correct to the nearest tenth of a unit. (HINT: Apothem $a = 3$)

b) Use $h$ from part (a) to find the volume of the pyramid.

# 10.4 Applications with Acute Triangles

**KEY CONCEPTS**

Area of a Triangle:

$A = \frac{1}{2}bc \sin \alpha$

$A = \frac{1}{2}ac \sin \beta$

$A = \frac{1}{2}ab \sin \gamma$

Law of Sines:

$\frac{\sin \alpha}{a} = \frac{\sin \beta}{b} = \frac{\sin \gamma}{c}$

Law of Cosines:

$c^2 = a^2 + b^2 - 2ab \cos \gamma$

$b^2 = a^2 + c^2 - 2ac \cos \beta$

$a^2 = b^2 + c^2 - 2bc \cos \alpha$

In Sections 10.1 through 10.3, our focus was strictly on right triangles. Thus, the sides of every triangle that we considered were two legs and a hypotenuse. We now turn our attention to some relationships that we will prove for, and apply with, *acute* triangles. The first relationship provides a formula for the area of a triangle in which $\alpha$, $\beta$, and $\gamma$ are all acute angles.

## AREA OF A TRIANGLE

THEOREM 10.4.1: The area of an acute triangle equals one-half the product of the lengths of two sides and the sine of the included angle.

*Given:* Acute $\triangle ABC$, as shown in Figure 10.37(a) on page 506

*Prove:* $A = \frac{1}{2}bc \sin \alpha$

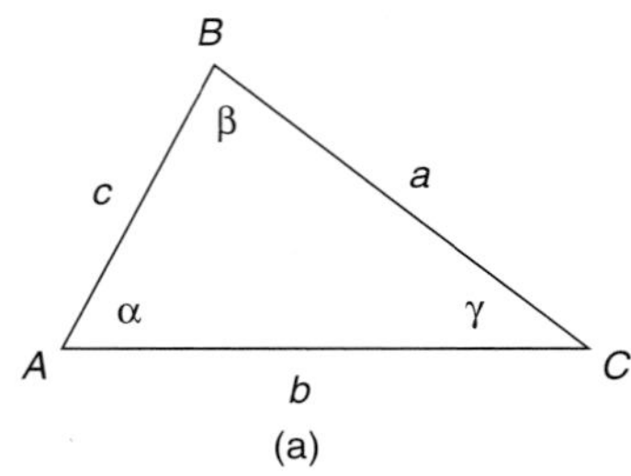

(a)

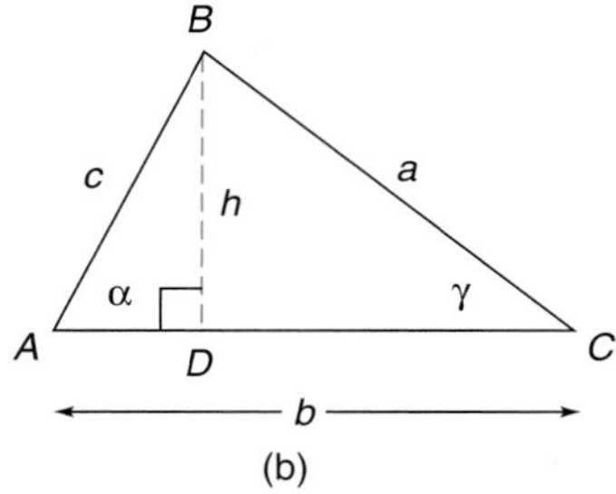

(b)

**FIGURE 10.37**

*Proof:* The area of the triangle is given by $A = \frac{1}{2}bh$. With the altitude $\overline{BD}$ [see Figure 10.37(b)], we see that $\sin \alpha = \frac{h}{c}$ in right $\triangle ABD$. Then $h = c \sin \alpha$. Consequently, $A = \frac{1}{2}bh$ becomes

$$A = \frac{1}{2}b(c \sin \alpha), \quad \text{so} \quad A = \frac{1}{2}bc \sin \alpha$$

Theorem 10.4.1 has three equivalent forms, as shown in the following box.

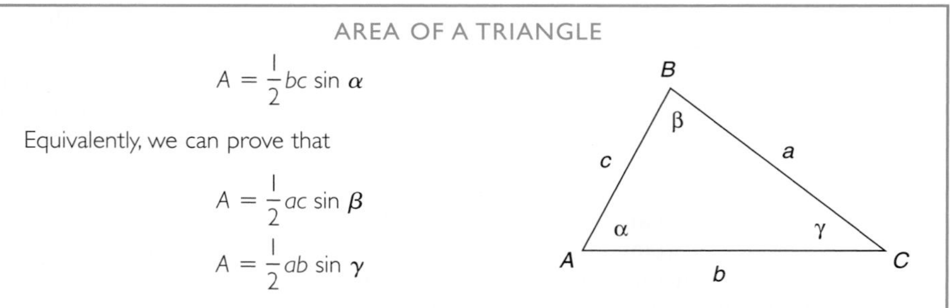
AREA OF A TRIANGLE

$$A = \frac{1}{2}bc \sin \alpha$$

Equivalently, we can prove that

$$A = \frac{1}{2}ac \sin \beta$$

$$A = \frac{1}{2}ab \sin \gamma$$



In the more advanced course called trigonometry, this area formula can also be proved for obtuse triangles. If the triangle is a right triangle with $\gamma = 90°$, then $A = \frac{1}{2}ab \sin \gamma$ becomes $A = \frac{1}{2}ab$ since $\sin \gamma = 1$.

*Technology Exploration*

If you have a graphing calculator, you can evaluate many results rather easily. For Example 1, evaluate $\left(\frac{1}{2}\right) \cdot 6 \cdot 10 \cdot \sin(33)$. Use degree mode.

EXAMPLE 1

In Figure 10.38, find the area of $\triangle ABC$.

***Solution***

We use $A = \frac{1}{2}bc \sin \alpha$ since $\alpha$, $b$, and $c$ are known.

$$A = \frac{1}{2} \cdot 6 \cdot 10 \cdot \sin 33°$$
$$= 30 \cdot \sin 33°$$
$$\approx 16.3 \text{ in}^2$$

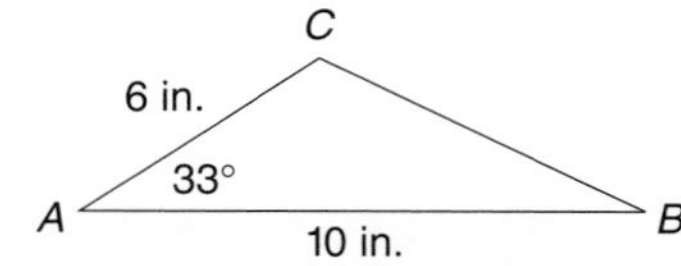

**FIGURE 10.38**

EXAMPLE 2

In $\triangle ABC$, $a = 10.2$ and $c = 7.6$. If the area of a $\triangle ABC$ is approximately 38.3 square units, find $\beta$ to the nearest degree.

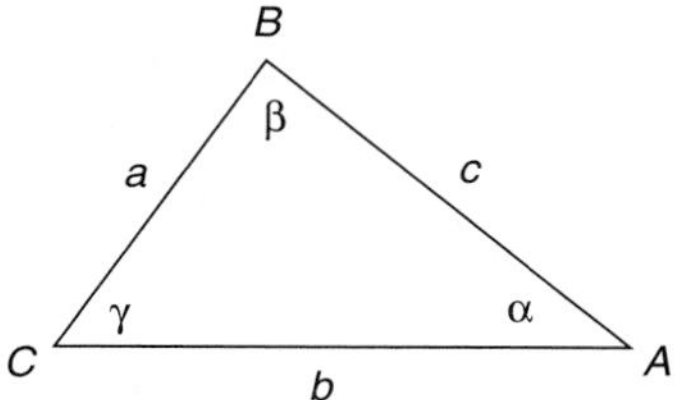

**FIGURE 10.39**

***Solution***

Using the area formula, $A = \frac{1}{2}ac \sin \beta$, we have $38.3 = (0.5)(10.2)(7.6) \sin \beta$, or $38.3 = 38.76 \sin \beta$.

Thus $\sin \beta = \frac{38.3}{38.76}$ and $\beta = \sin^{-1}\left(\frac{38.3}{38.76}\right)$. Then $\beta \approx 81°$ (rounded from 81.16).

*Exs. 1–4*

## LAW OF SINES

Because the area of a triangle is unique, we can equate the three area expressions characterized by Theorem 10.4.1 as follows:

$$\frac{1}{2}bc \sin \alpha = \frac{1}{2}ac \sin \beta = \frac{1}{2}ab \sin \gamma$$

Dividing each part of this equality by $\frac{1}{2}abc$, we find

$$\frac{\frac{1}{2}bc \sin \alpha}{\frac{1}{2}bca} = \frac{\frac{1}{2}ac \sin \beta}{\frac{1}{2}acb} = \frac{\frac{1}{2}ab \sin \gamma}{\frac{1}{2}abc}$$

$$\frac{\sin \alpha}{a} = \frac{\sin \beta}{b} = \frac{\sin \gamma}{c}$$

This relationship between the lengths of the sides of an acute triangle and the sines of their opposite angles is known as the Law of Sines. In trigonometry, it is shown that the Law of Sines is true for right triangles and obtuse triangles as well.

> THEOREM 10.4.2: **(Law of Sines)** In any acute triangle, the three ratios between the sines of the angles and the lengths of the opposite sides are equal. That is,
>
> $$\frac{\sin \alpha}{a} = \frac{\sin \beta}{b} = \frac{\sin \gamma}{c} \quad \text{or} \quad \frac{a}{\sin \alpha} = \frac{b}{\sin \beta} = \frac{c}{\sin \gamma}$$

When solving a problem, we equate only two of the equal ratios described in Theorem 10.4.2. For instance, we can use

$$\frac{\sin \alpha}{a} = \frac{\sin \beta}{b} \quad \text{or} \quad \frac{\sin \alpha}{a} = \frac{\sin \gamma}{c} \quad \text{or} \quad \frac{\sin \beta}{b} = \frac{\sin \gamma}{c}$$

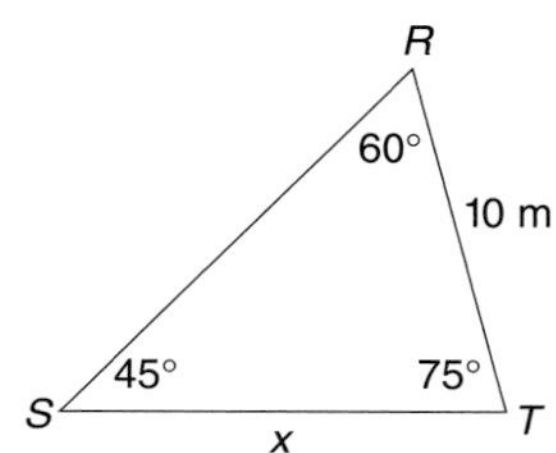


**FIGURE 10.40**

EXAMPLE 3

Use the Law of Sines to find the exact length $ST$ in Figure 10.40.

***Solution***

Because we know $RT$ and the measures of angles $S$ and $R$, we use $\frac{\sin S}{RT} = \frac{\sin R}{ST}$.

$$\frac{\sin 45°}{10} = \frac{\sin 60°}{x}$$

Because $\sin 45° = \frac{\sqrt{2}}{2}$ and $\sin 60° = \frac{\sqrt{3}}{2}$, we have

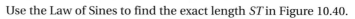

$$\frac{\frac{\sqrt{2}}{2}}{10} = \frac{\frac{\sqrt{3}}{2}}{x}$$

By the Means-Extremes Property,

$$\frac{\sqrt{2}}{2} \cdot x = \frac{\sqrt{3}}{2} \cdot 10$$

Multiplying by $\frac{2}{\sqrt{2}}$, we have

$$\frac{2}{\sqrt{2}} \cdot \frac{\sqrt{2}}{2} \cdot x = \frac{2}{\sqrt{2}} \cdot \frac{\sqrt{3}}{2} \cdot 10$$

$$x = \frac{10\sqrt{3}}{\sqrt{2}} = \frac{10\sqrt{3}}{\sqrt{2}} \cdot \frac{\sqrt{2}}{\sqrt{2}} = \frac{10\sqrt{6}}{2} = 5\sqrt{6}$$

Then $ST = 5\sqrt{6}$ m.

EXAMPLE 4

In $\triangle ABC$, $b = 12$, $c = 10$, and $\beta = 83°$. Find $\gamma$ to the nearest degree.

***Solution***

Knowing values of $b$, $c$, and $\beta$, we use the following form of the Law of Sines to find $\gamma$: $\frac{\sin \beta}{b} = \frac{\sin \gamma}{c}$.

B
β
a
c
γ
α
C
A
b

FIGURE 10.41

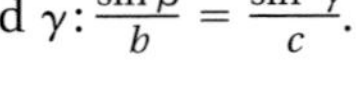

$$\frac{\sin 83°}{12} = \frac{\sin \gamma}{10}, \quad \text{so} \quad 12 \sin \gamma = 10 \sin 83°$$

Exs. 5–7

Then $\sin \gamma = \frac{10 \sin 83°}{12} \approx 0.8271$, so $\gamma = \sin^{-1}(0.8271) \approx 56°$.

## LAW OF COSINES

The final relationship that we consider is again proved only for an acute triangle. Like the Law of Sines, this relationship (known as the Law of Cosines) can be used to find unknown measures in a triangle. The Law of Cosines (which can also be established for obtuse triangles in a more advanced course) can be stated in words as "The square of one side of a triangle equals the sum of squares of the two remaining sides decreased by twice the product of those two sides and the cosine of their included angle." See Figure 10.42(a) as you read Theorem 10.4.3.

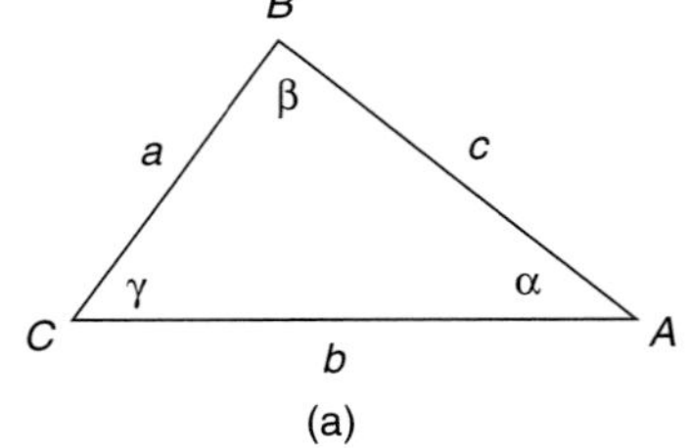


(a)

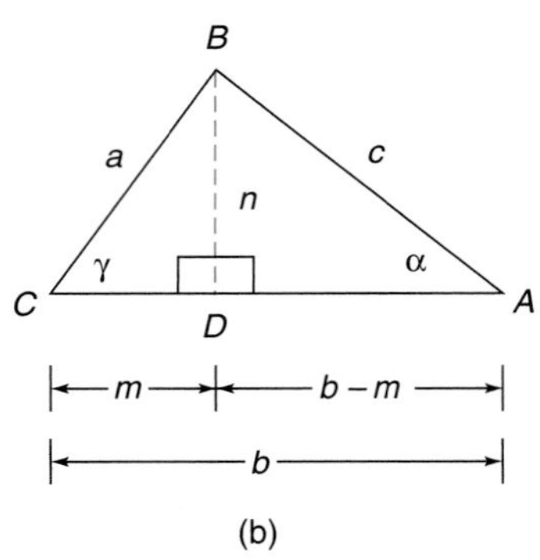


(b)

FIGURE 10.42

THEOREM 10.4.3: **(Law of Cosines)** In acute $\triangle ABC$,

$$c^2 = a^2 + b^2 - 2ab \cos \gamma$$
$$b^2 = a^2 + c^2 - 2ac \cos \beta$$
$$a^2 = b^2 + c^2 - 2bc \cos \alpha$$

The proof of the first form of the Law of Cosines follows.

*Given:* Acute $\triangle ABC$ in Figure 10.42(a)

*Prove:* $c^2 = a^2 + b^2 - 2ab \cos \gamma$

*Proof:* In Figure 10.42(a), draw the altitude $\overline{BD}$ from $B$ to $\overline{AC}$. We designate lengths of the line segments as shown in Figure 10.42(b). Now

$$(b - m)^2 + n^2 = c^2 \quad \text{and} \quad m^2 + n^2 = a^2$$

by the Pythagorean Theorem.

The second statement is equivalent to $m^2 = a^2 - n^2$. After we expand $(b - m)^2$, the first equation becomes

$$b^2 - 2bm + m^2 + n^2 = c^2$$

Then we replace $m^2$ by $(a^2 - n^2)$ to obtain

$$b^2 - 2bm + (a^2 - n^2) + n^2 = c^2$$

Simplifying yields

$$c^2 = a^2 + b^2 - 2bm$$

In right $\triangle CDB$,

$$\cos \gamma = \frac{m}{a} \qquad \text{so} \qquad m = a \cos \gamma$$

Hence

$$c^2 = a^2 + b^2 - 2bm$$
$$c^2 = a^2 + b^2 - 2b(a \cos \gamma)$$
$$c^2 = a^2 + b^2 - 2ab \cos \gamma$$

Arguments similar to the preceding proof can be provided for both remaining forms of the Law of Cosines. Although the Law of Cosines holds true for right triangles, the statement $c^2 = a^2 + b^2 - 2ab \cos \gamma$ reduces to the Pythagorean Theorem when $\gamma = 90°$ because $\cos 90° = 0$.

EXAMPLE 5

Find the length of $\overline{AB}$ in the triangle in Figure 10.43.

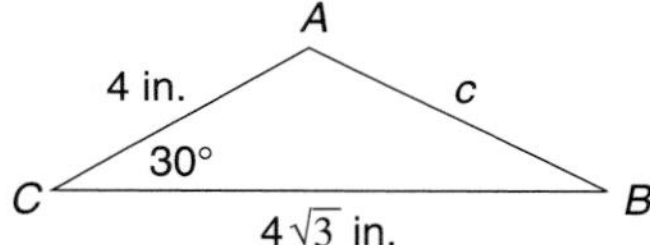

**FIGURE 10.43**

***Solution***

Referring to the 30° angle as $\gamma$, we use the form

$$c^2 = a^2 + b^2 - 2ab \cos \gamma$$
$$c^2 = (4\sqrt{3})^2 + 4^2 - 2 \cdot 4\sqrt{3} \cdot 4 \cdot \cos 30°$$
$$c^2 = 48 + 16 - 2 \cdot 4\sqrt{3} \cdot 4 \cdot \frac{\sqrt{3}}{2}$$
$$c^2 = 48 + 16 - 48$$
$$c^2 = 16$$
$$c = 4$$

Therefore, $AB = 4$ in.

The Law of Cosines can also be used to find the measure of an angle of a triangle when the lengths of its three sides are known.

EXAMPLE 6

In acute $\triangle ABC$ in Figure 10.44, find $\beta$ to the nearest degree.

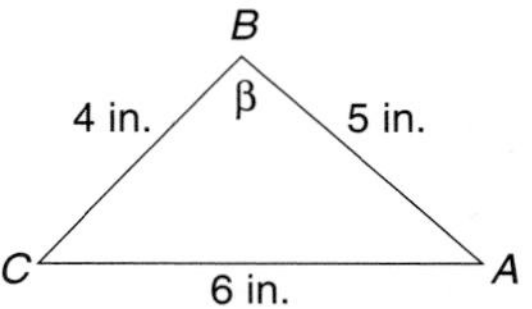

**FIGURE 10.44**

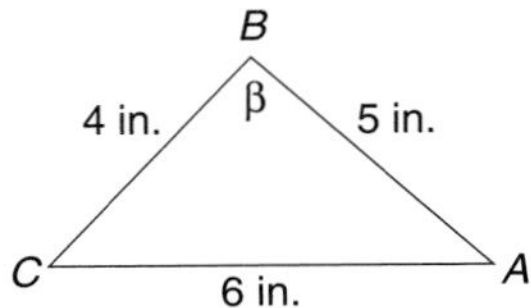


FIGURE 10.44

***Solution***

The form of the Law of Cosines involving $\beta$ is $b^2 = a^2 + c^2 - 2ac\cos\beta$. Because $b = 6$ is the length of the side opposite $\beta$, we have

$$6^2 = 4^2 + 5^2 - 2 \cdot 4 \cdot 5 \cdot \cos\beta$$
$$36 = 16 + 25 - 40\cos\beta$$
$$36 = 41 - 40\cos\beta$$

Therefore, $$40\cos\beta = 5$$

$$\cos\beta = \frac{5}{40} = \frac{1}{8} = 0.1250$$

With $\beta = \cos^{-1}(0.1250)$, we use a calculator to show that $\beta \approx 83°$.

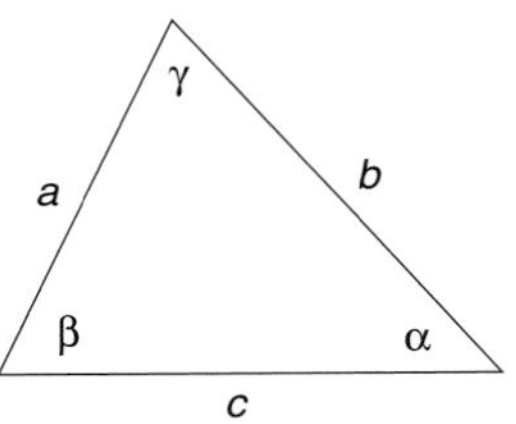


FIGURE 10.45

To find the measure of a side or an angle of an acute triangle, we must often decide which form of the Law of Sines or of the Law of Cosines should be applied. Table 10.4 deals with that question and is based on the acute triangle shown in the accompanying drawing. Note that $a$, $b$, and $c$ represent the lengths of the sides and that $\alpha$, $\beta$, and $\gamma$ represent the measures of the opposite angles, respectively (see Figure 10.45).

Warning (!)
If we know *only* the measures of the three angles of the triangle, then no length of side can be determined.

**TABLE 10.4**

**When to Use the Law of Sines/Law of Cosines**

1. *Three sides are known:* Use the Law of Cosines to find *any* angle.
   *Known measures: a, b, and c*
   *Desired measure: α*
   ∴ *Use* $a^2 = b^2 + c^2 - 2bc\cos\alpha$

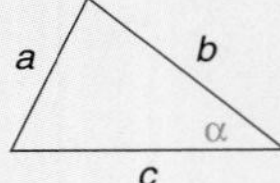


2. *Two sides and a non-included angle are known:* Use the Law of Sines to find the remaining non-included angle.
   *Known measures: a, b,* and $\alpha$
   *Desired measure:* $\beta$
   ∴ *Use* $\frac{\sin\alpha}{a} = \frac{\sin\beta}{b}$

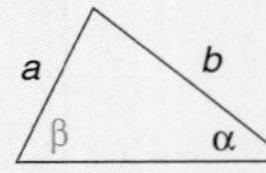


3. *Two sides and an included angle are known:* Use the Law of Cosines to find the remaining side.
   *Known measures: a, b,* and $\gamma$
   *Desired measure: c*
   ∴ *Use* $c^2 = a^2 + b^2 - 2ab\cos\gamma$

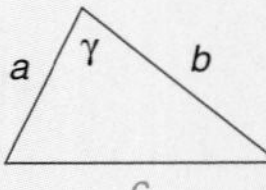


4. *Two angles and a non-included side are known:* Use the Law of Sines to find the other non-included side.
   *Known measures: a,* $\alpha$, and $\beta$
   *Desired measure: b*
   ∴ *Use* $\frac{\sin\alpha}{a} = \frac{\sin\beta}{b}$

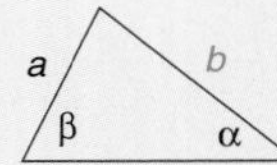

EXAMPLE 7

In the design of a child's swing set, each of the two metal posts that support the top bar measures 8 ft. At ground level, the posts are to be 6 ft apart (see Figure 10.46). At what angle should the two metal posts be secured? Give the answer to the nearest degree.

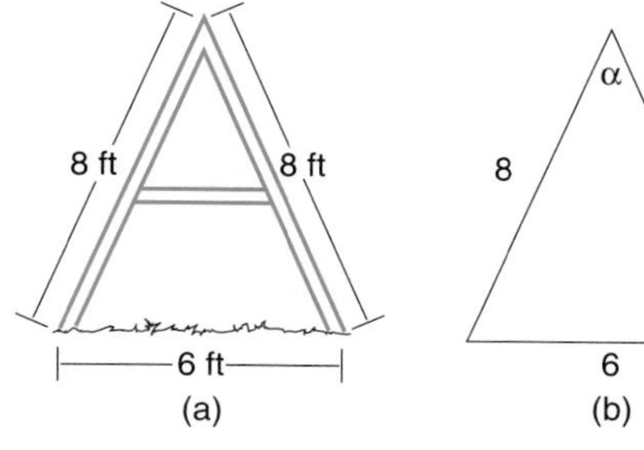


**FIGURE 10.46**

***Solution***

Call the desired angle measure $\alpha$. Because the three sides of the triangle are known, we use the Law of Cosines of the form $a^2 = b^2 + c^2 - 2bc\cos\alpha$.

Because $a$ represents the length of the side opposite the angle $\alpha$, $a = 6$ while $b = 8$ and $c = 8$. Consequently, we have

$$\begin{aligned} 6^2 &= 8^2 + 8^2 - 2 \cdot 8 \cdot 8 \cdot \cos\alpha \\ 36 &= 64 + 64 - 128\cos\alpha \\ 36 &= 128 - 128\cos\alpha \\ -92 &= -128\cos\alpha \\ \cos\alpha &= \frac{-92}{-128} \\ \cos\alpha &\approx 0.7188 \end{aligned}$$

Use of a calculator yields $\alpha \approx 44°$.

*Exs. 8–10*

## 10.4 Exercises

*In Exercises 1 and 2, use the given information to find an expression for the area of $\triangle ABC$. Give the answer in a form like $A = \frac{1}{2}(3)(4)\sin 32°$. See the figure at right.*

1. a) $a = 5$, $b = 6$, and $\gamma = 78°$
   b) $a = 5$, $b = 7$, $\alpha = 36°$, and $\beta = 88°$

2. a) $b = 7.3$, $c = 8.6$, and $\alpha = 38°$
   b) $a = 5.3$, $c = 8.4$, $\alpha = 36°$, and $\gamma = 87°$

*In Exercises 3 and 4, state the form of the Law of Sines used to solve the problem. Give the answer in a form like $\frac{\sin 72°}{6.3} = \frac{\sin 55°}{a}$.*

3. a) Find $\beta$ if it is known that $a = 5$, $b = 8$, and $\alpha = 40°$.
   b) Find $c$ if it is known that $a = 5.3$, $\alpha = 41°$, and $\gamma = 87°$.

B
β
a
c
γ
α
C
b
A

*Exercises 1–8*

4. a) Find $\beta$ if it is known that $b = 8.1$, $c = 8.4$, and $\gamma = 86°$.
   b) Find $c$ if it is known that $a = 5.3$, $\alpha = 40°$, and $\beta = 80°$.

*In Exercises 5 and 6, state the form of the Law of Cosines used to solve the problem. Using the values provided, give the answer in a form like $a^2 = b^2 + c^2 - 2bc\cos\alpha$. See the figure on page 511.*

5. a) Find $c$ if it is known that $a = 5.2$, $b = 7.9$, and $\gamma = 83°$.
   b) Find $\alpha$ if it is known that $a = 6$, $b = 9$, and $c = 10$.

6. a) Find $b$ if it is known that $a = 5.7$, $c = 8.2$, and $\beta = 79°$.
   b) Find $\beta$ if it is known that $a = 6$, $b = 8$, and $c = 9$.

*In Exercises 7 and 8, state the form of the Law of Sines or the Law of Cosines that you would use to solve the problem. See the figure on page 511.*

7. a) Find $\alpha$ if you know the values of $a$, $b$, and $\beta$.
   b) Find $\alpha$ if you know the values of $a$, $b$, and $c$.

8. a) Find $b$ if you know the values of $a$, $c$, and $\beta$.
   b) Find $b$ if you know the values of $a$, $\alpha$, and $\beta$.

9. For $\triangle ABC$ (not shown), suppose you know that $a = 3$, $b = 4$, and $c = 5$.
   a) Explain why you do *not* need to apply the Law of Sines or the Law of Cosines to find the measure of $\gamma$.
   b) Find $\gamma$.

10. For $\triangle ABC$ (not shown). suppose you know that $a = 3$, $\alpha = 57°$, and $\beta = 84°$.
    a) Explain why you do *not* need to apply the Law of Sines or the Law of Cosines to find the measure of $\gamma$.
    b) Find $\gamma$.

*In Exercises 11 to 14, find the area of each triangle shown. Give the answer to the nearest tenth of a square unit.*

11.

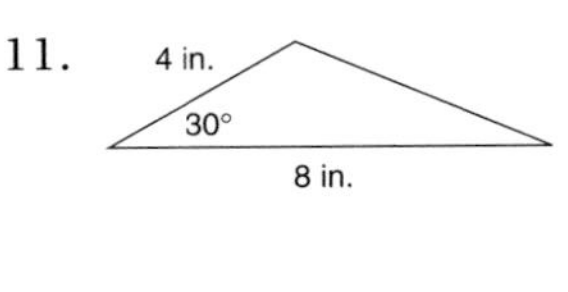


12.

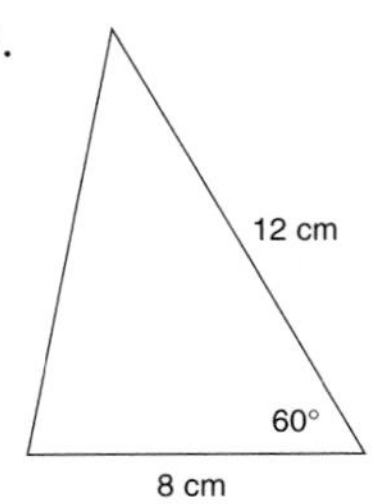


13.

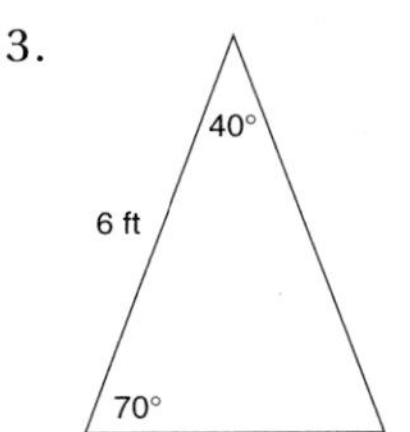


14.

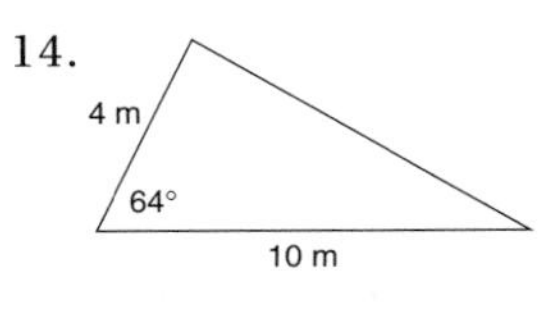


*In Exercises 15 and 16, find the area of the given figure. Give the answer to the nearest tenth of a square unit.*

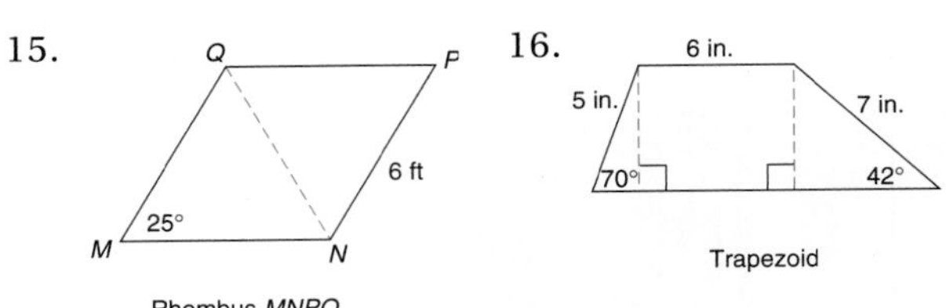


*In Exercises 17 to 22, use a form of the Law of Sines to find the measure of the indicated side or angle. Angle measures should be found to the nearest degree and lengths of sides to the nearest tenth of a unit.*

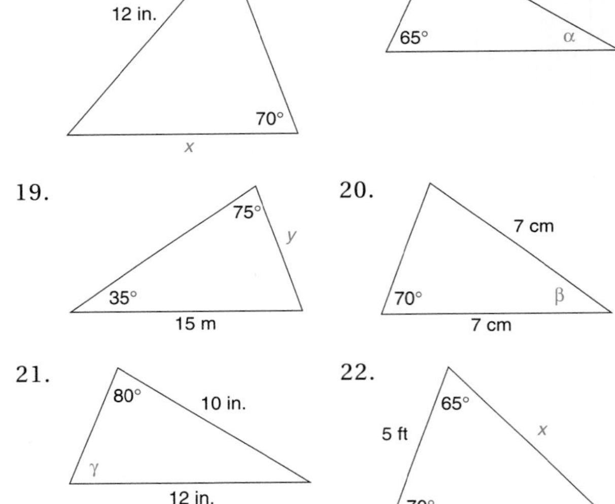


*In Exercises 23 to 28, use a form of the Law of Cosines to find the measure of the indicated side or angle. Angle measures should be found to the nearest degree and lengths of sides to the nearest tenth of a unit.*

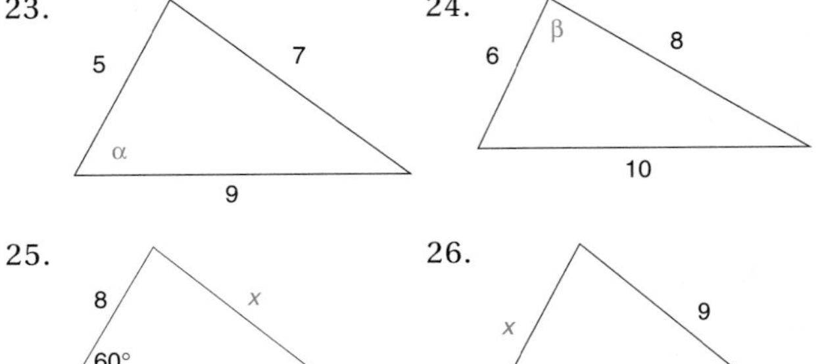

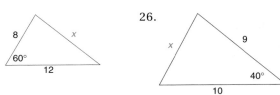

27.

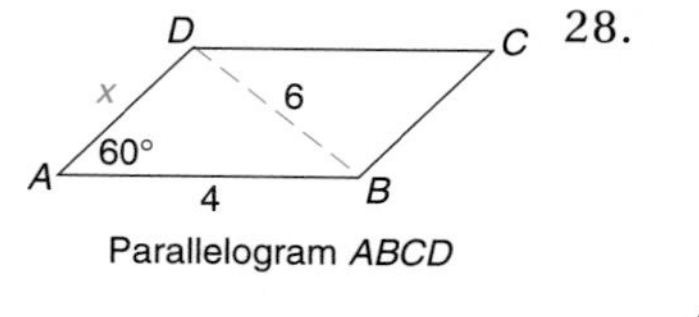

Parallelogram $ABCD$

28.

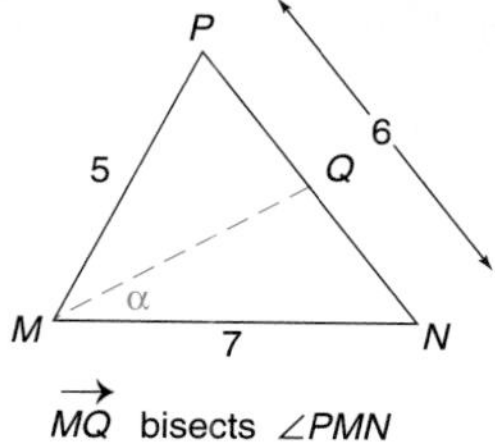

$\overrightarrow{MQ}$ bisects $\angle PMN$

*In Exercises 29 to 34, use the Law of Sines or the Law of Cosines to solve each problem. Angle measures should be found to the nearest degree and areas and distances to the nearest tenth of a unit.*

29. A triangular lot has street dimensions of 150 ft and 180 ft and an included angle of 80° for these two sides.

 a) Find the length of the remaining side of the lot.
 b) Find the area of the lot in square feet.

30. Two people observe a balloon. They are 500 ft apart, and their angles of observation are 47° and 65° as shown. Find the distance $x$ from the second observer to the balloon.

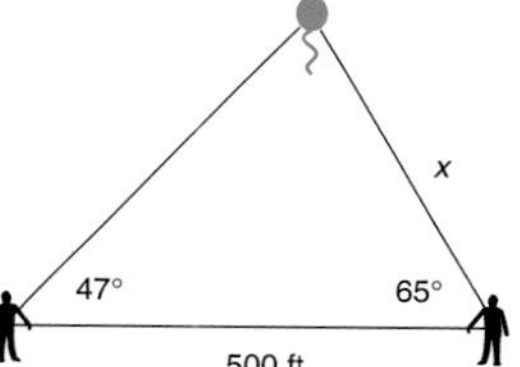


31. A surveillance aircraft at point $C$ sights an ammunition warehouse at $A$ and enemy headquarters at $B$ through the angles indicated. If points $A$ and $B$ are 10,000 m apart, what is the distance from the aircraft to enemy headquarters?

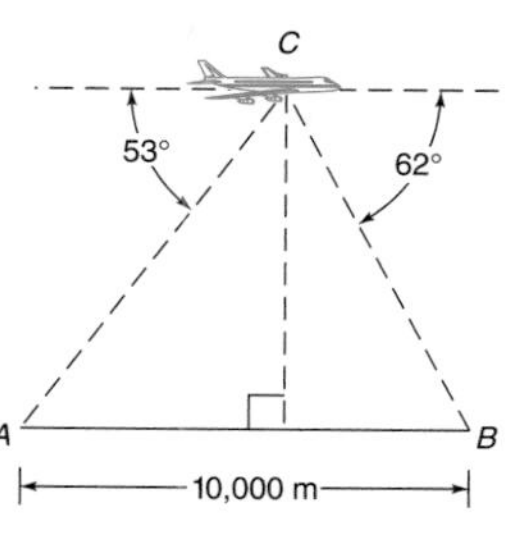


32. Above one room of a house the rafters meet as shown. What is the measure of the angle $\alpha$ at which they meet?

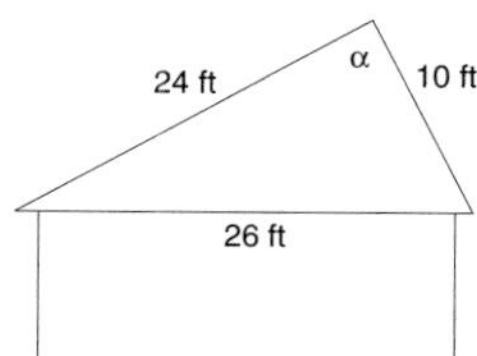


33. In an A-frame house, a bullet is found embedded at a point 8 ft up the sloped wall. If it was fired at a 30° angle with the horizontal, how far from the base of the wall was the gun fired?

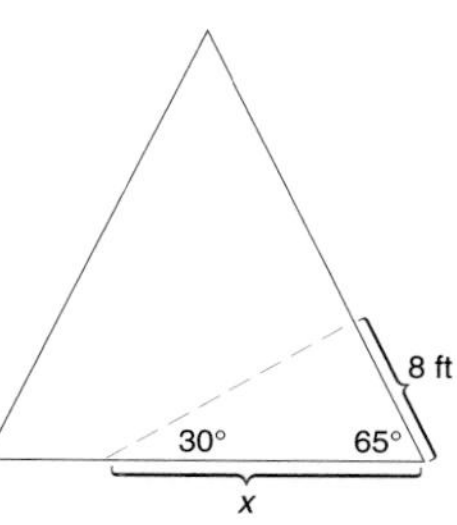


34. Clay pigeons are released at an angle of 30° with the horizontal. A sharpshooter hits one of the clay pigeons when shooting through an angle of elevation of 70°. If the point of release is 120 m from the sharpshooter, how far ($x$) is the sharpshooter from the target when it is hit?

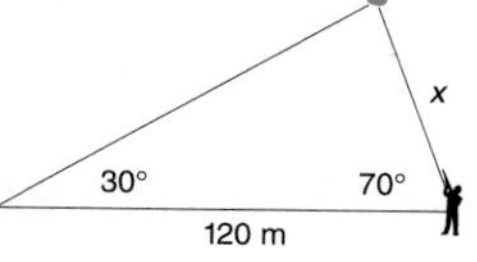


35. For the triangle shown, the area is exactly $18\sqrt{3}$ units². Determine the value of $x$.

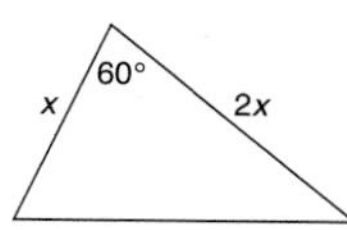


36. For the triangle shown, use the Law of Cosines to determine $b$.

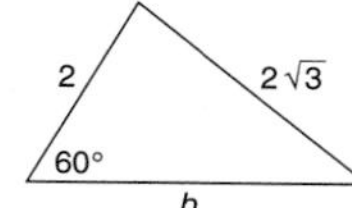


37. In the support structure for the Ferris wheel, $m\angle CAB = 30°$. If $AB = AC = 27$ ft, find $BC$.

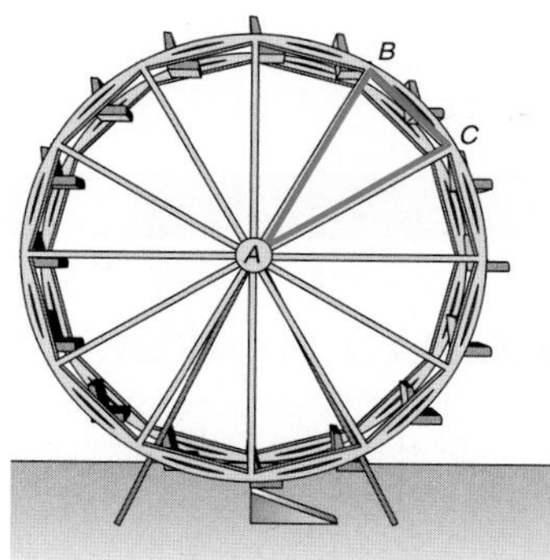


38. Show that the form of the Law of Cosines written $c^2 = a^2 + b^2 - 2ab\cos\gamma$ reduces to the Pythagorean Theorem when $\gamma = 90°$.

39. Explain why the area of the parallelogram shown is given by the formula $A = ab\sin\gamma$. (HINT: You will need to use $\overline{QN}$.)

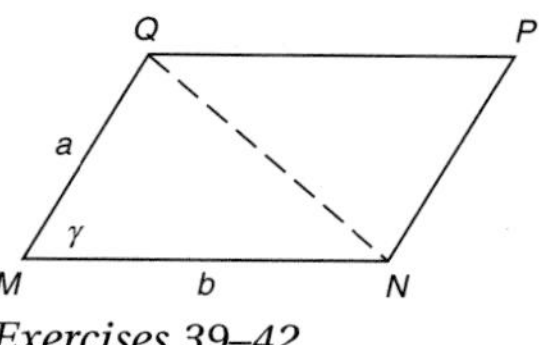

*Exercises 39–42*

40. Find the area of ▱$MNPQ$ if $a = 8$ cm, $b = 12$ cm, and $\gamma = 70°$. Answer to the nearest tenth of a square centimeter. (See Exercise 39.)

41. Find the area of ▱$MNPQ$ if $a = 6.3$ cm, $b = 8.9$ cm, and $\gamma = 67.5°$. Answer to the nearest *tenth* of a square centimeter. (See Exercise 39.)

42. The sides of a rhombus have length $a$. Two adjacent sides meet to form acute angle $\theta$. Use the formula from Exercise 39 to show that the area of the rhombus is given by $A = a^2\sin\theta$.

# Perspective on History

## SKETCH OF PLATO

Plato (428–348 B.C.) was a Greek philosopher who studied under Socrates in Athens until the time of Socrates' death. Because his master had been forced to drink poison, Plato feared for his own life and left Greece to travel. His journey began around 400 B.C. and lasted for 12 years, taking Plato to Egypt, Sicily, and Italy, where he became familiar with the Pythagoreans (see page 380.)

Plato eventually returned to Athens where he formed his own school, the Academy. Though primarily a philosopher, Plato held that the study of mathematical reasoning provided the most perfect training for the mind. So insistent was Plato that his students have some background in geometry that he placed above the door to the Academy a sign that read, "Let no man ignorant of geometry enter here."

Plato was the first to insist that all constructions be performed by using only two instruments, the compass and the straightedge. Given a line segment of length 1, Plato constructed line segments of lengths $\sqrt{2}$, $\sqrt{3}$, and so on. Unlike Archimedes (see page 168), Plato had no interest in applied mathematics. In fact, Plato's methodology was quite strict and required accurate definitions, precise hypotheses, and logical reasoning. Without doubt, his methods paved the way for the compilation of geometric knowledge in the form of *The Elements* by Euclid (see page 118).

Commenting on the life of Plato, Proclus stated the Plato caused mathematics (and geometry in particular) to make great advances. At that time, many of the discoveries in mathematics were made by Plato's students and by those who studied at the Academy after the death of Plato. It is ironic that although Plato was not himself a great mathematician, he was largely responsible for its development in his time.

# Perspective on Application

## RADIAN MEASURE OF ANGLES

In much of this textbook, we have considered angle measures from 0° to 180°. As you apply geometry, you will find that two things are true:

1. Angle measures do not have to be limited to degree measures from 0° to 180°.
2. The degree is not the only unit used in measuring angles.

We will address the first of these issues in Examples 1, 2, and 3.

### EXAMPLE 1

As the time changes from 1 P.M. to 1:45 P.M., through what angle does the minute hand rotate? See Figure 10.47.

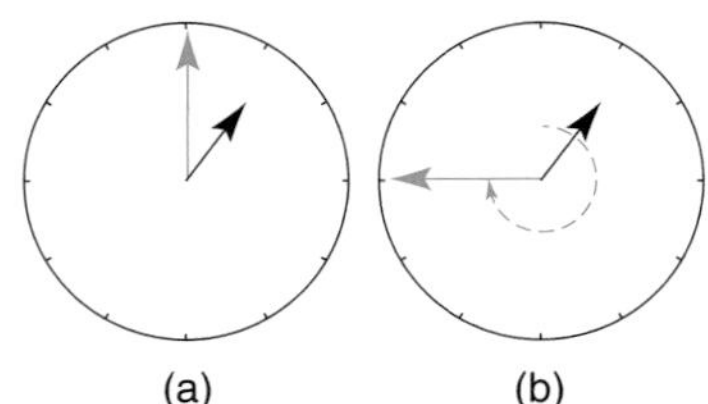

(a) (b)

**FIGURE 10.47**

***Solution***

Because the rotation is $\frac{3}{4}$ of a complete circle (360°), the result is $\frac{3}{4}$ (360°), or 270°.

### EXAMPLE 2

An airplane pilot is instructed to circle the control tower twice during a holding pattern before receiving clearance to land. Through what angle does the airplane move? See Figure 10.48.

***Solution***

Two circular rotations give 2(360°), or 720°.

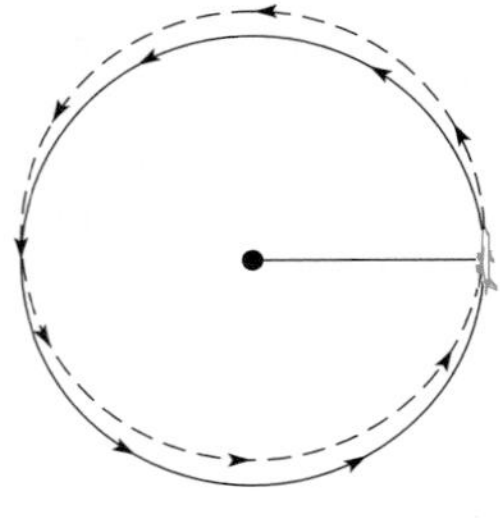

**FIGURE 10.48**

In trigonometry, negative measures for angles are used to distinguish the direction of rotation. A counterclockwise rotation is measured as positive, a clockwise rotation as negative. The arcs with arrows in Figure 10.49 are used to indicate the direction of rotation.

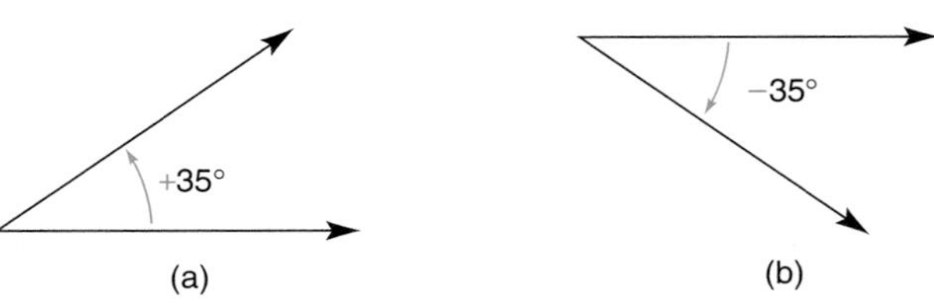


**FIGURE 10.49**

EXAMPLE 3

To tighten a hex bolt, a mechanic applies rotations of 45° several times. What is the measure of each rotation? See Figure 10.50.

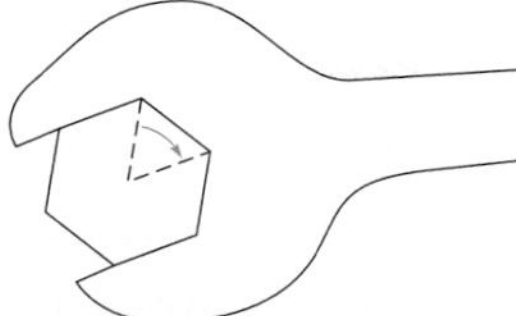

**FIGURE 10.50**

*Solution*

Tightening occurs if the angle is $-45°$.

NOTE: If the angle of rotation is 45° (that is, $+45°$), the bolt is loosened.

Our second concern is with an alternative unit for measuring angles, a unit often used in the study of trigonometry and calculus.

DEFINITION: In a circle, a **radian** (rad) is the measure of a central angle that intercepts an arc whose length is equal to the radius of the circle.

In Figure 10.51, the length of each radius and the intercepted arc are all equal to $r$. Thus the central angle shown measures 1 rad. A complete rotation about the circle corresponds to 360° and to $2\pi r$. Thus the arc length of 1 radius corresponds to the central angle measure of 1 rad, and the circumference of $2\pi$ radii corresponds to the complete rotation of $2\pi$ rad.

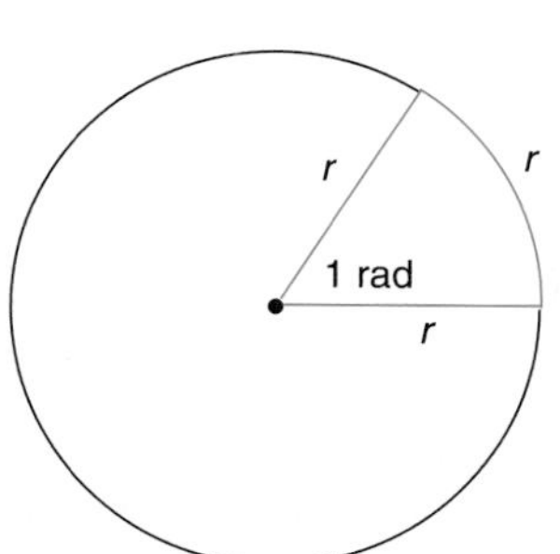


**FIGURE 10.51**

This relationship for the complete rotation allows us to equate 360° and $2\pi$ radians. As suggested by Figure 10.52, there are approximately 6.28 rad (or exactly $2\pi$ radians) about the circle. The exact result leads to an important relationship.

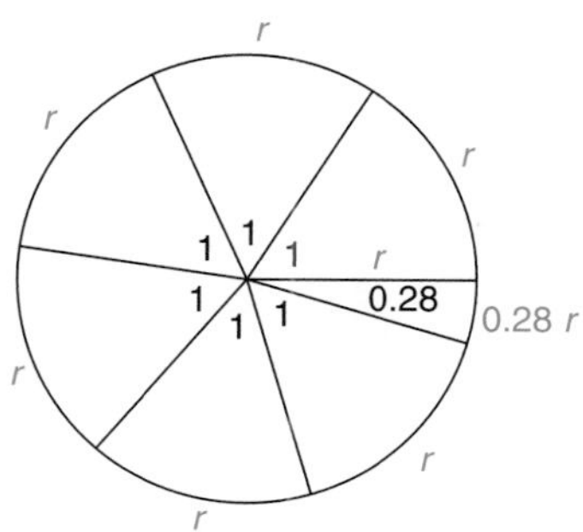


**FIGURE 10.52**

$$2\pi \text{ rad} = 360°$$
or
$$360° = 2\pi \text{ rad}$$

Through division by 2, the relationship is often restated as follows:

$$\pi \text{ rad} = 180°$$
or
$$180° = \pi \text{ rad}$$

With $\pi$ rad $= 180°$, we divide each side of this equation by $\pi$ to obtain the following relationship:

$$1 \text{ rad} = \frac{180°}{\pi} \approx 57.3°$$

To compare angle measures, we can also divide each side of the equation $180° = \pi$ rad by 180 to get the following relationship:

$$1° = \frac{\pi}{180} \text{ rad}$$

EXAMPLE 4

Using the fact that $1° = \frac{\pi}{180}$ rad, find the radian equivalencies for:

a) 30° b) 45° c) 60° d) $-90°$

*Solution*

a) $30° = 30(1°) = 30\left(\frac{\pi}{180}\right) \text{ rad} = \frac{\pi}{6} \text{ rad}$

b) $45° = 45(1°) = 45\left(\frac{\pi}{180}\right) \text{ rad} = \frac{\pi}{4} \text{ rad}$

c) $60° = 60(1°) = 60\left(\frac{\pi}{180}\right) \text{ rad} = \frac{\pi}{3} \text{ rad}$

d) $-90° = -90(1°) = -90\left(\frac{\pi}{180}\right) \text{ rad} = -\frac{\pi}{2} \text{ rad}$

EXAMPLE 5

Using the fact that $\pi$ rad $= 180°$, find the degree equivalencies for the following angles measured in radians:

a) $\frac{\pi}{6}$ b) $\frac{2\pi}{5}$ c) $\frac{-3\pi}{4}$ d) $\frac{\pi}{2}$

*Solution*

a) $\frac{\pi}{6} = \frac{180°}{6} = 30°$

b) $\frac{2\pi}{5} = \frac{2}{5} \cdot \pi = \frac{2}{5} \cdot 180° = 72°$

c) $\frac{-3\pi}{4} = \frac{-3}{4} \cdot \pi = \frac{-3}{4} \cdot 180° = -135°$

d) $\frac{\pi}{2} = \frac{180°}{2} = 90°$

Although we did not use this method of measuring angles in this textbook, you may need to use this method of angle measurement in a more advanced course.

# Summary

### A LOOK BACK AT CHAPTER 10

One goal of this chapter was to define the sine, cosine, and tangent ratios in terms of the sides of a right triangle. We derived a formula for finding the area of a triangle, given two sides and the included angle. We also proved the Law of Sines and the Law of Cosines for acute triangles. Another unit, called the radian, was introduced for the purpose of measuring angles.

### KEY CONCEPTS

**10.1** Greek Letters: $\alpha$, $\beta$, $\gamma$, $\theta$ • Opposite Side (Leg) • Hypotenuse • Sine Ratio: $\sin\theta = \dfrac{\text{opposite}}{\text{hypotenuse}}$ • Angle of Elevation • Angle of Depression

**10.2** Adjacent Side (Leg) • Cosine Ratio: $\cos\theta = \dfrac{\text{adjacent}}{\text{hypotenuse}}$ • Identity: $\sin^2\theta + \cos^2\theta = 1$

**10.3** Tangent Ratio: $\tan\theta = \dfrac{\text{opposite}}{\text{adjacent}}$ • Cotangent • Secant • Cosecant

**10.4** Area of a Triangle: $A = \frac{1}{2}bc\sin\alpha$
$A = \frac{1}{2}ac\sin\beta$
$A = \frac{1}{2}ab\sin\gamma$ •

Law of Sines: $\dfrac{\sin\alpha}{a} = \dfrac{\sin\beta}{b} = \dfrac{\sin\gamma}{c}$ •

Law of Cosines: $c^2 = a^2 + b^2 - 2ab\cos\gamma$
$b^2 = a^2 + c^2 - 2ac\cos\beta$
$a^2 = b^2 + c^2 - 2bc\cos\alpha$

# Chapter 10 Review Exercises

*In Exercises 1 to 4, state the ratio needed, and use it to find the measure of the indicated line segment to the nearest tenth of a unit.*

1.

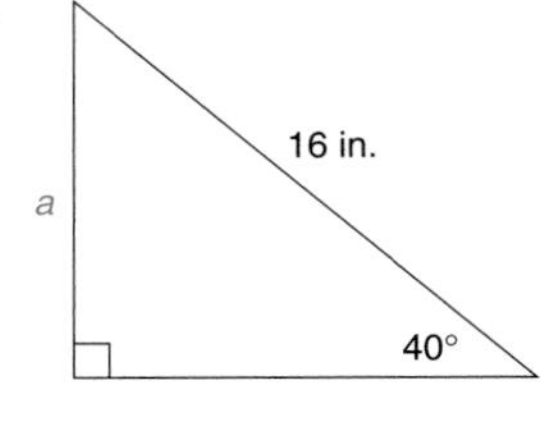


2.

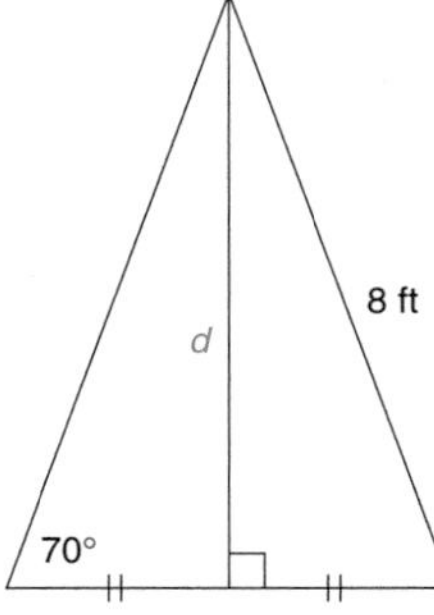


3.

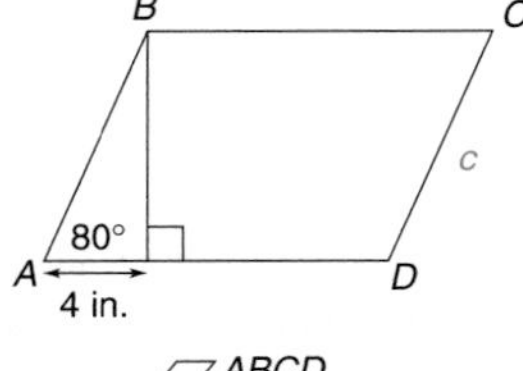


▱ABCD

4.

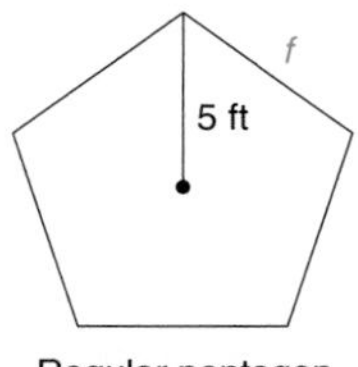


Regular pentagon with radius = 5 ft

*In Exercises 5 to 8, state the ratio needed, and use it to find the measure of the angle to the nearest degree.*

5.

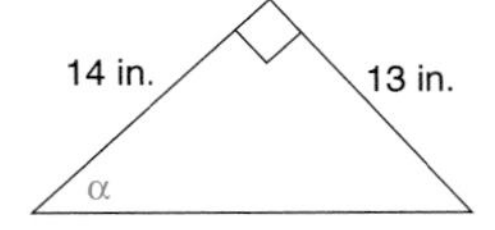


6.

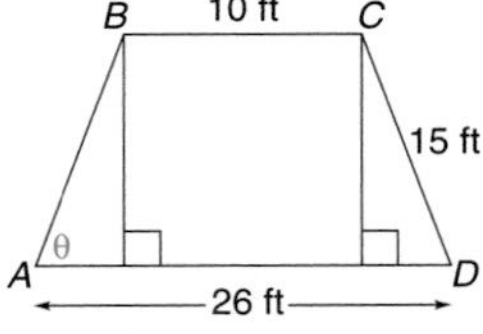


Isosceles trapezoid ABCD

7.

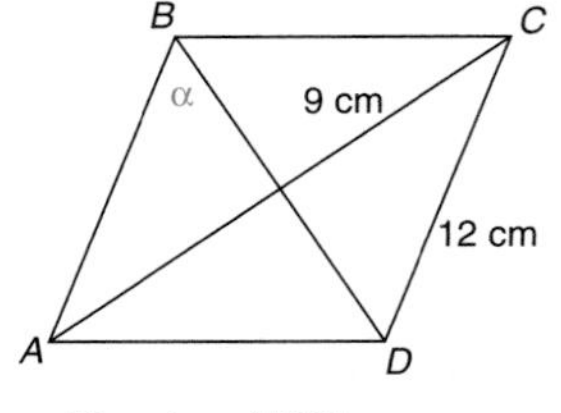


Rhombus ABCD

8.

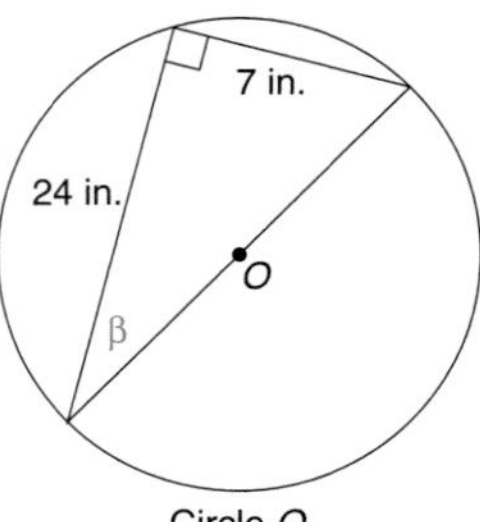


Circle O

*In Exercises 9 to 12, use the Law of Sines or the Law of Cosines to solve each triangle for the indicated length of side or angle measure. Angle measures should be found to the nearest degree; distances should be found to the nearest tenth of a unit.*

9.

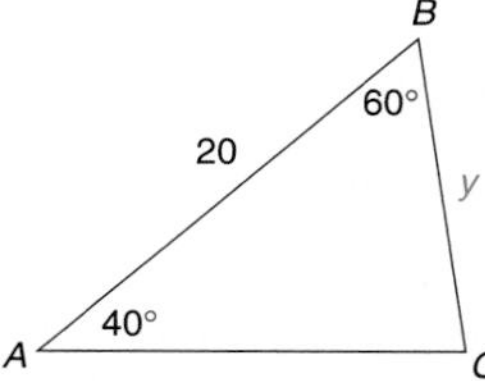

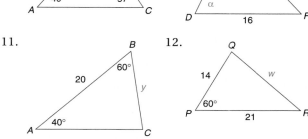

10.

11.

12.

*In Exercises 13 to 17, use the Law of Sines or the Law of Cosines to solve each problem. Angle measures should be found to the nearest degree; distances should be found to the nearest tenth of a unit.*

13. A building 50 ft tall is on a hillside. A surveyor at a point on the hill observes that the angle of elevation to the top of the building measures 43° and the angle of depression to the base of the building measures 16°. How far is the surveyor from the base of the building?

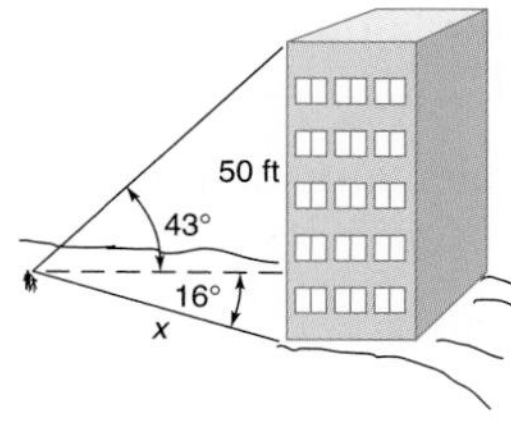


14. Two sides of a parallelogram are 50 cm and 70 cm long. Find the length of the shorter diagonal if a larger angle of the parallelogram measures 105°.

15. The sides of a rhombus are 6 in. each and the longer diagonal is 11 in. Find the measure of each of the acute angles in the rhombus.

16. The area of $\triangle ABC$ is 9.7 in$^2$. If $a = 6$ in. and $c = 4$ in., find the measure of angle $B$.

17. Find the area of the rhombus in Exercise 15.

*In Exercises 18 to 20, prove each statement without using a table or a calculator. Draw an appropriate right triangle.*

18. If $m\angle R = 45°$, then $\tan R = 1$.

19. If $m\angle S = 30°$, then $\sin S = \frac{1}{2}$.

20. If $m\angle T = 60°$, then $\sin T = \frac{\sqrt{3}}{2}$.

*In Exercises 21 to 30, use the drawings where provided to solve each problem. Angle measures should be found to the nearest degree; lengths should be found to the nearest tenth of a unit.*

21. In the evening, a tree that stands 12 ft tall casts a long shadow. If the angle of depression from the top of the tree to the tip of the shadow is 55°, what is the length of the shadow?

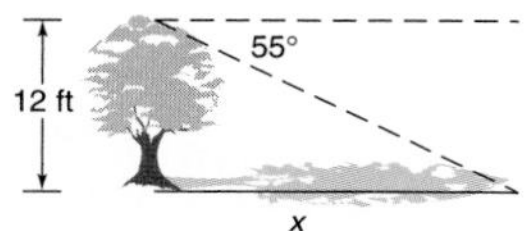


22. A rocket is shot into the air at an angle of 60°. If it is traveling at 200 ft per second, how high in the air is it after 5 seconds? (Ignoring gravity, assume that the path of the rocket is a straight line.)

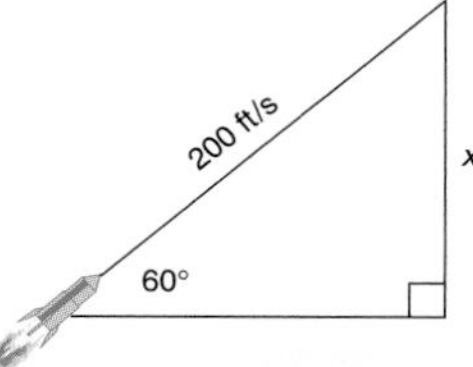


23. A 4-m beam is used to brace a wall. If the bottom of the beam is 3 m from the base of the wall, what is the angle of elevation to the top of the wall?

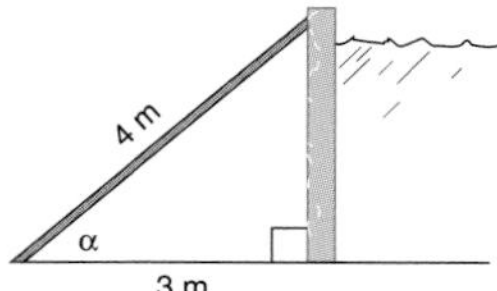


24. A hot-air balloon is 300 ft high. The pilot of the balloon sights a stadium 2200 ft away. What is the angle of depression?

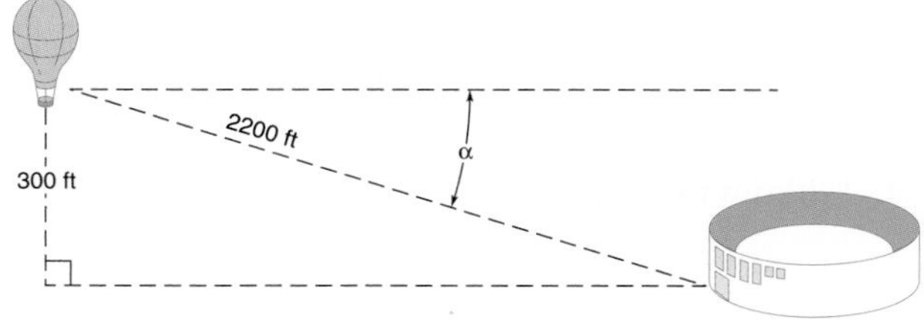


25. The apothem of a regular pentagon is approximately 3.44 cm. What is the approximate length of each side of the pentagon?

26. What is the approximate length of the radius of the pentagon in Exercise 25?

27. Each of the legs of an isosceles triangle is 40 cm in length. The base is 30 cm in length. Find the measure of a base angle.

28. The diagonals of a rhombus measure 12 in. and 16 in. Find the measure of the obtuse angle in the rhombus.

29. The unit used for measuring the steepness of a hill is the **grade.** A grade of $a$ to $b$ means the hill rises $a$ vertical units for every $b$ horizontal units. If, at some point, the hill is 3 ft above the horizontal and the angle of elevation to that point is 23°, what is the grade of this hill?

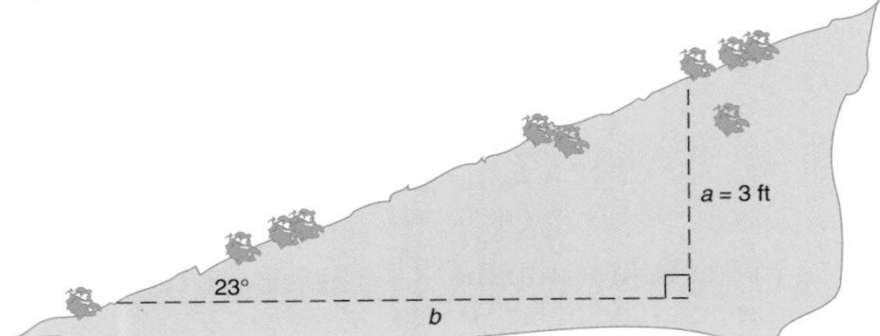


30. An observer in a plane 2500 m high sights two ships below. The angle of depression to one ship is 32°, and the angle of depression to the other ship is 44°. How far apart are the ships?

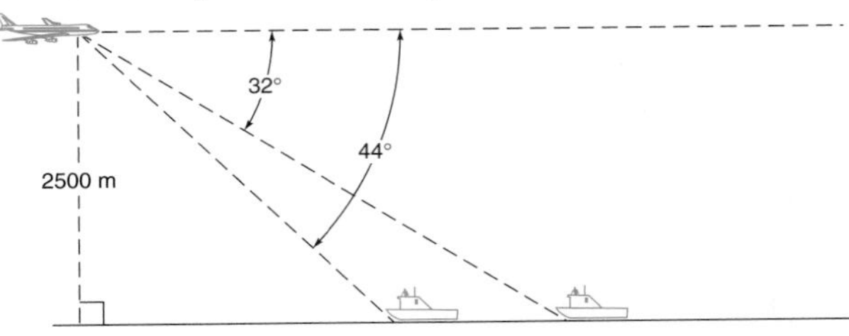


31. If $\sin \theta = \frac{7}{25}$, find $\cos \theta$ and $\sec \theta$.

32. If $\tan \theta = \frac{11}{60}$, find $\sec \theta$ and $\cot \theta$.

33. If $\cot \theta = \frac{21}{20}$, find $\csc \theta$ and $\sin \theta$.

34. In a right circular cone, the radius of the base is 3.2 ft in length and the angle formed by the radius and slant height measures $\theta = 65°$. To the nearest tenth of a foot, find the length of the altitude of the cone. Then use this length of altitude to find the volume of the cone to the nearest tenth of a cubic foot.

## Chapter 10 Test

1. For the right triangle shown, express each of the following in terms of $a$, $b$, and $c$:

   a) $\sin \alpha$ ________
   b) $\tan \beta$ ________

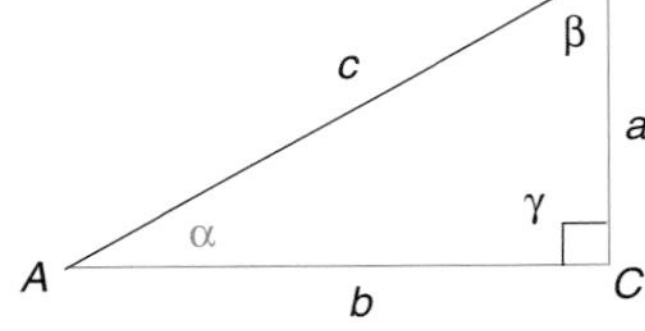


2. For the right triangle shown, express each fraction in lowest terms:

   a) $\cos \beta$ ________
   b) $\sin \alpha$ ________

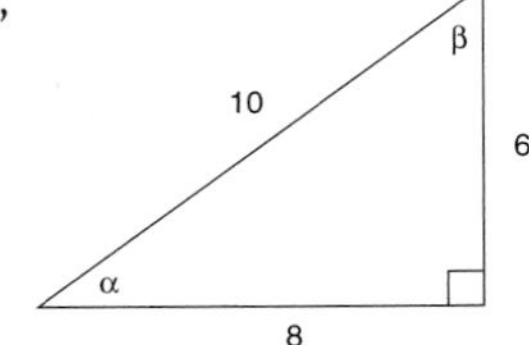


3. Without using a calculator, find the exact value of:

   a) tan 45° ________ b) sin 60° ________

4. Use your calculator to find each number correct to four decimal places.

   a) sin 23° ________ b) cos 79° ________

5. Using your calculator, find $\theta$ to the nearest degree if $\sin \theta = 0.6691$. ________

6. Without the calculator, determine which number is larger:

   a) tan 25° or tan 26° ________
   b) cos 47° or cos 48° ________

7. In the drawing provided, find the value of $a$ to the nearest whole number. ________

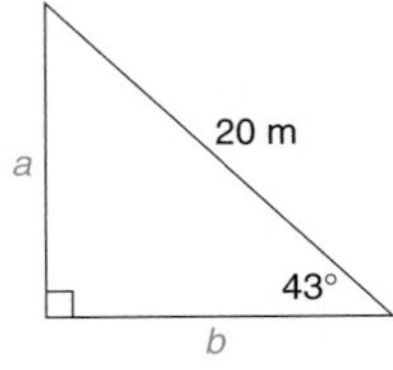


8. In the drawing provided, find the value of $y$ to the nearest whole number.

   ________

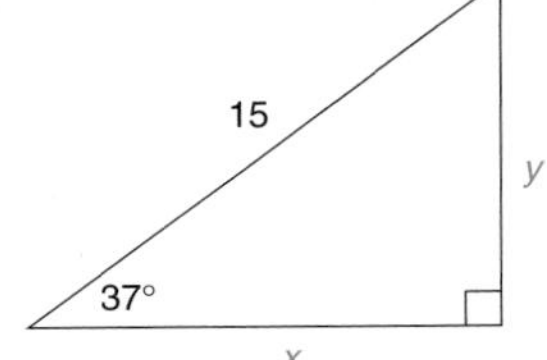

9. In the drawing provided, find the measure of $\theta$ to the nearest degree. ________

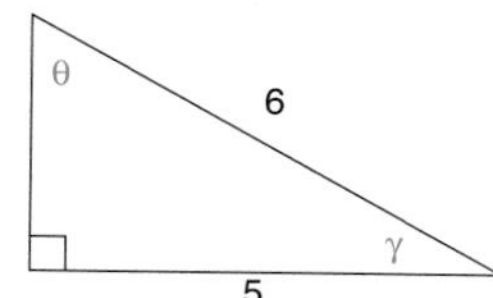


10. Using the drawing below, classify each statement as true or false:

a) $\cos\beta = \sin\alpha$ ________
b) $\sin^2\alpha + \cos^2\alpha = 1$ ________

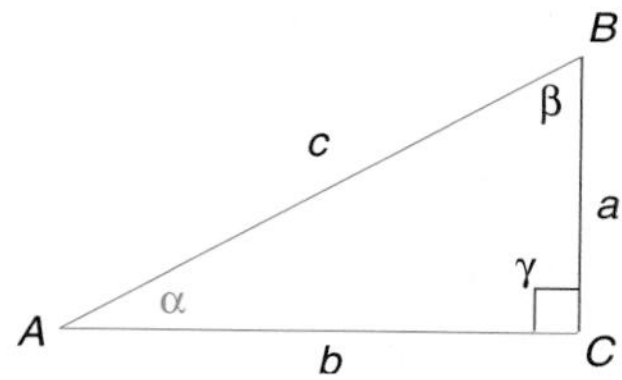


11. A kite is flying at an angle of elevation of 67° with the ground. If 100 feet of string have been paid out to the kite, how far is the kite above the ground? Answer to the nearest foot. ________

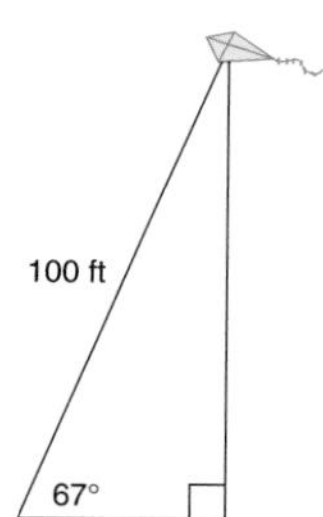


12. A roofline shows a span of 12 feet across a sloped roof and this span is accompanied by a 2 foot rise. To the nearest degree, find the measure of $\theta$. ________

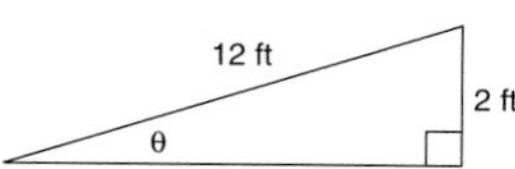


13. If $\sin\alpha = \frac{1}{2}$, find:

a) $\csc\alpha$ ________ b) $\alpha$ ________

14. In a right triangle with acute angles of measures $\alpha$ and $\beta$, $\cos\beta = \frac{a}{c}$. Find the following values in terms of the lengths of sides $a$, $b$, and $c$:

a) $\sin\alpha$ ________ b) $\sec\beta$ ________

15. Use one of the three forms for area $\left(\text{such as the form } A = \frac{1}{2}bc\sin\alpha\right)$ to find the area of the triangle shown. Answer to the nearest whole number. ________

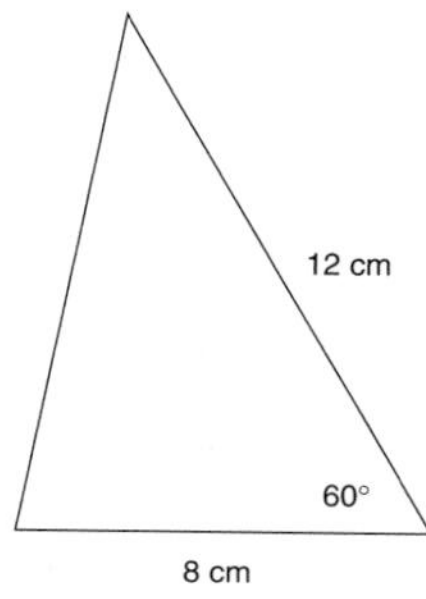


16. On the basis of the drawing provided, complete the Law of Sines.

$$\frac{\sin\alpha}{a} = \underline{\qquad\qquad} = \underline{\qquad\qquad}$$

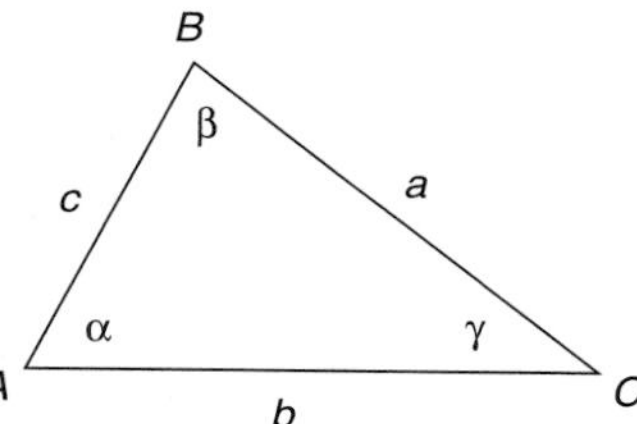


*Exercises 16, 17*

17. On the basis of the drawing provided, complete this form of the Law of Cosines.

$$a^2 = \underline{\qquad\qquad}$$

18. Use the Law of Sines or the Law of Cosines to find $\alpha$ to the nearest degree. ________

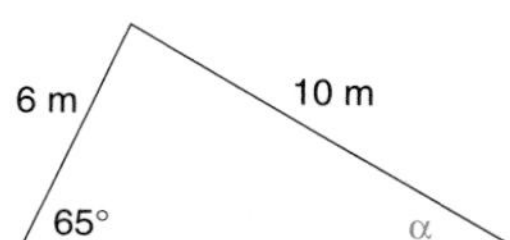


19. Use the Law of Sines or the Law of Cosines to find length $x$ to the nearest whole number. ________

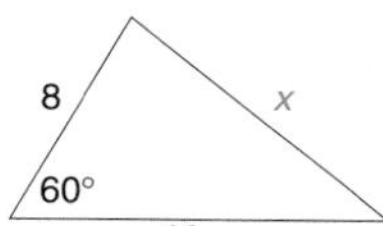

# Appendix A

## Algebra Review

## A.1 Algebraic Expressions

Algebra and geometry are both examples of mathematical systems. In algebra, we generally accept but do not define terms such as *addition, multiplication, number, positive,* and *equality.* For convenience, a real number is one that has a position on the number line, as shown in Figure A.1.

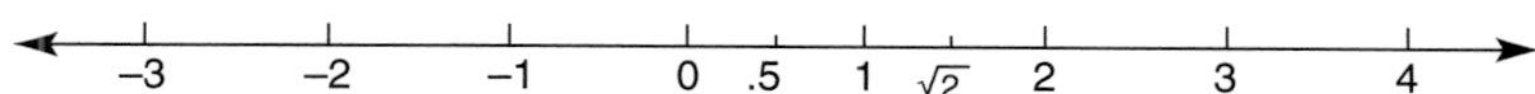


**FIGURE A.1**

Any real number positioned to the right of another real number is larger than that other number. For example, 4 is larger than $-2$; equivalently, $-2$ is less than 4 (smaller numbers are to the left). Numbers such as 3 and $-3$ are **opposites** or **additive inverses.** Two numerical expressions are **equal** if and only if they have the same value; for example, $2 + 3 = 5$. The axioms of equality are listed in the following box; they are also listed in Section 1.6.

> AXIOMS OF EQUALITY
>
> *Reflexive ($a = a$):* Any number equals itself.
>
> *Symmetric (if $a = b$, then $b = a$):* Two equal numbers are equal in either order.
>
> *Transitive* (if $a = b$ and $b = c$, then $a = c$): If a first number equals a second number and if the second number equals a third number, then the first number equals the third number.
>
> *Substitution:* If one numerical expression equals a second, then it may replace the second.

EXAMPLE 1

Name the axiom of equality illustrated in each case.

a) If $AB$ is the numerical length of the line segment $\overline{AB}$, then $AB = AB$.
b) If $17 = 2x - 3$, then $2x - 3 = 17$.
c) Given that $2x + 3x = 5x$, the statement $2x + 3x = 30$ can be replaced by $5x = 30$.

***Solution***

a) Reflexive  b) Symmetric  c) Substitution

To add two real numbers, think of the numbers as gains if positive and as losses if negative. For instance, $13 + (-5)$ represents the result of combining a gain of \$13 with a loss (or debt) of \$5. Therefore,

$$13 + (-5) = 8$$

The answer in addition is the **sum.** Three more examples of addition are

$$13 + 5 = 18 \qquad (-13) + 5 = -8 \qquad \text{and} \qquad (-13) + (-5) = -18$$

If you multiply two real numbers, the **product** (answer) will be positive if the two numbers have the same sign, negative if the two numbers have different signs, and 0 if either number is 0 or both numbers are 0.

EXAMPLE 2

Simplify each expression:

a) $5 + (-4)$ b) $5(-4)$ c) $(-7)(-6)$ d) $[5 + (-4)] + 8$

***Solution***

a) $5 + (-4) = 1$
b) $5(-4) = -20$
c) $(-7)(-6) = 42$
d) $[5 + (-4)] + 8 = 1 + 8 = 9$

What happens when you change the order in an addition problem? $(-3) + 9 = 6$ and $9 + (-3) = 6$. That the sums are equal when the order of the numbers added is reversed is often expressed by writing $a + b = b + a$. This property of real numbers is known as the Commutative Axiom for Addition. There is also a Commutative Axiom for Multiplication, which is illustrated by the fact that $(6)(-4) = (-4)(6)$; both products are $-24$.

In a numerical expression, grouping symbols indicate which operation should be performed first. However, $[5 + (-4)] + 8$ equals $5 + [(-4) + 8]$ because $1 + 8$ equals $5 + 4$. In general, the fact that $(a + b) + c$ equals $a + (b + c)$ is known as the Associative Axiom for Addition. There is also an Associative Axiom for Multiplication, which is illustrated below:

$$\begin{aligned} (3 \cdot 5)(-2) &= 3[5(-2)] \\ (15)(-2) &= 3(-10) \\ -30 &= -30 \end{aligned}$$

SELECTED AXIOMS OF REAL NUMBERS

*Commutative Axiom for Addition:* $a + b = b + a$

*Commutative Axiom for Multiplication:* $a \cdot b = b \cdot a$

*Associative Axiom for Addition:* $a + (b + c) = (a + b) + c$

*Associative Axiom for Multiplication:* $a \cdot (b \cdot c) = (a \cdot b) \cdot c$

To subtract $b$ from $a$ (that is, to find $a - b$), change the subtraction problem to an addition problem. The answer is the **difference** between $a$ and $b$.

DEFINITION OF SUBTRACTION:

$$a - b = a + (-b)$$

where $-b$ is the additive inverse (or opposite) of $b$.

For $b = 5$, we have $-b = -5$, and for $b = -2$, we have $-b = 2$. For the subtraction $a - (b + c)$, we use the additive inverse of $b + c$, which is $(-b) + (-c)$. That is,

$$a - (b + c) = a + [(-b) + (-c)]$$

EXAMPLE 3

Simplify each expression:

a) $5 - (-2)$ b) $(-7) - (-3)$ c) $12 - [3 + (-2)]$

***Solution***

a) $5 - (-2) = 5 + 2 = 7$
b) $(-7) - (-3) = (-7) + 3 = -4$
c) $12 - [3 + (-2)] = 12 + [(-3) + 2] = 12 + (-1) = 11$

Division can be replaced by multiplication just as subtraction was replaced by addition. We cannot divide by 0! Two numbers whose product is 1 are called **multiplicative inverses** (or **reciprocals**); $-\frac{3}{4}$ and $-\frac{4}{3}$ are multiplicative inverses because $-\frac{3}{4} \cdot -\frac{4}{3} = 1$. The answer in division is the **quotient.**

DEFINITION OF DIVISION: For $b \neq 0$,

$$a \div b = a \cdot \frac{1}{b}$$

where $\frac{1}{b}$ is the multiplicative inverse of $b$.

NOTE: $a \div b$ is also indicated by $a/b$ or $\frac{a}{b}$.

For $b = 5$ $\left(\text{that is, } b = \frac{5}{1}\right)$, we have $\frac{1}{b} = \frac{1}{5}$, and for $b = -\frac{2}{3}$, we have $\frac{1}{b} = -\frac{3}{2}$.

EXAMPLE 4

Simplify each expression:

a) $12 \div 2$ b) $(-5) \div \left(-\frac{2}{3}\right)$

***Solution***

a) $12 \div 2 = 12 \div \frac{2}{1}$

$= \frac{12}{1} \cdot \frac{1}{2}$ product of two positive numbers is a positive number

$= 6$

b) $(-5) \div \left(-\frac{2}{3}\right) = \left(-\frac{5}{1}\right) \div \left(-\frac{2}{3}\right)$

$= \left(-\frac{5}{1}\right) \cdot \left(-\frac{3}{2}\right)$

$= \frac{15}{2}$ product of two negative numbers is a positive number

EXAMPLE 5

Mason works at the grocery store for 3 hours on Friday after school and then works for 8 hours on Saturday. If he is paid $7 per hour, how much will he be paid in all?

***Solution***

Method I: Find the total number of hours worked and multiply by 7.

$$7(3 + 8) = 7 \cdot 11 = \$77$$

Method II: Figure the daily wages and add them.

$$\underbrace{(7 \cdot 3)}_{\text{Friday's wages}} + \underbrace{(7 \cdot 8)}_{\text{Saturday's wages}} = 21 + 56 = \$77$$

NOTE: We see that $7(3 + 8) = 7 \cdot 3 + 7 \cdot 8$, where the multiplications on the right are performed before the addition is completed.

The Distributive Axiom was illustrated in Example 5. Because multiplications are performed before additions, we write

$$\begin{aligned} a(b + c) &= a \cdot b + a \cdot c \\ 2(3 + 4) &= 2 \cdot 3 + 2 \cdot 4 \\ 2(7) &= 6 + 8 \end{aligned}$$

The "symmetric" form of the Distributive Axiom is

$$a \cdot b + a \cdot c = a(b + c)$$

This form can be used to combine *like terms* (expressions that contain the same variable factors). A **variable** is a letter that represents a number.

| | |
|---|---|
| $4x + 5x = x \cdot 4 + x \cdot 5$ | Commutative Axiom for Multiplication |
| $= x(4 + 5)$ | Symmetric form of Distributive Axiom |
| $= x(9)$ | Substitution |
| $= 9x$ | Commutative Axiom for Multiplication |
| $\therefore 4x + 5x = 9x$ | |

The Distributive Axiom also distributes multiplication over subtraction.

FORMS OF THE DISTRIBUTIVE AXIOM

$$\begin{aligned} a(b + c) &= a \cdot b + a \cdot c \\ a \cdot b + a \cdot c &= a(b + c) \\ a(b - c) &= a \cdot b - a \cdot c \\ a \cdot b - a \cdot c &= a(b - c) \end{aligned}$$

EXAMPLE 6

Combine like terms:

a) $7x + 3x$  b) $7x - 3x$  c) $3x^2y + 4x^2y + 6x^2y$
d) $3x^2y + 4xy^2 + 6xy^2$  e) $7x + 5y$

***Solution***

a) $7x + 3x = 10x$
b) $7x - 3x = 4x$
c) $3x^2y + 4x^2y + 6x^2y = (3x^2y + 4x^2y) + 6x^2y = 7x^2y + 6x^2y = 13x^2y$
d) $3x^2y + 4xy^2 + 6xy^2 = 3x^2y + (4xy^2 + 6xy^2) = 3x^2y + 10xy^2$
e) $7x + 5y$; cannot combine unlike terms

NOTE: In part (d), $3x^2y$ and $10xy^2$ are not like terms because $x^2y \neq xy^2$.

The statement $4x + 5x = 9x$ says that "the sum of 4 times a number and 5 times the same number equals 9 times the same number." Thus when $x = 3$, we are saying that "4 times 3 plus 5 times 3 equals 9 times 3." Because $x$ can be any real number, we may also write

$$4\pi + 5\pi = 9\pi$$

in which $\pi$ is the real number that equals approximately 3.14. Similarly,

$$4\sqrt{3} + 5\sqrt{3} = 9\sqrt{3}$$

in which $\sqrt{3}$ (read "the positive square root of 3") is equal to approximately 1.73.

You may recall the "order of operations" from a previous class; this order is used when simplifying more complicated expressions.

ORDER OF OPERATIONS

1. Simplify expressions within symbols such as parentheses ( ) or brackets [ ], beginning with the innermost symbols of inclusion.

NOTE: The presence of a fraction bar —— requires that you simplify a numerator or denominator before dividing.

2. Perform all calculations with exponents.
3. Do all multiplications and/or divisions, in order, from left to right.
4. Finally, do all additions and/or subtractions, in order, from left to right.

EXAMPLE 7

Simplify each numerical expression:

a) $3^2 + 4^2$ b) $4 \cdot 7 \div 2$ c) $2 \cdot 3 \cdot 5^2$

d) $\dfrac{8 - 6 \div (-3)}{4 + 3(2 + 5)}$ e) $2 + [3 + 4(5 - 1)]$

***Solution***

a) $3^2 + 4^2 = 9 + 16 = 25$

b) $4 \cdot 7 \div 2 = 28 \div 2 = 14$

c) $2 \cdot 3 \cdot 5^2 = 2 \cdot 3 \cdot 25$
$= (2 \cdot 3) \cdot 25 = 6 \cdot 25 = 150$

d) $\dfrac{8 - [6 \div (-3)]}{4 + 3(2 + 5)} = \dfrac{8 - (-2)}{4 + 3(7)}$
$= \dfrac{10}{4 + 21} = \dfrac{10}{25} = \dfrac{2}{5}$

e) $2 + [3 + 4(5 - 1)] = 2 + [3 + 4(4)]$
$= 2 + (3 + 16)$
$= 2 + 19 = 21$

An expression such as $(2 + 5)(6 + 4)$ can be simplified by two different methods. By following the rules of order, we have $(7)(10)$, or 70. An alternative method is described as the FOIL method: First, Outside, Inside, and Last terms are multiplied and then added. This is how it works:

$$\begin{aligned}(2 + 5)(6 + 4) &= 2 \cdot 6 + 2 \cdot 4 + 5 \cdot 6 + 5 \cdot 4\\ &= 12 + 8 + 30 + 20\\ &= 70\end{aligned}$$

FOIL is the Distributive Axiom in disguise. We would not generally use FOIL to find the product of $(2 + 5)$ and $(6 + 4)$, but we will use it to find the products in Example 8. Also see Example 2 in Section A.2.

### EXAMPLE 8

Use the FOIL method to find the products.

a) $(3x + 4)(2x - 3)$ b) $(5x + 2y)(6x - 5y)$

***Solution***

a) $(3x + 4)(2x - 3) = 3x \cdot 2x + 3x(-3) + 4(2x) + 4(-3)$
$= 6x^2 + (-9x) + 8x + (-12)$
$= 6x^2 - 1x - 12$
$= 6x^2 - x - 12$

b) $(5x + 2y)(6x - 5y) = 5x \cdot 6x + 5x(-5y) + 2y(6x) + 2y(-5y)$
$= 30x^2 + (-25xy) + 12xy + (-10y^2)$
$= 30x^2 - 13xy - 10y^2$

### EXAMPLE 9

A rectangular garden plot has been subdivided as shown in Figure A.2 into smaller rectangles. Find the area of:

a) Rectangle I
b) Rectangle II
c) Rectangle III
d) Rectangle IV
e) The large plot composed of I, II, III, and IV [HINT: Add parts (a), (b), (c), and (d).]
f) The large plot determined by multiplying the dimensions $(a + b)(c + d)$

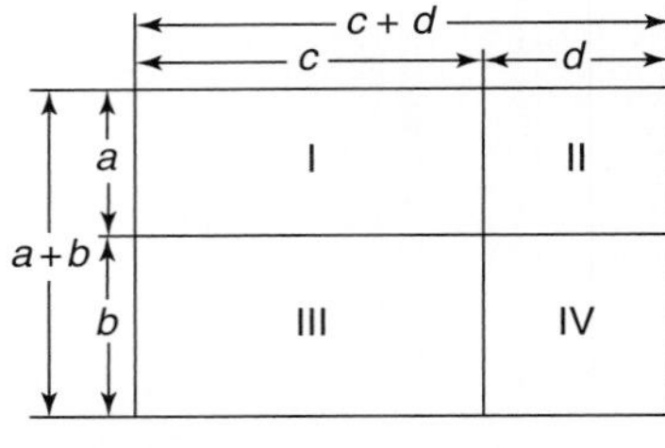


**FIGURE A.2**

***Solution***

a) $a \cdot c$ b) $a \cdot d$ c) $b \cdot c$ d) $b \cdot d$
e) $a \cdot c + a \cdot d + b \cdot c + b \cdot d$
f) $a \cdot c + a \cdot d + b \cdot c + b \cdot d$

NOTE: Results in parts (e) and (f) are necessarily identical.

## A.1 Exercises

1. Name the four parts of a mathematical system. (HINT: See Chapter 1, page 21.)
2. Name two examples of mathematical systems.
3. Which axiom of equality is illustrated in each of the following?
   a) $5 = 5$
   b) If $\frac{1}{2} = 0.5$ and $0.5 = 50\%$, then $\frac{1}{2} = 50\%$.
   c) Because $2 + 3 = 5$, we may replace $x + (2 + 3)$ by $x + 5$.
   d) If $7 = 2x - 3$, then $2x - 3 = 7$.

4. Give an example to illustrate each axiom of equality:

   a) Reflexive c) Transitive
   b) Symmetric d) Substitution

5. Find each sum:

   a) $5 + 7$ c) $(-5) + 7$
   b) $5 + (-7)$ d) $(-5) + (-7)$

6. Find each sum:

   a) $(-7) + 15$ c) $(-7) + (-15)$
   b) $7 + (-15)$ d) $(-7) + [(-7) + 15]$

7. Find each product:

   a) $5 \cdot 7$ c) $(-5)7$
   b) $5(-7)$ d) $(-5)(-7)$

8. Find each product:

   a) $(-7)(12)$ c) $(-7)[(3)(4)]$
   b) $(-7)(-12)$ d) $(-7)[(3)(-4)]$

9. The area (the number of squares) of the rectangle in the accompanying drawing can be determined by multiplying the measures of the two dimensions. Will the order of multiplication change the answer? Which axiom is illustrated?

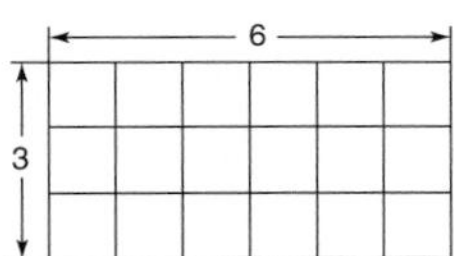


10. Identify the axiom of real numbers illustrated. Give a complete answer, such as Commutative Axiom for Multiplication.

    a) $7(5) = 5(7)$
    b) $(3 + 4) + 5 = 3 + (4 + 5)$
    c) $(-2) + 3 = 3 + (-2)$
    d) $(2 \cdot 3) \cdot 5 = 2 \cdot (3 \cdot 5)$

11. Perform each subtraction:

    a) $7 - (-2)$ c) $10 - 2$
    b) $(-7) - (+2)$ d) $(-10) - (-2)$

12. The temperature changes from $-3°$F at 2 A.M. to $7°$F at 7 A.M. Which expression represents the difference in temperatures from 2 A.M. to 7 A.M., $7 - (-3)$ or $(-3) - 7$?

13. Complete each division:

    a) $12 \div (-3)$ c) $(-12) \div \left(-\frac{2}{3}\right)$
    b) $12 \div \left(-\frac{1}{3}\right)$ d) $\left(-\frac{1}{12}\right) \div \left(\frac{1}{3}\right)$

14. Nine pegs are evenly spaced on a board so that the distance from each end to a peg equals the distance between any two pegs. If the board is 5 feet long, how far apart are the pegs?

15. The four owners of a shop realize a loss of \$240 in February. If the loss is shared equally, what number represents the profit for each owner for that month?

16. Bill works at a weekend convention by selling copies of a book. He receives a \$2 commission for each copy sold. If he sells 25 copies on Saturday and 30 copies on Sunday, what is Bill's total commission?

17. Use the Distributive Axiom to simplify each expression:

    a) $5(6 + 7)$ c) $\frac{1}{2}(7 + 11)$
    b) $4(7 - 3)$ d) $5x + 3x$

18. Use the Distributive Axiom to simplify each expression:

    a) $6(9 - 4)$ c) $7y - 2y$
    b) $\left(\frac{1}{2}\right) \cdot 6(4 + 8)$ d) $16x + 8x$

19. Simplify each expression:

    a) $6\pi + 4\pi$ c) $16x^2y - 9x^2y$
    b) $8\sqrt{2} + 3\sqrt{2}$ d) $9\sqrt{3} - 2\sqrt{3}$

20. Simplify each expression:

    a) $\pi r^2 + 2\pi r^2$ c) $7x^2y + 3xy^2$
    b) $7xy + 3xy$ d) $x + x + y$

21. Simplify each expression:

    a) $2 + 3 \cdot 4$ c) $2 + 3 \cdot 2^2$
    b) $(2 + 3) \cdot 4$ d) $2 + (3 \cdot 2)^2$

22. Simplify each expression:

    a) $3^2 + 4^2$ c) $3^2 + (8 - 2) \div 3$
    b) $(3 + 4)^2$ d) $[3^2 + (8 - 2)] \div 3$

23. Simplify each expression:

    a) $\dfrac{8 - 2}{2 - 8}$ c) $\dfrac{5 \cdot 2 - 6 \cdot 3}{7 - (-2)}$
    b) $\dfrac{8 - 2 \cdot 3}{(8 - 2) \cdot 3}$ d) $\dfrac{5 - 2 \cdot 6 + (-3)}{(-2)^2 + 4^2}$

24. Use the FOIL method to complete each multiplication:

    a) $(2 + 3)(4 + 5)$ b) $(7 - 2)(6 + 1)$

25. Use the FOIL method to complete each multiplication:

    a) $(3 - 1)(5 - 2)$ b) $(3x + 2)(4x - 5)$

26. Use the FOIL method to complete each multiplication:

    a) $(5x + 3)(2x - 7)$ b) $(2x + y)(3x - 5y)$

27. Using $x$ and $y$, find an expression for the length of the pegged board shown in the accompanying figure.

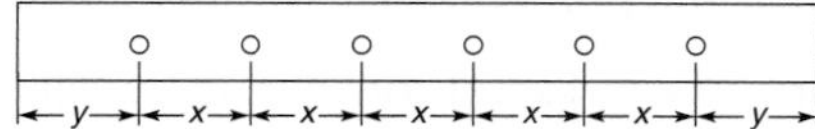


28. The cardboard used in the construction of the box shown in the accompanying figure has an area of $xy + yz + xz + xz + yz + xy$. Simplify this expression for the total area of the cardboard.

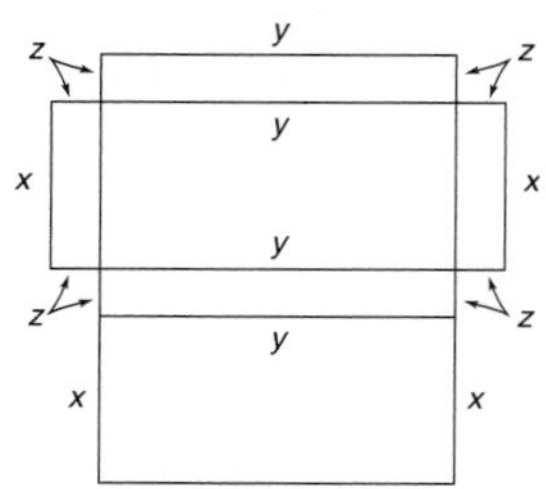


29. A large star is to be constructed, with lengths as shown in the accompanying figure. Give an expression for the total length of the wood strips used in the construction.

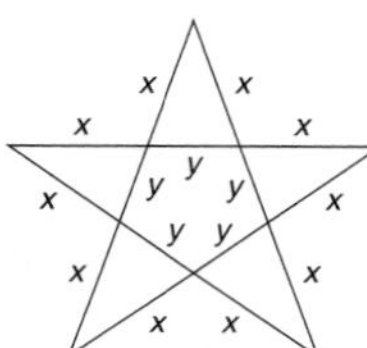


30. The area of an enclosed plot of ground that a farmer has subdivided can be found by multiplying $(x + y)$ times $(y + z)$. Use FOIL to complete the multiplication. How does this product compare with the total of the areas of the four smaller plots?

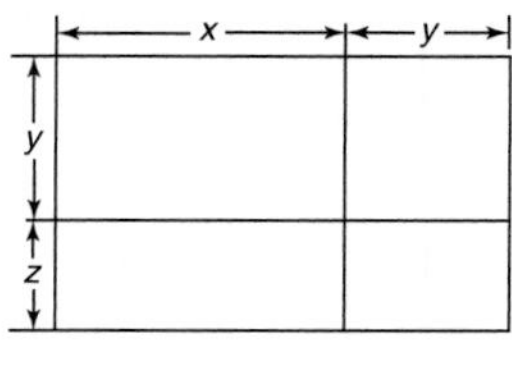


31. The degree measures of the angles of a triangle are $3x$, $5x$, and $2x$. Find an expression for the sum of the measures of these angles in terms of $x$.

32. The right circular cylinder shown in the accompanying figure has circular bases that have areas of $9\pi$ square units. The side has an area of $48\pi$ square units. Find an expression for the total surface area.

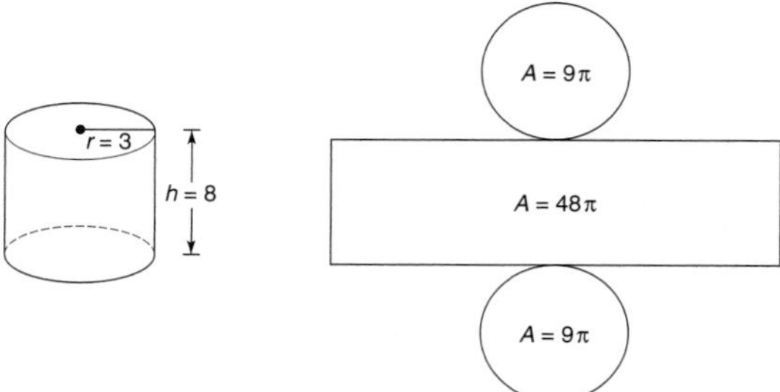


# A.2 Formulas and Equations

As you saw in the previous section, a **variable** is a letter used to represent an unknown number. The Greek letter $\pi$ is known as a **constant** because it always equals the same number (approximately 3.14). Although we will often use the letters $x$, $y$, and $z$ as variables, it is convenient to choose $r$ for radius, $h$ for height, $b$ for base, and so on.

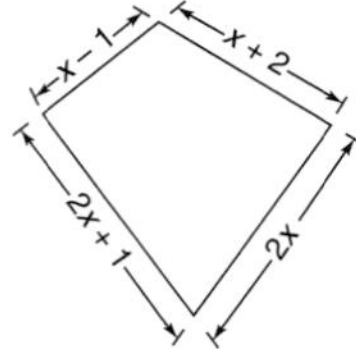


**FIGURE A.3**

EXAMPLE 1

For Figure A.3, combine like terms to find the perimeter (sum of the lengths of all sides) of the figure.

***Solution***

$$\begin{aligned} &(x - 1) + (x + 2) + 2x + (2x + 1) \\ &= x + (-1) + x + 2 + 2x + 2x + 1 \\ &= 1x + 1x + 2x + 2x + (-1) + 2 + 1 \\ &= 6x + 2 \end{aligned}$$

When the FOIL method is used with variable expressions, we combine like terms in the simplification.

EXAMPLE 2

Find a simplified expression for the area of a rectangle (not shown) with length $x + 5$ and width $x + 2$ by multiplying $(x + 5)$ times $(x + 2)$.

***Solution***

$$(x + 5)(x + 2)$$

(F, O, I, L)

$$= x \cdot x + 2 \cdot x + 5 \cdot x + 10$$
$$= x^2 + 7x + 10$$

In Example 2, we multiplied by the FOIL method before adding like terms in accordance with rules of order. When evaluating a variable expression, we must also follow that order. For instance, the value of $a^2 + b^2$ when $a = 3$ and $b = 4$ is given by

$$3^2 + 4^2 \quad \text{or} \quad 9 + 16 \quad \text{or} \quad 25$$

because exponential expressions must be simplified before addition occurs.

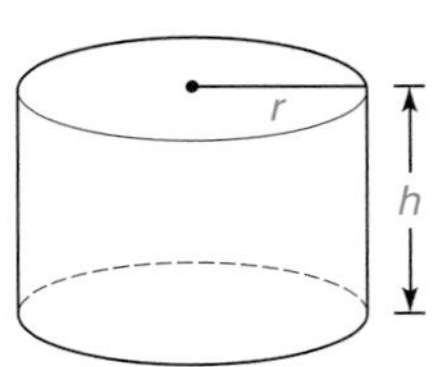


**FIGURE A.4**

EXAMPLE 3

Find the value of the following expressions.

a) $\pi r^2 h$, if $r = 3$ and $h = 4$ (leave $\pi$ in the answer)
b) $\frac{1}{2}h(b + B)$, if $h = 10$, $b = 7$, and $B = 13$

***Solution***

a) $\pi r^2 h = \pi \cdot 3^2 \cdot 4$
$= \pi \cdot 9 \cdot 4 = \pi(36) = 36\pi$

b) $\frac{1}{2}h(b + B) = \frac{1}{2} \cdot 10(7 + 13)$
$= \frac{1}{2} \cdot 10(20)$
$= \frac{1}{2} \cdot 10 \cdot 20$
$= 5 \cdot 20 = 100$

Many of the variable expressions that you will encounter are found in formulas. A **formula** is an equation that expresses a rule. For example, $V = \pi r^2 h$ is a formula for the volume $V$ of a right circular cylinder whose altitude has length $h$ and for which the circular base has a radius of length $r$. (See Figure A.4.)

EXAMPLE 4

Given the formula $P = 2\ell + 2w$, find the value of $P$ when $\ell = 7$ and $w = 3$.

***Solution***

By substitution, $P = 2\ell + 2w$ becomes

$$P = (2 \cdot 7) + (2 \cdot 3)$$
$$= 14 + 6$$
$$= 20$$

An **equation** is a statement that two expressions are equal. Although formulas are special types of equations, most equations are not formulas. Consider the following four examples of equations:

$$x + (x + 1) = 7$$
$$2(x + 1) = 8 - 2x$$
$$x^2 - 6x + 8 = 0$$
$$P = 2\ell + 2w \qquad \text{(a formula)}$$

The phrase *solving an equation* means finding the values of the variable that make the equation true when the variable is replaced by those values. These values are known as **solutions** for the equation. For example, 3 is a solution (in fact, the only solution) for the equation $x + (x + 1) = 7$ because $3 + (3 + 1) = 7$ is true.

When each side of an equation is transformed (changed) without having its solutions changed, we say that an **equivalent equation** is produced. Some of the properties that are used for equation solving are listed in the following box. Because these properties can be proved, they could be called theorems. However, we prefer to call an algebraic theorem by the name *property* in this book. These properties are also found in Section 1.5.

Warning (!)
We cannot multiply by 0 in solving an equation because the equation (say $2x - 1 = 7$) collapses to $0 = 0$. Division by 0 is likewise excluded.

PROPERTIES FOR EQUATION SOLVING

*Addition Property of Equality* (if $a = b$, then $a + c = b + c$): An equivalent equation results when the same number is added to each side of an equation.

*Subtraction Property of Equality* (if $a = b$, then $a - c = b - c$): An equivalent equation results when the same number is subtracted from each side of an equation.

*Multiplication Property of Equality* (if $a = b$, then $a \cdot c = b \cdot c$ for $c \neq 0$): An equivalent equation results when each side of an equation is multiplied by the same non-zero number.

*Division Property of Equality* $\left(\text{if } a = b, \text{ then } \frac{a}{c} = \frac{b}{c} \text{ for } c \neq 0\right)$: An equivalent equation results when each side of an equation is divided by the same non-zero number.

Addition and subtraction are **inverse operations,** as are multiplication and division. In problems that involve equation solving, we will utilize inverse operations. Specifically, we will

| | | |
|---|---|---|
| Add | to eliminate | a subtraction. |
| Subtract | to eliminate | an addition. |
| Multiply | to eliminate | a division. |
| Divide | to eliminate | a multiplication. |

EXAMPLE 5

Solve the equation $2x - 3 = 7$.

***Solution***

First add 3 (to eliminate the subtraction of 3 from 2x):

$$2x - 3 + 3 = 7 + 3$$
$$2x = 10 \qquad \text{simplifying}$$

Now divide by 2 (to eliminate the multiplication of 2 with x):

$$\frac{2x}{2} = \frac{10}{2}$$
$$x = 5 \qquad \text{simplifying}$$

In Example 5, the number 5 is the solution for the original equation. Replacing $x$ with 5, we can confirm this:

$$2x - 3 = 7$$
$$2(5) - 3 = 7$$
$$10 - 3 = 7$$

Some steps shown in this particular solution will be eliminated in the future. We may simplify as operations are performed so that the work has this appearance:

$$2x - 3 = 7$$
$$2x = 10 \quad \text{by addition}$$
$$x = 5 \quad \text{by division}$$

An equation that can be written in the form $ax + b = c$ for constants $a$, $b$, and $c$ is a **linear equation.** Our plan for solving such an equation involves getting variable terms together on one side of the equation and numerical terms together on the other side.

SOLVING A LINEAR EQUATION

1. Simplify each side of the equation; that is, combine like terms.
2. Eliminate additions and/or subtractions.
3. Eliminate multiplications and/or divisions.

EXAMPLE 6

Solve the equation $2(x - 3) + 5 = 13$.

***Solution***

$$2(x - 3) + 5 = 13$$
$$2x - 6 + 5 = 13 \quad \text{Distributive Axiom}$$
$$2x - 1 = 13 \quad \text{substitution}$$
$$2x = 14 \quad \text{addition}$$
$$x = 7 \quad \text{division}$$

NOTE: To check the solution, 7, we have

$$2(7 - 3) + 5 = 2(4) + 5 = 8 + 5 = 13$$

Some equations involve fractions. To avoid some of the difficulties that fractions bring, we often multiply each side of such equations by the **least common denominator (LCD)** of the fractions involved.

EXAMPLE 7

Solve the equation $\frac{x}{3} + \frac{x}{4} = 14$.

***Solution***

For the denominators 3 and 4, the LCD is 12. We must therefore multiply each side by 12 and use the Distributive Axiom on the left side.

$$12\left(\frac{x}{3} + \frac{x}{4}\right) = 12 \cdot 14$$
$$\frac{12}{1} \cdot \frac{x}{3} + \frac{12}{1} \cdot \frac{x}{4} = 168$$
$$4x + 3x = 168$$
$$7x = 168$$
$$x = 24$$

To check this result, we have

$$\frac{24}{3} + \frac{24}{4} = 14$$
$$8 + 6 = 14$$

It may happen that the variable appears in the denominator of the only fraction in an equation. In such cases, our method does not change! We continue to clear an equation of fractions through multiplication by the LCD, that denominator.

EXAMPLE 8

Solve the following equation for $n$:

$$\frac{360}{n} + 120 = 180$$

***Solution***

Multiplying by $n$ (the LCD), we have

$$n\left(\frac{360}{n} + 120\right) = 180 \cdot n$$
$$\frac{n}{1} \cdot \frac{360}{n} + 120 \cdot n = 180n$$
$$360 + 120n = 180n$$
$$360 = 60n$$
$$6 = n$$

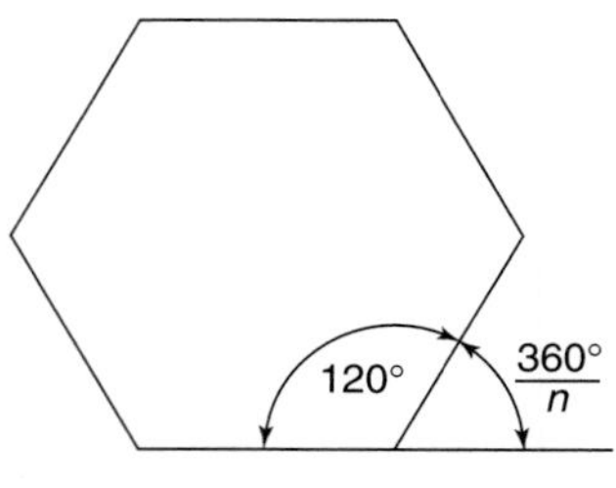


**FIGURE A.5**

NOTE: $n$ represents the number of sides possessed by the polygon in Figure A.5; $\frac{360}{n}$ and 120 represent the measures of angles in the figure.

Our final example combines many of the ideas introduced in this and the previous section. Example 9 is based on the formula for the area of a trapezoid.

EXAMPLE 9

See Figure A.6. For the formula $A = \frac{1}{2} \cdot h \cdot (b + B)$, suppose that $A = 77$, $b = 4$, and $B = 7$. Find the value of $h$.

***Solution***

Substitution leads to the equation

$$77 = \frac{1}{2} \cdot h \cdot (4 + 7)$$
$$77 = \frac{1}{2} \cdot h \cdot 11$$
$$2(77) = 2 \cdot \frac{1}{2} \cdot h \cdot 11 \quad \text{multiplying by 2}$$
$$154 = 11h \quad \text{simplifying}$$
$$14 = h \quad \text{dividing by 11}$$

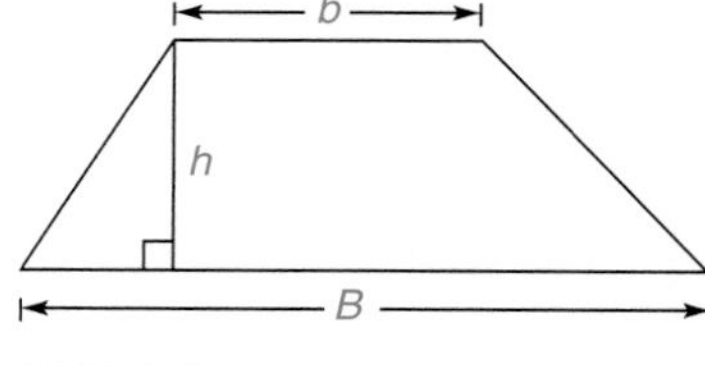


**FIGURE A.6**

In closing, we recall that the Addition Property of Equality (if $a = b$, then $a + c = b + c$) is really a theorem. We can show that $a + c = b + c$ must be true whenever $a = b$. This proof is sketched as follows:

*Because $a + c$ is the sum of two real numbers, $a + c$ is a real number also.*

| | | |
|---|---|---|
| | $\therefore a + c = a + c$ | Reflexive Axiom |
| But | $a = b$ | Hypothesis (Given) |
| | $\therefore a + c = b + c$ | Substitution (of $b$ for $a$) |

A more generalized form of the Addition Property of Equality can be stated as follows:

*If $a = b$ and $c = d$, then $a + c = b + d$.*

In this restatement, $d$ is substituted for $c$ on the right-hand side of the earlier equation $a + c = b + c$ to obtain $a + c = b + d$.

## A.2 Exercises

*In Exercises 1 to 6, simplify by combining similar terms.*

1. $(2x + 3) + (3x + 5)$
2. $(2x + 3) - (3x - 5)$
3. $x + (3x + 2) - (2x + 4)$
4. $(3x + 2) + (2x - 3) - (x + 1)$
5. $2(x + 1) + 3(x + 2)$ (HINT: Multiply before adding.)
6. $3(2x + 5) - 2(3x - 1)$

*In Exercises 7 to 12, simplify by using the FOIL method of multiplication.*

7. $(x + 3)(x + 4)$
8. $(x - 5)(x - 7)$
9. $(2x + 5)(3x - 2)$
10. $(3x + 7)(2x + 3)$
11. $(a + b)^2 + (a - b)^2$
12. $(x + 2)^2 - (x - 2)^2$

*In Exercises 13 to 16, evaluate each expression.*

13. $\ell \cdot w \cdot h$, if $\ell = 4$, $w = 3$, and $h = 5$
14. $a^2 + b^2$, if $a = 5$ and $b = 7$
15. $2 \cdot \ell + 2 \cdot w$, if $\ell = 13$ and $w = 7$
16. $a \cdot b \div c$, if $a = 6$, $b = 16$, and $c = 4$

*In Exercises 17 to 20, find the value of the variable named in each formula. Leave $\pi$ in the answers for Exercises 19 and 20.*

17. $S$, if $S = 2 \cdot \ell \cdot w + 2 \cdot w \cdot h + 2 \cdot \ell \cdot h$, $\ell = 6$, $w = 4$, and $h = 5$
18. $A$, if $A = \frac{1}{2}a(b + c + d)$, $a = 2$, $b = 6$, $c = 8$, and $d = 10$
19. $V$, if $V = \frac{1}{3}\pi \cdot r^2 \cdot h$, $r = 3$, and $h = 4$
20. $S$, if $S = 4\pi r^2$ and $r = 2$

*In Exercises 21 to 32, solve each equation.*

21. $2x + 3 = 17$
22. $3x - 3 = -6$
23. $-\frac{y}{3} + 2 = 6$
24. $3y = -21 - 4y$
25. $a + (a + 2) = 26$
26. $b = 27 - \frac{b}{2}$
27. $2(x + 1) = 30 - 6(x - 2)$
28. $2(x + 1) + 3(x + 2) = 22 + 4(10 - x)$
29. $\frac{x}{3} - \frac{x}{2} = -5$
30. $\frac{x}{2} + \frac{x}{3} + \frac{x}{4} = 26$
31. $\frac{360}{n} + 135 = 180$
32. $\frac{(n - 2) \cdot 180}{n} = 150$

*In Exercises 33 to 36, find the value of the indicated variable for each given formula.*

33. $w$, if $S = 2 \cdot \ell \cdot w + 2 \cdot w \cdot h + 2 \cdot \ell \cdot h$, $S = 148$, $\ell = 5$, and $h = 6$
34. $b$, if $A = \frac{1}{2} \cdot h \cdot (b + B)$, $A = 156$, $h = 12$, and $B = 11$
35. $y$, if $m = \frac{1}{2}(x - y)$, $m = 23$, and $x = 78$
36. $Y$, if $m = \frac{Y - y}{X - x}$, $m = \frac{-3}{2}$, $y = 1$, $X = 2$, and $x = -2$

*In Exercises 37 and 38, write a proof like the one in the final paragraph of this section. Assume that a, b, and c are real numbers.*

37. If $a = b$, then $a \cdot c = b \cdot c$.

38. If $a = b$, then $a - c = b - c$.

*In Exercises 39 and 40, the statements are not always true. Cite a counterexample for each claim.*

39. If $a \cdot c = b \cdot c$, then $a = b$. (NOTE: Name values of $a$, $b$, and $c$ for which $a \cdot c = b \cdot c$ but $a \neq b$.)

40. If $a^2 = b^2$, then $a = b$.

# A.3 Inequalities

In geometry, we sometimes need to work with or form inequalities. **Inequalities** are statements that involve one of the following relationships:

| | | |
|---|---|---|
| $<$ | means | "is less than" |
| $>$ | means | "is greater than" |
| $\leq$ | means | "is less than or equal to" |
| $\geq$ | means | "is greater than or equal to" |
| $\neq$ | means | "is not equal to" |

The statement $-4 < 7$ is true because negative 4 is less than positive 7. On the number line, the smaller number is always found to the left of the larger number. An equivalent claim is $7 > -4$, which means positive 7 is greater than negative 4.

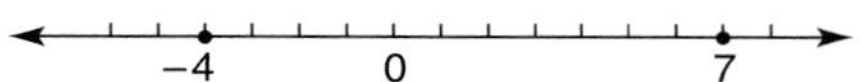


Both statements $6 \leq 6$ and $4 \leq 6$ are true. The statement $6 \leq 6$ could also be expressed by the statement $6 < 6$ or $6 = 6$, which is true because $6 = 6$ is true. Because $4 < 6$ is true, the statement $4 \leq 6$ is also true. A statement of the form *P or Q* is called a *disjunction*. See Section 1.1 for more information.

EXAMPLE 1

Give two true statements that involve the symbol $\geq$.

***Solution***

| | | |
|---|---|---|
| $5 \geq 5$ | because | $5 = 5$ is true |
| $12 \geq 5$ | because | $12 > 5$ is true |

The symbol $\neq$ is used to join any two numerical expressions that do not have the same value; for example, $2 + 3 \neq 7$. The following definition is also found in Section 3.5.

DEFINITION: $a$ is less than $b$ (that is, $a < b$) if and only if there is a positive number $p$ for which $a + p = b$; $a$ is greater than $b$ (that is, $a > b$) if and only if $b < a$.

EXAMPLE 2

Find, if possible, the following:

a) Any number $a$ for which "$a < a$" is true.
b) Any numbers $a$ and $b$ for which "$a < b$ and $b < a$" is true.

***Solution***

a) There is no such number. If $a < a$, then $a + p = a$ for some positive number $p$. Subtracting $a$ from each side of the equation gives $p = 0$. This statement ($p = 0$) contradicts the fact that $p$ is positive.

b) There are no such numbers. If $a < b$, then $a$ is to the left of $b$ on the number line. Therefore, $b < a$ is false, because this statement claims that $b$ is to the left of $a$.

EXAMPLE 3

What can you conclude for the numbers $x$, $y$, and $z$ if $x < y$ and $x > z$?

***Solution***

$x < y$ means that $x$ is to the left of $y$, as in Figure A.7. Similarly, $x > z$ (equivalently, $z < x$) means that $z$ is to the left of $x$. With $z$ to the left of $x$, which is itself to the left of $y$, we clearly have $z$ to the left of $y$; thus $z < y$.

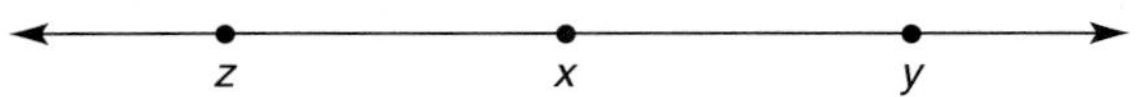


**FIGURE A.7**

Example 3 suggests a transitive relationship for the inequality "is less than," and this is stated in the following property. The Transitive Property of Inequality can also be stated using $>$, $\leq$, or $\geq$.

> TRANSITIVE PROPERTY OF INEQUALITY
>
> For numbers $a$, $b$, and $c$, if $a < b$ and $b < c$, then $a < c$.

This property can be proved as follows:

1. $a < b$ means that $a + p_1 = b$ for some positive number $p_1$.
2. $b < c$ means that $b + p_2 = c$ for some positive number $p_2$.
3. Substituting $a + p_1$ for $b$ (from statement 1) into the statement $b + p_2 = c$, we have $(a + p_1) + p_2 = c$.
4. Now $a + (p_1 + p_2) = c$.
5. But the sum of two positive numbers is also positive; that is, $p_1 + p_2 = p$, so statement 4 becomes $a + p = c$.
6. If $a + p = c$, then $a < c$, by the definition of "is less than."

Therefore, $a < b$ and $b < c$ implies that $a < c$.

The Transitive Property of Inequality can be extended to a series of unequal expressions. When a first value is less than a second, the second is less than a third, and so on, then the first is less than the last.

EXAMPLE 4

Two angles are complementary if the sum of their measures is exactly 90°. If the measure of the first of two complementary angles is more than 27°, what must you conclude about the measure of the second angle?

***Solution***

The second angle must measure less than 63°. The statements needed to establish this result are as follows:

1. If $x$ and $y$ are the angle measures, then $x + y = 90$.
2. If $x > 27$, then $27 < x$ and $27 + p = x$ for some positive number $p$.
3. $x + y = 90$ becomes $(27 + p) + y = 90$.
4. Restated, this equation is $27 + (p + y) = 90$.
5. Then $p + y = 63$.
6. Therefore, $y < 63$.

We now turn our attention to solving inequalities such as

$$x + (x + 1) < 7 \qquad \text{and} \qquad 2(x - 3) + 5 \geq 3$$

The method here is almost the same as the one used for equation solving, but there are some very important differences.

EXAMPLE 5

For the statement $-6 < 9$, determine the statement that results when each side is changed as follows:

a) Has 4 added to it
b) Has 2 subtracted from it
c) Is multiplied by 3
d) Is divided by $-3$

***Solution***

a) $-6 + 4 \,?\, 9 + 4$
   $-2 \,?\, 13 \rightarrow -2 < 13$
b) $-6 - 2 \,?\, 9 - 2$
   $-8 \,?\, 7 \rightarrow -8 < 7$
c) $(-6)(3) \,?\, 9(3)$
   $-18 \,?\, 27 \rightarrow -18 < 27$
d) $(-6) \div (-3) \,?\, 9 \div (-3)$
   $2 \,?\, -3 \rightarrow 2 > -3$

As Example 5 suggests, addition and subtraction preserve the inequality symbol. Multiplication and division by a *positive* number preserve the inequality symbol, but multiplication and division by a *negative* number reverse the inequality symbol.

PROPERTIES OF INEQUALITIES

Stated for $<$, these properties have counterparts involving $>$, $\leq$, and $\geq$.

*Addition:* If $a < b$, then $a + c < b + c$.

*Subtraction:* If $a < b$, then $a - c < b - c$.

*Multiplication:*
i) If $a < b$ and $c > 0$ ($c$ is positive), then $a \cdot c < b \cdot c$.
ii) If $a < b$ and $c < 0$ ($c$ is negative), then $a \cdot c > b \cdot c$.

*Division:*
i) If $a < b$ and $c > 0$ ($c$ is positive), then $\frac{a}{c} < \frac{b}{c}$.
ii) If $a < b$ and $c < 0$ ($c$ is negative), then $\frac{a}{c} > \frac{b}{c}$.

The plan for solving inequalities is similar to the one used for equation solving.

**Warning** (!)
Be sure to reverse the inequality symbol upon multiplying or dividing by a *negative* number.

**SOLVING AN INEQUALITY**

1. Simplify each side of the inequality; that is, combine like terms.
2. Eliminate additions and subtractions.
3. Eliminate multiplications and divisions.

## EXAMPLE 6

Solve $2x - 3 \le 7$.

***Solution***

$$\begin{aligned} 2x - 3 + 3 &\le 7 + 3 && \text{adding 3 preserves } \le \\ 2x &\le 10 && \text{simplify} \\ \frac{2x}{2} &\le \frac{10}{2} && \text{division by 2 preserves } \le \\ x &\le 5 && \text{simplify} \end{aligned}$$

The possible values of $x$ are shown on a number line in Figure A.8; this picture is the **graph** of the solutions. Note that the point above the 5 is shown solid in order to indicate that 5 is included as a solution.

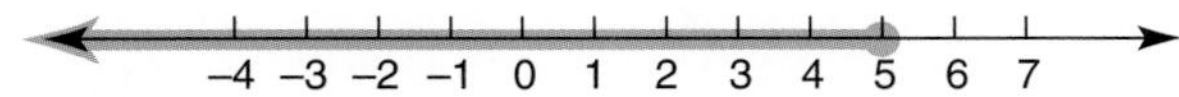


**FIGURE A.8**

## EXAMPLE 7

Solve $x(x - 2) - (x + 1)(x + 3) < 9$.

***Solution***

Using the Distributive Axiom and FOIL, we simplify the left side to get

$$(x^2 - 2x) - (x^2 + 4x + 3) < 9$$

Subtraction is performed by adding the additive inverse of each term in $(x^2 + 4x + 3)$.

$$\begin{aligned} \therefore (x^2 - 2x) + (-x^2 - 4x - 3) &< 9 \\ -6x - 3 &< 9 && \text{simplify} \\ -6x &< 12 && \text{add 3} \\ \frac{-6x}{-6} &> \frac{12}{-6} && \text{divide by } -6 \text{ and reverse the inequality symbol} \\ x &> -2 \end{aligned}$$

The graph of the solution is shown in Figure A.9. Notice that the circle above the $-2$ is shown open in order to indicate that $-2$ is not included as a solution.

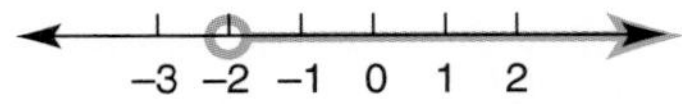


**FIGURE A.9**

## A.3 Exercises

1. If line segment $AB$ and line segment $CD$ in the accompanying drawing are drawn to scale, what does intuition tell you about the lengths of these segments?

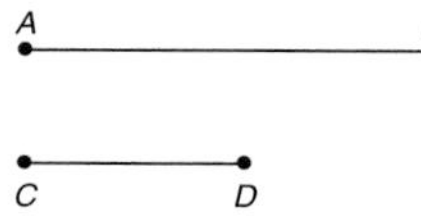


2. Using the number line shown, write two statements that relate the values of $e$ and $f$.

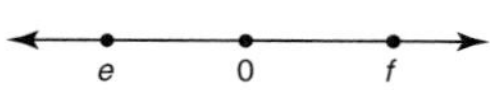


3. If angles $ABC$ and $DEF$ in the accompanying drawing were measured with a protractor, what does intuition tell you about the degree measures of these angles?

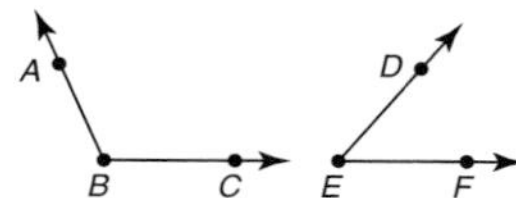


4. Consider the statement $x \geq 6$. Which of the following choices of $x$ below will make this a true statement?

   $x = -3$ $\quad x = 0$ $\quad x = 6$ $\quad x = 9$ $\quad x = 12$

5. According to the definition of $a < b$, there is a positive number $p$ for which $a + p = b$. Find the value of $p$ for the statement given.

   a) $3 < 7$ b) $-3 < 7$

6. Does the Transitive Property of Inequality hold true for four real numbers $a$, $b$, $c$, and $d$? That is, is the following statement true?

   If $a < b$, $b < c$, and $c < d$, then $a < d$.

7. Of several line segments, $AB > CD$ (the length of segment $AB$ is greater than that of segment $CD$), $CD > EF$, $EF > GH$, and $GH > IJ$. What conclusion does the Transitive Property of Inequality allow regarding $IJ$ and $AB$?

8. Of several angles, the degree measures are related in this way: $m\angle JKL > m\angle GHI$ (the measure of angle $JKL$ is greater than that of angle $GHI$), $m\angle GHI > m\angle DEF$, and $m\angle DEF > m\angle ABC$. What conclusion does the Transitive Property of Inequality allow regarding $m\angle ABC$ and $m\angle JKL$?

9. Classify as true or false.

   a) $5 \leq 4$ c) $5 \leq 5$
   b) $4 \leq 5$ d) $5 < 5$

10. Classify as true or false.

    a) $-5 \leq 4$ c) $-5 \leq -5$
    b) $5 \leq -4$ d) $5 \leq -5$

11. Two angles are supplementary if the sum of their measures is 180°. If the measure of the first of two supplementary angles is less than 32°, what must you conclude about the measure of the second angle?

12. Two trim boards need to be used together to cover a 12-ft length along one wall. If Jim recalls that one board is more than 7 ft long, what length must the second board be to span the 12-ft length?

13. Consider the inequality $-3 \leq 5$. Write the statement that results when

    a) each side is multiplied by 4.
    b) $-7$ is added to each side.
    c) each side is multiplied by $-6$.
    d) each side is divided by $-1$.

14. Consider the inequality $-6 > -9$. Write the statement that results when

    a) 8 is added to each side.
    b) each side is multiplied by $-2$.
    c) each side is multiplied by 2.
    d) each side is divided by $-3$.

15. Suppose that you are solving an inequality. Complete this chart for your work by indicating whether the inequality symbol should be reversed or kept by writing "change" or "no change."

| | *Positive* | *Negative* |
|---|---|---|
| Add | | |
| Subtract | | |
| Multiply | | |
| Divide | | |

*In Exercises 16 to 26, first solve each inequality. Then draw a number line graph of the solutions.*

16. $5x - 1 \leq 29$

17. $2x + 3 \leq 17$

18. $5 + 4x > 25$

19. $5 - 4x > 25$

20. $5(2 - x) \leq 30$

21. $2x + 3x < 200 - 5x$

22. $5(x + 2) < 6(9 - x)$

23. $\frac{x}{3} - \frac{x}{2} \leq 4$

24. $\frac{2x - 3}{-5} > 7$

25. $x^2 + 4x \leq x(x - 5) - 18$

26. $x(x + 2) < x(2 - x) + 2x^2$

*In Exercises 27 to 30, the claims made are not always true. Cite a counterexample to show why each claim fails.*

27. If $a < b$, then $a \cdot c < b \cdot c$.

28. If $a < b$, then $a \cdot c \neq b \cdot c$.

29. If $a < b$, then $a^2 < b^2$.

30. If $a \neq b$ and $b \neq c$, then $a \neq c$.

*In Exercises 31 and 32, the statements are true and can be called theorems. Use the definition of* $a < b$ *to prove each statement.*

31. If $a < b$ and $c < d$, then $a + c < b + d$.

32. If $a < b$, then $c - a > c - b$.

# A.4 Quadratic Equations

An equation that can be written in the form $ax^2 + bx + c = 0$ $(a \neq 0)$ is a **quadratic equation.** For example, $x^2 - 7x + 12 = 0$ and $6x^2 = 7x + 3$ are quadratic. Many quadratic equations can be solved by a factoring method that depends on the Zero Product Property.

ZERO PRODUCT PROPERTY

If $a \cdot b = 0$, then $a = 0$ or $b = 0$.

When this property is stated in words, it reads, "If the product of two expressions equals 0, then at least one of the factors must equal 0."

EXAMPLE 1

Solve $x^2 - 7x + 12 = 0$.

***Solution***

First you must factor the polynomial; then check the factors by using the FOIL method of multiplication.

| | | | |
|---|---|---|---|
| $(x - 3)(x - 4) = 0$ | | | factoring |
| $x - 3 = 0$ | or | $x - 4 = 0$ | Zero Product Property |
| $x = 3$ | or | $x = 4$ | Addition Property |

To check $x = 3$, substitute into the given equation:

$$3^2 - (7 \cdot 3) + 12 = 9 - 21 + 12 = 0$$

Similarly, to check $x = 4$, substitute again:

$$4^2 - (7 \cdot 4) + 12 = 16 - 28 + 12 = 0$$

Checks for later problems will not be provided. The solutions are usually expressed as a set—in this case, {3, 4}.

If you were asked to use factoring to solve the quadratic equation

$$6x^2 = 7x + 3,$$

it would be necessary to change the equation so that one side would be equal to 0. The form $ax^2 + bx + c = 0$ is the **standard form** of a quadratic equation.

**SOLVING A QUADRATIC EQUATION BY THE FACTORING METHOD**

1. Be sure the equation is in standard form (one side = 0).
2. Factor the polynomial side of the equation.
3. Set each factor containing the variable equal to 0.
4. Solve each equation found in step 3.
5. Check solutions by substituting into the original equation.

Step 5, which was shown in Example 1, is omitted in Example 2.

**EXAMPLE 2**

Solve $6x^2 = 7x + 3$.

***Solution***

First changing to standard form, we have

$$6x^2 - 7x - 3 = 0 \quad \text{standard form}$$

$$(2x - 3)(3x + 1) = 0 \quad \text{factoring}$$

$$2x - 3 = 0 \quad \text{or} \quad 3x + 1 = 0 \quad \text{Zero Product Property}$$

$$2x = 3 \quad \text{or} \quad 3x = -1 \quad \text{Addition-Subtraction Property}$$

$$x = \frac{3}{2} \quad \text{or} \quad x = \frac{-1}{3} \quad \text{division}$$

Therefore, $\left\{\frac{3}{2}, -\frac{1}{3}\right\}$ is the solution set.

In some instances, a common factor can be extracted from each term in the factoring step. In the equation $2x^2 + 10x - 48 = 0$, the left side of the equation has the common factor 2. Factoring leads to $2(x^2 + 5x - 24) = 0$ and then to $2(x + 8)(x - 3) = 0$. Of course, only the factors containing variables can equal 0, so the solutions to this equation are $-8$ and 3.

Equations such as $4x^2 = 9$ and $4x^2 - 12x = 0$ are **incomplete quadratic equations** because one term is missing from the standard form; the linear term (having exponent 1) is missing from the first equation, and the constant term is omitted in the second. Either equation can, however, be solved by factoring; in particular, the factoring is given by

$$4x^2 - 9 = (2x + 3)(2x - 3)$$

and

$$4x^2 - 12x = 4x(x - 3)$$

When solutions to $ax^2 + bx + c = 0$ cannot be found by factoring, they may be determined by the following formula, in which $a$ is the number multiplied by $x^2$, $b$ is the number multiplied by $x$, and $c$ is the constant term. The $\pm$ symbol tells us that there are generally two solutions, one found by adding and one by subtracting. The symbol $\sqrt{a}$ is read "square root of $a$."

**QUADRATIC FORMULA**

$x = \dfrac{-b \pm \sqrt{b^2 - 4ac}}{2a}$ are solutions for $ax^2 + bx + c = 0$, where $a \neq 0$.

Although the formula may provide two solutions for the equation, an application problem in geometry may have a single positive solution representing a segment (or angle) measure. Recall that for $a > 0$, $\sqrt{a}$ represents the principal square root of $a$.

> DEFINITION: Where $a > 0$, the number $\sqrt{a}$ is the positive number for which $(\sqrt{a})^2 = a$.

### EXAMPLE 3

a) Explain why $\sqrt{25}$ is equal to 5.
b) Without a calculator, find the value of $\sqrt{3} \cdot \sqrt{3}$.
c) Use a calculator to show that $\sqrt{5} \approx 2.236$.

***Solution***

a) We see that $\sqrt{25}$ must equal 5 because $5^2 = 25$.
b) By definition, $\sqrt{3}$ is the number for which $(\sqrt{3})^2 = 3$.
c) By using a calculator, we see that $2.236^2 \approx 5$.

### EXAMPLE 4

Simplify each expression, if possible.

a) $\sqrt{16}$ b) $\sqrt{0}$ c) $\sqrt{7}$ d) $\sqrt{400}$ e) $\sqrt{-4}$

***Solution***

a) $\sqrt{16} = 4$ because $4^2 = 16$.
b) $\sqrt{0} = 0$ because $0^2 = 0$.
c) $\sqrt{7}$ cannot be simplified; however, $\sqrt{7} \approx 2.646$.
d) $\sqrt{400} = 20$ because $20^2 = 400$; calculator can be used.
e) $\sqrt{-4}$ is not a real number; calculator gives an "ERROR" message.

Whereas $\sqrt{25}$ represents the principal square root of 25 (namely 5), the expression $-\sqrt{25}$ can be interpreted as "the negative number whose square is 25"; thus, $-\sqrt{25} = -5$ because $(-5)^2 = 25$. In expressions such as $\sqrt{9 + 16}$ and $\sqrt{4 + 9}$, we first simplify the radicand (the expression under the bar of the square root); thus $\sqrt{9 + 16} = \sqrt{25} = 5$ and $\sqrt{4 + 9} = \sqrt{13} \approx 3.606$.

Just as fractions are reduced to lower terms $\left(\frac{6}{8}\text{ is replaced by }\frac{3}{4}\right)$, it is also customary to reduce the size of the radicand when possible. To accomplish this, we use the Product Property of Square Roots.

> PRODUCT PROPERTY OF SQUARE ROOTS
>
> For $a \geq 0$ and $b \geq 0$, $\sqrt{a \cdot b} = \sqrt{a} \cdot \sqrt{b}$.

When simplifying, we replace the radicand by a product in which the largest possible number from the list of perfect squares below is selected as one of the factors:

4, 9, 16, 25, 36, 49, 64, 81, 100, 121, . . .

For example,

$$\begin{aligned}\sqrt{45} &= \sqrt{9 \cdot 5}\\ &= \sqrt{9} \cdot \sqrt{5}\\ &= 3\sqrt{5}\end{aligned}$$

The radicand has now been reduced from 45 to 5. Using a calculator, we see that $\sqrt{45} \approx 6.708$. Also, $3\sqrt{5}$ means 3 times $\sqrt{5}$, and with the calculator we see that $3\sqrt{5} \approx 6.708$.

> Leave the smallest possible integer under the square root symbol.

EXAMPLE 5

Simplify the following radicals:

a) $\sqrt{27}$ b) $\sqrt{125}$

***Solution***

a) 9 is the largest perfect square factor of 27. Therefore,

$$\sqrt{27} = \sqrt{9 \cdot 3} = \sqrt{9} \cdot \sqrt{3} = 3\sqrt{3}$$

b) 25 is the largest perfect square factor of 125. Therefore,

$$\sqrt{125} = \sqrt{25 \cdot 5} = \sqrt{25} \cdot \sqrt{5} = 5\sqrt{5}$$

The Product Property of Square Roots has a symmetric form that reads $\sqrt{a} \cdot \sqrt{b} = \sqrt{ab}$; for example, $\sqrt{2} \cdot \sqrt{3} = \sqrt{6}$ and $\sqrt{5} \cdot \sqrt{5} = \sqrt{25} = 5$.

> When calculator answers are requested or provided, the answers in this textbook will generally be rounded to two decimal places. For instance, $\sqrt{125} \approx 11.18$ (rounded from 11.1803).

The expression $ax^2 + bx + c$ may be **prime** (meaning "not factorable"). Because $x^2 - 5x + 3$ is prime, we solve the equation $x^2 - 5x + 3 = 0$ by using the Quadratic Formula $x = \frac{-b \pm \sqrt{b^2 - 4ac}}{2a}$; see Example 6.

NOTE: When square root radicals are left in an answer, the answer is considered to be exact. Once we use the calculator, the solutions are only approximate.

EXAMPLE 6

Find exact solutions for $x^2 - 5x + 3 = 0$. Then use a calculator to approximate these solutions correct to two decimal places.

***Solution***

With the equation in standard form, we see that $a = 1$, $b = -5$, and $c = 3$. So

$$x = \frac{-(-5) \pm \sqrt{(-5)^2 - 4(1)(3)}}{2(1)}$$

$$x = \frac{5 \pm \sqrt{25 - 12}}{2} \quad \text{or} \quad x = \frac{5 \pm \sqrt{13}}{2}$$

The exact solutions are $\frac{5 + \sqrt{13}}{2}$ and $\frac{5 - \sqrt{13}}{2}$. Using a calculator, we find that the approximate solutions are 4.30 and 0.70.

Using the Quadratic Formula to solve the equation $x^2 - 6x + 7 = 0$ yields $x = \frac{6 \pm \sqrt{8}}{2}$. In Example 7, we focus on the simplification of such an expression.

EXAMPLE 7

Simplify $\dfrac{6 \pm \sqrt{8}}{2}$.

***Solution***

Because $\sqrt{8} = \sqrt{4} \cdot \sqrt{2}$ or $2\sqrt{2}$, we simplify the expression as follows:

$$\frac{6 \pm \sqrt{8}}{2} = \frac{6 \pm 2\sqrt{2}}{2} = \frac{\cancel{2}(3 \pm \sqrt{2})}{\cancel{2}} = 3 \pm \sqrt{2}$$

NOTE 1: The number 2 was a common factor for the numerator and the denominator. We then reduced the fraction to lowest terms.

NOTE 2: The approximate values of $3 \pm \sqrt{2}$ are 4.41 and 1.59. Use your calculator to show that these values are the approximate solutions of the equation $x^2 - 6x + 7 = 0$.

Our final method for solving quadratic equations is used if the equation has the form $ax^2 + c = 0$.

SQUARE ROOTS PROPERTY

If $x^2 = p$ where $p \geq 0$, then $x = \pm\sqrt{p}$.

According to the Square Roots Property, the equation $x^2 = 6$ has the solutions $\pm\sqrt{6}$.

EXAMPLE 8

Use the Square Roots Property to solve the equation $2x^2 - 56 = 0$.

***Solution***

$$2x^2 - 56 = 0 \rightarrow 2x^2 = 56 \rightarrow x^2 = 28$$

Then

$$x = \pm\sqrt{28} = \pm\sqrt{4} \cdot \sqrt{7} = \pm 2\sqrt{7}$$

The exact solutions are $2\sqrt{7}$ and $-2\sqrt{7}$; the approximate solutions are 5.29 and $-5.29$.

In Example 10, the solutions for the quadratic equation will involve fractions. For this reason, we consider the Quotient Property of Square Roots. The Quotient Property enables us to replace the square root of a fraction by the square root of its numerator divided by the square root of its denominator.

QUOTIENT PROPERTY OF SQUARE ROOTS

For $a \geq 0$ and $b > 0$, $\sqrt{\frac{a}{b}} = \frac{\sqrt{a}}{\sqrt{b}}$.

EXAMPLE 9

Simplify the following square root expressions:

a) $\sqrt{\frac{16}{9}}$ b) $\sqrt{\frac{3}{4}}$

***Solution***

a) $\sqrt{\frac{16}{9}} = \frac{\sqrt{16}}{\sqrt{9}} = \frac{4}{3}$ b) $\sqrt{\frac{3}{4}} = \frac{\sqrt{3}}{\sqrt{4}} = \frac{\sqrt{3}}{2}$

EXAMPLE 10

Solve the equation $4x^2 - 9 = 0$.

***Solution***

$$4x^2 - 9 = 0 \rightarrow 4x^2 = 9 \rightarrow x^2 = \frac{9}{4}$$

Then

$$x = \pm\sqrt{\frac{9}{4}} = \pm\frac{\sqrt{9}}{\sqrt{4}} = \pm\frac{3}{2}$$

In summary, quadratic equations have the form $ax^2 + bx + c = 0$ and are solved by one of the following methods:

1. Factoring, when $ax^2 + bx + c$ is easily factored
2. The Quadratic Formula
$$x = \frac{-b \pm \sqrt{b^2 - 4ac}}{2a}$$
when $ax^2 + bx + c$ is not easily factored or cannot be factored
3. The Square Roots Property, when $b = 0$

## A.4 Exercises

1. Use your calculator to find the approximate value of each number, correct to two decimal places:

   a) $\sqrt{13}$ b) $\sqrt{8}$ c) $-\sqrt{29}$ d) $\sqrt{\frac{3}{5}}$

2. Use your calculator to find the approximate value of each number, correct to two decimal places:

   a) $\sqrt{17}$ b) $\sqrt{400}$ c) $-\sqrt{7}$ d) $\sqrt{1.6}$

3. Which equations are quadratic?

   a) $2x^2 - 5x + 3 = 0$ d) $\frac{1}{2}x^2 - \frac{1}{4}x - \frac{1}{8} = 0$
   b) $x^2 = x^2 + 4$ e) $\sqrt{2x - 1} = 3$
   c) $x^2 = 4$ f) $(x + 1)(x - 1) = 15$

4. Which equations are incomplete quadratic equations?

   a) $x^2 - 4 = 0$ d) $2x^2 - 4 = 2x^2 + 8x$
   b) $x^2 - 4x = 0$ e) $x^2 = \frac{9}{4}$
   c) $3x^2 = 2x$ f) $x^2 - 2x - 3 = 0$

5. Simplify each expression by using the Product Property of Square Roots:

   a) $\sqrt{8}$ c) $\sqrt{900}$
   b) $\sqrt{45}$ d) $(\sqrt{3})^2$

6. Simplify each expression by using the Product Property of Square Roots:

   a) $\sqrt{28}$ c) $\sqrt{54}$
   b) $\sqrt{32}$ d) $\sqrt{200}$

7. Simplify each expression by using the Quotient Property of Square Roots:

   a) $\sqrt{\frac{9}{16}}$ c) $\sqrt{\frac{7}{16}}$
   b) $\sqrt{\frac{25}{49}}$ d) $\sqrt{\frac{6}{9}}$

8. Simplify each expression by using the Quotient Property of Square Roots:

   a) $\sqrt{\frac{1}{4}}$ c) $\sqrt{\frac{5}{36}}$
   b) $\sqrt{\frac{16}{9}}$ d) $\sqrt{\frac{3}{16}}$

9. Use your calculator to verify that the following expressions are equivalent:

   a) $\sqrt{54}$ and $3\sqrt{6}$ b) $\sqrt{\frac{5}{16}}$ and $\frac{\sqrt{5}}{4}$

10. Use your calculator to verify that the following expressions are equivalent:

    a) $\sqrt{48}$ and $4\sqrt{3}$ b) $\sqrt{\frac{7}{9}}$ and $\frac{\sqrt{7}}{3}$

*In Exercises 11 to 18, solve each quadratic equation by factoring.*

11. $x^2 - 6x + 8 = 0$
12. $x^2 + 4x = 21$
13. $3x^2 - 51x + 180 = 0$
    (HINT: There is a common factor.)
14. $2x^2 + x - 6 = 0$
15. $3x^2 = 10x + 8$
16. $8x^2 + 40x - 112 = 0$
17. $6x^2 = 5x - 1$
18. $12x^2 + 10x = 12$

*In Exercises 19 to 26, solve each equation by using the Quadratic Formula. Give exact solutions in simplified form. When answers contain square roots, approximate the solutions rounded to two decimal places.*

19. $x^2 - 7x + 10 = 0$
20. $x^2 + 7x + 12 = 0$
21. $x^2 + 9 = 7x$
22. $2x^2 + 3x = 6$
23. $x^2 - 4x - 8 = 0$
24. $x^2 - 6x - 2 = 0$
25. $5x^2 = 3x + 7$
26. $2x^2 = 8x - 1$

*In Exercises 27 to 32, solve each incomplete quadratic equation. Use the Square Roots Property as needed.*

27. $2x^2 = 14$
28. $2x^2 = 14x$
29. $4x^2 - 25 = 0$
30. $4x^2 - 25x = 0$
31. $ax^2 - bx = 0$
32. $ax^2 - b = 0$
33. The length of a rectangle is 3 more than its width. If the area of the rectangle is 40, the dimensions $x$ and $x + 3$ can be found by solving the equation $x(x + 3) = 40$. Find these dimensions.
34. To find the lengths of $\overline{CP}$ (which is $x$), $\overline{PD}$, $\overline{AP}$, and $\overline{PB}$ in the circle, one must solve the equation

    $$x \cdot (x + 5) = (x + 1) \cdot 4$$

    Find the length of $\overline{CP}$.

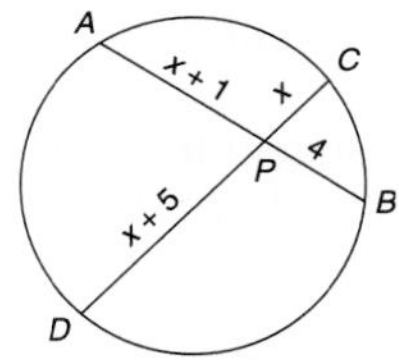


*In Exercises 35 and 36, use Theorem 2.5.1 to solve the problem. According to this theorem, the number of diagonals in a polygon of n sides is given by* $D = \frac{n(n-3)}{2}$.

35. Find the number of sides in a polygon that has 9 diagonals.
36. Find the number of sides in a polygon that has the same number of diagonals as it has sides.
37. In the right triangle, find $c$ if $a = 3$ and $b = 4$.
    (HINT: $c^2 = a^2 + b^2$)
38. In the right triangle, find $b$ if $a = 6$ and $c = 10$.
    (HINT: $c^2 = a^2 + b^2$)

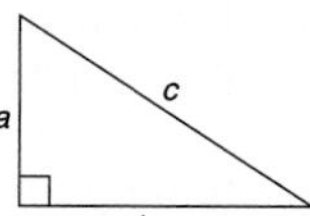

*Exercises 37, 38*

# Appendix B

## Summary of Constructions, Postulates, and Theorems and Corollaries

### Constructions

SECTION 1.2

1. To construct a segment congruent to a given segment.
2. To construct the midpoint $M$ of a given line segment $AB$.

SECTION 1.4

3. To construct an angle congruent to a given angle.
4. To construct the angle bisector of a given angle.

SECTION 1.6

5. To construct the line perpendicular to a given line at a specified point on the given line.

SECTION 2.1

6. To construct the line that is perpendicular to a given line from a point not on the given line.

SECTION 2.3

7. To construct the line parallel to a given line from a point not on that line.

SECTION 6.4

8. To construct a tangent to a circle at a point on the circle.
9. To construct a tangent to a circle from an external point.

### Postulates

SECTION 1.3

1. Through two distinct points, there is exactly one line.
2. (Ruler Postulate) The measure of any line segment is a unique positive number.
3. (Segment-Addition Postulate) If $X$ is a point on $\overline{AB}$ and $A$-$X$-$B$, then $AX + XB = AB$.
4. If two lines intersect, they intersect at a point.
5. Through three noncollinear points, there is exactly one plane.
6. If two distinct planes intersect, then their intersection is a line.
7. Given two distinct points in a plane, the line containing these points also lies in the plane.

SECTION 1.4

8. (Protractor Postulate) The measure of an angle is a unique positive number.
9. (Angle-Addition Postulate) If a point $D$ lies in the interior of angle $ABC$, then $m\angle ABD + m\angle DBC = m\angle ABC$.

SECTION 2.1

10. (Parallel Postulate) Through a point not on a line, exactly one line is parallel to the given line.
11. If two parallel lines are cut by a transversal, then the corresponding angles are congruent.

SECTION 3.1

12. If the three sides of one triangle are congruent to the three sides of a second triangle, then the triangles are congruent (SSS).
13. If two sides and the included angle of one triangle are congruent to two sides and the included angle of a second triangle, then the triangles are congruent (SAS).
14. If two angles and the included side of one triangle are congruent to two angles and the included side of a second triangle, then the triangles are congruent (ASA).

SECTION 5.2

15. If the three angles of one triangle are congruent to the three angles of a second triangle, then the triangles are similar (AAA).

SECTION 6.1

16. (Central Angle Postulate) In a circle, the degree measure of a central angle is equal to the degree measure of its intercepted arc.
17. (Arc-Addition Postulate) If $B$ lies between $A$ and $C$ on a circle, then $m\widehat{AB} + m\widehat{BC} = m\widehat{ABC}$.

SECTION 7.1

18. (Area Postulate) Corresponding to every bounded region is a unique positive number $A$, known as the area of that region.
19. If two closed plane figures are congruent, then their areas are equal.
20. (Area-Addition Postulate) Let $R$ and $S$ be two enclosed regions that do not overlap. Then $A_{R \cup S} = A_R + A_S$.
21. The area $A$ of a rectangle whose base has length $b$ and whose altitude has length $h$ is given by $A = bh$.

SECTION 7.4

22. The ratio of the circumference of a circle to the length of its diameter is a unique positive constant.

SECTION 7.5

23. The ratio of the degree measure $m$ of the central angle of a sector to 360° is the same as the ratio of the area of the sector to the area of the circle; that is, $\frac{\text{area of sector}}{\text{area of circle}} = \frac{m}{360°}$.

SECTION 8.1

24. (Volume Postulate) Corresponding to every solid is a unique positive number $V$ known as the volume of that solid.
25. The volume of a right rectangular prism is given by

$$V = \ell wh$$

where $\ell$ measures the length, $w$ the width, and $h$ the altitude of the prism.
26. The volume of a right prism is given by

$$V = Bh$$

where $B$ is the area of a base and $h$ is the altitude of the prism.

## Theorems and Corollaries

1.3.1 The midpoint of a line segment is unique.
1.4.1 There is one and only one angle bisector for a given angle.
1.6.1 If two lines are perpendicular, then they meet to form right angles.
1.6.2 If two lines intersect, then the vertical angles formed are congruent.
1.6.3 There is exactly one line perpendicular to a given line at any point on the line.
1.6.4 The perpendicular bisector of a line segment is unique.
1.7.1 If two lines meet to form a right angle, then these lines are perpendicular.
1.7.2 If two angles are complementary to the same angle (or to congruent angles), then these angles are congruent.
1.7.3 If two angles are supplementary to the same angle (or to congruent angles), then these angles are congruent.
1.7.4 Any two right angles are congruent.
1.7.5 If the exterior sides of two adjacent acute angles form perpendicular rays, then these angles are complementary.
1.7.6 If the exterior sides of two adjacent angles form a straight line, then these angles are supplementary.
1.7.7 If two segments are congruent, then their midpoints separate these segments into four congruent segments.
1.7.8 If two angles are congruent, then their bisectors separate these angles into four congruent angles.
2.1.1 From a point not on a given line, there is exactly one line perpendicular to the given line.
2.1.2 If two parallel lines are cut by a transversal, then the alternate interior angles are congruent.
2.1.3 If two parallel lines are cut by a transversal, then the alternate exterior angles are congruent.
2.1.4 If two parallel lines are cut by a transversal, then the interior angles on the same side of the transversal are supplementary.
2.1.5 If two parallel lines are cut by a transversal, then the exterior angles on the same side of the transversal are supplementary.
2.3.1 If two lines are cut by a transversal so that the corresponding angles are congruent, then these lines are parallel.
2.3.2 If two lines are cut by a transversal so that the alternate interior angles are congruent, then these lines are parallel.
2.3.3 If two lines are cut by a transversal so that the alternate exterior angles are congruent, then these lines are parallel.
2.3.4 If two lines are cut by a transversal so that the interior angles on the same side of the transversal are supplementary, then these lines are parallel.
2.3.5 If two lines are cut by a transversal so that the exterior angles on the same side of the transversal are supplementary, then these lines are parallel.
2.3.6 If two lines are both parallel to a third line, then these lines are parallel to each other.
2.3.7 If two coplanar lines are both perpendicular to a third line, then these lines are parallel to each other.
2.4.1 In a triangle, the sum of the measures of the interior angles is 180°.
2.4.2 Each angle of an equiangular triangle measures 60°.
2.4.3 The acute angles of a right triangle are complementary.
2.4.4 If two angles of one triangle are congruent to two angles of another triangle, then the third angles are also congruent.
2.4.5 The measure of an exterior angle of a triangle equals the sum of the measures of the two nonadjacent interior angles.
2.5.1 The total number of diagonals $D$ in a polygon of $n$ sides is given by the formula $D = \frac{n(n-3)}{2}$.
2.5.2 The sum $S$ of the measures of the interior angles of a polygon with $n$ sides is given by $S = (n-2) \cdot 180°$. Note that $n > 2$ for any polygon.
2.5.3 The measure $I$ of each interior angle of a regular or equiangular polygon of $n$ sides is $I = \frac{(n-2) \cdot 180°}{n}$.
2.5.4 The sum of the measures of the four interior angles of a quadrilateral is 360°.

2.5.5 The sum of the measures of the exterior angles, one at each vertex, of a polygon is 360°.

2.5.6 The measure $E$ of each exterior angle of a regular or equiangular polygon of $n$ sides is $E = \frac{360°}{n}$.

3.1.1 If two angles and a nonincluded side of one triangle are congruent to two angles and a nonincluded side of a second triangle, then the triangles are congruent (AAS).

3.2.1 If the hypotenuse and a leg of one right triangle are congruent to the hypotenuse and a leg of a second right triangle, then the triangles are congruent (HL).

3.3.1 Corresponding altitudes of congruent triangles are congruent.

3.3.2 The bisector of the vertex angle of an isosceles triangle separates the triangle into two congruent triangles.

3.3.3 If two sides of a triangle are congruent, then the angles opposite these sides are also congruent.

3.3.4 If two angles of a triangle are congruent, then the sides opposite these angles are also congruent.

3.3.5 An equilateral triangle is also equiangular.

3.3.6 An equiangular triangle is also equilateral.

3.5.1 The measure of a line segment is greater than the measure of any of its parts.

3.5.2 The measure of an angle is greater than the measure of any of its parts.

3.5.3 The measure of an exterior angle of a triangle is greater than the measure of either nonadjacent interior angle.

3.5.4 If a triangle contains a right or an obtuse angle, then the measure of this angle is greater than the measure of either of the remaining angles.

3.5.5 (Addition Property of Inequality): If $a > b$ and $c > d$, then $a + c > b + d$.

3.5.6 If one side of a triangle is longer than a second side, then the measure of the angle opposite the first side is greater than the measure of the angle opposite the second side.

3.5.7 If the measure of one angle of a triangle is greater than the measure of a second angle, then the side opposite the larger angle is longer than the side opposite the smaller angle.

3.5.8 The perpendicular segment from a point to a line is the shortest segment that can be drawn from the point to the line.

3.5.9 The perpendicular segment from a point to a plane is the shortest segment that can be drawn from the point to the plane.

3.5.10 (Triangle Inequality) The sum of the lengths of any two sides of a triangle is greater than the length of the third side.

3.5.10 *(Alternative)* The length of one side of a triangle must be between the sum and the difference of the lengths of the other two sides.

4.1.1 A diagonal of a parallelogram separates it into two congruent triangles.

4.1.2 The opposite angles of a parallelogram are congruent.

4.1.3 The opposite sides of a parallelogram are congruent.

4.1.4 The diagonals of a parallelogram bisect each other.

4.1.5 Two consecutive angles of a parallelogram are supplementary.

4.1.6 Two parallel lines are everywhere equidistant.

4.1.7 If two sides of one triangle are congruent to two sides of a second triangle and the included angle of the first triangle is greater than the included angle of the second, then the length of the side opposite the included angle of the first triangle is greater than the length of the side opposite the included angle of the second.

4.1.8 In a parallelogram with unequal pairs of consecutive angles, the longer diagonal lies opposite the obtuse angle.

4.2.1 If two sides of a quadrilateral are both congruent and parallel, then the quadrilateral is a parallelogram.

4.2.2 If both pairs of opposite sides of a quadrilateral are congruent, then it is a parallelogram.

4.2.3 If the diagonals of a quadrilateral bisect each other, then the quadrilateral is a parallelogram.

4.2.4 In a kite, one pair of opposite angles are congruent.

4.2.5 The segment that joins the midpoints of two sides of a triangle is parallel to the third side and has a length equal to one-half the length of the third side.

4.3.1 All angles of a rectangle are right angles.

4.3.2 The diagonals of a rectangle are congruent.

4.3.3 All sides of a square are congruent.

4.3.4 All sides of a rhombus are congruent.

4.3.5 The diagonals of a rhombus are perpendicular.

4.4.1 The base angles of an isosceles trapezoid are congruent.

4.4.2 The diagonals of an isosceles trapezoid are congruent.

4.4.3 The length of the median of a trapezoid equals one-half the sum of the lengths of the two bases.

4.4.4 The median of a trapezoid is parallel to each base.

4.4.5 If two base angles of a trapezoid are congruent, the trapezoid is an isosceles trapezoid.

4.4.6 If the diagonals of a trapezoid are congruent, the trapezoid is an isosceles trapezoid.

4.4.7 If three (or more) parallel lines intercept congruent segments on one transversal, then they intercept congruent segments on any transversal.

5.3.1 If two angles of one triangle are congruent to two angles of another triangle, then the triangles are similar (AA).

5.3.2 The lengths of the corresponding altitudes of similar triangles have the same ratio as the lengths of any pair of corresponding sides.

5.3.3 If an angle of one triangle is congruent to an angle of a second triangle and the pairs of sides including these angles are proportional (in length), then the triangles are similar (SAS~).

5.3.4 If the three sides of one triangle are proportional (in length) to the three corresponding sides of a second triangle, then the triangles are similar (SSS~).

5.3.5 If a line segment divides two sides of a triangle proportionally, then it is parallel to the third side.

5.4.1 The altitude drawn to the hypotenuse of a right triangle separates the right triangle into two right triangles that are similar to each other and to the original right triangle.

5.4.2 The length of the altitude to the hypotenuse of a right triangle is the geometric mean of the lengths of the segments of the hypotenuse.

5.4.3 The length of each leg of a right triangle is the geometric mean of the length of the hypotenuse and the length of the segment of the hypotenuse adjacent to that leg.

5.4.4 (Pythagorean Theorem) The square of the length of the hypotenuse of a right triangle is equal to the sum of the squares of the lengths of the legs.

5.4.5 (Converse of Pythagorean Theorem) If $a$, $b$, and $c$ are the lengths of the three sides of a triangle, with $c$ the length of the longest side, and if $c^2 = a^2 + b^2$, then the triangle is a right triangle with the right angle opposite the side of length $c$.

5.4.6 If the hypotenuse and a leg of one right triangle are congruent to the hypotenuse and a leg of a second right triangle, then the triangles are congruent (HL).

5.4.7 Let $a$, $b$, and $c$ represent the lengths of the three sides of a triangle with $c$ the length of the longest side.

1. If $c^2 > a^2 + b^2$, then the triangle is obtuse and the obtuse angle lies opposite the side of length $c$.
2. If $c^2 < a^2 + b^2$, then the triangle is acute.

5.5.1 (45-45-90 Theorem) In a triangle whose angles measure 45°, 45°, and 90°, the hypotenuse has a length equal to the product of $\sqrt{2}$ and the length of either leg.

5.5.2 (30-60-90 Theorem) In a triangle whose angles measure 30°, 60°, and 90°, the hypotenuse has a length equal to twice the length of the shorter leg, and the length of the longer leg is the product of $\sqrt{3}$ and the length of the shorter leg.

5.5.3 If the length of the hypotenuse of a right triangle equals the product of $\sqrt{2}$ and the length of either leg, then the angles of the triangle measure 45°, 45°, and 90°.

5.5.4 If the length of the hypotenuse of a right triangle is twice the length of one leg of the triangle, then the angle of the triangle opposite that leg measures 30°.

5.6.1 If a line is parallel to one side of a triangle and intersects the other two sides, then it divides these sides proportionally.

5.6.2 When three (or more) parallel lines are cut by a pair of transversals, the transversals are divided proportionally by the parallel lines.

5.6.3 (The Angle-Bisector Theorem) If a ray bisects one angle of a triangle, then it divides the opposite side into segments whose lengths are proportional to the two sides that form the bisected angle.

5.6.4 (Ceva's Theorem) Let $D$ be any point in the interior of $\triangle ABC$ and let $\overline{BE}$, $\overline{AF}$, and $\overline{CG}$ be determined by $D$ and the vertices of $\triangle ABC$. Then the product of the ratios of lengths of segments of the sides (taken in order) equals 1; that is, $\frac{AG}{GB} \cdot \frac{BF}{FC} \cdot \frac{CE}{EA} = 1$.

6.1.1 A radius that is perpendicular to a chord bisects the chord.

6.1.2 The measure of an inscribed angle of a circle is one-half the measure of its intercepted arc.

6.1.3 In a circle (or in congruent circles), congruent minor arcs have congruent central angles.

6.1.4 In a circle (or in congruent circles), congruent central angles have congruent arcs.

6.1.5 In a circle (or in congruent circles), congruent chords have congruent minor (major) arcs.

6.1.6 In a circle (or in congruent circles), congruent arcs have congruent chords.

6.1.7 Chords that are at the same distance from the center of a circle are congruent.

6.1.8 Congruent chords are located at the same distance from the center of a circle.

6.1.9 An angle inscribed in a semicircle is a right angle.

6.1.10 If two inscribed angles intercept the same arc, then these angles are congruent.

6.2.1 If a quadrilateral is inscribed in a circle, the opposite angles are supplementary.
*(Alternative)* The opposite angles of a cyclic quadrilateral are supplementary.

6.2.2 The measure of an angle formed by two chords that intersect within a circle is one-half the sum of the measures of the arcs intercepted by the angle and its vertical angle.

6.2.3 The radius (or any other line through the center of a circle) drawn to a tangent at the point of tangency is perpendicular to the tangent at that point.

6.2.4 The measure of an angle formed by a tangent and a chord drawn to the point of tangency is one-half the measure of the intercepted arc.

6.2.5 The measure of an angle formed when two secants intersect at a point outside the circle is one-half the difference of the measures of the two intercepted arcs.

6.2.6 If an angle is formed by a secant and tangent that intersect in the exterior of a circle, then the measure of the angle is one-half the difference of the measures of its intercepted arcs.

6.2.7 If an angle is formed by two intersecting tangents, then the measure of the angle is one-half the difference of the measures of the intercepted arcs.

6.2.8 If two parallel lines intersect a circle, the intercepted arcs between these lines are congruent.

6.3.1 If a line is drawn through the center of a circle perpendicular to a chord, then it bisects the chord and its arc.

6.3.2 If a line through the center of a circle bisects a chord other than a diameter, then it is perpendicular to the chord.

6.3.3 The perpendicular bisector of a chord contains the center of the circle.

6.3.4 The tangent segments to a circle from an external point are congruent.

6.3.5 If two chords intersect within a circle, then the product of the lengths of the segments (parts) of one chord is equal to the product of the lengths of the segments of the other chord.

6.3.6 If two secant segments are drawn to a circle from an external point, then the products of the lengths of each secant with its external segment are equal.

6.3.7 If a tangent segment and a secant segment are drawn to a circle from an external point, then the square of the length of the tangent equals the product of the length of the secant with the length of its external segment.

6.4.1 The line that is perpendicular to the radius of a circle at its endpoint on the circle is a tangent to the circle.

6.4.2 In a circle (or in congruent circles) containing two unequal central angles, the larger angle corresponds to the larger intercepted arc.

6.4.3 In a circle (or in congruent circles) containing two unequal arcs, the larger arc corresponds to the larger central angle.

6.4.4 In a circle (or in congruent circles) containing two unequal chords, the shorter chord is at the greater distance from the center of the circle.

6.4.5 In a circle (or in congruent circles) containing two unequal chords, the chord nearer the center of the circle has the greater length.

6.4.6 In a circle (or in congruent circles) containing two unequal chords, the longer chord corresponds to the greater minor arc.

6.4.7 In a circle (or in congruent circles) containing two unequal minor arcs, the greater minor arc corresponds to the longer of the chords related to these arcs.

6.5.1 The locus of points in a plane and equidistant from the sides of an angle is the angle bisector.

6.5.2 The locus of points in a plane that are equidistant from the endpoints of a line segment is the perpendicular bisector of that line segment.

6.6.1 The three angle bisectors of the angles of a triangle are concurrent.

6.6.2 The three perpendicular bisectors of the sides of a triangle are concurrent.

6.6.3 The three altitudes of a triangle are concurrent.

6.6.4 The three medians of a triangle are concurrent at a point that is two-thirds the distance from any vertex to the midpoint of the opposite side.

7.1.1 The area $A$ of a square whose sides are each of length $s$ is given by $A = s^2$.

7.1.2 The area $A$ of a parallelogram with a base of length $b$ and with corresponding altitude of length $h$ is given by

$$A = bh$$

7.1.3 The area $A$ of a triangle whose base has length $b$ and whose corresponding altitude has length $h$ is given by

$$A = \frac{1}{2}bh$$

7.1.4 The area $A$ of a right triangle with legs of lengths $a$ and $b$ is given $A = \frac{1}{2}ab$.

7.2.1 (Heron's Formula) If the three sides of a triangle have lengths $a$, $b$, and $c$, then the area $A$ of the triangle is given by

$$A = \sqrt{s(s-a)(s-b)(s-c)}$$

where the semiperimeter of the triangle is

$$s = \frac{1}{2}(a+b+c)$$

7.2.2 (Brahmagupta's Formula) For a cyclic quadrilateral with sides of lengths $a$, $b$, $c$, and $d$, the area $A$ is given by

$$A = \sqrt{(s-a)(s-b)(s-c)(s-d)}$$

where the semiperimeter of the quadrilateral is

$$s = \frac{1}{2}(a+b+c+d)$$

7.2.3 The area $A$ of a trapezoid whose bases have lengths $b_1$ and $b_2$ and whose altitude has length $h$ is given by

$$A = \frac{1}{2}h(b_1 + b_2)$$

7.2.4 The area $A$ of any quadrilateral with perpendicular diagonals of lengths $d_1$ and $d_2$ is given by

$$A = \frac{1}{2}d_1d_2$$

7.2.5 The area $A$ of a rhombus whose diagonals have lengths $d_1$ and $d_2$ is given by

$$A = \frac{1}{2}d_1d_2$$

7.2.6 The area $A$ of a kite whose diagonals have lengths $d_1$ and $d_2$ is given by

$$A = \frac{1}{2}d_1d_2$$

7.2.7 The ratio of the areas of two similar triangles equals the square of the ratio of the lengths of any two corresponding sides; that is,

$$\frac{A_1}{A_2} = \left(\frac{a_1}{a_2}\right)^2$$

7.3.1 A circle can be circumscribed about (or inscribed in) any regular polygon.

7.3.2 The measure of the central angle of a regular polygon of $n$ sides is given by $c = \frac{360}{n}$.

7.3.3 Any radius of a regular polygon bisects the angle at the vertex to which it is drawn.

7.3.4 Any apothem of a regular polygon bisects the side of the polygon to which it is drawn.

7.3.5 The area $A$ of a regular polygon whose apothem has length $a$ and whose perimeter is $P$ is given by

$$A = \frac{1}{2}aP$$

7.4.1 The circumference $C$ of a circle is given by the formula

$$C = \pi d \qquad \text{or} \qquad C = 2\pi r$$

7.4.2 In a circle whose circumference is $C$, the length $\ell$ of an arc whose degree measure is $m$ is given by

$$\ell = \frac{m}{360} \cdot C$$

7.4.3 The area $A$ of a circle whose radius is of length $r$ is given by $A = \pi r^2$.

7.5.1 In a circle of radius $r$, the area $A$ of a sector whose arc has degree measure $m$ is given by

$$A = \frac{m}{360}\pi r^2$$

7.5.2 The area of a semicircular region of radius $r$ is $A = \frac{1}{2}\pi r^2$.

7.5.3 Where $P$ represents the perimeter of a triangle and $r$ represents the length of the radius of its inscribed circle, the area $A$ of the triangle is given by

$$A = \frac{1}{2}rP$$

8.1.1 The lateral area $L$ of any prism whose altitude has measure $h$ and whose base has perimeter $P$ is given by $L = hP$.

8.1.2 The total area $T$ of any prism with lateral area $L$ and base area $B$ is given by $T = L + 2B$.

8.2.1 In a regular pyramid, the length $a$ of the apothem of the base, the altitude $h$, and the slant height $\ell$ satisfy the Pythagorean Theorem; that is, $\ell^2 = a^2 + h^2$ in every regular pyramid.

8.2.2 The lateral area $L$ of a regular pyramid with slant height of length $\ell$ and perimeter $P$ of the base is given by

$$L = \frac{1}{2}\ell P$$

8.2.3 The total area (surface area) $T$ of a pyramid with lateral area $L$ and base area $B$ is given by $T = L + B$.

8.2.4 The volume $V$ of a pyramid having a base area $B$ and an altitude of length $h$ is given by

$$V = \frac{1}{3}Bh$$

8.2.5 In a regular pyramid, the lengths of altitude $h$, radius $r$ of the base, and lateral edge $e$ satisfy the Pythagorean Theorem; that is, $e^2 = h^2 + r^2$.

8.3.1 The lateral area $L$ of a right circular cylinder with altitude of length $h$ and circumference $C$ of the base is given by $L = hC$.
*(Alternative)* Where $r$ is the length of the radius of the base, $L = 2\pi rh$.

8.3.2 The total area $T$ of a right circular cylinder with base area $B$ and lateral area $L$ is given by $T = L + 2B$.
*(Alternative)* Where $r$ is the length of the radius of the base and $h$ is the length of the altitude, $T = 2\pi rh + 2\pi r^2$.

8.3.3 The volume $V$ of a right circular cylinder with base area $B$ and altitude of length $h$ is given by $V = Bh$.
*(Alternative)* Where $r$ is the length of the radius of the base, $V = \pi r^2h$.

8.3.4 The lateral area $L$ of a right circular cone with slant height of length $\ell$ and circumference $C$ of the base is given by $L = \frac{1}{2}\ell C$.
*(Alternative)* Where $r$ is the length of the radius of the base, $L = \pi r\ell$.

8.3.5 The total area $T$ of a right circular cone with base area $B$ and lateral area $L$ is given by $T = B + L$.
*(Alternative)* Where $r$ is the length of the radius of the base and $\ell$ is the length of the slant height, $T = \pi r^2 + \pi r\ell$.

8.3.6 In a right circular cone, the lengths of the radius $r$ (of the base), the altitude $h$, and the slant height $\ell$ satisfy the Pythagorean Theorem; that is, $\ell^2 = r^2 + h^2$ in every right circular cone.

8.3.7 The volume $V$ of a right circular cone with base area $B$ and altitude of length $h$ is given by $V = \frac{1}{3}Bh$.
*(Alternative)* Where $r$ is the length of the radius of the base, $V = \frac{1}{3}\pi r^2 h$.

8.4.1 (Euler's Equation) The number of vertices $V$, the number of edges $E$, and the number of faces $F$ of a polyhedron are related by the equation $V + F = E + 2$.

8.4.2 The surface area $S$ of a sphere whose radius has length $r$ is given by $S = 4\pi r^2$.

8.4.3 The volume $V$ of a sphere with radius of length $r$ is given by $V = \frac{4}{3}\pi r^3$.

9.1.1 (Distance Formula) The distance $d$ between two points $(x_1, y_1)$ and $(x_2, y_2)$ is given by the formula

$$d = \sqrt{(x_2 - x_1)^2 + (y_2 - y_1)^2}$$

9.1.2 (Midpoint Formula) The midpoint $M$ of the line segment joining $(x_1, y_1)$ and $(x_2, y_2)$ has coordinates $x_M$ and $y_M$, where

$$(x_M, y_M) = \left(\frac{x_1 + x_2}{2}, \frac{y_1 + y_2}{2}\right)$$

That is, $$M = \left(\frac{x_1 + x_2}{2}, \frac{y_1 + y_2}{2}\right)$$

9.2.1 If two nonvertical lines are parallel, then their slopes are equal.
*(Alternative)* If $\ell_1 \parallel \ell_2$, then $m_1 = m_2$

9.2.2 If two lines (neither horizontal nor vertical) are perpendicular, then the product of their slopes is $-1$.
*(Alternative)* If $\ell_1 \perp \ell_2$, then $m_1 \cdot m_2 = -1$

9.4.1 The line segment determined by the midpoints of two sides of a triangle is parallel to the third side.

9.4.2 The diagonals of a parallelogram bisect each other.

9.4.3 The diagonals of a rhombus are perpendicular.

9.4.4 If the diagonals of a parallelogram are equal in length, then the parallelogram is a rectangle.

9.5.1 (Slope-Intercept Form of a Line) The line whose slope is $m$ and whose $y$ intercept is $b$ has the equation $y = mx + b$.

9.5.2 (Point-Slope Form of a Line) The line that has slope $m$ and contains the point $(x_1, y_1)$ has the equation

$$y - y_1 = m(x - x_1)$$

9.5.3 The three medians of a triangle are concurrent at a point that is two-thirds the distance from any vertex to the midpoint of the opposite side.

10.2.1 In any right triangle in which $\alpha$ is the measure of an acute angle,

$$\sin^2\alpha + \cos^2\alpha = 1$$

10.4.1 The area of any acute triangle equals one-half the product of the lengths of two sides and the sine of the included angle. That is,

$$A = \frac{1}{2}ab \sin \gamma$$
$$A = \frac{1}{2}ac \sin \beta$$
$$A = \frac{1}{2}bc \sin \alpha$$

10.4.2 (Law of Sines) In any acute triangle, the three ratios between the sines of the angles and the lengths of the opposite sides are equal. That is,

$$\frac{\sin \alpha}{a} = \frac{\sin \beta}{b} = \frac{\sin \gamma}{c}$$

10.4.3 (Law of Cosines) In acute triangle $ABC$,

$$c^2 = a^2 + b^2 - 2ab \cos \gamma$$
$$b^2 = a^2 + c^2 - 2ac \cos \beta$$
$$a^2 = b^2 + c^2 - 2bc \cos \alpha$$

# Answers

## Selected Exercises and Proofs

## Chapter 1

### 1.1 EXERCISES

**1.** (a) Not a statement (b) Statement; true (c) Statement; true (d) Statement; false
**3.** (a) Christopher Columbus did not cross the Atlantic Ocean. (b) Some jokes are not funny.
**5.** Conditional **7.** Simple **9.** Simple
**11.** H: You go to the game. C: You will have a great time.
**13.** H: The diagonals of a parallelogram are perpendicular. C: The parallelogram is a rhombus.
**15.** H: Two parallel lines are cut by a transversal. C: Corresponding angles are congruent.
**17.** First write the statement in "If, then" form: If a figure is a square, then it is a rectangle. H: A figure is a square. C: It is a rectangle.
**19.** True **21.** True **23.** False **25.** Induction
**27.** Deduction **29.** Intuition **31.** None
**33.** Angle 1 looks equal in measure to angle 2.
**35.** The three angles in one triangle are equal in measure to the corresponding three angles in the other triangle.
**37.** *A Prisoner of Society* might be nominated for an Academy Award. **39.** The instructor is a math teacher.
**41.** Angles 1 and 2 are complementary.
**43.** Alex has a strange sense of humor. **45.** None
**47.** June Jesse will be in the public eye.
**49.** Marilyn is a happy person. **51.** Valid **53.** Not valid **55.** (a) True (b) True (c) False

### 1.2 EXERCISES

**1.** $AB < CD$ **3.** Two; one **5.** One; none
**7.** $\angle ABC$, $\angle ABD$, $\angle DBC$ **9.** Yes; no; yes
**11.** $\angle ABC$, $\angle CBA$ **13.** Yes; no **15.** a, d
**17.** (a) 3 (b) $2\frac{1}{2}$ **19.** (a) 40° (b) 50°
**21.** Congruent; congruent **23.** Equal **25.** No
**27.** Yes **29.** Congruent **31.** $\overline{MN}$ and $\overline{QP}$ **33.** $\overline{AB}$
**35.** 22 **37.** $x = 9$ **39.** 124° **41.** 71° **43.** $x = 23$
**45.** 10.9 **47.** $x = 102$; $y = 78$ **49.** N 22° E

### 1.3 EXERCISES

**1.** $AC$ **3.** 75 in. **5.** 1.64 ft **7.** 3 mi
**9.** (a) $A$-$C$-$D$ (b) $A, B, C$ or $B, C, D$ or $A, B, D$
**11.** $\overleftrightarrow{CD}$ means line $CD$; $\overline{CD}$ means segment $CD$; $CD$ means the measure or length of $\overline{CD}$; $\overrightarrow{CD}$ means ray $CD$ with endpoint $C$.
**13.** (a) $m$ and $t$ (b) $m$ and $\overleftrightarrow{AD}$ or $\overleftrightarrow{AD}$ and $t$
**15.** $x = 3$; $AM = 7$ **17.** $x = 7$; $AB = 38$
**19.** (a) $\overrightarrow{OA}$ and $\overrightarrow{OD}$ (b) $\overrightarrow{OA}$ and $\overrightarrow{OB}$ (There are other possible answers.) **23.** Planes $M$ and $N$ intersect at $\overleftrightarrow{AB}$.
**25.** $A$ **27.** (a) $C$ (b) $C$ (c) $H$
**33.** (a) No (b) Yes (c) No (d) Yes
**35.** Six **37.** Nothing

### 1.4 EXERCISES

**1.** (a) Acute (b) Right (c) Obtuse
**3.** (a) Complementary (b) Supplementary
**5.** Adjacent angles
**7.** Complementary angles (also adjacent)
**9.** Yes; no
**11.** (a) Obtuse (b) Straight (c) Acute (d) Obtuse
**13.** $m\angle FAC + m\angle CAD = 180$; $\angle FAC$ and $\angle CAD$ are supplementary. **15.** (a) $x + y = 90$ (b) $x = y$
**17.** 42° **19.** $x = 20$; $m\angle RSV = 56°$
**21.** $x = 24$; $y = 8$ **23.** $\angle CAB \cong \angle DAB$
**25.** $\angle$s measure 128° and 52°.
**27.** (a) $(180 - x)°$ (b) $(192 - 3x)°$ (c) $(180 - 2x - 5y)°$
**29.** $x = 143$
**35.** It appears that the angle bisectors meet at one point.
**37.** It appears that the two sides opposite $\angle$s $A$ and $B$ are congruent. **39.** (a) 90° (b) 90° (c) Equal **41.** 135°

### 1.5 EXERCISES

**1.** Division (or Multiplication) Prop. of Equality
**3.** Subtraction Prop. of Equality
**5.** Multiplication Prop. of Equality
**7.** If 2$\angle$s are supp., the sum of their measures is 180°.
**9.** Angle-Addition Postulate
**11.** $AM + MB = AB$ **13.** $\overrightarrow{EG}$ bisects $\angle DEF$.
**15.** $m\angle 1 + m\angle 2 = 90°$ **17.** $2x = 10$ **19.** $7x + 2 = 30$
**21.** $6x - 3 = 27$ **23.** 1. Given 2. Distributive Prop. 3. Addition Prop. of Equality 4. Division Prop. of Equality
**25.** 1. $2(x + 3) - 7 = 11$ 2. $2x + 6 - 7 = 11$ 3. $2x - 1 = 11$ 4. $2x = 12$ 5. $x = 6$
**27.** 1. Given 2. Segment-Addition Postulate 3. Subtraction Prop. of Equality
**29.** 1. Given 2. Definition of angle bisector 3. Angle-Addition Postulate 4. Substitution 5. Substitution (Distribution) 6. Multiplication Prop. of Equality

**31.** S1. *M-N-P-Q* on $\overline{MQ}$ R1. Given 2. Segment-Addition Postulate 3. Segment-Addition Postulate 4. $MN + NP + PQ = MQ$ **33.** $5(x + y)$
**35.** $(-7)(-2) > 5(-2)$ or $14 > -10$

## 1.6 EXERCISES

**1.** 1. Given 2. If two ∠s are ≅ , then they are equal in measure. 3. Angle-Addition Postulate 4. Addition Property of Equality 5. Substitution 6. If two ∠s are equal in measure, then they are ≅.
**3.** 1. ∠1 ≅ ∠2 and ∠2 ≅ ∠3 2. ∠1 ≅ ∠3
**11.** 1. Given 3. Substitution 4. m∠1 = m∠2 5. ∠1 ≅ ∠2 **13.** No; yes; no **15.** No; yes; no
**17.** No; yes; yes **19.** (a) Perpendicular (b) Angles (c) Supplementary (d) Right (e) Measure of angle 1 (f) Adjacent (g) Complementary (h) Ray AB (i) Is congruent to (j) Vertical

**21.** PROOF

| Statements | Reasons |
|---|---|
| 1. *M-N-P-Q* on $\overline{MQ}$ | 1. Given |
| 2. $MN + NQ = MQ$ | 2. Segment-Addition Postulate |
| 3. $NP + PQ = NQ$ | 3. Segment-Addition Postulate |
| 4. $MN + NP + PQ = MQ$ | 4. Substitution |

**23.** PROOF

| Statements | Reasons |
|---|---|
| 1. ∠*TSW* with $\overrightarrow{SU}$ and $\overrightarrow{SV}$ | 1. Given |
| 2. m∠*TSW* = m∠*TSU* + m∠*USW* | 2. Angle-Addition Postulate |
| 3. m∠*USW* = m∠*USV* + m∠*VSW* | 3. Angle-Addition Postulate |
| 4. m∠*TSW* = m∠*TSU* + m∠*USV* + m∠*VSW* | 4. Substitution |

**25.** In space, there are an infinite number of lines that perpendicularly bisect a given line segment at its midpoint.

## 1.7 EXERCISES

**1.** H: A line segment is bisected. C: Each of the equal segments has half the length of the original segment.
**3.** First write the statement in "If, then" form. If a figure is a square, then it is a quadrilateral. H: A figure is a square. C: It is a quadrilateral.
**5.** H: Each angle is a right angle. C: Two angles are congruent. **7.** Statement, Drawing, Given, Prove, Proof

**9.** *Given:* $\overleftrightarrow{AB} \perp \overleftrightarrow{CD}$
*Prove:* ∠*AEC* is a right angle

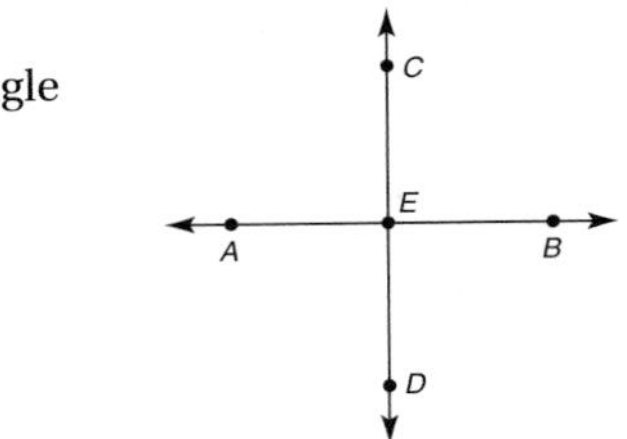

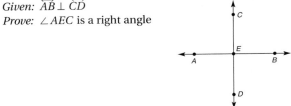

**11.** *Given:* ∠1 is comp. to ∠3; ∠2 is comp. to ∠3
*Prove:* ∠1 ≅ ∠2

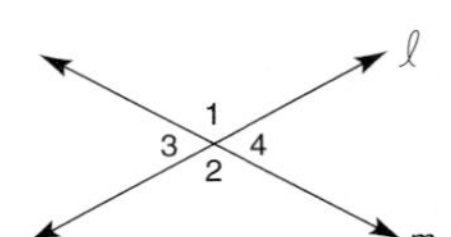


**13.** *Given:* Lines ℓ and *m* intersect as shown
*Prove:* ∠1 ≅ ∠2 and ∠3 ≅ ∠4

**15.** m∠2 = 55°; m∠3 = 125°; m∠4 = 55°
**17.** $x = 40$; m∠1 = 130°
**19.** $x = 60$; m∠1 = 120°
**21.** 1. Given 2. If two ∠s are complementary, the sum of their measures is 90. 3. Substitution 4. Subtraction Property of Equality 5. If two ∠s are equal in measure, then they are ≅. **25.** 1. Given 2. ∠*ABC* is a right ∠. 3. The measure of a rt. ∠ = 90. 4. Angle-Addition Postulate 6. ∠1 is comp. to ∠2.

## 1.7 SELECTED PROOF

**27.** PROOF

| Statements | Reasons |
|---|---|
| 1. ∠*ABC* ≅ ∠*EFG* | 1. Given |
| 2. m∠*ABC* = m∠*EFG* | 2. If two ∠s are ≅, their measures are = |
| 3. m∠*ABC* = m∠1 + m∠2<br>m∠*EFG* = m∠3 + m∠4 | 3. Angle-Addition Postulate |
| 4. m∠1 + m∠2 = m∠3 + m∠4 | 4. Substitution |
| 5. $\overrightarrow{BD}$ bisects ∠*ABC*<br>$\overrightarrow{FH}$ bisects ∠*EFG* | 5. Given |
| 6. m∠1 = m∠2 and m∠3 = m∠4 | 6. If a ray bisects an ∠, then two ∠s of equal measure are formed |
| 7. m∠1 + m∠1 = m∠3 + m∠3 or 2 · m∠1 = 2 · m∠3 | 7. Substitution |
| 8. m∠1 = m∠3 | 8. Division Prop. of Equality |
| 9. m∠1 = m∠2 = m∠3 = m∠4 | 9. Substitution (or Transitive) |
| 10. ∠1 ≅ ∠2 ≅ ∠3 ≅ ∠4 | 10. If ∠s are = in measure, then they are ≅ |

## CHAPTER 1 REVIEW EXERCISES

**1.** Undefined terms, defined terms, axioms or postulates, theorems **2.** Induction, deduction, intuition
**3.** 1. Names the term being defined 2. Places the term into a set or category 3. Distinguishes the term from other terms in the same category 4. Reversible
**4.** Intuition **5.** Induction **6.** Deduction
**7.** H: The diagonals of a trapezoid are equal in length. C: The trapezoid is isosceles.
**8.** H: The parallelogram is a rectangle. C: The diagonals of a parallelogram are congruent.
**9.** No conclusion **10.** Jody Smithers has a college degree.
**11.** Angle *A* is a right angle. **12.** *C*
**13.** $\angle RST$; $\angle S$; greater than 90° **14.** Perpendicular
**18.** 98° **19.** 47° **20.** 22 **21.** 17 **22.** 34
**23.** 152° **24.** 39°
**25.** (a) Point *M* (b) $\angle JMH$ (c) $\overrightarrow{MJ}$ (d) $\overleftrightarrow{KH}$
**26.** $67\frac{1}{2}°$ **27.** 28° and 152°
**28.** (a) $6x + 8$ (b) $x = 4$ (c) 11; 10; 11
**29.** The measure of angle 3 is less than 50°. **30.** 10 pegs
**31.** S **32.** S **33.** A **34.** S **35.** N
**36.** 2. $\angle 4 \cong \angle P$ 3. $\angle 1 \cong \angle 4$ 4. If two $\angle$s are $\cong$, then their measures are =. 5. Given 6. $m\angle 2 = m\angle 3$ 7. $m\angle 1 + m\angle 2 = m\angle 4 + m\angle 3$ 8. Angle-Addition Postulate 9. Substitution 10. $\angle TVP \cong \angle MVP$
**50.** 270°

## CHAPTER 1 REVIEW EXERCISES SELECTED PROOFS

**37.** PROOF

| Statements | Reasons |
|---|---|
| 1. $\overline{KF} \perp \overline{FH}$ | 1. Given |
| 2. $\angle KFH$ is a rt. $\angle$ | 2. If two segments are $\perp$, then they form a rt. $\angle$ |
| 3. $\angle JHF$ is a rt. $\angle$ | 3. Given |
| 4. $\angle KFH \cong \angle JHF$ | 4. Any two rt. $\angle$s are $\cong$ |

**38.** PROOF

| Statements | Reasons |
|---|---|
| 1. $\overline{KH} \cong \overline{FJ}$; *G* is the midpoint of both $\overline{KH}$ and $\overline{FJ}$ | 1. Given |
| 2. $\overline{KG} \cong \overline{GJ}$ | 2. If two segments are $\cong$, then their midpoints separate these segments into four $\cong$ segments |

**39.** PROOF

| Statements | Reasons |
|---|---|
| 1. $\overline{KF} \perp \overline{FH}$ | 1. Given |
| 2. $\angle KFJ$ is comp. to $\angle JFH$ | 2. If the exterior sides of two adjacent $\angle$s form $\perp$ rays, then these $\angle$s are comp. |

**40.** PROOF

| Statements | Reasons |
|---|---|
| 1. $\angle 1$ is comp. to $\angle M$ | 1. Given |
| 2. $\angle 2$ is comp. to $\angle M$ | 2. Given |
| 3. $\angle 1 \cong \angle 2$ | 3. If two $\angle$s are comp. to the same $\angle$, then these angles are $\cong$ |

**41.** PROOF

| Statements | Reasons |
|---|---|
| 1. $\angle MOP \cong \angle MPO$ | 1. Given |
| 2. $\overrightarrow{OR}$ bisects $\angle MOP$; $\overrightarrow{PR}$ bisects $\angle MPO$ | 2. Given |
| 3. $\angle 1 \cong \angle 2$ | 3. If two $\angle$s are $\cong$, then their bisectors separate these $\angle$s into four $\cong$ $\angle$s |

**42.** PROOF

| Statements | Reasons |
|---|---|
| 1. $\angle 4 \cong \angle 6$ | 1. Given |
| 2. $\angle 4 \cong \angle 5$ | 2. If two angles are vertical $\angle$s, then they are $\cong$ |
| 3. $\angle 5 \cong \angle 6$ | 3. Transitive Property |

**43.** PROOF

| Statements | Reasons |
|---|---|
| 1. Figure as shown | 1. Given |
| 2. $\angle 4$ is supp. to $\angle 2$ | 2. If the exterior sides of two adjacent $\angle$s form a line, then the $\angle$s are supp. |

**44.** PROOF

| Statements | Reasons |
|---|---|
| 1. $\angle 3$ is supp. to $\angle 5$<br>$\angle 4$ is supp. to $\angle 6$ | 1. Given |
| 2. $\angle 4 \cong \angle 5$ | 2. If two lines intersect, the vertical angles formed are $\cong$ |
| 3. $\angle 3 \cong \angle 6$ | 3. If two $\angle$s are supp. to congruent angles, then these angles are $\cong$ |

### CHAPTER 1 TEST

**1.** Induction [1.1] **2.** $\angle CBA$ or $\angle B$ [1.4] **3.** $AP + PB = AB$ [1.3] **4.** (a) Point (b) Line [1.3] **5.** (a) Right (b) Obtuse [1.4] **6.** (a) Supplementary (b) Congruent [1.4] **7.** $m\angle MNP = m\angle PNQ$ [1.4] **8.** (a) Right (b) Supplementary [1.7] **9.** Kianna will develop reasoning skills. [1.1] **10.** 10.4 in. [1.2] **11.** (a) 11 (b) 16 [1.3] **12.** 35° [1.4] **13.** (a) 24° (b) 45° [1.4] **14.** (a) 137° (b) 43° [1.4] **15.** (a) 25° (b) 47° [1.7] **16.** (a) 23° (b) 137° [1.7] **17.** $x + y = 90$ [1.4] **20.** 1. Given 2. Segment-Addition Postulate 3. Segment-Addition Postulate 4. Substitution [1.5] **21.** 1. $2x - 3 = 17$ 2. $2x = 20$ 3. $x = 10$ [1.5] **22.** 1. Given 2. 90° 3. Angle-Addition Postulate 4. 90° 5. Given 6. Definition of Angle-Bisector 7. Substitution 8. $m\angle 1 = 45°$ [1.7]

# Chapter 2

### 2.1 EXERCISES

**1.** (a) 108° (b) 72° **3.** (a) 68.3° (b) 68.3° **5.** (a) No (b) Yes (c) Yes **7.** Angle 9 appears to be a right angle. **9.** (a) $m\angle 3 = 87°$ (b) $m\angle 6 = 87°$ (c) $m\angle 1 = 93°$ (d) $m\angle 7 = 87°$ **11.** (a) $\angle 5$ (b) $\angle 5$ (c) $\angle 8$ (d) $\angle 5$ **13.** (a) $m\angle 2 = 68°$ (b) $m\angle 4 = 112°$ (c) $m\angle 5 = 112°$ (d) $m\angle MOQ = 34°$ **15.** $x = 10$; $m\angle 4 = 110°$ **17.** $x = 12$; $y = 4$; $m\angle 7 = 76°$ **19.** 1. Given 2. If two parallel lines are cut by a transversal, then the corresponding angles are $\cong$ 3. If two lines intersect, then the vertical angles are $\cong$ 4. $\angle 3 \cong \angle 4$ 5. $\angle 1 \cong \angle 4$ **25.** (a) $\angle 4 \cong \angle 2$ and $\angle 5 \cong \angle 3$ (b) 180° (c) 180° **29.** No

### 2.1 SELECTED PROOF

**21.** PROOF

| Statements | Reasons |
|---|---|
| 1. $\overleftrightarrow{CE} \parallel \overleftrightarrow{DF}$; transversal $\overleftrightarrow{AB}$ | 1. Given |
| 2. $\angle ACE \cong \angle ADF$ | 2. If two $\parallel$ lines are cut by a transversal, then the corresponding $\angle$s are $\cong$ |
| 3. $\overrightarrow{CX}$ bisects $\angle ACE$<br>$\overrightarrow{DE}$ bisects $\angle CDF$ | 3. Given |
| 4. $\angle 1 \cong \angle 3$ | 4. If two $\angle$s are $\cong$, then their bisectors separate these $\angle$s into four $\cong \angle$s |

### 2.2 EXERCISES

**1.** *Converse:* If Juan is rich, then he won the state lottery. FALSE.
*Inverse:* If Juan does not win the state lottery, then he will not be rich. FALSE.
*Contrapositive:* If Juan is not rich, then he did not win the state lottery. TRUE.
**3.** *Converse:* If two angles are complementary, then the sum of their measures is 90°. TRUE.
*Inverse:* If the sum of the measures of two angles is not 90°, then the two angles are not complementary. TRUE.
*Contrapositive:* If two angles are not complementary, then the sum of their measures is not 90°. TRUE.
**5.** No conclusion **7.** $x = 5$ **9.** (a), (b), and (e) **11.** Parallel

### 2.2 SELECTED PROOFS

**13.** Assume that $r \parallel s$. Then $\angle 1 \cong \angle 5$ because they are corresponding angles. But it is given that $\angle 1 \not\cong \angle 5$, which leads to a contradiction. Thus, the assumption that $r \parallel s$ is false and it follows that $r \nparallel s$.
**15.** Assume that $\overrightarrow{FH} \perp \overleftrightarrow{EG}$. Then $\angle 3 \cong \angle 4$ and $m\angle 3 = m\angle 4$. But it is given that $m\angle 3 > m\angle 4$, which leads to a contradiction. Then the assumption that $\overrightarrow{FH} \perp \overleftrightarrow{EG}$ must be false and it follows that $\overrightarrow{FH}$ is not perpendicular to $\overleftrightarrow{EG}$.
**17.** Assume that the angles are vertical angles. If they are vertical angles, then they are congruent. But this contradicts the hypothesis that the two angles are not congruent. Hence, our assumption must be false, and the angles are not vertical angles.
**21.** If $M$ is a midpoint of $\overline{AB}$, then $AM = \frac{1}{2}(AB)$. Assume that $N$ is also a midpoint of $\overline{AB}$ so that $AN = \frac{1}{2}(AB)$. By substitution, $AM = AN$. By the Segment-Addition Postulate, $AM = AN + NM$. Using substitution again, $AN + NM = AN$. Subtracting gives $NM = 0$. But this contradicts the Ruler Postulate, which states that the measure of a line segment is a positive number. Therefore, our assumption is wrong and $M$ is the only midpoint for $\overline{AB}$.

**2.3 EXERCISES**

**1.** $\ell \parallel m$ **3.** $\ell \nparallel m$ **5.** $\ell \nparallel m$ **7.** $p \parallel q$ 9. None **11.** $\ell \parallel n$ **13.** None **15.** $\ell \parallel n$ **17.** 1. Given 2. If two ∠s are comp. to the same ∠, then they are ≅ 3. $\overline{BC} \parallel \overline{DE}$ **23.** $x = 20$ **25.** $x = 9$ **27.** $x = 6$

**2.3 SELECTED PROOF**

**19.** PROOF

| Statements | Reasons |
|---|---|
| 1. $\overline{AD} \perp \overline{DC}$ and $\overline{BC} \perp \overline{DC}$ | 1. Given |
| 2. $\overline{AD} \parallel \overline{BC}$ | 2. If two lines are each ⊥ to a third line, then these lines are ∥ to each other |

**2.4 EXERCISES**

**1.** $m\angle C = 75°$ **3.** $m\angle B = 46°$ **5.** (a) Underdetermined (b) Determined (c) Overdetermined **7.** (a) Equilateral (b) Isosceles **9.** (a) Equiangular (b) Right **11.** If two ∠s of one triangle are ≅ to two ∠s of another triangle, then the third ∠s of the triangles are ≅. **13.** $m\angle 1 = 122°$; $m\angle 2 = 58°$; $m\angle 5 = 72°$ **15.** $m\angle 2 = 57.7°$; $m\angle 3 = 80.8°$; $m\angle 4 = 41.5°$ **17.** 35° **19.** 40° **21.** 360° **23.** $x = 45°$; $y = 45°$ **25.** $y = 20°$; $x = 100°$; $m\angle 5 = 60°$ **31.** 44° **33.** $m\angle N = 49°$; $m\angle P = 98°$ **35.** 35° **37.** 75° **45.** $m\angle M = 84°$

**2.5 EXERCISES**

**1.** Increase **3.** $x = 113°$; $y = 67°$; $z = 36°$ **5.** (a) 5 (b) 35 **7.** (a) 540° (b) 1440° **9.** (a) 90° (b) 150° **11.** (a) 90° (b) 30° **13.** (a) 7 (b) 9 **15.** (a) $n = 5$ (b) $n = 10$ **17.** (a) 15 (b) 20 **19.** 135°

**21.** 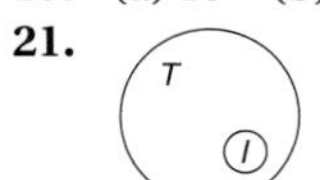


**23.** 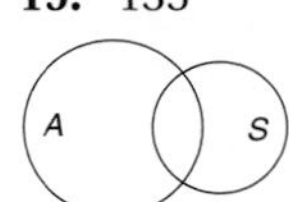


**25.** 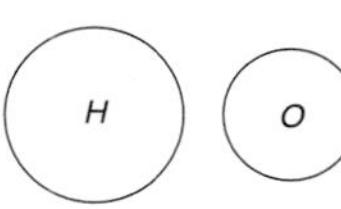


**31.** Figure (a): 90°, 90°, 120°, 120°, 120° Figure (b): 90°, 90°, 90°, 135°, 135° **33.** 36° **35.** The resulting polygon is also a regular polygon. **37.** 150° **39.** (a) $n - 3$ (b) $\frac{n(n-3)}{2}$ **41.** 221°

**2.5 SELECTED PROOF**

**29.** PROOF

| Statements | Reasons |
|---|---|
| 1. Quad. $RSTV$ with diagonals $\overline{RT}$ and $\overline{SV}$ intersecting at $W$ | 1. Given |
| 2. $m\angle RWS = m\angle 1 + m\angle 2$ | 2. The measure of an exterior ∠ of a △ equals the sum of the measures of the nonadjacent interior ∠s of the △ |
| 3. $m\angle RWS = m\angle 3 + m\angle 4$ | 3. Same as 2 |
| 4. $m\angle 1 + m\angle 2 = m\angle 3 + m\angle 4$ | 4. Substitution |

**2.6 EXERCISES**

**1.** M, T, X **3.** N, X **5.** (a), (c) **7.** (a), (b) **9.** MOM **11.** (a)  (b) 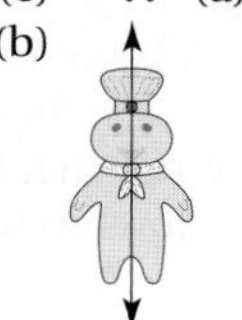

**13.** (a) 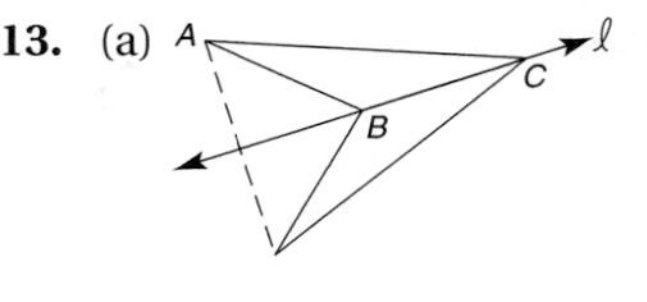


(b) 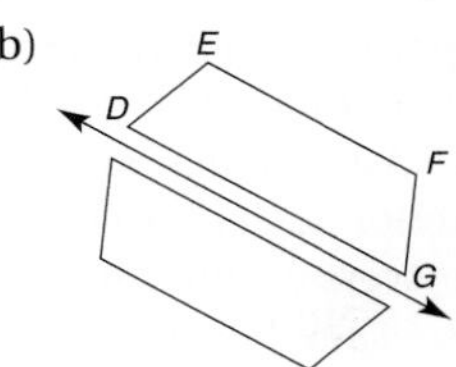


**15.** (a) 63° (b) Yes (c) Yes **17.** WHIM **19.** SIX **21.** WOW **23.** (a) Clockwise (b) Counterclockwise **25.** 62,365 kilowatt hours **27.** (a) Line (b) None (c) Line **29.** (b), (c) **31.** (a) 12 (b) 6 (c) 4 (d) 3

**CHAPTER 2 REVIEW EXERCISES**

**1.** (a) $\overline{BC} \parallel \overline{AD}$ (b) $\overline{AB} \parallel \overline{CD}$ **2.** 110° **3.** $x = 37$ **4.** $m\angle D = 75°$; $m\angle DEF = 125°$ **5.** $x = 20$; $y = 10$ **6.** $x = 30$; $y = 35$ **7.** $\overline{AE} \parallel \overline{BF}$ **8.** None **9.** $\overline{BE} \parallel \overline{CF}$ **10.** $\overline{BE} \parallel \overline{CF}$ **11.** $\overline{AC} \parallel \overline{DF}$ and $\overline{AE} \parallel \overline{BF}$ **12.** $x = 120°$; $y = 70°$ **13.** $x = 32°$; $y = 30°$ **14.** $y = -8$; $x = 24$ **15.** $x = 140°$ **16.** $x = 6$ **17.** $m\angle 3 = 69°$; $m\angle 4 = 67°$; $m\angle 5 = 44°$ **18.** 110° **19.** S **20.** N **21.** N **22.** S **23.** S **24.** A

**25.**

| Number of sides | 8 | 12 | 20 | 15 | 10 | 16 | 180 |
|---|---|---|---|---|---|---|---|
| Measure of each ext. ∠ | 45 | 30 | 18 | 24 | 36 | 22.5 | 2 |
| Measure of each int. ∠ | 135 | 150 | 162 | 156 | 144 | 157.5 | 178 |
| Number of diagonals | 20 | 54 | 170 | 90 | 35 | 104 | 15,930 |

**28.** Not possible
**30.** *Statement:* If two angles are right angles, then the angles are congruent.
*Converse:* If two angles are congruent, then the angles are right angles.
*Inverse:* If two angles are not right angles, then the angles are not congruent.
*Contrapositive:* If two angles are not congruent, then the angles are not right angles.
**31.** *Statement:* If it is not raining, then I am happy.
*Converse:* If I am happy, then it is not raining.
*Inverse:* If it is raining, then I am not happy.
*Contrapositive:* If I am not happy, then it is raining.
**32.** *Contrapositive* **37.** Assume $x = -3$.
**38.** Assume the sides opposite these angles are ≅ .
**39.** Assume that $\angle 1 \cong \angle 2$. Then $m \parallel n$ because congruent corresponding angles are formed. But this contradicts our hypothesis. Therefore, our assumption must be false, and it follows that $\angle 1 \not\cong \angle 2$.
**40.** Assume that $m \parallel n$. Then $\angle 1 \cong \angle 3$ because alternate exterior angles are congruent when parallel lines are cut by a transversal. But this contradicts the given fact that $\angle 1 \not\cong \angle 3$. Therefore, our assumption must be false, and it follows that m $\nparallel$ n.
**43.** (a) B, H, W (b) H, S **44.** (a) Isosceles triangle, Circle, Regular pentagon (b) Circle **45.** Congruent
**46.** (a)

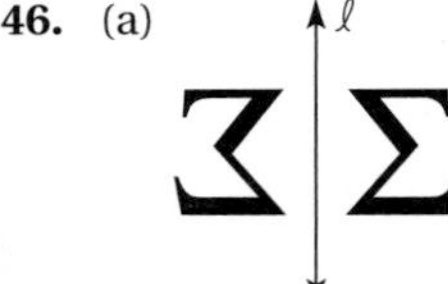

(b)

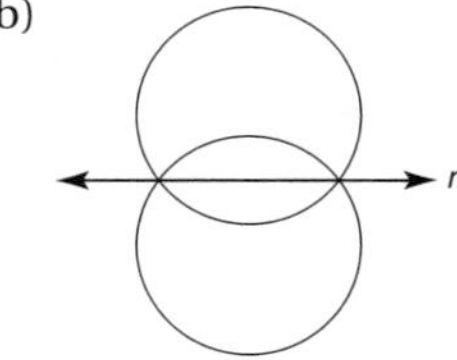

**47.** 90°

## CHAPTER 2 REVIEW EXERCISES SELECTED PROOFS

**33.** PROOF

| Statements | Reasons |
|---|---|
| 1. $\overline{AB} \parallel \overline{CF}$ | 1. Given |
| 2. $\angle 1 \cong \angle 2$ | 2. If two ∥ lines are cut by a transversal, then corresponding ∠s are ≅ |
| *(continued)* | |
| 3. $\angle 2 \cong \angle 3$ | 3. Given |
| 4. $\angle 1 \cong \angle 3$ | 4. Transitive Prop. of Congruence |

**34.** PROOF

| Statements | Reasons |
|---|---|
| 1 $\angle 1$ is comp. to $\angle 2$; $\angle 2$ is comp. to $\angle 3$ | 1. Given |
| 2. $\angle 1 \cong \angle 3$ | 2. If two ∠s are comp. to the same ∠, then these ∠s are ≅ |
| 3. $\overline{BD} \parallel \overline{AE}$ | 3. If two lines are cut by a transversal so that corresponding ∠s are ≅, then these lines are ∥ |

**35.** PROOF

| Statements | Reasons |
|---|---|
| 1. $\overline{BE} \perp \overline{DA}$; $\overline{CD} \perp \overline{DA}$ | 1. Given |
| 2. $\overline{BE} \parallel \overline{CD}$ | 2. If two lines are each ⊥ to a third line, then these lines are parallel to each other |
| 3. $\angle 1 \cong \angle 2$ | 3. If two ∥ lines are cut by a transversal, then the alternate interior ∠s are ≅ |

**36.** PROOF

| Statements | Reasons |
|---|---|
| 1. $\angle A \cong \angle C$ | 1. Given |
| 2. $\overline{DC} \parallel \overrightarrow{AB}$ | 2. Given |
| 3. $\angle C \cong \angle 1$ | 3. If two ∥ lines are cut by a transversal, the alt. int. ∠s are ≅ |
| 4. $\angle A \cong \angle 1$ | 4. Transitive Prop. of Congruence |
| 5. $\overline{DA} \parallel \overline{CD}$ | 5. If two lines are cut by a transversal so that corr. ∠s are ≅, then these lines are ∥ |

### CHAPTER 2 TEST

**1.** (a) $\angle 5$ (b) $\angle 3$ [2.1]
**2.** (a) $r$ and $s$ (b) $\ell$ and $m$ [2.3] **3.** "not $Q$" [2.2]
**4.** $\angle R$ and $\angle S$ are not both right angles. [2.2]
**5.** (a) $r \parallel t$ (b) $a \parallel c$ [2.3]
**7.** (a) 36° (b) 33° [2.4]
**8.** (a) Pentagon (b) Five [2.5]
**9.** (a) Equiangular hexagon (b) 120° [2.5]
**10.** A: line; D: line; N: point; O: both; X: both [2.6]
**11.** (a) Reflection (b) Slide (c) Rotation [2.6]
**12.** 61° [2.1] **13.** 54 [2.3] **14.** 50° [2.4]
**15.** 78° [2.4] **16.** 1. Given 2. $\angle 2 \cong \angle 3$
3. Transitive Prop. of Congruence 4. $\ell \parallel n$ [2.3]
**17.** Assume that $\angle M$ and $\angle Q$ are complementary. By definition, $m\angle M + m\angle Q = 90°$. Also, $m\angle M + m\angle Q + m\angle N = 180°$ because these are the three angles of $\triangle MNQ$. By substitution, $90° + m\angle N = 180°$, so it follows that $m\angle N = 90°$. But this leads to a contradiction because it is given that $m\angle N = 120°$. The assumption must be false, and it follows that $\angle M$ and $\angle Q$ are *not* complementary. [2.2]
**18.** 1. Given 2. 180° 3. $m\angle 1 + m\angle 2 + 90° = 180°$
4. 90° S5. $\angle 1$ and $\angle 2$ are complementary.
R5. Definition of complementary angles [2.4]

# Chapter 3

### 3.1 EXERCISES

**1.** $\angle A$; $\overline{AB}$; No; No **3.** $m\angle A = 72°$ **5.** SAS
**7.** $\triangle AED \cong \triangle FDE$ **9.** SSS **11.** AAS **13.** ASA
**15.** ASA **17.** SSS **19.** (a) $\angle A \cong \angle A$ (b) ASA
**21.** $\overline{AD} \cong \overline{EC}$ **23.** $\overline{MO} \cong \overline{MO}$ **25.** 1. Given
2. $\overline{AC} \cong \overline{AC}$ 3. SSS **33.** Yes; SAS or SSS **35.** No
**37.** (a) $\triangle CBE$, $\triangle ADE$, $\triangle CDE$ (b) $\triangle ADC$ (c) $\triangle CBD$

### 3.1 SELECTED PROOFS

**27.** PROOF

| Statements | Reasons |
|---|---|
| 1. $\overrightarrow{PQ}$ bisects $\angle MPN$ | 1. Given |
| 2. $\angle 1 \cong \angle 2$ | 2. If a ray bisects an $\angle$, it forms two $\cong$ $\angle$s |
| 3. $\overline{MP} \cong \overline{NP}$ | 3. Given |
| 4. $\overline{PQ} \cong \overline{PQ}$ | 4. Identity |
| 5. $\triangle MQP \cong \triangle NQP$ | 5. SAS |

**31.** PROOF

| Statements | Reasons |
|---|---|
| 1. $\angle VRS \cong \angle TSR$ and $\overline{RV} \cong \overline{TS}$ | 1. Given |
| 2. $\overline{RS} \cong \overline{RS}$ | 2. Identity |
| 3. $\triangle RST \cong \triangle SRV$ | 3. SAS |

### 3.2 EXERCISES

**9.** $m\angle 2 = 48°$; $m\angle 3 = 48°$; $m\angle 5 = 42°$; $m\angle 6 = 42°$
**11.** 1. Given 2. If two lines are $\perp$, then they form right $\angle$s
3. Identity 4. $\triangle HJK \cong \triangle HJL$ 5. $\overline{KJ} \cong \overline{JL}$
**17.** $c = 5$ **19.** $b = 8$ **21.** $c = \sqrt{41}$
**31.** (a) 8 (b) 37° (c) 53° **33.** 751 feet

### 3.2 SELECTED PROOFS

**1.** PROOF

| Statements | Reasons |
|---|---|
| 1. $\angle 1$ and $\angle 2$ are right $\angle$s $\overline{CA} \cong \overline{DA}$ | 1. Given |
| 2. $\overline{AB} \cong \overline{AB}$ | 2. Identity |
| 3. $\triangle ABC \cong \triangle ABD$ | 3. HL |

**5.** PROOF

| Statements | Reasons |
|---|---|
| 1. $\angle R$ and $\angle V$ are right $\angle$s $\angle 1 \cong \angle 2$ | 1. Given |
| 2. $\angle R \cong \angle V$ | 2. All right $\angle$s are $\cong$ |
| 3. $\overline{ST} \cong \overline{ST}$ | 3. Identity |
| 4. $\triangle RST \cong \triangle VST$ | 4. AAS |

**13.** PROOF

| Statements | Reasons |
|---|---|
| 1. $\angle$s $P$ and $R$ are right $\angle$s | 1. Given |
| 2. $\angle P \cong \angle R$ | 2. All right $\angle$s are $\cong$ |
| 3. $M$ is the midpoint of $\overline{PR}$ | 3. Given |
| 4. $\overline{PM} \cong \overline{MR}$ | 4. The midpoint of a segment forms two $\cong$ segments |
| 5. $\angle NMP \cong \angle QMR$ | 5. If two lines intersect, the vertical angles formed are $\cong$ |
| 6. $\triangle NPM \cong \triangle QRM$ | 6. ASA |
| 7. $\angle N \cong \angle Q$ | 7. CPCTC |

**23.** PROOF

| Statements | Reasons |
|---|---|
| 1. $\overline{DF} \cong \overline{DG}$ and $\overline{FE} \cong \overline{EG}$ | 1. Given |
| 2. $\overline{DE} \cong \overline{DE}$ | 2. Identity |
| 3. $\triangle FDE \cong \triangle GDE$ | 3. SSS |
| 4. $\angle FDE \cong \angle GDE$ | 4. CPCTC |
| 5. $\overrightarrow{DE}$ bisects $\angle FDG$ | 5. If a ray divides an $\angle$ into two $\cong \angle$s, then the ray bisects the angle |

**27.**

PROOF

| Statements | Reasons |
|---|---|
| 1. $\angle 1 \cong \angle 2$ and $\overline{MN} \cong \overline{QP}$ | 1. Given |
| 2. $\overline{MP} \cong \overline{MP}$ | 2. Identity |
| 3. $\triangle NMP \cong \triangle QPM$ | 3. SAS |
| 4. $\angle 3 \cong \angle 4$ | 4. CPCTC |
| 5. $\overline{MQ} \parallel \overline{NP}$ | 5. If two lines are cut by a transversal so that the alt. int. $\angle$s are $\cong$, then the lines are $\parallel$ |

### 3.3 EXERCISES

**1.** Isosceles **3.** $\overline{VT} \cong \overline{VU}$ **5.** $m\angle U = 69°$
**7.** $m\angle V = 36°$ **9.** $L = E$ (equivalent) **11.** $R$ and $S$ are disjoint; so $R \cap S = \varnothing$. **13.** Underdetermined
**15.** Overdetermined **17.** Determined **19.** 55°
**21.** $m\angle 2 = 68°$; $m\angle 1 = 44°$ **23.** $m\angle 5 = 124°$
**25.** $m\angle A = 52°$; $m\angle B = 64°$; $m\angle C = 64°$
**27.** 26 **29.** 12 **31.** Yes
**33.** 1. Given 2. $\angle 3 \cong \angle 2$ 3. $\angle 1 \cong \angle 2$
4. If two $\angle$s of a $\triangle$ are $\cong$, then the opposite sides are $\cong$
**39.** (a) 80° (b) 100° (c) 40° **41.** 75° each

### 3.3 SELECTED PROOF

**35.**

PROOF

| Statements | Reasons |
|---|---|
| 1. $\angle 1 \cong \angle 3$ | 1. Given |
| 2. $\overline{RU} \cong \overline{VU}$ | 2. Given |
| 3. $\angle R \cong \angle V$ | 3. If two sides of a $\triangle$ are $\cong$, then the $\angle$s opposite these sides are also $\cong$ |
| 4. $\triangle RUS \cong \triangle VUT$ | 4. ASA |
| 5. $\overline{SU} \cong \overline{TU}$ | 5. CPCTC |
| 6. $\triangle STU$ is isosceles | 6. If a $\triangle$ has two $\cong$ sides, it is an isosceles $\triangle$ |

### 3.4 EXERCISES

**19.** Construct a 90° angle; bisect it to form two 45° $\angle$s. Bisect one of the 45° angles to get a 22.5° $\angle$.
**31.** 120° **33.** 150° **39.** $D$ is on the bisector of $\angle A$.

### 3.5 EXERCISES

**1.** False **3.** True **5.** True **7.** False **9.** True
**11.** (a) Not possible ($100° + 100° + 60° \neq 180°$)
(b) Possible ($45° + 45° + 90° = 180°$)
**13.** (a) Possible (b) Not possible ($8 + 9 = 17$)
(c) Not possible ($8 + 9 < 18$)
**15.** Scalene right triangle ($m\angle Z = 90°$)
**17.** Isosceles obtuse triangle ($m\angle Z = 100°$)
**19.** 4 cm **21.** 72° (two such angles); 36° (one angle only) **23.** Nashville
**25.** 1. $m\angle ABC > m\angle DBE$ and $m\angle CBD > m\angle EBF$
3. Angle-Addition Postulate 4. $m\angle ABD > m\angle DBF$
**29.** $BC < EF$ **31.** $2 < x < 10$
**33.** $x + 2 < y < 5x + 12$
**35.** Proof: Assume $PM = PN$. Then $\triangle MPN$ is isosceles. But that contradicts the hypothesis; thus our assumption must be wrong, and $PM \neq PN$.

### 3.5 SELECTED PROOF

**27.**

PROOF

| Statements | Reasons |
|---|---|
| 1. Quad. $RSTU$ with diagonal $\overline{US}$; $\angle R$ and $\angle TUS$ are right $\angle$s | 1. Given |
| 2. $TS > US$ | 2. The shortest distance from a point to a line is the $\perp$ distance |
| 3. $US > UR$ | 3. Same as (2) |
| 4. $TS > UR$ | 4. Transitive Prop. of Inequality |

### CHAPTER 3 REVIEW EXERCISES

**15.** (a) $\overline{PR}$ (b) $\overline{PQ}$ **16.** $\overline{BC}, \overline{AC}, \overline{AB}$
**17.** $\angle R, \angle Q, \angle P$ **18.** $\overline{DA}$ **19.** (b) **20.** 5, 35
**21.** 20° **22.** 115° **23.** $m\angle C = 64°$ **24.** Isosceles
**25.** The triangle is also equilateral. **26.** 60°

### CHAPTER 3 REVIEW EXERCISES SELECTED PROOFS

**1.**

PROOF

| Statements | Reasons |
|---|---|
| 1. $\angle AEB \cong \angle DEC$ | 1. Given |
| 2. $\overline{AE} \cong \overline{ED}$ | 2. Given |
| 3. $\angle A \cong \angle D$ | 3. If two sides of a $\triangle$ are $\cong$, then the $\angle$s opposite these sides are also $\cong$ |
| 4. $\triangle AEB \cong \triangle DEC$ | 4. ASA |

**5.**

PROOF

| Statements | Reasons |
|---|---|
| 1. $\overline{AB} \cong \overline{DE}$ and $\overline{AB} \parallel \overline{DE}$ | 1. Given |
| 2. $\angle A \cong \angle D$ | 2. If two $\parallel$ lines are cut by a transversal, then the alt. int. $\angle$s are $\cong$ |
| 3. $\overline{AC} \cong \overline{DF}$ | 3. Given |
| 4. $\triangle BAC \cong \triangle EDF$ | 4. SAS |

*(continued)*

*(continued)*

| | |
|---|---|
| 5. $\angle BCA \cong \angle EFD$ | 5. CPCTC |
| 6. $\overline{BC} \parallel \overline{FE}$ | 6. If two lines are cut by a transversal so that alt. int. $\angle$s are $\cong$, then the lines are $\parallel$ |

**9.** PROOF

| Statements | Reasons |
|---|---|
| 1. $\overline{YZ}$ is the base of an isosceles triangle | 1. Given |
| 2. $\angle Y \cong \angle Z$ | 2. Base $\angle$s of an isosceles $\triangle$ are $\cong$ |
| 3. $\overrightarrow{XA} \parallel \overline{YZ}$ | 3. Given |
| 4. $\angle 1 \cong \angle Y$ | 4. If two $\parallel$ lines are cut by a transversal, then the corresponding $\angle$s are $\cong$ |
| 5. $\angle 2 \cong \angle Z$ | 5. If two $\parallel$ lines are cut by a transversal, then the alt. int. $\angle$s are $\cong$ |
| 6. $\angle 1 \cong \angle 2$ | 6. Transitive Prop. for Congruence |

**13.** PROOF

| Statements | Reasons |
|---|---|
| 1. $\overline{AB} \cong \overline{CD}$ | 1. Given |
| 2. $\angle BAD \cong \angle CDA$ | 2. Given |
| 3. $\overline{AD} \cong \overline{AD}$ | 3. Identity |
| 4. $\triangle BAD \cong \triangle CDA$ | 4. SAS |
| 5. $\angle CAD \cong \angle BDA$ | 5. CPCTC |
| 6. $\overline{AE} \cong \overline{ED}$ | 6. If two $\angle$s of a $\triangle$ are $\cong$, then the sides opposite these $\angle$s are also $\cong$ |
| 7. $\triangle AED$ is isosceles | 7. If a $\triangle$ has two $\cong$ sides, then it is an isosceles $\triangle$ |

### CHAPTER 3 TEST

**1.** (a) 75° (b) 4.7 cm [3.1]
**2.** (a) $\overline{XY}$ (b) $\angle Y$ [3.1]
**3.** (a) SAS (b) ASA [3.1]
**4.** Corresponding parts of congruent triangles are congruent. [3.2]
**5.** (a) No (b) Yes [3.2] **6.** Yes [3.2]
**7.** (a) $c = 10$ (b) $\sqrt{28}$ (or $2\sqrt{7}$) [3.2]
**8.** (a) $\overline{AM} \cong \overline{MB}$ (b) No [3.3]
**9.** (a) 38° (b) 36° [3.3]
**10.** (a) 7.6 inches (b) 57 [3.3]
**13.** (a) $\overline{BC}$ (b) $\overline{CA}$ [3.5]
**14.** $m\angle V > m\angle U > m\angle T$ [3.5]
**15.** $EB > DC$ since $EB = \sqrt{74}$ and $DC = \sqrt{65}$ [3.2]
**16.** $\overline{DA}$ [3.1]
**17.**

| Statements | Reasons |
|---|---|
| 1. $\angle R$ and $\angle V$ are rt $\angle$s | 1. Given |
| 2. $\angle R \cong \angle V$ | 2. All rt $\angle$s are $\cong$ |
| 3. $\angle 1 \cong \angle 2$ | 3. Given |
| 4. $\overline{ST} \cong \overline{ST}$ | 4. Identity |
| 5. $\triangle RST \cong \triangle VST$ | 5. AAS |

[3.1]

**18.** R1. Given R2. If 2 $\angle$s of a $\triangle$ are $\cong$, the opposite sides are $\cong$ S3. $\angle 1 \cong \angle 3$ R4. ASA S5. $\overline{US} \cong \overline{UT}$ S6. $\triangle STU$ is an isosceles triangle [3.3]

# Chapter 4

### 4.1 EXERCISES

**1.** (a) $AB = DC$ (b) $AD = BC$
**3.** (a) 8 (b) 5 (c) 70° (d) 110°
**5.** $AB = DC = 8$; $BC = AD = 9$
**7.** $m\angle A = m\angle C = 83°$; $m\angle B = m\angle D = 97°$
**9.** $\overline{AC}$ **11.** (a) $\overline{VY}$ (b) 16 **13.** True **15.** True
**17.** Parallelogram **19.** Parallelogram
**21.** 1. Given 2. $\overline{RV} \perp \overline{VT}$ and $\overline{ST} \perp \overline{VT}$ 3. $\overline{RV} \parallel \overline{ST}$ 4. $RSTV$ is a parallelogram **29.** $\angle P$ is a right angle
**31.** $\overline{RT}$ **33.** 255 mph **35.** $\overline{AC}$

### 4.1 SELECTED PROOF

**23.** PROOF

| Statements | Reasons |
|---|---|
| 1. Parallelogram $RSTV$ | 1. Given |
| 2. $\overline{RS} \parallel \overline{VT}$ | 2. Opposite sides of a parallelogram are $\parallel$ |
| 3. $\overline{XY} \parallel \overline{VT}$ | 3. Given |
| 4. $\overline{RS} \parallel \overline{XY}$ | 4. If two lines are each $\parallel$ to a third line, then the lines are $\parallel$ |
| 5. $RSYX$ is a parallelogram | 5. If a quadrilateral has opposite sides $\parallel$, then the quadrilateral is a parallelogram |
| 6. $\angle 1 \cong \angle S$ | 6. Opposite angles of a parallelogram are $\cong$ |

### 4.2 EXERCISES

**1.** (a) Yes (b) No **3.** Parallelogram
**5.** (a) Kite (b) Parallelogram **7.** $\overline{AC}$ **9.** 6.18
**11.** (a) 8 (b) 7 (c) 6 **13.** 10

**15.** (a) Yes; diagonal separating kite into two ≅ △ (b) No
**17.** Congruent **19.** 1. Given 2. Identity
3. $\triangle NMQ \cong \triangle NPQ$ 4. CPCTC 5. *MNPQ* is a kite
**29.** $y = 6$; $MN = 9$; $ST = 18$
**31.** $x = 5$; $RM = 11$; $ST = 22$ **33.** 270°

## 4.2 SELECTED PROOFS

**21.** PROOF

| Statements | Reasons |
|---|---|
| 1. *M-Q-T* and *P-Q-R* so that *MNPQ* and *QRST* are parallelograms | 1. Given |
| 2. $\angle N \cong \angle MQP$ | 2. Opposite ∠s in a parallelogram are ≅ |
| 3. $\angle MQP \cong \angle RQT$ | 3. If two lines intersect, the vertical ∠s formed are ≅ |
| 4. $\angle RQT \cong \angle S$ | 4. Same as (2) |
| 5. $\angle N \cong \angle S$ | 5. Transitive Prop. for Congruence |

**23.** PROOF

| Statements | Reasons |
|---|---|
| 1. Kite *HJKL* with diagonal $\overline{HK}$ | 1. Given |
| 2. $\overline{LH} \cong \overline{HJ}$ and $\overline{LK} \cong \overline{JK}$ | 2. A kite is a quadrilateral with two distinct pairs of ≅ adjacent sides |
| 3. $\overline{HK} \cong \overline{HK}$ | 3. Identity |
| 4. $\triangle LHK \cong \triangle JHK$ | 4. SSS |
| 5. $\angle LHK \cong \angle JHK$ | 5. CPCTC |
| 6. $\overrightarrow{HK}$ bisects $\angle LHJ$ | 6. If a ray divides an ∠ into two ≅ ∠s, then the ray bisects the ∠ |

## 4.3 EXERCISES

**1.** $m\angle A = 60°$; $m\angle ABC = 120°$
**3.** The parallelogram is a rectangle.
**5.** The quadrilateral is a rhombus.
**7.** $\overline{MN} \parallel$ to both $\overline{AB}$ and $\overline{DC}$; $MN = AB = DC$
**9.** $x = 5$; $DA = 19$ **11.** $NQ = 10$; $MP = 10$
**13.** $QP = \sqrt{72}$ or $6\sqrt{2}$; $MN = \sqrt{72}$ or $6\sqrt{2}$
**15.** $\sqrt{41}$ **17.** $\sqrt{34}$ **19.** 5 **21.** True
**23.** 1. Given 4. Same as (3) 5. If two lines are each ∥ to a third line, then the two lines are ∥ 6. Same as (2) 7. Same as (3) 8. Same as (3) 9. Same as (5) 10. *ABCD* is a parallelogram **25.** (a) **27.** 176
**39.** 20.4 ft **41.** Rhombus

## 4.4 EXERCISES

**1.** $m\angle D = 122°$; $m\angle B = 55°$
**3.** The trapezoid is an isosceles trapezoid.
**5.** The quadrilateral is a rhombus. **7.** Trapezoid
**9.** (a) Yes (b) No **11.** 9.7 **13.** 10.8
**15.** $7x + 2$ **19.** $h = 8$ **21.** 12 **23.** 22 ft
**25.** 14 **35.** (a) 7.0 (b) 14.2 (c) 10.6 (d) Yes
**37.** (a) 3 ft (b) 12 ft (c) 13 ft (d) $\sqrt{73}$ ft **39.** 8 ft

## CHAPTER 4 REVIEW EXERCISES

**1.** A **2.** S **3.** N **4.** S **5.** S **6.** A **7.** A
**8.** A **9.** A **10.** N **11.** S **12.** N
**13.** $AB = DC = 17$; $AD = BC = 31$
**14.** 106° **15.** 52 **16.** $m\angle M = 100°$; $m\angle P = 80°$
**17.** $\overline{PN}$ **18.** Kite
**19.** $m\angle G = m\angle F = 72°$; $m\angle E = 108°$ **20.** 14.9 cm
**21.** $MN = 23$; $PO = 7$ **22.** 26
**23.** $MN = 6$; $m\angle FMN = 80°$; $m\angle FNM = 40°$
**24.** $x = 3$; $MN = 15$; $JH = 30$
**32.** (a) Perpendicular (b) 13
**33.** (a) Perpendicular (b) 30
**34.** (a) Kites, rectangles, squares, rhombi, isosceles trapezoids (b) Parallelograms, rectangles, squares, rhombi
**35.** (a) Rhombus (b) Kite

## CHAPTER 4 REVIEW EXERCISES SELECTED PROOFS

**25.** PROOF

| Statements | Reasons |
|---|---|
| 1. *ABCD* is a parallelogram | 1. Given |
| 2. $\overline{AD} \cong \overline{CB}$ | 2. Opposite sides of a parallelogram are ≅ |
| 3. $\overline{AD} \parallel \overline{CB}$ | 3. Opposite sides of a parallelogram are ∥ |
| 4. $\angle 1 \cong \angle 2$ | 4. If two ∥ lines are cut by a transversal, then the alt. int. ∠s are ≅ |
| 5. $\overline{AF} \cong \overline{CE}$ | 5. Given |
| 6. $\triangle DAF \cong \triangle BCE$ | 6. SAS |
| 7. $\angle DFA \cong \angle BEC$ | 7. CPCTC |
| 8. $\overline{DF} \parallel \overline{EB}$ | 8. If two lines are cut by a transversal so that alt. ext. ∠s are ≅, then the lines are ∥ |

**26.** PROOF

| Statements | Reasons |
|---|---|
| 1. *ABEF* is a rectangle | 1. Given |
| 2. *ABEF* is a parallelogram | 2. A rectangle is a parallelogram with a rt. ∠ |

*(continued)*

*(continued)*

| Statements | Reasons |
|---|---|
| 3. $\overline{AF} \cong \overline{BE}$ | 3. Opposite sides of a parallelogram are ≅ |
| 4. *BCDE* is a rectangle | 4. Given |
| 5. $\angle F$ and $\angle BED$ are rt. ∠s | 5. All angles of a rectangle are rt. ∠s |
| 6. $\angle F \cong \angle BED$ | 6. Any two rt. ∠s are ≅ |
| 7. $\overline{FE} \cong \overline{ED}$ | 7. Given |
| 8. $\triangle AFE \cong \triangle BED$ | 8. SAS |
| 9. $\overline{AE} \cong \overline{BD}$ | 9. CPCTC |
| 10. $\angle AEF \cong \angle BDE$ | 10. CPCTC |
| 11. $\overline{AE} \parallel \overline{BD}$ | 11. If lines are cut by a transversal so that the corresponding ∠s are ≅, then the lines are ∥ |

**27.** PROOF

| Statements | Reasons |
|---|---|
| 1. $\overline{DE}$ is a median of $\triangle ADC$ | 1. Given |
| 2. *E* is the midpoint of $\overline{AC}$ | 2. A median of a △ is a line segment drawn from a vertex to the midpoint of the opposite side |
| 3. $\overline{AE} \cong \overline{EC}$ | 3. Midpoint of a segment forms two ≅ segments |
| 4. $\overline{BE} \cong \overline{FD}$ and $\overline{EF} \cong \overline{FD}$ | 4. Given |
| 5. $\overline{BE} \cong \overline{EF}$ | 5. Transitive Prop. for Congruence |
| 6. *ABCF* is a parallelogram | 6. If the diagonals of a quadrilateral bisect each other, then the quad. is a parallelogram |

**28.** PROOF

| Statements | Reasons |
|---|---|
| 1. $\triangle FAB \cong \triangle HCD$ | 1. Given |
| 2. $\overline{AB} \cong \overline{DC}$ | 2. CPCTC |
| 3. $\triangle EAD \cong \triangle GCB$ | 3. Given |
| 4. $\overline{AD} \cong \overline{BC}$ | 4. CPCTC |
| 5. *ABCD* is a parallelogram | 5. If a quadrilateral has both pairs of opposite sides ≅, then the quad. is a parallelogram |

**29.** PROOF

| Statements | Reasons |
|---|---|
| 1. *ABCD* is a parallelogram | 1. Given |
| 2. $\overline{DC} \cong \overline{BN}$ | 2. Given |
| 3. $\angle 3 \cong \angle 4$ | 3. Given |
| 4. $\overline{BN} \cong \overline{BC}$ | 4. If two ∠s of a △ are ≅, then the sides opposite these ∠s are also ≅ |
| 5. $\overline{DC} \cong \overline{BC}$ | 5. Transitive Prop. for Congruence |
| 6. *ABCD* is a rhombus | 6. If a parallelogram has two ≅ adjacent sides, then the parallelogram is a rhombus |

**30.** PROOF

| Statements | Reasons |
|---|---|
| 1. $\triangle TWX$ is isosceles with base $\overline{WX}$ | 1. Given |
| 2. $\angle W \cong \angle X$ | 2. Base ∠s of an isosceles △ are ≅ |
| 3. $\overline{RY} \parallel \overline{WX}$ | 3. Given |
| 4. $\angle TRY \cong \angle W$ and $\angle TYR \cong \angle X$ | 4. If two ∥ lines are cut by a transversal, then the corresp. ∠s are ≅ |
| 5. $\angle TRY \cong \angle TYR$ | 5. Transitive Prop. for Congruence |
| 6. $\overline{TR} \cong \overline{TY}$ | 6. If two ∠s of a △ are ≅, then the sides opposite these ∠s are also ≅ |
| 7. $\overline{TW} \cong \overline{TX}$ | 7. An isosceles △ has two ≅ sides |
| 8. $TR = TY$ and $TW = TX$ | 8. If two segments are ≅, then they are equal in length |
| 9. $TW = TR + RW$ and $TX = TY + YX$ | 9. Segment-Addition Postulate |
| 10. $TR + RW = TY + YX$ | 10. Substitution |
| 11. $RW = YX$ | 11. Subtraction Prop. of Equality |
| 12. $\overline{RW} \cong \overline{YX}$ | 12. If segments are = in length, then they are ≅ |
| 13. *RWXY* is an isosceles trapezoid | 13. If a quadrilateral has one pair of ∥ sides and the non-parallel sides are ≅, then the quad. is an isosceles trapezoid |

### CHAPTER 4 TEST

**1.** (a) Congruent (b) Supplementary [4.1]
**2.** 18.8 cm [4.1] **3.** $EB = 6$ [4.1]
**4.** $\overline{VS}$ [4.1] **5.** $x = 7$ [4.1]
**6.** (a) Kite (b) Parallelogram [4.2]
**7.** (a) Altitude (b) Rhombus [4.1]
**8.** (a) The line segments are parallel.
(b) $MN = \frac{1}{2}(BC)$ [4.2] **9.** 15.2 cm [4.2]
**10.** $x = 23$ [4.2] **11.** $AC = 13$ [4.3]
**12.** (a) $\overline{RV}, \overline{ST}$ (b) $\angle R$ and $\angle V$ (or $\angle S$ and $\angle T$) [4.4]
**13.** $MN = 14.3$ in. [4.4] **14.** $x = 5$ [4.4]
**15.** S1. Kite $ABCD$; $\overline{AB} \cong \overline{AD}$ and $\overline{BC} \cong \overline{DC}$
R1. Given S3. $\overline{AC} \cong \overline{AC}$
R4. SSS S5. $\angle B \cong \angle D$ R5. CPCTC [4.3]
**16.** S1. Trap. $ABCD$ with $\overline{AB} \parallel \overline{DC}$ and $\overline{AD} \cong \overline{BC}$
R1. Given R2. Congruent R3. Identity
R4. SAS S5. $\overline{AC} \cong \overline{DB}$ [4.4]

# Chapter 5

### 5.1 EXERCISES

**1.** (a) $\frac{4}{5}$ (b) $\frac{4}{5}$ (c) $\frac{2}{3}$ (d) Incommensurable
**3.** (a) $\frac{5}{8}$ (b) $\frac{1}{3}$ (c) $\frac{4}{3}$ (d) Incommensurable
**5.** (a) 3 (b) 8 **7.** (a) 6 (b) 4
**9.** (a) $\pm 2\sqrt{7} \approx \pm 5.29$ (b) $\pm 3\sqrt{2} \approx \pm 4.24$
**11.** (a) 4 (b) $-\frac{5}{6}$ or 3 **13.** (a) $\frac{3 \pm \sqrt{33}}{4} \approx 2.19$ or $-0.69$
(b) $\frac{7 \pm \sqrt{89}}{4} \approx 4.11$ or $-0.61$ **15.** 6.3 m/sec
**17.** $10\frac{1}{2}$ **19.** $\approx 24$ outlets **21.** (a) $4\sqrt{3} \approx 6.93$ (b) $4\frac{1}{2}$
**23.** Secretary's salary is \$24,900; salesperson's salary is \$37,350; vice-president's salary is \$62,250.
**25.** 40° and 50° **27.** 30.48 cm **29.** $2\frac{4}{7} \approx 2.57$
**31.** $a = 12$; $b = 16$ **33.** 4 in. by $4\frac{2}{3}$ in.
**35.** (a) $\frac{5 + 5\sqrt{5}}{2}$ (b) 8.1

### 5.2 EXERCISES

**1.** (a) Congruent (b) Proportional **3.** (a) Yes (b) No
**5.** (a) $\triangle ABC \sim \triangle XTN$ (b) $\triangle ACB \sim \triangle NXT$
**7.** Yes; Yes; Spheres have the same shape; one is an enlargement of the other unless they are congruent.
**9.** (a) 82° (b) 42° (c) $10\frac{1}{2}$ (d) 8
**11.** (a) Yes (b) Yes (c) Yes **13.** $5\frac{1}{3}$
**15.** 79° **17.** $n = 3$ **19.** 90° **21.** 12
**23.** $10 + 2\sqrt{5}$ or $10 - 2\sqrt{5}$; $\approx 14.47$ or 5.53
**25.** 75 **27.** 2.5 in. **29.** 3 ft, 9 in. **31.** 74 ft
**33.** No **35.** (a) Yes (b) Yes

### 5.3 EXERCISES

**1.** CASTC **3.** (a) True (b) True **5.** SSS ~
**7.** SAS ~ **9.** SAS ~ **11.** 1. Given
2. If 2 lines are ⊥, they form right angles.
3. All right angles are ≅. 4. Opposite ∠s of a ▱ are ≅
5. AA **13.** 1. Given 2. Definition of midpoint
3. If a line segment joins the midpoints of two sides of a △, its length is $\frac{1}{2}$ the length of the third side
4. Division Prop. of Eq. 5. Substitution 6. SSS ~
**15.** 1. $\overline{MN} \perp \overline{NP}$ and $\overline{QR} \perp \overline{RP}$
2. If two lines are ⊥, then they form a rt. ∠
3. $\angle N \cong \angle QRP$ 4. Identity S5. $\triangle MNP \sim \triangle QRP$
R5. AA **17.** 1. $\angle H \cong \angle F$ 2. If two ∠s are vertical ∠s, then they are ≅ S3. $\triangle HJK \sim \triangle FGK$ R3. AA
**19.** 1. $\frac{RQ}{NM} = \frac{RS}{NP} = \frac{QS}{MP}$ 2. $\triangle RQS \sim NMP$ 3. $\angle N \cong \angle R$
**21.** S1. $\overline{RS} \parallel \overline{UV}$ R1. Given 2. If 2 ∥ lines are cut by a transversal, alternate interior ∠s are ≅
3. $\triangle RST \sim \triangle VUT$ S4. $\frac{RT}{VT} = \frac{RS}{VU}$ R4. CSSTP
**23.** $4\frac{1}{2}$ **25.** 16 **27.** $EB = 24$ **29.** 27° **35.** $QS = 8$
**37.** 150 ft

### 5.3 SELECTED PROOFS

**31.** PROOF

| Statements | Reasons |
|---|---|
| 1. $\overline{AB} \parallel \overline{DF}$ and $\overline{BD} \parallel \overline{FG}$ | 1. Given |
| 2. $\angle A \cong \angle FEG$ and $\angle BCA \cong \angle G$ | 2. If two ∥ lines are cut by a transversal, then the corresponding ∠s are ≅ |
| 3. $\triangle ABC \sim \triangle EFG$ | 3. AA |

**33.** PROOF

| Statements | Reasons |
|---|---|
| 1. $\triangle DEF \sim \triangle MNP$; $\overline{DG}$ and $\overline{MQ}$ are altitudes | 1. Given |
| 2. $\overline{DG} \perp \overline{EF}$ and $\overline{MQ} \perp \overline{NP}$ | 2. An altitude is a segment drawn from a vertex ⊥ to the opposite side |
| 3. $\angle DGE$ and $\angle MQN$ are rt. ∠s | 3. ⊥ lines form a rt. ∠ |
| 4. $\angle DGE \cong \angle MQN$ | 4. Right ∠s are ≅ |
| 5. $\angle E \cong \angle N$ | 5. If two △s are ~ then the corresponding ∠s are ≅ |
| 6. $\triangle DGE \sim \triangle MQN$ | 6. AA |
| 7. $\frac{DG}{MQ} = \frac{DE}{MN}$ | 7. Corresponding sides of ~ △s are proportional |

### 5.4 EXERCISES

**1.** $\triangle RST \sim \triangle RVS \sim \triangle SVT$ **3.** $\frac{RT}{RS} = \frac{RS}{RV}$ or $\frac{RV}{RS} = \frac{RS}{RT}$
**5.** 4.5 **7.** (a) 10 (b) $\sqrt{34} \approx 5.83$ **9.** (a) 8 (b) 4
**11.** (a) Yes (b) No (c) Yes (d) No
**13.** (a) Right (b) Acute (c) Right (d) No △

**15.** 15 ft **17.** $6\sqrt{5} \approx 13.4$ m **19.** 20 ft **21.** 12 cm
**23.** The base is 8; the altitude is 6; the diagonals are 10.
**25.** $6\sqrt{7} \approx 15.87$ in **27.** 12 in **29.** 4 **31.** $9\frac{3}{13}$ in
**33.** $5\sqrt{5} \approx 11.18$ **39.** 60°
**41.** $TS = 13$; $RT = 13\sqrt{2} \approx 18.38$

## 5.5 EXERCISES

**1.** (a) $a$ (b) $a\sqrt{2}$ **3.** (a) $a\sqrt{3}$ (b) $2a$
**5.** $YZ = 8$; $XY = 8\sqrt{2} \approx 11.31$ **7.** $XZ = 10$; $YZ = 10$
**9.** $DF = 5\sqrt{3} \approx 8.66$; $FE = 10$ **11.** $DE = 12$; $FE = 24$
**13.** $HL = 6$; $HK = 12$; $MK = 6$ **15.** $AC = 6$; $AB = 6\sqrt{2} \approx 8.49$ **17.** $RS = 6$; $RT = 6\sqrt{3} \approx 10.39$
**19.** $DB = 5\sqrt{6} \approx 12.25$ **21.** $6\sqrt{3} + 6 \approx 16.39$
**23.** 45° **25.** 60°; 146 ft further
**27.** $DC = 2\sqrt{3} \approx 3.46$; $DB = 4\sqrt{3} \approx 6.93$
**29.** $6\sqrt{3} \approx 10.39$ **31.** $4\sqrt{3} \approx 6.93$
**33.** $6 + 6\sqrt{3} \approx 16.39$ **35.** (a) $6\sqrt{3}$ inches (b) 12 inches

## 5.6 EXERCISES

**1.** 30 oz of ingredient A; 24 oz of ingredient B; 36 oz of ingredient C
**3.** (a) Yes (b) Yes **5.** $EF = 4\frac{1}{6}$, $FG = 3\frac{1}{3}$, $GH = 2\frac{1}{2}$
**7.** $x = 5\frac{1}{3}$, $DE = 5\frac{1}{3}$, $EF = 6\frac{2}{3}$ **9.** $EC = 16\frac{4}{5}$
**11.** $a = 5$; $AD = 4$ **13.** (a) No (b) Yes
**15.** 9 **17.** $4\sqrt{6} \approx 9.80$ **19.** $\frac{AC}{CE} = \frac{AD}{DE}$, $\frac{DC}{CB} = \frac{DE}{EB}$
**21.** $SV = 2\sqrt{3} \approx 3.46$; $VT = 4\sqrt{3} \approx 6.93$
**23.** $x = \frac{1 + \sqrt{73}}{2}$ or $x = \frac{1 - \sqrt{73}}{2}$; reject both because each will give a negative number for the length of a side.
**25.** (a) True (b) True **27.** $RK = 1.8$
**29.** 1. Given 2. Means-Extremes Property 3. Addition Property of Equality 4. Distributive Property 6. Substitution
**35.** $\frac{-1 + \sqrt{5}}{2} \approx 0.62$ **37.** (a) $CD = 2$; $DB = 3$ (b) $CE = \frac{20}{11}$; $EA = \frac{24}{11}$ (c) $BF = \frac{10}{3}$; $FA = \frac{8}{3}$ (d) $\frac{3}{2} \cdot \frac{5}{6} \cdot \frac{4}{5} = 1$

## 5.6 SELECTED PROOF

**31.** PROOF

| Statements | Reasons |
|---|---|
| 1. $\triangle RST$ with $M$ the midpoint of $\overline{RS}$; $\overleftrightarrow{MN} \parallel \overline{ST}$ | 1. Given |
| 2. $RM = MS$ | 2. The midpoint of a segment divides the segment into two segments of equal measure |
| 3. $\frac{RM}{MS} = \frac{RN}{NT}$ | 3. If a line is $\parallel$ to one side of a $\triangle$ and intersects the other two sides, then it divides these sides proportionally |
| *(continued)* | |
| 4. $\frac{MS}{MS} = 1 = \frac{RN}{NT}$ | 4. Substitution |
| 5. $RN = NT$ | 5. Means-Extremes Property |
| 6. $N$ is the midpoint of $\overline{RT}$ | 6. If a point divides a segment into two segments of equal measure, then the point is a midpoint |

## CHAPTER 5 REVIEW EXERCISES

**1.** False **2.** True **3.** False **4.** True **5.** True
**6.** False **7.** True **8.** (a) $\pm\ 3\sqrt{2} \approx \pm\ 4.24$ (b) 26 (c) $-1$ (d) 2 (e) 7 or $-1$ (f) $-\frac{9}{5}$ or 4 (g) 6 or $-1$ (h) $-6$ or 3
**9.** \$3.78 **10.** Six packages **11.** \$79.20
**12.** The lengths of the sides are 8, 12, 20, and 28.
**13.** 18 **14.** 20 and $22\frac{1}{2}$ **15.** 150° **16.** (a) SSS ~ (b) AA (c) SAS ~ (d) SSS ~ **19.** $x = 5$; $m\angle F = 97°$
**20.** $AB = 6$; $BC = 12$ **21.** 3 **22.** $4\frac{1}{2}$ **23.** $6\frac{1}{4}$
**24.** $5\frac{3}{5}$ **25.** 10 **26.** 6 **27.** $EO = 1\frac{1}{5}$; $EK = 9$
**30.** (a) $8\frac{1}{3}$ (b) 21 (c) $2\sqrt{3} \approx 3.46$ (d) 3
**31.** (a) 16 (b) 40 (c) $2\sqrt{5} \approx 4.47$ (d) 4
**32.** (a) 30° (b) 24 (c) 20 (d) 16
**33.** $AE = 20$; $EF = 15$; $AF = 25$ **34.** $4\sqrt{2} \approx 5.66$ in.
**35.** $3\sqrt{2} \approx 4.24$ cm **36.** 25 cm **37.** $5\sqrt{3} \approx 8.66$ in.
**38.** $4\sqrt{3} \approx 6.93$ in. **39.** 12 cm
**40.** (a) $x = 9\sqrt{2} \approx 12.73$; $y = 9$ (b) $x = 4\frac{1}{2}$; $y = 6$ (c) $x = 12$; $y = 3$ (d) $x = 2\sqrt{14} \approx 7.48$; $y = 13$
**41.** 11 km **42.** (a) Acute (b) No $\triangle$ (c) Obtuse (d) Right (e) No $\triangle$ (f) Acute (g) Obtuse (h) Obtuse

## CHAPTER 5 REVIEW EXERCISES SELECTED PROOFS

**17.** PROOF

| Statements | Reasons |
|---|---|
| 1. $ABCD$ is a parallelogram; $\overline{DB}$ intersects $\overline{AE}$ at point $F$ | 1. Given |
| 2. $\overline{DC} \parallel \overline{AB}$ | 2. Opposite sides of a parallelogram are $\parallel$ |
| 3. $\angle CDB \cong \angle ABD$ | 3. If two $\parallel$ lines are cut by a transversal, then the alt. int. $\angle$s are $\cong$ |
| 4. $\angle DEF \cong \angle BAF$ | 4. Same as (3) |
| 5. $\triangle DFE \sim \triangle BFA$ | 5. AA |
| 6. $\frac{AF}{EF} = \frac{AB}{DE}$ | 6. Corresponding sides of $\sim \triangle$s are proportional |

**18.** PROOF

| Statements | Reasons |
|---|---|
| 1. $\angle 1 \cong \angle 2$ | 1. Given |
| 2. $\angle ADC \cong \angle 2$ | 2. If two lines intersect, then the vertical $\angle$s formed are $\cong$ |
| 3. $\angle ADC \cong \angle 1$ | 3. Transitive Prop. for Congruence |
| 4. $\angle A \cong \angle A$ | 4. Identity |
| 5. $\triangle BAE \sim \triangle CAD$ | 5. AA |
| 6. $\frac{AB}{AC} = \frac{BE}{CD}$ | 6. Corresponding sides of $\sim \triangle$s are proportional |

## CHAPTER 5 TEST

**1.** (a) 3:5 $\left(\text{or } \frac{3}{5}\right)$ (b) $\frac{25 \text{ mi}}{\text{gal}}$ [5.1]
**2.** (a) $\frac{40}{13}$ (b) 9, −9 [5.1] **3.** 15°; 75° [5.1]
**4.** (a) 92° (b) 12 [5.2] **5.** (a) SAS ~ (b) AA [5.3]
**6.** $\triangle ABC \sim \triangle ACD \sim \triangle CBD$ [5.4]
**7.** (a) $c = \sqrt{41}$ (b) $a = \sqrt{28} = 2\sqrt{7}$ [5.4]
**8.** (a) Yes (b) No [5.4] **9.** $DA = \sqrt{89}$ [5.4]
**10.** (a) $10\sqrt{2}$ in. (b) 8 cm [5.5]
**11.** (a) 5 m (b) 12 ft [5.5] **12.** $EC = 12$ [5.6]
**13.** $PQ = 4$; $QM = 6$ [5.6] **14.** 1 [5.6]
**15.** S1. $\overline{MN} \parallel \overline{QR}$ R1. Given
2. Corresponding $\angle$s are $\cong$ 3. $\angle P \cong \angle P$
4. AA [5.3] **16.** 1. Given 2. Identity 3. Given
5. Substitution 6. SAS ~ 7. $\angle PRC \cong \angle B$ [5.3]

# Chapter 6

## 6.1 EXERCISES

**1.** 29° **3.** 47.6° **5.** 56.6° **7.** 313°
**9.** (a) 90° (b) 270° (c) 135° (d) 135°
**11.** (a) 80° (b) 120° (c) 160° (d) 80° (e) 120° (f) 160° (g) 10° (h) 50° (i) 30°
**13.** (a) 72° (b) 144° (c) 36° (d) 72° (e) 18°
**15.** (a) 12 (b) $6\sqrt{2}$ **17.** 3 **19.** $\sqrt{7} + 3\sqrt{3}$
**21.** 90°; square
**23.** (a) The measure of an arc equals the measure of its corresponding central angle. Therefore, congruent arcs have congruent central angles. (b) The measure of a central angle equals the measure of its intercepted arc. Therefore, congruent central angles have congruent arcs. (c) Draw the radii to the endpoints of the congruent chords. The two triangles formed are congruent by SSS. The central angles of each triangle are congruent by CPCTC. Therefore, the arcs corresponding to the central angles are also congruent. Hence, congruent chords have congruent arcs. (d) Draw the four radii to the endpoints of the congruent arcs. Also draw the chords corresponding to the congruent arcs. The central angles corresponding to the congruent arcs are also congruent. Therefore, the triangles are congruent by SAS. The chords are congruent by CPCTC. Hence, congruent arcs have congruent chords. (e) Congruent central angles have congruent arcs (from b). Congruent arcs have congruent chords (from d). Hence, congruent central angles have congruent chords. (f) Congruent chords have congruent arcs (from c). Congruent arcs have congruent central angles (from a). Therefore, congruent chords have congruent central angles.
**25.** (a) 15° (b) 70° **27.** 72° **29.** 45°
**31.** 1. $\overline{MN} \parallel \overline{OP}$ in $\odot O$ 2. If two $\parallel$ lines are cut by a transversal, then the alt. int. $\angle$s are $\cong$ 3. If two $\angle$s are $\cong$, then their measures are = 4. The measure of an inscribed $\angle$ equals $\frac{1}{2}$ the measure of its intercepted arc
5. The measure of a central $\angle$ equals the measure of its arc
6. Substitution
**39.** If $\widehat{ST} \cong \widehat{TV}$, then $\overline{ST} \cong \overline{TV}$ ($\cong$ arcs in a circle have $\cong$ chords). $\triangle STV$ is an isosceles $\triangle$ because it has two $\cong$ sides.

## 6.1 SELECTED PROOF

**33.** *Proof:* Using the chords $\overline{AB}$, $\overline{BC}$, $\overline{CD}$, and $\overline{AD}$ in $\odot O$ as sides of inscribed angles, $\angle B \cong \angle D$ and $\angle A \cong \angle C$ because they are inscribed angles intercepting the same arc. $\triangle ABE \sim \triangle CDE$ by AA.

## 6.2 EXERCISES

**1.** (a) 8° (b) 46° (c) 38° (d) 54° (e) 126°
**3.** (a) 90° (b) 13° (c) 103° **5.** 18°
**7.** (a) 22° (b) 7° (c) 15°
**9.** (a) 136° (b) 224° (c) 68° (d) 44°
**11.** (a) 96° (b) 60°
**13.** (a) 120° (b) 240° (c) 60°
**15.** 28° **17.** $m\widehat{CE} = 88°$; $m\widehat{BD} = 36°$
**19.** (a) Supplementary (b) 107°
**21.** 1. $\overline{AB}$ and $\overline{AC}$ are tangents to $\odot O$ from $A$ 2. The measure of an $\angle$ formed by a tangent and a chord equals $\frac{1}{2}$ the arc measure 3. Substitution 4. If two $\angle$s are = in measure, they are $\cong$ 5. $\overline{AB} \cong \overline{AC}$ 6. $\triangle ABC$ is isosceles **27.** ≈ 154.95 mi **29.** $m\angle 1 = 36°$; $m\angle 2 = 108°$ **31.** (a) 30° (b) 60° (c) 150°
**33.** $\angle X \cong \angle X$; $\angle R \cong \angle W$; also, $\angle RVX \cong \angle WSX$ **35.** 10

## 6.2 SELECTED PROOF

**23.** *Given:* Tangent $\overline{AB}$ to $\odot O$ at point $B$; $m\angle A = m\angle B$
*Prove:* $m\widehat{BD} = 2 \cdot m\widehat{BC}$
*Proof:* $m\angle BCD = m\angle A + m\angle B$; but because $m\angle A = m\angle B$, $m\angle BCD = m\angle B + m\angle B$ or $m\angle BCD = 2 \cdot m\angle B$. $m\angle BCD$ also equals $\frac{1}{2}m\widehat{BD}$ because it is an inscribed $\angle$. Therefore, $\frac{1}{2}m\widehat{BD} = 2 \cdot m\angle B$ or $m\widehat{BD} = 4 \cdot m\angle B$. But if $\overline{AB}$ is a tangent to $\odot O$ at $B$, then $m\angle B = \frac{1}{2}m\widehat{BC}$. By substitution, $m\widehat{BD} = 4(\frac{1}{2}m\widehat{BC})$ or $m\widehat{BD} = 2 \cdot m\widehat{BC}$.

**6.3 EXERCISES**

**1.** 30° **3.** $6\sqrt{5}$ **7.** 3
**9.** $DE = 4$ and $EC = 12$ or $DE = 12$ and $EC = 4$ **11.** 4
**13.** $DE = 12$; $EC = 6$ **15.** $9\frac{2}{5}$ **17.** 9 **19.** $5\frac{1}{3}$
**21.** $3 + 3\sqrt{5}$ **23.** (a) None (b) One (c) 4
**29.** Yes; $\overline{AE} \cong \overline{CE}$; $\overline{DE} \cong \overline{EB}$ **31.** 20°
**33.** $AM = 5$; $PC = 7$; $BN = 9$ **35.** 12
**37.** 8.7 inches **39.** (a) Obtuse (b) Equilateral **41.** 45°

**6.3 SELECTED PROOFS**

**25.** If $\overline{AF}$ is a tangent to $\odot O$ and $\overline{AC}$ is a secant to $\odot O$, then $(AF)^2 = AC \cdot AB$. If $\overline{AF}$ is a tangent to $\odot Q$ and $\overline{AE}$ is a secant to $\odot Q$, then $(AF)^2 = AE \cdot AD$. By substitution, $AC \cdot AB = AE \cdot AD$.

**27.** *Proof:* Let $M, N, P$, and $Q$ be the points of tangency for $\overline{DC}$, $\overline{DA}$, $\overline{AB}$, and $\overline{BC}$, respectively. Because the tangent segments from an external point are congruent, $AP = AN$, $PB = BQ$, $CM = CQ$, and $MD = DN$. Thus $AP + PB + CM + MD = AN + BQ + CQ + DN$.
Reordering and associating, $(AP + PB) + (CM + MD) = (AN + DN) + (BQ + CQ)$ or $AB + CD = DA + BC$.

**43.** *Given:* $\overleftrightarrow{AB}$ contains $O$, the center of the circle, and $\overleftrightarrow{AB}$ contains $M$, the midpoint of $\overline{RS}$ (See Figure 6.38.)
*Prove:* $\overleftrightarrow{AB} \perp \overline{RS}$
*Proof:* If $M$ is the midpoint of $\overline{RS}$ in $\odot O$, then $\overline{RM} \cong \overline{MS}$. Draw $\overline{RO}$ and $\overline{OS}$, which are $\cong$ because they are radii in the same circle. Using $\overline{OM} \cong \overline{OM}$, $\triangle ROM \cong \triangle SOM$ by SSS. By CPCTC, $\angle OMS \cong \angle OMR$, and hence $\overleftrightarrow{AB} \perp \overline{RS}$.

**6.4 EXERCISES**

**1.** $m\angle CQD < m\angle AQB$ **3.** $QM < QN$ **5.** $CD < AB$
**7.** $QM > QN$ **11.** No; angles are not congruent.
**15.** $\overline{AB}$; $\overline{GH}$; for a circle containing unequal chords, the chord nearest the center has the greatest length and the chord at the greatest distance from the center has the least length.
**17.** (a) $\overline{OT}$ (b) $\overline{OD}$
**19.** (a) $m\widehat{MN} > m\widehat{QP}$ (b) $m\widehat{MPN} < m\widehat{PMQ}$
**21.** Obtuse
**23.** (a) $m\angle AOB > m\angle BOC$ (b) $AB > BC$
**25.** (a) $m\widehat{AB} > m\widehat{BC}$ (b) $AB > BC$
**27.** (a) $\angle C$ (b) $\overline{AC}$ **29.** (a) $\angle B$ (b) $\overline{AC}$
**31.** $\overline{AB}$ is $(4\sqrt{3} - 4\sqrt{2})$ closer than $\overline{CD}$. **37.** 7

**6.5 EXERCISES**

**1.** $A, C, E$ **11.** The locus of points at a given distance from a fixed line is two parallel lines on either side of the fixed line at the same (given) distance from the fixed line.
**13.** The locus of points at a distance of 3 in. from point $O$ is a circle with center $O$ and radius 3 in.
**15.** The locus of points equidistant from points $D$, $E$, and $F$ is the point $G$ for which $DG = EG = FG$.
**17.** The locus of the midpoints of the chords in $\odot Q$ parallel to diameter $\overline{PR}$ is the perpendicular bisector of $\overline{PR}$.
**19.** The locus of points equidistant from two given intersecting lines is two perpendicular lines that bisect the angles formed by the two intersecting lines.
**25.** The locus of points at a distance of 2 cm from a sphere whose radius is 5 cm is two concentric spheres with the same center. The radius of one sphere is 3 cm, and the radius of the other sphere is 7 cm.
**27.** The locus is another sphere with the same center and a radius of length 2.5 m.
**29.** The locus of points equidistant from an 8-ft ceiling and the floor is a plane parallel to the ceiling and the floor and midway between them.

**6.6 EXERCISES**

**1.** Yes **3.** Incenter **5.** Circumcenter **7.** (a) Angle bisectors (b) Perpendicular bisectors of sides (c) Altitudes (d) Medians **9.** No (need 2)
**11.** Equilateral triangle **13.** Midpoint of the hypotenuse
**23.** No **25.** $\frac{10\sqrt{3}}{3}$ **27.** $RQ = 10$; $SQ = \sqrt{89}$
**29.** (a) 4 (b) 6 (c) 10.5 **33.** Equilateral
**35.** (a) Yes (b) Yes **37.** (a) Yes (b) No

**CHAPTER 6 REVIEW EXERCISES**

**1.** 9 mm **2.** 30 cm **3.** $\sqrt{41}$ in. **4.** $6\sqrt{2}$ cm
**5.** 130° **6.** 35° **7.** 80° **8.** 35°
**9.** $m\widehat{AC} = m\widehat{DC} = 93\frac{1}{3}°$; $m\widehat{AD} = 173\frac{1}{3}°$
**10.** $m\widehat{AC} = 110°$ and $m\widehat{AD} = 180°$
**11.** $m\angle 2 = 44°$; $m\angle 3 = 90°$; $m\angle 4 = 46°$; $m\angle 5 = 44°$
**12.** $m\angle 1 = 50°$; $m\angle 2 = 40°$; $m\angle 3 = 90°$; $m\angle 4 = 50°$
**13.** 24 **14.** 10 **15.** A **16.** S **17.** N **18.** S
**19.** A **20.** N **21.** A **22.** N
**23.** (a) 70° (b) 28° (c) 64° (d) $m\angle P = 21°$ (e) $m\widehat{AB} = 90°$; $m\widehat{CD} = 40°$ (f) 260° **24.** (a) 3 (b) 8 (c) 16 (d) 4 (e) 4 (f) 8 or 1 (g) $3\sqrt{5}$ (h) 3 (i) $4\sqrt{3}$ (j) 3 **25.** 29
**26.** If $x = 7$, then $AC = 35$; $DE = 17\frac{1}{2}$. If $x = -4$, then $AC = 24$; $DE = 12$. **30.** $m\angle 1 = 93°$; $m\angle 2 = 25°$; $m\angle 3 = 43°$; $m\angle 4 = 68°$; $m\angle 5 = 90°$; $m\angle 6 = 22°$; $m\angle 7 = 68°$; $m\angle 8 = 22°$; $m\angle 9 = 50°$; $m\angle 10 = 112°$
**31.** $24\sqrt{2}$ cm **32.** $15 + 5\sqrt{3}$ cm
**33.** 14 cm and 15 cm **34.** $AD = 3$; $BE = 6$; $FC = 7$
**35.** (a) $AB > CD$ (b) $QP < QR$ (c) $m\angle A < m\angle C$
**39.** The locus of the midpoints of the radii of a circle is a concentric circle with radius half the length of the given radius.
**40.** The locus of the centers of all circles passing through two given points is the perpendicular bisector of the segment joining the two given points.
**41.** The locus of the center of a penny that rolls around a half-dollar is a circle.
**42.** The locus of points in space less than 3 units from a given point is the interior of a sphere.
**43.** The locus of points equidistant from two parallel planes is a parallel plane midway between the two planes.
**50.** (a) 12 (b) 2 (c) $2\sqrt{3}$ **51.** $BF = 6$; $AE = 9$

### CHAPTER 6 REVIEW EXERCISES SELECTED PROOFS

**27.** *Proof:* If $\overline{DC}$ is tangent to circles $B$ and $A$ at points $D$ and $C$, then $\overline{BD} \perp \overline{DC}$ and $\overline{AC} \perp \overline{DC}$. $\angle$s $D$ and $C$ are congruent because they are right angles. $\angle DEB \cong \angle CEA$ because of vertical angles. $\triangle BDE \sim \triangle ACE$ by AA. It follows that $\frac{AC}{CE} = \frac{BD}{ED}$ because corresponding sides are proportional. Hence, $AC \cdot ED = CE \cdot BD$.

**28.** *Proof:* In $\odot O$, if $\overline{EO} \perp \overline{BC}$, $\overline{DO} \perp \overline{BA}$, and $\overline{EO} \cong \overline{OD}$, then $\overline{BC} \cong \overline{BA}$. (Chords equidistant from the center of the circle are congruent.) It follows then that $\widehat{BC} \cong \widehat{BA}$.

**29.** *Proof:* If $\overline{AP}$ and $\overline{BP}$ are tangent to $\odot Q$ at $A$ and $B$, then $\overline{AP} \cong \overline{BP}$. $\widehat{AC} \cong \widehat{BC}$ because C is the midpoint of $\widehat{AB}$. It follows that $\overline{AC} \cong \overline{BC}$ and, using $\overline{CP} \cong \overline{CP}$, we have $\triangle ACP \cong \triangle BCP$ by SSS. $\angle APC \cong \angle BPC$ by CPCTC and hence $\overrightarrow{PC}$ bisects $\angle APB$.

### CHAPTER 6 TEST

**1.** (a) 272° (b) 134° [6.1]
**2.** (a) 69° (b) 32° [6.1]
**3.** (a) 48° (b) Isosceles [6.1]
**4.** (a) Right (b) Congruent [6.2]
**5.** (a) 69° (b) 37° [6.2]
**6.** (a) 214° (b) 34° [6.2]
**7.** (a) Concentric b) 8 [6.1]
**8.** (a) 1 (b) 2 [6.3] **9.** (a) 10 (b) 5 [6.3]
**11.** (a) $m\angle AQB > m\angle CQD$ (b) $AB > CD$ [6.4]
**12.** (a) The circle with center $P$ and radius length 3 centimeters (b) The sphere with center $P$ and radius length 3 centimeters [6.4]
**13.** (a) The line that is the perpendicular bisector of $\overline{AB}$ (b) The plane that is the perpendicular bisector of $\overline{AB}$ [6.5]
**14.** (a) Incenter (b) Centroid [6.6]
**15.** (a) 18 (b) $x = 4$ [6.6]
**16.** S1. In $\odot O$, chords $\overline{AD}$ and $\overline{BC}$ intersect at $E$
R1. Given 2. Vertical angles are congruent
4. AA 5. $\frac{AE}{CE} = \frac{BE}{DE}$ [6.3]

# Chapter 7

### 7.1 EXERCISES

**1.** Two triangles with equal areas are not necessarily congruent. Two squares with equal areas must be congruent because the sides are congruent. **3.** 37 units$^2$ **5.** The altitudes to $\overline{PN}$ and to $\overline{MN}$ are congruent. This is because $\triangle$s $QMN$ and $QPN$ are congruent; corresponding altitudes of $\cong$ $\triangle$s are $\cong$. **7.** Equal **9.** 54 cm$^2$ **11.** 18 m$^2$ **13.** 72 in$^2$ **15.** 100 in$^2$ **17.** 126 in$^2$ **19.** 264 units$^2$ **21.** 144 units$^2$ **23.** 192 ft$^2$ **25.** (a) 300 ft$^2$ (b) 3 gallons (c) \$46.50 **27.** $156 + 24\sqrt{10}$ ft$^2$ **29.** (a) 9 sq ft = 1 sq yd (b) 1296 sq in. = 1 sq yd **31.** 24 cm$^2$ **33.** $\overline{MN}$ joins the midpoints of $\overline{CA}$ and $\overline{CB}$, so $MN = \frac{1}{2}(AB)$. Therefore $\overline{AP} \cong \overline{PB} \cong \overline{MN}$. $\overline{PN}$ joins the midpoints of $\overline{CB}$ and $\overline{AB}$, so $PN = \frac{1}{2}(AC)$. Therefore $\overline{AM} \cong \overline{MC} \cong \overline{PN}$. $\overline{MP}$ joins the midpoints of $\overline{AB}$ and $\overline{AC}$, so $MP = \frac{1}{2}(BC)$. Therefore $\overline{CN} \cong \overline{NB} \cong \overline{MP}$. The four triangles are all $\cong$ by SSS. Therefore, the areas of all these triangles are the same. Hence, the area of the big triangle is equal to four times the area of one of the smaller triangles.
**37.** 8 in. **39.** (a) 12 in. (b) 84 in$^2$ **41.** 56 percent
**43.** By the Area-Addition Postulate, $A_{R \cup S} = A_R + A_S$. Now $A_{R \cup S}$, $A_R$, and $A_S$ are all positive numbers. Let $p$ represent the area of region $S$, so that $A_{R \cup S} = A_R + p$. By the definition of inequality, $A_R < A_{R \cup S}$, or $A_{R \cup S} > A_R$.
**45.** $(a + b)(c + d) = ac + ad + bc + bd$ **47.** $4\frac{8}{13}$ in.
**49.** 8 **51.** (a) 10 (b) 26 (c) 18 (d) No

### 7.1 SELECTED PROOF

**35.** *Proof:* $A = (LH)(HJ) = s^2$. By the Pythagorean Theorem, $s^2 + s^2 = d^2$.

$$2s^2 = d^2$$
$$s^2 = \frac{d^2}{2}$$
Thus
$$A = \frac{d^2}{2}$$

### 7.2 EXERCISES

**1.** 30 in. **3.** $4\sqrt{29}$ m **5.** 30 ft **7.** 38 **9.** 84 in$^2$
**11.** 1764 mm$^2$ **13.** 40 ft$^2$ **15.** 80 units$^2$
**17.** $36 + 36\sqrt{3}$ units$^2$ **19.** 16 in., 32 in., and 28 in.
**21.** 15 cm **23.** (a) $\frac{9}{4}$ (b) $\frac{4}{1}$ **27.** $24 + 4\sqrt{21}$ units$^2$
**29.** 96 units$^2$ **31.** 6 yd by 8 yd **33.** (a) 770 ft (b) \$454.30 **35.** 624 ft$^2$ **37.** Square with sides of length 10 in. **39.** (a) 52 units (b) 169 units$^2$ **41.** 60 in$^2$
**43.** (a) No (b) Yes **47.** 12 ft$^2$ **49.** 5 in$^2$
**51.** $h = 2.4$

### 7.2 SELECTED PROOFS

**25.** Using Heron's Formula, the semiperimeter is $\frac{1}{2}(3s)$, or $\frac{3s}{2}$. Then

$$A = \sqrt{\frac{3s}{2}\left(\frac{3s}{2} - s\right)\left(\frac{3s}{2} - s\right)\left(\frac{3s}{2} - s\right)}$$
$$A = \sqrt{\frac{3s}{2}\left(\frac{s}{2}\right)\left(\frac{s}{2}\right)\left(\frac{s}{2}\right)}$$
$$A = \sqrt{\frac{3s^4}{16}} = \frac{\sqrt{3} \cdot \sqrt{s^4}}{\sqrt{16}}$$
$$A = \frac{s^2\sqrt{3}}{4}$$

**45.** The area of a trapezoid $= \frac{1}{2}h(b_1 + b_2) = h \cdot \frac{1}{2}(b_1 + b_2)$. The length of the median of a trapezoid is $m = \frac{1}{2}(b_1 + b_2)$. By substitution, the area of a trapezoid is $A = hm$.

### 7.3 EXERCISES

**1.** First, construct the angle bisectors of two consecutive angles, say $A$ and $B$. The point of intersection, $O$, is the center of the inscribed circle.

Second, construct the line segment $\overline{OM}$ perpendicular to $\overline{AB}$. Then, using the radius $r = OM$, construct the inscribed circle with center $O$.
**3.** Draw the diagonals (angle bisectors) $\overline{JL}$ and $\overline{MK}$. These determine center $O$ of the inscribed circle. Now construct the line segment $\overline{OR} \perp \overline{MJ}$. Use $OR$ as the length of the radius of the inscribed circle. **9.** 27.2 in. **11.** 8.3 cm **13.** $a = 5$ in.; $r = 5\sqrt{2}$ in. **15.** $16\sqrt{3}$ ft; 16 ft **17.** (a) 120° (b) 90° (c) 72° (d) 60° **19.** (a) 4 (b) 8 (c) 6 (d) 15 **21.** 97.5 cm² **23.** 317.52 in² **25.** $54\sqrt{3}$ cm² **27.** $75\sqrt{3}$ in² **29.** $(24 + 12\sqrt{3})$ in² **31.** $\frac{2}{1}$ **33.** $4 + 4\sqrt{2}$ **35.** 168°

### 7.4 EXERCISES

**1.** $C = 16\pi$ cm; $A = 64\pi$ cm² **3.** $C = 66$ in.; $A = 346\frac{1}{2}$ in² **5.** (a) $r = 22$ in.; $d = 44$ in. (b) $r = 30$ ft; $d = 60$ ft **7.** (a) $r = 5$ in.; $d = 10$ in. (b) $r = 1.5$ cm; $d = 3.0$ cm **9.** $\frac{8}{3}\pi$ in. **11.** $C \approx 77.79$ in. **13.** $r \approx 6.7$ cm **15.** $\ell \approx 7.33$ in. **17.** 16 in² **19.** $5 < AN < 13$ **21.** $(32\pi - 64)$ in.² **23.** $(600 - 144\pi)$ ft² **25.** $\approx 7$ cm **27.** 8 in.
**29.** $A = A_{\text{LARGER CIRCLE}} - A_{\text{SMALLER CIRCLE}}$
$A = \pi R^2 - \pi r^2$
$A = \pi(R^2 - r^2)$
But $R^2 - r^2$ is a difference of two squares, so
$A = \pi(R + r)(R - r)$.
**31.** 3 in. and 4 in. **33.** (a) $\approx 201.06$ ft² (b) 2.87 pints. Thus 3 pints must be purchased. (c) \$8.85 **35.** (a) $\approx 1256$ ft² (b) 20.93. Thus 21 lb of seed is needed. (c) \$34.65 **37.** $\approx 43.98$ cm **39.** $\approx 14.43$ in. **41.** $\approx 27{,}488.94$ mi **43.** $12\pi$ cm²

### 7.5 EXERCISES

**1.** 34 in. **3.** 150 cm² **5.** $\frac{3}{2}rs$ **7.** 54 mm **9.** 24 in² **11.** 1 in. **13.** $P = \left(16 + \frac{8}{3}\pi\right)$ in. and $A = \frac{32}{3}\pi$ in² **15.** $\approx 30.57$ in. **17.** $P = (12 + 4\pi)$ in.; $A = (24\pi - 36\sqrt{3})$ in² **19.** $\left(25\sqrt{3} - \frac{25}{2}\pi\right)$ cm² **21.** $\frac{9}{2}$ cm **23.** $36\pi$ **25.** 90° **27.** Cut the pizza into 8 slices. **29.** $A = \left(\frac{\pi}{2}\right)s^2 - s^2$ **31.** $r = 3\frac{1}{3}$ ft or 3 ft 4 in. **35.** (a) 3 (b) 2 **37.** $\frac{308\pi}{3} \approx 322.54$ in² **39.** $1875\pi \approx 5890$ ft²

### CHAPTER 7 REVIEW EXERCISES

**1.** 480 **2.** (a) 40 (b) $40\sqrt{3}$ (c) $40\sqrt{2}$ **3.** 50 **4.** 204 **5.** 336 **6.** 36 **7.** (a) $24\sqrt{2} + 18$ (b) $24 + 9\sqrt{3}$ (c) $33\sqrt{3}$ **8.** $A = 216$ in²; $P = 60$ in. **9.** (a) 19,000 ft² (b) 4 bags (c) \$72 **10.** (a) 3 double rolls (b) 3 rolls **11.** (a) $\frac{289}{4}\sqrt{3} + 8\sqrt{33}$ (b) $50 + \sqrt{33}$ **12.** 168 **13.** 5 cm by 7 cm **14.** (a) 15 cm, 25 cm, and 20 cm (b) 150 cm² **15.** 36 **16.** $36\sqrt{3}$ cm² **17.** 20 **18.** (a) 72° (b) 108° (c) 72° **19.** $96\sqrt{3}$ ft² **20.** 6 in. **21.** $162\sqrt{3}$ in² **22.** (a) 8 (b) $\approx 120$ cm² **23.** (a) No. ⊥ bisectors of sides of a parallelogram are not necessarily concurrent. (b) No. ⊥ bisectors of sides of a rhombus are not necessarily concurrent. (c) Yes. ⊥ bisectors of sides of a rectangle are concurrent. (d) Yes. ⊥ bisectors of sides of a square are concurrent. **24.** (a) No. ∠ bisectors of a parallelogram are not necessarily concurrent. (b) Yes. ∠ bisectors of a rhombus are concurrent. (c) No. ∠ bisectors of a rectangle are not necessarily concurrent. (d) Yes. ∠ bisectors of a square are concurrent. **25.** $147\sqrt{3} \approx 254.61$ in² **26.** (a) 312 ft² (b) 35 yd² (c) \$348.95 **27.** $64 - 16\pi$ **28.** $\frac{49}{2}\pi - \frac{49}{2}\sqrt{3}$ **29.** $\frac{8}{3}\pi - 4\sqrt{3}$ **30.** $288 - 72\pi$ **31.** $25\sqrt{3} - \frac{25}{3}\pi$ **32.** $\ell = \frac{2\pi\sqrt{5}}{3}$ cm; $A = 5\pi$ cm² **33.** (a) 21 ft (b) $\approx 346\frac{1}{2}$ ft² **34.** (a) $6\pi$ ft² (b) $\left(6\sqrt{3} + \frac{4\pi}{3}\sqrt{3}\right)$ ft **35.** $(9\pi - 18)$ in² **39.** (a) $\approx 28$ yd² (b) $\approx 21.2$ ft² **40.** (a) $\approx 905$ ft² (b) \$162.90 (c) Approximately 151 flowers

### CHAPTER 7 REVIEW EXERCISES SELECTED PROOF

**36.** *Proof:* By an earlier theorem,

$$\begin{aligned} A_{\text{RING}} &= \pi R^2 - \pi r^2 \\ &= \pi(OC)^2 - \pi(OB)^2 \\ &= \pi[(OC)^2 - (OB)^2] \end{aligned}$$

In rt. △$OBC$,

$$(OB)^2 + (BC)^2 = (OC)^2$$

Thus $(OC)^2 - (OB)^2 = (BC)^2$

In turn, $A_{\text{RING}} = \pi(BC)^2$.

### CHAPTER 7 TEST

**1.** (a) Square inches (b) Equal [7.1] **2.** (a) $A = s^2$ (b) $C = 2\pi r$ [7.4] **3.** (a) True (b) False [7.2] **4.** 23 cm² [7.1] **5.** 120 ft² [7.1] **6.** 24 ft² [7.2] **7.** 24 cm² [7.2] **8.** 6 ft [7.2] **9.** (a) 29 in. (b) 58 in² [7.3] **10.** (a) $10\pi$ in. (b) $25\pi$ in² [7.4] **11.** $\approx 5\frac{1}{2}$ in. [7.4] **12.** 314 cm² [7.4] **13.** $(16\pi - 32)$ in² [7.5] **14.** $54\pi$ cm² [7.5] **15.** $(36\pi - 72)$ in² [7.5] **16.** $r = 2$ in. [7.3]

# Chapter 8

### 8.1 EXERCISES

**1.** (a) Yes (b) Oblique (c) Hexagon (d) Oblique hexagonal prism (e) Parallelogram **3.** (a) 12 (b) 18 (c) 8 **5.** (a) cm² (b) cm³ **7.** 132 cm² **9.** 120 cm³ **11.** (a) 16 (b) 8 (c) 16 **13.** (a) $2n$ (b) $n$ (c) $2n$ (d) $3n$ (e) $n$ (f) 2 (g) $n + 2$ **15.** (a) 671.6 cm² (b) 961.4 cm² (c) 2115.54 cm³ **17.** (a) 72 ft² (b) 84 ft² (c) 36 ft³ **19.** 1728 in³ **21.** 6 in. by 6 in. by 3 in. **23.** $x = 3$ **25.** \$4.44 **27.** 640 ft³

**29.** (a) $T = L + 2B$, $T = hP + 2(e \cdot e)$, $T = e(4e) + 2e^2$, $T = 4e^2 + 2e^2$, $T = 6e^2$ (b) 96 cm$^2$ (c) $V = Bh$, $V = e^2 \cdot e$, $V = e^3$ (d) 64 cm$^3$ **33.** \$128 **35.** 864 in$^3$ **37.** 10 gal **39.** 720 cm$^2$ **41.** 2952 cm$^3$

### 8.2 EXERCISES

**1.** (a) Right pentagonal prism (b) Oblique pentagonal prism **3.** (a) Regular square pyramid (b) Oblique square pyramid **5.** (a) Pyramid (b) $E$ (c) $\overline{EA}, \overline{EB}, \overline{EC}, \overline{ED}$ (d) $\triangle EAB, \triangle EBC, \triangle ECD, \triangle EAD$ (e) No **7.** (a) 5 (b) 8 (c) 5 **9.** 66 in$^2$ **11.** 32 cm$^3$ **13.** (a) $n + 1$ (b) $n$ (c) $n$ (d) $2n$ (e) $n$ (f) $n + 1$ **15.** 3a, 4a **17.** (a) Slant height (b) Lateral edge **19.** 4 in. **21.** (a) 144.9 cm$^2$ (b) 705.18 cm$^3$ **23.** (a) 60 ft$^2$ (b) 96 ft$^2$ (c) 48 ft$^3$ **25.** $36\sqrt{5} + 36 \approx 116.5$ in$^2$ **27.** 480 ft$^2$ **29.** 900 ft$^3$ **31.** $\approx 24$ ft **33.** 336 in$^3$ **37.** 96 in$^2$ **39.** $\frac{8}{1}$ or 8:1 **41.** 39.4 in$^3$

### 8.3 EXERCISES

**1.** (a) Yes (b) Yes (c) Yes **3.** (a) $60\pi \approx 188.50$ in$^2$ (b) $110\pi \approx 345.58$ in$^2$ (c) $150\pi \approx 471.24$ in$^3$ **5.** $\approx 54.19$ in$^2$ **7.** 5 cm **9.** The radius has a length of 2 in. and the altitude has a length of 3 in. **11.** $32\pi \approx 100.53$ in$^3$ **13.** $2\sqrt{13} \approx 7.21$ cm **15.** 2 m **17.** $4\sqrt{3} \approx 6.93$ in. **19.** $3\sqrt{5} \approx 6.71$ cm **21.** (a) $6\pi\sqrt{85} \approx 173.78$ in$^2$ (b) $6\pi\sqrt{85} + 36\pi \approx 286.88$ in$^2$ (c) $84\pi \approx 263.89$ in$^3$ **23.** $54\pi$ in$^3$ **25.** $2000\pi$ cm$^3$ **27.** $1200\pi$ cm$^3$ **29.** $65\pi \approx 204.2$ cm$^2$ **31.** $192\pi \approx 603.19$ in$^3$ **35.** $60\pi \approx 188.5$ in$^2$ **37.** $\frac{4}{1}$ or 4:1 **39.** $\approx 471.24$ gal **43.** $\approx 290.60$ cm$^3$ **45.** $\approx 318$ gal **47.** $\approx 38$ ft$^2$

### 8.4 EXERCISES

**1.** Polyhedron *EFGHIJK* is concave.
**3.** Polyhedron *EFGHIJK* has nine faces ($F$), seven vertices ($V$), and 14 edges ($E$). $V + F = E + 2$ becomes
$$7 + 9 = 14 + 2$$
**5.** A regular hexahedron has six faces ($F$), eight vertices ($V$), and 12 edges ($E$). $V + F = E + 2$ becomes
$$8 + 6 = 12 + 2$$
**7.** (a) 8 faces (b) Regular octahedron **9.** 9 faces **11.** (a) $\frac{1}{2}$ (b) $\frac{5}{12}$ (c) $\frac{5}{6}$ **13.** (a) $6\sqrt{2} \approx 8.49$ in. (b) $6\sqrt{3} \approx 10.39$ in. **15.** 44 in$^2$ **17.** 105.84 cm$^2$ **19.** (a) 17.64 m$^2$ (b) 4.2 m **21.** (a) 1468.8 cm$^2$ (b) \$8.81 **23.** (a) $\frac{3}{2}$ or 3:2 (b) $\frac{3}{2}$ or 3:2 **25.** $r = 3\sqrt{2} \approx 4.24$ in.; $h = 6\sqrt{2} \approx 8.49$ in. **27.** (a) $3\sqrt{3} \approx 5.20$ in. (b) 9 in. **29.** (a) $36\pi \approx 113.1$ m$^2$ (b) $36\pi \approx 113.1$ m$^3$ **31.** 1.5 in. **33.** 113.1 ft$^2$; $\approx 3$ pints **35.** $7.4\pi \approx 23.24$ in$^3$ **37.** (a) Yes (b) Yes **39.** Parallel **41.** Congruent **43.** $S = 36\pi$ units$^2$; $V = 36\pi$ units$^3$

### CHAPTER 8 REVIEW EXERCISES

**1.** 672 in$^2$ **2.** 297 cm$^2$
**3.** Dimensions are 6 in. by 6 in. by 20 in.; $V = 720$ in$^3$
**4.** $T = 468$ cm$^2$; $V = 648$ cm$^3$ **5.** (a) 360 in$^2$ (b) 468 in$^2$ (c) 540 in$^3$ **6.** (a) 624 cm$^2$ (b) $624 + 192\sqrt{3} \approx 956.55$ cm$^2$ (c) $1248\sqrt{3} \approx 2161.6$ cm$^3$ **7.** $\sqrt{89} \approx 9.43$ cm **8.** $3\sqrt{7} \approx 7.94$ in. **9.** $\sqrt{74} \approx 8.60$ in. **10.** $2\sqrt{3} \approx 3.46$ cm **11.** (a) 540 in$^2$ (b) 864 in$^2$ (c) 1296 in$^3$ **12.** (a) $36\sqrt{19} \approx 156.92$ cm$^2$ (b) $36\sqrt{19} + 36\sqrt{3} \approx 219.27$ cm$^2$ (c) $96\sqrt{3} \approx 166.28$ cm$^3$ **13.** (a) $120\pi$ in$^2$ (b) $192\pi$ in$^2$ (c) $360\pi$ in$^3$ **14.** (a) $\approx 351.68$ ft$^3$ (b) $\approx 452.16$ ft$^2$ **15.** (a) $72\pi \approx 226.19$ cm$^2$ (b) $108\pi \approx 339.29$ cm$^2$ (c) $72\pi\sqrt{3} \approx 391.78$ cm$^3$ **16.** $\ell = 10$ in. **17.** $\approx 616$ in$^2$ **18.** $\approx 904.32$ cm$^3$ **19.** $120\pi$ units$^3$ **20.** $\frac{\text{surface area of smaller}}{\text{surface area of larger}} = \frac{1}{9}$; $\frac{\text{volume of smaller}}{\text{volume of larger}} = \frac{1}{27}$ **21.** $\approx 183\frac{1}{3}$ in$^3$ **22.** $288\pi$ cm$^3$ **23.** $\frac{32\pi}{3}$ in$^3$ **24.** $\approx 1017.36$ in$^3$ **25.** $(2744 - \frac{1372}{3}\pi)$ in$^3$ **26.** (a) 8 (b) 4 (c) 12 **27.** $40\pi$ mm$^3$
**28.** (a) $V = 16$, $E = 24$, $F = 10$   $V + F = E + 2$ becomes $16 + 10 = 24 + 2$
(b) $V = 4$, $E = 6$, $F = 4$   $V + F = E + 2$ becomes $4 + 4 = 6 + 2$
(c) $V = 6$, $E = 12$, $F = 8$   $V + F = E + 2$ becomes $6 + 8 = 12 + 2$
**29.** 114 in$^3$ **30.** (a) $\frac{1}{2}$ (b) $\frac{5}{8}$ **31.** (a) 78 in$^2$ (b) $16\sqrt{3}$ cm$^2$ **32.** Right triangle (3-4-5)

### CHAPTER 8 TEST

**1.** (a) 15 (b) 7 [8.1] **2.** (a) 16 cm$^2$ (b) 112 cm$^2$ (c) 80 cm$^3$ [8.1] **3.** (a) 5 (b) 4 [8.2] **4.** (a) $32\sqrt{2}$ ft$^2$ (b) $(16 + 32\sqrt{2})$ft$^2$ [8.2] **5.** 15 ft [8.2] **6.** 3 in. [8.2] **7.** 50 ft$^3$ [8.2] **8.** (a) False (b) True [8.3] **9.** (a) True (b) True [8.3, 8.4] **10.** 12 [8.4] **11.** $3\sqrt{5}$ cm [8.3] **12.** (a) $48\pi$ cm$^2$ (b) $96\pi$ cm$^3$ [8.3] **13.** $h = 6$ in. [8.3] **14.** (a) $\frac{1}{2}$ (b) $\frac{3}{8}$ [8.4] **15.** (a) 1256.6 ft$^2$ (b) 4188.8 ft$^3$ [8.4]

# Chapter 9

### 9.1 EXERCISES

**3.** (a) 4 (b) 8 (c) 5 (d) 9 **5.** $b = 3.5$ or $b = 10.5$ **7.** (a) 5 (b) 10 (c) $2\sqrt{5}$ (d) $\sqrt{a^2 + b^2}$ **9.** (a) $\left(2, \frac{-3}{2}\right)$ (b) (1, 1) (c) (4, 0) (d) $\left(\frac{a}{2}, \frac{b}{2}\right)$ **11.** (a) $(-3, 4)$ (b) $(0, -2)$ (c) $(-a, 0)$ (d) $(-b, -c)$ **13.** (a) $\left(4, -\frac{5}{2}\right)$ (b) (0, 4) (c) $\left(\frac{7}{2}, -1\right)$ (d) $(a, b)$ **15.** (a) $(5, -1)$ (b) $(0, -5)$ (c) $(2, -a)$ (d) $(b, -c)$ **17.** (a) $(-3, -4)$ (b) $(-2, 0)$ (c) $(-a, 0)$ (d) $(-b, c)$ **19.** (a) $x = 4$ (b) $y = b$ (c) $x = 2$ (d) $y = 3$ **21.** $(2.5, -13.7)$ **23.** (2, 3); 16 **25.** (a) Isosceles (b) Equilateral (c) Isosceles right triangle **27.** $x + y = 6$ **29.** $(a, a\sqrt{3})$ or $(a, -a\sqrt{3})$ **31.** $(0, 1 + 3\sqrt{3})$ and $(0, 1 - 3\sqrt{3})$ **33.** 17 **35.** 9 **37.** (a) $135\pi$ units$^3$ (b) $75\pi$ units$^3$

**39.** (a) $96\pi$ units$^3$ (b) $144\pi$ units$^3$
**41.** (a) $90\pi$ units$^2$ (b) $90\pi$ units$^2$
**43.** (a) $(-3, -1)$ (b) $(1, -3)$ (c) $(3, 1)$
**45.** (a) $(5, 4)$ (b) $(5, 8)$ (c) $(3, 2)$

## 9.2 EXERCISES

**1.** (4, 0) and (0, 3) **3.** (5, 0) and $\left(0, -\frac{5}{2}\right)$ **5.** $(-3, 0)$
**7.** (6, 0) and (0, 3) **9.** (a) 4 (b) Undefined (c) $-1$
(d) 0 (e) $\frac{d-b}{c-a}$ (f) $-\frac{b}{a}$ **11.** (a) 10 (b) 15
**13.** (a) Collinear (b) Noncollinear **15.** (a) $\frac{3}{4}$ (b) $-\frac{5}{3}$
(c) $-2$ (d) $\frac{a-b}{c}$ **17.** (a) 2 (b) $-\frac{4}{3}$ (c) $-\frac{1}{3}$ (d) $-\frac{h+j}{f+g}$
**19.** None of these **21.** Perpendicular **23.** $\frac{3}{2}$
**25.** 23 **33.** Right triangle **35.** (4, 7); (0, $-1$); (10, $-3$)
**39.** $m_{\overline{EH}} = \frac{2c-0}{2b-0} = \frac{2c}{2b} = \frac{c}{b}$
$m_{\overline{FG}} = \frac{c-0}{(a+b)-a} = \frac{c}{b}$
Due to equal slopes, $\overline{EH} \parallel \overline{FG}$. Thus *EFGH* is a trapezoid.

## 9.2 SELECTED PROOF

**37.** $m_{\overline{VT}} = \frac{e-e}{(c-d)-(a+d)} = \frac{0}{c-a-2d} = 0$
$m_{\overline{RS}} = \frac{b-b}{c-a} = \frac{0}{c-a} = 0$
$\therefore \overline{VT} \parallel \overline{RS}$
$RV = \sqrt{[(a+d)-a]^2 + (e-b)^2}$
$= \sqrt{d^2 + (e-b)^2} = \sqrt{d^2 + e^2 - 2be + b^2}$
$ST = \sqrt{[c-(c-d)]^2 + (b-e)^2}$
$= \sqrt{(d)^2 + (b-e)^2}$
$= \sqrt{d^2 + b^2 - 2be + e^2}$
$\therefore RV = ST$
*RSTV* is an isosceles trapezoid.

## 9.3 EXERCISES

**1.** (a) $a\sqrt{2}$ if $a > 0$ (b) $\frac{d-b}{c-a}$ **3.** (a) $-1$ (b) $-\frac{b}{a}$
**5.** $\overline{AB}$ is horizontal and $\overline{BC}$ is vertical; $\therefore \overline{AB} \perp \overline{BC}$. Hence, $\angle B$ is a right $\angle$ and $\triangle ABC$ is a right triangle.
**7.** $m_{\overline{QM}} = \frac{c-0}{b-0} = \frac{c}{b}$
$m_{\overline{PN}} = \frac{c-0}{(a+b)-a} = \frac{c}{b}$
$\therefore \overline{QM} \parallel \overline{PN}$
$m_{\overline{QP}} = \frac{c-c}{(a+b)-b} = \frac{0}{a} = 0$
$m_{\overline{MN}} = \frac{0-0}{a-0} = \frac{0}{a} = 0$
$\therefore \overline{QP} \parallel \overline{MN}$
Because both pairs of opposite sides are parallel, *MNPQ* is a parallelogram.
**9.** $m_{\overline{MN}} = 0$ and $m_{\overline{QP}} = 0$; $\therefore \overline{MN} \parallel \overline{QP}$. $\overline{QM}$ and $\overline{PN}$ are both vertical; $\therefore \overline{QM} \parallel \overline{PN}$. Hence, *MQPN* is a parallelogram. Because $\overline{QM}$ is vertical and $\overline{MN}$ is horizontal, $\angle QMN$ is a right angle. Because parallelogram *MQPN* has a right $\angle$, it is also a rectangle.
**11.** $A = (0, 0)$; $B = (a, 0)$; $C = (a, b)$
**13.** $M = (0, 0)$; $N = (r, 0)$; $P = (r + s, t)$
**15.** $A = (0, 0)$; $B = (a, 0)$; $C = (a - c, d)$
**17.** (a) Square
$A = (0, 0)$; $B = (a, 0)$; $C = (a, a)$; $D = (0, a)$
(b) Square (with midpoints of sides)
$A = (0, 0)$; $B = (2a, 0)$; $C = (2a, 2a)$; $D = (0, 2a)$
**19.** (a) Parallelogram
$A = (0, 0)$; $B = (a, 0)$; $C = (a + b, c)$; $D = (b, c)$
(NOTE: *D* chosen before *C*)
(b) Parallelogram (with midpoints of sides)
$A = (0, 0)$; $B = (2a, 0)$; $C = (2a + 2b, 2c)$; $D = (2b, 2c)$
**21.** (a) Isosceles triangle
$R = (0, 0)$; $S = (2a, 0)$; $T = (a, b)$
(b) Isosceles triangle (with midpoints of sides)
$R = (0, 0)$; $S = (4a, 0)$; $T = (2a, 2b)$
**23.** $r^2 = s^2 + t^2$ **25.** $c^2 = a^2 - b^2$ **27.** $b^2 = 3a^2$
**29.** (a) Positive (b) Negative (c) $2a$
**31.** (a) Slope Formula (b) Distance Formula
(c) Midpoint Formula (d) Slope Formula

## 9.4 EXERCISES

**21.** $m_1 = -\frac{A}{B}$; $m_2 = \frac{B}{A}$; $m_1 \cdot m_2 = -1$, so $\ell_1 \perp \ell_2$.
**23.** $3x - 2y = 2$ **25.** $x^2 + y^2 = 9$
**27.** True. The quadrilateral that results is a parallelogram.

## 9.4 SELECTED PROOFS

**3.** The diagonals of a square are perpendicular bisectors of each other.

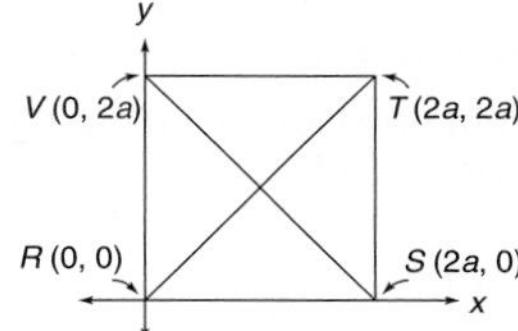


*Proof:* Let square *RSTV* have the vertices shown. Then the midpoints of the diagonals are $M_{\overline{RT}} = (a, a)$ and $M_{\overline{VS}} = (a, a)$. Also, $m_{\overline{RT}} = 1$ and $m_{\overline{VS}} = -1$. Because the two diagonals share the midpoint $(a, a)$ and the product of their slopes is $-1$, they are perpendicular bisectors of each other.
**7.** The segments that join the midpoints of the consecutive sides of a quadrilateral form a parallelogram.
*Proof:* The midpoints, as shown, of the sides of quadrilateral *ABCD* are

$$R = \left(\frac{0 + 2a}{2}, \frac{0 + 0}{2}\right) = (a, 0)$$
$$S = \left(\frac{2a + 2b}{2}, \frac{0 + 2c}{2}\right) = (a + b, c)$$
$$T = \left(\frac{2d + 2b}{2}, \frac{2e + 2c}{2}\right) = (d + b, e + c)$$
$$V = \left(\frac{0 + 2d}{2}, \frac{0 + 2e}{2}\right) = (d, e)$$

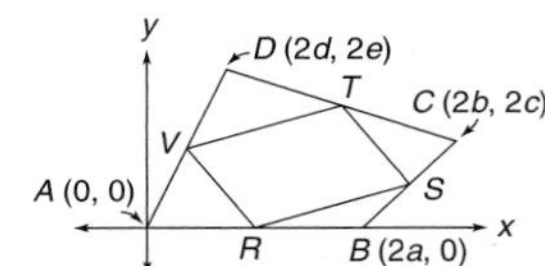

Now we determine slopes as follows:

$$m_{\overline{RS}} = \frac{c - 0}{(a + b) - a} = \frac{c}{b}$$
$$m_{\overline{ST}} = \frac{(e + c) - c}{(d + b) - (a + b)} = \frac{e}{d - a}$$
$$m_{\overline{TV}} = \frac{(e + c) - e}{(d + b) - d} = \frac{c}{b}$$
$$m_{\overline{VR}} = \frac{e - 0}{d - a} = \frac{e}{d - a}$$

Because $m_{\overline{RS}} = m_{\overline{TV}}$, $\overline{RS} \parallel \overline{TV}$. Also $m_{\overline{ST}} = m_{\overline{VR}}$, so $\overline{ST} \parallel \overline{VR}$. Then $RSTV$ is a parallelogram.

**11.** The midpoint of the hypotenuse of a right triangle is equidistant from the three vertices of the triangle.

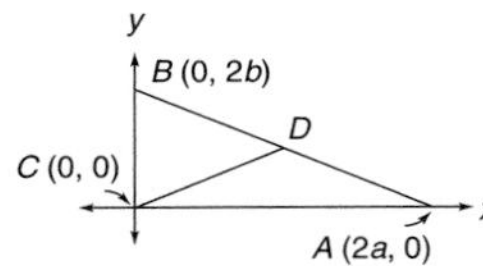


*Proof:* Let rt. $\triangle ABC$ have vertices as shown. Then $D$, the midpoint of the hypotenuse, is given by

$$D = \left(\frac{0 + 2a}{2}, \frac{2b + 0}{2}\right) = (a, b)$$

Now $BD = DA = \sqrt{(2a - a)^2 + (0 - b)^2} = \sqrt{a^2 + (-b)^2} = \sqrt{a^2 + b^2}$

Also, $CD = \sqrt{(a - 0)^2 + (b - 0)^2} = \sqrt{a^2 + b^2}$

Then $D$ is equidistant from $A$, $B$, and $C$.

**15.** If the midpoint of one side of a rectangle is joined to the endpoints of the opposite side, an isosceles triangle is formed.

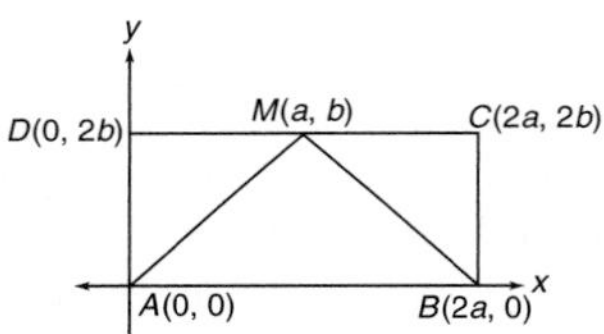


*Proof:* Let rectangle $ABCD$ have endpoints as shown above. With $M$ the midpoint of $\overline{DC}$,

$$M = \left(\frac{0 + 2a}{2}, \frac{2b + 2b}{2}\right) = (a, b)$$
$$MA = \sqrt{(a - 0)^2 + (b - 0)^2}$$
$$MA = \sqrt{a^2 + b^2}$$
$$MB = \sqrt{(a - 2a)^2 + (b - 0)^2}$$
$$MB = \sqrt{a^2 + b^2}$$

Because $MA = MB$, $\triangle AMB$ is isosceles.

### 9.5 EXERCISES

**1.** $x + 2y = 6$; $y = -\frac{1}{2}x + 3$
**3.** $-x + 3y = -40$; $y = \frac{1}{3}x - \frac{40}{3}$ **9.** $2x + 3y = 15$
**11.** $x + y = 6$ **13.** $-2x + 3y = -3$ **15.** $bx + ay = ab$
**17.** $-x + y = -2$ **19.** $5x + 2y = 5$
**21.** $4x + 3y = -12$ **23.** $-x + 3y = 2$
**25.** $y = -\frac{b}{a}x + \frac{bg + ha}{a}$ **27.** $(6, 0)$ **29.** $(5, -4)$
**31.** $(6, -2)$ **33.** $(3, 2)$ **35.** $(6, -1)$ **37.** $(5, -2)$
**39.** $(6, 0)$ **41.** $\left(\frac{a - c}{b}, a\right)$ **43.** $\left(b, \frac{ab - b^2}{c}\right)$

### 9.5 SELECTED PROOFS

**45.** The altitudes of a triangle are concurrent.

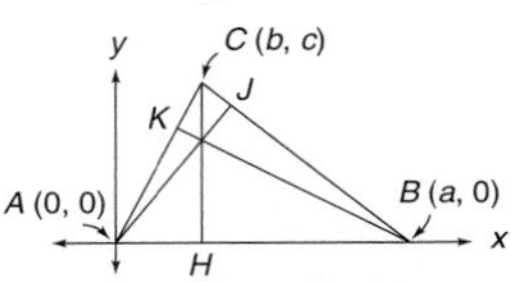


*Proof:* For $\triangle ABC$, let $\overline{CH}$, $\overline{AJ}$, and $\overline{BK}$ name the altitudes. Because $\overline{AB}$ is horizontal ($m_{\overline{AB}} = 0$), $\overline{CH}$ is vertical and has the equation $x = b$.

Because $m_{\overline{BC}} = \frac{c - 0}{b - a} = \frac{c}{b - a}$, the slope of altitude $\overline{AJ}$ is $m_{\overline{AJ}} = -\frac{b - a}{c} = \frac{a - b}{c}$. Since $\overline{AJ}$ contains (0, 0), its equation is $y = \frac{a - b}{c}x$.

The intersection of altitudes $\overline{CH}$ ($x = b$) and $\overline{AJ}$ $\left(y = \frac{a - b}{c}x\right)$ is at $x = b$, so $y = \frac{a - b}{c} \cdot b = \frac{b(a - b)}{c} = \frac{ab - b^2}{c}$. That is, $\overline{CH}$ and $\overline{AJ}$ intersect at $\left(b, \frac{ab - b^2}{c}\right)$. The remaining altitude is $\overline{BK}$. Since $m_{\overline{AC}} = \frac{c - 0}{b - 0} = \frac{c}{b}$, $m_{\overline{BK}} = -\frac{b}{c}$. Because $\overline{BK}$ contains (a, 0), its equation is $y - 0 = -\frac{b}{c}(x - a)$ or $y = \frac{-b}{c}(x - a)$.

For the three altitudes to be concurrent, $\left(b, \frac{ab - b^2}{c}\right)$ must lie on the line $y = \frac{-b}{c}(x - a)$. Substitution leads to

$$\frac{ab - b^2}{c} = \frac{-b}{c}(b - a) = \frac{-b(b - a)}{c} = \frac{-b^2 + ab}{c}$$

which is true. Then the three altitudes are concurrent.

**47.** First, find the equation of the line through $P$ perpendicular to $Ax + By = C$. Second, find the point of intersection $D$ of the two lines. Finally, use the Distance Formula to find the length of $\overline{PD}$.

### CHAPTER 9 REVIEW EXERCISES

**1.** (a) 7 (b) 6 (c) 13 (d) 5
**2.** (a) 8 (b) 10 (c) $4\sqrt{5}$ (d) 10
**3.** (a) $\left(6, \frac{1}{2}\right)$ (b) $(-2, 4)$ (c) $\left(1, -\frac{1}{2}\right)$ (d) $\left(\frac{2x - 3}{2}, y\right)$
**4.** (a) (2, 1) (b) $(-2, -2)$ (c) (0, 3) (d) $(x + 1, y + 1)$
**5.** (a) Undefined (b) 0 (c) $-\frac{5}{12}$ (d) $-\frac{4}{3}$
**6.** (a) Undefined (b) 0 (c) $\frac{1}{2}$ (d) $\frac{4}{3}$
**7.** $(-4, -8)$ **8.** (3, 7) **9.** $x = \frac{4}{3}$ **10.** $y = -4$
**11.** (a) Perpendicular (b) Parallel (c) Neither (d) Perpendicular **12.** Noncollinear
**13.** $x = 4$ **14.** (7, 0) and (0, 3)
**17.** (a) $3x + 5y = 21$ (b) $3x + y = -7$ (c) $-2x + y = -8$ (d) $y = 5$

**18.** $m_{\overline{AB}} = \frac{4}{3}$; $m_{\overline{BC}} = \frac{1}{2}$; $m_{\overline{AC}} = -2$. Because $m_{\overline{AC}} \cdot m_{\overline{BC}} = -1$, $\overline{AC} \perp \overline{BC}$ and $\angle C$ is a rt. $\angle$; the triangle is a rt. $\triangle$.
**19.** $AB = \sqrt{85}$; $BC = \sqrt{85}$. Because $AB = BC$, the triangle is isosceles.
**20.** $m_{\overline{RS}} = \frac{-4}{3}$; $m_{\overline{ST}} = \frac{5}{6}$; $m_{\overline{TV}} = \frac{-4}{3}$; $m_{\overline{RV}} = \frac{5}{6}$. Therefore $\overline{RS} \parallel \overline{VT}$ and $\overline{RV} \parallel \overline{ST}$ and $RSTV$ is a parallelogram.
**21.** (3, 5) **22.** (1, 4) **23.** (3, 5)
**24.** (1, 4) **25.** (16, 11), (4, −9), (−4, 5)
**26.** (a) $\sqrt{53}$ (b) −4 (c) $\frac{1}{4}$
**27.** $A = (-a, 0)$; $B = (0, b)$; $C = (a, 0)$
**28.** $D = (0, 0)$; $E = (a, 0)$; $F = (a, 2a)$; $G = (0, 2a)$
**29.** $R = (0, 0)$; $U = (0, a)$; $T = (a, a + b)$
**30.** $M = (0, 0)$; $N = (a, 0)$; $Q = (a + b, c)$; $P = (b, c)$
**31.** (a) $\sqrt{(a + c)^2 + (b + d - 2e)^2}$ (b) $-\frac{a}{b - e}$ or $\frac{a}{e - b}$
(c) $y - 2d = \frac{a}{e - b}(x - 2c)$

### CHAPTER 9 TEST

**1.** (a) (5, −3) (b) (0, −4) [9.1] **3.** $CD = 10$ [9.1]
**4.** (−3, 5) [9.1]
**5.**

| $x$ | 0 | 3 | **0** | 9 |
|---|---|---|---|---|
| $y$ | **4** | **2** | 4 | **−2** |

[9.2]

**6.** 

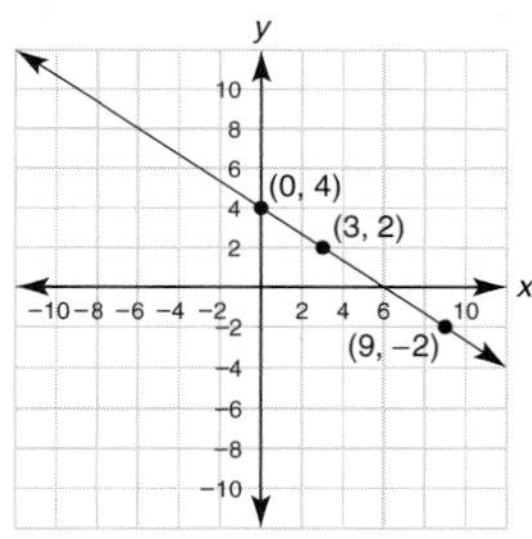


[9.2]

**7.** (a) −3 (b) $\frac{d - b}{c - a}$ [9.2] **8.** (a) $\frac{2}{3}$ (b) $-\frac{3}{2}$ [9.2]
**9.** Parallelogram [9.3]
**10.** $a = \sqrt{b^2 + c^2}$ or $a^2 = b^2 + c^2$ [9.3]
**11.** (a) Isosceles triangle (b) Trapezoid [9.3]
**12.** (a) Slope Formula (b) Distance Formula [9.3]
**13.** $D(0, 0)$, $E(2a, 0)$, $F(a, b)$ [9.3] **14.** (b) [9.4]
**15.** $m_{\overline{VR}} = m_{\overline{TS}}$, so $-\frac{v}{r} = \frac{v}{t - s}$.
Possible answers: $r = s - t$ or $s = r + t$ or equivalent [9.4]
**16.** (a) $y = x + 4$ (b) $y = \frac{3}{4}x - 3$ [9.5]
**17.** $y = cx + (b - ac)$ [9.5] **18.** (4, 1) [9.5]
**19.** (−1, 4) [9.5] **20.** $M = (a + b, c)$ and $N = (a, 0)$.
Then $m_{\overline{AC}} = \frac{2c - 0}{2b - 0} = \frac{c}{b}$ and $m_{\overline{MN}} = \frac{c - 0}{a + b - a} = \frac{c}{b}$.
With $m_{\overline{AC}} = m_{\overline{MN}}$, it follows that $\overline{AC} \parallel \overline{MN}$. [9.4]

# Chapter 10

### 10.1 EXERCISES

**1.** $\sin\alpha = \frac{5}{13}$; $\sin\beta = \frac{12}{13}$ **3.** $\sin\alpha = \frac{8}{17}$; $\sin\beta = \frac{15}{17}$
**5.** $\sin\alpha = \frac{\sqrt{15}}{5}$; $\sin\beta = \frac{\sqrt{10}}{5}$ **7.** 1 **9.** 0.2924
**11.** 0.9903 **13.** 0.9511 **15.** $a \approx 6.9$ in.; $b \approx 9.8$ in.
**17.** $a \approx 10.9$ ft; $b \approx 11.7$ ft **19.** $c \approx 8.8$ cm; $d \approx 28.7$ cm
**21.** $\alpha \approx 29°$; $\beta \approx 61°$ **23.** $\alpha \approx 17°$; $\beta \approx 73°$
**25.** $\alpha \approx 19°$; $\beta \approx 71°$ **27.** $\alpha \approx 23°$ **29.** $d \approx 103.5$ ft
**31.** $d \approx 128.0$ ft **33.** $\alpha \approx 24°$
**35.** (a) ≈ 5.4 ft (b) ≈ 54 ft² **37.** $\theta \approx 50°$

### 10.2 EXERCISES

**1.** $\cos\alpha = \frac{12}{13}$; $\cos\beta = \frac{5}{13}$ **3.** $\cos\alpha = \frac{3}{5}$; $\cos\beta = \frac{4}{5}$
**5.** $\cos\alpha = \frac{\sqrt{10}}{5}$; $\cos\beta = \frac{\sqrt{15}}{5}$ **7.** (a) $\sin\alpha = \frac{a}{c}$; $\cos\beta = \frac{a}{c}$.
Thus $\sin\alpha = \cos\beta$. (b) $\cos\alpha = \frac{b}{c}$; $\sin\beta = \frac{b}{c}$.
Thus $\cos\alpha = \sin\beta$. **9.** 0.9205 **11.** 0.9563
**13.** 0 **15.** 0.1392 **17.** $a \approx 84.8$ ft; $b \approx 53.0$ ft
**19.** $a = b = 5$ cm **21.** $c \approx 19.1$ in.; $d \approx 14.8$ in.
**23.** $\alpha = 60°$; $\beta = 30°$ **25.** $\alpha \approx 51°$; $\beta \approx 39°$
**27.** $\alpha \approx 65°$; $\beta \approx 25°$ **29.** $\theta \approx 34°$ **31.** $x \approx 1147.4$ ft
**33.** ≈ 8.1 in. **35.** ≈ 13.1 cm **37.** $\alpha \approx 55°$
**39.** (a) m∠$A$ = 68° (b) m∠$B$ = 112°

### 10.3 EXERCISES

**1.** $\tan\alpha = \frac{3}{4}$; $\tan\beta = \frac{4}{3}$ **3.** $\tan\alpha = \frac{\sqrt{5}}{2}$; $\tan\beta = \frac{2\sqrt{5}}{5}$
**5.** $\sin\alpha = \frac{5}{13}$; $\cos\alpha = \frac{12}{13}$; $\tan\alpha = \frac{5}{12}$; $\cot\alpha = \frac{12}{5}$;
$\sec\alpha = \frac{13}{12}$; $\csc\alpha = \frac{13}{5}$ **7.** $\sin\alpha = \frac{a}{c}$; $\cos\alpha = \frac{b}{c}$;
$\tan\alpha = \frac{a}{b}$; $\cot\alpha = \frac{b}{a}$; $\sec\alpha = \frac{c}{b}$; $\csc\alpha = \frac{c}{a}$
**9.** $\sin\alpha = \frac{x\sqrt{x^2 + 1}}{x^2 + 1}$; $\cos\alpha = \frac{\sqrt{x^2 + 1}}{x^2 + 1}$; $\tan\alpha = \frac{x}{1}$; $\cot\alpha = \frac{1}{x}$;
$\sec\alpha = \sqrt{x^2 + 1}$; $\csc\alpha = \frac{\sqrt{x^2 + 1}}{x}$ **11.** 0.2679
**13.** 1.5399 **15.** $x \approx 7.5$; $z \approx 14.2$
**17.** $y \approx 5.3$; $z \approx 8.5$ **19.** $d \approx 8.1$
**21.** $\alpha \approx 37°$; $\beta \approx 53°$ **23.** $\theta \approx 56°$; $\gamma \approx 34°$
**25.** $\alpha \approx 29°$; $\beta \approx 61°$ **27.** 1.4826 **29.** 2.0000
**31.** 1.3456 **33.** (b) ≈ 0.4245 **35.** (b) ≈ 7.1853
**37.** ≈ 1376.8 ft **39.** ≈ 4.1 in. **41.** ≈ 72°
**43.** $\alpha \approx 47°$. The heading may be described as N 47° W.
**45.** ≈ 26,730 ft
**47.** (a) $h \approx 9.2$ ft (b) $V \approx 110.4$ ft³

### 10.4 EXERCISES

**1.** (a) $\frac{1}{2} \cdot 5 \cdot 6 \cdot \sin 78°$ (b) $\frac{1}{2} \cdot 5 \cdot 7 \cdot \sin 56°$
**3.** (a) $\frac{\sin 40°}{5} = \frac{\sin\beta}{8}$ (b) $\frac{\sin 41°}{5.3} = \frac{\sin 87°}{c}$
**5.** (a) $c^2 = 5.2^2 + 7.9^2 - 2(5.2)(7.9)\cos 83°$
(b) $6^2 = 9^2 + 10^2 - 2 \cdot 9 \cdot 10 \cdot \cos\alpha$
**7.** (a) $\frac{\sin\alpha}{a} = \frac{\sin\beta}{b}$ (b) $a^2 = b^2 + c^2 - 2bc\cos\alpha$
**9.** (a) (3, 4, 5) is a Pythagorean Triple; $\gamma$ lies opposite the longest side and must be a right angle. (b) 90°
**11.** 8 in² **13.** ≈ 11.6 ft² **15.** ≈ 15.2 ft²
**17.** ≈ 11.1 in. **19.** ≈ 8.9 m **21.** ≈ 55°
**23.** ≈ 51° **25.** ≈ 10.6 **27.** ≈ 6.9
**29.** (a) ≈ 213.4 ft (b) ≈ 13,294.9 ft² **31.** ≈ 8812 m
**33.** ≈ 15.9 ft **35.** 6 **37.** ≈ 14.0 ft **41.** 51.8 cm²

### CHAPTER 10 REVIEW EXERCISES

**1.** sine; ≈ 10.3 in. **2.** sine; ≈ 7.5 ft **3.** cosine; ≈ 23.0 in. **4.** sine; ≈ 5.9 ft **5.** tangent; ≈ 43° **6.** cosine; ≈ 58° **7.** sine; ≈ 49° **8.** tangent; ≈ 16° **9.** ≈ 8.9 units **10.** ≈ 60° **11.** ≈ 13.1 units **12.** ≈ 18.5 units **13.** ≈ 42.7 ft **14.** ≈ 74.8 cm **15.** ≈ 47° **16.** ≈ 54° **17.** ≈ 26.3 in² **19.** If $m\angle S = 30°$ and $m\angle Q = 90°$, then the sides of $\triangle RQS$ can be represented by $RQ = x$, $RS = 2x$, and $SQ = x\sqrt{3}$. $\sin S = \sin 30° = \frac{x}{2x} = \frac{1}{2}$. **21.** ≈ 8.4 ft **22.** ≈ 866 ft **23.** ≈ 41° **24.** ≈ 8° **25.** ≈ 5.0 cm **26.** ≈ 4.3 cm **27.** ≈ 68° **28.** ≈ 106° **29.** 3 to 7 (or 3:7) **30.** ≈ 1412.0 m **31.** $\cos\theta = \frac{24}{25}$; $\sec\theta = \frac{25}{24}$ **32.** $\sec\theta = \frac{61}{60}$; $\cot\theta = \frac{60}{11}$ **33.** $\csc\theta = \frac{29}{20}$; $\sin\theta = \frac{20}{29}$ **34.** $h \approx 6.9$ ft; $V \approx 74.0$ ft³

### CHAPTER 10 TEST

**1.** (a) $\frac{a}{c}$ (b) $\frac{b}{a}$ [10.1, 10.3] **2.** (a) $\frac{3}{5}$ (b) $\frac{3}{5}$ [10.1, 10.2] **3.** (a) 1 (b) $\frac{\sqrt{3}}{2}$ [10.1, 10.3] **4.** (a) 0.3907 (b) 0.1908 [10.1, 10.2] **5.** $\theta \approx 42°$ [10.1] **6.** (a) tan 26° (b) cos 47° [10.2, 10.3] **7.** $a \approx 14$ [10.1] **8.** $y \approx 9$ [10.1] **9.** $\theta \approx 56°$ [10.1] **10.** (a) True (b) True [10.2] **11.** 92 ft [10.1] **12.** 10° [10.1] **13.** (a) $\csc\alpha = 2$ (b) $\alpha = 30°$ [10.1, 10.3] **14.** (a) $\frac{a}{c}$ (b) $\frac{c}{a}$ [10.3] **15.** ≈ 42 cm² [10.4] **16.** $\frac{\sin\alpha}{a} = \frac{\sin\beta}{b} = \frac{\sin\gamma}{c}$ [10.4] **17.** $a^2 = b^2 + c^2 - 2bc\cos\alpha$ [10.4] **18.** $\alpha \approx 33°$ [10.4] **19.** $x \approx 11$ [10.4]

## Appendix A

### A.1 ALGEBRAIC EXPRESSIONS

**1.** Undefined terms, definitions, axioms or postulates, and theorems **3.** (a) Reflexive (b) Transitive (c) Substitution (d) Symmetric **5.** (a) 12 (b) −2 (c) 2 (d) −12 **7.** (a) 35 (b) −35 (c) −35 (d) 35 **9.** No; Commutative Axiom for Multiplication **11.** (a) 9 (b) −9 (c) 8 (d) −8 **13.** (a) −4 (b) −36 (c) 18 (d) $-\frac{1}{4}$ **15.** −$60 **17.** (a) 65 (b) 16 (c) 9 (d) $8x$ **19.** (a) $10\pi$ (b) $11\sqrt{2}$ (c) $7x^2y$ (d) $7\sqrt{3}$ **21.** (a) 14 (b) 20 (c) 14 (d) 38 **23.** (a) −1 (b) $\frac{1}{9}$ (c) $-\frac{8}{9}$ (d) $-\frac{1}{2}$ **25.** (a) 6 (b) $12x^2 - 7x - 10$ **27.** $5x + 2y$ **29.** $10x + 5y$ **31.** $10x$

### A.2 FORMULAS AND EQUATIONS

**1.** $5x + 8$ **3.** $2x - 2$ **5.** $5x + 8$ **7.** $x^2 + 7x + 12$ **9.** $6x^2 + 11x - 10$ **11.** $2a^2 + 2b^2$ **13.** 60 **15.** 40 **17.** 148 **19.** $12\pi$ **21.** 7 **23.** −12 **25.** 12 **27.** 5 **29.** 30 **31.** 8 **33.** 4 **35.** 32

### A.3 INEQUALITIES

**1.** The length of $\overline{AB}$ is greater than the length of $\overline{CD}$. **3.** The measure of angle $ABC$ is greater than the measure of angle $DEF$. **5.** (a) 4 (b) 10 **7.** $IJ < AB$ **9.** (a) False (b) True (c) True (d) False **11.** The measure of the second angle must be greater than 148° and less than 180°. **13.** (a) $-12 \le 20$ (b) $-10 \le -2$ (c) $18 \ge -30$ (d) $3 \ge -5$

**15.**

| | |
|---|---|
| No change | No change |
| No change | No change |
| No change | Change |
| No change | Change |

**17.** $x \le 7$ **19.** $x < -5$ **21.** $x < 20$ **23.** $x \ge -24$ **25.** $x \le -2$ **27.** Not true if $c < 0$ **29.** Not true if $a = -3$ and $b = -2$ **31.** If $a < b$ and $c < d$, then $a + p_1 = b$ and $c + p_2 = d$, where $p_1$ and $p_2$ are both positive. Use the Addition Property to get

$$a + p_1 + c + p_2 = b + d$$
$$a + c + p_1 + p_2 = b + d$$

But because $p_1$ and $p_2$ are both positive, $p_1 + p_2$ must also be positive. $\therefore a + c < b + d$

### A.4 QUADRATIC EQUATIONS

**1.** (a) 3.61 (b) 2.83 (c) −5.39 (d) 0.77 **3.** a, c, d, f **5.** (a) $2\sqrt{2}$ (b) $3\sqrt{5}$ (c) 30 (d) 3 **7.** (a) $\frac{3}{4}$ (b) $\frac{5}{7}$ (c) $\frac{\sqrt{7}}{4}$ (d) $\frac{\sqrt{6}}{3}$ **9.** (a) $\sqrt{54} \approx 7.35$ and $3\sqrt{6} \approx 7.35$ (b) $\sqrt{\frac{5}{16}} \approx 0.56$ and $\frac{\sqrt{5}}{4} \approx 0.56$ **11.** $x = 4$ or $x = 2$ **13.** $x = 12$ or $x = 5$ **15.** $x = -\frac{2}{3}$ or $x = 4$ **17.** $x = \frac{1}{3}$ or $x = \frac{1}{2}$ **19.** $x = 5$ or $x = 2$ **21.** $x = \frac{7 \pm \sqrt{13}}{2} \approx 5.30$ or 1.70 **23.** $x = 2 \pm 2\sqrt{3} \approx 5.46$ or −1.46 **25.** $x = \frac{3 \pm \sqrt{149}}{10} \approx 1.52$ or −0.92 **27.** $x = \pm\sqrt{7} \approx \pm 2.65$ **29.** $x = \pm\frac{5}{2}$ **31.** $x = 0$ or $x = \frac{b}{a}$ **33.** 5 by 8 **35.** $n = 6$ **37.** $c = 5$

# Glossary

**acute angle.** an angle whose measure is between 0° and 90°

**acute triangle.** a triangle whose three interior angles are all acute

**adjacent angles.** two angles that have a common vertex and a common side between them

**altitude of cone (pyramid).** the line segment from the vertex perpendicular to the plane of the base

**altitude of cylinder (prism).** a line segment between and perpendicular to each of the two bases

**altitude of parallelogram.** a line segment drawn perpendicularly from a vertex to a nonadjacent side (known as the related base)

**altitude of trapezoid.** a line segment drawn perpendicularly from a vertex to the remaining parallel side

**altitude of triangle.** a line segment drawn perpendicularly from a vertex of the triangle to the opposite side of the triangle; the length of the altitude is the height of the triangle

**angle.** the plane figure formed by two rays that share a common endpoint

**angle bisector.** *see* bisector of angle

**angle of depression (elevation).** acute angle formed by a horizontal ray and a ray determined by a downward (an upward) rotation

**apothem of regular polygon.** any line segment drawn from the center of the regular polygon perpendicular to one of its sides

**arc.** the segment (part) of a circle determined by two points on the circle and all points between them

**area.** the measurement, in square units, of the amount of region within an enclosed plane figure

**auxiliary line.** a line (or part of a line) added to a drawing to help complete a proof or solve a problem

**axiom.** *see* postulate

**base.** a side (of a plane figure) or face (of a solid figure) to which an altitude is drawn

**base angles of isosceles triangle.** the two congruent angles of the isosceles triangle

**base of isosceles triangle.** the side of the triangle whose length is unique

**bases of trapezoid.** the two parallel sides of the trapezoid

**bisector of angle.** a ray that separates the given angle into two smaller, congruent angles

**center of circle.** the interior point of the circle whose distance from all points on the circle is the same

**center of regular polygon.** the common center of the inscribed and circumscribed circles of the regular polygon

**center of sphere.** the interior point of the sphere whose distance from all points on the sphere is the same

**central angle of circle.** an angle whose vertex is at the center of the circle and whose sides are radii of the circle

**central angle of regular polygon.** an angle whose vertex is at the center of the regular polygon and whose sides are two consecutive radii of the polygon

**centroid of triangle.** the point determined by the intersection of the three medians of the triangle

**chord of circle.** any line segment that joins two points on the circle

**circle.** the set of points in a plane that are at a fixed distance from a point (the center of the circle) in the plane

**circumcenter of triangle.** the center of the circumscribed circle of a triangle; the point determined by the intersection of the perpendicular bisectors of the three sides of the triangle

**circumference.** the linear measure of the distance around a circle

**circumscribed circle.** a circle that contains all vertices of a polygon such that the sides of the polygon are chords of the circle

**circumscribed polygon.** a polygon whose sides are all tangent to a circle in the interior of the polygon

**collinear points.** points that lie on the same line

**common tangent.** a line (or segment) that is tangent to more than one circle; can be a common external tangent or a common internal tangent

**complementary angles.** two angles whose sum of measures is 90°

**concave polygon.** a polygon in which at least one diagonal lies in the exterior of the polygon

**concentric circles (spheres).** two or more circles (spheres) that have the same center

**conclusion.** the "then" clause of an "If, then" statement; the part of a theorem indicating the claim to be proved

**concurrent lines.** three or more lines that contain the same point

**congruent.** used to describe figures (such as angles) that can be made to coincide

**converse.** relative to the statement "If *P*, then *Q*," this statement has the form "If *Q*, then *P*"

**convex polygon.** a polygon in which all diagonals lie in the interior of the polygon

**coplanar points.** points that lie in the same plane

**corollary.** a theorem that follows from another theorem as a "by-product"; a theorem that is easily proved as the consequence of another theorem

**cosecant.** in a right triangle, the ratio $\frac{\text{hypotenuse}}{\text{opposite}}$

**cosine.** in a right triangle, the ratio $\frac{\text{adjacent}}{\text{hypotenuse}}$

**cotangent.** in a right triangle, the ratio $\frac{\text{adjacent}}{\text{opposite}}$

**cube.** a right square prism whose edges are congruent

**cyclic polygon.** a polygon inscribed in a circle

**cylinder (circular).** the solid generated by using parallel line segments to join each point of one circle to each point of a second circle that lies in a plane parallel to that of the first circle

**decagon.** a polygon with exactly 10 sides

**deduction.** a form of reasoning in which specific conclusions are reached through the use of established principles

**degree.** the unit of measure that corresponds to $\frac{1}{360}$ of a complete revolution; used with angles and arcs

**diagonal of polygon.** a line segment that joins two nonconsecutive vertices of a polygon

**diameter.** any line segment that joins two points on a circle (or sphere) and contains the center of the circle (or sphere)

**dodecagon.** a polygon that has exactly 12 sides

**dodecahedron (regular).** a polyhedron that has exactly 12 faces that are congruent regular pentagons

**edge of polyhedron.** any line segment that joins two consecutive vertices of the polyhedron (includes prisms and pyramids)

**equiangular polygon.** a type of polygon whose angles are congruent (equal)

**equilateral polygon.** a type of polygon whose sides are congruent (equal)

**equivalent equations.** equations for which the solutions are the same

**extended proportion.** a proportion that has three or more members, such as $\frac{a}{b} = \frac{c}{d} = \frac{e}{f}$

**extended ratio.** a ratio that compares three or more numbers, such as $a{:}b{:}c$

**exterior.** refers to all points that lie outside an enclosed (bounded) plane or solid figure

**exterior angle of polygon.** an angle formed by one side of a polygon and an extension of a second side that has a common endpoint with the first side

**extremes of a proportion.** the first and last terms of a proportion; in $\frac{a}{b} = \frac{c}{d}$, $a$ and $d$ are the extremes

**face of polyhedron.** any one of the polygons that lies in a plane determined by the vertices of the polyhedron; includes base(s) and lateral faces of prisms and pyramids

**geometric mean.** the repeated second and third terms of certain proportions; in $\frac{a}{b} = \frac{b}{c}$, $b$ is the geometric mean of $a$ and $c$

**height.** *see* altitude

**heptagon.** a polygon that has exactly seven sides

**hexagon.** a polygon that has exactly six sides

**hexahedron (regular).** a polyhedron that has six congruent square faces; also called a cube

**hypotenuse of right triangle.** the side of a right triangle that lies opposite the right angle

**hypothesis.** the "if" clause of an "If, then" statement; the part of a theorem providing the given information

**icosahedron (regular).** a polyhedron with 20 faces that are equilateral triangles

**incenter of triangle.** the center of the inscribed circle of a triangle; the point determined by the intersection of the three angle bisectors of the angles of the triangle

**induction.** a form of reasoning in which a number of specific observations are used to draw a general conclusion

**inscribed angle of circle.** an angle whose vertex is on a circle and whose sides are chords of the circle

**inscribed circle.** a circle that lies inside a polygon such that the sides of the polygon are tangents of the circle

**inscribed polygon.** a polygon whose vertices all lie on a circle such that the sides of the polygon are chords of the circle

**intercepted arc.** the arc (an arc) of a circle that is cut off in the interior of an angle

**intercepts.** the points at which the graph of an equation intersects the axes

**interior.** refers to all points that lie inside an enclosed (bounded) plane or solid figure

**interior angle of polygon.** any angle formed by two consecutive sides of the polygon such that the angle lies in the interior of the polygon

**intersection.** the points that two geometric figures share

**intuition.** drawing a conclusion through insight

**inverse.** relative to the statement "If *P*, then *Q*," this statement has the form "If not *P*, then not *Q*"

**isosceles trapezoid.** a trapezoid that has two congruent legs (its nonparallel sides)

**isosceles triangle.** a triangle that has two congruent sides

**kite.** a quadrilateral that has two distinct pairs of congruent adjacent sides

**lateral area.** the sum of areas of the faces of a solid or the area of the curved surface, excluding the base area(s) (as in prisms, pyramids, cylinders, and cones)

**legs of isosceles triangle.** the two congruent sides of the triangle

**legs of right triangle.** the two sides that form the right angle of the triangle

**legs of trapezoid.** the two nonparallel sides of the trapezoid

**lemma.** a theorem that is introduced and proved so that a later theorem can be proved

**line of centers.** the line (or line segment) that joins the centers of two circles

**line segment.** the part of a line determined by two points and all points on the line that lie between those two points

**locus.** the set of all points that satisfy a given condition or conditions

**major arc.** an arc whose measure is between 180° and 360°

**mean proportional.** *see* geometric mean

**means of a proportion.** the second and third terms of a proportion; in $\frac{a}{b} = \frac{c}{d}$, $b$ and $c$ are the means

**median of trapezoid.** the line segment that joins the midpoints of the two legs (nonparallel sides) of the trapezoid

**median of triangle.** the line segment joining a vertex of the triangle to the midpoint of the opposite side

**midpoint.** the point on a line segment (or arc) that separates the line segment (arc) into two congruent parts

**minor arc.** an arc whose measure is between 0° and 180°

**nonagon.** a polygon that has exactly nine sides
**noncollinear points.** three or more points that do not lie on the same line
**noncoplanar points.** four or more points that do not lie in the same plane

**obtuse angle.** an angle whose measure is between 90° and 180°
**obtuse triangle.** a triangle that has exactly one interior obtuse angle
**octagon.** a polygon that has exactly eight sides
**octahedron (regular).** a polyhedron with eight congruent faces that are equilateral triangles
**opposite rays.** two rays having a common endpoint that together form a line
**orthocenter of triangle.** the point determined by the intersection of the three altitudes of the triangle

**parallel lines (planes).** two lines in a plane (or two planes) that do not intersect
**parallelogram.** a quadrilateral that has two pairs of parallel sides
**parallelepiped.** a right rectangular prism; a box
**pentagon.** a polygon that has exactly five sides
**perimeter of polygon.** the sum of the lengths of the sides of the polygon
**perpendicular bisector of line segment.** a line (or part of a line) that is both perpendicular to a given line segment and bisects that line segment
**perpendicular lines.** two lines that intersect to form congruent adjacent angles
**pi ($\pi$).** the constant ratio of the circumference of a circle to the length of its diameter; this ratio is commonly approximated by the fraction $\frac{22}{7}$ or the decimal 3.1416
**point of tangency (contact).** the point at which a tangent to a circle touches the circle
**polygon.** a plane figure whose sides are line segments that intersect only at their endpoints
**polyhedron.** a solid figure whose faces are polygons that intersect other faces along common sides of the polygons
**postulate.** a statement that is assumed true but is not proved

**Quadratic Formula.** the formula $x = \frac{-b \pm \sqrt{b^2 - 4ac}}{2a}$, which provides solutions for the equation $ax^2 + bx + c = 0$, where $a$, $b$, and $c$ are real numbers and $a \neq 0$
**quadrilateral.** a polygon that has exactly four sides

**radian.** the measure of a central angle of a circle whose intercepted arc has a length equal to the radius of the circle
**radius.** the line segment that joins the center of a circle (or sphere) to any point on the circle (or sphere)
**ratio.** a comparison between two quantities $a$ and $b$, generally written $\frac{a}{b}$ or $a{:}b$
**ray.** the part of a line that begins at a point and extends infinitely far in one direction
**rectangle.** a parallelogram that contains a right angle
**reflex angle.** an angle whose measure is greater than 180°, but less than 360°
**regular polygon.** a polygon whose sides are congruent and whose interior angles are congruent
**regular polyhedron.** a polyhedron whose edges are congruent and whose faces are congruent
**regular prism.** a right prism whose bases are regular polygons
**regular pyramid.** a pyramid whose base is a regular polygon and whose lateral faces are congruent isosceles triangles
**rhombus.** a parallelogram with two congruent adjacent sides
**right angle.** an angle whose measure is exactly 90°
**right circular cone.** a cone in which the line segment joining the vertex to the center of the circular base is perpendicular to the base
**right circular cylinder.** a cylinder in which the line segment joining the centers of the circular bases is perpendicular to the plane of each base
**right prism.** a prism in which lateral edges are perpendicular to the base edges they intersect
**right triangle.** a triangle in which exactly one of the interior angles is a right angle

**scalene triangle.** a triangle in which no two sides are congruent
**secant.** in a right triangle, the ratio $\frac{\text{hypotenuse}}{\text{adjacent}}$
**secant of circle.** a line (or part of a line) that intersects a circle at two points
**sector of circle.** the plane region bounded by two radii of the circle and the arc that is intercepted by the central angle formed by those radii
**segment of circle.** the plane region bounded by a chord and a minor arc (major arc) that has the same endpoints as that chord
**semicircle.** the arc of a circle determined by a diameter; an arc of a circle whose measure is exactly 180°
**set.** any collection of objects, numbers, or points
**similar polygons.** polygons that have the same shape
**sine.** in a right triangle, the ratio $\frac{\text{opposite}}{\text{hypotenuse}}$
**skew quadrilateral.** a quadrilateral whose sides do not all lie in one plane
**slant height of cone.** any line segment joining the vertex of the cone to a point on the circular base
**slant height of regular pyramid.** a line segment joining the vertex of the pyramid to the midpoint of a base edge of the pyramid
**slope.** a measure of the steepness of a line; in the rectangular coordinate system, the slope $m$ of the line through $(x_1, y_1)$ and $(x_2, y_2)$ is $m = \frac{y_2 - y_1}{x_2 - x_1}$
**sphere.** the set of points in space that are at a fixed distance from a point (the center of the sphere)
**straight angle.** an angle whose measure is exactly 180°; an angle whose sides are opposite rays
**straightedge.** an idealized instrument used to construct parts of lines
**supplementary angles.** two angles whose sum of measures is 180°
**surface area.** the measure of the total area (lateral area plus base area) of any solid figure
**symmetry with respect to a line ($\ell$).** figure for which every point $A$ has a second point $B$ on the figure for which $\ell$ is the perpendicular bisector of $\overline{AB}$
**symmetry with respect to a point ($P$).** figure for which every point $A$ has a second point $C$ on the figure for which $P$ is the midpoint of $\overline{AC}$.

**tangent.** in a right triangle, the ratio $\frac{\text{opposite}}{\text{adjacent}}$

**tangent circles.** two circles that have one point in common; the circles may be externally tangent or internally tangent

**tangent of circle.** a line (or part of a line) that touches a circle at only one point

**tetrahedron (regular).** a four-faced solid in which the faces are congruent equilateral triangles

**theorem.** a statement that follows logically from previous definitions and principles; a statement that can be proved

**torus.** a three-dimensional solid that has a "doughnut" shape

**transversal.** a line that intersects two or more lines, intersecting each at one point

**trapezoid.** a quadrilateral having exactly two parallel sides

**triangle.** a polygon that has exactly three sides

**triangle inequality.** a statement that the sum of the lengths of two sides of a triangle cannot be greater than the length of the third side

**union.** the joining together of any two sets, such as geometric figures

**valid argument.** an argument in which the conclusion follows logically from previously stated (and accepted) premises or assumptions

**vertex angle of isosceles triangle.** the angle formed by the two congruent sides of the triangle

**vertex of angle.** the point at which the two sides of the angle meet

**vertex of isosceles triangle.** the point at which the two congruent sides of the triangle meet

**vertex of polygon.** any point at which two sides of the polygon meet

**vertex of polyhedron.** any point at which two edges of the polyhedron meet

**vertical angles.** a pair of angles that lie in opposite positions when formed by two intersecting lines

**volume.** the measurement, in cubic units, of the amount of space within a bounded region of space

# Index

# Abbreviations

AA angle-angle (proves △s similar)
ASA angle-side-angle (proves △s congruent)
AAS angle-angle-side (proves △s congruent)
add. addition
adj. adjacent
alt. altitude, alternate
ax. axiom
cm centimeters
$cm^2$ square centimeters
$cm^3$ cubic centimeters
comp. complementary
corr. corresponding
cos cosine
cot cotangent
CPCTC Corresponding parts of congruent triangles are congruent.
csc cosecant
CSSTP Corresponding sides of similar triangles are proportional.
diag. diagonal
eq. equality
exs. exercises
ext. exterior
ft foot (or feet)
gal gallon
HL hypotenuse-leg (proves △s congruent)
hr hour
in. inch (or inches)
ineq. inequality
int. interior
isos. isosceles
km kilometers
m meters
mi miles
mm millimeters
$n$-gon polygon of $n$ sides
opp. opposite
pent. pentagon
post. postulate
prop. property
pt. point
quad. quadrilateral
rect. rectangle
rt. right
SAS side-angle-side (proves △s congruent)
SAS~ side-angle-side (proves △s similar)
sec, sec. secant, section
sin sine
SSS side-side-side (proves △s congruent)
SSS~ side-side-side (proves △s similar)
st. straight
supp. supplementary
tan tangent
trans. transversal
trap. trapezoid
vert. vertical (angle or line)
yd yards